Geometry

Solutions Manual

New York, New York
Columbus, Ohio
Chicago, Illinois
Peoria, Illinois
Woodland Hills, California

This booklet is provided in *Glencoe Geometry Answer Key Maker* (0-07-860264-5). Also provided are solutions for problems in the Prerequisite Skills, Extra Practice, and Mixed Problem Solving sections.

The McGraw-Hill Companies

Send all inquiries to:
Glencoe/McGraw-Hill
8787 Orion Place
Columbus, OH 43240-4027

ISBN: 0-07-860204-1

Geometry
Solutions Manual

5 6 7 8 9 10 009 12 11 10 09 08 07 06

CONTENTS

Chapter 1 Points, Lines, Planes, and Angles

Page 5 Getting Started

1–4.

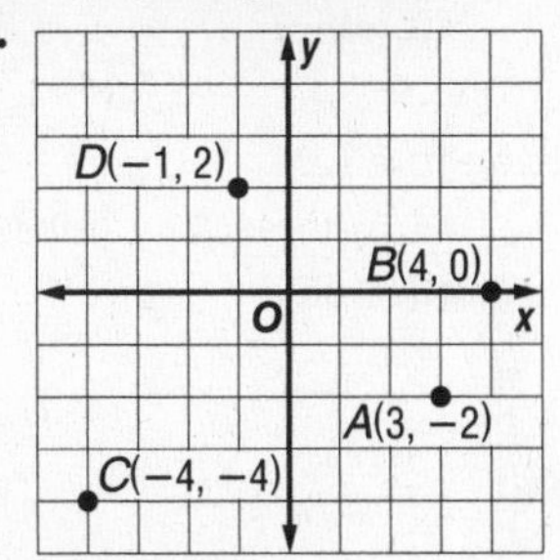

5. $\frac{3}{4} + \frac{3}{8} = \frac{6}{8} + \frac{3}{8}$
$= \frac{9}{8} = 1\frac{1}{8}$

6. $2\frac{5}{16} + 5\frac{1}{8} = \frac{37}{16} + \frac{41}{8}$
$= \frac{37}{16} + \frac{82}{16}$
$= \frac{119}{16} = 7\frac{7}{16}$

7. $\frac{7}{8} - \frac{9}{16} = \frac{14}{16} - \frac{9}{16}$
$= \frac{5}{16}$

8. $11\frac{1}{2} - 9\frac{7}{16} = \frac{23}{2} - \frac{151}{16}$
$= \frac{184}{16} - \frac{151}{16}$
$= \frac{33}{16} = 2\frac{1}{16}$

9. $2 - 17 = -15$

10. $23 - (-14) = 23 + 14$
$= 37$

11. $[-7 - (-2)]^2 = (-7 + 2)^2$
$= (-5)^2 = 25$

12. $9^2 + 13^2 = 81 + 169$
$= 250$

13. $P = 4s$
$= 4(5) = 20$
The perimeter is 20 in.

14. $P = 2\ell + 2w$
$= 2(6) + 2\left(2\frac{1}{2}\right)$
$= 12 + 5 = 17$
The perimeter is 17 ft.

15. $P = 2\ell + 2w$
$= 2(4.8) + 2(7.5)$
$= 9.6 + 15 = 24.6$
The perimeter is 24.6 m.

1-1 Points, Lines, and Planes

Page 8 Geometry Activity

1. no **2.** no

3. On $\overleftrightarrow{CD}$; see students' work.

4. See students' work.

Page 9 Check for Understanding

1. point, line, plane

2. See students' work; sample answer: Two lines intersect at a point.

3. Micha; the points must be noncollinear to determine a plane.

4. Sample answers: line p; plane R

5. Sample answer:

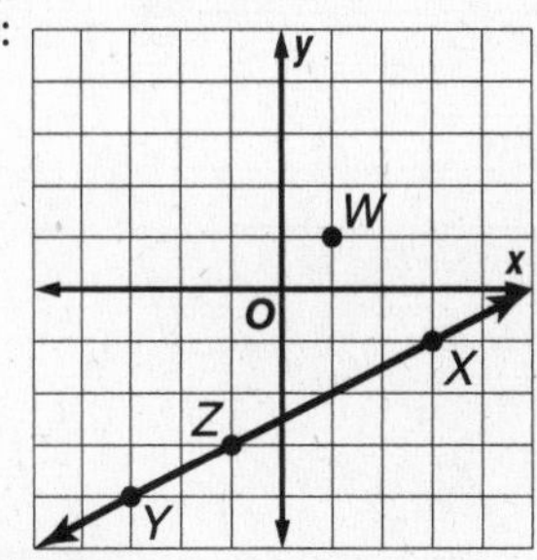

6.

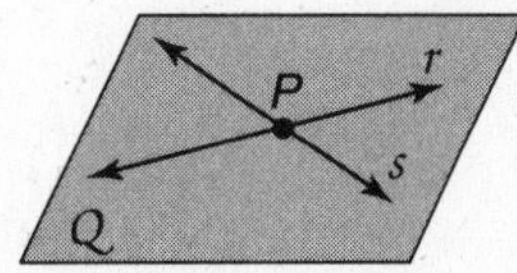

7. There are six planes: plane ABC, plane AGE, plane CDE, plane BCD, plane FAB, and plane DEF.

8. A, K, B or B, J, C

9. No; A, C, and J lie in plane ABC, but D does not.

10. line **11.** point **12.** plane

Pages 9–11 Practice and Apply

13. n **14.** F

15. R **16.** W

17. Sample answer: $\overleftrightarrow{PR}$

18. Yes, it intersects both m and n when all three lines are extended.

19. $(D, 9)$ **20.** Charlotte

21.

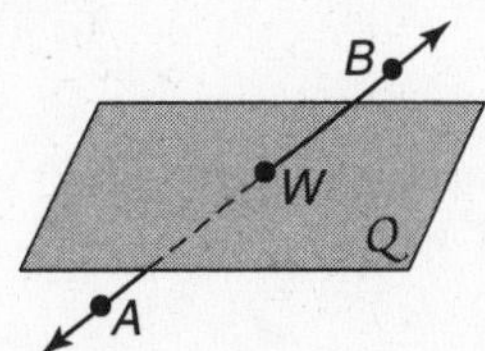

22.

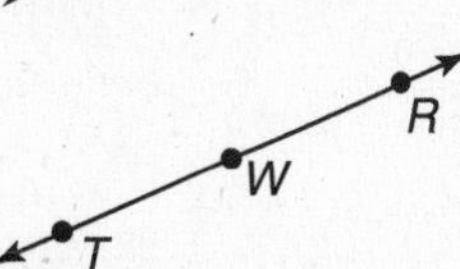

23. Sample answer:

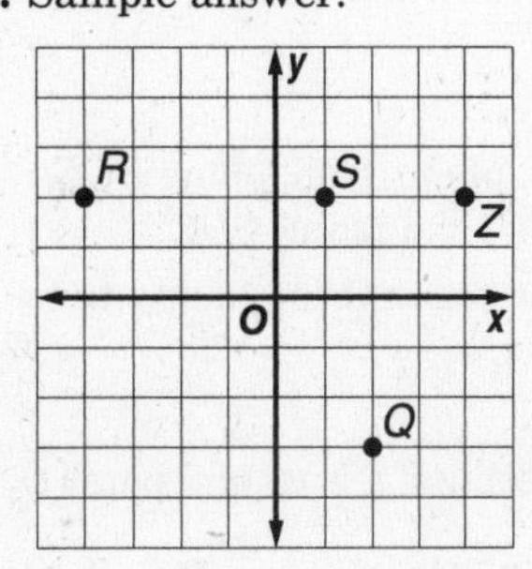

24. Sample answer:

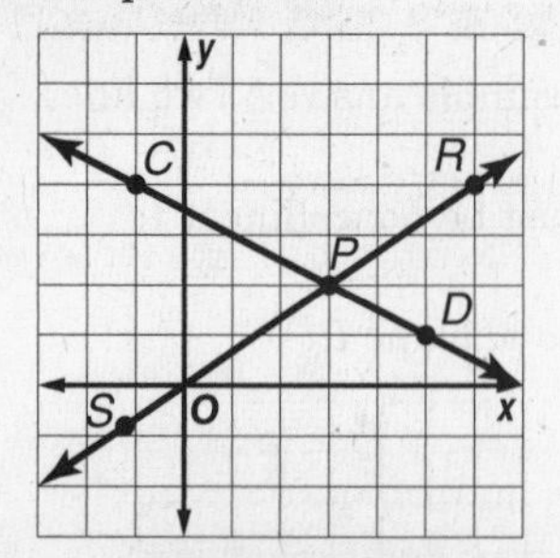

25.

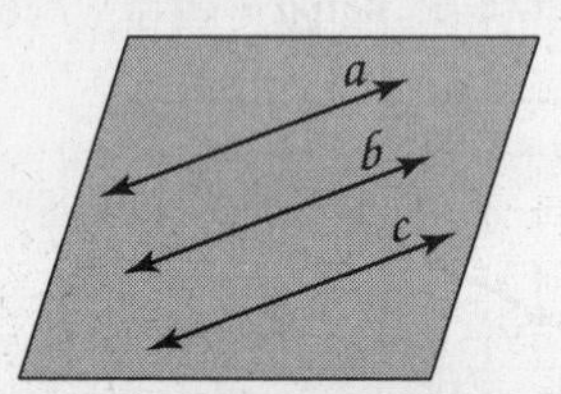

26.

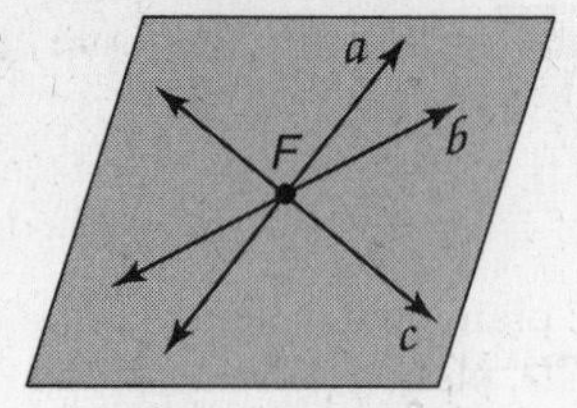

27.

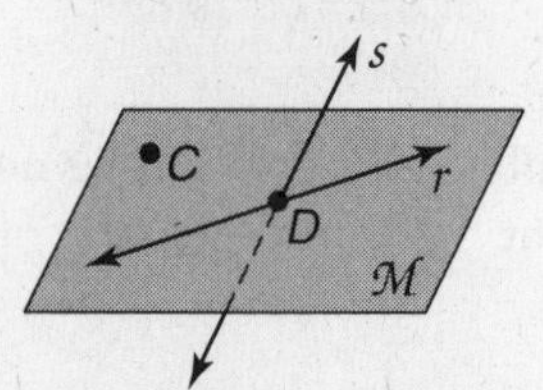

28.

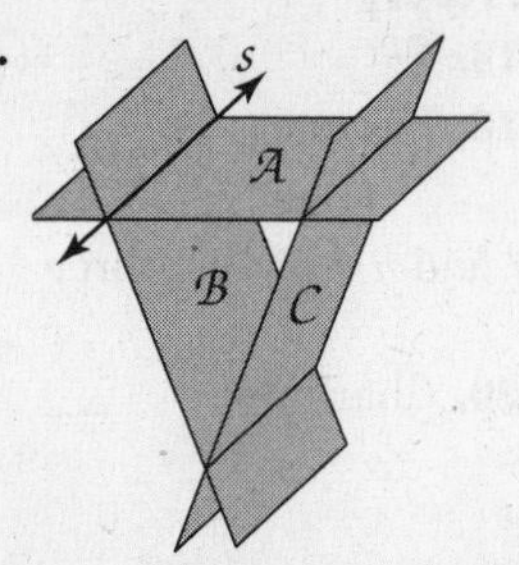

29. points that seem collinear; sample answer: (0, −2), (1, −3), (2, −4), (3, −5)

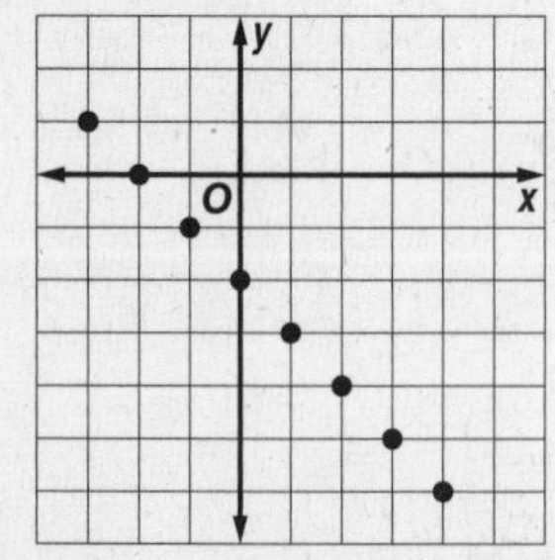

30. There are five planes: plane *ABC*, plane *BCE*, plane *ABE*, plane *ADE*, and plane *CDE*.

31. 1; There is exactly one plane through any three noncollinear points.

32. *E*, *F*, *C*

33. Because *A* and *B* determine a line, add point *G* anywhere on $\overleftrightarrow{AB}$.

34. *E*, *F*

35. *A*, *B*, *C*, *D* or *E*, *F*, *C*, *B*

36. Sample answer: points *E*, *A*, and *B* are coplanar, but points *E*, *A*, *B*, and *C* are not.

37. $\overleftrightarrow{AC}$

38. point

39. lines

40. plane

41. plane

42. two planes intersecting in a line

43. point

44. intersecting lines

45. point

46. line

47.

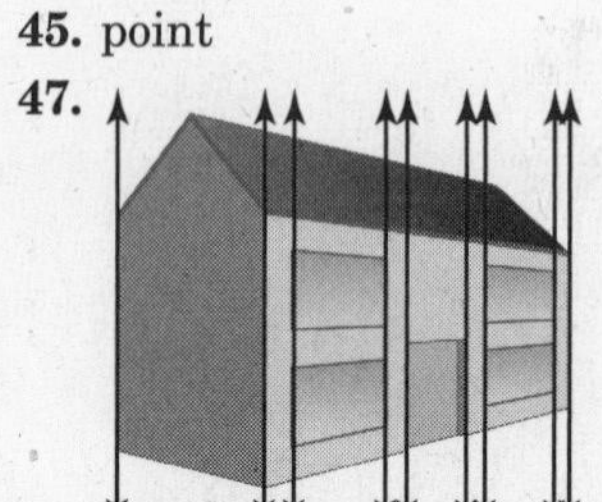

48.

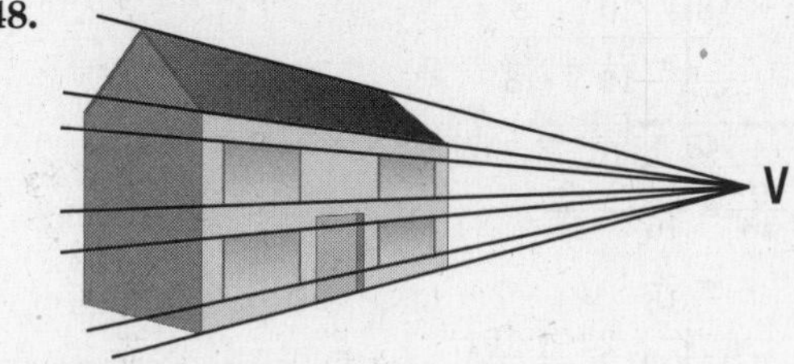

49. See students' work.

50. Sample answer: the image is rotated so that the front or back plane is not angled.

51. Sample answer:

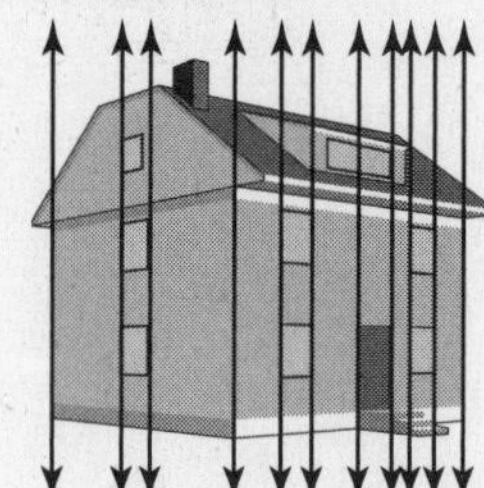

52. See picture.

53. vertical

54. Sample answer: the paths flown by airplanes flying in formation

55. Sample answer: Chairs wobble because all four legs do not touch the floor at the same time. Answers should include the following.

- The ends of the legs represent points. If all points lie in the same plane, the chair will not wobble.
- Because it only takes three points to determine a plane, a chair with three legs will never wobble.

56. C; Three lines intersect in a maximum of three points. The fourth line can cross the other three lines only one time each, adding three points to the figure. For example,

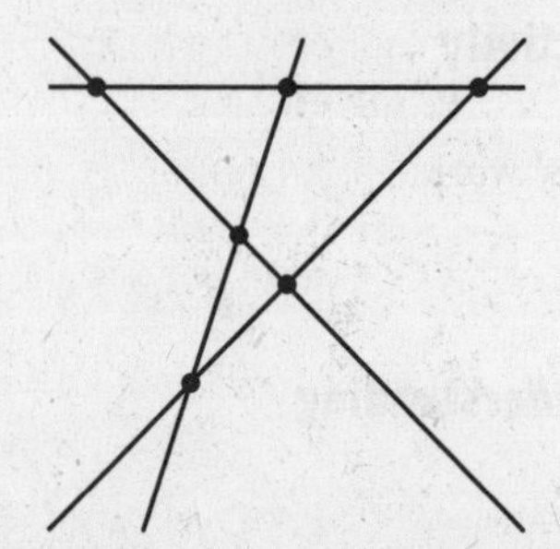

57. B; $2 + x = 2 - x$

$$2 + x - 2 = 2 - x - 2$$
$$x = -x$$
$$x + x = -x + x$$
$$2x = 0$$

Thus, x must be 0.

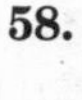

58.

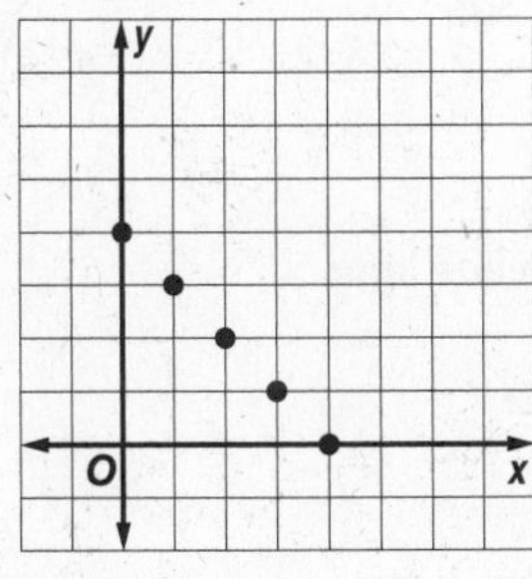

a line

59.

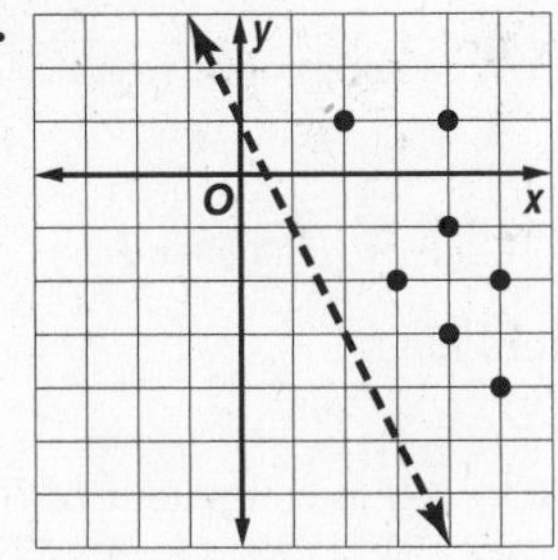

part of the coordinate plane above the line $y = -2x + 1$

60. $\frac{1}{2} = \frac{4}{8}$, so $\frac{1}{2}$ in. $> \frac{3}{8}$ in.

61. $\frac{1}{4} = \frac{4}{16}$, so $\frac{4}{16}$ in. $= \frac{1}{4}$ in.

62. $\frac{4}{5} = \frac{8}{10}$, so $\frac{4}{5}$ in. $> \frac{6}{10}$ in.

63. 10 mm = 1 cm

64. 2.5 cm = 25 mm, so 2.5 cm < 28 mm

65. 0.025 cm = 0.25 mm, so 0.025 cm < 25 mm

Page 12 Reading Mathematics

1. Points P, Q, and R lie on ℓ. Point T is not collinear with P, Q, and R.
2. Planes F, G, and H intersect at line j.
3. The intersection of planes W, X, Y, and Z is point P.

4.

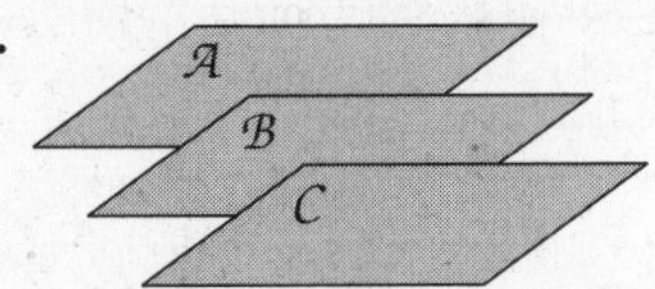

1-2 Linear Measure and Precision

Page 16 Check for Understanding

1. Align the 0 point on the ruler with the leftmost endpoint of the segment. Align the edge of the ruler along the segment. Note where the rightmost endpoint falls on the scale and read the closest eighth of an inch measurement.
2. Sample answers: rectangle, square, equilateral triangle
3. Each inch on the ruler is divided into eighths. Point Q is closer to the $1\frac{6}{8}$-inch mark. Thus, $\overline{PQ}$ is about $1\frac{6}{8}$ or $1\frac{3}{4}$ inches long.
4. The long marks on the ruler are centimeters, and the shorter marks are millimeters. There are 10 millimeters for each centimeter. Thus, the bee is 13 millimeters or 1.3 centimeters long.
5. The measurement is precise to within 0.5 meter. So, a measurement of 14 meters could be 13.5 to 14.5 meters.
6. The measuring tool is divided into $\frac{1}{4}$-inch increments. Thus, the measurement is precise to within $\frac{1}{2}\left(\frac{1}{4}\right)$ or $\frac{1}{8}$ inch. Therefore, the measurement could be between $3\frac{1}{4} - \frac{1}{8} = 3\frac{1}{8}$ inches and $3\frac{1}{4} + \frac{1}{8} = 3\frac{3}{8}$ inches.
7. $EG = EF + FG$
   $EG = 2.4 + 1.3$
   $EG = 3.7$
   So, $\overline{EG}$ is 3.7 centimeters long.
8. $$XY + YZ = XZ$$
   $$XY + 1\frac{5}{8} = 3$$
   $$XY + 1\frac{5}{8} - 1\frac{5}{8} = 3 - 1\frac{5}{8}$$
   $$XY = 1\frac{3}{8}$$
   So, $\overline{XY}$ is $1\frac{3}{8}$ inches long.

9.

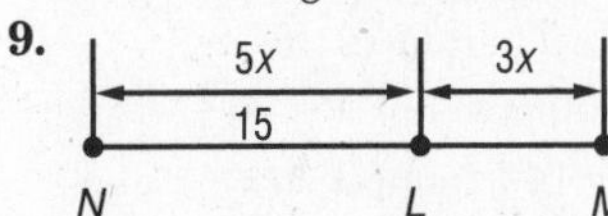

$NL = 5x = 15$
$$\frac{5x}{5} = \frac{15}{5}$$
$$x = 3$$
$LM = 3x$
$LM = 3(3)$
$LM = 9$

10.

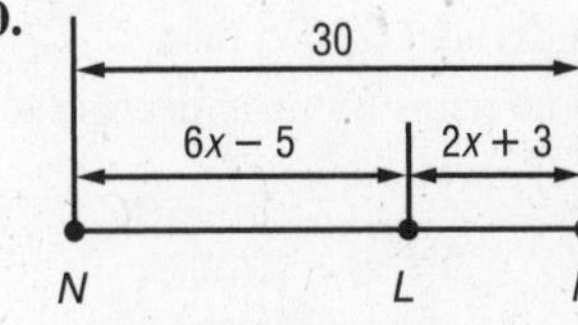

$$NM = NL + LM$$
$$30 = 6x - 5 + 2x + 3$$
$$30 = 8x - 2$$
$$30 + 2 = 8x - 2 + 2$$
$$32 = 8x$$
$$\frac{32}{8} = \frac{8x}{8}$$

$4 = x$
$LM = 2x + 3$
$LM = 2(4) + 3 = 8 + 3$
$LM = 11$

11. $\overline{BC} \cong \overline{CD}$ because they both have length 10 inches.
$\overline{BE} \cong \overline{ED}$ because they both have length 8 inches.
$\overline{BA} \cong \overline{DA}$ because they both have length 14.4 inches.

Pages 17–19 Practice and Apply

12. Each inch on the ruler is divided into sixteenths. Point B is closer to the $1\frac{5}{16}$-inch mark. Thus, $\overline{AB}$ is about $1\frac{5}{16}$ inches long.

13. The long marks on the ruler are centimeters, and the shorter marks are millimeters. Point D is closer to the 45-millimeter mark. Thus, $\overline{CD}$ is about 45 millimeters or 4.5 centimeters long.

14. The long marks on the ruler are centimeters, and the shorter marks are millimeters. The right end of the key is closer to the 33-millimeter mark. Thus, the key is about 33 millimeters or 3.3 centimeters long.

15. Each inch on the ruler is divided into sixteenths. The right tip of the paperclip is closer to the $1\frac{4}{16}$-inch mark. Thus, the paperclip is about $1\frac{4}{16}$ or $1\frac{1}{4}$ inches long.

16. The measurement is precise to within $\frac{1}{2}$ inch. So, a measurement of 80 inches could be $79\frac{1}{2}$ to $80\frac{1}{2}$ inches.

17. The measurement is precise to within 0.5 millimeter. So, a measurement of 22 millimeters could be 21.5 to 22.5 millimeters.

18. The measuring tool is divided into $\frac{1}{2}$-inch increments. Thus, the measurement is precise to within $\frac{1}{2}\left(\frac{1}{2}\right)$ or $\frac{1}{4}$ inch. Therefore, the measurement could be between $16\frac{1}{2} - \frac{1}{4} = 16\frac{1}{4}$ inches and $16\frac{1}{2} + \frac{1}{4} = 16\frac{3}{4}$ inches.

19. The measurement is precise to within 0.5 centimeter. So, a measurement of 308 centimeters could be between 307.5 and 308.5 centimeters.

20. The measurement is precise to within 0.005 meter or 5 millimeters. So, a measurement of 3.75 meters = 3750 millimeters could be between 3745 and 3755 millimeters.

21. The measuring tool is divided into $\frac{1}{4}$-foot increments. Thus, the measurement is precise to within $\frac{1}{2}\left(\frac{1}{4}\right)$ or $\frac{1}{8}$ foot. Therefore, the measurement could be between $3\frac{1}{4} - \frac{1}{8} = 3\frac{1}{8}$ feet and $3\frac{1}{4} + \frac{1}{8} = 3\frac{3}{8}$ feet.

22. $AC = AB + BC$
$AC = 16.7 + 12.8 = 29.5$
So, $\overline{AC}$ is 29.5 millimeters long.

23. $XZ = XY + YZ$
$XZ = \frac{1}{2} + \frac{3}{4} = \frac{2}{4} + \frac{3}{4}$
$XZ = \frac{5}{4}$
So, $\overline{XZ}$ is $\frac{5}{4}$ or $1\frac{1}{4}$ inches long.

24. $PR = PQ + QR$
$2\frac{1}{4} = \frac{5}{16} + QR$
$\frac{9}{4} - \frac{5}{16} = \frac{5}{16} + QR - \frac{5}{16}$
$\frac{36}{16} - \frac{5}{16} = QR$
$\frac{31}{16} = QR$
So, $\overline{QR}$ is $\frac{31}{16}$ or $1\frac{15}{16}$ inches long.

25. $RT = RS + ST$
$4.0 = 1.2 + ST$
$4.0 - 1.2 = 1.2 + ST - 1.2$
$2.8 = ST$
So, $\overline{ST}$ is 2.8 centimeters long.

26. $WY = WX + XY$
$4.8 = WX + WX$ $\quad$ $\overline{WX}$ is congruent to $\overline{XY}$.
$4.8 = 2WX$
$\frac{4.8}{2} = \frac{2WX}{2}$
$2.4 = WX$
So, $\overline{WX}$ is 2.4 centimeters long.

27. $AD = AB + BC + CD$
$3\frac{3}{4} = BC + BC + BC$ $\quad$ $\overline{AB}$ is congruent to $\overline{BC}$ and $\overline{CD}$ is congruent to $\overline{BC}$.
$3\frac{3}{4} = 3BC$
$\frac{1}{3}\left(\frac{15}{4}\right) = \frac{1}{3}(3BC)$
$\frac{15}{12} = BC$
So, $\overline{BC}$ is $1\frac{3}{12}$ or $1\frac{1}{4}$ inches long.

28.

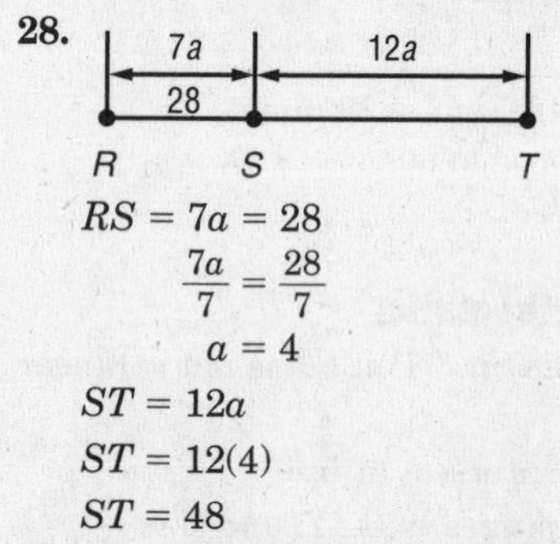

$RS = 7a = 28$
$\frac{7a}{7} = \frac{28}{7}$
$a = 4$
$ST = 12a$
$ST = 12(4)$
$ST = 48$

29.

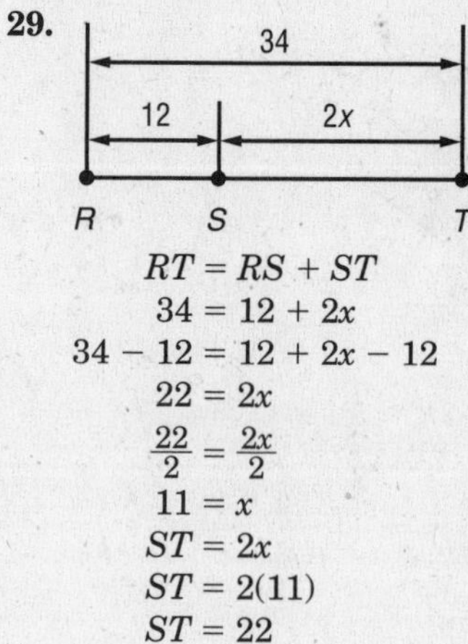

$RT = RS + ST$
$34 = 12 + 2x$
$34 - 12 = 12 + 2x - 12$
$22 = 2x$
$\frac{22}{2} = \frac{2x}{2}$
$11 = x$
$ST = 2x$
$ST = 2(11)$
$ST = 22$

30.

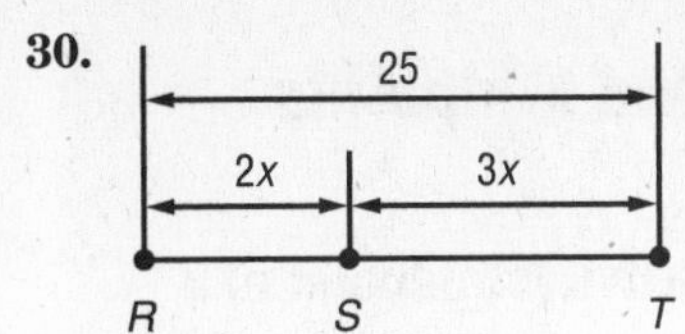

$RT = RS + ST$
$25 = 2x + 3x$
$25 = 5x$
$\frac{25}{5} = \frac{5x}{5}$
$5 = x$
$ST = 3x$
$ST = 3(5)$
$ST = 15$

31.

5x + 10
16
2x
R
S
T

$RT = RS + ST$
$5x + 10 = 16 + 2x$
$5x + 10 - 10 = 16 + 2x - 10$
$5x = 6 + 2x$
$5x - 2x = 6 + 2x - 2x$
$3x = 6$
$\frac{3x}{3} = \frac{6}{3}$
$x = 2$
$ST = 2x$
$ST = 2(2)$
$ST = 4$

32.

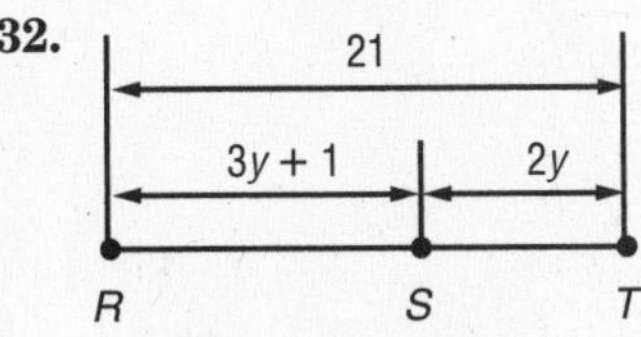

$RT = RS + ST$
$21 = 3y + 1 + 2y$
$21 = 5y + 1$
$21 - 1 = 5y + 1 - 1$
$20 = 5y$
$\frac{20}{5} = \frac{5y}{5}$
$4 = y$
$ST = 2y$
$ST = 2(4)$
$ST = 8$

33.

5y
4y − 1
2y − 1
R
S
T

$RT = RS + ST$
$5y = 4y - 1 + 2y - 1$
$5y = 6y - 2$
$5y + 2 = 6y - 2 + 2$
$5y + 2 = 6y$
$5y + 2 - 5y = 6y - 5y$
$2 = y$
$ST = 2y - 1$
$ST = 2(2) - 1$
$ST = 3$

34. yes; $AB = CD = 3$ cm

35. no; $EF = 6$ ft and $FG = 8$ ft

36. no; $NP = 1.75$ in. and $LM = 0.75$ in.

37. yes; $WX = XY = 6$ m

38. not from the information given

39. yes; $TR = 3(a + b) + 3c = 3a + 3b + 3c$
$SU = 3a + 3(b + c) = 3a + 3b + 3c$

40. The width of a music CD is 12 centimeters.

41. $\overline{CF} \cong \overline{DG}, \overline{AB} \cong \overline{HI}, \overline{CE} \cong \overline{ED} \cong \overline{EF} \cong \overline{EG}$

42. 144 cm^3; 343 mL could be actually as much as 343.5 mL and 200 mL as little as 199.5 mL; $343.5 - 199.5 = 144$.

43. The lengths of the bars are given in tenths of millions, and 0.1 million = 100,000. So the graph is precise to within 50,000 visitors.

44. 50,000 = 0.05 million, so a measurement of 98.5 million could be 98.45 million to 98.55 million visitors.

45. No; the number of visitors to Washington state parks could be as low as 46.35 million or as high as 46.45 million. The visitors to Illinois state parks could be as low as 44.45 million or as high as 44.55 million visitors. The difference in visitors could be as high as 2.0 million.

46. 12.5 cm; Each measurement is accurate within 0.5 cm, so the least perimeter is 2.5 cm + 4.5 cm + 5.5 cm.

47. 15.5 cm; Each measurement is accurate within 0.5 cm, so the greatest perimeter is 3.5 cm + 5.5 cm + 6.5 cm.

48.

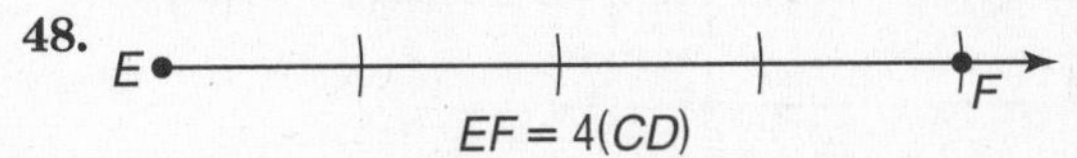

49.

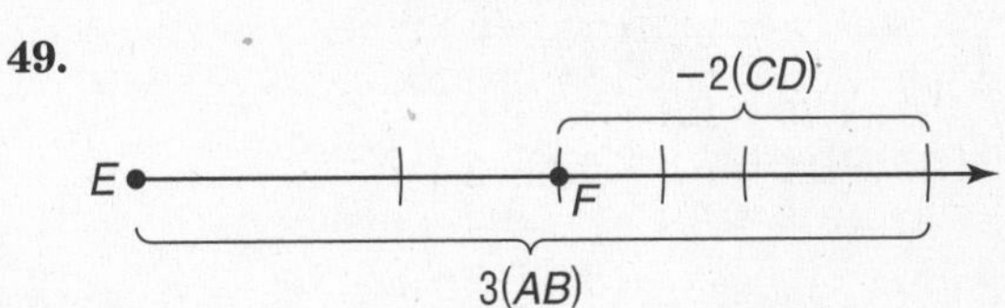

50a. 2 **50b.** 5 **50c.** 7

51. Sample answer: Units of measure are used to differentiate between size and distance, as well as for accuracy. Answers should include the following.

- When a measurement is stated, you do not know the precision of the instrument used to make the measure. Therefore, the actual measure could be greater or less than that stated.
- You can assume equal measures when segments are shown to be congruent.

52. $\frac{\text{allowable error}}{\text{measure}} = \frac{0.5 \text{ ft}}{27 \text{ ft}} \approx 0.019$ or 1.9%

53. $\frac{\text{allowable error}}{\text{measure}} = \frac{0.25 \text{ in.}}{14.5 \text{ in.}} \approx 0.017$ or 1.7%

54. $\frac{\text{allowable error}}{\text{measure}} = \frac{0.05 \text{ cm}}{42.3 \text{ cm}} \approx 0.001$ or 0.1%

55. $\frac{\text{allowable error}}{\text{measure}} = \frac{0.05 \text{ km}}{63.7 \text{ km}} \approx 0.0008$ or 0.08%

56. B; 5(12 in.) = 60 in. or 5 ft

57. D; forty percent are jazz tapes so sixty percent are blues tapes; $0.60(80) = 48$.

Page 19 Maintain Your Skills

58. B, G, E

59. Sample answer: planes ABC and BCD

60. C

61. There are five planes shown: plane ABC, plane $BCDE$, plane DEF, plane $ACDF$, and plane $ABEF$.

62. $2a + 2b = 2(3) + 2(8)$
$= 6 + 16 = 22$

63. $ac + bc = (3)(2) + (8)(2)$
$= 6 + 16 = 22$

64. $\frac{a - c}{2} = \frac{3 - 2}{2} = \frac{1}{2}$

65. $\sqrt{(c - a)^2} = \sqrt{(2 - 3)^2}$
$= \sqrt{(-1)^2}$
$= \sqrt{1} = 1$

Page 19 Practice Quiz 1

1. $\overleftrightarrow{PR}$ **2.** T **3.** $\overrightarrow{PR}$

4.

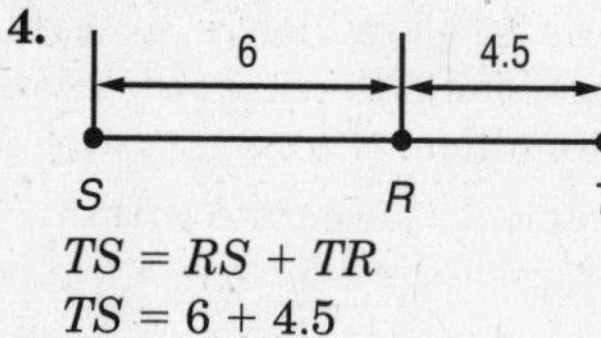

$TS = RS + TR$
$TS = 6 + 4.5$
$TS = 10.5$

5.

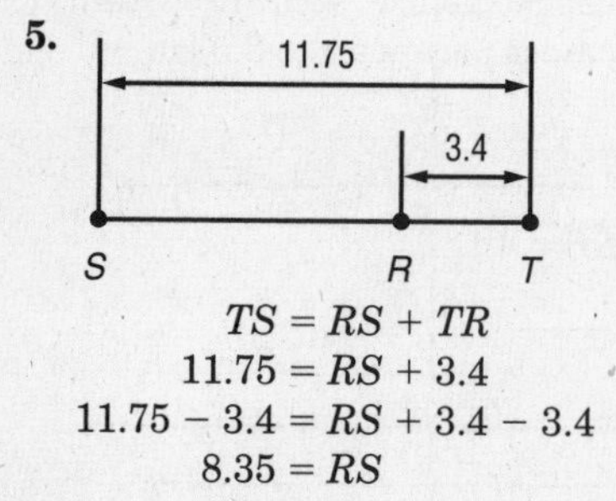

$TS = RS + TR$
$11.75 = RS + 3.4$
$11.75 - 3.4 = RS + 3.4 - 3.4$
$8.35 = RS$

Page 20 Geometry Activity: Probability and Segment Measure

1. $WZ = WX + XY + YZ$
$WZ = 2 + 1 + 3 = 6$
$P(J \text{ lies in } XY) = \frac{XY}{WZ}$
$= \frac{1}{6}$

2. $WZ = WX + XY + YZ$
$WZ = 2 + 1 + 3 = 6$
$P(R \text{ lies in } YZ) = \frac{YZ}{WZ}$
$= \frac{3}{6} = \frac{1}{2}$

3. $WY = WX + XY$
$WY = 2 + 1$
$WY = 3$
$P(S \text{ lies in } XY) = \frac{XY}{WY}$
$= \frac{1}{3}$

4. 1; $\overline{XY}$ contains all points that lie on both $\overline{WY}$ and $\overline{XZ}$.

5. 0; if point U lies on $\overline{WX}$, it cannot lie on $\overline{YZ}$.

1-3 Distance and Midpoints

Page 22 Geometry Activity: Midpoint of a Segment

1.

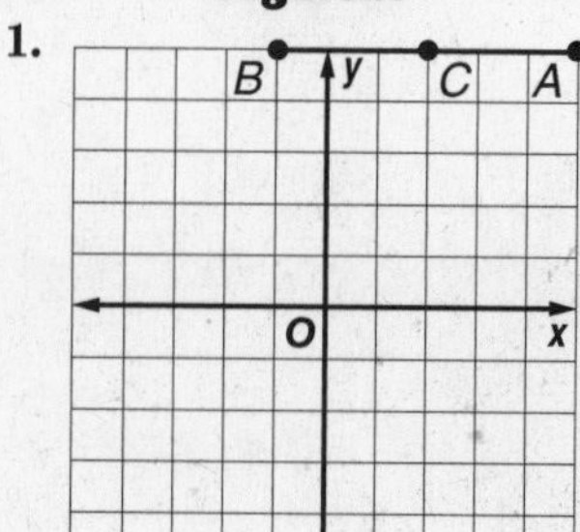

(2, 5)

2. $d = \sqrt{(x_2 - x_1)^2 + (y_2 - y_1)^2}$
$AC = \sqrt{(2 - 5)^2 + (5 - 5)^2}$
$AC = \sqrt{(-3)^2 + 0^2}$
$AC = \sqrt{9}$
$AC = 3$
$\overline{AC} \cong \overline{CB}$, so both are 3 units long.

3.

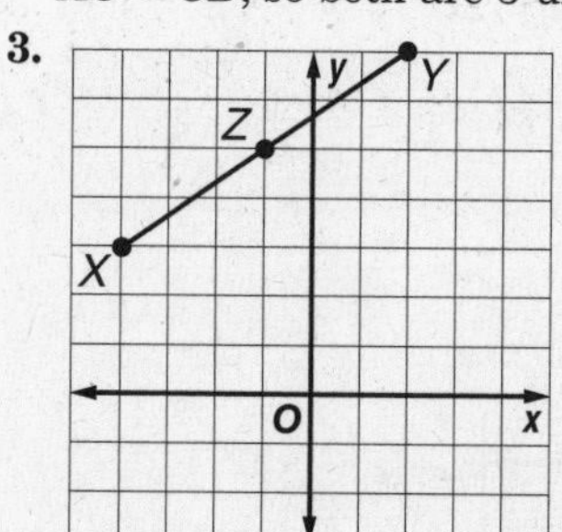

(−1, 5)

4. $d = \sqrt{(x_2 - x_1)^2 + (y_2 - y_1)^2}$
$XZ = \sqrt{[-1 - (-4)]^2 + (5 - 3)^2}$
$XZ = \sqrt{3^2 + 2^2}$
$XZ = \sqrt{13}$
$\overline{XZ} \cong \overline{ZY}$, so both are $\sqrt{13}$ units or about 3.6 units long.

5. Sample answer: The x-coordinate of the midpoint is one half the sum of the x-coordinates of the endpoints. The y-coordinate of the midpoint is one half the sum of the y-coordinates of the endpoints.

Page 25 Check for Understanding

1. Sample answers: (1) Use one of the Midpoint Formulas if you know the coordinates of the endpoints. (2) Draw a segment and fold the paper so that the endpoints match to locate the middle of the segment. (3) Use a compass and straightedge to construct the bisector of the segment.

2. Sample answer:

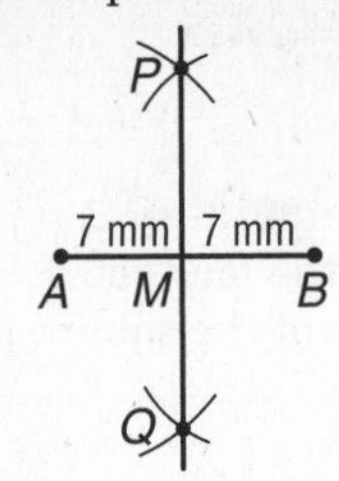

3. $AB = |2 - 10|$
$= |-8|$ or 8

4. $CD = |-3 - 4|$
$= |-7|$ or 7

5.

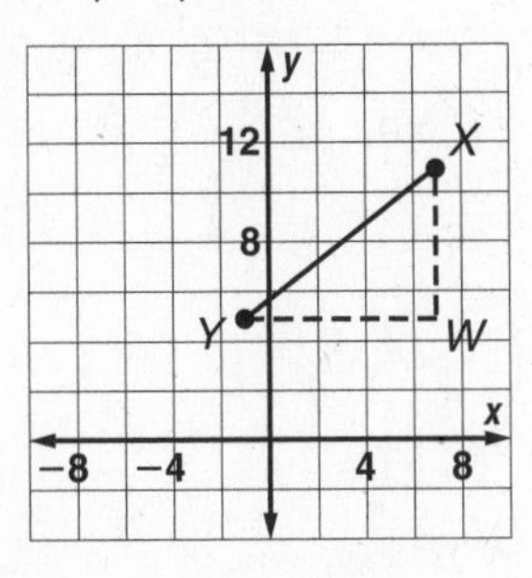

$(XY)^2 = (XW)^2 + (YW)^2$
$(XY)^2 = (6)^2 + (8)^2$
$(XY)^2 = 36 + 64$
$(XY)^2 = 100$
$XY = 10$

6. $d = \sqrt{(x_2 - x_1)^2 + (y_2 - y_1)^2}$
$DE = \sqrt{(8 - 2)^2 + (6 - 0)^2} = \sqrt{6^2 + 6^2}$
$DE = \sqrt{36 + 36}$
$DE = \sqrt{72}$
$DE \approx 8.49$

7. $M = \frac{-10 + (-2)}{2}$
$= \frac{-12}{2}$
$= -6$

8. $M = \frac{-3 + 6}{2}$
$= \frac{3}{2}$ or 1.5

9. $\left(\frac{x_1 + x_2}{2}, \frac{y_1 + y_2}{2}\right) = \left(\frac{-4 + (-1)}{2}, \frac{3 + 5}{2}\right)$
$= \left(-\frac{5}{2}, \frac{8}{2}\right)$ or $(-2.5, 4)$

10. $\left(\frac{x_1 + x_2}{2}, \frac{y_1 + y_2}{2}\right) = \left(\frac{2 + (-2)}{2}, \frac{8 + 2}{2}\right)$
$= (0, 5)$

11. $B(0, 5.5) = B\left(\frac{x_1 + (-3)}{2}, \frac{y_1 + 6}{2}\right)$

$0 = \frac{x_1 + (-3)}{2}$ $\quad$ $5.5 = \frac{y_1 + 6}{2}$
$0 = x_1 + (-3)$ $\quad$ $11 = y_1 + 6$
$3 = x_1$ $\quad$ $5 = y_1$

The coordinates of A are (3, 5)

12. B; $M(7, 8) = M\left(\frac{4 + 2x}{2}, \frac{6 + 2x}{2}\right)$

$7 = \frac{4 + 2x}{2}$ $\quad$ $8 = \frac{6 + 2x}{2}$
$14 = 4 + 2x$ $\quad$ $16 = 6 + 2x$
$10 = 2x$ $\quad$ $10 = 2x$
$5 = x$ $\quad$ $5 = x$

Pages 25–27 Practice and Apply

13. $DE = |2 - 4|$
$= |-2|$ or 2

14. $CF = |0 - 7|$
$= |-7|$ or 7

15. $AB = |-4 - (-1)|$
$= |-3|$ or 3

16. $AC = |-4 - 0|$
$= |-4|$ or 4

17. $AF = |-4 - 7|$
$= |-11|$ or 11

18. $BE = |-1 - 4|$
$= |-5|$ or 5

19.

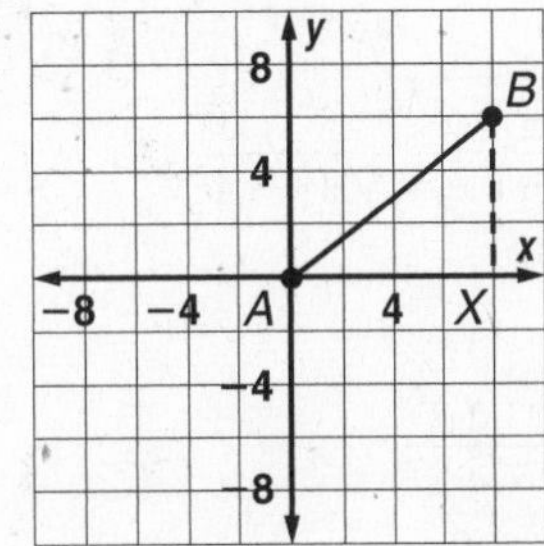

$(AB)^2 = (AX)^2 + (BX)^2$
$(AB)^2 = (8)^2 + (6)^2$
$(AB)^2 = 100$
$AB = 10$

20.

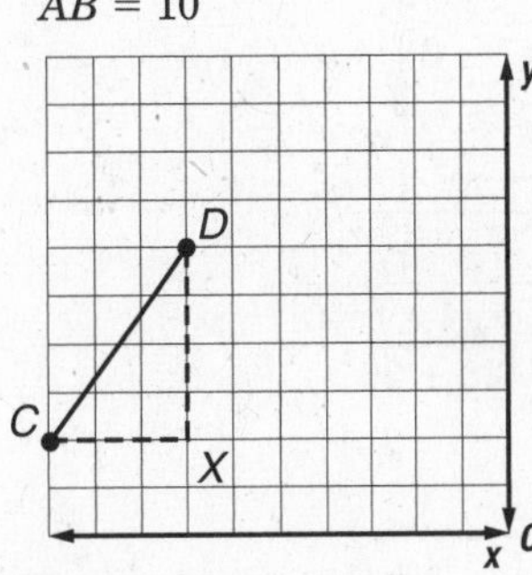

$(CD)^2 = (CX)^2 + (DX)^2$
$(CD)^2 = (3)^2 + (4)^2$
$(CD)^2 = 25$
$CD = 5$

21.

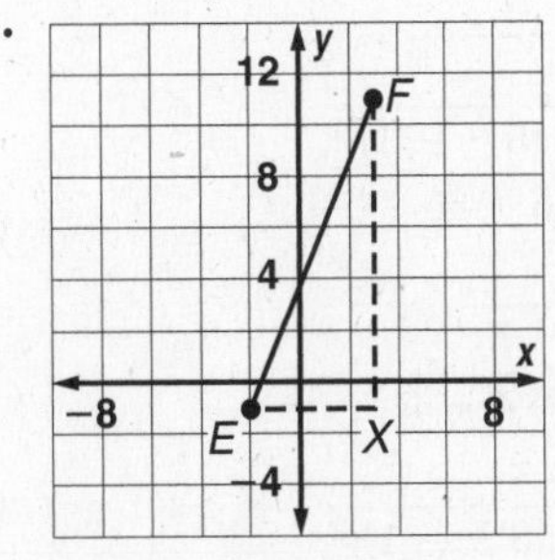

$(EF)^2 = (EX)^2 + (FX)^2$
$(EF)^2 = (5)^2 + (12)^2$
$(EF)^2 = 169$
$EF = 13$

22.

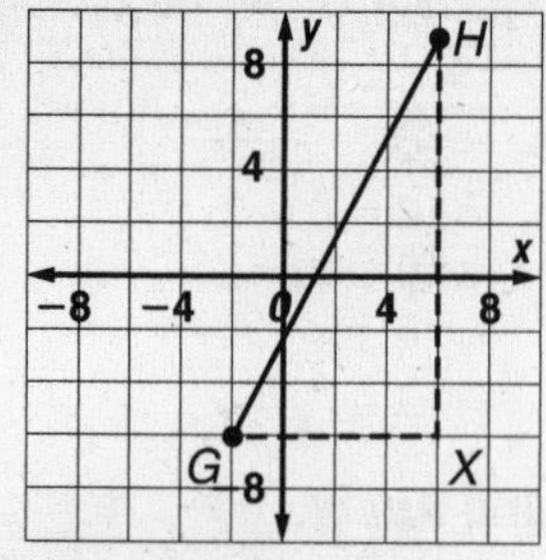

$(GH)^2 = (GX)^2 + (HX)^2$
$(GH)^2 = (8)^2 + (15)^2$
$(GH)^2 = 289$
$GH = 17$

23. $d = \sqrt{(x_2 - x_1)^2 + (y_2 - y_1)^2}$
$JK = \sqrt{(12 - 0)^2 + (9 - 0)^2} = \sqrt{12^2 + 9^2}$
$JK = \sqrt{225} = 15$

24. $d = \sqrt{(x_2 - x_1)^2 + (y_2 - y_1)^2}$
$LM = \sqrt{(7 - 3)^2 + (9 - 5)^2} = \sqrt{4^2 + 4^2}$
$LM = \sqrt{16 + 16} = \sqrt{32}$
$LM \approx 5.7$

25. $d = \sqrt{(x_2 - x_1)^2 + (y_2 - y_1)^2}$
$ST = \sqrt{[6 - (-3)]^2 + (5 - 2)^2}$
$ST = \sqrt{9^2 + 3^2} = \sqrt{81 + 9}$
$ST = \sqrt{90}$
$ST \approx 9.5$

26. $d = \sqrt{(x_2 - x_1)^2 + (y_2 - y_1)^2}$
$UV = \sqrt{(5 - 2)^2 + (7 - 3)^2} = \sqrt{3^2 + 4^2}$
$UV = \sqrt{9 + 16} = \sqrt{25}$
$UV = 5$

27. $d = \sqrt{(x_2 - x_1)^2 + (y_2 - y_1)^2}$
$NP = \sqrt{[3 - (-2)]^2 + [4 - (-2)]^2}$
$NP = \sqrt{5^2 + 6^2} = \sqrt{25 + 36}$
$NP = \sqrt{61}$
$NP \approx 7.8$

28. $d = \sqrt{(x_2 - x_1)^2 + (y_2 - y_1)^2}$
$QR = \sqrt{[1 - (-5)]^2 + (5 - 3)^2}$
$QR = \sqrt{6^2 + 2^2} = \sqrt{36 + 4}$
$QR = \sqrt{40}$
$QR \approx 6.3$

29. $d = \sqrt{(x_2 - x_1)^2 + (y_2 - y_1)^2}$
$XY = \sqrt{[2 - (-2)]^2 + [5 - (-1)]^2}$
$XY = \sqrt{4^2 + 6^2} = \sqrt{16 + 36}$
$XY = \sqrt{52}$
$YZ = \sqrt{(4 - 2)^2 + (3 - 5)^2}$
$YZ = \sqrt{2^2 + (-2)^2} = \sqrt{4 + 4}$
$YZ = \sqrt{8}$
$XZ = \sqrt{[4 - (-2)]^2 + [3 - (-1)]^2}$
$XZ = \sqrt{6^2 + 4^2} = \sqrt{36 + 16}$
$XZ = \sqrt{52}$
$XY + YZ + XZ = \sqrt{52} + \sqrt{8} + \sqrt{52}$
≈ 17.3 units

30. $d = \sqrt{(x_2 - x_1)^2 + (y_2 - y_1)^2}$
$AB = \sqrt{[-5 - (-4)]^2 + [1 - (-3)]^2}$
$AB = \sqrt{(-1)^2 + 4^2} = \sqrt{1 + 16}$
$AB = \sqrt{17}$
Because the figure is a square, the four sides are congruent. So the perimeter of the square is $4\sqrt{17} \approx 16.5$ units.

31. $M = \frac{-6 + 0}{2} = \frac{-6}{2} = -3$

32. $M = \frac{2 + 8}{2} = \frac{10}{2} = 5$

33. $M = \frac{0 + 5}{2} = \frac{5}{2}$ or 2.5

34. $M = \frac{-3 + 2}{2} = \frac{-1}{2}$ or −0.5

35. $M = \frac{-6 + 8}{2} = \frac{2}{2} = 1$

36. $M = \frac{-3 + 5}{2} = \frac{2}{2} = 1$

37. $\left(\frac{x_1 + x_2}{2}, \frac{y_1 + y_2}{2}\right) = \left(\frac{8 + 12}{2}, \frac{4 + 2}{2}\right) = \left(\frac{20}{2}, \frac{6}{2}\right) = (10, 3)$

38. $\left(\frac{x_1 + x_2}{2}, \frac{y_1 + y_2}{2}\right) = \left(\frac{9 + 17}{2}, \frac{5 + 4}{2}\right) = \left(\frac{26}{2}, \frac{9}{2}\right) = (13, 4.5)$

39. $\left(\frac{x_1 + x_2}{2}, \frac{y_1 + y_2}{2}\right) = \left(\frac{-11 + (-9)}{2}, \frac{-4 + (-2)}{2}\right) = \left(\frac{-20}{2}, \frac{-6}{2}\right) = (-10, -3)$

40. $\left(\frac{x_1 + x_2}{2}, \frac{y_1 + y_2}{2}\right) = \left(\frac{4 + 8}{2}, \frac{2 + (-6)}{2}\right) = \left(\frac{12}{2}, \frac{-4}{2}\right) = (6, -2)$

41. $\left(\frac{x_1 + x_2}{2}, \frac{y_1 + y_2}{2}\right) = \left(\frac{3.4 + 7.8}{2}, \frac{2.1 + 3.6}{2}\right) = \left(\frac{11.2}{2}, \frac{5.7}{2}\right) = (5.6, 2.85)$

42. $\left(\frac{x_1 + x_2}{2}, \frac{y_1 + y_2}{2}\right) = \left(\frac{-1.4 + 2.6}{2}, \frac{3.2 + (-5.4)}{2}\right) = \left(\frac{1.2}{2}, \frac{-2.2}{2}\right) = (0.6, -1.1)$

43. $S(-1, 5) = \left(\frac{x_1 + (-4)}{2}, \frac{y_1 + 3}{2}\right)$
$-1 = \frac{x_1 + (-4)}{2}$ $\quad 5 = \frac{y_1 + 3}{2}$
$-2 = x_1 + (-4)$ $\quad 10 = y_1 + 3$
$2 = x_1$ $\quad 7 = y_1$
The coordinates of R are (2, 7).

44. $S(-2, 2) = \left(\frac{x_1 + 2}{2}, \frac{y_1 + 8}{2}\right)$
$-2 = \frac{x_1 + 2}{2}$ $\quad 2 = \frac{y_1 + 8}{2}$
$-4 = x_1 + 2$ $\quad 4 = y_1 + 8$
$-6 = x_1$ $\quad -4 = y_1$
The coordinates of R are (−6, −4).

45. $S\left(\frac{5}{3}, 3\right) = \left(\frac{\frac{2}{3} + x_2}{2}, \frac{-5 + y_2}{2}\right)$

$\frac{5}{3} = \frac{\frac{2}{3} + x_2}{2}$ $\quad$ $3 = \frac{-5 + y_2}{2}$

$\frac{10}{3} = \frac{2}{3} + x_2$ $\quad$ $6 = -5 + y_2$

$\frac{8}{3} = x_2$ $\quad$ $11 = y_2$

The coordinates of T are $\left(\frac{8}{3}, 11\right)$.

46. $M(31.1, 99.3) = \left(\frac{x_1 + 31.8}{2}, \frac{y_1 + 106.4}{2}\right)$

$31.1 = \frac{x_1 + 31.8}{2}$ $\quad$ $99.3 = \frac{y_1 + 106.4}{2}$

$62.2 = x_1 + 31.8$ $\quad$ $198.6 = y_1 + 106.4$

$30.4 = x_1$ $\quad$ $92.2 = y_1$

The other endpoint is at (30.4°, 92.2°).

47. LaFayette, LA is near (30.4°, 92.2°).

48. Sample answer: = SQRT((A2 – C2) ∧ 2 + (B2 – D2) ∧ 2)

49a. $\sqrt{(54 - 113)^2 + (120 - 215)^2} \approx 111.8$

49b. $\sqrt{(68 - 175)^2 + (153 - 336)^2} \approx 212.0$

49c. $\sqrt{(421 - 502)^2 + (454 - 798)^2} \approx 353.4$

49d. $\sqrt{(837 - 612)^2 + (980 - 625)^2} \approx 420.3$

49e. $\sqrt{(1967 - 1998)^2 + (3 - 24)^2} \approx 37.4$

49f. $\sqrt{(4173.5 - 2080.6)^2 + (34.9 - 22.4)^2} \approx 2092.9$

50. $d = \sqrt{(x_2 - x_1)^2 + (y_2 - y_1)^2}$

$AB = \sqrt{(9-1)^2 + (10 - 3)^2}$

$AB = \sqrt{8^2 + 7^2} = \sqrt{64 + 49}$

$AB = \sqrt{113}$

$BC = \sqrt{(11 - 9)^2 + (18 - 10)^2}$

$BC = \sqrt{2^2 + 8^2} = \sqrt{4 + 64}$

$BC = \sqrt{68}$

$AC = \sqrt{(11 - 1)^2 + (18 - 3)^2}$

$AC = \sqrt{10^2 + 15^2} = \sqrt{100 + 225}$

$AC = \sqrt{325}$

The perimeter of $\triangle ABC$ is $\sqrt{113} + \sqrt{68} + \sqrt{325}$, which is approximately 36.9 units.

51. The new coordinates are $A'(2, 6)$, $B'(18, 20)$, $C'(22, 36)$.

$d = \sqrt{(x_2 - x_1)^2 + (y_2 - y_1)^2}$

$A'B' = \sqrt{(18 - 2)^2 + (20 - 6)^2}$

$A'B' = \sqrt{16^2 + 14^2} = \sqrt{256 + 196}$

$A'B' = \sqrt{452}$

$B'C' = \sqrt{(22 - 18)^2 + (36 - 20)^2}$

$B'C' = \sqrt{4^2 + 16^2} = \sqrt{16 + 256}$

$B'C' = \sqrt{272}$

$A'C' = \sqrt{(22 - 2)^2 + (36 - 6)^2}$

$A'C' = \sqrt{20^2 + 30^2} = \sqrt{400 + 900}$

$A'C' = \sqrt{1300}$

The perimeter of $\triangle A'B'C'$ is $\sqrt{452} + \sqrt{272} + \sqrt{1300}$, which is approximately 73.8 units.

52. The new coordinates are $A''(3, 9)$, $B''(27, 30)$, and $C''(33, 54)$.

$d = \sqrt{(x_2 - x_1)^2 + (y_2 - y_1)^2}$

$A''B'' = \sqrt{(27 - 3)^2 + (30 - 9)^2}$

$A''B'' = \sqrt{24^2 + 21^2} = \sqrt{576 + 441}$

$A''B'' = \sqrt{1017}$

$B''C'' = \sqrt{(33 - 27)^2 + (54 - 30)^2}$

$B''C'' = \sqrt{6^2 + 24^2} = \sqrt{36 + 576}$

$B''C'' = \sqrt{612}$

$A''C'' = \sqrt{(33 - 3)^2 + (54 - 9)^2}$

$A''C'' = \sqrt{30^2 + 45^2} = \sqrt{900 + 2025}$

$A''C'' = \sqrt{2925}$

The perimeter of $\triangle A''B''C''$ is $\sqrt{1017} + \sqrt{612} + \sqrt{2925}$, which is approximately 110.7 units.

53. Sample answer: The perimeter increases by the same factor.

54a. $F\left(\frac{x_1 + x_2}{2}, \frac{y_1 + y_2}{2}\right) = F\left(\frac{2 + 6}{2}, \frac{6 + 6}{2}\right)$

$= F\left(\frac{8}{2}, \frac{12}{2}\right)$ or $F(4, 6)$

$E\left(\frac{x_1 + x_2}{2}, \frac{y_1 + y_2}{2}\right) = E\left(\frac{6 + 6}{2}, \frac{6 + 2}{2}\right)$

$= E\left(\frac{12}{2}, \frac{8}{2}\right)$ or $E(6, 4)$

54b. $G(4, 4)$; it has the same x-coordinate as F and the same y-coordinate as E.

54c. $\overline{DG} \cong \overline{GB}$; use the Distance Formula to show $DG = GB$.

$d = \sqrt{(x_2 - x_1)^2 + (y_2 - y_1)^2}$

$DG = \sqrt{(4 - 2)^2 + (4 - 2)^2}$

$DG = \sqrt{2^2 + 2^2} = \sqrt{4 + 4}$

$DG = \sqrt{8}$

$GB = \sqrt{(6 - 4)^2 + (6 - 4)^2}$

$GB = \sqrt{2^2 + 2^2} = \sqrt{4 + 4}$

$GB = \sqrt{8}$

Thus, $DG = GB$.

55. $\frac{1}{4}(x_2 - x_1) = \frac{1}{4}[5 - (-3)]$

$= \frac{1}{4}(8) = 2$

$\frac{1}{4}(y_2 - y_1) = \frac{1}{4}[12 - (-8)]$

$= \frac{1}{4}(20) = 5$

$(-3 + 2, -8 + 5) = (-1, -3)$

Verify that the coordinates of X are $(-1, -3)$:

$WX = \sqrt{[-1 - (-3)]^2 + [-3 - (-8)]^2}$

$WX = \sqrt{2^2 + 5^2}$

$WX = \sqrt{4 + 25}$

$WX = \sqrt{29}$

$WZ = \sqrt{[5 - (-3)]^2 + [12 - (-8)]^2}$

$WZ = \sqrt{8^2 + 20^2}$

$WZ = \sqrt{64 + 400}$

$WZ = \sqrt{464}$

$\sqrt{29} \approx 5.385$

$\sqrt{464} \approx 21.540$

$5.385 = \frac{1}{4}(21.540)$, so $WX = \frac{1}{4}WZ$.

56. Sample answer: You can copy the segment onto a coordinate plane and then use either the Pythagorean Theorem or the Distance Formula to find its length. Answers should include the following.

- To use the Pythagorean Theorem, draw a vertical segment from one endpoint and a horizontal segment from the other endpoint to form a triangle. Use the measures of these segments as a and b in the formula $a^2 + b^2 = c^2$. Then solve for c. To use the Distance Formula, assign the coordinates of the endpoints of the segment as (x_1, y_1) and (x_2, y_2). Then use them in $d = \sqrt{(x_2 - x_1)^2 + (y_2 - y_1)^2}$ to find the length of the segment.
- $\sqrt{61} \approx 7.8$ units

57. B; $d = \sqrt{(x_2 - x_1)^2 + (y_2 - y_1)^2}$
$d = \sqrt{(-2 - 6)^2 + (-4 - 11)^2}$
$d = \sqrt{(-8)^2 + (-15)^2}$
$d = \sqrt{64 + 225}$
$d = \sqrt{289}$
$d = 17$

58. A

Page 27 Maintain Your Skills

59. $WY = WX + XY$
$WY = 1\frac{3}{4} + 2\frac{1}{2}$
$WY = 4\frac{1}{4}$
$\overline{WY}$ is $4\frac{1}{4}$ in. long.

60. $AC = AB + BC$
$8.5 = 3 + BC$
$5.5 = BC$
$\overline{BC}$ is 5.5 cm long.

61. Sample answer:

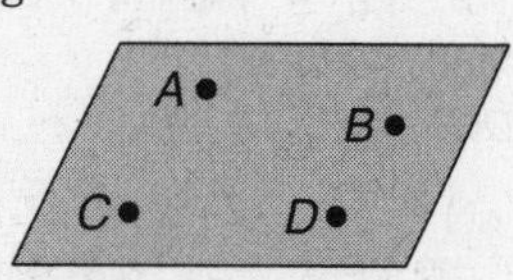

62.

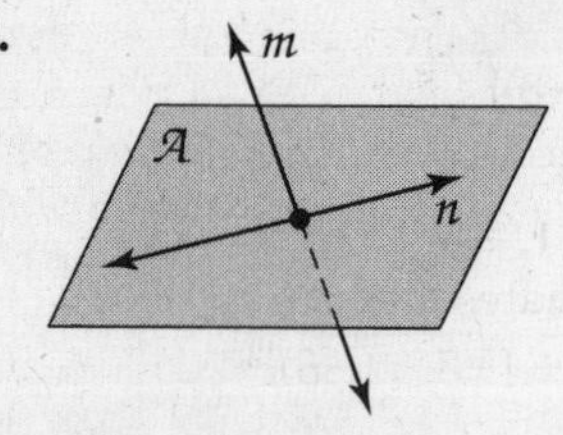

63. $2k = 5k - 30$
$-3k = -30$
$k = 10$

64. $14x - 31 = 12x + 8$
$14x = 12x + 39$
$2x = 39$
$x = \frac{39}{2}$ or 19.5

65. $180 - 8t = 90 + 2t$
$90 - 8t = 2t$
$90 = 10t$
$9 = t$

66. $12m + 7 = 3m + 52$
$12m = 3m + 45$
$9m = 45$
$m = 5$

67. $8x + 7 = 5x + 20$
$8x = 5x + 13$
$3x = 13$
$x = \frac{13}{3}$

68. $13n - 18 = 5n + 32$
$13n = 5n + 50$
$8n = 50$
$n = \frac{50}{8}$ or 6.25

Page 28 Geometry Activity: Modeling the Pythagorean Theorem

1. 25, 144, 169
2. 25 + 144 = 169
3. a^2, b^2, c^2
4. The formula for the Pythagorean Theorem can be expressed as $a^2 + b^2 = c^2$.
5. All of these fit the $a^2 + b^2 = c^2$ pattern.
6. The number of grid squares is $5^2 + 5^2$, which is 50 grid squares.

1-4 Angle Measure

Page 32 Geometry Activity: Bisect an Angle

1. They are congruent.
2. See students' work.
3. A segment bisector separates a segment into two congruent segments; an angle bisector separates an angle into two congruent angles.

Page 33 Check for Understanding

1. Yes; they all have the same measure.
2. Sample answer:

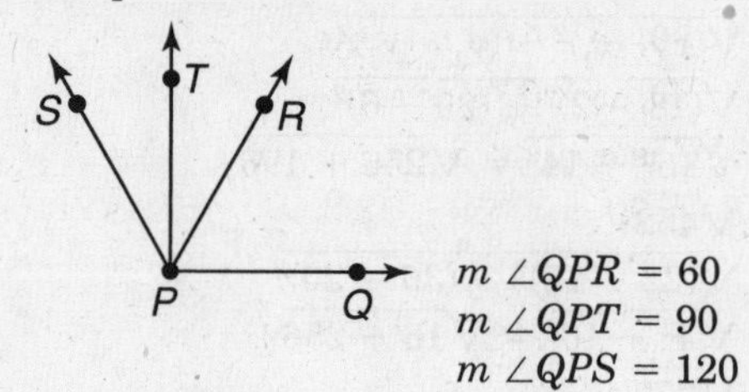

$m\angle QPR = 60$
$m\angle QPT = 90$
$m\angle QPS = 120$

3. $m\angle A = m\angle Z$
4. C
5. $\overrightarrow{BA}, \overrightarrow{BC}$
6. $\angle CDB, \angle 1$
7. 135°; $135 > 90$ and $135 < 180$ so $\angle WXY$ is obtuse.
8. 45°; $45 < 90$ so $\angle WXZ$ is acute.

9. $\overrightarrow{QT}$ bisects $\angle RQS$, so $\angle RQT \cong \angle SQT$.

$$\begin{aligned} m\angle RQT &= m\angle SQT \\ 6x + 5 &= 7x - 2 \\ 6x + 7 &= 7x \\ 7 &= x \\ m\angle RQT &= 6x + 5 \\ &= 6(7) + 5 \\ &= 42 + 5 \text{ or } 47 \end{aligned}$$

10. $\overrightarrow{QT}$ bisects $\angle RQS$, so $\angle RQT \cong \angle TQS$ and $m\angle RQS = 2 \cdot m\angle RQT$.

$$\begin{aligned} 22a - 11 &= 2(12a - 8) \\ 22a - 11 &= 24a - 16 \\ 22a + 5 &= 24a \\ 5 &= 2a \\ \tfrac{5}{2} &= a \\ m\angle TQS &= m\angle RQT \\ &= 12a - 8 \\ &= 12\left(\tfrac{5}{2}\right) - 8 \\ &= 30 - 8 \text{ or } 22 \end{aligned}$$

11. $\angle 1$, right; $\angle 2$, acute; $\angle 3$, obtuse

Pages 34–35 Practice and Apply

12. E **13.** B **14.** A

15. A **16.** $\overline{DA}$, $\overrightarrow{DB}$ **17.** $\overline{AB}$, $\overrightarrow{AD}$

18. $\overrightarrow{ED}$, $\overrightarrow{EG}$ **19.** $\overrightarrow{AD}$, $\overrightarrow{AE}$ **20.** $\angle ABC$, $\angle CBA$

21. $\angle FEA$, $\angle 4$

22. $\angle 2$, $\angle DBA$, $\angle EBA$, $\angle ABE$, $\angle FBA$, $\angle ABF$

23. $\angle AED$, $\angle DEA$, $\angle AEB$, $\angle BEA$, $\angle AEC$, $\angle CEA$

24. D, H **25.** $\angle 2$

26. Sample answer: $\angle 4$, $\angle 3$

27. $\overrightarrow{AD}$ bisects $\angle EAB$ so $\angle 5 \cong \angle 6$.

$$\begin{aligned} m\angle 5 &= \tfrac{1}{2} m\angle EAB \\ &= \tfrac{1}{2}(60) \\ &= 30 \\ m\angle 6 &= m\angle 5 \\ &= 30 \end{aligned}$$

28. $\angle BFD$ is marked with a right angle symbol, so $m\angle BFD = 90$; $\angle BFD$ is a right angle.

29. 60°; $60 < 90$, so $\angle AFB$ is acute.

30. 30°; $30 < 90$, so $\angle DFE$ is acute.

31. 90°; $\angle EFC$ is a right angle.

32. 150°; $150 > 90$ and $150 < 180$, so $\angle AFD$ is obtuse.

33. 120°; $120 > 90$ and $120 < 180$, so $\angle EFB$ is obtuse.

34. $\overrightarrow{YU}$ bisects $\angle ZYW$, so $\angle ZYU \cong \angle UYW$.

$$\begin{aligned} m\angle ZYU &= m\angle UYW \\ 8p - 10 &= 10p - 20 \\ 8p + 10 &= 10p \\ 10 &= 2p \\ 5 &= p \\ m\angle ZYU &= 8p - 10 \\ &= 8(5) - 10 \\ &= 40 - 10 \text{ or } 30 \end{aligned}$$

35. $\overrightarrow{YT}$ bisects $\angle XYW$, so $\angle 1 \cong \angle 2$.

$$\begin{aligned} m\angle 1 &= m\angle 2 \\ 5x + 10 &= 8x - 23 \\ 5x + 33 &= 8x \\ 33 &= 3x \\ 11 &= x \\ m\angle 2 &= 8x - 23 \\ &= 8(11) - 23 \\ &= 88 - 23 \\ &= 65 \end{aligned}$$

36. $\overrightarrow{YT}$ bisects $\angle XYW$, so $m\angle XYW = 2 \cdot m\angle 1$.

$$\begin{aligned} 6y - 24 &= 2y \\ -24 &= -4y \\ 6 &= y \end{aligned}$$

37. $\overrightarrow{YU}$ bisects $\angle WYZ$, so $m\angle WYZ = 2 \cdot m\angle ZYU$.

$$\begin{aligned} 82 &= 2(4r + 25) \\ 82 &= 8r + 50 \\ 32 &= 8r \\ 4 &= r \end{aligned}$$

38. $\overrightarrow{YU}$ bisects $\angle ZYW$, so $\angle ZYU \cong \angle UYW$, and $m\angle ZYU = m\angle UYW$.

$\overrightarrow{YX}$ and $\overrightarrow{YZ}$ are opposite rays, so $m\angle WYX + m\angle UYW + m\angle ZYU = 180$.

$$\begin{aligned} 2(12b + 7) + 9b - 1 + 9b - 1 &= 180 \\ 24b + 14 + 18b - 2 &= 180 \\ 42b + 12 &= 180 \\ 42b &= 168 \\ b &= 4 \end{aligned}$$

$$\begin{aligned} m\angle UYW &= m\angle ZYU \\ &= 9b - 1 \\ &= 9(4) - 1 \\ &= 36 - 1 \text{ or } 35 \end{aligned}$$

39. $\overrightarrow{YU}$ bisects $\angle ZYW$, so $m\angle ZYU = \frac{1}{2} \cdot m\angle ZYW$.

$$\begin{aligned} m\angle ZYU &= \tfrac{1}{2} \cdot m\angle ZYW \\ 13a - 7 &= \tfrac{1}{2}(90) \\ 13a - 7 &= 45 \\ 13a &= 52 \\ a &= 4 \end{aligned}$$

40. The angle at which the dogs must turn to get the scent of the article they wish to find is an acute angle.

41. Sample answer: *Acute* can mean something that is sharp or having a very fine tip like a pen, a knife, or a needle. *Obtuse* means not pointed or blunt, so something that is obtuse would be wide.

42.

$$\begin{aligned} m\angle 1 &= \tfrac{360}{6} = 60 \\ m\angle 2 &= \tfrac{360}{12} = 30 \\ m\angle 3 &= \tfrac{360}{4} = 90 \\ m\angle 4 &= \tfrac{360}{6} = 60 \\ m\angle 5 &= \tfrac{360}{3} = 120 \\ m\angle 6 &= \tfrac{360}{6} = 60 \end{aligned}$$

43. $m(\text{angle of reflection}) = \frac{1}{2} \cdot m\angle IBR$
$= \frac{1}{2}(62)$
$= 31$
$\overrightarrow{BN}$ is at a right angle to the barrier.
So $m\angle IBA = 90 - 31$ or 59.

44. You can only compare the measures of the angles. The arcs indicate both measures are the same regardless of the length of the rays.

45. 1, 3, 6, 10, 15

46. 3 rays: $(3 \times 2) \div 2 = 3$ angles;
4 rays: $(4 \times 3) \div 2 = 6$ angles;
5 rays: $(5 \times 4) \div 2 = 10$ angles;
6 rays: $(6 \times 5) \div 2 = 15$ angles

47. 7 rays: $(7 \times 6) \div 2 = 21$ angles
10 rays: $(10 \times 9) \div 2 = 45$ angles

48. $a = \frac{n(n-1)}{2}$, for a = number of angles and n = number of rays

49. Sample answer: A degree is $\frac{1}{360}$ of a circle. Answers should include the following.
- Place one side of the angle to coincide with 0 on the protractor and the vertex of the angle at the center point of the protractor. Observe the point at which the other side of the angle intersects the scale of the protractor.
- See students' work.

50. D

51. C; $5n + 4 = 7(n + 1) - 2n$
$5n + 4 = 7n + 7 - 2n$
$5n + 4 = 5n + 7$
$4 \neq 7$

Page 36 Maintain Your Skills

52. $d = \sqrt{(x_2 - x_1)^2 + (y_2 - y_1)^2}$
$AB = \sqrt{(5-2)^2 + (7-3)^2}$
$AB = \sqrt{3^2 + 4^2}$
$AB = \sqrt{9 + 16} = \sqrt{25}$
$AB = 5$
$M = \left(\frac{x_1 + x_2}{2}, \frac{y_1 + y_2}{2}\right)$
$= \left(\frac{2+5}{2}, \frac{3+7}{2}\right)$
$= \left(\frac{7}{2}, \frac{10}{2}\right)$ or (3.5, 5)

53. $d = \sqrt{(x_2 - x_1)^2 + (y_2 - y_1)^2}$
$CD = \sqrt{[6-(-2)]^2 + (4-0)^2}$
$CD = \sqrt{8^2 + 4^2} = \sqrt{64 + 16}$
$CD = \sqrt{80}$
$CD \approx 8.9$
$M = \left(\frac{x_1 + x_2}{2}, \frac{y_1 + y_2}{2}\right)$
$= \left(\frac{-2+6}{2}, \frac{0+4}{2}\right)$
$= \left(\frac{4}{2}, \frac{4}{2}\right)$ or (2, 2)

54. $d = \sqrt{(x_2 - x_1)^2 + (y_2 - y_1)^2}$
$EF = \sqrt{[5-(-3)]^2 + [8-(-2)]^2}$
$EF = \sqrt{8^2 + 10^2} = \sqrt{64 + 100}$
$EF = \sqrt{164}$
$EF \approx 12.8$
$M = \left(\frac{x_1 + x_2}{2}, \frac{y_1 + y_2}{2}\right)$
$= \left(\frac{-3+5}{2}, \frac{-2+8}{2}\right)$
$= \left(\frac{2}{2}, \frac{6}{2}\right)$ or (1, 3)

55. $WX = WR + RX$
$WX = 3\frac{5}{12} + 6\frac{1}{4}$
$WX = 9\frac{8}{12}$ or $9\frac{2}{3}$
$\overline{WX}$ is $9\frac{2}{3}$ ft long.

56. $XZ = XY + YZ$
$15.1 = 3.7 + YZ$
$11.4 = YZ$
$\overline{YZ}$ is 11.4 mm long.

57.
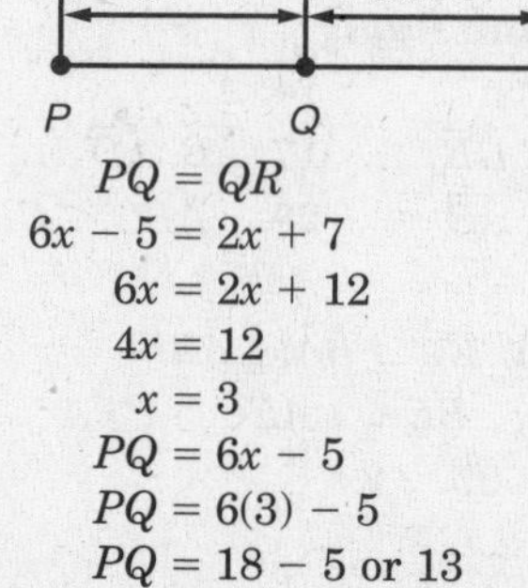

$PQ = QR$
$6x - 5 = 2x + 7$
$6x = 2x + 12$
$4x = 12$
$x = 3$
$PQ = 6x - 5$
$PQ = 6(3) - 5$
$PQ = 18 - 5$ or 13

58. Five planes are shown: plane *FJK*, plane *HJK*, plane *GHK*, plane *FGK*, and plane *FGH*.

59. *F*, *L*, *J*

60. *G* or *L*

61. $14x + (6x - 10) = 90$
$20x - 10 = 90$
$20x = 100$
$x = 5$

62. $2k + 30 = 180$
$2k = 150$
$k = 75$

63. $180 - 5y = 90 - 7y$
$90 - 5y = -7y$
$90 = -2y$
$-45 = y$

64. $90 - 4t = \frac{1}{4}(180 - t)$
$4(90 - 4t) = 180 - t$
$360 - 16t = 180 - t$
$180 - 16t = -t$
$180 = 15t$
$12 = t$

65. $(6m + 8) + (3m + 10) = 90$
$9m + 18 = 90$
$9m = 72$
$m = 8$

66. $(7n - 9) + (5n + 45) = 180$
$12n + 36 = 180$
$12n = 144$
$n = 12$

Page 36 Practice Quiz 2

1. $M = \left(\frac{x_1 + x_2}{2}, \frac{y_1 + y_2}{2}\right)$

$= \left(\frac{3 + (-4)}{2}, \frac{-1 + 3}{2}\right)$

$= \left(\frac{-1}{2}, \frac{2}{2}\right)$ or $\left(-\frac{1}{2}, 1\right)$

$d = \sqrt{(x_2 - x_1)^2 + (y_2 - y_1)^2}$

$AB = \sqrt{(-4 - 3)^2 + [3 - (-1)]^2}$

$AB = \sqrt{(-7)^2 + 4^2} = \sqrt{49 + 16}$

$AB = \sqrt{65}$

$AB \approx 8.1$

2. $M = \left(\frac{x_1 + x_2}{2}, \frac{y_1 + y_2}{2}\right)$

$= \left(\frac{6 + 2}{2}, \frac{4 + (-8)}{2}\right)$

$= \left(\frac{8}{2}, \frac{-4}{2}\right)$ or $(4, -2)$

$d = \sqrt{(x_2 - x_1)^2 + (y_2 - y_1)^2}$

$CD = \sqrt{(2 - 6)^2 + (-8 - 4)^2}$

$CD = \sqrt{(-4)^2 + (-12)^2} = \sqrt{16 + 144}$

$CD = \sqrt{160}$

$CD \approx 12.6$

3. $M = \left(\frac{x_1 + x_2}{2}, \frac{y_1 + y_2}{2}\right)$

$= \left(\frac{10 + (-10)}{2}, \frac{20 + (-20)}{2}\right)$

$= (0, 0)$

$d = \sqrt{(x_2 - x_1)^2 + (y_2 - y_1)^2}$

$EF = \sqrt{(-10 - 10)^2 + (-20 - 20)^2}$

$EF = \sqrt{(-20)^2 + (-40)^2}$

$EF = \sqrt{400 + 1600}$

$EF = \sqrt{2000}$

$EF \approx 44.7$

4. $m\angle RXT = m\angle SXT + m\angle RXS$

$111 = 3a - 4 + 2a + 5$

$111 = 5a + 1$

$110 = 5a$

$22 = a$

$m\angle RXS = 2a + 5$

$= 2(22) + 5$

$= 44 + 5$ or 49

5. $m\angle QXS = m\angle QXR + m\angle RXS$

$4a - 1 = a + 10 + 91$

$4a - 1 = a + 101$

$4a = a + 102$

$3a = 102$

$a = 34$

$m\angle QXS = 4a - 1$

$= 4(34) - 1$

$= 136 - 1$ or 135

1-5 Angle Relationships

Page 38 Geometry Activity: Angle Relationships

1. $\angle BCE \cong \angle DCA$
2. $\angle DCB \cong \angle ACE$
3. See students' work.
4. $\angle ACD$ and $\angle ECB$, $\angle DCB$ and $\angle ACE$; measures for each pair of vertical angles should be the same.
5. $\angle ACD$ and $\angle DCB$, $\angle DCB$ and $\angle BCE$, $\angle BCE$ and $\angle ECA$, $\angle ECA$ and $\angle ACD$; measures for each linear pair should add to 180.
6. Sample answers: The measures of vertical angles are equal or vertical angles are congruent.
   The sum of the measures of a linear pair is 180 or angles that form a linear pair are supplementary.

Page 41 Check for Understanding

1.

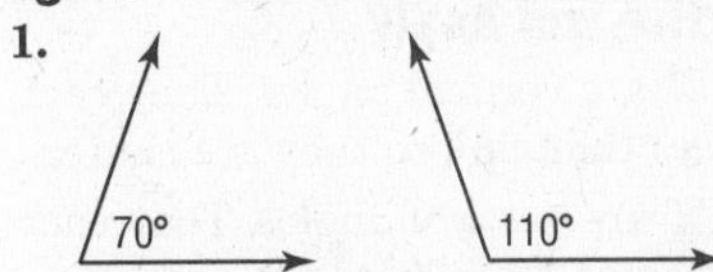

2. Sample answer: When two angles form a linear pair, then their noncommon sides form a straight angle, which measures 180. When the sum of the measures of two angles is 180, then the angles are supplementary.
3. Sample answer: The noncommon sides of a linear pair of angles form a straight line.
4. Sample answer: $\angle ABF$ and $\angle CBD$ are vertical angles. They each have measures less than 90°, so they are acute.
5. Sample answer: $\angle ABC$ and $\angle CBE$ are adjacent angles. They each have measures greater than 90°, so they are obtuse.
6. **Explore:** The problem involves three angles: an angle, its supplement, and its complement.

   Plan: Let $\angle A$ be the given angle, $\angle B$ its supplement and $\angle C$ the complement. Then $m\angle A + m\angle B = 180$ or $m\angle B = 180 - m\angle A$, and $m\angle A + m\angle C = 90$ or $m\angle C = 90 - m\angle A$. The problem states that $m\angle B = 3 \cdot m\angle C - 60$, so substitute for $m\angle B$ and $m\angle C$ and solve for $m\angle A$.

   Solve: $m\angle B = 3 \cdot m\angle C - 60$

   $180 - m\angle A = 3(90 - m\angle A) - 60$

   $180 - m\angle A = 270 - 3 \cdot m\angle A - 60$

   $180 + 2 \cdot m\angle A = 210$

   $2 \cdot m\angle A = 30$

   $m\angle A = 15$

   Examine: Check to see if the answer satisfies the problem.

   The measure of the complement of an angle with a measure of 15° is 75°. The measure of the supplement of the original angle is 165°.

   $3 \cdot 75 - 60 \stackrel{?}{=} 165$

   $165 = 165$

   The answer checks.

7. Lines p and q are perpendicular if angles 1 and 2 are both right angles. Then $m\angle 1 = m\angle 2 = 90$. So

$3x + 18 = 90$ $\quad$ $-8y - 70 = 90$
$3x = 72$ $\quad$ $-8y = 160$
$x = 24$ $\quad$ $y = -20$

8. No; while $\angle SRT$ appears to be a right angle, no information verifies this.
9. Yes; they share a common side and vertex, so they are adjacent. Since $\overline{PR}$ falls between $\overline{PQ}$ and $\overline{PS}$, $m\angle QPR < 90$, so the two angles cannot be complementary or supplementary.
10. $m\angle 4 = 60$ and $\angle 2$ and $\angle 4$ are vertical angles, so $m\angle 2 = 60$.
$\angle 1$ and $\angle 4$ are supplementary angles.
$m\angle 1 + m\angle 4 = 180$
$m\angle 1 + 60 = 180$
$m\angle 1 = 120$
$\angle 1$ and $\angle 3$ are vertical angles so $m\angle 3 = 120$.

Pages 42–43 Practice and Apply

11. $\angle WUT$ and $\angle VUX$ are vertical angles. They each have measures less than 90°, so they are acute.
12. $\angle WUV$ and $\angle XUT$ are vertical angles. They each have measures greater than 90°, so they are obtuse.
13. $\angle ZWU$ is a right angle and $\angle ZWU$ and $\angle YWU$ are supplementary so $\angle YWU$ is a right angle. Then $\angle UWT$ and $\angle TWY$ are adjacent angles that are complementary because $m\angle UWT + m\angle TWY = m\angle YWU$.
14. $\angle VXU$ and $\angle WYT$ are nonadjacent angles that are complementary because $m\angle VXU + m\angle WYT = 60 + 30$ or 90.
15. $\angle WTY$ and $\angle WTU$ is a linear pair with vertex T.
16. $\angle UVX$
17. **Explore:** The problem relates the measures formed by perpendicular rays.
Plan: $\overrightarrow{QP} \perp \overrightarrow{QR}$, so $\angle PQS$ and $\angle SQR$ are complementary.
$m\angle PQS + m\angle SQR = 90$

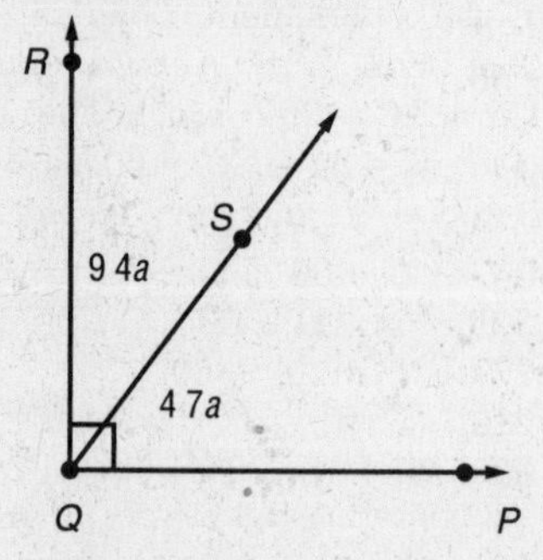

Solve: $4 + 7a + 9 + 4a = 90$
$13 + 11a = 90$
$11a = 77$
$a = 7$
$m\angle PQS = 4 + 7a$
$= 4 + 7(7)$
$= 4 + 49$ or 53
$m\angle SQR = 9 + 4a$
$= 9 + 4(7)$
$= 9 + 28$ or 37
Examine: Add the angle measures to verify their sum is 90.
$53 + 37 = 90$

18. **Explore:** The problem relates the measures of two complementary angles. You know that the sum of the measures of complementary angles is 90.
Plan: The angles are complementary, so the sum of the measures of the angles is 90.
Solve: $16z - 9 + 4z + 3 = 90$
$20z - 6 = 90$
$20z = 96$
$z = 4.8$
$16z - 9 = 16(4.8) - 9$
$= 76.8 - 9$ or 67.8
$4z + 3 = 4(4.8) + 3$
$= 19.2 + 3$ or 22.2
Examine: Add the angle measures to verify that the angles are complementary.
$67.8 + 22.2 = 90$
19. **Explore:** The problem involves an angle and its supplement. You know that the sum of the measures of two supplementary angles is 180.
Plan: Let $m\angle T = x$. Its supplement has measure $180 - x$. The problem also states that $m\angle T$ is 20 more than four times its supplement.
Solve: $x = 4(180 - x) + 20$
$x = 720 - 4x + 20$
$5x = 740$
$x = 148$
So $m\angle T = 148$.
Examine: Check to see if the answer satisfies the problem.
If $m\angle T = 148$, its supplement has a measure of 32.
$4 \times 32 + 20 = 148$
The answer checks.
20. **Explore:** The problem involves an angle and its supplement.
Plan: Let the measure of one angle be x. Its supplement has measure $180 - x$. The problem states that the measure of the angle's supplement is 44 less than x.
Solve: $180 - x = x - 44$
$224 - x = x$
$224 = 2x$
$112 = x$
$180 - x = 180 - 112$ or 68
The measures of the angle and its supplement are 112 and 68.
Examine: $112 - 44 = 68$
21. **Explore:** The problem involves an angle and its supplement.
Plan: Let the measure of one angle be x. Its supplement has measure $180 - x$. The problem states that one angle measures 12° more than the other.

Solve: $x + 12 = 180 - x$

$x = 168 - x$

$2x = 168$

$x = 84$

$180 - x = 180 - 84$ or 96

The measures of the angle and its supplement are 84 and 96.

Examine: $96 = 84 + 12$

22. **Explore:** The problem states that $m\angle 1$ is five less than $4 \cdot m\angle 2$ and $\angle 1$ and $\angle 2$ form a linear pair. So $m\angle 1 + m\angle 2 = 180$.

Plan: Let $m\angle 1 = x$.

Solve: $m\angle 1 + m\angle 2 = 180$

$x + m\angle 2 = 180$

$4 \cdot m\angle 2 - 5 + m\angle 2 = 180$

$5 \cdot m\angle 2 = 185$

$m\angle 2 = 37$

$m\angle 1 = 4 \cdot m\angle 2 - 5$

$= 4(37) - 5$

$= 148 - 5$ or 143

Examine: $4(37) - 5 \stackrel{?}{=} 143$

$143 = 143$

$\angle 1$ and $\angle 2$ are supplementary because $143 + 37 = 180$.

23. Always; the sum of two angles that each measure less than 90° can never equal 180°, so if one angle is acute the other must be obtuse.

24. Always; complementary angles are angles whose measures have a sum of 90°, so each angle must measure less than 90°.

25. Sometimes; for example, consider the following:

$m\angle A = 90$

$m\angle B = 90$

$m\angle C = 90$

Then $m\angle A + m\angle B = 180$ and $m\angle B + m\angle C = 180$ and $m\angle A + m\angle C = 180$.

However, now consider

$m\angle A = 100$

$m\angle B = 80$

$m\angle C = 100$

$m\angle A + m\angle B = 180$ and $m\angle B + m\angle C = 180$ but $m\angle A + m\angle C = 200$, so $\angle A$ is not supplementary to $\angle C$.

26. Never; $\overline{PN}$ and $\overline{PQ}$ have point P in common and form a right angle. $\angle NPQ$ is formed by $\overline{PN}$ and $\overline{PQ}$. Since P is the vertex, $\angle NPQ$ is a right angle.

27. $\overrightarrow{CF} \perp \overrightarrow{FD}$ if $\angle CFD$ is a right angle. So find a so that $m\angle CFD = 12a + 45$ is equal to 90.

$12a + 45 = 90$

$12a = 45$

$a = 3.75$

28. $m\angle AFC = m\angle AFB + m\angle BFC$

$90 = 8x - 6 + 14x + 8$

$90 = 22x + 2$

$88 = 22x$

$4 = x$

29. $\angle BFA$ and $\angle DFE$ are vertical angles so they have the same measure.

$3r + 12 = -8r + 210$

$3r = -8r + 198$

$11r = 198$

$r = 18$

$m\angle BFA = 3r + 12$

$= 3(18) + 12$

$= 54 + 12$ or 66

$\angle BFA$ and $\angle AFE$ are adjacent supplementary angles, so $m\angle BFA + m\angle AFE = 180$.

$66 + m\angle AFE = 180$

$m\angle AFE = 114$

30. $\angle L$ and $\angle M$ are complementary.

$m\angle L + m\angle M = 90$

$y - 2 + 2x + 3 = 90$

$y + 2x + 1 = 90$

$y + 2x = 89$

$y = 89 - 2x$

$\angle N$ and $\angle P$ are complementary.

$m\angle N + m\angle P = 90$

$2x - y + x - 1 = 90$

$3x - y - 1 = 90$

$3x - y = 91$

$-y = 91 - 3x$

$y = 3x - 91$

Equate the two expressions for y and solve for x.

$89 - 2x = 3x - 91$

$180 - 2x = 3x$

$180 = 5x$

$36 = x$

Now substitute the value of x into either expression for y and solve for y.

$y = 3x - 91$

$y = 3(36) - 91$

$y = 108 - 91$ or 17

$m\angle L = y - 2$

$m\angle L = 17 - 2$ or 15

$m\angle M = 2x + 3$

$m\angle M = 2(36) + 3$

$m\angle M = 72 + 3$ or 75

$m\angle N = 2x - y$

$m\angle N = 2(36) - 17$

$m\angle N = 72 - 17$ or 55

$m\angle P = x - 1$

$m\angle P = 36 - 1$ or 35

31. Yes; the symbol denotes that $\angle DAB$ is a right angle.

32. Yes; they are vertical angles.

33. Yes; the sum of their measures is $m\angle ADC$, which is 90.

34. No; there is no indication of the measures of these angles.

35. No; we do not know $m\angle ABC$.

36. Sample answer: *Complementary* means serving to fill out or complete, while *complimentary* means given as a courtesy or favor. *Complementary* has the mathematical meaning of an angle completing the measure to make 90°.

37. Sample answer:

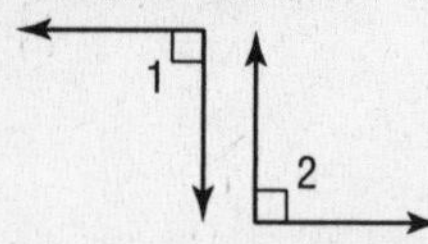

38. $m\angle AKB + m\angle BKC + m\angle CKD = 20 + 25 + 45$ or 90, so $\angle AKD$ is a right angle and $\overline{AK} \perp \overline{KD}$. $m\angle DKE + m\angle EKF = 60 + 30$ or 90, so $\angle DKF$ is a right angle and $\overline{KD} \perp \overline{KF}$. $m\angle EKF + m\angle FKG = 30 + 60$ or 90, so $\angle EKG$ is a right angle and $\overline{KE} \perp \overline{KG}$.

39. Because $\angle WUT$ and $\angle TUV$ are supplementary, let $m\angle WUT = x$ and $m\angle TUV = 180 - x$. A bisector creates measures that are half of the original angle, so $m\angle YUT = \frac{1}{2}m\angle WUT$ or $\frac{x}{2}$ and $m\angle TUZ = \frac{1}{2}m\angle TUV$ or $\frac{180-x}{2}$. Then $m\angle YUZ = m\angle YUT + m\angle TUZ$ or $\frac{x}{2} + \frac{180-x}{2}$. This sum simplifies to $\frac{180}{2}$ or 90. Because $m\angle YUZ = 90$, $\overline{YU} \perp \overline{UZ}$.

40. Sample answer: The types of angles formed depends on how the streets intersect. There may be as few as two angles or many more if there are more than two lines intersecting. Answers should include the following.

- linear pairs, vertical angles, adjacent angles
- See students' work.

41. A; $m\angle y = 89$ and $m\angle x + m\angle y = 180$.
$m\angle x = 180 - m\angle y$
$m\angle x = 180 - 89$ or 91

42. Let x be the number.
$4 \cdot 5 \cdot 6 = 2(10 + x)$
$120 = 20 + 2x$
$100 = 2x$
$50 = x$

43. Lines ℓ, m, and n are in plane E, so $\ell \perp \overleftrightarrow{AB}$, $m \perp \overleftrightarrow{AB}$, and $n \perp \overleftrightarrow{AB}$.

Page 43 Maintain Your Skills

44. $m\angle KFG < 90$, so $\angle KFG$ is acute.

45. $m\angle HFG > 90$, so $\angle HFG$ is obtuse.

46. $m\angle HFK = 90$, so $\angle HFK$ is a right angle.

47. $\angle JFE$ is marked with a right angle symbol, so $\angle JFE$ is a right angle.

48. $m\angle HFJ < 90$, so $\angle HFJ$ is acute.

49. $m\angle EFK > 90$, so $\angle EFK$ is obtuse.

50. $d = \sqrt{(x_2 - x_1)^2 + (y_2 - y_1)^2}$
$AB = \sqrt{(0-3)^2 + (1-5)^2}$
$AB = \sqrt{(-3)^2 + (-4)^2}$
$AB = \sqrt{9 + 16} = \sqrt{25}$
$AB = 5$

51. $d = \sqrt{(x_2 - x_1)^2 + (y_2 - y_1)^2}$
$CD = \sqrt{(5-5)^2 + (9-1)^2}$
$CD = \sqrt{0^2 + 8^2} = \sqrt{64}$
$CD = 8$

52. $d = \sqrt{(x_2 - x_1)^2 + (y_2 - y_1)^2}$
$EF = \sqrt{[-4 - (-2)]^2 + [10 - (-10)]^2}$
$EF = \sqrt{(-2)^2 + 20^2}$
$EF = \sqrt{4 + 400} = \sqrt{404}$
$EF \approx 20.1$

53. $d = \sqrt{(x_2 - x_1)^2 + (y_2 - y_1)^2}$
$GH = \sqrt{(-6-7)^2 + (0-2)^2}$
$GH = \sqrt{(-13)^2 + (-2)^2}$
$GH = \sqrt{169 + 4} = \sqrt{173}$
$GH \approx 13.2$

54. $d = \sqrt{(x_2 - x_1)^2 + (y_2 - y_1)^2}$
$JK = \sqrt{[4 - (-8)]^2 + (7-9)^2}$
$JK = \sqrt{12^2 + (-2)^2}$
$JK = \sqrt{144 + 4} = \sqrt{148}$
$JK \approx 12.2$

55. $d = \sqrt{(x_2 - x_1)^2 + (y_2 - y_1)^2}$
$LM = \sqrt{(3-1)^2 + (-1-3)^2}$
$LM = \sqrt{2^2 + (-4)^2} = \sqrt{4 + 16}$
$LM = \sqrt{20}$
$LM \approx 4.5$

56.

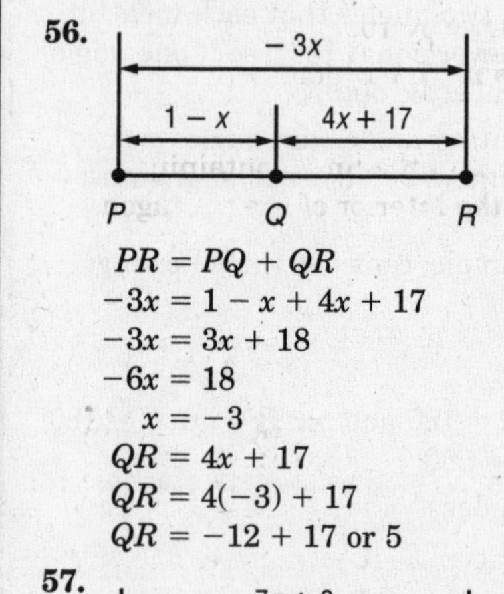

$PR = PQ + QR$
$-3x = 1 - x + 4x + 17$
$-3x = 3x + 18$
$-6x = 18$
$x = -3$
$QR = 4x + 17$
$QR = 4(-3) + 17$
$QR = -12 + 17$ or 5

57.

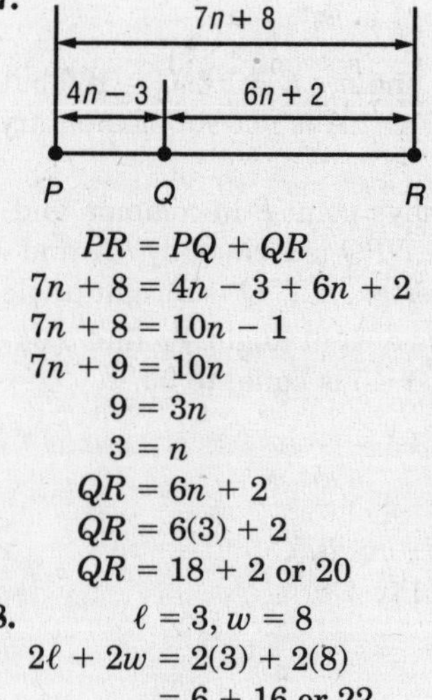

$PR = PQ + QR$
$7n + 8 = 4n - 3 + 6n + 2$
$7n + 8 = 10n - 1$
$7n + 9 = 10n$
$9 = 3n$
$3 = n$
$QR = 6n + 2$
$QR = 6(3) + 2$
$QR = 18 + 2$ or 20

58. $\ell = 3, w = 8$
$2\ell + 2w = 2(3) + 2(8)$
$= 6 + 16$ or 22

59. $\ell = 3, w = 8$
$\ell w = 3 \cdot 8$
$= 24$

60. $s = 2$
$4s = 4(2)$
$= 8$

61. $\ell = 3, w = 8, s = 2$
$\ell w + ws = 3 \cdot 8 + 8 \cdot 2$
$= 24 + 16$ or 40

62. $\ell = 3, w = 8, s = 2$
$s(\ell + w) = 2(3 + 8)$
$= 2(11)$ or 22

Page 44 Geometry Activity: Constructing Perpendiculars

1. See students' work.
2. The first step of the construction locates two points on the line. Then the process is very similar to the construction through a point on a line.

1-6 Polygons

Page 48 Check for Understanding

1. A regular decagon has 10 congruent sides, so divide the perimeter by 10.
2. Saul; Tiki's figure is not a polygon.
3. $P = 3s$
4. Sample answer: Some of the lines containing the sides pass through the interior of the pentagon.

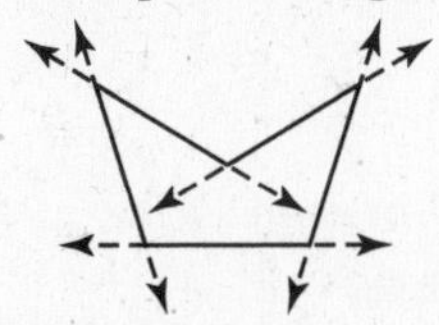

5. There are 5 sides, so the polygon is a pentagon. A line containing side $\overline{NM}$ will pass through the interior of the pentagon, so it is concave. The sides are not congruent, so it is irregular.
6. There are 6 sides, so the polygon is a hexagon. No line containing any of the sides will pass through the interior of the hexagon, so it is convex. The sides are congruent, and the angles are congruent, so it is regular.
7. Perimeter $= 8 + 8 + 6 + 5 + 6$
$= 33$ ft
8. The new side lengths would be 16 ft, 16 ft, 12 ft, 10 ft, and 12 ft so the new perimeter would be $16 + 16 + 12 + 10 + 12 = 66$ ft. The perimeter doubles.
9. $d = \sqrt{(x_2 - x_1)^2 + (y_2 - y_1)^2}$
$PQ = \sqrt{[0 - (-3)]^2 + (8 - 4)^2}$
$= \sqrt{3^2 + 4^2}$
$= \sqrt{25} = 5$
$QR = \sqrt{(3 - 0)^2 + (8 - 8)^2}$
$= \sqrt{3^2 + 0^2}$
$= \sqrt{9}$ or 3
$RS = \sqrt{(0 - 3)^2 + (4 - 8)^2}$
$= \sqrt{(-3)^2 + (-4)^2}$
$= \sqrt{25} = 5$
$PS = \sqrt{[0 - (-3)]^2 + (4 - 4)^2}$
$= \sqrt{3^2 + 0^2}$
$= \sqrt{9} = 3$
Perimeter $= PQ + QR + RS + PS$
$= 5 + 3 + 5 + 3$ or 16 units
10. $AB + BC + CD + AD = 95$
$3a + 2 + 2(a - 1) + 6a + 4 + 5a - 5 = 95$
$3a + 2 + 2a - 2 + 6a + 4 + 5a - 5 = 95$
$16a - 1 = 95$
$16a = 96$
$a = 6$
$AB = 3a + 2$
$= 3(6) + 2$
$= 18 + 2 = 20$
$BC = 2(a - 1)$
$= 2(6 - 1)$
$= 2(5) = 10$
$CD = 6a + 4$
$= 6(6) + 4$
$= 36 + 4 = 40$
$AD = 5a - 5$
$= 5(6) - 5$
$= 30 - 5 = 25$
11. $P = 5s$
$= 5(921)$
$= 4605$
The perimeter of the outside of the Pentagon is 4605 feet.

Pages 49–50 Practice and Apply

12. There are 4 sides, so the polygon is a quadrilateral. No line containing any of the sides will pass through the interior of the quadrilateral, so it is convex. The sides are congruent but the angles are not so it is irregular.
13. There are 8 sides, so the polygon is an octagon. No line containing any of the sides will pass through the interior of the octagon, so it is convex. The sides are congruent, and the angles are congruent, so it is regular.
14. There are 10 sides, so the polygon is a decagon. A line containing any of the sides will pass through the interior of the decagon, so it is concave. The sides are congruent, but the decagon is concave so it cannot be regular. It is irregular.
15. There are 5 sides, so the sign is a pentagon.
16. There are 4 sides, so the sign is a quadrilateral.
17. There are 3 sides, so the sign is a triangle.
18. There are 12 sides, so the sign is a dodecagon.
19. $P = 2\ell + 2w$
$= 2(28) + 2(13)$
$= 56 + 26 = 82$ ft
20. $P = 6 + 12 + 8 + 15 + 15$
$= 56$ m
21. $P = 6 + 2 + 2 + 6 + 2 + 2 + 6 + 2 + 2 + 6 + 2 + 2$
$= 40$ units
22. The new length would be 4(28) or 112 feet. The new width would be 4(13) or 52 feet.
$P = 2\ell + 2w$
$= 2(112) + 2(52) = 328$
The perimeter is multiplied by 4.
23. The new side lengths are 18 m, 36 m, 24 m, 45 m, and 45 m. The new perimeter is $18 + 36 + 24 + 45 + 45$ or 168 m. The perimeter is tripled.

24. The new side lengths are 3 and 1 units. The new perimeter is $3 + 1 + 1 + 3 + 1 + 1 + 3 + 1 + 1 + 3 + 1 + 1$ or 20 units. The perimeter is divided by 2.

25. The length of each side is multiplied by 10, so 10 can be factored out of the sum of the sides. Thus the new perimeter is multiplied by 10 so the perimeter is $12.5(10) = 125$ m.

26. $d = \sqrt{(x_2 - x_1)^2 + (y_2 - y_1)^2}$

$AB = \sqrt{[3 - (-1)]^2 + (4 - 1)^2}$
$= \sqrt{4^2 + 3^2}$
$= \sqrt{25} = 5$

$BC = \sqrt{(6 - 3)^2 + (0 - 4)^2}$
$= \sqrt{3^2 + (-4)^2}$
$= \sqrt{25} = 5$

$CD = \sqrt{(2 - 6)^2 + (-3 - 0)^2}$
$= \sqrt{(-4)^2 + (-3)^2}$
$= \sqrt{25} = 5$

$AD = \sqrt{[2 - (-1)]^2 + (-3 - 1)^2}$
$= \sqrt{3^2 + (-4)^2}$
$= \sqrt{25} = 5$

The perimeter is $AB + BC + CD + AD = 5 + 5 + 5 + 5$ or 20 units.

27. $d = \sqrt{(x_2 - x_1)^2 + (y_2 - y_1)^2}$

$PQ = \sqrt{[3 - (-2)]^2 + (3 - 3)^2}$
$= \sqrt{5^2 + 0^2}$
$= \sqrt{25} = 5$

$QR = \sqrt{(7 - 3)^2 + (0 - 3)^2}$
$= \sqrt{4^2 + (-3)^2}$
$= \sqrt{25} = 5$

$RS = \sqrt{(3 - 7)^2 + (-3 - 0)^2}$
$= \sqrt{(-4)^2 + (-3)^2}$
$= \sqrt{25} = 5$

$ST = \sqrt{(-2 - 3)^2 + [-3 - (-3)]^2}$
$= \sqrt{(-5)^2 + 0^2}$
$= \sqrt{25} = 5$

$TU = \sqrt{[-6 - (-2)]^2 + [0 - (-3)]^2}$
$= \sqrt{(-4)^2 + 3^2}$
$= \sqrt{25} = 5$

$PU = \sqrt{[-6 - (-2)]^2 + (0 - 3)^2}$
$= \sqrt{(-4)^2 + (-3)^2}$
$= \sqrt{25} = 5$

The perimeter is $PQ + QR + RS + ST + TU + PU$ or $5 + 5 + 5 + 5 + 5 + 5 = 30$ units.

28. $d = \sqrt{(x_2 - x_1)^2 + (y_2 - y_1)^2}$

$VW = \sqrt{(-2 - 3)^2 + (12 - 0)^2}$
$= \sqrt{(-5)^2 + 12^2}$
$= \sqrt{169} = 13$

$WX = \sqrt{[-10 - (-2)]^2 + (-3 - 12)^2}$
$= \sqrt{(-8)^2 + (-15)^2}$
$= \sqrt{289} = 17$

$XY = \sqrt{[-8 - (-10)]^2 + [-12 - (-3)]^2}$
$= \sqrt{2^2 + (-9)^2}$
$= \sqrt{85}$

$YZ = \sqrt{[-2 - (-8)]^2 + [-12 - (-12)]^2}$
$= \sqrt{6^2 + 0^2}$
$= \sqrt{36} = 6$

$VZ = \sqrt{(-2 - 3)^2 + (-12 - 0)^2}$
$= \sqrt{(-5)^2 + (-12)^2}$
$= \sqrt{169} = 13$

The perimeter is $VW + WX + XY + YZ + VZ$ or $13 + 17 + \sqrt{85} + 6 + 13 \approx 58.2$ units.

29. There are 6 sides and all sides are congruent. All sides are $\frac{90}{6}$ or 15 cm.

30. There are 4 sides and all sides are congruent. All sides are $\frac{14}{4}$ or 3.5 mi.

31. $P = x - 1 + x + 7 + 3x - 5$
$31 = 5x + 1$
$30 = 5x$
$6 = x$
$x - 1 = 6 - 1 = 5$
$x + 7 = 6 + 7 = 13$
$3x - 5 = 3(6) - 5 = 13$

The sides are 13 units, 13 units, and 5 units.

32. $P = 6x - 3 + 8x + 3 + 6x + 4$
$84 = 20x + 4$
$80 = 20x$
$4 = x$
$6x - 3 = 6(4) - 3 = 21$
$8x + 3 = 8(4) + 3 = 35$
$6x + 4 = 6(4) + 4 = 28$

The sides are 21 m, 35 m, and 28 m.

33. $P = 2\ell + 2w$
$42 = 2(3n + 2) + 2(n - 1)$
$42 = 6n + 4 + 2n - 2$
$42 = 8n + 2$
$40 = 8n$
$5 = n$
$3n + 2 = 3(5) + 2 = 17$
$n - 1 = 5 - 1 = 4$

The sides are 4 in., 4 in., 17 in., and 17 in.

34. $P = 2x - 1 + 2x + x + 2x$
$41 = 7x - 1$
$42 = 7x$
$6 = x$
$2x - 1 = 2(6) - 1 = 11$
$2x = 2(6) = 12$

The sides are 6 yd, 11 yd, 12 yd, and 12 yd.

35. 52 units; count the units in the figure.

36.

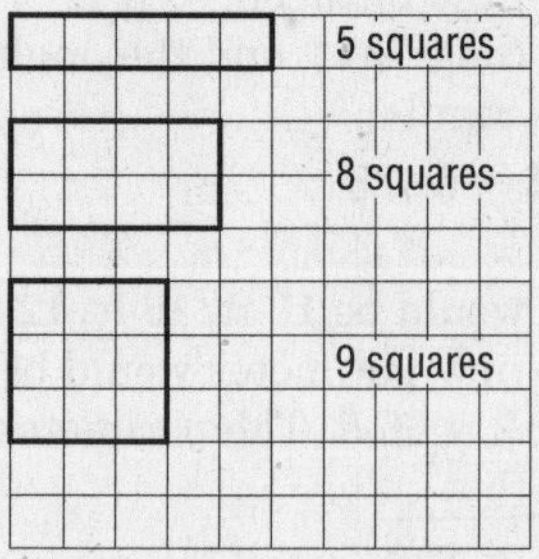

5 squares
8 squares
9 squares

36a. It is a square with side length of 3 units.

36b. In Part **a**, the rectangle with the greatest number of squares was a square with side of 3 units. So if a rectangle has perimeter of 36 units, a square would have the largest area. The side of this square would be $\frac{36}{4} = 9$ units.

37. Sample answer: Some toys use pieces to form polygons. Others have polygon-shaped pieces that connect together. Answers should include the following.

- triangles, quadrilaterals, pentagons
- 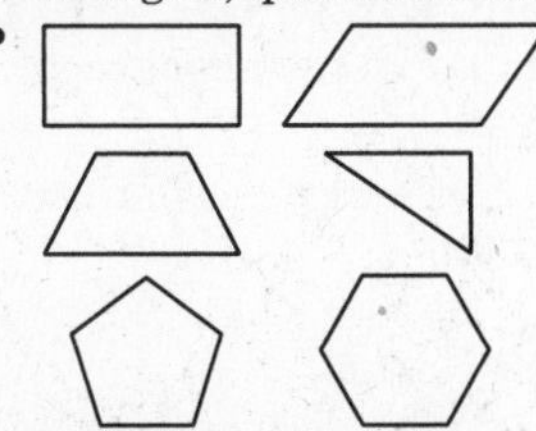

38. A square has four congruent sides. If one side remains to be fenced, three sides have been fenced. So each side has length $\frac{1}{3}(3x)$ or x meters.

39. D; $5n + 5 = 10$
$5n = 5$
$n = 1$
$11 - n = 11 - 1 = 10$

Page 50 Maintain Your Skills

40. Always; true by the definitions of linear pairs and supplementary angles.

41. Sometimes; angles with measures 100 and 80 are supplementary but two right angles are also supplementary.

42. $\overrightarrow{AM}$ bisects $\angle RAL$, so $\angle MAR \cong \angle MAL$.
$m\angle MAR = m\angle MAL$
$2x + 13 = 4x - 3$
$2x + 16 = 4x$
$16 = 2x$
$8 = x$
$m\angle RAL = m\angle MAR + m\angle MAL$
$= 2x + 13 + 4x - 3$
$= 2(8) + 13 + 4(8) - 3$ or 58

43. $\overrightarrow{AM}$ bisects $\angle RAL$, so $\angle MAR \cong \angle LAM$ and $m\angle RAL = 2 \cdot m\angle MAR$.
$x + 32 = 2(x - 31)$
$x + 32 = 2x - 62$
$x + 94 = 2x$
$94 = x$
$m\angle LAM = m\angle MAR$
$= x - 31$
$= 94 - 31 = 63$

44. $\overrightarrow{AS}$ bisects $\angle MAR$, so $\angle RAS \cong \angle SAM$.
$\overrightarrow{AM}$ bisects $\angle LAR$, so $m\angle LAR = 2 \cdot m\angle MAR$.
$m\angle RAS = m\angle SAM$
$25 - 2x = 3x + 5$
$20 - 2x = 3x$
$20 = 5x$
$4 = x$
$m\angle LAR = 2 \cdot m\angle MAR$
$= 2(m\angle SAM + m\angle RAS)$
$= 2(3x + 5 + 25 - 2x)$
$= 2(x + 30)$
$= 2x + 60$
$= 2(4) + 60 = 68$

Page 52 Geometry Software Investigation: Measuring Polygons

1. The sum of the side measures equals the perimeter.
2. $35.53 + 90.38 + 54.09 = 180$
3. See students' work.
4. Sample answer: When the lengths of the sides are doubled, the perimeter is doubled.
5. See student's work.
6. Sample answer: The sum of the measures of the angles of a triangle is 180.
7. Sample answer: The sum of the measures of the angles of a quadrilateral is 360; pentagon = 540; hexagon = 720.
8. Sample answer: The sum of the measures of the angles of polygons increases by 180 for each additional side.
9. yes; sample answer: triangle: 3 sides, angle measure sum: 180; quadrilateral: 4 sides, angle measure sum: $180 + 180 = 360$; pentagon: 5 sides, angle measure sum: $360 + 180 = 540$; hexagon: 6 sides, angle measure sum: $540 + 180 = 720$
10. Yes; sample answer: If the sides of a polygon are a, b, c, and d, then its perimeter is $a + b + c + d$. If each of the sides are increased by a factor of n then the sides measure na, nb, nc, and nd, and the perimeter is $na + nb + nc + nd$. By factoring, the perimeter is $n(a + b + c + d)$, which is the original perimeter increased by the same factor as the sides.

Chapter 1 Study Guide and Review

Pages 53–56

1. d; $\overline{AB} \cong \overline{BC}$, so B is the midpoint of $\overline{AC}$.

2. h	**3.** f	**4.** e
5. b	**6.** g	**7.** p or m
8. K or L	**9.** F	**10.** S

11.

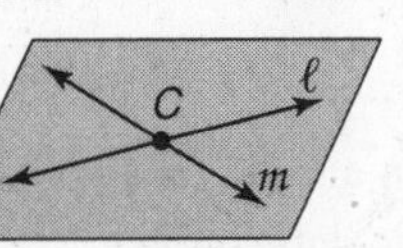

12.

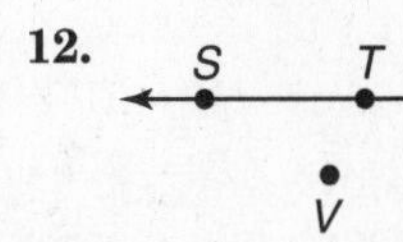

13.

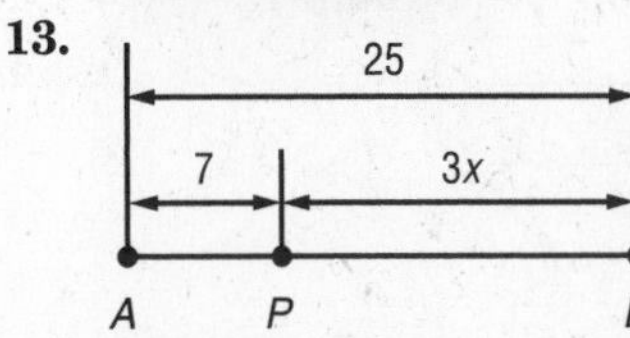

$AB = AP + PB$
$25 = 7 + 3x$
$18 = 3x$
$6 = x$
$PB = 3x$
$= 3(6)$
$= 18$

14.

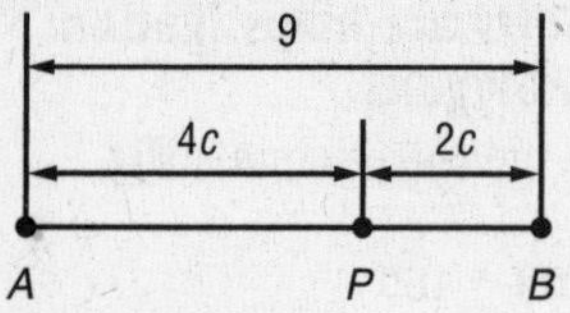

9

4c 2c

A P B

$AB = AP + PB$
$9 = 4c + 2c$
$9 = 6c$
$\frac{3}{2} = c$ or $c = 1.5$
$PB = 2c$
$= 2\left(\frac{3}{2}\right)$
$= 3$

15.

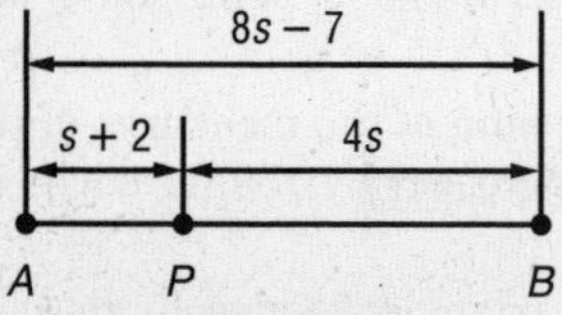

8s − 7

s + 2 4s

A P B

$AB = AP + PB$
$8s - 7 = s + 2 + 4s$
$8s - 7 = 5s + 2$
$8s = 5s + 9$
$3s = 9$
$s = 3$
$PB = 4s$
$= 4(3)$
$= 12$

16.

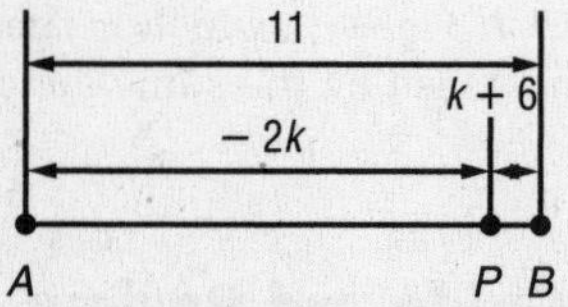

11

k + 6

− 2k

A P B

$AB = AP + PB$
$11 = -2k + k + 6$
$11 = -k + 6$
$5 = -k$
$-5 = k$
$PB = k + 6$
$= -5 + 6$
$= 1$

17. yes, because $HI = KJ = 9$ m

18. no, because $AB = 13.6$ cm and $AC = 19.3$ cm

19. not enough information to determine if $5x - 1 = 4x + 3$

20. $d = \sqrt{(x_2 - x_1)^2 + (y_2 - y_1)^2}$
$AB = \sqrt{(-3 - 1)^2 + (2 - 0)^2}$
$= \sqrt{(-4)^2 + 2^2} = \sqrt{20}$
≈ 4.5

21. $d = \sqrt{(x_2 - x_1)^2 + (y_2 - y_1)^2}$
$GL = \sqrt{[3 - (-7)]^2 + (3 - 4)^2}$
$= \sqrt{10^2 + (-1)^2} = \sqrt{101}$
≈ 10.0

22. $d = \sqrt{(x_2 - x_1)^2 + (y_2 - y_1)^2}$
$JK = \sqrt{(4 - 0)^2 + (-1 - 0)^2}$
$= \sqrt{4^2 + (-1)^2} = \sqrt{17}$
≈ 4.1

23. $d = \sqrt{(x_2 - x_1)^2 + (y_2 - y_1)^2}$
$MP = \sqrt{[-6 - (-4)]^2 + (19 - 16)^2}$
$= \sqrt{(-2)^2 + 3^2} = \sqrt{13}$
≈ 3.6

24. $M = \left(\frac{x_1 + x_2}{2}, \frac{y_1 + y_2}{2}\right)$
$= \left(\frac{0 + 22}{2}, \frac{0 + (-18)}{2}\right)$
$= \left(\frac{22}{2}, \frac{-18}{2}\right)$ or $(11, -9)$

25. $M = \left(\frac{x_1 + x_2}{2}, \frac{y_1 + y_2}{2}\right)$
$= \left(\frac{-6 + 12}{2}, \frac{-3 + (-7)}{2}\right)$
$= \left(\frac{6}{2}, \frac{-10}{2}\right)$ or $(3, -5)$

26. $M = \left(\frac{x_1 + x_2}{2}, \frac{y_1 + y_2}{2}\right)$
$= \left(\frac{2 + (-1)}{2}, \frac{5 + (-1)}{2}\right)$
$= \left(\frac{1}{2}, \frac{4}{2}\right)$ or $(0.5, 2)$

27. $M = \left(\frac{x_1 + x_2}{2}, \frac{y_1 + y_2}{2}\right)$
$= \left(\frac{3.4 + (-2.2)}{2}, \frac{-7.3 + (-5.4)}{2}\right)$
$= \left(\frac{1.2}{2}, \frac{-12.7}{2}\right)$ or $(0.6, -6.35)$

28. D **29.** $\overrightarrow{FE}, \overrightarrow{FG}$ **30.** $\angle DEH$

31. 70°; $70 < 90$, so $\angle SQT$ is acute.

32. 110°; $110 > 90$ and $110 < 180$, so $\angle PQT$ is obtuse.

33. 50°; $50 < 90$, so $\angle T$ is acute.

34. 70°; $70 < 90$, so $\angle PRT$ is acute.

35. $\overrightarrow{XV}$ bisects $\angle YXW$, so $\angle YXV \cong \angle VXW$.
$m\angle YXV = m\angle VXW$
$3x = 2x + 6$
$x = 6$
$m\angle YXW = m\angle YXV + m\angle VXW$
$= 3x + 2x + 6$
$= 3(6) + 2(6) + 6$
$= 36$

36. $\overrightarrow{XW}$ bisects $\angle YXZ$, so $\angle YXW \cong \angle WXZ$.
$m\angle YXW = m\angle WXZ$
$12x - 10 = 8(x + 1)$
$12x - 10 = 8x + 8$
$12x = 8x + 18$
$4x = 18$
$x = \frac{9}{2}$
$m\angle YXZ = m\angle YXW + m\angle WXZ$
$= 12x - 10 + 8(x + 1)$
$= 12\left(\frac{9}{2}\right) - 10 + 8\left(\frac{9}{2} + 1\right)$
$= 54 - 10 + 44$
$= 88$

37. $\overrightarrow{XW}$ bisects $\angle YXZ$, so $\angle YXW \cong \angle WXZ$ and $m\angle WXZ = \frac{1}{2} \cdot m\angle YXZ$.

$$m\angle WXZ = \frac{1}{2} \cdot m\angle YXZ$$
$$7x - 9 = \frac{1}{2}(9x + 17)$$
$$7x - 9 = \frac{9}{2}x + \frac{17}{2}$$
$$7x = \frac{9}{2}x + \frac{35}{2}$$
$$\frac{5}{2}x = \frac{35}{2}$$
$$x = 7$$
$$m\angle YXW = m\angle WXZ$$
$$= 7x - 9$$
$$= 7(7) - 9$$
$$= 40$$

38. $\angle TWY$, $\angle WYX$

39. $\angle TWY$, $\angle XWY$

40. $\overline{TW} \perp \overline{WZ}$ if $m\angle TWZ = 90$.

$$m\angle TWZ = 2c + 36$$
$$90 = 2c + 36$$
$$54 = 2c$$
$$27 = c$$

41. $\angle ZWX = \angle ZWY + \angle YWX$

$$90 = 4k - 2 + 5k + 11$$
$$90 = 9k + 9$$
$$81 = 9k$$
$$9 = k$$

42. There are 4 sides, so the polygon is a quadrilateral. No line containing any of the sides will pass through the interior of the quadrilateral, so it is convex. The sides are congruent, and the angles are congruent, so it is regular.

43. The figure is not a polygon because there are sides that intersect more than two other sides.

44. There are 8 sides, so the polygon is an octagon. There is a side such that a line containing that side will pass through the interior of the octagon, so it is concave. The sides are not congruent, so the octagon is irregular.

45. $d = \sqrt{(x_2 - x_1)^2 + (y_2 - y_1)^2}$

$$AB = \sqrt{(5-1)^2 + (1-2)^2}$$
$$= \sqrt{4^2 + (-1)^2} = \sqrt{17}$$
$$BC = \sqrt{(9-5)^2 + (2-1)^2}$$
$$= \sqrt{4^2 + 1^2} = \sqrt{17}$$
$$CD = \sqrt{(9-9)^2 + (5-2)^2}$$
$$= \sqrt{0^2 + 3^2} = 3$$
$$DE = \sqrt{(5-9)^2 + (6-5)^2}$$
$$= \sqrt{(-4)^2 + 1^2} = \sqrt{17}$$
$$EF = \sqrt{(1-5)^2 + (5-6)^2}$$
$$= \sqrt{(-4)^2 + (-1)^2} = \sqrt{17}$$
$$AF = \sqrt{(1-1)^2 + (5-2)^2}$$
$$= \sqrt{0^2 + 3^2} = 3$$

Perimeter $= AB + BC + CD + DE + EF + AF$

$$= \sqrt{17} + \sqrt{17} + 3 + \sqrt{17} + \sqrt{17} + 3$$
$$= 6 + 4\sqrt{17}$$
$$\approx 22.5 \text{ units}$$

46. $d = \sqrt{(x_2 - x_1)^2 + (y_2 - y_1)^2}$

$$WX = \sqrt{[7 - (-3)]^2 + (1-5)^2}$$
$$= \sqrt{10^2 + (-4)^2} = \sqrt{116}$$
$$XY = \sqrt{(5-7)^2 + (-4-1)^2}$$
$$= \sqrt{(-2)^2 + (-5)^2} = \sqrt{29}$$
$$YZ = \sqrt{(-5-5)^2 + [0 - (-4)]^2}$$
$$= \sqrt{(-10)^2 + 4^2} = \sqrt{116}$$
$$WZ = \sqrt{[-5 - (-3)]^2 + (0-5)^2}$$
$$= \sqrt{(-2)^2 + (-5)^2} = \sqrt{29}$$

Perimeter $= WX + XY + YZ + WZ$

$$= \sqrt{116} + \sqrt{29} + \sqrt{116} + \sqrt{29}$$
$$\approx 32.3 \text{ units}$$

Chapter 1 Practice Test

Page 57

1. true

2. true

3. False; the sum of two supplementary angles is 180.

4. true

5. m

6. D

7. C

8.

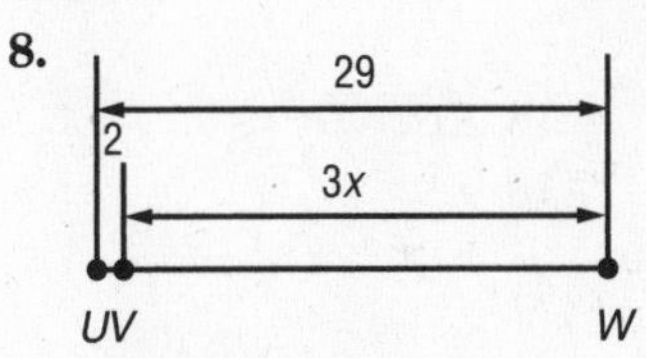

$$UW = UV + VW$$
$$29 = 2 + 3x$$
$$27 = 3x$$
$$9 = x$$
$$VW = 3x$$
$$= 3(9) = 27$$

9.

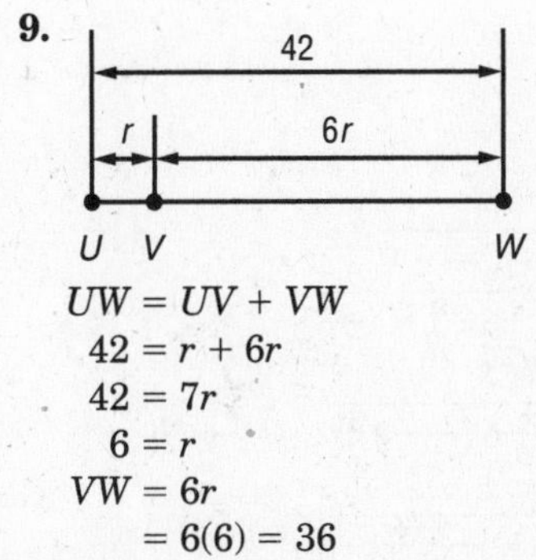

$$UW = UV + VW$$
$$42 = r + 6r$$
$$42 = 7r$$
$$6 = r$$
$$VW = 6r$$
$$= 6(6) = 36$$

10.

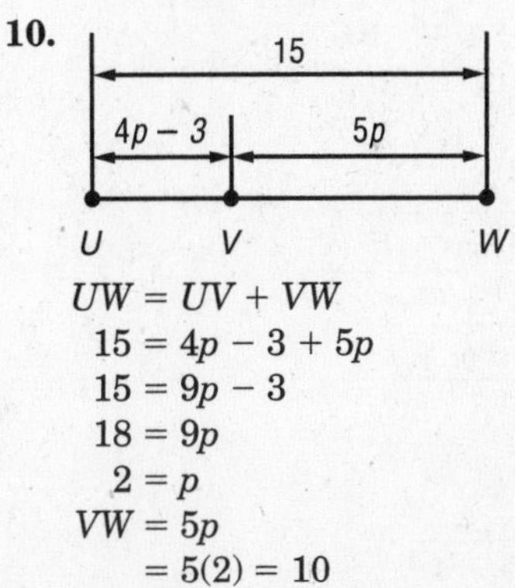

$$UW = UV + VW$$
$$15 = 4p - 3 + 5p$$
$$15 = 9p - 3$$
$$18 = 9p$$
$$2 = p$$
$$VW = 5p$$
$$= 5(2) = 10$$

11.

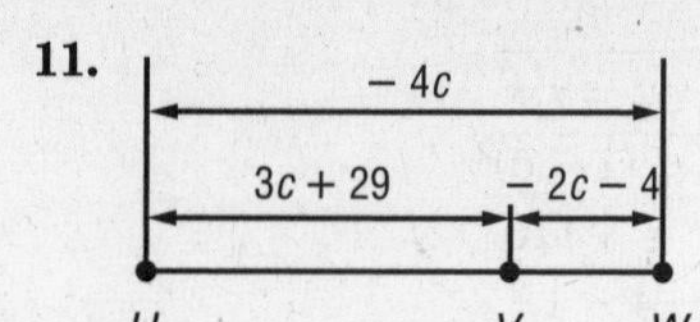

$UW = UV + VW$
$-4c = 3c + 29 + (-2c - 4)$
$-4c = c + 25$
$-5c = 25$
$c = -5$
$VW = -2c - 4$
$= -2(-5) - 4 = 6$

12. $d = \sqrt{(x_2 - x_1)^2 + (y_2 - y_1)^2}$
$GH = \sqrt{(-3 - 0)^2 + (4 - 0)^2}$
$= \sqrt{(-3)^2 + 4^2} = \sqrt{25}$
$= 5$

13. $d = \sqrt{(x_2 - x_1)^2 + (y_2 - y_1)^2}$
$NK = \sqrt{(-2 - 5)^2 + (8 - 2)^2}$
$= \sqrt{(-7)^2 + 6^2} = \sqrt{85}$
≈ 9.2

14. $d = \sqrt{(x_2 - x_1)^2 + (y_2 - y_1)^2}$
$AW = \sqrt{[-2 - (-4)]^2 + [2 - (-4)]^2}$
$= \sqrt{2^2 + 6^2} = \sqrt{40}$
≈ 6.3

15. C

16. $\overrightarrow{EC}, \overrightarrow{ED}$

17. $\angle ABD$ or $\angle ABE$

18. $\angle 9$

19. $4r + 7 + r - 2 = 180$
$5r + 5 = 180$
$5r = 175$
$r = 35$
$4r + 7 = 4(35) + 7$ or 147
$r - 2 = 35 - 2$ or 33

20. Let x be the measure of one angle. Then the other angle has measure $x + 26$ and $x + x + 26 = 90$.
$x + x + 26 = 90$
$2x + 26 = 90$
$2x = 64$
$x = 32$
$x + 26 = 32 + 26 = 58$

21. $d = \sqrt{(x_2 - x_1)^2 + (y_2 - y_1)^2}$
$PQ = \sqrt{[1 - (-6)]^2 + [-1 - (-3)]^2}$
$= \sqrt{7^2 + 2^2} = \sqrt{53}$
$QR = \sqrt{(1 - 1)^2 + [-5 - (-1)]^2}$
$= \sqrt{0^2 + (-4)^2} = 4$
$PR = \sqrt{[1 - (-6)]^2 + [-5 - (-3)]^2}$
$= \sqrt{7^2 + (-2)^2} = \sqrt{53}$
Perimeter $= PQ + QR + PR$
$= \sqrt{53} + 4 + \sqrt{53}$
≈ 18.6 units

22. $d = \sqrt{(x_2 - x_1)^2 + (y_2 - y_1)^2}$
$AB = \sqrt{[-4 - (-6)]^2 + (7 - 2)^2}$
$= \sqrt{2^2 + 5^2} = \sqrt{29}$
$BC = \sqrt{[0 - (-4)]^2 + (4 - 7)^2}$
$= \sqrt{4^2 + (-3)^2}$
$= \sqrt{25} = 5$
$CD = \sqrt{(0 - 0)^2 + (0 - 4)^2}$
$= \sqrt{0^2 + (-4)^2} = 4$
$DE = \sqrt{(-4 - 0)^2 + (-3 - 0)^2}$
$= \sqrt{(-4)^2 + (-3)^2} = \sqrt{25} = 5$
$AE = \sqrt{[-4 - (-6)]^2 + (-3 - 2)^2}$
$= \sqrt{2^2 + (-5)^2} = \sqrt{29}$
Perimeter $= AB + BC + CD + DE + AE$
$= \sqrt{29} + 5 + 4 + 5 + \sqrt{29}$
≈ 24.8 units

23.

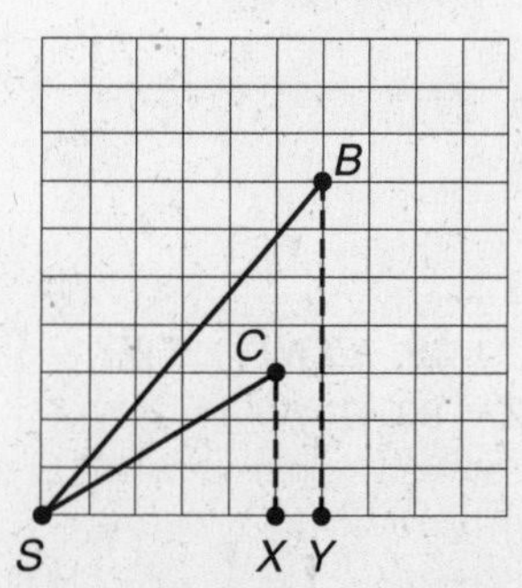

Use the information known about the cities to place them on a coordinate plane. Then use the gridlines to form a triangle using the points S for Springfield, B for Brighton, and Y. Use the Pythagorean Theorem.
$(SB)^2 = (SY)^2 + (BY)^2$
$(SB)^2 = 6^2 + 7^2$
$(SB)^2 = 85$
$SB \approx 9.2$
Highway 1 is approximately 9.2 miles long.

24.

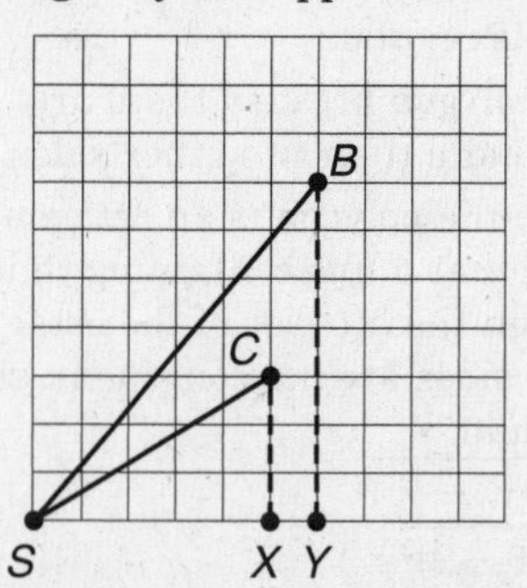

Use the information known about the cities to place them on a coordinate plane. Then use the gridlines to form a triangle using the points S for Springfield, C for Capital City, and X. Use the Pythagorean Theorem.
$(SC)^2 = (SX)^2 + (CX)^2$
$(SC)^2 = 5^2 + 3^2$
$(SC)^2 = 34$
$SC \approx 5.8$
Highway 4 is approximately 5.8 miles long.

25. C; the figure has sides with a common endpoint that are collinear.

Chapter 1 Standardized Test Practice

Pages 58–59

1. C; there are 15 hours between 7 A.M. and 10 P.M., so there are $15 \cdot 60 = 900$ minutes during that time. Then Juanita blinks $11 \cdot 900$ or 9900 times during the time she is awake.
2. A
3. B; $\frac{2x^2 + 12x + 16}{2x + 4} = \frac{2(x^2 + 6x + 8)}{2(x + 2)}$
$$= \frac{2(x + 4)(x + 2)}{2(x + 2)}$$
$$= x + 4$$
4. A; three noncollinear points determine a plane, so if the two planes are distinct, their intersection is not a plane. Two planes intersect in only one line.
5. B; 1 fathom = 6 feet so 55 fathoms = 55(6) or 330 feet, which is 110 yards.
6. C; use the Pythagorean Theorem, where x is the height up the side of the house that the ladder reaches.
$$(18)^2 = (6)^2 + x^2$$
$$324 = 36 + x^2$$
$$288 = x^2$$
$$17.0 \approx x$$
7. C; $m\angle ABD = m\angle CBD$
$$2x + 14 = 5x - 10$$
$$2x + 14 + 10 = 5x - 10 + 10$$
$$2x + 24 = 5x$$
$$2x + 24 - 2x = 5x - 2x$$
$$24 = 3x$$
$$8 = x$$
$$m\angle ABD = 2x + 14$$
$$= 2(8) + 14$$
$$= 30$$
8. D; let x be the measure of $\angle FEG$.
$$m\angle DEG + m\angle FEG = 180$$
$$\left(6\tfrac{1}{2}\right)x + x = 180$$
$$\tfrac{15}{2}x = 180$$
$$x = 24$$
$$m\angle DEG = 6\tfrac{1}{2}(24)$$
$$= 156$$
9. D;

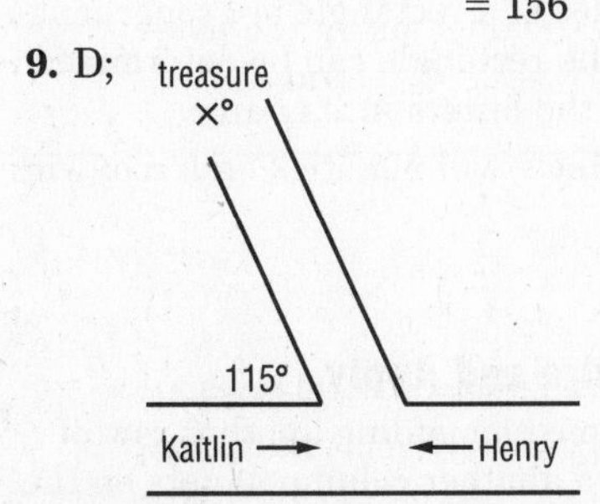

If Kaitlin turns 115°, Henry turns 180° − 115° or 65°.

10. $-2x + 6 + 4x^2 + x + x^2 - 5 = 5x^2 - x + 1$
11.
$$2y = 3x + 8$$
$$y = 2x + 3$$
$$2(2x + 3) = 3x + 8$$
$$4x + 6 = 3x + 8$$
$$x = 2$$
$$y = 2x + 3$$
$$y = 2(2) + 3 \text{ or } 7$$
The solution is (2, 7).

12.

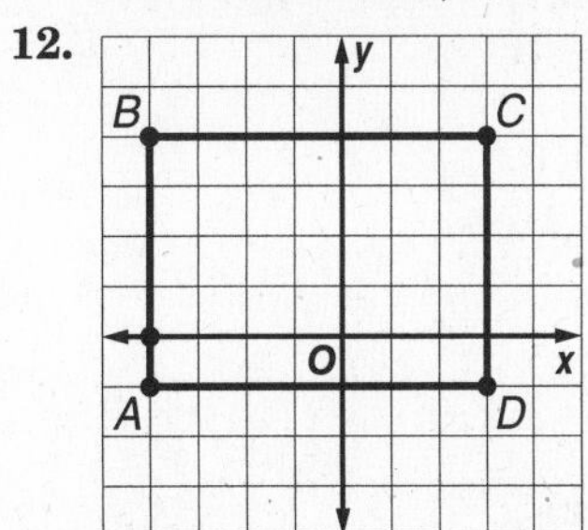

$ABCD$ is a rectangle, so D must have coordinates $D(3, -1)$.

13. $M = \left(\frac{x_1 + x_2}{2}, \frac{y_1 + y_2}{2}\right)$
$$= \left(\frac{2 + (-4)}{2}, \frac{-1 + 3}{2}\right)$$
$$= \left(\frac{-2}{2}, \frac{2}{2}\right) = (-1, 1)$$
14. $(AB)^2 = 60^2 + 110^2$
$$(AB)^2 = 3600 + 12{,}100$$
$$(AB)^2 = 15{,}700$$
$$AB \approx 125 \text{ m}$$
15. $P = 2\ell + 2w$
$$= 2(16) + 2(24)$$
$$= 80$$
The perimeter of the basement is 80 feet. If the pieces of plasterboard are 4 feet wide, it takes $\frac{80}{4}$ or 20 pieces of plasterboard to cover the walls.

16a.

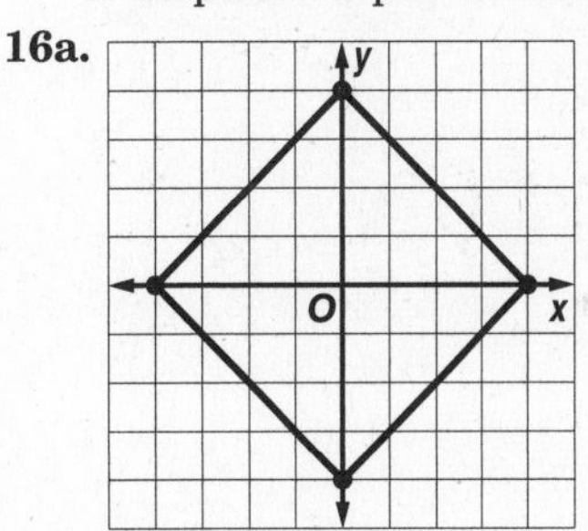

16b. $M = \left(\frac{x_1 + x_2}{2}, \frac{y_1 + y_2}{2}\right)$
$$\left(\frac{-4 + 0}{2}, \frac{0 + 4}{2}\right) = (-2, 2)$$
$$\left(\frac{-4 + 0}{2}, \frac{0 + (-4)}{2}\right) = (-2, -2)$$
$$\left(\frac{4 + 0}{2}, \frac{0 + (-4)}{2}\right) = (2, -2)$$
$$\left(\frac{4 + 0}{2}, \frac{0 + 4}{2}\right) = (2, 2)$$

17a. You are given that $a = 25$. The measure of the other angle marked with a single arc is also 25 because the arcs tell us that the angles are congruent. Both of the angles marked with double arcs have a measure of $a + 10$, which is 25 + 10 or 35. In the remaining angle, $b = 180 - (25 + 35 + 25 + 35)$ or 60 because the angles can be combined to form linear pairs, which are supplementary.

17b. All are acute.

Chapter 2 Reasoning and Proof

Page 61 Getting Started

1. $3n - 2 = 3(4) - 2$
   $= 12 - 2 = 10$
2. $(n + 1) + n = (6 + 1) + 6$
   $= 7 + 6$
   $= 13$
3. $n^2 - 3n = (3)^2 - 3(3)$
   $= 9 - 3(3)$
   $= 9 - 9 = 0$
4. $180(n - 2) = 180(5 - 2)$
   $= 180(3) = 540$
5. $n\left(\frac{n}{2}\right) = 10\left(\frac{10}{2}\right)$
   $= 10(5) = 50$
6. $\frac{n(n-3)}{2} = \frac{8(8-3)}{2}$
   $= \frac{8(5)}{2}$
   $= \frac{40}{2} = 20$
7. $6x - 42 = 4x$
   $-42 = -2x$
   $21 = x$
8. $8 - 3n = -2 + 2n$
   $10 - 3n = 2n$
   $10 = 5n$
   $2 = n$
9. $3(y + 2) = -12 + y$
   $3y + 6 = -12 + y$
   $3y = -18 + y$
   $2y = -18$
   $y = -9$
10. $12 + 7x = x - 18$
    $7x = x - 30$
    $6x = -30$
    $x = -5$
11. $3x + 4 = \frac{1}{2}x - 5$
    $3x = \frac{1}{2}x - 9$
    $\frac{5}{2}x = -9$
    $x = -\frac{18}{5}$
12. $2 - 2x = \frac{2}{3}x - 2$
    $4 - 2x = \frac{2}{3}x$
    $4 = \frac{8}{3}x$
    $\frac{3}{2} = x$
13. $\angle AGB$ and $\angle EGD$ are vertical angles, so $m\angle AGB = m\angle EGD$.
    $4x + 7 = 71$
    $4x = 64$
    $x = 16$
14. $m\angle BGC + m\angle CGD + m\angle DGE = 180$
    $45 + 8x + 4 + 15x - 7 = 180$
    $23x + 42 = 180$
    $23x = 138$
    $x = 6$

2-1 Inductive Reasoning and Conjecture

Pages 63–64 Check for Understanding

1. Sample answer: After the news is over, it's time for dinner.
2. Sometimes; the conjecture is true when E is between D and F; otherwise it is false.
3. Sample answer: When it is cloudy, it rains. Counterexample: It is often cloudy and it does not rain.
4. There is one of each shape in the first figure. There are two of each shape in the second figure and three of each shape in the third figure. So the next figure will have four of each shape.

5. 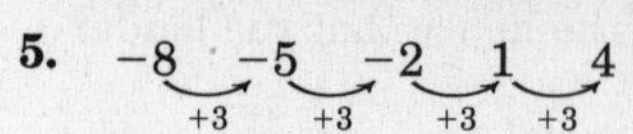

The numbers in the sequence increase by 3. The next number will increase by 3. So, it will be 4 + 3 or 7.

6. $PQ = TU$

P Q
R S
T U

7. $\overleftrightarrow{AB}$ and $\overleftrightarrow{CD}$ intersect at a single point P, so the lines are distinct. Thus the points A, B, C, and D are not all on the same line. So, A, B, C, and D are noncollinear.

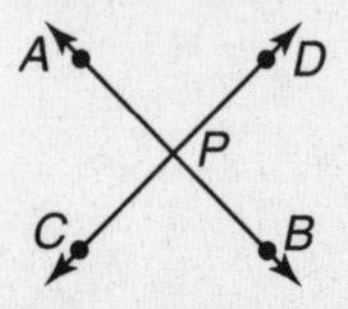

8. False; if $x = -2$, then $-x = -(-2)$ or 2.
9. True; opposite sides of a rectangle are congruent, and the sides of the rectangle can be determined from the order of the letters in its name.
10. Sample answer: Snow will not stick on a roof with a steep angle.

Pages 64–66 Practice and Apply

11. Each figure is formed by adding another row of dots to the top and another column of dots on the side. The number of dots in each figure is 2, 6, 12, 20.

The numbers increase by 4, 6, and 8. The next number will increase by 10. So, it will be 30.

• • • • • •
• • • • • •
• • • • • •
• • • • • •
• • • • • •

12. Each figure adds a triangle and changes the orientation of the triangles. The next figure will have five triangles with the same orientation as the three triangles in the second figure.

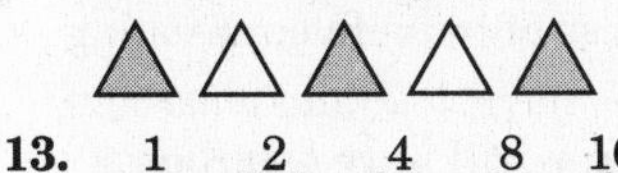

13. $1 \xrightarrow{+1} 2 \xrightarrow{+2} 4 \xrightarrow{+4} 8 \xrightarrow{+8} 16$

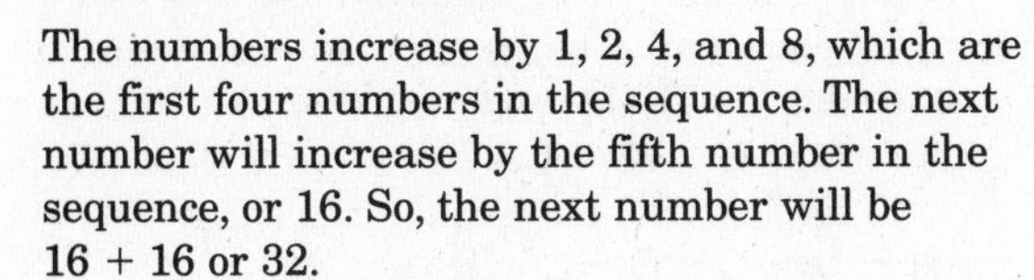

The numbers increase by 1, 2, 4, and 8, which are the first four numbers in the sequence. The next number will increase by the fifth number in the sequence, or 16. So, the next number will be $16 + 16$ or 32.

14. $4 \xrightarrow{+2} 6 \xrightarrow{+3} 9 \xrightarrow{+4} 13 \xrightarrow{+5} 18$

The numbers increase by 2, 3, 4, and 5. The next number will increase by 6. So, it will be $18 + 6$ or 24.

15. $\frac{1}{3} \xrightarrow{+\frac{2}{3}} 1 \xrightarrow{+\frac{2}{3}} \frac{5}{3} \xrightarrow{+\frac{2}{3}} \frac{7}{3} \xrightarrow{+\frac{2}{3}} 3$

The numbers increase by $\frac{2}{3}$. The next number will be $3 + \frac{2}{3}$ or $\frac{11}{3}$.

16. $1 \xrightarrow{\times\frac{1}{2}} \frac{1}{2} \xrightarrow{\times\frac{1}{2}} \frac{1}{4} \xrightarrow{\times\frac{1}{2}} \frac{1}{8} \xrightarrow{\times\frac{1}{2}} \frac{1}{16}$

The numbers are multiplied by $\frac{1}{2}$. The next number will be $\frac{1}{16} \times \frac{1}{2}$ or $\frac{1}{32}$.

17. $2 \xrightarrow{\times(-3)} -6 \xrightarrow{\times(-3)} 18 \xrightarrow{\times(-3)} -54$

The numbers are multiplied by -3. The next number will be $-54 \times (-3)$ or 162.

18. $-5 \xrightarrow{\times(-5)} 25 \xrightarrow{\times(-5)} -125 \xrightarrow{\times(-5)} 625$

The numbers are multiplied by -5. The next number will be $625 \times (-5)$ or -3125.

19. Each arrangement of blocks is formed by adding a level of blocks to the bottom. The numbers of blocks in the sequence are 1, 5, and 14.

$1 \xrightarrow{+4} 5 \xrightarrow{+9} 14$

The numbers increase by 4 and 9, which are squares of 2 and 3, respectively. The next number will increase by the square of 4, or 16. So it will be $14 + 16$ or 30.

20. Each arrangement of blocks is formed by adding a level of blocks to the bottom. The number of blocks on the bottom level of the figures is 1, 3, and 6.

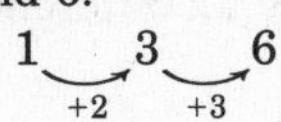

$1 \xrightarrow{+2} 3 \xrightarrow{+3} 6$

The number of blocks on the bottom increase by 2 and 3. The next increase will be 4. So there will be 10 blocks on the bottom of the fourth figure. The upper levels of the figure have a total of 10 blocks. The total number of blocks in this figure is 20.

21. Perpendicular lines form four right angles, so lines ℓ and m form four right angles.

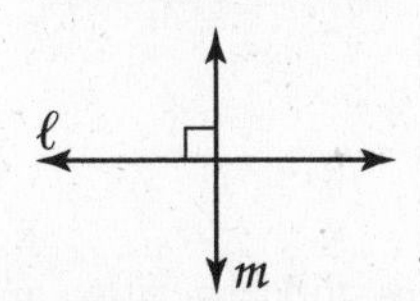

22. Graph A, B, and C. Connect the points to see that they lie on the same line. Thus A, B, and C are collinear.

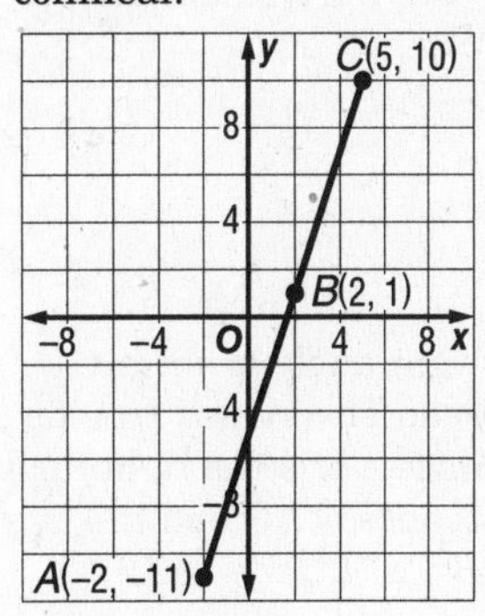

23. Linear pairs of angles are supplementary, so $\angle 3$ and $\angle 4$ are supplementary.

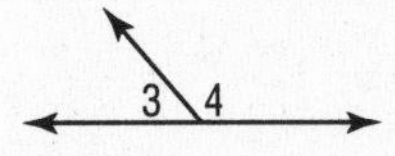

24. $\overrightarrow{BD}$ bisects $\angle ABC$ so the two angles formed are congruent: $\angle ABD \cong \angle DBC$.

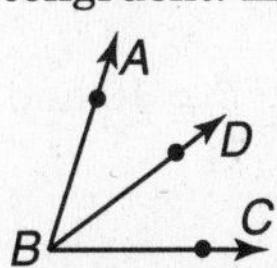

25. Graph P, Q, and R. The points form a triangle. Find the distance between each pair of points to determine the type of triangle.

$$d = \sqrt{(x_2 - x_1)^2 + (y_2 - y_1)^2}$$

$$PQ = \sqrt{[6 - (-1)]^2 + (-2 - 7)^2}$$

$$= \sqrt{7^2 + (-9)^2}$$

$$= \sqrt{130}$$

$$QR = \sqrt{(6 - 6)^2 + [5 - (-2)]^2}$$

$$= \sqrt{0^2 + 7^2}$$

$$= \sqrt{49} \text{ or } 7$$

$PR = \sqrt{[6 - (-1)]^2 + (5 - 7)^2}$
$= \sqrt{7^2 + (-2)^2}$
$= \sqrt{53}$

The lengths PQ, QR, and PR are all different, so $\triangle PQR$ is a scalene triangle.

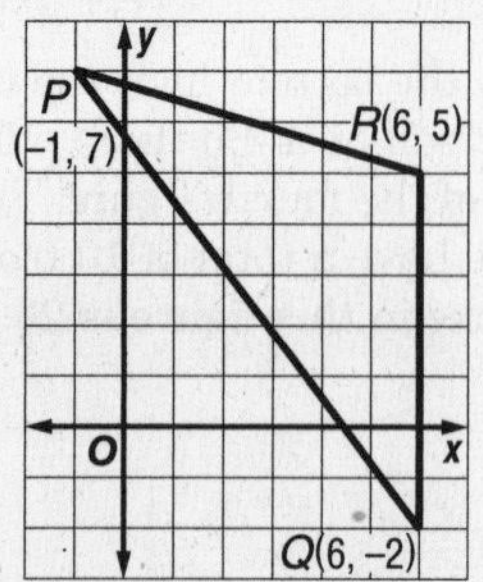

26. A square has 4 congruent sides. From the name of the square the sides are $\overline{HI}$, $\overline{IJ}$, $\overline{JK}$, and $\overline{KH}$. Thus, $HI = IJ = JK = KH$.

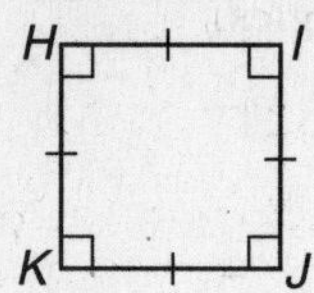

27. A rectangle has 4 sides where opposite sides are congruent. From the name of the rectangle the sides are $\overline{PQ}$, $\overline{QR}$, $\overline{SR}$, and $\overline{PS}$. The pairs of opposite sides are $\overline{PQ}$, $\overline{SR}$ and $\overline{QR}$, $\overline{PS}$. Thus, $PQ = SR$ and $QR = PS$.

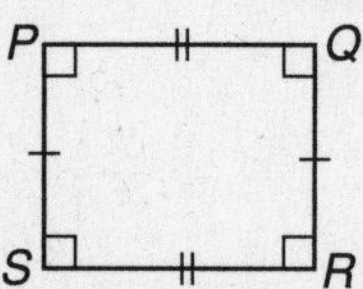

28. A triangle that has a right angle is a right triangle. The Pythagorean Theorem is true for every right triangle. Since $\angle B$ is the right angle of the triangle, the hypotenuse is $\overline{AC}$ and the legs are $\overline{AB}$ and $\overline{BC}$. So $(AB)^2 + (BC)^2 = (AC)^2$.

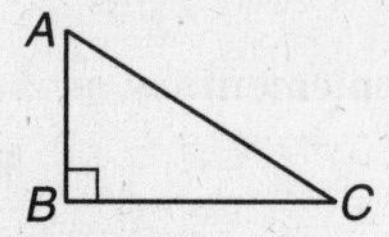

29. False;

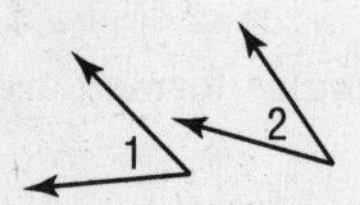

30. False; if $y = 7$ and $m = 5$, then $7 + 5 \geq 10$ and $5 \geq 4$, but $7 \not\geq 6$.

31. False;

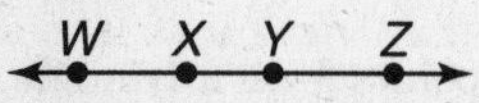

32. True;

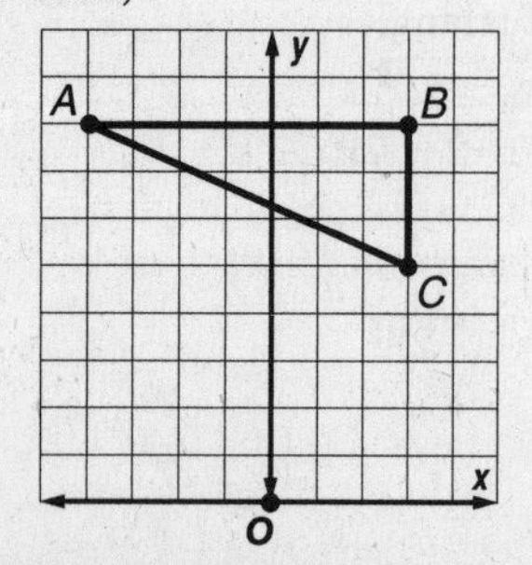

$\overline{AB}$ is horizontal, and $\overline{BC}$ is vertical. So the segments are perpendicular to each other. So $\angle B$ is a right angle. Therefore, $\triangle ABC$ is a right triangle.

33. True; the square of any real number is a nonnegative number.

34. False; D, E, and F do not have to be collinear.

35. False; $JKLM$ may not have a right angle.

36. True; any three noncollinear points form a triangle.

37. trial and error, a process of inductive reasoning

38. The first three alkanes have 1, 2, and 3 carbon atoms. The fourth alkane will have 4 carbon atoms. The first three alkanes have 4, 6, and 8 hydrogen atoms. The fourth alkane, butane, will have 8 + 2 or 10 hydrogen atoms.

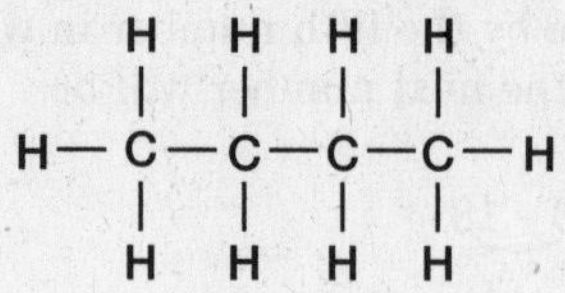

39. The 1st alkane has formula CH_4 (or C_1H_4).
The 2nd alkane has formula C_2H_6.
The 3rd alkane has formula C_3H_8.
The subscripts for C increase by 1 and are the same number as the number of the alkane in the series. The subscripts for H increase by 2. So the 7th alkane in the series has formula $C_7H_{(8+2+2+2+2)}$ or C_7H_{16}.

40. The nth alkane has n carbon atoms so the subscript for C is n. The number of hydrogen atoms in the nth alkane is 2 more than twice the numbers of the alkane in the series, or $2n + 2$. So the subscript for H is $2n + 2$. The formula is C_nH_{2n+2}.

41. False; if $n = 41$, then $n^2 - n + 41 = (41)^2 - 41 + 41$ or 41^2, which is not prime.

42. Sample answer: By past experience, when dark clouds appear, there is a chance of rain. Answers should include the following.

- When there is precipitation in the summer, it is usually rain because the temperature is above freezing. When the temperature is below freezing, as in the winter, ice or snow forms.
- See students' work.

43. C; 1, 1, 2, 3, 5, 8
$1 + 1 = 2$
$1 + 2 = 3$
$2 + 3 = 5$
$3 + 5 = 8$
Each number in the sequence is found by adding the two numbers before it. $5 + 8 = 13$, so the next number in the sequence is 13.

44. D; Let x be the sum of the six numbers and y be the sum of the three numbers whose average is 15.
$\frac{x}{6} = 18$
$x = 18(6)$ or 108

$\frac{y}{3} = 15$
$y = 15(3)$ or 45
The sum of the remaining three numbers is 108 − 45 or 63.

Page 66 Maintain Your Skills

45. There are 6 sides, so the polygon is a hexagon. No line containing any of the sides will pass through the interior of the hexagon, so it is convex. Not all sides are congruent, so the hexagon is irregular.

46. There are 5 sides, so the polygon is a pentagon. No line containing any of the sides will pass through the interior of the pentagon, so it is convex. The sides and angles are congruent, so the pentagon is regular.

47. There are 7 sides, so the polygon is a heptagon. There is a side such that a line containing the side will pass through the interior of the heptagon, so it is concave. Not all sides are congruent, so the heptagon is irregular.

48. Yes; the symbol denotes that $\angle KJN$ is a right angle.

49. No; we do not know anything about the angle measures.

50. No; we do not know whether $\angle MNP$ is a right angle.

51. Yes; they form a linear pair.

52. Yes; since the other three angles in rectangle *KLPJ* are right angles, $\angle KLP$ must also be a right angle.

53. $M = \left(\frac{x_1 + x_2}{2}, \frac{y_1 + y_2}{2}\right)$
$= \left(\frac{-1+5}{2}, \frac{3+(-5)}{2}\right)$
$= \left(\frac{4}{2}, \frac{-2}{2}\right)$ or $(2, -1)$

54. $M = \left(\frac{x_1 + x_2}{2}, \frac{y_1 + y_2}{2}\right)$
$= \left(\frac{4+(-3)}{2}, \frac{1+7}{2}\right)$
$= \left(\frac{1}{2}, \frac{8}{2}\right)$ or $(0.5, 4)$

55. $M = \left(\frac{x_1 + x_2}{2}, \frac{y_1 + y_2}{2}\right)$
$= \left(\frac{4+(-2)}{2}, \frac{-9+(-15)}{2}\right)$
$= \left(\frac{2}{2}, \frac{-24}{2}\right)$ or $(1, -12)$

56. $M = \left(\frac{x_1 + x_2}{2}, \frac{y_1 + y_2}{2}\right)$
$= \left(\frac{-5+7}{2}, \frac{-2+4}{2}\right)$
$= \left(\frac{2}{2}, \frac{2}{2}\right)$ or $(1, 1)$

57. $M = \left(\frac{x_1 + x_2}{2}, \frac{y_1 + y_2}{2}\right)$
$= \left(\frac{8+3}{2}, \frac{-1.8+6.2}{2}\right)$
$= \left(\frac{11}{2}, \frac{4.4}{2}\right)$ or $(5.5, 2.2)$

58. $M = \left(\frac{x_1 + x_2}{2}, \frac{y_1 + y_2}{2}\right)$
$= \left(\frac{-1.5+(-4)}{2}, \frac{-6+3}{2}\right)$
$= \left(\frac{-5.5}{2}, \frac{-3}{2}\right)$ or $(-2.75, -1.5)$

59. $PN = 3x$ and $PN = 24$.
$3x = 24$
$x = 8$
$MP = 7x$
$= 7(8)$ or 56

60. $PN = 9c$ and $PN = 63$.
$9c = 63$
$c = 7$
$MP = 2c$
$= 2(7)$ or 14

61. $MN = MP + PN$
$36 = 4x + 5x$
$36 = 9x$
$4 = x$
$MP = 4x$
$= 4(4)$ or 16

62. $MN = MP + PN$
$60 = 6q + 6q$
$60 = 12q$
$5 = q$
$MP = 6q$
$= 6(5)$ or 30

63. $MN = MP + PN$
$63 = 4y + 3 + 2y$
$63 = 6y + 3$
$60 = 6y$
$10 = y$
$MP = 4y + 3$
$= 4(10) + 3$
$= 43$

64. $MN = MP + PN$
$43 = 2b - 7 + 8b$
$43 = 10b - 7$
$50 = 10b$
$5 = b$
$MP = 2b - 7$
$= 2(5) - 7$
$= 3$

65. $x + 2 > 5$
$2 + 2 = 4$. $4 < 5$. So $2 + 2 \not> 5$.
$3 + 2 = 5$. $5 = 5$. So $3 + 2 \not> 5$.
$4 + 2 = 6$. $6 > 5$. So $4 + 2 > 5$.
$5 + 2 = 7$. $7 > 5$. So $5 + 2 > 5$.
The values in the replacement set that make the inequality true are 4 and 5.

66. $12 - x < 0$
$12 - 11 = 1$. $1 > 0$. So $12 - 11 \not< 0$.
$12 - 12 = 0$. $0 = 0$. So $12 - 12 \not< 0$.
$12 - 13 = -1$. $-1 < 0$. So $12 - 13 < 0$.
$12 - 14 = -2$. $-2 < 0$. So $12 - 14 < 0$.
The values in the replacement set that make the inequality true are 13 and 14.

67. $5x + 1 > 25$
$5(4) + 1 = 21$. $21 < 25$. So $5(4) + 1 \not> 25$.
$5(5) + 1 = 26$. $26 > 25$. So $5(5) + 1 > 25$.
$5(6) + 1 = 31$. $31 > 25$. So $5(6) + 1 > 25$.
$5(7) + 1 = 36$. $36 > 25$. So $5(7) + 1 > 25$.
The values in the replacement set that make the inequality true are 5, 6, and 7.

2-2 Logic

Pages 71–72 Check for Understanding

1. The conjunction (p and q) is represented by the intersection of the two circles.

2a. Sample answer: October has 31 days or $-5 + 3 = -8$.

2b. Sample answer: A square has five right angles and the Postal Service does not deliver mail on Sundays.

2c. Sample answer: July 5th is not a national holiday.

3. A conjunction is a compound statement using the word *and*, while a disjunction is a compound statement using the word *or*.

4. $9 + 5 = 14$ and February has 30 days.; false, because p is true and q is false

5. $9 + 5 = 14$ and a square has four sides.; true, because p is true and r is true

6. February has 30 days and a square has four sides.; false, because q is false and r is true

7. $9 + 5 = 14$ or February does not have 30 days.; true, because p is true and $\sim q$ is true

8. February has 30 days or a square has four sides.; true, because r is true

9. $9 + 5 \neq 14$ or a square does not have four sides.; false, because $\sim p$ is false and $\sim r$ is false

10.

p	q	$\sim q$	$p \wedge \sim q$
T	T	F	F
T	F	T	T
F	T	F	F
F	F	T	F

11.

p	q	$p \wedge q$
T	T	T
T	F	F
F	T	F
F	F	F

12.

q	r	$q \vee r$
T	T	T
T	F	T
F	T	T
F	F	F

13.

p	r	$\sim p$	$\sim p \wedge r$
T	T	F	F
T	F	F	F
F	T	T	T
F	F	T	F

14.

p	q	r	$p \vee q$	$(p \vee q) \vee r$
T	T	T	T	T
T	T	F	T	T
T	F	T	T	T
T	F	F	T	T
F	T	T	T	T
F	T	F	T	T
F	F	T	F	T
F	F	F	F	F

15. The states that produce more than 100 million bushels of corn are represented by the set labeled Corn. There are 14 states that produce more than 100 million bushels of corn.

16. The states that produce more than 100 million bushels of wheat are represented by the set labeled Wheat. There are 7 states that produce more than 100 million bushels of wheat.

17. The states that produce more than 100 million bushels of corn and wheat are represented by the intersection of the sets. There are 3 states that produce more than 100 million bushels of corn and wheat.

Pages 72–74 Practice and Apply

18. $\sqrt{-64} = 8$ and an equilateral triangle has three congruent sides; false, because p is false and q is true.

19. $\sqrt{-64} = 8$ or an equilateral triangle has three congruent sides; true, because q is true.

20. $\sqrt{-64} = 8$ and $0 < 0$; false, because p is false and r is false.

21. $0 < 0$ and an obtuse angle measures greater than 90° and less than 180°; false, because r is false and s is true.

22. An equilateral triangle has three congruent sides or $0 < 0$; true, because q is true.

23. An equilateral triangle has three congruent sides and an obtuse angle measures greater than 90° and less than 180°; true, because q is true and s is true.

24. $\sqrt{-64} = 8$ and an obtuse angle measures greater than 90° and less than 180°; false, because p is false and s is true.

25. An equilateral triangle has three congruent sides and $0 < 0$; false, because q is true and r is false.

26. $0 < 0$ or $\sqrt{-64} = 8$; false, because r is false and p is false.

27. An obtuse angle measures greater than 90° and less than 180° or an equilateral triangle has three congruent sides; true, because s is true and q is true.

28. $\sqrt{-64} = 8$ and an equilateral triangle has three congruent side, or an obtuse angle measures greater than 90° and less than 180°; true, because s is true.

29. An obtuse angle measures greater than 90° and less than 180°, or an equilateral triangle has three congruent sides and $0 < 0$; true, because s is true.

30.

p	q	$\sim p$	$\sim p \vee q$
T	T	F	T
T	F	F	F
F	T	T	T
F	F	T	T

31.

p	q	$\sim p$	$\sim q$	$\sim p \wedge \sim q$
T	T	F	F	F
T	F	F	T	F
F	T	T	F	F
F	F	T	T	T

32.

p	q	r	$p \vee q$	$(p \vee q) \wedge r$
T	T	T	T	T
T	T	F	T	F
T	F	T	T	T
T	F	F	T	F
F	T	T	T	T
F	T	F	T	F
F	F	T	F	F
F	F	F	F	F

33.

q	r	q and r
T	T	T
T	F	F
F	T	F
F	F	F

34.

p	q	p or q
T	T	T
T	F	T
F	T	T
F	F	F

35.

p	r	p or r
T	T	T
T	F	T
F	T	T
F	F	F

36.

p	q	p and q
T	T	T
T	F	F
F	T	F
F	F	F

37.

q	r	$\sim r$	$q \wedge \sim r$
T	T	F	F
T	F	T	T
F	T	F	F
F	F	T	F

38.

p	q	$\sim p$	$\sim q$	$\sim p \wedge \sim q$
T	T	F	F	F
T	F	F	T	F
F	T	T	F	F
F	F	T	T	T

39.

p	q	r	$\sim p$	$\sim r$	$q \wedge \sim r$	$\sim p \vee (q \wedge \sim r)$
T	T	T	F	F	F	F
T	T	F	F	T	T	T
T	F	T	F	F	F	F
T	F	F	F	T	F	F
F	T	T	T	F	F	T
F	T	F	T	T	T	T
F	F	T	T	F	F	T
F	F	F	T	T	F	T

40.

p	q	r	$\sim q$	$\sim r$	$\sim q \vee \sim r$	$p \wedge (\sim q \vee \sim r)$
T	T	T	F	F	F	F
T	T	F	F	T	T	T
T	F	T	T	F	T	T
T	F	F	T	T	T	T
F	T	T	F	F	F	F
F	T	F	F	T	T	F
F	F	T	T	F	T	F
F	F	F	T	T	T	F

41. The teens that said they listened to none of these types of music are represented by the region outside the Pop, Country, and Rap sets. There are 42 teens that said they listened to none of these types of music.

42. The teens that said they listened to all three types of music are represented by the intersection of three sets. There are 7 teens that said they listened to all three types of music.

43. The teens that said they listened to only pop and rap music are represented by the intersection of the Pop and Rap sets excluding the teens that also listen to country music (hence listen to all three types of music). There are 25 teens that said they listened to only pop and rap music.

44. The teens that said they listened to pop, rap, or country music are represented by the union of the Pop, Rap, and Country sets. There are $175 + 25 + 7 + 34 + 45 + 10 + 62$ or 358 teens that said they listened to pop, rap, or country music.

45. Level of Participation Among 310 Students

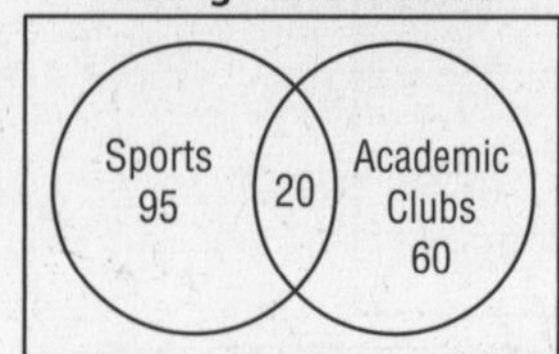

46. The students that participate in either clubs or sports are represented by the union of the sets. There are 60 + 20 + 95 or 175 students that participate in clubs or sports.

47. The students that do not participate in either clubs or sports are the students outside the union of the sets. There are 310 – 175 or 135 students that do not participate in either clubs or sports.

48. false

49. True; Rochester is located on Lake Ontario but Syracuse is not. The statement is a disjunction, so it is true.

50. False; Buffalo is located on Lake Erie, so the negation of the statement is false.

51.

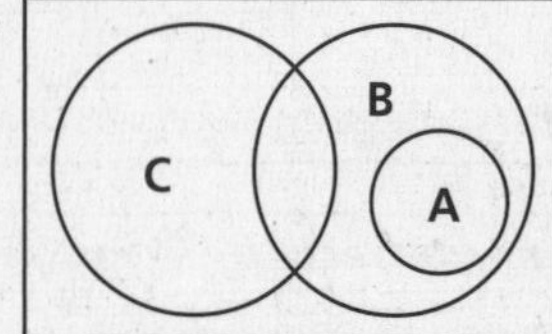

52. b; the relationship between Teams A and C is not known, so statements a and c might not be true. It is known that every member of Team A is also a member of Team B so b is true.

53. Sample answer: Logic can be used to eliminate false choices on a multiple choice test. Answers should include the following.

- Math is my favorite subject and drama club is my favorite activity.
- See students' work.

54. A; the marks on $\overline{AB}$ and $\overline{BC}$ indicate that $\overline{AB} \cong \overline{BC}$, so $AB = BC$ is true. Statement A is the only true statement about $\triangle ABC$.

55. C; let x be the first of the integers. Then $x + 2$ is the second integer. Their sum is 78.

$$x + x + 2 = 78$$
$$2x + 2 = 78$$
$$2x = 76$$
$$x = 38, x + 2 = 40$$

The two integers are 38 and 40, the greater of which is 40.

Page 74 Maintain Your Skills

56. 3, 5, 7, 9 (+2, +2, +2)

The numbers in the sequence increase by 2. The next number will increase by 2. So, it will be 9 + 2 or 11.

57. 1, 3, 9, 27 (×3, ×3, ×3)

The numbers in the sequence are multiplied by 3. The next number will be multiplied by 3. So, it will be 27 × 3 or 81.

58. $6, 3, \frac{3}{2}, \frac{3}{4}$ ($\times\frac{1}{2}$, $\times\frac{1}{2}$, $\times\frac{1}{2}$)

The numbers in the sequence are multiplied by $\frac{1}{2}$. The next number will be multiplied by $\frac{1}{2}$. So, it will be $\frac{3}{4} \times \frac{1}{2}$ or $\frac{3}{8}$.

59. 17, 13, 9, 5 (−4, −4, −4)

The numbers in the sequence decrease by 4. The next number will decrease by 4. So, it will be 5 − 4 or 1.

60. 64, 16, 4, 1 ($\times\frac{1}{4}$, $\times\frac{1}{4}$, $\times\frac{1}{4}$)

The numbers in the sequence are multiplied by $\frac{1}{4}$. The next number will be multiplied by $\frac{1}{4}$. So, it will be $1 \times \frac{1}{4}$ or $\frac{1}{4}$.

61. 5, 15, 45, 135 (×3, ×3, ×3)

The numbers in the sequence are multiplied by 3. The next number will be multiplied by 3. So, it will be 135 × 3 or 405.

62.
$$d = \sqrt{(x_2 - x_1)^2 + (y_2 - y_1)^2}$$
$$AB = \sqrt{[1 - (-6)]^2 + (3 - 7)^2} = \sqrt{7^2 + (-4)^2} = \sqrt{65}$$
$$BC = \sqrt{(-2 - 1)^2 + (-7 - 3)^2} = \sqrt{(-3)^2 + (-10)^2} = \sqrt{109}$$
$$AC = \sqrt{[-2 - (-6)]^2 + (-7 - 7)^2} = \sqrt{4^2 + (-14)^2} = \sqrt{212}$$
$$\text{Perimeter} = AB + BC + AC = \sqrt{65} + \sqrt{109} + \sqrt{212} \approx 33.1$$

63.
$$d = \sqrt{(x_2 - x_1)^2 + (y_2 - y_1)^2}$$
$$DE = \sqrt{[-5 - (-10)]^2 + [-2 - (-9)]^2} = \sqrt{5^2 + 7^2} = \sqrt{74}$$
$$P = 4s = 4 \cdot DE = 4\sqrt{74} \approx 34.4$$

64.
$$d = \sqrt{(x_2 - x_1)^2 + (y_2 - y_1)^2}$$
$$HI = \sqrt{(-8 - 5)^2 + [-9 - (-10)]^2} = \sqrt{(-13)^2 + 1^2} = \sqrt{170}$$
$$IJ = \sqrt{[-5 - (-8)]^2 + [-5 - (-9)]^2} = \sqrt{3^2 + 4^2} = \sqrt{25} = 5$$
$$JK = \sqrt{[-2 - (-5)]^2 + [-4 - (-5)]^2} = \sqrt{3^2 + 1^2} = \sqrt{10}$$
$$HK = \sqrt{(-2 - 5)^2 + [-4 - (-10)]^2} = \sqrt{(-7)^2 + 6^2} = \sqrt{85}$$
$$\text{Perimeter} = HI + IJ + JK + HK = \sqrt{170} + 5 + \sqrt{10} + \sqrt{85} \approx 30.4$$

65. $d = \sqrt{(x_2 - x_1)^2 + (y_2 - y_1)^2}$

$LM = \sqrt{(4-2)^2 + (5-1)^2}$
$= \sqrt{2^2 + 4^2}$
$= \sqrt{20} = 2\sqrt{5}$

$MN = \sqrt{(6-4)^2 + (4-5)^2}$
$= \sqrt{2^2 + (-1)^2} = \sqrt{5}$

$NP = \sqrt{(7-6)^2 + (-4-4)^2}$
$= \sqrt{1^2 + (-8)^2} = \sqrt{65}$

$PQ = \sqrt{(5-7)^2 + [-8-(-4)]^2}$
$= \sqrt{(-2)^2 + (-4)^2}$
$= \sqrt{20} = 2\sqrt{5}$

$QR = \sqrt{(3-5)^2 + [-7-(-8)]^2}$
$= \sqrt{(-2)^2 + 1^2}$
$= \sqrt{5}$

$LR = \sqrt{(3-2)^2 + (-7-1)^2}$
$= \sqrt{1^2 + (-8)^2}$
$= \sqrt{65}$

Perimeter
$= LM + MN + NP + PQ + QR + LR$
$= 2\sqrt{5} + \sqrt{5} + \sqrt{65} + 2\sqrt{5} + \sqrt{5} + \sqrt{65}$
$= 6\sqrt{5} + 2\sqrt{65}$
≈ 29.5

66. 145°; $90 < 145 < 180$ so $\angle ABC$ is obtuse.
67. 55°; $55 < 90$ so $\angle DBC$ is acute.
68. 90°; right
69. The front and back could be as much as 35.5 feet each and the sides could be as much as 75.5 feet each.
$P = 2\ell + 2w$
$= 2(35.5) + 2(75.5)$
$= 71 + 151 = 222$
Michelle should buy 222 feet of fencing.
70. $5a - 2b = 5(4) - 2(3)$
$= 20 - 6$ or 14
71. $4cd + 2d = 4(5)(2) + 2(2)$
$= 40 + 4$ or 44
72. $4e + 3f = 4(-1) + 3(-2)$
$= -4 + (-6)$ or -10
73. $3g^2 + h = 3(8)^2 + (-8)$
$= 3(64) + (-8)$
$= 192 + (-8)$ or 184

2-3 Conditional Statements

Page 78 Check for Understanding

1. Writing a conditional in if-then form is helpful so that the hypothesis and conclusion are easily recognizable.
2. Sample answer: If you eat your peas, then you will have dessert.
3. In the inverse, you negate both the hypothesis and the conclusion of the conditional. In the contrapositive, you negate the hypothesis and the conclusion of the converse.
4. Hypothesis: it rains on Monday;
Conclusion: I will stay home
5. Hypothesis: $x - 3 = 7$;
Conclusion: $x = 10$
6. Hypothesis: a polygon has six sides;
Conclusion: it is a hexagon
7. Sample answer: If a pitcher is a 32-ounce pitcher, then it holds a quart of liquid.
8. Sample answer: If two angles are supplementary, then the sum of the measures of the angles is 180.
9. Sample answer: If an angle is formed by perpendicular lines, then it is a right angle.
10. The hypothesis is true because you drove 70 miles per hour, and the conclusion is true because you received a speeding ticket. Since the promised result is true, the conditional statement is true.
11. The hypothesis is false, and the statement does not say what happens if you drive 65 miles per hour or less. You could still get a speeding ticket if you are driving in a zone where the posted speed limit is less than 65 miles per hour. In this case, we cannot say that the statement is false so the statement is true.
12. The hypothesis is true, but the conclusion is false. Because the result is not what was promised, the conditional statement is false.
13. Converse: If plants grow, then they have water; true.
Inverse: If plants do not have water, then they will not grow; true.
Contrapositive: If plants do not grow, then they do not have water. False; they may have been killed by overwatering.
14. Conditional in if-then form: If you are flying in an airplane, then you are safer than riding in a car.
Converse: If you are safer than riding in a car, then you are flying in an airplane. False; there are other places that are safer than riding in a car.
Inverse: If you are not flying in an airplane, then you are not safer than riding in a car. False; there are other places that are safer than riding in a car.
Contrapositive: If you are not safer than riding in a car, then you are not flying in an airplane; true.
15. Sample answer: If you are in Colorado, then aspen trees cover high areas of the mountains.
If you are in Florida, then cypress trees rise from the swamps.
If you are in Vermont, then maple trees are prevalent.

Pages 78–80 Practice and Apply

16. Hypothesis: $2x + 6 = 10$;
Conclusion: $x = 2$
17. Hypothesis: you are a teenager;
Conclusion: you are at least 13 years old
18. Hypothesis: you have a driver's license;
Conclusion: you are at least 16 years old
19. Hypothesis: three points lie on a line;
Conclusion: the points are collinear

20. Hypothesis: a man hasn't discovered something he will die for;
Conclusion: he isn't fit to live

21. Hypothesis: the measure of an angle is between 0 and 90;
Conclusion: the angle is acute

22. Sample answer: If you buy a 1-year fitness plan, then you get a free visit.

23. Sample answer: If you are a math teacher, then you love to solve problems.

24. Sample answer: If I think, then I am.

25. Sample answer: If two angles are adjacent, then they have a common side.

26. Sample answer: If two angles are vertical, then they are congruent.

27. Sample answer: If two triangles are equiangular, then they are equilateral.

28. The hypothesis is true because you are 19 years old, and the conclusion is true because you vote. Since the predicted result is true, the conditional statement is true.

29. The hypothesis is false, and the conclusion is true. The statement doesn't say what happens if you are younger than 18 years old. It is possible that you vote in a school election. In this case, we cannot say that the statement is false. Thus, the statement is true.

30. The hypothesis is true, and the conclusion is false. Because the result is not what was predicted, the conditional statement is false.

31. The hypothesis is false, and the conclusion is false. The statement doesn't say what happens if you are younger than 18 years old. In this case, we cannot say that the statement is false. Thus, the statement is true.

32. The hypothesis is true because your sister is 21 years old, and the conslusion is true because she votes. Since the predicted result is true, the conditional statement is true.

33. The hypothesis is true, and the conclusion is false. Because the result is not what was predicted, the conditional statement is false.

34. True; P, Q, and R are collinear, and P is in plane $\mathcal{M}$ and Q is in plane $\mathcal{N}$. The line containing P and Q is the intersection of $\mathcal{M}$ and $\mathcal{N}$, so the line that is the intersection of these planes is the line through P and Q, and thus R. So P, Q, and R are in $\mathcal{M}$.

35. True; points Q and B lie in plane $\mathcal{N}$, so the line that connects them also lies in plane $\mathcal{N}$.

36. True; Q is on the line through P that is the intersection of planes $\mathcal{M}$ and $\mathcal{N}$. The line is in $\mathcal{M}$, so Q is in $\mathcal{M}$.

37. False; P and A lie in plane $\mathcal{M}$ and Q and B lie in plane $\mathcal{N}$. $\mathcal{M}$ and $\mathcal{N}$ are distinct planes, so P, Q, A, and B are not coplanar.

38. false

39. True; line RQ is the same as line PQ since P, Q, and R are collinear. P is in $\mathcal{M}$ and Q is in $\mathcal{N}$ so $\mathcal{M}$ and $\mathcal{N}$ intersect at line PQ and hence line RQ.

40. Converse: If you live in Texas, then you live in Dallas. False; you could live in Austin.
Inverse: If you don't live in Dallas, then you don't live in Texas. False; you could live in Austin.
Contrapositive: If you don't live in Texas, then you don't live in Dallas; true.

41. Converse: If you are in good shape, then you exercise regularly; true.
Inverse: If you do not exercise regularly, then you are not in good shape; true.
Contrapositive: If you are not in good shape, then you do not exercise regularly. False; an ill person may exercise a lot, but still not be in good shape.

42. Conditional: If two angles are complementary, then their sum is 90.
Converse: If the sum of two angles is 90, then they are complementary; true.
Inverse: If two angles are not complementary, then their sum is not 90; true.
Contrapositive: If the sum of two angles is not 90, then they are not complementary; true.

43. Conditional: If a figure is a rectangle, then it is a quadrilateral.
Converse: If a figure is a quadrilateral, then it is a rectangle. False; it could be a rhombus.
Inverse: If a figure is not a rectangle, then it is not a quadrilateral. False; it could be a rhombus.
Contrapositive: If a figure is not a quadrilateral, then it is not a rectangle; true.

44. Conditional: If an angle is a right angle, then its measure is 90.
Converse: If an angle has a measure of 90, then it is a right angle; true.
Inverse: If an angle is not a right angle, then its measure is not 90; true.
Contrapositive: If an angle does not have a measure of 90, then it is not a right angle; true.

45. Conditional: If an angle is acute, then its measure is less than 90.
Converse: If an angle has measure less than 90, then it is acute; true.
Inverse: If an angle is not acute, then its measure is not less than 90; true.
Contrapositive: If an angle's measure is not less than 90, then it is not acute; true.

46. Sample answer: In Alaska, if it is summer, then there are more hours of daylight than darkness. In Alaska, if it is winter, then there are more hours of darkness than daylight.

47. Sample answer: In Alaska, if there are more hours of daylight than darkness, then it is summer. In Alaska, if there are more hours of darkness than daylight, then it is winter.

48. Sample answer: If I am exercising, then I am asleep. If I am exercising, then I am not asleep.

49. Conditional statements can be used to describe how to get a discount, rebate, or refund. Sample answers should include the following.

- If you are not 100% satisfied, then return the product for a full refund.
- Wearing a seatbelt reduces the risk of injuries.

50. C; The contrapositive of a statement always has the same truth value as the statement.

51. B; let x be the number of girls in class. Then there are $32 - x$ boys in class.

$$\frac{x}{32 - x} = \frac{5}{3}$$
$$3x = 5(32 - x)$$
$$3x = 160 - 5x$$
$$8x = 160$$
$$x = 20$$
$$32 - x = 12$$

Thus, there are 20 girls and 12 boys in class, so there are $20 - 12$ or 8 more girls than boys.

Page 80 Maintain Your Skills

52. George Washington was the first president of the United States and a hexagon has 5 sides.
$p \wedge q$ is false because p is true and q is false.

53. A hexagon has five sides or $60 \times 3 = 18$.
$q \vee r$ is false because q is false and r is false.

54. George Washington was the first president of the United States or a hexagon has five sides.
$p \vee q$ is true because p is true.

55. A hexagon doesn't have five sides or $60 \times 3 = 18$.
$\sim q \vee r$ is true because $\sim q$ is true.

56. George Washington was the first president of the United States and a hexagon doesn't have five sides.
$p \wedge \sim q$ is true because p is true and $\sim q$ is true.

57. George Washington was not the first president of the United States and $60 \times 3 \neq 18$.
$\sim p \wedge \sim r$ is false because $\sim p$ is false and $\sim r$ is true.

58. $AB = CD$; $AD = BC$

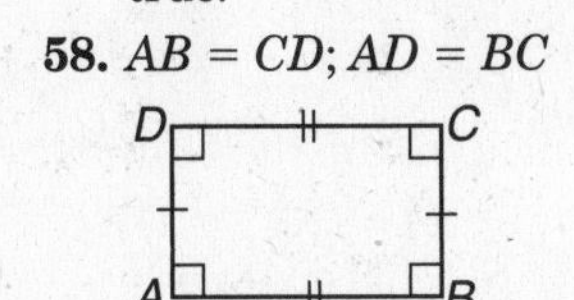

59. The sum of the measures of the angles in a triangle is 180.

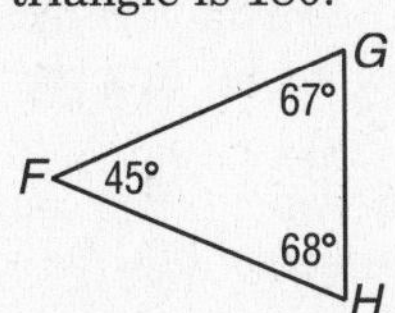

60. $\triangle JKL$ has two sides congruent.

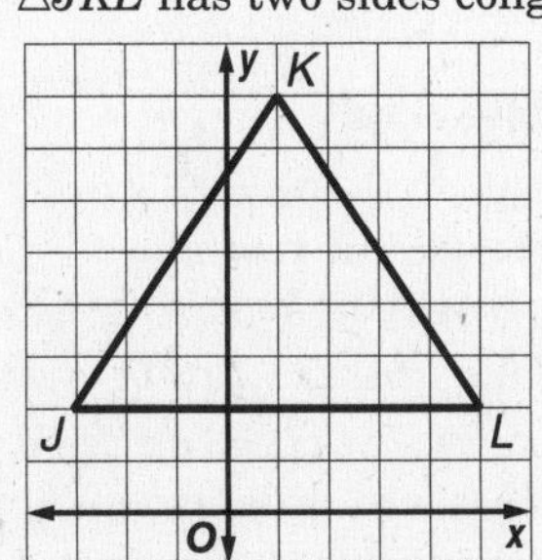

61. $\angle PQR$ is a right angle.

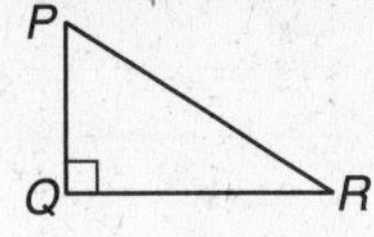

62. $d = \sqrt{(x_2 - x_1)^2 + (y_2 - y_1)^2}$
$CD = \sqrt{[0 - (-2)]^2 + [3 - (-1)]^2}$
$= \sqrt{2^2 + 4^2} = \sqrt{20}$
≈ 4.5

63. $d = \sqrt{(x_2 - x_1)^2 + (y_2 - y_1)^2}$
$JK = \sqrt{[1 - (-3)]^2 + (0 - 5)^2}$
$= \sqrt{4^2 + (-5)^2} = \sqrt{41}$
≈ 6.4

64. $d = \sqrt{(x_2 - x_1)^2 + (y_2 - y_1)^2}$
$PQ = \sqrt{[2 - (-3)]^2 + [-3 - (-1)]^2}$
$= \sqrt{5^2 + (-2)^2} = \sqrt{29}$
≈ 5.4

65. $d = \sqrt{(x_2 - x_1)^2 + (y_2 - y_1)^2}$
$RS = \sqrt{(-4 - 1)^2 + [3 - (-7)]^2}$
$= \sqrt{(-5)^2 + 10^2} = \sqrt{125}$
≈ 11.2

66. Subtract 4 from each side.

67. Multiply each side by 2.

68. Divide each side by 8.

Page 80 Practice Quiz 1

1. False

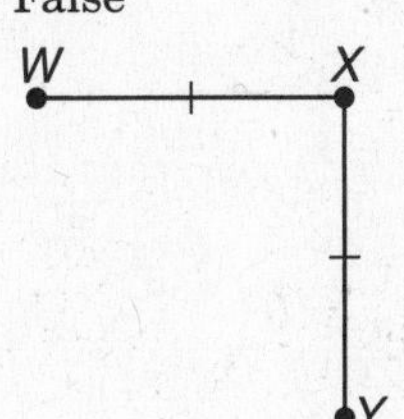

2. True; $m\angle 1 + m\angle 2 = 90$
$m\angle 2 = 90 - m\angle 1$
$m\angle 2 + m\angle 3 = 90$
$(90 - m\angle 1) + m\angle 3 = 90$
$-m\angle 1 + m\angle 3 = 0$
$m\angle 3 = m\angle 1$

3.

p	q	$\sim p$	$\sim p \wedge q$
T	T	F	F
T	F	F	F
F	T	T	T
F	F	T	F

4.

p	q	r	$q \wedge r$	$p \vee (q \wedge r)$
T	T	T	T	T
T	T	F	F	T
T	F	T	F	T
T	F	F	F	T
F	T	T	T	T
F	T	F	F	F
F	F	T	F	F
F	F	F	F	F

5. Converse: If two angles have a common vertex, then the angles are adjacent. False; $\angle ABD$ is not adjacent to $\angle ABC$.

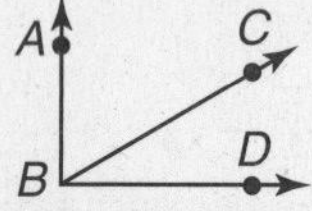

Inverse: If two angles are not adjacent, then they do not have a common vertex; False, $\angle ABC$ and $\angle DBE$ have a common vertex and are not adjacent.

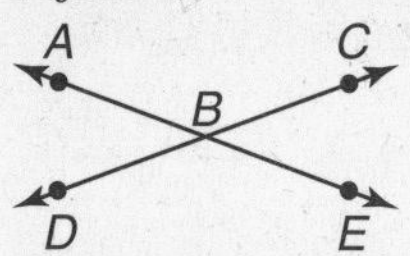

Contrapositive: If two angles do not have a common vertex, then they are not adjacent; true.

Page 81 Reading Mathematics

1. Conditional: If a calculator runs, then it has batteries.
   Converse: If a calculator has batteries, then it will run.
   False; a calculator may be solar powered.
2. Conditional: If two lines intersect, then they are not vertical.
   Converse: If two lines are not vertical, then they intersect.
   False; two parallel horizontal lines will not intersect.
3. Conditional: If two angles are congruent, then they have the same measure.
   Converse: If two angles have the same measure, then they are congruent.
   true
4. Conditional: If $3x - 4 = 20$, then $x = 7$.
   Converse: If $x = 7$, then $3x - 4 = 20$.
   False; $3x - 4 = 17$ when $x = 7$.
5. Conditional: If a line is a segment bisector, then it intersects the segment at its midpoint.
   Converse: If a line intersects a segment at its midpoint, then it is a segment bisector.
   true

2-4 Deductive Reasoning

Page 84 Check for Understanding

1. Sample answer: a: If it is rainy, the game will be cancelled.
   b: It is rainy.
   c: The game will be cancelled.
2. Transitive Property of Equality: $a = b$ and $b = c$ implies $a = c$. Law of Syllogism. a implies b and b implies c implies a implies c. Each statement establishes a relationship between a and c through their relationships to b.
3. Lakeisha; if you are dizzy, that does not necessarily mean that you are seasick and thus have an upset stomach.
4. Valid; the conditional is true and the hypothesis is true, so the conclusion is true.
5. Invalid; congruent angles do not have to be vertical.
6. no conclusion
7. Let p, q, and r represent the parts of the statement.
   p: the midpoint divides a segment
   q: two segments are congruent
   r: two segments have equal measures
   The given statements are true,
   Statement (1): $p \to q$
   Statement (2): $q \to r$
   So by the Law of Syllogism $p \to r$. Thus, the midpoint of a segment divides it into two segments with equal measures.
8. p: Molly arrives at school at 7:30 A.M.
   q: she will get help in math
   r: she will pass her math test
   Statement (3) is a valid conclusion by the Law of Syllogism.
9. Invalid; not all angles that are congruent are right angles.
10. A 35-year old female pays $14.35 per month for $30,000 of insurance, and Ann is a 35-year old female, so by the Law of Detachment, Ann pays $14.35 per month.
11. No; Terry could be a man or a woman. She could be 45 and have purchased $30,000 of life insurance.

Pages 85–87 Practice and Apply

12. invalid; $10 + 12 = 22$
13. Valid; since 5 and 7 are odd, the Law of Detachment indicates that their sum is even.
14. Valid; since 11 and 23 are odd, the Law of Detachment indicates that their sum is even.
15. Invalid; the sum is even.
16. Valid; A, B, and C are noncollinear, and by definition three noncollinear points determine a plane.

17. Invalid; E, F, and G are not necessarily noncollinear.

18. Invalid; the hypothesis is false as there are only two points.

19. Valid; the vertices of a triangle are noncollinear, and therefore determine a plane.

20. no conclusion

21. If the measure of an angle is less than 90, then it is not obtuse.

22. If X is the midpoint of $\overline{YZ}$, then $\overline{YX} \cong \overline{XZ}$.

23. no conclusion

24. p: you are an in-line skater
q: you live dangerously
r: you like to dance
Yes, statement (3) follows from (1) and (2) by the Law of Syllogism.

25. p: the measure of an angle is greater than 90
q: the angle is obtuse
Yes, statement (3) follows from (1) and (2) by the Law of Detachment.

26. Invalid; statement (1) is true, but statement (3) does not follow from (2). Not all congruent angles are vertical angles.

27. p: an angle is obtuse
q: the angle cannot be acute
Yes, statement (3) follows from (1) and (2) by the Law of Detachment.

28. p: you drive safely
q: you can avoid accidents
Yes, statement (3) follows from (1) and (2) by the Law of Detachment.

29. p: you are a customer
q: you are always right
r: you are a teenager
$r \rightarrow p$ does not follow from $(p \rightarrow q) \wedge (r \rightarrow q)$; invalid

30. p: John Steinbeck lived in Monterey
q: during the 1940s, one could hear the grating noise of the fish canneries.
If John Steinbeck lived in Monterey in 1941, then he could hear the grating noise of the fish canneries.

31. p: Catriona Le May Doan skated her second 500 meters in 37.45 seconds
q: She beat the time of Monique Garbrecht-Enfeldt
r: She would win the race
By the Law of Syllogism, if Catriona Le May Doan skated her second 500 meters in 37.45 seconds, then she would win the race.

32. Sample answer: Stacey assumed that the conditional statement was true.

33. Sample answer: Doctors and nurses use charts to assist in determining medications and their doses for patients. Answers should include the following.
- Doctors need to note a patient's symptoms to determine which medication to prescribe, then determine how much to prescribe based on weight, age, severity of the illness, and so on.
- Doctors use what is known to be true about diseases and when symptoms appear, then deduce that the patient has a particular illness.

34. C; if A were true, then Yasahiro would be a professional athlete by I, contradicting II. If B were true, then either Yasahiro is a professional athlete (contradicting II), or Yasahiro gets paid, which, together with III and I, contradicts II. D contradicts III. If C was not true, then II would be contradicted. Therefore C must be true.

35. B; 15% off the \$16 meal means the diner's meal cost (0.85)(\$16) or \$13.60. 20% of \$13.60 means the diner left a tip of (0.20)(\$13.60) or \$2.72. Thus, the diner paid a total of \$13.60 + \$2.72 or \$16.32.

Page 87 Maintain Your Skills

36. If you try Casa Fiesta, then you're looking for a fast, easy way to add some fun to your family's menu.

37. They are a fast, easy way to add fun to your family's menu.

38. No; the conclusion is implied.

39.

q	r	$q \wedge r$
T	T	T
T	F	F
F	T	F
F	F	F

40.

p	r	$\sim p$	$\sim p \vee r$
T	T	F	T
T	F	F	F
F	T	T	T
F	F	T	T

41.

p	q	r	$q \vee r$	$p \wedge (q \vee r)$
T	T	T	T	T
T	T	F	T	T
T	F	T	T	T
T	F	F	F	F
F	T	T	T	F
F	T	F	T	F
F	F	T	T	F
F	F	F	F	F

42.

p	q	r	$\sim q$	$\sim q \wedge r$	$p \vee (\sim q \wedge r)$
T	T	T	F	F	T
T	T	F	F	F	T
T	F	T	T	T	T
T	F	F	T	F	T
F	T	T	F	F	F
F	T	F	F	F	F
F	F	T	T	T	T
F	F	F	T	F	F

43. $\angle HDC$ is a right angle and $\angle HDC$ and $\angle HDF$ are a linear pair so $\angle HDF$ is a right angle. Thus, $\angle HDG$ is complementary to $\angle FDG$ because $m\angle HDG + m\angle FDG = 90$

44. Sample answer: $\angle KHJ$ and $\angle DHG$

45. Sample answer: $\angle JHK$ and $\angle DHK$

46. Congruent, adjacent, supplementary, linear pair

47. Yes, slashes on the segments indicate that they are congruent.

48.

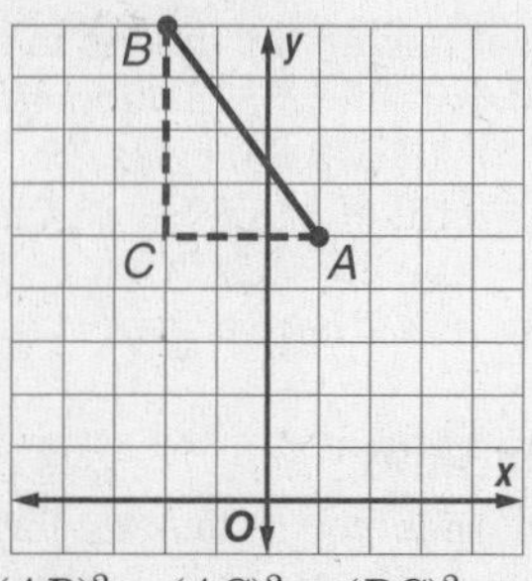

$(AB)^2 = (AC)^2 + (BC)^2$
$(AB)^2 = 3^2 + 4^2 = 25$
$AB = 5$

49.

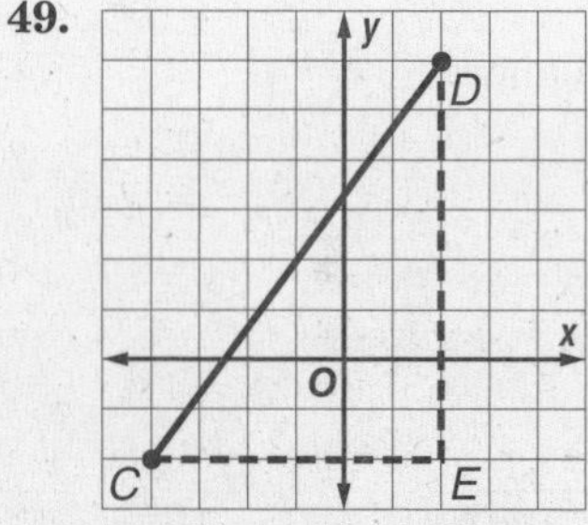

$(CD)^2 = (CE)^2 + (DE)^2$
$(CD)^2 = 6^2 + 8^2 = 100$
$CD = 10$

50.

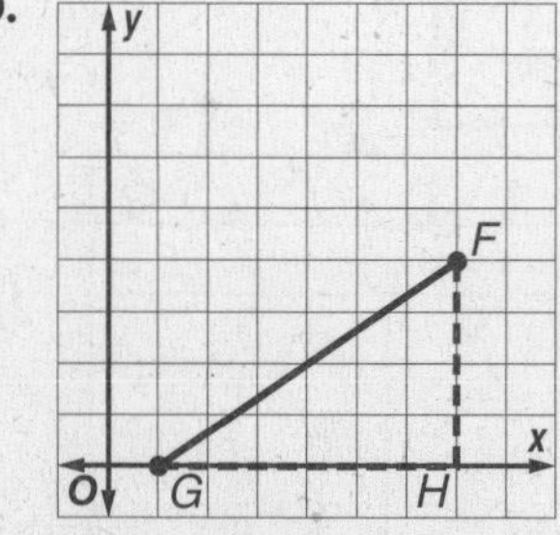

$(FG)^2 = (GH)^2 + (FH)^2$
$(FG)^2 = 6^2 + 4^2 = 52$
$FG = \sqrt{52} \approx 7.2$

51.

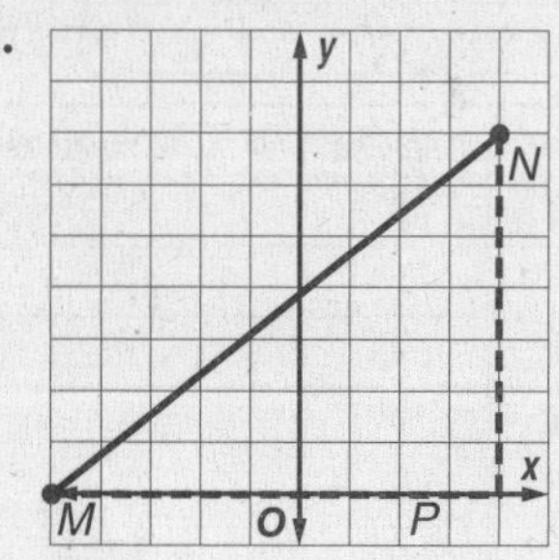

$(MN)^2 = (MP)^2 + (NP)^2$
$(MN)^2 = 9^2 + 7^2 = 130$
$MN = \sqrt{130} \approx 11.4$

52.

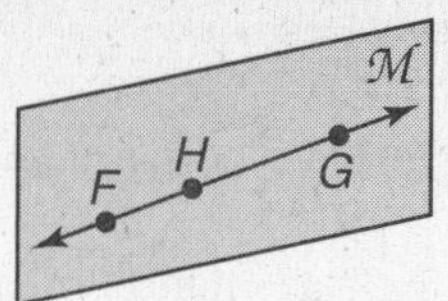

53.

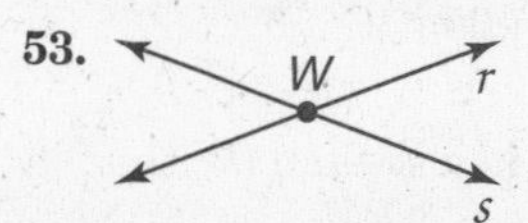

54.

55.

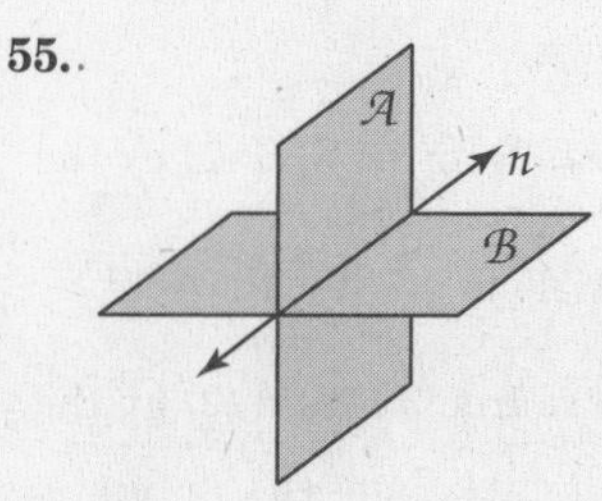

56. Sample answer: $\overline{AM} \cong \overline{CM}$, $\overline{CN} \cong \overline{BN}$, $AM = CM$, $CN = BN$, M is midpoint of $\overline{AC}$, N is midpoint of $\overline{BC}$.

57. Sample answer: $\angle 1$ and $\angle 2$ are complementary, $m\angle 1 + m\angle 2 = 90$.

58. Sample answer: $\angle 4$ and $\angle 5$ are supplementary, $m\angle 4 + m\angle 5 = 180$, $\angle 5$ and $\angle 6$ are supplementary, $m\angle 5 + m\angle 6 = 180$, $\angle 4 \cong \angle 6$, $m\angle 4 = m\angle 6$.

Page 88 Geometry Activity: Matrix Logic

1.

Job	Nate	John	Nick
Veterinarian's office	✓	✗	✗
Computer store	✗	✗	✓
Restaurant	✗	✓	✗

Nate works at the veterinarian's office, John works at the restaurant, and Nick works at the computer store.

2.

Apartment	Anita	Kelli	Scott	Eric	Ava	Roberto
A	✗	✗	✗	✗	✗	✓
B	✗	✗	✗	✓	✗	✗
C	✓	✗	✗	✗	✗	✗
D	✗	✗	✗	✗	✓	✗
E	✗	✗	✓	✗	✗	✗
F	✗	✓	✗	✗	✗	✗

Roberto lives in A, Eric lives in B, Anita lives in C, Ava lives in D, Scott lives in E, and Kelli lives in F.

2-5 Postulates and Paragraph Proofs

Page 91 Check for Understanding

1. Deductive reasoning is used to support claims that are made in a proof.
2. 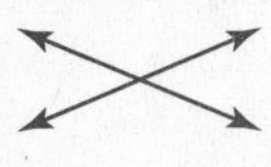

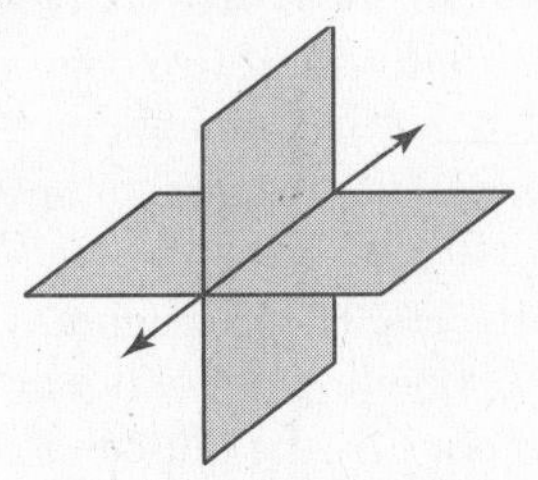

3. postulates, theorems, algebraic properties, definitions
4. **Explore:** There are four points, and each pair is to be connected by a segment.
Plan: Draw a diagram to illustrate the solution.

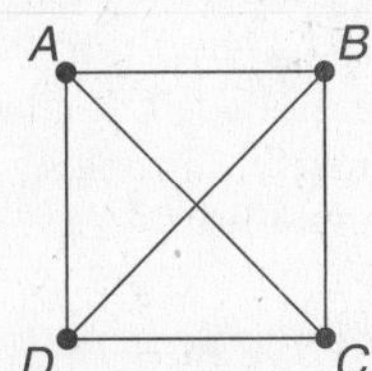

Solve: Connect each point with every other point. Then, count the number of segments. Between every two points there is exactly one segment. For the four points, six segments can be drawn.
Examine: The six segments that can be drawn are $\overline{AB}$, $\overline{AC}$, $\overline{AD}$, $\overline{BC}$, $\overline{BD}$, and $\overline{CD}$.

5. **Explore:** There are six points, and each pair is to be connected by a segment.
Plan: Draw a diagram to illustrate the solution

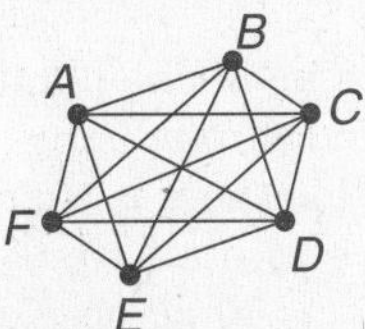

Solve: Connect each point with every other point. Then, count the number of segments. Between every two points there is exactly one segment. For the 6 points, 15 segments can be drawn.
Examine: The 15 segments that can be drawn are $\overline{AB}$, $\overline{AC}$, $\overline{AD}$, $\overline{AE}$, $\overline{AF}$, $\overline{BC}$, $\overline{BD}$, $\overline{BE}$, $\overline{BF}$, $\overline{CD}$, $\overline{CE}$, $\overline{CF}$, $\overline{DE}$, $\overline{DF}$, and $\overline{EF}$.

6. Sometimes; if the planes have a common intersection, then their intersection is one line.
7. Definition of collinear
8. Through any three points not on the same line, there is exactly one plane.
9. Through any two points, there is exactly one line.
10. **Given:** P is the midpoint of QR and ST, and $QR \cong ST$.
Prove: $PQ = PT$

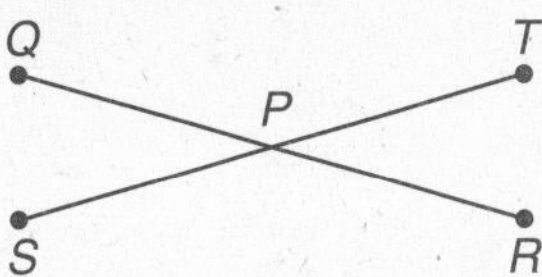

Proof: Since P is the midpoint of $\overline{QR}$ and $\overline{ST}$, $PQ = PR = \frac{1}{2}QR$ and $PS = PT = \frac{1}{2}ST$ by the definition of midpoint. We are given $\overline{QR} \cong \overline{ST}$ so $QR = ST$ by the definition of congruent segments. By the Multiplication Property, $\frac{1}{2}QR = \frac{1}{2}ST$. So, by substitution, $PQ = PT$.

11. **Explore:** There are six students, and each student is connected to five other students with ribbons.
Plan: Draw a diagram to illustrate the solution.

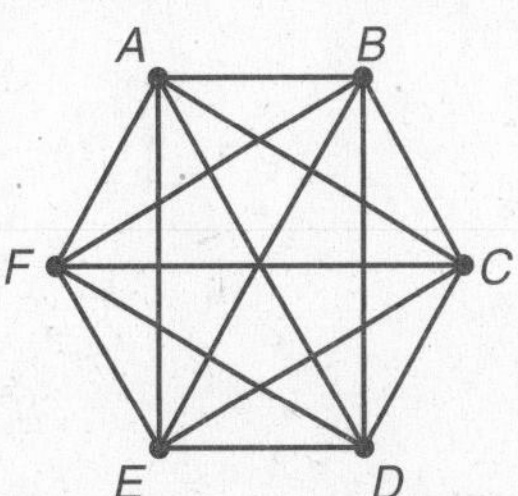

Solve: Let noncollinear points A, B, C, D, E, and F represent the six students. Connect each point with every other point. Then, count the number of segments. Between every two points there is exactly one segment. For the 6 points, 15 segments can be drawn.
Examine: In the figure, $\overline{AB}$, $\overline{AC}$, $\overline{AD}$, $\overline{AE}$, $\overline{AF}$, $\overline{BC}$, $\overline{BD}$, $\overline{BE}$, $\overline{BF}$, $\overline{CD}$, $\overline{CE}$, $\overline{CF}$, $\overline{DE}$, $\overline{DF}$, and $\overline{EF}$ each represent a ribbon between two students. There are 15 segments, so 15 ribbons are needed.

12. **Explore:** There are four points, and each pair is to be connected by a segment.
Plan: Draw a diagram.

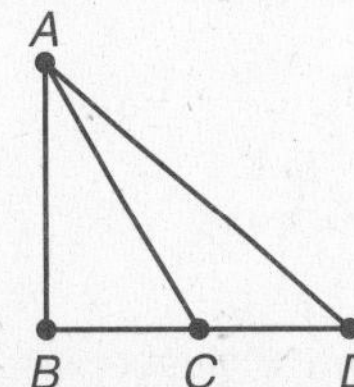

Solve: Connect each point with every other point. Then, count the number of segments. Between every two points there is exactly one segment. For the four points, six segments can be drawn.
Examine: The six segments that can be drawn are $\overline{AB}$, $\overline{AC}$, $\overline{AD}$, $\overline{BC}$, $\overline{BD}$, and $\overline{CD}$.

13. Explore: There are five points, and each pair is to be connected by a segment.
Plan: Draw a diagram.

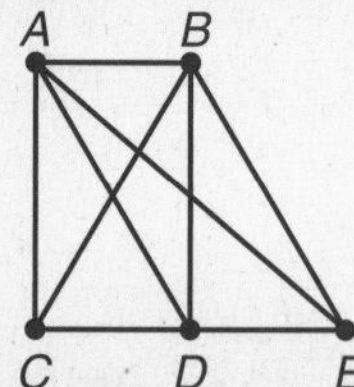

Solve: Connect each point with every other point. Then, count the number of segments. Between every two points there is exactly one segment. For the five points, ten segments can be drawn.
Examine: The ten segments that can be drawn are $\overline{AB}$, $\overline{AC}$, $\overline{AD}$, $\overline{AE}$, $\overline{BC}$, $\overline{BD}$, $\overline{BE}$, $\overline{CD}$, $\overline{CE}$, and $\overline{DE}$.

14. Explore: There are six points, and each pair is to be connected by a segment.
Plan: Draw a diagram.

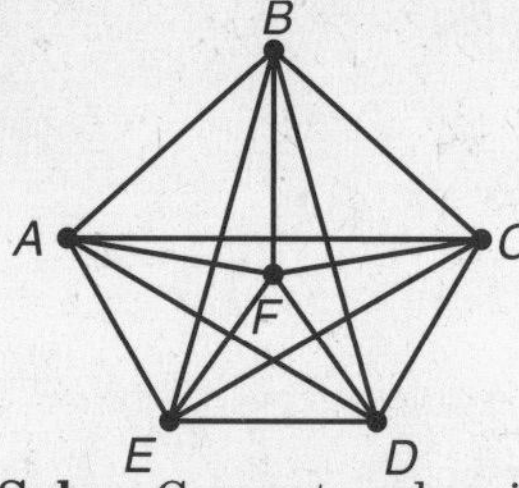

Solve: Connect each point with every other point. Then, count the number of segments. Between every two points there is exactly one segment. For the 6 points, 15 segments can be drawn.
Examine: The 15 segments that can be drawn are $\overline{AB}$, $\overline{AC}$, $\overline{AD}$, $\overline{AE}$, $\overline{AF}$, $\overline{BC}$, $\overline{BD}$, $\overline{BE}$, $\overline{BF}$, $\overline{CD}$, $\overline{CE}$, $\overline{CF}$, $\overline{DE}$, $\overline{DF}$, and $\overline{EF}$.

15. Explore: There are seven points, and each pair is to be connected by a segment.
Plan: Draw a diagram

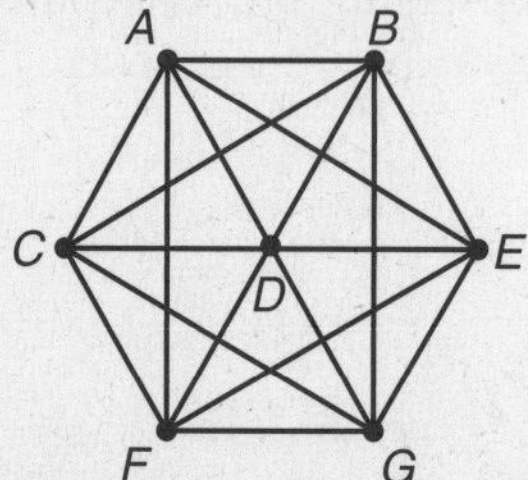

Solve: Connect each point with every other point. Then, count the number of segments. Between every two points there is exactly one segment. For the 7 points, 21 segments can be drawn.
Examine: The 21 segments that can be drawn are $\overline{AB}$, $\overline{AC}$, $\overline{AD}$, $\overline{AE}$, $\overline{AF}$, $\overline{AG}$, $\overline{BC}$, $\overline{BD}$, $\overline{BE}$, $\overline{BF}$, $\overline{BG}$, $\overline{CD}$, $\overline{CE}$, $\overline{CF}$, $\overline{CG}$, $\overline{DE}$, $\overline{DF}$, $\overline{DG}$, $\overline{EF}$, $\overline{EG}$, and $\overline{FG}$.

16. Sometimes; the three points cannot be on the same line.

17. Always; if two points lie in a plane, then the entire line containing those points lies in that plane.

18. Never; the intersection of a line and a plane can be a point, but the intersection of two planes is a line.

19. Sometimes; the three points cannot be on the same line.

20. Always; one plane contains at least three points, so it must contain two.

21. Sometimes; ℓ and m could be skew, so they would not lie in the same plane.

22. Postulate 2.1: Through any two points, there is exactly one line.

23. Postulate 2.5: If two points lie in a plane, then the entire line containing those points lies in that plane.

24. Postulate 2.2: Through any three points not on the same line, there is exactly one plane.

25. Postulate 2.5: If two points lie in a plane, then the entire line containing those points lies in the plane.

26. Postulate 2.1: Through any two points, there is exactly one line.

27. Postulate 2.2: Through any three points not on the same line, there is exactly one plane.

28. Given: C is the midpoint of $\overline{AB}$.
B is the midpoint of $\overline{CD}$.
Prove: $\overline{AC} \cong \overline{BD}$

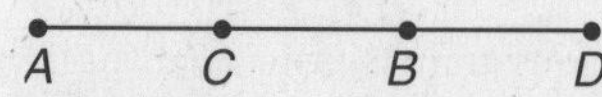

Proof: We are given that C is the midpoint of $\overline{AB}$, and B is the midpoint of $\overline{CD}$. By the definition of midpoint $\overline{AC} \cong \overline{CB}$ and $\overline{CB} \cong \overline{BD}$. Using the definition of congruent segments, $AC = CB$, and $CB = BD$. $AC = BD$ by the Transitive Property of Equality. Thus, $\overline{AC} \cong \overline{BD}$ by the definition of congruent segments.

29. There are 4 points, call them A, B, C, and D. Then there is exactly one line between each pair of points, so there are 6 lines: $\overleftrightarrow{AB}$, $\overleftrightarrow{AC}$, $\overleftrightarrow{AD}$, $\overleftrightarrow{BC}$, $\overleftrightarrow{BD}$, and $\overleftrightarrow{CD}$. The points are noncollinear and noncoplanar, and through any three points not on the same line there is exactly one plane. So there are 4 different planes: plane ABC, plane ACD, plane BCD, and plane ABD.

30. Sample answer: Lawyers make final arguments, which is a speech that uses deductive reasoning, in court cases.

31. It's possible that all five points lie in one plane. The points are noncollinear, and through any three points not on the same line there is exactly one plane. If the five points are points A, B, C, D, and E, then there are as many as 10 planes defined by these points: plane ABC, plane ABD, plane ABE, plane ACD, plane ACE, plane ADE, plane BCD, plane BCE, plane BDE, and plane CDE.

32. Sample answer: The forms and structures of different types of writing are accepted as true, such as the structure of a poem. Answers should include the following.

- The Declaration of Independence, "We hold these truths to be self-evident, ..."
- Through any two points, there is exactly one line.

33. C; A is true because Postulate 2.2 states that through any 3 points not on the same line, there is exactly one plane.
B is true because Postulate 2.6 states that if 2 lines intersect, then their intersection is one point.
D is true by the Midpoint Theorem.
C is not true because it contradicts Postulate 2.1 which states that through any two points, there is exactly one line.

34. A;
$(8x^4 - 2x^2 + 3x - 5) - (2x^4 + x^3 + 3x + 5)$
$= 8x^4 - 2x^2 + 3x - 5 - 2x^4 - x^3 - 3x - 5$
$= 6x^4 - x^3 - 2x^2 - 10$

Page 93 Maintain Your Skills

35. p: one has a part-time job
q: one must work 20 hours per week
Statement (3) is a valid conclusion by the Law of Detachment.

36. Converse: If you have a computer, then you have access to the Internet at your house. False; you can have a computer and not have access to the Internet.
Inverse: If you do not have access to the Internet at your house, then you do not have a computer. False; it is possible to not have access to the Internet and still have a computer.
Contrapositive: If you do not have a computer, then you do not have access to the Internet at your house. False; you could have Internet access through your television or wireless phone.

37. Converse: If $\triangle ABC$ has an angle measure greater than 90, then $\triangle ABC$ is a right triangle. False; the triangle would be obtuse.
Inverse: If $\triangle ABC$ is not a right triangle, none of its angle measures are greater than 90. False; it could be an obtuse triangle.
Contrapositive: If $\triangle ABC$ does not have an angle measure greater than 90, $\triangle ABC$ is not a right triangle. False; $m\angle ABC$ could still be 90 and $\triangle ABC$ be a right triangle.

38.

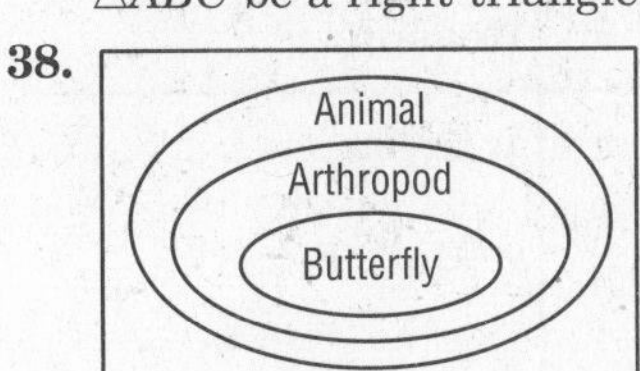

39. $d = \sqrt{(x_2 - x_1)^2 + (y_2 - y_1)^2}$
$DF = \sqrt{(4 - 3)^2 + (-1 - 3)^2}$
$= \sqrt{1^2 + (-4)^2} = \sqrt{17}$
≈ 4.1

40. $d = \sqrt{(x_2 - x_1)^2 + (y_2 - y_1)^2}$
$MN = \sqrt{(-5 - 0)^2 + (5 - 2)^2}$
$= \sqrt{(-5)^2 + 3^2} = \sqrt{34}$
≈ 5.8

41. $d = \sqrt{(x_2 - x_1)^2 + (y_2 - y_1)^2}$
$PQ = \sqrt{[1 - (-8)]^2 + (-3 - 2)^2}$
$= \sqrt{9^2 + (-5)^2} = \sqrt{106}$
≈ 10.3

42. $d = \sqrt{(x_2 - x_1)^2 + (y_2 - y_1)^2}$
$RS = \sqrt{[2 - (-5)]^2 + (1 - 12)^2}$
$= \sqrt{7^2 + (-11)^2} = \sqrt{170}$
≈ 13.0

43. $m - 17 = 8$
$m = 25$

44. $3y = 57$
$y = 19$

45. $\frac{y}{6} + 12 = 14$
$\frac{y}{6} = 2$
$y = 12$

46. $-t + 3 = 27$
$-t = 24$
$t = -24$

47. $8n - 39 = 41$
$8n = 80$
$n = 10$

48. $-6x + 33 = 0$
$-6x = -33$
$x = \frac{11}{2}$

2-6 Algebraic Proof

Page 97 Check for Understanding

1. Sample answer: If $x = 2$ and $x + y = 6$, then $2 + y = 6$.
2. given and prove statements and two columns, one of statements and one of reasons.
3. hypothesis; conclusion
4. Division Property
5. Multiplication Property
6. Substitution
7. Addition Property
8. **Given:** $\frac{x}{2} + 4x - 7 = 11$
   Prove: $x = 4$
   Proof:

Statements	Reasons
1. $\frac{x}{2} + 4x - 7 = 11$	1. Given
2. $2\left(\frac{x}{2} + 4x - 7\right) = 2(11)$	2. Mult. Prop.
3. $x + 8x - 14 = 22$	3. Distributive Prop.
4. $9x - 14 = 22$	4. Substitution
5. $9x = 36$	5. Add. Prop.
6. $x = 4$	6. Div. Prop.

9. **Given:** $5 - \frac{2}{3}x = 1$
   Prove: $x = 6$
   Proof:

Statements	Reasons
a. _?_ $5 - \frac{2}{3}x = 1$	a. Given
b. $3\left(5 - \frac{2}{3}x\right) = 3(1)$	b. _?_ Mult. Prop.
c. $15 - 2x = 3$	c. _?_ Dist. Prop.
d. _?_ $-2x = -12$	d. Subt. Prop.
e. $x = 6$	e. _?_ Div. Prop.

10. Given: $25 = -7(y - 3) + 5y$
Prove: $-2 = y$
Proof:

Statements	Reasons
1. $25 = -7(y - 3) + 5y$	1. Given
2. $25 = -7y + 21 + 5y$	2. Dist. Prop.
3. $25 = -2y + 21$	3. Substitution
4. $4 = -2y$	4. Subt. Prop.
5. $-2 = y$	5. Div. Prop.

11. Given: Rectangle $ABCD$, $AD = 3$, $AB = 10$
Prove: $AC = BD$

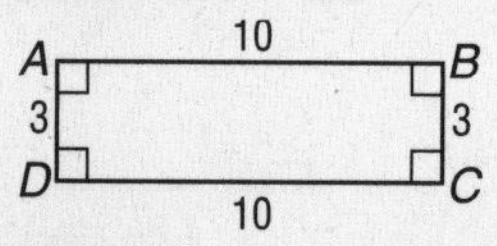

Proof:

Statements	Reasons
1. Rectangle $ABCD$, $AD = 3$, $AB = 10$	1. Given
2. Draw segments AC and DB.	2. Two points determine a line.
3. $\triangle ABC$ and $\triangle BCD$ are right triangles.	3. Def. of rt. $\triangle$
4. $AC = \sqrt{3^2 + 10^2}$, $DB = \sqrt{3^2 + 10^2}$	4. Pythag. Thm.
5. $AC = BD$	5. Substitution

12. Given: $c^2 = a^2 + b^2$
Prove: $a = \sqrt{c^2 - b^2}$
Proof:

Statements	Reasons
1. $c^2 = a^2 + b^2$	1. Given
2. $c^2 - b^2 = a^2$	2. Subt. Prop.
3. $a^2 = c^2 - b^2$	3. Symmetric Prop.
4. $\sqrt{a^2} = \sqrt{c^2 - b^2}$	4. Square Root Prop.
5. $a = \sqrt{c^2 - b^2}$	5. Square Root Prop.

13. C; $8 + x = 12$
$x = 4$
$4 - x = 4 - 4$ or 0

Pages 97–99 Practice and Apply

14. Transitive Property
15. Subtraction Property
16. Substitution
17. Substitution
18. Division or Multiplication Property
19. Reflexive Property
20. Distributive Property
21. Substitution
22. Division or Multiplication Property
23. Transitive Property

24. Given: $\frac{3x + 5}{2} = 7$
Prove: $x = 3$
Proof:

Statements	Reasons
a. $\frac{3x + 5}{2} = 7$	a. __?__ Given
b. __?__ $2\left(\frac{3x + 5}{2}\right) = 2(7)$	b. Mult. Prop.
c. $3x + 5 = 14$	c. __?__ Substitution
d. $3x = 9$	d. __?__ Subt. Prop.
e. __?__ $x = 3$	e. Div. Prop.

25. Given: $2x - 7 = \frac{1}{3}x - 2$
Prove: $x = 3$
Proof:

Statements	Reasons
a. __?__ $2x - 7 = \frac{1}{3}x - 2$	a. Given
b. __?__ $3(2x - 7) = 3\left(\frac{1}{3}x - 2\right)$	b. Mult. Prop.
c. $6x - 21 = x - 6$	c. __?__ Dist. Prop.
d. __?__ $5x - 21 = -6$	d. Subt. Prop.
e. $5x = 15$	e. __?__ Add. Prop.
f. __?__ $x = 3$	f. Div. Prop.

26. Given: $4 - \frac{1}{2}a = \frac{7}{2} - a$
Prove: $a = -1$
Proof:

Statements	Reasons
1. $4 - \frac{1}{2}a = \frac{7}{2} - a$	1. Given
2. $2\left(4 - \frac{1}{2}a\right) = 2\left(\frac{7}{2} - a\right)$	2. Mult. Prop.
3. $8 - a = 7 - 2a$	3. Dist. Prop.
4. $1 - a = -2a$	4. Subt. Prop.
5. $1 = -1a$	5. Add. Prop.
6. $-1 = a$	6. Div. Prop.
7. $a = -1$	7. Symmetric Prop.

27. Given: $-2y + \frac{3}{2} = 8$
Prove: $y = -\frac{13}{4}$
Proof:

Statements	Reasons
1. $-2y + \frac{3}{2} = 8$	1. Given
2. $2\left(-2y + \frac{3}{2}\right) = 2(8)$	2. Mult. Prop.
3. $-4y + 3 = 16$	3. Dist. Prop.
4. $-4y = 13$	4. Subt. Prop.
5. $y = -\frac{13}{4}$	5. Div. Prop.

28. Given: $-\frac{1}{2}m = 9$
Prove: $m = -18$
Proof:

Statements	Reasons
1. $-\frac{1}{2}m = 9$	1. Given
2. $-2\left(-\frac{1}{2}m\right) = -2(9)$	2. Mult. Prop.
3. $m = -18$	3. Substitution

29. Given: $5 - \frac{2}{3}z = 1$
Prove: $z = 6$
Proof:

Statements	Reasons
1. $5 - \frac{2}{3}z = 1$	1. Given
2. $3\left(5 - \frac{2}{3}z\right) = 3(1)$	2. Mult. Prop.
3. $15 - 2x = 3$	3. Dist. Prop.
4. $15 - 2x - 15 = 3 - 15$	4. Subt. Prop.
5. $-2x = -12$	5. Substitution
6. $\frac{-2x}{-2} = \frac{-12}{-2}$	6. Div. Prop.
7. $x = 6$	7. Substitution

30. Given: $XZ = ZY$,
$XZ = 4x + 1$, and
$ZY = 6x - 13$
Prove: $x = 7$

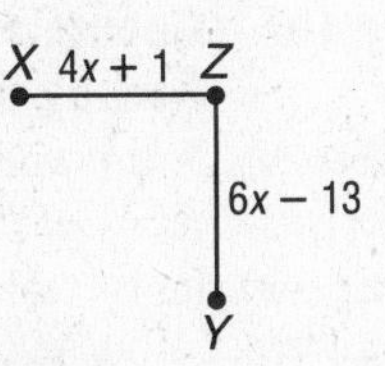

Proof:

Statements	Reasons
1. $XZ = ZY$, $XZ = 4x + 1$, and $ZY = 6x - 13$	1. Given
2. $4x + 1 = 6x - 13$	2. Substitution
3. $4x + 1 - 4x = 6x - 13 - 4x$	3. Subt. Prop.
4. $1 = 2x - 13$	4. Substitution
5. $1 + 13 = 2x - 13 + 13$	5. Add. Prop.
6. $14 = 2x$	6. Substitution
7. $\frac{14}{2} = \frac{2x}{2}$	7. Div. Prop.
8. $7 = x$	8. Substitution
9. $x = 7$	9. Symmetric

31. Given: $m\angle ACB = m\angle ABC$
Prove: $m\angle XCA = m\angle YBA$
Proof:

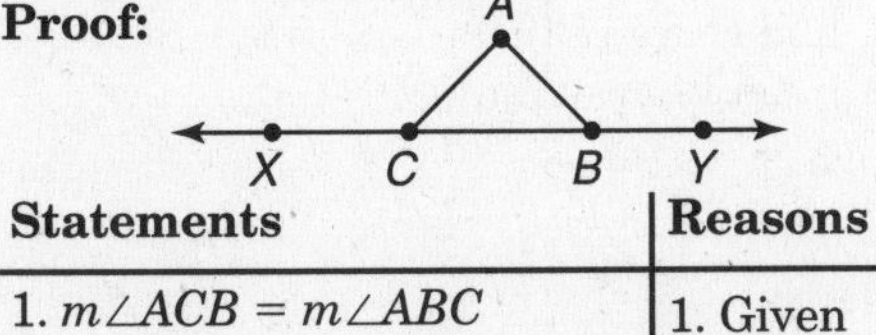

Statements	Reasons
1. $m\angle ACB = m\angle ABC$	1. Given
2. $m\angle XCA + m\angle ACB = 180$ $m\angle YBA + m\angle ABC = 180$	2. Def. of supp. ∠s
3. $m\angle XCA + m\angle ACB = m\angle YBA + m\angle ABC$	3. Substitution
4. $m\angle XCA + m\angle ACB = m\angle YBA + m\angle ACB$	4. Substitution
5. $m\angle XCA = m\angle YBA$	5. Subt. Prop.

32. Given: $E_k = hf + W$
Prove: $f = \frac{E_k - W}{h}$
Proof:

Statements	Reasons
1. $E_k = hf + W$	1. Given
2. $E_k - W = hf$	2. Subt. Prop.
3. $\frac{E_k - W}{h} = f$	3. Div. Prop.
4. $f = \frac{E_k - W}{h}$	4. Symmetric Prop.

33. Given: $m\angle ACB = m\angle DCE$
Prove: $m\angle ACB = m\angle ACG$
$m\angle DCE = m\angle ECF$
Proof:

Statements	Reasons
1. $m\angle ACB = m\angle DCE$	1. Given
2. $m\angle ACB = m\angle ECF$ $m\angle DCE = m\angle ACG$	2. Def. of vert. ∠s
3. $m\angle ACB = m\angle ACG$ $m\angle DCE = m\angle ECF$	3. Transitive Prop.

Thus, all of the angle measures would be equal.

34. Sample answer: Michael has a symmetric relationship of first cousin with Chris, Kevin, Diane, Dierdre, and Steven. Diane, Dierdre, and Steven have a symmetric and transitive relationship of sibling. Any direct line from bottom to top has a transitive descendent relationship.

35. See students' work.

36. Sample answer: Lawyers use evidence and testimony as reasons for justifying statements and actions. All of the evidence and testimony is linked together to prove a lawyer's case similar to a proof in mathematics.
Answers should include the following.

- Evidence is used to verify facts from witnesses or materials.
- Postulates, theorems, definitions, and properties can be used to justify statements made in mathematics.

37. B; $m\angle P + m\angle Q + m\angle R = 180$

$$m\angle Q + m\angle Q + 2(m\angle Q) = 180$$
$$4(m\angle Q) = 180$$
$$m\angle Q = 45$$

$$m\angle P = m\angle Q$$
$$= 45$$

38. B; $4 + x = y - 5$

$$x = y - 9$$

Page 100 Maintain Your Skills

39. Let the four buildings be named A, B, C, and D. In order to have exactly one sidewalk between each building, there should be 6 sidewalks. If $\overline{AB}$ is the sidewalk between buildings A and B, then the 6 sidewalks are $\overline{AB}$, $\overline{AC}$, $\overline{AD}$, $\overline{BC}$, $\overline{BD}$, and $\overline{CD}$.

40. Valid; since 24 is divisible by 6, the Law of Detachment says it is divisible by 3.

41. Invalid; $27 \div 6 = 4.5$, which is not an integer.

42. Valid; since 85 is not divisible by 3, the contrapositive of the statement and the Law of Detachment say that 85 is not divisible by 6.

43. Sample answer: If people are happy, then they rarely correct their faults.

44. Sample answer: If you don't know where you are going, then you will probably end up somewhere else.

45. Sample answer: If a person is a champion, then the person is afraid of losing.

46. Sample answer: If we would have new knowledge, then we must get a whole new world of questions.

47. The measurement is precise to within $\frac{1}{2}$ foot. So, a measurement of 13 feet could be $12\frac{1}{2}$ feet to $13\frac{1}{2}$ feet.

48. The measurement is precise to within 0.05 meter. So, a measurement of 5.9 meters could be 5.85 meters to 5.95 meters.

49. The measurement is precise to within 0.5 inch. So, a measurement of 74 inches could be 73.5 inches to 74.5 inches.

50. The measurement is precise to within 0.05 kilometer. So, a measurement of 3.1 kilometers could be 3.05 kilometers to 3.15 kilometers.

51. $JL = JK + KL$

$$25 = 14 + KL$$
$$11 = KL$$

52. $PS = PQ + QS$

$$51 = 23 + QS$$
$$28 = QS$$

53. $WZ = WY + YZ$

$$WZ = 38 + 9$$
$$WZ = 47$$

Page 100 Practice Quiz 2

1. Invalid; not all real numbers are integers.

2. Through any three points not on the same line, there is exactly one plane.

3. If two lines intersect, then their intersection is exactly one point.

4. If two points lie in a plane, then the entire line containing those points lies in that plane.

5. **Given:** $2(n - 3) + 5 = 3(n - 1)$
Prove: $n = 2$
Proof:

Statements	Reasons
1. $2(n - 3) + 5 = 3(n - 1)$	1. Given
2. $2n - 6 + 5 = 3n - 3$	2. Dist. Prop.
3. $2n - 1 = 3n - 3$	3. Substitution
4. $2n - 1 - 2n = 3n - 3 - 2n$	4. Subt. Prop.
5. $-1 = n - 3$	5. Substitution
6. $-1 + 3 = n - 3 + 3$	6. Add. Prop.
7. $2 = n$	7. Substitution
8. $n = 2$	8. Symmetric Prop.

2-7 Proving Segment Relationships

Page 101 Geometry Software Investigation: Adding Segment Measures

1. See students' work. The sum $AB + BC$ should always equal AC.

2. See students' work. The sum $AB + BC$ should always equal AC.

3. See students' work. The sum $AB + BC$ should always equal AC.

4. $AB + BC = AC$

5. no

Pages 103–104 Check for Understanding

1. Sample answer: The distance from Cleveland to Chicago is the same as the distance from Cleveland to Chicago.

2. Sample answer: If $\overline{AB} \cong \overline{XY}$ and $\overline{XY} \cong \overline{PQ}$, then $\overline{AB} \cong \overline{PQ}$.

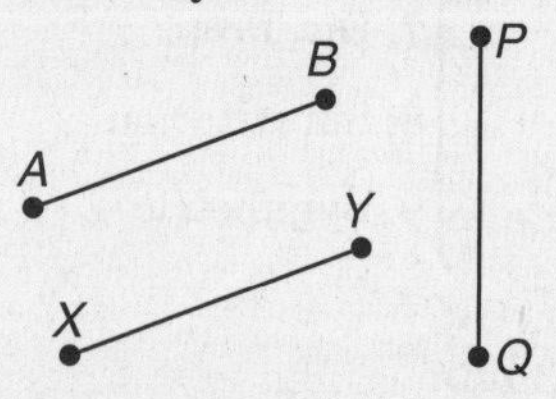

3. If A, B, and C are collinear and $AB + BC = AC$, then B is between A and C.

4. Reflexive

5. Symmetric

6. Subtraction

7. Given: $\overline{PQ} \cong \overline{RS}$, $\overline{QS} \cong \overline{ST}$
Prove: $\overline{PS} \cong \overline{RT}$

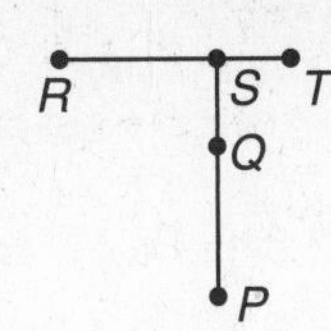

Proof:

Statements	Reasons
a. __?__, __?__ $\overline{PQ} \cong \overline{RS}$, $\overline{QS} \cong \overline{ST}$	a. Given
b. $PQ = RS$, $QS = ST$	b. __?__ Def. of ≅ segments
c. $PS = PQ + QS$, $RT = RS + ST$	c. __?__ Segment Addition Post.
d. __?__ $PQ + QS = RS + ST$	d. Addition Property
e. __?__ $PS = RT$	e. Substitution
f. $\overline{PS} \cong \overline{RT}$	f. __?__ Def. of ≅ segments

8. Given: $\overline{AP} \cong \overline{CP}$
$\overline{BP} \cong \overline{DP}$
Prove: $\overline{AB} \cong \overline{CD}$

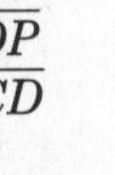

Proof:

Statements	Reasons
1. $\overline{AP} \cong \overline{CP}$ and $\overline{BP} \cong \overline{DP}$	1. Given
2. $AP = CP$ and $BP = DP$	2. Def. of ≅ segs.
3. $AP + PB = AB$	3. Seg. Add. Post.
4. $CP + DP = AB$	4. Substitution
5. $CP + PD = CD$	5. Seg. Add. Post.
6. $AB = CD$	6. Transitive Prop.
7. $\overline{AB} \cong \overline{CD}$	7. Def. of ≅ segs.

9. Given: $\overline{HI} \cong \overline{TU}$
and $\overline{HJ} \cong \overline{TV}$
Prove: $\overline{IJ} \cong \overline{UV}$

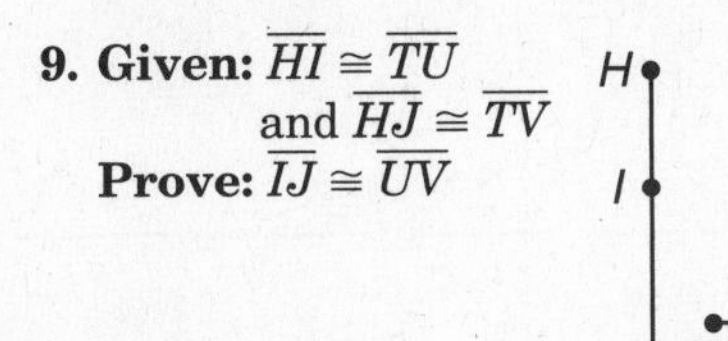

Proof:

Statements	Reasons
1. $\overline{HI} \cong \overline{TU}$ and $\overline{HJ} \cong \overline{TV}$	1. Given
2. $HI = TU$ and $HJ = TV$	2. Def. of ≅ segs.
3. $HI + IJ = HJ$	3. Seg. Add. Post.
4. $TU + IJ = TV$	4. Substitution
5. $TU + UV = TV$	5. Seg. Add. Post.
6. $TU + IJ = TU + UV$	6. Substitution
7. $TU = TU$	7. Reflexive Prop.
8. $IJ = UV$	8. Subt. Prop.
9. $\overline{IJ} \cong \overline{UV}$	9. Def. of ≅ segs.

10. Given: $\overline{AB} \cong \overline{CD}$
Prove: $\overline{CD} \cong \overline{AB}$

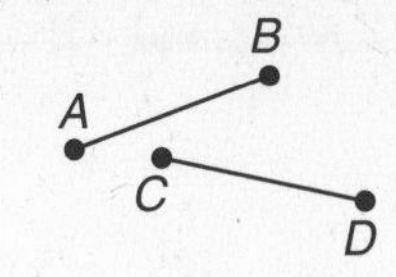

Proof:

Statements	Reasons
1. $\overline{AB} \cong \overline{CD}$	1. Given
2. $AB = CD$	2. Def. of ≅ segs.
3. $CD = AB$	3. Symmetric Prop.
4. $\overline{CD} \cong \overline{AB}$	4. Def. of ≅ segs.

11. Since Aberdeen is in South Dakota while Helena, Miles City, and Missoula are in Montana, Aberdeen is at one end of the line segment along which the four cities lie. Miles City is closest to Aberdeen (473 miles), Helena is next closest to Aberdeen (860 miles), and Missoula is farthest from Aberdeen (972 miles). Thus, Helena is between Missoula and Miles City.

Pages 104–106 Practice and Apply

12. Symmetric
13. Substitution
14. Segment Addition
15. Transitive
16. Addition
17. Subtraction
18. Given: $\overline{AD} \cong \overline{CE}$, $\overline{DB} \cong \overline{EB}$
Prove: $\overline{AB} \cong \overline{CB}$

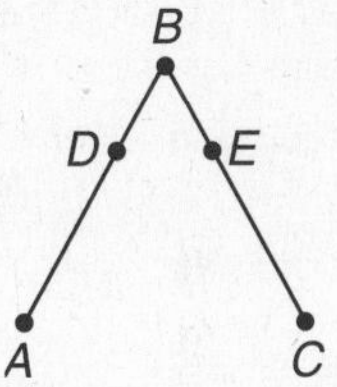

Proof:

Statements:	Reasons:
a. __?__ $\overline{AD} \cong \overline{CE}$, $\overline{DB} \cong \overline{EB}$	a. Given
b. $AD = CE$, $DB = EB$	b. __?__ Def. of ≅ segs.
c. $AD + DB = CE + EB$	c. __?__ Add. Prop.
d. __?__ $AB = AD + DB$, $CB = CE + EB$	d. Segment Addition Postulate
e. $AB = CB$	e. __?__ Substitution
f. $\overline{AB} \cong \overline{CB}$	f. __?__ Def. of ≅ segs.

19. Given: $\overline{XY} \cong \overline{WZ}$ and $\overline{WZ} \cong \overline{AB}$
Prove: $\overline{XY} \cong \overline{AB}$

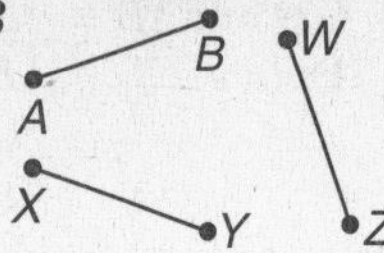

Proof:

Statements	Reasons
1. $\overline{XY} \cong \overline{WZ}$ and $\overline{WZ} \cong \overline{AB}$	1. Given
2. $XY = WZ$ and $WZ = AB$	2. Def. of ≅ segs.
3. $XY = AB$	3. Transitive Prop.
4. $\overline{XY} \cong \overline{AB}$	4. Def. of ≅ segs.

20. Given: $\overline{AB} \cong \overline{AC}$ and $\overline{PC} \cong \overline{QB}$
Prove: $\overline{AP} \cong \overline{AQ}$

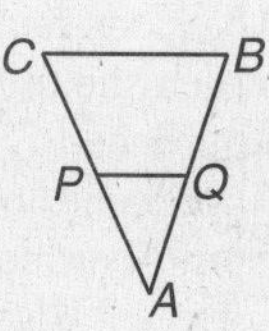

Proof:

Statements	Reasons
1. $\overline{AB} \cong \overline{AC}$ and $\overline{PC} \cong \overline{QB}$	1. Given
2. $AB = AC$, $PC = QB$	2. Def. of ≅ segs.
3. $AB = AQ + QB$, $AC = AP + PC$	3. Seg. Add. Post.
4. $AQ + QB = AP + PC$	4. Substitution
5. $AQ + QB = AP + QB$	5. Substitution
6. $QB = QB$	6. Reflexive Prop.
7. $AP = AQ$	7. Subt. Prop.
8. $\overline{AP} \cong \overline{AQ}$	8. Def. of ≅ segs.

21. Given: $\overline{WY} \cong \overline{ZX}$
A is the midpoint of $\overline{WY}$.
A is the midpoint of $\overline{ZX}$.
Prove: $\overline{WA} \cong \overline{ZA}$

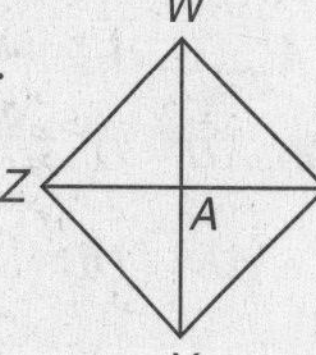

Proof:

Statements	Reasons
a. $\overline{WY} \cong \overline{ZX}$ A is the midpoint of $\overline{WY}$. A is the midpoint of $\overline{ZX}$.	a. __?__ Given
b. $WY = ZX$	b. __?__ Def. of ≅ segs.
c. __?__ $WA = AY$, $ZA = AX$	c. Definition of midpoint
d. $WY = WA + AY$, $ZX = ZA + AX$	d. __?__ Segment Addition Post.
e. $WA + AY = ZA + AX$	e. __?__ Substitution
f. $WA + WA = ZA + ZA$	f. __?__ Substitution
g. $2WA = 2ZA$	g. __?__ Substitution
h. __?__ $WA = ZA$	h. Division Property
i. $\overline{WA} \cong \overline{ZA}$	i. __?__ Def. of ≅ segs.

22. Given: $\overline{LM} \cong \overline{PN}$
and $\overline{XM} \cong \overline{XN}$
Prove: $\overline{LX} \cong \overline{PX}$

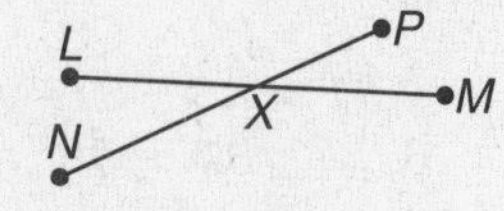

Proof:

Statements	Reasons
1. $\overline{LM} \cong \overline{PN}$ and $\overline{XM} \cong \overline{XN}$	1. Given
2. $LM = PN$ and $XM = XN$	2. Def. of ≅ segs.
3. $LM = LX + XM$, $PN = PX + XN$	3. Seg. Add. Post.
4. $LX + XM = PX + XN$	4. Substitution
5. $LX + XN = PX + XN$	5. Substitution
6. $XN = XN$	6. Reflexive Prop.
7. $LX = PX$	7. Subt. Prop.
8. $\overline{LM} \cong \overline{PX}$	8. Def. of ≅ segs.

23. Given: $AB = BC$
Prove: $AC = 2BC$
Proof:

A B C

Statements	Reasons
1. $AB = BC$	1. Given
2. $AC = AB + BC$	2. Seg. Add. Post.
3. $AC = BC + BC$	3. Substitution
4. $AC = 2BC$	4. Substitution

24. Given: $\overline{AB}$
Prove: $\overline{AB} \cong \overline{AB}$
Proof:

A B

Statements	Reasons
1. $\overline{AB}$	1. Given
2. $AB = AB$	2. Reflexive Prop.
3. $\overline{AB} \cong \overline{AB}$	3. Def. of ≅ segs.

25. Given: $\overline{AB} \cong \overline{DE}$
C is the midpoint of $\overline{BD}$.
Prove: $\overline{AC} \cong \overline{CE}$

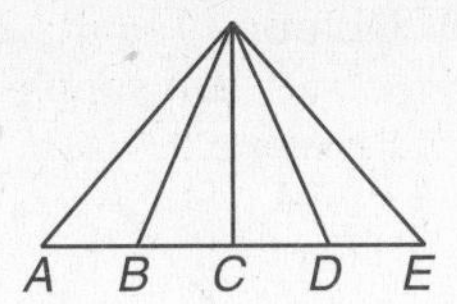

Proof:

Statements	Reasons
1. $\overline{AB} \cong \overline{DE}$, C is the midpoint of $\overline{BD}$	1. Given
2. $BC = CD$	2. Def. of midpoint
3. $AB = DE$	3. Def. of ≅ segs.
4. $AB + BC = CD + DE$	4. Add. Prop.
5. $AB + BC = AC$ $CD + DE = CE$	5. Seg. Add. Post.
6. $AC = CE$	6. Substitution
7. $\overline{AC} \cong \overline{CE}$	7. Def. of ≅ segs.

26. Given: $\overline{AB} \cong \overline{EF}$ and $\overline{BC} \cong \overline{DE}$
Prove: $\overline{AC} \cong \overline{DF}$

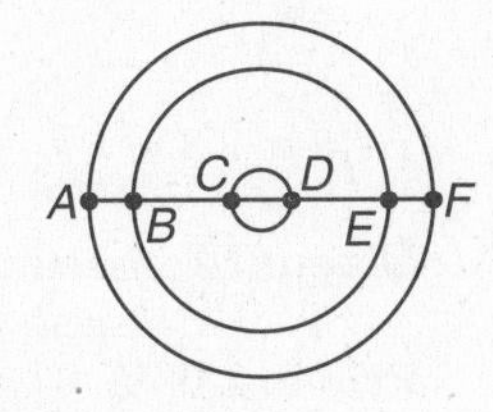

Proof:

Statements	Reasons
1. $\overline{AB} \cong \overline{EF}$ and $\overline{BC} \cong \overline{DE}$	1. Given
2. $AB = EF$ and $BC = DE$	2. Def. of ≅ segs.
3. $AB + BC = DE + EF$	3. Add. Prop.
4. $AC = AB + BC$, $DF = DE + EF$	4. Seg. Add. Post.
5. $AC = DF$	5. Substitution
6. $\overline{AC} \cong \overline{DF}$	6. Def. of ≅ segs.

27. Sample answers: $\overline{LN} \cong \overline{QO}$ and $\overline{LM} \cong \overline{MN} \cong \overline{RS} \cong \overline{ST} \cong \overline{QP} \cong \overline{PO}$

28. Sample answer: You can use segment addition to find the total distance between two destinations by adding the distances of various points in between. Answers should include the following.

- A passenger can add the distance from San Diego to Phoenix and the distance from Phoenix to Dallas to find the distance from San Diego to Dallas.
- The Segment Addition Postulate can be useful if you are traveling in a straight line.

29. B; $AD = AB + BC + CD$
$= 14\frac{1}{4} + 12\frac{3}{4} + 12\frac{1}{4}$
$= 39\frac{1}{4}$

$AQ = \frac{1}{2}AD$
$= \frac{1}{2}\left(39\frac{1}{4}\right)$
$= 19\frac{5}{8}$

$BP = \frac{1}{2}BC$
$= \frac{1}{2}\left(12\frac{3}{4}\right)$
$= 6\frac{3}{8}$

$AP = AB + BP$
$= 14\frac{1}{4} + 6\frac{3}{8}$
$= 20\frac{5}{8}$

$AP = AQ + QP$
$20\frac{5}{8} = 19\frac{5}{8} + QP$
$1 = QP$

30. Let x be the price of a box of popcorn. Then a tub of popcorn costs $2x$.
$60(2x) = 150$
$120x = 150$
$x = 1.25$
Thus, a box of popcorn costs $1.25 and a tub costs 2(1.25) or $2.50.
Total popcorn sales were $275 and $150 of that was for tubs, so boxes account for $275 − $150 or $125. So the number of boxes sold was $\frac{125}{1.25}$ or 100.

Page 106 Maintain Your Skills

31. Substitution
32. Distributive Property
33. Addition Property
34. Transitive Property
35. Never; the midpoint of a segment divides it into two congruent segments.
36. Sometimes; if the lines have a common intersection point, then it is a single point.
37. Always; if two planes intersect, they intersect in a line.
38. Sometimes; if the points are noncollinear, then they lie on three distinct lines.

39. $P = 2\ell + 2w$
$44 = 2(2x + 7) + 2(x + 6)$
$44 = 4x + 14 + 2x + 12$
$44 = 6x + 26$
$18 = 6x$
$3 = x$
$2x + 7 = 2(3) + 7$ or 13
$x + 6 = 3 + 6$ or 9
The dimensions of the rectangle are 9 cm by 13 cm.

40. $2x + x = 90$
$3x = 90$
$x = 30$

41. $2x + 4x = 90$
$6x = 90$
$x = 15$

42. $3x + 2 + x = 90$
$4x + 2 = 90$
$4x = 88$
$x = 22$

43. $x + 3x = 180$
$4x = 180$
$x = 45$

44. $26x + 10x = 180$
$36x = 180$
$x = 5$

45. $4x + 10 + 3x - 5 = 180$
$7x + 5 = 180$
$7x = 175$
$x = 25$

2-8 Proving Angle Relationships

Page 110 Geometry Activity: Right Angles

1. The lines are perpendicular.
2. They are congruent and they form linear pairs.
3. 90
4. They form right angles.
5. They all measure 90° and are congruent.

Pages 111–112 Check for Understanding

1. Tomas; Jacob's answer left out the part of $\angle ABC$ represented by $\angle EBF$.
2. Sample Answer: If $\angle 1 \cong \angle 2$ and $\angle 2 \cong \angle 3$, then $\angle 1 \cong \angle 3$.

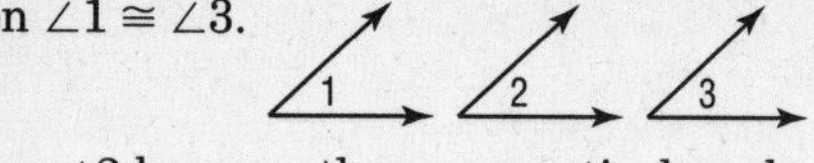

3. $\angle 1 \cong \angle 2$ because they are vertical angles.
$m\angle 1 = m\angle 2$
$65 = m\angle 2$

4. $m\angle 6 + m\angle 8 = 90$
$m\angle 6 + 47 = 90$
$m\angle 6 = 43$
$m\angle 6 + m\angle 7 + m\angle 8 = 180$
$43 + m\angle 7 + 47 = 180$
$m\angle 7 + 90 = 180$
$m\angle 7 = 90$

5. $m\angle 11 + m\angle 12 = 180$
$x - 4 + 2x - 5 = 180$
$3x - 9 = 180$
$3x = 189$
$x = 63$
$m\angle 11 = x - 4$
$= 63 - 4$ or 59
$m\angle 12 = 2x - 5$
$= 2(63) - 5$ or 121

6. **Given:** $\angle 1$ and $\angle 2$ are supplementary, $\angle 3$ and $\angle 4$ are supplementary, $\angle 1 \cong \angle 4$
Prove: $\angle 2 \cong \angle 3$

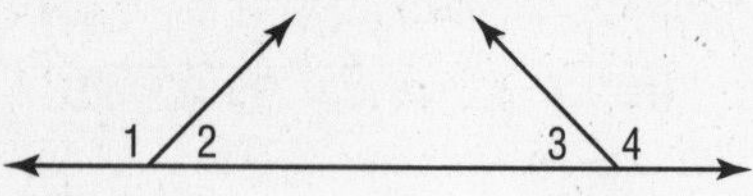

Proof:

Statements	Reasons
a. $\angle 1$ and $\angle 2$ are supplementary. $\angle 3$ and $\angle 4$ are supplementary. $\angle 1 \cong \angle 4$	a. __?__ Given
b. $m\angle 1 + m\angle 2 = 180$ $m\angle 3 + m\angle 4 = 180$	b. __?__ Def. of supp. ⦞s
c. $m\angle 1 + m\angle 2 = m\angle 3 + m\angle 4$	c. __?__ Substitution
d. $m\angle 1 = m\angle 4$	d. __?__ Def. of ≅ ⦞s
e. $m\angle 2 = m\angle 3$	e. __?__ Subt. Prob.
f. $\angle 2 \cong \angle 3$	f. __?__ Def. of ≅ ⦞s

7. **Given:** $\overrightarrow{VX}$ bisects $\angle WVY$, $\overrightarrow{VY}$ bisects $\angle XVZ$.
Prove: $\angle WVX \cong \angle YVZ$

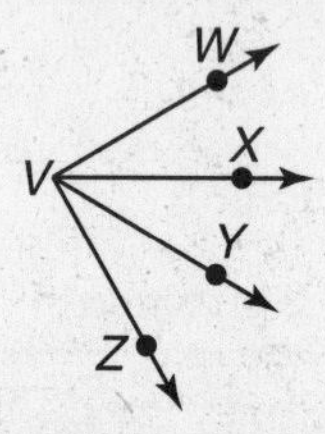

Proof:

Statements	Reasons
1. $\overrightarrow{VX}$ bisects $\angle WVY$. $\overrightarrow{VY}$ bisects $\angle XVZ$.	1. Given
2. $\angle WVX \cong \angle XVY$	2. Def. of $\angle$bisector
3. $\angle XVY \cong \angle YVZ$	3. Def. of $\angle$bisector
4. $\angle WVX \cong \angle YVZ$	4. Trans. Prop.

8. sometimes
9. sometimes
10. **Given:** Two angles form a linear pair.
Prove: The angles are supplementary.

Paragraph Proof: When two angles form a linear pair, the resulting angle is a straight angle whose measure is 180. By definition, two angles are supplementary if the sum of their measures is 180. By the Angle Addition Postulate, $m\angle 1 + m\angle 2 = 180$. Thus, if two angles form a linear pair, then the angles are supplementary.

11. Given: $\angle ABC$ is a right angle.
Prove: $\angle 1$ and $\angle 2$ are complementary angles.

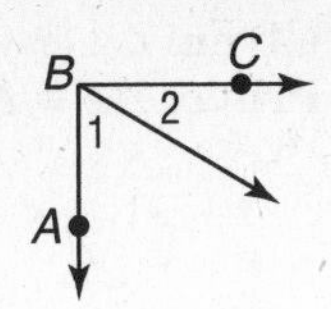

Proof:

Statements	Reasons
1. $\angle ABC$ is a right angle.	1. Given
2. $m\angle ABC = 90$	2. Def. of rt. $\angle$
3. $m\angle ABC = m\angle 1 + m\angle 2$	3. Angle Add. Post.
4. $m\angle 1 + m\angle 2 = 90$	4. Substitution
5. $\angle 1$ and $\angle 2$ are complementary angles.	5. Def. of comp. $\angle$s

12. $\angle 1$ and $\angle 2$ are complementary to $\angle X$, so $\angle 1 \cong \angle 2$.

$m\angle 1 = m\angle 2$
$2n + 2 = n + 32$
$2n = n + 30$
$n = 30$

13. $m\angle 1 = 2n + 2$
$= 2(30) + 2$
$= 62$

14. $m\angle 2 = n + 32$
$= 30 + 32$
$= 62$

15. $m\angle X + m\angle 1 = 90$
$m\angle X + 62 = 90$
$m\angle X = 28$

Pages 112–114 Practice and Apply

16. $m\angle 1 + m\angle 2 = 180$
$m\angle 1 + 67 = 180$
$m\angle 1 = 113$

17. $m\angle 3 + m\angle 4 = 90$
$38 + m\angle 4 = 90$
$m\angle 4 = 52$

18. $\angle 7$ and $\angle 8$ are complementary, so $m\angle 7 + m\angle 8 = 90$. Also, $m\angle 5 + m\angle 6 + m\angle 7 + m\angle 8 = 180$, so by substitution $m\angle 5 + m\angle 6 = 90$. $m\angle 6 = 29$, so $m\angle 5 = 61$. $\angle 5 \cong \angle 8$ so $m\angle 8 = m\angle 5 = 61$. Finally, $m\angle 7 + m\angle 8 = 90$ so $m\angle 7 = 90 - 61$ or 29.

19. $m\angle 9 + m\angle 10 = 180$
$2x - 4 + 2x + 4 = 180$
$4x = 180$
$x = 45$

$m\angle 9 = 2x - 4$
$= 2(45) - 4$ or 86

$m\angle 10 = 2x + 4$
$= 2(45) + 4$ or 94

20. $m\angle 11 + m\angle 12 = 180$
$4x + 2x - 6 = 180$
$6x - 6 = 180$
$6x = 186$
$x = 31$

$m\angle 11 = 4x$
$= 4(31)$ or 124

$m\angle 12 = 2x - 6$
$= 2(31) - 6$ or 56

21. $\angle 13 \cong \angle 14$
$m\angle 13 = m\angle 14$
$2x + 94 = 7x + 49$
$2x + 45 = 7x$
$45 = 5x$
$9 = x$

$m\angle 13 = 2x + 94$
$= 2(9) + 94$ or 112

$m\angle 14 = 7x + 49$
$= 7(9) + 49$ or 112

22. $\angle 15 \cong \angle 16$
$m\angle 15 = m\angle 16$
$x = 6x - 290$
$-5x = -290$
$x = 58$
$m\angle 15 = m\angle 16 = 58$

23. $\angle 17 \cong \angle 18$
$m\angle 17 = m\angle 18$
$2x + 7 = x + 30$
$2x = x + 23$
$x = 23$

$m\angle 17 = 2x + 7$
$= 2(23) + 7$
$= 53$

$m\angle 18 = x + 30$
$= 23 + 30$ or 53

24. $m\angle 19 + m\angle 20 = 180$
$100 + 20x + 20x = 180$
$100 + 40x = 180$
$40x = 80$
$x = 2$

$m\angle 19 = 100 + 20x$
$= 100 + 20(2)$ or 140

$m\angle 20 = 20x$
$= 20(2)$ or 40

25. Given: $\angle A$
Prove: $\angle A \cong \angle A$

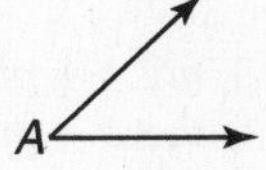

Proof:

Statements	Reasons
1. $\angle A$ is an angle.	1. Given
2. $m\angle A = m\angle A$	2. Reflexive Prop.
3. $\angle A \cong \angle A$	3. Def. of $\cong$ angles

26. Given: $\angle 1 \cong \angle 2$, $\angle 2 \cong \angle 3$
Prove: $\angle 1 \cong \angle 3$

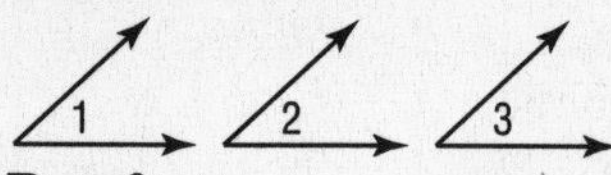

Proof:

Statements	Reasons
1. $\angle 1 \cong \angle 2$, $\angle 2 \cong \angle 3$	1. Given
2. $m\angle 1 = m\angle 2$, $m\angle 2 = m\angle 3$	2. Def. of $\cong$ angles
3. $m\angle 1 = m\angle 3$	3. Trans. Prop.
4. $\angle 1 \cong \angle 3$	4. Def. of $\cong$ angles

27. sometimes
28. always
29. always
30. sometimes
31. sometimes
32. always
33. Given: $\ell \perp m$
Prove: $\angle 2$, $\angle 3$, $\angle 4$ are rt. $\angle$s

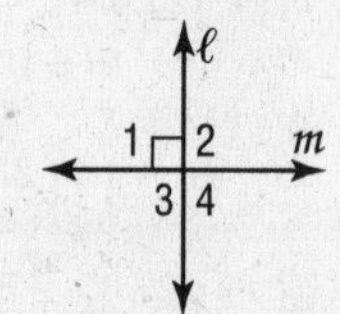

Proof:

Statements	Reasons
1. $\ell \perp m$	1. Given
2. $\angle 1$ is a right angle.	2. Def. of $\perp$
3. $m\angle 1 = 90$	3. Def. of rt. $\angle$s
4. $\angle 1 \cong \angle 4$	4. Vert. $\angle$s are $\cong$
5. $m\angle 1 = m\angle 4$	5. Def. of $\cong$ $\angle$s
6. $m\angle 4 = 90$	6. Substitution
7. $\angle 1$ and $\angle 2$ form a linear pair. $\angle 3$ and $\angle 4$ form a linear pair.	7. Def. of linear pair
8. $m\angle 1 + m\angle 2 = 180$, $m\angle 4 + m\angle 3 = 180$	8. Linear pairs are supplementary.
9. $90 + m\angle 2 = 180$, $90 + m\angle 3 = 180$	9. Substitution
10. $m\angle 2 = 90$, $m\angle 3 = 90$	10. Subt. Prop.
11. $\angle 2$, $\angle 3$, $\angle 4$, are rt. $\angle$s.	11. Def. of rt. $\angle$s (steps 6, 10)

34. Given: $\angle 1$ and $\angle 2$ are rt. $\angle$s.
Prove: $\angle 1 \cong \angle 2$

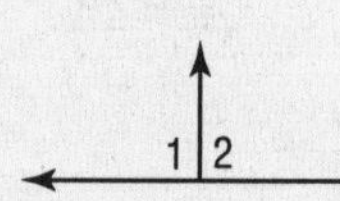

Proof:

Statements	Reasons
1. $\angle 1$ and $\angle 2$ are rt. $\angle$s.	1. Given
2. $m\angle 1 = 90$, $m\angle 2 = 90$	2. Def. of rt. $\angle$s
3. $m\angle 1 = m \angle 2$	3. Substitution
4. $\angle 1 \cong \angle 2$	4. Def. of $\cong$ angles

35. Given: $\ell \perp m$
Prove: $\angle 1 \cong \angle 2$

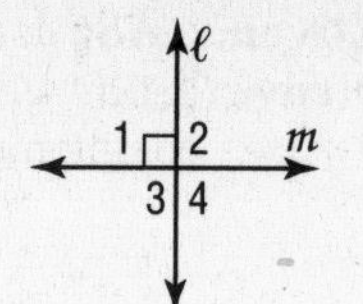

Proof:

Statements	Reasons
1. $\ell \perp m$	1. Given
2. $\angle 1$ and $\angle 2$ are rt. $\angle$s	2. $\perp$ lines intersect to form 4 rt. $\angle$s.
3. $\angle 1 \cong \angle 2$	3. All rt. $\angle$s are $\cong$.

36. Given: $\angle 1 \cong \angle 2$, $\angle 1$ and $\angle 2$ are supplementary.
Prove: $\angle 1$ and $\angle 2$ are rt. $\angle$s.
Proof:

Statements	Reasons
1. $\angle 1 \cong \angle 2$, $\angle 1$ and $\angle 2$ are supplementary.	1. Given
2. $m\angle 1 + m\angle 2 = 180$	2. Def. of supp. $\angle$s
3. $m\angle 1 = m\angle 2$	3. Def. of $\cong$ angle
4. $m\angle 1 + m\angle 1 = 180$	4. Substitution
5. $2(m\angle 1) = 180$	5. Add. Prop.
6. $m\angle 1 = 90$	6. Div. Prop.
7. $m\angle 2 = 90$	7. Substitution (steps 3, 6)
8. $\angle 1$ and $\angle 2$ are rt. $\angle$s.	8. Def. of rt. $\angle$s

37. Given: $\angle ABD \cong \angle CBD$, $\angle ABD$ and $\angle CBD$ form a linear pair
Prove: $\angle ABD$ and $\angle CBD$ are rt. $\angle$s.

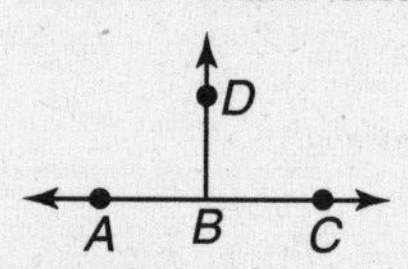

Proof:

Statements	Reasons
1. $\angle ABD \cong CBD$, $\angle ABD$ and $\angle CBD$ form a linear pair.	1. Given
2. $\angle ABD$ and $\angle CBD$ are supplementary.	2. Linear pairs are supplementary.
3. $\angle ABD$ and $\angle CBD$ are rt. $\angle$s.	3. If $\angle$s are $\cong$ and supp., they are rt. $\angle$s.

38. Given: $\angle ABD \cong \angle YXZ$
Prove: $\angle CBD \cong \angle WXZ$

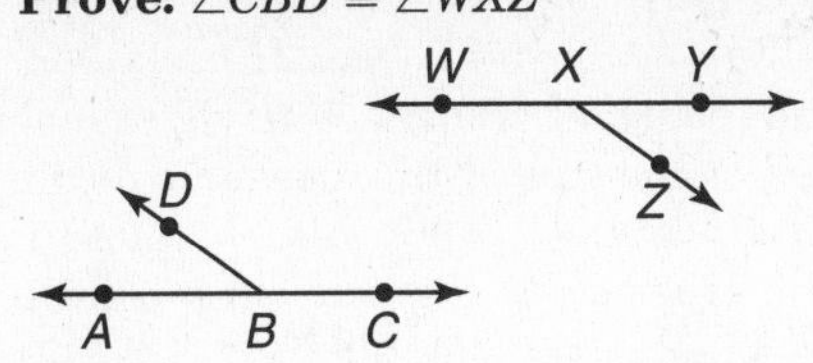

Proof:

Statements	Reasons
1. $\angle ABD \cong \angle YXZ$, $\angle ABD$ and $\angle CBD$ form a linear pair. $\angle YXZ$ and $\angle WXZ$ form a linear pair.	1. Given; from the figure
2. $m\angle ABD + m\angle CBD = 180$, $m\angle YXZ + m\angle WXZ = 180$	2. Linear pairs are supplementary.
3. $m\angle ABD + m\angle CBD = m\angle YXZ + m\angle WXZ$	3. Substitution
4. $m\angle ABD = m\angle YXZ$	4. Def. of $\cong$ $\angle$s
5. $m\angle YXZ + m\angle CBD = m\angle YXZ + m\angle WXZ$	5. Substitution
6. $m\angle YXZ = m\angle YXZ$	6. Reflexive Prop.
7. $m\angle CBD = m\angle WXZ$	7. Subt. Prop.
8. $\angle CBD \cong \angle WXZ$	8. Def. of $\cong$ $\angle$s

39. Given: $m\angle RSW = m\angle TSU$
Prove: $m\angle RST = m\angle WSU$

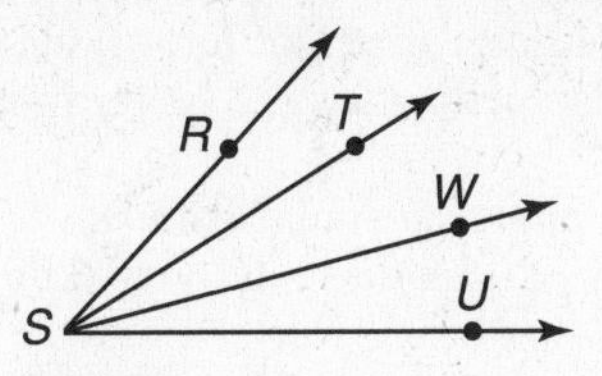

Proof:

Statements	Reasons
1. $m\angle RSW = m\angle TSU$	1. Given
2. $m\angle RSW = m\angle RST + m\angle TSW$, $m\angle TSU = m\angle TSW + m\angle WSU$	2. Angle Addition Postulate
3. $m\angle RST + m\angle TSW = m\angle TSW + m\angle WSU$	3. Substitution
4. $m\angle TSW = m\angle TSW$	4. Reflexive Prop.
5. $m\angle RST = m\angle WSU$	5. Subt. Prop.

40. $m\angle 1 + m\angle 2 = 180$
$28 + m\angle 2 = 180$
$m\angle 2 = 152$

41. Because the lines are perpendicular, the angles formed are right angles. All right angles are congruent. Therefore, $\angle 1$ is congruent to $\angle 2$.

42. $m\angle 1 + m\angle 4 = 90$;
$m\angle 1 + m\angle 2 + m\angle 3 + m\angle 4 = 180$
$m\angle 1 + m\angle 1 + m\angle 4 + m\angle 4 = 180$
$2(m\angle 1) + 2(m\angle 4) = 180$
$2(m\angle 1 + m\angle 4) = 180$
$m\angle 1 + m\angle 4 = 90$

43. Two angles that are supplementary to the same angle are congruent. Answers should include the following.

- $\angle 1$ and $\angle 2$ are supplementary; $\angle 2$ and $\angle 3$ are supplementary.
- $\angle 1$ and $\angle 3$ are vertical angles, and are therefore congruent.
- If two angles are complementary to the same angle, then the angles are congruent.

44. B; Let x be the measure of one angle. Then the measure of the other angle is $90 - x$.
$\frac{x}{90 - x} = \frac{4}{1}$
$x = 4(90 - x)$
$x = 360 - 4x$
$5x = 360$
$x = 72$
The other angle has mesure $90 - 72$ or 18.

45. B; The members of set T are 1, 4, 9, 16, 25, 36, and 49. The median of these numbers is 16.

Page 114 Maintain Your Skills

46. Given: G is between F and H.
H is between G and J.
Prove: $FG + GJ = FH + HJ$

Proof:

Statements	Reasons
1. G is between F and H; H is between G and J.	1. Given
2. $FG + GJ = FJ$, $FH + HJ = FJ$	2. Segment Addition Postulate
3. $FJ = FH + HJ$	3. Symmetric Prop.
4. $FG + GJ = FH + HJ$	4. Transitive Prop.

47. Given: X is the midpoint of $\overline{WY}$.
Prove: $WX + YZ = XZ$

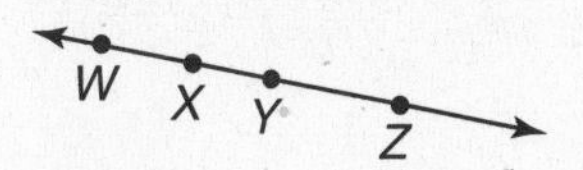

Proof:

Statements	Reasons
1. X is the midpoint of $\overline{WY}$.	1. Given
2. $WX = XY$	2. Def. of midpoint
3. $XY + YZ = XZ$	3. Segment Addition Postulate
4. $WX + YZ = XZ$	4. Substitution

48. Given: $AC = BD$
Prove: $AB = CD$
Proof:

Statements	Reasons
1. $AC = BD$	1. Given
2. $AB + BC = AC$, $BC + CD = BD$	2. Segment Addition Postulate
3. $BC = BC$	3. Reflexive Prop.
4. $AB + BC = BC + CD$	4. Substitution (2 and 3)
5. $AB = CD$	5. Subt. Prop.

49. $\angle ONM$, $\angle MNR$
50. $\angle PMQ \cong \angle QMN$
51. N or R
52. $\angle POQ$, $\angle QON$, $\angle NOM$, $\angle MOP$
53. obtuse
54. Sample answer: $\overrightarrow{NR}$ and $\overrightarrow{NP}$
55. $\angle NML$, $\angle NMP$, $\angle NMO$, $\angle RNM$, $\angle ONM$

Chapter 2 Study Guide and Review

Page 115 Vocabulary and Concept Check

1. conjecture
2. truth value
3. compound
4. and
5. hypothesis
6. converse
7. Postulates
8. informal proof

Pages 115–120 Lesson-by-Lesson Review

9. $m\angle A + m\angle B = 180$

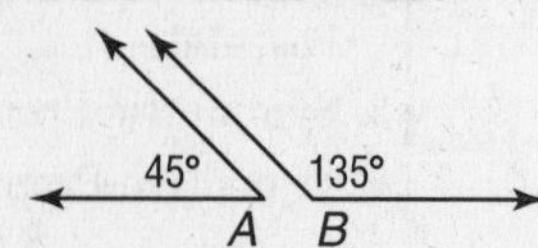

10. Y is the midpoint of $\overline{XZ}$.

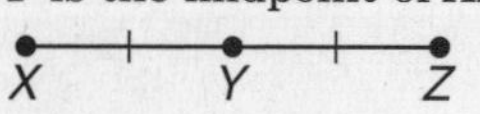

11. $LMNO$ is a square.

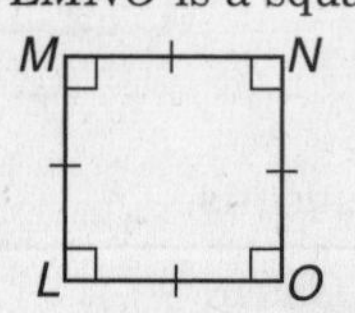

12. $-1 > 0$ and in a right triangle with right angle C, $a^2 + b^2 = c^2$.; false, because p is false and q is true.

13. In a right triangle with right angle C, $a^2 + b^2 = c^2$ or the sum of the measures of two supplementary angles is 180.; true, because q is true and r is true.

14. The sum of the measures of two supplementary angles is 180 and $-1 > 0$.; false, because r is true and p is false.

15. $-1 > 0$, and in a right triangle with right angle C, $a^2 + b^2 = c^2$, or the sum of the measures of two supplementary angles is 180.; false, because q is true and r is true so $q \vee r$ is true but p is false.

16. In a right triangle with right angle C, $a^2 + b^2 = c^2$, or $-1 > 0$ or the sum of the measures of two supplementary angles is 180.; true, because p is false and r is true so $p \vee r$ is true but q is true.

17. In a right triangle with right angle C, $a^2 + b^2 = c^2$ and the sum of the measures of two supplementary angles is 180, and $-1 > 0$.; false, because q is true and r is true so $q \wedge r$ is true but p is false.

18. Converse: If an angle is obtuse, then it measures 120. False; the measure could be any value between 90 and 180.
Inverse: If an angle measure does not equal 120, then it is not obtuse. False; the measure could be any value other than 120 between 90 and 180.
Contrapositive: If an angle is not obtuse, then its measure does not equal 120; true.

19. Converse: If a month has 31 days, then it is March. False; July has 31 days.
Inverse: If a month is not March, then it does not have 31 days. False; July has 31 days.
Contrapositive: If a month does not have 31 days, then it is not March; true.

20. Converse: If a point lies on the y-axis, then its ordered pair has 0 for its x-coordinate; true.
Inverse: If an ordered pair does not have 0 for its x-coordinate, then the point does not lie on the y-axis; true.
Contrapositive: If a point does not lie on the y-axis, then its ordered pair does not have 0 for its x-coordinate; true.

21. true, because the hypothesis is satisfied and the conclusion follows

22. true, because the hypothesis is not satisfied and we cannot say the statement is false

23. false, because the hypothesis is satisfied yet the conclusion does not follow

24. true, because the hypothesis is not satisfied and we cannot say the statement is false

25. Valid; by definition, adjacent angles have a common vertex.

26. Invalid; vertical angles also have a common vertex.

27. yes; Law of Detachment
p: a student attends North High School
q: a student has an ID number

28. Invalid; Statements (1) and (2) are true, but (3) does not follow from (1) and (2).

29. yes; Law of Syllogism
p: you like pizza with everything
q: you like Cardo's Pizza
r: you are a pizza connoisseur

30. Never; the intersection of two lines is a point.

31. Always; if P is the midpoint of $\overline{XY}$, then $\overline{XP} \cong \overline{PY}$. By definition of congruent segments, $XP = PY$.

32. sometimes; if M, X, and Y are collinear

33. sometimes; if the points are collinear

34. Always; there is exactly one line through Q and R. The line lies in at least one plane.

35. sometimes; if the right angles form a linear pair

36. Always; the Reflexive Property states that $\angle 1 \cong \angle 1$.

37. Never; adjacent angles must share a common side, and vertical angles do not.

38. Given: M is the midpoint of $\overline{AB}$ and Q is the midpoint of $\overline{AM}$.

Prove: $AQ = \frac{1}{4}AB$

A Q M B

Proof: If M is the midpoint of $\overline{AB}$, then $AM = \frac{1}{2}(AB)$. Since Q is the midpoint of $\overline{AM}$, $AQ = \frac{1}{2}AM$ or $\frac{1}{2}\left(\frac{1}{2}(AB)\right) = \frac{1}{4}AB$.

39. Distributive Property

40. Division Property

41. Subtraction Property

42. Transitive Property

43. Given: $5 = 2 - \frac{1}{2}x$

Prove: $x = -6$

Proof:

Statements	Reasons
1. $5 = 2 - \frac{1}{2}x$	1. Given
2. $5 - 2 = 2 - \frac{1}{2}x - 2$	2. Subt. Prop.
3. $3 = -\frac{1}{2}x$	3. Substitution
4. $-2(3) = -2\left(-\frac{1}{2}x\right)$	4. Mult. Prop.
5. $-6 = x$	5. Substitution
6. $x = -6$	6. Symmetric Prop.

44. Given: $x - 1 = \frac{x - 10}{-2}$

Prove: $x = 4$

Proof:

Statements	Reasons
1. $x - 1 = \frac{x - 10}{-2}$	1. Given
2. $-2(x - 1) = -2\left(\frac{x - 10}{-2}\right)$	2. Mult. Prop.
3. $-2x + 2 = x - 10$	3. Dist. Prop.
4. $-2x + 2 - 2 = x - 10 - 2$	4. Subt. Prop.
5. $-2x = x - 12$	5. Substitution
6. $-2x - x = x - 12 - x$	6. Subt. Prop.
7. $-3x = -12$	7. Substitution
8. $\frac{-3x}{-3} = \frac{-12}{-3}$	8. Div. Prop.
9. $x = 4$	9. Substitution

45. Given: $AC = AB$, $AC = 4x + 1$, $AB = 6x - 13$

Prove: $x = 7$

A 6x − 13 B

4x + 1 C

Proof:

Statements	Reasons
1. $AC = AB$, $AC = 4x + 1$, $AB = 6x - 13$	1. Given
2. $4x + 1 = 6x - 13$	2. Substitution
3. $4x + 1 - 1 = 6x - 13 - 1$	3. Subt. Prop.
4. $4x = 6x - 14$	4. Substitution
5. $4x - 6x = 6x - 14 - 6x$	5. Subt. Prop.
6. $-2x = -14$	6. Substitution
7. $\frac{-2x}{-2} = \frac{-14}{-2}$	7. Div. Prop.
8. $x = 7$	8. Substitution

46. Given: $MN = PQ$, $PQ = RS$

Prove: $MN = RS$

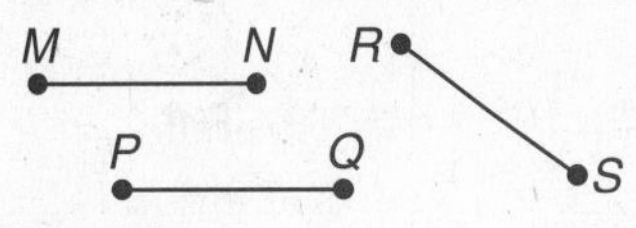

Proof:

Statements	Reasons
1. $MN = PQ$, $PQ = RS$	1. Given
2. $MN = RS$	2. Transitive Prop.

47. Reflexive Property

48. Symmetric Property

49. Addition Property

50. Transitive Property

51. Division or Multiplication Property

52. Addition Property

53. Given: $BC = EC$, $CA = CD$

Prove: $BA = DE$

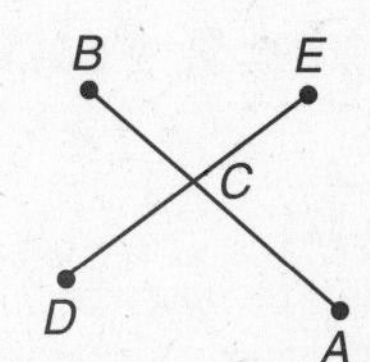

Proof:

Statements	Reasons
1. $BC = EC$, $CA = CD$	1. Given
2. $BC + CA = EC + CA$	2. Add. Prop.
3. $BC + CA = EC + CD$	3. Substitution
4. $BC + CA = BA$ $EC + CD = DE$	4. Seg. Add. Post.
5. $BA = DE$	5. Substitution

54. Given: $AB = CD$
Prove: $AC = BD$

Proof:

Statements	Reasons
1. $AB = CD$	1. Given
2. $BC = BC$	2. Reflexive Prop.
3. $AB + BC = CD + BC$	3. Add. Prop.
4. $AB + BC = AC$ $CD + BC = BD$	4. Seg. Add. Post.
5. $AC = BD$	5. Substitution

55. $m\angle 6 = 180 - 35$ or 145

56. $m\angle 7 = 180 - 157$ or 23

57. $m\angle 8 = 180 - 90$ or 90

58. Given: $\angle 1$ and $\angle 2$ form a linear pair. $m\angle 2 = 2(m\angle 1)$
Prove: $m\angle 1 = 60$

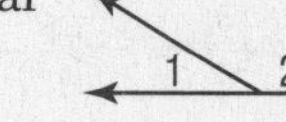

Proof:

Statements	Reasons
a. $\angle 1$ and $\angle 2$ form a linear pair.	**a.** ___?___ Given
b. $\angle 1$ and $\angle 2$ are supplementary.	**b.** ___?___ Supplement Theorem
c. ___?___ $m\angle 1 + m\angle 2 = 180$	**c.** Definition of supplementary angles
d. $m\angle 2 = 2(m\angle 1)$	**d.** ___?___ Given
e. ___?___ $m\angle 1 + 2(m\angle 1) = 180$	**e.** Substitution
f. ___?___ $3(m\angle 1) = 180$	**f.** Substitution
g. $\frac{3(m\angle 1)}{3} = \frac{180}{3}$	**g.** ___?___ Division Property
h. ___?___ $m\angle 1 = 60$	**h.** Substitution

Chapter 2 Practice Test

Page 121

1. Sample answer: Formal is the two-column proof, informal can be paragraph proofs.
2. Sample answer: You can use a counterexample.
3. Sample answer: statements and reasons to justify statements
4. true; Symmetric Prop.
5. false; $y = 2$
6. false; $a = -4$
7. $-3 > 2$ and $3x = 12$ when $x = 4$.; false, because p is false and q is true
8. $-3 > 2$ or $3x = 12$ when $x = 4$.; true, because p is false and q is true
9. $-3 > 2$, or $3x = 12$ when $x = 4$ and an equilateral triangle is also equiangular.; true, because q is true and r is true so $q \wedge r$ is true and p is false
10. Hypothesis: you eat an apple a day; Conclusion: the doctor will stay away; If you eat an apple a day, then the doctor will stay away.
    Converse: If the doctor stays away, then you eat an apple a day.
    Inverse: If you do not eat an apple a day, then the doctor will not stay away.
    Contrapositive: If the doctor does not stay away, then you do not eat an apple a day.
11. Hypothesis: a stone is rolling; Conclusion: it gathers no moss; If a stone is rolling, then it gathers no moss.
    Converse: If a stone gathers no moss, then it is rolling.
    Inverse: If a stone is not rolling, then it gathers moss.
    Contrapositive: If a stone gathers moss, then it is not rolling.
12. valid; Law of Detachment
    p: two lines are perpendicular
    q: the lines intersect
13. $m\angle 1 + 73 = 95$
    $m\angle 1 = 22$
14. $m\angle 2 = 180 - (m\angle 1 + 73)$
    $= 180 - (22 + 73)$
    $= 85$
15. $m\angle 3 = m\angle 2$
    $= 85$
16. **Given:** $y = 4x + 9$; $x = 2$
    Prove: $y = 17$
    Proof:

Statements	Reasons
1. $y = 4x + 9$; $x = 2$	1. Given
2. $y = 4(2) + 9$	2. Substitution
3. $y = 8 + 9$	3. Substitution
4. $y = 17$	4. Substitution

17. **Given:** $AM = CN$, $MB = ND$
    Prove: $AB = CD$

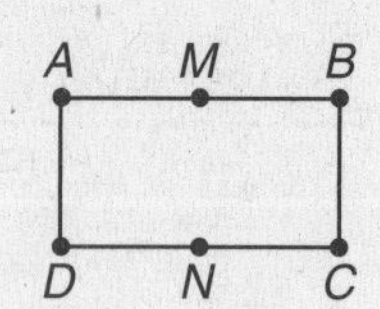

Proof:
We are given that $AM = CN$, $MB = ND$. By the Addition Property, $AM + MB = CN + MB$. Then by Substitution, $AM + MB = CN + ND$. Using the Segment Addition Postulate, $AB = AM + MB$, and $CD = CN + ND$. Then, by Substitution $AB = CD$.

18. Hypothesis: you are a hard-working person; Conclusion: you deserve a great vacation; If you are a hard-working person, then you deserve a great vacation.
19. A

Chapter 2 Standardized Test Practice

Pages 122–123

1. D; $-49 < 0.143 < 2.646 < 49$
2. C; $2(7) - 3 = 11$, so (7, 11) is on the line.
   $2(4) - 3 = 5$, so (4, 5) is on the line.
   $2(-2) - 3 = -7$, so $(-2, -10)$ is not on the line.
   $2(-5) - 3 = -13$, so $(-5, -13)$ is on the line.
3. A; a protractor is used to measure angles, not lengths. A calculator is not a measuring tool. A centimeter ruler is more accurate than a yardstick because its unit of measurement (centimeters) is smaller.
4. B; $\overline{DE} \cong \overline{EF}$ so $DE = EF$.
   $8x - 3 = 3x + 7$
   $8x = 3x + 10$
   $5x = 10$
   $x = 2$
5. A; $\angle ACF + \angle DCF = \angle ACD$ by the Angle Addition Postulate. So $m\angle ACF + m\angle DCF = 90$ since $\angle ACD$ is a right angle. Then $\angle ACF$ and $\angle DCF$ are complementary angles.
6. C; inductive reasoning uses specific examples to make a conjecture.
7. A
8. A; divide both sides by 3.
9. The shortest distance is the length of the hypotenuse of the right triangle whose legs have lengths 120 yd and $53\frac{1}{3}$ yd. Use the Pythagorean Theorem to find the length of the hypotenuse. Call this length d.
   $d^2 = (120)^2 + \left(53\frac{1}{3}\right)^2$
   $d^2 = 14{,}400 + \frac{25{,}600}{9}$
   $d \approx \sqrt{17{,}244.4}$
   $d \approx 131$ yd
10. inverse
11. $(p \rightarrow q) \wedge (q \rightarrow r) \rightarrow (p \rightarrow r)$
    Martina drank 300 mg of calcium.
12. Segment Addition Postulate
13. Sample answer: Marti can measure a third distance c, the distance between the ends of the two sides, and make sure it satisfies the equation $a^2 + b^2 = c^2$.

14a.

Possible lengths and widths where area is 100 sq ft	Perimeter of rectangle with given length and width
1 ft by 100 ft	202 ft
2 ft by 50 ft	104 ft
4 ft by 25 ft	58 ft
5 ft by 20 ft	50 ft
10 ft by 10 ft	40 ft

The dimensions that require the least amount of fencing are 10 ft by 10 ft.

14b. Sample answer: Make a list of all possible whole-number lengths and widths that will form a 100-square-foot area. Then find the perimeter of each rectangle. Choose the length and width combination that has the smallest perimeter.

14c. As the length and width get closer to having the same measure as one another, the amount of fencing required decreases.

15. Given: $\angle 1$ and $\angle 3$ are vertical angles.
$m\angle 1 = 3x + 5$, $m\angle 3 = 2x + 8$
Prove: $m\angle 1 = 14$

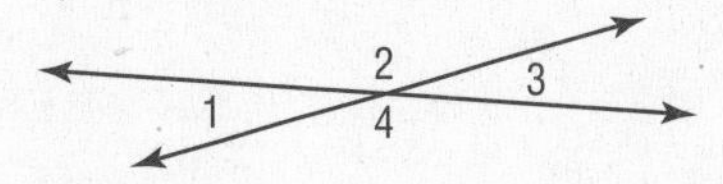

Proof:

Statements	Reasons
a. $\angle 1$ and $\angle 3$ are vertical angles. $m\angle 1 = 3x + 5$, $m\angle 3 = 2x + 8$	a. Given
b. $\angle 1 \cong \angle 3$	b. Vert. $\angle$s are $\cong$
c. $m\angle 1 = m\angle 3$	c. Def. of $\cong$ $\angle$s
d. $3x + 5 = 2x + 8$	d. Substitution
e. $x + 5 = 8$	e. Subt. Prop.
f. $x = 3$	f. Subt. Prop.
g. $m\angle 1 = 3(3) + 5$	g. Substitution
h. $m\angle 1 = 14$	h. Substitution

Chapter 3 Parallel and Perpendicular Lines

Page 125 Getting Started

1. $\overleftrightarrow{PQ}$
2. $\overleftrightarrow{PR}$ or $\overleftrightarrow{RS}$
3. $\overleftrightarrow{ST}$
4. $\overleftrightarrow{TR}$ or $\overleftrightarrow{TP}$
5. The arcs in the figure indicate that $\angle 2$ is congruent to $\angle 4$, $\angle 6$, and $\angle 8$.
6. The arcs in the figure indicate that $\angle 5$ is congruent to $\angle 1$, $\angle 3$, and $\angle 7$.
7. The arcs in the figure indicate that $\angle 3$ is congruent to $\angle 1$, $\angle 5$, and $\angle 7$.
8. The arcs in the figure indicate that $\angle 8$ is congruent to $\angle 2$, $\angle 4$, and $\angle 6$.
9. $y = 7x - 12$
   $= 7(3) - 12$
   $= 21 - 12 = 9$
10. $y = -\frac{2}{3}x + 4$
    $= -\frac{2}{3}(8) + 4$
    $= -\frac{16}{3} + \frac{12}{3} = -\frac{4}{3}$
11. $2x - 4y = 18$
    $2(6) - 4y = 18$
    $12 - 4y = 18$
    $-4y = 6$
    $y = -\frac{6}{4}$ or $-\frac{3}{2}$

3-1 Parallel Lines and Transversals

Page 126 Geometry Activity: Draw a Rectangular Prism

1. Planes ABC and EFG are parallel because those are the planes given as parallel. Planes BCG and ADH are parallel, as well as planes ABF and DCG because opposite sides of a rectangular prism are parallel the same way opposite sides of a rectangle are parallel.
2. Plane ABF intersects plane ABC at $\overleftrightarrow{AB}$; plane DCG intersects plane ABC at $\overleftrightarrow{DC}$; plane ADH intersects plane ABC at $\overleftrightarrow{AD}$; plane BCG intersects plane ABC at $\overleftrightarrow{BC}$.
3. $\overline{AE}$, $\overline{CG}$, and $\overline{DH}$ are parallel to $\overline{BF}$ because planes BCG and ADH are parallel and planes ABF and DCG are parallel.

Pages 128–129 Check for Understanding

1. Sample answer: The bottom and top of a cylinder are contained in parallel planes.

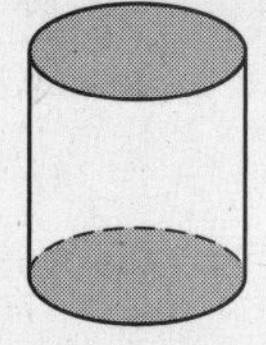

2. Juanita; Eric has listed interior angles, but they are not alternate interior angles.
3. Sample answer: looking down railroad tracks
4. ABC, JKL, ABK, CDM
5. $\overline{AB}$, $\overline{JK}$, $\overline{LM}$
6. $\overline{BK}$, $\overline{CL}$, $\overline{JK}$, $\overline{LM}$, $\overline{BL}$, $\overline{KM}$
7. q and r, q and t, r and t
8. p and q, p and t, q and t
9. p and r, p and t, r and t
10. p and q, p and r, q and r
11. alternate interior
12. corresponding
13. consecutive interior
14. alternate exterior
15. p; consecutive interior
16. p; alternate exterior
17. q; alternate interior
18. Sample answer: The pillars form parallel lines.
19. Sample answer: The roof and the floor are parallel planes.
20. Sample answer: One of the west pillars and the base on the east side form skew lines.
21. Sample answer: The top of the memorial "cuts" the pillars.

Pages 129–131 Practice and Apply

22. $\overline{DE}$, $\overline{PQ}$, $\overline{ST}$
23. ABC, ABQ, PQR, CDS, APU, DET
24. $\overline{BC}$, $\overline{EF}$, $\overline{QR}$
25. $\overline{AP}$, $\overline{BQ}$, $\overline{CR}$, $\overline{FU}$, $\overline{PU}$, $\overline{QR}$, $\overline{RS}$, $\overline{TU}$
26. ABC, AFU, BCR, CDS, EFU, PQR
27. $\overline{BC}$, $\overline{CD}$, $\overline{DE}$, $\overline{EF}$, $\overline{QR}$, $\overline{RS}$, $\overline{ST}$, $\overline{TU}$
28. b and c, b and r, r and c
29. a and c, a and r, r and c
30. a and b, a and r, b and r
31. a and b, a and c, b and c
32. corresponding
33. alternate exterior
34. alternate interior
35. corresponding
36. alternate exterior
37. alternate interior
38. consecutive interior
39. consecutive interior
40. p; corresponding
41. p; alternate interior
42. m; alternate exterior
43. ℓ; alternate exterior
44. m; corresponding
45. q; alternate interior
46. ℓ; corresponding
47. m; consecutive interior

48. Skew lines; the planes are flying in different directions and at different altitudes.

49. $\overline{CG}, \overline{DH}, \overline{EI}$

50. $\overline{DE}, \overline{FG}, \overline{HI}, \overline{GH}, \overline{BF}, \overline{DH}, \overline{EI}$

51. No; plane *ADE* will intersect all the planes if they are extended.

52. Sample answers: parallel bars in gymnastics, parallel port on a computer, parallel events, parallel voices in a choir, latitude parallels on a map

53. Infinite number; consider any line through *P* in any plane that does not contain ℓ.

54. 1

55. Sample answer: Parallel lines and planes are used in architecture to make structures that will be stable. Answers should include the following.

- Opposite walls should form parallel planes; the floor may be parallel to the ceiling.
- The plane that forms a stairway will not be parallel to some of the walls.

56. A

57. The elements of M are 15, 18, 21, 24, 27, and 30. The elements of P are 16, 20, 24, and 28. The numbers that are in P but not in M are 16, 20, and 28. Select any one of these three numbers.

Page 131 Maintain Your Skills

58. Given: $m\angle ABC = m\angle DFE$

$m\angle 1 = m\angle 4$

Prove: $m\angle 2 = m\angle 3$

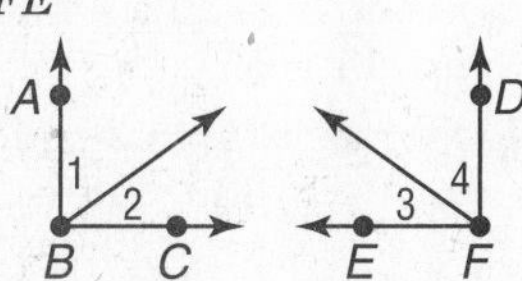

Proof:

Statements	Reasons
1. $m\angle ABC = m\angle DFE$ $m\angle 1 = m\angle 4$	1. Given
2. $m\angle ABC = m\angle 1 + m\angle 2$ $m\angle DFE = m\angle 3 + m\angle 4$	2. Angle Addition Post.
3. $m\angle 1 + m\angle 2 = m\angle 3 + m\angle 4$	3. Substitution Prop.
4. $m\angle 4 + m\angle 2 = m\angle 3 + m\angle 4$	4. Substitution Prop.
5. $m\angle 2 = m\angle 3$	5. Subt. Prop.

59. P Q R

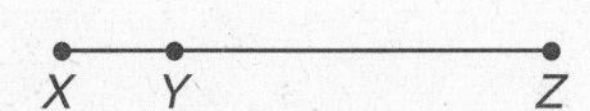

Since $\overline{PQ} \cong \overline{ZY}$ and $\overline{QR} \cong \overline{XY}$, $PQ = ZY$ and $QR = XY$ by the definition of congruent segments. By the Addition Property, $PQ + QR = ZY + XY$. Using the Segment Addition Postulate, $PR = PQ + QR$ and $XZ = XY + YZ$. By substitution, $PR = XZ$. Because the measures are equal, $\overline{PR} \cong \overline{XZ}$ by the definition of congruent segments.

60. no conclusion

61. $m\angle EFG$ is less than 90; Law of Detachment
p: an angle is acute
q: its measure is less than 90

62.
$$d = \sqrt{(x_2 - x_1)^2 + (y_2 - y_1)^2}$$
$$AB = \sqrt{[3 - (-1)]^2 + [4 - (-8)]^2} = \sqrt{4^2 + 12^2} = \sqrt{160} \approx 12.65$$

63.
$$d = \sqrt{(x_2 - x_1)^2 + (y_2 - y_1)^2}$$
$$CD = \sqrt{(-2 - 0)^2 + (9 - 1)^2} = \sqrt{(-2)^2 + 8^2} = \sqrt{68} \approx 8.25$$

64.
$$d = \sqrt{(x_2 - x_1)^2 + (y_2 - y_1)^2}$$
$$EF = \sqrt{[5 - (-3)]^2 + [4 - (-12)]^2} = \sqrt{8^2 + 16^2} = \sqrt{320} \approx 17.89$$

65.
$$d = \sqrt{(x_2 - x_1)^2 + (y_2 - y_1)^2}$$
$$GH = \sqrt{(9 - 4)^2 + [-25 - (-10)]^2} = \sqrt{5^2 + (-15)^2} = \sqrt{250} \approx 15.81$$

66.
$$d = \sqrt{(x_2 - x_1)^2 + (y_2 - y_1)^2}$$
$$JK = \sqrt{(-3 - 1)^2 + \left(-\frac{7}{4} - \frac{1}{4}\right)^2} = \sqrt{(-4)^2 + (-2)^2} = \sqrt{20} \approx 4.47$$

67.
$$d = \sqrt{(x_2 - x_1)^2 + (y_2 - y_1)^2}$$
$$LM = \sqrt{[5 - (-5)]^2 + \left(-\frac{2}{5} - \frac{8}{5}\right)^2} = \sqrt{10^2 + (-2)^2} = \sqrt{104} \approx 10.20$$

68. M, A, P, B, N

69. T, R, S, ℓ

70. 50, 180 − 50 or 130
The measures of the angles are 50 and 130.

71. 90, 180 − 90 or 90
The measures of the angles are 90 and 90.

72.
$$x + 2x = 180$$
$$3x = 180$$
$$x = 60$$
60, 2(60) or 120
The measures of the angles are 60 and 120.

73.
$$2y + 3y = 180$$
$$5y = 180$$
$$y = 36$$
$$2y = 2(36) \text{ or } 72$$
$$3y = 3(36) \text{ or } 108$$
The measures of the angles are 72 and 108.

74. $x + 2x + 3x = 180$

$6x = 180$

$x = 30$

There are two linear pairs in the figure. In one linear pair the angles measure x and $3x + 2x = 5x$, or 30 and 150. In the other pair the angles measure $3x$ and $x + 2x = 3x$, or 90 and 90.

75. $3x - 1 + 2x + 6 = 180$

$5x + 5 = 180$

$5x = 175$

$x = 35$

$3x - 1 = 3(35) - 1$

$= 104$

$2x + 6 = 2(35) + 6$

$= 76$

The measures of the angles are 76 and 104.

Page 132 Geometry Software Investigation: Angles and Parallel Lines

1. The pairs of corresponding angles are $\angle AEG$ and $\angle CFE$, $\angle AEF$ and $\angle CFH$, $\angle BEG$ and $\angle DFE$, $\angle BEF$ and $\angle DFH$. The pairs of alternate interior angles are $\angle AEF$ and $\angle DFE$, $\angle BEF$ and $\angle CFE$. The pairs of alternate exterior angles are $\angle AEG$ and $\angle DFH$, $\angle BEG$ and $\angle CFH$. The pairs of consecutive interior angles are $\angle AEF$ and $\angle CFE$, $\angle BEF$ and $\angle DFE$.

2. The pairs of corresponding angles in Exercise 1 that have the same measure are $\angle AEG$ and $\angle CFE$, $\angle AEF$ and $\angle CFH$, $\angle BEG$ and $\angle DFE$, $\angle BEF$ and $\angle DFH$. The pairs of alternate interior angles that have the same measure are $\angle AEF$ and $\angle DFE$, $\angle BEF$ and $\angle CFE$. The pairs of alternate exterior angles that have the same measure are $\angle AEG$ and $\angle DFH$, $\angle BEG$ and $\angle CFH$.

3. They are supplementary.

4a. If two parallel lines are cut by a transversal, then corresponding angles are congruent.

4b. If two parallel lines are cut by a transversal, then alternate interior angles are congruent.

4c. If two parallel lines are cut by a transversal, then alternate exterior angles are congruent.

4d. If two parallel lines are cut by a transversal, then consecutive interior angles are supplementary.

5. Yes; the angle pairs show the same relationships.

6. See students' work.

7a. Sample answer: All of the angles measure 90°.

7b. Sample answer: If two parallel lines are cut by a transversal so that it is perpendicular to one of the lines, then the transversal is perpendicular to the other line.

3-2 Angles and Parallel Lines

Page 136 Check for Understanding

1. Sometimes; if the transversal is perpendicular to the parallel lines, then $\angle 1$ and $\angle 2$ are right angles and are congruent.

2.

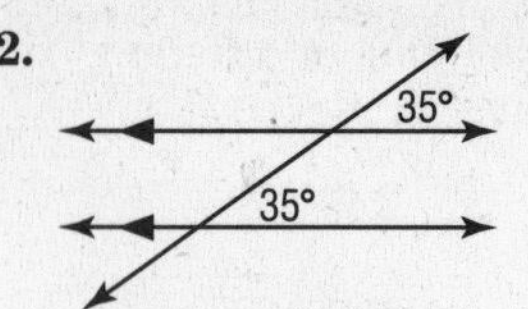

3. 1; all other angles can be determined using the Corresponding Angles Postulate, the Alternate Interior Angles Theorem, the Consecutive Interior Angles Theorem, and the Alternate Exterior Angles Theorem.

4. Alternate Interior Angles Theorem

5. $\angle 1 \cong \angle 3$

$m\angle 1 = m\angle 3$

$m\angle 1 = 110$

6. $\angle 6 \cong \angle 3$

$m\angle 6 = m\angle 3$

$m\angle 6 = 110$

7. $\angle 2$ and $\angle 3$ are supplementary.

$m\angle 2 + m\angle 3 = 180$

$m\angle 2 + 110 = 180$

$m\angle 2 = 70$

8. $\angle 10 \cong \angle 12$

$m\angle 10 = m\angle 12$

$m\angle 10 = 55$

9. $\angle 13 \cong \angle 12$

$m\angle 13 = m\angle 12$

$m\angle 13 = 55$

10. $\angle 15 \cong \angle 12$

$m\angle 15 = m\angle 12$

$m\angle 15 = 55$

11. $8y + 2 + 25y - 20 = 180$

$33y - 18 = 180$

$33y = 198$

$y = 6$

$10x + 8y + 2 = 180$

$10x + 8(6) + 2 = 180$

$10x + 50 = 180$

$10x = 130$

$x = 13$

12. $4x - 5 = 3x + 11$

$4x = 3x + 16$

$x = 16$

$4x - 5 + 3y + 1 = 180$

$4(16) - 5 + 3y + 1 = 180$

$60 + 3y = 180$

$3y = 120$

$y = 40$

13.

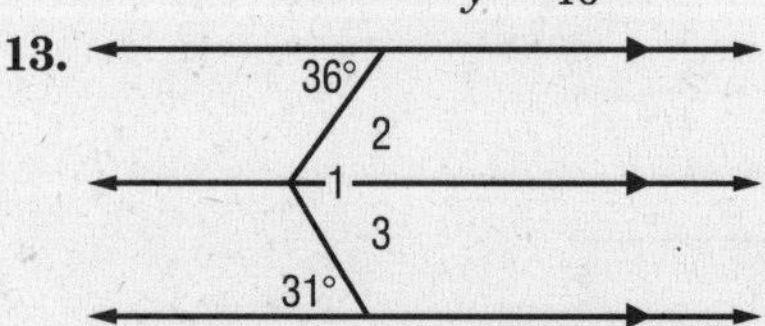

Draw a third line through the vertex of $\angle 1$ parallel to the two given lines.

$\angle 2$ is congruent to the angle whose measure is labeled 36° by the Alternate Interior Angles Theorem. So $m\angle 2 = 36$.

$\angle 3$ is congruent to the angle whose measure is labeled 31° by the Alternate Interior Angles

Theorem. So $m\angle 3 = 31$.
$m\angle 1 = m\angle 2 + m\angle 3$
$= 36 + 31$ or 67

Pages 136–138 Practice and Apply

14. $\angle 3 \cong \angle 9$
$m\angle 3 = m\angle 9$
$m\angle 3 = 75$

15. $\angle 5 \cong \angle 9$
$m\angle 5 = m\angle 9$
$m\angle 5 = 75$

16. $\angle 6$ and $\angle 9$ are supplementary.
$m\angle 6 + m\angle 9 = 180$
$m\angle 6 + 75 = 180$
$m\angle 6 = 105$

17. $\angle 7 \cong \angle 9$
$m\angle 7 = 75$
$\angle 7$ and $\angle 8$ are supplementary.
$m\angle 7 + m\angle 8 = 180$
$75 + m\angle 8 = 180$
$m\angle 8 = 105$

18. $\angle 11 \cong \angle 9$
$m\angle 11 = m\angle 9$
$m\angle 11 = 75$

19. $\angle 12$ and $\angle 9$ are supplementary.
$m\angle 12 + m\angle 9 = 180$
$m\angle 12 + 75 = 180$
$m\angle 12 = 105$

20. $\angle 2$ and $\angle 3$ are supplementary.
$m\angle 2 + m\angle 3 = 180$
$m\angle 2 + 43 = 180$
$m\angle 2 = 137$

21. $\angle 7 \cong \angle 3$
$m\angle 7 = m\angle 3$
$m\angle 7 = 43$

22. $\angle 10 \cong \angle 2$
$m\angle 10 = m\angle 2$
$m\angle 10 = 137$

23. $\angle 11 \cong \angle 3$
$m\angle 11 = m\angle 3$
$m\angle 11 = 43$

24. $\angle 13 \cong \angle 3$
$m\angle 13 = m\angle 3$
$m\angle 13 = 43$

25. $\angle 16 \cong \angle 2$
$m\angle 16 = m\angle 2$
$m\angle 16 = 137$

26.

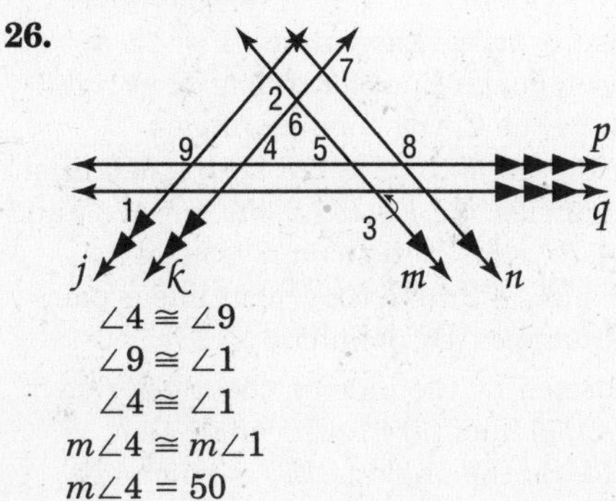

$\angle 4 \cong \angle 9$
$\angle 9 \cong \angle 1$
$\angle 4 \cong \angle 1$
$m\angle 4 \cong m\angle 1$
$m\angle 4 = 50$

27. $\angle 5 \cong \angle 3$
$m\angle 5 = m\angle 3$
$m\angle 5 = 60$

28. $\angle 2$ and $\angle 6$ are supplementary.
$m\angle 2 + m\angle 6 = 180$
$m\angle 4 + m\angle 5 + m\angle 6 = 180$
$m\angle 6 = 180 - (m\angle 4 + m\angle 5)$
$m\angle 2 + 180 - (m\angle 4 + m\angle 5) = 180$
$m\angle 2 = m\angle 4 + m\angle 5$
$\angle 5 \cong \angle 3$
$m\angle 5 = m\angle 3$
From Exercise 26,
$m\angle 4 = m\angle 1 = 50$
$m\angle 2 = m\angle 1 + m\angle 3$
$m\angle 2 = 50 + 60$
$m\angle 2 = 110$

29. $m\angle 6 + m\angle 4 + m\angle 5 = 180$
$m\angle 6 + 50 + 60 = 180$
$m\angle 6 = 70$

30. $\angle 7 \cong \angle 2$
$m\angle 7 = m\angle 2$
$m\angle 2 = 110$ (from Exercise 28)
So $m\angle 7 = 110$.

31. $\angle 8$ is congruent to an angle that forms a linear pair with $\angle 3$.
$m\angle 8 + m\angle 3 = 180$
$m\angle 8 + 60 = 180$
$m\angle 8 = 120$

32. $4x + 56 = 180$
$4x = 124$
$x = 31$
$3y - 11 + 56 = 180$
$3y + 45 = 180$
$3y = 135$
$y = 45$

33. $2x = 68$
$x = 34$
$68 + 3x - 15 + y^2 = 180$
$68 + 3(34) - 15 + y^2 = 180$
$155 + y^2 = 180$
$y^2 = 25$
$y = \pm 5$

34.

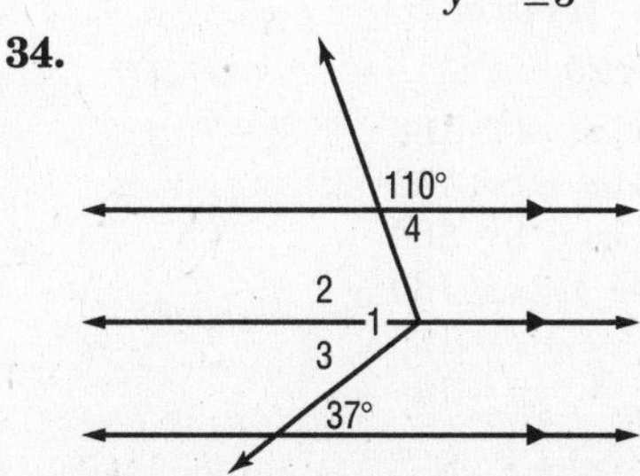

Draw a third line through the vertex of $\angle 1$ parallel to the two given lines.
$\angle 2 \cong \angle 4$ by the Alternate Interior Angles Theorem. $m\angle 4 + 110 = 180$, so $m\angle 4 = 70$ and hence $m\angle 2 = 70$.
$\angle 3$ is congruent to the angle whose measure is labeled 37° by the Alternate Interior Angles Theorem. So $m\angle 3 = 37$.
$m\angle 1 = m\angle 2 + m\angle 3$
$= 70 + 37$ or 107

35.

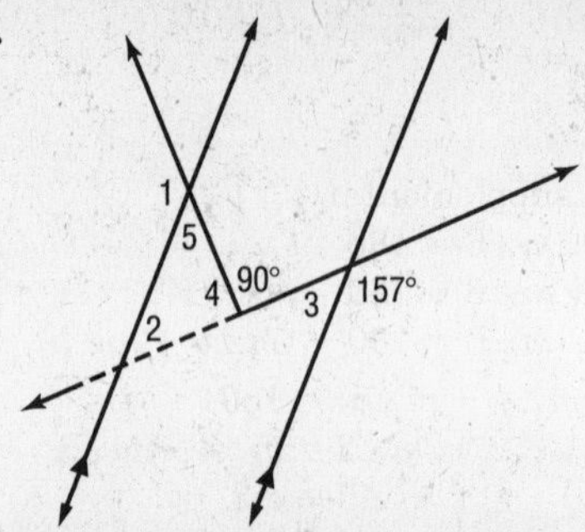

Extend the ray that forms the 157° angle in the opposite direction so that the line crosses the left line of the pair of parallel lines.
Then $m\angle 4 = 90$.

$m\angle 5 + m\angle 4 + m\angle 2 = 180$
$m\angle 5 = 180 - m\angle 4 - m\angle 2$
$m\angle 5 = 180 - 90 - m\angle 2$
$m\angle 5 = 90 - m\angle 2$
$\angle 2 \cong \angle 3$
$m\angle 2 = m\angle 3$
$m\angle 3 + 157 = 180$
$m\angle 3 = 23$
$m\angle 2 = 23$
$m\angle 1 + m\angle 5 = 180$
$m\angle 1 + 90 - m\angle 2 = 180$
$m\angle 1 + 90 - 23 = 180$
$m\angle 1 + 67 = 180$
$m\angle 1 = 113$

36. $x = 90$
$3y - 11 = y + 19$
$3y = y + 30$
$2y = 30$
$y = 15$
$3y - 11 + 4z + 2 + x = 180$
$3(15) - 11 + 4z + 2 + 90 = 180$
$4z + 126 = 180$
$4z = 54$
$z = 13.5$

37. $7y - 4 + 7x + 9 = 180$
$7y + 7x + 5 = 180$
$7y + 7x = 175$
$7y = 175 - 7x$
$y = \frac{1}{7}(175 - 7x)$
$y = 25 - x$
$2y + 5 + 11x - 1 = 180$
$2(25 - x) + 5 + 11x - 1 = 180$
$50 - 2x + 5 + 11x - 1 = 180$
$54 + 9x = 180$
$9x = 126$
$x = 14$
$y = 25 - x$
$= 25 - 14$ or 11
$z + 7x + 9 = 180$
$z + 7(14) + 9 = 180$
$z + 107 = 180$
$z = 73$

38. The angle with measure 40° is congruent to an angle that forms a linear pair with the angle whose measure is x°. So $40 + x = 180$. Then $x = 140$.

39. Given: $\ell \parallel m$
Prove: $\angle 1 \cong \angle 8$
$\angle 2 \cong \angle 7$

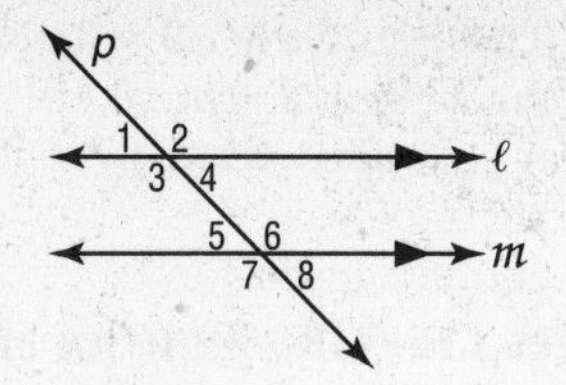

Proof:

Statements	Reasons
1. $\ell \parallel m$	1. Given
2. $\angle 1 \cong \angle 5$, $\angle 2 \cong \angle 6$	2. Corresponding Angles Postulate
3. $\angle 5 \cong \angle 8$, $\angle 6 \cong \angle 7$	3. Vertical Angles Theorem
4. $\angle 1 \cong \angle 8$, $\angle 2 \cong \angle 7$	4. Transitive Property

40. Given: $m \parallel n$, ℓ is a transversal.
Prove: $\angle 1$ and $\angle 2$ are supplementary; $\angle 3$ and $\angle 4$ are supplementary.

Proof:

Statements	Reasons
1. $m \parallel n$, ℓ is a transversal	1. Given
2. $\angle 1$ and $\angle 3$ form a linear pair; $\angle 2$ and $\angle 4$ form a linear pair.	2. Def. of linear pair
3. $\angle 1$ and $\angle 3$ are supplementary; $\angle 2$ and $\angle 4$ are supplementary.	3. If two angles form a linear pair, then they are supplementary.
4. $\angle 1 \cong \angle 4$, $\angle 2 \cong \angle 3$	4. Alt. int. ∠s ≅
5. $\angle 1$ and $\angle 2$ are supplementary; $\angle 3$ and $\angle 4$ are supplementary.	5. Substitution

41. Given: $\ell \perp m$, $m \parallel n$
Prove: $\ell \perp n$

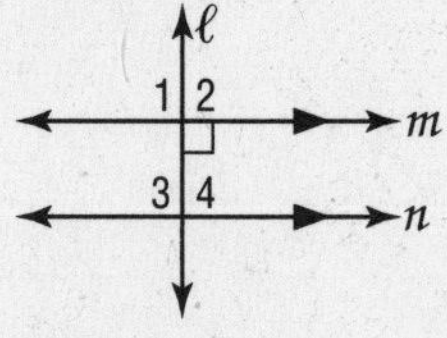

Proof: Since $\ell \perp m$, we know that $\angle 1 \cong \angle 2$, because perpendicular lines form congruent right angles. Then by the Corresponding Angles Postulate, $\angle 1 \cong \angle 3$ and $\angle 2 \cong \angle 4$. By the definition of congruent angles, $m\angle 1 = m\angle 2$, $m\angle 1 = m\angle 3$ and $m\angle 2 = m\angle 4$. By substitution, $m\angle 3 = m\angle 4$. Because $\angle 3$ and $\angle 4$ form a congruent linear pair, they are right angles. By definition, $\ell \perp n$.

42. The angle formed by the pipe on the other side of the road is supplementary to the angle that measures 65°. So the angle is $180 - 65$ or 115.

43. $\angle 2$ and $\angle 6$ are consecutive interior angles for the same transversal, which makes them supplementary because $\overline{WX} \parallel \overline{YZ}$. $\angle 4$ and $\angle 6$ are not necessarily supplementary because $\overline{WZ}$ may not be parallel to $\overline{XY}$.

44. Sample answer: Angles and lines are used in art to show depth, and to create realistic objects. Answers should include the following.
   - Rectangular shapes are made by drawing parallel lines and perpendiculars.
   - M. C. Escher and Pablo Picasso use lines and angles in their art.

45. C; Let $y°$ be the measure of the third angle of the right triangle in the figure.

$$160 = y + 120$$
$$40 = y$$
$$x + y + 90 = 180$$
$$x + 40 + 90 = 180$$
$$x = 50$$

46. C;
$$ax = bx + c$$
$$ax - bx = c$$
$$(a - b)x = c$$
$$x = \frac{c}{a - b}$$

Page 138 Maintain Your Skills

47. $\overline{FG}$

48. $\overline{AB}, \overline{DE}, \overline{FG}, \overline{IJ}, \overline{AE}, \overline{FJ}$

49. CDH

50. $\overline{BG}, \overline{CH}, \overline{FG}, \overline{HI}$

51. $m\angle 1 + 124 = 180$
$m\angle 1 = 56$

52. $m\angle 2 = 53$

53. Hypothesis: it rains this evening
Conclusion: I will mow the lawn tomorrow

54. Hypothesis: you eat a balanced diet
Conclusion: it will keep you healthy

55. $\frac{7 - 9}{8 - 5} = \frac{-2}{3}$ or $-\frac{2}{3}$

56. $\frac{-3 - 6}{2 - 8} = \frac{-9}{-6}$
$= \frac{3}{2}$

57. $\frac{14 - 11}{23 - 15} = \frac{3}{8}$

58. $\frac{15 - 23}{14 - 11} = \frac{-8}{3}$ or $-\frac{8}{3}$

59. $\frac{2}{9} \cdot \left(-\frac{18}{5}\right) = -\frac{36}{45}$
$= -\frac{4}{5}$

Page 138 Practice Quiz 1

1. p; alternate exterior
2. ℓ; consecutive interior
3. q; alternate interior
4. $\angle 6 \cong \angle 1$
$m\angle 6 = m\angle 1$
$m\angle 6 = 105$
5. $\angle 4 \cong \angle 2$
$m\angle 4 = m\angle 2$
$m\angle 2 + m\angle 1 = 180$
$m\angle 2 + 105 = 180$
$m\angle 2 = 75$
$m\angle 4 = 75$

3-3 Slopes of Lines

Page 142 Check for Understanding

1. horizontal; vertical
2. Curtis; Lori added the coordinates instead of finding the difference.
3. horizontal line, vertical line
4. $m = \frac{(y_2 - y_1)}{(x_2 - x_1)}$
$= \frac{-1 - 3}{-2 - (-4)}$
$= \frac{-4}{2}$ or -2
5. Line ℓ goes through $P(0, 4)$ and $Q(4, 2)$.
$m = \frac{(y_2 - y_1)}{(x_2 - x_1)}$
$= \frac{2 - 4}{4 - 0}$
$= \frac{-2}{4}$ or $-\frac{1}{2}$
6. Line m goes through $C(0, -3)$ and $D(3, -1)$.
$m = \frac{(y_2 - y_1)}{(x_2 - x_1)}$
$= \frac{-1 - (-3)}{3 - 0}$
$= \frac{2}{3}$
7. Line ℓ has slope $-\frac{1}{2}$ (from Exercise 5). Any line perpendicular to ℓ has a slope that is the opposite reciprocal of $-\frac{1}{2}$, or 2.
8. slope of $\overleftrightarrow{GH} = \frac{0 - 13}{-11 - 14}$
$= \frac{-13}{-25}$ or $\frac{13}{25}$
slope of $\overleftrightarrow{RS} = \frac{-5 - 7}{-4 - (-3)}$
$= \frac{-12}{-1}$ or 12
The slopes are not the same, so $\overleftrightarrow{GH}$ and $\overleftrightarrow{RS}$ are not parallel. The product of the slopes is $\frac{156}{25}$, so $\overleftrightarrow{GH}$ and $\overleftrightarrow{RS}$ are not perpendicular. Therefore, $\overleftrightarrow{GH}$ and $\overleftrightarrow{RS}$ are neither parallel nor perpendicular.
9. slope of $\overleftrightarrow{GH} = \frac{-9 - (-9)}{9 - 15}$
$= \frac{0}{-6}$ or 0
slope of $\overleftrightarrow{RS} = \frac{-1 - (-1)}{3 - (-4)}$
$= \frac{0}{7}$ or 0
The slopes are the same so $\overleftrightarrow{GH}$ and $\overleftrightarrow{RS}$ are parallel.

10. Start at (1, 2). Move up 2 units and then move right 1 unit. Draw the line through this point and (1, 2).

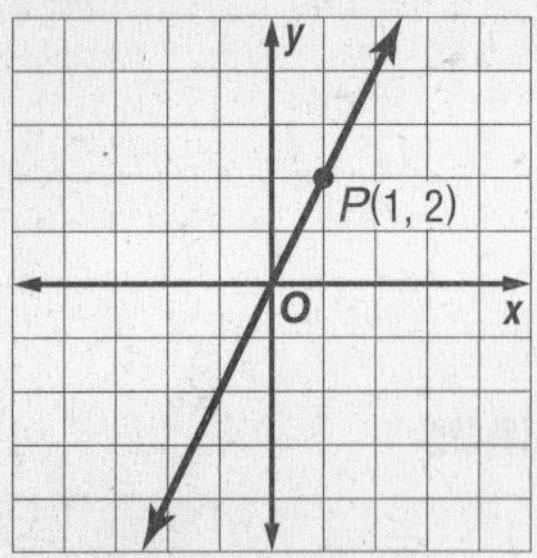

11. Find the slope of $\overleftrightarrow{MN}$.

$m = \frac{(y_2 - y_1)}{(x_2 - x_1)}$
$= \frac{2 - 0}{1 - 5}$
$= -\frac{2}{4}$ or $-\frac{1}{2}$

Since $\left(-\frac{1}{2}\right)(2) = -1$, the slope of the line perpendicular to $\overleftrightarrow{MN}$ through $A(6, 4)$ is 2. Graph the line. Start at (6, 4). Move up 2 units and then move right 1 unit. Draw the line through this point and (6, 4).

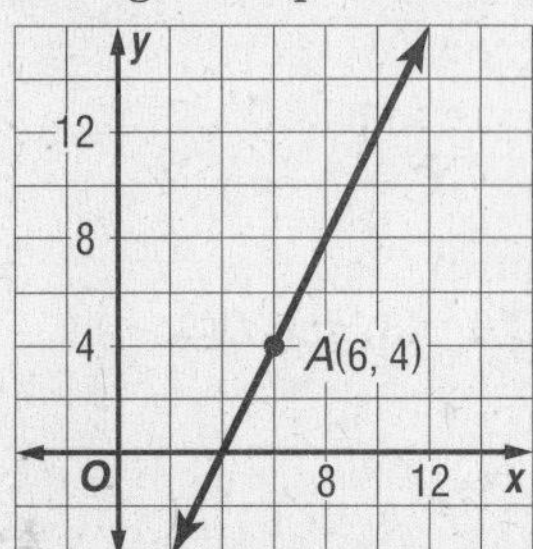

12. The hill has an 8% grade, so the road will rise or fall 8 units vertically with every 100 horizontal units traveled. So the slope is either $\frac{8}{100} = \frac{2}{25}$ or $-\frac{8}{100} = -\frac{2}{25}$.

13. Let $(x_1, y_1) = (0, 0)$ and $m = -\frac{2}{25}$. Then $y_2 = -120$ because the biker is 120 meters below her starting position.

$m = \frac{(y_2 - y_1)}{(x_2 - x_1)}$
$-\frac{2}{25} = \frac{-120 - 0}{x_2 - 0}$
$-\frac{2}{25} = \frac{-120}{x_2}$
$x_2 = 1500$

If $m = \frac{2}{25}$ then $x_2 = -1500$. So the current position of the biker is represented by (1500, −120) or (−1500, −120).

14. The distance is the same no matter which coordinates are used for the biker's current position.

$d = \sqrt{(x_2 - x_1)^2 + (y_2 - y_1)^2}$
$= \sqrt{(1500 - 0)^2 + (-120 - 0)^2}$
$= \sqrt{1500^2 + (-120)^2}$
$= \sqrt{2{,}264{,}400}$
≈ 1505

The biker has traveled 1505 meters down the hill.

Pages 142–144 Practice and Apply

15. $m = \frac{(y_2 - y_1)}{(x_2 - x_1)}$
$= \frac{3 - 2}{7 - 0}$ or $\frac{1}{7}$

16. $m = \frac{(y_2 - y_1)}{(x_2 - x_1)}$
$= \frac{-5 - (-3)}{-6 - (-2)}$
$= \frac{-2}{-4}$ or $\frac{1}{2}$

17. $m = \frac{(y_2 - y_1)}{(x_2 - x_1)}$
$= \frac{-3 - 2}{4 - 3}$
$= \frac{-5}{1}$ or -5

18. $m = \frac{(y_2 - y_1)}{(x_2 - x_1)}$
$= \frac{3 - 7}{4 - 1}$
$= -\frac{4}{3}$

19. slope of $\overleftrightarrow{PQ} = \frac{1 - (-2)}{9 - (-3)}$
$= \frac{3}{12}$ or $\frac{1}{4}$

slope of $\overleftrightarrow{UV} = \frac{-2 - 6}{5 - 3}$
$= \frac{-8}{2}$ or -4

The product of the slopes is $\frac{1}{4}(-4)$ or -1. So, $\overleftrightarrow{PQ}$ is perpendicular to $\overleftrightarrow{UV}$.

20. slope of $\overleftrightarrow{PQ} = \frac{3 - 0}{0 - (-4)}$
$= \frac{3}{4}$

slope of $\overleftrightarrow{UV} = \frac{6 - (-3)}{8 - (-4)}$
$= \frac{9}{12}$ or $\frac{3}{4}$

The slopes are the same, so $\overleftrightarrow{PQ}$ is parallel to $\overleftrightarrow{UV}$.

21. slope of $\overleftrightarrow{PQ} = \frac{1 - 7}{2 - (-10)}$
$= \frac{-6}{12}$ or $-\frac{1}{2}$

slope of $\overleftrightarrow{UV} = \frac{1 - 0}{6 - 4}$
$= \frac{1}{2}$

The slopes are not the same, so $\overleftrightarrow{PQ}$ and $\overleftrightarrow{UV}$ are not parallel. The product of the slopes is $\left(-\frac{1}{2}\right)\left(\frac{1}{2}\right)$ or $-\frac{1}{4}$, so $\overleftrightarrow{PQ}$ and $\overleftrightarrow{UV}$ are not perpendicular. Therefore, $\overleftrightarrow{PQ}$ and $\overleftrightarrow{UV}$ are neither parallel nor perpendicular.

22. slope of $\overleftrightarrow{PQ} = \frac{1 - 2}{0 - (-9)}$
$= -\frac{1}{9}$

slope of $\overleftrightarrow{UV} = \frac{-1 - 8}{-2 - (-1)}$
$= \frac{-9}{-1}$ or 9

The product of the slopes is $\left(-\frac{1}{9}\right)(9)$ or -1. So, $\overleftrightarrow{PQ}$ is perpendicular to $\overleftrightarrow{UV}$.

23. slope of $\overleftrightarrow{PQ} = \frac{8 - 1}{9 - 1}$
$= \frac{7}{8}$

slope of $\overleftrightarrow{UV} = \frac{8 - 1}{2 - (-6)}$
$= \frac{7}{8}$

The slopes are the same, so $\overleftrightarrow{PQ}$ is parallel to $\overleftrightarrow{UV}$.

24. slope of $\overleftrightarrow{PQ} = \frac{0-(-4)}{10-5}$
$= \frac{4}{5}$
slope of $\overleftrightarrow{UV} = \frac{-13-(-8)}{5-9}$
$= \frac{5}{4}$
The slopes are not the same, so $\overleftrightarrow{PQ}$ and $\overleftrightarrow{UV}$ are not parallel. The product of the slopes is $\left(\frac{4}{5}\right)\left(\frac{5}{4}\right)$ or 1, so $\overleftrightarrow{PQ}$ and $\overleftrightarrow{UV}$ are not perpendicular. Therefore, $\overleftrightarrow{PQ}$ and $\overleftrightarrow{UV}$ are neither parallel nor perpendicular.

25. $m = \frac{(y_2 - y_1)}{(x_2 - x_1)}$
$= \frac{-1-2}{0-(-1)}$
$= \frac{-3}{1}$ or -3

26. $m = \frac{(y_2 - y_1)}{(x_2 - x_1)}$
$= \frac{4-(-5)}{1-(-4)}$
$= \frac{9}{5}$

27. $m = \frac{(y_2 - y_1)}{(x_2 - x_1)}$
$= \frac{-1-5}{-3-(-2)}$
$= \frac{-6}{-1}$ or 6

28. $m = \frac{(y_2 - y_1)}{(x_2 - x_1)}$
$= \frac{-4-(-4)}{4-(-2)}$
$= \frac{0}{6}$ or 0

29. The slope of $\overleftrightarrow{LM}$ is 6 (from Exercise 27). A line parallel to $\overleftrightarrow{LM}$ has the same slope as $\overleftrightarrow{LM}$, thus has slope 6.

30. The slope of $\overleftrightarrow{PQ}$ is $\frac{9}{5}$ (from Exercise 26). A line perpendicular to $\overleftrightarrow{PQ}$ has slope that is the opposite reciprocal of $\frac{9}{5}$, or $-\frac{5}{9}$.

31. The slope of $\overleftrightarrow{EF}$ is 0 (from Exercise 28). $\overleftrightarrow{EF}$ is horizontal, so a line perpendicular to $\overleftrightarrow{EF}$ is vertical and has undefined slope.

32. The slope of $\overleftrightarrow{AB}$ is -3 (from Exercise 25). A line parallel to $\overleftrightarrow{AB}$ has the same slope as $\overleftrightarrow{AB}$, thus has slope -3.

33. Start at $(-2, 1)$. Move down 4 units and then move right 1 unit. Draw the line through this point and $(-2, 1)$.

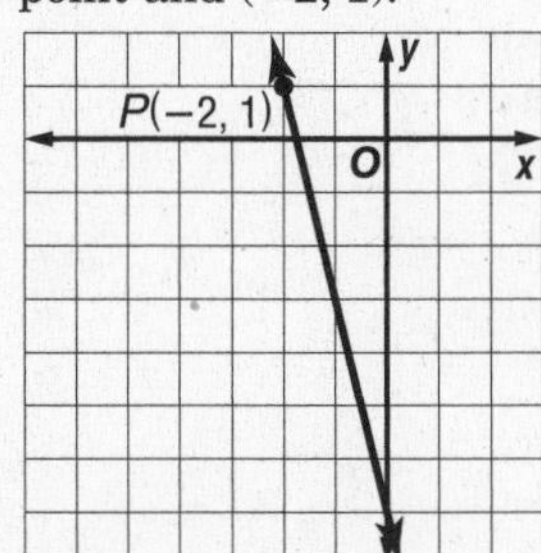

34. Find the slope of $\overleftrightarrow{CD}$.
$m = \frac{(y_2 - y_1)}{(x_2 - x_1)}$
$= \frac{1-7}{5-(-1)}$
$= \frac{-6}{6}$ or -1

The line to be graphed is parallel to $\overleftrightarrow{CD}$ so the line has slope -1.
Start at $(-1, -3)$. Move down 1 unit and then move right 1 unit. Draw the line through this point and $(-1, -3)$.

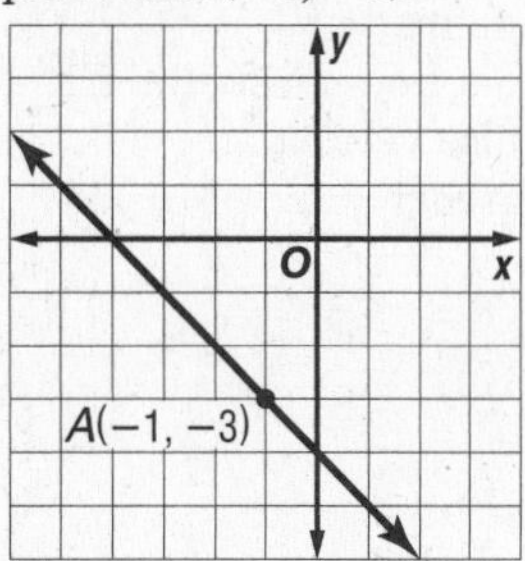

35. Find the slope of $\overleftrightarrow{GH}$.
$m = \frac{(y_2 - y_1)}{(x_2 - x_1)}$
$= \frac{0-3}{-3-0}$
$= \frac{-3}{-3}$ or 1
Since $1(-1) = -1$, the slope of the line perpendicular to $\overleftrightarrow{GH}$ through $M(4, 1)$, is -1.
Start at $(4, 1)$. Move down 1 unit and then move right 1 unit. Draw the line through this point and $(4, 1)$.

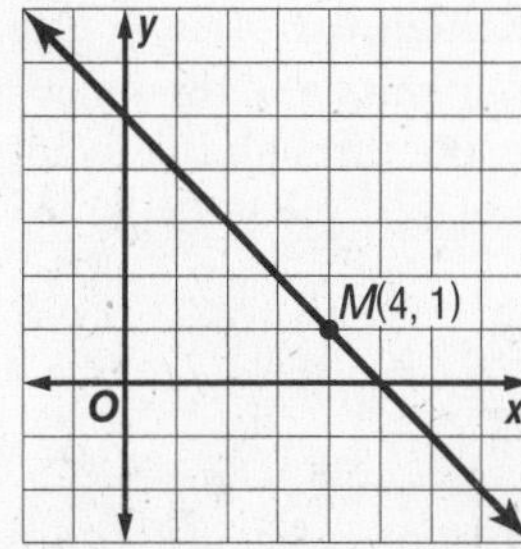

36. Start at $(-7, -1)$. Move up 2 units and then move right 5 units. Draw the line through this point and $(-7, -1)$.

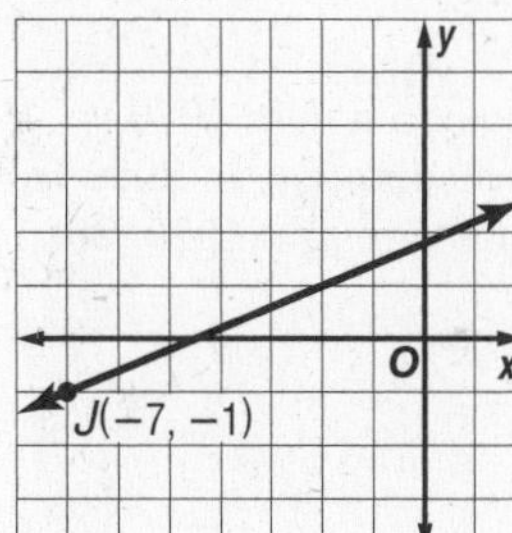

37. Find the slope of $\overleftrightarrow{KL}$.

$m = \frac{(y_2 - y_1)}{(x_2 - x_1)}$

$= \frac{-12 - 7}{2 - 2}$

$= \frac{-19}{0}$ which is undefined

$\overleftrightarrow{KL}$ is a vertical line, so a line parallel to $\overleftrightarrow{KL}$ through $Q(-2, -4)$ is also vertical.

Draw a vertical line through $(-2, -4)$.

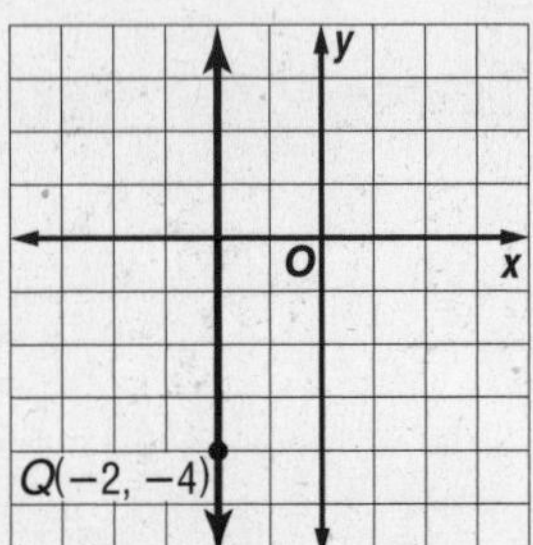

38. Find the slope of $\overleftrightarrow{DE}$.

$m = \frac{(y_2 - y_1)}{(x_2 - x_1)}$

$= \frac{0 - 2}{5 - 0}$

$= \frac{-2}{5}$

Since $\left(\frac{-2}{5}\right)\left(\frac{5}{2}\right) = -1$, the slope of the line perpendicular to $\overleftrightarrow{DE}$ through $W(6, 4)$ is $\frac{5}{2}$.

Start at (6, 4). Move up 5 units and then move right 2 units. Draw the line through this point and (6, 4).

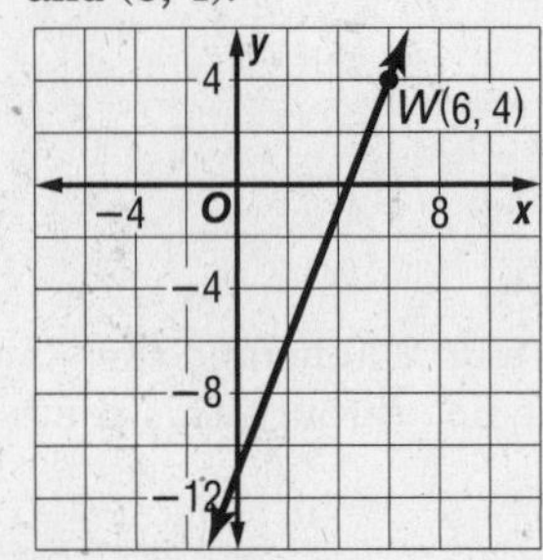

39. Sample answer: The median age in 1970 is approximately 28. The median age in 2000 is 35.3. Find the slope of the line through (1970, 28) and (2000, 35.3).

$m = \frac{(y_2 - y_1)}{(x_2 - x_1)}$

$= \frac{35.3 - 28}{2000 - 1970}$

$= \frac{7.3}{30}$

≈ 0.24

The annual rate of change is approximately 0.24 year per year.

40. Sample answer: Let $(x_1, y_1) = (2000, 35.3)$ and $m = 0.24$.

$m = \frac{(y_2 - y_1)}{(x_2 - x_1)}$

$0.24 = \frac{y_2 - 35.3}{2010 - 2000}$

$0.24 = \frac{y_2 - 35.3}{10}$

$2.4 = y_2 - 35.3$

$37.7 = y_2$

The median age will be 37.7 in 2010.

41. Let $(x_1, y_1) = (2000, 35.3)$ and $m = \frac{1}{3}$.

$m = \frac{(y_2 - y_1)}{(x_2 - x_1)}$

$\frac{1}{3} = \frac{40.6 - 35.3}{x_2 - 2000}$

$x_2 - 2000 = 3(5.3)$

$x_2 = 2015.9$

The median age will be 40.6 in 2016.

42. $m = \frac{(y_2 - y_1)}{(x_2 - x_1)}$

$-\frac{3}{7} = \frac{-1 - 2}{x - 6}$

$-\frac{3}{7} = \frac{-3}{x - 6}$

$7 = x - 6$

$13 = x$

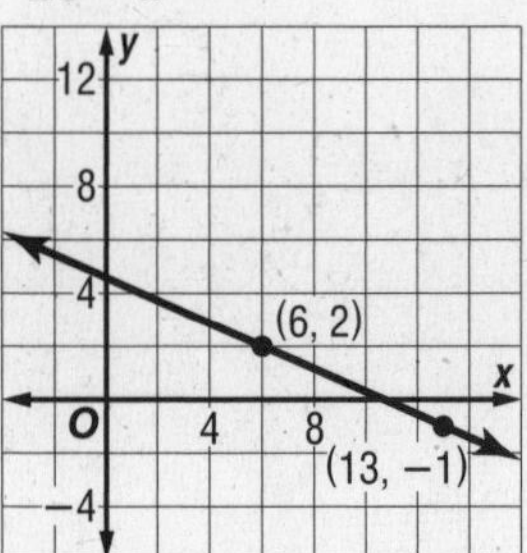

43. $m_1 = \frac{(y_2 - y_1)}{(x_2 - x_1)}$

$= \frac{-1 - 8}{2 - 4}$

$= \frac{9}{2}$

$\left(\frac{9}{2}\right)\left(-\frac{2}{9}\right) = -1$, so the line containing $(x, 2)$ and $(-4, 5)$ perpendicular to the line containing (4, 8) and $(2, -1)$ has slope $-\frac{2}{9}$.

$m_2 = \frac{(y_2 - y_1)}{(x_2 - x_1)}$

$-\frac{2}{9} = \frac{5 - 2}{-4 - x}$

$-\frac{2}{9} = \frac{3}{-4 - x}$

$-2(-4 - x) = 27$

$8 + 2x = 27$

$2x = 19$

$x = \frac{19}{2}$

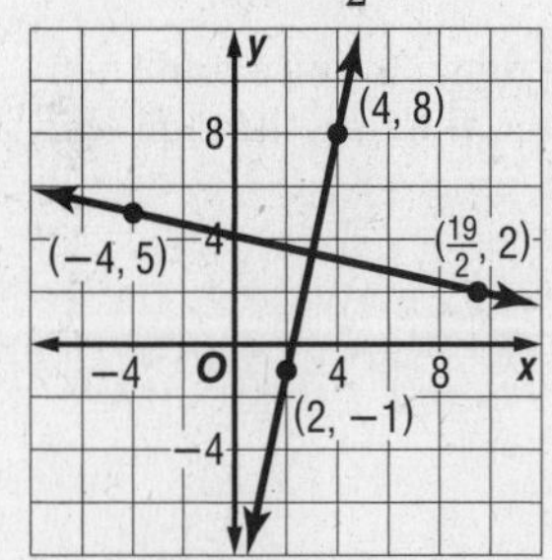

44. $m = \frac{(y_2 - y_1)}{(x_2 - x_1)}$

$= \frac{77 - 51}{2000 - 1998}$

$= \frac{26}{2}$ or 13

The percent changes at a rate of 13% per year.

45. $m = \frac{(y_2 - y_1)}{(x_2 - x_1)}$

$= \frac{77 - 64}{2000 - 1999}$

$= 13$

$$13 = \frac{90 - 77}{x - 2000}$$
$$13(x - 2000) = 13$$
$$x - 2000 = 1$$
$$x = 2001$$
90% of classrooms will have Internet access in 2001.

46. No; the graph can only rise until it reaches 100%.

47. The y-intercept can be found by setting the equation for x equal to zero, solving for t, then using this value in the equation for y.
$$x = 5 + 2t$$
$$0 = 5 + 2t$$
$$-5 = 2t$$
$$-\frac{5}{2} = t$$
$$y = -3 + t$$
$$y = -3 + \left(-\frac{5}{2}\right)$$
$$y = -\frac{11}{2}$$
Find the x-intercept in a similar manner.
$$y = -3 + t$$
$$0 = -3 + t$$
$$3 = t$$
$$x = 5 + 2t$$
$$x = 5 + 2(3)$$
$$x = 11$$
So two points on the line are $(0, -\frac{11}{2})$ and $(11, 0)$.
$$m = \frac{(y_2 - y_1)}{(x_2 - x_1)}$$
$$= \frac{0 - \left(-\frac{11}{2}\right)}{11 - 0}$$
$$= \frac{\frac{11}{2}}{11} \text{ or } \frac{1}{2}$$
The slope-intercept form of the equation of the line is $y = \frac{1}{2}x - \frac{11}{2}$.

48. Sample answer: Slope is used when driving through hills to determine how fast to go. Answers should include the following.
- Drivers should be notified of the grade so that they can adjust their speed accordingly. A positive slope indicates that the driver must speed up, while a negative slope indicates that the driver should slow down.
- An escalator must be at a steep enough slope to be efficient, but also must be gradual enough to ensure comfort.

49. C; $m = \frac{(y_2 - y_1)}{(x_2 - x_1)}$
$$= \frac{-2 - 1}{-3 - (-5)}$$
$$= \frac{-3}{2}$$
$\left(\frac{-3}{2}\right)\left(\frac{2}{3}\right) = -1$, so the slope of the line perpendicular to the line containing $(-5, 1)$ and $(-3, -2)$ is $\frac{2}{3}$.

50. A; the winning sailboat completed the race in $\frac{24}{9}$ hours, or $2\frac{2}{3}$ hours. The second-place boat completed the race in $\frac{24}{8}$ hours, or 3 hours. The difference in times is $\frac{1}{3}$ hour, or 20 minutes.

Page 144 Maintain Your Skills

51. $\angle 6 \cong \angle 1$
$$m\angle 6 = m\angle 1$$
$$m\angle 6 = 131$$

52. $\angle 7$ is supplementary to $\angle 6$.
$$m\angle 7 + m\angle 6 = 180$$
$$m\angle 7 + 131 = 180$$
$$m\angle 7 = 49$$

53. $\angle 4 \cong \angle 7$
$$m\angle 4 = m\angle 7$$
$$m\angle 4 = 49$$

54. $m\angle 2 + m\angle 1 = 180$
$$m\angle 2 + 131 = 180$$
$$m\angle 2 = 49$$

55. $m\angle 5 + m\angle 1 = 180$
$$m\angle 5 + 131 = 180$$
$$m\angle 5 = 49$$

56. $\angle 8 \cong \angle 6$
$$\angle 6 \cong \angle 1$$
$$\angle 8 \cong \angle 1$$
$$m\angle 8 = m\angle 1$$
$$m\angle 8 = 131$$

57. ℓ; alternate exterior

58. ℓ; corresponding

59. p; alternate interior

60. q; consecutive interior

61. m; alternate interior

62. q; corresponding

63. H, I, and J are noncollinear.

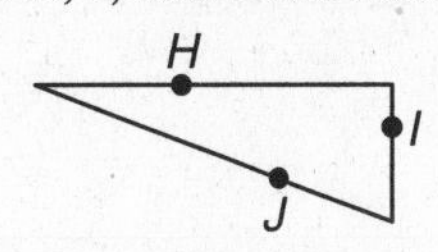

64. $XZ + ZY = XY$.

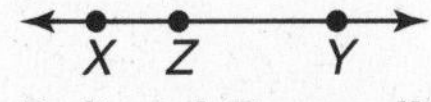

65. R, S, and T are collinear.

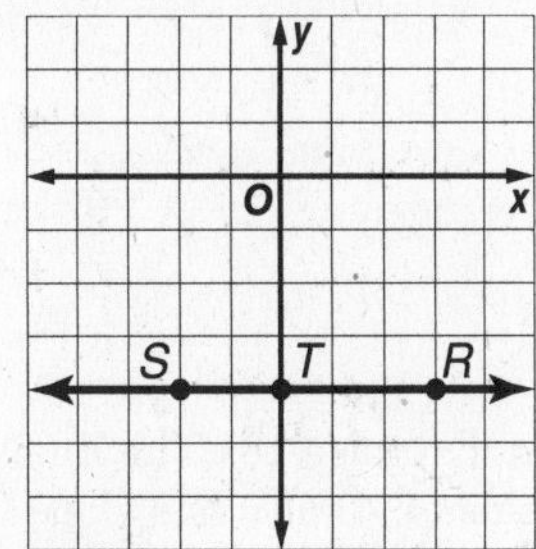

66. acute

67. obtuse

68. right

69. obtuse

70. $2x + y = 7$
$$y = -2x + 7$$

71. $2x + 4y = -5$
$$4y = -2x - 5$$
$$y = -\frac{1}{2}x - \frac{5}{4}$$

72. $5x - 2y + 4 = 0$
$$5x + 4 = 2y$$
$$\frac{5}{2}x + 2 = y$$
$$y = \frac{5}{2}x + 2$$

3-4 Equations of Lines

Pages 147–148 Check for Understanding

1. Sample answer: Use the point-slope form where $(x_1, y_1) = (-2, 8)$ and $m = -\frac{2}{5}$.
2. Sample answer: $y = 2x - 3, y = -x - 6$
3. Sample answer: $y = x$

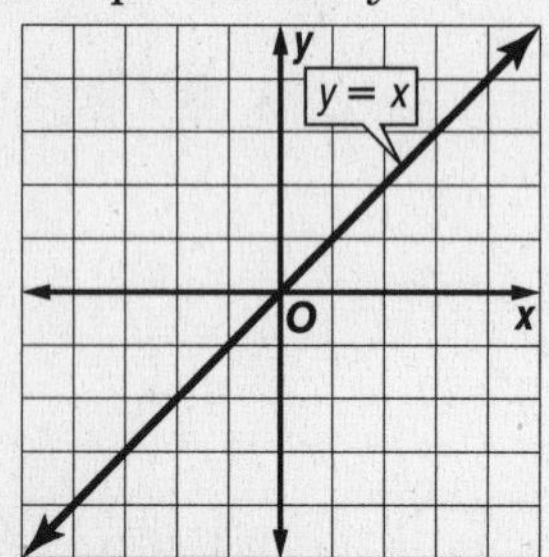

4. $y = mx + b$
   $y = \frac{1}{2}x + 4$
5. $y = mx + b$
   $y = -\frac{3}{5}x - 2$
6. $y = mx + b$
   $y = 3x - 4$
7. $y - y_1 = m(x - x_1)$
   $y - (-1) = \frac{3}{2}(x - 4)$
   $y + 1 = \frac{3}{2}(x - 4)$
8. $y - y_1 = m(x - x_1)$
   $y - 5 = 3(x - 7)$
9. $y - y_1 = m(x - x_1)$
   $y - 137.5 = 1.25(x - 20)$
10. $m = \frac{(y_2 - y_1)}{(x_2 - x_1)}$
    $= \frac{5 - 3}{0 - (-1)}$
    $= 2$
    $y = 2x + 5$
11. $m = \frac{(y_2 - y_1)}{(x_2 - x_1)}$
    $= \frac{3 - 2}{-1 - 0}$
    $= -1$
    $y = -x + 2$
12. The slope of ℓ is 2 (from Exercise 10). The line parallel to ℓ that contains (4, 4) also has slope 2.
    $y - y_1 = m(x - x_1)$
    $y - 4 = 2(x - 4)$
    $y - 4 = 2x - 8$
    $y = 2x - 4$
13. The total monthly cost of Justin's current plan is $y = 39.95$. For the other provider, the cost increases \$0.95 for each hour of connection so the slope is 0.95. The y-intercept is where 0 hours are used, or \$4.95.
    $y = mx + b$
    $y = 0.95x + 4.95$
14. Current plan: $y = 39.95$ (no matter how many hours Justin uses)
    Alternate plan: $y = 0.95x + 4.95$
    $= 0.95(60) + 4.95$
    $= 61.95$
    He should keep his current plan, based on his average usage.

Pages 148–149 Practice and Apply

15. $y = mx + b$
    $y = \frac{1}{6}x - 4$
16. $y = mx + b$
    $y = \frac{2}{3}x + 8$
17. $y = mx + b$
    $y = \frac{5}{8}x - 6$
18. $y = mx + b$
    $y = \frac{2}{9}x + \frac{1}{3}$
19. $y = mx + b$
    $y = -x - 3$
20. $y = mx + b$
    $y = -\frac{1}{12}x + 1$
21. $y - y_1 = m(x - x_1)$
    $y - 1 = 2(x - 3)$
22. $y - y_1 = m(x - x_1)$
    $y - 7 = -5(x - 4)$
23. $y - y_1 = m(x - x_1)$
    $y - (-5) = -\frac{4}{5}[x - (-12)]$
    $y + 5 = -\frac{4}{5}(x + 12)$
24. $y - y_1 = m(x - x_1)$
    $y - 11 = \frac{1}{16}(x - 3)$
25. $y - y_1 = m(x - x_1)$
    $y - 17.12 = 0.48(x - 5)$
26. $y - y_1 = m(x - x_1)$
    $y - 87.5 = -1.3\ (x - 10)$
27. Find the slope of k using (0, −2) and (−1, 1).
    $m = \frac{(y_2 - y_1)}{(x_2 - x_1)}$
    $= \frac{1 - (-2)}{-1 - 0}$
    $= -3$
    $y = mx + b$
    $y = -3x - 2$
28. Find the slope of ℓ using (0, 5) and (−1, 4).
    $m = \frac{(y_2 - y_1)}{(x_2 - x_1)}$
    $= \frac{4 - 5}{-1 - 0}$
    $= 1$
    $y = mx + b$
    $y = x + 5$
29. Find the slope of m using (2, 0) and (1, −2).
    $m = \frac{(y_2 - y_1)}{(x_2 - x_1)}$
    $= \frac{-2 - 0}{1 - 2}$
    $= 2$
    $y - y_1 = m(x - x_1)$
    $y - 0 = 2(x - 2)$
    $y = 2x - 4$

30. Find the slope of n using (0, 6) and (8, 5).

$m = \frac{(y_2 - y_1)}{(x_2 - x_1)}$

$= \frac{5 - 6}{8 - 0}$

$= -\frac{1}{8}$

$y = mx + b$

$y = -\frac{1}{8}x + 6$

31. Since the slope of line ℓ is 1 (from Exercise 28), the slope of a line perpendicular to it is -1.

$y - y_1 = m(x - x_1)$

$y - 6 = -1[x - (-1)]$

$y - 6 = -x - 1$

$y = -x + 5$

32. Since the slope of line k is -3 (from Exercise 27), the slope of a line parallel to it is -3.

$y - y_1 = m(x - x_1)$

$y - 0 = -3(x - 7)$

$y = -3x + 21$

33. Since the slope of line n is $-\frac{1}{8}$ (from Exercise 30), the slope of a line parallel to it is $-\frac{1}{8}$.

$y - y_1 = m(x - x_1)$

$y - 0 = -\frac{1}{8}(x - 0)$

$y = -\frac{1}{8}x$

34. Since the slope of line m is 2 (from Exercise 29), the slope of a line perpendicular to it is $-\frac{1}{2}$.

$y - y_1 = m(x - x_1)$

$y - (-3) = -\frac{1}{2}[x - (-3)]$

$y + 3 = \frac{-1}{2}x - \frac{3}{2}$

$y = -\frac{1}{2}x - \frac{9}{2}$

35. $y = mx + b$

$y = -3x + 5$

36. $y = mx + b$

$y = 0 \cdot x + 6$

$y = 6$

37. $m = \frac{(y_2 - y_1)}{(x_2 - x_1)}$

$= \frac{3 - 0}{0 - 5}$

$= -\frac{3}{5}$

$y = mx + b$

$y = -\frac{3}{5}x + 3$

38. $m = \frac{(y_2 - y_1)}{(x_2 - x_1)}$

$= \frac{-1 - (-1)}{-2 - 4}$

$= \frac{0}{-6}$ or 0

$y - y_1 = m(x - x_1)$

$y - (-1) = 0(x - 4)$

$y + 1 = 0$

$y = -1$

39. $m = \frac{(y_2 - y_1)}{(x_2 - x_1)}$

$= \frac{-6 - (-3)}{10 - (-5)}$

$= \frac{-3}{15}$ or $-\frac{1}{5}$

$y - y_1 = m(x - x_1)$

$y - (-3) = -\frac{1}{5}[x - (-5)]$

$y + 3 = -\frac{1}{5}x - 1$

$y = -\frac{1}{5}x - 4$

40. $m = \frac{(y_2 - y_1)}{(x_2 - x_1)}$

$= \frac{-1 - 0}{0 - 5}$

$= \frac{1}{5}$

$y = mx + b$

$y = \frac{1}{5}x - 1$

41. $m = \frac{(y_2 - y_1)}{(x_2 - x_1)}$

$= \frac{-4 - 8}{-6 - (-6)}$

$= \frac{-12}{0}$ which is undefined

There is no slope-intercept form for this line. An equation for the line is $x = -6$.

42. $m = \frac{(y_2 - y_1)}{(x_2 - x_1)}$

$= \frac{-5 - (-1)}{-8 - (-4)}$

$= \frac{-4}{-4}$ or 1

$y - y_1 = m(x - x_1)$

$y - (-1) = 1[x - (-4)]$

$y + 1 = x + 4$

$y = x + 3$

43. $2x - 5y = 8$

$-5y = -2x + 8$

$y = \frac{2}{5}x - \frac{8}{5}$

The slope of the line is $\frac{2}{5}$, so the slope of a line parallel to it is $\frac{2}{5}$.

$y - y_1 = m(x - x_1)$

$y - (-2) = \frac{2}{5}(x - 7)$

$y + 2 = \frac{2}{5}x - \frac{14}{5}$

$y = \frac{2}{5}x - \frac{24}{5}$

44. $2y + 2 = -\frac{7}{4}(x - 7)$

$2y + 2 = -\frac{7}{4}x + \frac{49}{4}$

$2y = -\frac{7}{4}x + \frac{41}{4}$

$y = -\frac{7}{8}x + \frac{41}{8}$

The slope of the line is $-\frac{7}{8}$, so the slope of a line perpendicular to it is $\frac{8}{7}$.

$y - y_1 = m(x - x_1)$

$y - (-3) = \frac{8}{7}[x - (-2)]$

$y + 3 = \frac{8}{7}x + \frac{16}{7}$

$y = \frac{8}{7}x - \frac{5}{7}$

45. For each appliance Ann sells she earns \$50, so Ann earns 15(\$50) or \$750 plus commission. If the total price of the appliances Ann sells is x dollars, her commission is $0.05x$. So, Ann earned $y = 0.05x + 750$ dollars in a week in which she sold 15 appliances.

46. $750x$

47. The number of gallons of paint in stock decreases at a rate of 750 gallons per day.

$y = -750x + 10{,}800$

48.

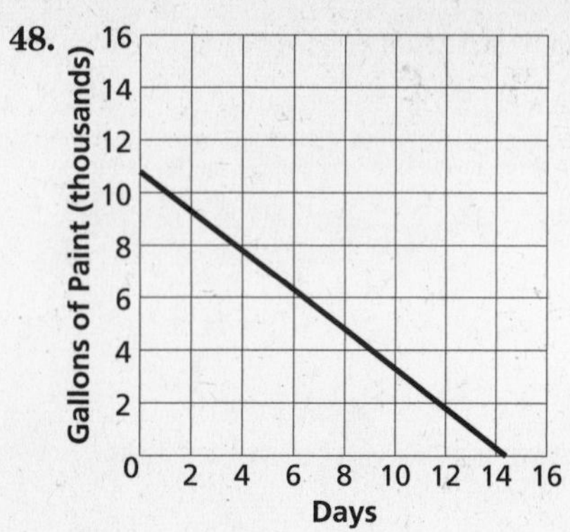

49. Find x when $y = 0$.

$y = -750x + 10{,}800$
$0 = -750x + 10{,}800$
$750x = 10{,}800$
$x \approx 14$

The store will run out of paint after 14 days, so the manager should order more paint after 14 − 4 or 10 days.

50. Two points on the line are (120, −60) and (130, −70).

$m = \frac{(y_2 - y_1)}{(x_2 - x_1)}$
$= \frac{-70 - (-60)}{130 - 120}$
$= \frac{-10}{10}$ or -1
$y - y_1 = m(x - x_1)$
$y - (-60) = -1(x - 120)$
$y + 60 = -x + 120$
$y = -x + 60$

51. The slope of the Jeff Davis/Reeves County line is −1 (from Exercise 50), so the slope of the Reeves/Pecos County line is 1.

$y - y_1 = m(x - x_1)$
$y - (-60) = 1(x - 120)$
$y + 60 = x - 120$
$y = x - 180$

52. The equation $y = mx$ is the special case of $y = m(x - x_1) + y_1$ when $x_1 = y_1 = 0$.

53. Sample answer: In the equation of a line, the b value indicates the fixed rate, while the mx value indicates charges based on usage. Answers should include the following.

- The fee for air time can be considered the slope of the equation.
- We can find where the equations intersect to see where the plans would be equal.

54. A; $2x - 8y = 16$

$-8y = -2x + 16$
$y = \frac{1}{4}x - 16$
$\left(\frac{1}{4}\right)(-4) = -1$

55. B; $y^2 < 1$, so $y < 1$ and $-y < 1$, thus $y < 1$ and $y > -1$.

Page 150 Maintain Your Skills

56. $m = \frac{(y_2 - y_1)}{(x_2 - x_1)}$
$= \frac{0 - 6}{4 - 0}$
$= \frac{-6}{4}$ or $-\frac{3}{2}$

57. $m = \frac{(y_2 - y_1)}{(x_2 - x_1)}$
$= \frac{-6 - 1}{8 - 8}$
$= \frac{-7}{0}$, which is undefined

58. $m = \frac{(y_2 - y_1)}{(x_2 - x_1)}$
$= \frac{3 - 3}{-6 - 6}$
$= \frac{0}{-12}$ or 0

59. $\angle 7 \cong \angle 1$
$m\angle 7 = m\angle 1$
$m\angle 7 = 58$

60. $\angle 5 \cong \angle 2$
$m\angle 5 = m\angle 2$
$m\angle 5 = 47$

61. $m\angle 1 + m\angle 2 + m\angle 6 = 180$
$58 + 47 + m\angle 6 = 180$
$m\angle 6 = 75$

62. $m\angle 4 + m\angle 2 + m\angle 3 = 180$
$m\angle 4 + 47 + 26 = 180$
$m\angle 4 = 107$

63. $m\angle 8 = m\angle 2 + m\angle 3$
$m\angle 8 = 47 + 26$
$m\angle 8 = 73$

64. From Exercise 59, $m\angle 7 = 58$.
From Exercise 63, $m\angle 8 = 73$.
$m\angle 7 + m\angle 8 + m\angle 9 = 180$
$58 + 73 + m\angle 9 = 180$
$m\angle 9 = 49$

65. Given: $AC = DF$
$AB = DE$
Prove: $BC = EF$

A B C
D E F

Proof:

Statements	Reasons
1. $AC = DF$, $AB = DE$	1. Given
2. $AC = AB + BC$, $DF = DE + EF$	2. Segment Addition Postulate
3. $AB + BC = DE + EF$	3. Substitution Property
4. $BC = EF$	4. Subtraction Property

66. $d = \sqrt{(x_2 - x_1)^2 + (y_2 - y_1)^2}$
$AB = \sqrt{(-2 - 10)^2 + [-8 - (-6)]^2}$
$= \sqrt{(-12)^2 + (-2)^2}$
$= \sqrt{148}$
$BC = \sqrt{[-5 - (-2)]^2 + [-7 - (-8)]^2}$
$= \sqrt{(-3)^2 + 1^2}$
$= \sqrt{10}$

$AC = \sqrt{(-5 - 10)^2 + [7 - (-6)]^2}$
$= \sqrt{(-15)^2 + (-1)^2} = \sqrt{226}$
$AB + BC + AC = \sqrt{148} + \sqrt{10} + \sqrt{226}$
≈ 30.36

67. $d = \sqrt{(x_2 - x_1)^2 + (y_2 - y_1)^2}$
$AB = \sqrt{[2 - (-3)]^2 + (-9 - 2)^2}$
$= \sqrt{5^2 + (-11)^2} = \sqrt{146}$
$BC = \sqrt{(0 - 2)^2 + [-10 - (-9)]^2}$
$= \sqrt{(-2)^2 + (-1)^2} = \sqrt{5}$
$AC = \sqrt{[0 - (-3)]^2 + (-10 - 2)^2}$
$= \sqrt{3^2 + (-12)^2} = \sqrt{153}$
$AB + BC + AC = \sqrt{146} + \sqrt{5} + \sqrt{153}$
≈ 26.69

68. $\angle 2$ and $\angle 5$, $\angle 3$ and $\angle 8$
69. $\angle 1$ and $\angle 5$, $\angle 2$ and $\angle 6$, $\angle 4$ and $\angle 8$, $\angle 3$ and $\angle 7$
70. $\angle 1$ and $\angle 7$, $\angle 4$ and $\angle 6$
71. $\angle 2$ and $\angle 8$, $\angle 3$ and $\angle 5$

Page 150 Practice Quiz 2

1. slope of $\overleftrightarrow{AB} = \frac{1 - (-1)}{6 - 3}$
$= \frac{2}{3}$
slope of $\overleftrightarrow{CD} = \frac{4 - (-2)}{2 - (-2)}$
$= \frac{6}{4}$ or $\frac{3}{2}$

The slopes are not the same, so $\overleftrightarrow{AB}$ and $\overleftrightarrow{CD}$ are not parallel. The product of the slopes is $\left(\frac{2}{3}\right)\left(\frac{3}{2}\right)$ or 1, so $\overleftrightarrow{AB}$ and $\overleftrightarrow{CD}$ are not perpendicular. Therefore, $\overleftrightarrow{AB}$ and $\overleftrightarrow{CD}$ are neither parallel nor perpendicular.

2. slope of $\overleftrightarrow{AB} = \frac{13 - (-11)}{3 - (-3)}$
$= \frac{24}{6}$ or 4
slope of $\overleftrightarrow{CD} = \frac{-8 - (-6)}{8 - 0}$
$= \frac{-2}{8}$ or $-\frac{1}{4}$

The product of the slopes is $4\left(-\frac{1}{4}\right)$ or -1. So $\overleftrightarrow{AB}$ is perpendicular to $\overleftrightarrow{CD}$.

3. Line p contains $(0, -4)$ and $(2, 3)$.
$m = \frac{(y_2 - y_1)}{(x_2 - x_1)}$
$= \frac{3 - (-4)}{2 - 0} = \frac{7}{2}$

4. A line parallel to q has the same slope as q. Line q contains $(0, 2)$ and $(2, 3)$.
$m = \frac{(y_2 - y_1)}{(x_2 - x_1)}$
$= \frac{3 - 2}{2 - 0} = \frac{1}{2}$

5. Line r contains $(1, -1)$ and $(-4, 3)$.
$m = \frac{(y_2 - y_1)}{(x_2 - x_1)}$
$= \frac{3 - (-1)}{-4 - 1} = -\frac{4}{5}$

$\left(-\frac{4}{5}\right)\left(\frac{5}{4}\right) = -1$, so a line perpendicular to r has slope $\frac{5}{4}$.

6. From Exercise 4, q has slope $\frac{1}{2}$ and has y-intercept 2.
$y = mx + b$
$y = \frac{1}{2}x + 2$

7. From Exercise 5, r has slope $-\frac{4}{5}$. A line parallel to r also has slope $-\frac{4}{5}$.
$y - y_1 = m(x - x_1)$
$y - 4 = -\frac{4}{5}[x - (-1)]$
$y - 4 = -\frac{4}{5}x - \frac{4}{5}$
$y = -\frac{4}{5}x + \frac{16}{5}$

8. From Exercise 3, p has slope $\frac{7}{2}$. A line perpendicular to p has slope $-\frac{2}{7}$.
$y = mx + b$
$y = -\frac{2}{7}x + 0$
$y = -\frac{2}{7}x$

9. The lines are parallel so they have the same slope: $-\frac{1}{4}$.
$y - y_1 = m(x - x_1)$
$y - (-8) = -\frac{1}{4}(x - 5)$
$y + 8 = -\frac{1}{4}(x - 5)$

10. The line $y = -3$ is a horizontal line, so a line perpendicular to this line is vertical. An equation of the vertical line through $(-4, -4)$ is $0 = x + 4$.

3-5 Proving Lines Parallel

Page 154 Check for Understanding

1. Sample answer: Use a pair of alternate exterior $\angle$s that are congruent and cut by a transversal; show that a pair of consecutive interior angles are supplementary; show that alternate interior $\angle$s are $\cong$; show two lines are $\perp$ to same line; show corresponding $\angle$s are $\cong$.

2. Sample answer:

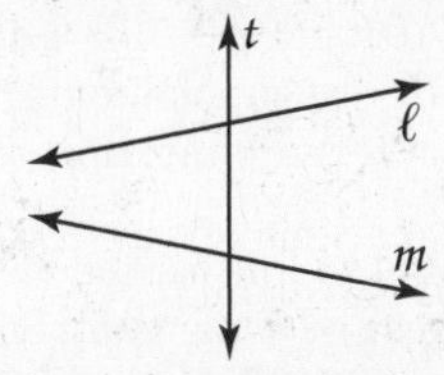

3. Sample answer: A basketball court has parallel lines, as does a newspaper. The edges should be equidistant along the entire line.
4. $\ell \parallel m$; corresponding angles
5. $\ell \parallel m$; alternate interior angles
6. $p \parallel q$; consecutive interior angles
7. $p \parallel q$; alternate exterior angles

8.

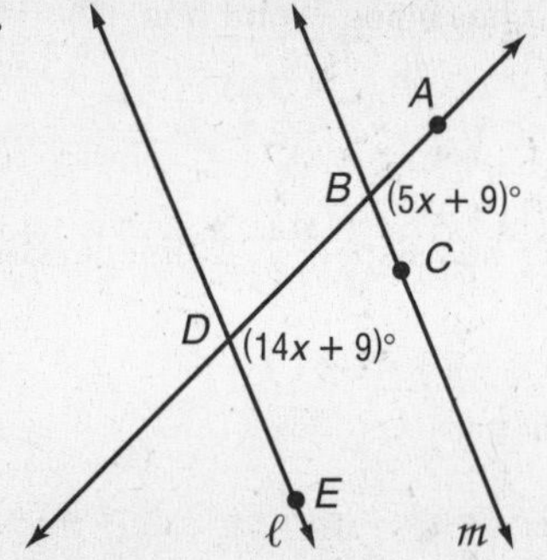

Explore: From the figure, you know that $m\angle ABC = 5x + 90$ and $m\angle ADE = 14x + 9$. You also know that $\angle ABC$ and $\angle ADE$ are corresponding angles.

Plan: For line ℓ to be parallel to line m, the corresponding angles must be congruent. So, $m\angle ABC = m\angle ADE$. Substitute the given angle measures into this equation and solve for x.

Solve: $m\angle ABC = m\angle ADE$

$$\begin{aligned} 5x + 90 &= 14x + 9 \\ 90 &= 9x + 9 \\ 81 &= 9x \\ 9 &= x \end{aligned}$$

Examine: Now use the value of x to find $m\angle ABC$.

$$\begin{aligned} m\angle ABC &= 5x + 90 \\ &= 5(9) + 90 = 135 \end{aligned}$$

Verify the angle measure by using the value of x to find $m\angle ADE$. That is, $14x + 9 = 14(9) + 9$ or 135. Since $m\angle ABC = m\angle ADE$, $\angle ABC \cong \angle ADE$ and $\ell \parallel m$.

9.

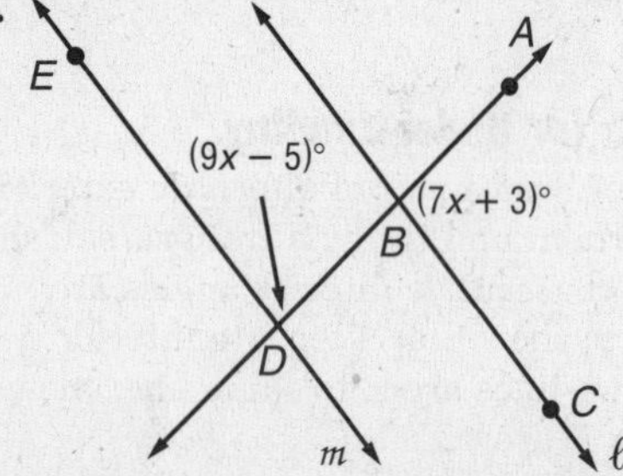

Explore: From the figure, you know that $m\angle ABC = 7x + 3$ and $m\angle ADE = 9x - 5$. You also know that $\angle ABC$ and $\angle CBD$ are supplementary, and $\angle CBD$ and $\angle ADE$ are alternate interior angles.

Plan: For line ℓ to be parallel to line m, the alternate interior angles must be congruent. So, $m\angle CBD = m\angle ADE$, where $m\angle CBD + m\angle ABC = 180$. Thus $m\angle CBD = 180 - m\angle ABC$, so $m\angle ADE = 180 - m\angle ABC$. Substitute the given angle measures into this equation and solve for x.

Solve: $m\angle ADE = 180 - m\angle ABC$

$$\begin{aligned} 9x - 5 &= 180 - (7x + 3) \\ 9x - 5 &= 180 - 7x - 3 \\ 16x - 5 &= 180 - 3 \\ 16x &= 182 \\ x &= 11.375 \end{aligned}$$

Examine: Now use the value of x to find $m\angle ADE$.

$$\begin{aligned} m\angle ADE &= 9x - 5 \\ &= 9(11.375) - 5 \\ &= 97.375 \end{aligned}$$

Verify the angle measure by using the value of x to find $m\angle ABC$.

$$\begin{aligned} m\angle ABC &= 7x + 3 \\ &= 7(11.375) + 3 \\ &= 82.625 \end{aligned}$$

Now $m\angle CBD = 180 - m\angle ABC = 180 - 82.625$ or 97.375. Since $m\angle CBD = m\angle ADE$, $\angle CBD \cong \angle ADE$ and $\ell \parallel m$.

10. Given: $\angle 1 \cong \angle 2$

Prove: $\ell \parallel m$

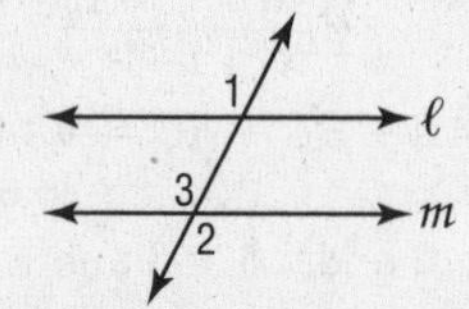

Proof:

Statements	Reasons
1. $\angle 1 \cong \angle 2$	1. Given
2. $\angle 2 \cong \angle 3$	2. Vertical angles are congruent.
3. $\angle 1 \cong \angle 3$	3. Trans. Prop. of $\cong$
4. $\ell \parallel m$	4. If corr. ∠s are $\cong$, then lines are $\parallel$.

11. slope of $\overleftrightarrow{AB} = \frac{-2 - (-3)}{0 - (-7)}$ or $\frac{1}{7}$

slope of $\overleftrightarrow{CD} = \frac{\frac{7}{4} - \frac{1}{2}}{6 - (-4)}$

$= \frac{\frac{5}{4}}{10}$ or $\frac{1}{8}$

The slope of $\overleftrightarrow{CD}$ is $\frac{1}{8}$, and the slope of line $\overleftrightarrow{AB}$ is $\frac{1}{7}$. The slopes are not equal, so the lines are not parallel.

12. Yes; sample answer: Pairs of alternate interior angles are congruent.

Pages 155–157 Practice and Apply

13. $a \parallel b$; alternate interior ∠s

14. none

15. $\ell \parallel m$; corresponding ∠s

16. none

17. $\overleftrightarrow{AE}$ and $\overleftrightarrow{BF}$; $\cong$ corresponding ∠s

18. $\overleftrightarrow{AE}$ and $\overleftrightarrow{BF}$; $\cong$ corresponding ∠s

19. $\overleftrightarrow{AC}$ and $\overleftrightarrow{EG}$; $\cong$ alternate interior ∠s

20. $\overleftrightarrow{AC}$ and $\overleftrightarrow{EG}$; supplementary consecutive interior ∠s

21. $\overleftrightarrow{HS}$ and $\overleftrightarrow{JT}$; $\cong$ corresponding ∠s

22. $\overleftrightarrow{HS}$ and $\overleftrightarrow{JT}$; $\cong$ alternate interior ∠s

23. $\overleftrightarrow{KN}$ and $\overleftrightarrow{PR}$; supplementary consecutive interior ∠s

24. $\overleftrightarrow{HS}$ and $\overleftrightarrow{JT}$; 2 lines $\perp$ the same line

25. Given: $\ell \perp t$
$m \perp t$
Prove: $\ell \parallel m$

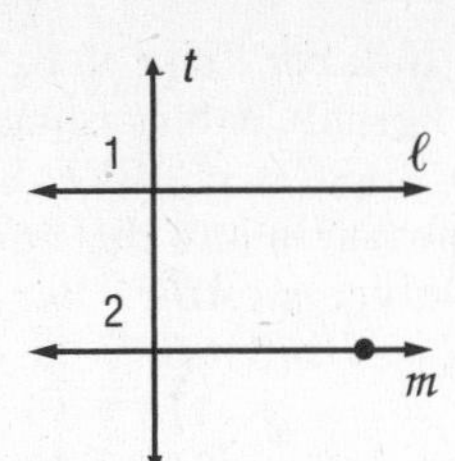

Proof:

Statements	Reasons
1. $\ell \perp t, m \perp t$	1. Given
2. $\angle 1$ and $\angle 2$ are right angles.	2. Definition of perpendicular
3. $\angle 1 \cong \angle 2$	3. All rt. ∠s are ≅.
4. $\ell \parallel m$	4. If corr. ∠s are ≅, then lines are ∥.

26.

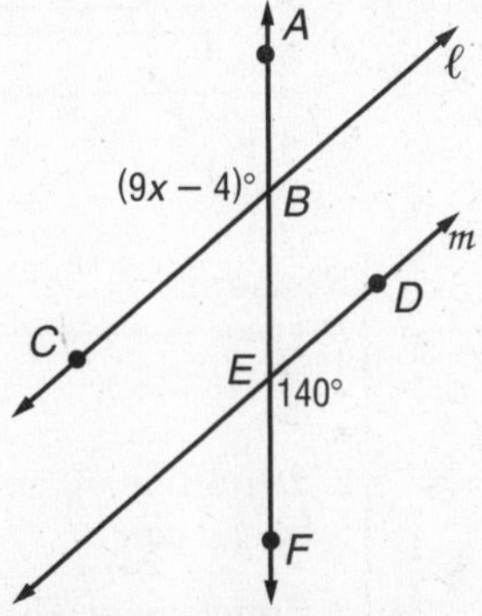

Explore: From the figure, you know that $m\angle ABC = 9x - 4$ and $m\angle DEF = 140$. You also know that $\angle ABC$ and $\angle DEF$ are alternate exterior angles.
Plan: For line ℓ to be parallel to line m, the alternate exterior angles must be congruent. So, $m\angle ABC = m\angle DEF$. Substitute the given angle measures into this equation and solve for x.
Solve: $m\angle ABC = m\angle DEF$
$9x - 4 = 140$
$9x = 144$
$x = 16$
Examine: Verify the angle measure by using the value of x to find $m\angle ABC$. That is, $9x - 4 = 9(16) - 4$ or 140. Since $m\angle ABC = m\angle DEF$, $\angle ABC \cong \angle DEF$ and $\ell \parallel m$.

27.

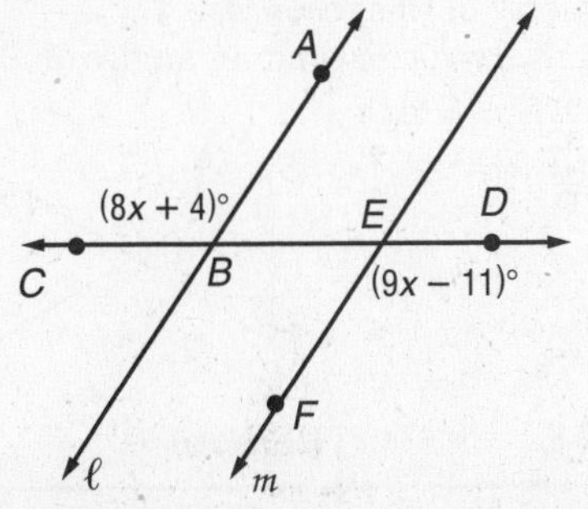

Explore: From the figure, you know that $m\angle ABC = 8x + 4$ and $m\angle DEF = 9x - 11$. You also know that $\angle ABC$ and $\angle DEF$ are alternate exterior angles.
Plan: For line ℓ to be parallel to line m, the alternate exterior angles must be congruent. So, $m\angle ABC = m\angle DEF$. Substitute the given angle measures into this equation and solve for x.

Solve: $m\angle ABC = m\angle DEF$
$8x + 4 = 9x - 11$
$4 = x - 11$
$15 = x$
Examine: Verify the angle measures by using the value of x to find $m\angle ABC$ and $m\angle DEF$.
$m\angle ABC = 8x + 4$
$= 8(15) + 4$ or 124
$m\angle DEF = 9x - 11$
$= 9(15) - 11$ or 124
Since $m\angle ABC = m\angle DEF$, $\angle ABC \cong \angle DEF$ and $\ell \parallel m$.

28.

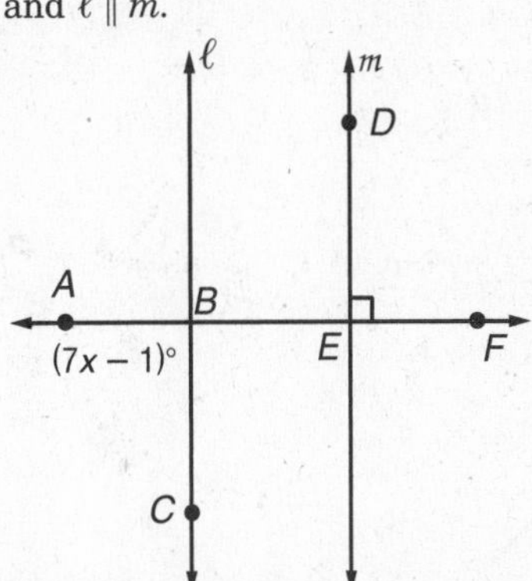

Explore: From the figure, you know that $m\angle ABC = 7x - 1$ and $\angle DEF$ is a right angle and hence has measure 90. You also know that $\angle ABC$ and $\angle DEF$ are alternate exterior angles.
Plan: For line ℓ to be parallel to line m, the alternate exterior angles must be congruent. So, $m\angle ABC = m\angle DEF$. Substitute the given angle measures into this equation and solve for x.
Solve: $m\angle ABC = m\angle DEF$
$7x - 1 = 90$
$7x = 91$
$x = 13$
Examine: Verify the angle measure by using the value of x to find $m\angle ABC$. That is, $7x - 1 = 7(13) - 1$ or 90. Since $m\angle ABC = m\angle DEF$, $\angle ABC \cong \angle DEF$ and $\ell \parallel m$.

29.

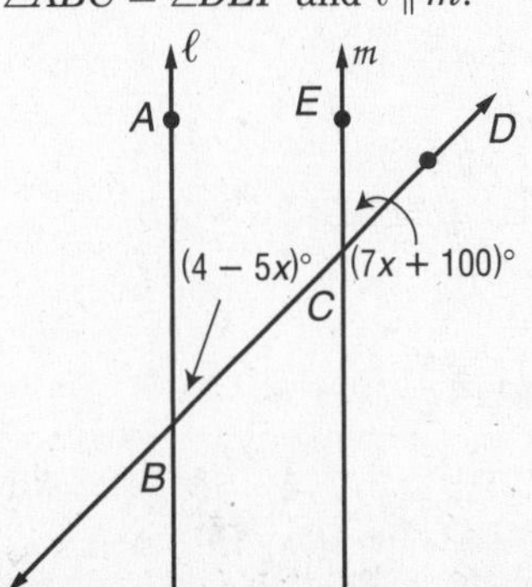

Explore: From the figure, you know that $m\angle ABC = 4 - 5x$ and $m\angle ECD = 7x + 100$. You also know that $\angle ABC$ and $\angle ECD$ are corresponding angles.
Plan: For line ℓ to be parallel to line m, the corresponding angles must be congruent. So, $m\angle ABC = m\angle ECD$. Substitute the given angle measures into this equation and solve for x.

Solve: $m\angle ABC = m\angle ECD$

$$4 - 5x = 7x + 100$$
$$4 = 12x + 100$$
$$-96 = 12x$$
$$-8 = x$$

Examine: Verify the angle measures by using the value of x to find $m\angle ABC$ and $m\angle ECD$.

$m\angle ABC = 4 - 5x$
$= 4 - 5(-8)$ or 44

$m\angle ECD = 7x + 100$
$= 7(-8) + 100$ or 44

Since $m\angle ABC = m\angle ECD$, $\angle ABC \cong \angle ECD$ and $\ell \parallel m$.

30.

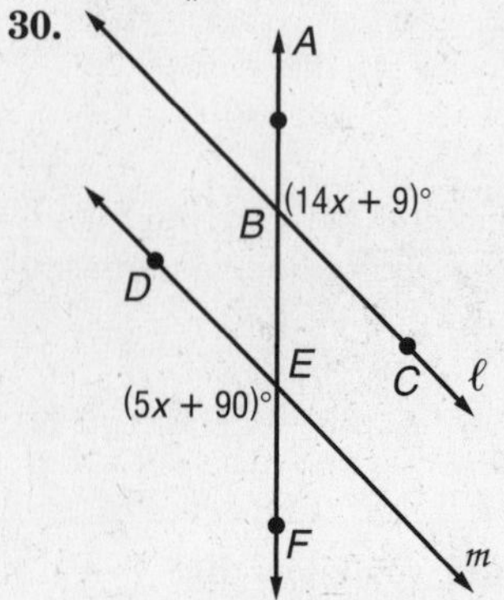

Explore: From the figure, you know that $m\angle ABC = 14x + 9$ and $m\angle DEF = 5x + 90$. You also know that $\angle ABC$ and $\angle DEF$ are alternate exterior angles.

Plan: For line ℓ to be parallel to line m, the alternate exterior angles must be congruent. So, $m\angle ABC = m\angle DEF$. Substitute the given angle measures into this equation and solve for x.

Solve: $m\angle ABC = m\angle DEF$

$$14x + 9 = 5x + 90$$
$$9x + 9 = 90$$
$$9x = 81$$
$$x = 9$$

Examine: Verify the angle measures by using the value of x to find $m\angle ABC$ and $m\angle DEF$.

$m\angle ABC = 14x + 9$
$= 14(9) + 9$ or 135

$m\angle DEF = 5x + 90$
$= 5(9) + 90$ or 135

Since $m\angle ABC = m\angle DEF$, $\angle ABC \cong \angle DEF$ and $\ell \parallel m$.

31.

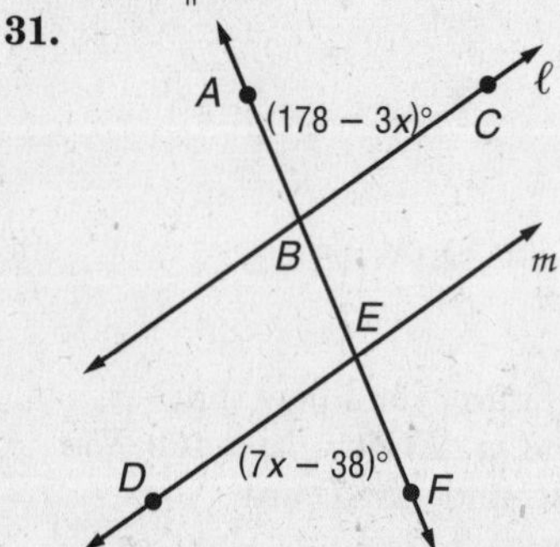

Explore: From the figure, you know that $m\angle ABC = 178 - 3x$ and $m\angle DEF = 7x - 38$. You also know that $\angle ABC$ and $\angle DEF$ are alternate exterior angles.

Plan: For line ℓ to be parallel to line m, the alternate exterior angles must be congruent. So, $m\angle ABC = m\angle DEF$. Substitute the given angle measures into this equation and solve for x.

Solve: $m\angle ABC = m\angle DEF$

$$178 - 3x = 7x - 38$$
$$178 = 10x - 38$$
$$216 = 10x$$
$$21.6 = x$$

Examine: Verify the angle measures by using the value of x to find $m\angle ABC$ and $m\angle DEF$.

$m\angle ABC = 178 - 3x$
$= 178 - 3(21.6)$ or 113.2

$m\angle DEF = 7x - 38$
$= 7(21.6) - 38$ or 113.2

Since $m\angle ABC = m\angle DEF$, $\angle ABC \cong \angle DEF$ and $\ell \parallel m$.

32. Given: $\angle 1$ and $\angle 2$ are supplementary.
Prove: $\ell \parallel m$

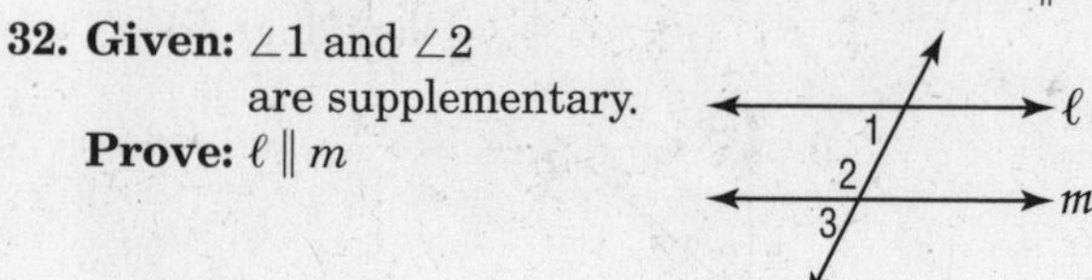

Proof:

Statements	Reasons
1. $\angle 1$ and $\angle 2$ are supplementary.	1. Given
2. $\angle 2$ and $\angle 3$ form a linear pair.	2. Definition of linear pair
3. $\angle 2$ and $\angle 3$ are supplementary.	3. Supplement Th.
4. $\angle 1 \cong \angle 3$	4. ∠s suppl. to same ∠ or ≅ ∠s are ≅.
5. $\ell \parallel m$	5. If corr. ∠s are ≅, then lines are ∥.

33. Given: $\angle 4 \cong \angle 6$
Prove: $\ell \parallel m$

Proof: We know that $\angle 4 \cong \angle 6$. Because $\angle 6$ and $\angle 7$ are vertical angles they are congruent. By the Transitive Property of Congruence, $\angle 4 \cong \angle 7$. Since $\angle 4$ and $\angle 7$ are corresponding angles, and they are congruent, $\ell \parallel m$.

34. Given: $\angle 2 \cong \angle 1$, $\angle 1 \cong \angle 3$
Prove: $\overline{ST} \parallel \overline{UV}$

Proof:

Statements	Reasons
1. $\angle 2 \cong \angle 1$, $\angle 1 \cong \angle 3$	1. Given
2. $\angle 2 \cong \angle 3$	2. Trans. Prop.
3. $\overline{ST} \parallel \overline{UV}$	3. If alt. int. ∠s are ≅, lines are ∥.

35. Given: $\overline{AD} \perp \overline{CD}$, $\angle 1 \cong \angle 2$
Prove: $\overline{BC} \perp \overline{CD}$

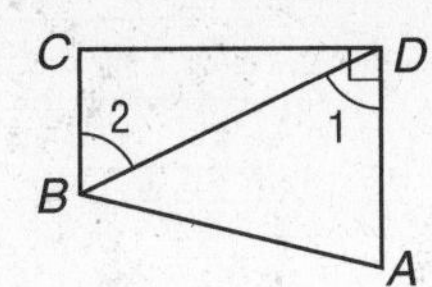

Proof:

Statements	Reasons
1. $\overline{AD} \perp \overline{CD}$, $\angle 1 \cong \angle 2$	1. Given
2. $\overline{AD} \parallel \overline{BC}$	2. If alt. int. $\angle$s are $\cong$, lines are $\parallel$.
3. $\overline{BC} \perp \overline{CD}$	3. Perpendicular Transversal Th.

36. Given: $\overline{JM} \parallel \overline{KN}$, $\angle 1 \cong \angle 2$, $\angle 3 \cong \angle 4$
Prove: $\overline{KM} \parallel \overline{LN}$

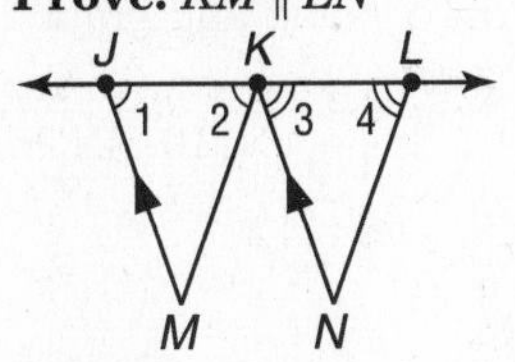

Proof:

Statements	Reasons
1. $\overline{JM} \parallel \overline{KN}$, $\angle 1 \cong \angle 2$, $\angle 3 \cong \angle 4$	1. Given
2. $\angle 1 \cong \angle 3$	2. If lines are $\parallel$, corr. $\angle$s are $\cong$.
3. $\angle 2 \cong \angle 4$	3. Substitution
4. $\overline{KM} \parallel \overline{LN}$	4. If corr. $\angle$s are $\cong$, lines are $\parallel$.

37. Given: $\angle RSP \cong \angle PQR$, $\angle QRS$ and $\angle PQR$ are supplementary
Prove: $\overline{PS} \parallel \overline{QR}$

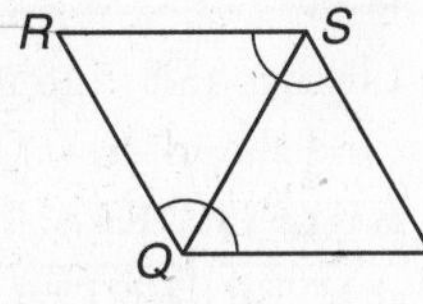

Proof:

Statements	Reasons
1. $\angle RSP \cong \angle PQR$, $\angle QRS$ and $\angle PQR$ are supplementary	1. Given
2. $m\angle RSP = m\angle PQR$	2. Def. of $\cong$ $\angle$s
3. $m\angle QRS + m\angle PQR = 180$	3. Definition of suppl. $\angle$s
4. $m\angle QRS + m\angle RSP = 180$	4. Substitution
5. $\angle QRS$ and $\angle RSP$ are supplementary.	5. Def. of suppl. $\angle$s
6. $\overline{PS} \parallel \overline{QR}$	6. If cons. int. $\angle$s are suppl., lines are $\parallel$.

38. slope of $\overleftrightarrow{AB} = \frac{4-2}{4-(-4)}$ or $\frac{1}{4}$

slope of $\overleftrightarrow{CD} = \frac{1-0}{4-0}$ or $\frac{1}{4}$

Yes, the lines are parallel since the slopes are the same.

39. slope of $\overleftrightarrow{AB} = \frac{-1.5-(-0.75)}{2-(-1)}$

$= \frac{-0.75}{3}$ or -0.25

slope of $\overleftrightarrow{CD} = \frac{1.5-1.8}{0-(-1.5)}$

$= \frac{-0.3}{1.5}$ or -0.2

No, the lines are not parallel since the slopes are not the same.

40. When he measures the angle that each picket makes with the 2 by 4, he is measuring corresponding angles. When all of the corresponding angles are congruent, the pickets must be parallel.

41. The 10-yard lines will be parallel because they are all perpendicular to the sideline and two or more lines perpendicular to the same line are parallel.

42. Consecutive angles are supplementary; opposite angles are congruent; the sum of the measures of the angles is 360.

43. See students' work.

44. Sample answer: They should appear to have the same slope. Answers should include the following.
- The corresponding angles must be equal in order for the lines to be parallel.
- The parking lot spaces have right angles.

45. B; $\angle 2$ and $\angle 3$ are supplementary to $\angle 1$ because they each form a linear pair with $\angle 1$. $\angle 4$ is not necessarily supplementary to $\angle 1$ because $\angle 1$ and $\angle 4$ are vertical angles and hence are congruent. $\angle 1 \cong \angle 5$ because $\ell \parallel m$ and $\angle 1$ and $\angle 5$ are corresponding angles. $\angle 7$ and $\angle 6$ are supplementary to $\angle 5$ because they each form a linear pair with $\angle 5$, and so by substitution $\angle 7$ and $\angle 6$ are supplementary to $\angle 1$. $\angle 8$ is not necessarily supplementary to $\angle 1$ because $\angle 8$ and $\angle 5$ are vertical angles, so $\angle 8 \cong \angle 5$ and $\angle 8 \cong \angle 1$.

46. D; let p, n, d, and q represent the numbers of pennies, nickels, dimes, and quarters Kendra has, respectively. Then we are given that $p = 3n$, $n = d$, and $d = 2q$. So we can see that $n = 2q$ so $p = 3(2q)$ or $6q$, and thus all the numbers of coins depend on the number of quarters. Kendra has at least one quarter, so if $q = 1$ then $d = 2(1)$ or 2, $n = 2$, and $p = 6(1)$ or 6.

Then $0.01p + 0.05n + 0.10d + 0.25q$
$= 0.01(6) + 0.05(2) + 0.10(2) + 0.25(1)$
$= 0.61$

Page 157 Maintain Your Skills

47. $y = mx + b$
$y = 0.3x - 6$

48. $y - y_1 = m(x - x_1)$
$y - (-15) = \frac{1}{3}[x - (-3)]$
$y + 15 = \frac{1}{3}x + 1$
$y = \frac{1}{3}x - 14$

49. $m = \frac{(y_2 - y_1)}{(x_2 - x_1)}$

$= \frac{11 - 7}{-3 - 5}$

$= -\frac{4}{8}$ or $-\frac{1}{2}$

$y - y_1 = m(x - x_1)$

$y - 7 = -\frac{1}{2}(x - 5)$

$y - 7 = -\frac{1}{2}x + \frac{5}{2}$

$y = -\frac{1}{2}x + \frac{19}{2}$

50. The slope of $y = \frac{1}{2}x - 4$ is $\frac{1}{2}$, so a line perpendicular to $y = \frac{1}{2}x - 4$ has slope -2.

$y - y_1 = m(x - x_1)$

$y - 1 = -2(x - 4)$

$y - 1 = -2x + 8$

$y = -2x + 9$

51. slope of $\overleftrightarrow{BD} = \frac{-3 - 2}{4 - 0}$ or $\frac{-5}{4}$

52. slope of $\overleftrightarrow{CD} = \frac{-3 - (-3)}{4 - (-1)}$ or 0

53. slope of $\overleftrightarrow{AB} = \frac{2 - (-2)}{0 - (-4)}$ or 1

54. slope of $\overleftrightarrow{EO} = \frac{0 - 2}{0 - 4}$ or $\frac{1}{2}$

55. slope of $\overleftrightarrow{DE} = \frac{2 - (-3)}{4 - 4}$

$= \frac{5}{0}$, which is undefined

Any line parallel to $\overleftrightarrow{DE}$ also has undefined slope.

56. slope of $\overleftrightarrow{BD} = \frac{-5}{4}$ (from Exercise 51)

The slope of any line perpendicular to $\overleftrightarrow{BD}$ is $\frac{4}{5}$.

57.

p	q	p and q
T	T	T
T	F	F
F	T	F
F	F	F

58.

p	q	$\sim q$	p or $\sim q$
T	T	F	T
T	F	T	T
F	T	F	F
F	F	T	T

59.

p	q	$\sim p$	$\sim p \wedge q$
T	T	F	F
T	F	F	F
F	T	T	T
F	F	T	F

60.

p	q	$\sim p$	$\sim q$	$\sim p \wedge \sim q$
T	T	F	F	F
T	F	F	T	F
F	T	T	F	F
F	F	T	T	T

61. A picture frame has right angles, so the angles must be complementary angles.

62. $d = \sqrt{(x_2 - x_1)^2 + (y_2 - y_1)^2}$

$= \sqrt{(7 - 2)^2 + (19 - 7)^2}$

$= \sqrt{5^2 + 12^2}$

$= \sqrt{169}$ or 13

63. $d = \sqrt{(x_2 - x_1)^2 + (y_2 - y_1)^2}$

$= \sqrt{(-1 - 8)^2 + (2 - 0)^2}$

$= \sqrt{(-9)^2 + 2^2}$

$= \sqrt{85}$

≈ 9.22

64. $d = \sqrt{(x_2 - x_1)^2 + (y_2 - y_1)^2}$

$= \sqrt{[-8 - (-6)]^2 + [-2 - (-4)]^2}$

$= \sqrt{(-2)^2 + 2^2}$

$= \sqrt{8}$

≈ 2.83

Page 158 Graphing Calculator Investigation: Points of Intersection

1. Enter the equations in the Y = list and graph in the standard viewing window.

KEYSTROKES: [Y =] 2 [X, T, θ, n] [–] 10 [ENTER] 2 [X, T, θ, n] [–] 2 [ENTER] [(–)] 1 [÷] 2 [X, T, θ, N] [+] 4 [ZOOM] 6

Use the CALC menu to find the points of intersection.

Find the intersection of a and t.

KEYSTROKES: [2nd] [CALC] 5 [ENTER] [▼] [ENTER] [ENTER]

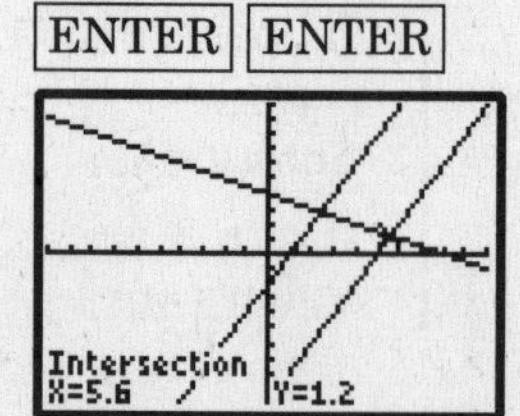

Lines a and t intersect at (5.6, 1.2).

Find the intersection of b and t.

KEYSTROKES: [2nd] [CALC] 5 [▼] [ENTER] [ENTER] [ENTER]

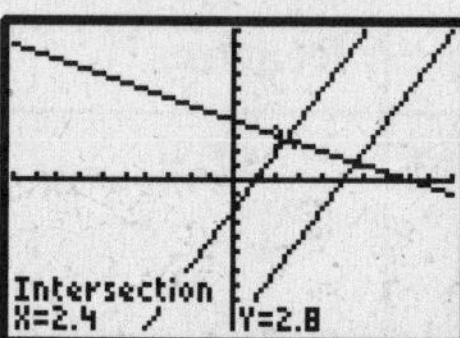

Lines b and t intersect at (2.4, 2.8).

2. Enter the equations in the Y = list and graph in the standard viewing window.

KEYSTROKES: [Y =] [(–)] [X, T, θ, n] [–] 3 [ENTER] [(–)] [X, T, θ, n] [+] 5 [ENTER] [X, T, θ, n] [–] 6 [ZOOM] 6

Use the CALC menu to find the points of intersection.

Find the intersection of a and t.

KEYSTROKES: 2nd [CALC] 5 ENTER ▼ ENTER ENTER

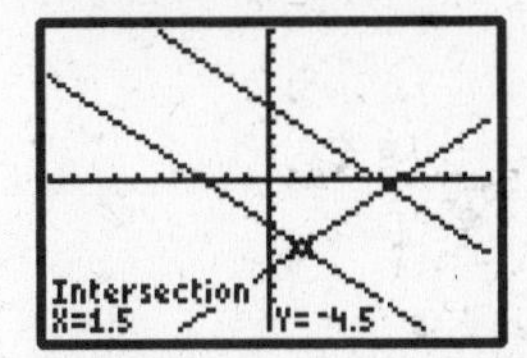

Lines a and t intersect at (1.5, −4.5).

Find the intersection of b and t.

KEYSTROKES: 2nd [CALC] 5 ▼ ENTER ENTER ENTER

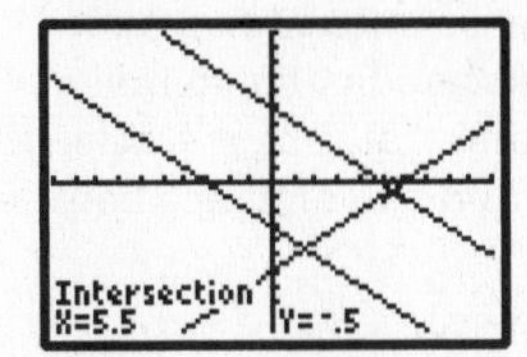

Lines b and t intersect at (5.5, −0.5).

3. Enter the equations for a and b in the Y = list. Use the DRAW menu to graph t, since it is a vertical line.

KEYSTROKES: Y = 6 ENTER 0 ENTER 2nd [DRAW] 4 (−) 2 ENTER

Use the TRACE function to find the points of intersection.

Find the intersection of a and t.

KEYSTROKES: TRACE (−) 2 ENTER

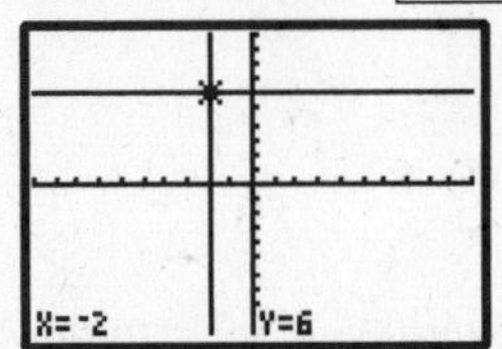

Lines a and t intersect at (−2, 6).

Find the intersection of b and t.

KEYSTROKES: TRACE ▼ (−) 2 ENTER

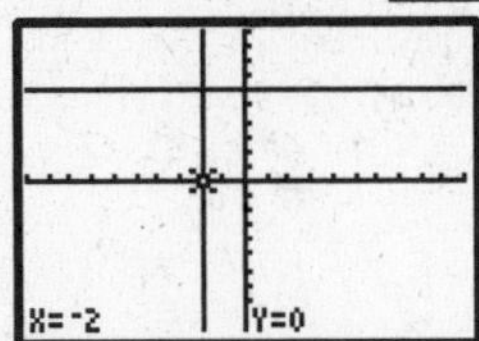

Lines b and t intersect at (−2, 0).

4. Enter the equations in the Y = list and graph in the standard viewing window.

KEYSTROKES: Y = (−) 3 X, T, θ, n + 1 ENTER

(−) 3 X, T, θ, n − 3 ENTER 1

÷ 3 X, T, θ, n + 8 ZOOM 6

Use the CALC menu to find the points of intersection.

Find the intersection of a and t.

KEYSTROKES: 2nd [CALC] 5 ENTER ▼ ENTER ENTER

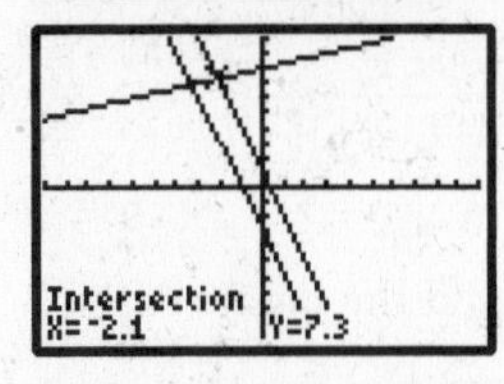

Lines a and t intersect at (−2.1, 7.3)

Find the intersection of b and t.

KEYSTROKES: 2nd [CALC] 5 ▼ ENTER ENTER ENTER

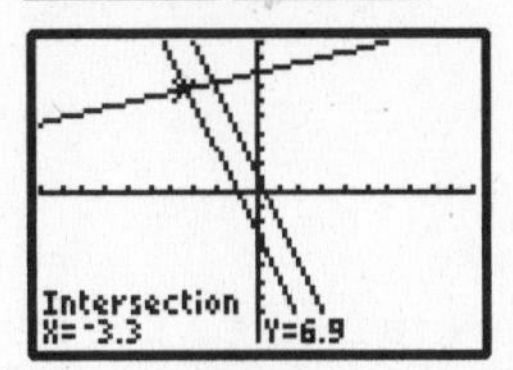

Lines b and t intersect at (−3.3, 6.9)

5. Enter the equations in the Y = list and graph in the standard viewing window.

KEYSTROKES: Y = 4 ÷ 5 X, T, θ, n − 2 ENTER 4 ÷ 5 X, T, θ, n − 7 ENTER (−) 5 ÷ 4 X, T, θ, n ZOOM 6

Use the CALC menu to find the points of intersection.

Find the intersection of a and t.

KEYSTROKES: 2nd [CALC] 5 ENTER ▼ ENTER ENTER

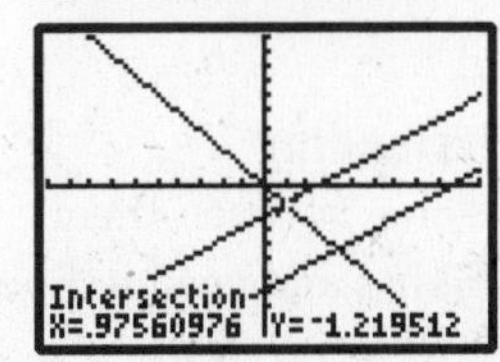

Lines a and t intersect at (1.0, −1.2).

Find the intersection of b and t.

KEYSTROKES: 2nd [CALC] 5 ▼ ENTER ENTER ENTER

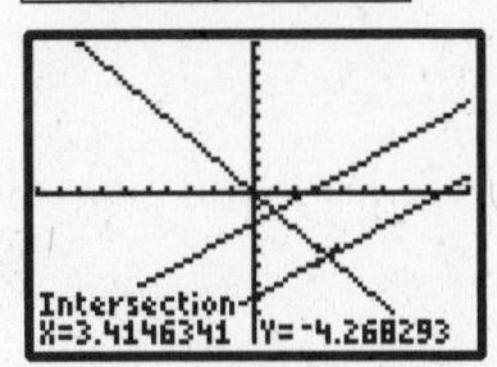

Lines b and t intersect at (3.4, −4.3).

6. Enter the equations in the Y = list and graph in the decimal viewing window.
KEYSTROKES: [Y =] [(−)] 1 [÷] 6 [X, T, θ, n] [+] 2 [÷] 3 [ENTER] [(−)] 1 [÷] 6 [X, T, θ, n] [+] 5 [÷] 12 [ENTER] 6 [X, T, θ, n] [+] 2 [ZOOM] 4

Use the CALC menu to find the points of intersection.
Find the intersection of a and t.
KEYSTROKES: [2nd] [CALC] 5 [ENTER] [▼] [ENTER] [ENTER]

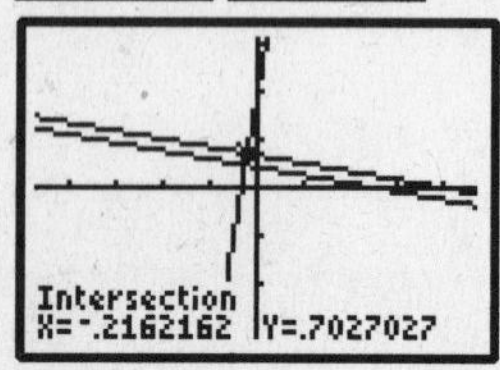

Lines a and t intersect at $(-0.2, 0.7)$.
Find the intersection of b and t.
KEYSTROKES: [2nd] [CALC] 5 [▼] [ENTER] [ENTER] [ENTER]

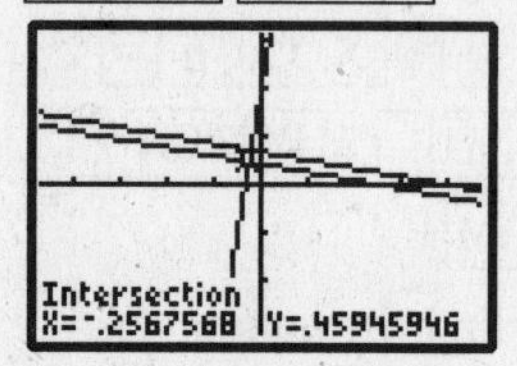

Lines b and t intersect at $(-0.3, 0.5)$.

3-6 Perpendiculars and Distance

Page 162 Check for Understanding

1. Construct a perpendicular line between them.
2. Sample answer: You are hiking and need to find the shortest path to a shelter.
3. Sample answer: Measure distances at different parts; compare slopes; measure angles. Finding slopes is the most readily available method.
4. L M K N

5.

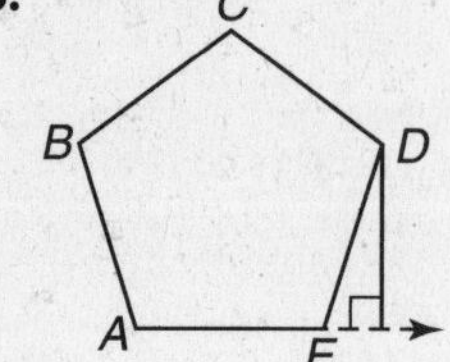

6.

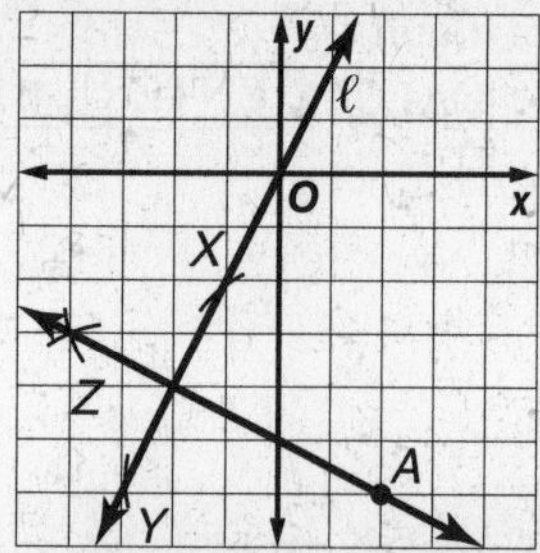

1. Graph line ℓ and point A. Place the compass point at point A. Make the setting wide enough so that when an arc is drawn, it intersects ℓ in two places. Label these points X and Y.
2. Put the compass at point X and draw an arc above line ℓ.
4. Draw $\overleftrightarrow{AZ}$. $\overleftrightarrow{AZ} \perp \ell$. The segment constructed from point $A(2, -6)$ perpendicular to line ℓ appears to intersect line ℓ at $(-2, -4)$. Use the Distance Formula to find the distance between point A and ℓ.

$$d = \sqrt{(x_2 - x_1)^2 + (y_2 - y_1)^2}$$
$$= \sqrt{(-2 - 2)^2 + [-4 - (-6)]^2}$$
$$= \sqrt{(-4)^2 + (2)^2} = \sqrt{16 + 4} = \sqrt{20} \approx 4.47$$

The distance between A and ℓ is about 4.47 units.

7. First, write an equation of a line p perpendicular to the given lines. The slope of p is the opposite reciprocal of $\frac{3}{4}$, or $-\frac{4}{3}$. Use the y-intercept of $y = \frac{3}{4}x - 1$, $(0, -1)$, as one of the endpoints of the perpendicular segment.

$$y - y_1 = m(x - x_1)$$
$$y - (-1) = -\frac{4}{3}(x - 0)$$
$$y + 1 = -\frac{4}{3}x$$
$$y = -\frac{4}{3}x - 1$$

Next, use a system of equations to determine the point of intersection of $y = \frac{3}{4}x + \frac{1}{8}$ and p.

$$\frac{3}{4}x + \frac{1}{8} = -\frac{4}{3}x - 1$$
$$\frac{3}{4}x + \frac{4}{3}x = -1 - \frac{1}{8}$$
$$\frac{25}{12}x = -\frac{9}{8}$$
$$x = -\frac{27}{50}$$
$$y = \frac{3}{4}\left(-\frac{27}{50}\right) + \frac{1}{8}$$
$$y = -\frac{7}{25}$$

The point of intersection is $\left(-\frac{27}{50}, -\frac{7}{25}\right)$.
Then, use the Distance Formula to determine the distance between $(0, -1)$ and $\left(-\frac{27}{50}, -\frac{7}{25}\right)$.

$$d = \sqrt{(x_2 - x_1)^2 + (y_2 - y_1)^2}$$
$$= \sqrt{\left(-\frac{27}{50} - 0\right)^2 + \left[-\frac{7}{25} - (-1)\right]^2}$$
$$= \sqrt{.81}$$
$$= 0.9$$

The distance between the lines is 0.9 unit.

8. $x + 3y = 6$

$3y = -x + 6$

$y = -\frac{1}{3}x + 2$

$x + 3y = -14$

$3y = -x - 14$

$y = -\frac{1}{3}x - \frac{14}{3}$

First, write an equation of a line p perpendicular to the given lines. The slope of p is the opposite reciprocal of $-\frac{1}{3}$, or 3. Use the y-intercept of $y = -\frac{1}{3}x + 2$, (0, 2), as one of the endpoints of the perpendicular segment.

$y - y_1 = m(x - x_1)$

$y - 2 = 3(x - 0)$

$y - 2 = 3x$

$y = 3x + 2$

Next, use a system of equations to determine the point of intersection of $y = -\frac{1}{3}x - \frac{14}{3}$ and p.

$-\frac{1}{3}x - \frac{14}{3} = 3x + 2$

$-\frac{1}{3}x - 3x = 2 + \frac{14}{3}$

$-\frac{10}{3}x = \frac{20}{3}$

$x = -2$

$y = 3(-2) + 2$

$y = -4$

The point of intersection is (−2, −4).

Then, use the Distance Formula to determine the distance between (0, 2) and (−2, −4).

$d = \sqrt{(x_2 - x_1)^2 + (y_2 - y_1)^2}$

$= \sqrt{(-2 - 0)^2 + (-4 - 2)^2}$

$= \sqrt{40}$

$= 2\sqrt{10}$

The distance between the lines is $2\sqrt{10}$, or approximately 6.32 units.

9.

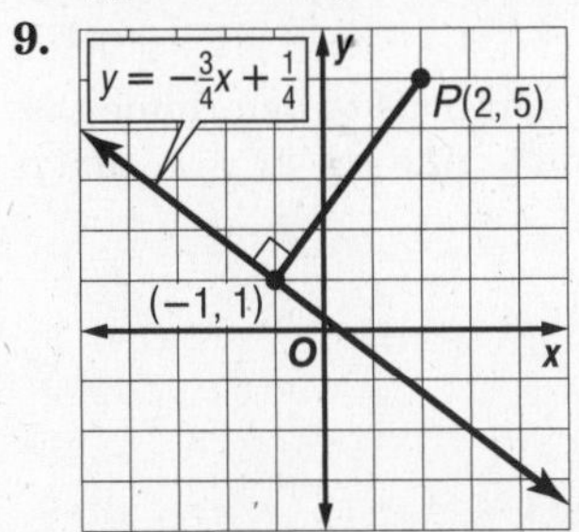

1. Graph $y = -\frac{3}{4}x + \frac{1}{4}$ and point P. Place the compass at point P. Make the setting wide enough so that when an arc is drawn, it intersects $y = -\frac{3}{4}x + \frac{1}{4}$ in two places. Label these points of intersection A and B.
2. Put the compass at point A and draw an arc below the line.
3. Using the same compass setting, put the compass at point B and draw an arc to intersect the one drawn in step 2. Label the point of intersection Q.
4. Draw $\overleftrightarrow{PQ}$. $\overleftrightarrow{PQ}$ is perpendicular to $y = -\frac{3}{4}x + \frac{1}{4}$. Label point R at the intersection of the lines. The segment constructed from $P(2, 5)$ perpendicular to $y = -\frac{3}{4}x + \frac{1}{4}$ appears to intersect $y = -\frac{3}{4}x + \frac{1}{4}$ at $R(-1, 1)$. Use the Distance Formula to find the distance between P and $y = -\frac{3}{4}x + \frac{1}{4}$.

$d = \sqrt{(x_2 - x_1)^2 + (y_2 - y_1)^2}$

$= \sqrt{[2 - (-1)]^2 + (5 - 1)^2}$

$= \sqrt{3^2 + 4^2}$

$= \sqrt{25} = 5$

The distance between P and $y = -\frac{3}{4}x + \frac{1}{4}$ is 5 units.

10.

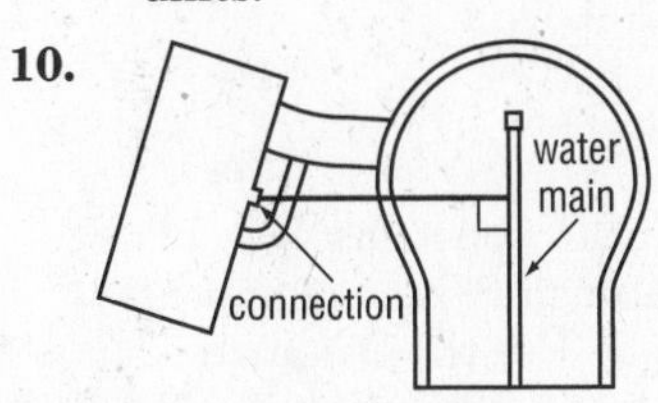

Pages 162–164 Practice and Apply

11.

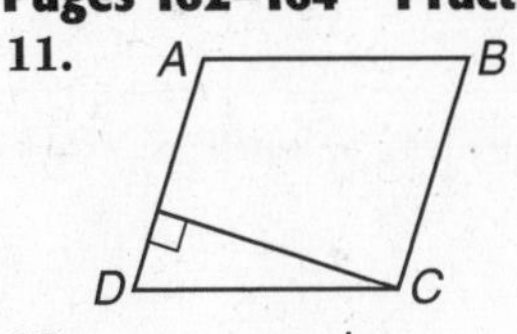

12.

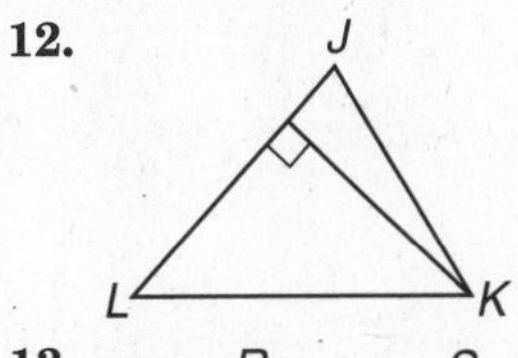

13.

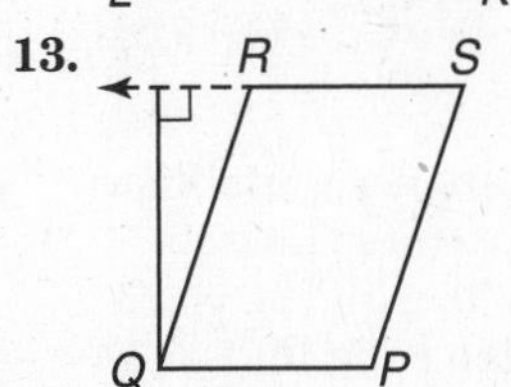

14.

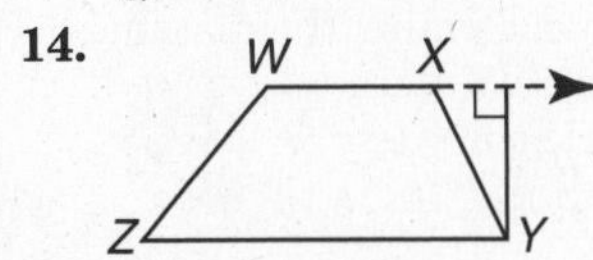

15.

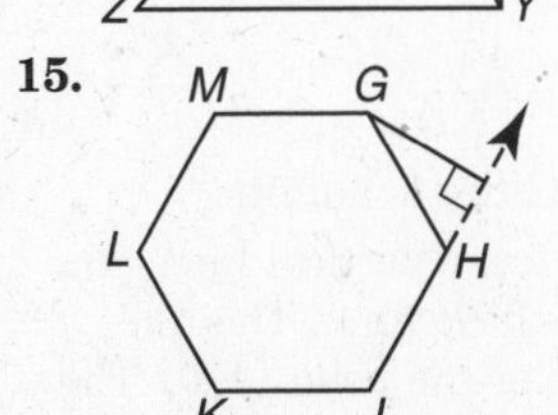

16.

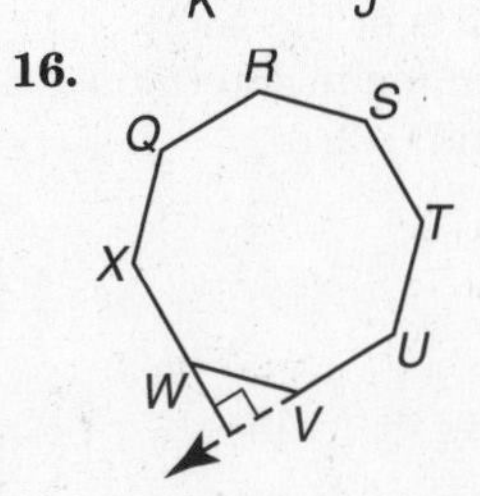

17. Graph line ℓ and point P. Construct m perpendicular to ℓ through P. Line m appears to intersect line ℓ at (4, 0). Use the Distance Formula to find the distance between P and ℓ.

$d = \sqrt{(x_2 - x_1)^2 + (y_2 - y_1)^2}$
$= \sqrt{(4 - 4)^2 + (3 - 0)^2}$
$= \sqrt{9} = 3$

The distance between P and ℓ is 3 units.

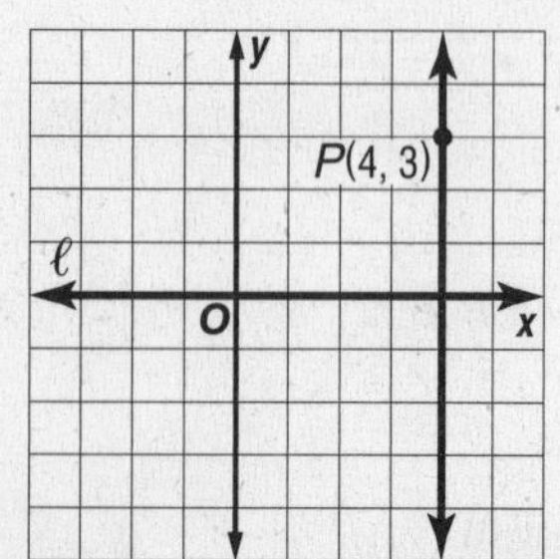

18. Graph line ℓ and point P. Construct m perpendicular to ℓ through P. Line m appears to intersect line ℓ at (1, 3). Use the Distance Formula to find the distance between P and ℓ.

$d = \sqrt{(x_2 - x_1)^2 + (y_2 - y_1)^2}$
$= \sqrt{(-4 - 1)^2 + (4 - 3)^2}$
$= \sqrt{(-5)^2 + 1^2} = \sqrt{26}$

The distance between P and ℓ is $\sqrt{26}$ units.

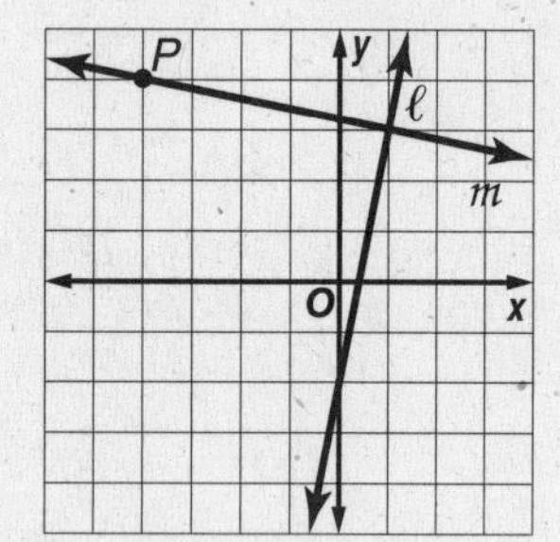

19. The lines $y = -3$ and $y = 1$ are horizontal lines. The y-axis is perpendicular to these lines. The y-intercept of $y = -3$ is $(0, -3)$. The y-axis intersects the line $y = 1$ at the point (0, 1). Use the Distance Formula to determine the distance between $(0, -3)$ and (0, 1).

$d = \sqrt{(x_2 - x_1)^2 + (y_2 - y_1)^2}$
$= \sqrt{(0 - 0)^2 + [1 - (-3)]^2}$
$= \sqrt{0^2 + 4^2}$
$= \sqrt{16}$ or 4

The distance between the lines is 4 units.

20. The lines $x = 4$ and $x = -2$ are vertical lines. The x-axis is perpendicular to these lines. The line $x = 4$ intersects the x-axis at the point (4, 0). The line $x = -2$ intersects the x-axis at the point $(-2, 0)$. Use the Distance Formula to determine the distance between (4, 0) and $(-2, 0)$.

$d = \sqrt{(x_2 - x_1)^2 + (y_2 - y_1)^2}$
$= \sqrt{(-2 - 4)^2 + (0 - 0)^2}$
$= \sqrt{(-6)^2 + 0^2} = \sqrt{36}$ or 6

The distance between the lines is 6 units.

21. First, write an equation of a line p perpendicular to the lines $y = 2x + 2$ and $y = 2x - 3$. The slope of p is the opposite reciprocal of 2, or $-\frac{1}{2}$. Use the y-intercept of the line $y = 2x + 2$, (0, 2), as one of the endpoints of the perpendicular segment.

$y - y_1 = m(x - x_1)$
$y - 2 = -\frac{1}{2}(x - 0)$
$y - 2 = -\frac{1}{2}x$
$y = -\frac{1}{2}x + 2$

Next, use a system of equations to determine the point of intersection of the line $y = 2x - 3$ and p.

$2x - 3 = -\frac{1}{2}x + 2$
$2x + \frac{1}{2}x = 2 + 3$
$\frac{5}{2}x = 5$
$x = 2$
$y = 2(2) - 3$
$y = 1$

The point of intersection is (2, 1).

Then, use the Distance Formula to determine the distance between (0, 2) and (2, 1).

$d = \sqrt{(x_2 - x_1)^2 + (y_2 - y_1)^2}$
$= \sqrt{(2 - 0)^2 + (1 - 2)^2}$
$= \sqrt{(2)^2 + (-1)^2}$
$= \sqrt{5}$

The distance between the lines is $\sqrt{5}$ units.

22. First, write an equation of a line p perpendicular to the lines $y = 4x$ and $y = 4x - 17$. The slope of p is the opposite reciprocal of 4, or $-\frac{1}{4}$. Use the y-intercept of the line $y = 4x$, (0, 0), as one of the endpoints of the perpendicular segment.

$y - y_1 = m(x - x_1)$
$y - 0 = -\frac{1}{4}(x - 0)$
$y = -\frac{1}{4}x$

Next, use a system of equations to determine the point of intersection of the line $y = 4x - 17$ and p.

$4x - 17 = -\frac{1}{4}x$
$-17 = -\frac{17}{4}x$
$4 = x$
$y = 4(4) - 17$
$y = -1$

The point of intersection is $(4, -1)$.

Then, use the Distance Formula to determine the distance between (0, 0) and $(4, -1)$.

$d = \sqrt{(x_2 - x_1)^2 + (y_2 - y_1)^2}$
$= \sqrt{(-1 - 0)^2 + (4 - 0)^2}$
$= \sqrt{(-1)^2 + (4)^2}$
$= \sqrt{17}$

The distance between the lines is $\sqrt{17}$ units.

23. $2x - y = -4$

$-y = -2x - 4$

$y = 2x + 4$

First, write an equation of a line p perpendicular to the lines $y = 2x - 3$ and $y = 2x + 4$. The slope of p is the opposite reciprocal of 2, or $-\frac{1}{2}$. Use the y-intercept of the line $y = 2x - 3$, $(0, -3)$, as one of the endpoints of the perpendicular segment.

$y - y_1 = m(x - x_1)$

$y - (-3) = -\frac{1}{2}(x - 0)$

$y + 3 = -\frac{1}{2}x$

$y = -\frac{1}{2}x - 3$

Next, use a system of equations to determine the point of intersection of the line $y = 2x + 4$ and p.

$2x + 4 = -\frac{1}{2}x - 3$

$2x + \frac{1}{2}x = -3 - 4$

$\frac{5}{2}x = -7$

$x = -\frac{14}{5}$

$y = 2\left(-\frac{14}{5}\right) + 4$

$y = -\frac{8}{5}$

The point of intersection is $\left(-\frac{14}{5}, -\frac{8}{5}\right)$.

Then, use the Distance Formula to determine the distance between $(0, -3)$ and $\left(-\frac{14}{5}, -\frac{8}{5}\right)$.

$d = \sqrt{(x_2 - x_1)^2 + (y_2 - y_1)^2}$

$= \sqrt{\left(-\frac{14}{5} - 0\right)^2 + \left[-\frac{8}{5} - (-3)\right]^2}$

$= \sqrt{\left(-\frac{14}{5}\right)^2 + \left(\frac{7}{5}\right)^2}$

$= \sqrt{\frac{196}{25} + \frac{49}{25}}$

$= \sqrt{\frac{245}{25}}$ or $\sqrt{9.8}$

The distance between the lines is $\sqrt{9.8}$ units.

24. $3x + 4y = 20$

$4y = -3x + 20$

$y = -\frac{3}{4}x + 5$

First, write an equation of a line p perpendicular to the lines $y = -\frac{3}{4}x - 1$ and $y = -\frac{3}{4}x + 5$. The slope of p is the opposite reciprocal of $-\frac{3}{4}$, or $\frac{4}{3}$. Use the y-intercept of the line $y = -\frac{3}{4}x - 1$, $(0, -1)$, as one of the endpoints of the perpendicular segment.

$y - y_1 = m(x - x_1)$

$y - (-1) = \frac{4}{3}(x - 0)$

$y + 1 = \frac{4}{3}x$

$y = \frac{4}{3}x - 1$

Next, use a system of equations to determine the point of intersection of the line $y = -\frac{3}{4}x + 5$ and p.

$-\frac{3}{4}x + 5 = \frac{4}{3}x - 1$

$-\frac{3}{4}x - \frac{4}{3}x = -1 - 5$

$-\frac{25}{12}x = -6$

$x = \frac{72}{25}$

$y = -\frac{3}{4}\left(\frac{72}{25}\right) + 5$

$y = \frac{71}{25}$

The point of intersection is $\left(\frac{72}{25}, \frac{71}{25}\right)$.

Then, use the Distance Formula to determine the distance between $(0, -1)$ and $\left(\frac{72}{25}, \frac{71}{25}\right)$.

$d = \sqrt{(x_2 - x_1)^2 + (y_2 - y_1)^2}$

$= \sqrt{\left(\frac{72}{25} - 0\right)^2 + \left[\frac{71}{25} - (-1)\right]^2}$

$= \sqrt{\left(\frac{72}{25}\right)^2 + \left(\frac{96}{25}\right)^2}$

$= \sqrt{\frac{576}{25}}$ or $\frac{24}{5}$

The distance between the lines is $\frac{24}{5}$ units.

25.

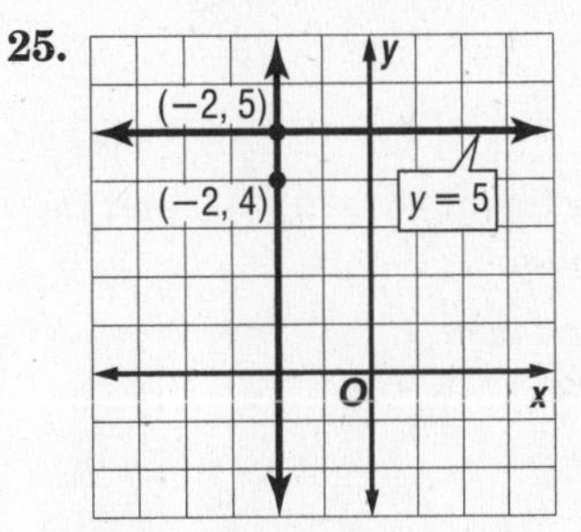

The distance from a line to a point not on the line is the length of the segment perpendicular to the line from the point. From the figure, this distance is 1.

26.

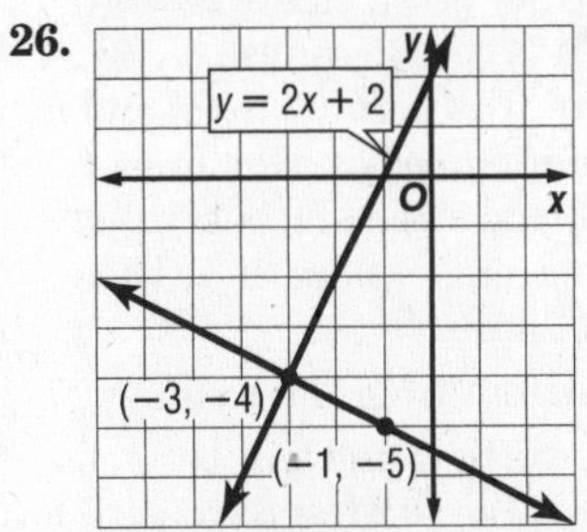

The perpendicular segment from the point $(-1, -5)$ to the line $y = 2x + 2$ appears to intersect the line $y = 2x + 2$ at $(-3, -4)$. Use the Distance Formula to find the distance between $(-1, -5)$ and $y = 2x + 2$.

$d = \sqrt{(x_2 - x_1)^2 + (y_2 - y_1)^2}$

$= \sqrt{[-1 - (-3)]^2 + [-5 - (-4)]^2}$

$= \sqrt{2^2 + (-1)^2}$

$= \sqrt{5}$

The distance between the line $y = 2x + 2$ and the point $(-1, -5)$ is $\sqrt{5}$ units.

27. $2x - 3y = -9$

$-3y = -2x - 9$

$y = \frac{2}{3}x + 3$

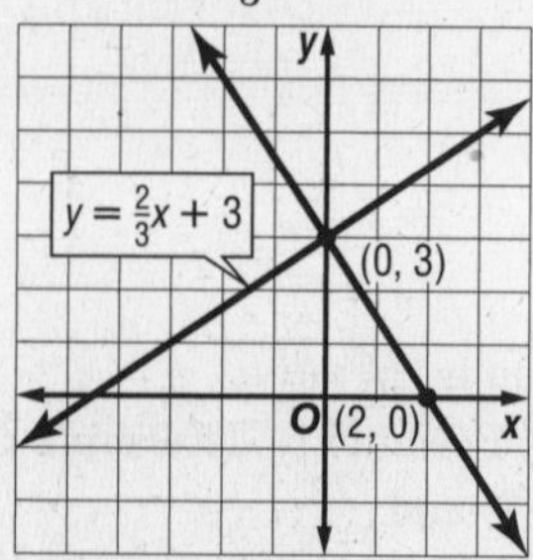

The perpendicular segment from the point (2, 0) to the line $y = \frac{2}{3}x + 3$ appears to intersect the line $y = \frac{2}{3}x + 3$ at (0, 3). Use the Distance Formula to find the distance between (2, 0) and $y = \frac{2}{3}x + 3$.

$d = \sqrt{(x_2 - x_1)^2 + (y_2 - y_1)^2}$

$= \sqrt{(2 - 0)^2 + (0 - 3)^2}$

$= \sqrt{2^2 + (-3)^2}$

$= \sqrt{13}$

The distance between the line $y = \frac{2}{3}x + 3$ and the point (2, 0) is $\sqrt{13}$ units.

28. Given: ℓ is equidistant to m.
n is equidistant to m.

Prove: $\ell \parallel n$

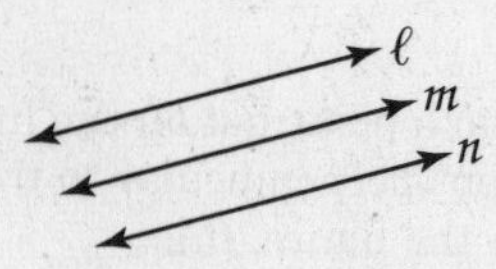

Paragraph proof: We are given that ℓ is equidistant to m, and n is equidistant to m. By definition of equidistant, ℓ is parallel to m, and n is parallel to m. By definition of parallel lines, slope of ℓ = slope of m, and slope of n = slope of m. By substitution, slope of ℓ = slope of n. Then, by definition of parallel lines, $\ell \parallel n$.

29. It is everywhere equidistant from the ceiling.

30. The plumb line will be vertical and will be perpendicular to the floor. The shortest distance from a point to the floor will be along the plumb line.

31. $a = 3, b = 4, c = 6, (x_1, y_1) = (4, 6)$

$$\frac{|ax_1 + by_1 - c|}{\sqrt{a^2 + b^2}} = \frac{|3(4) + 4(6) - 6|}{\sqrt{3^2 + 4^2}}$$

$$= \frac{|30|}{\sqrt{25}}$$

$$= \frac{30}{5} \text{ or } 6$$

32a.

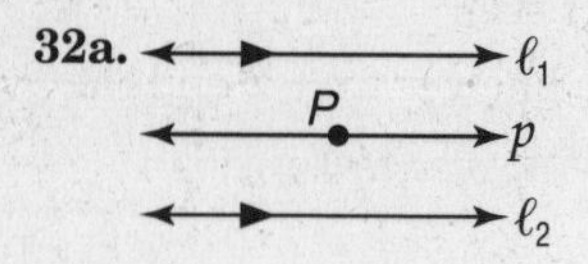

32b.

32c.

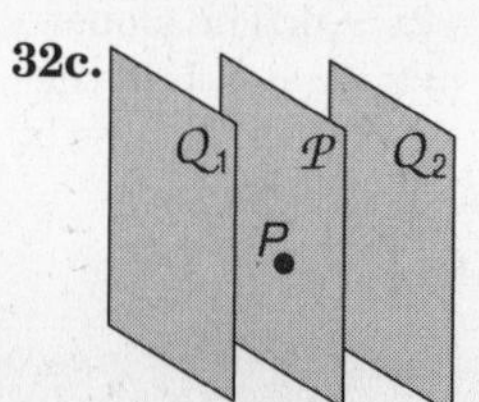

32d.

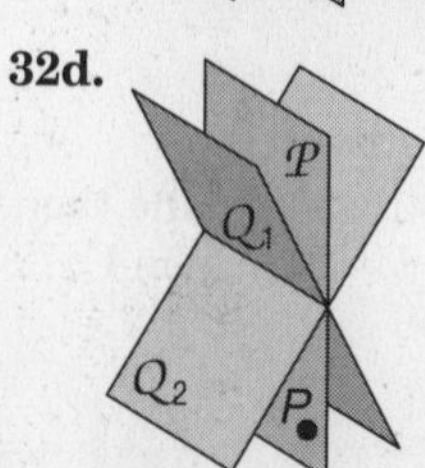

32e.

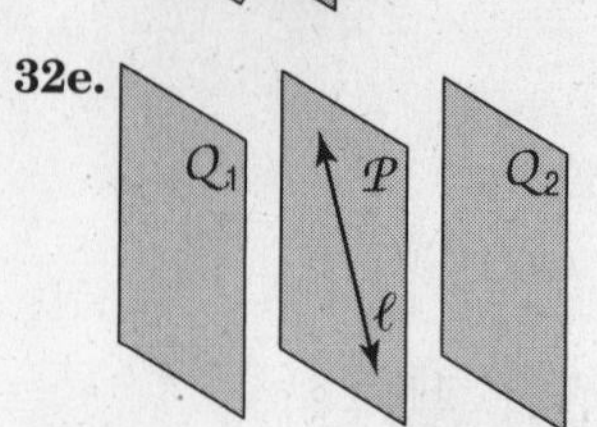

32f.

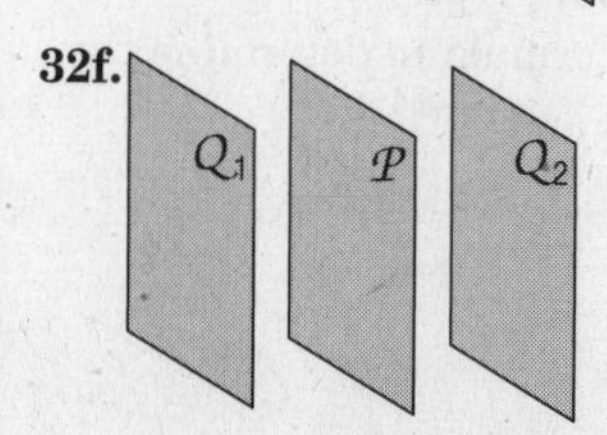

33. Sample answer: We want new shelves to be parallel so they will line up.

Answers should include the following.

- After marking several points, a slope can be calculated, which should be the same slope as the original brace.
- Building walls requires parallel lines.

34.

$AB = 16$ and X is the midpoint of $\overline{AB}$, so $XB = 8$. $CD = 20$ and X is the midpoint of $\overline{CD}$, so $XD = 10$. $\overline{BD} \perp \overline{AB}$, so XBD is a right triangle. Use the Pythagorean Theorem to find BD.

$(XD)^2 = (XB)^2 + (BD)^2$

$10^2 = 8^2 + (BD)^2$

$36 = (BD)^2$

$6 = BD$

35. D; The coin came up heads 14 times, but since the first and last flips were both heads and there were 24 total flips, it's not possible to have all 14 times heads came up be consecutive. But it is possible that the first 13 flips were heads, or the last 13 flips were heads.

Page 164 Maintain Your Skills

36. $\overleftrightarrow{DE} \parallel \overleftrightarrow{CF}$; alternate interior ∠s

37. $\overleftrightarrow{DA} \parallel \overleftrightarrow{EF}$; corresponding ∠s

38. $\overleftrightarrow{DA} \parallel \overleftrightarrow{EF}$; $\angle 1 \cong \angle 4$ and consecutive interior ∠s are supplementary

39. Find the slope of a using (0, 3) and (2, 4).

$m = \frac{(y_2 - y_1)}{(x_2 - x_1)}$

$= \frac{4-3}{2-0}$ or $\frac{1}{2}$

$y = mx + b$

$y = \frac{1}{2}x + 3$

40. Find the slope of b using (0, 5) and (1, 4).

$m = \frac{(y_2 - y_1)}{(x_2 - x_1)}$

$= \frac{4-5}{1-0}$ or -1

$y = mx + b$

$y = -x + 5$

41. Find the slope of c using (0, −2) and (3, 0).

$m = \frac{(y_2 - y_1)}{(x_2 - x_1)}$

$= \frac{0-(-2)}{3-0}$ or $\frac{2}{3}$

$y = mx + b$

$y = \frac{2}{3}x - 2$

42. Line a has slope $\frac{1}{2}$ (from Exercise 39), so a line perpendicular to a has slope −2.

$y - y_1 = m(x - x_1)$

$y - (-4) = -2[x - (-1)]$

$y + 4 = -2x - 2$

$y = -2x - 6$

43. Line c has slope $\frac{2}{3}$ (from Exercise 41), so a line parallel to c has slope $\frac{2}{3}$.

$y - y_1 = m(x - x_1)$

$y - 5 = \frac{2}{3}(x - 2)$

$y - 5 = \frac{2}{3}x - \frac{4}{3}$

$y = \frac{2}{3}x + \frac{11}{3}$

44. Given: $NL = NM, AL = BM$
Prove: $NA = NB$

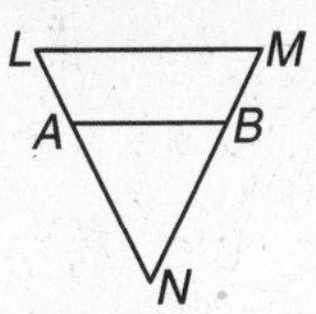

Proof:

Statements	Reasons
1. $NL = NM$ $AL = BM$	1. Given
2. $NL = NA + AL$ $NM = NB + BM$	2. Segment Addition Post.
3. $NA + AL = NB + BM$	3. Substitution
4. $NA + BM = NB + BM$	4. Substitution
5. $NA = NB$	5. Subt. Prop.

Pages 165–166 Geometry Activity: Non-Euclidean Geometry

1. The great circle is finite.

2. A curved path on the great circle passing through two points is the shortest distance between the two points.

3. There exist no parallel lines.

4. Two distinct great circles intersect in exactly two points.

5. A pair of perpendicular great circles divides the sphere into four finite congruent regions.

6. There exist no parallel lines.

7. There are two distances between two points.

8. true

9. False; in spherical geometry, if three points are collinear, any point can be between the other two.

10. False; in spherical geometry, there are no parallel lines.

Chapter 3 Study Guide and Review

Page 167 Vocabulary and Concept Check

1. alternate
2. perpendicular
3. parallel
4. transversal
5. alternate exterior
6. congruent
7. consecutive

Pages 167-170 Lesson-by-Lesson Review

8. corresponding
9. alternate exterior
10. consecutive interior
11. corresponding
12. alternate interior

13. consecutive interior

14. alternate exterior

15. alternate interior

16. $\angle 1$ and $\angle 2$ are supplementary.

$m\angle 1 + m\angle 2 = 180$
$53 + m\angle 2 = 180$
$m\angle 2 = 127$

17. $\angle 3 \cong \angle 6$
$\angle 6 \cong \angle 1$
$\angle 3 \cong \angle 1$
$m\angle 3 = m\angle 1$
$m\angle 3 = 53$

18. $m\angle 4 + m\angle 3 = 180$
$m\angle 4 + 53 = 180$
$m\angle 4 = 127$

19. $m\angle 5 + m\angle 6 = 180$
$\angle 6 \cong \angle 1$
$m\angle 6 = m\angle 1$
$m\angle 5 + m\angle 1 = 180$
$m\angle 5 + 53 = 180$
$m\angle 5 = 127$

20. $\angle 6 \cong \angle 1$
$m\angle 6 = m\angle 1$
$m\angle 6 = 53$

21. $m\angle 7 + m\angle 6 = 180$
$m\angle 7 + 53 = 180$
$m\angle 7 = 127$

22. Find a.

$\angle 1$ and $\angle 2$ are supplementary, so
$m\angle 1 + m\angle 2 = 180.$
$3a + 40 + 2a + 25 = 180$
$5a + 65 = 180$
$5a = 115$
$a = 23$

Find b.

$\angle 2 \cong \angle 4$ and $m\angle 4 + m\angle 3 = 180$, so
$m\angle 2 + m\angle 3 = 180.$
$2a + 25 + 5b - 26 = 180$
$2(23) + 25 + 5b - 26 = 180$
$5b + 45 = 180$
$5b = 135$
$b = 27$

23. slope of $\overleftrightarrow{AB} = \frac{-1-1}{3-(-4)}$
$= -\frac{2}{7}$

slope of $\overleftrightarrow{CD} = \frac{9-2}{0-2}$
$= -\frac{7}{2}$

The slopes are not the same, so $\overleftrightarrow{AB}$ and $\overleftrightarrow{CD}$ are not parallel. The product of the slopes is $\left(-\frac{2}{7}\right)\left(-\frac{7}{2}\right)$ or 1, so $\overleftrightarrow{AB}$ and $\overleftrightarrow{CD}$ are not perpendicular. Therefore, $\overleftrightarrow{AB}$ and $\overleftrightarrow{CD}$ are neither parallel nor perpendicular.

24. slope of $\overleftrightarrow{AB} = \frac{-2-2}{2-6}$
$= \frac{-4}{-4}$ or 1

slope of $\overleftrightarrow{CD} = \frac{2-(-4)}{5-(-1)}$
$= \frac{6}{6}$ or 1

The slopes are the same, so $\overleftrightarrow{AB}$ and $\overleftrightarrow{CD}$ are parallel.

25. slope of $\overleftrightarrow{AB} = \frac{5-(-3)}{4-1}$
$= \frac{8}{3}$

slope of $\overleftrightarrow{CD} = \frac{2-(-1)}{-7-1}$
$= -\frac{3}{8}$

The product of the slopes is $\left(\frac{8}{3}\right)\left(-\frac{8}{3}\right)$ or -1. So, $\overleftrightarrow{AB}$ is perpendicular to $\overleftrightarrow{CD}$.

26. slope of $\overleftrightarrow{AB} = \frac{3-0}{6-2}$
$= \frac{3}{4}$

slope of $CD = \frac{-1-(-4)}{3-(-1)}$
$= \frac{3}{4}$

The slopes are the same, so $\overleftrightarrow{AB}$ and $\overleftrightarrow{CD}$ are parallel.

27. First, find the slope of $\overleftrightarrow{AB}$.

$m = \frac{(y_2 - y_1)}{(x_2 - x_1)}$
$= \frac{6-2}{1-(-1)}$
$= \frac{4}{2}$ or 2

Parallel lines have the same slope, so the slope of the line to be drawn is 2.

Graph the line. Start at (2, 3). Move up 2 units and then move right 1 unit. Draw the line through this point and (2, 3).

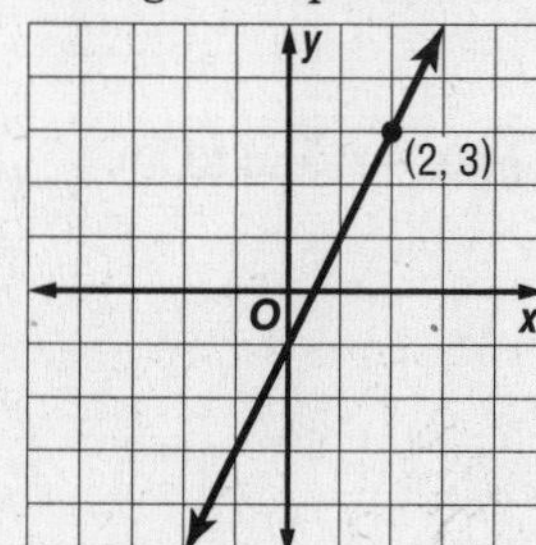

28. First, find the slope of $\overleftrightarrow{PQ}$.

$m = \frac{(y_2 - y_1)}{(x_2 - x_1)}$
$= \frac{-4-2}{3-5}$
$= \frac{-6}{-2}$ or 3

Since $3\left(-\frac{1}{3}\right) = -1$, the slope of the line to be drawn is $-\frac{1}{3}$. Graph the line. Start at $(-2, -2)$. Move down 1 unit and then move right 3 units. Draw the line through this point and $(-2, -2)$.

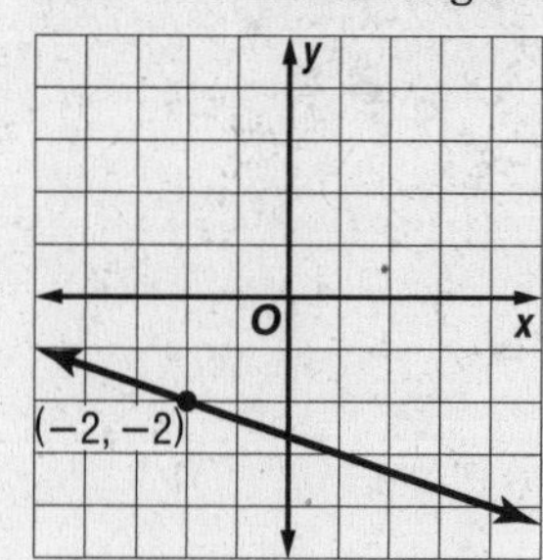

29. $y - y_1 = m(x - x_1)$
$y - (-5) = 2(x - 1)$
$y + 5 = 2x - 2$
$y = 2x - 7$

30. $m = \frac{(y_2 - y_1)}{(x_2 - x_1)}$
$= \frac{-1 - 5}{-2 - 2}$
$= \frac{-6}{-4}$ or $\frac{3}{2}$
$y - y_1 = m\ (x - x_1)$
$y - 5 = \frac{3}{2}\ (x - 2)$
$y - 5 = \frac{3}{2}x - 3$
$y = \frac{3}{2}x + 2$

31. $y = mx + b$
$y = -\frac{2}{7}x + 4$

32. $y - y_1 = m(x - x_1)$
$y - (-4) = -\frac{3}{2}\ (x - 2)$
$y + 4 = -\frac{3}{2}x + 3$
$y = -\frac{3}{2}x - 1$

33. $y = mx + b$
$y = 5x - 3$

34. $m = \frac{(y_2 - y_1)}{(x_2 - x_1)}$
$= \frac{6 - (-1)}{-4 - 3}$
$= \frac{7}{-7}$ or -1
$y - y_1 = m(x - x_1)$
$y - (-1) = -1(x - 3)$
$y + 1 = -x + 3$
$y = -x + 2$

35. $\overleftrightarrow{AL}$ and $\overleftrightarrow{BJ}$, alternate exterior $\angle$s are $\cong$

36. $\overleftrightarrow{AL}$ and $\overleftrightarrow{BJ}$, consecutive interior $\angle$s are supplementary

37. $\overleftrightarrow{CF}$ and $\overleftrightarrow{GK}$, 2 lines $\perp$ to the same line

38. $\overleftrightarrow{AL}$ and $\overleftrightarrow{BJ}$, alternate interior $\angle$s are $\cong$

39. $\overleftrightarrow{CF}$ and $\overleftrightarrow{GK}$, consecutive interior $\angle$s are supplementary

40. $\overleftrightarrow{CF}$ and $\overleftrightarrow{GK}$, corresponding $\angle$s are $\cong$

41. First, write an equation of a line p perpendicular to $y = 2x - 4$ and $y = 2x + 1$. The slope of p is the opposite reciprocal of 2, or $-\frac{1}{2}$. Use the y-intrecept of $y = 2x - 4$, $(0, -4)$, as one of the endpoints of the perpendicular segment.
$y - y_1 = m(x - x_1)$
$y - (-4) = -\frac{1}{2}(x - 0)$
$y + 4 = -\frac{1}{2}x$
$y = -\frac{1}{2}x - 4$
Next, use a system of equations to determine the point of intersection of the line $y = 2x + 1$ and p.
$2x + 1 = -\frac{1}{2}x - 4$
$2x + \frac{1}{2}x = -4 - 1$
$\frac{5}{2}x = -5$
$x = -2$
$y = 2(-2) + 1$
$y = -3$
The point of intersection is $(-2, -3)$.
Then use the Distance Formula to determine the distance between $(0, -4)$ and $(-2, -3)$.
$d = \sqrt{(x_2 - x_1)^2 + (y_2 - y_1)^2}$
$= \sqrt{(-2 - 0)^2 + [-3 - (-4)]^2}$
$= \sqrt{(-2)^2 + 1^2}$
$= \sqrt{5}$
The distance between the lines is $\sqrt{5}$ units.

42. First, write an equation of a line p perpendicular to $y = \frac{1}{2}x$ and $y = \frac{1}{2}x + 5$. The slope of p is the opposite reciprocal of $\frac{1}{2}$, or -2. Use the y-intercept of $y = \frac{1}{2}x$, $(0, 0)$, as one of the endpoints of the perpendicular segment.
$y - y_1 = m(x - x_1)$
$y - 0 = -2\ (x - 0)$
$y = -2x$
Next, use a system of equations to determine the point of intersection of the line $y = \frac{1}{2}x + 5$ and p.
$\frac{1}{2}x + 5 = -2x$
$5 = -2x - \frac{1}{2}x$
$5 = -\frac{5}{2}x$
$-2 = x$
$y = \frac{1}{2}(-2) + 5$
$y = 4$
The point of intersection is $(-2, 4)$.
Then, use the Distance Formula to determine the distance between $(0, 0)$ and $(-2, 4)$.
$d = \sqrt{(x_2 - x_1)^2 + (y_2 - y_1)^2}$
$= \sqrt{(-2 - 0)^2 + (4 - 0)^2}$
$= \sqrt{(-2)^2 + 4^2}$
$= \sqrt{20} = 2\sqrt{5}$
The distance between the lines is $\sqrt{20}$ units.

Chapter 3 Practice Test

Page 171

1. The slope of a line perpendicular to $y = 3x - \frac{2}{7}$ is the opposite reciprocal of 3, or $-\frac{1}{3}$.
Sample answer: $y = -\frac{1}{3}x + 1$

2. Sample answer: If alternate interior $\angle$s are $\cong$, then lines are $\parallel$.

3. $\angle 2$ and $\angle 6$ each form a linear pair with $\angle 1$, so $\angle 2$ and $\angle 6$ are supplementary to $\angle 1$.
$\angle 3 \cong \angle 1$, so $\angle 3$ is not necessarily supplementary to $\angle 1$. $m\angle 3 + m\angle 4 + m\angle 5 = 180$, so $m\angle 1 + m\angle 4 + m\angle 5 = 180$. So $\angle 4$ and $\angle 5$ are not supplementary to $\angle 1$.

4. $m\angle 8 + m\angle 12 = 180$
$m\angle 8 + 64 = 180$
$m\angle 8 = 116$

5. $\angle 13 \cong \angle 12$
$m\angle 13 = m\angle 12$
$m\angle 13 = 64$

6. $\angle 7 \cong \angle 12$
$m\angle 7 = m\angle 12$
$m\angle 7 = 64$

7. $m\angle 11 + m\angle 12 = 180$
$m\angle 11 + 64 = 180$
$m\angle 11 = 116$

8. $m\angle 3 + m\angle 4 = 180$
$\angle 4 \cong \angle 12$
$m\angle 4 = m\angle 12$
$m\angle 3 + m\angle 12 = 180$
$m\angle 3 + 64 = 180$
$m\angle 3 = 116$

9. $\angle 4 \cong \angle 12$
$m\angle 4 = m\angle 12$
$m\angle 4 = 64$

10. $m\angle 9 + m\angle 10 = 180$
$\angle 10 \cong \angle 12$
$m\angle 10 = m\angle 12$
$m\angle 9 + m\angle 12 = 180$
$m\angle 9 + 64 = 180$
$m\angle 9 = 116$

11. $\angle 5 \cong \angle 7$
$\angle 7 \cong \angle 12$
$\angle 5 \cong \angle 12$
$m\angle 5 = m\angle 12$
$m\angle 5 = 64$

12. Start at $(-2, 1)$. Move 1 unit down and then move 1 unit right. Draw the line through this point and $(-2, 1)$.

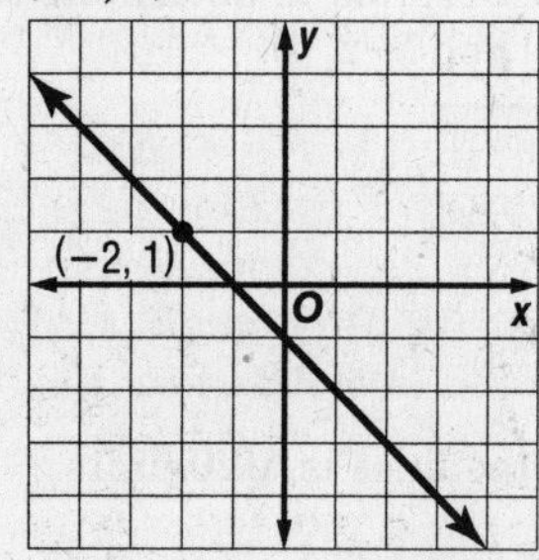

13. slope of $\overleftrightarrow{AB} = \frac{3 - 0}{4 - (-2)}$
$= \frac{3}{6}$ or $\frac{1}{2}$

$\frac{1}{2}(-2) = -1$, so the slope of the line to be graphed is -2.
Start at $(-1, 3)$. Move 2 units down and then move 1 unit right. Draw the line through this point and $(-1, 3)$.

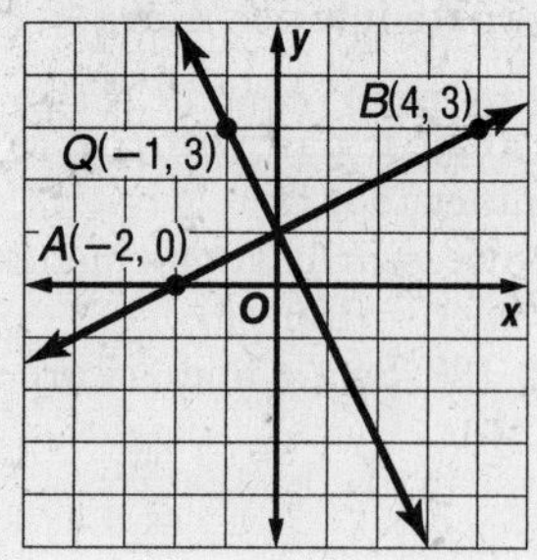

14. slope of $\overleftrightarrow{FG} = \frac{-1 - 5}{-3 - 3}$
$= \frac{-6}{-6}$ or 1

Start at $(1, -1)$. Move 1 unit up and then move 1 unit right. Draw the line through this point and $(1, -1)$.

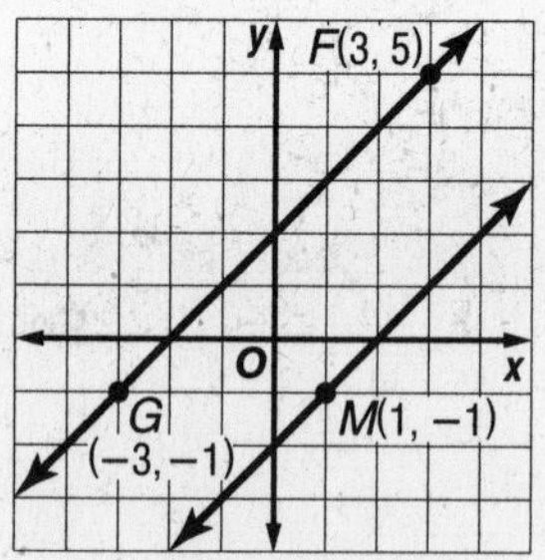

15. Start at $(3, -2)$. Move 4 units up and then move 3 units left. Draw the line through this point and $(3, -2)$.

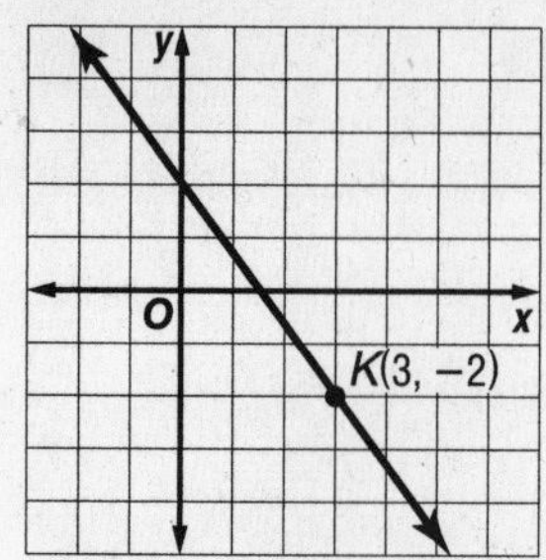

16. $\angle ABD \cong \angle ACE$
$m\angle ACE + m\angle ECF = 180$
$m\angle ABD + m\angle ECF = 180$
$3x - 60 + 2x + 15 = 180$
$5x - 45 = 180$
$5x = 225$
$x = 45$

17. $\angle DBC \cong \angle ECF$
$m\angle DBC = m\angle ECF$
$y = 2(45) + 15$
$y = 105$

18. $m\angle FCF = 2x + 15$
$= 2(45) + 15$
$= 105$

19. $m\angle ABD = 3x - 60$
$= 3(45) - 60$
$= 75$

20. $m\angle BCE + m\angle FCE = 180$
$m\angle BCE + 105 = 180$
$m\angle BCE = 75$

21. $m\angle CBD = y$
$= 105$

22. The slope of a line perpendicular to $y = 2x - 1$ and $y = 2x + 9$ is $-\frac{1}{2}$. Use the y-intercept of $y = 2x - 1$, $(0, -1)$, as one of the endpoints of the perpendicular segment.

$$y - y_1 = m(x - x_1)$$
$$y - (-1) = -\frac{1}{2}(x - 0)$$
$$y + 1 = -\frac{1}{2}x$$
$$y = -\frac{1}{2}x - 1$$

Use a system of equations to determine the point of intersection of the line $y = 2x + 9$ and the perpendicular segment.

$$2x + 9 = -\frac{1}{2}x - 1$$
$$2x + \frac{1}{2}x = -1 - 9$$
$$\frac{5}{2}x = -10$$
$$x = -4$$
$$y = 2(-4) + 9$$
$$y = 1$$

The point of intersection is $(-4, 1)$.
Use the Distance Formula to determine the distance between $(0, -1)$ and $(-4, 1)$.

$$d = \sqrt{(x_2 - x_1)^2 + (y_2 - y_1)^2}$$
$$= \sqrt{(-4 - 0)^2 + [1 - (-1)]^2}$$
$$= \sqrt{(-4)^2 + 2^2}$$
$$= \sqrt{20}$$

The distance between the lines is $\sqrt{20}$ or about 4.47 units.

23. The slope of a line perpendicular to $y = -x + 4$ and $y = -x - 2$ is 1. Use the y-intercept of $y = -x + 4$, $(0, 4)$, as one of the endpoints of the perpendicular segment.

$$y - y_1 = m(x - x_1)$$
$$y - 4 = 1(x - 0)$$
$$y - 4 = x$$
$$y = x + 4$$

Use a system of equations to determine the point of intersection of the line $y = -x - 2$ and the perpendicular segment.

$$-x - 2 = x + 4$$
$$-2x = 6$$
$$x = -3$$
$$y = -(-3) - 2$$
$$y = 1$$

The point of intersection is $(-3, 1)$.
Use the Distance Formula to determine the distance between $(0, 4)$ and $(-3, 1)$.

$$d = \sqrt{(x_2 - x_1)^2 + (y_2 - y_1)^2}$$
$$= \sqrt{(-3 - 0)^2 + (1 - 4)^2}$$
$$= \sqrt{(-3)^2 + (-3)^2}$$
$$= \sqrt{18}$$

The distance between the lines is $\sqrt{18}$ or about 4.24 units.

24. The slope of a line perpendicular to $y = -x - 4$ and $y = -x$ is 1. Use the y-intercept of $y = -x$, $(0, 0)$, as one of the endpoints of the perpendicular segment.

$$y - y_1 = m(x - x_1)$$
$$y - 0 = 1(x - 0)$$
$$y = x$$

Use a system of equations to determine the point of intersection of the line $y = -x - 4$ and the perpendicular segment.

$$-x - 4 = x$$
$$-4 = 2x$$
$$-2 = x$$
$$y = -(-2) - 4$$
$$y = -2$$

The point of intersection is $(-2, -2)$.
Use the Distance Formula to determine the distance between $(0, 0)$ and $(-2, -2)$.

$$d = \sqrt{(x_2 - x_1)^2 + (y_2 - y_1)^2}$$
$$= \sqrt{(-2 - 0)^2 + (-2 - 0)^2}$$
$$= \sqrt{(-2)^2 + (-2)^2} = \sqrt{8}$$
$$\approx 2.83$$

The distance between the lines $y = -x$ and $y = -x - 4$ is about 2.83 units. So, the distance between Lorain Road and Detroit Road is about 2.83 miles.

25. B; $\angle 1 \cong \angle 3$, and $\angle 3$ and $\angle 4$ are supplementary so $\angle 1$ and $\angle 4$ are supplementary. $m\angle 4 = 180 - m\angle 1$ so $m\angle 4 = 107$ and hence $m\angle 4 > 73$. Furthermore, $\angle 1$ is not congruent to $\angle 4$. Lines m and ℓ are parallel so consecutive interior angles are supplementary. Hence, $m\angle 2 + m\angle 3 = 180$. So the only statement that cannot be true is B.

Chapter 3 Standardized Test Practice

Pages 172–173

1. B; 2 m = 200 cm
2. C; $d = \sqrt{(x_2 - x_1)^2 + (y_2 - y_1)^2}$
   $= \sqrt{(-2 - 2)^2 + (-3 - 4)^2}$
   $= \sqrt{(-4)^2 + (-7)^2} = \sqrt{65}$
3. A; statement A is true by definition of angle bisector. There is not enough information to establish that any of the other statements are true.
4. D; $180 - 72 = 108$
5. D; $4x + 4 = 6x - 8$
   $4 + 8 = 6x - 4x$
   $12 = 2x$
   $6 = x$
6. C
7. B; $\angle 1$ and $\angle 3$ are corresponding angles.

8. C; $4y - x = 8$

$4y = x + 8$

$y = \frac{1}{4}x + 2$

$\left(\frac{1}{4}\right)(-4) = -1$, so the slope of the perpendicular line is -4. $y = -4x - 15$ is the only choice with slope -4.

9. C; the number 2 in $y = 2x - 5$ is the slope, which determines the steepness of the line.

10. $3\left(\frac{4x - 6}{3}\right) = 3(10)$

11. $m\angle FHC + m\angle FHD = 180$

$\angle FHD \cong \angle HGB$

$m\angle FHD = m\angle HGB$

$m\angle FHC + m\angle HGB = 180$

$m\angle FHC + 70 = 180$

$m\angle FHC = 110$

The flag holder rotates 110°.

12. $\angle CHG \cong \angle HGB$

$m\angle CHG = m\angle HGB$

$m\angle CHG = 70$

13. $m = \frac{(y_2 - y_1)}{(x_2 - x_1)}$

$= \frac{6 - 4}{9 - 3}$

$= \frac{2}{6}$ or $\frac{1}{3}$

14. The ball did not reach home plate. The distance between second base and home plate forms the hypotenuse of a right triangle, with second base to third base as one leg, and third base to home plate as the other leg. The Pythagorean Theorem is used to find the distance between second base and home plate.

$90^2 + 90^2 = c^2$

$8100 + 8100 = c^2$

$16{,}200 = c^2$

$\sqrt{16{,}200} = c$

$127.3 \approx c$

Since the ball traveled 120 feet and the distance from second base to home plate is about 127.3 feet, the ball did not make it to home plate.

15a. $m = \frac{(y_2 - y_1)}{(x_2 - x_1)}$

$= \frac{48 - 44}{2 - 1}$

$= 4$

15b. The slope represents the increase in the average monthly cable bill each year.

15c. The slope of the line through the points is 4. One point on the line is (1, 44). Find the equation of the line.

$y - 44 = 4(x - 1)$

$y - 44 = 4x - 4$

$y = 4x + 40$

After 10 years the cable bill will be $y = 4(10) + 40$, or \$80.

Chapter 4 Congruent Triangles

Page 177 Getting Started

1. $2x + 18 = 5$
$2x = -13$
$x = -\frac{13}{2}$ or $-6\frac{1}{2}$

2. $3m - 16 = 12$
$3m = 28$
$m = \frac{28}{3}$ or $9\frac{1}{3}$

3. $4y + 12 = 16$
$4y = 4$
$y = 1$

4. $10 = 8 - 3z$
$2 = -3z$
$-\frac{2}{3} = z$

5. $6 = 2a + \frac{1}{2}$
$\frac{11}{2} = 2a$
$\frac{11}{4} = a$
$2\frac{3}{4} = a$

6. $\frac{2}{3}b + 9 = -15$
$\frac{2}{3}b = -24$
$b = -36$

7.

$\angle 8 \cong \angle 2$	vertical $\angle$s
$\angle 8 \cong \angle 12$	corresponding $\angle$s
$\angle 8 \cong \angle 15$	alternate exterior $\angle$s
$\angle 8 \cong \angle 6$	corresponding $\angle$s
$\angle 8 \cong \angle 9$	alternate exterior $\angle$s
$\angle 8 \cong \angle 3$	$\angle 8 \cong \angle 15$ alternate exterior $\angle$s, $\angle 15 \cong \angle 3$ alternate interior $\angle$s, transitivity
$\angle 8 \cong \angle 13$	$\angle 8 \cong \angle 15$ alternate exterior $\angle$s, $\angle 15 \cong \angle 13$ corresponding $\angle$s, transitivity

8.

$\angle 13 \cong \angle 2$	$\angle 13 \cong \angle 15$ corresponding $\angle$s, $\angle 15 \cong \angle 2$ corresponding $\angle$s, transitivity
$\angle 13 \cong \angle 12$	alternate exterior $\angle$s
$\angle 13 \cong \angle 15$	corresponding $\angle$s
$\angle 13 \cong \angle 6$	alternate exterior $\angle$s
$\angle 13 \cong \angle 9$	corresponding $\angle$s
$\angle 13 \cong \angle 3$	vertical $\angle$s
$\angle 13 \cong \angle 8$	$\angle 13 \cong \angle 15$ corresponding $\angle$s, $\angle 15 \cong \angle 8$ alternate exterior $\angle$s, transitivity

9.

$\angle 1$ is supplementary to $\angle 6$	linear pair
$\angle 1$ is supplementary to $\angle 9$	linear pair
$\angle 1$ is supplementary to $\angle 3$	consecutive interior $\angle$s
$\angle 1$ is supplementary to $\angle 13$	$\angle 1$ is supplementary to $\angle 3$, $\angle 3 \cong \angle 13$ vertical $\angle$s
$\angle 1$ is supplementary to $\angle 2$	consecutive interior $\angle$s
$\angle 1$ is supplementary to $\angle 8$	$\angle 1$ is supplementary to $\angle 2$, $\angle 2 \cong \angle 8$ vertical $\angle$s
$\angle 1$ is supplementary to $\angle 12$	$\angle 1 \cong \angle 10$ corresponding $\angle$s, $\angle 10$ is supplementary to $\angle 12$ consecutive interior $\angle$s
$\angle 1$ is supplementary to $\angle 15$	$\angle 1 \cong \angle 14$ corresponding $\angle$s, $\angle 14$ supplementary to $\angle 15$ consecutive interior $\angle$s

10.

$\angle 12$ is supplementary to $\angle 4$	linear pair
$\angle 12$ is supplementary to $\angle 16$	linear pair
$\angle 12$ is supplementary to $\angle 11$	$\angle 12 \cong \angle 3$ corresponding $\angle$s, $\angle 3$ is supplementary to $\angle 11$ linear pair
$\angle 12$ is supplementary to $\angle 14$	$\angle 12$ is supplementary to $\angle 16$ linear pair, $\angle 16 \cong \angle 14$ corresponding $\angle$s
$\angle 12$ is supplementary to $\angle 5$	$\angle 12$ is supplementary to $\angle 10$ consecutive interior $\angle$s, $\angle 10 \cong \angle 5$ alternate exterior $\angle$s
$\angle 12$ is supplementary to $\angle 1$	$\angle 12$ is supplementary to $\angle 10$ consecutive interior $\angle$s, $\angle 10 \cong \angle 1$ corresponding $\angle$s
$\angle 12$ is supplementary to $\angle 7$	$\angle 12 \cong \angle 8$ corresponding $\angle$s, $\angle 8$ is supplementary to $\angle 7$ linear pair
$\angle 12$ is supplementary to $\angle 10$	consecutive interior $\angle$s

11.
$$d = \sqrt{(x_2 - x_1)^2 + (y_2 - y_1)^2}$$
$$= \sqrt{(-4 - 6)^2 + (3 - 8)^2}$$
$$= \sqrt{(-10)^2 + (-5)^2} = \sqrt{125}$$
$$\approx 11.2$$

12.
$$d = \sqrt{(x_2 - x_1)^2 + (y_2 - y_1)^2}$$
$$= \sqrt{[6 - (-15)]^2 + (18 - 12)^2}$$
$$= \sqrt{21^2 + 6^2} = \sqrt{477}$$
$$\approx 21.8$$

13.
$$d = \sqrt{(x_2 - x_1)^2 + (y_2 - y_1)^2}$$
$$= \sqrt{(-3 - 11)^2 + [-4 - (-8)]^2}$$
$$= \sqrt{(-14)^2 + 4^2} = \sqrt{212}$$
$$\approx 14.6$$

4-1 Classifying Triangles

Page 179 Geometry Activity: Equilateral Triangles

1. Yes, all edges of the paper are an equal length.
2. See students' work.
3. See students' work.

Pages 180–181 Check for Understanding

1. Triangles are classified by sides and angles. For example, a triangle can have a right angle and have no two sides congruent.
2. Sample answer: 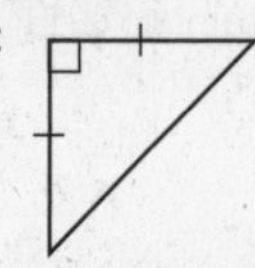
3. Always; equiangular triangles have three acute angles.
4. Never; right triangles have one right angle and acute triangles have all acute angles.
5. The triangle has one angle with measure greater than 90, so it is an obtuse triangle.
6. The triangle has three congruent angles, so it is an equiangular triangle.
7. $\triangle MJK$ is obtuse because $m\angle MJK > 90$.
   $\triangle KLM$ is obtuse because $m\angle KLM = m\angle MJK$ and $m\angle MJK > 90$.
   $\triangle JKN$ is obtuse because $m\angle JNK = 180 - m\angle JNM = 180 - 52$ or 128.
   $\triangle LMN$ is obtuse because $m\angle LNM = 180 - m\angle JNM = 180 - 52$ or 128.
8. $\triangle GHD$ is a right triangle because $\overline{GH} \perp \overline{DF}$ so $\angle GHD$ is a right angle.
   $\triangle GHJ$ is a right triangle because $\overline{GH} \perp \overline{DF}$ so $\angle GHJ$ is a right angle.
   $\triangle IJF$ is a right triangle because $\overline{IJ} \parallel \overline{GH}$, $\overline{GH} \perp \overline{DF}$ so $\overline{IJ} \perp \overline{DF}$ and $\angle IJF$ is a right angle.
   $\triangle EIG$ is a right triangle because $\overline{GI} \perp \overline{EF}$ so $\angle EIG$ is a right angle.
9. $\overline{JM} \cong \overline{MN}$ so $JM = MN$.
   $2x - 5 = 3x - 9$
   $-5 + 9 = 3x - 2x$
   $4 = x$
   $JM = 2x - 5$
   $= 2(4) - 5 = 3$
   $MN = 3x - 9$
   $= 3(4) - 9 = 3$
   $JN = x - 2$
   $= 4 - 2 = 2$
10. Since $\triangle QRS$ is equilateral, each side has the same length. So $QR = RS$.
    $4x = 2x + 1$
    $2x = 1$
    $x = \frac{1}{2}$
    $QR = 4x$
    $= 4\left(\frac{1}{2}\right) = 2$
    $RS = 2x + 1$
    $= 2\left(\frac{1}{2}\right) + 1 = 2$
    $QS = 6x - 1$
    $= 6\left(\frac{1}{2}\right) - 1 = 2$
11. $TW = \sqrt{(4-2)^2 + (-5-6)^2}$
    $= \sqrt{4 + 121} = \sqrt{125}$
    $WZ = \sqrt{(-3-4)^2 + [0-(-5)]^2}$
    $= \sqrt{49 + 25} = \sqrt{74}$
    $TZ = \sqrt{(-3-2)^2 + (0-6)^2}$
    $= \sqrt{25 + 36} = \sqrt{61}$
    $\triangle TWZ$ is scalene because no two sides are congruent.
12. 8 scalene triangles (green), 8 isosceles triangles in the middle (blue), 4 isosceles triangles around the middle (yellow), and 4 isosceles triangles at the corners of the square (purple).

Pages 181–183 Practice and Apply

13. The triangle has a right angle, so it is a right triangle.
14. The triangle has three acute angles, so it is an acute triangle.
15. The triangle has three acute angles, so it is an acute triangle.
16. The triangle has one angle with measure greater than 90, so it is an obtuse triangle.
17. The triangle has one angle with measure greater than 90, so it is an obtuse triangle.
18. The triangle has a right angle, so it is a right triangle.
19. The triangle has three congruent sides and three congruent angles, so it is equilateral and equiangular.
20. See students' work.
21. The triangles have two congruent sides, so they are isosceles. The triangles have three acute angles, so they are acute.
22. $\angle BGD$ is a right angle, so $\overline{AD} \perp \overline{BC}$. Then $\angle AGB$, $\angle AGC$, $\angle DGB$, and $\angle DGC$ are right angles, so $\triangle AGB$, $\triangle AGC$, $\triangle DGB$, and $\triangle DGC$ are right triangles.
23. $\angle BAC$ and $\angle CDB$ are obtuse, so $\triangle BAC$ and $\triangle CDB$ are obtuse triangles.
24. $\triangle AGB$, $\triangle AGC$, $\triangle DGB$, and $\triangle DGC$ are scalene because they each have no congruent sides.
25. $\triangle ABD$, $\triangle ACD$, $\triangle BAC$, and $\triangle CDB$ are isosceles triangles because they each have two congruent sides.
26. $\overline{HG} \cong \overline{JG}$ so $HG = JG$.
    $x + 7 = 3x - 5$
    $7 + 5 = 3x - x$
    $12 = 2x$
    $6 = x$
    $GH = x + 7$
    $= 6 + 7 = 13$
    $GJ = 3x - 5$
    $= 3(6) - 5 = 13$
    $HJ = x - 1$
    $= 6 - 1 = 5$
27. $\triangle MPN$ is equilateral so all three sides are congruent.
    $MN = MP$
    $3x - 6 = x + 4$
    $3x - x = 4 + 6$
    $2x = 10$
    $x = 5$
    $MN = 3x - 6$
    $= 3(5) - 6 = 9$
    $MP = x + 4$
    $= 5 + 4 = 9$
    $NP = 2x - 1$
    $= 2(5) - 1 = 9$

28. $\triangle QRS$ is equilateral so all three sides are congruent.

$QR = 2x - 2$
$RS = x + 6$
$QS = 3x - 10$
$QR = RS$
$2x - 2 = x + 6$
$2x - x = 6 + 2$
$x = 8$
$QR = 2x - 2$
$= 2(8) - 2 = 14$
$RS = x + 6$
$= 8 + 6 = 14$
$QS = 3x - 10$
$= 3(8) - 10 = 14$

29. $JL = 2x - 5$
$JK = x + 3$
$KL = x - 1$
$\overline{KJ} \cong \overline{LJ}$
$KJ = LJ$
$x + 3 = 2x - 5$
$3 + 5 = 2x - x$
$8 = x$
$JL = 2x - 5$
$= 2(8) - 5 = 11$
$JK = x + 3$
$= 8 + 3 = 11$
$KL = x - 1$
$= 8 - 1 = 7$

30. P is the midpoint of $\overline{MN}$, so $MP = PN = \frac{1}{2}(24)$ or 12. $\overline{OP} \perp \overline{MN}$, so $\triangle MPO$ and $\triangle NPO$ are right triangles. Use the Pythagorean Theorem to find MO and NO.

$(MO)^2 = (MP)^2 + (OP)^2$
$(MO)^2 = 12^2 + 12^2$
$(MO)^2 = 288$
$MO = \sqrt{288}$
$(NO)^2 = (PN)^2 + (OP)^2$
$(NO)^2 = 12^2 + 12^2$
$(NO)^2 = 288$
$NO = \sqrt{288}$

$\triangle MPO$ and $\triangle NPO$ are not equilateral because $MO = NO = \sqrt{288}$.

31. Let x be the distance from Lexington to Nashville. Then the distance from Cairo to Nashville is $x - 40$, the distance from Cairo to Lexington is $x + 81$, and $(x - 40) + (x + 81) + x = 593$

$3x + 41 = 593$
$3x = 552$
$x = 184$
$x - 40 = 184 - 40$ or 144
$x + 81 = 184 + 81$ or 265

The triangle formed is scalene; it is 184 miles from Lexington to Nashville, 265 miles from Cairo to Lexington, and 144 miles from Cairo to Nashville.

32. $AB = \sqrt{(3-5)^2 + (-1-4)^2}$
$= \sqrt{4 + 25} = \sqrt{29}$
$BC = \sqrt{(7-3)^2 + [-1-(-1)]^2}$
$= \sqrt{16 + 0} = \sqrt{16}$ or 4
$AC = \sqrt{(7-5)^2 + (-1-4)^2}$
$= \sqrt{4 + 25} = \sqrt{29}$

Since $\overline{AB}$ and $\overline{AC}$ have the same length, $\triangle ABC$ is isosceles.

33. $AB = \sqrt{[5-(-4)]^2 + (6-1)^2}$
$= \sqrt{81 + 25} = \sqrt{106}$
$BC = \sqrt{(-3-5)^2 + (-7-6)^2}$
$= \sqrt{64 + 169} = \sqrt{233}$
$AC = \sqrt{[-3-(-4)]^2 + (-7-1)^2}$
$= \sqrt{1 + 64} = \sqrt{65}$

$\triangle ABC$ is scalene because no two sides are congruent.

34. $AB = \sqrt{[-7-(-7)]^2 + (-1-9)^2}$
$= \sqrt{0 + 100} = \sqrt{100}$ or 10
$BC = \sqrt{[4-(-7)]^2 + [-1-(-1)]^2}$
$= \sqrt{121 + 0} = \sqrt{121}$ or 11
$AC = \sqrt{[4-(-7)]^2 + (-1-9)^2}$
$= \sqrt{121 + 100} = \sqrt{221}$

$\triangle ABC$ is scalene because no two sides are congruent.

35. $AB = \sqrt{[2-(-3)]^2 + [1-(-1)]^2}$
$= \sqrt{25 + 4} = \sqrt{29}$
$BC = \sqrt{(2-2)^2 + (-3-1)^2}$
$= \sqrt{0 + 16} = \sqrt{16}$ or 4
$AC = \sqrt{[2-(-3)]^2 + [-3-(-1)]^2}$
$= \sqrt{25 + 4} = \sqrt{29}$

Since $\overline{AB}$ and $\overline{AC}$ have the same length, $\triangle ABC$ is isosceles.

36. $AB = \sqrt{(5\sqrt{3} - 0)^2 + (2-5)^2}$
$= \sqrt{75 + 9} = \sqrt{84}$
$BC = \sqrt{(0 - 5\sqrt{3})^2 + (-1-2)^2}$
$= \sqrt{75 + 9} = \sqrt{84}$
$AC = \sqrt{(0-0)^2 + (-1-5)^2}$
$= \sqrt{0 + 36} = \sqrt{36}$ or 6

Since $\overline{AB}$ and $\overline{BC}$ have the same length, $\triangle ABC$ is isosceles.

37. $AB = \sqrt{[-5-(-9)]^2 + (6\sqrt{3} - 0)^2}$
$= \sqrt{16 + 108} = \sqrt{124}$
$BC = \sqrt{[-1-(-5)]^2 + (0 - 6\sqrt{3})^2}$
$= \sqrt{16 + 108} = \sqrt{124}$
$AC = \sqrt{[-1-(-9)]^2 + (0-0)^2}$
$= \sqrt{64 + 0} = \sqrt{64}$ or 8

Since $\overline{AB}$ and $\overline{BC}$ have the same length, $\triangle ABC$ is isosceles.

38. Given: $\triangle EUI$ is equiangular.
$\overline{QL} \parallel \overline{UI}$
Prove: $\triangle EQL$ is equiangular.

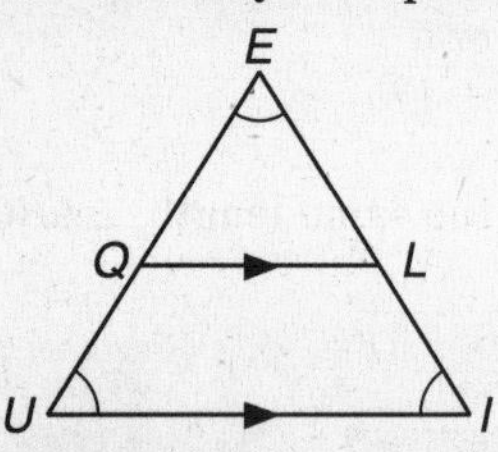

Proof:

Statements	Reasons
1. $\triangle EUI$ is equiangular. $\overline{QL} \parallel \overline{UI}$	1. Given
2. $\angle E \cong \angle EUI \cong \angle EIU$	2. Definition of equiangular triangle
3. $\angle EUI \cong \angle EQL$ $\angle EIU \cong \angle ELQ$	3. Corresponding ∡ are ≅.
4. $\angle E \cong \angle E$	4. Reflexive Property
5. $\angle E \cong \angle EQL \cong \angle ELQ$	5. Transitive Property
6. $\triangle EQL$ is equiangular.	6. Definition of equiangular triangles

39. Given: $m\angle NPM = 33$
Prove: $\triangle RPM$ is obtuse.

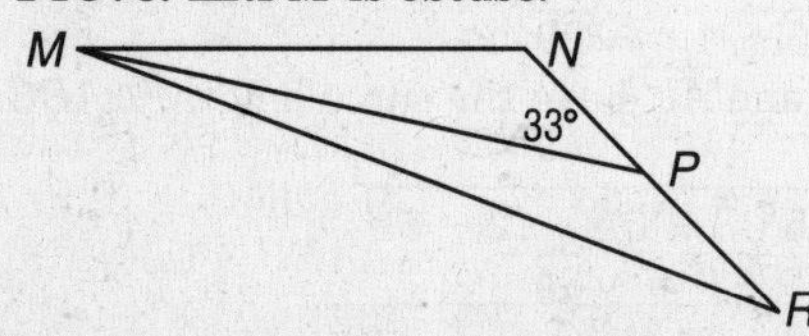

Proof: $\angle NPM$ and $\angle RPM$ form a linear pair. $\angle NPM$ and $\angle RPM$ are supplementary because if two angles form a linear pair, then they are supplementary. So, $m\angle NPM + m\angle RPM = 180$. It is given that $m\angle NPM = 33$. By substitution, $33 + m\angle RPM = 180$. Subtract to find that $m\angle RPM = 147$. $\angle RPM$ is obtuse by definition. $\triangle RPM$ is obtuse by definition.

40. $TS = \sqrt{[-7-(-4)]^2 + (8-14)^2}$
$= \sqrt{9+36}$ or $\sqrt{45}$
$SR = \sqrt{[-10-(-7)]^2 + (2-8)^2}$
$= \sqrt{9+36}$ or $\sqrt{45}$
S is the midpoint of $\overline{RT}$.
$UT = \sqrt{[0-(-4)]^2 + (8-14)^2}$
$= \sqrt{16+36}$ or $\sqrt{52}$
$VU = \sqrt{(4-0)^2 + (2-8)^2}$
$= \sqrt{16+36}$ or $\sqrt{52}$
U is the midpoint of $\overline{TV}$.

41. $AD = \sqrt{\left(0-\frac{a}{2}\right)^2 + (0-b)^2}$
$= \sqrt{\left(-\frac{a}{2}\right)^2 + (-b)^2} = \sqrt{\frac{a^2}{4} + b^2}$
$CD = \sqrt{\left(a-\frac{a}{2}\right)^2 + (0-b)^2}$
$= \sqrt{\left(\frac{a}{2}\right)^2 + (-b)^2} = \sqrt{\frac{a^2}{4} + b^2}$
$AD = CD$, so $\overline{AD} \cong \overline{CD}$. $\triangle ADC$ is isosceles by definition.

42. Use the Distance Formula and Slope Formula.
$KL = \sqrt{(4-2)^2 + (2-6)^2}$
$= \sqrt{4+16}$
$= \sqrt{20}$
$\overline{KL} \cong \overline{LM}$, so $LM = \sqrt{20}$.
slope of $\overline{KL} = \frac{2-6}{4-2}$
$= \frac{-4}{2}$ or -2
$\angle KLM$ is a right angle, so $\overline{KL} \perp \overline{LM}$ and $\overline{LM}$ has slope $\frac{1}{2}$.
Let (x_2, y_2) be the coordinates of M.
slope of $\overline{LM} = \frac{y_2-2}{x_2-4}$
$\frac{1}{2} = \frac{y_2-2}{x_2-4}$
$x_2 - 4 = 2(y_2 - 2)$
$LM = \sqrt{(x_2-4)^2 + (y_2-2)^2}$
$\sqrt{20} = \sqrt{[2(y_2-2)]^2 + (y_2-2)^2}$
$\sqrt{20} = \sqrt{4(y_2-2)^2 + (y_2-2)^2}$
$\sqrt{20} = \sqrt{5(y_2-2)^2}$
$20 = 5(y_2-2)^2$
$4 = (y_2-2)^2$
$2 = y_2 - 2$ or $-2 = y_2 - 2$
$4 = y_2$ or $0 = y_2$
$x_2 = 2(y_2-2) + 4$
$x_2 = 2(4-2) + 4$ or $x_2 = 2(0-2) + 4$
$x_2 = 8$ or $x_2 = 0$
M has coordinates (8, 4) or (0, 0).

43. Sample answer: Triangles are used in construction as structural support. Answers should include the following.

- Triangles can be classified by sides and angles. If the measure of each angle is less than 90, the triangle is acute. If the measure of one angle is greater than 90, the triangle is obtuse. If one angle equals 90°, the triangle is right. If each angle has the same measure, the triangle is equiangular. If no two sides are congruent, the triangle is scalene. If at least two sides are congruent, it is isosceles. If all of the sides are congruent, the triangle is equilateral.
- Isosceles triangles seem to be used more often in architecture and construction.

44. C; $AB = \sqrt{[1-(-1)]^2 + (3-1)^2}$
$= \sqrt{4+4} = \sqrt{8}$
$BC = \sqrt{(3-1)^2 + (-1-3)^2}$
$= \sqrt{4+16} = \sqrt{20}$
$AC = \sqrt{[3-(-1)]^2 + (-1-1)^2}$
$= \sqrt{16+4} = \sqrt{20}$
Since $\overline{AC}$ and $\overline{BC}$ have the same length, $\triangle ABC$ is isosceles.

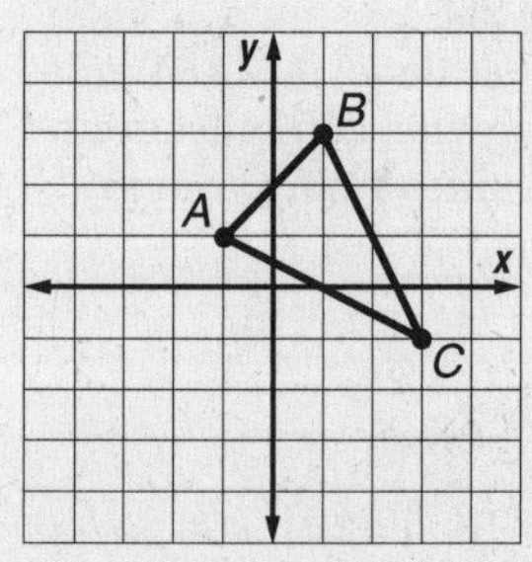

All three angles of $\triangle ABC$ are acute, so $\triangle ABC$ is acute.

45. B; $\frac{x + y + 15 + 35}{4} = 25$

$$x + y + 50 = 100$$
$$x + y = 50$$
$$\frac{x + 15 + 35}{3} = 27$$
$$x + 50 = 81$$
$$x = 31$$

Substituting 31 for x,
$31 + y = 50$
$y = 19$

Page 183 Maintain Your Skills

46.

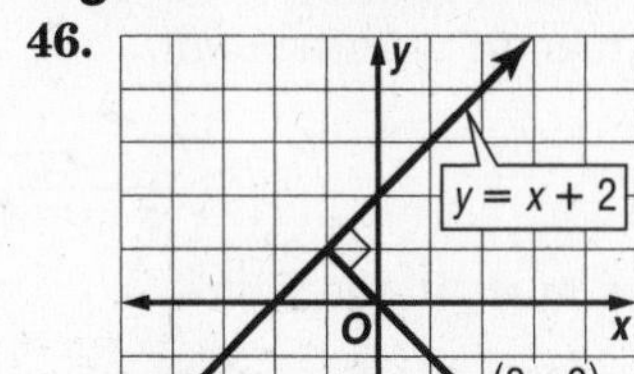

The perpendicular segment from the point $(2, -2)$ to the line $y = x + 2$ appears to intersect the line $y = x + 2$ at $(-1, 1)$. Use the Distance Formula to find the distance between $(2, -2)$ and $y = x + 2$.

$$d = \sqrt{(x_2 - x_1)^2 + (y_2 - y_1)^2}$$
$$= \sqrt{(-1 - 2)^2 + [1 - (-2)]^2}$$
$$= \sqrt{9 + 9}$$
$$= \sqrt{18}$$

The distance between the line $y = x + 2$ and the point $(2, -2)$ is $\sqrt{18}$ units.

47.

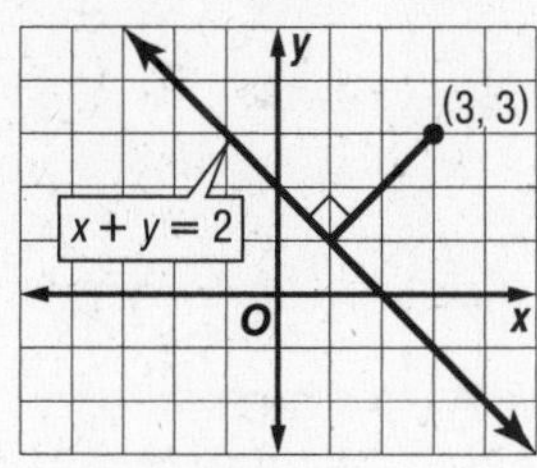

$x + y = 2$
$y = -x + 2$

The perpendicular segment from the point $(3, 3)$ to the line $x + y = 2$ appears to intersect the line $x + y = 2$ at $(1, 1)$. Use the Distance Formula to find the distance between $(3, 3)$ and $x + y = 2$.

$$d = \sqrt{(x_2 - x_1)^2 + (y_2 - y_1)^2}$$
$$= \sqrt{(3 - 1)^2 + (3 - 1)^2}$$
$$= \sqrt{4 + 4}$$
$$= \sqrt{8}$$

The distance between the line $x + y = 2$ and the point $(3, 3)$ is $\sqrt{8}$ units.

48.

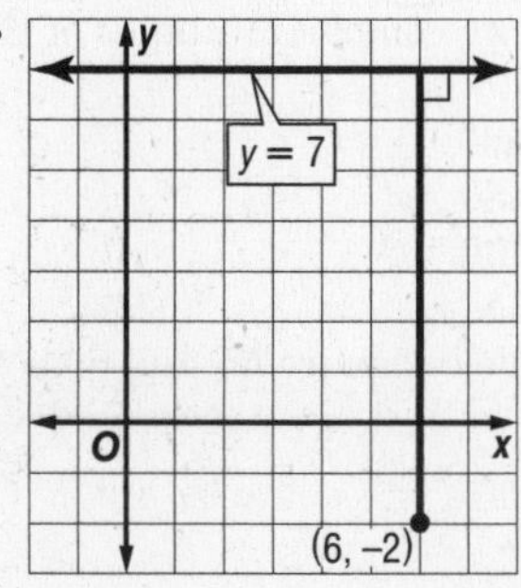

The perpendicular segment from the point $(6, -2)$ to the line $y = 7$ intersects the line $y = 7$ at $(6, 7)$. Use the Distance Formula to find the distance between $(6, -2)$ and $y = 7$.

$$d = \sqrt{(x_2 - x_1)^2 + (y_2 - y_1)^2}$$
$$= \sqrt{(6 - 6)^2 + (-2 - 7)^2}$$
$$= \sqrt{0 + 81}$$
$$= \sqrt{81} \text{ or } 9$$

The distance between the line $y = 7$ and the point $(6, -2)$ is 9 units.

49.

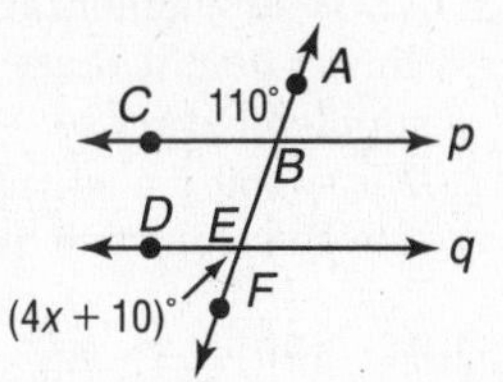

Explore: From the figure, you know that $m\angle ABC = 110$ and $m\angle DEF = 4x + 10$.
Plan: For line p to be parallel to line q, corresponding angles must be congruent, so $\angle ABC \cong \angle AED$. Since $\angle AED$ is supplementary to $\angle DEF$ because they are a linear pair, it must be true that $p \parallel q$ if $\angle ABC$ is supplementary to $\angle DEF$, or $m\angle ABC + m\angle DEF = 180$. Substitute the given angle measures into this equation and solve for x.
Solve: $m\angle ABC + m\angle DEF = 180$

$$110 + 4x + 10 = 180$$
$$4x = 60$$
$$x = 15$$

Examine: Verify the measure of $\angle DEF$ by using the value of x. That is, $4x + 10 = 4(15) + 10$ or 70, and $110 + 70 = 180$. Since $m\angle ABC + m\angle DEF = 180$, $\angle ABC$ is supplementary to $\angle DEF$ and $p \parallel q$.

50.

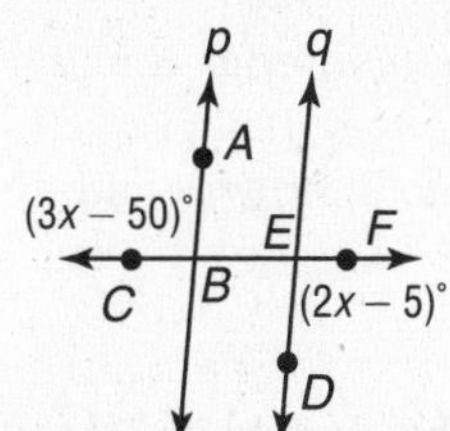

Explore: From the figure, you know that $m\angle ABC = 3x - 50$ and $m\angle DEF = 2x - 5$. You also know that $\angle ABC$ and $\angle DEF$ are alternate exterior angles.
Plan: For line p to be parallel to line q, the alternate exterior angles must be congruent. So $m\angle ABC = m\angle DEF$. Substitute the given angle measures into this equation and solve for x.
Solve: $m\angle ABC = m\angle DEF$

$$3x - 50 = 2x - 5$$
$$x = 45$$

Examine: Verify the angle measures by using the value of x to find $m\angle ABC$ and $m\angle DEF$.

$$m\angle ABC = 3x - 50$$
$$= 3(45) - 50$$
$$= 85$$
$$m\angle DEF = 2x - 5$$
$$= 2(45) - 5$$
$$= 85$$

Since $m\angle ABC = m\angle DEF$, $\angle ABC \cong \angle DEF$ and $p \parallel q$.

51.

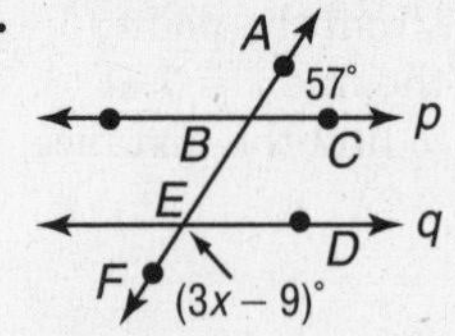

Explore: From the figure, you know that $m\angle ABC = 57$ and $m\angle DEF = 3x - 9$.
Plan: For line p to be parallel to line q, corresponding angles must be congruent, so $\angle ABC \cong \angle AED$. Since $\angle AED$ is supplementary to $\angle DEF$, because they are a linear pair, it must be true that $p \parallel q$ if $\angle ABC$ is supplementary to $\angle DEF$, or $m\angle ABC + m\angle DEF = 180$. Substitute the given angle measures into this equation and solve for x.
Solve: $m\angle ABC + m\angle DEF = 180$
$57 + 3x - 9 = 180$
$3x = 132$
$x = 44$
Examine: Verify the measure of $\angle DEF$ by using the value of x. That is, $3x - 9 = 3(44) - 9$ or 123, and $57 + 123 = 180$. Since $m\angle ABC + m\angle DEF = 180$, $m\angle ABC$ is supplementary to $\angle DEF$, and $p \parallel q$.

52. **1.** Given
2. Subtraction Property
3. Addition Property
4. Division Property

53. any three: $\angle 2$ and $\angle 11$, $\angle 3$ and $\angle 6$, $\angle 4$ and $\angle 7$, $\angle 3$ and $\angle 12$, $\angle 7$ and $\angle 10$, $\angle 8$ and $\angle 11$

54. $\angle 1$ and $\angle 4$, $\angle 1$ and $\angle 10$, $\angle 5$ and $\angle 2$, $\angle 5$ and $\angle 8$, $\angle 9$ and $\angle 6$, $\angle 9$ and $\angle 12$

55. $\angle 6$, $\angle 9$, and $\angle 12$ by alternate interior ∡ and transitivity

56. $\angle 1$, $\angle 4$, and $\angle 10$ by alternate interior ∡ and transitivity

57. $\angle 2$, $\angle 5$, and $\angle 8$ by alternate interior ∡ and transitivity

Page 184 Geometry Activity: Angles of Triangles

1. congruent
2. congruent
3. congruent
4. 180, because $\angle DFA + \angle DFE = \angle AFE$, and $\angle AFE$ and $\angle EFC$ form a linear pair.
5. $m\angle A + m\angle B + m\angle C = 180$ by substitution
6. The sum of the measures of the angles of any triangle is 180.
7. $m\angle A + m\angle B$ is the measure of the exterior angle at C.
8. See students' work.
9. yes
10. See students' work.
11. See students' work.
12. The measure of an exterior angle is equal to the sum of the measures of the two remote interior angles.

4-2 Angles of Triangles

Pages 188–189 Check for Understanding

1. Sample answer: $\angle 2$ and $\angle 3$ are the remote interior angles of exterior $\angle 1$.

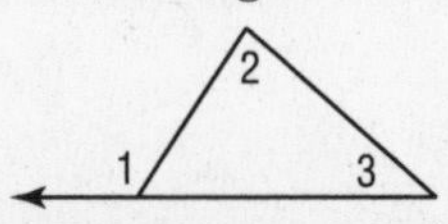

2. Najee; the sum of the measures of the remote interior angles is equal to the measure of the corresponding exterior angle.

3. Let $\angle P$ be the unknown angle at Pittsburgh.
$m\angle P + 85 + 52 = 180$
$m\angle P + 137 = 180$
$m\angle P = 43$

4. Let $\angle A$ be the unknown angle in the figure.
$m\angle A + 62 + 19 = 180$
$m\angle A + 81 = 180$
$m\angle A = 99$

5. $m\angle 1 = 23 + 32$
$= 55$

6. $m\angle 2 + 22 = m\angle 1$
$m\angle 2 + 22 = 55$
$m\angle 2 = 33$

7. $m\angle 3 = 22 + (180 - m\angle 1)$
$= 22 + 180 - 55$
$= 147$

8. $m\angle 1 + 25 = 90$
$m\angle 1 = 65$

9. $m\angle 2 + 65 = 90$
$m\angle 2 = 25$

10. $m\angle 1 + m\angle 2 = 90$
$m\angle 1 + 70 = 90$
$m\angle 1 = 20$

Pages 189–191 Practice and Apply

11. Let $\angle X$ be the third angle in the triangle.
$m\angle X + 40 + 47 = 180$
$m\angle X + 87 = 180$
$m\angle X = 93$

12. Let $\angle X$ be one of the two congruent angles of the triangle.
$m\angle X + m\angle X + 39 = 180$
$2m\angle X + 39 = 180$
$2m\angle X = 141$
$m\angle X = 70.5$
The missing angles have measure 70.5 and 70.5.

13. Let $\angle X$ be one of the two congruent angles of the triangle.
$m\angle X + m\angle X + 50 = 180$
$2m\angle X + 50 = 180$
$2m\angle X = 130$
$m\angle X = 65$
The missing angles have measure 65 and 65.

14. Let $\angle X$ be the unknown acute angle of the triangle.
$m\angle X + 27 = 90$
$m\angle X = 63$

15. $m\angle 1 + 47 + 57 = 180$
$m\angle 1 + 104 = 180$
$m\angle 1 = 76$

16. $m\angle 2 = m\angle 1$
$m\angle 2 = 76$

17. $m\angle 3 + m\angle 2 + 55 = 180$
$m\angle 3 + 76 + 55 = 180$
$m\angle 3 + 131 = 180$
$m\angle 3 = 49$

18. $m\angle 1 + 69 + 47 = 180$
$m\angle 1 + 116 = 180$
$m\angle 1 = 64$

19. $m\angle 1 + m\angle 2 + 63 = 180$
$64 + m\angle 2 + 63 = 180$
$m\angle 2 + 127 = 180$
$m\angle 2 = 53$

20. $m\angle 3 = m\angle 2 + 63$
$= 53 + 63$
$= 116$

21. $m\angle 4 + m\angle 5 + m\angle 3 = 180$
$2m\angle 4 + 116 = 180$
$2m\angle 4 = 64$
$m\angle 4 = 32$

22. $m\angle 5 = m\angle 4$
$m\angle 5 = 32$

23. $m\angle 6 + 136 = 180$
$m\angle 6 = 44$

24. $m\angle 7 + 47 = 136$
$m\angle 7 = 89$

25. $m\angle 1 + 33 + 24 = 180$
$m\angle 1 + 57 = 180$
$m\angle 1 = 123$

26. $m\angle 2 + 95 = m\angle 1$
$m\angle 2 + 95 = 123$
$m\angle 2 = 28$

27. $m\angle 3 + 109 = m\angle 1$
$m\angle 3 + 109 = 123$
$m\angle 3 = 14$

28. $m\angle 1 + 126 = 180$
$m\angle 1 = 54$

29. $m\angle 2 + 73 = 126$
$m\angle 2 = 53$

30. $m\angle 3 + 43 = 180$
$m\angle 3 = 137$

31. $m\angle 4 + 34 + 43 = 180$
$m\angle 4 + 77 = 180$
$m\angle 4 = 103$

32. $m\angle 1 + m\angle DGF = 90$
$m\angle 1 + 53 = 90$
$m\angle 1 = 37$

33. $m\angle 2 + m\angle AGC = 90$
$m\angle 2 + 40 = 90$
$m\angle 2 = 50$

34. $m\angle 3 + m\angle AGC = 90$
$m\angle 3 + 40 = 90$
$m\angle 3 = 50$

35. $m\angle 4 + m\angle 2 = 90$
$m\angle 4 + 50 = 90$
$m\angle 4 = 40$

36. $m\angle 1 + 26 + 101 = 180$
$m\angle 1 + 127 = 180$
$m\angle 1 = 53$

37. $m\angle 2 = 26 + 103$
$= 129$

38. $m\angle 3 = 101 + (180 - 128)$
$= 153$

39. Given: $\angle FGI \cong \angle IGH$
$\overline{GI} \perp \overline{FH}$
Prove: $\angle F \cong \angle H$

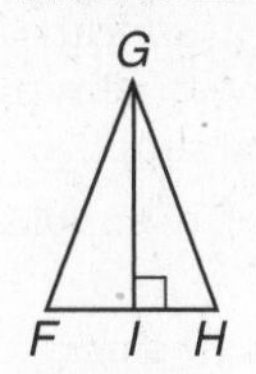

Proof:

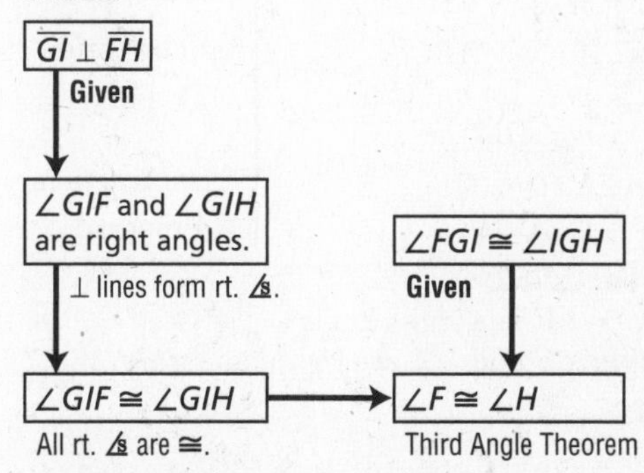

40. Given: $ABCD$ is a quadrilateral.
Prove: $m\angle DAB + m\angle B + m\angle BCD + m\angle D = 360$

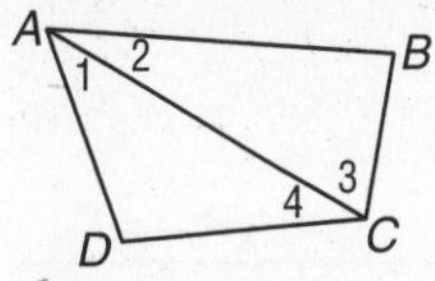

Proof:

Statements	**Reasons**
1. $ABCD$ is a quadrilateral.	1. Given
2. $m\angle 2 + m\angle 3 + m\angle B = 180$ $m\angle 1 + m\angle 4 + m\angle D = 180$	2. Angle Sum Theorem
3. $m\angle 2 + m\angle 3 + m\angle B + m\angle 1 + m\angle 4 + m\angle D = 360$	3. Addition Property
4. $m\angle DAB = m\angle 1 + m\angle 2$ $m\angle BCD = m\angle 3 + m\angle 4$	4. Angle addition
5. $m\angle DAB + m\angle B + m\angle BCD + m\angle D = 360$	5. Substitution

41. Given: $\triangle ABC$
Prove: $m\angle CBD = m\angle A + m\angle C$

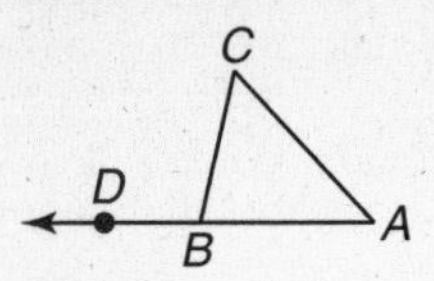

Proof:

Statements	Reasons
1. $\triangle ABC$	1. Given
2. $\angle CBD$ and $\angle ABC$ form a linear pair.	2. Def. of linear pair
3. $\angle CBD$ and $\angle ABC$ are supplementary.	3. If 2 $\angle$s form a linear pair, they are suppl.
4. $m\angle CBD + m\angle ABC = 180$	4. Def. of suppl.
5. $m\angle A + m\angle ABC + m\angle C = 180$	5. Angle Sum Theorem
6. $m\angle A + m\angle ABC + m\angle C = m\angle CBD + m\angle ABC$	6. Subsitution Property
7. $m\angle A + m\angle C = m\angle CBD$	7. Subtraction Property

42. Given: $\triangle RST$
$\angle R$ is a right angle
Prove: $\angle S$ and $\angle T$ are complementary

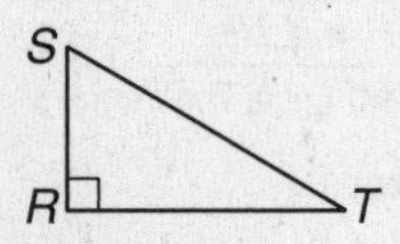

Proof:

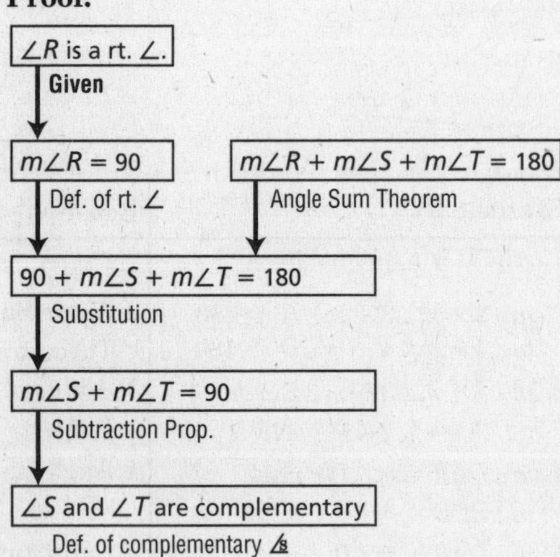

43. Given: $\triangle MNO$
$\angle M$ is a right angle.
Prove: There can be at most one right angle in a triangle.

Proof: In $\triangle MNO$, $\angle M$ is a right angle. $m\angle M + m\angle N + m\angle O = 180$. $m\angle M = 90$, so $m\angle N + m\angle O = 90$. If $\angle N$ were a right angle, then $m\angle O = 0$. But that is impossible, so there cannot be two right angles in a triangle.

Given: $\triangle PQR$
$\angle P$ is obtuse.
Prove: There can be at most one obtuse angle in a triangle.

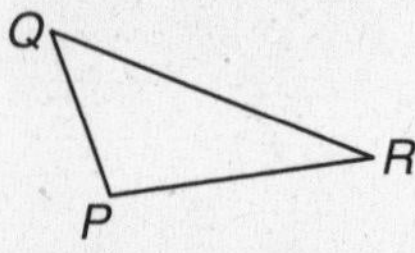

Proof: In $\triangle PQR$, $\angle P$ is obtuse. So $m\angle P > 90$. $m\angle P + m\angle Q + m\angle R = 180$. It must be that $m\angle Q + m\angle R < 90$. So, $\angle Q$ and $\angle R$ must be acute.

44. Given: $\angle A \cong \angle D$
$\angle B \cong \angle E$
Prove: $\angle C \cong \angle F$

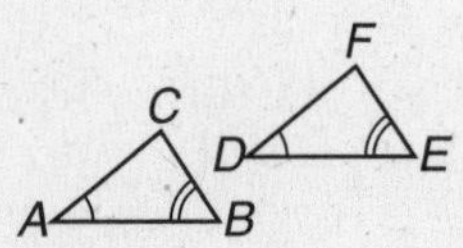

Proof:

Statements	Reasons
1. $\angle A \cong \angle D$ $\angle B \cong \angle E$	1. Given
2. $m\angle A = m\angle D$ $m\angle B = m\angle E$	2. Def. of $\cong$ $\angle$s
3. $m\angle A + m\angle B + m\angle C = 180$ $m\angle D + m\angle E + m\angle F = 180$	3. Angle Sum Theorem
4. $m\angle A + m\angle B + m\angle C = m\angle D + m\angle E + m\angle F$	4. Transitive Property
5. $m\angle D + m\angle E + m\angle C = m\angle D + m\angle E + m\angle F$	5. Substitution Property
6. $m\angle C = m\angle F$	6. Subtraction Property
7. $\angle C \cong \angle F$	7. Def. of $\cong$ $\angle$s

45. $m\angle 1 + m\angle 2 + m\angle 3 = 180$
$4x + 5x + 6x = 180$
$15x = 180$
$x = 12$
$m\angle 1 = 4(12)$ or 48
$m\angle 2 = 5(12)$ or 60
$m\angle 3 = 6(12)$ or 72

46. Sample answer: The shape of a kite is symmetric. If triangles are used on one side of the kite, congruent triangles are used on the opposite side. The wings of this kite are made from congruent right triangles. Answers should include the following.
- By the Third Angle Theorem, if two angles of two congruent triangles are congruent, then the third angles of each triangle are congruent.
- If one angle measures 90, the other two angles are both acute.

47. A; $m\angle Z + m\angle X = 90$
$\frac{a}{2} + 2a = 90$
$a + 4a = 180$
$5a = 180$
$a = 36$
$m\angle Z = \frac{a}{2}$
$= \frac{36}{2}$ or 18

48. B; let x be the measure of the first angle. Then the other angles have measure $3x$ and $x + 25$.

$x + 3x + x + 25 = 180$
$5x + 25 = 180$
$5x = 155$
$x = 31$
$3x = 3(31)$ or 93
$x + 25 = 31 + 25$ or 56

Page 191 Maintain Your Skills

49. $\triangle AED$ is scalene because no two sides are congruent.

50. $\triangle AED$ is obtuse because $m\angle AED > 90$. $m\angle AED = m\angle BEC$, so $\triangle BEC$ is obtuse.

51. $\triangle BEC$ is isosceles because $\overline{EB} \cong \overline{EC}$.

52. First, write an equation of a line p perpendicular to $y = x + 6$ and $y = x - 10$. The slope of p is the opposite reciprocal of 1, or -1. Use the y-intercept of $y = x + 6$, $(0, 6)$, as one of the endpoints of the perpendicular segment.

$y - y_1 = m(x - x_1)$
$y - 6 = -1(x - 0)$
$y - 6 = -x$
$y = -x + 6$

Next, use a system of equations to determine the point of intersection of line $y = x - 10$ and p.

$x - 10 = -x + 6$
$2x = 16$
$x = 8$
$y = 8 - 10$
$y = -2$

The point of intersection is $(8, -2)$.
Then, use the Distance Formula to determine the distance between $(0, 6)$ and $(8, -2)$.

$d = \sqrt{(x_2 - x_1)^2 + (y_2 - y_1)^2}$
$= \sqrt{(8 - 0)^2 + (-2 - 6)^2}$
$= \sqrt{64 + 64}$
$= \sqrt{128}$

The distance between the lines is $8\sqrt{2}$ units.

53. First, write an equation of a line p perpendicular to $y = -2x + 3$ and $y = -2x - 7$. The slope of p is the opposite reciprocal of -2, or $\frac{1}{2}$. Use the y-intercept of $y = -2x + 3$, $(0, 3)$, as one of the endpoints of the perpendicular segment.

$y - y_1 = m(x - x_1)$
$y - 3 = \frac{1}{2}(x - 0)$
$y - 3 = \frac{1}{2}x$
$y = \frac{1}{2}x + 3$

Next, use a system of equations to determine the point of intersection of line $y = -2x - 7$ and p.

$-2x - 7 = \frac{1}{2}x + 3$
$-\frac{5}{2}x = 10$
$x = -4$
$y = -2(-4) - 7$
$y = 1$

The point of intersection is $(-4, 1)$.
Then, use the Distance Formula to determine the distance between $(0, 3)$, and $(-4, 1)$.

$d = \sqrt{(x_2 - x_1)^2 + (y_2 - y_1)^2}$
$= \sqrt{(-4 - 0\ )^2 + (1 - 3)^2}$
$= \sqrt{16 + 4}$
$= \sqrt{20}$

The distance between the lines is $2\sqrt{5}$ units.

54. $4x - y = 20$
$-y = -4x + 20$
$y = 4x - 20$
$4x - y = 3$
$-y = -4x + 3$
$y = 4x - 3$

First, write an equation of a line p perpendicular to $4x - y = 20$ and $4x - y = 3$. The slope of p is the opposite reciprocal of 4, or $-\frac{1}{4}$. Use the y-intercept of $4x - y = 20$, $(0, -20)$, as one of the endpoints of the perpendicular segment.

$y - y_1 = m(x - x_1)$
$y - (-20) = -\frac{1}{4}(x - 0)$
$y + 20 = -\frac{1}{4}x$
$y = -\frac{1}{4}x - 20$

Next, use a system of equations to determine the point of intersection of line $4x - y = 3$ and p.

$4x - 3 = -\frac{1}{4}x - 20$
$\frac{17}{4}x = -17$
$x = -4$
$y = -\frac{1}{4}(-4) - 20$
$y = -19$

The point of intersection is $(-4, -19)$.
Then, use the Distance Formula to determine the distance between $(0, -20)$ and $(-4, -19)$.

$d = \sqrt{(x_2 - x_1)^2 + (y_2 - y_1)^2}$
$= \sqrt{(-4 - 0)^2 + [-19 - (-20)]^2}$
$= \sqrt{16 + 1}$
$= \sqrt{17}$

The distance between the lines is $\sqrt{17}$ units.

55. $2x - 3y = -9$
$-3y = -2x - 9$
$y = \frac{2}{3}x + 3$
$2x - 3y = -6$
$-3y = -2x - 6$
$y = \frac{2}{3}x + 2$

First, write an equation of a line p perpendicular to $2x - 3y = -9$ and $2x - 3y = -6$. The slope of p is the opposite reciprocal of $\frac{2}{3}$, or $-\frac{3}{2}$. Use the y-intercept of $2x - 3y = -9$, $(0, 3)$, as one of the endpoints of the perpendicular segment.

$y - y_1 = m(x - x_1)$
$y - 3 = -\frac{3}{2}(x - 0)$
$y - 3 = -\frac{3}{2}x$
$y = -\frac{3}{2}x + 3$

Next, use a system of equations to determine the point of intersection of line $2x - 3y = -6$ and p.

$\frac{2}{3}x + 2 = -\frac{3}{2}x + 3$
$\frac{13}{6}x = 1$
$x = \frac{6}{13}$

$y = -\frac{3}{2}\left(\frac{6}{13}\right) + 3$

$y = -\frac{9}{13} + 3$ or $\frac{30}{13}$

The point of intersection is $\left(\frac{6}{13}, \frac{30}{13}\right)$.

Then, use the Distance Formula to determine the distance between (0, 3) and $\left(\frac{6}{13}, \frac{30}{13}\right)$.

$d = \sqrt{(x_2 - x_1)^2 + (y_2 - y_1)^2}$

$= \sqrt{\left(\frac{6}{13} - 0\right)^2 + \left(\frac{30}{13} - 3\right)^2}$

$= \sqrt{\frac{36}{169} + \frac{81}{169}}$

$= \sqrt{\frac{117}{169}}$ or $\frac{\sqrt{117}}{13}$

The distance between the lines is $\frac{\sqrt{117}}{13}$ units.

56. $2y + 8 + 142 = 180$ linear pair
$2y + 150 = 180$
$2y = 30$
$y = 15$
$4x + 6 = 142$ corresponding angles
$4x = 136$
$x = 34$
$z = 4x + 6$ alternate exterior angles
$z = 4(34) + 6$
$z = 142$

57. $x + 68 = 180$ supplementary consecutive interior angles
$x = 112$
$4y + 68 = 180$ linear pair
$4y = 112$
$y = 28$
$5z + 2 = x$ alternate interior angles
$5z + 2 = 112$
$5z = 110$
$z = 22$

58. $3x = 48$ alternate interior angles
$x = 16$
$y + 42 + 48 = 180$ Angle Sum Theorem
$y + 90 = 180$
$y = 90$
$z = 42$ alternate interior angles

59. reflexive **60.** symmetric
61. symmetric **62.** transitive
63. transitive **64.** transitive

4-3 Congruent Triangles

Page 195 Check for Understanding

1. The sides and the angles of the triangle are not affected by a congruence transformation, so congruence is preserved.

2. Sample answer:

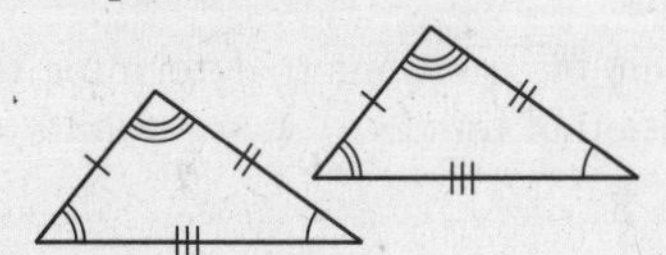

3. $\triangle AFC \cong \triangle DFB$ **4.** $\triangle HJT \cong \triangle TKH$

5. $\angle W \cong \angle S$, $\angle X \cong \angle T$, $\angle Z \cong \angle J$, $\overline{WX} \cong \overline{ST}$, $\overline{XZ} \cong \overline{TJ}$, $\overline{WZ} \cong \overline{SJ}$

6. The red triangles are congruent: $\triangle BME$, $\triangle ANG$, $\triangle DKH$, $\triangle CLF$. The blue triangles are congruent: $\triangle EMJ$, $\triangle GNJ$, $\triangle HKJ$, $\triangle FLJ$. The purple triangles are congruent to each other and to the triangles made up of a blue triangle and a red triangle: $\triangle BLJ$, $\triangle AMJ$, $\triangle JND$, $\triangle JKC$, $\triangle BMJ$, $\triangle ANJ$, $\triangle JKD$, $\triangle JLC$. Another set of congruent triangles consists of triangles made up of a red, a blue, and a purple triangle: $\triangle BAJ$, $\triangle ADJ$, $\triangle DCJ$, $\triangle CBJ$. Another set of congruent triangles consists of the triangles which are each half of the square: $\triangle BCD$, $\triangle ADC$, $\triangle CBA$, $\triangle DAB$.

7.

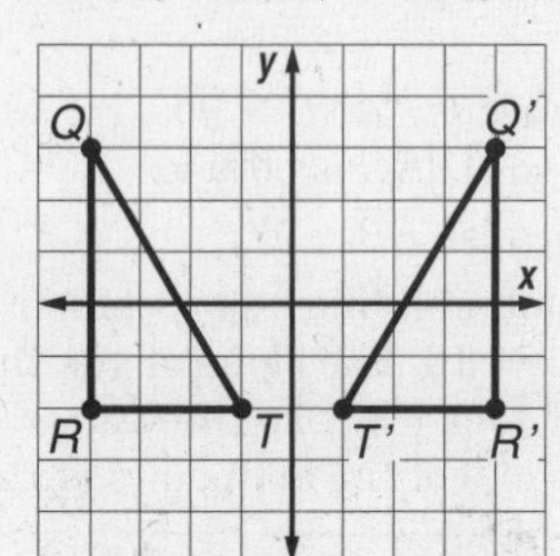

Use the Distance Formula to find the length of each side in the triangles.

$QR = \sqrt{[-4 - (-4)]^2 + (-2 - 3)^2}$
$= \sqrt{0 + 25}$ or 5
$Q'R' = \sqrt{(4 - 4)^2 + (-2 - 3)^2}$
$= \sqrt{0 + 25}$ or 5
$RT = \sqrt{[-1 - (-4)]^2 + [-2 - (-2)]^2}$
$= \sqrt{9 + 0}$ or 3
$R'T' = \sqrt{(1 - 4)^2 + [-2 - (-2)]^2}$
$= \sqrt{9 + 0}$ or 3
$QT = \sqrt{[-1 - (-4)]^2 + (-2 - 3)^2}$
$= \sqrt{9 + 25}$ or $\sqrt{34}$
$Q'T' = \sqrt{(1 - 4)^2 + (-2 - 3)^2}$
$= \sqrt{9 + 25}$ or $\sqrt{34}$

The lengths of the corresponding sides of two triangles are equal. Therefore, by the definition of congruence, $\overline{QR} \cong \overline{Q'R'}$, $\overline{RT} \cong \overline{R'T'}$, and $\overline{QT} \cong \overline{Q'T'}$. Use a protractor to measure the angles of the triangles. You will find that the measures are the same. In conclusion, because $\overline{QR} \cong \overline{Q'R'}$, $\overline{RT} \cong \overline{R'T'}$, $\overline{QT} \cong \overline{Q'T'}$, $\angle Q \cong \angle Q'$, $\angle R \cong \angle R'$, and $\angle T \cong \angle T'$, $\triangle QRT \cong \triangle Q'R'T'$. $\triangle Q'R'T'$ is a flip of $\triangle QRT$.

8. $\angle G \cong \angle K$, $\angle H \cong \angle L$, $\angle J \cong \angle P$, $\overline{GH} \cong \overline{KL}$, $\overline{HJ} \cong \overline{LP}$, $\overline{GJ} \cong \overline{KP}$

Pages 195–198 Practice and Apply

9. $\triangle CFH \cong \triangle JKL$ **10.** $\triangle RSV \cong \triangle TSV$
11. $\triangle WPZ \cong \triangle QVS$ **12.** $\triangle EFH \cong \triangle GHF$

13. $\angle T \cong \angle X$, $\angle U \cong \angle Y$, $\angle V \cong \angle Z$, $\overline{TU} \cong \overline{XY}$, $\overline{UV} \cong \overline{YZ}$, $\overline{TV} \cong \overline{XZ}$

14. $\angle C \cong \angle R$, $\angle D \cong \angle S$, $\angle G \cong \angle W$, $\overline{CD} \cong \overline{RS}$, $\overline{DG} \cong \overline{SW}$, $\overline{CG} \cong \overline{RW}$

15. $\angle B \cong \angle D$, $\angle C \cong \angle G$, $\angle F \cong \angle H$, $\overline{BC} \cong \overline{DG}$, $\overline{CF} \cong \overline{GH}$, $\overline{BF} \cong \overline{DH}$

16. $\angle A \cong \angle H$, $\angle D \cong \angle K$, $\angle G \cong \angle L$, $\overline{AD} \cong \overline{HK}$, $\overline{DG} \cong \overline{KL}$, $\overline{AG} \cong \overline{HL}$

17. $\triangle 1 \cong \triangle 10$, $\triangle 2 \cong \triangle 9$, $\triangle 3 \cong \triangle 8$, $\triangle 4 \cong \triangle 7$, $\triangle 5 \cong \triangle 6$

18. $\triangle$s 1–4, $\triangle$s 5–12, $\triangle$s 13–20

19. $\triangle$s 1, 5, 6, and 11, $\triangle$s 3, 8, 10, and 12, $\triangle$s 2, 4, 7, and 9

20. $\triangle UFS$, $\triangle TDV$, $\triangle ACB$

21. We need to know that all the corresponding angles are congruent and that the other corresponding sides are congruent.

22. Use the figure and the Distance Formula to find the length of each side in the triangles.

$PQ = 2$
$P'Q' = 2$
$QV = 4$
$Q'V' = 4$

$$PV = \sqrt{[-2-(-4)]^2 + (4-8)^2} = \sqrt{4+16} \text{ or } \sqrt{20}$$
$$P'V' = \sqrt{(2-4)^2 + (4-8)^2} = \sqrt{4+16} \text{ or } \sqrt{20}$$

The lengths of the corresponding sides of the two triangles are equal. Therefore, by the definition of congruence, $\overline{PQ} \cong \overline{P'Q'}$, $\overline{QV} \cong \overline{Q'V'}$, and $\overline{PV} \cong \overline{P'V'}$.
Use a protractor to confirm that the corresponding angles are congruent.
Therefore, $\triangle PQV \cong \triangle P'Q'V'$.
$\triangle P'Q'V'$ is a flip of $\triangle PQV$.

23. Use the figure and the Distance Formula to find the length of each side in the triangles.

$MN = 8$
$M'N' = 8$
$NP = 2$
$N'P' = 2$

$$MP = \sqrt{[2-(-6)]^2 + (2-4)^2} = \sqrt{64+4} \text{ or } \sqrt{68}$$
$$M'P' = \sqrt{[2-(-6)]^2 + [-2-(-4)]^2} = \sqrt{64+4} \text{ or } \sqrt{68}$$

The lengths of the corresponding sides of the two triangles are equal. Therefore, by the definition of congruence, $\overline{MN} \cong \overline{M'N'}$, $\overline{NP} \cong \overline{N'P'}$, and $\overline{MP} \cong \overline{M'P'}$.
Use a protractor to confirm that the corresponding angles are congruent.
Therefore, $\triangle MNP \cong \triangle M'N'P'$.
$\triangle M'N'P'$ is a flip of $\triangle MNP$.

24. Use the figure and the Distance Formula to find the length of each side in the triangles.

$$GF = \sqrt{(5-2)^2 + (3-2)^2} = \sqrt{9+1} \text{ or } \sqrt{10}$$
$$G'F' = \sqrt{(12-9)^2 + (3-2)^2} = \sqrt{9+1} \text{ or } \sqrt{10}$$
$$GH = \sqrt{(3-2)^2 + (5-2)^2} = \sqrt{1+9} \text{ or } \sqrt{10}$$
$$G'H' = \sqrt{(10-9)^2 + (5-2)^2} = \sqrt{1+9} \text{ or } \sqrt{10}$$
$$HF = \sqrt{(5-3)^2 + (3-5)^2} = \sqrt{4+4} \text{ or } \sqrt{8}$$
$$H'F' = \sqrt{(12-10)^2 + (3-5)^2} = \sqrt{4+4} \text{ or } \sqrt{8}$$

The lengths of the corresponding sides of the two triangles are equal. Therefore, by the definition of congruence, $\overline{GF} \cong \overline{G'F'}$, $\overline{GH} \cong \overline{G'H'}$, and $\overline{HF} \cong \overline{H'F'}$.
Use a protractor to confirm that the corresponding angles are congruent.
Therefore, $\triangle GHF \cong \triangle G'H'F'$.
$\triangle G'H'F'$ is a slide of $\triangle GHF$.

25. Use the figure and the Distance Formula to find the length of each side in the triangles.

$$JK = \sqrt{[-2-(-4)]^2 + (-3-3)^2} = \sqrt{4+36} \text{ or } \sqrt{40}$$
$$J'K' = \sqrt{(8-2)^2 + [-1-(-3)]^2} = \sqrt{36+4} \text{ or } \sqrt{40}$$
$$KL = \sqrt{[0-(-2)]^2 + [2-(-3)]^2} = \sqrt{4+25} \text{ or } \sqrt{29}$$
$$K'L' = \sqrt{(3-8)^2 + [1-(-1)]^2} = \sqrt{25+4} \text{ or } \sqrt{29}$$
$$JL = \sqrt{[0-(-4)]^2 + (2-3)^2} = \sqrt{16+1} \text{ or } \sqrt{17}$$
$$J'L' = \sqrt{(3-2)^2 + [1-(-3)]^2} = \sqrt{1+16} \text{ or } \sqrt{17}$$

The lengths of the corresponding sides of the two triangles are equal. Therefore, by the definition of congruence, $\overline{JK} \cong \overline{J'K'}$, $\overline{KL} \cong \overline{K'L'}$, and $\overline{JL} \cong \overline{J'L'}$.
Use a protractor to confirm that the corresponding angles are congruent.
Therefore, $\triangle JKL \cong \triangle J'K'L'$.
$\triangle J'K'L'$ is a turn of $\triangle JKL$.

26. False; $\angle A \cong \angle X$, $\angle B \cong \angle Y$, and $\angle C \cong \angle Z$ but the corresponding sides are not congruent.

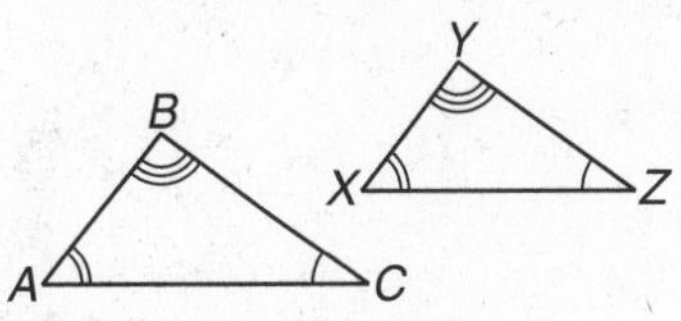

27. True;

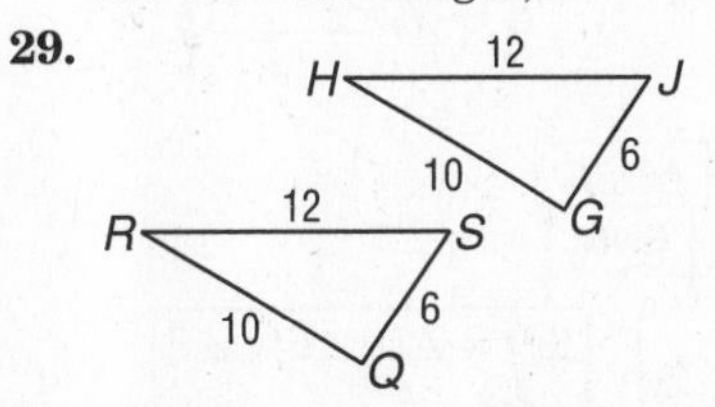

28. Both statements are correct because the spokes are the same length, $\overline{EA} \cong \overline{IA}$, and $\overline{AE} \cong \overline{AI}$.

29.

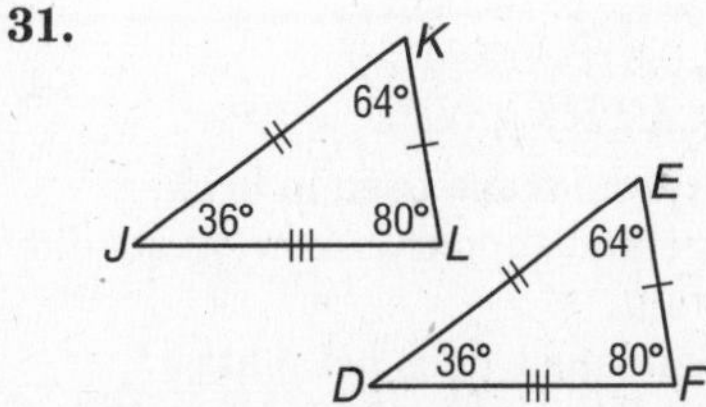

30. $\overline{HJ}$ corresponds to $\overline{RS}$, so $\overline{HJ} \cong \overline{RS}$.

$$2x - 4 = 12$$
$$2x = 16$$
$$x = 8$$

31.

K
64°
J 36° 80° L
E
64°
D 36° 80° F

32. $\angle D \cong \angle J$, so $m\angle D = m\angle J = 36$.

$$m\angle D + m\angle E + m\angle F = 180$$
$$36 + 64 + 3x + 52 = 180$$
$$3x + 152 = 180$$
$$3x = 28$$
$$x = \frac{28}{3}$$

33. Given: $\triangle RST \cong \triangle XYZ$
Prove: $\triangle XYZ \cong \triangle RST$

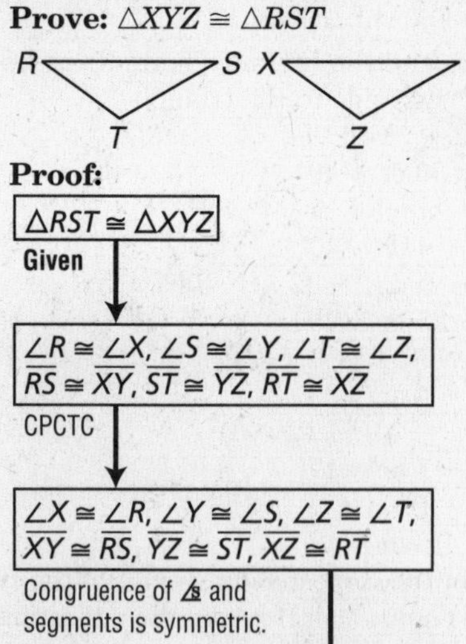

Proof:

34. a. Given
b. Given
c. Congruence of segments is reflexive.
d. Given
e. Def. of ⊥ lines
f. Given
g. Def. of ⊥ lines
h. All right ∡ are ≅.
i. Given
j. Alt. int. ∡ are ≅.
k. Given
l. Alt. int. ∡ are ≅.
m. Def. of ≅ Δs

35. Given: $\triangle DEF$
Prove: $\triangle DEF \cong \triangle DEF$

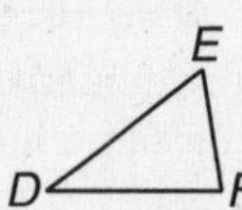

Proof:

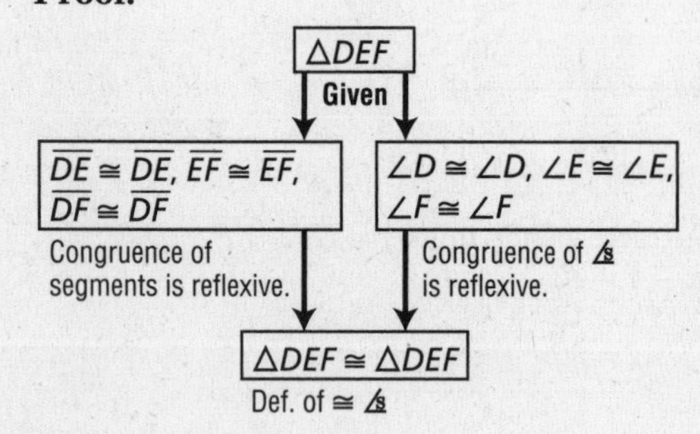

36. $\angle SMP \cong \angle TNP$, $\angle MPS \cong \angle NPT$

37. Sample answer: Triangles are used in bridge design for structure and support. Answers should include the following.
- The shape of the triangle does not matter.
- Some of the triangles used in the bridge supports seem to be congruent.

38. B; by the order of the vertices in the triangle names, $\overline{AC} \cong \overline{XZ}$

39. D; $DF = \sqrt{[3-(-5)]^2 + (-7-4)^2}$
$= \sqrt{64 + 121}$ or $\sqrt{185}$

Page 198 Maintain Your Skills

40. $x + 40 = 115$
$x = 75$

41. $x + 42 = 100$
$x = 58$

42. $x + x + 30 = 180$
$2x = 150$
$x = 75$

43. $\overline{BC} \cong \overline{CD}$, so $BC = CD$.
$2x + 4 = 10$
$2x = 6$
$x = 3$
$BC = 2x + 4$
$= 2(3) + 4$ or 10
$CD = 10$
$BD = x + 2$
$= 3 + 2$ or 5

44. $\triangle HKT$ is equilateral, so $\overline{HK} \cong \overline{HT}$ hence $HK = HT$.
$x + 7 = 4x - 8$
$15 = 3x$
$5 = x$
$HK = KT = HT = 4x - 8$
$= 4(5) - 8$ or 12

45. $m = \frac{(y_2 - y_1)}{(x_2 - x_1)}$
$= \frac{-3-3}{4-0}$ or $-\frac{3}{2}$
$y = mx + b$
$y = -\frac{3}{2}x + 3$

46. $y = mx + b$
$y = \frac{3}{4}x + 8$

47. $m = -4$
$y - y_1 = m(x - x_1)$
$y - 1 = -4[x - (-3)]$
$y - 1 = -4x - 12$
$y = -4x - 11$

48. $y - y_1 = m(x - x_1)$
$y - 2 = -4[x - (-3)]$
$y - 2 = -4x - 12$
$y = -4x - 10$

49. $d = \sqrt{(x_2 - x_1)^2 + (y_2 - y_1)^2}$
$= \sqrt{[1-(-1)]^2 + (6-7)^2}$
$= \sqrt{4+1}$ or $\sqrt{5}$

50. $d = \sqrt{(x_2 - x_1)^2 + (y_2 - y_1)^2}$
$= \sqrt{(4-8)^2 + (-2-2)^2}$
$= \sqrt{16+16}$ or $\sqrt{32}$

51. $d = \sqrt{(x_2 - x_1)^2 + (y_2 - y_1)^2}$
$= \sqrt{(5-3)^2 + (2-5)^2}$
$= \sqrt{4+9}$ or $\sqrt{13}$

Page 198 Practice Quiz 1

1. The segments $\overline{FJ}$, $\overline{GJ}$, $\overline{HJ}$, and $\overline{DJ}$ are all congruent. So $\triangle DFJ$, $\triangle GJF$, $\triangle HJG$, and $\triangle DJH$ are isosceles triangles because they each have a pair of congruent sides.

2. $\triangle ABC$ is equilateral, so all sides are congruent.
$2x = 4x - 7$
$-2x = -7$
$x = 3.5$

3. $AB = 2x$
$= 2(3.5)$ or 7
$BC = 4x - 7$
$= 4(3.5) - 7$ or 7
$AC = x + 3.5$
$= 3.5 + 3.5$ or 7

4. $m\angle 1 + 50 + 70 = 180$
$m\angle 1 + 120 = 180$
$m\angle 1 = 60$
$m\angle 2 = m\angle 1 + 50$
$= 60 + 50$ or 110
$m\angle 3 + m\angle 2 + 21 = 180$
$m\angle 3 + 110 + 21 = 180$
$m\angle 3 = 49$

5. $\angle M \cong \angle J, \angle N \cong \angle K, \angle P \cong \angle L;$
$\overline{MN} \cong \overline{JK}, \overline{NP} \cong \overline{KL}, \overline{MP} \cong \overline{JL}$

Page 199 Reading Mathematics

1. Sample answer: If side lengths are given, determine the number of congruent sides and name the triangle. Some isosceles triangles are equilateral triangles.
2. $\triangle ABC$ is obtuse because $m\angle C > 90$.
3. equiangular or equilateral

4-4 Proving Congruence—SSS, SAS

Pages 203–204 Check for Understanding

1. Sample answer: In $\triangle QRS$, $\angle R$ is the included angle of the sides $\overline{QR}$ and $\overline{RS}$.

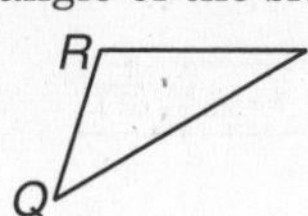

2. Jonathan; the measure of $\angle DEF$ is needed to use SAS.

3. $EG = \sqrt{[-2 - (-4)]^2 + [-3 - (-3)]^2}$
$= \sqrt{4 + 0}$ or 2
$MP = \sqrt{(2 - 4)^2 + [-3 - (-3)]^2}$
$= \sqrt{4 + 0}$ or 2
$FG = \sqrt{[-2 - (-2)]^2 + (-3 - 1)^2}$
$= \sqrt{0 + 16}$ or 4
$NP = \sqrt{(2 - 2)^2 + (-3 - 1)^2}$
$= \sqrt{0 + 16}$ or 4
$EF = \sqrt{[-2 - (-4)]^2 + [1 - (-3)]^2}$
$= \sqrt{4 + 16}$ or $\sqrt{20}$
$MN = \sqrt{(2 - 4)^2 + [1 - (-3)]^2}$
$= \sqrt{4 + 16}$ or $\sqrt{20}$
$EG = MP$, $FG = NP$, and $EF = MN$. The corresponding sides have the same measure and are congruent. $\triangle EFG \cong \triangle MNP$ by SSS.

4. $EG = \sqrt{[-3 - (-2)]^2 + [1 - (-2)]^2}$
$= \sqrt{1 + 9}$ or $\sqrt{10}$
$MP = \sqrt{(3 - 2)^2 + (1 - 2)^2}$
$= \sqrt{1 + 1}$ or $\sqrt{2}$
$FG = \sqrt{[-3 - (-4)]^2 + (1 - 6)^2}$
$= \sqrt{1 + 25}$ or $\sqrt{26}$
$NP = \sqrt{(3 - 4)^2 + (1 - 6)^2}$
$= \sqrt{1 + 25}$ or $\sqrt{26}$
$EF = \sqrt{[-4 - (-2)]^2 + [6 - (-2)]^2}$
$= \sqrt{4 + 64}$ or $\sqrt{68}$
$MN = \sqrt{(4 - 2)^2 + (6 - 2)^2}$
$= \sqrt{4 + 16}$ or $\sqrt{20}$
The corresponding sides are not congruent, so the triangles are not congruent.

5. **Given:** $\overline{DE}$ and $\overline{BC}$ bisect each other.
Prove: $\triangle DGB \cong \triangle EGC$

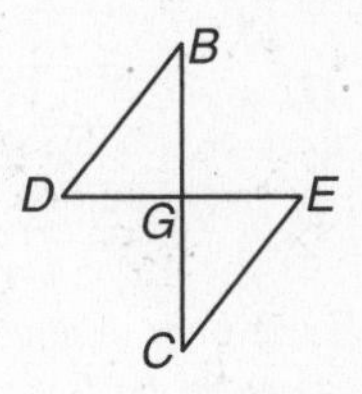

Proof:

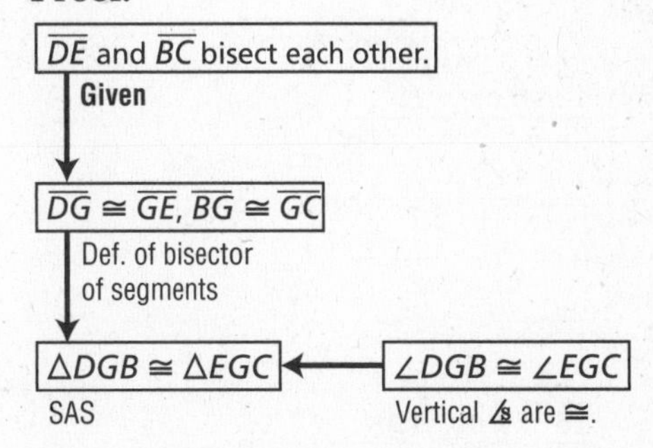

6. **Given:** $\overline{KM} \parallel \overline{JL}, \overline{KM} \cong \overline{JL}$
Prove: $\triangle JKM \cong \triangle MLJ$

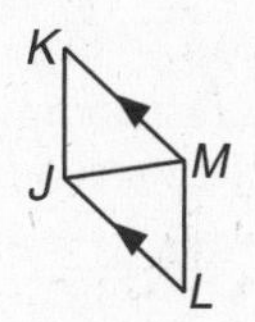

Proof:

Statements	Reasons
1. $\overline{KM} \parallel \overline{JL}, \overline{KM} \cong \overline{JL}$	1. Given
2. $\angle KMJ \cong \angle LJM$	2. Alt. int. ∠s are ≅.
3. $\overline{JM} \cong \overline{JM}$	3. Reflexive Property
4. $\triangle JKM \cong \triangle MLJ$	4. SAS

7. The triangles have two pairs of sides and the included angles congruent, so the triangles are congruent by the SAS postulate.
8. Each pair of corresponding sides are congruent, so the triangles are congruent by the SSS postulate.
9. **Given:** T is the midpoint of $\overline{SQ}$.
$\overline{SR} \cong \overline{QR}$
Prove: $\triangle SRT \cong \triangle QRT$

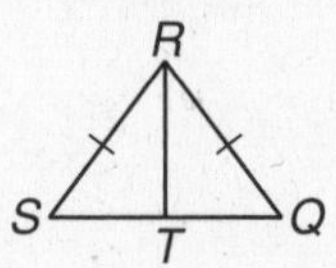

Proof:

Statements	Reasons
1. T is the midpoint of $\overline{SQ}$.	1. Given
2. $\overline{ST} \cong \overline{TQ}$	2. Midpoint Theorem
3. $\overline{SR} \cong \overline{QR}$	3. Given
4. $\overline{RT} \cong \overline{RT}$	4. Reflexive Property
5. $\triangle SRT \cong \triangle QRT$	5. SSS

Pages 204–206 Practice and Apply

10. $JK = \sqrt{[-7-(-3)]^2 + (4-2)^2}$
$= \sqrt{16+4}$ or $\sqrt{20}$
$FG = \sqrt{(4-2)^2 + (7-3)^2}$
$= \sqrt{4+16}$ or $\sqrt{20}$
$KL = \sqrt{[-1-(-7)]^2 + (9-4)^2}$
$= \sqrt{36+25}$ or $\sqrt{61}$
$GH = \sqrt{(9-4)^2 + (1-7)^2}$
$= \sqrt{25+36}$ or $\sqrt{61}$
$JL = \sqrt{[-1-(-3)]^2 + (9-2)^2}$
$= \sqrt{4+49}$ or $\sqrt{53}$
$FH = \sqrt{(9-2)^2 + (1-3)^2}$
$= \sqrt{49+4}$ or $\sqrt{53}$
Each pair of corresponding sides has the same measure so they are congruent. $\triangle JKL \cong \triangle FGH$ by SSS.

11. $JK = \sqrt{[-2-(-1)]^2 + (-2-1)^2}$
$= \sqrt{1+9}$ or $\sqrt{10}$
$FG = \sqrt{(3-2)^2 + [-2-(-1)]^2}$
$= \sqrt{1+1}$ or $\sqrt{2}$
$KL = \sqrt{[-5-(-2)]^2 + [-1-(-2)]^2}$
$= \sqrt{9+1}$ or $\sqrt{10}$
$GH = \sqrt{(2-3)^2 + [5-(-2)]^2}$
$= \sqrt{1+49}$ or $\sqrt{50}$
$JL = \sqrt{[-5-(-1)]^2 + (-1-1)^2}$
$= \sqrt{16+4}$ or $\sqrt{20}$
$FH = \sqrt{(2-2)^2 + [5-(-1)]^2}$
$= \sqrt{0+36}$ or 6
The corresponding sides are not congruent, so $\triangle JKL$ is not congruent to $\triangle FGH$.

12. $JK = \sqrt{[0-(-1)]^2 + [6-(-1)]^2}$
$= \sqrt{1+49}$ or $\sqrt{50}$
$FG = \sqrt{(5-3)^2 + (3-1)^2}$
$= \sqrt{4+4}$ or $\sqrt{8}$
$KL = \sqrt{(2-0)^2 + (3-6)^2}$
$= \sqrt{4+9}$ or $\sqrt{13}$
$GH = \sqrt{(8-5)^2 + (1-3)^2}$
$= \sqrt{9+4}$ or $\sqrt{13}$
$JL = \sqrt{[2-(-1)]^2 + [3-(-1)]^2}$
$= \sqrt{9+16}$ or 5
$FH = \sqrt{(8-3)^2 + (1-1)^2}$
$= \sqrt{25+0}$ or 5
The corresponding sides are not congruent, so $\triangle JKL$ is not congruent to $\triangle FGH$.

13. $JK = \sqrt{(4-3)^2 + (6-9)^2}$
$= \sqrt{1+9}$ or $\sqrt{10}$
$FG = \sqrt{(2-1)^2 + (4-7)^2}$
$= \sqrt{1+9}$ or $\sqrt{10}$
$KL = \sqrt{(1-4)^2 + (5-6)^2}$
$= \sqrt{9+1}$ or $\sqrt{10}$
$GH = \sqrt{(-1-2)^2 + (3-4)^2}$
$= \sqrt{9+1}$ or $\sqrt{10}$
$JL = \sqrt{(1-3)^2 + (5-9)^2}$
$= \sqrt{4+16}$ or $\sqrt{20}$
$FH = \sqrt{(-1-1)^2 + (3-7)^2}$
$= \sqrt{4+16}$ or $\sqrt{20}$
The corresponding sides have the same measure so they are congruent. $\triangle JKL \cong \triangle FGH$ by SSS.

14. Given: $\overline{AE} \cong \overline{FC}, \overline{AB} \cong \overline{BC}, \overline{BE} \cong \overline{BF}$
Prove: $\triangle AFB \cong \triangle CEB$

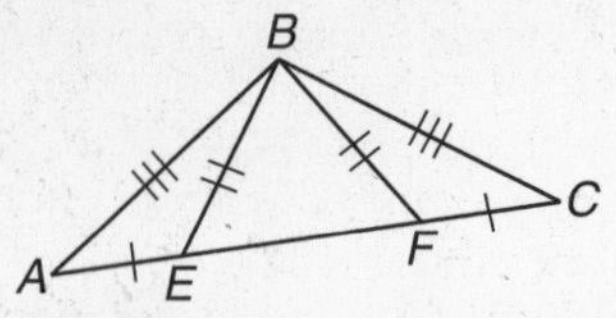

Proof:

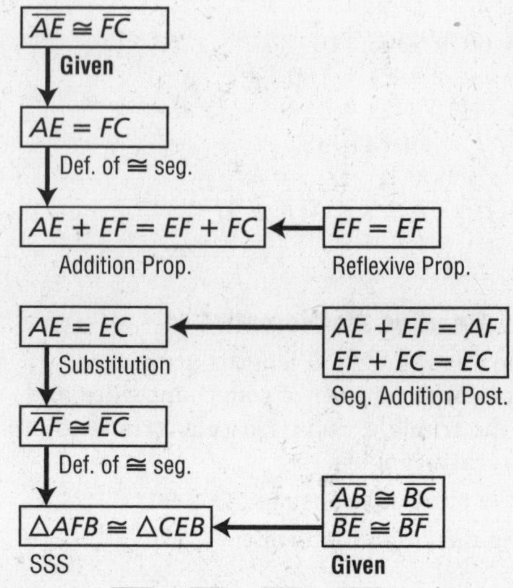

15. Given: $\overline{RQ} \cong \overline{TQ} \cong \overline{YQ} \cong \overline{WQ}$
$\angle RQY \cong \angle WQT$
Prove: $\triangle QWT \cong \triangle QYR$

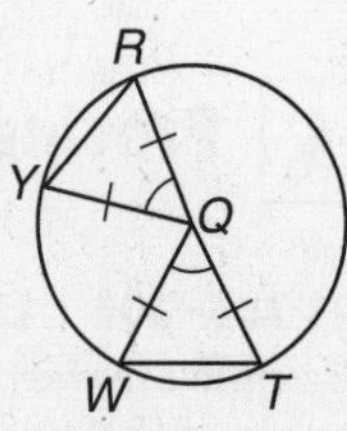

Proof:

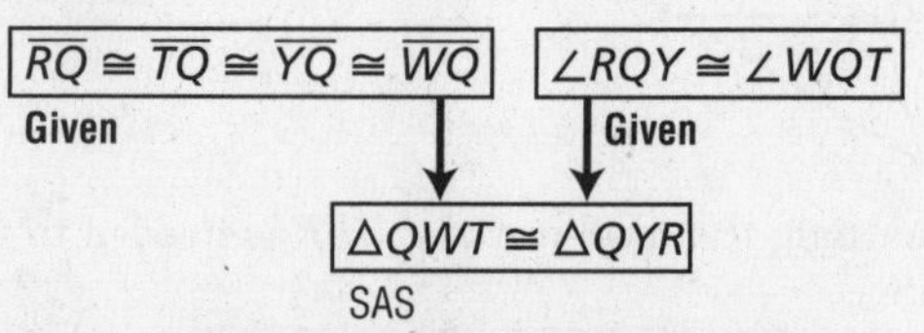

16. Given: $\triangle CDE$ is an isosceles triangle.
G is the midpoint of $\overline{CE}$.
Prove: $\triangle CDG \cong \triangle EDG$

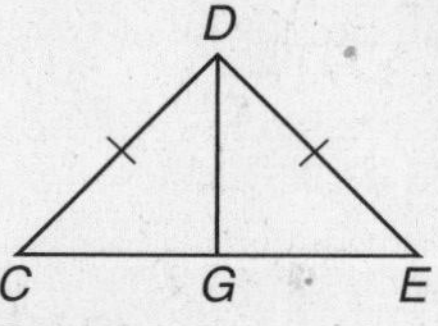

Proof:

Statements	Reasons
1. $\triangle CDE$ is an isosceles triangle, G is the midpoint of $\overline{CE}$.	1. Given
2. $\overline{CD} \cong \overline{DE}$	2. Def. of isos. $\triangle$
3. $\overline{CG} \cong \overline{GE}$	3. Midpoint Th.
4. $\overline{DG} \cong \overline{DG}$	4. Reflexive Property
5. $\triangle CDG \cong \triangle EDG$	5. SSS

17. Given: $\triangle MRN \cong \triangle QRP$, $\angle MNP \cong \angle QPN$
Prove: $\triangle MNP \cong \triangle QPN$

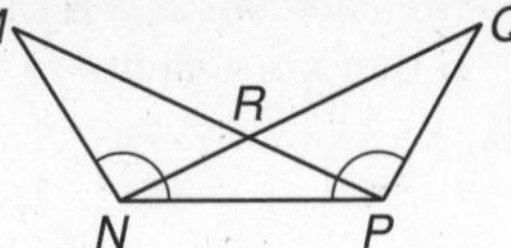

Proof:

Statements	Reasons
1. $\triangle MRN \cong \triangle QRP$, $\angle MNP \cong \angle QPN$	1. Given
2. $\overline{MN} \cong \overline{QP}$	2. CPCTC
3. $\overline{NP} \cong \overline{NP}$	3. Reflexive Property
4. $\triangle MNP \cong \triangle QPN$	4. SAS

18. Given: $\overline{AC} \cong \overline{GC}$
$\overline{EC}$ bisects $\overline{AG}$.
Prove: $\triangle GEC \cong \triangle AEC$

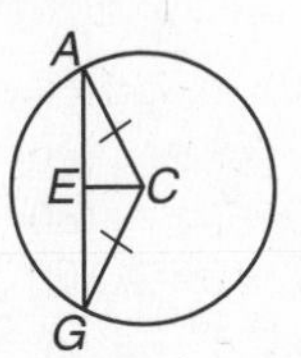

Proof:

Statements	Reasons
1. $\overline{AC} \cong \overline{GC}, \overline{EC}$ bisects $\overline{AG}$.	1. Given
2. $\overline{AE} \cong \overline{EG}$	2. Def. of segment bisector
3. $\overline{EC} \cong \overline{EC}$	3. Reflexive Property
4. $\triangle GEC \cong \triangle AEC$	4. SSS

19. Given: $\triangle GHJ \cong \triangle LKJ$
Prove: $\triangle GHL \cong \triangle LKG$

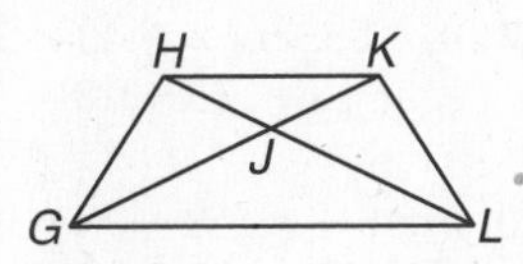

Proof:

Statements	Reasons
1. $\triangle GHJ \cong \triangle LKJ$	1. Given
2. $\overline{HJ} \cong \overline{KJ}, \overline{GJ} \cong \overline{LJ}, \overline{GH} \cong \overline{LK}$	2. CPCTC
3. $HJ = KJ, GJ = LJ$	3. Def. of $\cong$ segments
4. $HJ + LJ = KJ + JG$	4. Addition Property
5. $KJ + GJ = KG$; $HJ + LJ = HL$	5. Segment Addition
6. $KG = HL$	6. Substitution
7. $\overline{KG} \cong \overline{HL}$	7. Def. of $\cong$ segments
8. $\overline{GL} \cong \overline{GL}$	8. Reflexive Property
9. $\triangle GHL \cong \triangle LKG$	9. SSS

20. Given: $\overline{RS} \cong \overline{PN}$
$\overline{RT} \cong \overline{MP}$
$\angle S \cong \angle N$
$\angle T \cong \angle M$
Prove: $\triangle RST \cong \triangle PNM$

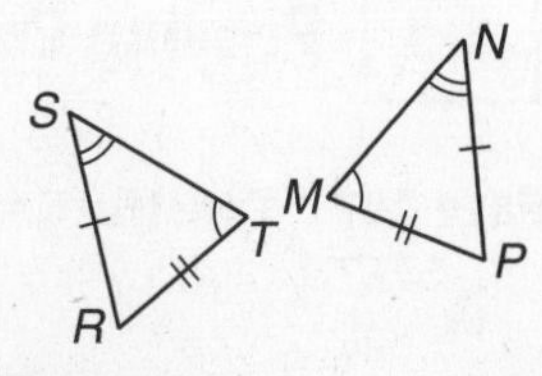

Proof:

Statements	Reasons
1. $\overline{RS} \cong \overline{PN}, \overline{RT} \cong \overline{MP}$	1. Given
2. $\angle S \cong \angle N$, and $\angle T \cong \angle M$	2. Given
3. $\angle R \cong \angle P$	3. Third Angle Theorem
4. $\triangle RST \cong \triangle PNM$	4. SAS

21. Given: $\overline{EF} \cong \overline{HF}$
G is the midpoint of $\overline{EH}$.
Prove: $\triangle EFG \cong \triangle HFG$

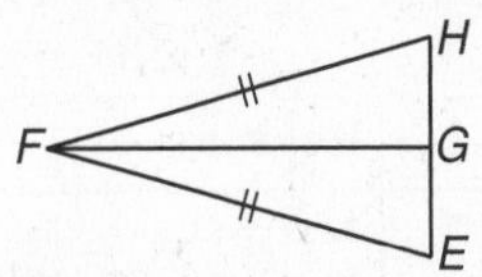

Proof:

Statements	Reasons
1. $\overline{EF} \cong \overline{HF}$; G is the midpoint of $\overline{EH}$.	1. Given
2. $\overline{EG} \cong \overline{GH}$	2. Midpoint Theorem
3. $\overline{FG} \cong \overline{FG}$	3. Reflexive Property
4. $\triangle EFG \cong \triangle HFG$	4. SSS

22. Each pair of corresponding sides is congruent. The triangles are congruent by the SSS Postulate.

23. The triangles have two pairs of corresponding sides congruent and one pair of angles congruent but what is needed is the pair of included angles to be congruent. It is *not* possible to prove the triangles are congruent.

24. The triangles have one pair of angles congruent and one pair of sides (the shared side) congruent. It is *not* possible to prove the triangles are congruent.

25. The triangles have three pairs of corresponding sides congruent and one pair of corresponding angles congruent. The triangles are congruent by the SSS or SAS Postulates.

26. Given: $\overline{TS} \cong \overline{SF} \cong \overline{FH} \cong \overline{HT}$
$\angle TSF$, $\angle SFH$, $\angle FHT$, and $\angle HTS$ are right angles.
Prove: $\overline{HS} \cong \overline{TF}$

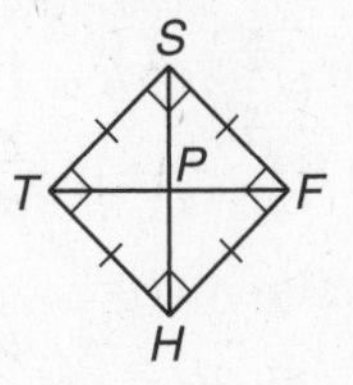

Proof:

Statements	Reasons
1. $\overline{TS} \cong \overline{SF} \cong \overline{FH} \cong \overline{HT}$	1. Given
2. $\angle TSF$, $\angle SFH$, $\angle FHT$, and $\angle HTS$ are right	2. Given
3. $\angle STH \cong \angle THF$	3. All right $\angle$s are $\cong$.
4. $\triangle STH \cong \triangle THF$	4. SAS
5. $\overline{HS} \cong \overline{TF}$	5. CPCTC

27. Given: $\overline{TS} \cong \overline{SF} \cong \overline{FH} \cong \overline{HT}$
$\angle TSF$, $\angle SFH$, $\angle FHT$, and $\angle HTS$ are right angles.
Prove: $\angle SHT \cong \angle SHF$

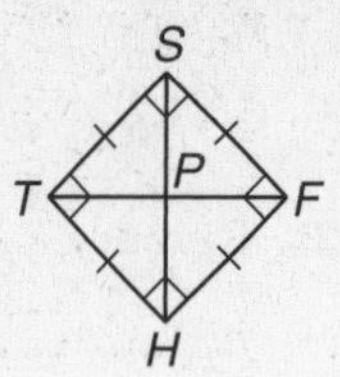

Proof:

Statements	Reasons
1. $\overline{TS} \cong \overline{SF} \cong \overline{FH} \cong \overline{HT}$	1. Given
2. $\angle TSF$, $\angle SFH$, $\angle FHT$, and $\angle HTS$ are right angles.	2. Given
3. $\angle STH \cong \angle SFH$	3. All rt. ∠s are ≅.
4. $\triangle STH \cong \triangle SFH$	4. SAS
5. $\angle SHT \cong \angle SHF$	5. CPCTC

28. Given: $\overline{DE} \cong \overline{FB}$, $\overline{AE} \cong \overline{FC}$, $\overline{AE} \perp \overline{DB}$, $\overline{CF} \perp \overline{DB}$
Prove: $\triangle ABD \cong \triangle CDB$

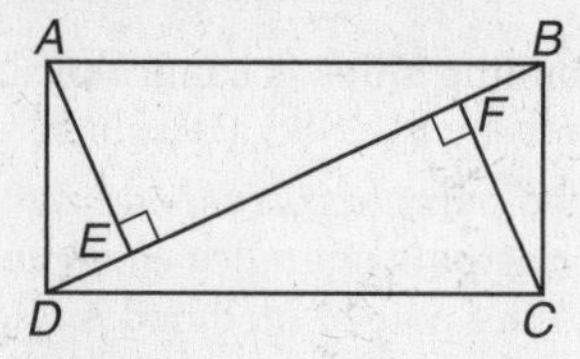

Plan: First use SAS to show that $\triangle ADE \cong \triangle CBF$. Next use CPCTC and Reflexive Property for segments to show $\triangle ABD \cong \triangle CDB$.
Proof:

Statements	Reasons
1. $\overline{DE} \cong \overline{FB}$, $\overline{AE} \cong \overline{FC}$	1. Given
2. $\overline{AE} \perp \overline{DB}$, $\overline{CF} \perp \overline{DB}$	2. Given
3. $\angle AED$ is a right angle. $\angle CFB$ is a right angle.	3. ⊥ lines form right ∠s.
4. $\angle AED \cong \angle CFB$	4. All right angles are ≅.
5. $\triangle ADE \cong \triangle CBF$	5. SAS
6. $\overline{AD} \cong \overline{BC}$	6. CPCTC
7. $\overline{DB} \cong \overline{DB}$	7. Reflexive Property for Segments
8. $\angle CBD \cong \angle ADB$	8. CPCTC
9. $\triangle ABD \cong \triangle CDB$	9. SAS

29. Sample answer: The properties of congruent triangles help land surveyors double check measurements. Answers should include the following.
- If each pair of corresponding angles and sides are congruent, the triangles are congruent by definition. If two pairs of corresponding sides and the included angle are congruent, the triangles are congruent by SAS. If each pair of corresponding sides are congruent, the triangles are congruent by SSS.
- Sample answer: Architects also use congruent triangles when designing buildings.

30. C; using vertical angles and exterior angles, $a + b = 90$. Given this fact, it is impossible for the other statements to be true.

31. B; $3x + 6x + 7x = 180$
$16x = 180$
$x = 11.25$
$3x = 3(11.25)$ or 33.75
$6x = 6(11.25)$ or 67.5
$7x = 7(11.25)$ or 78.75
Because the angles are all less than 90°, the triangle is acute.

Page 206 Maintain Your Skills

32. $\triangle ACB \cong \triangle DCE$
33. $\triangle WXZ \cong \triangle YXZ$
34. $\triangle LMP \cong \triangle NPM$
35. $m\angle 2 = 78$
36. $m\angle 3 + m\angle 2 = 180$
$m\angle 3 + 78 = 180$
$m\angle 3 = 102$
37. $m\angle 4 + m\angle 5 = 90$
$m\angle 5 = 90 - m\angle 4$
$m\angle 4 + m\angle 3 + 56 = 180$
$m\angle 4 + 102 + 56 = 180$
$m\angle 4 = 22$
$m\angle 5 = 90 - m\angle 4$
$= 90 - 22$ or 68
38. $m\angle 4 + m\angle 3 + 56 = 180$
$m\angle 4 + 102 + 56 = 180$
$m\angle 4 = 22$
39. $m\angle 1 + m\angle 2 + 43 = 180$
$m\angle 1 + 78 + 43 = 180$
$m\angle 1 = 59$
40. $m\angle 6 + m\angle 5 + 78 = 180$
$m\angle 6 + 68 + 78 = 180$
$m\angle 6 = 34$
41. rate of change $= \frac{0.3 - 1.3}{2 - 1}$
$= -1$
42. rate of change $= \frac{-1.1 - 0.3}{3 - 2}$
$= -1.4$
43. There is a steeper rate of decline from the second quarter to the third.
44. $\overline{BE}$, since $\overrightarrow{AE}$ bisects $\overline{BC}$
45. $\angle CBD$, since $\overrightarrow{BD}$ bisects $\angle ABC$
46. $\angle BDA$, since $\angle BDC$ is a right angle and forms a linear pair with $\angle BDA$
47. $\overline{CD}$, since $\overrightarrow{BD}$ bisects $\overline{AC}$

4-5 Proving Congruence—ASA, AAS

Page 207 Construction

5. $\triangle JKL \cong \triangle ABC$

Page 208 Geometry Activity

1. They are congruent.
2. The triangles are congruent.

Page 210 Check for Understanding

1. Two triangles can have corresponding congruent angles without corresponding congruent sides. $\angle A \cong \angle D$, $\angle B \cong \angle E$, and $\angle C \cong \angle F$. However, $\overline{AB} \ncong \overline{DE}$, so $\triangle ABC \ncong \triangle DEF$.

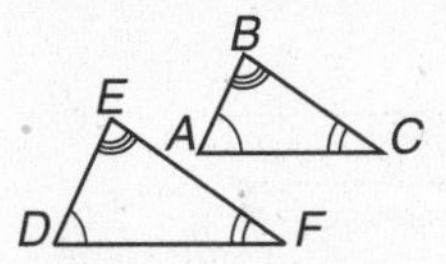

2. Sample answer: In $\triangle ABC$, $\overline{AB}$ is the included side of $\angle A$ and $\angle B$.

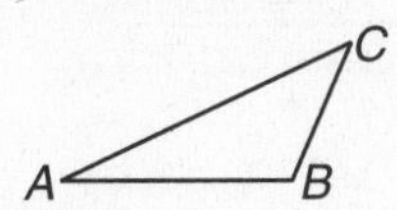

3. AAS can be proven using the Third Angle Theorem. Postulates are accepted as true without proof.

4. **Given:** $\overline{GH} \parallel \overline{KJ}$, $\overline{GK} \parallel \overline{HJ}$
Prove: $\triangle GJK \cong \triangle JGH$

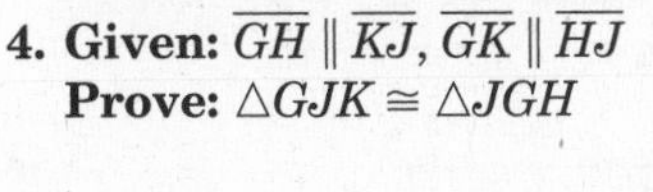

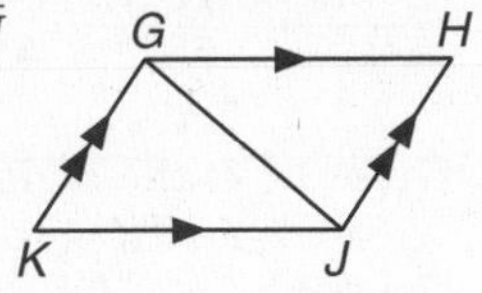

Proof:

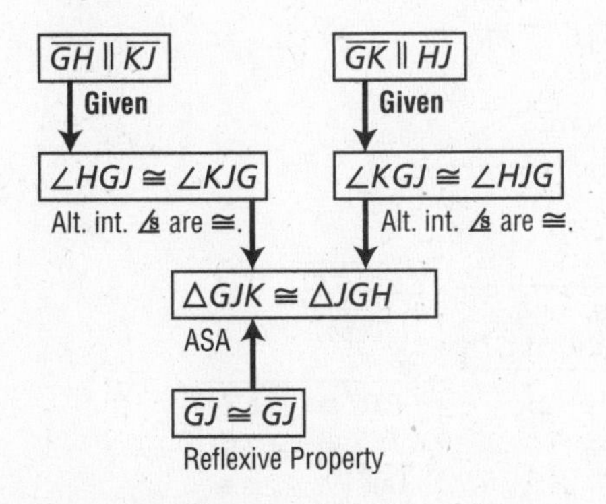

5. **Given:** $\overline{XW} \parallel \overline{YZ}$, $\angle X \cong \angle Z$
Prove: $\triangle WXY \cong \triangle YZW$

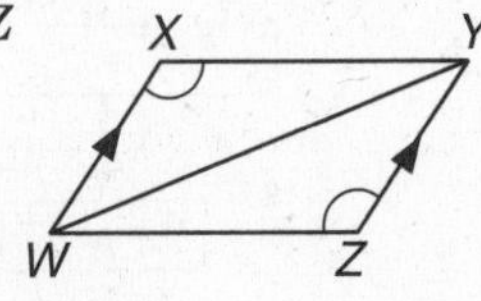

Proof:

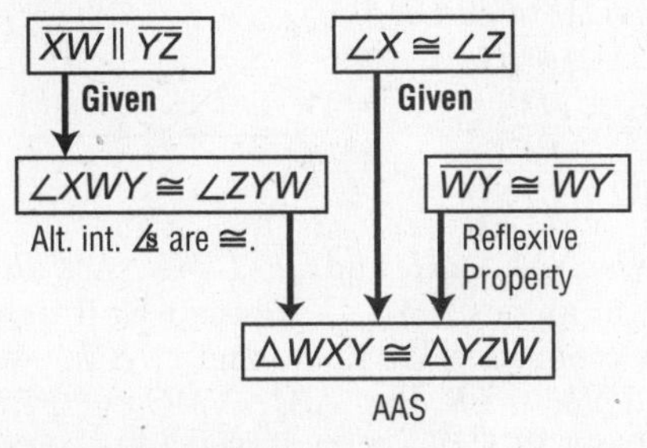

6. **Given:** $\overline{QS}$ bisects $\angle RST$; $\angle R \cong \angle T$.
Prove: $\triangle QRS \cong \triangle QTS$

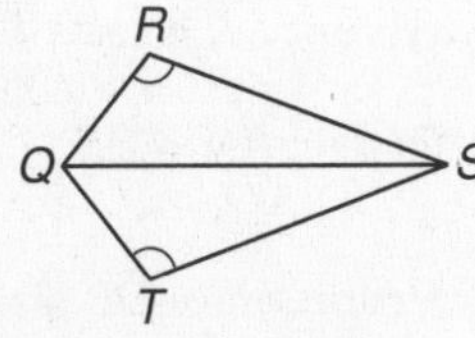

Proof: We are given that $\angle R \cong \angle T$ and $\overline{QS}$ bisects $\angle RST$, so by definition of angle bisector, $\angle RSQ \cong \angle TSQ$. By the Reflexive Property, $\overline{QS} \cong \overline{QS}$. $\triangle QRS \cong \triangle QTS$ by AAS.

7. **Given:** $\angle E \cong \angle K$, $\angle DGH \cong \angle DHG$, $\overline{EG} \cong \overline{KH}$
Prove: $\triangle EGD \cong \triangle KHD$

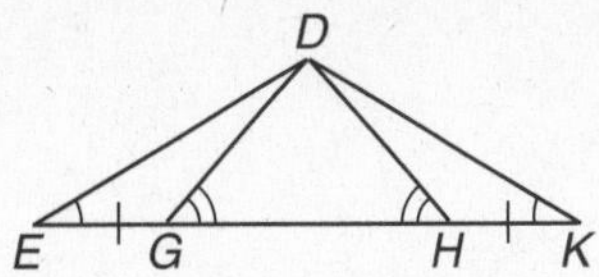

Proof: Since $\angle EGD$ and $\angle DGH$ are linear pairs, the angles are supplementary. Likewise, $\angle KHD$ and $\angle DHG$ are supplementary. We are given that $\angle DGH \cong \angle DHG$. Angles supplementary to congruent angles are congruent so $\angle EGD \cong \angle KHD$. Since we are given that $\angle E \cong \angle K$ and $\overline{EG} \cong \overline{KH}$, $\triangle EGD \cong \triangle KHD$ by ASA.

8. This cannot be determined. The information given cannot be used with any of the triangle congruence postulates, theorems or the definition of congruent triangles. By construction, two different triangles can be shown with the given information. Therefore, it cannot be determined if $\triangle SRT \cong \triangle MKL$.

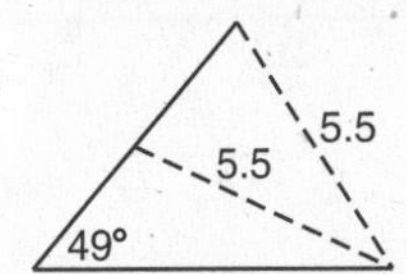

Pages 211–213 Practice and Apply

9. **Given:** $\overline{EF} \parallel \overline{GH}$, $\overline{EF} \cong \overline{GH}$
Prove: $\overline{EK} \cong \overline{KH}$

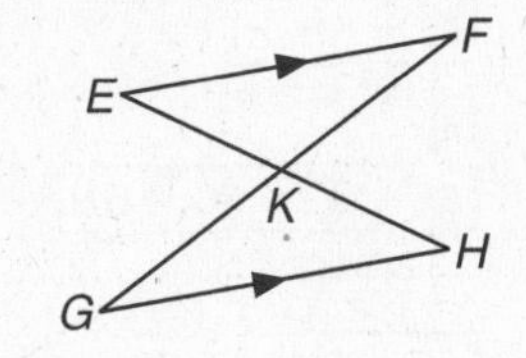

Proof:

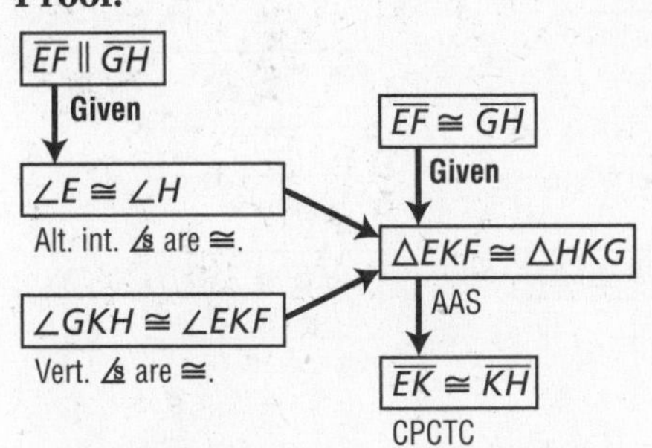

10. **Given:** $\overline{DE} \parallel \overline{JK}$, $\overline{DK}$ bisects $\overline{JE}$.
Prove: $\triangle EGD \cong \triangle JGK$

J D G K E

Proof:

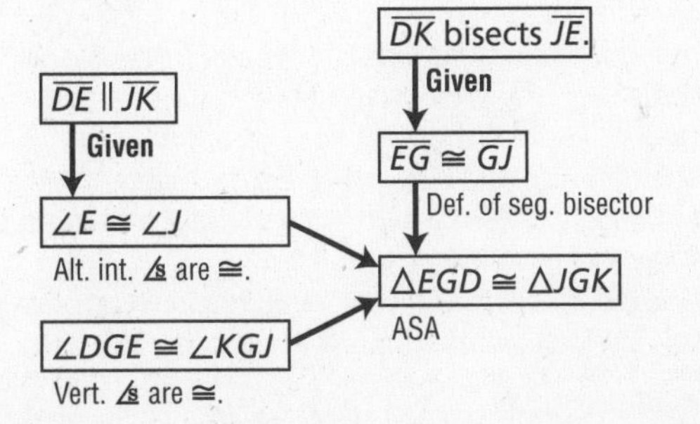

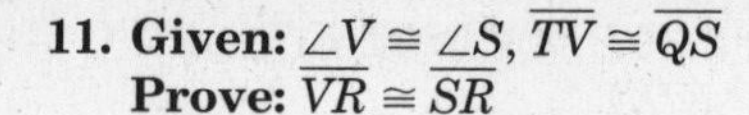

11. Given: $\angle V \cong \angle S$, $\overline{TV} \cong \overline{QS}$
Prove: $\overline{VR} \cong \overline{SR}$

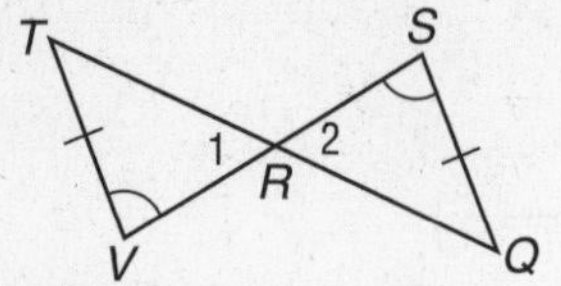

Proof:

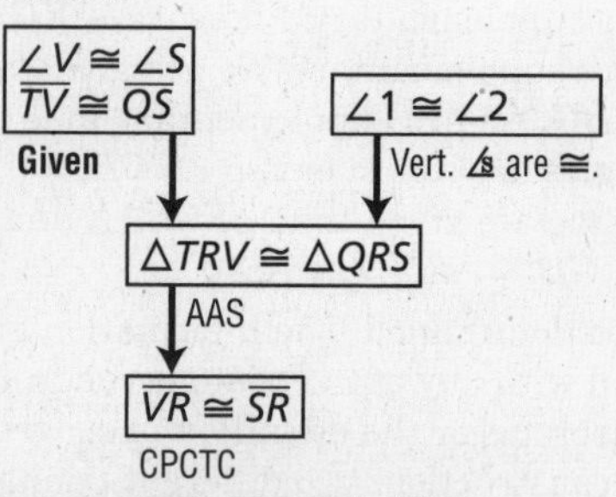

12. Given: $\overline{EJ} \parallel \overline{FK}$, $\overline{JG} \parallel \overline{KH}$
$\overline{EF} \cong \overline{GH}$
Prove: $\triangle EJG \cong \triangle FKH$

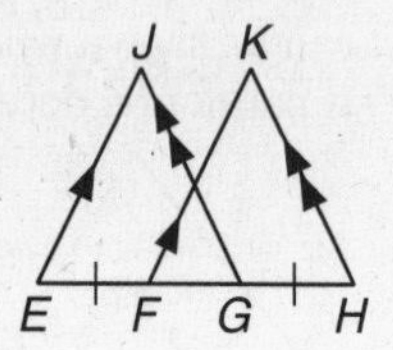

Proof:

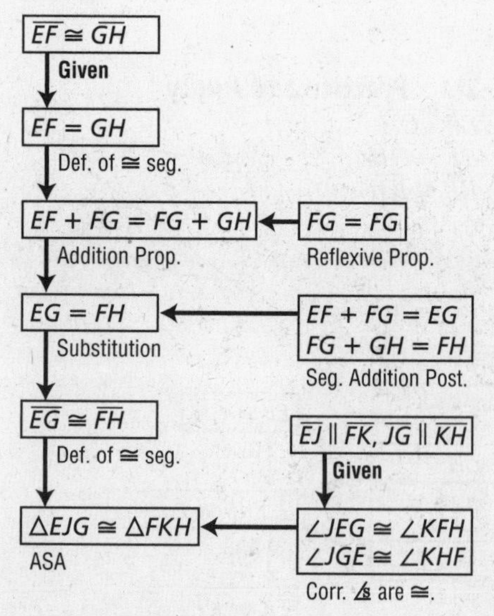

13. Given: $\overline{MN} \cong \overline{PQ}$,
$\angle M \cong \angle Q$
$\angle 2 \cong \angle 3$
Prove: $\triangle MLP \cong \triangle QLN$

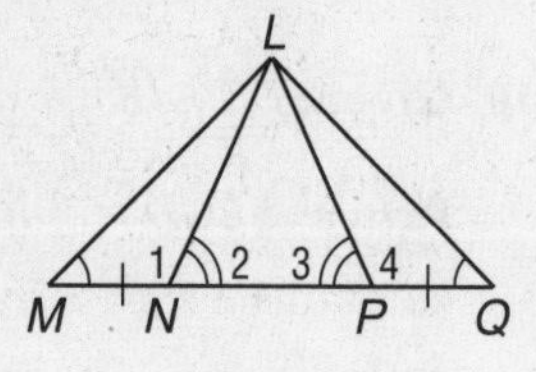

Proof:

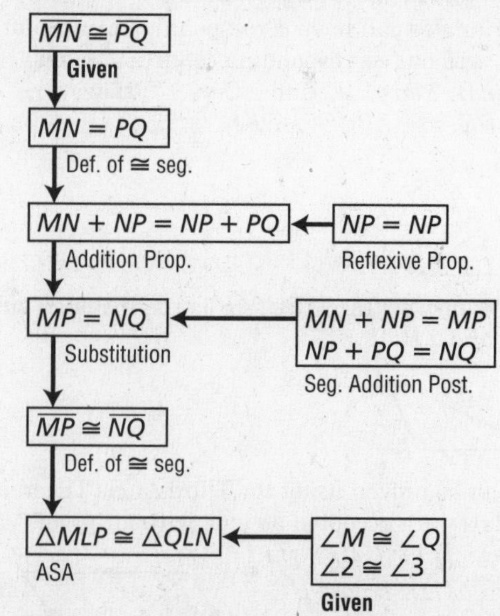

14. Given: Z is the midpoint of $\overline{CT}$.
$\overline{CY} \parallel \overline{TE}$
Prove: $\overline{YZ} \cong \overline{EZ}$

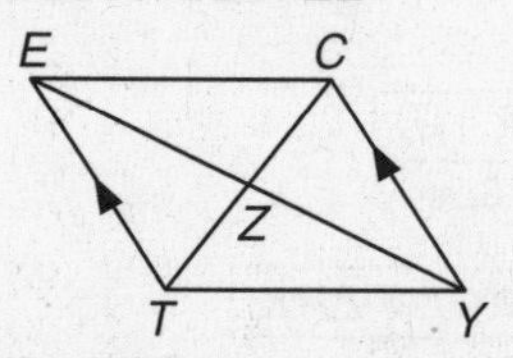

Proof:

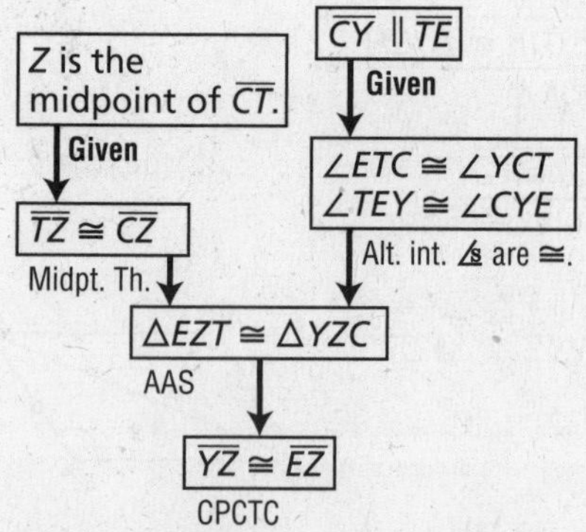

15. Given: $\angle NOM \cong \angle POR$,
$\overline{NM} \perp \overline{MR}$,
$\overline{PR} \perp \overline{MR}$,
$\overline{NM} \cong \overline{PR}$
Prove: $\overline{MO} \cong \overline{OR}$

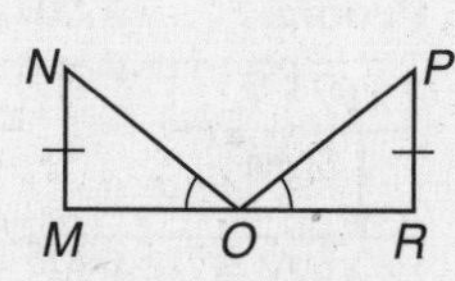

Proof: Since $\overline{NM} \perp \overline{MR}$ and $\overline{PR} \perp \overline{MR}$, $\angle M$ and $\angle R$ are right angles. $\angle M \cong \angle R$ because all right angles are congruent. We know that $\angle NOM \cong \angle POR$ and $\overline{NM} \cong \overline{PR}$. By AAS, $\triangle NMO \cong \triangle PRO$. $\overline{MO} \cong \overline{OR}$ by CPCTC.

16. Given: $\overline{DL}$ bisects $\overline{BN}$.
$\angle XLN \cong \angle XDB$
Prove: $\overline{LN} \cong \overline{DB}$

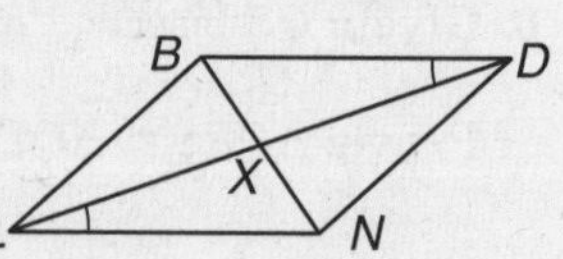

Proof: Since $\overline{DL}$ bisects $\overline{BN}$, $\overline{BX} \cong \overline{XN}$. $\angle XLN \cong \angle XDB$. $\angle LXN \cong \angle DXB$ because vertical angles are congruent. $\triangle LXN \cong \triangle DXB$ by AAS. $\overline{LN} \cong \overline{DB}$ by CPCTC.

17. Given: $\angle F \cong \angle J$, $\angle E \cong \angle H$, $\overline{EC} \cong \overline{GH}$
Prove: $\overline{EF} \cong \overline{HJ}$

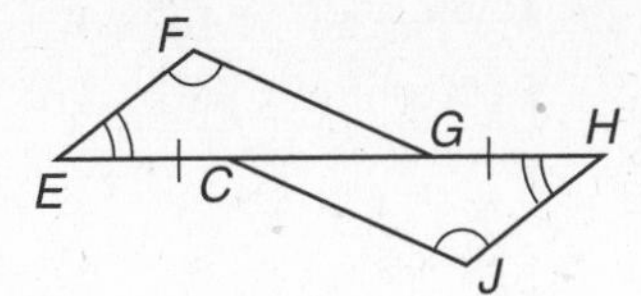

Proof: We are given that $\angle F \cong \angle J$, $\angle E \cong \angle H$, and $\overline{EC} \cong \overline{GH}$. By the Reflexive Property, $\overline{CG} \cong \overline{CG}$. Segment addition results in $EG = EC + CG$ and $CH = CG + GH$. By the definition of congruence, $EC = GH$ and $CG = CG$. Substitute to find $EG = CH$. By AAS, $\triangle EFG \cong \triangle HJC$. By CPCTC, $\overline{EF} \cong \overline{HJ}$.

18. Given: $\overline{TX} \parallel \overline{SY}$, $\angle TXY \cong \angle TSY$
Prove: $\triangle TSY \cong \triangle YXT$

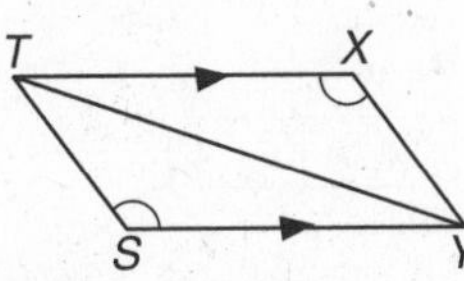

Proof: Since $\overline{TX} \parallel \overline{SY}$, $\angle YTX \cong \angle TYS$ by Alternate Interior Angles Theorem. $\overline{TY} \cong \overline{TY}$ by the Reflexive Property. Given $\angle TXY \cong \angle TSY$, $\triangle TSY \cong \triangle YXT$ by AAS.

19. Given: $\angle MYT \cong \angle NYT$, $\angle MTY \cong \angle NTY$
Prove: $\triangle RYM \cong \triangle RYN$

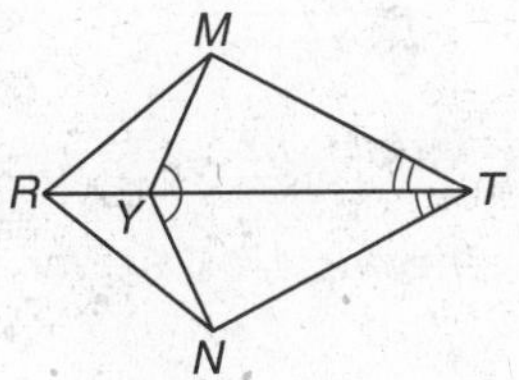

Proof:

Statements	Reasons
1. $\angle MYT \cong \angle NYT$, $\angle MTY \cong \angle NTY$	1. Given
2. $\overline{YT} \cong \overline{YT}$, $\overline{RY} \cong \overline{RY}$	2. Reflexive Property
3. $\triangle MYT \cong \triangle NYT$	3. ASA
4. $\overline{MY} \cong \overline{NY}$	4. CPCTC
5. $\angle RYM$ and $\angle MYT$ are a linear pair; $\angle RYN$ and $\angle NYT$ are a linear pair	5. Def. of linear pair
6. $\angle RYM$ and $\angle MYT$ are supplementary and $\angle RYN$ and $\angle NYT$ are supplementary.	6. Supplement Theorem
7. $\angle RYM \cong \angle RYN$	7. $\angle$s suppl. to $\cong$ $\angle$s are $\cong$.
8. $\triangle RYM \cong \triangle RYN$	8. SAS

20. Given: $\triangle BMI \cong \triangle KMT$, $\overline{IP} \cong \overline{PT}$
Prove: $\triangle IPK \cong \triangle TPB$

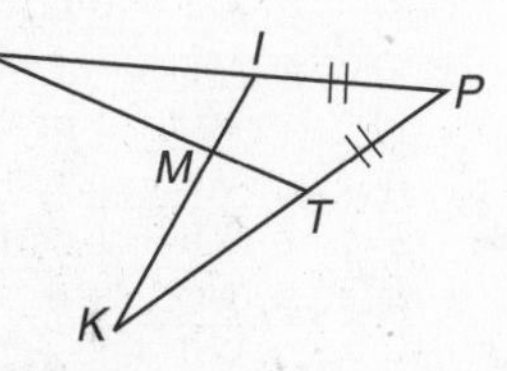

Proof:

Statements	Reasons
1. $\triangle BMI \cong \triangle KMT$	1. Given
2. $\angle B \cong \angle K$	2. CPCTC
3. $\overline{IP} \cong \overline{PT}$	3. Given
4. $\angle P \cong \angle P$	4. Reflexive Property
5. $\triangle IPK \cong \triangle TPB$	5. AAS

21. Explore: We are given the measurement of one side of each triangle. We need to determine whether two triangles are congruent.
Plan: $\overline{CD} \cong \overline{GH}$, because the segments have the same measure. $\angle CFD \cong \angle HFG$ because vertical angles are congruent. Since F is the midpoint of $\overline{DG}$, $\overline{DF} \cong \overline{FG}$.
Solve: We are given information about side-side-angle (SSA). This is not a method to prove two triangles congruent.
Examine: Use a compass, protractor, and ruler to draw a triangle with the given measurements. For simplicity of measurement, use 1 inch instead of 4 feet and 2 inches instead of 8 feet, so the measurements of the construction and those of the garden will be proportional.

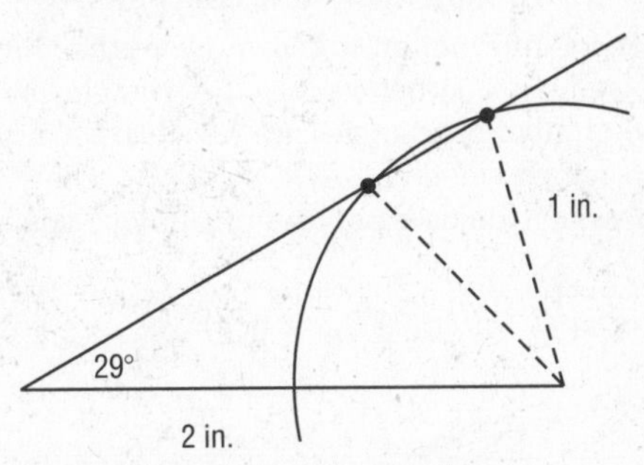

- Draw a segment 2 inches long.
- At one end, draw an angle of 29°. Extend the line longer than 2 inches.
- At the other end of the segment, draw an arc with a radius of 1 inch such that it intersects the line.

Notice that there are two possible segments that could determine the triangle. It cannot be determined whether $\triangle CFD \cong \triangle HFG$. The information given does not lead to a unique triangle.

22. Explore: We need to determine whether two triangles are congruent.
Plan: Since F is the midpoint of $\overline{DG}$, $\overline{DF} \cong \overline{FG}$. F is also the midpoint of $\overline{CH}$, so $\overline{CF} \cong \overline{FH}$. Since $\overline{DG} \cong \overline{CH}$, $\overline{DF} \cong \overline{CF}$ and $\overline{FG} \cong \overline{FH}$. $\angle CFD \cong \angle HFG$ because vertical angles are congruent.
Solve: $\triangle CFD \cong \triangle HFG$ by SAS.
Examine: The corresponding sides and angles used to determine the triangles are congruent by SAS are $\overline{DF} \cong \overline{FG}$, $\angle CFD \cong \angle HFG$, and $\overline{CF} \cong \overline{FH}$.

23. Explore: We need to determine whether two triangles are congruent.
Plan: Since N is the midpoint of $\overline{JL}$, $\overline{JN} \cong \overline{NL}$. $\angle JNK \cong \angle LNK$ because perpendicular lines form right angles and right angles are congruent. By the Reflexive Property, $\overline{KN} \cong \overline{KN}$.
Solve: $\triangle JKN \cong \triangle LKN$ by SAS.
Examine: The corresponding sides and angles used to determine the triangles are congruent by SAS are $\overline{JN} \cong \overline{NL}$, $\angle JNK \cong \angle LNK$, and $\overline{KN} \cong \overline{KN}$.

24. Explore: We are given the measurements of one side and one angle of each triangle. We need to determine whether the two triangles are congruent.
Plan: It is given that $\overline{JM} \cong \overline{LM}$ and $\angle NJM \cong \angle NLM$. By the Reflexive Property, $\overline{NM} \cong \overline{NM}$.
Solve: We are given information about side-side-angle (SSA). This is not a method to prove two triangles congruent.
Examine: Use a compass, protractor, and ruler to draw a triangle with the given measurements. For simplicity of measurement, we will use centimeters instead of feet, so the measurements of the construction and those of the kite will be proportional.

- Draw a segment 2.7 centimeters long.
- At one end, draw an angle of 68°. Extend the line to exactly 2 centimeters.
- At the other end of the segment, draw a seg-ment that intersects the 2 centimeter segment at any place other than either of its endpoints.

Because no information is known about the length of the segment that determines the triangle, an infinite number of triangles are possible. It cannot be determined whether$\triangle JNM \cong \triangle LNM$. The information given does not lead to a unique triangle.

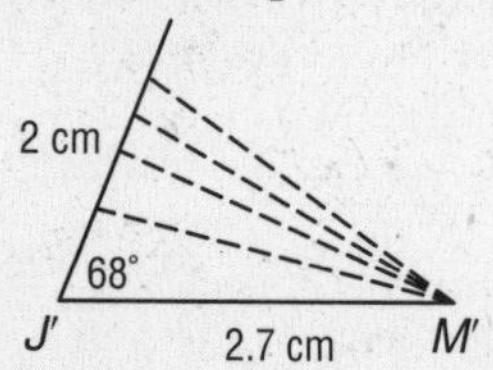

25. *VNR*, AAS or ASA

26. *VMN*, ASA or AAS

27. *MIN*, SAS

28. *RMI*, AAS or ASA

29. Since Aiko is perpendicular to the ground, two right angles are formed and right angles are congruent. The angles of sight are the same and her height is the same for each triangle. The triangles are congruent by ASA. By CPCTC, the distances are the same. The method is valid.

30. Sample answer: The triangular trusses support the structure. Answers should include the following.

- To determine whether two triangles are congruent, information is needed about consecutive side-angle-side, side-side-side, angle-side-angle, angle-angle-side, or about each angle and each side.
- Triangles that are congruent will support weight better because the pressure will be evenly divided.

31. D; $m\angle B + m\angle BAC + m\angle BCA = 180$
$76 + m\angle BAC + m\angle BCA = 180$
$m\angle BAC + m\angle BCA = 104$

$\overline{AD}$ bisects $\angle BAC$, so $m\angle DAC = \frac{1}{2}m\angle BAC$.

$\overline{DC}$ bisects $\angle BCA$, so $m\angle DCA = \frac{1}{2}m\angle BCA$.

$m\angle DAC + m\angle DCA = \frac{1}{2}m\angle BAC + \frac{1}{2}m\angle BCA$
$= \frac{1}{2}(m\angle BAC + m\angle BCA)$
$= \frac{1}{2}(104)$ or 52

$m\angle ADC + m\angle DAC + m\angle DCA = 180$
$m\angle ADC + 52 = 180$
$m\angle ADC = 128$

32. A; x percent of 10,000 $= \frac{x}{100}(10{,}000)$
1 percent of x percent of 10,000
$= \frac{1}{100}\left(\frac{x}{100}\right)(10{,}000)$
$= \frac{x}{10{,}000}(10{,}000)$
$= x$

Page 213 Maintain Your Skills

33. Given: $\overline{BA} \cong \overline{DE}$, $\overline{DA} \cong \overline{BE}$
Prove: $\triangle BEA \cong \triangle DAE$

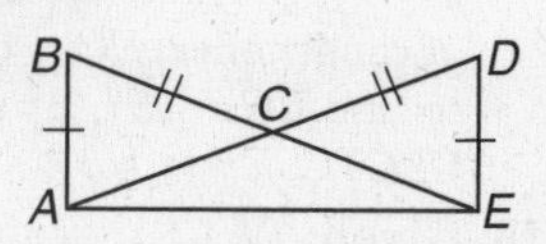

Proof:

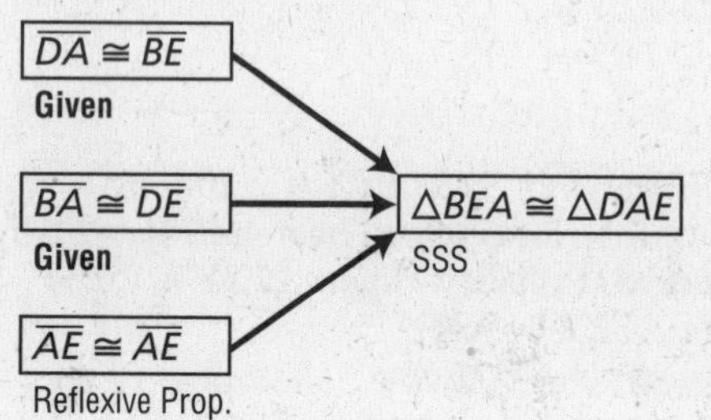

34. Given: $\overline{XZ} \perp \overline{WY}$, $\overline{XZ}$ bisects $\overline{WY}$.
Prove: $\triangle WZX \cong \triangle YZX$

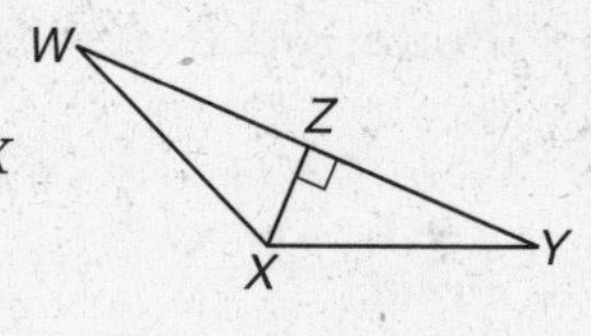

Proof:

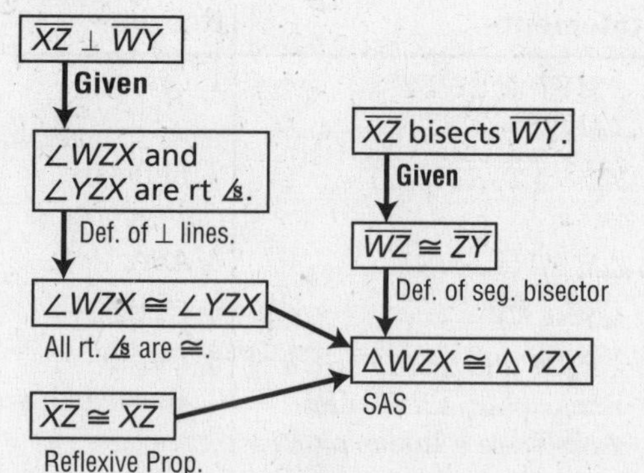

35. $RS = \sqrt{[-1-(-2)]^2 + (1-2)^2}$
$= \sqrt{1+1}$ or $\sqrt{2}$
$R'S' = \sqrt{(1-2)^2 + [-1-(-2)]^2}$
$= \sqrt{1+1}$ or $\sqrt{2}$
$ST = \sqrt{[-2-(-1)]^2 + (1-1)^2}$
$= \sqrt{1+0}$ or 1
$S'T' = \sqrt{(2-1)^2 + [-1-(-1)]^2}$
$= \sqrt{1+0}$ or 1
$RT = \sqrt{[-2-(-2)]^2 + (1-2)^2}$
$= \sqrt{0+1}$ or 1
$R'T' = \sqrt{(2-2)^2 + [-1-(-2)]^2}$
$= \sqrt{0+1}$ or 1
$R'T' = \sqrt{(2-2)^2 + [-1-(-2)]^2}$
$= \sqrt{0+1}$ or 1

Each pair of corresponding sides has the same measure, so they are congruent. Use a protractor to confirm that the corresponding angles are congruent. Therefore, $\triangle RST \cong \triangle R'S'T'$. $\triangle R'S'T'$ is a turn of $\triangle RST$.

36. $MP = \sqrt{[-1-(-3)]^2 + (1-1)^2}$
$= \sqrt{4+0}$ or 2
$M'P' = \sqrt{(0-2)^2 + (1-1)^2}$
$= \sqrt{4+0}$ or 2
$MN = \sqrt{[-3-(-3)]^2 + (4-1)^2}$
$= \sqrt{0+9}$ or 3
$M'N' = \sqrt{(2-2)^2 + (4-1)^2}$
$= \sqrt{0+9}$ or 3
$NP = \sqrt{[-1-(-3)]^2 + (1-4)^2}$
$= \sqrt{4+9}$ or $\sqrt{13}$
$N'P' = \sqrt{(2-0)^2 + (4-1)^2}$
$= \sqrt{4+9}$ or $\sqrt{13}$
Each pair of corresponding sides has the same measure, so they are congruent. Use a protractor to confirm that the corresponding angles are congruent. Therefore, $\triangle MPN \cong \triangle M'P'N'$. $\triangle M'P'N'$ is a flip of $\triangle MPN$.

37. If people are happy, then they rarely correct their faults.

38. If a person is a champion, then he or she is afraid of losing.

39. Since two sides are marked congruent, the triangle is isosceles.

40. Since three sides are marked congruent, the triangle is equilateral.

41. Since two sides are marked congruent, the triangle is isosceles.

Pages 214–215 Geometry Activity: Congruence in Right Triangles

1. yes; **a.** SAS, **b.** ASA, **c.** AAS

2a. LL

2b. LA

2c. HA

3. None; two pairs of legs congruent is sufficient for proving right triangles congruent.

4. yes **5.** yes

6. SSA is a valid test of congruence for right triangles.

7. Given: $\triangle DEF$ and $\triangle RST$ are right triangles.
$\angle E$ and $\angle S$ are right angles.
$\overline{EF} \cong \overline{ST}$
$\overline{ED} \cong \overline{SR}$
Prove: $\triangle DEF \cong \triangle RST$

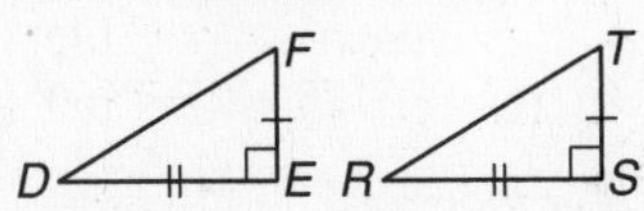

Proof: We are given that $\overline{EF} \cong \overline{ST}$, $\overline{ED} \cong \overline{SR}$, and $\angle E$ and $\angle S$ are right angles. Since all right angles are congruent, $\angle E \cong \angle S$. Therefore, by SAS, $\triangle DEF \cong \triangle RST$.

8. Given: $\triangle ABC$ and $\triangle XYZ$ are right triangles.
$\angle A$ and $\angle X$ are right angles.
$\overline{BC} \cong \overline{YZ}$
$\angle B \cong \angle Y$
Prove: $\triangle ABC \cong \triangle XYZ$

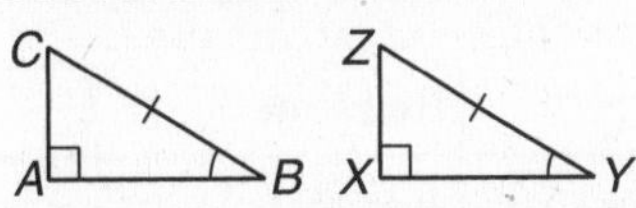

Proof: We are given that $\triangle ABC$ and $\triangle XZY$ are right triangles with right angles $\angle A$ and $\angle X$, $\overline{BC} \cong \overline{YZ}$, and $\angle B \cong \angle Y$. Since all right angles are congruent, $\angle A \cong \angle X$. Therefore, $\triangle ABC \cong \triangle XYZ$ by AAS.

9. Case 1:
Given: $\triangle ABC$ and $\triangle DEF$ are right triangles.
$\overline{AC} \cong \overline{DF}$
$\angle C \cong \angle F$
Prove: $\triangle ABC \cong \triangle DEF$

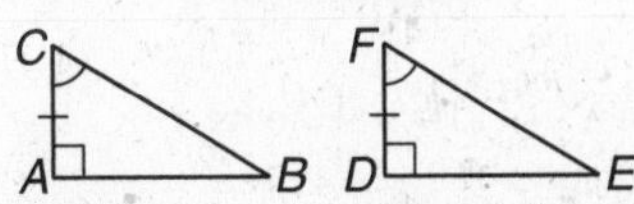

Proof: It is given that $\triangle ABC$ are $\triangle DEF$ are right triangles, $\overline{AC} \cong \overline{DF}$, $\angle C \cong \angle F$. By the definition of right triangles, $\angle A$ and $\angle D$ are right angles. Thus, $\angle A \cong \angle D$ since all right angles are congruent. $\triangle ABC \cong \triangle DEF$ by ASA.
Case 2:
Given: $\triangle ABC$ and $\triangle DEF$ are right triangles.
$\overline{AC} \cong \overline{DF}$
$\angle B \cong \angle E$
Prove: $\triangle ABC \cong \triangle DEF$

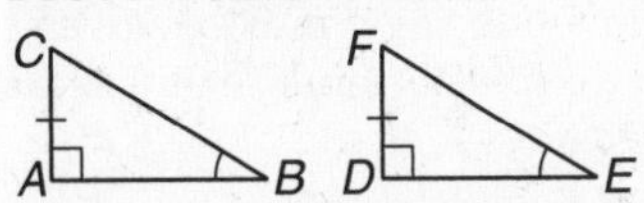

Proof: If is given that $\triangle ABC$ and $\triangle DEF$ are right triangles, $\overline{AC} \cong \overline{DF}$, and $\angle B \cong \angle E$. By the definition of right triangle, $\angle A$ and $\angle D$ are right angles. Thus, $\angle A \cong \angle D$ since all right angles are congruent. $\triangle ABC \cong \triangle DEF$ by AAS.

10. Given: $\overline{ML} \perp \overline{MK}$, $\overline{JK} \perp \overline{KM}$, $\angle J \cong \angle L$
Prove: $\overline{JM} \cong \overline{KL}$

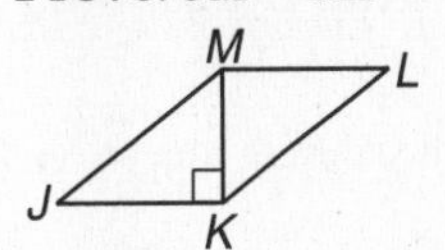

Proof:

Statements	Reasons
1. $\overline{ML} \perp \overline{MK}$, $\overline{JK} \perp \overline{KM}$, $\angle J \cong \angle L$	1. Given
2. $\angle LMK$ and $\angle JKM$ are rt. $\angle$s	2. $\perp$ lines form $\cong$ rt. $\angle$s.
3. $\triangle LMK$ and $\triangle JKM$ are rt. $\triangle$s	3. Def. of rt. $\triangle$
4. $\overline{MK} \cong \overline{MK}$	4. Reflexive Property
5. $\triangle LMK \cong \triangle JKM$	5. LA
6. $\overline{JM} \cong \overline{KL}$	6. CPCTC

11. **Given:** $\overline{JK} \perp \overline{KM}$, $\overline{JM} \cong \overline{KL}$, $\overline{ML} \parallel \overline{JK}$
Prove: $\overline{ML} \cong \overline{JK}$

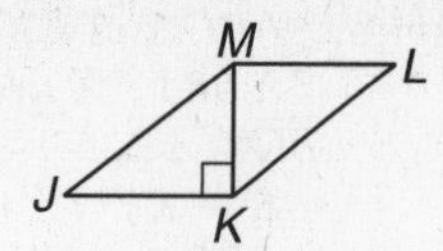

Proof:

Statements	Reasons
1. $\overline{JK} \perp \overline{KM}$, $\overline{JM} \cong \overline{KL}$, $\overline{ML} \parallel \overline{JK}$	1. Given
2. $\angle JKM$ is a rt. $\angle$	2. $\perp$ lines form rt. $\angle$s
3. $\overline{KM} \perp \overline{ML}$	3. Perpendicular Transversal Th.
4. $\angle LMK$ is a rt. $\angle$	4. $\perp$ lines form rt. $\angle$s
5. $\overline{MK} \cong \overline{MK}$	5. Reflexive Property
6. $\triangle JMK \cong \triangle LMK$	6. HL
7. $\overline{ML} \cong \overline{JK}$	7. CPCTC

4-6 Isosceles Triangles

Page 216 Geometry Activity: Isosceles Triangles

1. $\angle A \cong \angle B$
2. They are congruent.
3. They are congruent.

Page 219 Check for Understanding

1. The measure of only one angle must be given in an isosceles triangle to determine the measures of the other two angles.
2. $\overline{WX} \cong \overline{ZX}$; $\angle W \cong \angle Z$
3. Sample answer: Draw a line segment. Set your compass to the length of the line segment and draw an arc from each endpoint. Draw segments from the intersection of the arcs to each endpoint.
4. $\angle ADH$ is opposite $\overline{AH}$ and $\angle AHD$ is opposite $\overline{AD}$, so $\angle ADH \cong \angle AHD$.
5. $\overline{BH}$ is opposite $\angle BDH$ and $\overline{BD}$ is opposite $\angle BHD$, so $\overline{BH} \cong \overline{BD}$.
6. Each angle of an equilateral triangle measures 60°.
$m\angle F = 3x + 4$
$60 = 3x + 4$
$56 = 3x$
$\frac{56}{3} = x$
$m\angle G = 6y$
$60 = 6y$
$10 = y$
$m\angle H = 19z + 3$
$60 = 19z + 3$
$57 = 19z$
$3 = z$

7. **Given:** $\triangle CTE$ is isosceles with vertex $\angle C$. $m\angle T = 60$
Prove: $\triangle CTE$ is equilateral.

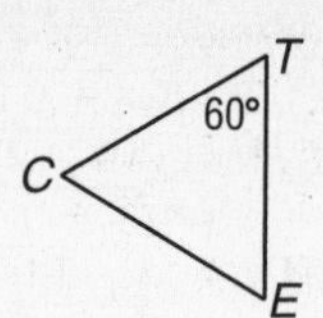

Proof:

Statements	Reasons
1. $\triangle CTE$ is isosceles with vertex $\angle C$.	1. Given
2. $\overline{CT} \cong \overline{CE}$	2. Def. of isosceles triangle
3. $\angle E \cong \angle T$	3. Isosceles Triangle Theorem
4. $m\angle E = m\angle T$	4. Def. of $\cong$ $\angle$s
5. $m\angle T = 60$	5. Given
6. $m\angle E = 60$	6. Substitution
7. $m\angle C + m\angle E + m\angle T = 180$	7. Angle Sum Theorem
8. $m\angle C + 60 + 60 = 180$	8. Substitution
9. $m\angle C = 60$	9. Subtraction
10. $\triangle CTE$ is equiangular.	10. Def. of equiangular $\triangle$
11. $\triangle CTE$ is equilateral.	11. Equiangular $\triangle$s are equilateral.

8. A;
Read the Test Item
$\triangle PQS$ is isosceles with base $\overline{PS}$. Likewise, $\triangle QRS$ is isosceles with base $\overline{QS}$.
Solve the Test Item
Step 1 The base angles of $\triangle QRS$ are congruent.
Let $x = m\angle RSQ = m\angle RQS$.
$m\angle PRS + m\angle RSQ + m\angle RQS = 180$
$72 + x + x = 180$
$72 + 2x = 180$
$2x = 108$
$x = 54$
So, $m\angle RSQ = m\angle RQS = 54$.
Step 2 $\angle RQS$ and $\angle PQS$ form a linear pair. Solve for $m\angle PQS$.
$m\angle RQS + m\angle PQS = 180$
$54 + m\angle PQS = 180$
$m\angle PQS = 126$
Step 3 The base angles of $\triangle PQS$ are congruent.
Let y represent $m\angle QPS$ and $m\angle PSQ$.
$m\angle PQS + m\angle QPS + m\angle PSQ = 180$
$126 + y + y = 180$
$126 + 2y = 180$
$2y = 54$
$y = 27$
The measure of $\angle QPS$ is 27. Choice A is correct.

Pages 219–221 Practice and Apply

9. $\angle LRT$ is opposite $\overline{LT}$ and $\angle LTR$ is opposite $\overline{LR}$, so $\angle LRT \cong \angle LTR$.
10. $\angle LXW$ is opposite $\overline{LW}$ and $\angle LWX$ is opposite $\overline{LX}$, so $\angle LXW \cong \angle LWX$.
11. $\angle LSQ$ is opposite $\overline{QL}$ and $\angle LQS$ is opposite $\overline{SL}$, so $\angle LSQ \cong \angle LQS$.

12. $\overline{LX}$ is opposite $\angle LYX$ and $\overline{LY}$ is opposite $\angle LXY$, so $\overline{LX} \cong \overline{LY}$.

13. $\overline{LS}$ is opposite $\angle LRS$ and $\overline{LR}$ is opposite $\angle LSR$, so $\overline{LS} \cong \overline{LR}$.

14. $\overline{LY}$ is opposite $\angle LWY$ and $\overline{LW}$ is opposite $\angle LYW$, so $\overline{LY} \cong \overline{LW}$.

15. The base angles of an isosceles triangle are congruent. So $\angle LNM \cong \angle MLN$. From the figure, $m\angle MLN = 20$. So $m\angle LNM = 20$.

16. $m\angle LNM + m\angle MLN + m\angle M = 180$
$20 + 20 + m\angle M = 180$
$m\angle M = 140$

17. The base angles of an isosceles triangle are congruent, so $m\angle LKN = m\angle LNK$.
$m\angle LKN + m\angle LNK + m\angle KLN = 180$
$m\angle LKN + m\angle LKN + 18 = 180$
$2m\angle LKN = 162$
$m\angle LKN = 81$

18. $m\angle JKN = m\angle JKL + m\angle LKN$
$130 = m\angle JKL + 81$
$49 = m\angle JKL$
$m\angle JKL + m\angle JLK + m\angle J = 180$
$49 + 25 + m\angle J = 180$
$m\angle J = 106$

19. The base angles of $\triangle DFG$ are congruent, so $\angle DFG \cong \angle D$.
$m\angle DFG = m\angle D$
$= 28$

20. $m\angle DGF + m\angle D + m\angle DFG = 180$
$28 + 28 + m\angle DGF = 180$
$m\angle DGF = 124$

21. $\angle FGH$ and $\angle DGF$ are a linear pair.
$m\angle DGF + m\angle FGH = 180$
$124 + m\angle FGH = 180$
$m\angle FGH = 56$

22. The base angles of $\triangle FGH$ are congruent, so $\angle FGH \cong \angle H$. From Exercise 21, $m\angle FGH = 56$.
$m\angle FGH + m\angle H + m\angle GFH = 180$
$56 + 56 + m\angle GFH = 180$
$m\angle GFH = 68$

23. $\triangle MLP$ is isosceles with base $\overline{MP}$. $\triangle JMP$ is isosceles with base $\overline{JP}$.

Step 1 The base angles of $\triangle MLP$ are congruent. Let x represent $m\angle PML$ and $m\angle MPL$.
$m\angle PML + m\angle MPL + m\angle PLJ = 180$
$x + x + 34 = 180$
$2x + 34 = 180$
$2x = 146$
$x = 73$
So, $m\angle PML = m\angle MPL = 73$.

Step 2 $\angle PML$ and $\angle JMP$ form a linear pair. Solve for $m\angle JMP$.
$m\angle PML + m\angle JMP = 180$
$73 + m\angle JMP = 180$
$m\angle JMP = 107$

Step 3 The base angles of $\triangle JMP$ are congruent. Let y represent $m\angle J$ and $m\angle JPM$.
$m\angle JPM + m\angle J + m\angle JMP = 180$
$y + y + 107 = 180$
$2y + 107 = 180$
$2y = 73$
$y = 36.5$
So, $m\angle JPM = m\angle J = 36.5$.

24. $\triangle MLP$ is isosceles with base $\overline{MP}$, $\triangle JMP$ is isosceles with base $\overline{JP}$.

Step 1 The base angles of $\triangle MLP$ are congruent. Let x represent $m\angle PML$ and $m\angle MPL$.
$m\angle PML + m\angle MPL + m\angle PLJ = 180$
$x + x + 58 = 180$
$2x + 58 = 180$
$2x = 122$
$x = 61$
So, $m\angle PML = m\angle MPL = 61$.

Step 2 $\angle PML$ and $\angle JMP$ form a linear pair. Solve for $m\angle JMP$.
$m\angle PML + m\angle JMP = 180$
$61 + m\angle JMP = 180$
$m\angle JMP = 119$

Step 3 The base angles of $\triangle JMP$ are congruent. Let y represent $m\angle PJL$ and $m\angle JPM$.
$m\angle JPM + m\angle PJL + m\angle JMP = 180$
$y + y + 119 = 180$
$2y + 119 = 180$
$2y = 61$
$y = 30.5$
So, $m\angle JPM = m\angle PJL = 30.5$.

25. $\triangle GKH$ is isosceles with base $\overline{HK}$. $\triangle JKH$ is isosceles with base $\overline{HJ}$.

Step 1 The base angles of $\triangle GKH$ are congruent. Let x represent $m\angle GHK$ and $m\angle GKH$.
$m\angle HGK + m\angle GHK + m\angle GKH = 180$
$28 + x + x = 180$
$2x = 152$
$x = 76$
So, $m\angle GHK = m\angle GKH = 76$.

Step 2 $\angle GKH$ and $\angle HKJ$ form a linear pair. Solve for $m\angle HKJ$.
$m\angle GKH + m\angle HKJ = 180$
$76 + m\angle HKJ = 180$
$m\angle HKJ = 104$

Step 3 The base angles of $\triangle JKH$ are congruent. Let x represent $m\angle HJK$ and $m\angle JHK$.
$m\angle HJK + m\angle JHK + m\angle HKJ = 180$
$x + x + 104 = 180$
$2x + 104 = 180$
$2x = 76$
$x = 38$
So, $m\angle HJK = m\angle JHK = 38$.

26. $\triangle GKH$ is isosceles with base $\overline{HK}$.

Step 1 The base angles of $\triangle GKH$ are congruent. Let x represent $m\angle GHK$ and $m\angle GKH$.
$m\angle HGK + m\angle GHK + m\angle GKH = 180$
$42 + x + x = 180$
$42 + 2x = 180$
$2x = 138$
$x = 69$
So, $m\angle GHK = m\angle GKH = 69$.

Step 2 $\angle GKH$ and $\angle HKJ$ form a linear pair. Solve for $m\angle HKJ$.
$m\angle GKH + m\angle HKJ = 180$
$69 + m\angle HKJ = 180$
$m\angle HKJ = 111$

27. $\triangle LMN$ is equilateral, so $\overline{LM} \cong \overline{MN} \cong \overline{LN}$. $\overline{MP}$ bisects $\overline{LN}$, so $\overline{MP}$ bisects $\angle LMN$. $\triangle LMN$ is equilateral, hence equiangular. So $m\angle LMN = m\angle MLP = 60$ and $m\angle PML = 30$.

$$LM = MN$$
$$3x + 1 = 4x - 2$$
$$3 = x$$
$$m\angle PLM + m\angle PML + m\angle MPL = 180$$
$$60 + 30 + 5y = 180$$
$$90 + 5y = 180$$
$$5y = 180$$
$$y = 18$$

28. $LM = 3x + 1$
$= 3(3) + 1$ or 10
$LM = MN = LN$, so all sides have measure 10.

29. Given: $\triangle XKF$ is equilateral.
$\overline{XJ}$ bisects $\angle X$.
Prove: J is the midpoint of $\overline{KF}$.

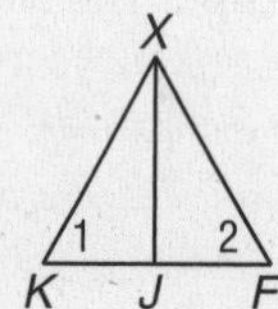

Proof:

Statements	Reasons
1. $\triangle XKF$ is equilateral.	1. Given
2. $\overline{KX} \cong \overline{FX}$	2. Definition of equilateral $\triangle$
3. $\angle 1 \cong \angle 2$	3. Isosceles Triangle Theorem
4. $\overline{XJ}$ bisects $\angle X$	4. Given
5. $\angle KXJ \cong \angle FXJ$	5. Def. of $\angle$ bisector
6. $\triangle KXJ \cong \triangle FXJ$	6. ASA
7. $\overline{KJ} \cong \overline{JF}$	7. CPCTC
8. J is the midpoint of $\overline{KF}$.	8. Def. of midpoint

30. Given: $\triangle MLP$ is isosceles.
N is the midpoint of $\overline{MP}$.
Prove: $\overline{LN} \perp \overline{MP}$

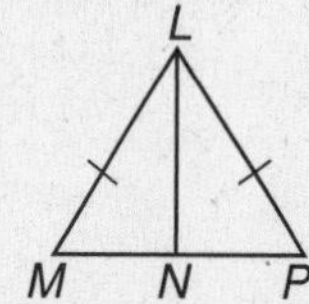

Proof:

Statements	Reasons
1. $\triangle MLP$ is isosceles.	1. Given
2. $\overline{ML} \cong \overline{LP}$	2. Definition of isosceles $\triangle$
3. $\angle M \cong \angle P$	3. Isosceles Triangle Theorem
4. N is the midpoint of $\overline{MP}$.	4. Given
5. $\overline{MN} \cong \overline{NP}$	5. Midpoint Theorem
6. $\triangle MNL \cong \triangle PNL$	6. SAS
7. $\angle LNM \cong \angle LNP$	7. CPCTC
8. $m\angle LNM = m\angle LNP$	8. Congruent angles have equal measures.
9. $\angle LNM$ and $\angle LNP$ are a linear pair	9. Definition of linear pair
10. $m\angle LNM + m\angle LNP = 180$	10. Sum of measures of a linear pair of angles is 180
11. $2m\angle LNM = 180$	11. Substitution
12. $m\angle LNM = 90$	12. Division
13. $\angle LNM$ is a right angle.	13. Definition of right angle
14. $\overline{LN} \perp \overline{MP}$	14. Definition of perpendicular

31. Case I:
Given: $\triangle ABC$ is an equilateral triangle.
Prove: $\triangle ABC$ is an equiangular triangle.

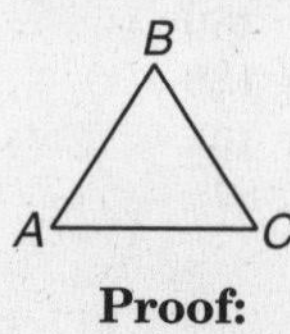

Proof:

Statements	Reasons
1. $\triangle ABC$ is an equilateral triangle.	1. Given
2. $\overline{AB} \cong \overline{AC} \cong \overline{BC}$	2. Def. of equilateral $\triangle$
3. $\angle A \cong \angle B$, $\angle B \cong \angle C$, $\angle A \cong \angle C$	3. Isosceles Triangle Theorem
4. $\angle A \cong \angle B \cong \angle C$	4. Substitution
5. $\triangle ABC$ is an equiangular triangle.	5. Def. of equiangular $\triangle$

Case II:
Given: $\triangle ABC$ is an equiangular triangle.
Prove: $\triangle ABC$ is an equilateral triangle.

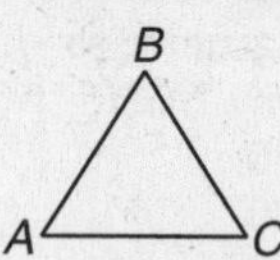

Proof:

Statements	Reasons
1. $\triangle ABC$ is an equiangular triangle.	1. Given
2. $\angle A \cong \angle B \cong \angle C$	2. Def. of equiangular $\triangle$
3. $\overline{AB} \cong \overline{AC}$, $\overline{AB} \cong \overline{BC}$, $\overline{AC} \cong \overline{BC}$	3. Conv. of Isos. $\triangle$ Th.
4. $\overline{AB} \cong \overline{AC} \cong \overline{BC}$	4. Substitution
5. $\triangle ABC$ is an equilateral triangle.	5. Def. of equilateral $\triangle$

32. Given: $\triangle MNO$ is an equilateral triangle.
Prove: $m\angle M = m\angle N = m\angle O = 60$

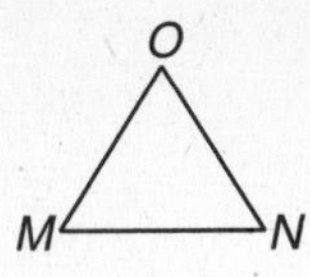

Proof:

Statements	Reasons
1. $\triangle MNO$ is an equilateral triangle.	1. Given
2. $\overline{MN} \cong \overline{MO} \cong \overline{NO}$	2. Def. of equilateral $\triangle$
3. $\angle M \cong \angle N \cong \angle O$	3. Isosceles $\triangle$ Thm.
4. $m\angle M = m\angle N = m\angle O$	4. Def. of $\cong$ $\triangle$s
5. $m\angle M + m\angle N + m\angle O = 180$	5. Angle Sum Theorem
6. $3m\angle M = 180$	6. Substitution
7. $m\angle M = 60$	7. Division Property
	8. Substitution

33. Given: $\triangle ABC$
$\angle A \cong \angle C$
Prove: $\overline{AB} \cong \overline{CB}$

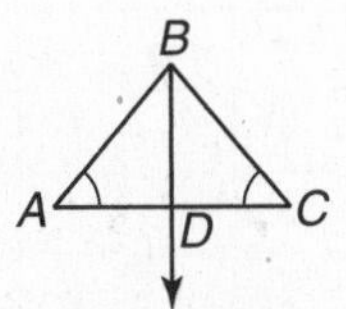

Proof:

Statements	Reasons
1. Let $\overrightarrow{BD}$ bisect $\angle ABC$.	1. Protractor Postulate
2. $\angle ABD \cong \angle CBD$	2. Def. of bisector
3. $\angle A \cong \angle C$	3. Given
4. $\overline{BD} \cong \overline{BD}$	4. Reflexive Property
5. $\triangle ABD \cong \triangle CBD$	5. AAS
6. $\overline{AB} \cong \overline{CB}$	6. CPCTC

34. The minimum requirement is that two angles measure 60°.

35. The front face of the figure has two congruent angles and one angle of measure 60. Let y represent the measure of each of the congruent angles.
$y + y + 60 = 180$
$2y + 60 = 180$
$2y = 120$
$y = 60$
Therefore all angles have measure 60, so the triangle is equiangular and equilateral. All sides have length $2x + 5$, in particular the edge between the front face and the side face showing. Because the side face has two congruent base angles it is isosceles and the sides opposite the congruent angles are congruent.
$2x + 5 = 3x - 13$
$2x + 18 = 3x$
$18 = x$

36. The triangle is isosceles with base angles having measure $3x + 8$.
$(3x + 8) + (3x + 8) + (2x + 20) = 180$
$8x + 36 = 180$
$8x = 144$
$x = 18$

37. The triangle on the bottom half of the figure is isosceles. The base angles are congruent.
$2x - 25 = x + 5$
$x - 25 = 5$
$x = 30$

38. There are two sets of 12 isosceles triangles. One black set forms a circle with their bases on the outside of the circle. Another black set encircles a circle in the middle.

39. The triangles in each set appear to be acute.

40. $\triangle DCE$ is equilateral, hence equiangular so $m\angle CDE = m\angle DEC = m\angle DCE = 60$. Then we know, $m\angle ACB > m\angle DCE$ so $m\angle ACB > 60$ and $m\angle FCG < m\angle DCE$ so $m\angle FCG < 60$. This means that in isosceles $\triangle ABC$, $\angle BAC$ and $\angle ABC$ are the congruent base angles, so $m\angle ABC = m\angle BAC = 42$. Also, in isosceles $\triangle FCG$, $\angle CFG$ and $\angle FGC$ are the congruent base angles, so $m\angle CFG = m\angle FGC = 77$.
$m\angle ACB + m\angle ABC + m\angle BAC = 180$
$m\angle ACB + 42 + 42 = 180$
$m\angle ACB = 96$
$m\angle FCG + m\angle CFG + m\angle FGC = 180$
$m\angle FCG + 77 + 77 = 180$
$m\angle FCG = 26$
So $m\angle 3 = 26$
$m\angle CFD + m\angle CFG = 180$
$m\angle CFD + 77 = 180$
$m\angle CFD = 103$
$m\angle CGF + m\angle CGE = 180$
$77 + m\angle CGE = 180$
$m\angle CGE = 103$
$m\angle 2 + m\angle CDF + m\angle CFD = 180$
$m\angle 2 + 60 + 103 = 180$
$m\angle 2 = 17$
$m\angle 4 + m\angle CEG + m\angle CGE = 180$
$m\angle 4 + 60 + 103 = 180$
$m\angle 4 = 17$
$\triangle CDA \cong \triangle CEB$ by AAS Postulate
So $\angle 1 \cong \angle 5$
$2m\angle 1 + 2(17) + 26 = m\angle ACB$
$2m\angle 1 + 60 = 96$
$2m\angle 1 = 36$
$m\angle 1 = 18$
So, $m\angle 1 = 18$, $m\angle 2 = 17$, $m\angle 3 = 26$, $m\angle 4 = 17$, and $m\angle 5 = 18$.

41. Sample answer: Artists use angles, lines, and shapes to create visual images. Answers should include the following.
- Rectangle, squares, rhombi, and other polygons are used in many works of art.
- There are two rows of isosceles triangles in the painting. One row has three congruent isosceles triangles. The other row has six congruent isosceles triangles.

42. A; $m\angle ZXY + m\angle PYZ = 90$

$$m\angle ZXY + 26 = 90$$
$$m\angle ZXY = 64$$

$\triangle PYZ$ is isosceles since $YP = YZ$, so $m\angle YPZ = m\angle YZP$.

$$m\angle YPZ + m\angle YZP + m\angle PYZ = 180$$
$$2(m\angle YPZ) + 26 = 180$$
$$2(m\angle YPZ) = 154$$
$$m\angle YPZ = 77$$
$$m\angle XPZ + m\angle YPZ = 180$$
$$m\angle XPZ + 77 = 180$$
$$m\angle XPZ = 103$$
$$m\angle XZP + m\angle XPZ + m\angle PXZ = 180$$
$$m\angle XZP + 103 + 64 = 180$$
$$m\angle XZP = 13$$

43. D; $\left(\frac{x_1 + x_2}{2}, \frac{y_1 + y_2}{2}\right) = \left(\frac{3 + 9}{2}, \frac{5 + 13}{2}\right) = \left(\frac{12}{2}, \frac{18}{2}\right)$ or $(6, 9)$

Page 221 Maintain Your Skills

44. Given: $\angle N \cong \angle D$, $\angle G \cong \angle I$, $\overline{AN} \cong \overline{SD}$

Prove: $\triangle ANG \cong \triangle SDI$

Proof: We are given $\angle N \cong \angle D$ and $\angle G \cong \angle I$. By the Third Angle Theorem, $\angle A \cong \angle S$. We are also given $\overline{AN} \cong \overline{SD}$. $\triangle ANG \cong \triangle SDI$ by ASA.

45. Given: $\overline{VR} \perp \overline{RS}$, $\overline{UT} \perp \overline{SU}$, $\overline{RS} \cong \overline{US}$

Prove: $\triangle VRS \cong \triangle TUS$

Proof: We are given that $\overline{VR} \perp \overline{RS}$, $\overline{UT} \perp \overline{SU}$ and $\overline{RS} \cong \overline{US}$. Perpendicular lines form four right angles so $\angle R$ and $\angle U$ are right angles. $\angle R \cong \angle U$ because all right angles are congruent. $\angle RSV \cong \angle UST$ since vertical angles are congruent. Therefore, $\triangle VRS \cong \triangle TUS$ by ASA.

46. $QR = \sqrt{[1 - (-3)]^2 + (2 - 1)^2} = \sqrt{16 + 1}$ or $\sqrt{17}$

$EG = \sqrt{(2 - 6)^2 + [-3 - (-2)]^2} = \sqrt{16 + 1}$ or $\sqrt{17}$

$RS = \sqrt{(-1 - 1)^2 + (-2 - 2)^2} = \sqrt{4 + 16}$ or $\sqrt{20}$

$GH = \sqrt{(4 - 2)^2 + [1 - (-3)]^2} = \sqrt{4 + 16}$ or $\sqrt{20}$

$QS = \sqrt{[-1 - (-3)]^2 + (-2 - 1)^2} = \sqrt{4 + 9}$ or $\sqrt{13}$

$EH = \sqrt{(4 - 6)^2 + [1 - (-2)]^2} = \sqrt{4 + 9}$ or $\sqrt{13}$

Each pair of corresponding sides has the same measure so they are congruent. $\triangle QRS \cong \triangle EGH$ by SSS.

47. $QR = \sqrt{(5 - 1)^2 + [1 - (-5)]^2} = \sqrt{16 + 36}$ or $\sqrt{52}$

$EG = \sqrt{[-1 - (-4)]^2 + [2 - (-3)]^2} = \sqrt{9 + 25}$ or $\sqrt{34}$

$RS = \sqrt{(4 - 5)^2 + (0 - 1)^2} = \sqrt{1 + 1}$ or $\sqrt{2}$

$GH = \sqrt{[2 - (-1)]^2 + (1 - 2)^2} = \sqrt{9 + 1}$ or $\sqrt{10}$

$QS = \sqrt{(4 - 1)^2 + [0 - (-5)]^2} = \sqrt{9 + 25}$ or $\sqrt{34}$

$EH = \sqrt{[2 - (-4)]^2 + [1 - (-3)]^2} = \sqrt{36 + 16}$ or $\sqrt{52}$

The corresponding sides are not congruent so $\triangle QRS$ is not congruent to $\triangle EGH$.

48.

a	*b*	*a* **and** *b*
T	T	T
T	F	F
F	T	F
F	F	F

49.

p	*q*	~*p*	~*q*	~*p* **or** ~*q*
T	T	F	F	F
T	F	F	T	T
F	T	T	F	T
F	F	T	T	T

50.

k	*m*	~*m*	*k* **and** ~*m*
T	T	F	F
T	F	T	T
F	T	F	F
F	F	T	F

51.

y	*z*	~*y*	~*y* **or** *z*
T	T	F	T
T	F	F	F
F	T	T	T
F	F	T	T

52. $\left(\frac{x_1 + x_2}{2}, \frac{y_1 + y_2}{2}\right) = \left(\frac{2 + 7}{2}, \frac{15 + 9}{2}\right) = (4.5, 12)$

53. $\left(\frac{x_1 + x_2}{2}, \frac{y_1 + y_2}{2}\right) = \left(\frac{-4 + 2}{2}, \frac{6 + (-12)}{2}\right) = (-1, -3)$

54. $\left(\frac{x_1 + x_2}{2}, \frac{y_1 + y_2}{2}\right) = \left(\frac{3 + 7.5}{2}, \frac{2.5 + 4}{2}\right) = (5.25, 3.25)$

Page 221 Practice Quiz 2

1. $JM = \sqrt{[-2-(-4)]^2 + (6-5)^2}$
$= \sqrt{4+1}$ or $\sqrt{5}$
$BD = \sqrt{[-4-(-3)]^2 + [-2-(-4)]^2}$
$= \sqrt{1+4}$ or $\sqrt{5}$
$ML = \sqrt{[-1-(-2)]^2 + (1-6)^2}$
$= \sqrt{1+25}$ or $\sqrt{26}$
$DG = \sqrt{[1-(-4)]^2 + [-1-(-2)]^2}$
$= \sqrt{25+1}$ or $\sqrt{26}$
$JL = \sqrt{[-1-(-4)]^2 + (1-5)^2}$
$= \sqrt{9+16}$ or 5
$BG = \sqrt{[1-(-3)]^2 + [-1-(-4)]^2}$
$= \sqrt{16+9}$ or 5
Each pair of corresponding sides has the same measure so they are congruent. $\triangle JML \cong \triangle BDG$ by SSS.

2. **Given:** $\angle A \cong \angle H$, $\angle AEJ \cong \angle HJE$
Prove: $\overline{AJ} \cong \overline{EH}$

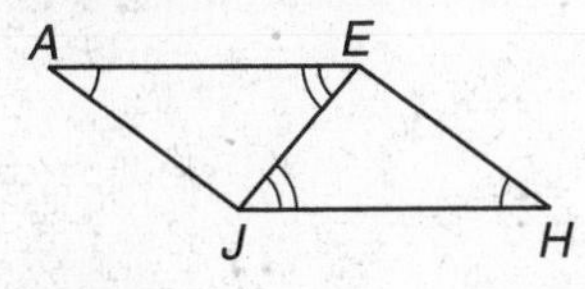

Proof:

Statements	Reasons
1. $\angle A \cong \angle H$, $\angle AEJ \cong \angle HJE$	1. Given
2. $\overline{EJ} \cong \overline{EJ}$	2. Reflexive Property
3. $\triangle AEJ \cong \triangle HJE$	3. AAS
4. $\overline{AJ} \cong \overline{EH}$	4. CPCTC

3. From the figure, $\overline{WX} \cong \overline{XY}$ so the angles opposite these sides are congruent. That is, $m\angle XWY = m\angle XYW$.
$m\angle XYW + m\angle XYZ = 180$
$m\angle XYW + 128 = 180$
$m\angle XYW = 52$
So, $m\angle XWY = 52$.

4. From Question 3, $m\angle XWY = 52$.
$m\angle WXY + m\angle XWY + m\angle XYW = 180$
$m\angle WXY + 52 + 52 = 180$
$m\angle WXY = 76$

5. $\triangle XYZ$ is isosceles with $\angle YZX \cong \angle YXZ$.
$m\angle YZX + m\angle YXZ + m\angle XYZ = 180$
$2(m\angle YZX) + 128 = 180$
$2(m\angle YZX) = 52$
$m\angle YZX = 26$

4-7 Triangles and Coordinate Proof

Page 224 Check for Understanding

1. Place one vertex at the origin, place one side of the triangle on the positive x-axis. Label the coordinates with expressions that will simplify the computations.

2. Sample answer:

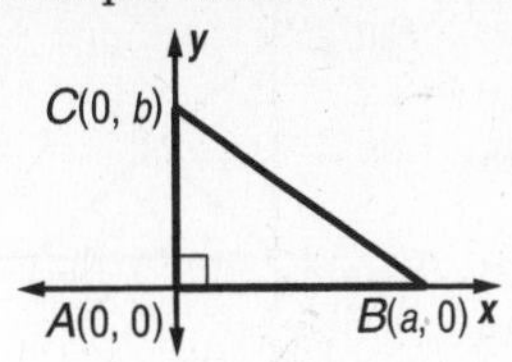

3. • Use the origin as vertex F of the triangle.
• Place the base of the triangle along the positive x-axis.
• Position the triangle in the first quadrant.
• Since H is on the x-axis, its y-coordinate is 0. Its x-coordinate is $2b$ because the base of the triangle is $2b$ units long.
• Since $\triangle FGH$ is isosceles, the x-coordinate of G is halfway between 0 and $2b$ or b. We cannot determine the y-coordinate in terms of b, so call it c.

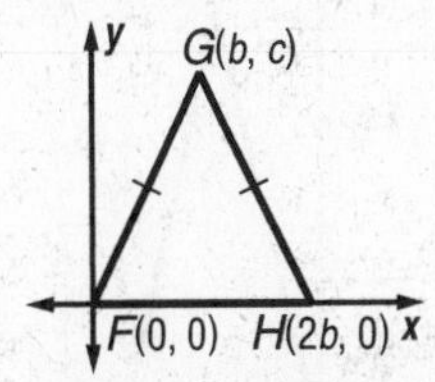

4. • Use the origin as vertex C of the triangle.
• Place side $\overline{CD}$ of the triangle along the positive x-axis.
• Position the triangle in the first quadrant.
• Since D is on the x-axis, its y-coordinate is 0. Its x-coordinate is a because each side of the triangle is a units long.
• Since $\triangle CDE$ is equilateral, the x-coordinate of E is halfway between 0 and a, or $\frac{a}{2}$. We cannot determine the y-coordinate in terms of a, so call it b.

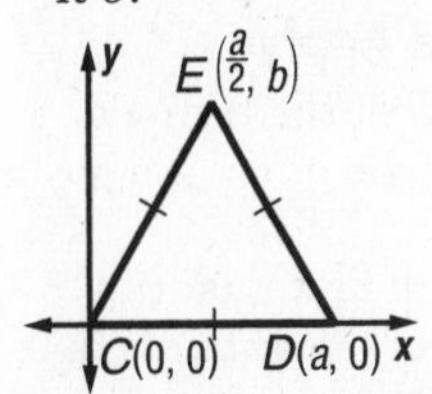

5. Vertex P is on the y-axis, so its x-coordinate is 0. $\angle R$ is a right angle and $\overline{RQ}$ has length a, but there is no other information given to determine the y-coordinate of P, so let the y-coordinate be b. So, the coordinates of P are $(0, b)$.

6. Vertex Q is on the x-axis, so its y-coordinate is 0. $\triangle PQN$ is equilateral and the y-axis bisects side $\overline{QN}$. The x-coordinate of N is $2a$, so the x-coordinate of Q is $-2a$. So, the coordinates of Q are $(-2a, 0)$.

7. Vertex N is on the y-axis and no information is given to determine its y-coordinate, so the coordinates of N are $(0, b)$ for some b. Vertex Q is on the x-axis, so its y-coordinate is 0. $\triangle NRQ$ is isosceles and the y-axis bisects its base $\overline{RQ}$. The x-coordinate of R is $-a$, so the x-coordinate of Q is a. So, the coordinates of Q are $(a, 0)$.

8. Given: $\triangle ABC$ is a right triangle with hypotenuse $\overline{BC}$. M is the midpoint of $\overline{BC}$.

Prove: M is equidistant from the vertices.

Proof: The coordinates of M, the midpoint of $\overline{BC}$, will be $\left(\frac{2c}{2}, \frac{2b}{2}\right) = (c, b)$.

The distance from M to each of the vertices can be found using the Distance Formula.

$MB = \sqrt{(c-0)^2 + (b-2b)^2} = \sqrt{c^2 + b^2}$

$MC = \sqrt{(c-2c)^2 + (b-0)^2} = \sqrt{c^2 + b^2}$

$MA = \sqrt{(c-0)^2 + (b-0)^2} = \sqrt{c^2 + b^2}$

Thus, $MB = MC = MA$, and M is equidistant from the vertices.

9. Given: $\triangle ABC$

Prove: $\triangle ABC$ is isosceles.

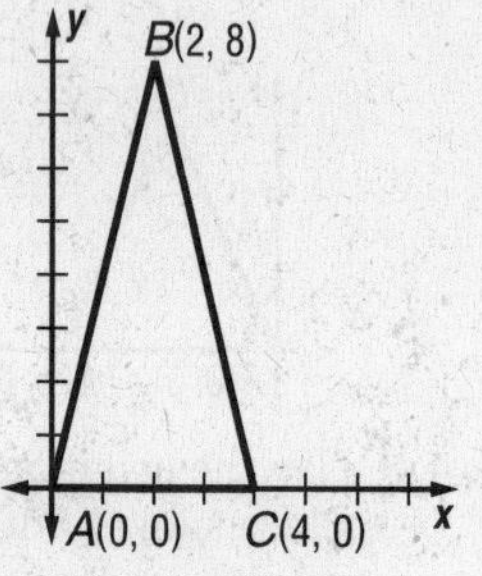

Proof: Use the Distance Formula to find AB and BC.

$AB = \sqrt{(2-0)^2 + (8-0)^2} = \sqrt{4+64}$ or $\sqrt{68}$

$BC = \sqrt{(4-2)^2 + (0-8)^2} = \sqrt{4+64}$ or $\sqrt{68}$

Since $AB = BC$, $\overline{AB} \cong \overline{BC}$. Since the legs are congruent, $\triangle ABC$ is isosceles.

Pages 224–226 Practice and Apply

10.
- Use the origin as vertex Q of the triangle.
- Place the base of the triangle along the positive x-axis.
- Position the triangle in the first quadrant.
- Since R is on the x-axis, its y-coordinate is 0. Its x-coordinate is b because the base $\overline{QR}$ is b units long.
- Since $\triangle QRT$ is isosceles, the x-coordinate of T is halfway between 0 and b or $\frac{b}{2}$. We cannot determine the y-coordinate in terms of b, so call it c.

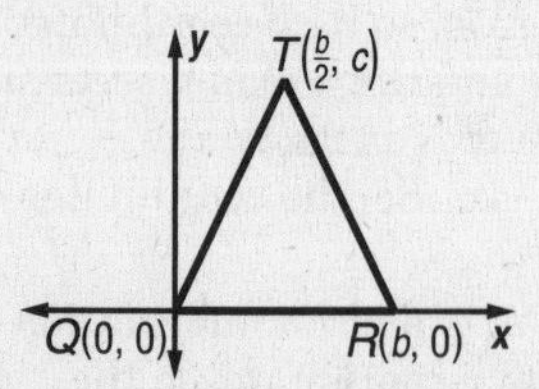

11.
- Use the origin as vertex M of the triangle.
- Place side $\overline{MN}$ of the triangle along the positive x-axis.
- Position the triangle in the first quadrant.
- Since N is on the x-axis, its y-coordinate is 0. Its x-coordinate is $2a$ because each side of the triangle is $2a$ units long.
- Since $\triangle MNP$ is equilateral, the x-coordinate of P is halfway between 0 and $2a$, or a. We cannot determine the y-coordinate in terms of a, so call it b.

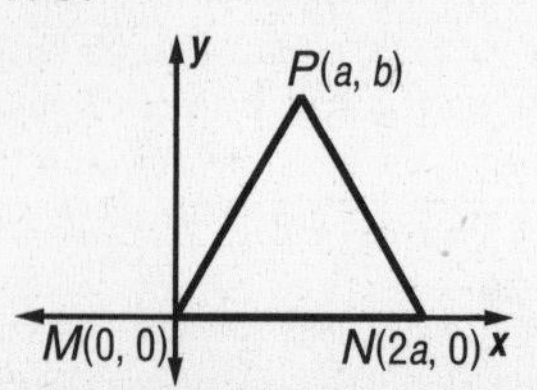

12.
- Use the origin as vertex L of the triangle.
- Place leg $\overline{LM}$ of the triangle along the positive x-axis.
- Position the triangle in the first quadrant.
- Since M is on the x-axis, its y-coordinate is 0. Its x-coordinate is c because each leg is c units long.
- Since J is on the y-axis, its x-coordinate is 0. Its y-coordinate is c because each leg is c units long.

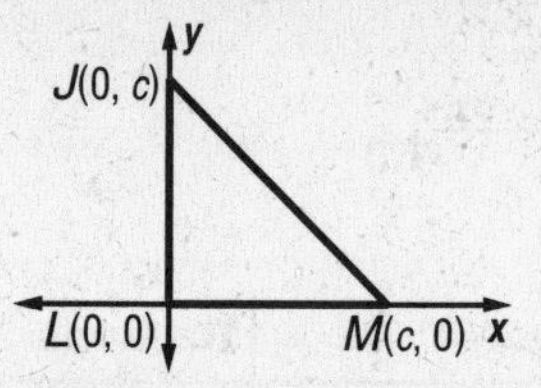

13.
- Use the origin as vertex W of the triangle.
- Place side $\overline{WZ}$ of the triangle along the positive x-axis.
- Position the triangle in the first quadrant.
- Since Z is on the x-axis, its y-coordinate is 0. Its x-coordinate is $\frac{b}{2}$ because each side of the triangle is $\frac{1}{2}b$ units long.
- Since $\triangle WXZ$ is equilateral, the x-coordinate of X is halfway between 0 and $\frac{b}{2}$, or $\frac{b}{4}$. We cannot determine the y-coordinate in terms of b, so call it c.

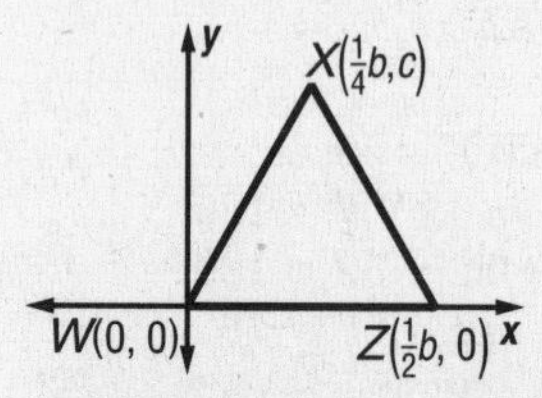

14.
- Use the origin as vertex P of the triangle.
- Place side $\overline{PW}$ of the triangle along the positive x-axis.
- Position the triangle in the first quadrant.
- Since W is on the x-axis, its y-coordinate is 0. Its x-coordinate is $a + b$ because the base is $a + b$ units long.
- Since $\triangle PWY$ is isosceles, the x-coordinate of Y is halfway between 0 and $a + b$ or $\frac{a+b}{2}$. We cannot determine the y-coordinate in terms of a and b, so call it c.

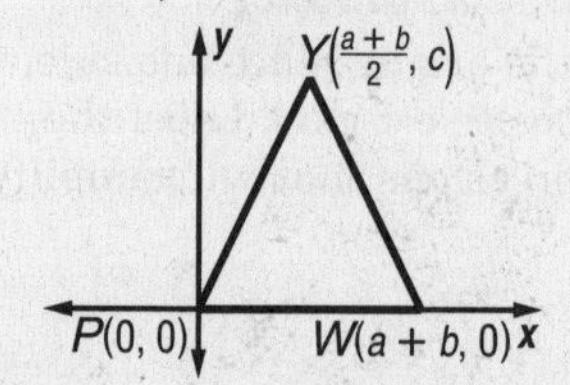

15. • Use the origin as vertex Y of the triangle.
• Place leg $\overline{YZ}$ of the triangle along the positive x-axis.
• Position the triangle in the first quadrant.
• Since X is on the y-axis, its x-coordinate is 0. Its y-coordinate is b because $\overline{XY}$ is b units long.
• Since Z is on the x-axis, its y-coordinate is 0. Its x-coordinate is $2b$ because $XY = b$ and $ZY = 2(XY)$ or $2b$.

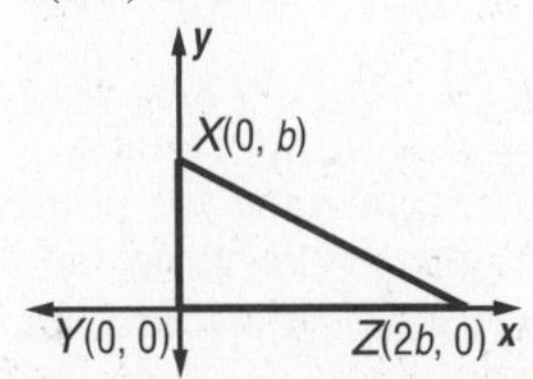

16. Since $\triangle PQR$ is equilateral, the x-coordinate of R is halfway between 0 and $2a$, or a. So, the coordinates of R are (a, b).

17. Since $\triangle LPQ$ is right isosceles, $\overline{LP} \cong \overline{PQ}$ and $\overline{LP} \perp \overline{PQ}$. Then Q and P have the same x-coordinate, so the x-coordinate of P is a. P is on the x-axis, so its y-coordinate is 0. The distance from L to P is a units. The distance from P to Q must be the same. So, the coordinates of Q are (a, a) and the coordinates of P are $(a, 0)$.

18. $\triangle JKN$ is isosceles, so $\overline{JK} \cong \overline{JN}$. The distance from J to K is $2a$ units. The distance from J to N must be the same. N is on the y-axis, so the coordinates of N are $(0, 2a)$.

19. $\triangle CDF$ is equilateral, so the x-coordinate of F is halfway between 0 and the x-coordinate of D. So the x-coordinate of D is $2b$. D is on the x-axis, so the coordinates of D are $(2b, 0)$.

20. $\triangle BCE$ is isosceles and the y-axis intersects side $\overline{BC}$ at its midpoint. The distance from the origin to C is the same as the distance from the origin to B. B is on the x-axis to the left of 0, so the coordinates of B are $(-a, 0)$. E is on the y-axis, and we cannot determine the y-coordinate in terms of a, so call it b. The coordinates of E are $(0, b)$.

21. $\triangle MNP$ is isosceles and the y-axis intersects side $\overline{MN}$ at its midpoint. The distance from the origin to M is the same as the distance from the origin to N. N is on the x-axis, so the coordinates of N are $(2b, 0)$. P is on the y-axis, and we cannot determine the y-coordinate in terms of b, so call it c. The coordinates of P are $(0, c)$.

22. $\triangle JHG$ is isosceles and the y-axis intersects side $\overline{JH}$ at its midpoint. The distance from the origin to J is the same as the distance from the origin to H. H is on the x-axis, so the coordinates of H are $(b, 0)$. G is on the y-axis, and we cannot determine the y-coordinate in terms of b, so call it c. The coordinates of G are $(0, c)$.

23. $\triangle JKL$ is isosceles, so the x-coordinate of J is halfway between 0 and $2c$, or c. We cannot determine the y-coordinate in terms of c, so call it b. The coordinates of J are (c, b).

24. Since $\triangle NPQ$ is right isosceles, $\overline{NQ} \cong \overline{PQ}$ and $\overline{NQ} \perp \overline{PQ}$. Then Q and P have the same x-coordinate, so the x-coordinate of Q is a. Q is on the x-axis, so the coordinates of Q are $(a, 0)$.

25. **Given:** isosceles $\triangle ABC$ with $\overline{AC} \cong \overline{BC}$; R and S are midpoints of legs $\overline{AC}$ and $\overline{BC}$.
Prove: $\overline{AS} \cong \overline{BR}$

Proof: The coordinates of R are $\left(\frac{2a+0}{2}, \frac{2b+0}{2}\right)$ or (a, b).
The coordinates of S are $\left(\frac{2a+4a}{2}, \frac{2b+0}{2}\right)$ or $(3a, b)$.

$BR = \sqrt{(4a-a)^2 + (0-b)^2}$
$= \sqrt{(3a)^2 + (-b)^2}$ or $\sqrt{9a^2 + b^2}$
$AS = \sqrt{(3a-0)^2 + (b-0)^2}$
$= \sqrt{(3a)^2 + (b)^2}$ or $\sqrt{9a^2 + b^2}$
Since $BR = AS$, $\overline{AS} \cong \overline{BR}$.

26. **Given:** isosceles $\triangle ABC$ with $\overline{BC} \cong \overline{AC}$; R, S, and T are midpoints of their respective sides.
Prove: $\triangle RST$ is isosceles.

Proof:
Midpoint R is $\left(\frac{a+0}{2}, \frac{b+0}{2}\right)$ or $\left(\frac{a}{2}, \frac{b}{2}\right)$.
Midpoint S is $\left(\frac{a+2a}{2}, \frac{b+0}{2}\right)$ or $\left(\frac{3a}{2}, \frac{b}{2}\right)$.
Midpoint T is $\left(\frac{2a+0}{2}, \frac{0+0}{2}\right)$ or $(a, 0)$.

$RT = \sqrt{\left(\frac{a}{2} - a\right)^2 + \left(\frac{b}{2} - 0\right)^2}$
$= \sqrt{\left(-\frac{a}{2}\right)^2 + \left(\frac{b}{2}\right)^2}$
$= \sqrt{\left(\frac{a}{2}\right)^2 + \left(\frac{b}{2}\right)^2}$
$ST = \sqrt{\left(\frac{3a}{2} - a\right)^2 + \left(\frac{b}{2} - 0\right)^2}$
$= \sqrt{\left(\frac{a}{2}\right)^2 + \left(\frac{b}{2}\right)^2}$
$RT = ST$ and $\overline{RT} \cong \overline{ST}$ and $\triangle RST$ is isosceles.

27. **Given:** $\triangle ABC$; S is the midpoint of $\overline{AC}$. T is the midpoint of $\overline{BC}$.
Prove: $\overline{ST} \parallel \overline{AB}$

Proof:
Midpoint S is $\left(\frac{b+0}{2}, \frac{c+0}{2}\right)$ or $\left(\frac{b}{2}, \frac{c}{2}\right)$.
Midpoint T is $\left(\frac{a+b}{2}, \frac{0+c}{2}\right)$ or $\left(\frac{a+b}{2}, \frac{c}{2}\right)$.

Slope of $\overline{ST} = \dfrac{\frac{c}{2} - \frac{c}{2}}{\frac{a+b}{2} - \frac{b}{2}} = \dfrac{0}{\frac{a}{2}}$ or 0.

Slope of $\overline{AB} = \dfrac{0-0}{a-0} = \dfrac{0}{a}$ or 0.
$\overline{ST}$ and $\overline{AB}$ have the same slope so $\overline{ST} \parallel \overline{AB}$.

28. Given: $\triangle ABC$
S is the midpoint of $\overline{AC}$.
T is the midpoint of $\overline{BC}$.
Prove: $ST = \frac{1}{2}AB$

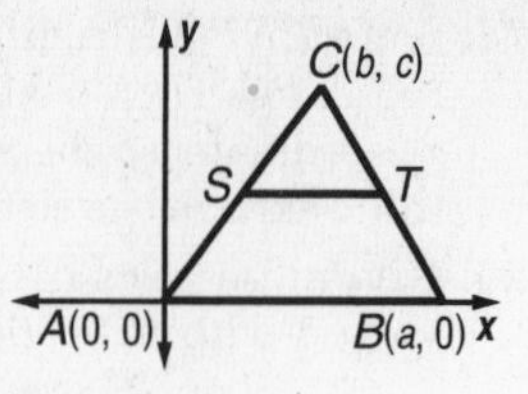

Midpoint S is $\left(\frac{b+0}{2}, \frac{c+0}{2}\right)$ or $\left(\frac{b}{2}, \frac{c}{2}\right)$.
Midpoint T is $\left(\frac{a+b}{2}, \frac{0+c}{2}\right)$ or $\left(\frac{a+b}{2}, \frac{c}{2}\right)$.

Proof: $ST = \sqrt{\left(\frac{a+b}{2} - \frac{b}{2}\right)^2 + \left(\frac{c}{2} - \frac{c}{2}\right)^2}$

$= \sqrt{\left(\frac{a}{2}\right)^2 + 0^2}$

$= \sqrt{\left(\frac{a}{2}\right)^2}$ or $\frac{a}{2}$

$AB = \sqrt{(a-0)^2 + (0-0)^2}$

$= \sqrt{a^2 + 0^2}$ or a

$ST = \frac{1}{2}AB$

29. Given: $\triangle ABD$, $\triangle FBD$
$AF = 6$, $BD = 3$
Prove: $\triangle ABD \cong \triangle FBD$

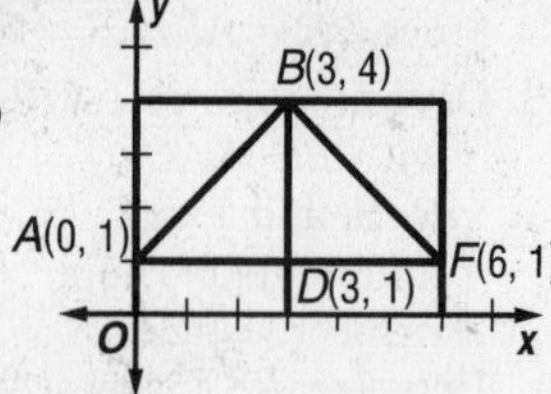

Proof: $\overline{BD} \cong \overline{BD}$ by the Reflexive Property.

$AD = \sqrt{(3-0)^2 + (1-1)^2} = \sqrt{9+0}$ or 3

$DF = \sqrt{(6-3)^2 + (1-1)^2} = \sqrt{9+0}$ or 3

Since $AD = DF$, $\overline{AD} \cong \overline{DF}$.

$AB = \sqrt{(3-0)^2 + (4-1)^2} = \sqrt{9+9}$ or $3\sqrt{2}$

$BF = \sqrt{(6-3)^2 + (1-4)^2} = \sqrt{9+9}$ or $3\sqrt{2}$

Since $AB = BF$, $\overline{AB} \cong \overline{BF}$.

$\triangle ABD \cong \triangle FBD$ by SSS.

30. Given: $\triangle BPR$
$PR = 800$,
$BR = 800$
Prove: $\triangle BPR$ is an isosceles right triangle.

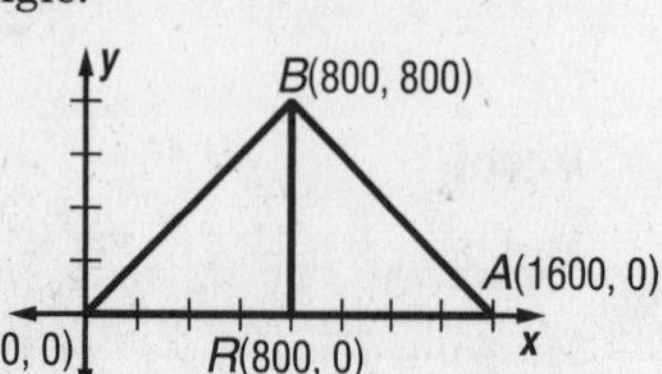

Proof: Since PR and BR have the same measure, $\overline{PR} \cong \overline{BR}$.

The slope of $PR = \frac{0-0}{800-0}$ or 0.

The slope of $BR = \frac{800-0}{800-800}$, which is undefined.

$\overline{PR} \perp \overline{BR}$, so $\angle PRB$ is a right angle. $\triangle BPR$ is an isosceles right triangle.

31. Given: $\triangle BPR$,
$\triangle BAR$
$PR = 800$,
$BR = 800$,
$RA = 800$
Prove: $\overline{PB} \cong \overline{BA}$

y
B(800, 800)
A(1600, 0)
P(0, 0)
R(800, 0)
x

Proof:

$PB = \sqrt{(800-0)^2 + (800-0)^2}$ or $\sqrt{1{,}280{,}000}$

$BA = \sqrt{(800-1600)^2 + (800-0)^2}$ or $\sqrt{1{,}280{,}000}$

$PB = BA$, so $\overline{PB} \cong \overline{BA}$.

32. Given: $\triangle JCT$
Prove: $\triangle JCT$ is a right triangle.

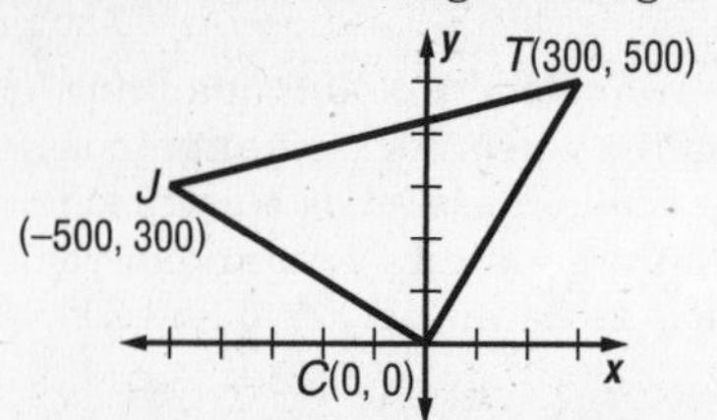

Proof: The slope of $\overline{JC} = \frac{300-0}{-500-0}$ or $-\frac{3}{5}$.

The slope of $\overline{TC} = \frac{500-0}{300-0}$ or $\frac{5}{3}$.

The slope of $\overline{TC}$ is the opposite reciprocal of the slope of $\overline{JC}$. $\overline{JC} \perp \overline{TC}$, so $\angle TCJ$ is a right angle. $\triangle JCT$ is a right triangle.

33. Use the Distance Formula to find the distance between $J(-500, 300)$ and $T(300, 500)$.

$JT = \sqrt{[300-(-500)]^2 + (500-300)^2}$

$= \sqrt{680{,}000}$

≈ 824.6

The distance between Tami and Juan is $\sqrt{680{,}000}$ or approximately 824.6 feet.

34. X is at the origin, so place $Y(a, b)$ in the first quadrant. Draw a perpendicular segment from Y to the x-axis, label the intersection point Z. Z has the same x-coordinate as Y, or a. Z is on the x-axis, so the coordinates of Z are $(a, 0)$.

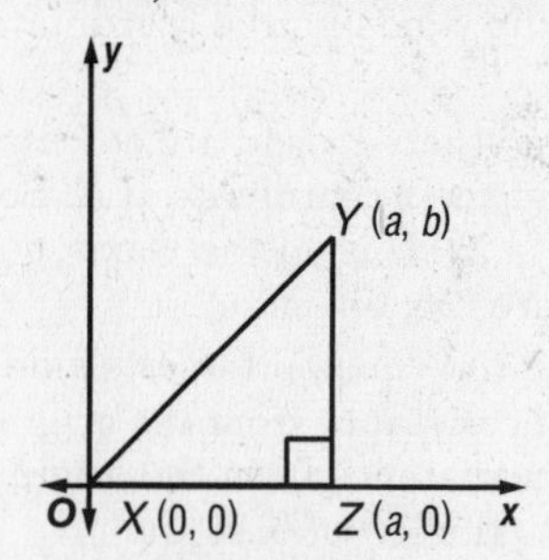

35. X is at the origin, so place $Y(a, b)$ in the first quadrant. $\triangle XYZ$ is isosceles, so place Z on the x-axis so that the x-coordinate of Y is halfway between 0 and the x-coordinate of Z. So, the coordinates of Z are $(2a, 0)$.

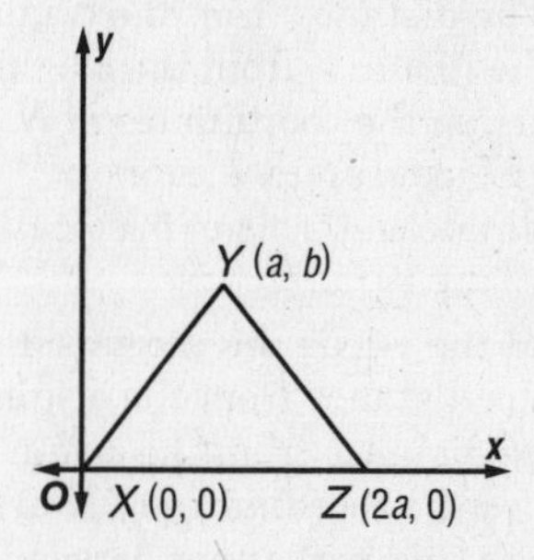

36. Sample answer: X is at the origin, so place $Y(a, b)$ in the first quadrant. $\triangle XYZ$ is scalene, so place Z on the x-axis so that the x-coordinate c is such that $XZ \neq YZ \neq XY$. So, the coordinates of Z are $(c, 0)$.

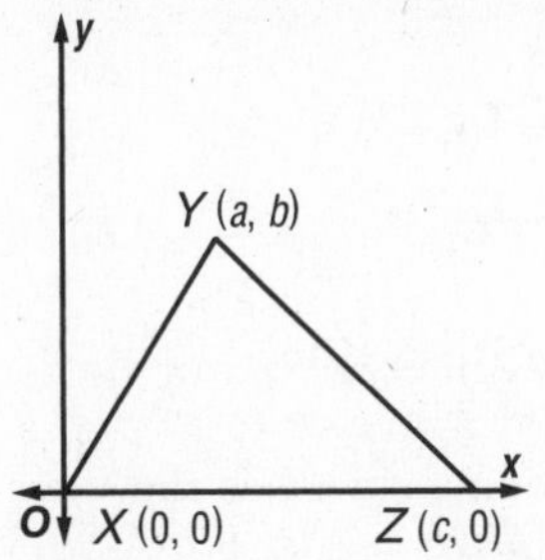

37. $AB = 4a$

$AC = \sqrt{(0 - (-2a))^2 + (2a - 0)^2}$
$= \sqrt{4a^2 + 4a^2}$ or $\sqrt{8a^2}$

$CB = \sqrt{(0 - 2a)^2 + (2a - 0)^2}$
$= \sqrt{4a^2 + 4a^2}$ or $\sqrt{8a^2}$

Slope of $\overline{AC} = \frac{2a - 0}{0 - (-2a)}$ or 1;

slope of $\overline{CB} = \frac{2a - 0}{0 - 2a}$ or -1.

$\overline{AC} \perp \overline{CB}$ and $\overline{AC} \cong \overline{CB}$, so $\triangle ABC$ is a right isosceles triangle.

38. Sample answer: Placing the figures on the coordinate plane is useful in proofs. We can use coordinate geometry to prove theorems and verify properties. Answers should include the following.

- flow proof, two-column proofs, paragraph proofs, informal proofs, and coordinate proofs
- Sample answer: The Isosceles Triangle Theorem can be proved using coordinate proof.

39. C; $d = \sqrt{(-3 - 1)^2 + [1 - (-2)]^2}$
$= \sqrt{16 + 9}$ or 5

40. B; $\left(\frac{x_1 + x_2}{2}, \frac{y_1 + y_2}{2}\right) = \left(\frac{-5 + (-2)}{2}, \frac{4 + (-1)}{2}\right)$
$= (-3.5, 1.5)$

Page 226 Maintain Your Skills

41. Given: $\angle 3 \cong \angle 4$
Prove: $\overline{QR} \cong \overline{QS}$

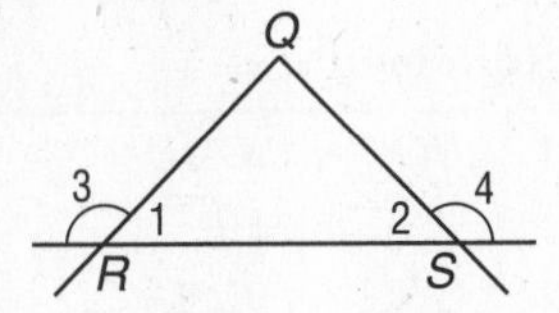

Proof:

Statements	Reasons
1. $\angle 3 \cong \angle 4$	1. Given
2. $\angle 2$ and $\angle 4$ form a linear pair. $\angle 1$ and $\angle 3$ form a linear pair.	2. Def. of linear pair
3. $\angle 2$ and $\angle 4$ are supplementary. $\angle 1$ and $\angle 3$ are supplementary.	3. If 2 ∠s form a linear pair, then they are suppl.
4. $\angle 2 \cong \angle 1$	4. Angles that are suppl. to ≅ ∠s are ≅.
5. $\overline{QR} \cong \overline{QS}$	5. Conv. of Isos. △ Th.

42. Given: isosceles triangle JKN with vertex $\angle N$, $\overline{JK} \parallel \overline{LM}$
Prove: $\triangle NML$ is isosceles

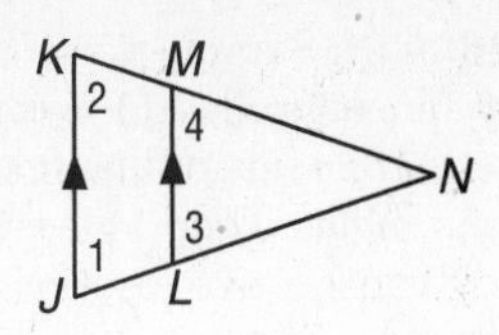

Proof:

Statements	Reasons
1. isosceles triangle JKN with vertex $\angle N$	1. Given
2. $\overline{NJ} \cong \overline{NK}$	2. Def. of isosceles triangle
3. $\angle 2 \cong \angle 1$	3. Isosceles Triangle Theorem
4. $\overline{JK} \parallel \overline{LM}$	4. Given
5. $\angle 1 \cong \angle 3$, $\angle 4 \cong \angle 2$	5. Corr. ∠s are ≅.
6. $\angle 2 \cong \angle 3$, $\angle 4 \cong \angle 1$	6. Congruence of ∠s is transitive. (Statements 3 and 5)
7. $\angle 4 \cong \angle 3$	7. Congruence of ∠s is transitive. (Statements 3 and 6)
8. $\overline{LN} \cong \overline{MN}$	8. If 2 ∠s of a △ are ≅, then the sides opp. those ∠s are ≅.
9. $\triangle NML$ is an isosceles triangle.	9. Def. of isosceles triangle

43. Given: $\overline{AD} \cong \overline{CE}$, $\overline{AD} \parallel \overline{CE}$
Prove: $\triangle ABD \cong \triangle EBC$

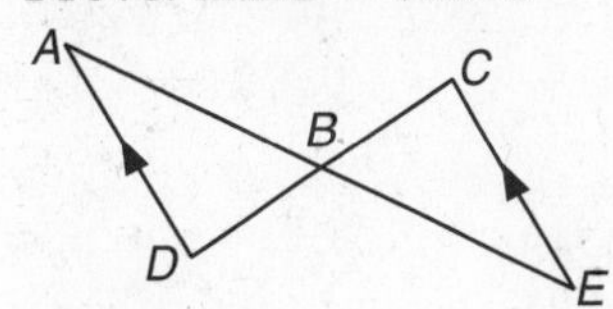

Proof:

Statements	Reasons
1. $\overline{AD} \parallel \overline{CE}$	1. Given
2. $\angle A \cong \angle E$, $\angle D \cong \angle C$	2. Alt. int. ∠s are ≅.
3. $\overline{AD} \cong \overline{CE}$	3. Given
4. $\triangle ABD \cong \triangle EBC$	4. ASA

44. Given: $\overline{WX} \cong \overline{XY}$, $\angle V \cong \angle Z$
Prove: $\overline{WV} \cong \overline{YZ}$

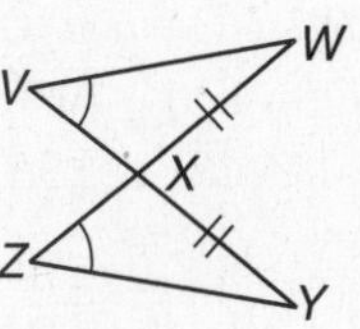

Proof:

Statements	Reasons
1. $\overline{WX} \cong \overline{XY}$, $\angle V \cong \angle Z$	1. Given
2. $\angle WXV \cong \angle YXZ$	2. Vert. ∠s are ≅.
3. $\triangle WXV \cong \triangle YXZ$	3. AAS
4. $\overline{WV} \cong \overline{YZ}$	4. CPCTC

45. $\angle BCA \cong \angle DAC$, so $\overline{BC} \parallel \overline{AD}$; if alternate interior ∠s are ≅, lines are ∥.

46. $h \parallel j$; Sample answer: The angle in the figure whose measure is 111 is congruent to the angle on the other side of line h such that the angles are vertical angles. Then $111 + 69 = 180$, so $h \parallel j$ because consecutive interior angles are supplementary.

47. $l \parallel m$; 2 lines $\perp$ to the same line are parallel

Chapter 4 Study Guide and Review

Page 227 Vocabulary and Concept Check

1. h	**2.** g	**3.** d
4. j	**5.** a	**6.** c
7. b	**8.** f	

Pages 228–230 Lesson-by-Lesson Review

9. $\triangle ABC$ has one angle with measure greater than 90, so it is an obtuse triangle. $\triangle ABC$ has two congruent sides, so it is isosceles.

10. $\triangle BDP$ has a right angle, so it is a right triangle. The measure of one of the acute angles is 60 so the measure of the other acute angle is $90 - 60$ or 30. $\triangle BDP$ cannot be isosceles or equilateral so it must be scalene.

11. $\triangle BPQ$ has at least two congruent sides, and one of the base angles is 60° so the other base angle must also be 60°. Then the third angle is $180 - (60 + 60)$ or 60 so $\triangle BPQ$ is equiangular and hence equilateral.

12. $m\angle 1 = 45 + 40$
$= 85$

13. $m\angle 2 + m\angle 1 + 70 = 180$
$m\angle 2 + 85 + 70 = 180$
$m\angle 2 = 25$

14. $m\angle 3 + m\angle 1 = 180$
$m\angle 3 + 85 = 180$
$m\angle 3 = 95$

15. $\angle E \cong \angle D$, $\angle F \cong \angle C$, $\angle G \cong \angle B$, $\overline{EF} \cong \overline{DC}$, $\overline{FG} \cong \overline{CB}$, $\overline{GE} \cong \overline{BD}$

16. $\angle FGC \cong \angle DLC$, $\angle GCF \cong \angle LCD$, $\angle GFC \cong \angle LDC$, $\overline{GC} \cong \overline{LC}$, $\overline{CF} \cong \overline{CD}$, $\overline{FG} \cong \overline{DL}$

17. $\angle KNC \cong \angle RKE$, $\angle NCK \cong \angle KER$, $\angle CKN \cong \angle ERK$, $\overline{NC} \cong \overline{KE}$, $\overline{CK} \cong \overline{ER}$, $\overline{KN} \cong \overline{RK}$

18. $MN = \sqrt{(-4 - 0)^2 + (3 - 3)^2}$
$= \sqrt{16 + 0}$ or 4
$QR = \sqrt{(2 - 5)^2 + (6 - 6)^2}$
$= \sqrt{9 + 0}$ or 3
$NP = \sqrt{[-4 - (-4)]^2 + (6 - 3)^2}$
$= \sqrt{0 + 9}$ or 3
$RS = \sqrt{(2 - 2)^2 + (2 - 6)^2}$
$= \sqrt{0 + 16}$ or 4
$MP = \sqrt{(-4 - 0)^2 + (6 - 3)^2}$
$= \sqrt{16 + 9}$ or 5
$QS = \sqrt{(2 - 5)^2 + (2 - 6)^2}$
$= \sqrt{9 + 16}$ or 5

Each pair of corresponding sides does not have the same measure. Therefore, $\triangle MNP$ is not congruent to $\triangle QRS$. $\triangle MNP$ is congruent to $\triangle SRQ$.

19. $MN = \sqrt{(7 - 3)^2 + (4 - 2)^2}$
$= \sqrt{16 + 4}$ or $\sqrt{20}$
$QR = \sqrt{[-4 - (-2)]^2 + (7 - 3)^2}$
$= \sqrt{4 + 16}$ or $\sqrt{20}$
$NP = \sqrt{(6 - 7)^2 + (6 - 4)^2}$
$= \sqrt{1 + 4}$ or $\sqrt{5}$
$RS = \sqrt{[-6 - (-4)]^2 + (6 - 7)^2}$
$= \sqrt{4 + 1}$ or $\sqrt{5}$
$MP = \sqrt{(6 - 3)^2 + (6 - 2)^2}$
$= \sqrt{9 + 16}$ or 5
$QS = \sqrt{[-6 - (-2)]^2 + (6 - 3)^2}$
$= \sqrt{16 + 9}$ or 5

Each pair of corresponding sides has the same measure. Therefore, $\triangle MNP \cong \triangle QRS$ by SSS.

20. Given: $\overline{DF}$ bisects $\angle CDE$, $\overline{CE} \perp \overline{DF}$.
Prove: $\triangle DGC \cong \triangle DGE$

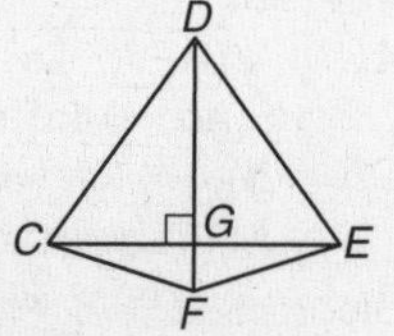

Proof:

Statements	Reasons
1. $\overline{DF}$ bisects $\angle CDE$, $\overline{CE} \perp \overline{DF}$.	1. Given
2. $\overline{DG} \cong \overline{DG}$	2. Reflexive Property
3. $\angle CDF \cong \angle EDF$	3. Def. of $\angle$ bisector
4. $\angle DGC$ is a rt. $\angle$; $\angle DGE$ is a rt. $\angle$	4. Def. of $\perp$ segments
5. $\angle DGC \cong \angle DGE$	5. All rt. $\angle$s are $\cong$.
6. $\triangle DGC \cong \triangle DGE$	6. ASA

21. Given: $\triangle DGC \cong \triangle DGE$, $\triangle GCF \cong \triangle GEF$
Prove: $\triangle DFC \cong \triangle DFE$

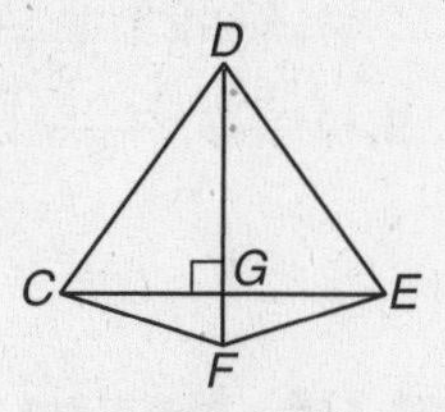

Proof:

Statements	Reasons
1. $\triangle DGC \cong \triangle DGE$, $\triangle GCF \cong \triangle GEF$	1. Given
2. $\angle CDG \cong \angle EDG$, $\overline{CD} \cong \overline{ED}$, and $\angle CFD \cong \angle EFD$	2. CPCTC
3. $\triangle DFC \cong \triangle DFE$	3. AAS

22. $\triangle PQU$ is isosceles with base $\overline{PU}$. The base angles of $\triangle PQU$ are congruent, so $\angle P \cong \angle PUQ$. We are given $m\angle P = 32$, so $m\angle PUQ = 32$.

23. $\triangle PQU$ is isosceles with base $\overline{PU}$. $\triangle PRT$ is isosceles with base $\overline{PT}$.

Step 1 The base angles of $\triangle PQU$ are congruent. Let x represent $m\angle P$ and $m\angle QUP$.

$$m\angle P + m\angle QUP + m\angle PQU = 180$$
$$x + x + 40 = 180$$
$$2x = 140$$
$$x = 70$$

So, $m\angle P = m\angle QUP = 70$.

Step 2 The base angles of $\triangle PRT$ are congruent. By Step 1 we know $m\angle P = 70$, so $m\angle T = 70$. Let y represent $m\angle R$.

$$m\angle P + m\angle T + m\angle R = 180$$
$$70 + 70 + y = 180$$
$$y = 40$$

So, $m\angle R = 40$.

24. $\triangle RQS$ is isosceles with base $\overline{QS}$. The base angles of $\triangle RQS$ are congruent, so $\angle RQS \cong \angle RSQ$. We are given $m\angle RQS = 75$, so $m\angle RSQ = 75$. Let y represent $m\angle R$.

$$m\angle RQS + m\angle RSQ + m\angle R = 180$$
$$75 + 75 + y = 180$$
$$y = 30$$

So, $m\angle R = 30$.

25. $\triangle RQS$ is isosceles with base $\overline{QS}$. $\triangle RPT$ is isosceles with base $\overline{PT}$.

Step 1 The base angles of $\triangle RQS$ are congruent, and $m\angle RQS = 80$ so $m\angle RSQ = 80$. Let x represent $m\angle R$.

$$m\angle RQS + m\angle RSQ + m\angle R = 180$$
$$80 + 80 + x = 180$$
$$x = 20$$

So, $m\angle R = 20$.

Step 2 The base angles of $\triangle RPT$ are congruent. Let y represent $m\angle P$ and $m\angle T$.

$$m\angle P + m\angle T + m\angle R = 180$$
$$y + y + 20 = 180$$
$$2y = 160$$
$$y = 80$$

So, $m\angle P = m\angle T = 80$.

26.
- Use the origin as vertex T of the triangle.
- Place the base of the triangle along the positive x-axis.
- Position the triangle in the first quadrant.
- Since I is on the x-axis, its y-coordinate is 0. Its x-coordinate is $4a$ because the base of the triangle is $4a$ units long.
- Since $\triangle TRI$ is isosceles, the x-coordinate of R is halfway between 0 and $4a$, or $2a$. We cannot determine the y-coordinate in terms of a, so call it b.

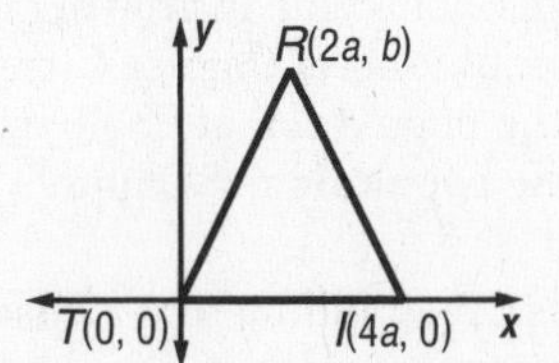

27.
- Use the origin as vertex B of the triangle.
- Place side $\overline{BD}$ along the positive x-axis.
- Position the triangle in the first quadrant.
- Since D is on the x-axis, its y-coordinate is 0. Its x-coordinate is $6m$ because each side length is $6m$ units long.
- Since $\triangle BCD$ is equilateral, the x-coordinate of C is halfway between 0 and $6m$, or $3m$. We cannot determine the y-coordinate in terms of m, so call it n.

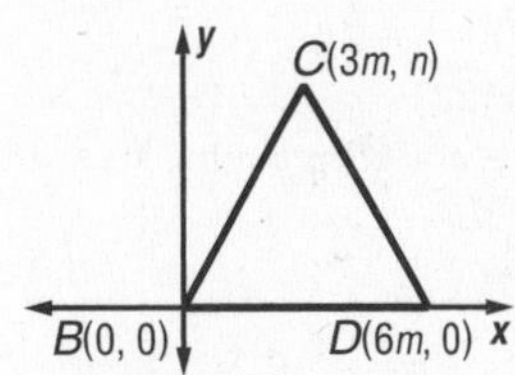

28.
- Use the origin as vertex K of the triangle, the right angle.
- Place leg $\overline{KL}$ along the positive x-axis.
- Place leg $\overline{JK}$ along the positive y-axis.
- Since L is on the x-axis, its y-coordinate is 0. Its x-coordinate is a because one of the leg lengths is a units.
- Since J is on the y-axis, its x-coordinate is 0. Its y-coordinate is b because the other leg length is b units.

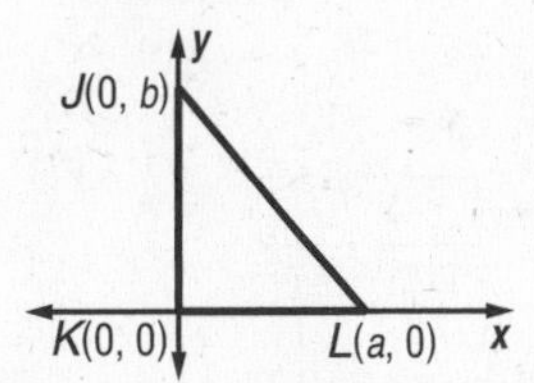

Chapter 4 Practice Test

Page 231

1. b **2.** a **3.** c

4. $\triangle PCD$ is obtuse because $\angle PCD$ has measure greater than 90.

5. $\triangle PAC$ is isosceles because $\overline{PA} \cong \overline{PC}$.

6. $\triangle PBA$, $\triangle PBC$, and $\triangle PBD$ are right triangles because $\overline{PB} \perp \overline{AD}$, so $\angle PBA$ and $\angle PBD$ are right angles.

7. $m\angle 1 + 100 = 180$
$m\angle 1 = 80$

8. $m\angle 2 + 75 = 180$
$m\angle 2 = 105$

9. $m\angle 3 + m\angle 1 = m\angle 2$
$m\angle 3 + 80 = 105$
$m\angle 3 = 25$

10. $\angle D \cong \angle P$, $\angle E \cong \angle Q$, $\angle F \cong \angle R$, $\overline{DE} \cong \overline{PQ}$, $\overline{EF} \cong \overline{QR}$, $\overline{DF} \cong \overline{PR}$

11. $\angle F \cong \angle H$, $\angle M \cong \angle N$, $\angle G \cong \angle J$, $\overline{FM} \cong \overline{HN}$, $\overline{MG} \cong \overline{NJ}$, $\overline{FG} \cong \overline{HJ}$

12. $\angle X \cong \angle Z$, $\angle Y \cong \angle Y$, $\angle Z \cong \angle X$, $\overline{XY} \cong \overline{ZY}$, $\overline{YZ} \cong \overline{YX}$, and $\overline{XZ} \cong \overline{ZX}$

13. $JK = \sqrt{[2 - (-1)]^2 + [-3 - (-2)]^2}$
$= \sqrt{9 + 1}$ or $\sqrt{10}$
$MN = \sqrt{[-2 - (-6)]^2 + [1 - (-7)]^2}$
$= \sqrt{16 + 64}$ or $\sqrt{80}$
$KL = \sqrt{(3 - 2)^2 + [1 - (-3)]^2}$
$= \sqrt{1 + 16}$ or $\sqrt{17}$
$NP = \sqrt{[5 - (-2)]^2 + (3 - 1)^2}$
$= \sqrt{49 + 4}$ or $\sqrt{53}$
$JL = \sqrt{[3 - (-1)]^2 + [1 - (-2)]^2}$
$= \sqrt{16 + 9}$ or 5
$MP = \sqrt{[5 - (-6)]^2 + [3 - (-7)]^2}$
$= \sqrt{121 + 100}$ or $\sqrt{221}$
Corresponding sides are not congruent, so $\triangle JKL$ is not congruent to $\triangle MNP$.

14. Given: $\triangle JKM \cong \triangle JNM$
Prove: $\triangle JKL \cong \triangle JNL$

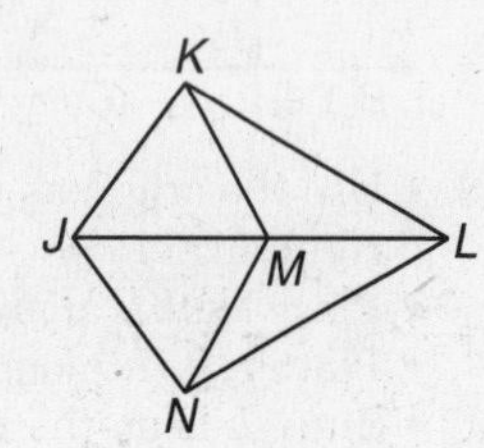

Proof:

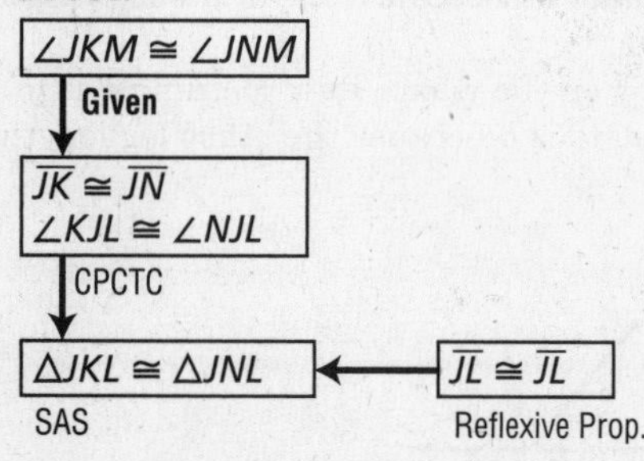

15. $\triangle FJH$ is isosceles with base $\overline{JH}$. The base angles are congruent, so $m\angle J = m\angle FHJ$. Let x represent $m\angle J$ and $m\angle FHJ$.
$m\angle J + m\angle FHJ + m\angle JFH = 180$
$x + x + 34 = 180$
$2x = 146$
$x = 73$
So, $m\angle J = m\angle FHJ = 73$.

16. $\triangle JFH$ is isosceles with base $\overline{JH}$. $\triangle FGH$ is isosceles with base $\overline{FH}$.
Step 1 The base angles of $\triangle FGH$ are congruent. Let x represent $m\angle GFH$ and $m\angle GHF$.
$m\angle GFH + m\angle GHF + m\angle G = 180$
$x + x + 32 = 180$
$2x = 148$
$x = 74$
So, $m\angle GFH = m\angle GHF = 74$.
Step 2 $\angle GHF + \angle FHJ = \angle GHJ$ by the Angle Addition Postulate.
$m\angle GHF + m\angle FHJ = m\angle GHJ$
$74 + m\angle FHJ = 152$
$m\angle FHJ = 78$
Step 3 The base angles of $\triangle JFH$ are congruent, so $\angle J \cong \angle FHJ$. Then $m\angle J = m\angle FHJ = 78$. Let y represent $m\angle JFH$.
$m\angle J + m\angle FHJ + m\angle JFH = 180$
$78 + 78 + y = 180$
$y = 24$
So, $m\angle JFH = 24$.

17. Given: $\triangle ABE$, $\triangle BCE$
$AB = 22$, $AC = 44$,
$AE = 36$, $CD = 36$
$\angle A$ and $\angle C$ are right angles.
Prove: $\triangle ABE \cong \triangle CBD$

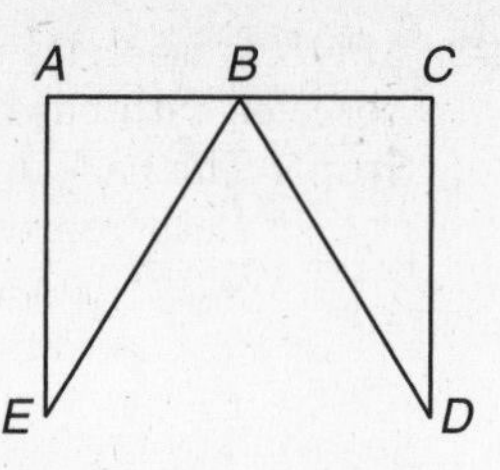

Proof: We are given that $AB = 22$ and $AC = 44$. By the Segment Addition Postulate,
$AB + BC = AC$
$22 + BC = 44$ Substitution
$BC = 22$ Subtract 22 from each side.
Since $AB = BC$, then by the definition of congruent segments, $\overline{AB} \cong \overline{BC}$.
We are given that $AE = 36$ and $CD = 36$. Then also by the definition of congruent segments, $\overline{AE} \cong \overline{CD}$.
We are additionally given that both $\angle A$ and $\angle C$ are right angles. Since all right angles are congruent, $\angle A \cong \angle C$.
Since $\overline{AB} \cong \overline{BC}$, $\angle A \cong \angle C$, and $\overline{AE} \cong \overline{CD}$, then by SAS, $\triangle ABE \cong \triangle CBD$.

18. C; $\triangle JGH$ is isosceles with base $\overline{JH}$.
Step 1 $\triangle FGH$ is a right triangle, so the sum of the measures of the two acute angles is 90.
$m\angle F + m\angle H = 90$
$28 + m\angle H = 90$
$m\angle H = 62$
Step 2 The base angles of $\triangle JGH$ are congruent, so $\angle H \cong \angle HJG$. Then $m\angle HJG = m\angle H = 62$. Let x represent $m\angle JGH$.
$m\angle H + m\angle HJG + m\angle JGH = 180$
$62 + 62 + x = 180$
$x = 56$
So, $m\angle JGH = 56$.

Chapter 4 Standardized Test Practice

Pages 232–233

1. B; find when the populations will be equal. Let x represent the number of years after 2002.
$2010 + 150x = 1040 + 340x$
$2010 = 1040 + 190x$
$970 = 190x$
$5.1 \approx x$
The populations will be equal about 5 years after 2002, or 2007.
Since the population of Shelbyville is growing at a faster rate than the population of Capitol City, the following year, 2008, the population of Shelbyville will be greater than the population of Capitol City.

2. C; grams measure mass, feet and meters measure length, and liters measure volumes of liquid.

3. B; let x represent the length of the shadow. Use the Pythagorean Theorem to solve for x to the nearest foot.
$12^2 = 9^2 + x^2$
$144 = 81 + x^2$
$63 = x^2$
$8 \approx x$
The shadow is about 8 feet long.

4. D

5. D; the slope of the line in Kris's graph can be found using points (0, 3) and (4, 11).
$\frac{(y_2 - y_1)}{(x_2 - x_1)} = \frac{11 - 3}{4 - 0}$ or 2
The slope of the line in Mitzi's graph is the same as the slope of the line in Kris's graph, or 2. So the line in Mitzi's graph has equation $y = 2x + b$, or $2x - y = -b$. The only answer of this form is $2x - y = 1$, so $-b = 1$ or $b = -1$.

6. B; $m\angle EFG = m\angle FDE + m\angle FED$
$9x + 7 = 5x + 5x$
$9x + 7 = 10x$
$7 = x$
$m\angle EFG = 9(7) + 7$
$= 70$

7. C; $\triangle ABD \cong \triangle CBD$, so $\triangle CBD$ is a flip of $\triangle ABD$ over the x-axis. Corresponding points have the same x-coordinate and opposite y-coordinates.

8. A; if we know $\overline{BC} \cong \overline{CE}$ then $\triangle ACB \cong \triangle DCE$ by SAS.

9. $3s^2(2s^3 - 7) = 6s^5 - 21s^2$

10. Brian's second statement was the converse of his original statement.

11. **Explore:** Creston (C) and Dixville (D) are endpoints of the base of an isosceles triangle formed by Creston, Dixville, and Milford. We are looking for the coordinates $(x, -1)$ to satisfy these conditions.

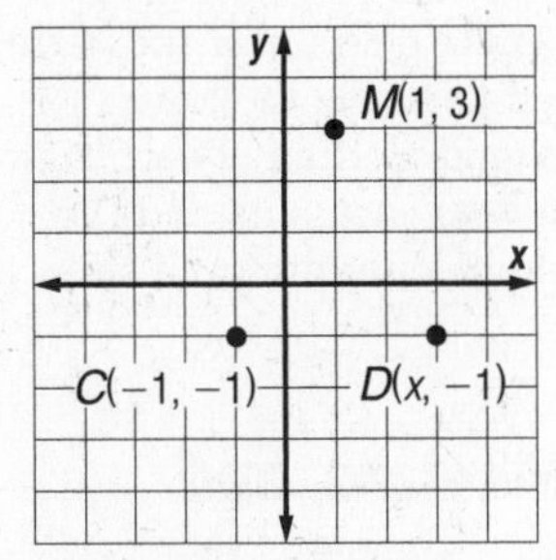

Solve: The x-coordinate of the vertex angle is halfway between the x-coordinates of C and D.
So, $1 = \frac{(-1 + x)}{2}$
$2 = -1 + x$
$3 = x$

Examine: C and D are the base vertices of the isosceles triangle formed by C, D, and M.

12. The angle adjacent to the 105° angle has measure 180 − 105 or 75. The tower is an isosceles triangle, so its base angles are congruent. Let x represent the measure of the angle at the top of the tower.
$x + 75 + 75 = 180$
$x = 30$
The measure of the angle at the top of the tower is 30.

13. $\angle BCA \cong \angle EFD$ since all right angles are congruent. $\overline{AC} \cong \overline{DF}$ since the planes are equidistant from the ground. $\angle CAB \cong \angle FDE$ since the planes descend at the same angle. So, $\triangle ABC \cong \triangle DEF$ by ASA.

14. Let x represent $m\angle A$. $\overline{AB} \cong \overline{BC}$ so $\angle A \cong \angle C$ and $m\angle C = m\angle A = x$. Also, $m\angle B = 3(m\angle A)$ or $3x$.
$x + x + 3x = 180$
$5x = 180$
$x = 36$
$m\angle C = x$
$= 36$

15a. The railroad ties that run across train tracks are parallel. So, $x = 90$ because the angle whose measure is x is supplementary to an angle that corresponds to the angle whose measure is known to be 90.

15b. Perpendicular lines, because the ties are parallel and the tracks are parallel so all angles are 90° angles.

15c. They are congruent. Sample answers: Both are right angles; they are supplementary angles; they are corresponding angles.

16a. Sample answer:

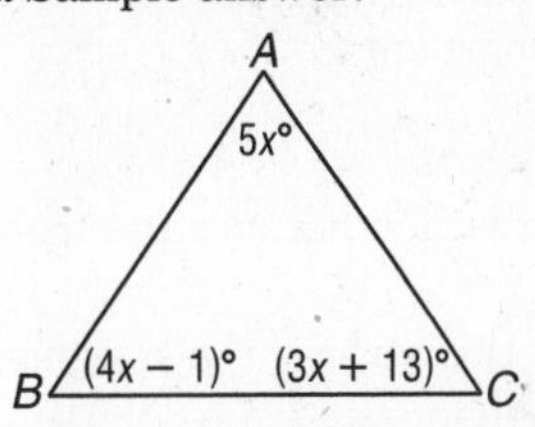

16b. From the Angle Sum Theorem, we know that $m\angle A + m\angle B + m\angle C = 180$. Substituting the given measures, $5x + 4x - 1 + 3x + 13 = 180$. Solve for x to find that $x = 14$. Substitute 14 for x to find the measures: $5x = 5(14)$ or 70, $4x - 1 = 4(14) - 1$ or 55, and $3x + 13 = 3(14) + 13$ or 55.

16c. If two angles of a triangle are congruent, then the sides opposite those angles are congruent (Converse of the Isosceles Triangle Theorem). Since two sides of this triangle are congruent, it is an isosceles triangle (Definition of Isosceles Triangle).

Chapter 5 Relationships in Triangles

Page 235 Getting Started

1. $\left(\frac{x_1 + x_2}{2}, \frac{y_1 + y_2}{2}\right) = \left(\frac{-12 + 4}{2}, \frac{-5 + 15}{2}\right)$
$= (-4, 5)$
2. $\left(\frac{x_1 + x_2}{2}, \frac{y_1 + y_2}{2}\right) = \left(\frac{-22 + 10}{2}, \frac{-25 + 10}{2}\right)$
$= (-6, -7.5)$
3. $\left(\frac{x_1 + x_2}{2}, \frac{y_1 + y_2}{2}\right) = \left(\frac{19 + (-20)}{2}, \frac{-7 + (-3)}{2}\right)$
$= (-0.5, -5)$
4. $m\angle 1 + 104 = 180$
$m\angle 1 = 76$
5. $m\angle 2 + 36 = 104$
$m\angle 2 = 68$
6. $m\angle 3 + 104 = 180$
$m\angle 3 = 76$
7. $m\angle 4 = 40$
8. $m\angle 5 + m\angle 4 = 104$
$m\angle 5 + 40 = 104$
$m\angle 5 = 64$
9. $m\angle 5 + m\angle 6 = 90$
$64 + m\angle 6 = 90$
$m\angle 6 = 26$
10. $m\angle 7 + 40 = 180$
$m\angle 7 = 140$
11. $m\angle 8 + m\angle 6 = m\angle 4$
$m\angle 8 + 26 = 40$
$m\angle 8 = 14$
12. Let p and q represent the parts of the statement.
p: the three sides of one triangle are congruent to the three sides of a second triangle
q: the triangles are congruent
Statement (1): $p \to q$
Statement (2): q
No conclusion can be reached because the truth of p is unknown.
13. Let p and q represent the parts of the statement.
p: a polygon is a triangle
q: the sum of the measures of the angles is 180
Statement (1): $p \to q$
Statement (2): p
Since the given statements are true, use the Law of Detachment to conclude that the sum of the measures of the angles of polygon JKL is 180.

Pages 236–237 Geometry Activity: Bisectors, Medians, and Altitudes

1.
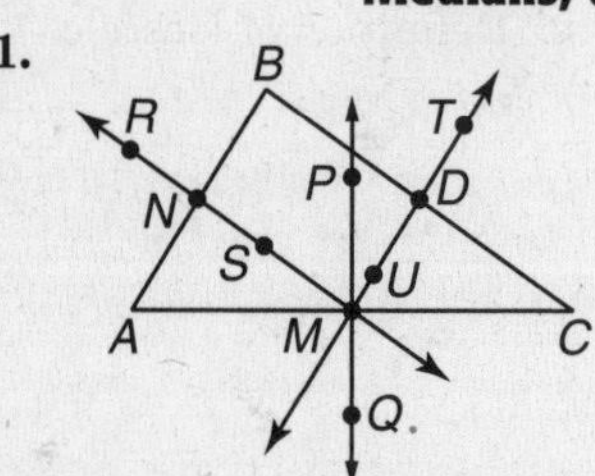

2. They intersect at the same point.

3.
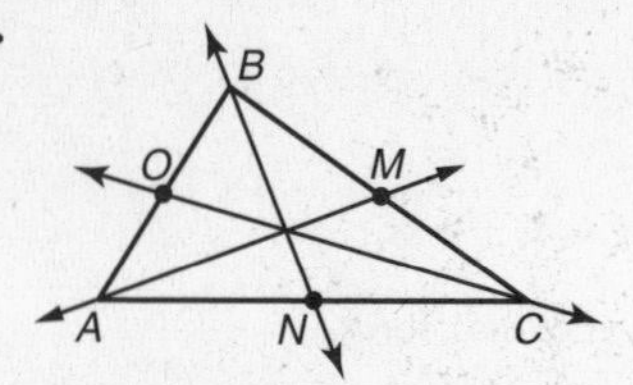

4. They intersect at the same point.

5.
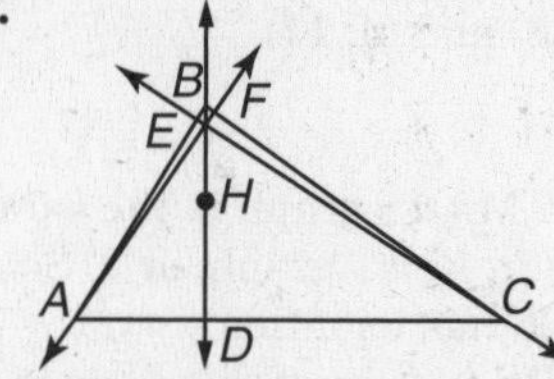

6. They intersect at the same point.

7.
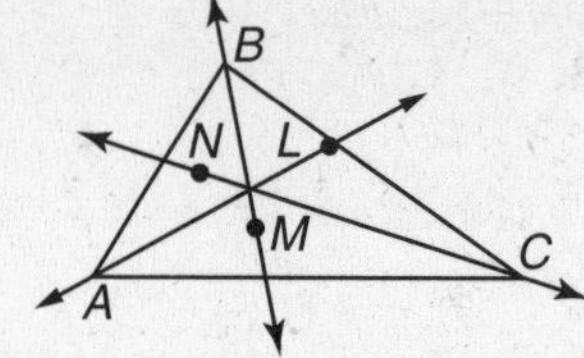

8. They intersect at the same point.
9. See students' work.
10. Acute: all intersect inside the triangle; obtuse: perpendicular bisectors and altitudes intersect outside the triangle; medians and angle bisectors intersect inside the triangle; right: perpendicular bisectors intersect on the hypotenuse, medians intersect inside the triangle, altitudes intersect on the vertex of the right angle, and angle bisectors intersect inside the triangle.
11. For an isosceles triangle, the perpendicular bisector and median of the side opposite the vertex are the same as the altitude from the vertex angle and the angle bisector of the vertex angle. In an equilateral triangle, the perpendicular bisector and median of each side is the same as the altitude to each side and the angle bisector of the angle opposite each side.

5-1 Bisectors, Medians, and Altitudes

Page 242 Check for Understanding

1. Sample answer: Both pass through the midpoint of a side. A perpendicular bisector is perpendicular to the side of a triangle, and does not necessarily pass through the vertex opposite the side, while a median does pass through the vertex and is not necessarily perpendicular to the side.

2. Sample answer:

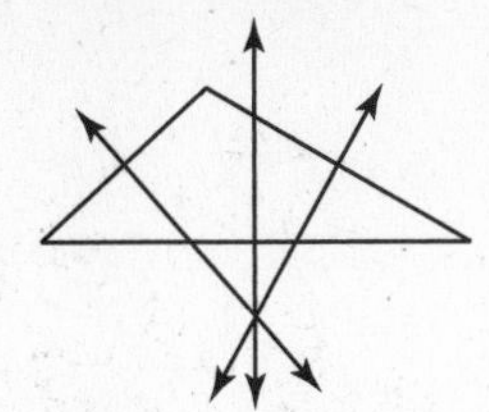

3. Sample answer: An altitude and angle bisector of a triangle are the same segment in an equilateral triangle.

4. Find an equation of the perpendicular bisector of $\overline{AB}$.

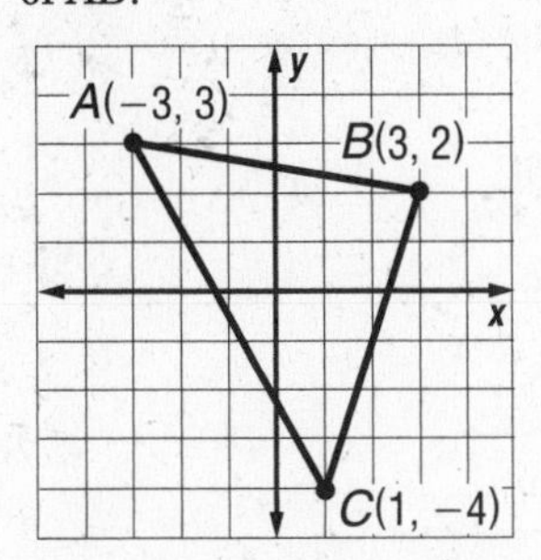

The slope of $\overline{AB}$ is $-\frac{1}{6}$, so the slope of the perpendicular bisector is 6. The midpoint of $\overline{AB}$ is $\left(0, \frac{5}{2}\right)$.

$y - y_1 = m(x - x_1)$

$y - \frac{5}{2} = 6(x - 0)$

$y - \frac{5}{2} = 6x$

$y = 6x + \frac{5}{2}$

Find an equation of the perpendicular bisector of $\overline{BC}$. The slope of $\overline{BC}$ is 3, so the slope of the perpendicular bisector of $\overline{BC}$ is $-\frac{1}{3}$. The midpoint of $\overline{BC}$ is $(2, -1)$.

$y - y_1 = m(x - x_1)$

$y - (-1) = -\frac{1}{3}(x - 2)$

$y + 1 = -\frac{1}{3}x + \frac{2}{3}$

$y = -\frac{1}{3}x - \frac{1}{3}$

Solve a system of equations to find the point of intersection of the perpendicular bisectors.

Find x.

$y = 6x + \frac{5}{2}$

$-\frac{1}{3}x - \frac{1}{3} = 6x + \frac{5}{2}$

$-2x - 2 = 36x + 15$

$-17 = 38x$

$-\frac{17}{38} = x$

Replace x with $-\frac{17}{38}$ in one of the equations to find the y-coordinate.

$y = 6\left(-\frac{17}{38}\right) + \frac{5}{2}$

$y = -\frac{7}{38}$

The coordinates of the circumcenter of $\triangle ABC$ are $\left(-\frac{17}{38}, -\frac{7}{38}\right)$.

5. Given: $\overline{XY} \cong \overline{XZ}$

$\overline{YM}$ and $\overline{ZN}$ are medians.

Prove: $\overline{YM} \cong \overline{ZN}$

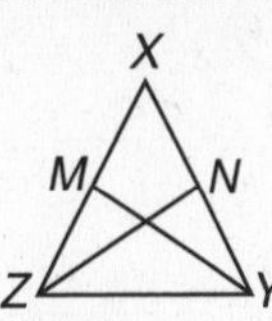

Proof:

Statements	Reasons
1. $\overline{XY} \cong \overline{XZ}$, $\overline{YM}$ and $\overline{ZN}$ are medians.	1. Given
2. M is the midpoint of $\overline{XZ}$. N is the midpoint of $\overline{XY}$.	2. Def. of median
3. $XY = XZ$	3. Def. of $\cong$
4. $\overline{XM} \cong \overline{MZ}$, $\overline{XN} \cong \overline{NY}$	4. Midpoint Theorem
5. $XM = MZ$, $XN = NY$	5. Def. of $\cong$
6. $XM + MZ = XZ$, $XN + NY = XY$	6. Segment Addition Postulate
7. $XM + MZ = XN + NY$	7. Substitution
8. $MZ + MZ = NY + NY$	8. Substitution
9. $2MZ = 2NY$	9. Addition Property
10. $MZ = NY$	10. Division Property
11. $\overline{MZ} \cong \overline{NY}$	11. Def. of $\cong$
12. $\angle XZY \cong \angle XYZ$	12. Isosceles Triangle Theorem
13. $\overline{YZ} \cong \overline{YZ}$	13. Reflexive Property
14. $\triangle MYZ \cong \triangle NZY$	14. SAS
15. $\overline{YM} \cong \overline{ZN}$	15. CPCTC

6. Find x.

$TQ = TR$

$2x = 8$

$x = 4$

Find y.

$PT = TR$

$3y - 1 = 8$

$3y = 9$

$y = 3$

Find z.

Line ℓ bisects $\overline{PR}$, so $z + 4 = 7$, or $z = 3$.

Pages 243–245 Practice and Apply

7. Find an equation of the median from $D(4, 0)$ to the midpoint of $\overline{EF}$. The midpoint of $\overline{EF}$ is $(-1, 5)$. The slope of the median is -1.

$y - y_1 = m(x - x_1)$

$y - 0 = -1(x - 4)$

$y = -x + 4$

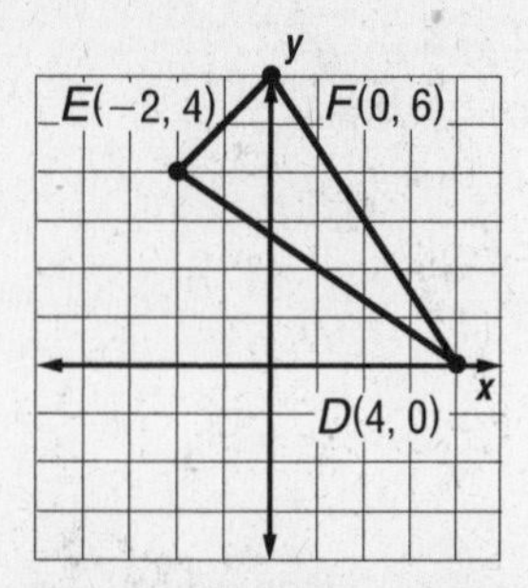

Find an equation of the median from $F(0, 6)$ to the midpoint of $\overline{DE}$. The midpoint of $\overline{DE}$ is (1, 2). The slope of the median is −4.

$y - y_1 = m(x - x_1)$
$y - 6 = -4(x - 0)$
$y - 6 = -4x$
$y = -4x + 6$

Solve a system of equations to find the point of intersection of the medians.

Find x.

$y = -x + 4$
$y = -4x + 6$
$-4x + 6 = -x + 4$
$6 = 3x + 4$
$2 = 3x$
$\frac{2}{3} = x$

Replace x with $\frac{2}{3}$ in one of the equations to find the y-coordinate.

$y = -\frac{2}{3} + 4$
$y = 3\frac{1}{3}$

The coordinates of the centroid are $\left(\frac{2}{3}, 3\frac{1}{3}\right)$.

8.

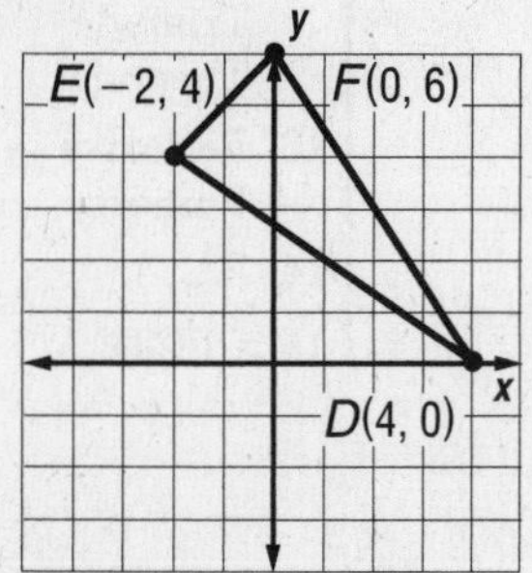

Find an equation of the altitude from $D(4, 0)$ to $\overline{EF}$. The slope of $\overline{EF}$ is 1, so the slope of the altitude is −1.

$y - y_1 = m(x - x_1)$
$y - 0 = -1(x - 4)$
$y = -x + 4$

Find an equation of the altitude from $F(0, 6)$ to $\overline{DE}$. The slope of $\overline{DE}$ is $-\frac{2}{3}$, so the slope of the altitude is $\frac{3}{2}$.

$y - y_1 = m(x - x_1)$
$y - 6 = \frac{3}{2}(x - 0)$
$y - 6 = \frac{3}{2}x$
$y = \frac{3}{2}x + 6$

Solve a system of equations to find the point of intersection of the altitudes.

Find x.

$y = -x + 4$
$y = \frac{3}{2}x + 6$
$\frac{3}{2}x + 6 = -x + 4$
$6 = -\frac{5}{2}x + 4$
$2 = -\frac{5}{2}x$
$-\frac{4}{5} = x$

Replace x with $-\frac{4}{5}$ in one of the equations to find the y-coordinate.

$y = -\left(-\frac{4}{5}\right) + 4$
$y = 4\frac{4}{5}$

The coordinates of the orthocenter are $\left(-\frac{4}{5}, 4\frac{4}{5}\right)$.

9.

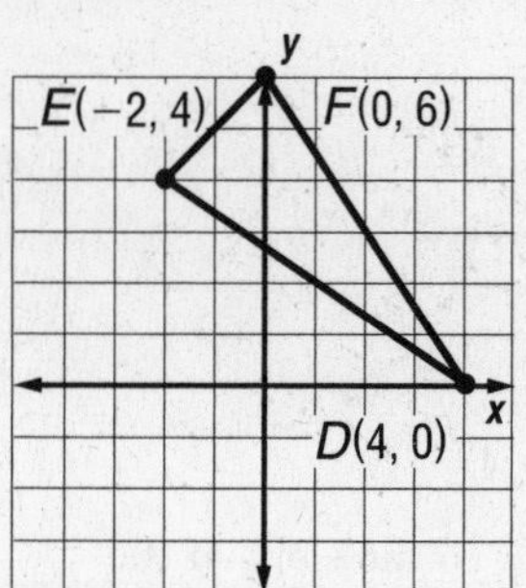

Find an equation of the perpendicular bisector of $\overline{DE}$. The slope of $\overline{DE}$ is $-\frac{2}{3}$, so the slope of the perpendicular bisector is $\frac{3}{2}$. The midpoint of $\overline{DE}$ is (1, 2).

$y - y_1 = m(x - x_1)$
$y - 2 = \frac{3}{2}(x - 1)$
$y - 2 = \frac{3}{2}x - \frac{3}{2}$
$y = \frac{3}{2}x + \frac{1}{2}$

Find an equation of the perpendicular bisector of $\overline{EF}$. The slope of $\overline{EF}$ is 1, so the slope of the perpendicular bisector is −1. The midpoint of $\overline{EF}$ is (−1, 5).

$y - y_1 = m(x - x_1)$
$y - 5 = -1[x - (-1)]$
$y - 5 = -x - 1$
$y = -x + 4$

Solve a system of equations to find the point of intersection of the perpendicular bisectors.

Find x.

$y = -x + 4$
$y = \frac{3}{2}x + \frac{1}{2}$
$-x + 4 = \frac{3}{2}x + \frac{1}{2}$
$\frac{7}{2} = \frac{5}{2}x$
$7 = 5x$
$\frac{7}{5} = x$

Replace x with $\frac{7}{5}$ in one of the equations to find the y-coordinate.

$y = -\frac{7}{5} + 4$
$y = \frac{13}{5}$

The coordinates of the circumcenter are $\left(\frac{7}{5}, \frac{13}{5}\right)$ or $\left(1\frac{2}{5}, 2\frac{3}{5}\right)$.

10. Given: $\overline{CD}$ is the perpendicular bisector of $\overline{AB}$.
E is a point on $\overline{CD}$.

Prove: $EB = EA$

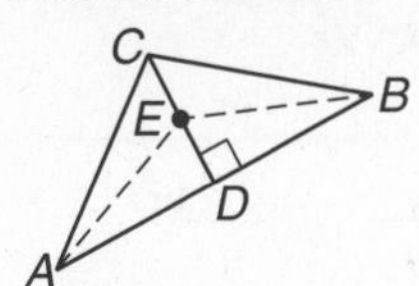

Proof:
$\overline{CD}$ is the perpendicular bisector of $\overline{AB}$. By definition of perpendicular bisector, D is the midpoint of $\overline{AB}$. Thus, $\overline{AD} \cong \overline{BD}$ by the Midpoint Theorem. $\angle CDA$ and $\angle CDB$ are right angles by definition of perpendicular. Since all right angles are congruent, $\angle CDA \cong \angle CDB$. Since E is a point on $\overline{CD}$, $\angle EDA$ and $\angle EDB$ are right angles and are congruent. By the Reflexive Property, $\overline{ED} \cong \overline{ED}$. Thus, $\triangle EDA \cong \triangle EDB$ by SAS. $\overline{EB} \cong \overline{EA}$ because CPCTC, and by definition of congruence, $EB = EA$.

11. Given: $\triangle UVW$ is isosceles with vertex angle UVW.
$\overline{YV}$ is the bisector of $\angle UVW$.

Prove: $\overline{YV}$ is a median.

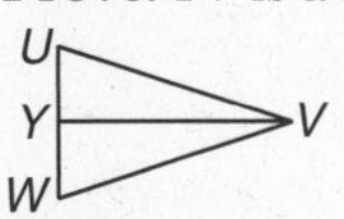

Proof:

Statements	Reasons
1. $\triangle UVW$ is an isosceles triangle with vertex angle UVW, $\overline{YV}$ is the bisector of $\angle UVW$.	1. Given
2. $\overline{UV} \cong \overline{WV}$	2. Def. of isosceles $\triangle$
3. $\angle UVY \cong \angle WVY$	3. Def. of angle bisector
4. $\overline{YV} \cong \overline{YV}$	4. Reflexive Property
5. $\triangle UVY \cong \triangle WVY$	5. SAS
6. $\overline{UY} \cong \overline{WY}$	6. CPCTC
7. Y is the midpoint of $\overline{UW}$.	7. Def. of midpoint
8. $\overline{YV}$ is a median.	8. Def. of median

12. Given: $\overline{GL}$ is a median of $\triangle EGH$.
$\overline{JM}$ is a median of $\triangle IJK$.
$\triangle EGH \cong \triangle IJK$

Prove: $\overline{GL} \cong \overline{JM}$

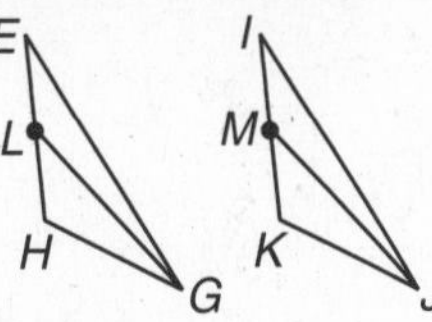

Proof:

Statements	Reasons
1. $\overline{GL}$ is a median of $\triangle EGH$, $\overline{JM}$ is a median of $\triangle IJK$, and $\triangle EGH \cong \triangle IJK$.	1. Given
2. $\overline{GH} \cong \overline{JK}$, $\angle GHL \cong \angle JKM$, $\overline{EH} \cong \overline{IK}$	2. CPCTC
3. $EH = IK$	3. Def. of $\cong$
4. $\overline{EL} \cong \overline{LH}$, $\overline{IM} \cong \overline{MK}$	4. Def. of median
5. $EL = LH$, $IM = MK$	5. Def. of $\cong$
6. $EL + LH = EH$, $IM + MK = IK$	6. Segment Addition Postulate
7. $EL + LH = IM + MK$	7. Substitution
8. $LH + LH = MK + MK$	8. Substitution
9. $2LH = 2MK$	9. Addition Property
10. $LH = MK$	10. Division Property
11. $\overline{LH} \cong \overline{MK}$	11. Def of $\cong$
12. $\triangle GHL \cong \triangle JKM$	12. SAS
13. $\overline{GL} \cong \overline{JM}$	13. CPCTC

13. $\overline{MS}$ is an altitude of $\triangle MNQ$, so $\angle MSQ$ is a right angle.

$m\angle 1 + m\angle 2 = 90$
$3x + 11 + 7x + 9 = 90$
$10x + 20 = 90$
$10x = 70$
$x = 7$
$m\angle 2 = 7(7) + 9$
$= 58$

14. $\overline{MS}$ is a median of $\triangle MNQ$, so $\overline{QS} \cong \overline{SN}$.

$3a - 14 = 2a + 1$
$a - 14 = 1$
$a = 15$
$m\angle MSQ = 7a + 1$
$= 7(15) + 1$
$= 106$

$\overline{MS}$ is not an altitude of $\triangle MNQ$ because $m\angle MSQ = 106$.

15. $\overline{WP}$ is an angle bisector, so $m\angle HWP = \frac{1}{2}(m\angle HWA)$.

$x + 12 = \frac{1}{2}(4x - 16)$
$x + 12 = 2x - 8$
$12 = x - 8$
$20 = x$

$\overline{WP}$ is a median, so $\overline{AP} \cong \overline{PH}$.

$3y + 11 = 7y - 5$
$11 = 4y - 5$
$16 = 4y$
$4 = y$

$m\angle PAW = 3x - 2$
$= 3(20) - 2$
$= 58$

$m\angle PWA = m\angle HWP$ because $\overline{WP}$ is an angle bisector.

So, $m\angle PWA = x + 12$
$= 20 + 12$ or 32

$m\angle WPA + m\angle PAW + m\angle PWA = 180$
$m\angle WPA + 58 + 32 = 180$
$m\angle WPA = 90$

$\angle WPA$ is a right angle, so $\overline{WP}$ is also an altitude.

16. $\overline{WP}$ bisects $\overline{AH}$, so $\overline{AP} \cong \overline{PH}$.

$6r + 4 = 22 + 3r$
$3r + 4 = 22$
$3r = 18$
$r = 6$

$\overline{WP}$ is perpendicular to $\overline{AH}$, so $m\angle HPW = 90$.

$m\angle WHA + m\angle HWP + m\angle HPW = 180$
$8q + 17 + 10 + q + 90 = 180$
$9q + 117 = 180$
$9q = 63$
$q = 7$

$m\angle HWP = 10 + q$
$= 10 + 7$
$= 17$

17. Always; each median is always completely contained in the interior of the triangle, so the intersection point must also be inside the triangle.

18. sometimes; true for a right triangle but false for an equilateral triangle

19. Never; an angle bisector lies between two sides of the triangle and is contained in the triangle up to the point where it intersects the opposite side. so, the intersection point of the three angle bisectors must also be inside the triangle.

20. sometimes; true for an obtuse triangle but false for an acute triangle

21. $\overline{PS}$ is a median of $\triangle PQR$, so $\overline{QS} \cong \overline{SR}$.

$10x - 7 = 5x + 3$
$5x - 7 = 3$
$5x = 10$
$x = 2$

22. $\overline{AD}$ is an altitude of $\triangle ABC$, so $\angle ADC$ is a right angle.

$m\angle ADC = 90$
$4x - 6 = 90$
$4x = 96$
$x = 24$

23. $\overline{PX}$ is an altitude of $\triangle PQR$, so $\angle PXR$ is a right angle.

$m\angle PXR = 90$
$2a + 10 = 90$
$2a = 80$
$a = 40$

24. $\overline{RZ}$ bisects $\angle PRQ$, so $m\angle PRZ = m\angle ZRQ$.

$4b - 17 = 3b - 4$
$b - 17 = -4$
$b = 13$

$m\angle PRZ = 4(13) - 17$
$= 35$

25. $\overline{QY}$ is a median, so $PY = YR$.

$2c - 1 = 4c - 11$
$-1 = 2c - 11$
$10 = 2c$
$5 = c$

$PR = PY + YR$
$= 2(5) - 1 + 4(5) - 11$
$= 10 - 1 + 20 - 11$
$= 18$

26. $\overleftrightarrow{QY}$ is perpendicular to $\overline{PR}$, so $\angle QYR$ is a right angle.

$m\angle QYR = 90$
$7b + 6 = 90$
$7b = 84$
$b = 12$

27. X is the midpoint of $\overline{ST}$.

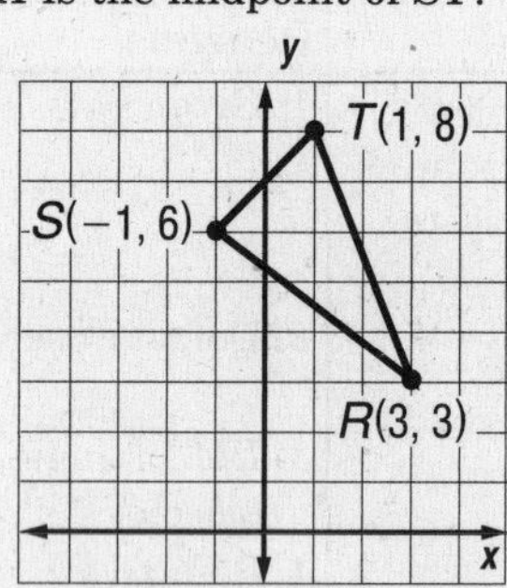

$\left(\frac{x_1 + x_2}{2}, \frac{y_1 + y_2}{2}\right) = \left(\frac{-1+1}{2}, \frac{6+8}{2}\right)$
$= (0, 7)$

28. $d = \sqrt{(x_2 - x_1)^2 + (y_2 - y_1)^2}$

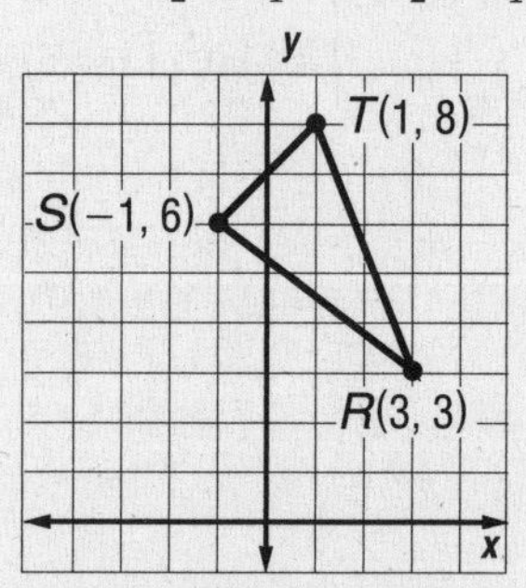

$RX = \sqrt{(0 - 3)^2 + (7 - 3)^2}$
$= \sqrt{9 + 16}$ or 5 units

29. $m = \frac{(y_2 - y_1)}{(x_2 - x_1)}$

$= \frac{7-3}{0-3}$ or $-\frac{4}{3}$

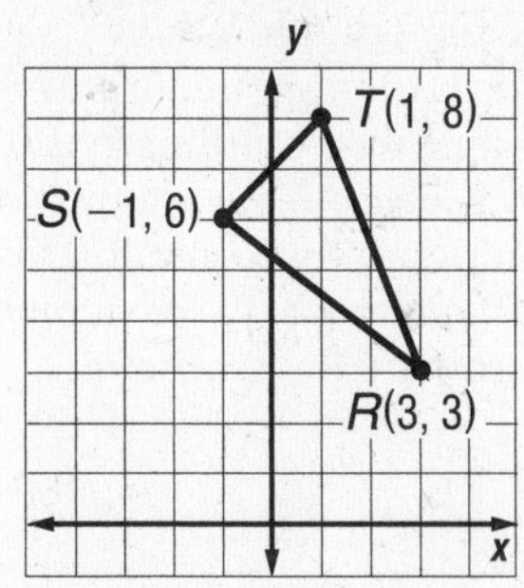

30. No, $\overline{RX}$ is not an altitude of $\triangle RST$. The slope of $\overline{ST}$ is 1. The product of the slopes of $\overline{ST}$ and $\overline{RX}$ is $-\frac{4}{3}$, not -1. Thus, the segments are not perpendicular.

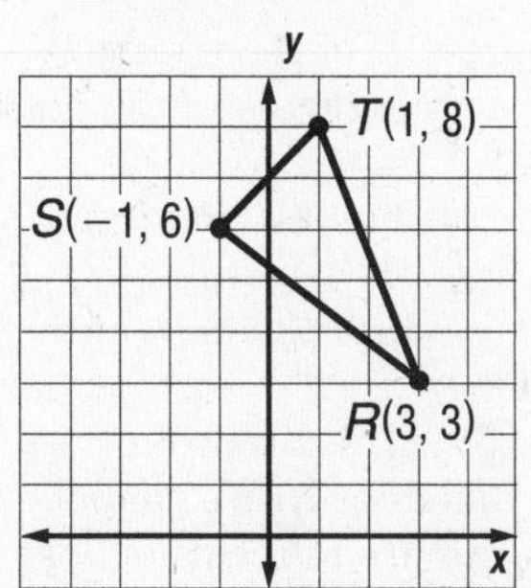

31. Given: $\overline{CA} \cong \overline{CB}, \overline{AD} \cong \overline{BD}$
Prove: C and D are on the perpendicular bisector of $\overline{AB}$.

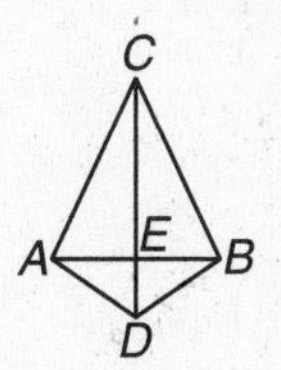

Proof:

Statements	Reasons
1. $\overline{CA} \cong \overline{CB}, \overline{AD} \cong \overline{BD}$	1. Given
2. $\overline{CD} \cong \overline{CD}$	2. Reflexive Property
3. $\triangle ACD \cong \triangle BCD$	3. SSS
4. $\angle ACD \cong \angle BCD$	4. CPCTC
5. $\overline{CE} \cong \overline{CE}$	5. Reflexive Property
6. $\triangle CEA \cong \triangle CEB$	6. SAS
7. $\overline{AE} \cong \overline{BE}$	7. CPCTC
8. E is the midpoint of $\overline{AB}$.	8. Def. of midpoint
9. $\angle CEA \cong \angle CEB$	9. CPCTC
10. $\angle CEA$ and $\angle CEB$ form a linear pair.	10. Def. of linear pair
11. $\angle CEA$ and $\angle CEB$ are supplementary.	11. Supplement Theorem
12. $m\angle CEA + m\angle CEB = 180$	12. Def. of suppl. ∠s
13. $m\angle CEA + m\angle CEA = 180$	13. Substitution
14. $2(m\angle CEA) = 180$	14. Substitution
15. $m\angle CEA = 90$	15. Division Property
16. $\angle CEA$ and $\angle CEB$ are rt. ∠s.	16. Def. of rt. ∠
17. $\overline{CD} \perp \overline{AB}$	17. Def. of ⊥
18. $\overline{CD}$ is the perpendicular bisector of $\overline{AB}$.	18. Def. of ⊥ bisector
19. C and D are on the perpendicular bisector of $\overline{AB}$.	19. Def. of points on a line

32. Given: $\angle BAC$, P is in the interior of $\angle BAC$, $PD = PE$
Prove: $\overrightarrow{AP}$ is the angle bisector of $\angle BAC$

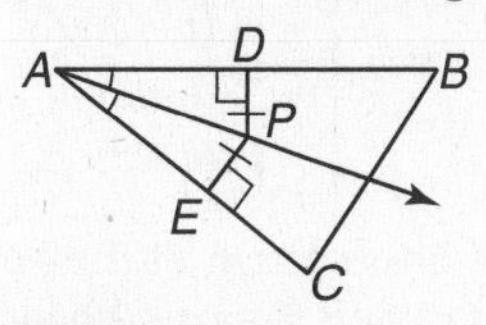

Proof:

Statements	Reasons
1. $\angle BAC$, P is in the interior of $\angle BAC$, $PD = PE$	1. Given
2. $\overline{PD} \cong \overline{PE}$	2. Def. of ≅
3. $\overline{PD} \perp \overline{AB}, \overline{PE} \perp \overline{AC}$	3. Distance from a point to a line is measured along ⊥ segment from the point to the line.
4. $\angle ADP$ and $\angle AEP$ are rt. ∠s	4. Def. of ⊥
5. $\triangle ADP$ and $\triangle AEP$ are rt. △s	5. Def. of rt. △
6. $\overline{AP} \cong \overline{AP}$	6. Reflexive Property
7. $\triangle ADP \cong \triangle AEP$	7. HL
8. $\angle DAP \cong \angle EAP$	8. CPCTC
9. $\overrightarrow{AP}$ is the angle bisector of $\angle BAC$	9. Def. of ∠ bisector

33. Given: $\triangle ABC$ with angle bisectors $\overrightarrow{AD}$, $\overrightarrow{BE}$, and $\overrightarrow{CF}$, $\overline{KP} \perp \overline{AB}$, $\overline{KQ} \perp \overline{BC}$, $\overline{KR} \perp \overline{AC}$

Prove: $KP = KQ = KR$

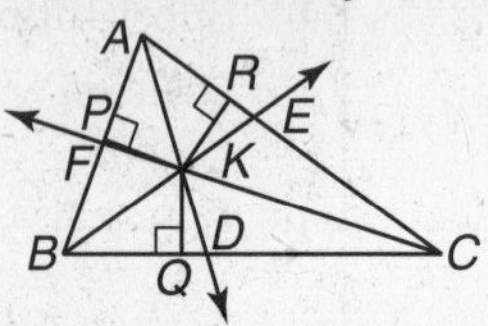

Proof:

Statements	Reasons
1. $\triangle ABC$ with angle bisectors $\overrightarrow{AD}$, $\overrightarrow{BE}$, and $\overrightarrow{CF}$, $\overline{KP} \perp \overline{AB}$, $\overline{KQ} \perp \overline{BC}$, $\overline{KR} \perp \overline{AC}$	1. Given
2. $KP = KQ$, $KQ = KR$, $KP = KR$	2. Any point on the $\angle$ bisector is equidistant from the sides of the angle.
3. $KP = KQ = KR$	3. Transitive Property

34. The flag is located at the intersection of the angle bisector between Amesbury and Stearns Roads and the perpendicular bisector of the segment joining Grand Tower and the park entrance.

35. $\frac{16 + 2 + (-6)}{3} = 4$

36. $\frac{8 + 4 + 12}{3} = 8$

37.

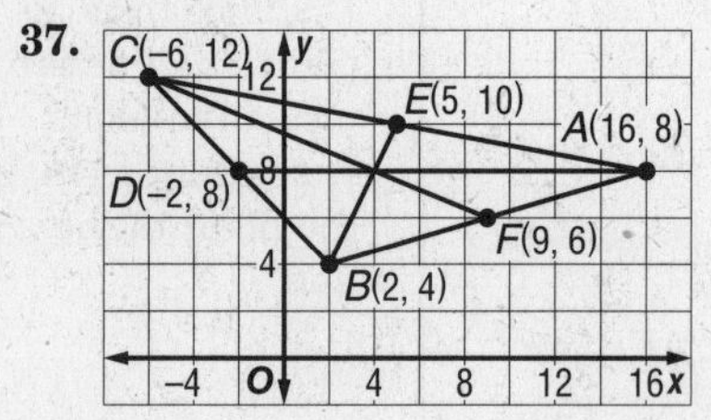

38. The centroid has the same coordinates as the means of the vertices' coordinates.

39. The altitude will be the same for both triangles, and the bases will be congruent, so the areas will be equal.

40. Sample answer: You can balance a triangle on a pencil point by locating the center of gravity of the triangle. Answers should include the following.

- centroid
-

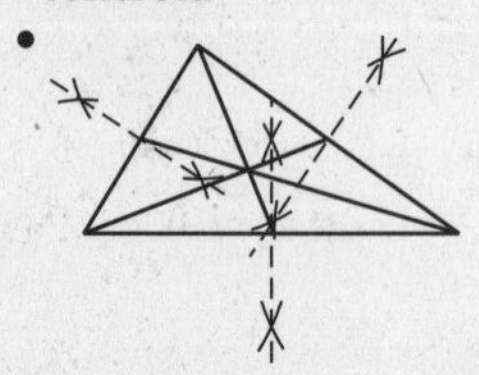

41. C; $GJ = JH$, so J is the midpoint of GH and FJ is a median of $\triangle FGH$.

42. D; $3x = 0.3y$
$10.0x = y$
$10.0 = \frac{y}{x}$

Page 245 Maintain Your Skills

43. Sample answer:
- Use the origin as vertex A of the triangle.
- Place side AB along the positive x-axis.
- Position the triangle in the first quadrant.
- Since B is on the x-axis, its y-coordinate is 0. Its x-coordinate is n because $\overline{AB}$ is n units long.
- Since $\triangle ABC$ is equilateral, the x-coordinate of C is halfway between 0 and n, or $\frac{n}{2}$. We cannot determine the y-coordinate in terms of n, so call it m.

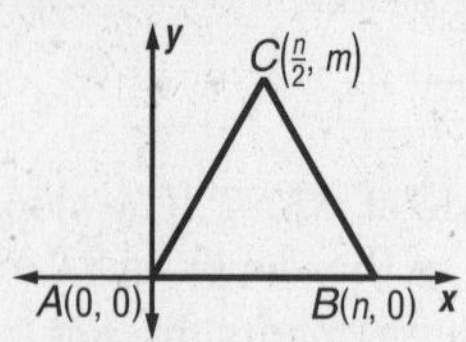

44. Sample answer:
- Use the origin as vertex D of the triangle.
- Place the base of the triangle along the positive x-axis.
- Position the triangle in the first quadrant.
- Since F is on the x-axis, its y-coordinate is 0. Its x-coordinate is a because the base of the triangle is a units long.
- Since $\triangle DEF$ is isosceles, the x-coordinate of E is halfway between 0 and a, or $\frac{a}{2}$. We cannot determine the y-coordinate in terms of a, so call it b.

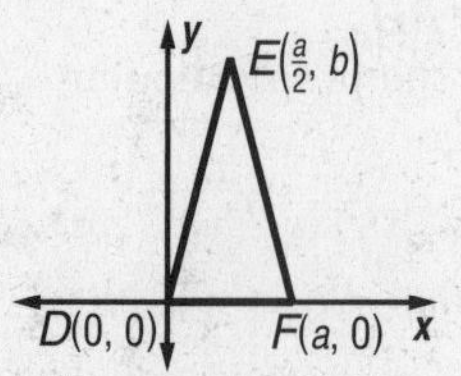

45. Sample answer:
- Use the origin as vertex H of the triangle.
- Place leg $\overline{HI}$ along the positive y-axis.
- $\overline{GI}$ is the hypotenuse so H is a right angle and leg $\overline{GH}$ is on the x-axis. Position $\overline{GH}$ on the positive x-axis.
- Since G is on the x-axis, its y-coordinate is 0. Its x-coordinate is x because $\overline{GH}$ is x units long.
- Since I is on the y-axis, its x-coordinate is 0. Its y-coordinate is $3x$ because $HI = 3GH$ or $3x$.

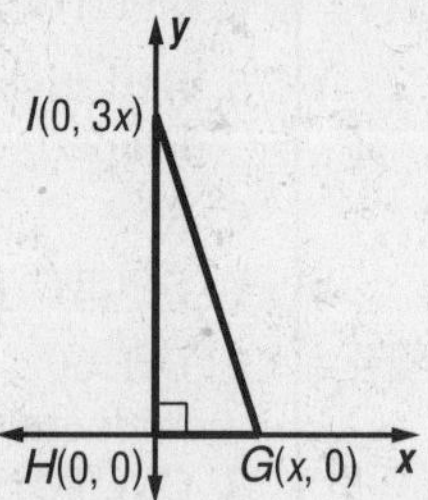

46. $\overline{MT} \cong \overline{MR}$ by the converse of the Isosceles Triangle Theorem.

47. $\angle 5 \cong \angle 11$ by the Isosceles Triangle Theorem.

48. $\angle 7 \cong \angle 10$ by the Isosceles Triangle Theorem.

49. $\overline{ML} \cong \overline{MN}$ by the converse of the Isosceles Triangle Theorem.

50. It is everywhere equidistant.

51. $\frac{3}{8} > \frac{5}{16}$ because $\frac{3}{8} = \frac{6}{16}$.

52. $2.7 > \frac{5}{3}$ because $\frac{5}{3} = 1.\overline{6}$.

53. $-4.25 > -\frac{19}{4}$ because $-\frac{19}{4} = -4.75$.

54. $-\frac{18}{25} < -\frac{19}{27}$ because $-\frac{18}{25} = -0.72$ and $-\frac{19}{27} = -0.\overline{703}$.

Page 246 Reading Mathematics: Math Words and Everyday Words

1. Sample answer: A median of a triangle is a segment that has one endpoint at a vertex and the other at the midpoint of the opposite side; the everyday meaning says it is a paved or planted strip in the middle of a highway.

2. Sample answer: the intersection of two or more lines or curves, the top of the head, the highest point

3. Sample answer: in a trapezoid, the segment joining the midpoints of the legs; the middle value of a set of data that has been arranged into an ordered sequence

4. Sample answer: a separate piece of something; a portion cut off from a geometric figure by one or more points, lines, or planes.

5-2 Inequalities and Triangles

Page 249 Geometry Activity: Inequalities for Sides and Angles of Triangles

1. Sample answer: It is the greatest measure.

2. Sample answer: It is the least measure.

3. See students' work.

4. Sample answer: The measures of the angles opposite the sides are in the same order as the lengths of the respective sides.

Page 251 Check for Understanding

1. never;

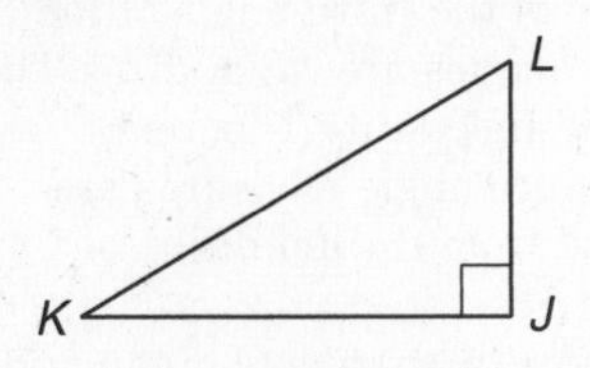

$\angle J$ is a right angle.

Since $m\angle J = 2 \cdot m\angle K$, $90 = 2 \cdot m\angle K$

$45 = m\angle K$

$m\angle L = 180 - 90 - 45 = 45$

$m\angle L = m\angle K$, so $\triangle LKJ$ is isosceles, and $KJ = LJ$.

Let $KJ = LJ = 1$. If the statement in the problem is true, then $LK = 2$. Since $\triangle LKJ$ is a right triangle, the Pythagorean Theorem applies.

$1^2 + 1^2 = 2^2$

$2 = 4$

This is a false statement, so the statement in the problem is never true.

2. A, C, B

Sample answer: $m\angle CAB$, $m\angle ACB$, $m\angle ABC$; $\overline{BC}$, $\overline{AB}$, $\overline{AC}$

3. Grace; she placed the shorter side with the smaller angle and the longer side with the larger angle.

4. **Explore:** Compare the measure of $\angle 2$ to the measures of $\angle 1$ and $\angle 4$

Plan: Use properties and theorems of real numbers to compare the angle measures.

Solve: Compare $m\angle 1$ to $m\angle 2$.

By the Exterior Angle Theorem, $m\angle 2 = m\angle 1 + m\angle 4$. Since angle measures are positive numbers and from the definition of inequality, $m\angle 2 > m\angle 1$.

Compare $m\angle 4$ to $m\angle 2$.

Again, by the Exterior Angle Theorem, $m\angle 2 = m\angle 1 + m\angle 4$. The definition of inequality states that if $m\angle 2 = m\angle 1 + m\angle 4$ then $m\angle 2 > m\angle 4$.

Examine: $m\angle 2$ is greater than $m\angle 1$ and $m\angle 4$. Therefore, $\angle 2$ has the greatest measure.

5. **Explore:** Compare the measure of $\angle 3$ to the measures of $\angle 2$ and $\angle 5$.

Plan: Use properties and theorems of real numbers to compare the angle measures.

Solve: Compare $m\angle 2$ to $m\angle 3$.

By the Exterior Angle Theorem, $m\angle 3 = m\angle 2 + m\angle 5$. Since angle measures are positive numbers and from the definition of inequality, $m\angle 3 > m\angle 2$.

Compare $m\angle 5$ to $m\angle 3$.

Again, by the Exterior Angle Theorem, $m\angle 3 = m\angle 2 + m\angle 5$. The definition of inequality states that if $m\angle 3 = m\angle 2 + m\angle 5$ then $m\angle 3 > m\angle 5$.

Examine: $m\angle 3$ is greater than $m\angle 2$ and $m\angle 5$. Therefore, $\angle 3$ has the greatest measure.

6. **Explore:** Compare the measure of $\angle 3$ to the measures of $\angle 1$, $\angle 2$, $\angle 4$, and $\angle 5$.

Plan: Use properties and theorems of real numbers to compare the angle measures.

Solve: From Exercise 4, $m\angle 2 > m\angle 1$ and $m\angle 2 > m\angle 4$.

From Exercise 5, $m\angle 3 > m\angle 2$ and $m\angle 3 > m\angle 5$.

Then by transitivity, $m\angle 3 > m\angle 1$ and $m\angle 3 > m\angle 4$.

Examine: $m\angle 3$ is greater than $m\angle 1$, $m\angle 2$, $m\angle 4$, and $m\angle 5$. Therefore, $\angle 3$ has the greatest measure.

7. By the Exterior Angle Inequality Theorem, $m\angle 1 > m\angle 4$ and $m\angle 1 > m\angle 5 + m\angle 6$ so $m\angle 1 > m\angle 5$ and $m\angle 6$. Thus, the measures of $\angle 4$, $\angle 5$, and $\angle 6$ are all less than $m\angle 1$.

8. By the Exterior Angle Inequality Theorem, $m\angle 1 > m\angle 5 + m\angle 6$ and $m\angle 7 > m\angle 5 + m\angle 6$ so $m\angle 1 > m\angle 6$ and $m\angle 7 > m\angle 6$. Thus, the measures of $\angle 1$ and $\angle 7$ are greater than $m\angle 6$.

9. By the Exterior Angle Inequality Theorem, $m\angle 7 > m\angle 2 + m\angle 3$ and $m\angle 7 > m\angle 5 + m\angle 6$, so $m\angle 7 > m\angle 2$, $m\angle 7 > m\angle 3$, $m\angle 7 > m\angle 5$, and $m\angle 7 > m\angle 6$. Thus, the measures of $\angle 2$, $\angle 3$, $\angle 5$, and $\angle 6$ are less than $m\angle 7$.
10. The side opposite $\angle WXY$ is longer than the side opposite $\angle XYW$, so $m\angle WXY > m\angle XYW$.
11. The side opposite $\angle XZY$ is shorter than the side opposite $\angle XYZ$, so $m\angle XZY < m\angle XYZ$.
12. The side opposite $\angle WYX$ is shorter than the side opposite $\angle XWY$, so $m\angle WYX < m\angle XWY$.
13. $\overline{AE}$ is opposite a 30° angle, and $\overline{EB}$ is opposite a 110° angle. If one angle of a triangle has a greater measure than another angle, then the side opposite the greater angle is longer than the side opposite the lesser angle, so $AE < EB$.
14. $\overline{CE}$ is opposite $\angle CDE$, and $m\angle CDE = 180 - (50 + 55)$ or 75. $\overline{CD}$ is opposite a 55° angle. If one angle of a triangle has a greater measure than another angle, then the side opposite the greater angle is longer than the side opposite the lesser angle, so $CE > CD$.
15. $\overline{BC}$ is opposite $\angle BEC$, and $m\angle BEC = 180 - (40 + 100)$ or 40. $\overline{EC}$ is opposite a 40° angle. Thus, $BC = EC$.
16. Second base; the angle opposite the side from third base to second base is smaller than the angle opposite the side from third to first. Therefore, the distance from third to second is shorter than the distance from third to first.

Pages 252–253 Practice and Apply

17. **Explore:** Compare the measure of $\angle 1$ to the measures of $\angle 2$ and $\angle 4$.
Plan: Use properties and theorems of real numbers to compare angle measures.
Solve: Compare $m\angle 1$ to $m\angle 2$.
By the Exterior Angle Inequality Theorem, $m\angle 1 > m\angle 2$.
Compare $m\angle 1$ to $m\angle 4$.
By the Exterior Angle Inequality Theorem, $m\angle 1 > m\angle 4$.
Examine: $m\angle 1$ is greater than $m\angle 2$ and $m\angle 4$. Therefore, $\angle 1$ has the greatest measure.
18. **Explore:** Compare the measure of $\angle 2$ to the measures of $\angle 4$ and $\angle 6$.
Plan: Use properties and theorems of real numbers to compare angle measures.
Solve: Compare $m\angle 2$ to $m\angle 4$.
By the Exterior Angle Inequality Theorem, $m\angle 2 > m\angle 4$.
Compare $m\angle 2$ to $m\angle 6$.
By the Exterior Angle Inequality Theorem, $m\angle 2 > m\angle 6$.
Examine: $m\angle 2$ is greater than $m\angle 4$ and $m\angle 6$. Therefore, $\angle 2$ has the greatest measure.
19. **Explore:** Compare the measure of $\angle 7$ to the measures of $\angle 3$ and $\angle 5$.
Plan: Use properties and theorems of real numbers to compare angle measures.
Solve: Compare $m\angle 7$ to $m\angle 3$.
By the Exterior Angle Inequality Theorem, $m\angle 7 > m\angle 3$.
Compare $m\angle 7$ to $m\angle 5$.
By the Exterior Angle Inequality Theorem, $m\angle 7 > m\angle 5$.
Examine: $m\angle 7$ is greater than $m\angle 3$ and $m\angle 5$. Therefore, $\angle 7$ has the greatest measure.
20. **Explore:** Compare the measure of $\angle 1$ to the measures of $\angle 2$ and $\angle 6$.
Plan: Use properties and theorems of real numbers to compare angle measures.
Solve: Compare $m\angle 1$ to $m\angle 2$.
By the Exterior Angle Inequality Theorem, $m\angle 1 > m\angle 2$.
Compare $m\angle 1$ to $m\angle 6$.
By the Exterior Angle Inequality Theorem, $m\angle 1 > m\angle 6$.
Examine: $m\angle 1$ is greater than $m\angle 2$ and $m\angle 6$. Therefore, $\angle 1$ has the greatest measure.
21. **Explore:** Compare the measure of $\angle 7$ to the measures of $\angle 5$ and $\angle 8$.
Plan: Use properties and theorems of real numbers to compare angle measures.
Solve: Compare $m\angle 7$ to $m\angle 5$.
By the Exterior Angle Inequality Theorem, $m\angle 7 > m\angle 5$.
Compare $m\angle 7$ to $m\angle 8$.
By the Exterior Angle Inequality Theorem, $m\angle 7 > m\angle 8$.
Examine: $m\angle 7$ is greater than $m\angle 5$ and $m\angle 8$. Therefore, $\angle 7$ has the greatest measure.
22. **Explore:** Compare the measure of $\angle 2$ to the measures of $\angle 6$ and $\angle 8$.
Plan: Use properties and theorems of real numbers to compare angle measures.
Solve: Compare $m\angle 2$ to $m\angle 6$.
By the Exterior Angle Inequality Theorem, $m\angle 2 > m\angle 6$.
Compare $m\angle 2$ to $m\angle 8$.
Let x be the measure of the third angle of the triangle whose other angles are $\angle 3$ and $\angle 4$. Then, by the Exterior Angle Inequality Theorem, $m\angle 2 > (x + m\angle 8)$. Since angle measures are positive numbers and from the definition of inequality, $m\angle 2 > m\angle 8$.
Examine: $m\angle 2$ is greater than $m\angle 6$ and $m\angle 8$. Therefore, $\angle 2$ has the greatest measure.
23. By the Exterior Angle Inequality Theorem, $m\angle 5 > m\angle 7$, $m\angle 5 > m\angle 10$, and $m\angle 5 > m\angle 2 + m\angle 8$ so $m\angle 5 > m\angle 2$ and $m\angle 5 > m\angle 8$. Thus, the measures of $\angle 2$, $\angle 7$, $\angle 8$, and $\angle 10$ are all less than $m\angle 5$.
24. By the Exterior Angle Inequality Theorem, $m\angle 4 > m\angle 6$, $m\angle 1 > m\angle 6 + m\angle 9$ so $m\angle 1 > m\angle 6$ and $m\angle 11 > m\angle 6 + m\angle 9$ so $m\angle 11 > m\angle 6$. Thus, the measures of $\angle 1$, $\angle 4$, and $\angle 11$ are all greater than $m\angle 6$.

25. By the Exterior Angle Inequality Theorem, $m\angle 3 > m\angle 10$ and $m\angle 5 > m\angle 10$. Thus, the measures of $\angle 3$ and $\angle 5$ are greater than $m\angle 10$.

26. By the Exterior Angle Inequality Theorem, $m\angle 1 > m\angle 3$, $m\angle 1 > m\angle 6$, and $m\angle 1 > m\angle 9$. Thus, the measures of $\angle 3$, $\angle 6$, and $\angle 9$ are all less than $m\angle 1$.

27. By the Exterior Angle Inequality Theorem, $m\angle 8 > m\angle 9$, $m\angle 7 > m\angle 9$, $m\angle 3 > m\angle 9$, and $m\angle 1 > m\angle 9$. Thus, the measures of $\angle 8$, $\angle 7$, $\angle 3$, and $\angle 1$ are all greater than $m\angle 9$.

28. By the Exterior Angle Inequality Theorem, $m\angle 8 > m\angle 2$, $m\angle 8 > m\angle 4$, $m\angle 8 > m\angle 5$, and $m\angle 8 > m\angle 9$. Thus, the measures of $\angle 2$, $\angle 4$, $\angle 5$, and $\angle 9$ are all less than $m\angle 8$.

29. The side opposite $\angle KAJ$ is shorter than the side opposite $\angle AJK$, so $m\angle KAJ < m\angle AJK$.

30. The side opposite $\angle MJY$ is longer than the side opposite $\angle JYM$, so $m\angle MJY > m\angle JYM$.

31. The side opposite $\angle SMJ$ is longer than the side opposite $\angle MJS$, so $m\angle SMJ > m\angle MJS$.

32. The side opposite $\angle AKJ$ is longer than the side opposite $\angle JAK$, so $m\angle AKJ > m\angle JAK$.

33. The side opposite $\angle MYJ$ is shorter than the side opposite $\angle JMY$, so $m\angle MYJ < m\angle JMY$.

34. The side opposite $\angle JSY$ is longer than the side opposite $\angle JYS$, so $m\angle JSY > m\angle JYS$.

35. **Given:** $\overline{JM} \cong \overline{JL}$
$\overline{JL} \cong \overline{KL}$
Prove: $m\angle 1 > m\angle 2$

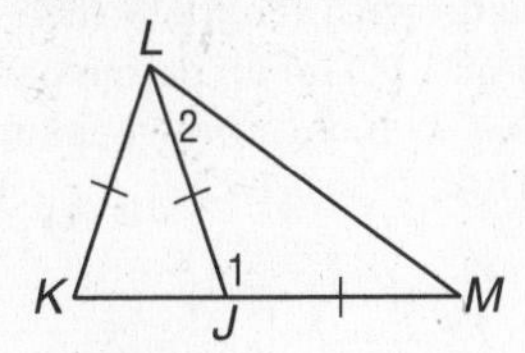

Proof:

Statements	Reasons
1. $\overline{JM} \cong \overline{JL}$, $\overline{JL} \cong \overline{KL}$	1. Given
2. $\angle LKJ \cong \angle LJK$	2. Isosceles △ Theorem
3. $m\angle LKJ = m\angle LJK$	3. Def. of ≅
4. $m\angle 1 > m\angle LKJ$	4. Ext. ∠ Inequality Theorem
5. $m\angle 1 > m\angle LJK$	5. Substitution
6. $m\angle LJK > m\angle 2$	6. Ext. ∠ Inequality Theorem
7. $m\angle 1 > m\angle 2$	7. Trans. Prop. of Inequality

36. **Given:** $\overline{PR} \cong \overline{PQ}$; $QR > QP$
Prove: $m\angle P > m\angle Q$

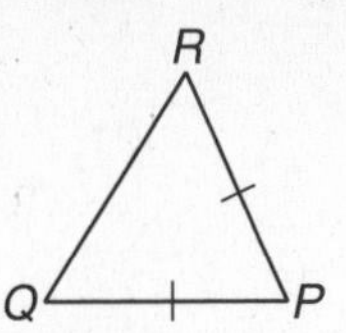

Proof:

Statements	Reasons
1. $QR > QP$	1. Given
2. $m\angle P > m\angle R$	2. If one side of a △ is longer than another, then the ∠ opp. the longer side is greater than the ∠ opposite the shorter side.
3. $\overline{PR} \cong \overline{PQ}$	3. Given
4. $\angle Q \cong \angle R$	4. Isosceles △ Theorem
5. $m\angle Q = m\angle R$	5. Def. of ≅
6. $m\angle P > m\angle Q$	6. Substitution

37. $\overline{ZY}$ is opposite a 45° angle. $\overline{YR}$ is opposite $\angle YZR$, and $m\angle YZR = 180 - (95 + 45)$ or 40. If one angle of a triangle has a greater measure than another angle, then the side opposite the greater angle is longer than the side opposite the lesser angle, so $ZY > YR$.

38. $\overline{SR}$ is opposite a 43° angle. $\overline{ZS}$ is opposite $\angle ZRS$, and $m\angle ZRS = 180 - (43 + 97)$ or 40. If one angle of a triangle has a greater measure than another angle, then the side opposite the greater angle is longer than the side opposite the lesser angle, so $SR > ZS$.

39. $\overline{RZ}$ is opposite a 97° angle. $\overline{SR}$ is opposite a 43° angle. If one angle of a triangle has a greater measure than another angle, then the side opposite the greater angle is longer than the side opposite the lesser angle, so $RZ > SR$.

40. $\overline{ZY}$ is opposite a 45° angle. $\overline{RZ}$ is opposite a 95° angle. If one angle of a triangle has a greater measure than another angle, then the side opposite the greater angle is longer than the side opposite the lesser angle, so $ZY < RZ$.

41. $\overline{TY}$ is opposite $\angle TZY$. $m\angle TZY + m\angle ZYT = 91$, so $m\angle TZY = 91 - 66$ or 25. $\overline{ZY}$ is opposite an 89° angle. If one angle of a triangle has a greater measure than another angle, then the side opposite the greater angle is longer than the side opposite the lesser angle, so $TY < ZY$.

42. $\overline{TY}$ is opposite $\angle TZY$. $m\angle TZY + m\angle ZYT = 91$, so $m\angle TZY = 91 - 66$ or 25. $\overline{ZT}$ is opposite a 66° angle. If one angle of a triangle has a greater measure than another angle, then the side opposite the greater angle is longer than the side opposite the lesser angle, so $TY < ZT$.

43. $KL = \sqrt{(-1 - 3)^2 + (5 - 2)^2}$
$= \sqrt{16 + 9}$ or 5
$LM = \sqrt{[-3 - (-1)]^2 + (-7 - 5)^2}$
$= \sqrt{4 + 144}$
$= \sqrt{148}$
≈ 12.2
$KM = \sqrt{(-3 - 3)^2 + (-7 - 2)^2}$
$= \sqrt{36 + 81}$
$= \sqrt{117}$
≈ 10.8
The side lengths in order from least to greatest are KL, KM, LM, so the angles in order from the least to the greatest measure are $\angle M$, $\angle L$, $\angle K$.

44. Sample answer: Draw a triangle that satisfies the given information. Then draw the medians and measure them to find their lengths.

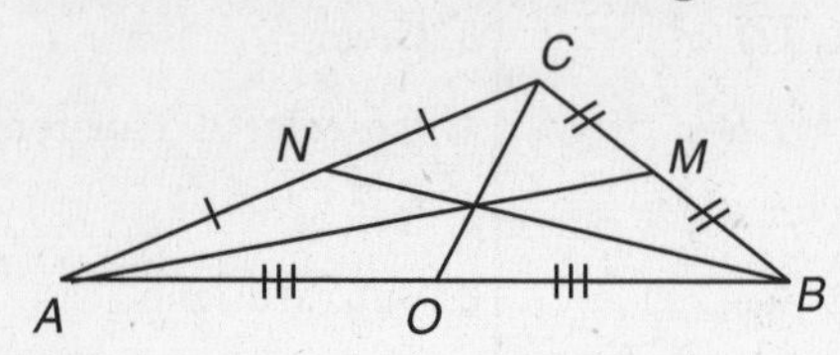

In $\triangle ABC$, $AB > AC > BC$. Measure $\overline{AM}$, $\overline{BN}$, and $\overline{CO}$. In this triangle $CO < BN < AM$.

45. $8x + 4 + 11x - 37 + 5x + 21 = 180$
$24x - 12 = 180$
$24x = 192$
$x = 8$

$8x + 4 = 8(8) + 4$ or 68
$11x - 37 = 11(8) - 37$ or 51
$5x + 21 = 5(8) + 21$ or 61
$68 > 61 > 51$, so the lengths of the legs of the trip in order from greatest to least are Phoenix to Atlanta, Des Moines to Phoenix, Atlanta to Des Moines.

46. $m\angle P + m\angle Q + m\angle R = 180$
$9n + 29 + 93 - 5n + 10n + 2 = 180$
$14n + 124 = 180$
$14n = 56$
$n = 4$

$m\angle P = 9(4) + 29$ or 65
$m\angle Q = 93 - 5(4)$ or 73
$m\angle R = 10(4) + 2$ or 42
$m\angle R < m\angle P < m\angle Q$, so the sides of $\triangle PQR$ in order from shortest to longest are $\overline{PQ}$, $\overline{QR}$, $\overline{PR}$.

47. $m\angle P + m\angle Q + m\angle R = 180$
$12n - 9 + 62 - 3n + 16n + 2 = 180$
$25n + 55 = 180$
$25n = 125$
$n = 5$

$m\angle P = 12(5) - 9$ or 51
$m\angle Q = 62 - 3(5)$ or 47
$m\angle R = 16(5) + 2$ or 82
$m\angle Q < m\angle P < m\angle R$, so the sides of $\triangle PQR$ in order from shortest to longest are $\overline{PR}$, $\overline{QR}$, $\overline{PQ}$.

48. $m\angle P + m\angle Q + m\angle R = 180$
$9n - 4 + 4n - 16 + 68 - 2n = 180$
$11n + 48 = 180$
$11n = 132$
$n = 12$

$m\angle P = 9(12) - 4$ or 104
$m\angle Q = 4(12) - 16$ or 32
$m\angle R = 68 - 2(12)$ or 44
$m\angle Q < m\angle R < m\angle P$, so the sides of $\triangle PQR$ in order from shortest to longest are $\overline{PR}$, $\overline{PQ}$, $\overline{QR}$.

49. $m\angle P + m\angle Q + m\angle R = 180$
$3n + 20 + 2n + 37 + 4n + 15 = 180$
$9n + 72 = 180$
$9n = 108$
$n = 12$

$m\angle P = 3(12) + 20$ or 56
$m\angle Q = 2(12) + 37$ or 61
$m\angle R = 4(12) + 15$ or 63
$m\angle P < m\angle Q < m\angle R$, so the sides of $\triangle PQR$ in order from shortest to longest are $\overline{QR}$, $\overline{PR}$, $\overline{PQ}$.

50. $m\angle P + m\angle Q + m\angle R = 180$
$4n + 61 + 67 - 3n + n + 74 = 180$
$2n + 202 = 180$
$2n = -22$
$n = -11$

$m\angle P = 4(-11) + 61$ or 17
$m\angle Q = 67 - 3(-11)$ or 100
$m\angle R = -11 + 74$ or 63
$m\angle P < m\angle R < m\angle Q$, so the sides of $\triangle PQR$ in order from shortest to longest are $\overline{QR}$, $\overline{PQ}$, $\overline{PR}$.

51. The angle opposite the side with length $\frac{x}{3}$ inches has measure $180 - 163$ or 17. The angle opposite the side with length $2(y + 1)$ inches has measure $180 - (17 + 75)$ or 88. Thus, $2(y + 1) > \frac{x}{3}$.
$2(y + 1) > \frac{x}{3}$
$y + 1 > \frac{x}{6}$
$y > \frac{x}{6} - 1$
$y > \frac{x - 6}{6}$

52.

A
B C
T M

Given: $\triangle ABC$ is scalene; $\overline{AM}$ is the median from A to $\overline{BC}$; $\overline{AT}$ is the altitude from A to $\overline{BC}$.
Prove: $AM > AT$
$\angle ATB$ and $\angle ATM$ are right angles by the definition of altitude and $m\angle ATB = m\angle ATM$ because all right angles are congruent. By the Exterior Angle Inequality Theorem, $m\angle ATB > m\angle AMT$. So, $m\angle ATM > m\angle AMT$ by Substitution. If one angle of a triangle has a greater measure than another angle, then the side opposite the greater angle is longer than the side opposite the lesser angle. Thus, $AM > AT$.

53. $m\angle A + m\angle B + m\angle C = 180$

$2y + 12 + y - 18 + 4y + 12 = 180$

$7y + 6 = 180$

$7y = 174$

$y \approx 25$

$m\angle A \approx 2(25) + 12$ or 62

$m\angle B \approx 25 - 18$ or 7

$m\angle C \approx 4(25) + 12$ or 112

$m\angle C > m\angle A$ so $3x + 15 > 4x + 7$

$15 > x + 7$

$8 > x$

$CB > 0$, so $4x + 7 > 0$

$4x > -7$

$x > -\frac{7}{4}$

Thus, $-\frac{7}{4} < x < 8$.

54. Sample answer: The largest corner is opposite the longest side. Answers should include the following.
- the Exterior Angle Inequality Theorem
- the angle opposite the side that is 51 feet long

55. A; $n = p + 180 - m$ so $m + n - 180 = p$

56. D; $\frac{1}{2}x - 3 = 2\left(\frac{x-1}{5}\right)$

$5x - 30 = 4(x - 1)$

$5x - 30 = 4x - 4$

$x - 30 = -4$

$x = 26$

Page 254 Maintain Your Skills

57. D is on $\overline{CB}$ with $\overline{CD} \cong \overline{DB}$ and the slope of $\overline{CD}$ equals the slope of $\overline{DB}$.

$DB = \sqrt{(12 - 9)^2 + (3 - 12)^2}$

$= \sqrt{9 + 81}$

$= \sqrt{90}$

slope of $\overline{DB} = \frac{3 - 12}{12 - 9}$

$= \frac{-9}{3}$ or -3

D is 9 units down and 3 units to the right from B. The point C that is 9 units down and 3 units to the right from D has coordinates $(12 + 3, 3 - 9)$ or $(15, -6)$. Check that $CD = DB$.

$CD = \sqrt{(15 - 12)^2 + (-6 - 3)^2}$

$= \sqrt{9 + 81}$

$= \sqrt{90}$

C has coordinates $(15, -6)$.

58. slope of $\overline{BC} = -3$

slope of $\overline{AD} = \frac{3 - 8}{12 - 3}$ or $-\frac{5}{9}$

No, $\overline{AD}$ is not an altitude of $\triangle ABC$ because $-\frac{5}{9}(-3) \neq -1$.

59. slope of $\overline{BD} = -3$ $\quad \frac{\sqrt{90}}{2}$

slope of $\overline{EF} = \frac{7\frac{1}{2} - 6}{10\frac{1}{2} - 6}$

$= \frac{1\frac{1}{2}}{}$

$= \frac{1}{3}$

$\overline{BD} \perp \overline{EF}$ because $\frac{1}{3}(-3) = -1$.

$BF = \sqrt{\left(10\frac{1}{2} - 9\right)^2 + \left(7\frac{1}{2} - 12\right)^2}$

$= \sqrt{\frac{9}{4} + \frac{81}{4}}$

$= \sqrt{\frac{90}{4}}$

$= \frac{\sqrt{90}}{2}$

$DF = \sqrt{\left(10\frac{1}{2} - 12\right)^2 + \left(7\frac{1}{2} - 3\right)^2}$

$= \sqrt{\frac{9}{4} + \frac{81}{4}}$

$= \sqrt{\frac{90}{4}}$

$= \frac{\sqrt{90}}{2}$

$BF = DF$, so $\overline{BF} \cong \overline{DF}$ and F is the midpoint of $\overline{BD}$. Therefore, $\overline{EF}$ is a perpendicular bisector of $\overline{BD}$.

60. $x = \frac{0 + a + b}{3} = \frac{a + b}{3}$

$y = \frac{0 + 0 + c}{3} = \frac{c}{3}$

D has coordinates $D\left(\frac{a + b}{3}, \frac{c}{3}\right)$.

61. Label the midpoints of $\overline{AB}$, $\overline{BC}$, and $\overline{CA}$ as E, F, and G respectively. Then the coordinates of E, F, and G are $\left(\frac{a}{2}, 0\right)$, $\left(\frac{a + b}{2}, \frac{c}{2}\right)$, and $\left(\frac{b}{2}, \frac{c}{2}\right)$ respectively. The slope of $\overline{AF} = \frac{c}{a + b}$, and the slope

of $\overline{AD} = \frac{c}{a + b}$, so D is on $\overline{AF}$. The slope of $\overline{BG} = \frac{c}{b - 2a}$ and the slope of $\overline{BD} = \frac{c}{b - 2a}$, so D is on $\overline{BG}$. The slope of $\overline{CE} = \frac{2c}{2b - a}$ and the slope of $\overline{CD} = \frac{2c}{2b - a}$, so D is on $\overline{CE}$. Since D is on $\overline{AF}$, $\overline{BG}$, and $\overline{CE}$, it is the intersection point of the three segments.

62. $\angle T \cong \angle X$, $\angle U \cong \angle Y$, $\angle V \cong \angle Z$, $\overline{TU} \cong \overline{XY}$, $\overline{UV} \cong \overline{YZ}$, $\overline{TV} \cong \overline{XZ}$

63. $\angle C \cong \angle R$, $\angle D \cong \angle S$, $\angle G \cong \angle W$, $\overline{CD} \cong \overline{RS}$, $\overline{DG} \cong \overline{SW}$, $\overline{CG} \cong \overline{RW}$

64. $\angle B \cong \angle D$, $\angle C \cong \angle G$, $\angle F \cong \angle H$, $\overline{BC} \cong \overline{DG}$, $\overline{CF} \cong \overline{GH}$, $\overline{BF} \cong \overline{DH}$

65. slope of line containing $(4, 8)$ and $(2, -1)$:

$\frac{-1 - 8}{2 - 4} = \frac{-9}{-2} = \frac{9}{2}$

slope of line containing $(x, 2)$ and $(-4, 5)$:

$\frac{5 - 2}{-4 - x} = \frac{3}{-4 - x}$

Solve for x. Since the lines are perpendicular,

$\frac{3}{-4 - x} = -\frac{2}{9}$

$3(9) = -2(-4 - x)$

$27 = 8 + 2x$

$19 = 2x$

$9.5 = x$

66. true; $2ab = 2(2)(5)$

$= 20$

67. false; $c(b - a) = 6(5 - 2)$

$= 6(3)$

$= 18$

68. true; $a + c = 2 + 6$ or 8
$a + b = 2 + 5$ or 7
$a + c > a + b$ since $8 > 7$

Page 254 Practice Quiz 1

1. $BD = DC$
$4x + 9 = 7x - 6$
$9 = 3x - 6$
$15 = 3x$
$5 = x$
2. $\overline{AD} \perp \overline{BC}$, so $m\angle ADC = 90$.
$2y - 6 = 90$
$2y = 96$
$y = 48$
3. Never; a median is a segment from one vertex to the side opposite the vertex and never intersects any other vertex.
4. Always; an angle bisector lies between two sides of the triangle and is contained in the triangle up to the point where it intersects the opposite side. So, the intersection point of the three angle bisectors must also be inside the triangle.
5. sometimes; true for an obtuse triangle but false for an acute triangle
6. sometimes; true for right triangles but false for other triangles
7. No triangle; by Exercise 4, the angle bisectors always intersect at a point in the interior of the triangle.
8. $m\angle T > m\angle S > m\angle U$, so the sides of $\triangle STU$ in order from longest to shortest are $\overline{SU}, \overline{TU}, \overline{ST}$.
9. $m\angle Q + m\angle R + m\angle S = 180$
$3x + 20 + 2x + 37 + 4x + 15 = 180$
$9x + 72 = 180$
$9x = 108$
$x = 12$

$m\angle Q = 3(12) + 20$ or 56
$m\angle R = 2(12) + 37$ or 61
$m\angle S = 4(12) + 15$ or 63
10. $m\angle Q < m\angle R < m\angle S$, so the sides of $\triangle QRS$ in order from shortest to longest are $\overline{RS}, \overline{QS}, \overline{QR}$.

5-3 Indirect Proof

Pages 257–258 Check for Understanding

1. If a statement is shown to be false, then its opposite must be true.
2. Sample answer: Indirect proofs are proved using the contrapositive, showing $\sim Q \to \sim P$. In a direct proof, it would be shown that $P \to Q$. For example, indirect reasoning can be used to prove that a person is not guilty of a crime by assuming the person is guilty, then contradicting evidence to show that the person could not have committed the crime.
3. Sample answer: $\triangle ABC$ is scalene.
Given: $\triangle ABC$; $AB \neq BC$; $BC \neq AC$; $AB \neq AC$
Prove: $\triangle ABC$ is scalene.

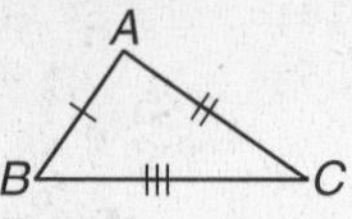

Proof:
Step 1 Assume $\triangle ABC$ is not scalene.
Case 1: $\triangle ABC$ is isoceles.
If $\triangle ABC$ is isosceles, then $AB = BC$, $BC = AC$, or $AB = AC$.
This contradicts the given information, so $\triangle ABC$ is not isosceles.
Case 2: $\triangle ABC$ is equilateral.
In order for a triangle to be equilateral, it must also be isosceles, and Case 1 proved that $\triangle ABC$ is not isosceles. Thus, $\triangle ABC$ is not equilateral.
Therefore, $\triangle ABC$ is scalene.
4. $x \geq 5$
5. The lines are not parallel.
6. The lines are not parallel.
7. **Given:** $a > 0$
Prove: $\frac{1}{a} > 0$
Proof:
Step 1 Assume $\frac{1}{a} \leq 0$.
Step 2 $\frac{1}{a} \leq 0$; $a \cdot \frac{1}{a} \leq 0 \cdot a$, $1 \leq 0$
Step 3 The conclusion that $1 \leq 0$ is false, so the assumption that $\frac{1}{a} \leq 0$ must be false. Therefore, $\frac{1}{a} > 0$.
8. **Given:** n is odd.
Prove: n^2 is odd.
Proof:
Step 1 Assume n^2 is even.
Step 2 n is odd, so n can be expressed as $2a + 1$.
$n^2 = (2a + 1)^2$ Substitution
$= (2a + 1)(2a + 1)$ Multiply.
$= 4a^2 + 4a + 1$ Simplify.
$= 2(2a^2 + 2a) + 1$ Distributive Property
Step 3 $2(2a^2 + 2a) + 1$ is an odd number. This contradicts the assumption, so the assumption must be false. Thus n^2 is odd.
9. **Given:** $\triangle ABC$
Prove: There can be no more than one obtuse angle in $\triangle ABC$.

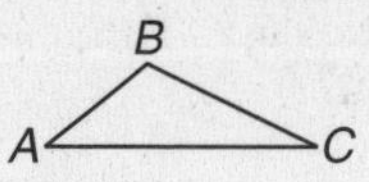

Proof:
Step 1 Assume that there can be more than one obtuse angle in $\triangle ABC$.
Step 2 An obtuse angle has a measure greater than 90. Suppose $\angle A$ and $\angle B$ are obtuse angles. Then $m\angle A + m\angle B > 180$ and $m\angle A + m\angle B + m\angle C > 180$.
Step 3 The conclusion contradicts the fact that the sum of the measures of the angles of a triangle equals 180. Thus, there can be at most one obtuse angle in $\triangle ABC$.

10. **Given:** $m \nparallel n$
Prove: Lines m and n intersect at exactly one point.

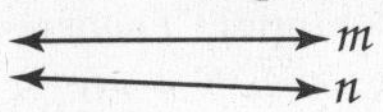

Proof:
Case 1: m and n intersect at more than one point.
Step 1 Assume that m and n intersect at more than one point.
Step 2 Lines m and n intersect at points P and Q. Both lines m and n contain P and Q.
Step 3 By postulate, there is exactly one line through any two points. Thus the assumption is false, and lines m and n intersect in no more than one point.
Case 2: m and n do not intersect.
Step 1 Assume that m and n do not intersect.
Step 2 If lines m and n do not intersect, then they are parallel.
Step 3 This conclusion contradicts the given information. Therefore the assumption is false, and lines m and n intersect in at least one point. Combining the two cases, lines m and n intersect in no more than one point and no less than one point. So lines m and n intersect in exactly one point.

11. **Given:** $\triangle ABC$ is a right triangle; $\angle C$ is a right angle.
Prove: $AB > BC$ and $AB > AC$

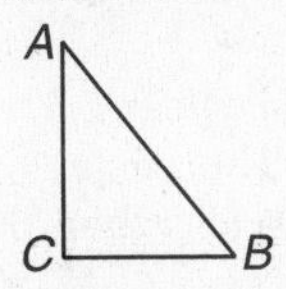

Proof:
Step 1 Assume that the hypotenuse of a right triangle is not the longest side. That is, $AB < BC$ or $AB < AC$.
Step 2 If $AB < BC$, then $m\angle C < m\angle A$. Since $m\angle C = 90$, $m\angle A > 90$. So, $m\angle C + m\angle A > 180$. By the same reasoning, if $AB < AC$, then $m\angle C + m\angle B > 180$.
Step 3 Both relationships contradict the fact that the sum of the measures of the angles of a triangle equals 180. Therefore, the hypotenuse must be the longest side of a right triangle.

12. **Given:** $x + y > 270$
Prove: $x > 135$ or $y > 135$
Proof:
Step 1 Assume $x \leq 135$ and $y \leq 135$.
Step 2 $x + y \leq 270$
Step 3 This contradicts the fact that $x + y > 270$. Therefore, at least one of the stages was longer than 135 miles.

Pages 258–260 Practice and Apply

13. $\overline{PQ} \ncong \overline{ST}$
14. $x \leq 4$
15. 6 cannot be expressed as $\frac{a}{b}$.
16. A median of an isosceles triangle is not an altitude.
17. Points P, Q, and R are noncollinear.
18. The angle bisector of the vertex angle of an isosceles triangle is not an altitude of the triangle.
19. **Given:** $\frac{1}{a} < 0$
Prove: a is negative.
Proof:
Step 1 Assume $a > 0$. $a \neq 0$ since that would make $\frac{1}{a}$ undefined.
Step 2 $\frac{1}{a} < 0$
$a\left(\frac{1}{a}\right) < (0)a$
$1 < 0$
Step 3 $1 > 0$, so the assumption must be false. Thus, a must be negative.

20. **Given:** n^2 is even.
Prove: n^2 is divisible by 4.
Proof:
Step 1 Assume n^2 is not divisible by 4. In other words, 4 is not a factor of n^2.
Step 2 If the square of a numbers is even, then the number is also even. So, if n^2 is even, then n must be even.
Let $n = 2a$.
$n = 2a$
$n^2 = (2a)^2$ or $4a^2$
Step 3 4 is a factor of n^2, which contradicts the assumption.

21. **Given:** $\overline{PQ} \cong \overline{PR}$
$\angle 1 \ncong \angle 2$
Prove: PZ is not a median of $\triangle PQR$.

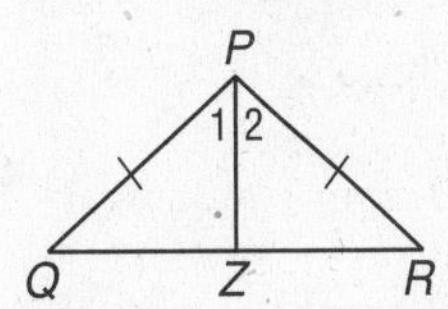

Proof:
Step 1 Assume $\overline{PZ}$ is a median of $\triangle PQR$.
Step 2 If $\overline{PZ}$ is a median of $\triangle PQR$, then Z is the midpoint of $\overline{QR}$, and $\overline{QZ} \cong \overline{RZ}$. $\overline{PZ} \cong \overline{PZ}$ by the Reflexive Property. $\triangle PZQ \cong \triangle PZR$ by SSS. $\angle 1 \cong \angle 2$ by CPCTC.
Step 3 This conclusion contradicts the given fact $\angle 1 \ncong \angle 2$. Thus, $\overline{PZ}$ is not a median of $\triangle PQR$.

22. **Given:** $m\angle 2 \neq m\angle 1$
Prove: $\ell \nparallel m$

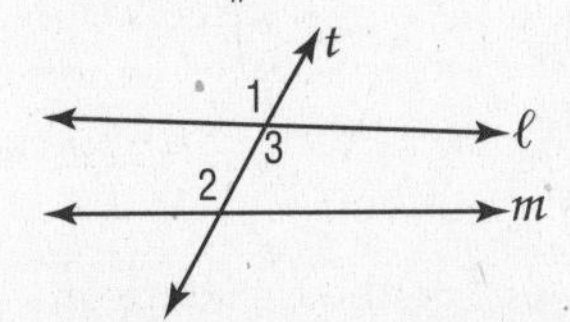

Proof:
Step 1 Assume that $\ell \parallel m$.
Step 2 If $\ell \parallel m$, then $\angle 1 \cong \angle 2$ because they are corresponding angles. Thus, $m\angle 1 = m\angle 2$.
Step 3 This contradicts the given fact that $m\angle 1 \neq m\angle 2$. Thus the assumption $\ell \parallel m$ is false. Therefore, $\ell \nparallel m$.

23. Given: $a > 0, b > 0$, and $a > b$

Prove: $\frac{a}{b} > 1$

Proof:

Step 1 Assume that $\frac{a}{b} \leq 1$.

Step 2

Case 1	**Case 2**
$\frac{a}{b} < 1$	$\frac{a}{b} = 1$
$a < b$	$a = b$

Step 3 The conclusion of both cases contradicts the given fact $a > b$.
Thus, $\frac{a}{b} > 1$.

24. Given: $\overline{AB} \not\cong \overline{AC}$

Prove: $\angle 1 \not\cong \angle 2$

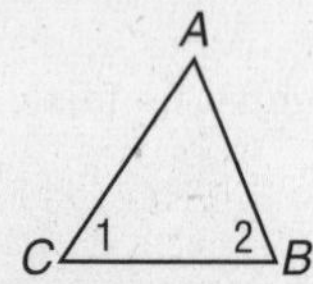

Proof:

Step 1 Assume that $\angle 1 \cong \angle 2$.

Step 2 If $\angle 1 \cong \angle 2$, then the sides opposite the angles are congruent.
Thus $\overline{AB} \cong \overline{AC}$.

Step 3 The conclusion contradicts the given information. Thus $\angle 1 \cong \angle 2$ is false.
Therefore, $\angle 1 \not\cong \angle 2$.

25. Given: $\triangle ABC$ and $\triangle ABD$ are equilateral.
$\triangle ACD$ is not equilateral.

Prove: $\triangle BCD$ is not equilateral.

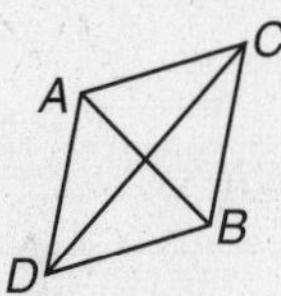

Proof:

Step 1 Assume that $\triangle BCD$ is an equilateral triangle.

Step 2 If $\triangle BCD$ is an equilateral triangle, then $\overline{BC} \cong \overline{CD} \cong \overline{DB}$. Since $\triangle ABC$ and $\triangle ABD$ are equilateral triangles, $\overline{AC} \cong \overline{AB} \cong \overline{BC}$ and $\overline{AD} \cong \overline{AB} \cong \overline{DB}$. By the Transitive Property, $\overline{AC} \cong \overline{AD} \cong \overline{CD}$. Therefore, $\triangle ACD$ is an equilateral triangle.

Step 3 This conclusion contradicts the given information. Thus, the assumption is false. Therefore, $\triangle BCD$ is not an equilateral triangle.

26. Given: $m\angle A > m\angle ABC$

Prove: $BC > AC$

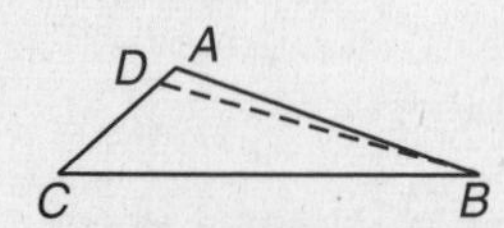

Proof: Assume $BC \not> AC$. By the Comparison Property, $BC = AC$ or $BC < AC$.

Case 1: If $BC = AC$, then $\angle ABC \cong \angle A$ by the Isosceles Triangle Theorem (If two sides of a triangle are congruent, then the angles opposite those sides are congruent.) But, $\angle ABC \cong \angle A$ contradicts the given statement that $m\angle A > m\angle ABC$. So, $BC \neq AC$.

Case 2: If $BC < AC$, then there must be a point D between A and C so that $\overline{DC} \cong \overline{BC}$. Draw the auxiliary segment $\overline{BD}$. Since $DC = BC$, by the Isosceles Triangle Theorem $\angle BDC \cong \angle DBC$. Now $\angle BDC$ is an exterior angle of $\triangle BAD$, and by the Exterior Angles Inequality Theorem (the measure of an exterior angle of a triangle is greater than the measure of either corresponding remote interior angle) $m\angle BDC > m\angle A$. By the Angle Addition Postulate, $m\angle ABC = m\angle ABD + m\angle DBC$. Then by the definition of inequality, $m\angle ABC > m\angle DBC$. By Substitution and the Transitive Property of Inequality, $m\angle ABC > m\angle A$. But this contradicts the given statement that $m\angle A > m\angle ABC$. In both cases, a contradiction was found, and hence our assumption must have been false. Therefore, $BC > AC$.

27. Use $r = \frac{d}{t}$, $t = 3$, and $d = 175$.

Proof:

Step 1 Assume that Ramon's average speed was greater than or equal to 60 miles per hour, $r \geq 60$.

Step 2

Case 1	**Case 2**
$r = 60$	$r > 60$
$60 \stackrel{?}{=} \frac{175}{3}$	$\frac{175}{3} \stackrel{?}{>} 60$
$60 \neq 58.3$	$58.3 \not> 60$

Step 3 The conclusions are false, so the assumption must be false. Therefore, Ramon's average speed was less than 60 miles per hour.

28. A majority is greater than half or 50%.

Proof:

Step 1 Assume that the percent of college-bound seniors receiving information from guidance counselors is less than 50%.

Step 2 By examining the graph, you can see that 56% of college-bound seniors received information from guidance counselors.

Step 3 Since 56% > 50%, the assumption is false. Therefore, a majority of college-bound seniors received information from guidance counselors.

29. $1500 \cdot 15\% \stackrel{?}{=} 225$
$1500 \cdot 0.15 \stackrel{?}{=} 225$
$225 = 225$

30. teachers and friends; 15% + 18% = 33%, 33% > 31%

31. Yes; if you assume the client was at the scene of the crime, it is contradicted by his presence in Chicago at that time. Thus, the assumption that he was present at the crime is false.

32. See students' work.

33. Proof:

Step 1 Assume that $\sqrt{2}$ is a rational number.

Step 2 If $\sqrt{2}$ is a rational number, it can be written as $\frac{a}{b}$, where a and b are integers with no common factors, and $b \neq 0$. If $\sqrt{2} = \frac{a}{b}$, then $2 = \frac{a^2}{b^2}$, and $2b^2 = a^2$. Thus a^2 is an even number, as is a. Because a is even it can be written as $2n$.

$2b^2 = a^2$
$2b^2 = (2n)^2$
$2b^2 = 4n^2$
$b^2 = 2n^2$

b^2 is an even number. So, b is also an even number.

Step 3 Because b and a are both even numbers, they have a common factor of 2. This contradicts the definition of rational numbers. Therefore, $\sqrt{2}$ is not rational.

34. Sample answer: Indirect proof is sometimes used in mystery novels. Answers should include the following.

- Sherlock Holmes would disprove all possibilities except the actual solution to a mystery.
- medical diagnosis, trials, scientific research

35. D; $x + 80 = 140$
$x = 60$
A, B, and C are true. D is not true.

36. A; $\frac{8}{16} \cdot \frac{7}{15} \cdot \frac{6}{14} = \frac{1}{10}$

Page 260 Maintain Your Skills

37. The angle in $\triangle MOP$ with the greatest measure is opposite the side with the greatest measure. Side $\overline{OM}$ has measure 9, which is greater than all other sides of the triangle. So, $\angle P$ has the greatest measure.

38. The angle in $\triangle LMN$ with the least measure is opposite the side with the least measure. Side $\overline{LM}$ has measure 6, which is less than all other sides of the triangle. So, $\angle N$ has the least measure.

39. Given: $\overline{CD}$ is an angle bisector. $\overline{CD}$ is an altitude.
Prove: $\triangle ABC$ is isosceles.

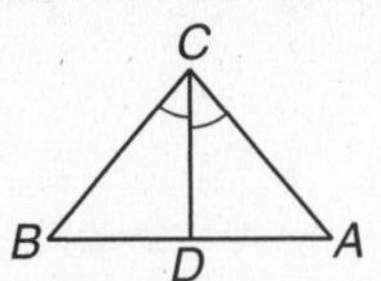

Proof:

Statements	Reasons
1. $\overline{CD}$ is an angle bisector. $\overline{CD}$ is an altitude.	1. Given
2. $\angle ACD \cong \angle BCD$	2. Def. of $\angle$ bisector
3. $\overline{CD} \perp \overline{AB}$	3. Def. of altitude
4. $\angle CDA$ and $\angle CDB$ are rt. $\angle$s	4. $\perp$ lines form 4 rt. $\angle$s.
5. $\angle CDA \cong \angle CDB$	5. All rt. $\angle$s are $\cong$.
6. $\overline{CD} \cong \overline{CD}$	6. Reflexive Prop.
7. $\triangle ACD \cong \triangle BCD$	7. ASA
8. $\overline{AC} \cong \overline{BC}$	8. CPCTC
9. $\triangle ABC$ is isosceles.	9. Def. of isosceles $\triangle$

40. Given: $\overline{QT}$ is a median. $\triangle QRS$ is isosceles with base $\overline{RS}$.
Prove: $\overline{QT}$ bisects $\angle SQR$

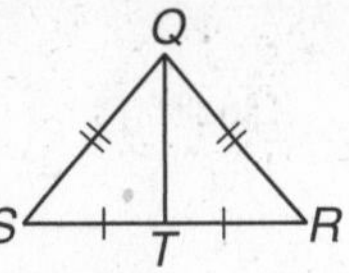

Proof:

Statements	Reasons
1. $\overline{QT}$ is a median. $\triangle QRS$ is isosceles with base $\overline{RS}$.	1. Given
2. $\overline{RT} \cong \overline{ST}$	2. Def. of median
3. $\overline{QR} \cong \overline{QS}$	3. Def. of isosceles $\triangle$
4. $\overline{QT} \cong \overline{QT}$	4. Reflexive Prop.
5. $\triangle QRT \cong \triangle QST$	5. SSS
6. $\angle SQT \cong \angle RQT$	6. CPCTC
7. $\overline{QT}$ bisects $\angle SQR$	7. Def. of $\angle$ bisector

41. Given: $\triangle ABC \cong \triangle DEF$; $\overline{BG}$ is an angle bisector of $\angle ABC$. $\overline{EH}$ is an angle bisector of $\angle DEF$.
Prove: $\overline{BG} \cong \overline{EH}$

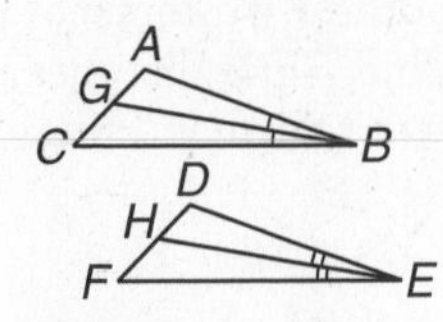

Proof:

Statements	Reasons
1. $\triangle ABC \cong \triangle DEF$	1. Given
2. $\angle A \cong \angle D$, $\overline{AB} \cong \overline{DE}$, $\angle ABC \cong \angle DEF$	2. CPCTC
3. $\overline{BG}$ is an angle bisector of $\angle ABC$. $\overline{EH}$ is an angle bisector of $\angle DEF$.	3. Given
4. $\angle ABG \cong \angle GBC$, $\angle DEH \cong \angle HEF$	4. Def. of $\angle$ bisector
5. $m\angle ABC = m\angle DEF$	5. Def. of $\cong$ $\angle$s
6. $m\angle ABG = m\angle GBC$, $m\angle DEH = m\angle HEF$	6. Def. of $\cong$ $\angle$s
7. $m\angle ABC = m\angle ABG + m\angle GBC$, $m\angle DEF = m\angle DEH + m\angle HEF$	7. Angle Addition Property
8. $m\angle ABC = m\angle ABG + m\angle ABG$, $m\angle DEF = m\angle DEH + m\angle DEH$	8. Substitution
9. $m\angle ABG + m\angle ABG = m\angle DEH + m\angle DEH$	9. Substitution
10. $2m\angle ABG = 2m\angle DEH$	10. Substitution
11. $m\angle ABG = m\angle DEH$	11. Division
12. $\angle ABG \cong m\angle DEH$	12. Def. of $\cong$ $\angle$s
13. $\triangle ABG \cong \triangle DEH$	13. ASA
14. $\overline{BG} \cong \overline{EH}$	14. CPCTC

42. $m\angle R + m\angle S + m\angle A = 180$
$41 + 109 + m\angle A = 180$
$150 + m\angle A = 180$
$m\angle A = 30$

43. $y - y_1 = m(x - x_1)$
$y - 3 = 2(x - 4)$

44. $y - y_1 = m(x - x_1)$
$y - (-2) = -3(x - 2)$
$y + 2 = -3(x - 2)$

45. $y - y_1 = m(x - x_1)$
$y - (-9) = 11[x - (-4)]$
$y + 9 = 11(x + 4)$

46. true; $19 - 10 \overset{?}{<} 11$
$9 < 11$

47. false; $31 - 17 \overset{?}{<} 12$
$14 \not< 12$

48. true; $38 + 76 \overset{?}{>} 109$
$114 > 109$

5-4 The Triangle Inequality

Pages 263–264 Check for Understanding

1. Sample answer: If the lines are not horizontal, then the segment connecting their y-intercepts is not perpendicular to either line. Since distance is measured along a perpendicular segment, this segment cannot be used.

2. Jameson; $5 + 10 > 13$ but $5 + 8 \not> 13$.

3. Sample answer: 2, 3, 4 and 1, 2, 3

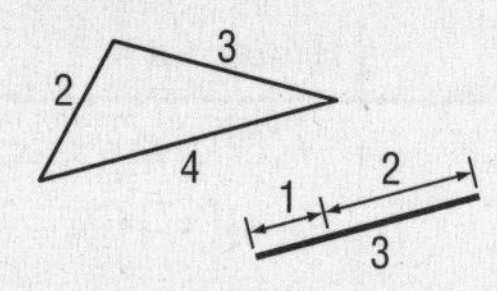

4. Check each inequality.

$5 + 4 \overset{?}{>} 3$ $\quad$ $4 + 3 \overset{?}{>} 5$ $\quad$ $5 + 3 \overset{?}{>} 4$
$9 > 3$ ✓ $\quad$ $7 > 5$ ✓ $\quad$ $8 > 4$ ✓

All of the inequalities are true, so 5, 4, and 3 can be the lengths of the sides of a triangle.

5. $5 + 10 \overset{?}{>} 15$
$15 \not> 15$
Because the sum of two measures equals the third measure, the sides cannot form a triangle.

6. $30.1 + 0.8 \overset{?}{>} 31$
$30.9 \not> 31$
Because the sum of two measures is less than the third measure, the sides cannot form a triangle.

7. Check each inequality.

$5.6 + 10.1 \overset{?}{>} 5.2$ $\quad$ $5.6 + 5.2 \overset{?}{>} 10.1$
$15.7 > 5.2$ ✓ $\quad$ $10.8 > 10.1$ ✓
$5.2 + 10.1 \overset{?}{>} 5.6$
$15.3 > 5.6$ ✓

All of the inequalities are true, so 5.6, 10.1, and 5.2 can be the lengths of the sides of a triangle.

8. Let the measure of the third side be n.

$7 + 12 > n$ $\quad$ $7 + n > 12$ $\quad$ $12 + n > 7$
$19 > n$ or $n < 19$ $\quad$ $n > 5$ $\quad$ $n > -5$

Graph the inequalities on the same number line.

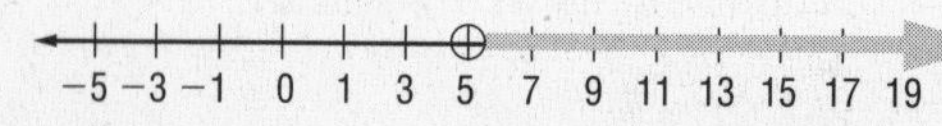

$n < 19$

−5 −3 −1 0 1 3 5 7 9 11 13 15 17 19

$n > 5$

−5 −3 −1 0 1 3 5 7 9 11 13 15 17 19

$n > -5$

−5 −3 −1 0 1 3 5 7 9 11 13 15 17 19

Find the intersection.
The range of values that fit all three inequalities is $5 < n < 19$.

9. Let the measure of the third side be n.

$14 + 23 > n$ $\quad$ $14 + n > 23$ $\quad$ $23 + n > 14$
$37 > n$ or $n < 37$ $\quad$ $n > 9$ $\quad$ $n > -9$

Graph the inequalities on the same number line.

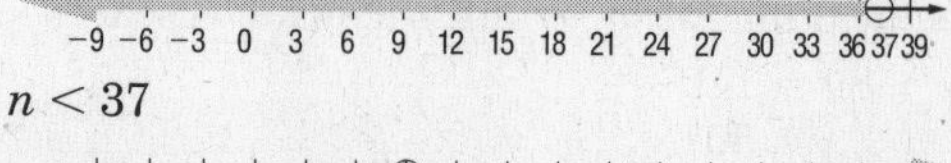

$n < 37$

−9 −6 −3 0 3 6 9 12 15 18 21 24 27 30 33 36 39

$n > 9$

−9 −6 −3 0 3 6 9 12 15 18 21 24 27 30 33 36 39

$n > -9$

−9 −6 −3 0 3 6 9 12 15 18 21 24 27 30 33 36 37 39

Find the intersection.
The range of values that fit all three inequalities is $9 < n < 37$.

10. Let the measure of the third side be n.

$22 + 34 > n$ $\quad$ $22 + n > 34$ $\quad$ $34 + n > 22$
$56 > n$ or $n < 56$ $\quad$ $n > 12$ $\quad$ $n > -12$

Graph the inequalities on the same number line.

−12 −6 0 6 12 18 24 30 36 42 48 54 56

$n < 56$

−12 −6 0 6 12 18 24 30 36 42 48 54

$n > 12$

−12 −6 0 6 12 18 24 30 36 42 48 54

$n > -12$

−12 −6 0 6 12 18 24 30 36 42 48 54 56

Find the intersection.
The range of values that fit all three inequalities is $12 < n < 56$.

11. Let the measure of the third side be n.

$15 + 18 > n$ $\quad$ $15 + n > 18$ $\quad$ $18 + n > 15$
$33 > n$ or $n < 33$ $\quad$ $n > 3$ $\quad$ $n > -3$

Graph the inequalities on the same number line.

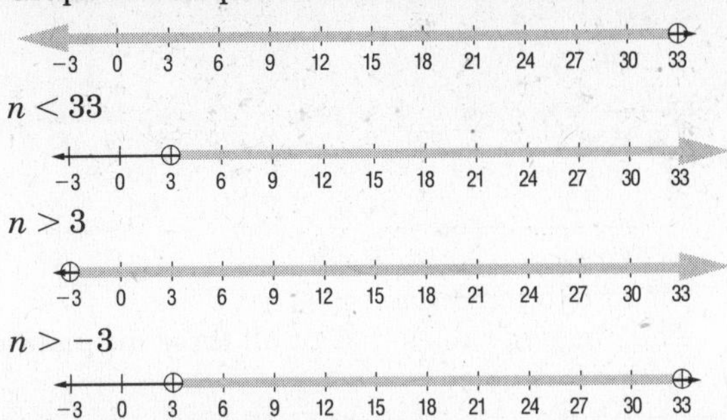

Find the intersection.
The range of values that fit all three inequalities is $3 < n < 33$.

12. Given: $\overline{PQ} \perp$ plane $\mathcal{M}$
Prove: $\overline{PQ}$ is the shortest segment from P to plane $\mathcal{M}$.

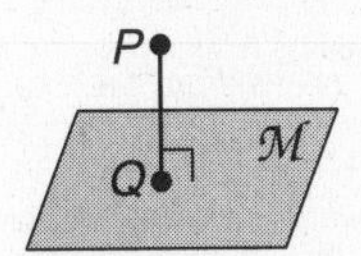

Proof: By definition, $\overline{PQ}$ is perpendicular to plane $\mathcal{M}$ if it is perpendicular to every line in $\mathcal{M}$ that intersects it. But since the perpendicular segment from a point to a line is the shortest segment from the point to the line, that perpendicular segment is the shortest segment from the point to each of these lines. Therefore, $\overline{PQ}$ is the shortest segment from P to $\mathcal{M}$.

13. B; Let x be the length of each of the congruent sides of the triangle.
$x + x > 10$ $\quad$ $x + 10 > x$
$2x > 10$ $\quad$ $10 > 0$ true for all x
$x > 5$
The side length x is a whole number greater than 5. The smallest number x for which this is true is 6. Thus, the answer is choice B.

Pages 264–265 Practice and Apply

14. $1 + 2 \overset{?}{>} 3$
$3 \not> 3$
Because the sum of two measures equals the third measure, the sides cannot form a triangle.

15. $2 + 6 \overset{?}{>} 11$
$8 \not> 11$
Because the sum of two measures is less than the third measure, the sides cannot form a triangle.

16. Check each inequality.
$8 + 8 \overset{?}{>} 15$ $\quad$ $8 + 15 \overset{?}{>} 8$
$16 > 15$ ✓ $\quad$ $23 > 8$ ✓
All of the inequalities are true, so 8, 8, and 15 can be the lengths of the sides of a triangle.

17. $13 + 16 \overset{?}{>} 29$
$29 \not> 29$
Because the sum of two measures equals the third measure, the sides cannot form a triangle.

18. Check each inequality.
$18 + 32 \overset{?}{>} 21$ $\quad$ $18 + 21 \overset{?}{>} 32$ $\quad$ $32 + 21 \overset{?}{>} 18$
$50 > 21$ ✓ $\quad$ $39 > 32$ ✓ $\quad$ $53 > 18$ ✓
All of the inequalities are true, so 18, 32, and 21 can be the lengths of the sides of a triangle.

19. Check each inequality.
$9 + 21 \overset{?}{>} 20$ $\quad$ $9 + 20 \overset{?}{>} 21$ $\quad$ $20 + 21 \overset{?}{>} 9$
$30 > 20$ ✓ $\quad$ $29 > 21$ ✓ $\quad$ $41 > 9$ ✓
All of the inequalities are true, so 9, 21, and 20 can be the lengths of the sides of a triangle.

20. $5 + 9 \overset{?}{>} 17$
$14 \not> 17$
Because the sum of two measures is less than the third measure, the sides cannot form a triangle.

21. Check each inequality.
$17 + 30 \overset{?}{>} 30$ $\quad$ $30 + 30 \overset{?}{>} 17$
$47 > 30$ ✓ $\quad$ $60 > 17$ ✓
All of the inequalities are true, so 17, 30, and 30 can be the lengths of the sides of a triangle.

22. Check each inequality.
$8.4 + 7.2 \overset{?}{>} 3.5$ $\quad$ $8.4 + 3.5 \overset{?}{>} 7.2$ $\quad$ $3.5 + 7.2 \overset{?}{>} 8.4$
$15.6 > 3.5$ ✓ $\quad$ $11.9 > 7.2$ ✓ $\quad$ $10.7 > 8.4$ ✓
All of the inequalities are true, so 8.4, 7.2, and 3.5 can be the lengths of the sides of a triangle.

23. Check each inequality.
$4 + 0.9 \overset{?}{>} 4.1$ $\quad$ $4 + 4.1 \overset{?}{>} 0.9$ $\quad$ $4.1 + 0.9 \overset{?}{>} 4$
$4.9 > 4.1$ ✓ $\quad$ $8.1 > 0.9$ ✓ $\quad$ $5 > 4$ ✓
All of the inequalities are true, so 4, 0.9, and 4.1 can be the lengths of the sides of a triangle.

24. $2.2 + 12 \overset{?}{>} 14.3$
$14.2 \not> 14.3$
Because the sum of two measures is less than the third measure, the sides cannot form a triangle.

25. $0.18 + 0.21 \overset{?}{>} 0.52$
$0.39 \not> 0.52$
Because the sum of two measures is less than the third measure, the sides cannot form a triangle.

26. Let the measure of the third side be n.
$5 + 11 > n$ $\quad$ $5 + n > 11$ $\quad$ $11 + n > 5$
$16 > n$ or $n < 16$ $\quad$ $n > 6$ $\quad$ $n > -6$
Graph the inequalities on the same number line.

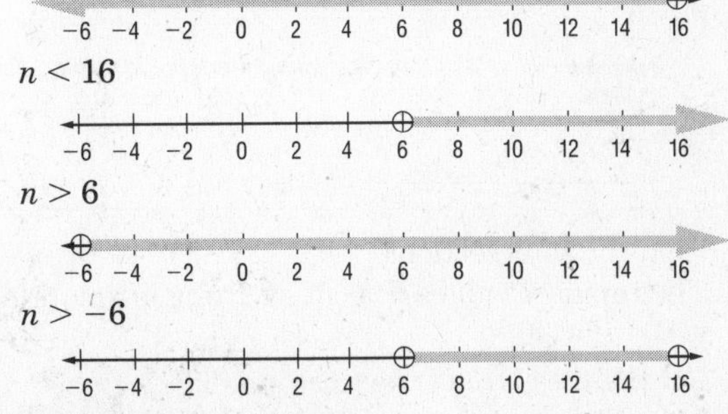

Find the intersection.
The range of values that fit all three inequalities is $6 < n < 16$.

27. Let the measure of the third side be n.

$7 + 9 > n$ $\quad$ $7 + n > 9$ $\quad$ $9 + n > 7$

$16 > n$ or $n < 16$ $\quad$ $n > 2$ $\quad$ $n > -2$

Graph the inequalities on the same number line.

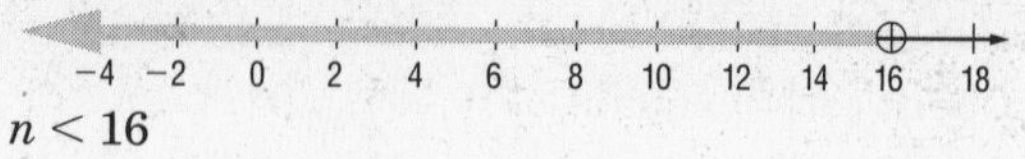

$n < 16$

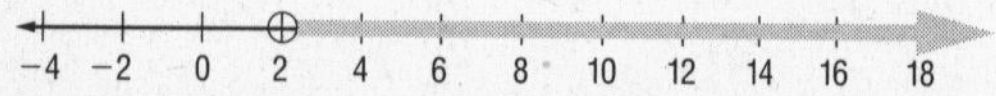

$n > 2$

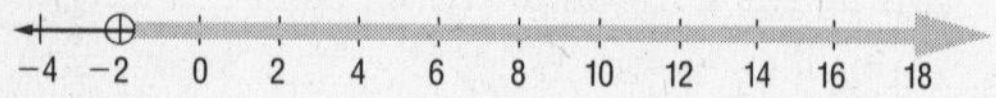

$n > -2$

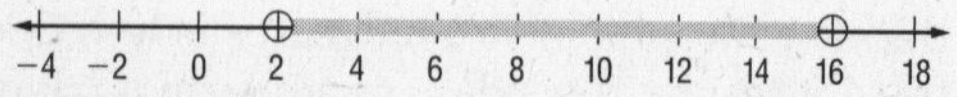

Find the intersection.

The range of values that fit all three inequalities is $2 < n < 16$.

28. Let the measure of the third side be n.

$10 + 15 > n$ $\quad$ $10 + n > 15$ $\quad$ $15 + n > 10$

$25 > n$ or $n < 25$ $\quad$ $n > 5$ $\quad$ $n > -5$

Graph the inequalities on the same number line.

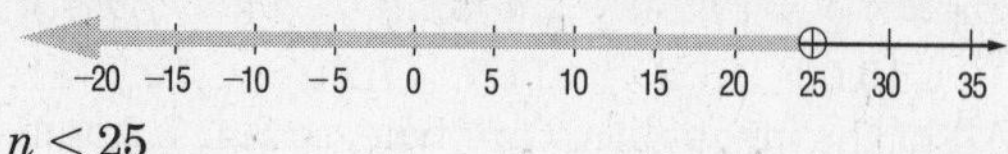

$n < 25$

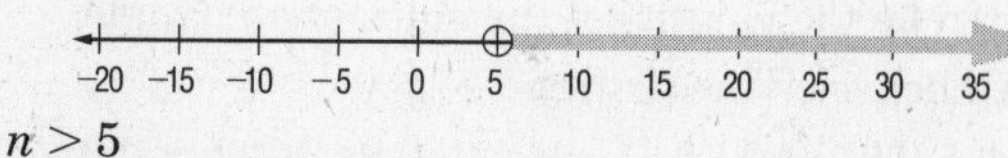

$n > 5$

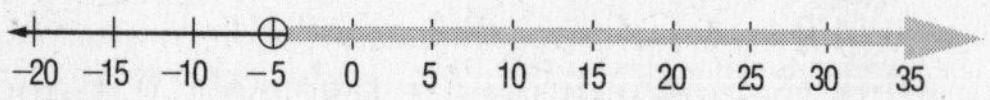

$n > -5$

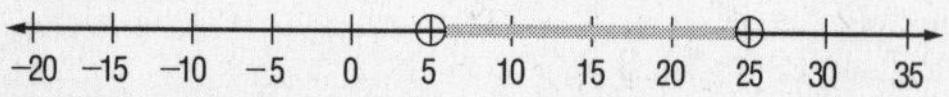

Find the intersection.

The range of values that fit all three inequalities is $5 < n < 25$.

29. Let the measure of the third side be n.

$12 + 18 > n$ $\quad$ $12 + n > 18$ $\quad$ $18 + n > 12$

$30 > n$ or $n < 30$ $\quad$ $n > 6$ $\quad$ $n > -6$

Graph the inequalities on the same number line.

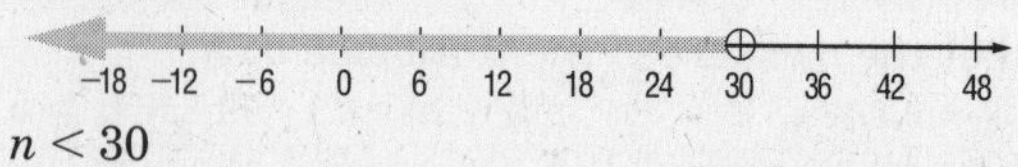

$n < 30$

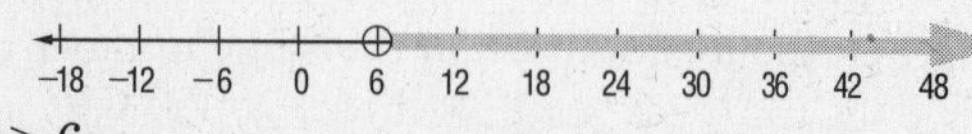

$n > 6$

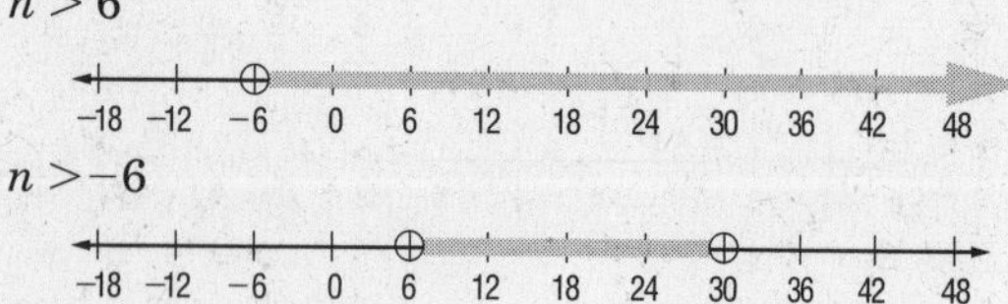

$n > -6$

Find the intersection.

The range of values that fit all three inequalities is $6 < n < 30$.

30. Let the measure of the third side be n.

$21 + 47 > n$ $\quad$ $21 + n > 47$ $\quad$ $47 + n > 21$

$68 > n$ or $n < 68$ $\quad$ $n > 26$ $\quad$ $n > -26$

Graph the inequalities on the same number line.

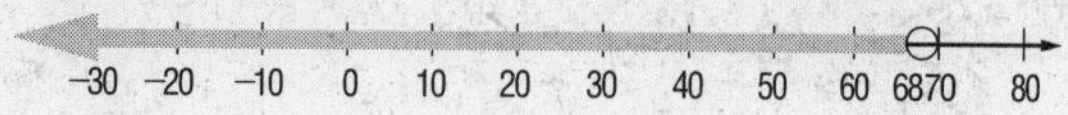

$n < 68$

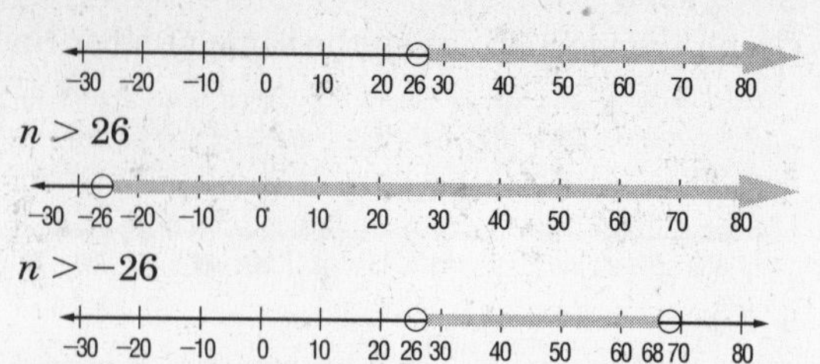

$n > 26$

$n > -26$

Find the intersection.

The range of values that fit all three inequalities is $26 < n < 68$.

31. Let the measure of the third side be n.

$32 + 61 > n$ $\quad$ $32 + n > 61$ $\quad$ $61 + n > 32$

$93 > n$ or $n < 93$ $\quad$ $n > 29$ $\quad$ $n > -29$

Graph the inequalities on the same number line.

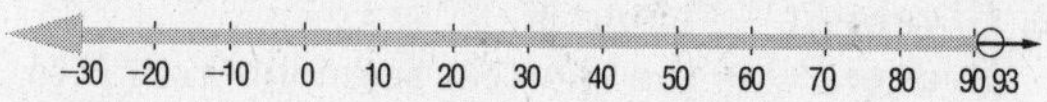

$n < 93$

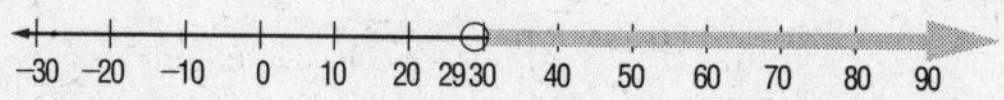

$n > 29$

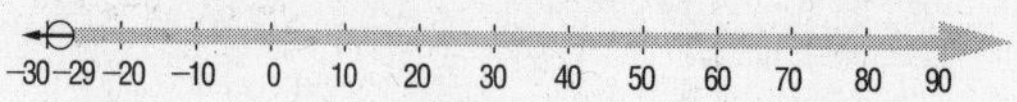

$n > -29$

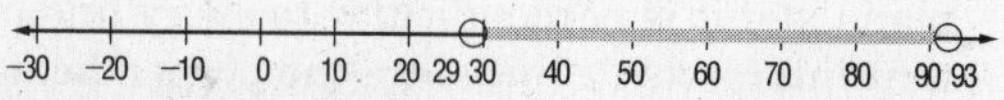

Find the intersection.

The range of values that fit all three inequalities is $29 < n < 93$.

32. Let the measure of the third side be n.

$30 + 30 > n$ $\quad$ $30 + n > 30$

$60 > n$ or $n < 60$ $\quad$ $n > 0$

Graph the inequalities on the same number line.

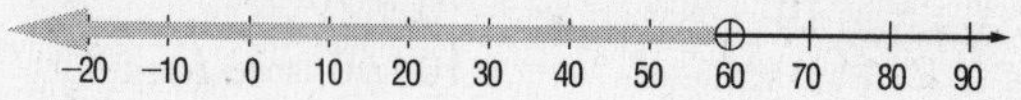

$n < 60$

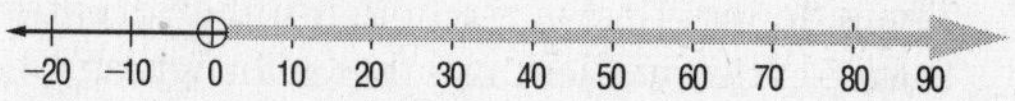

$n > 0$

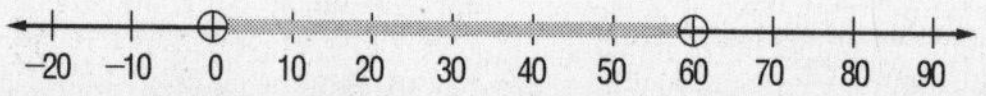

Find the intersection.

The range of values that fit all three inequalities is $0 < n < 60$.

33. Let the measure of the third side be n.

$64 + 88 > n$ $\quad$ $64 + n > 88$ $\quad$ $88 + n > 64$

$152 > n$ or $n < 152$ $\quad$ $n > 24$ $\quad$ $n > -24$

Graph the inequalities on the same number line.

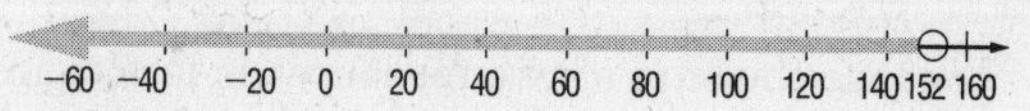

$n < 152$

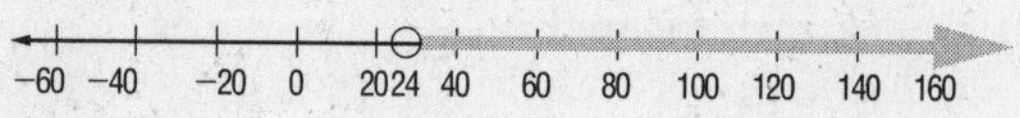

$n > 24$

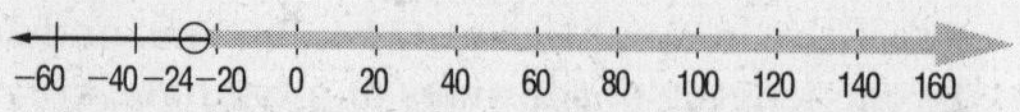

$n > -24$

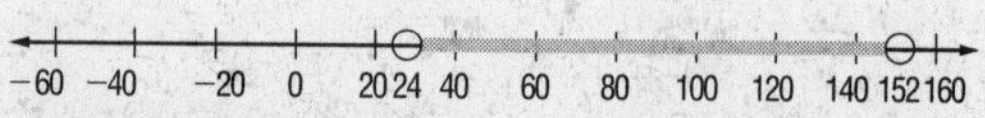

Find the intersection.

The range of values that fit all three inequalities is $24 < n < 152$.

34. Let the measure of the third side be n.

$57 + 55 > n$ $\quad$ $57 + n > 55$ $\quad$ $55 + n > 57$

$112 > n$ or $n < 112$ $\quad$ $n > -2$ $\quad$ $n > 2$

Graph the inequalities on the same number line.

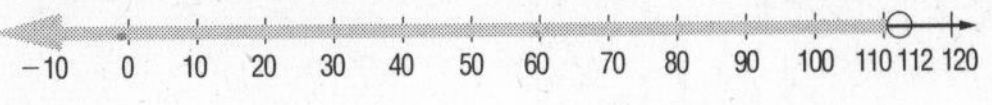

$n < 112$

$n > -2$

$n > 2$

Find the intersection.

The range of values that fit all three inequalities is $2 < n < 112$.

35. Let the measure of the third side be n.

$75 + 75 > n$ $\quad$ $75 + n > 75$

$150 > n$ or $n < 150$ $\quad$ $n > 0$

Graph the inequalities on the same number line.

$n < 150$

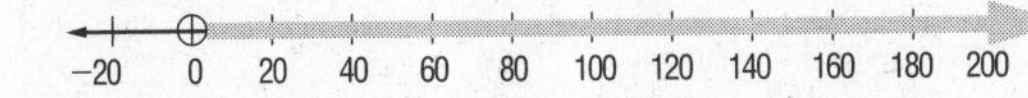

$n > 0$

Find the intersection.

The range of values that fit all three inequalities is $0 < n < 150$.

36. Let the measure of the third side be n.

$78 + 5 > n$ $\quad$ $78 + n > 5$ $\quad$ $5 + n > 78$

$83 > n$ or $n < 83$ $\quad$ $n > -73$ $\quad$ $n > 73$

Graph the inequalities on the same number line.

$n < 83$

$n > -73$

$n > 73$

Find the intersection.

The range of values that fit all three inequalities is $73 < n < 83$.

37. Let the measure of the third side be n.

$99 + 2 > n$ $\quad$ $99 + n > 2$ $\quad$ $2 + n > 99$

$101 > n$ or $n < 101$ $\quad$ $n > -97$ $\quad$ $n > 97$

Graph the inequalities on the same number line.

$n < 101$

$n > -97$

$n > 97$

Find the intersection.

The range of values that fit all three inequalities is $97 < n < 101$.

38. Given: $\angle B \cong \angle ACB$

Prove: $AD + AB > CD$

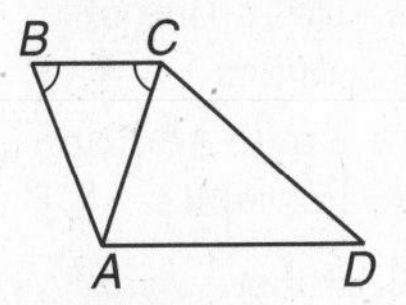

Proof:

Statements	**Reasons**
1. $\angle B \cong \angle ACB$	1. Given
2. $\overline{AB} \cong \overline{AC}$	2. If two ∠s are ≅, the sides opposite the two ∠s are ≅.
3. $AB = AC$	3. Def. of ≅ segments
4. $AD + AC > CD$	4. Triangle Inequality
5. $AD + AB > CD$	5. Substitution

39. Given: $\overline{HE} \cong \overline{EG}$

Prove: $HE + FG > EF$

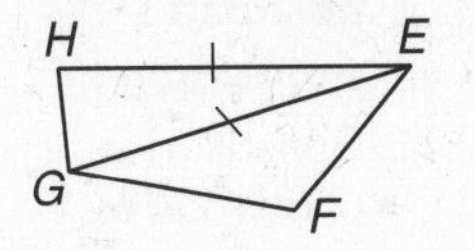

Proof:

Statements	**Reasons**
1. $\overline{HE} \cong \overline{EG}$	1. Given
2. $HE = EG$	2. Def. of ≅ segments
3. $EG + FG > EF$	3. Triangle Inequality
4. $HE + FG > EF$	4. Substitution

40. Given: $\triangle ABC$

Prove: $AC + BC > AB$

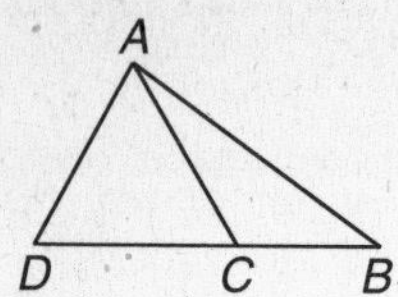

Proof:

Statements	Reasons
1. Construct $\overline{CD}$ so that C is between B and D and $\overline{CD} \cong \overline{AC}$.	1. Ruler Postulate
2. $CD = AC$	2. Definition of $\cong$
3. $\angle CAD \cong \angle ADC$	3. Isosceles Triangle Theorem
4. $m\angle CAD = m\angle ADC$	4. Definition of $\cong$ angles
5. $m\angle BAC + m\angle CAD = m\angle BAD$	5. Angle Addition Postulate
6. $m\angle BAC + m\angle ADC = m\angle BAD$	6. Substitution
7. $m\angle ADC < m\angle BAD$	7. Definition of inequality
8. $AB < BD$	8. If an angle of a triangle is greater than a second angle, then the side opposite the greater angle is longer than the side opposite the lesser angle.
9. $BD = BC + CD$	9. Segment Addition Postulate
10. $AB < BC + CD$	10. Substitution
11. $AB < BC + AC$	11. Substitution (Steps 2, 10)

41. $AB = \sqrt{(2-5)^2 + (-4-8)^2}$
$= \sqrt{9 + 144}$
$= \sqrt{153}$
≈ 12.4

$BC = \sqrt{(-3-2)^2 + [-1-(-4)]^2}$
$= \sqrt{25 + 9}$
$= \sqrt{34}$
≈ 5.8

$AC = \sqrt{(-3-5)^2 + (-1-8)^2}$
$= \sqrt{64 + 81}$
$= \sqrt{145}$
≈ 12.0

$AB + BC \overset{?}{>} AC$
$12.4 + 5.8 \overset{?}{>} 12.0$
$18.2 > 12.0$ ✓

$AB + AC \overset{?}{>} BC$
$12.4 + 12.0 \overset{?}{>} 5.8$
$24.4 > 5.8$ ✓

$AC + BC \overset{?}{>} AB$
$12.0 + 5.8 \overset{?}{>} 12.4$
$17.8 > 12.4$ ✓

All of the inequalities are true, so the coordinates can be the vertices of a triangle.

42. $LM = \sqrt{[-22-(-24)]^2 + [20-(-19)]^2}$
$= \sqrt{4 + 1521}$
$= \sqrt{1525}$
≈ 39.1

$MN = \sqrt{[-5-(-22)]^2 + (-7-20)^2}$
$= \sqrt{289 + 729}$
$= \sqrt{1018}$
≈ 31.9

$LN = \sqrt{[-5-(-24)]^2 + [-7-(-19)]^2}$
$= \sqrt{361 + 144}$
$= \sqrt{505}$
≈ 22.5

$LM + MN \overset{?}{>} LN$
$39.1 + 31.9 \overset{?}{>} 22.5$
$71 > 22.5$ ✓

$LM + LN \overset{?}{>} MN$
$39.1 + 22.5 \overset{?}{>} 31.9$
$61.6 > 31.9$ ✓

$LN + MN \overset{?}{>} LM$
$22.5 + 31.9 \overset{?}{>} 39.1$
$54.4 > 39.1$ ✓

All of the inequalities are true, so the coordinates can be the vertices of a triangle.

43. $XY = \sqrt{(16-0)^2 + [-12-(-8)]^2}$
$= \sqrt{256 + 16}$
$= \sqrt{272}$
≈ 16.49

$YZ = \sqrt{(28-16)^2 + [-15-(-12)]^2}$
$= \sqrt{144 + 9}$
$= \sqrt{153}$
≈ 12.37

$XZ = \sqrt{(28-0)^2 + [-15-(-8)]^2}$
$= \sqrt{784 + 49}$
$= \sqrt{833}$
≈ 28.86

$XY + YZ \overset{?}{>} XZ$
$16.49 + 12.37 \overset{?}{>} 28.86$
$28.86 = 28.86$

Because the sum of two measures equals the third measure, the sides cannot form a triangle and so the coordinates cannot be the vertices of a triangle.

44. $RS = \sqrt{(-3-1)^2 + [-20-(-4)]^2}$
$= \sqrt{16 + 256}$
$= \sqrt{272}$
≈ 16.5

$RT = \sqrt{(5-1)^2 + [12-(-4)]^2}$
$= \sqrt{16 + 256}$
$= \sqrt{272}$
≈ 16.5

$ST = \sqrt{[5-(-3)]^2 + [12-(-20)]^2}$
$= \sqrt{64 + 1024}$
$= \sqrt{1088}$
≈ 33

$RS + RT \stackrel{?}{>} ST$

$16.5 + 16.5 \stackrel{?}{>} 33$

$33 = 33$

Because the sum of two measures equals the third measure, the sides cannot form a triangle and so the coordinates cannot be the vertices of a triangle.

45. Consider all possible triples using the lengths 3, 4, 5, 6, and 12.

$3 + 4 \stackrel{?}{>} 5$ $\quad$ $4 + 5 \stackrel{?}{>} 3$ $\quad$ $3 + 5 \stackrel{?}{>} 4$

$7 > 5$ ✓ $\quad$ $9 > 3$ ✓ $\quad$ $8 > 4$ ✓

$3 + 4 \stackrel{?}{>} 6$ $\quad$ $4 + 6 \stackrel{?}{>} 3$ $\quad$ $3 + 6 \stackrel{?}{>} 4$

$7 > 6$ ✓ $\quad$ $10 > 3$ ✓ $\quad$ $9 > 4$ ✓

$3 + 4 > 12$

$7 \not> 12$

$4 + 5 \stackrel{?}{>} 6$ $\quad$ $4 + 6 \stackrel{?}{>} 5$ $\quad$ $5 + 6 \stackrel{?}{>} 4$

$9 > 6$ ✓ $\quad$ $10 > 5$ ✓ $\quad$ $11 > 4$ ✓

$4 + 5 \stackrel{?}{>} 12$

$9 \not> 12$

$3 + 5 \stackrel{?}{>} 6$ $\quad$ $5 + 6 \stackrel{?}{>} 3$ $\quad$ $3 + 6 \stackrel{?}{>} 5$

$8 > 6$ ✓ $\quad$ $11 > 3$ ✓ $\quad$ $9 > 5$ ✓

$3 + 5 \stackrel{?}{>} 12$

$8 \not> 12$

$3 + 6 \stackrel{?}{>} 12$

$9 \not> 12$

$4 + 6 \stackrel{?}{>} 12$

$10 \not> 12$

$5 + 6 \stackrel{?}{>} 12$

$11 \not> 12$

Of all possible triples, 4 of them satisfy the triangle inequality, so there are 4 possible triangles.

46. $3 + 4 + 5 = 12$, which is divisible by 3
$3 + 4 + 6 = 13$, which is not divisible by 3
$4 + 5 + 6 = 15$, which is divisible by 3
$3 + 5 + 6 = 14$, which is not divisible by 3
Carlota could make 2 different triangles with a perimeter that is divisible by 3.

47. The rope has 13 knots that determine 12 segments of the rope. We need to determine how many triangles there are whose perimeter is 12. First determine whether any triangle can have a side with 1 segment. If one side has length 1, then the possible triples of side lengths are 1, 1, 10; 1, 2, 9; 1, 3, 8; 1, 4, 7; and 1, 5, 6. Check each of these triples using the Triangle Inequality.

$1 + 1 \not> 10$
$1 + 2 \not> 9$
$1 + 3 \not> 8$
$1 + 4 \not> 7$
$1 + 5 \not> 6$

Therefore, there is no possible triangle with one side of length 1 and perimeter 12.

Determine whether any triangle can have a side with 2 segments. If one side has length 2, then the possible triples of side lengths are 2, 1, 9; 2, 2, 8; 2, 3, 7; 2, 4, 6; and 2, 5, 5. Check each of these triples using the Triangle Inequality.

$2 + 1 \not> 9$
$2 + 2 \not> 8$
$2 + 3 \not> 7$
$2 + 4 \not> 6$

$2 + 5 > 5$ and $5 + 5 > 2$, so there could be a triangle with side lengths 2, 5, and 5 units.

Determine whether any triangle can have a side with 3 segments. If one side has length 3, then the possible triples of side lengths are 3, 1, 8; 3, 2, 7; 3, 3, 6; and 3, 4, 5. Check each of these triples using the Triangle Inequality.

$3 + 1 \not> 8$
$3 + 2 \not> 7$
$3 + 3 \not> 6$

$3 + 4 > 5$ and $3 + 5 > 4$ and $4 + 5 > 3$, so there could be a triangle with side lengths 3, 4, and 5 units.

Determine whether any triangle can have a side with 4 segments. If one side has length 4, then the possible triples of side lengths are 4, 1, 7; 4, 2, 6; 4, 3, 5; and 4, 4, 4. Check each of these triples using the Triangle Inequality. Note that we have already shown that there can be a triangle with sides 4, 3, and 5 units.

$4 + 1 \not> 7$
$4 + 2 \not> 6$

$4 + 4 > 4$, so there could be a triangle with side lengths 4, 4, and 4 units.

By examining all of the triples we have considered to this point, we can see that all possible triples have been listed and checked. Therefore, there are 3 triangles that can be formed using the rope shown in the figure.

48. $14 < m < 17$, so m is either 15 or 16 feet.
$13 < n < 17$, so n is 14, 15, or 16 feet.
Check all possible triples using the triangle inequality.

$2 + 15 \stackrel{?}{>} 14$ $\quad$ $2 + 14 \stackrel{?}{>} 15$ $\quad$ $14 + 15 \stackrel{?}{>} 2$

$17 > 14$ ✓ $\quad$ $16 > 15$ ✓ $\quad$ $29 > 2$ ✓

$2 + 15 \stackrel{?}{>} 15$ $\quad$ $15 + 15 \stackrel{?}{>} 2$

$17 > 15$ ✓ $\quad$ $30 > 2$ ✓

$2 + 15 \stackrel{?}{>} 16$ $\quad$ $2 + 16 \stackrel{?}{>} 15$ $\quad$ $15 + 16 \stackrel{?}{>} 2$

$17 > 16$ ✓ $\quad$ $18 > 15$ ✓ $\quad$ $31 > 2$ ✓

$2 + 16 \stackrel{?}{>} 14$ $\quad$ $2 + 14 > 16$

$18 > 14$ ✓ $\quad$ $16 \not> 16$

$2 + 16 \stackrel{?}{>} 15$ $\quad$ $2 + 15 \stackrel{?}{>} 16$ $\quad$ $15 + 16 \stackrel{?}{>} 2$

$18 > 15$ ✓ $\quad$ $17 > 16$ ✓ $\quad$ $31 > 2$ ✓

$2 + 16 \stackrel{?}{>} 16$ $\quad$ $16 + 16 \stackrel{?}{>} 2$

$18 > 16$ ✓ $\quad$ $32 > 2$ ✓

The possible triangles that can be made from sides with measures 2 ft, m ft, and n ft are (2 ft, 15 ft, 14 ft), (2 ft, 15 ft, 15 ft), (2 ft, 15 ft, 16 ft), and (2 ft, 16 ft, 16 ft).

49. Of the 4 possible triangles listed in Exercise 48, 2 are isosceles, so the probability is $\frac{2}{4}$ or $\frac{1}{2}$.

50. Sample answer: The length of any side of a triangle is greater than the differences between the lengths of the other two sides.
Paragraph Proof:
By the Triangle Inequality Theorem, for $\triangle ABC$ with side measures a, b, and c, $a + b > c$, $b + c > a$, and $c + a > b$. Using the Subtraction Property of Inequality, $a > c - b$, $b > a - c$, and $c > b - a$.

51. Sample answer: You can use the Triangle Inequality Theorem to verify the shortest route between two locations. Answers should include the following.
- A longer route might be better if you want to collect frequent flier miles.
- A straight route might not always be available.

52. D; If the perimeter is 29, the measure of the third side is 10.

$7 + 10 \overset{?}{>} 12$ $7 + 12 \overset{?}{>} 10$ $10 + 12 \overset{?}{>} 7$
$17 > 12$ ✓ $19 > 10$ ✓ $22 > 7$ ✓

So, 7, 12, and 10 could be the sides of a triangle with perimeter 29.
If the perimeter is 34, the measure of the third side is 15.

$7 + 15 \overset{?}{>} 12$ $7 + 12 \overset{?}{>} 15$ $12 + 15 \overset{?}{>} 7$
$22 > 12$ ✓ $19 > 15$ ✓ $27 > 7$ ✓

So, 7, 12, and 15 could be the sides of a triangle with perimeter 34.
If the perimeter is 37, the measure of the third side is 18.

$7 + 18 \overset{?}{>} 12$ $7 + 12 \overset{?}{>} 18$ $12 + 18 \overset{?}{>} 7$
$25 > 12$ ✓ $19 > 18$ ✓ $30 > 7$ ✓

So, 7, 12, and 18 could be the sides of a triangle with perimeter 37.
If the perimeter is 38, the measure of the third side is 19.

$7 + 12 \overset{?}{>} 19$
$19 \not> 19$

So, 7, 12, and 19 cannot be the sides of a triangle.

53. A; If the graphs of the equations do intersect, then we can solve the system of equations and find the coordinates of any points of intersection. Substitute $-x$ for y in the equation $(x - 5)^2 + (y - 5)^2 = 4$ and solve for x.

$$(x - 5)^2 + [(-x) - 5]^2 = 4$$
$$x^2 - 10x + 25 + x^2 + 10x + 25 = 4$$
$$2x^2 + 50 = 4$$
$$2x^2 = -46$$
$$x^2 = -23$$

Because there is no real number x whose square is equal to -23, there are no points of intersection.

Page 266 Maintain Your Skills

54. Given: P is a point not on line ℓ.
Prove: $\overleftrightarrow{PQ}$ is the only line through P perpendicular to ℓ.

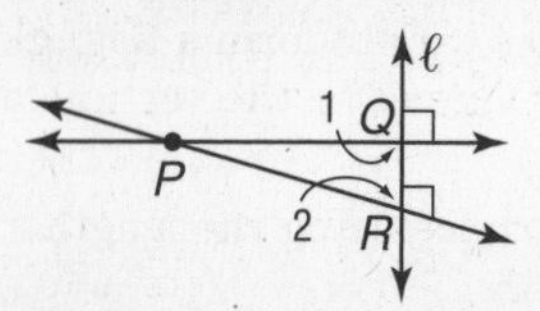

Proof:

Statements	Reasons
1. $\overleftrightarrow{PQ}$ is not the only line through P perpendicular to ℓ.	1. Assumption
2. $\angle 1$ and $\angle 2$ are right angles.	2. $\perp$ lines form 4 rt. $\angle$s.
3. $m\angle 1 = 90$, $m\angle 2 = 90$	3. Def. of rt. $\angle$
4. $m\angle 1 + m\angle 2 + m\angle QPR = 180$	4. The sum of $\angle$s in a $\triangle$ is 180.
5. $90 + 90 + m\angle QPR = 180$	5. Substitution
6. $m\angle QPR = 0$	6. Subtraction Property

This contradicts the fact that the measure of an angle is greater than 0. Thus, $\overleftrightarrow{PQ}$ is the only line through P perpendicular to ℓ.

55.
$$m\angle P + m\angle Q + m\angle R = 180$$
$$7x + 8 + 8x - 10 + 7x + 6 = 180$$
$$22x + 4 = 180$$
$$22x = 176$$
$$x = 8$$

$m\angle P = 7(8) + 8$ or 64
$m\angle Q = 8(8) - 10$ or 54
$m\angle R = 7(8) + 6$ or 62
$m\angle P > m\angle R > m\angle Q$, so the sides of $\triangle PQR$ in order from longest to shortest are $\overline{QR}$, $\overline{PQ}$, $\overline{PR}$.

56.
$$m\angle P + m\angle Q + m\angle R = 180$$
$$3x + 44 + 68 - 3x + x + 61 = 180$$
$$x + 173 = 180$$
$$x = 7$$

$m\angle P = 3(7) + 44$ or 65
$m\angle Q = 68 - 3(7)$ or 47
$m\angle R = 7 + 61$ or 68
$m\angle R > m\angle P > m\angle Q$, so the sides of $\triangle PQR$ in order from longest to shortest are $\overline{PQ}$, $\overline{QR}$, $\overline{PR}$.

57.
$JK = \sqrt{(0 - 0)^2 + (0 - 5)^2}$
$= \sqrt{0 + 25}$ or 5
$PQ = \sqrt{(4 - 4)^2 + (3 - 8)^2}$
$= \sqrt{0 + 25}$ or 5
$KL = \sqrt{(-2 - 0)^2 + (0 - 0)^2}$
$= \sqrt{4 + 0}$ or 2
$QR = \sqrt{(6 - 4)^2 + (3 - 3)^2}$
$= \sqrt{4 + 0}$ or 2
$JL = \sqrt{(-2 - 0)^2 + (0 - 5)^2}$
$= \sqrt{4 + 25}$ or $\sqrt{29}$
$PR = \sqrt{(6 - 4)^2 + (3 - 8)^2}$
$= \sqrt{4 + 25}$ or $\sqrt{29}$

The corresponding sides have the same measure and are congruent. $\triangle JKL \cong \triangle PQR$ by SSS.

58. $JK = \sqrt{(1-6)^2 + (-6-4)^2}$
$= \sqrt{25 + 100}$ or $\sqrt{125}$
$PQ = \sqrt{(5-0)^2 + (-3-7)^2}$
$= \sqrt{25 + 100}$ or $\sqrt{125}$
$KL = \sqrt{(-9-1)^2 + [5-(-6)]^2}$
$= \sqrt{100 + 121}$ or $\sqrt{221}$
$QR = \sqrt{(15-5)^2 + [8-(-3)]^2}$
$= \sqrt{100 + 121}$ or $\sqrt{221}$
$JL = \sqrt{(-9-6)^2 + (5-4)^2}$
$= \sqrt{225 + 1}$ or $\sqrt{226}$
$PR = \sqrt{(15-0)^2 + (8-7)^2}$
$= \sqrt{225 + 1}$ or $\sqrt{226}$
The corresponding sides have the same measure and are congruent. $\triangle JKL \cong \triangle PQR$ by SSS.

59. $JK = \sqrt{[1-(-6)]^2 + [5-(-3)]^2}$
$= \sqrt{49 + 64}$ or $\sqrt{113}$
$PQ = \sqrt{(5-2)^2 + [-4-(-11)]^2}$
$= \sqrt{9 + 49}$ or $\sqrt{58}$
$KL = \sqrt{(2-1)^2 + (-2-5)^2}$
$= \sqrt{1 + 49}$ or $\sqrt{50}$
$QR = \sqrt{(10-5)^2 + [-10-(-4)]^2}$
$= \sqrt{25 + 36}$ or $\sqrt{61}$
$JL = \sqrt{[2-(-6)]^2 + [-2-(-3)]^2}$
$= \sqrt{64 + 1}$ or $\sqrt{65}$
$PR = \sqrt{(10-2)^2 + [-10-(-11)]^2}$
$= \sqrt{64 + 1}$ or $\sqrt{65}$
The corresponding sides are not congruent, so the triangles are not congruent.

60. $3x + 54 < 90$
$3x < 36$
$x < 12$

61. $8x - 14 < 3x + 19$
$5x - 14 < 19$
$5x < 33$
$x < 6.6$

62. $4x + 7 < 180$
$4x < 173$
$x < 43.25$

Page 266 Practice Quiz 2

1. The number 117 is not divisible by 13.
2. $m\angle C \geq m\angle D$
3. **Step 1** Assume that $x \leq 8$.
Step 2 $7x > 56$
$x > 8$
Step 3 The solution of $7x > 56$ contradicts the assumption. Thus, $x \leq 8$ must be false. Therefore, $x > 8$.

4. **Given:** $\overline{MO} \cong \overline{ON}$
$\overline{MP} \not\cong \overline{NP}$
Prove: $\angle MOP \not\cong \angle NOP$

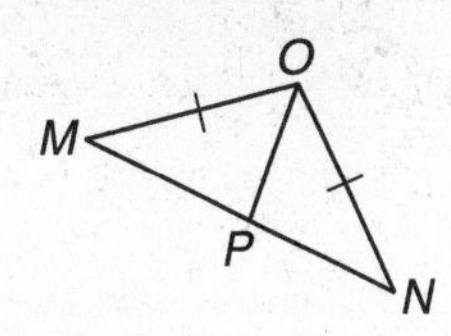

Proof:
Step 1 Assume that $\angle MOP \cong \angle NOP$.
Step 2 We know that $\overline{MO} \cong \overline{ON}$, and $\overline{OP} \cong \overline{OP}$ by the Reflexive Property. If $\angle MOP \cong \angle NOP$, then $\triangle MOP \cong \triangle NOP$ by SAS. Then, $\overline{MP} \cong \overline{NP}$ by CPCTC.
Step 3 The conclusion that $\overline{MP} \cong \overline{NP}$ contradicts the given information. Thus, the assumption is false. Therefore, $\angle MOP \not\cong \angle NOP$.

5. **Given:** $m\angle ADC \neq m\angle ADB$
Prove: $\overline{AD}$ is not an altitude of $\triangle ABC$.

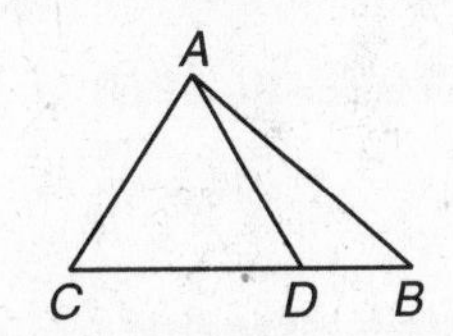

Proof:

Statements	Reasons
1. $\overline{AD}$ is an altitude of $\triangle ABC$.	1. Assumption
2. $\angle ADC$ and $\angle ADB$ are right angles.	2. Def. of altitude
3. $\angle ADC \cong \angle ADB$	3. All rt $\angle$s are $\cong$.
4. $m\angle ADC = m\angle ADB$	4. Def. of $\cong$ angles

This contradicts the given information that $m\angle ADC \neq m\angle ADB$. Thus, $\overline{AD}$ is not an altitude of $\triangle ABC$.

6. Check each inequality.
$7 + 24 \overset{?}{>} 25$ $7 + 25 \overset{?}{>} 24$ $24 + 25 \overset{?}{>} 7$
$31 > 25$ ✓ $32 > 24$ ✓ $49 > 7$ ✓
All of the inequalities are true, so 7, 24, and 25 can be the lengths of the sides of a triangle.

7. $25 + 35 \overset{?}{>} 60$
$60 \not> 60$
Because the sum of two measures equals the third measure, the sides cannot form a triangle.

8. Check each inequality.
$3 + 18 \overset{?}{>} 20$ $18 + 20 \overset{?}{>} 3$ $3 + 20 \overset{?}{>} 18$
$21 > 20$ ✓ $38 > 3$ ✓ $23 > 18$ ✓
All of the inequalities are true, so 20, 3, and 18 can be the lengths of the sides of a triangle.

9. Check each inequality.
$5 + 10 \overset{?}{>} 6$ $5 + 6 \overset{?}{>} 10$ $6 + 10 \overset{?}{>} 5$
$15 > 6$ ✓ $11 > 10$ ✓ $16 > 5$ ✓
All of the inequalities are true, so 5, 10, and 6 can be the lengths of the sides of a triangle.

10. Let the measure of the third side be n.

$57 + 32 > n$ $\quad$ $57 + n > 32$ $\quad$ $32 + n > 57$

$89 > n$ or $n < 89$ $\quad$ $n > -25$ $\quad$ $n > 25$

Graph the inequalities on the same number line.

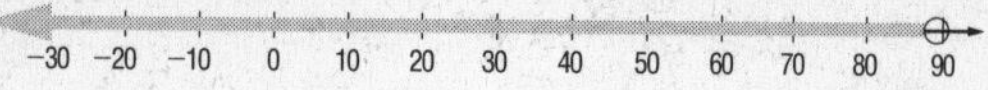

$n < 89$

$n > -25$

$n > 25$

Find the intersection.

The range of values that fit all three inequalities is $25 < n < 89$.

5-5 Inequalities Involving Two Triangles

Pages 270–271 Check for Understanding

1. Sample answer: A pair of scissors illustrates the SSS inequality. As the distance between the tips of the scissors decreases, the angle between the blades decreases, allowing the blades to cut.
2. The SSS Inequality Theorem compares the angle between two sides of a triangle for which the two sides are congruent and the third side is different. The SSS Postulate states that two triangles that have three sides congruent are congruent.
3. In $\triangle BDC$ and $\triangle BDA$, $\overline{BC} \cong \overline{AD}$, $\overline{BD} \cong \overline{BD}$, and $m\angle BDA > m\angle CBD$. The SAS Inequality Theorem allows us to conclude that $AB < CD$.
4. In $\triangle PQS$ and $\triangle RQS$, $\overline{RQ} \cong \overline{PQ}$, $\overline{QS} \cong \overline{QS}$, and $PS > RS$. The SSS Inequality Theorem allows us to conclude that $m\angle PQS > m\angle RQS$.
5. The upper triangle is equilateral and so all angles are 60 degrees. The SAS inequality allows us to conclude that $x + 5 > 3x - 7$.

$x + 5 > 3x - 7$

$5 > 2x - 7$

$12 > 2x$

$6 > x$

Also, the measure of any side is greater than 0.

$3x - 7 > 0$

$3x > 7$

$x > \frac{7}{3}$

The two inequalities can be written as the compound inequality $\frac{7}{3} < x < 6$.

6. Because $12 > 8$, the SSS Inequality allows us to conclude $140 > 7x + 4$.

$140 > 7x + 4$

$136 > 7x$

$\frac{136}{7} > x$

Also, the measure of any angle is always greater than 0.

$7x + 4 > 0$

$7x > -4$

$x > -\frac{4}{7}$

The two inequalities can be written as the compound inequality $-\frac{4}{7} < x < \frac{136}{7}$.

7. **Given:** $\overline{PQ} \cong \overline{SQ}$

Prove: $PR > SR$

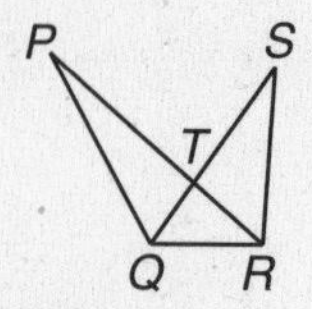

Proof:

Statements	Reasons
1. $\overline{PQ} \cong \overline{SQ}$	1. Given
2. $\overline{QR} \cong \overline{QR}$	2. Reflexive Property
3. $m\angle PQR = m\angle PQS + m\angle SQR$	3. Angle Addition Postulate
4. $m\angle PQR > m\angle SQR$	4. Def. of inequality
5. $PR > SR$	5. SAS Inequality

8. **Given:** $\overline{TU} \cong \overline{US}$; $\overline{US} \cong \overline{SV}$

Prove: $ST > UV$

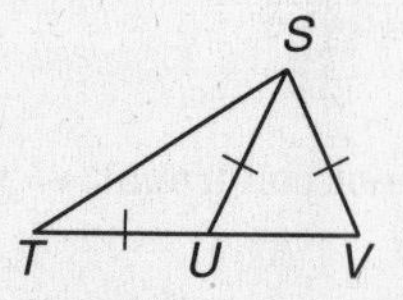

Proof:

Statements	Reasons
1. $\overline{TU} \cong \overline{US}$, $\overline{US} \cong \overline{SV}$	1. Given
2. $m\angle SUT > m\angle USV$	2. Ext. ∠ Inequality Theorem
3. $ST > UV$	3. SAS Inequality

9. Sample answer: The pliers are an example of the SAS inequality. As force is applied to the handles, the angle between them decreases causing the distance between them to decrease. As the distance between the ends of the pliers decreases, more force is applied to a smaller area.

Pages 271–273 Practice and Apply

10. From the figure, $AB = 9$ and $FD = 6$, so $AB > FD$.
11. In $\triangle BDC$ and $\triangle FDB$, $\overline{FD} \cong \overline{DC}$, $\overline{BD} \cong \overline{BD}$, and $BC < BF$. The SSS Inequality allows us to conclude that $m\angle BDC < m\angle FDB$.
12. In $\triangle BFA$ and $\triangle DBF$, $\overline{AB} \cong \overline{BD}$, $\overline{BF} \cong \overline{BF}$, and $AF > FD$. The SSS Inequality allows us to conclude that $m\angle FBA > m\angle DBF$.
13. In $\triangle ABD$ and $\triangle CBD$, $\overline{BC} \cong \overline{BA}$, $\overline{BD} \cong \overline{BD}$, and $m\angle ABD > m\angle CBD$. The SAS Inequality allows us to conclude that $AD > DC$.

14. In $\triangle ABO$ and $\triangle CBO$, $\overline{AB} \cong \overline{CB}$, $\overline{OB} \cong \overline{OB}$, and $m\angle CBO < m\angle ABO$. The SAS Inequality allows us to conclude that $OC < OA$.

15. In $\triangle ABC$, $\overline{AB} \cong \overline{CB}$ so $\triangle ABC$ is isosceles with base angles of measure $\frac{1}{2}[180 - (40 + 60)]$ or 40.

$m\angle AOB = 180 - (m\angle OAB + m\angle ABO)$
$= 180 - (40 + 60)$
$= 80$

$m\angle AOD = 180 - m\angle AOB$
$= 100$

Therefore, $m\angle AOD > m\angle AOB$.

16. By the SAS Inequality, $10 < 3x - 2$.

$10 < 3x - 2$
$12 < 3x$
$4 < x$ or $x > 4$

17. The triangle on the left is equilateral, so all angles have measure 60. Then by the SAS inequality, $x + 2 > 2x - 8$.

$x + 2 > 2x - 8$
$2 > x - 8$
$10 > x$

The measure of any side is always greater than 0.

$2x - 8 > 0$
$2x > 8$
$x > 4$

The two inequalities can be written as the compound inequality $4 < x < 10$.

18. $\overline{MC} \cong \overline{MC}$, so by the SSS Inequality $m\angle 1 > m\angle 2$.

$5x + 20 > 8x - 100$
$20 > 3x - 100$
$120 > 3x$
$40 > x$

The measure of any angle is always greater than 0.

$8x - 100 > 0$
$8x > 100$
$x > 12.5$

The two inequalities can be written as the compound inequality $12.5 < x < 40$.

19. $\angle RTV \cong \angle TRV$, so $\triangle RVT$ is isosceles and $\overline{RV} \cong \overline{TV}$. $\overline{SV} \cong \overline{SV}$ and $RS < ST$, so by the SSS Inequality, $m\angle RVS < m\angle SVT$.

$15 + 5x < 10x - 20$
$15 < 5x - 20$
$35 < 5x$
$7 < x$

The measure of $\angle SVT$ is less than 180.

$10x - 20 < 180$
$10x < 200$
$x < 20$

The two inequalities can be written as the compound inequality $7 < x < 20$.

20. Given: $\triangle ABC$, $\overline{AB} \cong \overline{CD}$
Prove: $BC > AD$

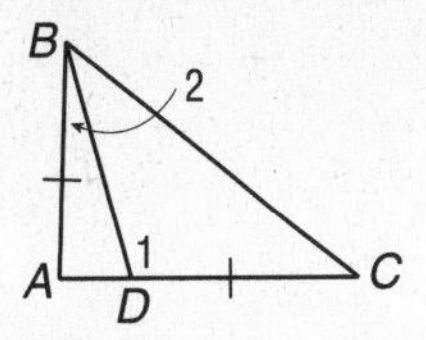

Proof:

Statements	Reasons
1. $\triangle ABC$, $\overline{AB} \cong \overline{CD}$	1. Given
2. $\overline{BD} \cong \overline{BD}$	2. Reflexive Property
3. $m\angle 1 > m\angle 2$	3. If an $\angle$ is an ext. $\angle$ of a $\triangle$, then its measure is greater than the measure of either remote int. $\angle$.
4. $BC > AD$	4. SAS Inequality

21. Given: $\overline{PQ} \cong \overline{RS}$, $QR < PS$
Prove: $m\angle 3 < m\angle 1$

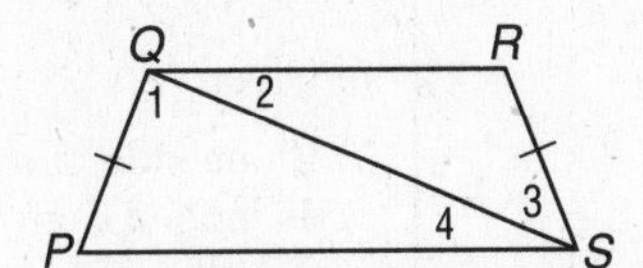

Proof:

Statements	Reasons
1. $\overline{PQ} \cong \overline{RS}$	1. Given
2. $\overline{QS} \cong \overline{QS}$	2. Reflexive Property
3. $QR < PS$	3. Given
4. $m\angle 3 < m\angle 1$	4. SSS Inequality

22. Given: $\overline{PR} \cong \overline{PQ}$, $SQ > SR$
Prove: $m\angle 1 < m\angle 2$

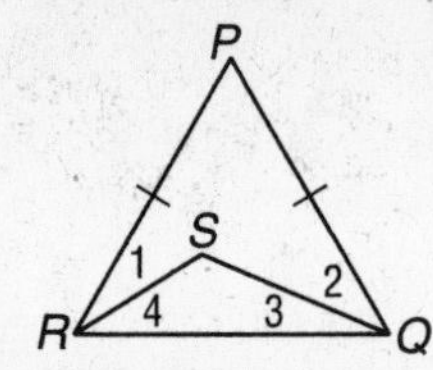

Proof:

Statements	Reasons
1. $\overline{PR} \cong \overline{PQ}$	1. Given
2. $\angle PRQ \cong \angle PQR$	2. If two sides of $\triangle$ are $\cong$, the angles opposite the sides are $\cong$.
3. $m\angle PRQ = m\angle 1 + m\angle 4$, $m\angle PQR = m\angle 2 + m\angle 3$	3. Angle Add. Post.
4. $m\angle PRQ = m\angle PQR$	4. Def. of $\cong$ angles
5. $m\angle 1 + m\angle 4 = m\angle 2 + m\angle 3$	5. Substitution
6. $SQ > SR$	6. Given
7. $m\angle 4 > m\angle 3$	7. If one side of a $\triangle$ is longer than another side, then the $\angle$ opposite the longer side is greater than the $\angle$ opposite the shorter side.
8. $m\angle 4 = m\angle 3 + x$	8. Def. of inequality
9. $m\angle 1 + m\angle 3 + x = m\angle 2 + m\angle 3$	9. Substitution
10. $m\angle 1 + x = m\angle 2$	10. Subtraction Prop.
11. $m\angle 1 < m\angle 2$	11. Def. of inequality

23. Given: $\overline{ED} \cong \overline{DF}$; $m\angle 1 > m\angle 2$; D is the midpoint of $\overline{CB}$; $\overline{AE} \cong \overline{AF}$.
Prove: $AC > AB$

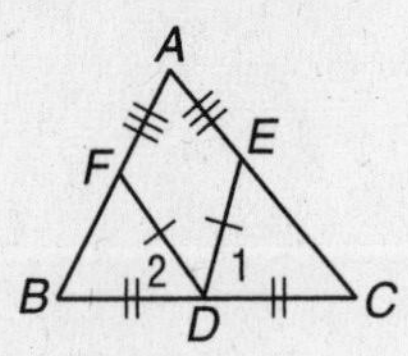

Proof:

Statements	Reasons
1. $\overline{ED} \cong \overline{DF}$; D is the midpoint of $\overline{DB}$.	1. Given
2. $CD = BD$	2. Def. of midpoint
3. $\overline{CD} \cong \overline{BD}$	3. Def. of $\cong$ segments
4. $m\angle 1 > m\angle 2$	4. Given
5. $EC > FB$	5. SAS Inequality
6. $\overline{AE} \cong \overline{AF}$	6. Given
7. $AE = AF$	7. Def. of $\cong$ segments
8. $AE + EC > AE + FB$	8. Add. Prop. of Inequality
9. $AE + EC > AF + FB$	9. Substitution
10. $AE + EC = AC$, $AF + FB = AB$	10. Segment Add. Post.
11. $AC > AB$	11. Substitution

24. Given: $\overline{RS} \cong \overline{UW}$
$\overline{ST} \cong \overline{WV}$
$RT > UV$
Prove: $m\angle S > m\angle W$

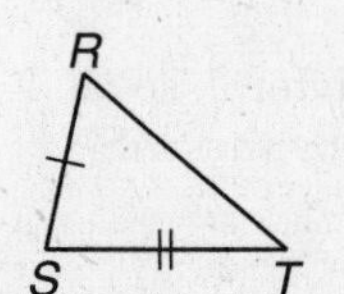

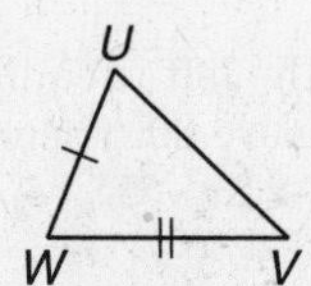

Indirect Proof:
Step 1 Assume $m\angle S \leq m\angle W$.
Step 2 If $m\angle S \leq m\angle W$, then either $m\angle S < m\angle W$ or $m\angle S = m\angle W$.
Case 1: If $m\angle S < m\angle W$, then $RT < UV$ by the SAS Inequality.
Case 2: If $m\angle S = m\angle W$, then $\triangle RST \cong \triangle UWV$ by SAS, and $\overline{RT} \cong \overline{UV}$ by CPCTC. Thus $RT = UV$.
Step 3 Both cases contradict the given $RT > UV$. Therefore, the assumption must be false, and the conclusion, $m\angle S > m\angle W$, must be true.

25. As the door is opened wider, the angle formed increases and the distance from the end of the door to the door frame increases.

26. By the SAS Inequality Theorem, if the tree started to lean, one of the angles of the triangle formed by the tree, the ground, and the stake would change, and the side opposite that angle would change as well. However, with the stake in the ground and fixed to the tree, none of the sides of the triangle can change length. Thus, none of the angles can change. This ensures that the tree will stay straight.

27. As the vertex angle increases, the base angles decrease. Thus, as the base angles decrease, the altitude of the triangle decreases.

28.
$$v = \frac{0.78s^{1.67}}{h^{1.17}} = \frac{0.78(1.0)^{1.67}}{0.85^{1.17}} \approx 0.94 \text{ m/s}$$
$$v = \frac{0.78s^{1.67}}{h^{1.17}} = \frac{0.78(1.2)^{1.67}}{0.85^{1.17}} \approx 1.28 \text{ m/s}$$

29.

Stride (m)	$v = \frac{0.78s^{1.67}}{h^{1.17}}$	Velocity (m/s)
0.25	$\frac{0.78(0.25)^{1.67}}{1.1^{1.17}}$	0.07
0.50	$\frac{0.78(0.50)^{1.67}}{1.1^{1.17}}$	0.22
0.75	$\frac{0.78(0.75)^{1.67}}{1.1^{1.17}}$	0.43
1.00	$\frac{0.78(1.00)^{1.67}}{1.1^{1.17}}$	0.70
1.25	$\frac{0.78(1.25)^{1.67}}{1.1^{1.17}}$	1.01
1.50	$\frac{0.78(1.50)^{1.67}}{1.1^{1.17}}$	1.37

30. As the length of the stride increases, the angle formed at the hip increases.

31. Sample answer: A backhoe digs when the angle between the two arms decreases and the shovel moves through the dirt. Answers should include the following.
- As the operator digs, the angle between the arms decreases.
- The distance between the ends of the arms increases as the angle between the arms increases, and decreases as the angle decreases.

32. B; $AD = BD$ by definition of a median.
$AC > BC$ since $m\angle 1 > m\angle 2$ and by the SAS Inequality.
$m\angle 1 > m\angle B$ because if an $\angle$ is an ext. $\angle$ of a $\triangle$, then its measure is greater than the measure of either remote int. $\angle$.
$m\angle ADC = m\angle 1$ and $m\angle BDC = m\angle 2$, and $m\angle 1 > m\angle 2$ is given, so $m\angle ADC > m\angle BDC$.
The correct answer is B.

33. B; $\frac{\frac{1}{2}(99.50 + 88.95 + 95.90 + 102.45)}{4} = 48.35$

Page 273 Maintain Your Skills

34. no; $1 + 21 \not> 25$

35. yes; $16 + 6 \overset{?}{>} 19$, $22 > 19$ ✓; $6 + 19 \overset{?}{>} 16$, $25 > 16$ ✓; $16 + 19 \overset{?}{>} 6$, $35 > 6$ ✓

36. no; $8 + 7 \not> 15$

37. $\overline{AD}$ is a not a median of $\triangle ABC$.

38. The triangle is not isosceles.

39. Given: $\overline{AD}$ bisects $\overline{BE}$; $\overline{AB} \parallel \overline{DE}$.
Prove: $\triangle ABC \cong \triangle DEC$

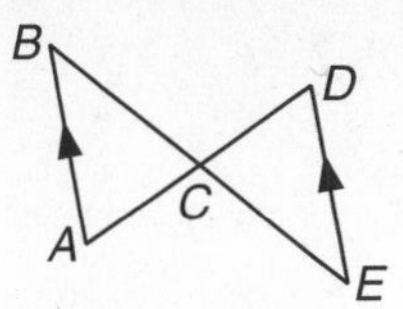

Proof:

Statements	Reasons
1. $\overline{AD}$ bisects $\overline{BE}$; $\overline{AB} \parallel \overline{DE}$.	1. Given
2. $\overline{BC} \cong \overline{EC}$	2. Def. of seg. bisector
3. $\angle B \cong \angle E$	3. Alt. int. $\angle$s Thm.
4. $\angle BCA \cong \angle ECD$	4. Vert. $\angle$s are $\cong$.
5. $\triangle ABC \cong \triangle DEC$	5. ASA

40. Given: $\overrightarrow{OM}$ bisects $\angle LMN$; $\overline{LM} \cong \overline{MN}$.
Prove: $\triangle MOL \cong \triangle MON$

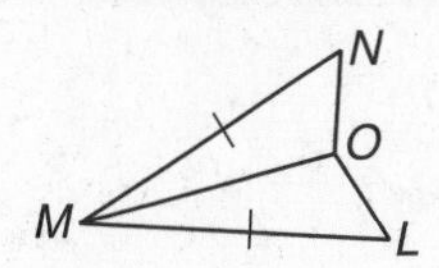

Proof:

Statements	Reasons
1. $\overrightarrow{OM}$ bisects $\angle LMN$; $\overline{LM} \cong \overline{MN}$.	1. Given
2. $\angle LMO \cong \angle NMO$	2. Def. of $\angle$ bisector
3. $\overline{OM} \cong \overline{OM}$	3. Reflexive Prop.
4. $\triangle MOL \cong \triangle MON$	4. SAS

41. $EF = \sqrt{(4-4)^2 + (11-6)^2}$
$= \sqrt{0 + 25}$ or 5
$FG = \sqrt{(9-4)^2 + (6-11)^2}$
$= \sqrt{25 + 25}$ or $\sqrt{50}$
$EG = \sqrt{(9-4)^2 + (6-6)^2}$
$= \sqrt{25 + 0}$ or 5
$\triangle EFG$ is isosceles.

42. $EF = \sqrt{[15 - (-7)]^2 + (0 - 10)^2}$
$= \sqrt{484 + 100}$ or $\sqrt{584}$
$FG = \sqrt{(-2 - 15)^2 + (-1 - 0)^2}$
$= \sqrt{289 + 1}$ or $\sqrt{290}$
$EG = \sqrt{[-2 - (-7)]^2 + (-1 - 10)^2}$
$= \sqrt{25 + 121}$ or $\sqrt{146}$
$\triangle EFG$ is scalene.

43. $EF = \sqrt{(7 - 16)^2 + (6 - 14)^2}$
$= \sqrt{81 + 64}$ or $\sqrt{145}$
$FG = \sqrt{(-5 - 7)^2 + (-14 - 6)^2}$
$= \sqrt{144 + 400}$ or $\sqrt{544}$
$EG = \sqrt{(-5 - 16)^2 + (-14 - 14)^2}$
$= \sqrt{441 + 784}$ or 35
$\triangle EFG$ is scalene.

44. $EF = \sqrt{(12-9)^2 + (14-9)^2}$
$= \sqrt{9+25}$ or $\sqrt{34}$
$FG = \sqrt{(14-12)^2 + (6-14)^2}$
$= \sqrt{4+64}$ or $\sqrt{68}$
$EG = \sqrt{(14-9)^2 + (6-9)^2}$
$= \sqrt{25+9}$ or $\sqrt{34}$
$\triangle EFG$ is isosceles.

45. Let p, q be the parts of the statement.
p: it has to be special
q: it has to be Wildflowers
The statement p is true, so it follows that Catalina should go to Wildflowers by the Law of Detachment.

Chapter 5 Study Guide and Review

Page 274 Vocabulary and Concept Check

1. incenter
2. median
3. Triangle Inequality Theorem
4. centroid
5. angle bisector
6. perpendicular bisectors
7. orthocenter

Pages 274–276 Lesson-by-Lesson Review

8. $m\angle ACQ = m\angle QCB$
$m\angle QCB = \frac{1}{2}m\angle ACB$
$42 + x = \frac{1}{2}(123 - x)$
$42 + x = 61.5 - \frac{x}{2}$
$42 + \frac{3}{2}x = 61.5$
$\frac{3}{2}x = 19.5$
$x = 13$
$m\angle ACQ = 42 + 13$ or 55

9. $AR = RB$
$3x + 6 = 5x - 14$
$6 = 2x - 14$
$20 = 2x$
$10 = x$
$AB = AR + AB$
$= 3(10) + 6 + 5(10) - 14$
$= 72$

10. $m\angle APC = 90$
$72 + x = 90$
$x = 18$

11. The side opposite $\angle DEF$ is longer than the side opposite $\angle DFE$, so $m\angle DEF > m\angle DFE$.

12. The side opposite $\angle GDF$ is shorter than the side opposite $\angle DGF$, so $m\angle GDF < m\angle DGF$.

13. The side opposite $\angle DEF$ is longer than the side opposite $\angle FDE$, so $m\angle DEF > m\angle FDE$.

14. The angle opposite $\overline{SR}$ has a greater measure than the angle opposite $\overline{SD}$, so $SR > SD$.

15. $m\angle QDR + m\angle RDS = 180$
$m\angle QDR + 110 = 180$
$m\angle QDR = 70$
$m\angle QRD + m\angle QDR + m\angle RQD = 180$
$m\angle QRD + 70 + 73 = 180$
$m\angle QRD = 37$
The angle opposite $\overline{DQ}$ has a smaller measure than the angle opposite $\overline{DR}$, so $DQ < DR$.

16. $m\angle QRP = 37$ (see exercise 15)
$m\angle QPR = 27$
In $\triangle PQR$, the angle opposite $\overline{PQ}$ has a greater measure than the angle opposite $\overline{QR}$, so $PQ > QR$.

17. $m\angle SRQ = m\angle SRD + m\angle DRQ$
$= 34 + 37$
$= 71$
The angle opposite $\overline{SR}$ has a greater measure than the angle opposite $\overline{SQ}$, so $SR > SQ$.

18. $\sqrt{2}$ is a rational number.

19. The triangles are not congruent.

20. Assume that Miguel completed at most 20 passes in each of the five games in which he played. If we let p_1, p_2, p_3, p_4, and p_5 be the number of passes Miguel completed in games 1, 2, 3, 4, and 5, respectively, then
$p_1 + p_2 + p_3 + p_4 + p_5 =$ the total number of passes Miguel completed $= 101$.
Because we have assumed that he completed at most 20 passes in each of the five games,
$p_1 \le 20$ and $p_2 \le 20$ and $p_3 \le 20$ and $p_4 \le 20$ and $p_5 \le 20$.
Then, by a property of inequalities,
$p_1 + p_2 + p_3 + p_4 + p_5 \le 20 + 20 + 20 + 20 + 20$ or 100 passes.
But this says that Miguel completed at most 100 passes this season, which contradicts the information we were given, that he completed 101 passes. So our assumption must be false. Thus, Miguel completed more than 20 passes in at least one game this season.

21. no; $7 + 5 \overset{?}{>} 20$
$12 \not> 20$
Because the sum of two measures is less than the third measure, the sides cannot form a triangle.

22. Check each inequality.
$16 + 20 \overset{?}{>} 5$ $\quad 16 + 5 \overset{?}{>} 20$ $\quad 20 + 5 \overset{?}{>} 16$
$36 > 5$ ✓ $\quad 21 > 20$ ✓ $\quad 25 > 16$ ✓
All of the inequalities are true, so 16, 20, and 5 can be the lengths of the sides of a triangle.

23. Check each inequality.
$18 + 20 \overset{?}{>} 6$ $\quad 18 + 6 \overset{?}{>} 20$ $\quad 20 + 6 \overset{?}{>} 18$
$38 > 6$ ✓ $\quad 24 > 20$ ✓ $\quad 26 > 18$ ✓
All of the inequalities are true, so 18, 20, and 6 can be the lengths of the sides of a triangle.

24. In $\triangle BAM$ and $\triangle DAM$, $\overline{AB} \cong \overline{AD}$, $\overline{AM} \cong \overline{AM}$, and $BM > DM$. The SSS Inequality allows us to conclude that $m\angle BAC > m\angle DAC$.

25. In $\triangle BMC$ and $\triangle DCM$, $\overline{BM} \cong \overline{CD}$, $\overline{MC} \cong \overline{MC}$, and $m\angle BMC > m\angle DCM$. The SAS Inequality allows us to conclude that $BC > MD$.

26. Using the SSS Inequality, $54 > 28$ so $41 > x + 20$ or $21 > x$. $x + 20 > 0$, so $x > -20$. The two inequalities can be written as the compound inequality $-20 < x < 21$.

27. In the upper triangle, the bottom angle has measure $90 - 60$ or 30, so the upper left angle has measure $180 - (95 + 30)$ or 55. Then by the SAS Inequality, $5x + 3 > 3x + 17$.
$5x + 3 > 3x + 17$
$2x + 3 > 17$
$2x > 14$
$x > 7$

Chapter 5 Practice Test

Page 277

1. b
2. c
3. a
4. $HP = PJ$
$5x - 16 = 3x + 8$
$2x - 16 = 8$
$2x = 24$
$x = 12$
$HJ = HP + PJ$
$= 5(12) - 16 + 3(12) + 8$
$= 88$
5. $m\angle GJN = m\angle NJH$
$6y - 3 = 4y + 23$
$2y - 3 = 23$
$2y = 26$
$y = 13$
$m\angle GJH = m\angle GJN + m\angle NJH$
$= 6(13) - 3 + 4(13) + 23$
$= 150$
6. $m\angle HMG = 90$
$4z + 14 = 90$
$4z = 76$
$z = 19$
7. By the Exterior Angle Theorem, $m\angle 8 > m\angle 7$ and $m\angle 5 > m\angle 8$. By transitivity, $m\angle 5 > m\angle 7$, so $\angle 5$ has the greatest measure.
8. By the Exterior Angle Theorem, $m\angle 8 > m\angle 7$ and $m\angle 8 > m\angle 6$, so $\angle 8$ has the greatest measure.
9. By the Exterior Angle Theorem, $m\angle 1 > m\angle 6$ and $m\angle 1 > m\angle 9$, so $\angle 1$ has the greatest measure.
10. $2^n + 1$ is even.
11. Alternate interior angles are not congruent.
12. Assume that Marcus spent less than one half hour on a teleconference every day. If we let t_1, t_2, and t_3 be the time spent on a teleconference on days 1, 2, and 3, respectively, then $t_1 + t_2 + t_3 =$ the total amount of time over the three days spent on the teleconference.
Because he spent less than a half hour every day on a teleconference, $t_1 < 0.5$ and $t_2 < 0.5$ and $t_3 < 0.5$.
Then, by a property of inequalities, $t_1 + t_2 + t_3 < 0.5 + 0.5 + 0.5$ or 1.5 hours.
But this says that Marcus spent less than one and one-half hours on a teleconference over the three days, which contradicts the information we were given. So we must abandon our assumption. Thus, Marcus spent at least one half-hour on a teleconference, on at least one of the three days.
13. Let the measure of the third side be n.
$1 + 14 > n$ $\quad$ $1 + n > 14$ $\quad$ $14 + n > 1$
$15 > n$ or $n < 15$ $\quad$ $n > 13$ $\quad$ $n > -13$
Graph the inequalities on the same number line.

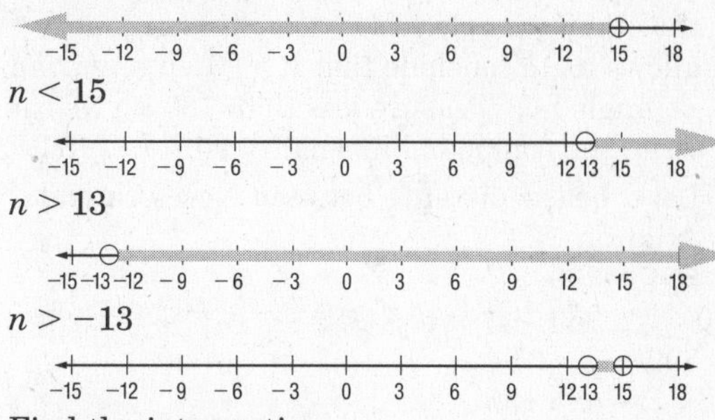

Find the intersection.
The range of values that fit all three inequalities is $13 < n < 15$.
14. Let the measure of the third side be n.
$14 + 11 > n$ $\quad$ $14 + n > 11$ $\quad$ $11 + n > 14$
$25 > n$ or $n < 25$ $\quad$ $n > -3$ $\quad$ $n > 3$
Graph the inequalities on the same number line.

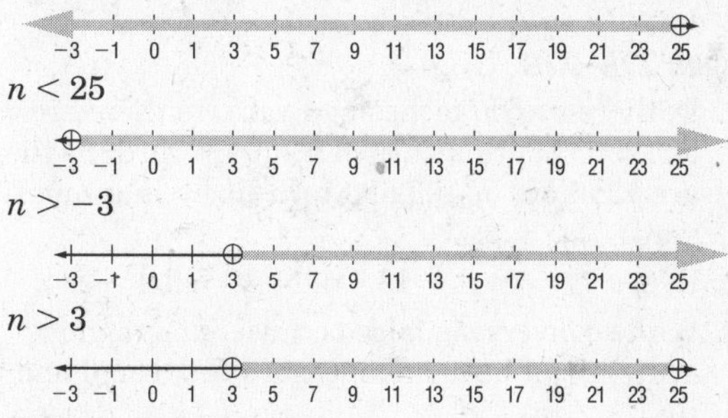

Find the intersection.
The range of values that fit all three inequalities is $3 < n < 25$.
15. Let the measure of the third side be n.
$13 + 19 > n$ $\quad$ $13 + n > 19$ $\quad$ $19 + n > 13$
$32 > n$ or $n < 32$ $\quad$ $n > 6$ $\quad$ $n > -6$
Graph the inequalities on the same number line.

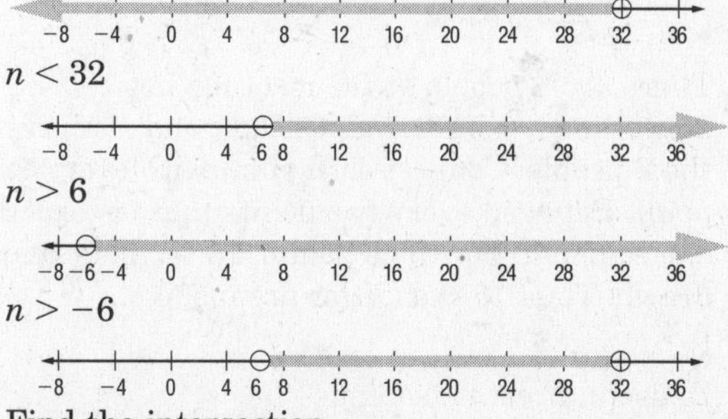

Find the intersection.
The range of values that fit all three inequalities is $6 < n < 32$.
16. $y > 0$, so $y + 1 > y$. Then $7 > x$ and $x > 0$, so $0 < x < 7$.

17. $x + 7 > 11$

$x > 4$

$7 + 11 > 2x$

$18 > 2x$

$9 > x$

$4 < x < 9$

18. In the two triangles, each has a side of measure 12 and the shared side is congruent to itself. The hypotenuse of the right triangle has a length of 13.86, which is shorter than the third length of the other triangle, 14. So the SSS Inequality allows us to conclude that $x > 90$. In any triangle, an angle has measure less than 180, so we can write the compound inequality $90 < x < 180$.

19. Let d be the distance between NewYork and Atlanta.

$554 + 399 > d$

$953 > d$ or $d < 953$

$399 + d > 554$

$d > 155$

Therefore, 155 mi $< d <$ 953 mi.

20. A; $3 + 8 = 11$, so 3, 8, 11 cannot be the sides of a triangle.

Chapter 5 Standardized Test Practice

Pages 278–279

1. D; there are 36 inches in a yard, so there are 36^2 or 1296 square inches in a square yard. So there are 1296(80) or 103,680 yarn fibers in a square yard.

2. B; $6 + 7 + 6 + 4 + 12 + 11 = 46$ units

3. B; the converse is false because if an angle is acute it can have any measure between 0 and 90.

4. B;

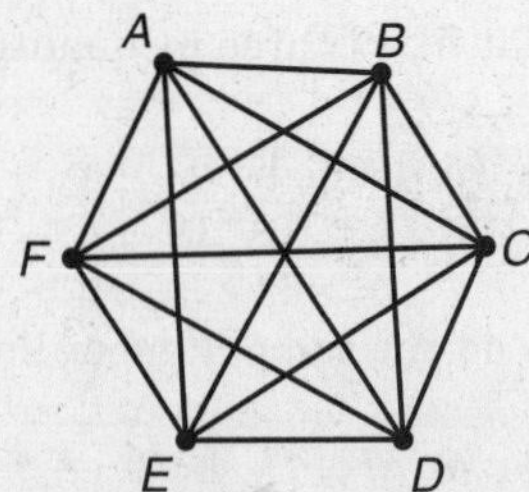

There are 6 people at the meeting. Let noncollinear points A, B, C, D, E, and F represent the 6 people. Connect each point with every other point. Between every two points there is exactly one segment. For the 6 points, 15 segments can be drawn. Thus 15 exchanges are made.

5. B

6. B; use indirect reasoning

7. C; the shortest distance from a point to a line segment is a perpendicular segment

8. A; $2 + 9 \overset{?}{>} 10$ $2 + 10 \overset{?}{>} 9$ $9 + 10 \overset{?}{>} 2$

$11 > 10$ ✓ $12 > 9$ ✓ $19 > 2$ ✓

$5 + 8 = 13$, so 5, 8, 13 cannot be the measures of the sides of the triangle.

$7 + 11 < 20$, so 7, 11, 20 cannot be the measures of the sides of the triangle.

$9 + 13 < 26$, so 9, 13, 26 cannot be the measures of the sides of the triangle.

9. The ramp rises 2 feet as it runs 24 feet, so the slope is $\frac{2}{24}$ or $\frac{1}{12}$.

10. $x + 55 = 90$

$x = 35$

11. The point P is the midpoint of $\overline{BC}$ with coordinates $\left(\frac{8+8}{2}, \frac{2+10}{2}\right) = (8, 6)$.

12. $\overline{BC}$ is vertical, so $\overline{AT}$ is horizontal. The y-coordinate of A is 4, and so the y-coordinate of T is also 4. T is on $\overline{BC}$, so the x-coordinate of T is 8. Therefore the coordinates of T are (8, 4).

13. SSS Inequality

14a. From the points (0, 200) and (4, 120), the slope of the line is $\frac{(200 - 120)}{(0 - 4)} = -20$. Find slope again using one of the given points and (10, 0); the slope is $\frac{(120 - 0)}{(4 - 10)} = -20$. Since the slope is the same, (10, 0) must be on the original line. Students may check by drawing an extension of the line and will see that it goes through (10, 0).

14b. The point (10, 0) shows that on the tenth payment Kendell's balance will be $0, so the amount will be paid in full.

15a.

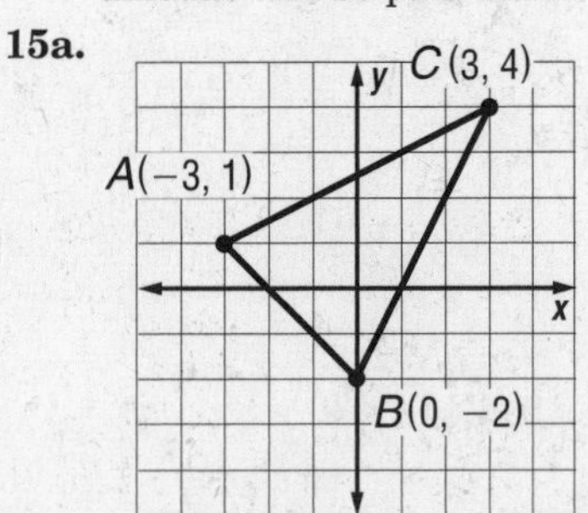

15b. $AB = \sqrt{[0 - (-3)]^2 + (-2 - 1)^2}$

$= \sqrt{9 + 9}$

$= \sqrt{18}$

≈ 4.2

$BC = \sqrt{(3 - 0)^2 + [4 - (-2)]^2}$

$= \sqrt{9 + 36}$

$= \sqrt{45}$

≈ 6.7

$AC = \sqrt{[3 - (-3)]^2 + (4 - 1)^2}$

$= \sqrt{36 + 9}$

$= \sqrt{45}$

≈ 6.7

15c. isosceles triangle because $\overline{BC}$ is congruent to $\overline{AC}$

15d. According to the Isosceles Triangle Theorem, if two sides of a triangle are congruent, then the angles opposite those sides are congruent. Since $\overline{BC} \cong \overline{AC}$, $\angle A \cong \angle B$.

15e. If one side of a triangle is longer than another side, the angle opposite the longer side has a greater measure than the angle opposite the shorter side. Since $\overline{BC}$ is longer than $\overline{AB}$, $m\angle A > m\angle C$.

Chapter 6 Proportions and Similarity

Page 281 Getting Started

1. $\frac{2}{3}y - 4 = 6$
   $\frac{2}{3}y = 10$
   $y = \frac{3}{2}(10)$ or 15
2. $\frac{5}{6} = \frac{x-4}{12}$
   $12 \cdot \frac{5}{6} = 12 \cdot \frac{x-4}{12}$
   $10 = x - 4$
   $14 = x$
3. $\frac{4}{3} = \frac{y+2}{y-1}$
   $4(y-1) = 3(y+2)$
   $4y - 4 = 3y + 6$
   $y - 4 = 6$
   $y = 10$
4. $\frac{2y}{4} = \frac{32}{y}$
   $2y^2 = 128$
   $y^2 = 64$
   $y = \pm 8$
5. $m = \frac{(y_2 - y_1)}{(x_2 - x_1)}$
   $= \frac{-1-5}{0-3}$
   $= \frac{-6}{-3}$
   $= 2$
6. $m = \frac{(y_2 - y_1)}{(x_2 - x_1)}$
   $= \frac{-3-(-3)}{2-(-6)}$
   $= \frac{0}{8}$
   $= 0$
7. $m = \frac{(y_2 - y_1)}{(x_2 - x_1)}$
   $= \frac{-2-4}{2-(-3)}$
   $= -\frac{6}{5}$
8. yes; congruent alternate exterior angles
9. yes; congruent alternate interior angles
10. No; $\angle 5$ and $\angle 3$ do not have a relationship that could be used to determine whether the lines are parallel.
11. $2^1 = 2$
    $2^2 = 4$
    $2^3 = 8$
    $2^4 = 16$
12. $1^2 - 2 = 1 - 2 = -1$
    $2^2 - 2 = 4 - 2 = 2$
    $3^2 - 2 = 9 - 2 = 7$
    $4^2 - 2 = 16 - 2 = 14$
13. $3^1 - 2 = 3 - 2 = 1$
    $3^2 - 2 = 9 - 2 = 7$
    $3^3 - 2 = 27 - 2 = 25$
    $3^4 - 2 = 81 - 2 = 79$

6-1 Proportions

Pages 284–285 Check for Understanding

1. Cross multiply and divide by 28.
2. Sample answer: $\frac{5}{4} = \frac{10}{8}$, $\frac{5}{10} = \frac{4}{8}$
3. Suki; Madeline did not find the cross products correctly.
4. $\frac{\text{number of goals}}{\text{number of games}} = \frac{9}{12}$ or 3 : 4
5. $\frac{\text{height of replica}}{\text{height of statue}} = \frac{10 \text{ inches}}{10 \text{ feet}}$
   $= \frac{10 \text{ inches}}{120 \text{ inches}}$
   $= \frac{1}{12}$
6. $\frac{x}{5} = \frac{11}{35}$
   $35x = 5(11)$
   $35x = 55$
   $x = \frac{11}{7}$
7. $\frac{2.3}{4} = \frac{x}{3.7}$
   $2.3(3.7) = 4x$
   $8.51 = 4x$
   $2.1275 = x$
8. $\frac{x-2}{2} = \frac{4}{5}$
   $5(x-2) = 2(4)$
   $5x - 10 = 8$
   $5x = 18$
   $x = 3.6$
9. Rewrite 9 : 8 : 7 as $9x : 8x : 7x$ and use those measures for the sides of the triangle. Write an equation to represent the perimeter of the triangle as the sum of the measures of its sides.
   $9x + 8x + 7x = 144$
   $24x = 144$
   $x = 6$
   Use this value of x to find the measures of the sides of the triangle.
   $9x = 9(6)$ or 54 units
   $8x = 8(6)$ or 48 units
   $7x = 7(6)$ or 42 units
10. Rewrite 5 : 7 : 8 as $5x : 7x : 8x$ and use those measures for the angles of the triangle. Write an equation to represent the sum of the angle measures of the triangle.
    $5x + 7x + 8x = 180$
    $20x = 180$
    $x = 9$
    Use this value of x to find the measures of the angles of the triangle.
    $5x = 5(9)$ or 45
    $7x = 7(9)$ or 63
    $8x = 8(9)$ or 72

11. $\frac{\text{scale on map (cm)}}{\text{distance represented (mi)}} = \frac{\text{distance on map (cm)}}{\text{actual distance (mi)}}$

$$\frac{1.5}{200} = \frac{2.4}{x}$$
$$1.5x = 200(2.4)$$
$$1.5x = 480$$
$$x = 320$$

The cities are 320 miles apart.

Pages 285–287 Practice and Apply

12. $\frac{\text{number of hits}}{\text{number of games}} = \frac{8}{10}$ or 4 : 5

13. $\frac{\text{number of boys}}{\text{number of girls}} = \frac{76}{165 - 76}$ or 76 : 89

14. $\frac{\text{number of rands}}{\text{number of dollars}} = \frac{208}{18}$ or 104 : 9

15. $\frac{\text{number of students}}{\text{number of teachers}} = \frac{44{,}125}{1747}$ or about 25.3 : 1

16. $\frac{AC}{BH} = \frac{20 - 0}{70 - 10}$

$= \frac{20}{60}$ or 1 : 3

17. Rewrite 3 : 4 as $3x : 4x$ and use those measures for the two lengths of cable.

$3x + 4x = 42$
$7x = 42$
$x = 6$

Use this value of x to find the measures of the two lengths of cable.

$3x = 3(6)$ or 18 ft
$4x = 4(6)$ or 24 ft

18. Rewrite 2 : 5 : 3 as $2x : 5x : 3x$ and use those measures for the angles of the triangle.

$2x + 5x + 3x = 180$
$10x = 180$
$x = 18$

Use this value of x to find the measures of the angles of the triangle.

$2x = 2(18)$ or 36
$5x = 5(18)$ or 90
$3x = 3(18)$ or 54

19. Rewrite 6 : 9 : 10 as $6x : 9x : 10x$ and use those measures for the angles of the triangle.

$6x + 9x + 10x = 180$
$25x = 180$
$x = 7.2$

Use this value of x to find the measures of the angles of the triangle.

$6x = 6(7.2)$ or 43.2
$9x = 9(7.2)$ or 64.8
$10x = 10(7.2)$ or 72

20. Rewrite 8 : 7 : 5 as $8x : 7x : 5x$ and use those measures for the sides of the triangle.

$8x + 7x + 5x = 240$
$20x = 240$
$x = 12$

Use this value of x to find the measures of the sides of the triangle.

$8x = 8(12)$ or 96 ft
$7x = 7(12)$ or 84 ft
$5x = 5(12)$ or 60 ft

21. Rewrite 3 : 4 : 5 as $3x : 4x : 5x$ and use those measures for the sides of the triangle.

$3x + 4x + 5x = 72$
$12x = 72$
$x = 6$

Use this value of x to find the measures of the sides of the triangle.

$3x = 3(6)$ or 18 in.
$4x = 4(6)$ or 24 in.
$5x = 5(6)$ or 30 in.

22. Rewrite $\frac{1}{2} : \frac{1}{3} : \frac{1}{5}$ as $\frac{x}{2} : \frac{x}{3} : \frac{x}{5}$ and use those measures for the sides of the triangle.

$$\frac{x}{2} + \frac{x}{3} + \frac{x}{5} = 6.2$$
$$30\left(\frac{x}{2} + \frac{x}{3} + \frac{x}{5}\right) = 30(6.2)$$
$$15x + 10x + 6x = 186$$
$$31x = 186$$
$$x = 6$$

Use this value of x to find the measures of the sides of the triangle.

$\frac{x}{2} = \frac{6}{2}$ or 3 cm

$\frac{x}{3} = \frac{6}{3}$ or 2 cm

$\frac{x}{5} = \frac{6}{5}$ or 1.2 cm

23. $\frac{\text{height of the door}}{\text{Alice's height in Wonderland}} = \frac{15}{10}$ or $\frac{3}{2}$

24. $\frac{\text{height of the door}}{\text{Alice's height in Wonderland}}$

$= \frac{\text{height of door in Alice's normal world}}{\text{Alice's normal height}}$

$\frac{15}{10} = \frac{x}{50}$
$15(50) = 10x$
$750 = 10x$
$75 = x$

The height of the door in Alice's normal world would be about 75 inches.

25. $\frac{\text{Lincoln's height in model}}{\text{Lincoln's height in theater}} = \frac{8 \text{ in.}}{6 \text{ ft } 4 \text{ in.}}$

$= \frac{8 \text{ in.}}{76 \text{ in.}}$

$= 2 : 19$

26. $\frac{\text{number of people in United States}}{\text{number of pounds of ice cream consumed}}$

$= \frac{\text{number of people in Raleigh, NC}}{\text{number of pounds of ice cream}}$

$\frac{255{,}082{,}000}{4{,}183{,}344{,}800} = \frac{276{,}000}{x}$

$255{,}082{,}000x = 1{,}154{,}603{,}165{,}000{,}000$
$x = 4{,}526{,}400$

The people of Raleigh, North Carolina, might consume 4,526,400 pounds of ice cream.

27. $\frac{\text{number of people in United States}}{\text{number of pounds of ice cream consumed}}$

$= \frac{\text{1 person}}{\text{number of pounds of ice cream}}$

$\frac{255{,}082{,}000}{4{,}183{,}344{,}800} = \frac{1}{x}$

$255{,}082{,}000x = 4{,}183{,}344{,}800$
$x = 16.4$

One person consumed about 16.4 pounds of ice cream.

28. $\frac{3}{8} = \frac{x}{5}$

$3(5) = 8x$

$15 = 8x$

$\frac{15}{8} = x$

29. $\frac{a}{5.18} = \frac{1}{4}$

$4a = 5.18(1)$

$a = 1.295$

30. $\frac{3x}{23} = \frac{48}{92}$

$3x(92) = 23(48)$

$276x = 1104$

$x = 4$

31. $\frac{13}{49} = \frac{26}{7x}$

$13(7x) = 49(26)$

$91x = 1274$

$x = 14$

32. $\frac{2x - 13}{28} = \frac{-4}{7}$

$(2x - 13)(7) = 28(-4)$

$14x - 91 = -112$

$14x = -21$

$x = -\frac{3}{2}$

33. $\frac{4x + 3}{12} = \frac{5}{4}$

$(4x + 3)(4) = 12(5)$

$16x + 12 = 60$

$16x = 48$

$x = 3$

34. $\frac{b + 1}{b - 1} = \frac{5}{6}$

$(b + 1)(6) = (b - 1)(5)$

$6b + 6 = 5b - 5$

$b + 6 = -5$

$b = -11$

35. $\frac{3x - 1}{2} = \frac{-2}{x + 2}$

$(3x - 1)(x + 2) = -4$

$3x^2 + 6x - x - 2 = -4$

$3x^2 + 5x + 2 = 0$

$(3x + 2)(x + 1) = 0$

$3x + 2 = 0$ or $x + 1 = 0$

$3x = -2$ $\quad x = -1$

$x = -\frac{2}{3}$

36. The larger dimension of the photograph is 27.5 cm, so reducing this dimension to 10 cm will give the maximum dimensions of the reduced photograph.

$\frac{27.5}{10} = \frac{21.3}{x}$

$27.5x = 10(21.3)$

$27.5x = 213$

$x \approx 7.75$

The maximum dimensions are 7.75 cm by 10 cm.

37. $\frac{\text{reduced length}}{\text{original length}} = \frac{7.75}{21.3}$

≈ 0.36 or 36%

38a. The ratio 2 : 2 : 3 indicates that there are three sides, two of which have the same measure. This description fits an isosceles triangle.

38b. The ratio 3 : 3 : 3 : 3 indicates that there are four sides with all the same measure. This description fits a square or a rhombus.

38c. The ratio 4 : 5 : 4 : 5 indicates that there are four sides, and opposite sides are congruent. This description fits a rectangle or a parallelogram.

39. Sample answer: It appears that Tiffany used rectangles with areas that were in proportion as a background for this artwork. Answers should include the following.

- The center column pieces are to the third column from the left pieces as the pieces from the third column are to the pieces in the outside column.
- The dimensions are approximately 24 inches by 34 inches.

40. $\frac{1.618}{1} = \frac{12}{x}$

$1.618x = 12$

$x \approx 7.4$ cm

41. D; the ratio of wheat to rice to oats is 3 : 1 : 2, so the ratio of wheat to oats is 3 : 2.

$\frac{3}{2} = \frac{x}{120}$

$3(120) = 2x$

$360 = 2x$

$180 = x$

180 pounds of wheat will be used. The answer is D.

Page 287 Maintain Your Skills

42. always; $m\angle 1 + m\angle 2 = 180$

$3x - 50 + x + 30 = 180$

$4x - 20 = 180$

$4x = 200$

$x = 50$

$m\angle 1 = 3(50) - 50$

$= 150 - 50$ or 100

$m\angle 2 = 50 + 30$ or 80

So, $m\angle 1 > m\angle 2$, and by the SAS Inequality $LS > SN$.

43. Always; $\angle PNO$ is an exterior angle of $\triangle SNO$, so $m\angle PNO > m\angle 2$. Then by the SAS Inequality, $OP > SN$.

44. never; $m\angle 1 + m\angle 2 = 180$

$3x - 50 + x + 30 = 180$

$4x - 20 = 180$

$4x = 200$

$x = 50$

45. Let the measure of the third side be x.

$16 + 31 > x$ $\quad 16 + x > 31$ $\quad 31 + x > 16$

$47 > x$ or $x < 47$ $\quad x > 15$ $\quad x > -15$

Graph the inequalities on the same number line.

$x < 47$

$x > 15$

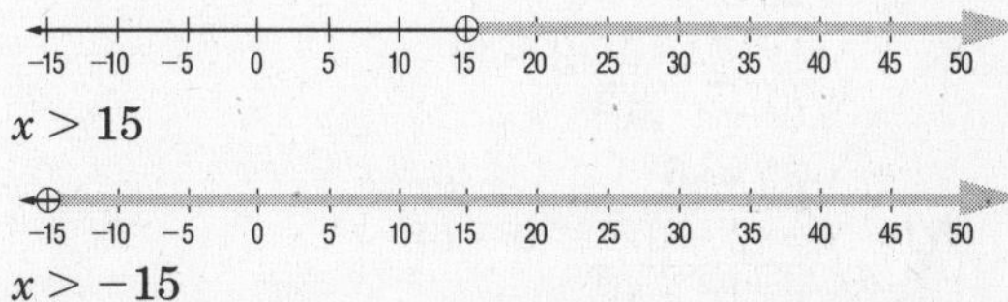

$x > -15$

Find the intersection.
The range of values that fit all three inequalities is $15 < x < 47$.

46. Let the measure of the third side be x.

$26 + 40 > x$ $26 + x > 40$ $40 + x > 26$

$66 > x$ or $x < 66$ $x > 14$ $x > -14$

Graph the inequalities on the same number line.

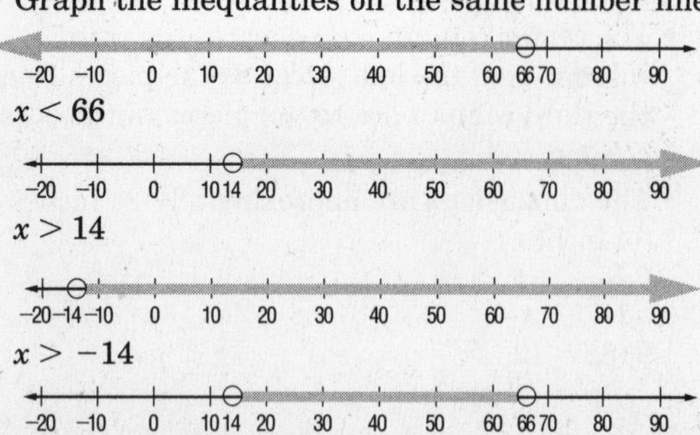

Find the intersection.
The range of values that fit all three inequalities is $14 < x < 66$.

47. Let the measure of the third side be x.

$11 + 23 > x$ $11 + x > 23$ $23 + x > 11$

$34 > x$ or $x < 34$ $x > 12$ $x > -12$

Graph the inequalities on the same numer line.

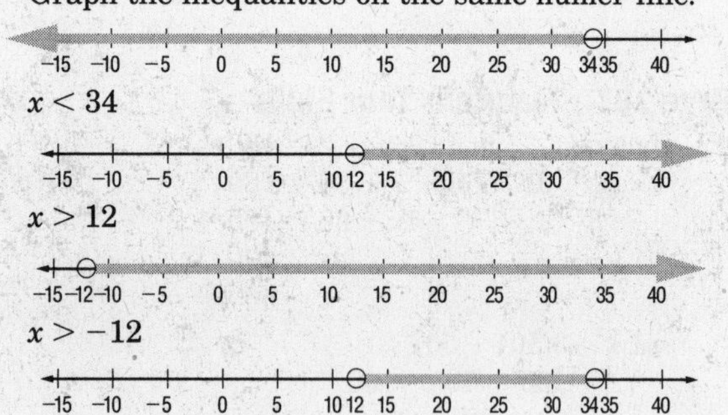

Find the intersection.
The range of values that fit all three inequalities is $12 < x < 34$.

48.

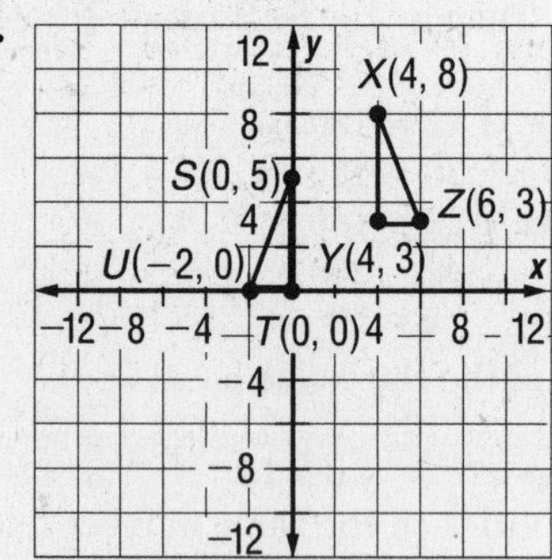

$ST = \sqrt{(0-0)^2 + (0-5)^2}$
$= \sqrt{0^2 + (-5)^2}$
$= \sqrt{25}$ or 5

$XY = \sqrt{(4-4)^2 + (3-8)^2}$
$= \sqrt{0^2 + (-5)^2}$
$= \sqrt{25}$ or 5

$TU = \sqrt{(-2-0)^2 + (0-0)^2}$
$= \sqrt{(-2)^2 + 0^2}$
$= \sqrt{4}$ or 2

$YZ = \sqrt{(6-4)^2 + (3-3)^2}$
$= \sqrt{2^2 + 0^2}$
$= \sqrt{4}$ or 2

$SU = \sqrt{(-2-0)^2 + (0-5)^2}$
$= \sqrt{(-2)^2 + (-5)^2}$
$= \sqrt{29}$

$XZ = \sqrt{(6-4)^2 + (3-8)^2}$
$= \sqrt{2^2 + (-5)^2}$
$= \sqrt{29}$

$\triangle STU \cong \triangle XYZ$ by SSS.

49. Start at $P(-3, -4)$. Move up 3 units and then move right 5 units. Draw the line through this point and P.

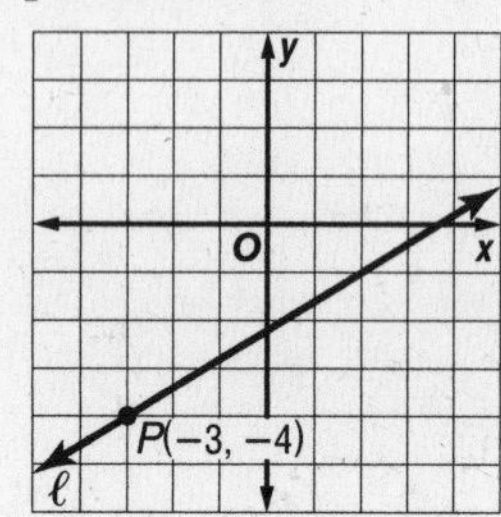

50. Plot points $A(5, 3)$ and $B(-1, 8)$. Draw line $\overleftrightarrow{AB}$.

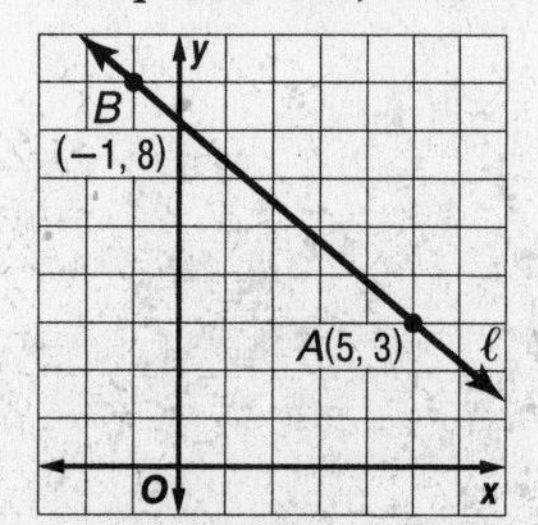

51. Find the slope of $\overleftrightarrow{JK}$.

$m = \frac{(y_2 - y_1)}{(x_2 - x_1)}$
$= \frac{3-5}{4-(-1)}$
$= -\frac{2}{5}$

The line through $E(2, 2)$ also has slope $-\frac{2}{5}$.
Start at $E(2, 2)$. Move down 2 units and then move right 5 units. Draw the line through this point and E.

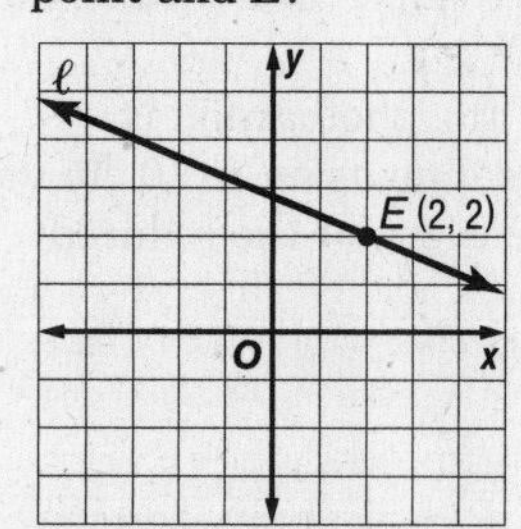

52. Find the slope of $\overleftrightarrow{QR}$.

$m = \frac{(y_2 - y_1)}{(x_2 - x_1)}$
$= \frac{-6-2}{-4-6}$
$= \frac{-8}{-10}$ or $\frac{4}{5}$

The product of the slopes of two perpendicular lines is -1.
Since $\frac{4}{5}\left(-\frac{5}{4}\right) = -1$, the slope of the line perpendicular to $\overleftrightarrow{QR}$ through $S(8, 1)$ is $-\frac{5}{4}$.

Start at $S(8, 1)$. Move down 5 units and then move right 4 units. Draw the line through this point and S.

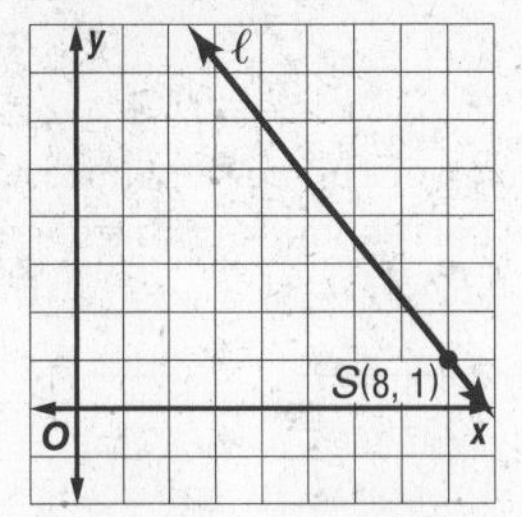

53. Yes; 100 km and 62 mi are the same length, so $AB = CD$. By the definition of congruent segments, $\overline{AB} \cong \overline{CD}$.

54. $d = \sqrt{(x_2 - x_1)^2 + (y_2 - y_1)^2}$

$AB = \sqrt{(-8 - 12)^2 + (3 - 3)^2}$

$= \sqrt{(-20)^2 + 0^2}$

$= \sqrt{400}$ or 20.0

55. $d = \sqrt{(x_2 - x_1)^2 + (y_2 - y_1)^2}$

$CD = \sqrt{(5 - 0)^2 + (12 - 0)^2}$

$= \sqrt{5^2 + 12^2}$

$= \sqrt{169}$ or 13.0

56. $d = \sqrt{(x_2 - x_1)^2 + (y_2 - y_1)^2}$

$EF = \sqrt{\left(2 - \frac{4}{5}\right)^2 + \left[\frac{-1}{2} - (-1)\right]^2}$

$= \sqrt{1.2^2 + 0.5^2}$

$= \sqrt{1.69}$ or 1.3

57. $d = \sqrt{(x_2 - x_1)^2 + (y_2 - y_1)^2}$

$GH = \sqrt{(4 - 3)^2 + \left(-\frac{2}{7} - \frac{3}{7}\right)^2}$

$= \sqrt{1^2 + \left(-\frac{5}{7}\right)^2}$

$= \sqrt{1 + \frac{25}{49}}$

$= \sqrt{\frac{74}{49}}$

≈ 1.2

Page 288 Spreadsheet Investigation: Fibonacci Sequence and Ratios

1. It increases also.
2. odd-odd-even
3. It approaches 1.618.
4. The increase in terms confirms the original observations.
5. As the number of terms increases, the ratio of each term to its preceding term approaches the golden ratio.

6-2 Similar Polygons

Pages 292–293 Check for Understanding

1. Both students are correct. One student has inverted the ratio and reversed the order of the comparison.
2. See students' drawings. Sample counterexample: A rectangle with consecutive sides of 4 in. and 12 in. would not have sides proportional to a rectangle with consecutive sides of 6 in. and 8 in. because $\frac{4}{6} \neq \frac{12}{8}$.
3. If two polygons are congruent, then they are similar. All of the corresponding angles are congruent, and the ratio of measures of the corresponding sides is 1. Two similar figures have congruent angles, and the sides are in proportion, but not always congruent. If the scale factor is 1, then the figures are congruent.
4. $\triangle PQR$ and $\triangle GHI$ each have two angles with measure 60, so the third angle of each triangle must also have measure 60. Thus, $\angle P \cong \angle Q \cong \angle R \cong \angle G \cong \angle H \cong \angle I$ and $\frac{PQ}{GH} = \frac{QR}{HI} = \frac{RP}{IG} = \frac{3}{7}$, so $\triangle PQR \sim \triangle GHI$.
5. From the diagram, $\angle A \cong \angle E$, $\angle B \cong \angle F$, $\angle C \cong \angle G$, and $\angle D \cong \angle H$.

   $\frac{AD}{EH} = \frac{CB}{GF} = \frac{4}{6}$ or $\frac{2}{3}$

   $\frac{DC}{HG} = \frac{BA}{FE} = \frac{3}{\frac{9}{2}}$ or $\frac{2}{3}$

   The ratios of the measures of the corresponding sides are equal, and the corresponding angles are congruent, so parallelogram $ABCD \sim$ parallelogram $EFGH$.
6. Use the congruent angles to write the corresponding vertices in order: $\triangle ACB \sim \triangle DFE$. Write a proportion to find x.

   $\frac{AC}{DF} = \frac{CB}{FE}$

   $\frac{21}{x} = \frac{27}{18}$

   $21(18) = 27x$

   $378 = 27x$

   $14 = x$

   So, $DF = 14$.

   The scale factor is $\frac{AC}{DF} = \frac{21}{14}$ or $\frac{3}{2}$.
7. Use the congruent angles to write the corresponding vertices in order: polygon $ABCD \sim$ polygon $EFGH$. Write a proportion to find x.

   $\frac{DA}{HE} = \frac{BA}{FE}$

   $\frac{10}{x - 3} = \frac{14}{x + 5}$

   $10(x + 5) = (x - 3)(14)$

   $10x + 50 = 14x - 42$

   $10x + 92 = 14x$

   $92 = 4x$

   $23 = x$

   $EF = x + 5$

   $= 23 + 5$ or 28

   $EH = x - 3$

   $= 23 - 3$ or 20

   $\frac{GF}{CB} = \frac{EF}{AB}$

   $\frac{GF}{16} = \frac{28}{14}$

   $\frac{GF}{16} = 2$, so $GF = 2(16)$ or 32.

   The scale factor is $\frac{DA}{HE} = \frac{10}{20}$ or $\frac{1}{2}$.

8. Write proportions for finding side measures.

$$\frac{\text{new length} \rightarrow x}{\text{original length} \rightarrow 60} = \frac{1}{4}$$
$$4x = 60$$
$$x = 15$$
$$\frac{\text{new height} \rightarrow y}{\text{original height} \rightarrow 40} = \frac{1}{4}$$
$$4y = 40$$
$$y = 10$$

The new length is 15 cm, and the new height is 10 cm.

9. Write proportions for finding side measures.

$$\frac{\text{new first side} \rightarrow x}{\text{original first side} \rightarrow 3} = 5$$
$$x = 15$$
$$\frac{\text{new second side} \rightarrow y}{\text{original second side} \rightarrow 5} = 5$$
$$y = 25$$
$$\frac{\text{new third side} \rightarrow z}{\text{original third side} \rightarrow 4} = 5$$
$$z = 20$$

The new side lengths are 15 m, 25 m, and 20 m, so the perimeter is 15 m + 25 m + 20 m = 60 m.

10. See students' drawings. The drawings will be similar since the measures of the corresponding sides will be proportional and the corresponding angles will be congruent.

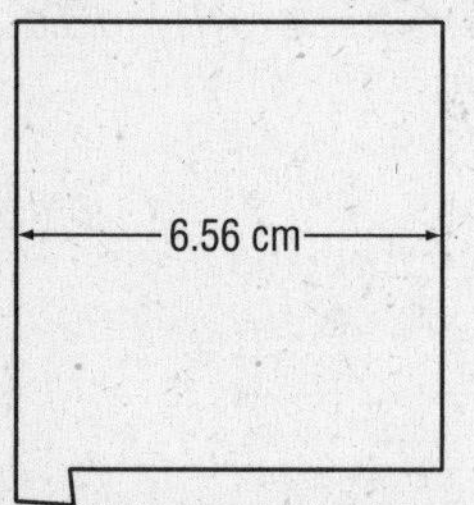

Figure is not shown actual size.

Pages 293–297 Practice and Apply

11. $\angle BCF \cong \angle DCF$, $\angle ABC \cong \angle EDC$, $\angle BAF \cong \angle DEF$, and $\angle AFC \cong \angle EFC$; $\overline{BC} \cong \overline{DC}$, $\overline{AB} \cong \overline{ED}$; $\overline{AF} \cong \overline{EF}$, and $\overline{CF} \cong \overline{CF}$. Therefore, $ABCF$ is similar to $EDCF$ since they are congruent.

12. $\angle 1 \cong \angle 2 \cong \angle 3 \cong 4$; $\angle 6 \cong \angle 5$ because if two angles of one triangle are congruent to two angles of a second triangle, then the third angles are congruent. $\overline{XY} \cong \overline{XW} \cong \overline{XZ}$, and $\overline{YW} \cong \overline{WZ}$, so the ratio of the corresponding sides is 1. Therefore, $\triangle XYW \sim \triangle XWZ$.

13. $\triangle ABC$ is not similar to $\triangle DEF$. From the lengths of the sides we can determine that $\angle A$ corresponds to $\angle D$, but $\angle A \not\cong \angle D$.

14. $\angle B \cong \angle P$, $\angle D \cong \angle M$, and $\angle C \cong \angle N$ because if two angles of one triangle are congruent to two angles of a second triangle, then the third angles are congruent.

$$\frac{DB}{MP} = \frac{2}{5\frac{1}{3}} \text{ or } \frac{3}{8}$$
$$\frac{BC}{PN} = \frac{4}{10\frac{2}{3}} \text{ or } \frac{3}{8}$$
$$\frac{CD}{NM} = \frac{3}{8}$$

The ratios of the measures of the corresponding sides are equal, and the corresponding angles are congruent, so $\triangle BCD \sim \triangle PNM$.

15.
$$\frac{\text{height of replica}}{\text{height of actual tower}} = \frac{350\frac{2}{3}\text{ feet}}{1052\text{ feet}} = \frac{1}{3}$$

The scale factor is $\frac{1}{3}$.

16. The first copy is 80% or $\frac{8}{10}$ of the original. The second copy must be 100% or 1 of the original.

$$\frac{\text{second copy}}{\text{first copy}} = \frac{1}{\frac{8}{10}} = \frac{10}{8} = 1.25$$

Use a scale factor of 1.25 or 125%.

17. Use the congruent angles to write the corresponding vertices in order: polygon $ABCD \sim$ polygon $EFGH$. Write a proportion to find x.

$$\frac{AB}{EF} = \frac{CD}{GH}$$
$$\frac{x+1}{8} = \frac{x-1}{5}$$
$$(x+1)(5) = 8(x-1)$$
$$5x + 5 = 8x - 8$$
$$5 = 3x - 8$$
$$13 = 3x$$
$$\frac{13}{3} = x$$
$$AB = x + 1 = \frac{13}{3} + 1 \text{ or } \frac{16}{3}$$
$$CD = x - 1 = \frac{13}{3} - 1 \text{ or } \frac{10}{3}$$

The scale factor is $\frac{AB}{EF} = \frac{\frac{16}{3}}{8}$ or $\frac{2}{3}$.

18. Use the congruent angles to write the corresponding vertices in order: $\triangle ABC \sim \triangle EDC$. Write a proportion to find x.

$$\frac{AC}{EC} = \frac{CB}{CD}$$
$$\frac{x+7}{12-x} = \frac{4}{6}$$
$$(x+7)(6) = (12-x)(4)$$
$$6x + 42 = 48 - 4x$$
$$10x + 42 = 48$$
$$10x = 6$$
$$x = \frac{3}{5}$$
$$AC = x + 7 = \frac{3}{5} + 7 \text{ or } 7\frac{3}{5}$$
$$CE = 12 - x = 12 - \frac{3}{5} \text{ or } 11\frac{2}{5}$$

The scale factor is $\frac{CB}{CD} = \frac{4}{6}$ or $\frac{2}{3}$.

19. Use the congruent angles to write the corresponding vertices in order: $\triangle ABE \sim \triangle ACD$. Write a proportion to find x.

$$\frac{AB}{AC} = \frac{AE}{AD}$$

$$\frac{10}{10 + x + 2} = \frac{6.25}{6.25 + x - 1}$$

$$\frac{10}{x + 12} = \frac{6.25}{x + 5.25}$$

$$10(x + 5.25) = (x + 12)(6.25)$$

$$10x + 52.5 = 6.25x + 75$$

$$3.75x + 52.5 = 75$$

$$3.75x = 22.5$$

$$x = 6$$

$BC = x + 2$

$= 6 + 2$ or 8

$ED = x - 1$

$= 6 - 1$ or 5

The scale factor is $\frac{AB}{AC} = \frac{AB}{AB + BC} = \frac{10}{10 + 8} = \frac{10}{18}$ or $\frac{5}{9}$.

20. Use the congruent angles to write the corresponding vertices in order: $\triangle RST \sim \triangle EGF$. Write a proportion to find x.

$$\frac{RT}{EF} = \frac{ST}{GF}$$

$$\frac{15}{11.25} = \frac{10}{x}$$

$$15x = 11.25(10)$$

$$15x = 112.5$$

$$x = 7.5$$

$GF = x$

$= 7.5$

$$\frac{EG}{SR} = \frac{GF}{ST}$$

$$\frac{EG}{20.7} = \frac{7.5}{10}$$

$$(EG)(10) = (20.7)(7.5)$$

$$(EG)(10) = 155.25$$

$$EG = 15.525$$

The scale factor is $\frac{ST}{GF} = \frac{10}{7.5}$ or $\frac{4}{3}$.

21. Write proportions for finding side measures.

$$\begin{array}{r} \text{first new length} \rightarrow \\ \text{original length} \rightarrow \end{array} \frac{x}{2.5} = \frac{5}{4}$$

$$4x = 2.5(5)$$

$$4x = 12.5$$

$$x = 3.125$$

$$\begin{array}{r} \text{second new length} \rightarrow \\ \text{first new length} \rightarrow \end{array} \frac{y}{3.125} = \frac{5}{4}$$

$$4y = 3.125(5)$$

$$4y = 15.625$$

$$y \approx 3.9$$

$$\begin{array}{r} \text{first new width} \rightarrow \\ \text{original width} \rightarrow \end{array} \frac{z}{4} = \frac{5}{4}$$

$$4z = 4(5)$$

$$4z = 20$$

$$z = 5$$

$$\begin{array}{r} \text{second new width} \rightarrow \\ \text{first new width} \rightarrow \end{array} \frac{w}{5} = \frac{5}{4}$$

$$4w = 5(5)$$

$$4w = 25$$

$$w = 6.25$$

After both enlargements the dimensions were about 3.9 inches by 6.25 inches.

22. The enlargement process E can be represented by the equation $E = \frac{5}{4}\left(\frac{5}{4}x\right)$.

23. $\frac{5}{4} \cdot \frac{5}{4} = \frac{25}{16}$

24. Explore: Every millimeter represents 1 meter. The dimensions of the field are about 69 meters by 105 meters.

Plan: Create a proportion relating each measurement to the scale to find the measurements in millimeters. Then make a scale drawing.

Solve:

$$\begin{array}{r} \text{millimeters} \rightarrow \\ \text{meters} \rightarrow \end{array} \frac{1}{1} = \frac{x}{69} \begin{array}{l} \leftarrow \text{millimeters} \\ \leftarrow \text{meters} \end{array}$$

$$69 = x$$

The width of the field should be 69 millimeters in the drawing.

$$\begin{array}{r} \text{millimeters} \rightarrow \\ \text{meters} \rightarrow \end{array} \frac{1}{1} = \frac{y}{105} \begin{array}{l} \leftarrow \text{millimeters} \\ \leftarrow \text{meters} \end{array}$$

$$105 = y$$

The length of the field should be 105 millimeters in the drawing.

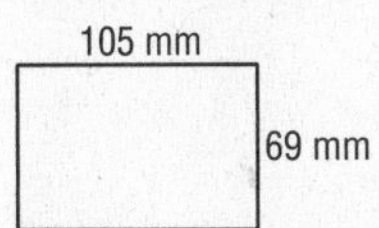

Figure is not shown actual size.

Examine: The scale is 1 : 1, so it is clear that the dimensions in the drawing are reasonable.

25. Explore: Every $\frac{1}{4}$ inch represents 4 feet. The dimensions of the basketball court are 84 feet by 50 feet.

Plan: Create a proportion relating each measurement to the scale to find the measurements in inches. Then make a scale drawing.

Solve:

$$\begin{array}{r} \text{inches} \rightarrow \\ \text{feet} \rightarrow \end{array} \frac{\frac{1}{4}}{4} = \frac{x}{84} \begin{array}{l} \leftarrow \text{inches} \\ \leftarrow \text{feet} \end{array}$$

$$\frac{1}{4}(84) = 4x$$

$$21 = 4x$$

$$5.25 = x$$

The length of the court should be 5.25 inches in the drawing.

$$\begin{array}{r} \text{inches} \\ \text{feet} \end{array} \frac{\frac{1}{4}}{4} = \frac{y}{50} \begin{array}{l} \leftarrow \text{inches} \\ \leftarrow \text{feet} \end{array}$$

$$\frac{1}{4}(50) = 4y$$

$$12.5 = 4y$$

$$3.125 = y$$

The width of the court should be 3.125 inches in the drawing.

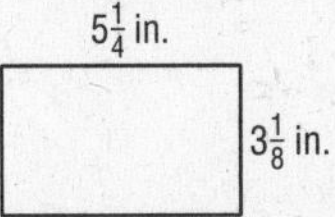

Figure is not shown actual size.

Examine: The scale is $\frac{1}{4}$: 4. The dimensions in the drawing are reasonable.

26. Explore: Every $\frac{1}{8}$ inch represents 1 foot. The dimensions of the tennis court are 36 feet by 78 feet.

Plan: Create a proportion relating each measurement to the scale to find the measurements in inches. Then make a scale drawing.

Solve:

$$\begin{matrix} \text{inches} \rightarrow \\ \text{feet} \rightarrow \end{matrix} \frac{\frac{1}{8}}{1} = \frac{x}{36} \begin{matrix} \leftarrow \text{inches} \\ \leftarrow \text{feet} \end{matrix}$$

$$\tfrac{1}{8}(36) = x$$

$$4.5 = x$$

The width of the court should be 4.5 inches in the drawing.

$$\begin{matrix} \text{inches} \rightarrow \\ \text{feet} \rightarrow \end{matrix} \frac{\frac{1}{8}}{1} = \frac{y}{78} \begin{matrix} \leftarrow \text{inches} \\ \leftarrow \text{feet} \end{matrix}$$

$$\tfrac{1}{8}(78) = y$$

$$9.75 = y$$

The length of the court should be 9.75 inches in the drawing.

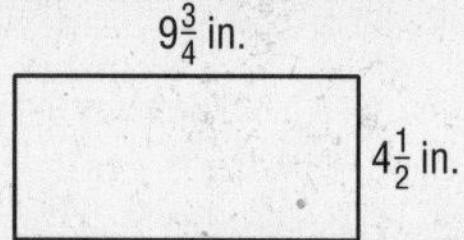

Figure is not shown actual size.

Examine: The scale is $\frac{1}{8}$: 1. The dimensions in the drawing are reasonable.

27. Always; the corresponding angles are congruent and the ratios of the measures of the corresponding sides are all 1.

28. Always; all angles are right angles and so all are congruent, and the ratios of the measures of the corresponding sides are all the same.

29. Never; the number of angles and sides of the figures must be the same for the figures to be compared.

30. sometimes; true when corresponding angles are congruent and ratios of measures of corresponding sides are equal, false when one of these does not hold

31. sometimes; true when the ratios of the measures of the corresponding sides are equal, false when they are not.

32. sometimes; true when corresponding angles are congruent and ratios of measures of corresponding sides are equal, false when one of these does not hold

33. Always; all angles have measure 60 and are congruent, and the ratios of the measures of the corresponding sides are all the same.

34.
$$\angle G \cong \angle L$$
$$m\angle G = m\angle L$$
$$87 = x - 4$$
$$91 = x$$
$$\angle J \cong \angle O$$
$$m\angle J = m\angle O$$
$$y + 30 = 60$$
$$y = 30$$

35.
$$\angle L \cong \angle S$$
$$m\angle L = m\angle S$$
$$30 = x$$
$$\angle K \cong \angle R$$
$$m\angle K = m\angle R$$
$$m\angle K = 180 - (m\angle Q + m\angle S)$$
$$y = 180 - (80 + 30)$$
$$y = 180 - 110$$
$$y = 70$$

36.
$$\frac{AB}{FE} = \frac{BC}{EH}$$
$$\frac{x+2}{15} = \frac{8}{10}$$
$$(x + 2)(10) = 15(8)$$
$$10x + 20 = 120$$
$$10x = 100$$
$$x = 10$$
$$\frac{DC}{GH} = \frac{BC}{EH}$$
$$\frac{y-3}{5} = \frac{8}{10}$$
$$(y - 3)(10) = 5(8)$$
$$10y - 30 = 40$$
$$10y = 70$$
$$y = 7$$

37.
$$\frac{x-3}{16} = \frac{12}{8}$$
$$(x - 3)(8) = 16(12)$$
$$8x - 24 = 192$$
$$8x = 216$$
$$x = 27$$
$$\frac{y+1}{10} = \frac{12}{8}$$
$$(y + 1)(8) = 10(12)$$
$$8y + 8 = 120$$
$$8y = 112$$
$$y = 14$$

38.
$$\frac{2x}{12} = \frac{20}{15}$$
$$2x(15) = 12(20)$$
$$30x = 240$$
$$x = 8$$
$$\frac{y+4}{12} = \frac{15}{20}$$
$$(y + 4)(20) = 12(15)$$
$$20y + 80 = 180$$
$$20y = 100$$
$$y = 5$$

39.
$$\frac{RS}{WU} = \frac{TS}{VW}$$
$$\frac{x}{29} = \frac{49}{20}$$
$$20x = 29(49)$$
$$20x = 1421$$
$$x = 71.05$$
$$\frac{RT}{UV} = \frac{TS}{VW}$$
$$\frac{y+3}{21} = \frac{49}{20}$$
$$(y + 3)(20) = 21(49)$$
$$20y + 60 = 1029$$
$$20y = 969$$
$$y = 48.45$$

40. $\frac{AD}{AG} = \frac{12}{12 - 4.5} = \frac{12}{7.5} = \frac{8}{5}$

41. $AG = AD - GD$
$= 12 - 4.5$
$= 7.5$

42.
$$\frac{DC}{GF} = \frac{AD}{AG}$$
$$\frac{DC}{14} = \frac{12}{7.5}$$
$$7.5(DC) = 14(12)$$
$$7.5(DC) = 168$$
$$DC = 22.4$$

43. $m\angle ADC = m\angle AGF$
$= 108$

44.
$$\frac{BC}{EF} = \frac{AD}{AG}$$
$$\frac{BC}{8} = \frac{12}{7.5}$$
$$7.5(BC) = 8(12)$$
$$7.5(BC) = 96$$
$$BC = 12.8$$

45. $AB + BC + CD + AD = 26 + 12.8 + 22.4 + 12 = 73.2$

46.
$$\frac{AE}{AB} = \frac{AG}{AD}$$
$$\frac{AE}{26} = \frac{7.5}{12}$$
$$12(AE) = 26(7.5)$$
$$12(AE) = 195$$
$$AE = 16.25$$

$AE + EF + FG + AG = 16.25 + 8 + 14 + 7.5$
$= 45.75$

47. $\frac{73.2}{45.75} = \frac{8}{5}$

48. $\triangle ABC \sim \triangle IHG \sim \triangle JLK$ and $\triangle NMO \sim \triangle PRS$;
$\angle A \cong \angle I \cong \angle J$ because each one measures 53°.
$\angle B \cong \angle H \cong \angle L$ because each one is a right angle.
$\angle C \cong \angle G \cong \angle K$ because each one measures 90°-53° or 37°.
So all corresponding angles are congruent. Now determine whether corresponding sides are proportional.
Sides opposite 90° angle.
$\frac{AC}{IG} = \frac{5}{10} = \frac{1}{2}$ $\quad \frac{AC}{JK} = \frac{5}{1.25} = 4$ $\quad \frac{IG}{JK} = \frac{10}{1.25} = 8$
Sides opposite 53° angle.
$\frac{BC}{HG} = \frac{4}{8} = \frac{1}{2}$ $\quad \frac{BC}{LK} = \frac{4}{1} = 4$ $\quad \frac{HG}{LK} = \frac{8}{1} = 8$
Sides opposite 37° angle.
$\frac{AB}{IH} = \frac{3}{6} = \frac{1}{2}$ $\quad \frac{AB}{JL} = \frac{3}{0.75} = 4$ $\quad \frac{IH}{JL} = \frac{6}{0.75} = 8$
The ratios of the measures of the corresponding sides are equal, and the corresponding angles are congruent, so $\triangle ABC \sim \triangle IHG$, $\triangle ABC \sim \triangle JLK$, and $\triangle IHG \sim \triangle JLK$.
$\angle N \cong \angle P$ because each one measures 67°.
$\angle M \cong \angle R$ because each one is a right angle.
$\angle O \cong \angle S$ because each one measures 90°-67° or 23°.
So all corresponding angles are congruent. Now determine whether corresponding sides are proportional.
Sides opposite 90° angle.
$\frac{PS}{NO} = \frac{32.5}{13} = 2.5$
Sides opposite 67° angle.
$\frac{RS}{MO} = \frac{30}{12} = 2.5$
Sides opposite 23° angle.
$\frac{PR}{NM} = \frac{12.5}{5} = 2.5$
The ratios of the measures of the corresponding sides are equal, and the corresponding angles are congruent, so $\triangle NMO \sim \triangle PRS$.

49.

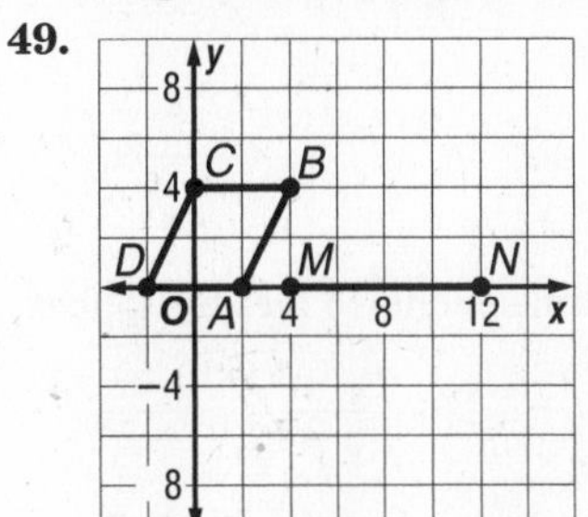

$DA = 4$ and $MN = 8$ so the scale factor is $\frac{8}{4}$ or 2.
To move from point A to point B, move up 4 units and then move 2 units to the right. Because the scale factor is 2, to move from N to L, move up 8 units and move 4 units to the right. So the coordinates of L could be (16, 8). Similarly the coordinates of P could be (8, 8)
Another similar polygon can be obtained by moving down and to the right. To move from N to L, move down 8 units and move 4 units to the right. So the coordinates of L could also be $L(16, -8)$. The coordinates of P could be $(8, -8)$.

50.

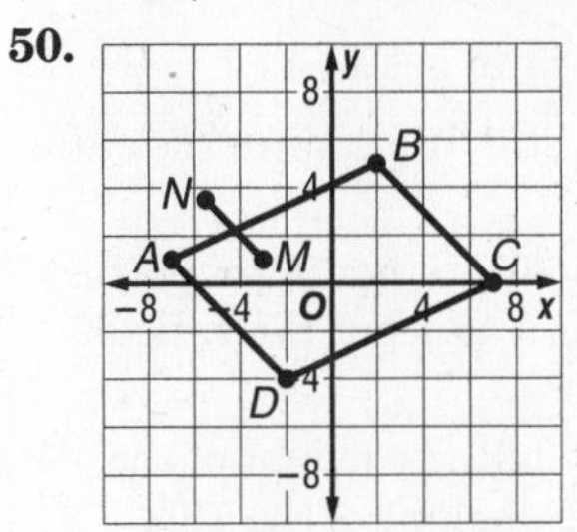

$$AD = \sqrt{[-2-(-7)]^2 + (-4-1)^2}$$
$$= \sqrt{5^2 + (-5)^2}$$
$$= \sqrt{50} \text{ or } 5\sqrt{2}$$

$$NM = \sqrt{\left[-\frac{11}{2} - (-3)\right]^2 + \left(\frac{7}{2} - 1\right)^2}$$
$$= \sqrt{\left(-\frac{5}{2}\right)^2 + \left(\frac{5}{2}\right)^2}$$
$$= \sqrt{\frac{50}{4}} \text{ or } \frac{5}{2}\sqrt{2}$$

The scale factor is $\frac{MN}{AD} = \frac{\frac{5}{2}\sqrt{2}}{5\sqrt{2}} = \frac{1}{2}$.

To move from point A to point B, move up 4 units and then move 9 units to the right. Because the scale factor is $\frac{1}{2}$, to obtain L from N move up 2 units and then move 4.5 units to the right. So the coordinates of L could be $\left(-1, \frac{11}{2}\right)$. Similarly, the coordinates of P could be $\left(\frac{3}{2}, 3\right)$.
Another similar polygon can be obtained by moving down and to the left. To move from N to L move down 2 units and then move 4.5 units to the left. So the coordinates of L could be $\left(-10, \frac{3}{2}\right)$.
Similarly, the coordinates of P could be $\left(-\frac{15}{2}, -1\right)$.

51. $\frac{1 \text{ inch}}{24 \text{ feet}} = \frac{\frac{3}{4} \text{ inch}}{x \text{ feet}}$

$x = 24\left(\frac{3}{4}\right)$

$x = 18$

$\frac{1 \text{ inch}}{24 \text{ feet}} = \frac{\frac{5}{8} \text{ inch}}{y \text{ feet}}$

$y = 24\left(\frac{5}{8}\right)$

$y = 15$

The living room has dimensions 18 feet by 15 feet.

52. $\frac{1 \text{ inch}}{24 \text{ feet}} = \frac{1\frac{1}{4} \text{ inches}}{x \text{ feet}}$

$x = 24(1.25)$

$x = 30$

$\frac{1 \text{ inch}}{24 \text{ feet}} = \frac{\frac{3}{8} \text{ inch}}{y \text{ feet}}$

$y = 24\left(\frac{3}{8}\right)$

$y = 9$

The deck has dimensions 30 feet by 9 feet.

53. The sides are in a ratio of 4 : 1, so if the length of *WXYZ* is x then the length of *ABCD* is $4x$, and if the width of *WXYZ* is y then the width of *ABCD* is $4y$. Then the areas are in a ratio of $\frac{(4x)(4y)}{xy} = \frac{16xy}{xy}$ or 16 : 1.

54. $\frac{4}{1} = \frac{4(3)}{1(3)} = \frac{12}{3}$ or $\frac{4}{1}$. The ratio is still 4 : 1.

55. The ratio of the areas is still 16 : 1 since the ratio of the sides is still 4 : 1.

56. No; the corresponding sides are not in proportion. The ratio of the widths is 1 to 1 but the ratio of the heights is 2 to 1.

57. The widths are the same but the height of the 36% rectangle is twice the height of the 18% rectangle, so the ratio of the areas is $\frac{2\ell w}{\ell w}$ or 2 : 1. The ratio of the percents is $\frac{36\%}{18\%} = \frac{0.36}{0.18}$ or 2 : 1, so the ratios are the same.

58. Sample answer: The difference between increase and decrease is 8%, so the level of courtesy is only slightly decreased.

59. $\frac{a}{3a} = \frac{b}{3b} = \frac{c}{3c} = \frac{1}{3}$

$\frac{a + b + c}{3a + 3b + 3c} = \frac{a + b + c}{3(a + b + c)} = \frac{1}{3}$

60. $\frac{a + 6}{3a + 6} \neq \frac{c + 6}{3c + 6}$

The sides are no longer proportional, so the new triangles are not similar.

61. Sample answer: Artists use geometric shapes in patterns to create another scene or picture. The included objects have the same shape but are different sizes. Answers should include the following.

- The objects are enclosed within a circle. The objects seem to go on and on.
- Each "ring" of figures has images that are approximately the same width, but vary in number and design.

62. B; rewrite 5 : 3 as $5x : 3x$.

$5x + 3x = 32$

$8x = 32$

$x = 4$

$5x = 5(4)$ or 20

$3x = 3(4)$ or 12

There are 20 − 12 or 8 more girls than boys. The answer is B.

63. D;

$180 - (51 + 85) = 44$

$180 - (51 + 44) = 85$

Corresponding angles are congruent.

$\frac{12}{4} = 3$ and $\frac{9.3}{3.1} = 3$, so the ratios of two pairs of corresponding sides are equal, so the triangles are similar.

$\frac{x}{2.8} = \frac{9.3}{3.1}$

$3.1x = 2.8(9.3)$

$3.1x = 26.04$

$x = 8.4$

The answer is D.

64. Multiply each coordinate by 2.

A′ has coordinates $(0 \cdot 2, 0 \cdot 2) = (0, 0)$.

B′ has coordinates $(8 \cdot 2, 0 \cdot 2) = (16, 0)$.

C′ has coordinates $(2 \cdot 2, 7 \cdot 2) = (4, 14)$.

65.

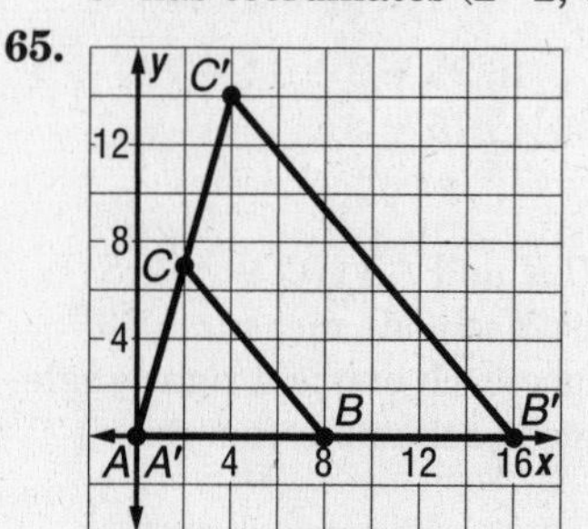

66. $AB = \sqrt{(8 - 0)^2 + (0 - 0)^2}$

$= \sqrt{8^2 + 0^2}$

$= \sqrt{64}$ or 8

$A'B' = \sqrt{(16 - 0)^2 + (0 - 0)^2}$

$= \sqrt{16^2 + 0^2}$

$= \sqrt{256}$ or 16

$BC = \sqrt{(2 - 8)^2 + (7 - 0)^2}$

$= \sqrt{(-6)^2 + 7^2}$

$= \sqrt{85}$

$B'C' = \sqrt{(4 - 16)^2 + (14 - 0)^2}$

$= \sqrt{(-12)^2 + 14^2}$

$= \sqrt{340}$ or $2\sqrt{85}$

$AC = \sqrt{(2 - 0)^2 + (7 - 0)^2}$

$= \sqrt{2^2 + 7^2}$

$= \sqrt{53}$

$A'C' = \sqrt{(4 - 0)^2 + (14 - 0)^2}$

$= \sqrt{4^2 + 14^2}$

$= \sqrt{212}$ or $2\sqrt{53}$

67. $\frac{AB}{A'B'} = \frac{8}{16}$ or $\frac{1}{2}$

$\frac{AC}{A'C'} = \frac{\sqrt{53}}{2\sqrt{53}}$ or $\frac{1}{2}$

$\frac{BC}{B'C'} = \frac{\sqrt{85}}{2\sqrt{85}}$ or $\frac{1}{2}$

68. You could use the slope formula to find that $\overline{BC} \parallel \overline{B'C'}$. Thus, $\angle ABC \cong \angle A'B'C'$ and $\angle ACB \cong \angle A'C'B'$ because of corresponding angles. $\angle A \cong \angle A'$ because of the Third Angle Theorem.

69. The sides are proportional and the angles are congruent, so the triangles are similar.

Page 297 Maintain Your Skills

70. $\frac{b}{7.8} = \frac{2}{3}$
$3b = 7.8(2)$
$3b = 15.6$
$b = 5.2$

71. $\frac{c-2}{c+3} = \frac{5}{4}$
$(c-2)(4) = (c+3)(5)$
$4c - 8 = 5c + 15$
$-c - 8 = 15$
$-c = 23$
$c = -23$

72. $\frac{2}{4y+5} = \frac{-4}{y}$
$2y = (4y+5)(-4)$
$2y = -16y - 20$
$18y = -20$
$y = -\frac{10}{9}$

73. $\overline{BC} \cong \overline{BA}$, $\overline{BO} \cong \overline{BO}$, and $m\angle OBC > m\angle OBA$. By the SAS Inequality, $OC > AO$.

74. $\triangle ABC$ is isosceles with base angles $\frac{1}{2}[180 - (68 + 40)] = 36$. Then $m\angle AOB = 180 - (40 + 36) = 104$ and $m\angle AOD = 180 - 104 = 76$ so, $m\angle AOD < m\angle AOB$.

75. $m\angle ABD > m\angle ADB$ because if one side of a triangle is longer than another side, then the angle opposite the longer side has a greater measure than the angle opposite the shorter side.

76. $x + 52 + 35 = 180$
$x = 93$

77. $x + 32 + 57 = 180$
$x = 91$

78. $x + 40 + 25 = 180$
$x = 115$

79. $m\angle 1 = m\angle 2$
$10x - 9 = 9x + 3$
$x - 9 = 3$
$x = 12$
$m\angle 1 = 10(12) - 9$
$= 120 - 9$ or 111
$m\angle 2 = 9(12) + 3$
$= 108 + 3$ or 111

80. $m\angle 1 = m\angle 4$ (≅ alternate interior ⦞)
$= 118)$

81. $m\angle 2 + m\angle 4 = 180$ (supplementary consecutive interior ⦞)
$m\angle 2 + 118 = 180$
$m\angle 2 = 62$

82. $m\angle 3 + m\angle 4 = 180$ (linear pair)
$m\angle 3 + 118 = 180$
$m\angle 3 = 62$

83. $m\angle 2 + m\angle 5 = 180$ (supplementary consecutive interior ⦞)
$62 + m\angle 5 = 180$
$m\angle 5 = 118$

84. $m\angle ABD = m\angle 1$ (≅ corresponding ⦞)
$= 118$

85. $m\angle 6 + m\angle 4 = 180$ (supplementary consecutive interior ⦞)
$m\angle 6 + 118 = 180$
$m\angle 6 = 62$

86. $m\angle 7 = m\angle 6$ (≅ alternate interior ⦞)
$= 62$

87. $m\angle 8 = m\angle 4$ (≅ corresponding ⦞)
$= 118$

6-3 Similar Triangles

Page 298 Geometry Activity: Similar Triangles

1. $\frac{FD}{ST} = \frac{4}{7} \approx 0.57$
$\frac{EF}{RS} = \frac{2.5}{4.4} \approx 0.57$
$\frac{ED}{RT} = \frac{4.3}{7.6} \approx 0.57$
All of the ratios equal about 0.57.

2. Yes, all sides are in the same ratio.

3. Sample answer: Either all sides proportional or two corresponding angles congruent.

Page 301 Check for Understanding

1. Sample answer: Two triangles are congruent by the SSS, SAS, and ASA Postulates and the AAS Theorem. In these triangles, corresponding parts must be congruent. Two triangles are similar by AA similarity, SSS Similarity, and SAS Similarity. In similar triangles, the sides are proportional and the angles are congruent. Congruent triangles are always similar triangles. Similar triangles are congruent only when the scale factor for the proportional sides is 1. SSS and SAS are common relationships for both congruence and similarity.

2. Yes; suppose $\triangle RST$ has angles that measure 46°, 54°, and 80°, $\triangle ABC$ has angles that measure 39°, 63°, and 78°, and $\triangle EFG$ has angles that measure 39°, 63°, and 78°. So $\triangle ABC$ is not similar to $\triangle RST$ and $\triangle RST$ is not similar to $\triangle EFG$, but $\triangle ABC$ is similar to $\triangle EFG$.

3. Alicia; while both have corresponding sides in a ratio, Alicia has them in proper order with the numerators from the same triangle.

4. $\angle A \cong \angle D$ and $\angle F \cong \angle C$, so by AA Similarity, $\triangle ABC \sim \triangle DEF$.
$\frac{DE}{AB} = \frac{FE}{CB}$
$\frac{x}{45} = \frac{3}{15}$
$15x = 45(3)$
$15x = 135$
$x = 9$
$DE = x = 9$

5. $\angle A \cong \angle D$ and $\angle B \cong \angle E$, so by AA Similarity, $\triangle ABC \sim \triangle DEF$.

$$\frac{AB}{DE} = \frac{BC}{EF}$$
$$\frac{x}{x-4} = \frac{5}{3}$$
$$3x = 5(x-4)$$
$$3x = 5x - 20$$
$$-2x = -20$$
$$x = 10$$
$$AB = x = 10$$
$$DE = x - 4$$
$$= 10 - 4 \text{ or } 6$$

6. Triangles EFD and BCA are right triangles. To determine whether corresponding sides are proportional, find AB.

$$(AB)^2 = (AC)^2 + (BC)^2$$
$$(AB)^2 = 10^2 + 5^2 = 100 + 25$$
$$AB = \sqrt{125} \text{ or } 5\sqrt{5}$$
$$\frac{DE}{AB} = \frac{8}{5\sqrt{5}} \text{ and } \frac{EF}{BC} = \frac{4}{5}$$

$\frac{8}{5\sqrt{5}} \neq \frac{4}{5}$, so the triangles are not similar because corresponding sides are not proportional.

7. If the measures of the corresponding sides are proportional, then the triangles are similar.

$\frac{DF}{AB} = \frac{9}{3}$ or 3, $\frac{DE}{AC} = \frac{25}{8\frac{1}{3}}$ or 3, $\frac{FE}{BC} = \frac{21}{7}$ or 3

$\frac{DF}{AB} = \frac{DE}{AC} = \frac{FE}{BC}$, so by SSS Similarity, $\triangle DEF \sim \triangle ACB$.

8. Triangles ABC and EDF are isosceles triangles because they each have a pair of congruent sides. Base angles of isosceles triangles are congruent, so $\angle B \cong \angle C$ and $\angle D \cong \angle F$. From the figure, $\angle B \cong \angle D$, so by transitivity $\angle B \cong \angle F$ and $\angle D \cong \angle C$. So $\triangle ABC \sim \triangle EDF$ by AA Similarity.

9.

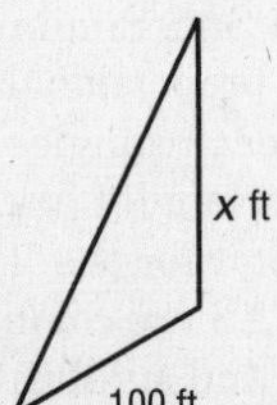

Assuming that the sun's rays form similar triangles, the following proportion can be written.

$$\frac{\text{height of tower (ft)}}{\text{height of post (ft)}} = \frac{\text{tower shadow (ft)}}{\text{post shadow (ft)}}$$

Substitute the known values and let x be the height of the cell phone tower.

$$\frac{x}{4\frac{1}{2}} = \frac{100}{3\frac{1}{3}}$$
$$\left(3\frac{1}{3}\right)x = \left(4\frac{1}{2}\right)(100)$$
$$\left(3\frac{1}{3}\right)x = 450$$
$$x = 135$$

The cellphone tower is 135 feet tall.

Pages 302–305 Practice and Apply

10. If the measures of the corresponding sides are proportional, then the triangles are similar.

$\frac{QR}{NO} = \frac{7}{21}$ or $\frac{1}{3}$, $\frac{QP}{NM} = \frac{10}{30}$ or $\frac{1}{3}$, $\frac{RP}{OM} = \frac{15}{45}$ or $\frac{1}{3}$

$\frac{QR}{NO} = \frac{QP}{NM} = \frac{RP}{OM}$, so by SSS Similarity, $\triangle MNO \sim \triangle PQR$.

11. If the measures of the corresponding sides are proportional, then the triangles are similar.

$\frac{QR}{TV} = \frac{SR}{UV} = \frac{7}{14}$ or $\frac{1}{2}$ and $\frac{QS}{TU} = \frac{3}{6}$ or $\frac{1}{2}$

$\frac{QR}{TV} = \frac{SR}{UV} = \frac{QS}{TU}$, so by SSS Similarity, $\triangle QRS \sim \triangle TVU$.

12. $\angle F \cong \angle J$ and $\frac{EF}{IJ} = \frac{EG}{IA} = \frac{4}{5}$, but SSA is not a valid justification for similarity. There is not enough information to determine whether the triangles are similar.

13.
$$m\angle R + m\angle S + m\angle T = 180$$
$$m\angle R + 120 + 20 = 180$$
$$m\angle R = 40$$
$$m\angle J + m\angle K + m\angle L = 180$$
$$40 + 120 + m\angle L = 180$$
$$m\angle L = 20$$

$\angle S \cong \angle K$ and $\angle R \cong \angle J$, so $\triangle RST \sim \triangle JKL$ by AA Similarity.

14. $ST = XV$, $UT = WV$, and $\angle T \cong \angle V$, so $\triangle STU \sim \triangle XVW$ by SAS Similarity.

15. $\frac{AB}{JK} = \frac{3}{9}$ or $\frac{1}{3}$ and $\frac{BC}{KL} = \frac{5}{15}$ or $\frac{1}{3}$, and $\angle B \cong \angle K$. $\triangle ABC \sim \triangle JKL$ by SAS Similarity.

16. Since $\overline{AE} \parallel \overline{BD}$, $\angle EAC \cong \angle DBC$ and $\angle AEC \cong \angle BDC$ because they are corresponding angles. By AA Similarity, $\triangle AEC \sim \triangle BDC$.

17. If the measures of the corresponding sides are proportional, then the triangles are similar.

$\frac{ST}{BA} = \frac{6}{20}$ or $\frac{3}{10}$ and $\frac{SR}{BC} = \frac{10.5}{30}$ or $\frac{3.5}{10}$

$\frac{ST}{BA} \neq \frac{SR}{BC}$, so $\triangle RST$ is not similar to $\triangle CBA$ because the sides are not proportional.

18. $\overline{AE} \parallel \overline{DC}$, so $\angle A \cong \angle C$ and $\angle E \cong \angle D$ because they are alternate interior angles. By AA Similarity, $\triangle ABE \sim \triangle CBD$.

$$\frac{AB}{CB} = \frac{EB}{DB}$$
$$\frac{x+3}{2x-8} = \frac{5}{3}$$
$$3(x+3) = 5(2x-8)$$
$$3x + 9 = 10x - 40$$
$$-7x + 9 = -40$$
$$-7x = -49$$
$$x = 7$$
$$AB = x + 3$$
$$= 7 + 3 \text{ or } 10$$
$$BC = 2x - 8$$
$$= 2(7) - 8 \text{ or } 6$$

19. $\overline{DC} \parallel \overline{EB}$, so $\angle ADC \cong \angle AEB$ and $\angle ACD \cong \angle ABE$ because they are corresponding angles. By AA Similarity, $\triangle ABE \sim \triangle ACD$.

$$\frac{AB}{AC} = \frac{AE}{AD}$$
$$\frac{x+2}{x+2+6} = \frac{8-5}{8}$$
$$8(x+2) = (x+8)(3)$$
$$8x + 16 = 3x + 24$$
$$5x + 16 = 24$$
$$5x = 8$$
$$x = \frac{8}{5}$$
$$AB = x + 2$$
$$= \frac{8}{5} + 2 \text{ or } 3\frac{3}{5}$$
$$AC = x + 2 + 6$$
$$= \frac{8}{5} + 2 + 6 \text{ or } 9\frac{3}{5}$$

20. $\triangle ABD$ and $\triangle FEC$ are right triangles with $\angle A \cong \angle F$ and right angles B and E. Because all right angles are congruent, $\triangle ABD \sim \triangle FEC$ by AA Similarity.

$$\frac{BD}{CE} = \frac{AB}{FE}$$
$$\frac{x-1}{x+2} = \frac{3}{8}$$
$$8(x-1) = 3(x+2)$$
$$8x - 8 = 3x + 6$$
$$5x - 8 = 6$$
$$5x = 14$$
$$x = \frac{14}{5}$$
$$BD = x - 1$$
$$= \frac{14}{5} - 1 \text{ or } \frac{9}{5}$$
$$EC = x + 2$$
$$= \frac{14}{5} + 2 \text{ or } \frac{24}{5}$$

21. $\triangle ABC$ and $\triangle ARS$ are right triangles with $\angle A \cong \angle A$ and right angles $\angle ASR$ and $\angle ACB$. Because all right angles are congruent, $\triangle ABC \sim \triangle ARS$ by AA Similarity.

$$\frac{AS}{AC} = \frac{SR}{CB}$$
$$\frac{x}{12} = \frac{6}{9}$$
$$9x = 72$$
$$x = 8$$
$$AB = x + 7$$
$$= 8 + 7 \text{ or } 15$$
$$AS = x$$
$$= 8$$

22.

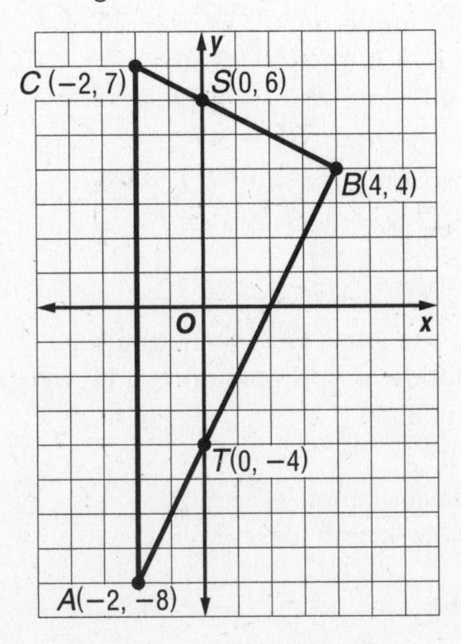

$$AB = \sqrt{6^2 + 12^2} = \sqrt{180} \text{ or } 6\sqrt{5}$$
$$BC = \sqrt{6^2 + 3^2} = \sqrt{45} \text{ or } 3\sqrt{5}$$
$$CA = |7 - (-8)| = 15$$
$$ST = |6 - (-4)| = 10$$
$$TB = \sqrt{8^2 + 4^2} = \sqrt{80} \text{ or } 4\sqrt{5}$$
$$BS = \sqrt{2^2 + 4^2} = \sqrt{20} \text{ or } 2\sqrt{5}$$

$\frac{CA}{ST} = \frac{15}{10}$ or $\frac{3}{2}$, $\frac{AB}{TB} = \frac{6\sqrt{5}}{4\sqrt{5}}$ or $\frac{3}{2}$, and $\frac{BC}{BS} = \frac{3\sqrt{5}}{2\sqrt{5}}$ or $\frac{3}{2}$.

Since $\frac{CA}{ST} = \frac{AB}{TB} = \frac{BC}{BS}$, $\triangle ABC \sim \triangle TBS$ by SSS Similarity.

23. The perimeter of $\triangle ABC$ is $6\sqrt{5} + 3\sqrt{5} + 15$ or $15 + 9\sqrt{5}$. The perimeter of $\triangle TBS$ is $4\sqrt{5} + 2\sqrt{5} + 10$ or $10 + 6\sqrt{5}$.

$$\frac{15 + 9\sqrt{5}}{10 + 6\sqrt{5}} = \frac{3(5 + 3\sqrt{5})}{2(5 + 3\sqrt{5})} \text{ or } \frac{3}{2}$$

24. False; this is not true for equilateral or isosceles triangles.

25. True; similarity of triangles is transitive.

26. $\angle QRS$ and $\angle STR$ are right angles, so $\angle QRS \cong \angle STR$. $\angle Q \cong \angle Q$, so $\triangle QRS \sim \triangle QTR$ by AA Similarity. $\angle S \cong \angle S$, so $\triangle QRS \sim \triangle RTS$ by AA Similarity. Therefore, $\triangle QTR \sim \triangle RTS$ by transitivity.

27. $AB \parallel FD$, so $\angle BAE \cong \angle AFC$ and $\angle ABE \cong \angle ECF$ because they are alternate interior angles. Then $\triangle EAB \sim \triangle EFC$ by AA Similarity. $\overline{AD} \parallel \overline{BC}$, so $\angle ADF \cong \angle ECF$ because they are corresponding angles. $\angle F \cong \angle F$, so $\triangle EFC \sim \triangle AFD$ by AA Similarity. Then $\triangle EAB \sim \triangle AFD$ by transitivity.

28. $\angle PSY$ and $\angle PQR$ are right angles and so they are congruent. $\angle P \cong \angle P$, so $\triangle PSY \sim \triangle PQR$ by AA Similarity. $\overline{PR} \parallel \overline{WX}$, so $\angle YWX \cong \angle WYP$ because they are alternate interior angles. $\angle WYX \cong \angle YSP$ because they are right angles, so $\triangle WYX \sim \triangle YSP$ by AA Similarity.

$$\frac{PY}{XW} = \frac{PS}{XY}$$
$$\frac{PY}{10} = \frac{3}{6}$$
$$6(PY) = 10(3)$$
$$6(PY) = 30$$
$$PY = 5$$
$$\frac{SY}{YW} = \frac{PS}{XY}$$
$$\frac{SY}{8} = \frac{3}{6}$$
$$6(SY) = 8(3)$$
$$6(SY) = 24$$
$$SY = 4$$
$$\frac{PQ}{PS} = \frac{PR}{PY}$$
$$\frac{PQ}{3} = \frac{5+5}{5}$$
$$5(PQ) = 3(10)$$
$$5(PQ) = 30$$
$$PQ = 6$$

29. $\overline{PR} \parallel \overline{KL}$, so $\angle RQM \cong \angle LNM$, $\angle PQM \cong \angle KNM$, $\angle L \cong \angle QRM$, and $\angle K \cong \angle QPM$, since these are all pairs of corresponding angles. $\angle LNM$, $\angle KNM$, $\angle RQM$, $\angle PQM$, and $\angle LMK$ are all right angles, so each is congruent to the others.

Since $\angle LMK \cong \angle MNK$ and $\angle K \cong \angle K$, $\triangle LMK \sim \triangle MNK$ by AA Similarity.

$$\frac{LK}{MK} = \frac{MK}{NK}$$
$$\frac{16 + 9}{2(KP) + KP} = \frac{2(KP) + KP}{9}$$
$$\frac{25}{3(KP)} = \frac{3(KP)}{9}$$
$$25(9) = 9(KP)^2$$
$$25 = (KP)^2$$
$$5 = KP$$

$KM = KP + PM$
$= KP + 2(KP)$
$= 3(KP)$
$= 3(5)$ or 15

$PM = 2(KP) = 2(5)$ or 10

Since $\angle L \cong \angle QRM$ and $\angle K \cong \angle QPM$, $\triangle LKM \sim \triangle RPM$ by AA Similarity.

$$\frac{LK}{RP} = \frac{KM}{PM}$$
$$\frac{25}{RP} = \frac{15}{10}$$
$$25(10) = 15(RP)$$
$$\frac{250}{15} = RP$$
$$\frac{50}{3} = RP$$

Since $\angle KNM \cong \angle PQM$ and $\angle K \cong \angle QPM$, $\triangle KNM \sim \triangle PQM$ by AA Similarity.

$$\frac{KN}{PQ} = \frac{KM}{PM}$$
$$\frac{9}{PQ} = \frac{15}{10}$$
$$9(10) = 15(PQ)$$
$$90 = 15(PQ)$$
$$6 = PQ$$
$$RP = RQ + PQ$$
$$\frac{50}{3} = RQ + 6$$
$$\frac{50}{3} - 6 = RQ$$
$$\frac{32}{3} = RQ$$

$m\angle MQP + m\angle QPM + m\angle PMQ = 180$
$90 + m\angle QPM + m\angle PMQ = 180$
$m\angle QPM + m\angle PMQ = 90$

Since $\angle RMP$ is a right angle,
$m\angle RMQ + m\angle PMQ = 90$
$m\angle QPM + m\angle PMQ = m\angle RMQ + m\angle PMQ$
$m\angle QPM = m\angle RMQ$, so $\angle QPM \cong \angle RMQ$.

Therefore, $\triangle RQM \sim \triangle MQP$ by AA Similarity.

$$\frac{RQ}{MQ} = \frac{MQ}{QP}$$
$$\frac{\frac{32}{3}}{MQ} = \frac{MQ}{6}$$
$$(MQ)^2 = \frac{32}{3} \cdot 6$$
$$(MQ)^2 = 64$$
$$MQ = 8$$
$$\frac{RM}{MP} = \frac{QM}{QP}$$
$$\frac{RM}{10} = \frac{8}{6}$$
$$6(RM) = 10(8)$$
$$RM = \frac{80}{6} \text{ or } \frac{40}{3}$$

Since $\triangle KNM \sim \triangle PQM$,

$$\frac{NM}{QM} = \frac{KN}{PQ}$$
$$\frac{NM}{8} = \frac{9}{6}$$
$$6(NM) = 8(9)$$
$$NM = \frac{72}{6} \text{ or } 12$$

Since $\angle L \cong \angle QRM$ and $\angle LNM \cong \angle RQM$, $\triangle QRM \sim \triangle NLM$ by AA Similarity.

$$\frac{QR}{NL} = \frac{RM}{LM}$$
$$\frac{\frac{32}{3}}{16} = \frac{\frac{40}{3}}{LM}$$
$$\frac{32}{3}(LM) = 16\left(\frac{40}{3}\right)$$
$$LM = 16 \cdot \frac{40}{3} \cdot \frac{3}{32}$$
$$LM = 20$$

30. $\frac{IJ}{XJ} = \frac{HJ}{YJ}$ and $\angle J \cong \angle J$, so $\triangle IJH \sim \triangle XJY$ by SAS Similarity.

$m\angle JXY = 180 - m\angle WXJ$
$= 180 - 130$
$= 50$

$m\angle JIH = m\angle JXY$ by corr. ∠s
$= 50$

$m\angle YIZ = m\angle JIH$ by vert. ∠s
$= 50$

$m\angle JYX = m\angle YIZ + m\angle WZG$
$= 50 + 20$
$= 70$

because exterior angle = sum of remote interior angles

$m\angle JHI = m\angle JYX = 70$ by corr. ∠s

$m\angle J + m\angle JXY + m\angle JYX = 180$
$m\angle J + 50 + 70 = 180$
$m\angle J = 60$

$m\angle JHG + m\angle JHI = 180$ Linear pair
$m\angle JHG + 70 = 180$
$m\angle JHG = 110$

31. $\angle RST$ is a right angle, so $m\angle RST = 90$.
$m\angle RTS = 47$.
$m\angle RTS + m\angle R + m\angle RST = 180$
$47 + m\angle R + 90 = 180$
$m\angle R = 43$

$\angle UVT$ is a right angle, so $m\angle UVT = 90$.
$m\angle UVT + m\angle TUV + m\angle UTV = 180$
$90 + m\angle TUV + 47 = 180$
$m\angle TUV = 43$

$\angle RUS$ is a right angle, so $m\angle RUS = 90$.
$m\angle RUS + m\angle R + m\angle RSU = 180$
$90 + 43 + m\angle RSU = 180$
$m\angle RSU = 47$

$\angle SUT$ is a right angle, so $m\angle SUT = 90$.
$m\angle SUV + m\angle TUV = 90$
$m\angle SUV + 43 = 90$
$m\angle SUV = 47$

32. Assuming that the sun's rays form similar triangles, the following proportion can be written.

$$\frac{\text{height of pyramid (ft)}}{\text{height of staff (paces} \cdot 3)} = \frac{\text{pyramid shadow length (paces} \cdot 3)}{\text{staff shadow length (paces} \cdot 3)}$$

Substitute the known values and let x be the height of the pyramid.

$\frac{x}{2(3)} = \frac{(125 + 114)(3)}{3(3)}$
$\frac{x}{6} = \frac{717}{9}$
$9x = 6(717)$
$9x = 4302$
$x = 478$

The pyramid was about 478 feet tall at that time.

33. x must equal y. If $\overline{BD} \parallel \overline{AE}$, then $\angle CBD \cong \angle CAE$ because they are corresponding angles and $\angle C \cong \angle C$ so $\triangle BCD \sim \triangle ACE$ by AA Similarity. Then $\frac{BC}{AC} = \frac{DC}{EC}$. Thus, $\frac{2}{4} = \frac{x}{x+y}$. Cross multiply and solve for y.

$\frac{2}{4} = \frac{x}{x+y}$
$2(x + y) = 4x$
$2x + 2y = 4x$
$2y = 2x$
$y = x$

34. Given: $\angle B \cong \angle E$; $\overline{QP} \parallel \overline{BC}$; $\overline{QP} \cong \overline{EF}$; $\frac{AB}{DE} = \frac{BC}{EF}$
Prove: $\triangle ABC \sim \triangle DEF$

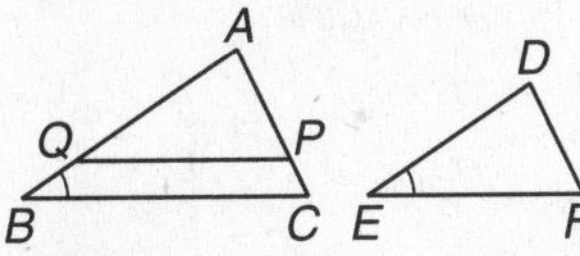

Proof:

Statements	Reasons
1. $\angle B \cong \angle E$, $\overline{QP} \parallel \overline{BC}$; $\overline{QP} \cong \overline{EF}$; $\frac{AB}{DE} = \frac{BC}{EF}$	1. Given
2. $\angle APQ \cong \angle C$ $\angle AQP \cong \angle B$	2. Corresponding ≅ Postulate
3. $\angle AQP \cong \angle E$	3. Transitive Prop. of ≅ ∠s
4. $\triangle ABC \sim \triangle AQP$	4. AA Similarity
5. $\frac{AB}{AQ} = \frac{BC}{QP}$	5. Def. of ~ △s
6. $AB \cdot QP = AQ \cdot BC$ $AB \cdot EF = DE \cdot BC$	6. Cross products
7. $QP = EF$	7. Def. of ≅ segments
8. $AB \cdot EF = AQ \cdot BC$	8. Substitution
9. $AQ \cdot BC = DE \cdot BC$	9. Substitution
10. $AQ = DE$	10. Div. Prop.
11. $\overline{AQ} \cong \overline{DE}$	11. Def. of ≅ segments
12. $\triangle AQP \cong \triangle DEF$	12. SAS
13. $\angle APQ \cong \angle F$	13. CPCTC
14. $\angle C \cong \angle F$	14. Transitive Prop.
15. $\triangle ABC \sim \triangle DEF$	15. AA Similarity

35. Given: $\overline{LP} \parallel \overline{MN}$
Prove: $\frac{LJ}{JN} = \frac{PJ}{JM}$
Proof:

Statements	Reasons
1. $\overline{LP} \parallel \overline{MN}$	1. Given
2. $\angle PLN \cong \angle LNM$, $\angle LPM \cong \angle PMN$	2. Alternate Interior ∠s Theorem
3. $\triangle LPJ \sim \triangle NMJ$	3. AA Similarity
4. $\frac{LJ}{JN} = \frac{PJ}{JM}$	4. Corr. sides of ~ △s are proportional.

36. Given: $\overline{EB} \perp \overline{AC}$, $\overline{BH} \perp \overline{AE}$, $\overline{CJ} \perp \overline{AE}$
a. Prove: $\triangle ABH \sim \triangle DCB$

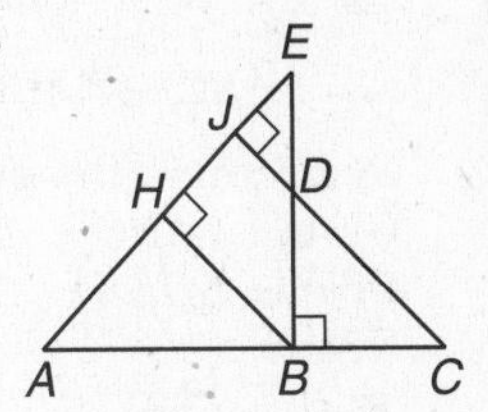

Proof:
$\angle AHB$, $\angle AJC$, and $\angle EBC$ are right angles because perpendicular lines form right angles. Since all right angles are congruent, $\angle AHB \cong \angle AJC \cong \angle EBC$. Since $\angle A \cong \angle A$ by the Reflexive Property, $\triangle ABH \sim \triangle ACJ$, by AA Similarity. Likewise, since $\angle C \cong \angle C$, $\triangle ACJ \sim \triangle DCB$. By the Transitive Property, $\triangle ABH \sim \triangle DCB$.

b. Prove: $\frac{BC}{BE} = \frac{BD}{BA}$
Proof:
From part a, $\angle A \cong \angle CDB$ by definition of similar triangles. $\angle ABE \cong \angle DBC$ because all right angles are congruent. Thus, $\triangle ABE \sim \triangle DBC$ by AA Similarity.
$\frac{BC}{BE} = \frac{BD}{BA}$ from definition of similar triangles.

37. Given: $\triangle BAC$ and $\triangle EDF$ are right triangles.
$\frac{AB}{DE} = \frac{AC}{DF}$
Prove: $\triangle ABC \sim \triangle DEF$

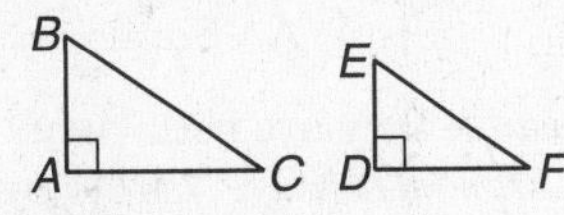

Proof:

Statements	Reasons
1. $\triangle BAC$ and $\triangle EDF$ are right triangles.	1. Given
2. $\angle BAC$ and $\angle EDF$ are right angles.	2. Def. of rt. △
3. $\angle BAC \cong \angle EDF$	3. All rt. ∠s are ≅.
4. $\frac{AB}{DE} = \frac{AC}{DF}$	4. Given
5. $\triangle ABC \sim \triangle DEF$	5. SAS Similarity

38. Reflexive Property of Similarity
Given: $\triangle ABC$
Prove: $\triangle ABC \sim \triangle ABC$

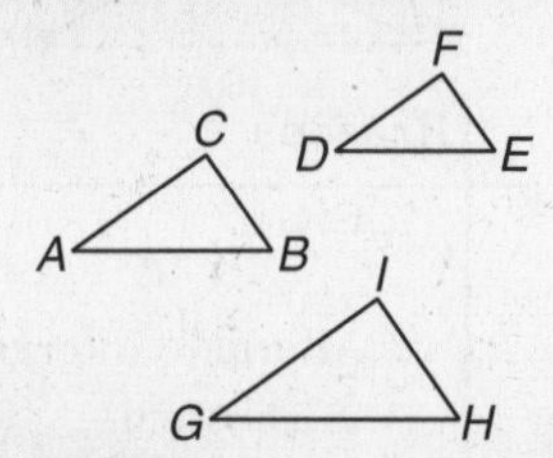

Proof:

Statements	Reasons
1. $\triangle ABC$	1. Given
2. $\angle A \cong \angle A$, $\angle B \cong \angle B$	2. Reflexive Prop.
3. $\triangle ABC \sim \triangle ABC$	3. AA Similarity

Symmetric Property of Similarity
Given: $\triangle ABC \sim \triangle DEF$
Prove: $\triangle DEF \sim \triangle ABC$
Proof:

Statements	Reasons
1. $\triangle ABC \sim \triangle DEF$	1. Given
2. $\angle A \cong \angle D$, $\angle B \cong \angle E$	2. Def. of ~ polygons
3. $\angle D \cong \angle A$, $\angle E \cong \angle B$	3. Symmetric Prop.
4. $\triangle DEF \sim \triangle ABC$	4. AA Similarity

Transitive Property of Similarity
Given: $\triangle ABC \sim \triangle DEF$ and $\triangle DEF \sim \triangle GHI$
Prove: $\triangle ABC \sim \triangle GHI$
Proof:

Statements	Reasons
1. $\triangle ABC \sim \triangle DEF$, $\triangle DEF \sim \triangle GHI$	1. Given
2. $\angle A \cong \angle D$, $\angle B \cong \angle E$, $\angle D \cong \angle G$, $\angle E \cong \angle H$	2. Def. of ~ polygons
3. $\angle A \cong \angle G$, $\angle B \cong \angle H$	3. Trans. Prop.
4. $\triangle ABC \sim \triangle GHI$	4. AA Similarity

39. $\angle MKO \cong \angle MOP$ because they are both right angles, and $\angle M \cong \angle M$, so $\triangle MKO \sim \triangle MOP$ by AA Similarity. $\angle OKP$ is a right angle because it forms a linear pair with right angle $\angle MKO$. $\angle OKP \cong \angle MOP$ and $\angle P \cong \angle P$, so $\triangle MOP \sim \triangle OKP$ by AA Similarity. Then $\triangle MKO \sim \triangle OKP$ by transitivity.

$\frac{MK}{OK} = \frac{OK}{KP}$

$\frac{1.5}{4.5} = \frac{4.5}{KP}$

$1.5(KP) = 4.5(4.5)$

$1.5(KP) = 20.25$

$KP = 13.5$

The distance KP is 13.5 feet.

40. If the side of $\triangle DEF$ that is 36 cm corresponds to the shortest side of $\triangle ABC$, then we can find the lengths of the other sides of $\triangle DEF$ using proportions.

$\frac{36}{4} = \frac{x}{6}$

$36(6) = 4x$

$216 = 4x$

$54 = x$

$\frac{36}{4} = \frac{y}{9}$

$36(9) = 4y$

$324 = 4y$

$81 = y$

The perimeter of $\triangle DEF$ is 36 + 54 + 81 or 171 cm.

41. Assume the lines of sight create similar triangles.

$\frac{x}{1.92} = \frac{87.6}{0.4}$

$0.4x = 1.92(87.6)$

$0.4x = 168.192$

$x = 420.48$

The tower is about 420.5 m tall.

42. It is difficult to measure shadows within a city.

43. Assume that $ADFE$ is a rectangle.

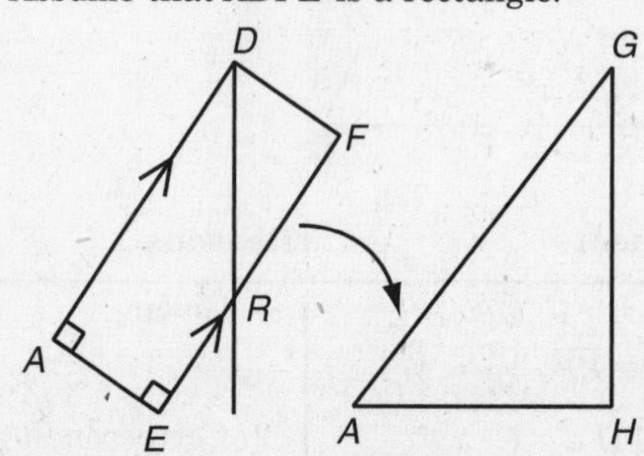

Let R be the point on $\overline{EF}$ where the vertical line from D crosses $\overline{EF}$. Then $\angle ADR \cong \angle DRF$ because $\overline{AD} \parallel \overline{EF}$ and alternate interior angles are congruent. $\angle DFR \cong \angle AHG$ because all right angles are congruent. So $\triangle AGH \sim \triangle DRF$ by AA Similarity.

$\frac{GH}{RF} = \frac{AH}{DF}$

$\frac{x}{6} = \frac{1500}{10}$

$10x = 6(1500)$

$10x = 9000$

$x = 900$ cm or 9 m

So, the height of the tree is 9 m + 1.75 m or 10.75 m.

44. No; the towns are on different latitudinal lines, so the sun is at a different angle to the two buildings.

45.

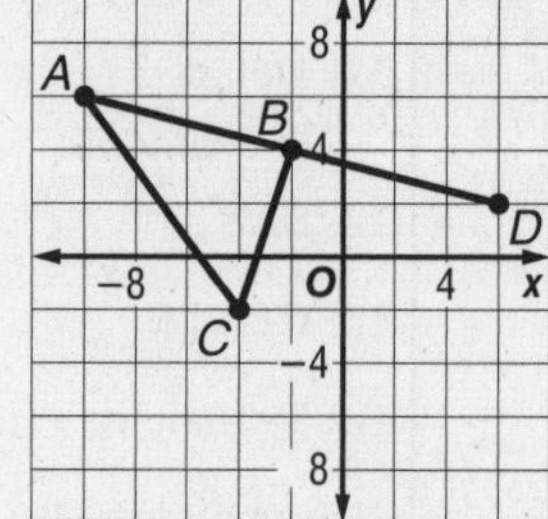

46. If $\triangle ABC \sim \triangle ADE$ then $\frac{AB}{AD} = \frac{BC}{DE}$. Plot point E so that $\overline{BC} \parallel \overline{DE}$ and the proportion is true.

$AB = \sqrt{[-2-(-10)]^2 + (4-6)^2}$
$= \sqrt{8^2 + (-2)^2}$
$= \sqrt{68}$ or $2\sqrt{17}$

$AD = \sqrt{[6-(-10)]^2 + (2-6)^2}$
$= \sqrt{16^2 + (-4)^2}$
$= \sqrt{272}$ or $4\sqrt{17}$

$\frac{AB}{AD} = \frac{2\sqrt{17}}{4\sqrt{17}}$ or $\frac{1}{2}$, so $DE = 2(BC)$. To get from point B to point C, move down 6 units and then move left 2 units. To get from point D to point E, move down 12 units and then move left 4 units. Locate point E at $(2, -10)$.

47. $\triangle ABC \sim \triangle ACD$ by AA Similarity.
$\triangle ABC \sim \triangle CBD$ by AA Similarity.
$\triangle ACD \sim \triangle CBD$ by AA Similarity.

48. Sample answer: Engineers use triangles, some the same shape, but different in size, to complete a project. Answers should include the following.
- Engineers use triangles in construction because they are rigid shapes.
- With the small ground pressure, the tower does not sink, shift, lean, or fall over.

49. A; $\triangle ABE \sim \triangle ACD$ by AA Similarity, so $\frac{AE}{AD} = \frac{AB}{AC}$.

$\frac{10-4}{10} = \frac{x-2}{x-2+5}$
$\frac{6}{10} = \frac{x-2}{x+3}$
$6(x+3) = 10(x-2)$
$6x + 18 = 10x - 20$
$18 = 4x - 20$
$38 = 4x$
$9.5 = x$

50. B;

$\frac{x+3}{6} = \frac{x}{x-2}$
$(x+3)(x-2) = 6x$
$x^2 - 2x + 3x - 6 = 6x$
$x^2 + x - 6 = 6x$
$x^2 - 5x - 6 = 0$
$(x-6)(x+1) = 0$
$x - 6 = 0$ or $x + 1 = 0$
$x = 6$ $\quad x = -1$

Page 306 Maintain Your Skills

51. Use the congruent angles to write the corresponding vertices in order.
$PQRS \sim ABCD$

$\frac{PQ}{AB} = \frac{SR}{DC}$
$\frac{x}{3.2} = \frac{0.7}{1.4}$
$1.4x = 3.2(0.7)$
$1.4x = 2.24$
$x = 1.6$

$\frac{BC}{QR} = \frac{CD}{RS}$
$\frac{BC}{0.7} = \frac{1.4}{0.7}$
$0.7(BC) = 0.7(1.4)$
$BC = 1.4$

$\frac{PS}{AD} = \frac{SR}{DC}$
$\frac{PS}{2.2} = \frac{0.7}{1.4}$
$1.4(PS) = 2.2(0.7)$
$1.4(PS) = 1.54$
$PS = 1.1$

The scale factor is $\frac{SR}{CD} = \frac{0.7}{1.4}$ or $\frac{1}{2}$.

52. Use the congruent angles to write the corresponding vertices in order. $\triangle EFG \sim \triangle XYZ$

$\frac{XY}{EF} = \frac{YZ}{FG}$
$\frac{22.5}{6x} = \frac{7.5}{10}$
$22.5(10) = 6x(7.5)$
$225 = 45x$
$5 = x$
$EF = 6x$
$= 6(5)$ or 30

$\frac{XZ}{EG} = \frac{YZ}{FG}$
$\frac{XZ}{25} = \frac{7.5}{10}$
$10(XZ) = 25(7.5)$
$10(XZ) = 187.5$
$XZ = 18.75$

The scale factor is $\frac{FG}{YZ} = \frac{10}{7.5}$ or $\frac{4}{3}$.

53. $\frac{1}{y} = \frac{3}{15}$
$15 = 3y$
$5 = y$

54. $\frac{6}{8} = \frac{7}{b}$
$6b = 8(7)$
$6b = 56$
$b = \frac{28}{3}$

55. $\frac{20}{28} = \frac{m}{21}$
$20(21) = 28m$
$420 = 28m$
$15 = m$

56. $\frac{16}{7} = \frac{9}{s}$
$16s = 7(9)$
$16s = 63$
$s = \frac{63}{16}$

57. Find the coordinates of T.

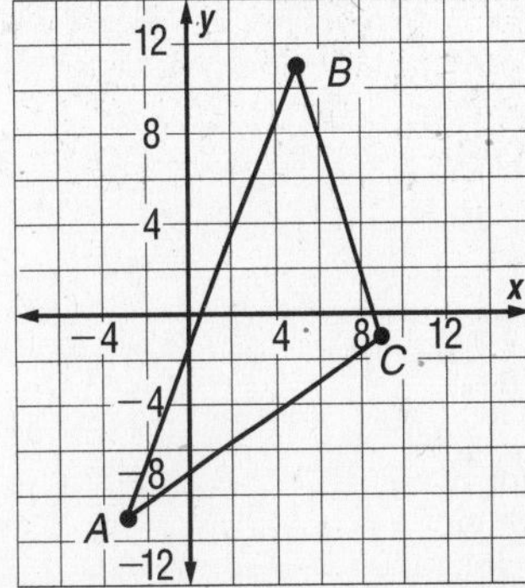

$\overline{AT}$ is a median from A to $\overline{BC}$, so T is the midpoint of $\overline{BC}$. Then T has coordinates

$\left(\frac{5+9}{2}, \frac{11+(-1)}{2}\right) = (7, 5)$.

The slope of $\overline{BC}$ is $\frac{-1-11}{9-5} = \frac{-12}{4}$ or -3.

The slope of $\overline{AT}$ is $\frac{5-(-9)}{7-(-3)} = \frac{14}{10}$ or $\frac{7}{5}$.

$-3\left(\frac{7}{5}\right) \neq -1$, so $\overline{AT}$ is not perpendicular to $\overline{BC}$. $\overline{AT}$ is not an altitude.

58. p: you are at least 54 inches tall
q: you may ride the roller coaster
Adam is 5 feet 8 inches tall, or $5 \cdot 12 + 8 =$ 68 inches, so he can ride the roller coaster by the Law of Detachment.

59. $\left(\frac{2+9}{2}, \frac{15+11}{2}\right) = \left(\frac{11}{2}, \frac{26}{2}\right)$
$= (5.5, 13)$

60. $\left(\frac{-4+2}{2}, \frac{4+(-12)}{2}\right) = \left(\frac{-2}{2}, \frac{-8}{2}\right)$
$= (-1, -4)$

61. $\left(\frac{0+7}{2}, \frac{8+(-13)}{2}\right) = \left(\frac{7}{2}, \frac{-5}{2}\right)$
$= (3.5, -2.5)$

Page 306 Practice Quiz 1

1. yes; $\angle A \cong \angle E$, $\angle B \cong \angle D$, $\angle 1 \cong \angle 3$, $\angle 2 \cong \angle 4$ and $\frac{AB}{ED} = \frac{BC}{DC} = \frac{AF}{EF} = \frac{FC}{FC} = 1$

2. no; $\frac{6.5}{6} \neq \frac{5}{5.5}$ and $\frac{6.5}{5.5} \neq \frac{5}{6}$

3. $\angle D \cong \angle B$ and $\angle AED \cong \angle CEB$, so $\triangle ADE \sim \triangle CBE$ by AA Similarity.

$\frac{DE}{BE} = \frac{AD}{CB}$

$\frac{3x-2}{6} = \frac{10}{15}$

$15(3x-2) = 6(10)$

$45x - 30 = 60$

$45x = 90$

$x = 2$

$\frac{AE}{CE} = \frac{AD}{CB}$

$\frac{AE}{12} = \frac{10}{15}$

$15(AE) = 12(10)$

$15(AE) = 120$

$AE = 8$

$DE = 3x - 2$
$= 3(2) - 2$ or 4

4. $\overline{ST} \parallel \overline{QP}$, so $\angle TSR \cong \angle PQR$. $\angle R \cong \angle R$, so $\triangle PQR \sim \triangle TSR$ by AA Similarity.

$\frac{PQ}{TS} = \frac{QR}{SR}$

$\frac{25}{TS} = \frac{10+5}{10}$

$25(10) = (10+5)(TS)$

$250 = 15(TS)$

$\frac{50}{3} = TS$

$(SR)^2 + (TR)^2 = (TS)^2$

$10^2 + x^2 = \left(\frac{50}{3}\right)^2$

$100 + x^2 = \frac{2500}{9}$

$x^2 = \frac{1600}{9}$

$x = \frac{40}{3}$

$\frac{PR}{TR} = \frac{QR}{SR}$

$\frac{PT+x}{x} = \frac{10+5}{10}$

$\frac{PT+\frac{40}{3}}{\frac{40}{3}} = \frac{15}{10}$

$10\left(PT + \frac{40}{3}\right) = \left(\frac{40}{3}\right)(15)$

$10(PT) + \frac{400}{3} = \frac{600}{3}$

$10(PT) = \frac{200}{3}$

$PT = \frac{20}{3}$

5. $\frac{\text{scale on map (cm)}}{\text{distance on map (mi)}} = \frac{\text{distance on map (cm)}}{\text{actual distance (mi)}}$

$\frac{1.5}{100} = \frac{29.2}{x}$

$1.5x = 100(29.2)$

$1.5x = 2920$

$x \approx 1947$

The cities are about 1947 miles apart.

6-4 Parallel Lines and Proportional Parts

Pages 311–312 Check for Understanding

1. Sample answer: If a line intersects two sides of a triangle and separates sides into corresponding segments of proportional lengths, then it is parallel to the third side.

2. Sample answer:

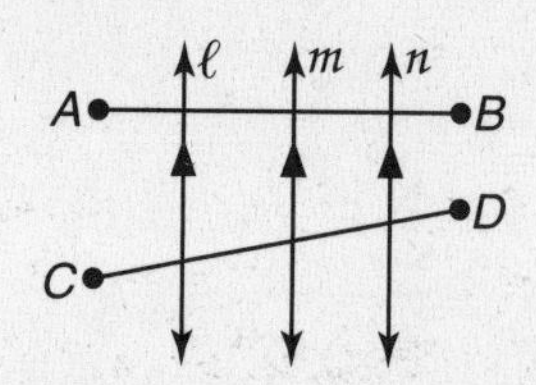

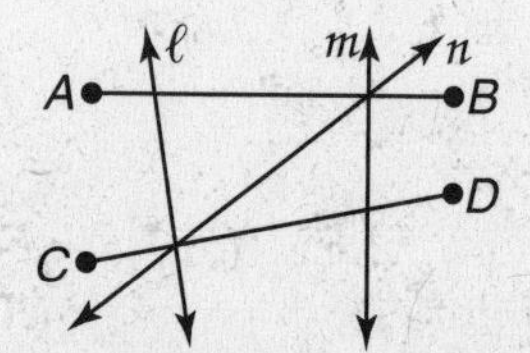

3. Given three or more parallel lines intersecting two transversals, Corollary 6.1 states that the parts of the transversals are proportional. Corollary 6.2 states that if the parts of one transversal are congruent, then the parts of every transversal are congruent.

4. $\overline{LW} \parallel \overline{TS}$, so by the Triangle Proportionality Theorem, $\frac{LT}{RL} = \frac{WS}{RW}$. Substitute the known measures.

$\frac{9-5}{5} = \frac{6}{RW}$

$4(RW) = 5(6)$

$4(RW) = 30$

$RW = 7.5$

5. $\overline{LW} \parallel \overline{TS}$, so by the Triangle Proportionality Theorem, $\frac{TL}{LR} = \frac{WS}{RW}$. Substitute the known measures.

$\frac{8-3}{3} = \frac{WS}{6}$
$6(5) = 3(WS)$
$30 = 3(WS)$
$10 = WS$

6. Use the Midpoint Formula to find the midpoints of $\overline{AB}$ and $\overline{AC}$.

$D\left(\frac{10+(-2)}{2}, \frac{0+6}{2}\right) = D(4, 3)$

$E\left(\frac{-4+(-2)}{2}, \frac{0+6}{2}\right) = E(-3, 3)$

7. If the slopes of $\overline{DE}$ and $\overline{BC}$ are equal, $\overline{DE} \parallel \overline{BC}$.

slope of $\overline{DE} = \frac{3-3}{-3-4}$ or 0

slope of $\overline{BC} = \frac{0-0}{10-(-4)}$ or 0

Because the slopes of $\overline{DE}$ and $\overline{BC}$ are equal, $\overline{DE} \parallel \overline{BC}$.

8. First, use the Distance Formula to find BC and DE.

$BC = \sqrt{[10-(-4)]^2 + (0-0)^2}$
$= \sqrt{196 + 0}$
$= 14$
$DE = \sqrt{(-3-4)^2 + (3-3)^2}$
$= \sqrt{49 + 0}$
$= 7$
$\frac{DE}{BC} = \frac{7}{14}$ or $\frac{1}{2}$
If $\frac{DE}{BC} = \frac{1}{2}$, then $DE = \frac{1}{2}BC$.

9. $MQ = MR + RQ$
$12.5 = 4.5 + RQ$
$8 = RQ$
$MP = MN + NP$
$25 = 9 + NP$
$16 = NP$

In order to show $\overline{RN} \parallel \overline{QP}$, we must show that $\frac{MR}{RQ} = \frac{MN}{NP}$.

$\frac{MR}{RQ} = \frac{4.5}{8}$ or $\frac{9}{16}$
$\frac{MN}{NP} = \frac{9}{16}$

Thus, $\frac{MR}{RQ} = \frac{MN}{NP} = \frac{9}{16}$. Since the sides have proportional lengths, $\overline{RN} \parallel \overline{QP}$.

10. In order to show $\overline{DB} \parallel \overline{AE}$, we must show that $\frac{ED}{DC} = \frac{AB}{BC}$.

$\frac{ED}{DC} = \frac{8}{20}$ or $\frac{2}{5}$
$\frac{AB}{BC} = \frac{12}{25}$

$\frac{ED}{DC} \neq \frac{AB}{BC}$, so the sides do not have proportional lengths and $\overline{DB}$ is not parallel to $\overline{AE}$.

11. To find x:
$20 - 5x = 2x + 6$
$20 = 7x + 6$
$14 = 7x$
$2 = x$

To find y:
$y = \frac{3}{5}y + 2$
$\frac{2}{5}y = 2$
$y = 5$

12. To find x:
$\frac{1}{3}x + 2 = \frac{2}{3}x - 4$
$-\frac{1}{3}x + 2 = -4$
$-\frac{1}{3}x = -6$
$x = 18$

To find y:
$5y = \frac{7}{3}y + 8$
$\frac{8}{3}y = 8$
$y = 3$

13. The streets form a triangle cut by a Walkthrough that is parallel to the bottom of the triangle. Use the Triangle Proportionality Theorem.

Talbot Rd. / Woodbury Ave.

$\frac{\text{Entrance to Walkthrough}}{\text{Walkthrough to Clay Rd.}} = \frac{\text{Entrance to Walkthrough}}{\text{Walkthrough to Clay Rd.}}$

$\frac{880}{1408} = \frac{x}{1760}$
$880(1760) = 1408x$
$1{,}548{,}800 = 1408x$
$1100 = x$

The distance from the entrance to the Walkthrough along Woodbury Avenue is 1100 yards.

Pages 312–315 Practice and Apply

14. $\overline{MN} \parallel \overline{YZ}$, so by the Triangle Proportionality Theorem, $\frac{MY}{XM} = \frac{NZ}{XN}$. Substitute the known measures.

$\frac{MY}{4} = \frac{9}{6}$
$6(MY) = 4(9)$
$6(MY) = 36$
$MY = 6$

$XY = XM + MY$
$= 4 + 6$ or 10

15. $\overline{MN} \parallel \overline{YZ}$, so by the Triangle Proportionality Theorem, $\frac{MY}{XM} = \frac{NZ}{XN}$. Substitute the known measures.

$\frac{10-2}{2} = \frac{t+1}{t-2}$
$8(t-2) = 2(t+1)$
$8t - 16 = 2t + 2$
$6t - 16 = 2$
$6t = 18$
$t = 3$

16. $\overline{DE} \parallel \overline{BC}$, so by the Triangle Proportionality Theorem, $\frac{DB}{AD} = \frac{EC}{AE}$. Substitute the known measures.

$\frac{24}{AD} = \frac{18}{3}$
$24(3) = 18(AD)$
$72 = 18(AD)$
$4 = AD$

17. $\overline{EB} \parallel \overline{DC}$, so by the Triangle Proportionality Theorem, $\frac{ED}{AE} = \frac{BC}{AB}$. Substitute the known measures.

$$\frac{2x-3}{3} = \frac{6}{2}$$
$$2(2x-3) = 3(6)$$
$$4x - 6 = 18$$
$$4x = 24$$
$$x = 6$$
$$ED = 2x - 3$$
$$= 2(6) - 3 \text{ or } 9$$

18. $\overline{BC} \parallel \overline{ED}$, so by the Triangle Proportionality Theorem, $\frac{BE}{AB} = \frac{CD}{AC}$. Substitute the known measures.

$$\frac{20}{16} = \frac{x+5}{x-3}$$
$$20(x-3) = 16(x+5)$$
$$20x - 60 = 16x + 80$$
$$4x - 60 = 80$$
$$4x = 140$$
$$x = 35$$
$$AC = x - 3$$
$$= 35 - 3 \text{ or } 32$$
$$CD = x + 5$$
$$= 35 + 5 \text{ or } 40$$

19. $\overline{BF} \parallel \overline{CE}$ and $\overline{AC} \parallel \overline{DF}$, so by the Triangle Proportionality Theorem, $\frac{BC}{AB} = \frac{FE}{AF}$ and $\frac{AF}{FE} = \frac{CD}{DE}$. Substitute the known measures.

$$\frac{x}{6} = \frac{x + \frac{10}{3}}{8}$$
$$8x = 6\left(x + \frac{10}{3}\right)$$
$$8x = 6x + 20$$
$$2x = 20$$
$$x = 10$$
$$\frac{8}{x + \frac{10}{3}} = \frac{y}{2y-3}$$
$$\frac{8}{10 + \frac{10}{3}} = \frac{y}{2y-3}$$
$$\frac{8}{\frac{40}{3}} = \frac{y}{2y-3}$$
$$8(2y-3) = \frac{40}{3}y$$
$$16y - 24 = \frac{40}{3}y$$
$$-24 = -\frac{8}{3}y$$
$$9 = y$$
$$BC = x$$
$$= 10$$
$$FE = x + \frac{10}{3}$$
$$= 10 + \frac{10}{3} \text{ or } 13\frac{1}{3}$$
$$CD = y$$
$$= 9$$
$$DE = 2y - 3$$
$$= 2(9) - 3 \text{ or } 15$$

20. In order to have $\overline{GJ} \parallel \overline{FK}$, it must be true that $\frac{HG}{GF} = \frac{HJ}{JK}$.

$$\frac{6}{12} = \frac{8}{x-4}$$
$$6(x-4) = 12(8)$$
$$6x - 24 = 96$$
$$6x = 120$$
$$x = 20$$

21. In order to have $\overline{GJ} \parallel \overline{FK}$, it must be true that $\frac{HG}{GF} = \frac{HJ}{JK}$.

$$\frac{x-4}{18} = \frac{x-5}{15}$$
$$15(x-4) = 18(x-5)$$
$$15x - 60 = 18x - 90$$
$$-3x - 60 = -90$$
$$-3x = -30$$
$$x = 10$$

22. In order to have $\overline{GJ} \parallel \overline{FK}$, it must be true that $\frac{HG}{GF} = \frac{HJ}{JK}$.

$$\frac{x+3.5}{21-(x+3.5)} = \frac{x-8.5}{7-(x-8.5)}$$
$$\frac{x+3.5}{17.5-x} = \frac{x-8.5}{15.5-x}$$
$$(x+3.5)(15.5-x) = (17.5-x)(x-8.5)$$
$$15.5x - x^2 + 54.25 - 3.5x = 17.5x - 148.75 - x^2 + 8.5x$$
$$-x^2 + 12x + 54.25 = -x^2 + 26x - 148.75$$
$$12x + 54.25 = 26x - 148.75$$
$$54.25 = 14x - 148.75$$
$$203 = 14x$$
$$14.5 = x$$

23. In order to show $\overline{QT} \parallel \overline{RS}$, we must show that $\frac{PQ}{QR} = \frac{PT}{TS}$.

$$\frac{PQ}{QR} = \frac{9}{30-9}$$
$$= \frac{9}{21} \text{ or } \frac{3}{7}$$
$$\frac{PT}{TS} = \frac{12}{18-12}$$
$$= \frac{12}{6} \text{ or } 2$$

Because $\frac{PQ}{QR} \neq \frac{PT}{TS}$, $\overline{QT}$ is not parallel to $\overline{RS}$.

24. In order to show $\overline{QT} \parallel \overline{RS}$, we must show that $\frac{PQ}{QR} = \frac{PT}{TS}$.

$$\frac{PQ}{QR} = \frac{65-22}{22}$$
$$= \frac{43}{22}$$

Let $TS = x$. Then $SP = 3x$ and $PT = 3x - x$ or $2x$.

$$\frac{PT}{TS} = \frac{2x}{x} = 2$$

Because $\frac{PQ}{QR} \neq \frac{PT}{TS}$, $\overline{QT}$ is not parallel to $\overline{RS}$.

25. In order to show $\overline{QT} \parallel \overline{RS}$, we must show that $\frac{PQ}{QR} = \frac{PT}{TS}$.

Let $RQ = x$. Then $PQ = \frac{x}{2}$.

$$\frac{PQ}{QR} = \frac{\frac{x}{2}}{x} \text{ or } \frac{1}{2}$$
$$\frac{PT}{TS} = \frac{12.9 - 8.6}{8.6}$$
$$= \frac{4.3}{8.6} \text{ or } \frac{1}{2}$$

Thus, $\frac{PQ}{QR} = \frac{PT}{TS} = \frac{1}{2}$. Since the sides have proportional lengths, $\overline{QT} \parallel \overline{RS}$.

26. In order to show $\overline{QT} \parallel \overline{RS}$, we must show that $\frac{PQ}{QR} = \frac{PT}{TS}$.

$\frac{PQ}{QR} = \frac{34.88}{18.32}$ or $\frac{436}{229}$

$\frac{PT}{TS} = \frac{33.25 - 11.45}{11.45}$

$= \frac{21.8}{11.45}$ or $\frac{436}{229}$

Thus, $\frac{PQ}{QR} = \frac{PT}{TS} = \frac{436}{229}$. Since the sides have proportional lengths, $\overline{QT} \parallel \overline{RS}$.

27. $\overline{DE}$ is a midsegment of $\triangle ABC$ and $\overline{DE} \parallel \overline{BC}$, so by the Triangle Midsegment Theorem, $DE = \frac{1}{2}BC$. Then $BC = 2DE$.

$DE = \sqrt{(4-1)^2 + (3-1)^2}$
$= \sqrt{9+4}$
$= \sqrt{13}$

$BC = 2DE$
$= 2\sqrt{13}$
$= \sqrt{4 \cdot 13}$ or $\sqrt{52}$

28. If the slopes of $\overline{WM}$ and $\overline{TS}$ are equal, $\overline{WM} \parallel \overline{TS}$.

slope of $WM = \frac{12-14}{5-3}$ or -1

slope of $TS = \frac{20-26}{17-11}$ or -1

Because the slopes of $\overline{WM}$ and $\overline{TS}$ are equal, $\overline{WM} \parallel \overline{TS}$.

$\overline{WM}$ is a midsegment of $\triangle RST$ if W is the midpoint of $\overline{RT}$ and M is the midpoint of RS.

The midpoint of $\overline{RT}$ is $\left(\frac{-1+11}{2}, \frac{8+26}{2}\right) = (5, 17)$.

These are not the coordinates of W.

The midpoint of $\overline{RS}$ is $\left(\frac{-1+17}{2}, \frac{8+20}{2}\right) = (8, 14)$.

These are not the coordinates of M.

$\overline{WM}$ is not a midsegment because W and M are not midpoints of their respective sides.

29. $\overline{DE}$ is a midsegment of $\triangle ABC$. Use the Midpoint Formula to find the coordinates of D and E.

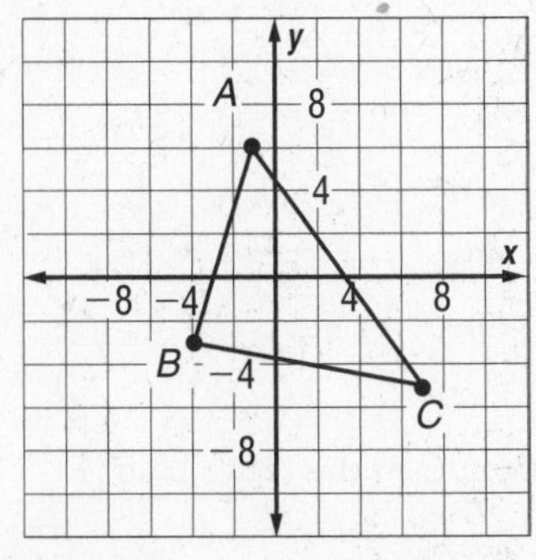

$D\left(\frac{-1+7}{2}, \frac{6+(-5)}{2}\right) = D\left(3, \frac{1}{2}\right)$

$E\left(\frac{-4+7}{2}, \frac{-3+(-5)}{2}\right) = E\left(\frac{3}{2}, -4\right)$

Find the slopes of $\overline{DE}$ and $\overline{AB}$.

slope of $\overline{DE} = \frac{-4 - \frac{1}{2}}{\frac{3}{2} - 3}$ or 3

slope of $\overline{AB} = \frac{-3-6}{-4-(-1)}$ or 3

Both $\overline{DE}$ and $\overline{AB}$ have slope 3, so $\overline{DE}$ is parallel to $\overline{AB}$.

30. Use the Distance Formula to find DE and AB.

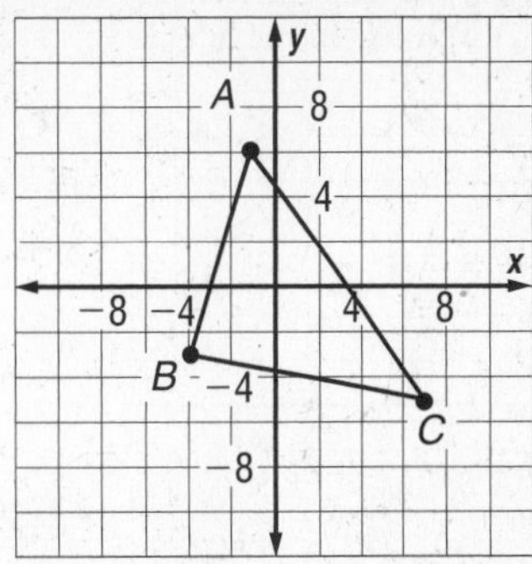

$DE = \sqrt{\left(\frac{3}{2} - 3\right)^2 + \left(-4 - \frac{1}{2}\right)^2}$
$= \sqrt{\frac{9}{4} + \frac{81}{4}}$
$= \frac{3}{2}\sqrt{10}$

$AB = \sqrt{[-4-(-1)]^2 + (-3-6)^2}$
$= \sqrt{9+81}$
$= 3\sqrt{10}$

So, $\frac{3}{2}\sqrt{10} = \frac{1}{2}(3\sqrt{10})$ and thus $DE = \frac{1}{2}AB$.

31.

Graph $\overleftrightarrow{AB}$. We can find segments of $\overleftrightarrow{AB}$ with lengths in a ratio of 2 to 1 by considering a second line and parallel lines that intersect this line and $\overleftrightarrow{AB}$.

Graph $C(0, 12)$ and $D(0, 0)$ and lines $\overleftrightarrow{CA}$ and $\overleftrightarrow{DB}$. $\overleftrightarrow{CA}$ and $\overleftrightarrow{DB}$ are horizontal lines and are parallel lines intersecting transversals $\overleftrightarrow{CD}$ (the y-axis) and $\overleftrightarrow{AB}$. We can find P by finding a third parallel line intersecting $\overleftrightarrow{CD}$ and $\overleftrightarrow{AB}$ so that this line separates $\overleftrightarrow{CD}$ into two parts with a ratio of 2 to 1. CD is 12 units, so if a horizontal line intersects $\overleftrightarrow{CD}$ at (0, 4) or (0, 8) then this line separates $\overleftrightarrow{CD}$ into two parts with a ratio of 2 to 1. These horizontal lines intersect $\overleftrightarrow{AB}$ at (4, 4) and (3, 8). These points cut off $\overleftrightarrow{AB}$ into parts with a ratio of 2 to 1, so P could have coordinates (4, 4) or (3, 8).

32.

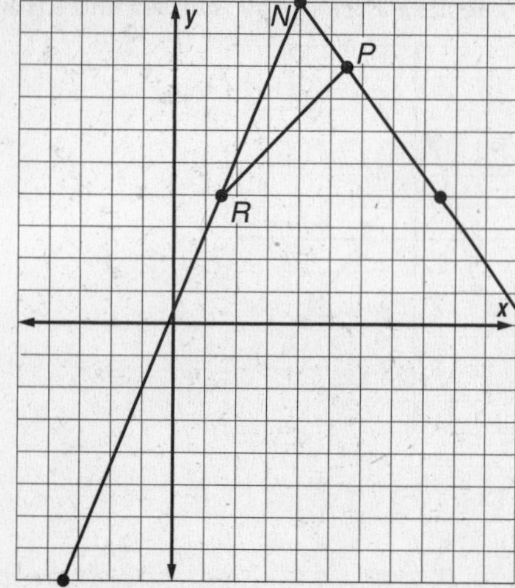

L is on $\overrightarrow{PN}$ and M is on $\overrightarrow{RN}$ so graph N, P, and R and extend $\overline{PN}$ and $\overline{RN}$ so that $\overline{PR}$ divides $\overline{NL}$ and $\overline{MN}$. $\frac{LP}{PN} = \frac{2}{1}$, and $\overline{PR}$ divides $\overline{NL}$ and $\overline{MN}$ proportionally so $\frac{MR}{RN} = \frac{2}{1}$. Then $LP = 2(PN)$ and $MR = 2(RN)$. Starting at $N(8, 20)$, move to $P(11, 16)$ by moving down 4 units and then right 3 units. Locate L by moving from P down 8 units and then right 6 units. The coordinates of L are $(17, 8)$. Now starting at $N(8, 20)$, move to $R(3, 8)$ by moving down 12 units and then left 5 units. Locate M by moving from R down 24 units and then left 10 units. The coordinates of M are $(-7, -16)$.

Verify that $LP = 2(PN)$ and $MR = 2(RN)$.

$PN = \sqrt{(11-8)^2 + (16-20)^2}$
$= \sqrt{9 + 16}$
$= \sqrt{25}$ or 5

$LP = \sqrt{(11-17)^2 + (16-8)^2}$
$= \sqrt{36 + 64}$
$= \sqrt{100}$ or 10

So, $LP = 2(PN)$.

$RN = \sqrt{(8-3)^2 + (20-8)^2}$
$= \sqrt{25 + 144}$
$= \sqrt{169}$ or 13

$MR = \sqrt{[3-(-7)]^2 + [8-(-16)]^2}$
$= \sqrt{100 + 576}$
$= \sqrt{676}$ or 26

So, $MR = 2(RN)$.

33. To find x:

The sides of the large triangle are cut in equal parts by the segment whose length is labeled $x + 2$, so this segment is a midsegment and its length is half the length of the segment whose length is labeled $\frac{5}{3}x + 11$.

$x + 2 = \frac{1}{2}\left(\frac{5}{3}x + 11\right)$
$x + 2 = \frac{5}{6}x + \frac{11}{2}$
$\frac{1}{6}x + 2 = \frac{11}{2}$
$\frac{1}{6}x = \frac{7}{2}$
$x = 21$

To find y:

$2y + 6 = 3y - 9$
$6 = y - 9$
$15 = y$

34. To find x:

$2x + 3 = 6 - x$
$3x + 3 = 6$
$3x = 3$
$x = 1$

To find y:

$\frac{4}{3}y + 1 = 2y$
$1 = \frac{2}{3}y$
$\frac{3}{2} = y$

35. The poles form parallel line segments and the wires are transversals cutting through the ends of the parallel segments.

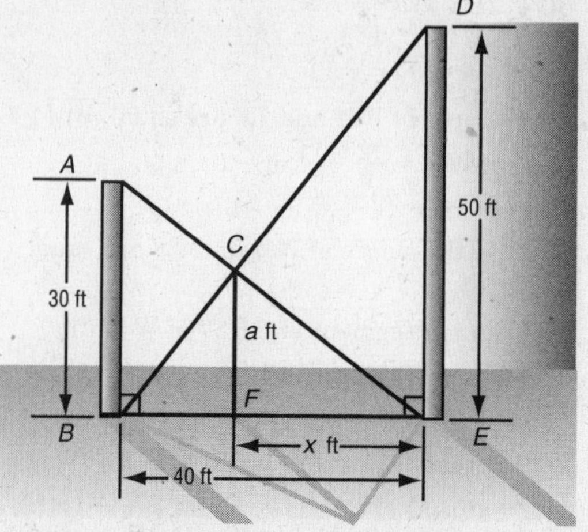

$\angle CFE$ is a right angle, so $\angle CFE \cong \angle ABE$ and $\angle CFB \cong \angle DEB$. Also, $\angle CEF \cong \angle AEB$, so $\triangle AEB \sim \triangle CEF$ by AA Similarity. $\angle CBF \cong \angle DBE$, so $\triangle CBF \sim \triangle DBE$ by AA Similarity.

$\frac{x}{40} = \frac{a}{30}$
$30x = 40a$
$\frac{30}{40}x = a$
$\frac{a}{50} = \frac{40 - x}{40}$
$40a = 50(40 - x)$
$40\left(\frac{30}{40}x\right) = 2000 - 50x$
$30x = 2000 - 50x$
$80x = 2000$
$x = 25$

So, the distance from C to the taller pole is 25 feet.

36.
$\frac{x}{a} = \frac{40}{30}$
$30x = 40a$
$30(25) = 40a$
$750 = 40a$
$18.75 = a$

The coupling is 18.75 feet above the ground.

37. Let y represent the length of the wire from the smaller pole.

$$30^2 + 40^2 = y^2$$
$$900 + 1600 = y^2$$
$$2500 = y^2$$
$$50 = y$$

Let z represent the length of the wire from the top of the smaller pole to the coupling.

$$\frac{z}{50 - z} = \frac{40 - 25}{25}$$
$$25z = 15(50 - z)$$
$$25z = 750 - 15z$$
$$40z = 750$$
$$z = 18.75$$

The coupling is 18.75 feet down the wire from the top of the smaller pole.

38. Given: $\frac{DB}{AD} = \frac{EC}{AE}$

Prove: $\overline{DE} \parallel \overline{BC}$

A, D, E, B, C

Proof:

Statements	Reasons
1. $\frac{DB}{AD} = \frac{EC}{AE}$	1. Given
2. $\frac{AD}{AD} + \frac{DB}{AD} = \frac{AE}{AE} + \frac{EC}{AE}$	2. Addition Prop.
3. $\frac{AD + DB}{AD} = \frac{AE + EC}{AE}$	3. Substitution
4. $AB = AD + DB$,	4. Segment Addition Postulate
5. $\frac{AB}{AD} = \frac{AC}{AE}$	5. Substitution
6. $\angle A \cong \angle A$	6. Reflexive Prop.
7. $\triangle ADE \sim \triangle ABC$	7. SAS Similarity
8. $\angle ADE \cong \angle ABC$	8. Def. of ~ polygons
9. $\overline{DE} \parallel \overline{BC}$	9. If corr. ∠s are ≅, then the lines are ∥.

39. Given: D is the midpoint of $\overline{AB}$. E is the midpoint of $\overline{AC}$.

Prove: $\overline{DE} \parallel \overline{BC}$; $DE = \frac{1}{2}BC$

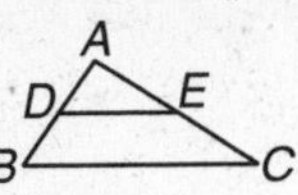

Proof:

Statements	Reasons
1. D is the midpoint of $\overline{AB}$. E is the midpoint of $\overline{AC}$.	1. Given
2. $\overline{AD} \cong \overline{DB}$, $\overline{AE} \cong \overline{EC}$	2. Midpoint Theorem
3. $AD = DB$, $AE = EC$	3. Def. of ≅ segments
4. $AB = AD + DB$, $AC = AE + EC$	4. Segment Addition Postulate
5. $AB = AD + AD$, $AC = AE + AE$	5. Substitution
6. $AB = 2AD$, $AC = 2AE$	6. Substitution
7. $\frac{AB}{AD} = 2$, $\frac{AC}{AE} = 2$	7. Division Prop.
8. $\frac{AB}{AD} = \frac{AC}{AE}$	8. Transitive Prop.
9. $\angle A \cong \angle A$	9. Reflexive Prop.
10. $\triangle ADE \sim \triangle ABC$	10. SAS Similarity
11. $\angle ADE \cong \angle ABC$	11. Def. of ~ polygons
12. $\overline{DE} \parallel \overline{BC}$	12. If corr. ∠s are ≅, then the lines are parallel.
13. $\frac{BC}{DE} = \frac{AB}{AD}$	13. Def. of ~ polygons
14. $\frac{BC}{DE} = 2$	14. Substitution Prop.
15. $2DE = BC$	15. Mult. Prop.
16. $DE = \frac{1}{2}BC$	16. Division Prop.

40. Figure is not shown actual size.

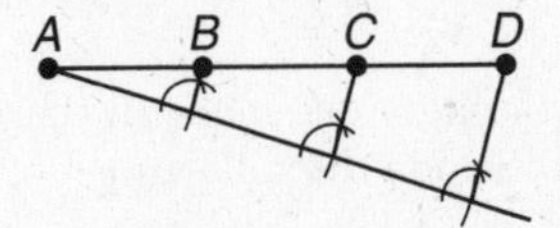

41.

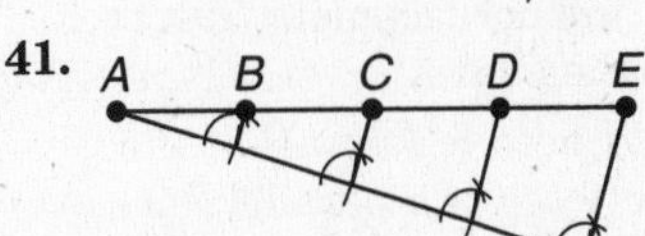

42.

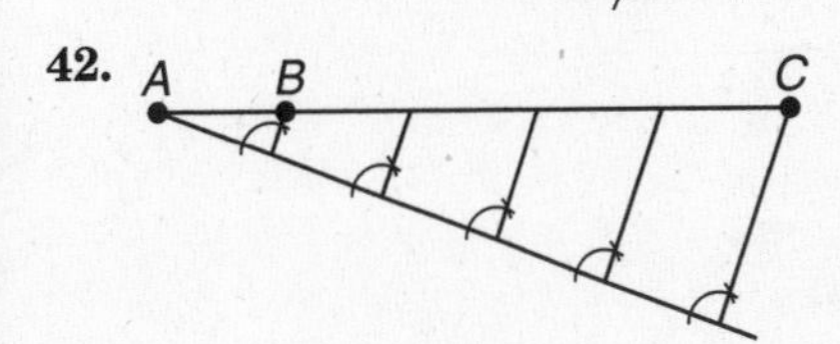

43. The total length of the lots along Lake Creek Drive is 20 + 22 + 25 + 18 + 28 or 113 meters. The lines dividing the lots are perpendicular to Lake Creek Drive, so they are all parallel. Write proportions to solve for each variable.

$$\frac{\text{total lake frontage}}{\text{total length of lots along street}} = \frac{\text{individual lake frontage}}{\text{individual length along street}}$$

$\frac{135.6}{113} = \frac{u}{20}$
$135.6(20) = 113u$
$2712 = 113u$
$24 = u$

$\frac{135.6}{113} = \frac{w}{22}$
$135.6(22) = 113w$
$2983.2 = 113w$
$26.4 = w$

$\frac{135.6}{113} = \frac{x}{25}$
$135.6(25) = 113x$
$3390 = 113x$
$30 = x$

$\frac{135.6}{113} = \frac{y}{18}$
$135.6(18) = 113y$
$2440.8 = 113y$
$21.6 = y$

$\frac{135.6}{113} = \frac{z}{28}$
$135.6(28) = 113z$
$3796.8 = 113z$
$33.6 = z$

44. Use the Triangle Proportionality Theorem in $\triangle DCA$.
$\overline{BG} \parallel \overline{AD}$, so $\frac{AB}{BC} = \frac{DG}{GC}$. Also, in $\triangle DCF$, $\frac{DG}{GC} = \frac{DE}{EF}$. Using the Transitive Property of Equality, $\frac{AB}{BC} = \frac{DE}{EF}$.

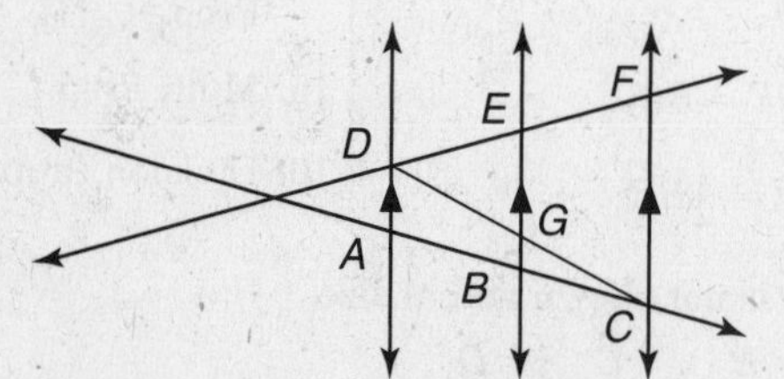

45. Sample answer: City planners use maps in their work. Answers should include the following.

- City planners need to know geometry facts when developing zoning laws.
- A city planner would need to know that the shortest distance between two parallel lines is the perpendicular distance.

46. B; $\frac{12}{18} = \frac{x}{42 - x}$
$12(42 - x) = 18x$
$504 - 12x = 18x$
$504 = 30x$
$16.8 = x$

47. $\frac{a + b}{2} = 18$
$a + b = 36$
$a = 36 - b$
$\frac{a}{b} = \frac{5}{4}$
$\frac{36 - b}{b} = \frac{5}{4}$
$4(36 - b) = 5b$
$144 - 4b = 5b$
$144 = 9b$
$16 = b$
$a = 36 - b$
$= 36 - 16$ or 20
$a - b = 20 - 16$ or 4

48a. See students' work. $\overline{EF} \parallel \overline{GH}$, $\overline{FG} \parallel \overline{EH}$, $\overline{EF} \cong \overline{GH}$, $\overline{FG} \cong \overline{EH}$

48b. No, there will be an odd number of sides so it is not possible to pair opposite sides.

Page 315 Maintain Your Skills

49. Yes, by AA Similarity; the parallel lines determine congruent corresponding angles.

50. Yes, by SSS Similarity; $\frac{6}{8} = \frac{9}{12} = \frac{12}{16} = \frac{3}{4}$

51. No; corresponding angles are not congruent. The third angles have measure $180 - (72 + 66) = 42$ and $180 - (66 + 38) = 76$.

52. $\frac{x}{20} = \frac{7}{14}$
$14x = 20(7)$
$14x = 140$
$x = 10$
$\frac{y}{9} = \frac{14}{7}$
$7y = 126$
$y = 18$

53. $\frac{x}{18} = \frac{14}{21}$
$21x = 252$
$x = 12$
$\frac{y}{9} = \frac{14}{21}$
$21y = 126$
$y = 6$

54. If one side of a triangle is longer than another side, then the angle opposite the longer side has a greater measure than the angle opposite the shorter side. In $\triangle ADB$, $\angle ADB$ is opposite a side whose measure is 15, and $\angle ABD$ is opposite a side whose measure is 12. $15 > 12$, so $m\angle ADB > m\angle ABD$.

55. If one side of a triangle is longer than another side, then the angle opposite the longer side has a greater measure than the angle opposite the shorter side. In $\triangle ADB$, $\angle ABD$ is opposite a side whose measure is 12, and $\angle BAD$ is opposite a side whose measure is 9. $12 > 9$, so $m\angle ABD > m\angle BAD$.

56. If one side of a triangle is longer than another side, then the angle opposite the longer side has a greater measure than the angle opposite the shorter side. In $\triangle BDC$, $\angle BCD$ is opposite a side whose measure is 9, and $\angle CDB$ is opposite a side whose measure is 13. $9 < 13$, so $m\angle BCD < m\angle CDB$.

57. If one side of a triangle is longer than another side, then the angle opposite the longer side has a greater measure than the angle opposite the shorter side. In $\triangle BDC$, $\angle CBD$ is opposite a side whose measure is 10, and $\angle BCD$ is opposite a side whose measure is 9. $10 > 9$, so $m\angle CBD > m\angle BCD$.

58. There are 6 equilateral triangles.

59. There are 18 obtuse triangles in the figure, 3 in each equilateral triangle.

60. True; the hypothesis is false, so we cannot say that the statement is false.

61. False; the hypothesis is true, but the conclusion is false.

62. True; the hypothesis is false, so we cannot say that the statement is false.

63. True; the hypothesis is true and the conclusion is true.

64. $\angle A \cong \angle D$, $\angle B \cong \angle E$, $\angle C \cong \angle F$, $\overline{AB} \cong \overline{DE}$, $\overline{BC} \cong \overline{EF}$, $\overline{AC} \cong \overline{DF}$

65. $\angle R \cong \angle X$, $\angle S \cong \angle Y$, $\angle T \cong \angle Z$, $\overline{RS} \cong \overline{XY}$, $\overline{ST} \cong \overline{YZ}$, $\overline{RT} \cong \overline{XZ}$

66. $\angle P \cong \angle K$, $\angle Q \cong \angle L$, $\angle R \cong \angle M$, $\overline{PQ} \cong \overline{KL}$, $\overline{QR} \cong \overline{LM}$, $\overline{PR} \cong \overline{KM}$

6-5 Parts of Similar Triangles

Page 319 Check for Understanding

1. $\triangle ABC \sim \triangle MNQ$ and $\overline{AD}$ and $\overline{MR}$ are altitudes, angle bisectors, or medians.

2. Sample answer: The perimeters are in the same proportion as the side measures, which is $\frac{24}{36}$ or $\frac{2}{3}$. So if the smaller triangle has side lengths 6, 8, and 10 (so that its perimeter is $6 + 8 + 10 = 24$), then the larger triangle has side lengths $\frac{3}{2}(6)$, $\frac{3}{2}(8)$, and $\frac{3}{2}(10)$ or 9, 12, and 15.

3. Let x represent the perimeter of $\triangle DEF$. The perimeter of $\triangle ABC = 5 + 6 + 7$ or 18.

$$\frac{DE}{AB} = \frac{\text{perimeter of } \triangle DEF}{\text{perimeter of } \triangle ABC}$$
$$\frac{3}{5} = \frac{x}{18}$$
$$54 = 5x$$
$$10.8 = x$$

The perimeter of $\triangle DEF$ is 10.8 units.

4. Let x represent the perimeter of $\triangle WZX$.

$$\frac{WX}{ST} = \frac{\text{perimeter of } \triangle WZX}{\text{perimeter of } \triangle SRT}$$
$$\frac{5}{6} = \frac{x}{15}$$
$$75 = 6x$$
$$12.5 = x$$

The perimeter of $\triangle WZX$ is 12.5 units.

5. $\frac{x}{12} = \frac{6.5}{13}$

$13x = 78$

$x = 6$

6. $\frac{20}{x} = \frac{16}{12}$

$240 = 16x$

$15 = x$

7. $\frac{x}{9} = \frac{18}{24}$

$24x = 162$

$x = 6.75$

8. **Given:** $\triangle ABC \sim \triangle DEF$ and $\frac{AB}{DE} = \frac{m}{n}$

Prove: $\frac{\text{perimeter of } \triangle ABC}{\text{perimeter of } \triangle DEF} = \frac{m}{n}$

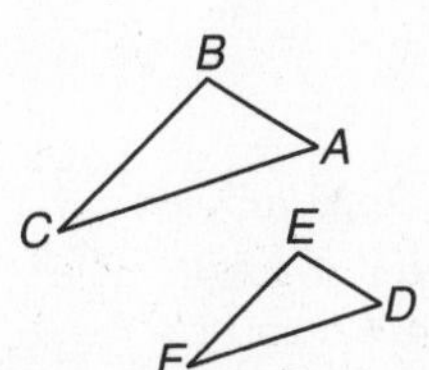

Proof: Because $\triangle ABC \sim \triangle DEF$, $\frac{AB}{DE} = \frac{BC}{EF} = \frac{AC}{DF}$. So $\frac{BC}{EF} = \frac{AC}{DF} = \frac{m}{n}$. Cross products yield $AB = DE\left(\frac{m}{n}\right)$, $BC = EF\left(\frac{m}{n}\right)$, and $AC = DF\left(\frac{m}{n}\right)$. Using substitution, the perimeter of $\triangle ABC = DE\left(\frac{m}{n}\right) + EF\left(\frac{m}{n}\right) + DF\left(\frac{m}{n}\right)$, or $\frac{m}{n}(DE + EF + DF)$. The ratio of the two perimeters

$$= \frac{\frac{m}{n}(DE + EF + DF)}{DE + EF + DF} \text{ or } \frac{m}{n}.$$

9. $$\frac{\text{height of Tamika}}{\text{height of image}} = \frac{\text{distance from camera}}{\text{length of camera}}$$
$$\frac{165}{5} = \frac{x}{10}$$
$$1650 = 5x$$
$$330 = x$$

Tamika should be 330 cm or 3.3 m from the camera.

Pages 320–322 Practice and Apply

10. Let x represent the perimeter of $\triangle BCD$. The perimeter of $\triangle FDE = 4 + 5 + 8$ or 17.

$$\frac{CD}{DE} = \frac{\text{perimeter of } \triangle BCD}{\text{perimeter of } \triangle FDE}$$
$$\frac{12}{8} = \frac{x}{17}$$
$$204 = 8x$$
$$25.5 = x$$

The perimeter of $\triangle BCD$ is 25.5 units.

11. Let x represent the perimeter of $\triangle ADF$. The perimeter of $\triangle BCE = 24 + 12 + 18$ or 54.

$\frac{DF}{CE} = \frac{\text{perimeter of } \triangle ADF}{\text{perimeter of } \triangle BCE}$

$\frac{21}{18} = \frac{x}{54}$

$1134 = 18x$

$63 = x$

The perimeter of $\triangle ADF$ is 63 units.

12. Let x represent the perimeter of $\triangle CBH$. The perimeter of $\triangle FEH = 11 + 6 + 10$ or 27. $ADEG$ is a parallelogram, so $\overline{AD} \parallel \overline{GE}$ and $\angle BCH \cong \angle HFE$. so $\overline{CH}$ corresponds to $\overline{FH}$.

$\frac{CH}{FH} = \frac{\text{perimeter of } \triangle CBH}{\text{perimeter of } \triangle FEH}$

$\frac{7}{10} = \frac{x}{27}$

$189 = 10x$

$18.9 = x$

The perimeter of $\triangle CBH$ is 18.9 units.

13. Let x represent the perimeter of $\triangle DEF$.

$\frac{DF}{FC} = \frac{\text{perimeter of } \triangle DEF}{\text{perimeter of } \triangle CBF}$

$\frac{6}{8} = \frac{x}{27}$

$162 = 8x$

$20.25 = x$

The perimeter of $\triangle DEF$ is 20.25 units.

14. Let x represent the perimeter of $\triangle ABC$. The perimeter of $\triangle CBD = 3 + 4 + 5$ or 12.

$\frac{CB}{DB} = \frac{\text{perimeter of } \triangle ABC}{\text{perimeter of } \triangle CBD}$

$\frac{5}{3} = \frac{x}{12}$

$60 = 3x$

$20 = x$

The perimeter of $\triangle ABC$ is 20 units.

15. Let x represent the perimeter of $\triangle ABC$. $\triangle CBD$ is a right triangle, so use the Pythagorean Theorem to find BD.

$(CD)^2 + (BD)^2 = (CB)^2$

$12^2 + (BD)^2 = 31.2^2$

$144 + (BD)^2 = 973.44$

$(BD)^2 = 829.44$

$BD = 28.8$

The perimeter of $\triangle CBD = 31.2 + 28.8 + 12$ or 72.

$\frac{AB}{CB} = \frac{\text{perimeter of } \triangle ABC}{\text{perimeter of } \triangle CBD}$

$\frac{5 + 28.8}{31.2} = \frac{x}{72}$

$33.8(72) = 31.2x$

$2433.6 = 31.2x$

$78 = x$

The perimeter of $\triangle ABC$ is 78 units.

16. The original picture is similar to the enlarged picture, so the dimensions and perimeters are proportional. The perimeter of the original picture is $2(18) + 2(24)$ or 84 cm. The perimeter of the enlarged picture is $0.30(84) + 84$ or 109.2 cm. The enlarged picture will take approximately 109.2 cm of cord, so 110 cm will be enough.

17. Yes, the perimeters are in the same proportion as the sides, $\frac{300}{600}$ or $\frac{1}{2}$.

18. Let x represent EG.

$\frac{AD}{EH} = \frac{AC}{EG}$

$\frac{15}{7.5} = \frac{17}{x}$

$15x = 127.5$

$x = 8.5$

Thus, $EG = 8.5$.

19. Let x represent EH.

$\frac{BG}{EH} = \frac{BC}{EF}$

$\frac{3}{x} = \frac{4+2}{1+2}$

$\frac{3}{x} = \frac{6}{3}$

$9 = 6x$

$\frac{3}{2} = x$

Thus, $EH = \frac{3}{2}$.

20. $\frac{FB}{SA} = \frac{FG}{ST}$

$\frac{7-x}{2} = \frac{5}{x}$

$7x - x^2 = 10$

$0 = x^2 - 7x + 10$

$0 = (x-5)(x-2)$

$x - 5 = 0$ or $x - 2 = 0$

$x = 5$ $\quad x = 2$

We must choose $x = 5$ because otherwise $FB = 5$ and so $FB = FG$, but the hypotenuse must be longer than either leg. Thus, $x = 5$ and $FB = 2$.

21. $\frac{DG}{JM} = \frac{DC}{JL}$

$\frac{2}{x} = \frac{6-x}{4}$

$8 = 6x - x^2$

$x^2 - 6x + 8 = 0$

$(x-4)(x-2) = 0$

$x - 4 = 0$ or $x - 2 = 0$

$x = 4$ $\quad x = 2$

We must choose $x = 2$ because otherwise $JM = 4$ and so $JM = JL$, but the hypotenuse must be longer than either leg. Thus, $x = 2$ and $DC = 4$.

22. $\frac{12}{32} = \frac{x-5}{2x-3}$

$24x - 36 = 32x - 160$

$-36 = 8x - 160$

$124 = 8x$

$15.5 = x$

23. $\frac{11}{14} = \frac{20-x}{x}$

$11x = 280 - 14x$

$25x = 280$

$x = 11\frac{1}{5}$

24. $\frac{x+3}{6} = \frac{x}{4}$

$4x + 12 = 6x$

$12 = 2x$

$6 = x$

25. $\frac{x}{8} = \frac{9}{2x}$

$2x^2 = 72$

$x^2 = 36$

$x = \pm 6$

x represents a length, which must be positive. So, $x = 6$.

26. $\overline{TA}$ and $\overline{WB}$ are medians, so $RA = AS$ and $UB = BV$. Then $RS = 2(RA)$ and $UV = 2(UB)$.

$\frac{TA}{WB} = \frac{RS}{UV}$

$\frac{8}{3x-6} = \frac{2(3)}{2(x+2)}$

$\frac{8}{3x-6} = \frac{6}{2x+4}$

$16x + 32 = 18x - 36$

$32 = 2x - 36$

$68 = 2x$

$34 = x$

$UB = x + 2$

$= 34 + 2$ or 36

27. $\overline{BF}$ bisects $\angle ABC$, so by the Angle Bisector Theorem, $\frac{AF}{CF} = \frac{BA}{BC}$. Let x represent CF. Then $AF = 9 - x$.

$\frac{9-x}{x} = \frac{6}{7.5}$

$67.5 - 7.5x = 6x$

$67.5 = 13.5x$

$5 = x$

Thus, $CF = 5$.

$\overline{AC} \parallel \overline{ED}$, so $\angle BED \cong \angle BFC$. $\angle FBC \cong \angle EBD$, so by AA Similarity, $\triangle EBD \sim \triangle FBC$. Let y represent BD.

$\frac{BD}{BC} = \frac{ED}{FC}$

$\frac{y}{7.5} = \frac{9}{5}$

$5y = 67.5$

$y = 13.5$

Thus, $BD = 13.5$

28. $\frac{\text{height of person}}{\text{height of image}} = \frac{\text{distance from camera}}{\text{length of camera}}$

$\frac{x}{12} = \frac{7 \cdot 12}{15}$

$15x = 1008$

$x = 67.2$

The person is 67.2 inches or about 5 feet 7 inches tall.

29. $xy = z^2$; $\triangle ACD \sim \triangle CBD$ by AA Similarity. Thus, $\frac{CD}{BD} = \frac{AD}{CD}$ or $\frac{z}{y} = \frac{x}{z}$. The cross products yield $xy = z^2$.

30. Given: $\triangle ABC \sim \triangle PQR$

Prove: $\frac{BD}{QS} = \frac{BA}{QP}$

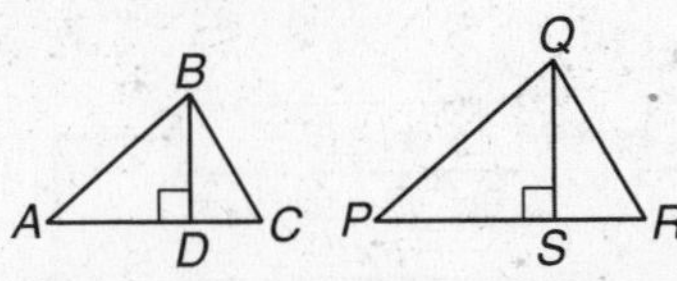

Proof: Since $\triangle ABC \sim \triangle PQR$, $\angle A \cong \angle P$. $\angle BDA \cong \angle QSP$ because they are both right angles created by the altitude drawn to the opposite side and all right angles are congruent. Thus $\triangle ABD \sim \triangle PQS$ by AA Similarity and $\frac{BD}{QS} = \frac{BA}{QP}$ by the definition of similar polygons.

31. Given: $\triangle ABC \sim \triangle RST$, $\overline{AD}$ is a median of $\triangle ABC$. $\overline{RU}$ is a median of $\triangle RST$.

Prove: $\frac{AD}{RU} = \frac{AB}{RS}$

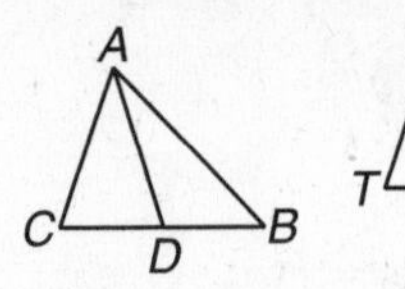

Proof:

Statements	Reasons
1. $\triangle ABC \sim \triangle RST$ $\overline{AD}$ is a median of $\triangle ABC$. $\overline{RU}$ is a median of $\triangle RST$.	1. Given
2. $CD = DB$; $TU = US$	2. Def. of median
3. $\frac{AB}{RS} = \frac{CB}{TS}$	3. Def. of ~ polygons
4. $CB = CD + DB$; $TS = TU + US$	4. Segment Addition Postulate
5. $\frac{AB}{RS} = \frac{CD + DB}{TU + US}$	5. Substitution
6. $\frac{AB}{RS} = \frac{DB + DB}{US + US}$ or $\frac{2(DB)}{2(US)}$	6. Substitution
7. $\frac{AB}{RS} = \frac{DB}{US}$	7. Substitution
8. $\angle B \cong \angle S$	8. Def. of ~ polygons
9. $\triangle ABD \sim \triangle RSU$	9. SAS Similarity
10. $\frac{AD}{RU} = \frac{AB}{RS}$	10. Def. of ~ polygons

32. Given: $\overline{CD}$ bisects $\angle ACB$. By construction $\overline{AE} \parallel \overline{CD}$.

Prove: $\frac{AD}{DB} = \frac{AC}{BC}$

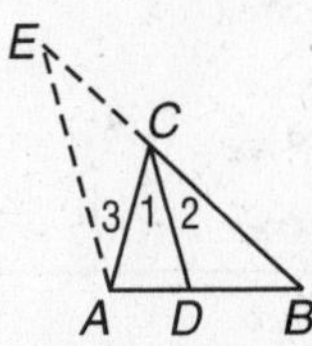

Proof:

Statements	Reasons
1. $\overline{CD}$ bisects $\angle ACB$. By construction, $\overline{AE} \parallel \overline{CD}$.	1. Given
2. $\frac{AD}{DB} = \frac{EC}{BC}$	2. Triangle Proportionality Theorem
3. $\angle 1 \cong \angle 2$	3. Definition of Angle Bisector
4. $\angle 3 \cong \angle 1$	4. Alternate Interior Angle Theorem
5. $\angle 2 \cong \angle E$	5. Corresponding Angle Postulate
6. $\angle 3 \cong \angle E$	6. Transitive Prop.
7. $\overline{EC} \cong \overline{AC}$	7. Isosceles $\triangle$ Th.
8. $EC = AC$	8. Def. of congruent segments
9. $\frac{AD}{DB} = \frac{AC}{BC}$	9. Substitution

33. Given: $\triangle ABC \sim \triangle PQR$, $\overline{BD}$ is an altitude of $\triangle ABC$. $\overline{QS}$ is an altitude of $\triangle PQR$.

Prove: $\frac{QP}{BA} = \frac{QS}{BD}$

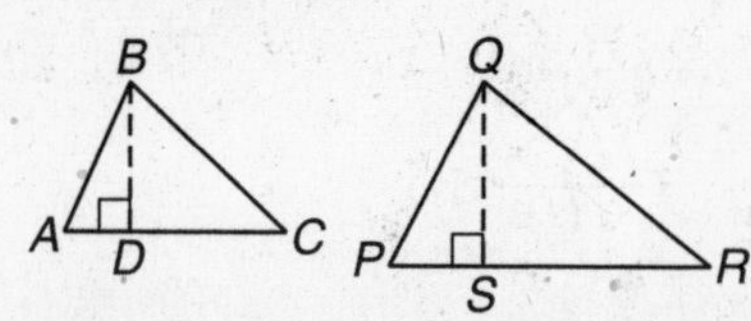

Proof: $\angle A \cong \angle P$ because of the definition of similar polygons. Since $\overline{BD}$ and $\overline{QS}$ are perpendicular to $\overline{AC}$ and $\overline{PR}$, $\angle BDA \cong \angle QSP$. So, $\triangle ABD \sim \triangle PQS$ by AA Similarity and $\frac{QP}{BA} = \frac{QS}{BD}$ by definition of similar polygons.

34. Given: $\angle C \cong \angle BDA$

Prove: $\frac{AC}{DA} = \frac{AD}{BA}$

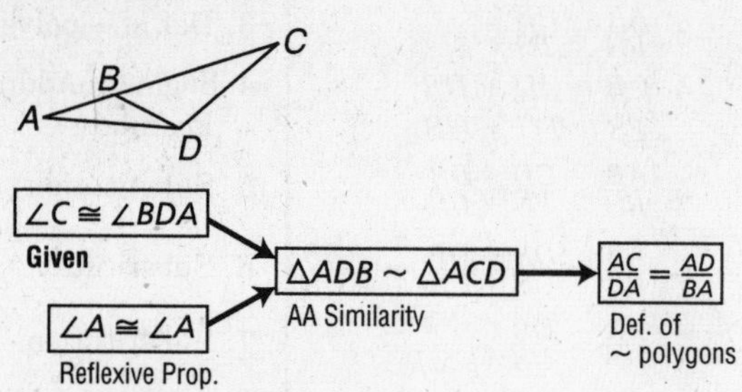

35. Given: $\overline{JF}$ bisects $\angle EFG$. $\overline{EH} \parallel \overline{FG}$, $\overline{EF} \parallel \overline{HG}$

Prove: $\frac{EK}{KF} = \frac{GJ}{JF}$

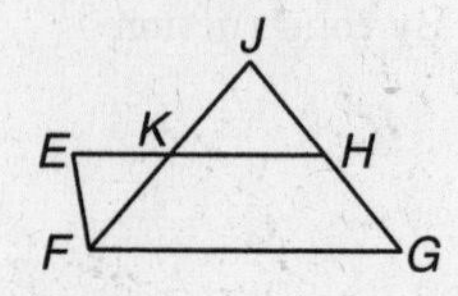

Proof:

Statements	Reasons
1. $\overline{JF}$ bisects $\angle EFG$. $\overline{EH} \parallel \overline{FG}$, $\overline{EF} \parallel \overline{HG}$	1. Given
2. $\angle EFK \cong \angle KFG$	2. Def. of $\angle$ bisector
3. $\angle KFG \cong \angle JKH$	3. Corresponding $\angle$s Postulate
4. $\angle JKH \cong \angle EKF$	4. Vertical $\angle$s are $\cong$.
5. $\angle EFK \cong \angle EKF$	5. Transitive Prop.
6. $\angle FJH \cong \angle EFK$	6. Alternate Interior $\angle$s Theorem
7. $\angle FJH \cong \angle EKF$	7. Transitive Prop.
8. $\triangle EKF \sim \triangle GJF$	8. AA Similarity
9. $\frac{EK}{KF} = \frac{GJ}{JF}$	9. Def. of $\sim \triangle$s

36. Given: $\overline{RU}$ bisects $\angle SRT$. $\overline{VU} \parallel \overline{RT}$

Prove: $\frac{SV}{VR} = \frac{SR}{RT}$

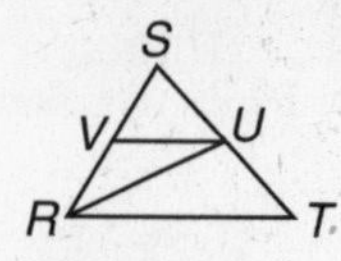

Proof:

Statements	Reasons
1. $\overline{RU}$ bisects $\angle SRT$. $\overline{VU} \parallel \overline{RT}$	1. Given
2. $\angle S \cong \angle S$	2. Reflexive Prop.
3. $\angle SUV \cong \angle STR$	3. Corresponding $\angle$s Postulate
4. $\triangle SUV \sim \triangle STR$	4. AA Similarity
5. $\frac{SV}{VU} = \frac{SR}{RT}$	5. Def. of $\sim \triangle$s
6. $\angle URT \cong \angle VUR$	6. Alternate Interior $\angle$s Theorem
7. $\angle VRU \cong \angle URT$	7. Def. of $\angle$ bisector
8. $\angle VUR \cong \angle VRU$	8. Transitive Prop.
9. $\overline{VU} \cong \overline{VR}$	9. If 2 $\angle$s of a $\triangle$ are $\cong$, the sides opp. these $\angle$s are $\cong$.
10. $VU = VR$	10. Def. of $\cong$ segments
11. $\frac{SV}{VR} = \frac{SR}{RT}$	11. Substitution

37. Given: $\triangle RST \sim \triangle ABC$, W and D are midpoints of $\overline{TS}$ and $\overline{CB}$, respectively.

Prove: $\triangle RWS \sim \triangle ADB$

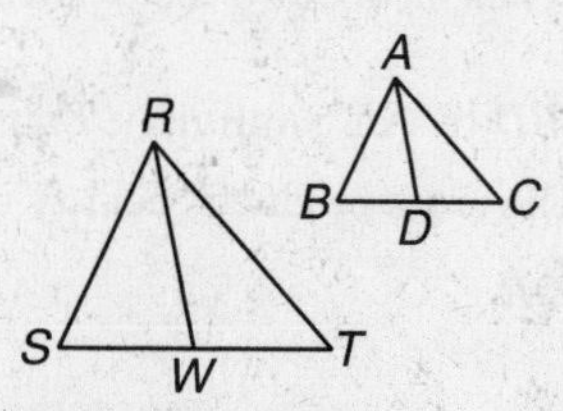

Proof:

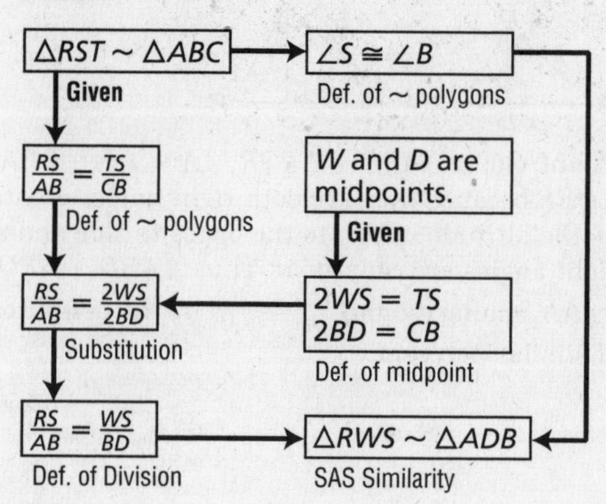

38. Sample answer: The geometry occurs inside the camera as the image is formed on the film. Answers should include the following.

- 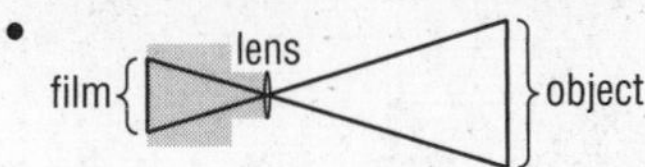

- The triangles are similar because the SAS Similarity Theorem holds. The congruent angles are the vertical angles and the corresponding sides are the congruent sides of the isosceles triangles.

39. Let x represent DF.

$\frac{AC}{DF} = \frac{AB}{DE}$

$\frac{10.5}{x} = \frac{6.5}{8}$

$84 = 6.5x$

$12.9 \approx x$

40. B; let x be one of the two numbers that are the same. Then $x + x + 3x = 180$, so $5x = 180$ or $x = 36$. So, the numbers are 36, 36, and 108. The answer is B.

Page 323 Maintain Your Skills

41. $LO = LM + MO$

$14 = 7 + MO$

$7 = MO$

$LP = LN + NP$

$16 = 9 + NP$

$7 = NP$

In order to show $\overline{MN} \parallel \overline{OP}$, we must show that $\frac{LM}{MO} = \frac{LN}{NP}$. $\frac{LM}{MO} = \frac{7}{7}$ or 1. $\frac{LN}{NP} = \frac{9}{7}$. Since $\frac{LM}{MO} \neq \frac{LN}{NP}$, the sides are not proportional and $\overline{MN}$ is not parallel to $\overline{OP}$.

42. There is not enough information given to determine whether $\overline{MN}$ is parallel to $\overline{OP}$.

43. In order to show $\overline{MN} \parallel \overline{OP}$, we must show that $\frac{LM}{MO} = \frac{LN}{NP}$. $\frac{LM}{MO} = \frac{15}{5}$ or 3. $\frac{LN}{NP} = \frac{12}{4}$ or 3. Thus, $\frac{LM}{MO} = \frac{LN}{NP}$. Since the sides have proportional lengths, $\overline{MN} \parallel \overline{OP}$.

44. $\overline{VZ} \parallel \overline{YX}$, so $\angle Z \cong \angle Y$ and $\angle V \cong \angle X$. So, $\triangle VZW \sim \triangle XYW$. Then $\frac{VW}{XW} = \frac{ZW}{YW}$.

$\frac{3x - 6}{x + 4} = \frac{6}{5}$

$15x - 30 = 6x + 24$

$9x - 30 = 24$

$9x = 54$

$x = 6$

$VW = 3x - 6$

$= 3(6) - 6$ or 12

$WX = x + 4$

$= 6 + 4$ or 10

45. $\overline{RS} \parallel \overline{QT}$, so $\angle R \cong \angle PQT$ and $\angle S \cong \angle QTP$. So, $\triangle PQT \sim \triangle PRS$. Then $\frac{PT}{PS} = \frac{PQ}{PR}$.

$\frac{10}{10 + 4} = \frac{2x + 1}{2x + 1 + 6}$

$\frac{10}{14} = \frac{2x + 1}{2x + 7}$

$20x + 70 = 28x + 14$

$70 = 8x + 14$

$56 = 8x$

$7 = x$

$PQ = 2x + 1$

$= 2(7) + 1$ or 15

46. The line goes through (3, 0) and (0, −3).

$m = \frac{-3 - 0}{0 - 3}$ or 1

$y = mx + b$

$y = 1x + (-3)$

$y = x - 3$

47. $y - y_1 = m(x - x_1)$

$y - (-1) = 2[x - (-1)]$

$y + 1 = 2x + 2$

$y = 2x + 1$

48. 5 12 19 26 33 (+7 +7 +7 +7)

The numbers increase by 7.

The next number will increase by 7. So, it will be 33 + 7 or 40. The next number will also increase by 7. So, it will be 40 + 7 or 47.

49. 10 20 40 80 160 (×2 ×2 ×2 ×2)

The numbers are multiplied by 2.

The next number will be multiplied by 2. So, it will be 160 × 2 or 320. The next number will also be multiplied by 2. So, it will be 320 × 2 or 640.

50. 0 5 4 9 8 13 (+5 −1 +5 −1 +5)

The pattern is to add 5 then subtract 1.

The last number was obtained by adding 5, so the next number is obtained by subtracting 1. So, it will be 13 − 1 or 12. The next number is obtained by adding 5, so it will be 12 + 5 or 17.

Page 323 Practice Quiz 2

1. $\overline{DE} \parallel \overline{BC}$, so $\frac{DB}{AD} = \frac{EC}{AE}$.

$\frac{DB}{8} = \frac{18}{12}$

$12(DB) = 144$

$DB = 12$

$AB = AD + DB$

$= 8 + 12$ or 20

2. $AB = AD + DB$

$20 = 4 + DB$

$16 = DB$

$\overline{DE} \parallel \overline{BC}$, so $\frac{DB}{AD} = \frac{EC}{AE}$.

$\frac{16}{4} = \frac{m + 4}{m - 2}$

$16m - 32 = 4m + 16$

$12m - 32 = 16$

$12m = 48$

$m = 4$

3. $XY = XV + VY$

$30 = 9 + VY$

$21 = VY$

$XZ = XW + WZ$

$18 = 12 + WZ$

$6 = WZ$

In order to show $\overline{YZ} \parallel \overline{VW}$, we must show that $\frac{XV}{VY} = \frac{XW}{WZ}$. $\frac{XV}{VY} = \frac{9}{21}$ or $\frac{3}{7}$. $\frac{XW}{WZ} = \frac{12}{6}$ or 2. Thus, $\frac{XV}{VY} \neq \frac{XW}{WZ}$ so the sides are not proportional. $\overline{YZ}$ is not parallel to $\overline{VW}$.

4. $XZ = XW + WZ$
$33.25 = XW + 11.45$
$21.8 = XW$
In order to show $\overline{YZ} \parallel \overline{VW}$, we must show that $\frac{XV}{VY} = \frac{XW}{WZ}$. $\frac{XV}{VY} = \frac{34.88}{18.32}$ or $\frac{436}{229}$. $\frac{XW}{WZ} = \frac{21.8}{11.45}$ or $\frac{436}{229}$.
Thus, $\frac{XV}{VY} = \frac{XW}{WZ} = \frac{436}{229}$. Since the sides have proportional lengths, $\overline{YZ} \parallel \overline{VW}$.

5. Let x represent the perimeter of $\triangle DEF$. The perimeter of $\triangle GFH = 2 + 2.5 + 4$ or 8.5.
$\frac{EF}{FH} = \frac{\text{perimeter of } \triangle DEF}{\text{perimeter of } \triangle GFH}$
$\frac{6}{4} = \frac{x}{8.5}$
$51 = 4x$
$12.75 = x$

6. Let x represent the perimeter of $\triangle RUW$. The perimeter of $\triangle STV = 12 + 18 + 24$ or 54.
$\frac{UW}{VT} = \frac{\text{perimeter of } \triangle RUW}{\text{perimeter of } \triangle STV}$
$\frac{21}{18} = \frac{x}{54}$
$1134 = 18x$
$63 = x$

7. $\frac{18 - x}{x} = \frac{10}{14}$
$252 - 14x = 10x$
$252 = 24x$
$10.5 = x$

8. $\frac{4}{x} = \frac{6}{x + 3}$
$4x + 12 = 6x$
$12 = 2x$
$6 = x$

9. $\frac{2x}{5} = \frac{10}{x}$
$2x^2 = 50$
$x^2 = 25$
$x = \pm 5$
Because x represents length, x cannot be negative. So, $x = 5$.

10. Let x represent the longest side of the second garden.
$\frac{53.5}{32.1} = \frac{25}{x}$
$53.5x = 802.5$
$x = 15$
The longest side of the second garden is 15 feet.

Page 324 Geometry Activity: Sierpinski Triangle

1. Stage 3:

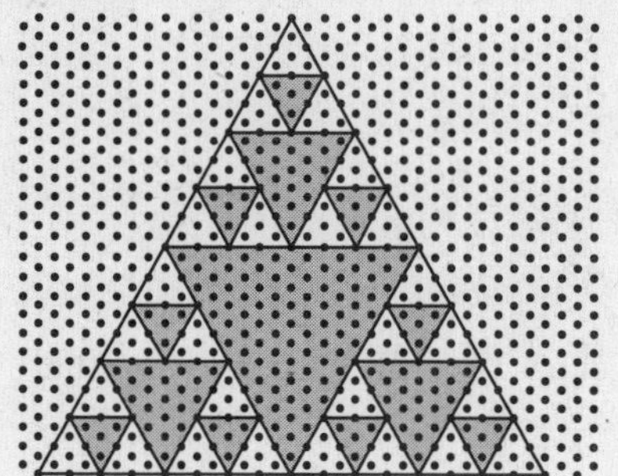

Stage 4:

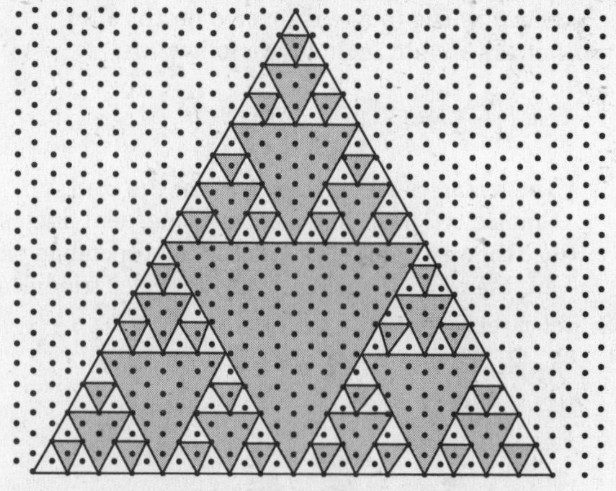

There are 81 nonshaded triangles at Stage 4.

2. Stage 0: 16 + 16 + 16 or 48 units
Stage 1: 8 + 8 + 8 or 24 units
Stage 2: 4 + 4 + 4 or 12 units
Stage 3: 2 + 2 + 2 or 6 units
Stage 4: 1 + 1 + 1 or 3 units
3. The perimeter will be divided by 2 from stage to stage, getting smaller and smaller and approaching zero.
4. Yes, $\triangle DFM$ is equilateral with 2 units on each side. Yes, $\triangle BCE$, $\triangle GHL$, and $\triangle IJN$ are equilateral with 1 unit on each side.
5. yes, by AA Similarity because each triangle has all angles of measure 60
6. 3, each sharing a side with the shaded triangle in Stage 1.
7.

8. Three copies of the Stage 4 Sierpinski triangle combine to form a Stage 5 triangle (just as three copies of a Stage 2 triangle combine to form a Stage 3 triangle).
9. 9 copies: 3 to make a Stage 5 triangle, and 3 copies of the Stage 5 triangle, so a total of $3 \cdot 3$ or 9 Stage 4 triangles

6-6 Fractals and Self-Similarity

Page 328 Check for Understanding

1. Sample answer: irregular shape formed by iteration of self-similar shapes
2. They can accurately calculate thousands of iterations.
3. Sample answer: icebergs, ferns, leaf veins

4. Stage 1: 2, Stage 2: 6, Stage 3: 14, Stage 4: 30

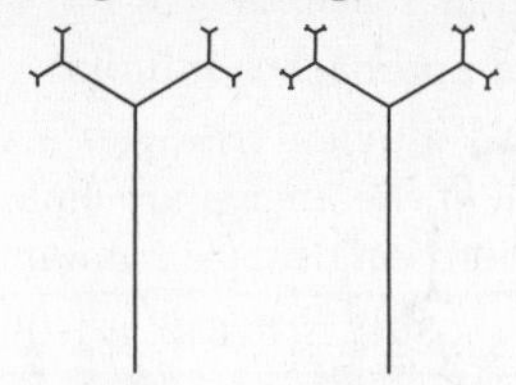

5.

Stage	Number of Branches	Pattern
1	2	$2(2^1 - 1)$
2	6	$2(2^2 - 1)$
3	14	$2(2^3 - 1)$
4	30	$2(2^4 - 1)$

The formula is twice the difference of 2 to a power that is the stage number and 1: $A_n = 2(2^n - 1)$.

6. No, the base of the tree or segment of a branch without an end does not contain a replica of the entire tree.

7. $\sqrt{2} \approx 1.4142...$
$\sqrt{\sqrt{2}} \approx 1.1892...$

8. $\sqrt{\sqrt{\sqrt{2}}} \approx 1.0905...$; the results are getting closer to 1, so the result after 100 repeats approaches 1.

9. Yes, the procedure is repeated over and over again.

10. First, write an equation to find the balance after one year.
current balance + (current balance × interest rate) = new balance

$4000 + (4000 \cdot 0.011) = 4044$
$4044 + (4044 \cdot 0.011) = 4088.48$
$4088.48 + (4088.48 \cdot 0.011) = 4133.45$
$4133.45 + (4133.45 \cdot 0.011) = 4178.92$

After compounding interest four times, Jamir will have $4178.92 in the account.

Pages 328–331 Practice and Apply

11. 9 holes

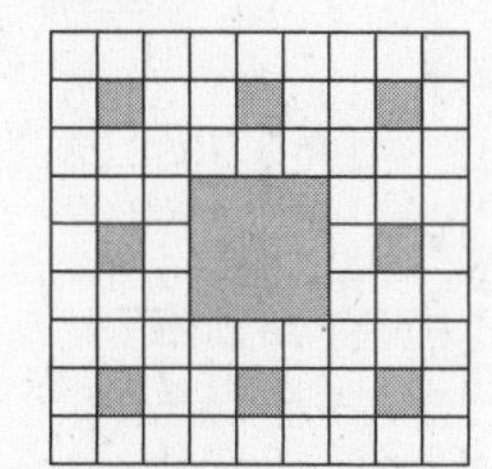

12. 73 holes

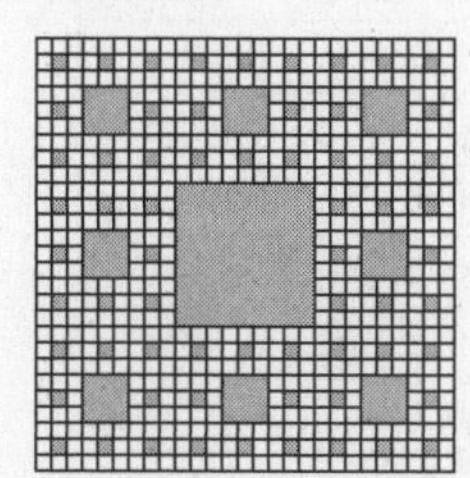

13. Yes, any part contains the same figure as the whole, 9 squares with the middle shaded.

14.

Stage	Number of Dots	Pattern
1	1	1 + 0
2	3	2 + 1
3	6	3 + 3
4	10	4 + 6

The formula is the stage number plus the number of dots in the previous stage: $A_n = n + A_{n-1}$.
So, in the seventh stage, the number of dots is $A_7 = 7 + A_6$, where $A_6 = 6 + A_5$ and $A_5 = 5 + A_4 = 5 + 10$ or 15. Then $A_6 = 6 + 15$ or 21 and $A_7 = 7 + 21$ or 28. So there are 28 dots.

15. 1, 3, 6, 10, 15...; Each difference is 1 more than the preceding difference.

16. The triangular numbers are the numbers in the diagonal.

17. The result is similar to a Stage 3 Sierpinski triangle.

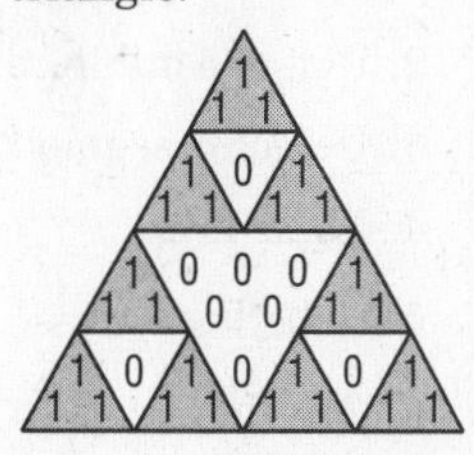

18. It is similar to a Stage 1 Sierpinski triangle.

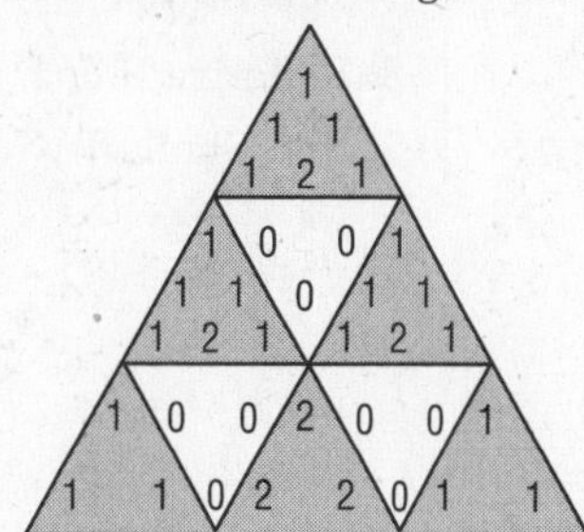

19. All of the numbers in the outside diagonal of Pascal's triangle are 1, so the sum of 25 1's is 25.

20. The second diagonal consists of the natural numbers, 1, 2, 3, 4, 5,..., 50. The sum is

$1 + 2 + 3 + 4 + \cdots + 46 + 47 + 48 + 49 + 50$
$= 50 + (1 + 49) + (2 + 48) + (3 + 47) + \cdots + (24 + 26) + 25$
$= 50 + 24(50) + 25$
$= 1275$

21. Given: $\triangle ABC$ is equilateral. $CD = \frac{1}{3}CB$ and $CE = \frac{1}{3}CA$

Prove: $\triangle CED \sim \triangle CAB$

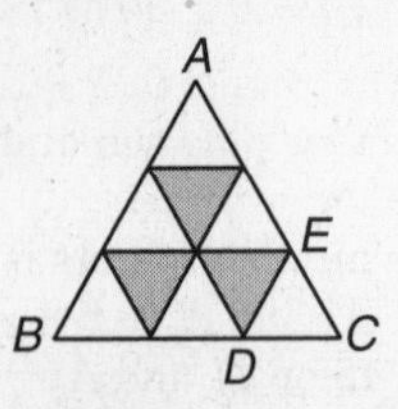

Proof:

Statements	Reasons
1. $\triangle ABC$ is equilateral. $CD = \frac{1}{3}CB$ $CE = \frac{1}{3}CA$	1. Given
2. $\overline{AC} \cong \overline{BC}$	2. Def. of equilateral $\triangle$
3. $AC = BC$	3. Def. of $\cong$ segments
4. $\frac{1}{3}AC = \frac{1}{3}CB$	4. Mult. Prop.
5. $CD = CE$	5. Substitution
6. $\frac{CD}{CB} = \frac{CE}{CB}$	6. Division Prop.
7. $\frac{CD}{CB} = \frac{CE}{CA}$	7. Substitution
8. $\angle C \cong \angle C$	8. Reflexive Prop.
9. $\triangle CED \sim \triangle CAB$	9. SAS Similarity

22.

23. Yes; the smaller and smaller details of the shape have the same geometric characteristics as the original form.

24. Stage 1: 6
Stage 2: 36

25.

Stage	Number of Segments	Pattern
0	1	4^0
1	4	4^1
2	16	4^2
3	64	4^3

The formula is 4 to the power of the stage number: $A_n = 4^n$.
So in Stage 8 there are $4^8 = 65{,}536$ segments.

26. Stage 0: 1 unit, Stage 1: $\frac{1}{3}$ unit, Stage 2: $\frac{1}{9}$ unit, Stage 3: $\frac{1}{27}$ unit; as the stages increase, the length of the segments will approach zero.

27. Stage 0: 3 units, Stage 1: $3 \cdot \frac{4}{3}$ or 4 units, Stage 2: $3\left(\frac{4}{3}\right)\left(\frac{4}{3}\right) = 3\left(\frac{4}{3}\right)^2$ or $5\frac{1}{3}$ units, Stage 3: $3\left(\frac{4}{3}\right)^3$ or $7\frac{1}{9}$ units

28. $P = 3\left(\frac{4}{3}\right)^n$; as the stages increase, the perimeter increases and approaches infinity.

29. The original triangle and the new triangles are equilateral and thus, all of the angles are equal to 60. By AA Similarity, the triangles are similar.

30.

x	12	3.46410...	1.8612...	1.3642...	1.168...
$\sqrt{x}$	3.46410...	1.8612...	1.3642...	1.168...	1.0807...

The numbers converge to 1.

31.

x	5	0.2	5	0.2	5
$\frac{1}{x}$	0.2	5	0.2	5	0.2

The numbers alternate between 0.2 and 5.

32.

x	0.3	0.6694...	0.8747...	0.9563...	0.9852...
$x^{\frac{1}{3}}$	0.6694...	0.8747...	0.9563...	0.9852...	0.9950...

The numbers converge to 1.

33.

x	0	1	2	4	16
2^x	1	2	4	16	65,536

The numbers approach positive infinity.

34.

x	1	3	7
$2x + 1$	3	7	15

35.

x	5	0	−5
$x - 5$	0	−5	−10

36.

x	2	3	8
$x^2 - 1$	3	8	63

37.

x	4	−6	24
$3(2 - x)$	−6	24	−66

38. Write an equation to find the balance each month.

$$\begin{pmatrix}\text{current}\\\text{balance}\end{pmatrix} + \begin{pmatrix}\text{current}\\\text{balance}\end{pmatrix} \times \begin{pmatrix}\text{interest}\\\text{rate}\end{pmatrix} - \begin{pmatrix}100\\\text{payment}\end{pmatrix} = \begin{pmatrix}\text{new}\\\text{balance}\end{pmatrix}$$

$1250 + (1250 \times 0.015) - 100 = 1168.75$
$1168.75 + (1168.75 \times 0.015) - 100 = 1086.28$
$1086.28 + (1086.28 \times 0.015) - 100 = 1002.57$

The balance after 3 months will be $1,002.57.

39.

0.200	0.64
0.64	0.9216...
0.9216...	0.2890...
0.2890...	0.8219...
0.8219...	0.5854...
0.5854...	0.9708...
0.9708...	0.1133...
0.1133...	0.4019...
0.4019...	0.9615...
0.9615...	0.1478...

0.201	0.6423...
0.6423...	0.9188...
0.9188...	0.2981...
0.2981...	0.8369...
0.8369...	0.5458...
0.5458...	0.9916...
0.9916...	0.0333...
0.0333...	0.1287...
0.1287...	0.4487...
0.4487...	0.9894...

Yes, the initial value affected the tenth value.

40. A small difference in initial data can have a large effect in later data.

41. Sample answer: The leaves in the tree and the branches of the trees are self-similar. These self-similar shapes are repeated throughout the painting.

42a. The flower and mountain are computer-generated; the feathers and moss are real.

42b. The fractals exhibit self-similarity and iteration.

43. See students' work.

44.

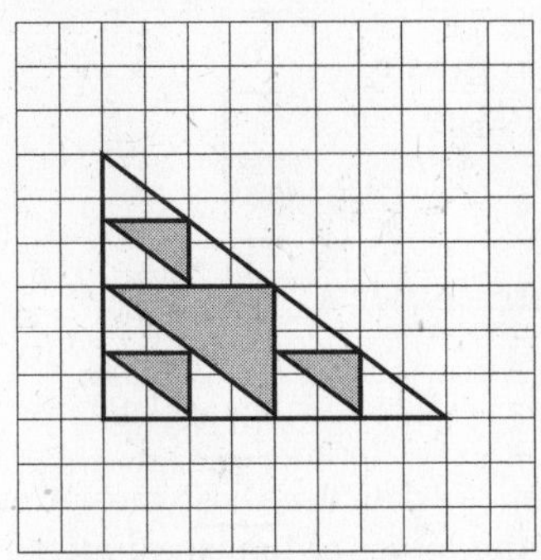

In Stage 1, the shaded triangle has legs 3 and 4 units. The hypotenuse has length c, where $c^2 = 3^2 + 4^2$, or $c = \sqrt{9 + 16}$ or 5. So, the perimeter of the triangle is 3 + 4 + 5 or 12 units. In Stage 2, there are three small shaded triangles and the larger shaded triangle from Stage 1. Each of the small shaded triangles has legs 2 and 1.5 units. The hypotenuse has length d, where $d^2 = 2^2 + 1.5^2$, or $d = \sqrt{4 + 2.25}$ or 2.5. So, the perimeter of all shaded triangles in Stage 2 is 3(2 + 1.5 + 2.5) + 12 or 30 units.

45. Sample answer: Fractal geometry can be found in the repeating patterns of nature. Answers should include the following.

- Broccoli is an example of fractal geometry because the shape of the florets is repeated throughout; one floret looks the same as the stalk.
- Sample answer: Scientists can use fractals to study the human body, rivers, and tributaries, and to model how landscapes change over time.

46. Suppose the 24-inch side of the larger triangle corresponds to the smallest side of the smaller triangle. Let P represent the perimeter of the larger triangle.

$\frac{24}{3} = \frac{P}{3 + 6 + 8}$

$24(3 + 6 + 8) = 3P$

$408 = 3P$

$136 = P$

The maximum perimeter is 136 inches.

47. C; Let x be the number of minutes the repair technician worked in excess of 30 minutes.

$170 = 80 + 2x$

$90 = 2x$

$45 = x$

The repair technician worked 30 + 45 or 75 minutes.

Page 331 Maintain Your Skills

48. $\frac{21}{14} = \frac{3x - 6}{x + 4}$

$21x + 84 = 42x - 84$

$84 = 21x - 84$

$168 = 21x$

$8 = x$

49. $\frac{x}{17} = \frac{16}{20}$

$20x = 272$

$x = 13\frac{3}{5}$

50. $\frac{3x}{6} = \frac{8}{x}$

$3x^2 = 48$

$x^2 = 16$

$x = \pm 4$

Reject $x = -4$ because x represents length, which is positive. So, $x = 4$.

51. $\frac{7}{x} = \frac{2x + 1}{15}$

$105 = 2x^2 + x$

$0 = 2x^2 + x - 105$

$(x - 7)(2x + 15) = 0$

$x - 7 = 0$ or $2x + 15 = 0$

$x = 7$ $\quad 2x = -15$

$x = -7.5$

Reject $x = -7.5$ because x represents length, which is positive. So, $x = 7$.

52. $\frac{AK}{JA} = \frac{BL}{JB}$

$\frac{18 - (JA)}{JA} = \frac{9}{27 - 9}$

$\frac{18 - (JA)}{JA} = \frac{9}{18}$

$324 - 18(JA) = 9(JA)$

$324 = 27(JA)$

$12 = JA$

53. $\frac{JB}{JL} = \frac{AB}{KL}$

$\frac{13}{JL} = \frac{8}{10}$

$130 = 8(JL)$

$16\frac{1}{4} = JL$

54. $\frac{AK}{JA} = \frac{BL}{JB}$

$\frac{10}{25} = \frac{14}{JB}$

$10(JB) = 350$

$JB = 35$

55. If one side of a triangle is longer than another side, then the angle opposite the longer side has a greater measure than the angle opposite the shorter side. The sides of the triangle in order from least to greatest are 965, 1038, and 1042. So, the angles opposite these sides in the same order are arranged from least to greatest. That order is Miami, Bermuda, and San Juan.

56. $P = 5s$

$60 = 5s$

$12 = s$

Each side has measure 12 cm.

57. $P = 2\ell + 2w$
$54 = 2(2x + 1) + 2(x + 2)$
$54 = 4x + 2 + 2x + 4$
$54 = 6x + 6$
$48 = 6x$
$8 = x$
$x + 2 = 8 + 2$ or 10
$2x + 1 = 2(8) + 1$ or 17
The sides of the polygon have measures 10 feet, 10 feet, 17 feet, and 17 feet.

58. $P = n + 2(n + 2) + 2n - 7$
$57 = n + 2n + 4 + 2n - 7$
$57 = 5n - 3$
$60 = 5n$
$12 = n$
$n + 2 = 12 + 2$ or 14
$2n - 7 = 2(12) - 7$ or 17
The sides of the polygon have measures 17, 14, 14, and 12 units.

Chapter 6 Study Guide and Review

Page 332 Vocabulary and Concept Check

1. true
2. false, proportional
3. true
4. false, sides
5. false, iteration
6. false, one-half
7. true
8. true
9. false, parallel to

Pages 332–336 Lesson-by-Lesson Review

10. $\frac{3}{4} = \frac{x}{12}$
$3(12) = 4x$
$36 = 4x$
$9 = x$

11. $\frac{7}{3} = \frac{28}{z}$
$7z = 3(28)$
$7z = 84$
$z = 12$

12. $\frac{x + 2}{5} = \frac{14}{10}$
$10(x + 2) = 5(14)$
$10x + 20 = 70$
$10x = 50$
$x = 5$

13. $\frac{3}{7} = \frac{7}{y - 3}$
$3(y - 3) = 7(7)$
$3y - 9 = 49$
$3y = 58$
$y = \frac{58}{3}$

14. $\frac{4 - x}{3 + x} = \frac{16}{25}$
$25(4 - x) = 16(3 + x)$
$100 - 25x = 48 + 16x$
$100 = 48 + 41x$
$52 = 41x$
$\frac{52}{41} = x$

15. $\frac{x - 12}{6} = \frac{x + 7}{-4}$
$-4(x - 12) = 6(x + 7)$
$-4x + 48 = 6x + 42$
$48 = 10x + 42$
$6 = 10x$
$\frac{3}{5} = x$

16. $\frac{\text{number of total bases from hits}}{\text{number of total at-bats}}$
$= \frac{263}{416}$
≈ 0.632

17. Rewrite 2 : 7 as $2x : 7x$ and use those measures as the lengths of the pieces of the board after cutting it.
$2x + 7x = 108$
$9x = 108$
$x = 12$
$2x = 2(12)$ or 24
$7x = 7(12)$ or 84
The two pieces have lengths 24 inches and 84 inches.

18. In similar polygons, corresponding sides are in proportion. $\frac{TU}{VW} = \frac{6}{9}$ or $\frac{2}{3}$. But $\frac{UV}{UV} = 1$, so two of the sides are in a 2 : 3 ratio while two others are equal in length. Thus, the triangles are not similar.

19. The figures are rectangles, so all angles are congruent. $\frac{LK}{RQ} = \frac{24}{16}$ or $\frac{3}{2}$ and $\frac{LM}{PQ} = \frac{30}{20}$ or $\frac{3}{2}$, so the sides are in a 3 : 2 ratio. Opposite sides are congruent, so all sides are in a 3 : 2 ratio. Thus, the figures are similar.

20. $ABCD \sim AEFG$
$\frac{AD}{AG} = \frac{AB}{AE}$
$\frac{x}{x + 7.5} = \frac{x - 2}{x - 2 + 5}$
$\frac{x}{x + 7.5} = \frac{x - 2}{x + 3}$
$x(x + 3) = (x + 7.5)(x - 2)$
$x^2 + 3x = x^2 - 2x + 7.5x - 15$
$x^2 + 3x = x^2 + 5.5x - 15$
$3x = 5.5x - 15$
$-2.5x = -15$
$x = 6$
$AB = x - 2$
$= 6 - 2$ or 4
$AG = x + 7.5$
$= 6 + 7.5$ or 13.5
The scale factor is $\frac{AD}{AG} = \frac{x}{x + 7.5} = \frac{6}{13.5}$ or $\frac{4}{9}$.

21. $\overline{PT} \parallel \overline{SR}$, so $\angle P \cong \angle R$ and $\angle T \cong \angle S$ because they are alternate interior angles. $\angle PQT \cong \angle SQR$ because they are vertical angles. Thus, $\triangle PQT \sim \triangle RQS$.

$\frac{PQ}{RQ} = \frac{TQ}{SQ}$

$\frac{6 - x}{6 + x} = \frac{3}{3 + x}$

$(6 - x)(3 + x) = (6 + x)(3)$

$18 + 6x - 3x - x^2 = 18 + 3x$

$18 + 3x - x^2 = 18 + 3x$

$18 - x^2 = 18$

$-x^2 = 0$

$x = 0$

$PQ = 6 - x$

$= 6 - 0$ or 6

$QS = 3 + x$

$= 3 + 0$ or 3

The scale factor is $\frac{TQ}{SQ} = \frac{3}{3 + x} = \frac{3}{3 + 0}$ or 1.

22. Triangles ABC and DFE are isosceles triangles. $\angle A \cong \angle D$, and $BA = CA$ and $FD = ED$ so $\frac{BA}{FD} = \frac{CA}{ED}$. Thus, $\triangle ABC \sim \triangle DFE$ by SAS Similarity.

23. $\overline{HI} \parallel \overline{JK}$, so $\angle GHI \cong \angle GJK$ and $\angle GIH \cong \angle GKJ$ because they are corresponding angles. Thus, $\triangle GHI \sim \triangle GJK$ by AA Similarity.

24. $m\angle L + m\angle Q + m\angle LMQ = 180$

$35 + 85 + m\angle LMQ = 180$

$m\angle LMQ = 60$

$\angle LMQ \cong \angle NMP$, so $m\angle NMP = 60$.

$m\angle N + m\angle P + m\angle NMP = 180$

$m\angle N + 40 + 60 = 180$

$m\angle N = 80$

$\triangle LMQ$ is not similar to $\triangle PMN$ because the angles of the triangles are not congruent.

25. Since $\overline{AB} \parallel \overline{DE}$, $\angle B \cong \angle E$ and $\angle A \cong \angle D$ because they are alternate interior angles. By AA Similarity, $\triangle ABC \sim \triangle DEC$. Using the definition of similar polygons, $\frac{BC}{EC} = \frac{AC}{DC}$.

$\frac{x + 3}{11x - 2} = \frac{1}{6}$

$6(x + 3) = 1(11x - 2)$

$6x + 18 = 11x - 2$

$18 = 5x - 2$

$20 = 5x$

$4 = x$

26. $\angle V \cong \angle UST$ and $\angle T \cong \angle T$, so by AA Similarity, $\triangle RVT \sim \triangle UST$. Using the definition of similar polygons, $\frac{RT}{UT} = \frac{VT}{ST}$.

$\frac{2x + 4}{x + 2} = \frac{3 + x + 2}{4}$

$\frac{2x + 4}{x + 2} = \frac{x + 5}{4}$

$4(2x + 4) = (x + 2)(x + 5)$

$8x + 16 = x^2 + 5x + 2x + 10$

$8x + 16 = x^2 + 7x + 10$

$16 = x^2 - x + 10$

$0 = x^2 - x - 6$

$0 = (x - 3)(x + 2)$

$x - 3 = 0$ or $x + 2 = 0$

$x = 3 \qquad x = -2$

Reject $x = -2$ because otherwise $UT = x + 2 = -2 + 2$ or 0. So $x = 3$.

27. In order to show that $\overline{GL} \parallel \overline{HK}$, we must show that $\frac{IH}{HG} = \frac{IK}{KL}$.

$\frac{IH}{HG} = \frac{21}{14}$ or $\frac{3}{2}$, and $\frac{IK}{KL} = \frac{15}{9}$ or $\frac{5}{3}$.

Because the side lengths are not proportional, $\overline{GL} \nparallel \overline{HK}$.

28. $IL = IK + KL$

$36 = 28 + KL$

$8 = KL$

In order to show that $\overline{GL} \parallel \overline{HK}$, we must show that $\frac{IH}{HG} = \frac{IK}{KL}$.

$\frac{IH}{HG} = \frac{35}{10}$ or $\frac{7}{2}$, and $\frac{IK}{KL} = \frac{28}{8}$ or $\frac{7}{2}$.

$\frac{IH}{HG} = \frac{IK}{KL}$, so $\overline{GL} \parallel \overline{HK}$.

29. In order to show that $\overline{GL} \parallel \overline{HK}$, we must show that $\frac{IH}{HG} = \frac{IK}{KL}$. Let $KL = x$. Then $IL = 3x$ and $IK = 3x - x$ or $2x$.

$\frac{IK}{KL} = \frac{2x}{x}$ or 2, and $\frac{IH}{HG} = \frac{22}{11}$ or 2. $\frac{IK}{KL} = \frac{IH}{HG}$. Thus, $\overline{GL} \parallel \overline{HK}$.

30. In order to show that $\overline{GL} \parallel \overline{HK}$, we must show that $\frac{IH}{HG} = \frac{IK}{KL}$. Let $HI = x$. Then $IG = 3x$, so $HG = 3x - x$ or $2x$.

$\frac{IH}{HG} = \frac{x}{2x}$ or $\frac{1}{2}$, and $\frac{IK}{KL} = \frac{18}{6}$ or 3. So the sides are not proportional and $\overline{GL} \nparallel \overline{HK}$.

31. From the Triangle Proportionality Theorem, $\frac{BC}{AB} = \frac{ED}{AE}$.

Substitute the known measures.

$\frac{4}{6} = \frac{ED}{9}$

$4(9) = 6(ED)$

$36 = 6(ED)$

$6 = ED$

32. From the Triangle Proportionality Theorem, $\frac{BC}{AB} = \frac{ED}{AE}$.

Substitute the known measures.

$\frac{16 - 12}{12} = \frac{5}{AE}$

$\frac{4}{12} = \frac{5}{AE}$

$4(AE) = 12(5)$

$4(AE) = 60$

$AE = 15$

33. Since $\overline{BE} \parallel \overline{CD}$, $\angle ABE \cong \angle ACD$ and $\angle AEB \cong \angle ADC$ by the corresponding angles postulate. Then $\triangle ABE \sim \triangle ACD$ by the AA Similarity. Using the definition of similar polygons, $\frac{CD}{BE} = \frac{AD}{AE}$.

Substitute the known measures.

$\frac{CD}{6} = \frac{8 + 4}{8}$

$\frac{CD}{6} = \frac{12}{8}$

$8(CD) = 6(12)$

$8(CD) = 72$

$CD = 9$

34. Since $\overline{BE} \parallel \overline{CD}$, $\angle ABE \cong \angle ACD$ and $\angle AEB \cong \angle ADC$ by the corresponding angles postulate. Then $\triangle ABE \sim \triangle ACD$ by the AA Similarity. Using the definition of similar polygons, $\frac{AC}{AB} = \frac{CD}{BE}$.
Substitute the known measures.
$\frac{33 + BC}{33} = \frac{32}{24}$
$24(33 + BC) = 33(32)$
$792 + 24(BC) = 1056$
$24(BC) = 264$
$BC = 11$

35. Let x represent the perimeter of $\triangle DEF$. The perimeter of $\triangle ABC = 7 + 6 + 3$ or 16.
$\frac{CB}{FE} = \frac{\text{perimeter of } \triangle ABC}{\text{perimeter of } \triangle DEF}$
$\frac{6}{9} = \frac{16}{x}$
$6x = 144$
$x = 24$
The perimeter of $\triangle DEF$ is 24 units.

36. Let x represent the perimeter of $\triangle QRS$. The perimeter of $\triangle QTP = 15 + 16 + 11$ or 42.
$\frac{TQ}{RQ} = \frac{\text{perimeter of } \triangle QTP}{\text{perimeter of } \triangle QRS}$
$\frac{15}{5} = \frac{42}{x}$
$15x = 210$
$x = 14$
The perimeter of $\triangle QRS$ is 14 units.

37. Let x represent the perimeter of $\triangle CPD$. $\angle CPD \cong \angle BPA$ and $\angle D \cong \angle A$, so $\triangle CPD \sim \triangle BPA$ by AA Similarity. Similar triangles have corresponding medians proportional to the corresponding sides, so the corresponding medians are also proportional to the perimeters.
$\frac{BM}{CN} = \frac{\text{perimeter of } \triangle BPA}{\text{perimeter of } \triangle CPD}$
$\frac{\sqrt{13}}{3\sqrt{13}} = \frac{12}{x}$
$x\sqrt{13} = 36\sqrt{13}$
$x = 36$

38. Use the Pythagorean Theorem in $\triangle PQM$ to find PM.
$(PM)^2 + (QM)^2 = (PQ)^2$
$(PM)^2 + 12^2 = 13^2$
$(PM)^2 = 169 - 144$
$(PM)^2 = 25$
$PM = 5$
Since $\triangle PQM \sim \triangle PRQ$,
$\frac{PQ}{PR} = \frac{PM}{PQ}$
$\frac{13}{PR} = \frac{5}{13}$
$5(PR) = 13(13)$
$5(PR) = 169$
$PR = 33.8$
$\frac{PM}{PQ} = \frac{QM}{RQ}$
$\frac{5}{13} = \frac{12}{RQ}$
$5(RQ) = 13(12)$
$5(RQ) = 156$
$RQ = 31.2$
The perimeter of $\triangle PQR = 13 + 33.8 + 31.2$ or 78.

39. Stage 2 is not similar to Stage 1.

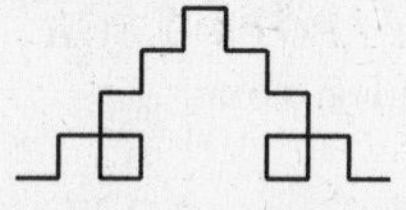

40.

x	2	4	60
$x^3 - 4$	4	60	215,996

41.

x	−4	−8	−20
$3x + 4$	−8	−20	−56

42.

x	10	0.1	10
$\frac{1}{x}$	0.1	10	0.1

43.

x	30	−6	−9.6
$\frac{x}{10} - 9$	−6	−9.6	−9.96

Chapter 6 Practice Test

Page 337

1. b
2. a
3. c

4. $\frac{x}{14} = \frac{1}{2}$
$2x = 14$
$x = 7$

5. $\frac{4x}{3} = \frac{108}{x}$
$4x^2 = 324$
$x^2 = 81$
$x = \pm 9$

6. $\frac{k + 2}{7} = \frac{k - 2}{3}$
$3(k + 2) = 7(k - 2)$
$3k + 6 = 7k - 14$
$6 = 4k - 14$
$20 = 4k$
$5 = k$

7. Using corresponding angles, pentagon $DCABE \sim$ pentagon $DGIHF$. The scale factor is $\frac{CD}{GD} = \frac{12}{18}$ or $2 : 3$.

8. Using corresponding angles, $\triangle PQR \sim \triangle PST$.
$\frac{PQ}{PS} = \frac{QR}{ST}$
$\frac{2x + 2}{2x + 2 + 6} = \frac{2x}{15}$
$\frac{2x + 2}{2x + 8} = \frac{2x}{15}$
$15(2x + 2) = 2x(2x + 8)$
$30x + 30 = 4x^2 + 16x$
$30 = 4x^2 - 14x$
$0 = 4x^2 - 14x - 30$
$0 = 2x^2 - 7x - 15$
$0 = (x - 5)(2x + 3)$
$x - 5 = 0$ or $2x + 3 = 0$
$x = 5$ $x = -\frac{3}{2}$
Reject $x = -\frac{3}{2}$, because otherwise $QR = 2x = -3$, and length must be positive. So, $x = 5$.
The scale factor is $\frac{QR}{ST} = \frac{2x}{15} = \frac{2(5)}{15} = \frac{10}{15}$ or $2 : 3$.

9. Using corresponding angles, $\triangle MAD \sim \triangle MCB$.

$\frac{AM}{CM} = \frac{DM}{BM}$

$\frac{25}{x + 20} = \frac{-3x}{12}$

$300 = -3x^2 - 60x$

$0 = -3x^2 - 60x - 300$

$0 = x^2 + 20x + 100$

$0 = (x + 10)(x + 10)$

$x + 10 = 0$

$x = -10$

The scale factor is $\frac{DM}{BM} = \frac{-3x}{12} = \frac{-3(-10)}{12} = \frac{30}{12}$ or 5 : 2.

10. $\frac{PQ}{ML} = \frac{5}{10}$ or $\frac{1}{2}$

$\frac{PR}{MN} = \frac{6}{12}$ or $\frac{1}{2}$

$\frac{QR}{LN} = \frac{3}{6}$ or $\frac{1}{2}$

Corresponding sides are proportional, so the triangles are similar by SSS Similarity.

11. $\angle PTS \cong \angle QTR$ since they are vertical angles.

$m\angle PTS = m\angle QTR = 90$.

$m\angle PTS + m\angle S + m\angle P = 180$

$90 + 62 + m\angle P = 180$

$m\angle P = 28$

$m\angle QTR + m\angle R + m\angle Q = 180$

$90 + 66 + m\angle Q = 180$

$m\angle Q = 24$

Since only one pair of corresponding angles is congruent, the triangles are not similar.

12. $\overline{ED} \parallel \overline{CB}$ so $\angle AED \cong \angle ACB$ and $\angle ADE \cong \angle ABC$ because corresponding angles are congruent. Then the triangles are similar by AA Similarity.

13. $\frac{GK}{KJ} = \frac{GH}{HI}$

$\frac{8 - KJ}{KJ} = \frac{12}{4}$

$4(8 - KJ) = 12(KJ)$

$32 - 4(KJ) = 12(KJ)$

$32 = 16(KJ)$

$2 = KJ$

14. $\frac{GK}{KJ} = \frac{GH}{HI}$

$\frac{GK}{6} = \frac{7}{14 - 7}$

$\frac{GK}{6} = \frac{7}{7}$

$7(GK) = 42$

$GK = 6$

15. $\frac{GK}{KJ} = \frac{GH}{HI}$

$\frac{6}{4} = \frac{9}{HI}$

$6(HI) = 36$

$HI = 6$

$GI = GH + HI$

$= 9 + 6$ or 15

16. Let x represent the perimeter of $\triangle DEF$.

$\frac{CB}{EF} = \frac{\text{perimeter of } \triangle ACB}{\text{perimeter of } \triangle DEF}$

$\frac{10}{14} = \frac{7 + 10 + 13}{x}$

$\frac{10}{14} = \frac{30}{x}$

$10x = 420$

$x = 42$

The perimeter of $\triangle DEF$ is 42 units.

17. $\angle CMA \cong \angle ACB$ and $\angle A \cong \angle A$, so $\triangle ABC \sim \triangle ACM$ by AA Similarity.

$(AM)^2 + 6^2 = 10^2$

$(AM)^2 + 36 = 100$

$(AM)^2 = 64$

$AM = 8$

$\frac{BC}{CM} = \frac{AC}{AM}$

$\frac{BC}{6} = \frac{10}{8}$

$8(BC) = 60$

$BC = 7.5$

$\frac{BC}{CM} = \frac{AB}{AC}$

$\frac{7.5}{6} = \frac{8 + MB}{10}$

$75 = 48 + 6(MB)$

$27 = 6(MB)$

$4.5 = MB$

The perimeter of $\triangle ABC$ is 10 + 7.5 + 8 + 4.5 or 30.

18.

x	−3	12	87
$5x + 27$	12	87	462

19. $\frac{\text{height of fence}}{\text{height of backboard}} = \frac{\text{shadow of fence}}{\text{shadow of backboard}}$

$\frac{4 \text{ ft}}{x} = \frac{20 \text{ in.}}{65 \text{ in.}}$

$\frac{48}{x} = \frac{20}{65}$

$3120 = 20x$

$156 = x$

The height of the top of the backboard is 156 inches, or 13 feet.

20. B; $X - Y$ is spent, which is $\frac{X - Y}{X}$ as part of the weekly salary.

Chapter 6 Standardized Test Practice

Pages 338–339

1. B; $|-8 + 2| = |-6| = 6$

2. B; $d = \sqrt{[-5 - (-3)]^2 + (2 - 17)^2}$

$= \sqrt{(-2)^2 + (-15)^2}$

$= \sqrt{4 + 225}$

$= \sqrt{229}$ miles

3. A

4. D

5. A; $\frac{1.5 \text{ cm}}{2 \text{ ft}} = \frac{11.25 \text{ cm}}{x \text{ ft}}$

$1.5x = 22.5$

$x = 15$ ft

6. B; $\frac{45}{63} = \frac{5}{7}$, so the dimensions could be 7 inches by 5 inches.

7. C; $\frac{s}{u} = \frac{p}{r}$ because side UT corresponds to side RQ and side ST corresponds to side PQ.

8. D

9. counterexample

10. $y - y_1 = m(x - x_1)$
$y - 2 = 3(x - 2)$
$y - 2 = 3x - 6$
$y = 3x - 4$

11. $PQ = \frac{1}{2}EF$
$20 = \frac{1}{2}(3x + 4)$
$20 = \frac{3}{2}x + 2$
$18 = \frac{3}{2}x$
$12 = x$

12. $\frac{20 \text{ cm}}{40 \text{ m}} = \frac{\text{perimeter of model}}{23 + 40 + 46}$
$\frac{20}{40} = \frac{x}{109}$
$2180 = 40x$
$54.5 = x$
The perimeter of the model is 54.5 cm.

13a. $m = \frac{(160 - 40)}{(4 - 0)}$
$= \frac{120}{4}$
$= 30$

13b. The slope represents the monthly flat rate, so the company charges a flat rate of $30 per month.

13c. $y - y_1 = m(x - x_1)$
$y - 40 = 30(x - 0)$
$y - 40 = 30x$
$y = 30x + 40$

13d. The new monthly rate will be $25 per month, so the equation will be $y = 25x + 40$. The graph will have a less steep slope.

14a. Using the Triangle Proportionality Theorem, if a line is parallel to one side of a triangle and intersects the other two sides in two distinct points, then it separates these sides into segments of proportional length, so $\frac{AE}{AC} = \frac{AD}{AB}$. Since the corresponding sides are proportional and the included angle, $\angle A$, is the same, by SAS Similarity Theorem we know that the triangles are similar.

14b. If B is between A and D and C is between A and E, then $\frac{AB}{BC} = \frac{AD}{DE}$.
$\frac{3500}{1400} = \frac{3500 + 1500}{DE}$
$\frac{3500}{1400} = \frac{5000}{DE}$
$3500(DE) = 7{,}000{,}000$
$DE = 2000$
If D is between A and B and E is between A and C, then $\frac{AB}{BC} = \frac{AD}{DE}$.
$\frac{3500}{1400} = \frac{3500 - 1500}{DE}$
$\frac{3500}{1400} = \frac{2000}{DE}$
$3500(DE) = 2{,}800{,}000$
$DE = 800$
So, DE is 2000 feet or 800 feet.

Chapter 7 Right Triangles and Trigonometry

Page 341 Getting Started

1. $\frac{3}{4} = \frac{12}{a}$
   $3a = 48$
   $a = 16$
2. $\frac{c}{5} = \frac{8}{3}$
   $3c = 40$
   $c \approx 13.33$
3. $\frac{e}{20} = \frac{6}{5}$
   $5e = 120$
   $e = 24$
   $\frac{6}{5} = \frac{f}{10}$
   $60 = 5f$
   $12 = f$
4. $\frac{4}{3} = \frac{6}{y}$
   $4y = 18$
   $y = 4.5$
   $\frac{4}{3} = \frac{1}{z}$
   $4z = 3$
   $z = 0.75$
5. $c^2 = 5^2 + 12^2$
   $c^2 = 25 + 144 = 169$
   $c = 13$
6. $c^2 = 6^2 + 8^2$
   $c^2 = 36 + 64 = 100$
   $c = 10$
7. $c^2 = 15^2 + 15^2$
   $c^2 = 225 + 225 = 450$
   $c \approx 21.21$
8. $c^2 = 14^2 + 27^2$
   $c^2 = 196 + 729 = 925$
   $c \approx 30.41$
9. $\sqrt{8} = \sqrt{4 \cdot 2}$
   $= \sqrt{4} \cdot \sqrt{2} = 2\sqrt{2}$
10. $\sqrt{10^2 - 5^2} = \sqrt{100 - 25}$
    $= \sqrt{75} = \sqrt{25 \cdot 3}$
    $= \sqrt{25} \cdot \sqrt{3} = 5\sqrt{3}$
11. $\sqrt{39^2 - 36^2} = \sqrt{1521 - 1296}$
    $= \sqrt{225} = 15$
12. $\frac{7}{\sqrt{2}} = \frac{7\sqrt{2}}{\sqrt{2} \cdot \sqrt{2}}$
    $= \frac{7\sqrt{2}}{\sqrt{4}} = \frac{7\sqrt{2}}{2}$
13. $x + 44 + 38 = 180$
    $x + 82 = 180$
    $x = 98$
14. $x + 40 = 155$
    $x = 115$
15. $x + 2x + 21 + 90 = 180$
    $3x + 111 = 180$
    $3x = 69$
    $x = 23$

7-1 Geometric Mean

Page 343 Geometry Software Investigation

1. See students' work.
2. They are equal.
3. They are equal.
4. They are similar.

Pages 345–346 Check for Understanding

1. Sample answer: 2 and 72.
   Let x represent the geometric mean.
   $\frac{2}{x} = \frac{x}{72}$
   $x^2 = 144$
   $x = \sqrt{144}$ or 12
2. 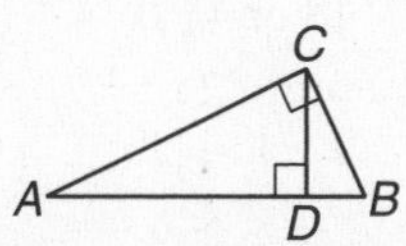

   For leg $\overline{CB}$, $\overline{DB}$ is the segment of the hypotenuse that shares an endpoint. Thus, it is the adjacent segment. The same is true for leg $\overline{AC}$ and segment $\overline{AD}$.
3. Ian; his proportion shows that the altitude is the geometric mean of the two segments of the hypotenuse.
4. Let x represent the geometric mean.
   $\frac{9}{x} = \frac{x}{4}$
   $x^2 = 36$
   $x = \sqrt{36}$ or 6
5. Let x represent the geometric mean.
   $\frac{36}{x} = \frac{x}{49}$
   $x^2 = 1764$
   $x = \sqrt{1764}$ or 42
6. Let x represent the geometric mean.
   $\frac{6}{x} = \frac{x}{8}$
   $x^2 = 48$
   $x = \sqrt{48}$ or $4\sqrt{3}$
   $x \approx 6.9$
7. Let x represent the geometric mean.
   $\frac{2\sqrt{2}}{x} = \frac{x}{3\sqrt{2}}$
   $x^2 = 12$
   $x = \sqrt{12}$ or $2\sqrt{3}$
   $x \approx 3.5$
8. Let $x = CD$.
   $\frac{AD}{CD} = \frac{CD}{BD}$
   $\frac{2}{x} = \frac{x}{6}$
   $x^2 = 12$
   $x = \sqrt{12}$ or $2\sqrt{3}$
   $x \approx 3.5$

9. Let $x = EH$.

$\frac{GH}{EH} = \frac{EH}{FH}$

$\frac{16 - 12}{x} = \frac{x}{12}$

$x^2 = 48$

$x = \sqrt{48}$ or $4\sqrt{3}$

$x \approx 6.9$

10. $\frac{BD}{CD} = \frac{CD}{AD}$

$\frac{8}{x} = \frac{x}{3}$

$x^2 = 24$

$x = \sqrt{24}$

$x = 2\sqrt{6}$

$\frac{BA}{CA} = \frac{CA}{DA}$

$\frac{8 + 3}{y} = \frac{y}{3}$

$y^2 = 33$

$y = \sqrt{33}$

11. $\frac{BD}{CD} = \frac{CD}{DA}$

$\frac{x}{2\sqrt{3}} = \frac{2\sqrt{3}}{2}$

$2x = 12$

$x = 6$

$\frac{BA}{BC} = \frac{BC}{BD}$

$\frac{x + 2}{y} = \frac{y}{x}$

$\frac{6 + 2}{y} = \frac{y}{6}$

$y^2 = 48$

$y = \sqrt{48}$ or $4\sqrt{3}$

12.

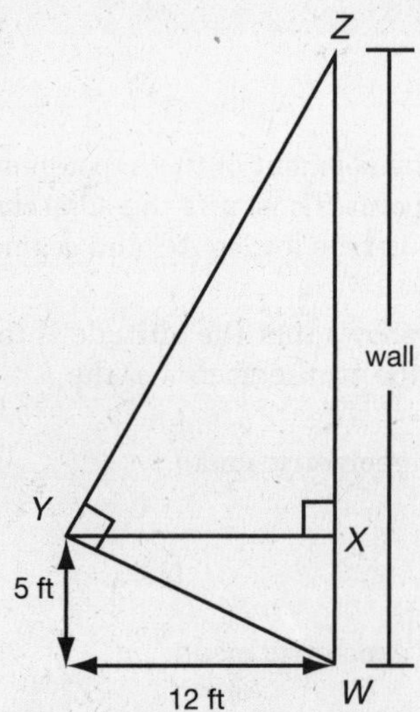

Draw a diagram. Let $\overline{YX}$ be the altitude drawn from the right angle of $\triangle WYZ$.

$\frac{WX}{YX} = \frac{YX}{ZX}$

$\frac{5}{12} = \frac{12}{ZX}$

$5ZX = 144$

$ZX \approx 28.8$

Khaliah estimates that the wall is about 5 + 28.8 or 33.8 feet high.

Pages 346–348 Practice and Apply

13. Let x represent the geometric mean.

$\frac{5}{x} = \frac{x}{6}$

$x^2 = 30$

$x = \sqrt{30}$

$x \approx 5.5$

14. Let x represent the geometric mean.

$\frac{24}{x} = \frac{x}{25}$

$x^2 = 600$

$x = \sqrt{600}$ or $10\sqrt{6}$

$x \approx 24.5$

15. Let x represent the geometric mean.

$\frac{\sqrt{45}}{x} = \frac{x}{\sqrt{80}}$

$x^2 = \sqrt{3600}$

$x^2 = 60$

$x = \sqrt{60}$ or $2\sqrt{15}$

$x \approx 7.7$

16. Let x represent the geometric mean.

$\frac{\sqrt{28}}{x} = \frac{x}{\sqrt{1372}}$

$x^2 = \sqrt{38{,}416}$

$x^2 = 196$

$x = \sqrt{196}$ or 14

17. Let x represent the geometric mean.

$\frac{\frac{3}{5}}{x} = \frac{x}{1}$

$x^2 = \frac{3}{5}$

$x = \sqrt{\frac{3}{5}}$

$x = \frac{\sqrt{3}}{\sqrt{5}}$

$x = \frac{\sqrt{15}}{5}$

$x \approx 0.8$

18. Let x represent the geometric mean.

$\frac{\frac{8\sqrt{3}}{5}}{x} = \frac{x}{\frac{6\sqrt{3}}{5}}$

$x^2 = \frac{144}{25}$

$x = \sqrt{\frac{144}{25}}$

$x = \frac{\sqrt{144}}{\sqrt{25}}$

$x = \frac{12}{5}$ or 2.4

19. Let x represent the geometric mean.

$\frac{\frac{2\sqrt{2}}{6}}{x} = \frac{x}{\frac{5\sqrt{2}}{6}}$

$x^2 = \frac{20}{36}$

$x = \sqrt{\frac{20}{36}}$

$x = \frac{\sqrt{20}}{\sqrt{36}}$

$x = \frac{2\sqrt{5}}{6}$

$x = \frac{\sqrt{5}}{3}$

$x \approx 0.7$

20. Let x represent the geometric mean.

$\frac{\frac{13}{7}}{x} = \frac{x}{\frac{5}{7}}$

$x^2 = \frac{65}{49}$

$x = \sqrt{\frac{65}{49}}$

$x = \frac{\sqrt{65}}{\sqrt{49}} = \frac{\sqrt{65}}{7}$

$x \approx 1.2$

21. Let $x = AD$.

$$\frac{BD}{AD} = \frac{AD}{CD}$$
$$\frac{5}{x} = \frac{x}{9}$$
$$x^2 = 45$$
$$x = \sqrt{45} \text{ or } 3\sqrt{5}$$
$$x \approx 6.7$$

22. Let $x = EH$.

$$\frac{FH}{EH} = \frac{EH}{GH}$$
$$\frac{12}{x} = \frac{x}{12}$$
$$x^2 = 144$$
$$x = \sqrt{144} \text{ or } 12$$

23. Let $x = LM$.

$$\frac{JM}{LM} = \frac{LM}{KM}$$
$$\frac{8}{x} = \frac{x}{16}$$
$$x^2 = 128$$
$$x = \sqrt{128} \text{ or } 8\sqrt{2}$$
$$x \approx 11.3$$

24. Let $x = QS$.

$$\frac{PS}{QS} = \frac{QS}{RS}$$
$$\frac{21}{x} = \frac{x}{7}$$
$$x^2 = 147$$
$$x = \sqrt{147}$$
$$x \approx 12.1$$

25. Let $x = UW$.

$$\frac{VW}{UW} = \frac{UW}{TW}$$
$$\frac{2}{x} = \frac{x}{13}$$
$$x^2 = 26$$
$$x = \sqrt{26}$$
$$x \approx 5.1$$

26. Let $x = ZN$.

$$\frac{YN}{ZN} = \frac{ZN}{XN}$$
$$\frac{2.5}{x} = \frac{x}{10}$$
$$x^2 = 25$$
$$x = \sqrt{25} \text{ or } 5$$

27. $\frac{3+8}{x} = \frac{x}{8}$

$$x^2 = 88$$
$$x = \sqrt{88} \text{ or } 2\sqrt{22}$$
$$x \approx 9.4$$
$$\frac{3+8}{y} = \frac{y}{3}$$
$$y^2 = 33$$
$$y = \sqrt{33}$$
$$y \approx 5.7$$
$$\frac{8}{z} = \frac{z}{3}$$
$$z^2 = 24$$
$$z = \sqrt{24} \text{ or } 2\sqrt{6}$$
$$z \approx 4.9$$

28. $\frac{6}{8} = \frac{8}{x-6}$

$$6x - 36 = 64$$
$$6x = 100$$
$$x = \frac{50}{3}$$
$$\frac{x}{y} = \frac{y}{6}$$
$$y^2 = 6x$$
$$y^2 = 6\left(\frac{50}{3}\right)$$
$$y^2 = 100$$
$$y = \sqrt{100}$$
$$y = 10$$
$$\frac{x}{z} = \frac{z}{x-6}$$
$$z^2 = x^2 - 6x$$
$$z^2 = \left(\frac{50}{3}\right)^2 - 6\left(\frac{50}{3}\right)$$
$$z^2 = \frac{2500}{9} - 100$$
$$z^2 = \frac{1600}{9}$$
$$z = \sqrt{\frac{1600}{9}} \text{ or } \frac{40}{3}$$

29. $z^2 + 5^2 = 15^2$

$$z^2 + 25 = 225$$
$$z^2 = 200$$
$$z = \sqrt{200} \text{ or } 10\sqrt{2}$$
$$z \approx 14.1$$
$$\frac{15}{z} = \frac{z}{x}$$
$$15x = z^2$$
$$15x = (10\sqrt{2})^2$$
$$15x = 200$$
$$x = \frac{40}{3}$$
$$\frac{15}{5} = \frac{5}{y}$$
$$15y = 25$$
$$y = \frac{5}{3}$$

30. $x^2 + 4^2 = 10^2$

$$x^2 + 16 = 100$$
$$x^2 = 84$$
$$x = \sqrt{84} \text{ or } 2\sqrt{21}$$
$$x \approx 9.2$$
$$\frac{y+4}{10} = \frac{10}{4}$$
$$4y + 16 = 100$$
$$4y = 84$$
$$y = 21$$
$$z = y + 4$$
$$z = 21 + 4 = 25$$

31. $\frac{6x}{36} = \frac{36}{x}$

$$6x^2 = 1296$$
$$x^2 = 216$$
$$x = \sqrt{216} \text{ or } 6\sqrt{6}$$
$$x \approx 14.7$$
$$\frac{6x + x}{y} = \frac{y}{x}$$
$$y^2 = 7x^2$$
$$y^2 = 7(6\sqrt{6})^2$$
$$y^2 = 1512$$
$$y = \sqrt{1512} \text{ or } 6\sqrt{42}$$
$$y \approx 38.9$$
$$\frac{6x + x}{z} = \frac{z}{6x}$$
$$z^2 = 42x^2$$
$$z^2 = 42(6\sqrt{6})^2$$
$$z^2 = 9072$$
$$z = \sqrt{9072} \text{ or } 36\sqrt{7}$$
$$z \approx 95.2$$

32. $\frac{12}{x} = \frac{x}{8}$

$x^2 = 96$

$x = \sqrt{96}$ or $4\sqrt{6}$

$x \approx 9.8$

$\frac{12 - 8}{y} = \frac{y}{8}$

$y^2 = 32$

$y = \sqrt{32}$ or $4\sqrt{2}$

$y \approx 5.7$

$\frac{12}{z} = \frac{z}{12 - 8}$

$z^2 = 48$

$z = \sqrt{48}$ or $4\sqrt{3}$

$z \approx 6.9$

33. $\frac{a}{\sqrt{17}} = \frac{\sqrt{17}}{b}$

$ab = 17$

$7b = 17$

$b = \frac{17}{7}$

34. $\frac{x}{\sqrt{12}} = \frac{\sqrt{12}}{y}$

$xy = 12$

$x\sqrt{3} = 12$

$x = \frac{12}{\sqrt{3}}$ or $4\sqrt{3}$

$x \approx 6.9$

35. Never; let x and $x + 1$ be consecutive positive integers. The average of the numbers is $\frac{x + (x + 1)}{2}$ or $\frac{2x + 1}{2}$

$\frac{x}{\frac{2x+1}{2}} \stackrel{?}{=} \frac{\frac{2x+1}{2}}{x + 1}$

$x^2 + x \stackrel{?}{=} \frac{4x^2 + 4x + 1}{4}$

$x^2 + x \stackrel{?}{=} x^2 + x + \frac{1}{4}$

$0 \neq \frac{1}{4}$

Since $0 \neq \frac{1}{4}$, the first equation is never true.

The average of the numbers is not the geometric mean of the numbers.

36. Always; let a and b be positive integers, and let x be the geometric mean of a^2 and b^2.

$\frac{a^2}{x} = \frac{x}{b^2}$

$a^2b^2 = x^2$

$(ab)^2 = x^2$

$ab = x$

Because a and b are positive integers, their product x is a positive integer.

37. Sometimes; let a and b be positive integers, and let x be their geometric mean.

$\frac{a}{x} = \frac{x}{b}$

$ab = x^2$

$\sqrt{ab} = x$

x is an integer when ab is a perfect square.

38. Sometimes; true when the triangle is a right triangle, but not necessarily true otherwise.

39. $\triangle FGH$ is a right triangle. $\overline{OG}$ is the altitude from the vertex of the right angle to the hypotenuse of that triangle. So, by Theorem 7.2, OG is the geometric mean between OF and OH, and so on.

40. Sample answer: The golden ratio occurs when the geometric mean is approximately 1.62.

41. Let x be the length of the brace. Let y be the segment of the hypotenuse adjacent to the leg with measure 3 yards.

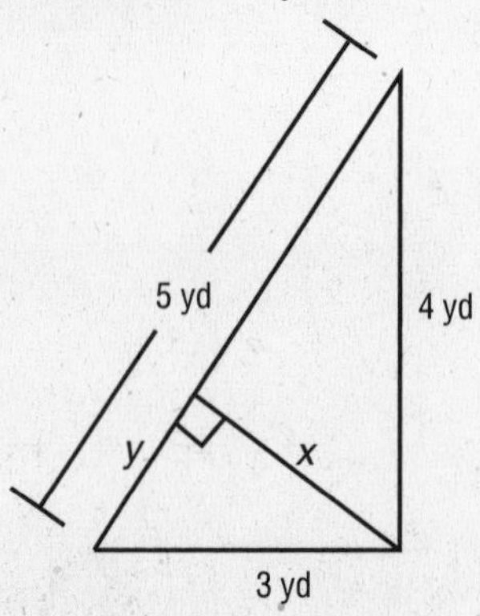

$\frac{5}{3} = \frac{3}{y}$

$5y = 9$

$y = \frac{9}{5}$

$\frac{y}{x} = \frac{x}{5 - y}$

$x^2 = 5y - y^2$

$x^2 = 5\left(\frac{9}{5}\right) - \left(\frac{9}{5}\right)^2$

$x^2 = 9 - \frac{81}{25}$

$x^2 = \frac{144}{25}$

$x = \sqrt{\frac{144}{25}}$ or $\frac{12}{5}$

$x = 2.4$

The brace is 2.4 yards long.

42. Let x be the geometric mean.

The number of players from Indiana is 10, and the number of players from North Carolina is 7.

$\frac{10}{x} = \frac{x}{7}$

$x^2 = 70$

$x = \sqrt{70}$

$x \approx 8.4$

43. Let x be the geometric mean between UCLA and Clemson.

$\frac{15}{x} = \frac{x}{6}$

$x^2 = 90$

$x = \sqrt{90}$

$x = 3\sqrt{10}$

Let y be the geometric mean between Indiana and Virginia.

$\frac{10}{y} = \frac{y}{9}$

$y^2 = 90$

$y = \sqrt{90}$

$y = 3\sqrt{10}$

The geometric mean between Indiana and Virginia is the same as for UCLA and Clemson.

44. $\frac{AD}{CD} = \frac{CD}{BD}$

$\frac{12}{CD} = \frac{CD}{4}$

$(CD)^2 = 48$

$CD = \sqrt{48}$ or $4\sqrt{3}$

$(CB)^2 = (CD)^2 + (DB)^2$

$(CB)^2 = (4\sqrt{3})^2 + 4^2$

$(CB)^2 = 48 + 16$

$(CB)^2 = 64$

$CB = \sqrt{64}$ or 8

$\triangle CED \sim \triangle ACB$, so $\frac{AB}{CD} = \frac{CB}{DE}$.

$\frac{12 + 4}{4\sqrt{3}} = \frac{8}{DE}$

$16DE = 32\sqrt{3}$

$DE = 2\sqrt{3}$

45. Given: $\angle PQR$ is a $\approx$ right angle. $\overline{QS}$ is an altitude of $\triangle PQR$.

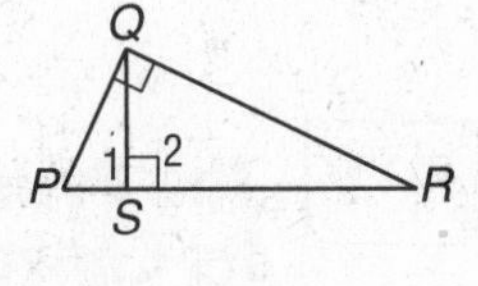

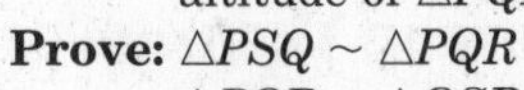

Prove: $\triangle PSQ \sim \triangle PQR$
$\triangle PQR \sim \triangle QSR$
$\triangle PSQ \sim \triangle QSR$

Proof:

Statements	Reasons
1. $\angle PQR$ is a right angle. $\overline{QS}$ is an altitude of $\triangle PQR$.	1. Given
2. $\overline{QS} \perp \overline{RP}$	2. Definition of altitude
3. $\angle 1$ and $\angle 2$ are right angles.	3. Definition of perpendicular lines
4. $\angle 1 \cong \angle PQR$ $\angle 2 \cong \angle PQR$	4. All right ∡ are ≅.
5. $\angle P \cong \angle P$ $\angle R \cong \angle R$	5. Congruence of angles is reflexive.
6. $\triangle PSQ \sim \triangle PQR$ $\triangle PQR \sim \triangle QSR$	6. AA Similarity (statements 4 and 5)
7. $\triangle PSQ \sim \triangle QSR$	7. Similarity of triangles is transitive.

46. Given: $\angle ADC$ is a right angle. $\overline{DB}$ is an altitude of $\triangle ADC$.

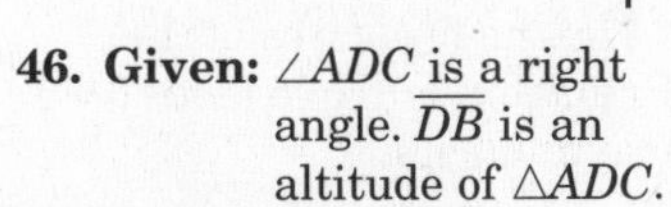

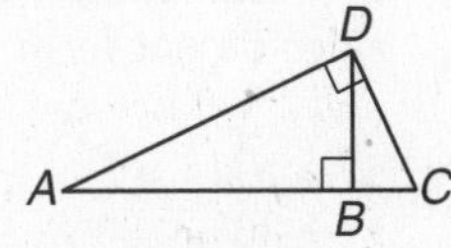

Prove: $\frac{AB}{DB} = \frac{DB}{CB}$

Proof: It is given that $\angle ADC$ is a right angle and $\overline{DB}$ is an altitude of $\triangle ADC$. $\triangle ADC$ is a right triangle by the definition of a right triangle. Therefore, $\triangle ADB \sim \triangle DCB$, because if the altitude is drawn from the vertex of the right angle to the hypotenuse of a right triangle, then the two triangles formed are similar to the given triangle and to each other. So $\frac{AB}{DB} = \frac{DB}{CB}$ by definition of similar polygons.

47. Given: $\angle ADC$ is a right angle. $\overline{DB}$ is an altitude of $\triangle ADC$.

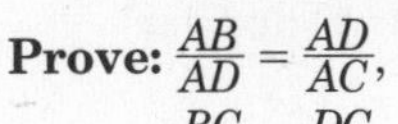

Prove: $\frac{AB}{AD} = \frac{AD}{AC}$, $\frac{BC}{DC} = \frac{DC}{AC}$

Proof:

Statements	Reasons
1. $\angle ADC$ is a right angle. $\overline{DB}$ is an altitude of $\triangle ADC$.	1. Given
2. $\triangle ADC$ is a right triangle.	2. Definition of right triangle
3. $\triangle ABD \sim \triangle ADC$ $\triangle DBC \sim \triangle ADC$	3. If the altitude is drawn from the vertex of the rt. ∠ to the hypotenuse of a rt. △, then the 2 △s formed are similar to the given △ and to each other.
4. $\frac{AB}{AD} = \frac{AD}{AC}$, $\frac{BC}{DC} = \frac{DC}{AC}$	4. Definition of similar polygons

48. Sample answer: The geometric mean can be used to help determine the optimum viewing distance. Answers should include the following.

- If you are too far from a painting, you may not be able to see fine details. If you are too close, you may not be able to see the entire painting.
- A curator can use the geometric mean to help determine how far from the painting the roping should be.

49. C; $\frac{x}{6} = \frac{6}{10}$

$10x = 36$

$x = 3.6$

$\frac{y}{8} = \frac{8}{10}$

$10y = 64$

$y = 6.4$

50. B; $5x^2 + 405 = 1125$

$5x^2 = 720$

$x^2 = 144$

$x = \pm\sqrt{144}$

$x = \pm 12$

Page 348 Maintain Your Skills

51.

x	12	15	18
$x + 3$	15	18	21

52.

x	4	14	44
$3x + 2$	14	44	134

53.

x	3	7	47
$x^2 - 2$	7	47	2207

54.

x	1	−4	−14
$2(x - 3)$	−4	−14	−34

55. The smallest angle is opposite the side with the smallest measure, 20. Let x and $20 - x$ be the measures of the segments formed by the angle bisector. By the Angle Bisector Theorem,

$$\frac{x}{20 - x} = \frac{24}{30}.$$
$$\frac{x}{20 - x} = \frac{24}{30}$$
$$30x = 480 - 24x$$
$$54x = 480$$
$$x = \frac{80}{9} \text{ or } 8\frac{8}{9}$$
$$20 - x = \frac{100}{9} \text{ or } 11\frac{1}{9}$$

The segments have measures $8\frac{8}{9}$ and $11\frac{1}{9}$.

56. By the Exterior Angle Inequality Theorem, $m\angle 8 > m\angle 6$, $m\angle 8 > m\angle 3 + m\angle 4$, and $m\angle 8 > m\angle 2$. Thus, the measures of $\angle 6$, $\angle 4$, $\angle 2$, and $\angle 3$ are all less than $m\angle 8$.

57. By the Exterior Angle Inequality Theorem, $m\angle 1 < m\angle 5$ and $m\angle 1 < m\angle 7$. Thus, the measures of $\angle 5$ and $\angle 7$ are greater than $m\angle 1$.

58. By the Exterior Angle Inequality Theorem, $m\angle 1 < m\angle 7$ and $m\angle 6 < m\angle 7$. Thus, the measures of $\angle 1$ and $\angle 6$ are less than $m\angle 7$.

59. By the Exterior Angle Inequality Theorem, $m\angle 2 > m\angle 6$, $m\angle 7 > m\angle 6$, and $m\angle 8 > m\angle 6$. Thus, the measures of $\angle 2$, $\angle 7$, and $\angle 8$ are all greater than $m\angle 6$.

60. $y = mx + b$
$y = 2x + 4$

61. $m = \frac{-8 - 0}{0 - 2}$
$= \frac{-8}{-2}$ or 4
$y = mx + b$
$y = 4x + (-8)$
$y = 4x - 8$

62. $m = \frac{0 - 6}{-1 - 2}$
$= \frac{-6}{-3}$ or 2
$y - y_1 = m(x - x_1)$
$y - 6 = 2(x - 2)$
$y - 6 = 2x - 4$
$y = 2x + 2$

63. $y - y_1 = m(x - x_1)$
$y - (-3) = -4[x - (-2)]$
$y + 3 = -4(x + 2)$
$y + 3 = -4x - 8$
$y = -4x - 11$

64. $c^2 = 3^2 + 4^2$
$c^2 = 9 + 16$
$c^2 = 25$
$c = \sqrt{25}$ or 5
The length of the hypotenuse is 5 cm.

65. $c^2 = 5^2 + 12^2$
$c^2 = 25 + 144$
$c^2 = 169$
$c = \sqrt{169}$ or 13
The length of hypotenuse is 13 feet.

66. $c^2 = 3^2 + 5^2$
$c^2 = 9 + 25$
$c^2 = 34$
$c = \sqrt{34}$
$c \approx 5.8$
The length of the hypotenuse is about 5.8 inches.

Page 349 Geometry Activity: The Pythagorean Theorem

1. yes
2. $a^2 + b^2 = c^2$
3. Sample answer: The sum of the areas of the two smaller squares is equal to the area of the largest square.

7-2 The Pythagorean Theorem and Its Converse

Pages 353–354 Check for Understanding

1. Maria; Colin does not have the longest side as the value of c.
2. Since the numbers in a Pythagorean triple satisfy the equation $a^2 + b^2 = c^2$, they represent the sides of a right triangle by the converse of the Pythagorean Theorem.
3.

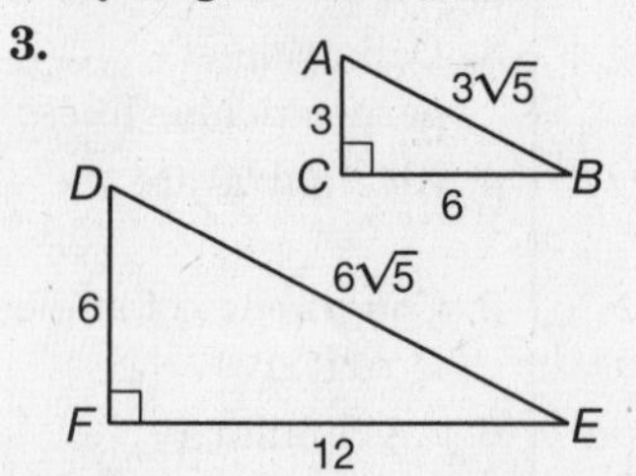

Sample answer: $\triangle ABC \sim \triangle DEF$, $\angle A \cong \angle D$, $\angle B \cong \angle E$, and $\angle C \cong \angle F$, $\overline{AB}$ corresponds to $\overline{DE}$, $\overline{BC}$ corresponds to $\overline{EF}$, $\overline{AC}$ corresponds to $\overline{DF}$. The scale factor is $\frac{2}{1}$. No; the measures of the sides do not form a Pythagorean triple since $6\sqrt{5}$ and $3\sqrt{5}$ are not whole numbers.

4. $x^2 + 6^2 = 10^2$
$x^2 + 36 = 100$
$x^2 = 64$
$x = \sqrt{64}$ or 8

5. $x^2 + \left(\frac{4}{7}\right)^2 = \left(\frac{5}{7}\right)^2$
$x^2 + \frac{16}{49} = \frac{25}{49}$
$x^2 = \frac{9}{49}$
$x = \sqrt{\frac{9}{49}}$ or $\frac{3}{7}$

6. $20^2 + 37.5^2 = x^2$
$400 + 1406.25 = x^2$
$1806.25 = x^2$
$\sqrt{1806.25} = x$
$42.5 = x$

7.

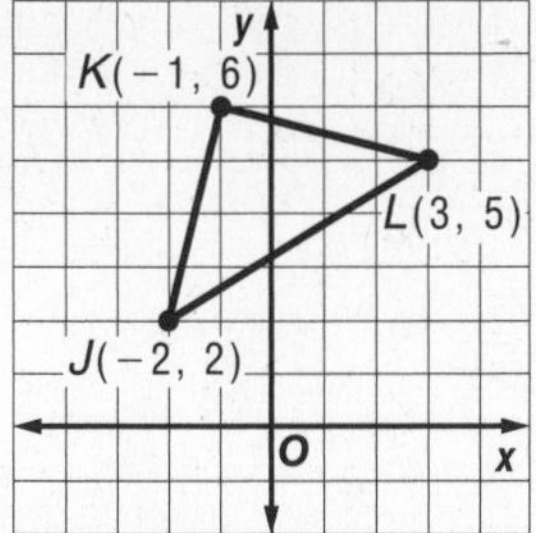

$JK = \sqrt{[-1 - (-2)]^2 + (6 - 2)^2}$
$= \sqrt{1^2 + 4^2}$ or $\sqrt{17}$
$KL = \sqrt{[3 - (-1)]^2 + (5 - 6)^2}$
$= \sqrt{4^2 + (-1)^2}$ or $\sqrt{17}$
$JL = \sqrt{[3 - (-2)]^2 + (5 - 2)^2}$
$= \sqrt{5^2 + 3^2}$ or $\sqrt{34}$

$JK^2 + KL^2 \stackrel{?}{=} JL^2$
$\left(\sqrt{17}\right)^2 + \left(\sqrt{17}\right)^2 \stackrel{?}{=} \left(\sqrt{34}\right)^2$
$17 + 17 \stackrel{?}{=} 34$
$34 = 34$

Yes; $\triangle JKL$ is a right triangle since the sum of the squares of two sides equals the square of the longest side.

8. Since the measure of the longest side is 39, 39 must be c, and a or b are 15 and 36.

$a^2 + b^2 = c^2$
$15^2 + 36^2 \stackrel{?}{=} 39^2$
$225 + 1296 \stackrel{?}{=} 1521$
$1521 = 1521$

These segments form the sides of a right triangle since they satisfy the Pythagorean Theorem. The measures are whole numbers and form a Pythagorean triple.

9. Since the measure of the longest side is 21, 21 must be c, and a or b are $\sqrt{40}$ and 20.

$a^2 + b^2 = c^2$
$\left(\sqrt{40}\right)^2 + 20^2 \stackrel{?}{=} 21^2$
$40 + 400 \stackrel{?}{=} 441$
$440 \neq 441$

Since $400 \neq 441$, segments with these measures cannot form a right triangle. Therefore, they do not form a Pythagorean triple.

10. Since the measure of the longest side is $\sqrt{108}$, $\sqrt{108}$ must be c, and a or b are $\sqrt{44}$ and 8.

$a^2 + b^2 = c^2$
$\left(\sqrt{44}\right)^2 + 8^2 \stackrel{?}{=} \left(\sqrt{108}\right)^2$
$44 + 64 \stackrel{?}{=} 108$
$108 = 108$

Since $108 = 108$, segments with these measures form a right triangle. However the three numbers are not all whole numbers. Therefore, they do not form a Pythagorean triple.

11. Let x represent the width of the screen.

$11.5^2 + x^2 = 19^2$
$132.25 + x^2 = 361$
$x^2 = 228.75$
$x = \sqrt{228.75}$
$x \approx 15.1$

The screen is about 15.1 inches wide.

Pages 354–356 Practice and Apply

12. The altitude divides the side of measure 14 into two congruent segments of measure 7 because the triangle is an isosceles triangle.

$x^2 + 7^2 = 8^2$
$x^2 + 49 = 64$
$x^2 = 15$
$x = \sqrt{15}$
$x \approx 3.9$

13. $x^2 + 4^2 = 8^2$
$x^2 + 16 = 64$
$x^2 = 48$
$x = \sqrt{48}$ or $4\sqrt{3}$
$x \approx 6.9$

14. $20^2 + 28^2 = x^2$
$400 + 784 = x^2$
$1184 = x^2$
$\sqrt{1184} = x$
$4\sqrt{74} = x$
$34.4 \approx x$

15. $40^2 + 32^2 = x^2$
$1600 + 1024 = x^2$
$2624 = x^2$
$\sqrt{2624} = x$
$8\sqrt{41} = x$
$51.2 \approx x$

16. $x^2 + 25^2 = 33^2$
$x^2 + 625 = 1089$
$x^2 = 464$
$x = \sqrt{464}$
$x = 4\sqrt{29}$
$x \approx 21.5$

17. $x^2 + 15^2 = 25^2$
$x^2 + 225 = 625$
$x^2 = 400$
$x = \sqrt{400}$ or 20

18.

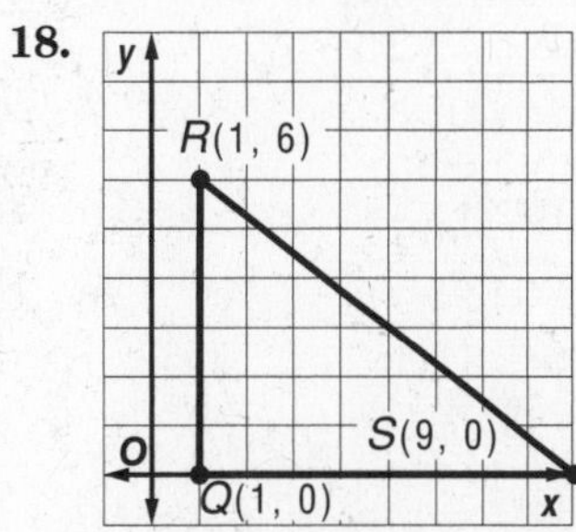

$QR = \sqrt{(1 - 1)^2 + (6 - 0)^2}$
$= \sqrt{0^2 + 6^2}$
$= \sqrt{36}$ or 6
$RS = \sqrt{(9 - 1)^2 + (0 - 6)^2}$
$= \sqrt{8^2 + (-6)^2}$
$= \sqrt{100}$ or 10
$QS = \sqrt{(9 - 1)^2 + (0 - 0)^2}$
$= \sqrt{8^2 + 0^2}$
$= \sqrt{64}$ or 8

$QR^2 + QS^2 \stackrel{?}{=} RS^2$
$6^2 + 8^2 \stackrel{?}{=} 10^2$
$36 + 64 \stackrel{?}{=} 100$
$100 = 100$

Since the sum of the squares of two sides equals the square of the longest side, $\triangle QRS$ is a right triangle.

19. 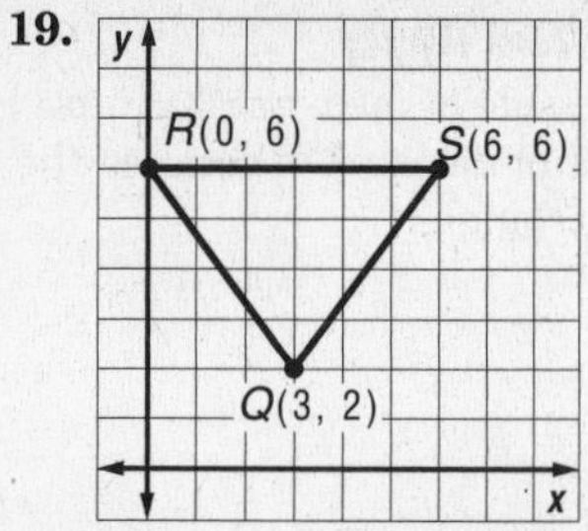

$QR = \sqrt{(0-3)^2 + (6-2)^2}$
$= \sqrt{(-3)^2 + 4^2}$
$= \sqrt{25}$ or 5

$RS = \sqrt{(6-0)^2 + (6-6)^2}$
$= \sqrt{6^2 + 0^2}$
$= \sqrt{36}$ or 6

$QS = \sqrt{(6-3)^2 + (6-2)^2}$
$= \sqrt{3^2 + 4^2}$
$= \sqrt{25}$ or 5

$QR^2 + QS^2 \stackrel{?}{=} RS^2$
$5^2 + 5^2 \stackrel{?}{=} 6^2$
$25 + 25 \stackrel{?}{=} 36$
$50 \neq 36$

Since the sum of the squares of two sides is not equal to the square of the longest side, $\triangle QRS$ is not a right triangle.

20. 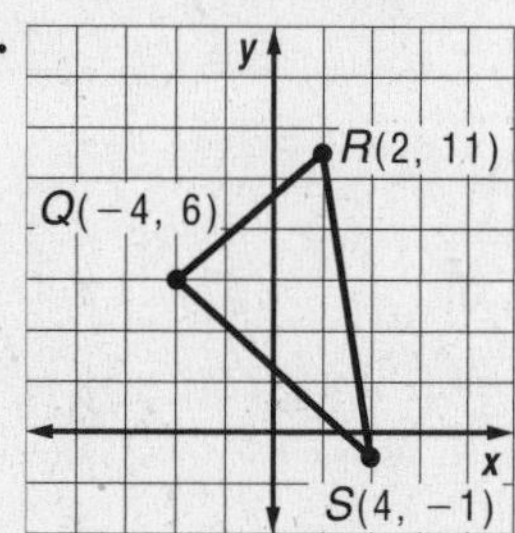

$QR = \sqrt{[2-(-4)]^2 + (11-6)^2}$
$= \sqrt{6^2 + 5^2}$ or $\sqrt{61}$

$RS = \sqrt{(4-2)^2 + (-1-11)^2}$
$= \sqrt{2^2 + (-12)^2}$ or $\sqrt{148}$

$QS = \sqrt{[4-(-4)]^2 + (-1-6)^2}$
$= \sqrt{8^2 + (-7)^2}$ or $\sqrt{113}$

$QR^2 + QS^2 \stackrel{?}{=} RS^2$
$\left(\sqrt{61}\right)^2 + \left(\sqrt{113}\right)^2 \stackrel{?}{=} \left(\sqrt{148}\right)^2$
$61 + 113 \stackrel{?}{=} 148$
$174 \neq 148$

Since the sum of the squares of two sides is not equal to the square of the longest side, $\triangle QRS$ is not a right triangle.

21. 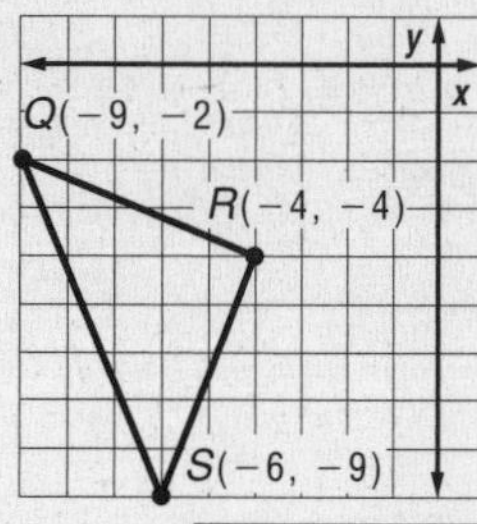

$QR = \sqrt{[-4-(-9)]^2 + [-4-(-2)]^2}$
$= \sqrt{5^2 + (-2)^2}$ or $\sqrt{29}$

$RS = \sqrt{[-6-(-4)]^2 + [-9-(-4)]^2}$
$= \sqrt{(-2)^2 + (-5)^2}$ or $\sqrt{29}$

$QS = \sqrt{[-6-(-9)]^2 + [-9-(-2)]^2}$
$= \sqrt{3^2 + (-7)^2}$ or $\sqrt{58}$

$QR^2 + RS^2 \stackrel{?}{=} QS^2$
$(\sqrt{29})^2 + (\sqrt{29})^2 \stackrel{?}{=} (\sqrt{58})^2$
$29 + 29 \stackrel{?}{=} 58$
$58 = 58$

Since the sum of the squares of two sides equals the square of the longest side, $\triangle QRS$ is a right triangle.

22. Since the measure of the longest side is 17, 17 must be c, and a or b are 8 and 15.
$a^2 + b^2 = c^2$
$8^2 + 15^2 \stackrel{?}{=} 17^2$
$64 + 225 \stackrel{?}{=} 289$
$289 = 289$

These segments form the sides of a right triangle since they satisfy the Pythagorean Theorem. The measures are whole numbers and form a Pythagorean triple.

23. Since the measure of the longest side is 25, 25 must be c, and a or b are 7 and 24.
$a^2 + b^2 = c^2$
$7^2 + 24^2 \stackrel{?}{=} 25^2$
$49 + 576 \stackrel{?}{=} 625$
$625 = 625$

These segments form the sides of a right triangle since they satisfy the Pythagorean Theorem. The measures are whole numbers and form a Pythagorean triple.

24. Since the measure of the longest side is 31, 31 must be c, and a or b are 20 and 21.
$a^2 + b^2 = c^2$
$20^2 + 21^2 \stackrel{?}{=} 31^2$
$400 + 441 \stackrel{?}{=} 961$
$841 \neq 961$

Since $841 \neq 961$, segments with these measures cannot form a right triangle. Therefore, they do not form a Pythagorean triple.

25. Since the measure of the longest side is 37, 37 must be c, and a or b are 12 and 34.
$a^2 + b^2 = c^2$
$12^2 + 34^2 \stackrel{?}{=} 37^2$
$144 + 1156 \stackrel{?}{=} 1369$
$1300 \neq 1369$

Since $1300 \neq 1369$, segments with these measures cannot form a right triangle. Therefore, they do not form a Pythagorean triple.

26. Since the measure of the longest side is $\frac{\sqrt{74}}{35}$, $\frac{\sqrt{74}}{35}$ must be c, and a or b are $\frac{1}{5}$ and $\frac{1}{7}$.

$$a^2 + b^2 = c^2$$
$$\left(\frac{1}{5}\right)^2 + \left(\frac{1}{7}\right)^2 \stackrel{?}{=} \left(\frac{\sqrt{74}}{35}\right)^2$$
$$\frac{1}{25} + \frac{1}{49} \stackrel{?}{=} \frac{74}{1225}$$
$$\frac{74}{1225} = \frac{74}{1225}$$

These segments form the sides of a right triangle since they satisfy the Pythagorean Theorem. However, the three numbers are not whole numbers. Therefore, they do not form a Pythagorean triple.

27. Since the measure of the longest side is $\frac{35}{36}$, $\frac{35}{36}$ must be c, and a or b are $\frac{\sqrt{3}}{2}$ and $\frac{\sqrt{2}}{3}$.

$$a^2 + b^2 = c^2$$
$$\left(\frac{\sqrt{3}}{2}\right)^2 + \left(\frac{\sqrt{2}}{3}\right)^2 \stackrel{?}{=} \left(\frac{35}{36}\right)^2$$
$$\frac{3}{4} + \frac{2}{9} \stackrel{?}{=} \frac{1225}{1296}$$
$$\frac{35}{36} \neq \frac{1225}{1296}$$

Since $\frac{35}{36} \neq \frac{1225}{1296}$, segments with these measures cannot form a right triangle. Therefore, they do not form a Pythagorean triple.

28. Since the measure of the longest side is 1, 1 must be c, and a or b are $\frac{3}{5}$ and $\frac{4}{5}$.

$$a^2 + b^2 = c^2$$
$$\left(\frac{3}{5}\right)^2 + \left(\frac{4}{5}\right)^2 \stackrel{?}{=} 1^2$$
$$\frac{9}{25} + \frac{16}{25} \stackrel{?}{=} 1$$
$$\frac{25}{25} = 1$$

These segments form the sides of a right triangle since they satisfy the Pythagorean Theorem. However, the three numbers are not all whole numbers. Therefore, they do not form a Pythagorean triple.

29. Since the measure of the longest side is $\frac{10}{7}$, $\frac{10}{7}$ must be c, and a or b are $\frac{6}{7}$ and $\frac{8}{7}$.

$$a^2 + b^2 = c^2$$
$$\left(\frac{6}{7}\right)^2 + \left(\frac{8}{7}\right)^2 \stackrel{?}{=} \left(\frac{10}{7}\right)^2$$
$$\frac{36}{49} + \frac{64}{49} \stackrel{?}{=} \frac{100}{49}$$
$$\frac{100}{49} = \frac{100}{49}$$

Since $\frac{100}{49} = \frac{100}{49}$, segments with these measures form a right triangle. However, the three numbers are not whole numbers. Therefore, they do not form a Pythagorean triple.

30.

5	12	13
10	24	**? 26**
15	**? 36**	39
? 20	48	52

31. 5-12-13

32. Sample answer: The triples are all multiples of the triple 5-12-13.

33. Sample answer: They consist of any number of similar triangles.

34. Yes; the measures of the sides are always multiples of 5, 12, and 13.

35a. Find multiples of the triple 8, 15, 17.

$$8 \cdot 2 = 16$$
$$15 \cdot 2 = 30$$
$$17 \cdot 2 = 34$$
$$16^2 + 30^2 \stackrel{?}{=} 34^2$$
$$256 + 900 \stackrel{?}{=} 1156$$
$$1156 = 1156$$
$$8 \cdot 3 = 24$$
$$15 \cdot 3 = 45$$
$$17 \cdot 3 = 51$$
$$24^2 + 45^2 \stackrel{?}{=} 51^2$$
$$576 + 2025 \stackrel{?}{=} 2601$$
$$2601 = 2601$$

Two triples are 16-30-34 and 24-45-51.

b. Find multiples of the triple 9, 40, 41.

$$9 \cdot 2 = 18$$
$$40 \cdot 2 = 80$$
$$41 \cdot 2 = 82$$
$$18^2 + 80^2 \stackrel{?}{=} 82^2$$
$$324 + 6400 \stackrel{?}{=} 6724$$
$$6724 = 6724$$
$$9 \cdot 3 = 27$$
$$40 \cdot 3 = 120$$
$$41 \cdot 3 = 123$$
$$27^2 + 120^2 \stackrel{?}{=} 123^2$$
$$729 + 14{,}400 \stackrel{?}{=} 15{,}129$$
$$15{,}129 = 15{,}129$$

Two triples are 18-80-82 and 27-120-123.

c. Find multiples of the triple 7, 24, 25.

$$7 \cdot 2 = 14$$
$$24 \cdot 2 = 48$$
$$25 \cdot 2 = 50$$
$$14^2 + 48^2 \stackrel{?}{=} 50^2$$
$$196 + 2304 \stackrel{?}{=} 2500$$
$$2500 = 2500$$
$$7 \cdot 3 = 21$$
$$24 \cdot 3 = 72$$
$$25 \cdot 3 = 75$$
$$21^2 + 72^2 \stackrel{?}{=} 75^2$$
$$441 + 5184 \stackrel{?}{=} 5625$$
$$5625 = 5625$$

Two triples are 14-48-50 and 21-72-75.

36.
$$d = \sqrt{(105 - 122)^2 + (40 - 38)^2}$$
$$= \sqrt{(-17)^2 + 2^2}$$
$$= \sqrt{289 + 4}$$
$$= \sqrt{293}$$
$$\approx 17.1$$

The distance from San Francisco to Denver is about 17.1 degrees.

37.
$$d = \sqrt{(105 - 115)^2 + (40 - 36)^2}$$
$$= \sqrt{(-10)^2 + 4^2}$$
$$= \sqrt{100 + 16}$$
$$= \sqrt{116}$$
$$\approx 10.8$$

The distance from Las Vegas to Denver is about 10.8 degrees.

38. Given: $\triangle ABC$ with sides of measure a, b, and c, where $c^2 = a^2 + b^2$

Prove: $\triangle ABC$ is a right triangle.

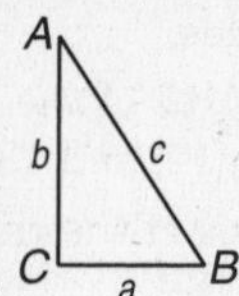

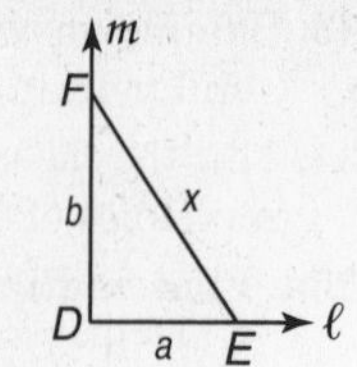

Proof: Draw $\overline{DE}$ on line ℓ with measure equal to a. At D, draw line $m \perp \overline{DE}$. Locate point F on m so that $DF = b$. Draw $\overline{FE}$ and call its measure x. Because $\triangle FED$ is a right triangle, $a^2 + b^2 = x^2$. But $a^2 + b^2 = c^2$, so $x^2 = c^2$ or $x = c$. Thus, $\triangle ABC \cong \triangle FED$ by SSS. This means $\angle C \cong \angle DE$. Therefore, $\angle C$ must be a right angle, making $\triangle ABC$ a right triangle.

39. Given: $\triangle ABC$ with right angle at C, $AB = d$

Prove: $d = \sqrt{(x_2 - x_1)^2 + (y_2 - y_1)^2}$

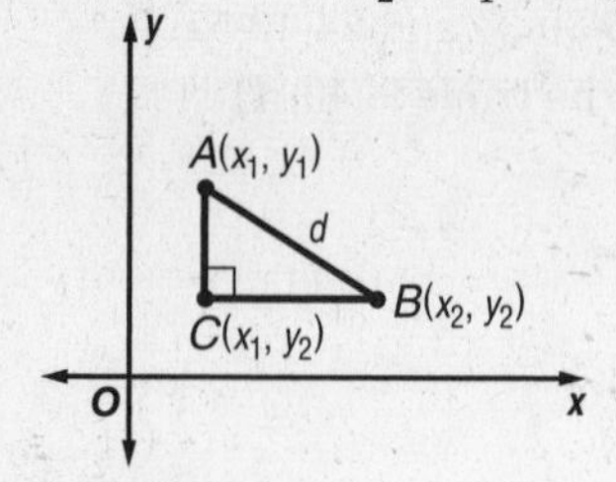

Proof:

Statements	Reasons
1. $\triangle ABC$ with right angle at C, $AB = d$	1. Given
2. $(CB)^2 + (AC)^2 = (AB)^2$	2. Pythagorean Theorem
3. $\lvert x_2 - x_1 \rvert = CB$ $\lvert y_2 - y_1 \rvert = AC$	3. Distance on a number line
4. $\lvert x_2 - x_1 \rvert^2 + \lvert y_2 - y_1 \rvert^2 = d^2$	4. Substitution
5. $(x_2 - x_1)^2 + (y_2 - y_1)^2 = d^2$	5. Substitution
6. $\sqrt{(x_2 - x_1)^2 + (y_2 - y_1)^2} = d$	6. Take the square root of each side.
7. $d = \sqrt{(x_2 - x_1)^2 + (y_2 - y_1)^2}$	7. Reflexive Property

40. First, use the Pythagorean Theorem to find the length of the ladder, represented by y.

$12^2 + 16^2 = y^2$
$144 + 256 = y^2$
$400 = y^2$
$\sqrt{400} = y$
$20 = y$

The ladder is 20 feet long.

$(2 + 12)^2 + x^2 = 20^2$
$14^2 + x^2 = 20^2$
$196 + x^2 = 400$
$x^2 = 204$
$x = \sqrt{204}$
$x = 2\sqrt{51}$
$x \approx 14.3$

The ladder reaches about 14.3 feet up the side of the house.

41. Let x be the hypotenuse of the triangle with height 26 + 12 or 38 and base $\frac{1}{2}(9)$ or 4.5.

$4.5^2 + 38^2 = x^2$
$20.25 + 1444 = x^2$
$1464.25 = x^2$
$\sqrt{1464.25} = x$

The length of wire needed is $2x = 2\sqrt{1464.25}$ or about 76.53 feet.

42. Let s represent the side of each square stone.

$s^2 + (2s)^2 = x^2$
$s^2 + 4s^2 = 15^2$
$5s^2 = 225$
$s^2 = 45$
$s = \sqrt{45}$ or $3\sqrt{5}$

The area of each square stone is $s^2 = 45$, so the area of the walkway is $6 \cdot 45$ or 270 in^2.

43. Let x represent the number of miles away from the starting point.

$6^2 + 12^2 = x^2$
$36 + 144 = x^2$
$180 = x^2$
$\sqrt{180} = x$
$13.4 \approx x$

The trawler traveled about 13.4 miles out of the way.

44. $AD = BC = 6$

$AD^2 + AB^2 = BD^2$
$6^2 + 8^2 = BD^2$
$36 + 64 = BD^2$
$100 = BD^2$
$\sqrt{100} = BD$
$10 = BD$

$HD = BF = 8$

$DM = \frac{1}{2}(BD)$
$= \frac{1}{2}(10)$
$= 5$

$HD^2 + DM^2 = HM^2$
$8^2 + 5^2 = HM^2$
$64 + 25 = HM^2$
$89 = HM^2$
$\sqrt{89} = HM$
$9.4 \approx HM$

$HM = EM = FM = GM$ because $HD = EA = FB = GC$ and $AC = DB$ so $AM = BM = CM = DM$.

45. Sample answer: The road, the tower that is perpendicular to the road, and the cables form the right triangles. Answers should include the following.

- Right triangles are formed by the bridge, the towers, and the cables.
- The cable is the hypotenuse in each triangle.

46. A; let $B(8, 0)$ be the vertex of right triangle ABE.

$AB^2 + BE^2 = AE^2$
$8^2 + h^2 = 10^2$
$64 + h^2 = 100$
$h^2 = 36$
$h = \sqrt{36}$
$h = 6$

47. C; $x^2 + 36 = (9 - x)^2$
$$x^2 + 36 = 81 - 18x + x^2$$
$$36 = 81 - 18x$$
$$-45 = -18x$$
$$2.5 = x$$

48. 3-4-5, 6-8-10, 12-16-20, 24-32-40, 27-36-45

49. $3 \cdot 4 \cdot 5 = 60$
$6 \cdot 8 \cdot 10 = 480$
$= 8 \cdot 60$
$12 \cdot 16 \cdot 20 = 3840$
$= 64 \cdot 60$
$24 \cdot 32 \cdot 40 = 30{,}720$
$= 512 \cdot 60$
$27 \cdot 36 \cdot 45 = 43{,}740$
$= 729 \cdot 60$
Yes, the conjecture holds true.

Page 356 Maintain Your Skills

50. Let x represent the geometric mean.
$\frac{3}{x} = \frac{x}{12}$
$x^2 = 36$
$x = \sqrt{36}$ or 6

51. Let x represent the geometric mean.
$\frac{9}{x} = \frac{x}{12}$
$x^2 = 108$
$x = \sqrt{108}$ or $6\sqrt{3}$
$x \approx 10.4$

52. Let x represent the geometric mean.
$\frac{11}{x} = \frac{x}{7}$
$x^2 = 77$
$x = \sqrt{77}$
$x \approx 8.8$

53. Let x represent the geometric mean.
$\frac{6}{x} = \frac{x}{9}$
$x^2 = 54$
$x = \sqrt{54}$ or $3\sqrt{6}$
$x \approx 7.3$

54. Let x represent the geometric mean.
$\frac{2}{x} = \frac{x}{7}$
$x^2 = 14$
$x = \sqrt{14}$
$x \approx 3.7$

55. Let x represent the geometric mean.
$\frac{2}{x} = \frac{x}{5}$
$x^2 = 10$
$x = \sqrt{10}$
$x \approx 3.2$

56.

x	5	$\sqrt{10}$	2.51...	2.24...	2.11...
$\sqrt{2x}$	$\sqrt{10}$	2.51...	2.24...	2.11...	2.05...

The sequence of numbers converges to 2.

57.

x	1	3	27
3^x	3	27	7.6×10^{12}

The sequence of numbers approaches positive infinity.

58.

x	4	2	1.41...	1.18...	1.09...	1.04...
$x^{\frac{1}{2}}$	2	1.41...	1.18...	1.09...	1.04...	1.02...

The sequence of numbers converges to 1.

59.

x	4	0.25	4	0.25
$\frac{1}{x}$	0.25	4	0.25	4

The sequence of numbers alternates between 0.25 and 4.

60. No; $12 + 13 \not> 25$, so the sides do not satisfy the Triangle Inequality.

61. $\frac{7}{\sqrt{3}} \cdot \frac{\sqrt{3}}{\sqrt{3}} = \frac{7\sqrt{3}}{3}$

62. $\frac{18}{\sqrt{2}} \cdot \frac{\sqrt{2}}{\sqrt{2}} = \frac{18\sqrt{2}}{2} = 9\sqrt{2}$

63. $\frac{14}{\sqrt{2}} \cdot \frac{\sqrt{2}}{\sqrt{2}} = \frac{\sqrt{28}}{2}$
$= \frac{2\sqrt{7}}{2} = \sqrt{7}$

64. $\frac{3\sqrt{11}}{\sqrt{3}} \cdot \frac{\sqrt{3}}{\sqrt{3}} = \frac{3\sqrt{33}}{3} = \sqrt{33}$

65. $\frac{24}{\sqrt{2}} \cdot \frac{\sqrt{2}}{\sqrt{2}} = \frac{24\sqrt{2}}{2} = 12\sqrt{2}$

66. $\frac{12}{\sqrt{3}} \cdot \frac{\sqrt{3}}{\sqrt{3}} = \frac{12\sqrt{3}}{3} = 4\sqrt{3}$

67. $\frac{2\sqrt{6}}{\sqrt{3}} \cdot \frac{\sqrt{3}}{\sqrt{3}} = \frac{2\sqrt{18}}{\sqrt{3}}$
$= \frac{6\sqrt{2}}{3} = 2\sqrt{2}$

68. $\frac{15}{\sqrt{3}} \cdot \frac{\sqrt{3}}{\sqrt{3}} = \frac{15\sqrt{3}}{3} = 5\sqrt{3}$

69. $\frac{2}{\sqrt{8}} \cdot \frac{\sqrt{8}}{\sqrt{8}} = \frac{2\sqrt{8}}{8}$
$= \frac{4\sqrt{2}}{8} = \frac{\sqrt{2}}{2}$

70. $\frac{25}{\sqrt{10}} \cdot \frac{\sqrt{10}}{\sqrt{10}} = \frac{25\sqrt{10}}{10} = \frac{5\sqrt{10}}{2}$

7-3 Special Right Triangles

Page 360 Check for Understanding

1. Sample answer: Construct two perpendicular lines. Use a ruler to measure 3 cm from the point of intersection on one ray. Use the compass to copy the 3 cm segment. Connect the two endpoints to form a 45°-45°-90° triangle with sides of 3 cm and a hypotenuse of $3\sqrt{2}$ cm.

2. Sample answer: Draw a line using a ruler. Then use a protractor to measure a 90° angle. On one ray mark a 2 cm length, and at that endpoint use the protractor to measure a 30° angle toward the other ray. Where this ray intersects the other ray should form a 60° angle, completing the 30°-60°-90° triangle with sides 2 cm, $2\sqrt{3}$cm, and a hypotenuse of 4 cm.

3. The diagonal is twice as long as its width w, so the diagonal forms a 30°-60°-90° triangle with the length ℓ and the width w. Then the length of the rectangle is $\sqrt{3}$ times the width, or $\ell = \sqrt{3}w$.

4. The triangle is a 45°-45°-90° triangle. The legs are congruent, so $x = 3$. The length of the hypotenuse is $\sqrt{2}$ times the length of the leg, so $y = 3\sqrt{2}$.

5. The triangle is a 45°-45°-90° triangle. The legs are congruent, so $x = y$. The length of the hypotenuse is $\sqrt{2}$ times the length of the leg.

$$10 = x\sqrt{2}$$
$$\frac{10}{\sqrt{2}} = x$$
$$\frac{10}{\sqrt{2}} \cdot \frac{\sqrt{2}}{\sqrt{2}} = x$$
$$\frac{10\sqrt{2}}{2} = x$$
$$5\sqrt{2} = x$$

So, $x = 5\sqrt{2}$ and $y = 5\sqrt{2}$.

6. The triangle is a 30°-60°-90° triangle. y is the measure of the hypotenuse and x is the measure of the longer leg.

$y = 2(8)$ or 16
$x = 8\sqrt{3}$

7. $\triangle ABC$ is a 30°-60°-90° triangle with hypotenuse c, shorter leg a and longer leg b.

$c = 2a$
$8 = 2a$
$4 = a$
$b = \sqrt{3}(a)$
$b = \sqrt{3}(4)$ or $4\sqrt{3}$

8. $\triangle ABC$ is a 30°-60°-90° triangle with hypotenuse c, shorter leg a and longer leg b.

$$b = \sqrt{3}(a)$$
$$18 = \sqrt{3}(a)$$
$$\frac{18}{\sqrt{3}} = a$$
$$\frac{18}{\sqrt{3}} \cdot \frac{\sqrt{3}}{\sqrt{3}} = a$$
$$\frac{18\sqrt{3}}{3} = a$$
$$6\sqrt{3} = a$$

$c = 2a$
$c = 2(6\sqrt{3})$
$c = 12\sqrt{3}$

9. 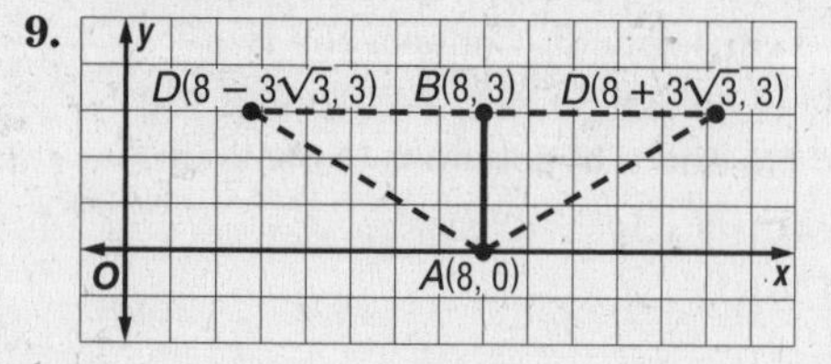

Graph A and B. $\overline{AB}$ lies on a vertical gridline of the coordinate plane. Since $\overline{BD}$ will be perpendicular to $\overline{AB}$, it lies on a horizontal gridline.

$AB = |3 - 0| = 3$
$\overline{AB}$ is the shorter leg. $\overline{BD}$ is the longer leg.
$BD = \sqrt{3}(AB)$
$BD = \sqrt{3}(3)$ or $3\sqrt{3}$

Point D has the same y-coordinate as B. D is located $3\sqrt{3}$ units to the right of B or to the left of B. So, the coordinates of D are $(8 + 3\sqrt{3}, 3)$ or about (13.20, 3) or $(8 - 3\sqrt{3}, 3)$ or about (2.80, 3).

10. 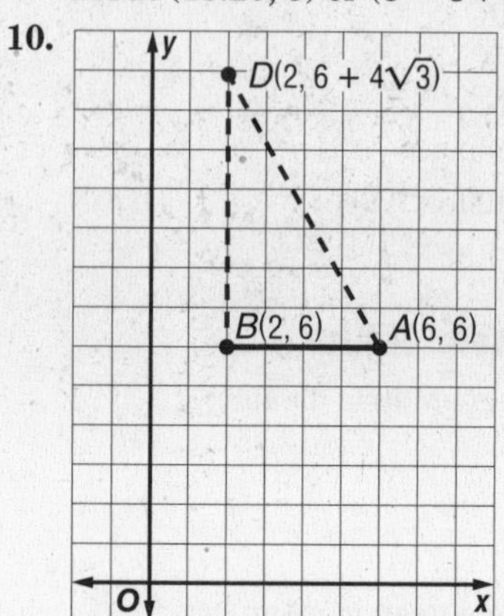

Graph A and B. $\overline{AB}$ lies on a horizontal gridline of the coordinate plane. Since $\overline{BD}$ will be perpendicular to $\overline{AB}$, it lies on a vertical gridline.

$AB = |2 - 6| = 4$
$\overline{AB}$ is the shorter leg. $\overline{BD}$ is the longer leg.
$BD = \sqrt{3}(AB)$
$BD = \sqrt{3}(4)$ or $4\sqrt{3}$

Point D has the same x-coordinate as B. D is located $4\sqrt{3}$ units above B. So, the coordinates of D are $(2, 6 + 4\sqrt{3})$ or about (2, 12.93).

11. The length of each leg of the 45°-45°-90° triangle formed by homeplate, first base, and second base is 90 feet. The distance d is the hypotenuse and is $\sqrt{2}$ times as long as a leg. Then $d = 90\sqrt{2}$ or about 127.28 feet.

Pages 360–362 Practice and Apply

12. The figure is a square, so each triangle is a 45°-45°-90° triangle. Thus, $x = 45$. The length of the hypotenuse is $\sqrt{2}$ times the length of a leg of the triangle. Therefore, $y = \sqrt{2}(9.6)$ or $9.6\sqrt{2}$.

13. The figure is a square, so each triangle is a 45°-45°-90° triangle. Thus, $y = 45$. the length of the hypotenuse is $\sqrt{2}$ times the length of a leg of the triangle.

$$17 = x\sqrt{2}$$
$$\frac{17}{\sqrt{2}} = x$$
$$\frac{17}{\sqrt{2}} \cdot \frac{\sqrt{2}}{\sqrt{2}} = x$$
$$\frac{17\sqrt{2}}{2} = x$$

14. The triangle is a 30°-60°-90° triangle where x is the measure of the shorter leg and y is the measure of the longer leg.

$18 = 2x$
$9 = x$
$y = \sqrt{3}(x)$
$y = \sqrt{3}(9)$ or $9\sqrt{3}$

15.

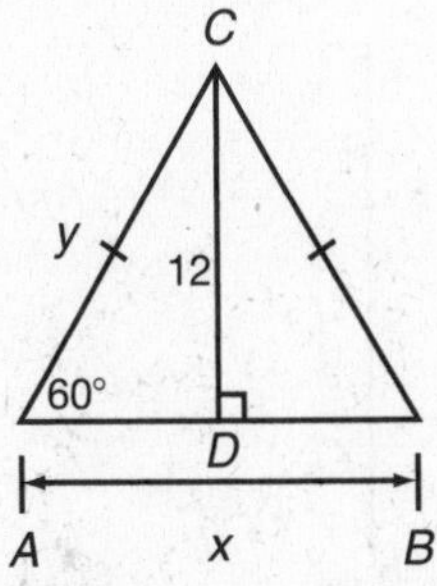

$\triangle ABC$ is equilateral because $\overline{CD}$ bisects the base and $\overline{CD}$ is an altitude. Therefore $\triangle ADC$ is a 30°-60°-90° triangle with hypotenuse y and shorter leg $\frac{x}{2}$.

$\frac{x}{2} \cdot \sqrt{3} = 12$
$x = 12 \cdot \frac{2}{\sqrt{3}}$
$x = \frac{24}{\sqrt{3}} \cdot \frac{\sqrt{3}}{\sqrt{3}}$
$x = \frac{24\sqrt{3}}{3}$
$x = 8\sqrt{3}$
$y = 2\left(\frac{x}{2}\right)$
$y = x$
$y = 8\sqrt{3}$

16. x is the shorter leg and y is the longer leg of the 30°-60°-90° triangle.

$2x = 11$
$x = \frac{11}{2}$ or 5.5
$y = \sqrt{3}(x)$
$y = \sqrt{3}(5.5)$ or $5.5\sqrt{3}$

17.

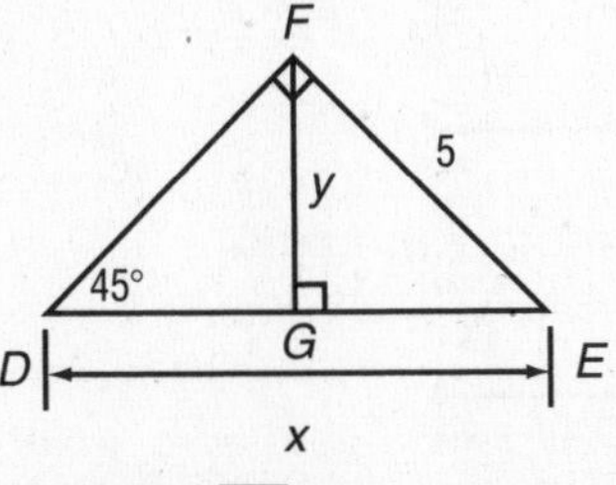

In $\triangle DEF$, $\overline{DE}$ is the hypotenuse so $x = 5\sqrt{2}$. In $\triangle FGE$, $\overline{FG}$ is a leg and $\overline{FE}$ is the hypotenuse.

$\sqrt{2}(y) = 5$
$y = \frac{5}{\sqrt{2}}$
$y = \frac{5}{\sqrt{2}} \cdot \frac{\sqrt{2}}{\sqrt{2}}$
$y = \frac{5\sqrt{2}}{2}$

18. In $\triangle BCE$, a is the measure of the hypotenuse and $\overline{CE}$ is the longer leg.

$a = 2(BE)$
$10\sqrt{3} = 2(BE)$
$5\sqrt{3} = BE$
$CE = \sqrt{3}(BE)$
$CE = \sqrt{3}(5\sqrt{3})$
$CE = 5 \cdot 3$ or 15

In $\triangle CEA$, $\overline{AE}$ is the longer leg.

$y = \sqrt{3}(CE)$
$y = \sqrt{3}(15)$ or $15\sqrt{3}$

19. In $\triangle BEC$, a is the measure of the hypotenuse and x is the measure of the shorter leg.

$a = 2x$
$a = 2(7\sqrt{3})$
$a = 14\sqrt{3}$
$CE = \sqrt{3}(x)$
$CE = \sqrt{3}(7\sqrt{3})$
$CE = 7 \cdot 3$ or 21

In $\triangle ACE$, y is the measure of the longer leg and b is the measure of the hypotenuse.

$y = \sqrt{3}(CE)$
$y = \sqrt{3}(21)$ or $21\sqrt{3}$
$b = 2(CE)$
$b = 2(21)$ or 42

20. The altitude is the longer leg of a 30°-60°-90° triangle. Let x represent the length of the shorter leg.

$12 = \sqrt{3}(x)$
$\frac{12}{\sqrt{3}} = x$
$\frac{12}{\sqrt{3}} \cdot \frac{\sqrt{3}}{\sqrt{3}} = x$
$\frac{12\sqrt{3}}{3} = x$
$4\sqrt{3} = x$

Then the hypotenuse, which is a side of the equilateral triangle, has measure $2x = 2(4\sqrt{3})$ or $8\sqrt{3} \approx 13.86$ feet.

21. The perimeter is 45, so each congruent side has measure $\frac{45}{3}$ or 15 cm. An altitude is the longer leg of a 30°-60°-90° triangle. Let x represent the length of the shorter leg. Then $x = \frac{15}{2}$ or 7.5. The altitude has measure $\sqrt{3}(x) = 7.5\sqrt{3}$ or about 12.99 cm.

22. Each side of the square is a leg of a 45°-45°-90° triangle. The length of the hypotenuse, $22\sqrt{2}$ mm, is $\sqrt{2}$ times as long as a leg. So each leg has measure 22. Then the perimeter of the square is 4(22) or 88 mm.

23. The altitude is the longer leg of a 30°-60°-90° triangle. Let x represent the length of the shorter leg.

$7.4 = \sqrt{3}(x)$
$\frac{7.4}{\sqrt{3}} = x$
$\frac{7.4}{\sqrt{3}} \cdot \frac{\sqrt{3}}{\sqrt{3}} = x$
$\frac{7.4\sqrt{3}}{3} = x$

Then the hypotenuse, which is a side of the equilateral triangle, has measure $2x = 2\left(\frac{7.4\sqrt{3}}{3}\right)$ or $\frac{14.8\sqrt{3}}{3}$. Thus, the perimeter of the equilateral triangle is $3\left(\frac{14.8\sqrt{3}}{3}\right) = 14.8\sqrt{3}$ or about 25.63 m.

24.

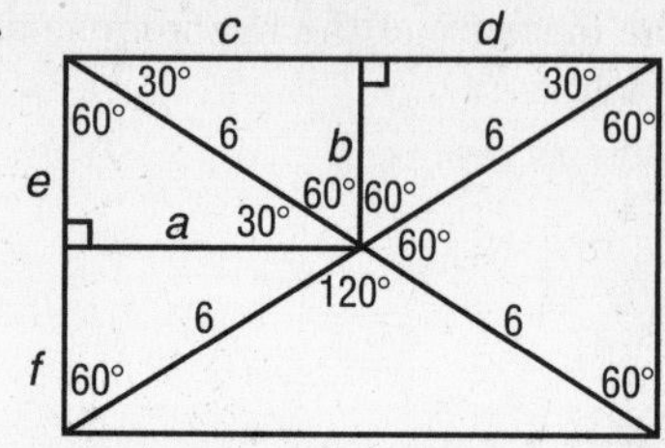

The diagonals determine equilateral triangles with sides equal to half the length of each diagonal, or 6 inches. So $e + f = 6$, and $e = f = 3$. Then $a = \sqrt{3}(e)$ or $3\sqrt{3}$, so $c = d = 3\sqrt{3}$. Then the perimeter of the rectangle is $2(6) + 2(3\sqrt{3} + 3\sqrt{3})$ or $12 + 12\sqrt{3} \approx 32.78$ inches.

25. Each side has measure $\sqrt{\frac{256}{4}}$ or 8. The length of a diagonal is $\sqrt{2}$ times as long as a side of the square. So the measure of a diagonal is $8\sqrt{2} \approx 11.31$.

26. $8 = 2x$

$4 = x$

$y = \sqrt{3}(x)$

$y = \sqrt{3}(4)$ or $4\sqrt{3}$

$z = 6$

Each leg of the 45°-45°-90° triangle has length x, or 4. Then $CB = 4\sqrt{2}$. So, the perimeter of $ABCD$ is $4\sqrt{3} + 6 + 4 + 4\sqrt{2} + 6 + 8 = 4\sqrt{3} + 4\sqrt{2} + 24$, or about 36.59 units.

27.

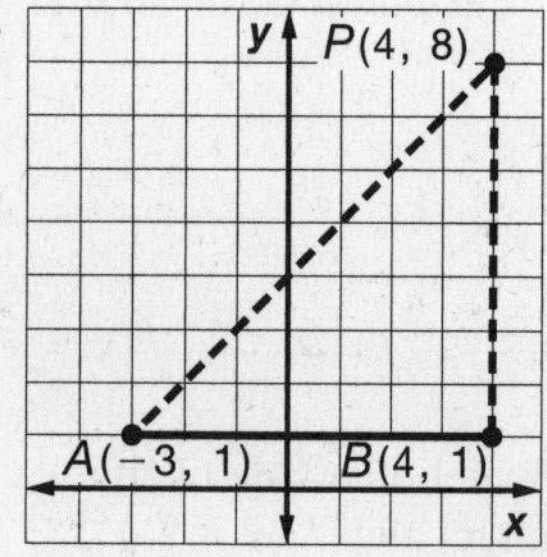

Graph A and B. $\overline{AB}$ lies on a horizontal gridline of the coordinate plane. Since $\overline{PB}$ will be perpendicular to $\overline{AB}$, it lies on a vertical gridline.

$AB = |4 - (-3)| = 7$

AB and PB are congruent, so $PB = 7$. Point P has the same x-coordinate as B. P is located 7 units above B. So, the coordinates of P are $(4, 1 + 7)$ or $(4, 8)$.

28.

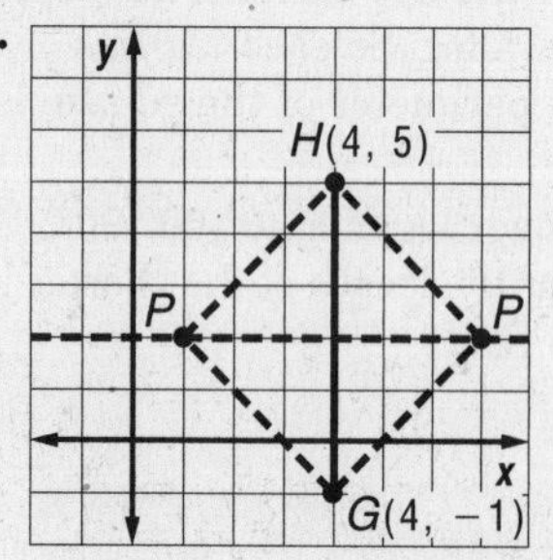

Graph G and H. If GH is the diagonal of a square with vertices P, G, and H then the other diagonal is perpendicular to and bisects $\overline{GH}$. $\overline{GH}$ lies on a vertical gridline of the coordinate plane, so the other diagonal through P is horizontal and goes through the midpoint of $\overline{GH}$, $\left(\frac{4+4}{2}, \frac{5+(-1)}{2}\right)$ or $(4, 2)$.

$GH = |5 - (-1)| = 6$

The diagonal through P also has measure 6. Point P has the same y-coordinate as the midpoint of $\overline{GH}$. P is located 3 units to the left or right of the midpoint of $\overline{GH}$. So, the coordinates of P are $(4 - 3, 2) = (1, 2)$ or $(4 + 3, 2) = (7, 2)$.

29.

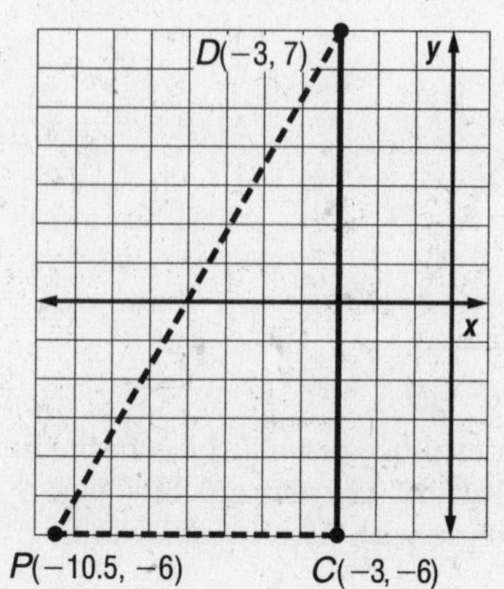

Graph C and D. $\overline{CD}$ lies on a vertical gridline of the coordinate plane. Since $\overline{PC}$ will be perpendicular to $\overline{CD}$, it lies on a horizontal gridline.

$CD = |7 - (-6)| = 13$

$\overline{CD}$ is the longer leg. $\overline{PC}$ is the shorter leg. So, $CD = \sqrt{3}(PC)$.

$13 = \sqrt{3}(PC)$

$\frac{13}{\sqrt{3}} = PC$

$\frac{13}{\sqrt{3}} \cdot \frac{\sqrt{3}}{\sqrt{3}} = PC$

$\frac{13\sqrt{3}}{3} = PC$

Point P has the same y-coordinate as C. P is located $\frac{13\sqrt{3}}{3}$ units to the left of C. So, the coordinates of P are $\left(-3 - \frac{13\sqrt{3}}{3}, -6\right)$ or about $(-10.51, -6)$.

30.

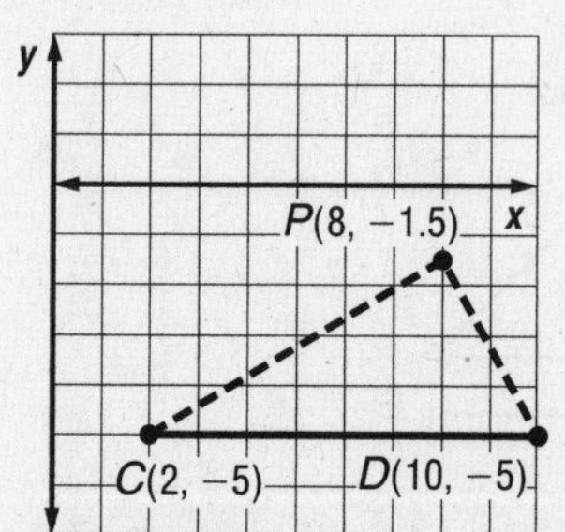

Graph C and D. $\overline{CD}$ lies on a horizontal gridline of the coordinate plane. $m\angle C = 30$, so $m\angle D = 60$. Let Q be the point on $\overline{CD}$ where the altitude from P intersects CD. Then $m\angle CPQ = 60$.

$CD = |10 - 2| = 8$

$\overline{CP}$ is the longer leg, and $\overline{PD}$ is the shorter leg.

$2(PD) = CD$

$2(PD) = 8$

$PD = 4$

$\sqrt{3}(PD) = CP$

$\sqrt{3}(4) = CP$

$4\sqrt{3} = CP$

$\overline{CP}$ is the hypotenuse of 30°-60°-90° triangle CPQ. $\overline{CQ}$ is the longer side and PQ is the shorter side.

$2(PQ) = CP$
$2(PQ) = 4\sqrt{3}$
$PQ = 2\sqrt{3}$
$\sqrt{3}(PQ) = CQ$
$\sqrt{3}(2\sqrt{3}) = CQ$
$2(3) = CQ$
$6 = CQ$

Q is on CD 6 units to the right of C, so Q has coordinates $(2 + 6, -5) = (8, -5)$. P is $2\sqrt{3}$ units above Q, so P has coordinates $(8, -5 + 2\sqrt{3})$.

31. $\overline{ST}$ is the shorter leg of a 30°-60°-90° triangle.
$SP = 2(ST)$
$6\sqrt{3} = 2(ST)$
$3\sqrt{3} = ST$
Then $c = 3\sqrt{3}$. PT is a vertical line segment, so $a = c = 3\sqrt{3}$.
$PT = \sqrt{3}(ST)$
$PT = \sqrt{3}(3\sqrt{3})$
$PT = 3 \cdot 3$ or 9
Then $b = 9$. $\overline{PQ} \parallel \overline{SR}$, so PQ is a horizontal line segment. Thus $d = b = 9$.

32. 12 triangles

33. The smaller angle is rotated, so it is the 30° angle.

34. There are no gaps because when a 30° angle is rotated 12 times, it rotates 360°.

35. Sample answer:

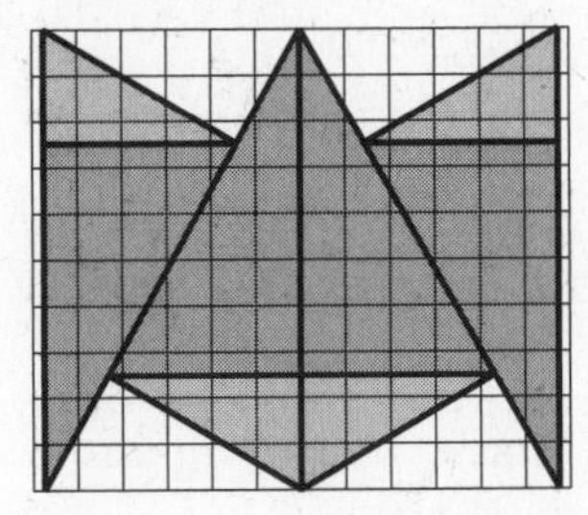

36.
$2x = 8$
$x = 4$
$y = \sqrt{3}(x)$
$y = \sqrt{3}(4)$ or $4\sqrt{3}$
$y = \sqrt{2}(z)$
$4\sqrt{3} = \sqrt{2}(z)$
$\frac{4\sqrt{3}}{\sqrt{2}} = z$
$\frac{4\sqrt{3}}{\sqrt{2}} \cdot \frac{\sqrt{2}}{\sqrt{2}} = z$
$\frac{4\sqrt{6}}{2} = z$
$2\sqrt{6} = z$

37.
$BD = \sqrt{3}(DH)$
$8\sqrt{3} = \sqrt{3}(DH)$
$8 = DH$
$BH = 2(DH)$
$BH = 2(8)$
$BH = 16$

38.
$4 = 2a$
$2 = a$
$b = \sqrt{3}(a)$
$b = 2\sqrt{3}$
$b = 2c$
$2\sqrt{3} = 2c$
$\sqrt{3} = c$
$d = \sqrt{3}(c)$
$d = \sqrt{3}(\sqrt{3})$
$d = 3$
$d = 2e$
$3 = 2e$
$1.5 = e$
$f = \sqrt{3}(e)$
$f = 1.5\sqrt{3}$
$f = 2g$
$1.5\sqrt{3} = 2g$
$0.75\sqrt{3} = g$
$x = \sqrt{3}(g)$
$x = \sqrt{3}(0.75\sqrt{3})$
$x = 0.75 \cdot 3 = 2.25$

39. The hexagon consists of six equilateral triangles. So, $m\angle UXY = 60$. WY bisects a 60° angle, so $m\angle XYW = 30$. WY is twice the length of the longer side of a 30°-60°-90° triangle.
$WY = 2\left(\frac{12}{2}\sqrt{3}\right)$
$= 12\sqrt{3}$
≈ 20.78 cm

40. Find CB.
$AB = \sqrt{2}(CB)$
$347 = \sqrt{2}(CB)$
$\frac{347}{\sqrt{2}} = CB$
$\frac{347}{\sqrt{2}} \cdot \frac{\sqrt{2}}{\sqrt{2}} = CB$
$\frac{347\sqrt{2}}{2} = CB$
The center fielder is standing $\frac{347\sqrt{2}}{2}$ or about 245.4 feet from home plate.

41.

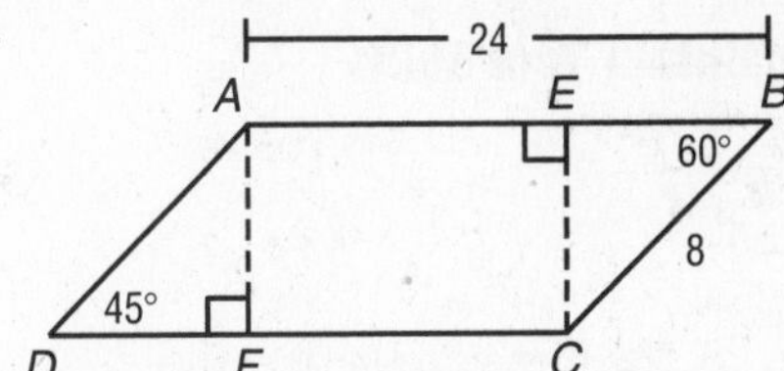

Draw altitudes $\overline{CE}$ and $\overline{AF}$. $\triangle BEC$ is a 30°-60°-90° triangle and $\triangle DAF$ is a 45°-45°-90° triangle.
In $\triangle BEC$, $\overline{CB}$ is the hypotenuse and $\overline{EC}$ is the longer leg.
$CB = 2(EB)$
$8 = 2(EB)$
$4 = EB$
$EC = \sqrt{3}(EB)$
$EC = 4\sqrt{3}$
$AF = EC$, so $AF = 4\sqrt{3}$.
$DF = AF$, so $DF = 4\sqrt{3}$.
$AD = \sqrt{2}(DF)$
$AD = \sqrt{2}(4\sqrt{3})$
$AD = 4\sqrt{6}$
$FC = AE$
$= AB - EB$
$= 24 - 4$
$= 20$
The perimeter of $ABCD$ is $AB + BC + DF + FC + AD = 24 + 8 + 4\sqrt{3} + 20 + 4\sqrt{6}$ or $52 + 4\sqrt{3} + 4\sqrt{6}$ units.

42. Sample answer: Congruent triangles of different color can be arranged to create patterns. Answers should include the following.

- 5, 9, 15, and 17; 3, 4, 6, 7, 10, 11, and 12
- Placing 45° angles next to one another forms 90° angles, which can be placed next to each other, leaving no holes.

43. C; $2x + 4x = 90$

$6x = 90$

$x = 15$

$m\angle A = 4x$

$= 4(15)$

$= 60$

$m\angle B = 2x$

$= 2(15)$

$= 30$

$BC = \sqrt{3}(AC)$

$6 = \sqrt{3}(AC)$

$\frac{6}{\sqrt{3}} = AC$

$\frac{6}{\sqrt{3}} \cdot \frac{\sqrt{3}}{\sqrt{3}} = AC$

$\frac{6\sqrt{3}}{3} = AC$

$2\sqrt{3} = AC$

$AB = 2(AC)$

$AB = 2(2\sqrt{3}) = 4\sqrt{3}$

44. $(3 \star 4)(5 \star 3) = \left(\frac{3^2}{4^2}\right)\left(\frac{5^2}{3^2}\right)$

$= \frac{5^2}{4^2}$

$= \frac{25}{16}$

Page 363 Maintain Your Skills

45. $a^2 + b^2 = c^2$

$3^2 + 4^2 \stackrel{?}{=} 5^2$

$9 + 16 \stackrel{?}{=} 25$

$25 = 25$

Since $25 = 25$, the measures satisfy the Pythagorean Theorem, so the sides can be the sides of a right triangle. All side lengths are whole numbers, so the measures form a Pythagorean triple.

46. $a^2 + b^2 = c^2$

$9^2 + 40^2 \stackrel{?}{=} 41^2$

$81 + 1600 \stackrel{?}{=} 1681$

$1681 = 1681$

Since $1681 = 1681$, the measures satisfy the Pythagorean Theorem. So the sides can be the sides of a right triangle. All side lengths are whole numbers, so the measures form a Pythagorean triple.

47. $a^2 + b^2 = c^2$

$20^2 + 21^2 \stackrel{?}{=} 31^2$

$400 + 441 \stackrel{?}{=} 961$

$841 \neq 961$

Since $841 \neq 961$, the measures do not satisfy the Pythagorean Theorem. So the sides cannot be the sides of a right triangle. The measures do not form a Pythagorean triple.

48. $a^2 + b^2 = c^2$

$20^2 + 48^2 \stackrel{?}{=} 52^2$

$400 + 2304 \stackrel{?}{=} 2704$

$2704 = 2704$

Since $2704 = 2704$, the measures satisfy the Pythagorean Theorem. So the sides can be the sides of a right triangle. All side lengths are whole numbers, so the measures form a Pythagorean triple.

49. $a^2 + b^2 = c^2$

$7^2 + 24^2 \stackrel{?}{=} 25^2$

$49 + 576 \stackrel{?}{=} 625$

$625 = 625$

Since $625 = 625$, the measures satisfy the Pythagorean Theorem. So the sides can be the sides of a right triangle. All side lengths are whole numbers, so the measures form a Pythagorean triple.

50. $a^2 + b^2 = c^2$

$12^2 + 34^2 \stackrel{?}{=} 37^2$

$144 + 1156 \stackrel{?}{=} 1369$

$1300 \neq 1369$

Since $1300 \neq 1369$, the measures do not satisfy the Pythagorean Theorem. So the sides cannot be the sides of a right triangle. The measures do not form a Pythagorean triple.

51. $\frac{z}{10} = \frac{10}{4}$

$4z = 100$

$z = 25$

$y = z - 4$

$y = 25 - 4$ or 21

$\frac{y}{x} = \frac{x}{4}$

$4y = x^2$

$4(21) = x^2$

$84 = x^2$

$\sqrt{84} = x$

$2\sqrt{21} = x$

$9.2 \approx x$

52. $\frac{12}{x} = \frac{x}{8}$

$96 = x^2$

$\sqrt{96} = x$

$4\sqrt{6} = x$

$9.8 \approx x$

$\frac{8}{y} = \frac{y}{12 - 8}$

$y^2 = 32$

$y = \sqrt{32}$

$y = 4\sqrt{2}$

$y \approx 5.7$

$\frac{12}{z} = \frac{z}{12 - 8}$

$z^2 = 48$

$z = \sqrt{48}$

$z = 4\sqrt{3}$

$z \approx 6.9$

53. $\frac{15}{5} = \frac{5}{y}$
$15y = 25$
$y = \frac{5}{3}$
$x = 15 - y$
$x = 15 - \frac{5}{3}$
$x = \frac{40}{3}$
$\frac{15}{z} = \frac{z}{x}$
$15x = z^2$
$15\left(\frac{40}{3}\right) = z^2$
$200 = z^2$
$\sqrt{200} = z$
$10\sqrt{2} = z$
$14.1 \approx z$

54. In $\triangle ALK$ and $\triangle ALN$, $\overline{AL} \cong \overline{AL}$, $\overline{KL} \cong \overline{NL}$, and $AK < AN$. Then $m\angle ALK < m\angle ALN$ by the SSS Inequality.

55. In $\triangle ALK$ and $\triangle NLO$, $\overline{AL} \cong \overline{OL}$, $\overline{KL} \cong \overline{NL}$, and $AK < NO$. So, $m\angle ALK < m\angle NLO$ by the SSS Inequality.

56. In $\triangle OLK$ and $\triangle NLO$, $\overline{LO} \cong \overline{LO}$, $\overline{KL} \cong \overline{NL}$, and $KO > NO$. So, $m\angle OLK > m\angle NLO$ by the SSS Inequality.

57. In $\triangle KLO$ and $\triangle ALN$, $\overline{KL} \cong \overline{NL}$, $\overline{LO} \cong \overline{AL}$, and $\overline{AN} \cong \overline{KO}$. $\triangle KLO \cong \triangle ALN$ by SSS Congruence. So $m\angle KLO = m\angle ALN$ by CPCTC and the definition of congruent angles.

58. $JK = \sqrt{[-1-(-3)]^2 + (5-2)^2}$
$= \sqrt{2^2 + 3^2}$
$= \sqrt{13}$
$RS = \sqrt{[-4-(-6)]^2 + (3-6)^2}$
$= \sqrt{2^2 + (-3)^2}$
$= \sqrt{13}$
$KL = \sqrt{[4-(-1)]^2 + (4-5)^2}$
$= \sqrt{5^2 + (-1)^2}$
$= \sqrt{26}$
$ST = \sqrt{[1-(-4)]^2 + (4-3)^2}$
$= \sqrt{5^2 + 1^2}$
$= \sqrt{26}$
$JL = \sqrt{[4-(-3)]^2 + (4-2)^2}$
$= \sqrt{7^2 + 2^2}$
$= \sqrt{53}$
$RT = \sqrt{[1-(-6)]^2 + (4-6)^2}$
$= \sqrt{7^2 + (-2)^2}$
$= \sqrt{53}$
$\triangle JKL \cong \triangle RST$ by SSS.

59. $5 = \frac{x}{3}$
$15 = x$

60. $\frac{x}{9} = 0.14$
$x = 1.26$

61. $0.5 = \frac{10}{k}$
$0.5k = 10$
$k = 20$

62. $0.2 = \frac{13}{g}$
$0.2g = 13$
$g = 65$

63. $\frac{7}{n} = 0.25$
$7 = 0.25n$
$28 = n$

64. $9 = \frac{m}{0.8}$
$7.2 = m$

65. $\frac{24}{x} = 0.4$
$24 = 0.4x$
$60 = x$

66. $\frac{35}{y} = 0.07$
$35 = 0.07y$
$500 = y$

Page 363 Practice Quiz 1

1. Let x represent the measure of the altitude.
$\frac{21}{x} = \frac{x}{7}$
$x^2 = 147$
$x = \sqrt{147}$ or $7\sqrt{3}$
$x \approx 12.1$

2. Let x represent the measure of the altitude.
$\frac{9}{x} = \frac{x}{5}$
$x^2 = 45$
$x = \sqrt{45}$ or $3\sqrt{5}$
$x \approx 6.7$

3. $AB = \sqrt{(4-2)^2 + (0-1)^2}$
$= \sqrt{2^2 + (-1)^2}$
$= \sqrt{5}$
$BC = \sqrt{(5-4)^2 + (7-0)^2}$
$= \sqrt{1^2 + 7^2}$
$= \sqrt{50}$
$AC = \sqrt{(5-2)^2 + (7-1)^2}$
$= \sqrt{3^2 + 6^2}$
$= \sqrt{45}$
$AB^2 + AC^2 \stackrel{?}{=} BC^2$
$(\sqrt{5})^2 + (\sqrt{45})^2 \stackrel{?}{=} (\sqrt{50})^2$
$5 + 45 \stackrel{?}{=} 50$
$50 = 50$
The triangle is a right triangle because the side lengths satisfy the Pythagorean Theorem.

4. $x = 3$
$y = (\sqrt{2})3$ or $3\sqrt{2}$

5. $x = 2(6)$ or 12
$y = \sqrt{3}(6)$ or $6\sqrt{3}$

7-4 Trigonometry

Page 365 Geometry Activity: Trigonometric Ratios

1. They are similar triangles because corresponding sides are proportional.

2.

	In $\triangle AED$	In $\triangle AGF$	In $\triangle ABC$
$\sin A$	$\frac{DE}{AD} \approx 0.6114$	$\frac{FG}{AF} \approx 0.6114$	$\frac{BC}{AC} \approx 0.6114$
$\cos A$	$\frac{AE}{AD} \approx 0.7913$	$\frac{AG}{AF} \approx 0.7913$	$\frac{AB}{AC} \approx 0.7913$
$\tan A$	$\frac{DE}{AE} \approx 0.7727$	$\frac{FG}{AG} \approx 0.7727$	$\frac{BC}{AB} \approx 0.7727$

3. Sample answer: Regardless of the side lengths, the trigonometric ratio is the same when comparing angles in similar triangles.

4. $m\angle A$ is the same in all triangles.

Pages 367–368 Check for Understanding

1. The triangles are similar, so the ratios remain the same.

2. Sample answer:

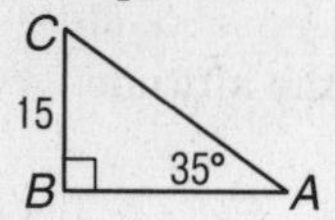

$m\angle B = 90, m\angle C = 55, b \approx 26.2, c \approx 21.4$

3. All three ratios involve two sides of a right triangle. The sine ratio is the measure of the opposite side divided by the measure of the hypotenuse. The cosine ratio is the measure of the adjacent side divided by the measure of the hypotenuse. The tangent ratio is the measure of the opposite side divided by the measure of the adjacent side.

4. The tan is the ratio of the measure of the opposite side divided by the measure of the adjacent side for a given angle in a right triangle. The $\tan^{-1}$ is the measure of the angle with a certain tangent ratio.

5. $\sin A = \frac{\text{opposite leg}}{\text{hypotenuse}} = \frac{a}{c} = \frac{14}{50} = 0.28$

$\cos A = \frac{\text{adjacent leg}}{\text{hypotenuse}} = \frac{b}{c} = \frac{48}{50} = 0.96$

$\tan A = \frac{\text{opposite leg}}{\text{adjacent leg}} = \frac{a}{b} = \frac{14}{48} \approx 0.29$

$\sin B = \frac{\text{opposite leg}}{\text{hypotenuse}} = \frac{b}{c} = \frac{48}{50} = 0.96$

$\cos B = \frac{\text{adjacent leg}}{\text{hypotenuse}} = \frac{a}{c} = \frac{14}{50} = 0.28$

$\tan B = \frac{\text{opposite leg}}{\text{adjacent leg}} = \frac{b}{a} = \frac{48}{14} \approx 3.43$

6. $\sin A = \frac{\text{opposite leg}}{\text{hypotenuse}} = \frac{a}{c} = \frac{8}{17} \approx 0.47$

$\cos A = \frac{\text{adjacent leg}}{\text{hypotenuse}} = \frac{b}{c} = \frac{15}{17} \approx 0.88$

$\tan A = \frac{\text{opposite leg}}{\text{adjacent leg}} = \frac{a}{b} = \frac{8}{15} \approx 0.53$

$\sin B = \frac{\text{opposite leg}}{\text{hypotenuse}} = \frac{b}{c} = \frac{15}{17} \approx 0.88$

$\cos B = \frac{\text{adjacent leg}}{\text{hypotenuse}} = \frac{a}{c} = \frac{8}{17} \approx 0.47$

$\tan B = \frac{\text{opposite leg}}{\text{adjacent leg}} = \frac{b}{a} = \frac{15}{8} \approx 1.88$

7. KEYSTROKES: [SIN] 57 [ENTER]

$\sin 57° \approx 0.8387$

8. KEYSTROKES: [COS] 60 [ENTER]

$\cos 60° = 0.5000$

9. KEYSTROKES: [COS] 33 [ENTER]

$\cos 33° \approx 0.8387$

10. KEYSTROKES: [TAN] 30 [ENTER]

$\tan 30° \approx 0.5774$

11. KEYSTROKES: [TAN] 45 [ENTER]

$\tan 45° = 1.0000$

12. KEYSTROKES: [SIN] 85 [ENTER]

$\sin 85° \approx 0.9962$

13. KEYSTROKES: [2nd] [TAN^{-1}] 1.4176 [ENTER]

$m\angle A \approx 54.8$

14. KEYSTROKES: [2nd] [SIN^{-1}] 0.6307 [ENTER]

$m\angle B \approx 39.1$

15.

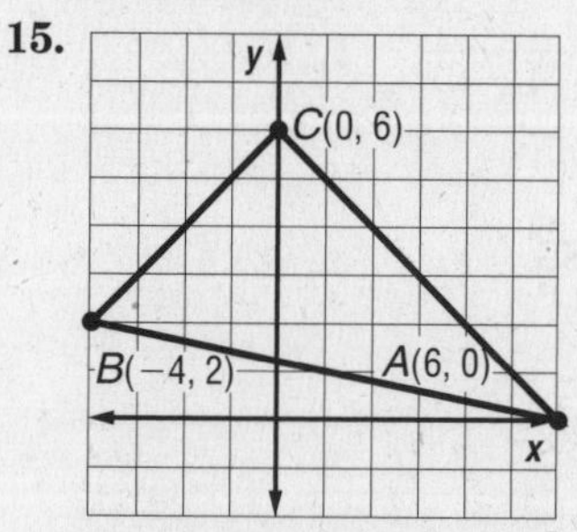

Explore: You know the coordinates of the vertices of a right triangle and that $\angle C$ is the right angle. You need to find the measure of one of the angles.

Plan: Use the Distance Formula to find the measure of each side. Then use one of the trigonometric ratios to write an equation. Use the inverse to find $m\angle A$.

Solve: $AB = \sqrt{(-4-6)^2 + (2-0)^2}$
$= \sqrt{100 + 4}$
$= \sqrt{104}$ or $2\sqrt{26}$
$BC = \sqrt{[0-(-4)]^2 + (6-2)^2}$
$= \sqrt{16 + 16}$
$= \sqrt{32}$ or $4\sqrt{2}$
$AC = \sqrt{(0-6)^2 + (6-0)^2}$
$= \sqrt{36 + 36}$
$= \sqrt{72}$ or $6\sqrt{2}$

Use the tangent ratio.

$\tan A = \frac{BC}{AC}$

$\tan A = \frac{4\sqrt{2}}{6\sqrt{2}}$ or $\frac{2}{3}$

$A = \tan^{-1}\left(\frac{2}{3}\right)$

KEYSTROKES: [2nd] [TAN^{-1}] 2 [÷] 3 [ENTER]

$m\angle A \approx 33.69006753$

The measure of $\angle A$ is about 33.7.

Examine: Use the sine ratio to check the answer.

$\sin A = \frac{BC}{AB}$

$\sin A = \frac{4\sqrt{2}}{2\sqrt{26}}$

$A = \sin^{-1}\left(\frac{4\sqrt{2}}{2\sqrt{26}}\right)$

KEYSTROKES: [2nd] [SIN^{-1}] 4 [2nd] [$\sqrt{\ }$] 2 [)] [÷] [(] 2 [2nd] [$\sqrt{\ }$] 26 [)] [)] [)]

$m\angle A \approx 33.69006753$

The answer is correct.

16.

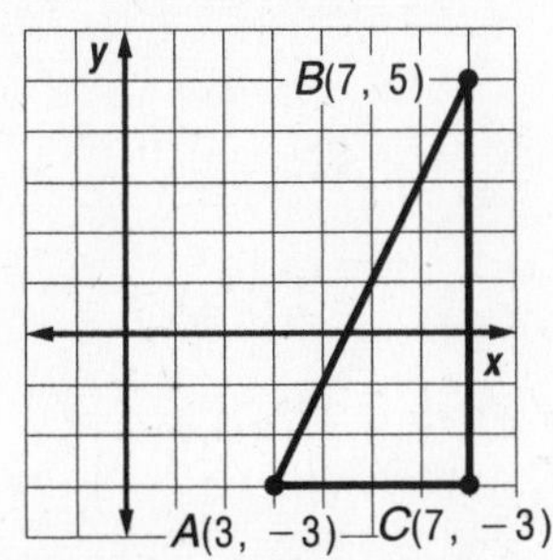

Explore: You know the coordinates of the vertices of a right triangle and that $\angle C$ is the right angle. You need to find the measure of one of the angles.
Plan: Use the Distance Formula to find the measure of each side. Then use one of the trigonometric ratios to write an equation. Use the inverse to find $m\angle B$.

Solve: $AB = \sqrt{(7-3)^2 + [5-(-3)]^2}$
$= \sqrt{16 + 64}$
$= \sqrt{80}$ or $4\sqrt{5}$
$BC = \sqrt{(7-7)^2 + (-3-5)^2}$
$= \sqrt{0 + 64}$ or 8
$AC = \sqrt{(7-3)^2 + [-3-(-3)]^2}$
$= \sqrt{16 + 0}$ or 4

Use the tangent ratio.

$\tan B = \frac{AC}{BC}$

$\tan B = \frac{4}{8}$ or $\frac{1}{2}$

$B = \tan^{-1}\left(\frac{1}{2}\right)$

KEYSTROKES: [2nd] [TAN^{-1}] 1 [÷] 2 [ENTER]

$m\angle B \approx 26.56505118$

The measure of $\angle B$ is about 26.6.

Examine: Use the sine ratio to check the answer.

$\sin B = \frac{AC}{AB}$

$\sin B = \frac{4}{4\sqrt{5}}$ or $\frac{1}{\sqrt{5}}$

$B = \sin^{-1}\left(\frac{1}{\sqrt{5}}\right)$

KEYSTROKES: [2nd] [SIN^{-1}] 1 [÷] [2nd] [$\sqrt{\ }$] 5 [ENTER]

$m\angle B \approx 26.56505118$

The answer is correct.

17. Let x be Maureen's distance from the tower in feet.

$\tan 31.2 = \frac{1815}{x}$

$x = \frac{1815}{\tan 31.2}$

KEYSTROKES: 1815 [÷] [TAN] 31.2 [ENTER]

Maureen is about 2997 feet from the tower.

Pages 368–370 Practice and Apply

18. $\sin P = \frac{\text{opposite leg}}{\text{hypotenuse}}$
$= \frac{p}{r}$
$= \frac{12}{37} \approx 0.32$

$\cos P = \frac{\text{adjacent leg}}{\text{hypotenuse}}$
$= \frac{q}{r}$
$= \frac{35}{37} \approx 0.95$

$\tan P = \frac{\text{opposite leg}}{\text{adjacent leg}}$
$= \frac{p}{q}$
$= \frac{12}{35} \approx 0.34$

$\sin Q = \frac{\text{opposite leg}}{\text{hypotenuse}}$
$= \frac{q}{r}$
$= \frac{35}{37} \approx 0.95$

$\cos Q = \frac{\text{adjacent leg}}{\text{hypotenuse}}$
$= \frac{p}{r}$
$= \frac{12}{37} \approx 0.32$

$\tan Q = \frac{\text{opposite leg}}{\text{adjacent leg}}$
$= \frac{q}{p}$
$= \frac{35}{12} \approx 2.92$

19. $\sin P = \frac{\text{opposite leg}}{\text{hypotenuse}}$
$= \frac{p}{r}$
$= \frac{\sqrt{6}}{3\sqrt{2}} \cdot \frac{\sqrt{2}}{\sqrt{2}}$
$= \frac{\sqrt{3}}{3} \approx 0.58$

$\cos P = \frac{\text{adjacent leg}}{\text{hypotenuse}}$
$= \frac{a}{r}$
$= \frac{2\sqrt{3}}{3\sqrt{2}} \cdot \frac{\sqrt{2}}{\sqrt{2}}$
$= \frac{\sqrt{6}}{3} \approx 0.82$

$\tan P = \frac{\text{opposite leg}}{\text{adjacent leg}}$
$= \frac{p}{q}$
$= \frac{\sqrt{6}}{2\sqrt{3}} \cdot \frac{\sqrt{3}}{\sqrt{3}}$
$= \frac{\sqrt{2}}{2} \approx 0.71$

$\sin Q = \frac{\text{opposite leg}}{\text{hypotenuse}}$
$= \frac{q}{r}$
$= \frac{2\sqrt{3}}{3\sqrt{2}} \cdot \frac{\sqrt{2}}{\sqrt{2}}$
$= \frac{\sqrt{6}}{3} \approx 0.82$

$\cos Q = \frac{\text{adjacent leg}}{\text{hypotenuse}}$
$= \frac{p}{r}$
$= \frac{\sqrt{6}}{3\sqrt{2}} \cdot \frac{\sqrt{2}}{\sqrt{2}}$
$= \frac{\sqrt{3}}{3} \approx 0.58$

$\tan Q = \frac{\text{opposite leg}}{\text{adjacent leg}}$
$= \frac{q}{p}$
$= \frac{2\sqrt{3}}{\sqrt{6}} \cdot \frac{\sqrt{6}}{\sqrt{6}}$
$= \sqrt{2} \approx 1.41$

20. $\sin P = \frac{\text{opposite leg}}{\text{hypotenuse}}$
$= \frac{p}{r}$
$= \frac{\frac{3}{2}}{3}$
$= \frac{1}{2} = 0.50$

$\cos P = \frac{\text{adjacent leg}}{\text{hypotenuse}}$
$= \frac{a}{r}$
$= \frac{\frac{3\sqrt{3}}{2}}{3}$
$= \frac{\sqrt{3}}{2} \approx 0.87$

$\tan P = \frac{\text{opposite leg}}{\text{adjacent leg}}$
$= \frac{p}{q}$
$= \frac{\frac{3}{2}}{\frac{3\sqrt{3}}{2}}$
$= \frac{\sqrt{3}}{3} \approx 0.58$

$\sin Q = \frac{\text{opposite leg}}{\text{hypotenuse}}$
$= \frac{q}{r}$
$= \frac{\frac{3\sqrt{3}}{2}}{3}$
$= \frac{\sqrt{3}}{2} \approx 0.87$

$\cos Q = \frac{\text{adjacent leg}}{\text{hypotenuse}}$
$= \frac{p}{r}$
$= \frac{\frac{3}{2}}{3}$
$= \frac{1}{2} = 0.50$

$\tan Q = \frac{\text{opposite leg}}{\text{adjacent leg}}$
$= \frac{q}{p}$
$= \frac{\frac{3\sqrt{3}}{2}}{\frac{3}{2}}$
$= \sqrt{3} \approx 1.73$

21. $\sin P = \frac{\text{opposite leg}}{\text{hypotenuse}}$
$= \frac{p}{r}$
$= \frac{2\sqrt{3}}{3\sqrt{3}}$
$= \frac{2}{3} \approx 0.67$

$\cos P = \frac{\text{adjacent leg}}{\text{hypotenuse}}$
$= \frac{q}{r}$
$= \frac{\sqrt{15}}{3\sqrt{3}} \cdot \frac{\sqrt{3}}{\sqrt{3}}$
$= \frac{\sqrt{5}}{3} \approx 0.75$

$\tan P = \frac{\text{opposite leg}}{\text{adjacent leg}}$
$= \frac{p}{q}$
$= \frac{2\sqrt{3}}{\sqrt{15}} \cdot \frac{\sqrt{15}}{\sqrt{15}}$
$= \frac{2\sqrt{5}}{5} \approx 0.89$

$\sin Q = \frac{\text{opposite leg}}{\text{hypotenuse}}$
$= \frac{q}{r}$
$= \frac{\sqrt{15}}{3\sqrt{3}} \cdot \frac{\sqrt{3}}{\sqrt{3}}$
$= \frac{\sqrt{5}}{3} \approx 0.75$

$\cos Q = \frac{\text{adjacent leg}}{\text{hypotenuse}}$
$= \frac{p}{r}$
$= \frac{2\sqrt{3}}{3\sqrt{3}}$
$= \frac{2}{3} \approx 0.67$

$\tan Q = \frac{\text{opposite leg}}{\text{adjacent leg}}$
$= \frac{q}{p}$
$= \frac{\sqrt{15}}{2\sqrt{3}} \cdot \frac{\sqrt{3}}{\sqrt{3}}$
$= \frac{\sqrt{5}}{2} \approx 1.12$

22. **KEYSTROKES:** SIN 6 ENTER
$\sin 6° \approx 0.1045$

23. **KEYSTROKES:** TAN 42.8 ENTER
$\tan 42.8° \approx 0.9260$

24. **KEYSTROKES:** COS 77 ENTER
$\cos 77° \approx 0.2250$

25. **KEYSTROKES:** SIN 85.9 ENTER
$\sin 85.9° \approx 0.9974$

26. **KEYSTROKES:** TAN 12.7 ENTER
$\tan 12.7° \approx 0.2254$

27. **KEYSTROKES:** COS 22.5 ENTER
$\cos 22.5° \approx 0.9239$

28. $\sin A = \frac{\text{opposite leg}}{\text{hypotenuse}}$
$= \frac{\sqrt{26}}{26}$
≈ 0.1961

29. $\tan B = \frac{\text{opposite leg}}{\text{adjacent leg}}$
$= \frac{5\sqrt{26}}{1\sqrt{26}}$ or $\frac{5}{1}$
$= 5.0000$

30. $\cos A = \frac{\text{adjacent leg}}{\text{hypotenuse}}$
$= \frac{5\sqrt{26}}{26}$
≈ 0.9806

31. $\sin x° = \frac{\text{opposite leg}}{\text{hypotenuse}}$
$= \frac{25}{5\sqrt{26}}$
$= \frac{25}{5\sqrt{26}} \cdot \frac{\sqrt{26}}{\sqrt{26}}$
$= \frac{25\sqrt{26}}{5(26)}$
$= \frac{5\sqrt{26}}{26}$
≈ 0.9806

32. $\cos x° = \frac{\text{adjacent leg}}{\text{hypotenuse}}$
$= \frac{5}{5\sqrt{26}}$
$= \frac{5}{5\sqrt{26}} \cdot \frac{\sqrt{26}}{\sqrt{26}}$
$= \frac{\sqrt{26}}{26}$
≈ 0.1961

33. $\tan A = \frac{\text{opposite leg}}{\text{adjacent leg}}$
$= \frac{\sqrt{26}}{5\sqrt{26}}$
$= \frac{1}{5}$
$= 0.2000$

34. $\cos B = \frac{\text{adjacent leg}}{\text{hypotenuse}}$
$= \frac{\sqrt{26}}{26}$
≈ 0.1961

35. $\sin y° = \frac{\text{opposite leg}}{\text{hypotenuse}}$
$= \frac{1}{\sqrt{26}}$
$= \frac{1}{\sqrt{26}} \cdot \frac{\sqrt{26}}{\sqrt{26}}$
$= \frac{\sqrt{26}}{26}$
≈ 0.1961

36. $\tan x° = \frac{\text{opposite leg}}{\text{adjacent leg}}$
$= \frac{25}{5}$
$= 5.0000$

37. **KEYSTROKES:** 2nd [SIN^{-1}] 0.7245 ENTER
$m\angle B \approx 46.42726961$
The measure of $\angle B$ is about 46.4.

38. **KEYSTROKES:** 2nd [COS^{-1}] 0.2493 ENTER
$m\angle C \approx 75.56390633$
The measure of $\angle C$ is about 75.6.

39. **KEYSTROKES:** 2nd [TAN^{-1}] 9.4618 ENTER
$m\angle E \approx 83.96691253$
The measure of $\angle E$ is about 84.0.

40. **KEYSTROKES:** 2nd [SIN^{-1}] 0.4567 ENTER
$m\angle A \approx 27.17436867$
The measure of $\angle A$ is about 27.2.

41. **KEYSTROKES:** 2nd [COS^{-1}] 0.1212 ENTER
$m\angle D \approx 83.03863696$
The measure of $\angle D$ is about 83.0.

42. **KEYSTROKES:** 2nd [TAN^{-1}] 0.4279 ENTER
$m\angle F \approx 23.16608208$
The measure of $\angle F$ is about 23.2.

43. $\tan 24° = \frac{x}{19}$
$19 \tan 24° = x$
KEYSTROKES: 19 TAN 24 ENTER
$x \approx 8.5$

44. $\sin x° = \frac{12}{17}$
$x° = \sin^{-1}\left(\frac{12}{17}\right)$
KEYSTROKES: 2nd [SIN^{-1}] 12 ÷ 17 ENTER
$x \approx 44.9$

45. $\cos 62° = \frac{x}{60}$
$60 \cos 62° = x$
KEYSTROKES: 60 COS 62 ENTER
$x \approx 28.2$

46. $\cos 31° = \frac{x}{34}$
$34 \cos 31° = x$
KEYSTROKES: 34 COS 31 ENTER
$x \approx 29.1$

47. $\sin 17° = \frac{6.6}{x}$
$x = \frac{6.6}{\sin 17°}$
KEYSTROKES: 6.6 ÷ SIN 17 ENTER
$x \approx 22.6$

48. $\tan x° = \frac{15}{18}$
$x° = \tan^{-1}\left(\frac{15}{18}\right)$
KEYSTROKES: 2nd [TAN^{-1}] 15 ÷ 18 ENTER
$x \approx 39.8$

49. Let x represent the vertical change of the plane after climbing at a constant angle of 3° for 60 ground miles.

$\tan 3° = \frac{x}{60}$

$60 \tan 3° = x$

KEYSTROKES: 60 [TAN] 3 [ENTER]

$x \approx 3.1$

The plane is about 3.1 + 1 or 4.1 miles above sea level.

50. Let x represent the maximum height.

$\sin 75° = \frac{x}{20}$

$20 \sin 75° = x$

KEYSTROKES: 20 [SIN] 75 [ENTER]

$x \approx 19.32$

The ladder can reach a maximum height of about 19.32 feet.

51. Let x represent the distance from the base of the ladder to the building.

$\cos 75° = \frac{x}{20}$

$20 \cos 75° = x$

KEYSTROKES: 20 [COS] 75 [ENTER]

$x \approx 5.18$

The base of the ladder is about 5.18 feet from the building.

52.

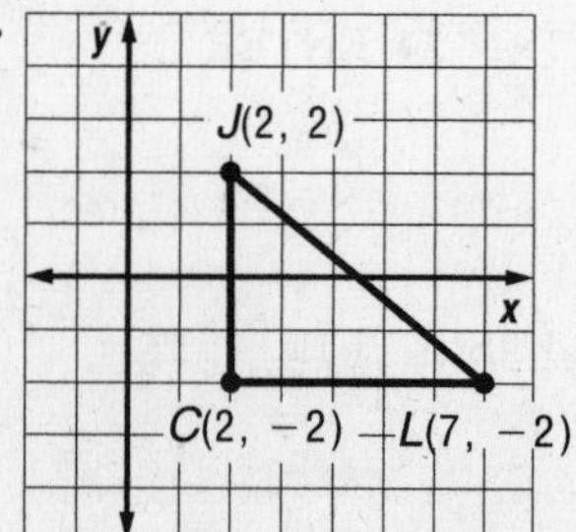

Explore: You know the coordinates of the vertices of a right triangle and that $\angle C$ is the right angle. You need to find the measure of one of the angles.

Plan: Use the Distance Formula to find the measure of each side. Then use one of the trigonometric ratios to write an equation. Use the inverse to find $m\angle J$.

Solve: $JC = \sqrt{(2-2)^2 + (-2-2)^2}$

$= \sqrt{0 + 16}$ or 4

$CL = \sqrt{(7-2)^2 + [-2-(-2)]^2}$

$= \sqrt{25 + 0}$ or 5

$JL = \sqrt{(7-2)^2 + (-2-2)^2}$

$= \sqrt{25 + 16}$

$= \sqrt{41}$

$\tan J = \frac{CL}{JC}$

$\tan J = \frac{5}{4}$

$J = \tan^{-1}\left(\frac{5}{4}\right)$

KEYSTROKES: [2nd] [TAN^{-1}] 5 [÷] 4 [ENTER]

$m\angle J \approx 51.34019175$

The measure of $\angle J$ is about 51.3.

Examine: Use the sine ratio to check the answer.

$\sin J = \frac{CL}{JL}$

$\sin J = \frac{5}{\sqrt{41}}$

$J = \sin^{-1}\left(\frac{5}{\sqrt{41}}\right)$

KEYSTROKES: [2nd] [SIN^{-1}] 5 [÷] [2nd] [$\sqrt{\ }$] 41 [ENTER]

$m\angle J \approx 51.34019175$

The answer is correct.

53.

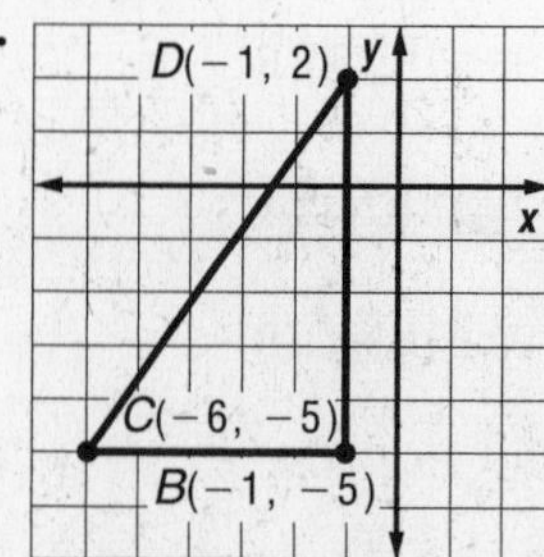

Explore: You know the coordinates of the vertices of a right triangle and that $\angle B$ is the right angle. You need to find the measure of one of the angles.

Plan: Use the Distance Formula to find the measure of each side. Then use one of the trigonometric ratios to write an equation. Use the inverse to find $m\angle C$.

Solve: $BC = \sqrt{[-6-(-1)]^2 + [-5-(-5)]^2}$

$= \sqrt{25 + 0}$ or 5

$BD = \sqrt{[-1-(-1)]^2 + [2-(-5)]^2}$

$= \sqrt{0 + 49}$ or 7

$CD = \sqrt{[-1-(-6)]^2 + [2-(-5)]^2}$

$= \sqrt{25 + 49}$

$= \sqrt{74}$

$\tan C = \frac{BD}{BC}$

$\tan C = \frac{7}{5}$

$C = \tan^{-1}\left(\frac{7}{5}\right)$

KEYSTROKES: [2nd] [TAN^{-1}] 7 [÷] 5 [ENTER]

$m\angle C \approx 54.46232221$

The measure of $\angle C$ is about 54.5.

Examine: Use the sine ratio to check the answer.

$\sin C = \frac{BD}{CD}$

$\sin C = \frac{7}{\sqrt{74}}$

$C = \sin^{-1}\left(\frac{7}{\sqrt{74}}\right)$

KEYSTROKES: [2nd] [SIN^{-1}] 7 [÷] [2nd] [$\sqrt{\ }$] 74 [ENTER]

$m\angle C \approx 54.46232221$

The answer is correct.

54.

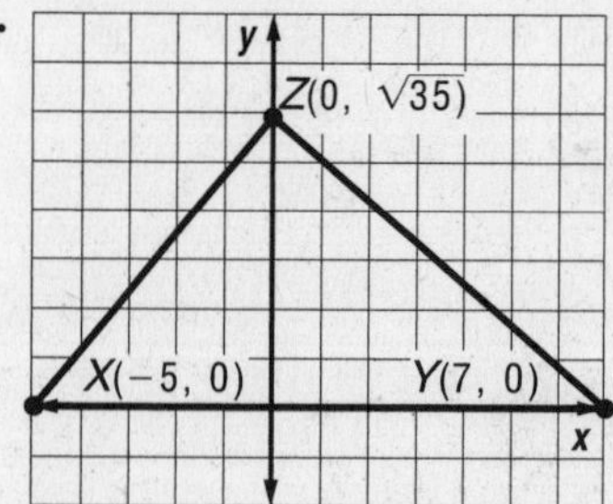

Explore: You know the coordinates of the vertices of a right triangle and that $\angle Z$ is the right angle. You need to find the measure of one of the angles.

Plan: Use the Distance Formula to find the measure of each side. Then use one of the trigonometric ratios to write an equation. Use the inverse to find $m\angle X$.

Solve: $XY = \sqrt{[7-(-5)]^2 + (0-0)^2}$
$= \sqrt{144 + 0}$ or 12
$YZ = \sqrt{(0-7)^2 + (\sqrt{35}-0)^2}$
$= \sqrt{49 + 35}$
$= \sqrt{84}$ or $2\sqrt{21}$
$XZ = \sqrt{[0-(-5)]^2 + (\sqrt{35}-0)^2}$
$= \sqrt{25 + 35}$
$= \sqrt{60}$ or $2\sqrt{15}$

$\cos X = \frac{XZ}{XY}$

$\cos X = \frac{2\sqrt{15}}{12}$ or $\frac{\sqrt{15}}{6}$

$X = \cos^{-1}\left(\frac{\sqrt{15}}{6}\right)$

KEYSTROKES: [2nd] [COS⁻¹] [2nd] [√] 15 [)] [÷] 6 [ENTER]

$m\angle X \approx 49.79703411$

The measure of $\angle X$ is about 49.8.

Examine: Use the sine ratio to check the answer.

$\sin X = \frac{YZ}{XY}$

$\sin X = \frac{2\sqrt{21}}{12}$ or $\frac{\sqrt{21}}{6}$

$X = \sin^{-1}\left(\frac{\sqrt{21}}{6}\right)$

KEYSTROKES: [2nd] [SIN⁻¹] [2nd] [√] 21 [)] [÷] 6 [ENTER]

$m\angle X \approx 49.79703411$

The answer is correct.

55. 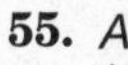

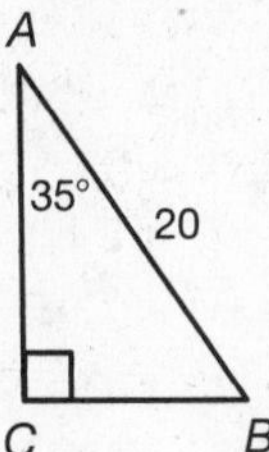

$\sin 35° = \frac{CB}{20}$

$20\sin 35° = CB$

KEYSTROKES: 20 [SIN] 35 [ENTER]

$CB \approx 11.5$

$\cos 35° = \frac{AC}{20}$

$20\cos 35° = AC$

KEYSTROKES: 20 [COS] 35 [ENTER]

$AC \approx 16.4$

The perimeter is about 20 + 11.5 + 16.4 or 47.9 inches.

56. $\sin x° = \frac{24}{36}$

$x° = \sin^{-1}\left(\frac{24}{36}\right)$

KEYSTROKES: [2nd] [SIN⁻¹] 24 [÷] 36 [ENTER]

$x \approx 41.8$

Let a represent the side of the smaller triangle opposite the angle of measure y.

$(2a)^2 + 24^2 = 36^2$

$4a^2 + 576 = 1296$

$4a^2 = 720$

$a^2 = 180$

$a = \sqrt{180}$ or $6\sqrt{5}$

$\tan y° = \frac{6\sqrt{5}}{24}$

$\tan y° = \frac{\sqrt{5}}{4}$

$y° = \tan^{-1}\left(\frac{\sqrt{5}}{4}\right)$

KEYSTROKES: [2nd] [TAN⁻¹] [2nd] [√] 5 [)] [÷] 4 [ENTER]

$y \approx 29.2$

57. $\tan 55° = \frac{x}{12}$

$12\tan 55° = x$

KEYSTROKES: 12 [TAN] 55 [ENTER]

$x \approx 17.1$

$\sin 47° = \frac{x}{y}$

$y = \frac{x}{\sin 47°}$

KEYSTROKES: 12 [TAN] 55 [)] [÷] [SIN] 47 [ENTER]

$y \approx 23.4$

58. $\tan 32° = \frac{24}{x}$

$x = \frac{24}{\tan 32°}$

KEYSTROKES: 24 [÷] [TAN] 32 [ENTER]

$x \approx 38.4$

$\cos 32° = \frac{y}{x}$

$x\cos 32° = y$

KEYSTROKES: 24 [÷] [TAN] 32 [)] [×] [COS] 32 [ENTER]

$y \approx 32.6$

59. Let d represent the distance between Alpha Centauri and the sun.

$\tan 0.00021 = \frac{1}{d}$

$d = \frac{1}{\tan 0.00021}$

KEYSTROKES: 1 [÷] [TAN] 0.00021 [ENTER]

$d \approx 272{,}837$

The distance is about 272,837 astronomical units.

60. The stellar parallax would be too small.

61. Let E be the point where $\overline{DB}$ intersects $\overline{AC}$. $\triangle EAB$ is a 45°-45°-90° triangle because it is an isosceles right triangle.

$AB = \sqrt{2}(AE)$

$8 = \sqrt{2}(AE)$

$\frac{8}{\sqrt{2}} = AE$

$\frac{8}{\sqrt{2}} \cdot \frac{\sqrt{2}}{\sqrt{2}} = AE$

$4\sqrt{2} = AE$

$\sin x° = \frac{AE}{AD}$

$\sin x° = \frac{4\sqrt{2}}{10}$

$\sin x° = \frac{2\sqrt{2}}{5}$

62. Sample answer: Surveyors use a theodolite to measure angles to determine distances and heights. Answers should include the following.

- Theodolites are used in surveying, navigation, and meteorology. They are used to measure angles.
- The angle measures from two points, which are a fixed distance apart, to a third point.

63. C; $(AC)^2 + 3^2 = 5^2$
$(AC)^2 + 9 = 25$
$(AC)^2 = 16$
$AC = \sqrt{16}$ or 4
$\cos C = \frac{AC}{BC}$
$\cos C = \frac{4}{5}$

64. B; $x^2 = 15^2 + 24^2 - 15(24)$
$x^2 = 225 + 576 - 360$
$x^2 = 441$
$x = \sqrt{441}$
$x = 21$

65. $\csc A = \frac{5}{3}$; $\sec A = \frac{5}{4}$; $\cot A = \frac{4}{3}$; $\csc B = \frac{5}{4}$;
$\sec B = \frac{5}{3}$; $\cot B = \frac{3}{4}$

66. $\csc A = \frac{13}{12}$; $\sec A = \frac{13}{5}$; $\cot A = \frac{5}{12}$; $\csc B = \frac{13}{5}$;
$\sec B = \frac{13}{12}$; $\cot B = \frac{12}{5}$

67. $\csc A = \frac{8}{4}$ or 2; $\sec A = \frac{8}{4\sqrt{3}}$ or $\frac{2\sqrt{3}}{3}$;
$\cot A = \frac{4\sqrt{3}}{4}$ or $\sqrt{3}$; $\csc B = \frac{8}{4\sqrt{3}}$ or $\frac{2\sqrt{3}}{3}$;
$\sec B = \frac{8}{4}$ or 2; $\cot B = \frac{4}{4\sqrt{3}}$ or $\frac{\sqrt{3}}{3}$

68. $\csc A = \frac{4}{2\sqrt{2}}$ or $\sqrt{2}$; $\sec A = \frac{4}{2\sqrt{2}}$ or $\sqrt{2}$;
$\cot A = \frac{2\sqrt{2}}{2\sqrt{2}}$ or 1; $\csc B = \frac{4}{2\sqrt{2}}$ or $\sqrt{2}$;
$\sec B = \frac{4}{2\sqrt{2}}$ or $\sqrt{2}$; $\cot B = \frac{2\sqrt{2}}{2\sqrt{2}}$ or 1

Page 370 Maintain Your Skills

69. $b = \sqrt{3}a$
$b = \sqrt{3}(4)$ or $4\sqrt{3}$
$c = 2a$
$c = 2(4)$ or 8

70. $b = \sqrt{3}a$
$3 = \sqrt{3}a$
$\frac{3}{\sqrt{3}} = a$
$\frac{3}{\sqrt{3}} \cdot \frac{\sqrt{3}}{\sqrt{3}} = a$
$\sqrt{3} = a$
$c = 2a$
$c = 2\sqrt{3}$

71. $c = 2a$
$5 = 2a$
$2.5 = a$
$b = \sqrt{3}a$
$b = \sqrt{3}(2.5)$ or $2.5\sqrt{3}$

72. $a^2 + b^2 = c^2$
$4^2 + 5^2 \stackrel{?}{=} 6^2$
$16 + 25 \stackrel{?}{=} 36$
$41 \neq 36$
Since $41 \neq 36$, the measures cannot be the sides of a right triangle because they do not satisfy the Pythagorean Theorem. The measures do not form a Pythagorean triple.

73. $a^2 + b^2 = c^2$
$5^2 + 12^2 \stackrel{?}{=} 13^2$
$25 + 144 \stackrel{?}{=} 169$
$169 = 169$
Since the side measures satisfy the Pythagorean Theorem, they can be the sides of a right triangle. The measures are all whole numbers, so they do form a Pythagorean triple.

74. $a^2 + b^2 = c^2$
$9^2 + 12^2 \stackrel{?}{=} 15^2$
$81 + 144 \stackrel{?}{=} 225$
$225 = 225$
Since the side measures satisfy the Pythagorean Theorem, they can be the sides of a right triangle. The measures are all whole numbers, so they do form a Pythagorean triple.

75. $a^2 + b^2 = c^2$
$8^2 + 12^2 \stackrel{?}{=} 16^2$
$64 + 144 \stackrel{?}{=} 256$
$208 \neq 256$
Since $208 \neq 256$, the measures cannot be the sides of a right triangle because they do not satisfy the Pythagorean Theorem. The measures do not form a Pythagorean triple.

76. Rewrite 4 : 11 as $4x : 11x$ and use those values for the number of minutes of commercials and actual show.
$4x + 11x = 30$
$15x = 30$
$x = 2$
$4x = 4(2)$ or 8
8 minutes are spent on commercials.

77. $m\angle 15 = 117$ vertical ⦞s

78. $m\angle 7 = 30$ corresponding ⦞s

79. $m\angle 3 + 30 = 180$ linear pair
$m\angle 3 = 150$

80. $m\angle 12 + 117 = 180$ linear pair
$m\angle 12 = 63$

81. $m\angle 11 = m\angle 12$ alternate interior ⦞s
$m\angle 11 = 63$

82. $m\angle 4 + 30 = 180$ linear pair
$m\angle 4 = 150$

7-5 Angles of Elevation and Depression

Page 373 Check for Understanding

1. Sample answer: $\angle ABC$

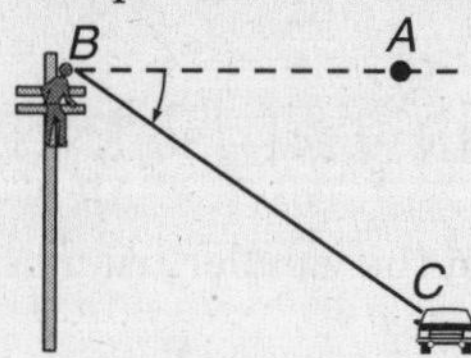

2. Sample answer: An angle of elevation is called that because the angle formed by a horizontal line and a segment joining two endpoints rises above the horizontal line.

3. The angle of depression is $\angle FPB$ and the angle of elevation is $\angle TBP$.

4.

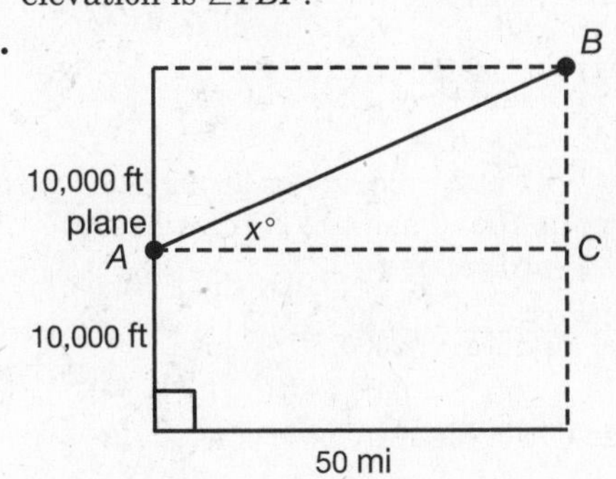

AC is 50 miles or 50(5280) = 264,000 feet.

Let x represent $m\angle CAB$.

$$\tan x° = \frac{CB}{AC}$$
$$\tan x° = \frac{10{,}000}{264{,}000}$$
$$x = \tan^{-1}\left(\frac{10{,}000}{264{,}000}\right)$$
$$x \approx 2.2$$

The angle of elevation should be about 2.2°.

5.

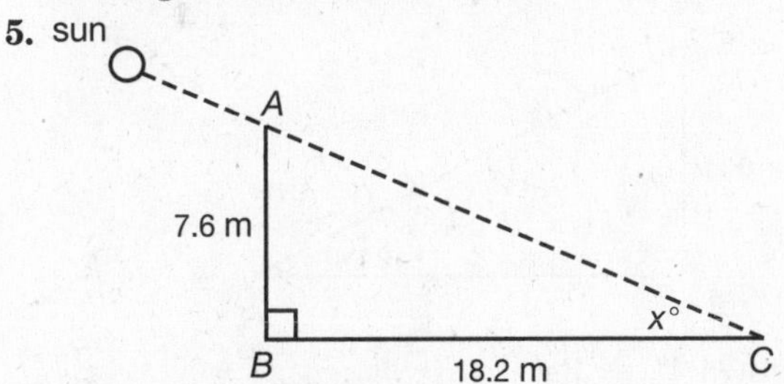

Let x represent $m\angle ACB$.

$$\tan x° = \frac{AB}{BC}$$
$$\tan x° = \frac{7.6}{18.2}$$
$$x = \tan^{-1}\left(\frac{7.6}{18.2}\right)$$
$$x \approx 22.7$$

The angle of elevation is 22.7°.

6. The angle of depression between the ship and the horizontal is 13.25°. Find the length along the ocean floor.

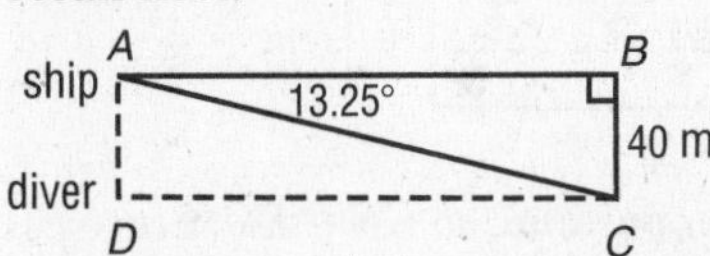

The ocean floor and the horizontal level with the ship are parallel, creating congruent alternate interior angles $\angle BAC$ and $\angle ACD$.

$$\tan 13.25° = \frac{40}{DC}$$
$$DC \tan 13.25° = 40$$
$$DC = \frac{40}{\tan 13.25°}$$
$$DC \approx 169.9$$

The diver must walk about 169.9 meters along the ocean floor.

7. The angle of depression between the top of the tower and the horizontal is 12°. Find the distance along the ground from the plane to the tower.

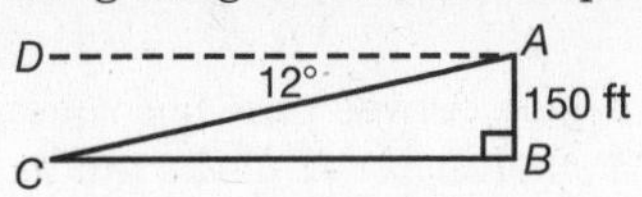

Because $\overline{DA}$ and $\overline{CB}$ are horizontal, they are parallel. Thus, $\angle DAC \cong \angle ACB$. So $m\angle ACB = 12$.

$$\tan 12° = \frac{150}{CB}$$
$$CB \tan 12° = 150$$
$$CB = \frac{150}{\tan 12°}$$
$$CB \approx 706$$

The plane is about 706 feet from the base of the tower.

Pages 374–376 Practice and Apply

8.

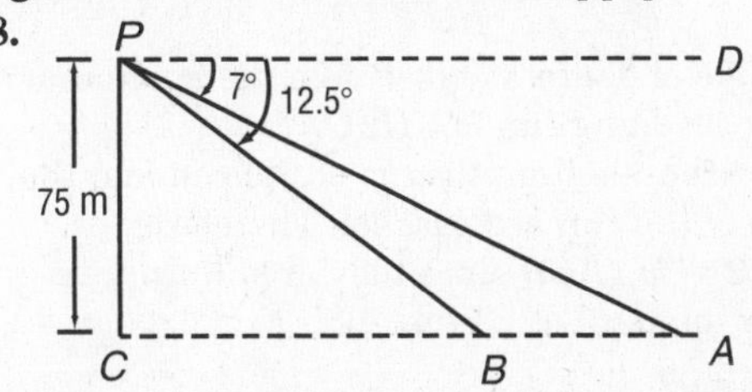

The parasailer is at P and the boats are at A and B. $\triangle PCB$ and $\triangle PCA$ are right triangles. The distance between the boats is AB or $AC - BC$. Because PD and CA are horizontal lines, they are parallel. Thus, $\angle DPA \cong \angle PAC$ and $\angle DPB \cong \angle PBC$ because they are alternate interior angles. This means that $m\angle PAC = 7$ and $m\angle PBC = 12.5$.

$$\tan 7° = \frac{75}{AC}$$
$$AC \tan 7° = 75$$
$$AC = \frac{75}{\tan 7°}$$
$$AC \approx 610.83$$
$$\tan 12.5° = \frac{75}{BC}$$
$$BC \tan 12.5° = 75$$
$$BC = \frac{75}{\tan 12.5°}$$
$$BC \approx 338.30$$
$$AC - BC \approx 610.83 - 338.30 \text{ or about } 273.$$

The distance between the boats is about 273 m.

9. The angle of elevation between the green and the horizontal is 12°.

Let d represent the distance from the tee to the hole.

$$\sin 12° = \frac{36}{d}$$
$$d \sin 12° = 36$$
$$d = \frac{36}{\sin 12°}$$
$$d \approx 173.2$$

The distance is about 173.2 yards.

10. 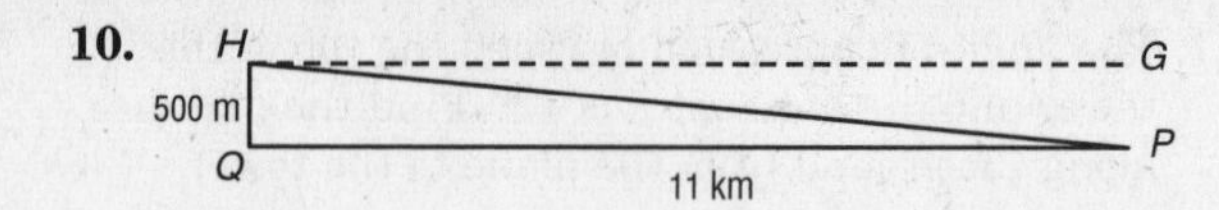

The angle of depression between the horizontal and the flight of the helicopter H to the landing pad P is $\angle GHP$.
The ground and the horizontal from the helicopter are parallel. Therefore, $m\angle GHP = m\angle HPQ$ since they are alternate interior angles. Let $x = m\angle HPQ$.

$\tan x° = \frac{0.5}{11}$

$x = \tan^{-1}\left(\frac{0.5}{11}\right)$

$x \approx 2.6$

The angle of depression is about 2.6°.

11. 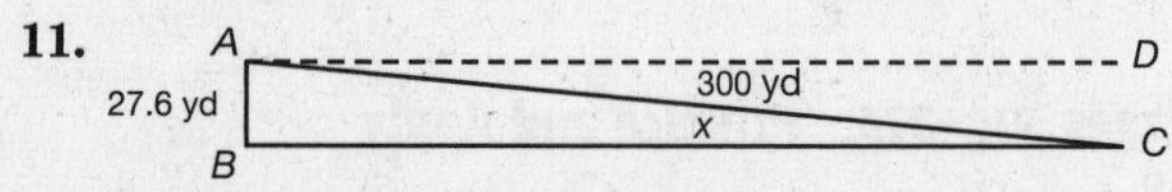

The angle of depression between the horizontal and the sledding run is $\angle DAC$.
The horizontals from the top of the run and the bottom of the run are parallel. Therefore, $m\angle DAC = m\angle ACB$ since they are alternate interior angles. Let $x = m\angle ACB$.

$\sin x° = \frac{27.6}{300}$

$x = \sin^{-1}\left(\frac{27.6}{300}\right)$

$x \approx 5.3$

The angle of depression is about 5.3°.

12. 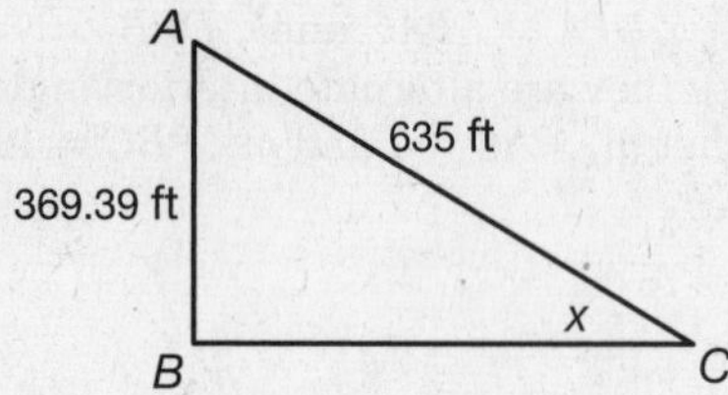

Let x represent $m\angle ACB$.

$\sin x° = \frac{369.39}{635}$

$x = \sin^{-1}\left(\frac{369.39}{635}\right)$

$x \approx 35.6$

The angle of elevation (incline) is about 35.6°.

13. 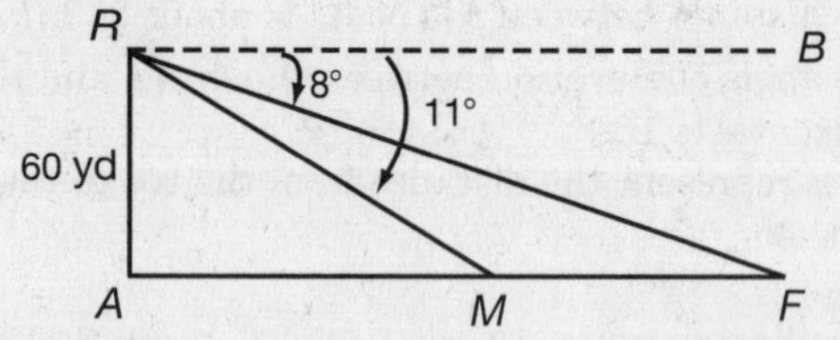

$\triangle RAM$ and $\triangle RAF$ are right triangles. The distance between the merry-go-round M and the Ferris wheel F is MF or $AF - AM$.
Because RB and AF are horizontal lines, they are parallel. Thus, $\angle BRF \cong \angle RFA$ and $\angle BRM \cong \angle RMA$ because they are alternate interior angles. This means that $m\angle RFA = 8$ and $m\angle RMA = 11$.

$\tan 8° = \frac{60}{AF}$

$AF \tan 8° = 60$

$AF = \frac{60}{\tan 8°}$

$AF \approx 426.92$

$\tan 11° = \frac{60}{AM}$

$AM \tan 11° = 60$

$AM = \frac{60}{\tan 11°}$

$AM \approx 308.67$

$AF - AM \approx 426.92 - 308.67$ or about 118.2 yards.
The merry-go-round and the Ferris wheel are about 76.4 yards apart.

14. $\frac{\text{vertical rise}}{\text{horizontal distance}} = \frac{140}{2000}$

$= 0.07$ or 7%

The grade of the highway is 7%.

15. 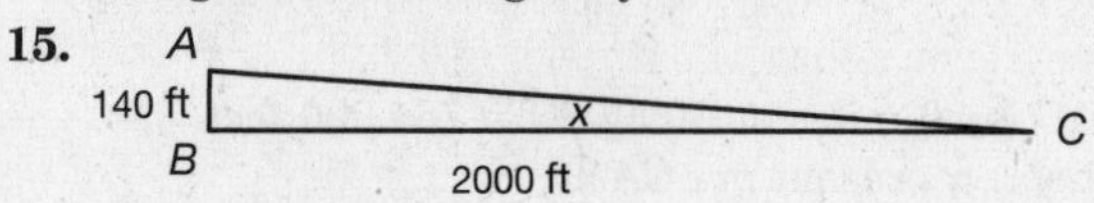

Let x represent $m\angle ACB$.

$\tan x° = \frac{140}{2000}$

$x = \tan^{-1}\left(\frac{140}{2000}\right)$

$x \approx 4.00$

The angle of elevation is about 4°.

16. 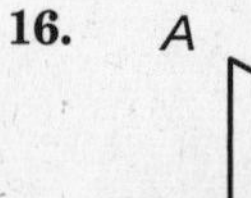
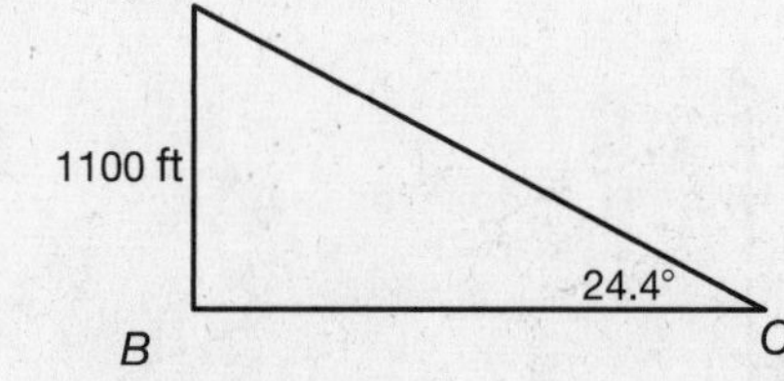

$\sin 24.4° = \frac{1100}{AC}$

$AC \sin 24.4° = 1100$

$AC = \frac{1100}{\sin 24.4°}$

$AC \approx 2663$

The ski run is about 2663 feet.

17. 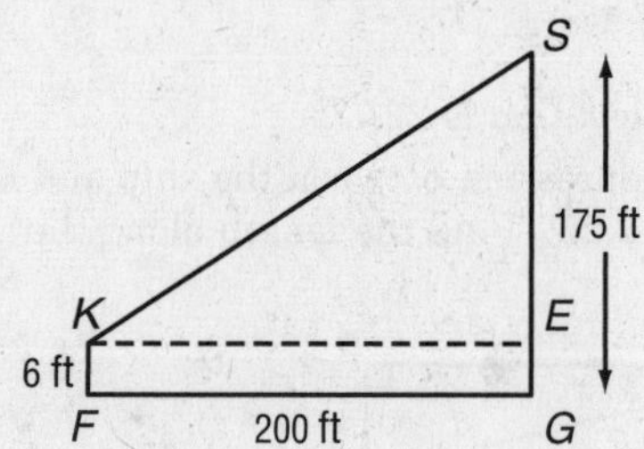

$\overline{KE}$ and $\overline{FG}$ are parallel, so $KF = EG$. Since SG is 175 feet and EG is 6 feet, SE is 169 feet. Let x represent $m\angle SKE$.

$\tan x° = \frac{169}{200}$

$x = \tan^{-1}\left(\frac{169}{200}\right)$

$x \approx 40.2$

The angle of elevation is about 40.2°.

18.

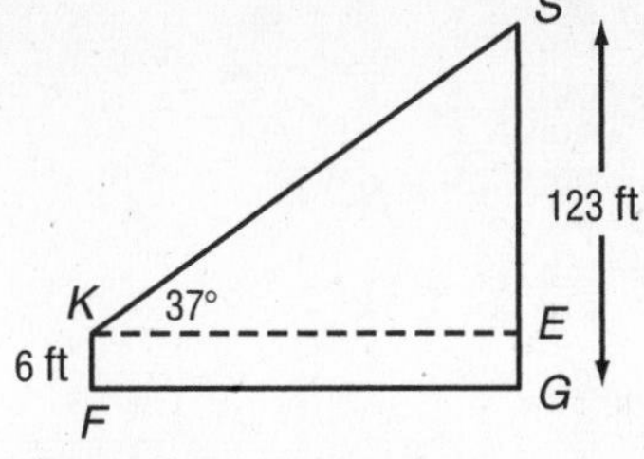

$\overline{KE}$ and $\overline{FG}$ are parallel, so $KF = EG$. Since SG is 123 feet and EG is 6 feet, SE is 117 feet.

$$\tan 37° = \frac{117}{KE}$$
$$KE \tan 37° = 117$$
$$KE = \frac{117}{\tan 37°}$$
$$KE \approx 155.3$$

$KE = FG$, so the distance between kirk and the geyser is about 155.3 feet.

19.

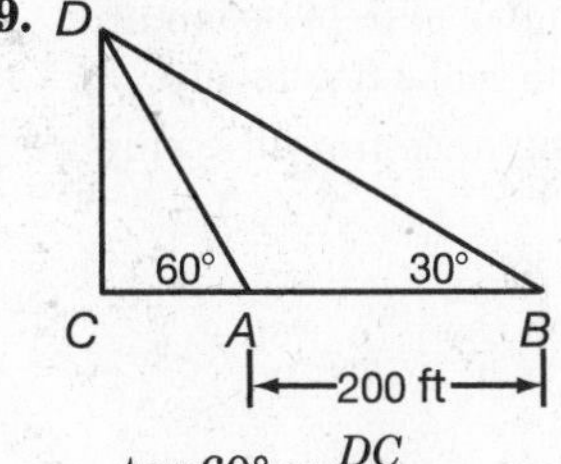

$$\tan 60° = \frac{DC}{AC}$$
$$AC \tan 60° = DC$$
$$\tan 30° = \frac{DC}{BC}$$
$$\tan 30° = \frac{AC \tan 60°}{BC}$$
$$BC \tan 30° = AC \tan 60°$$
$$BC = \frac{AC \tan 60°}{\tan 30°}$$
$$AC + 200 = AC\left(\frac{\tan 60°}{\tan 30°}\right)$$
$$AC + 200 = AC(3)$$
$$200 = 3AC - AC$$
$$200 = 2AC$$
$$100 = AC$$
$$BC = AC + 200$$
$$= 100 + 200 \text{ or } 300$$

The observers are 100 feet and 300 feet from the base of the tree.

20. Let x represent the distance from the spotlight to the base of the cloud formation.

$$\tan 62.7° = \frac{x}{83}$$
$$83 \tan 62.7° = x$$
$$160.8 \approx x$$

The ceiling is about 160.8 + 1.5 or 162.3 meters.

21.

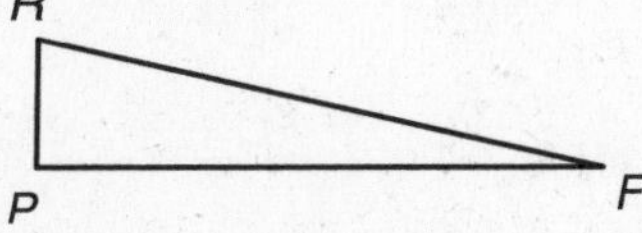

$RF = 48$ and $m\angle RFP = 10$. Find RP.

$$\sin 10° = \frac{RP}{RF}$$
$$\sin 10° = \frac{RP}{48}$$
$$48 \sin 10° = RP$$
$$8.3 \approx RP$$

The raised end of the treadmill is about 8.3 inches off the floor.

22.

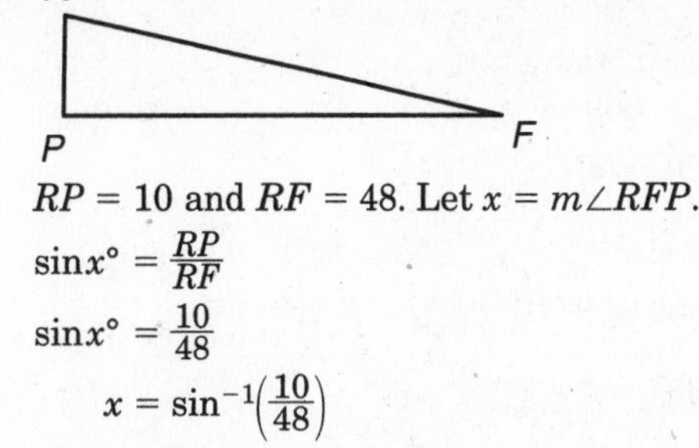

$RP = 10$ and $RF = 48$. Let $x = m\angle RFP$.

$$\sin x° = \frac{RP}{RF}$$
$$\sin x° = \frac{10}{48}$$
$$x = \sin^{-1}\left(\frac{10}{48}\right)$$
$$x \approx 12$$

The incline of the treadmill is about 12°.

23. Let x_i represent the rise at each stage i, $i = 1, 2, 3, 4, 5$. The length of the treadmill is 48 inches. Suppose the incline at the beginning of the exam is 10°.

Stage 1: $\sin 10° = \frac{x_1}{48}$
$$x_1 = 48 \sin 10°$$
$$x_1 \approx 8.3351$$

Stage 2: $\sin 12° = \frac{x_2}{48}$
$$x_2 = 48 \sin 12°$$
$$x_2 \approx 9.9798$$

Stage 3: $\sin 14° = \frac{x_3}{48}$
$$x_3 = 48 \sin 14°$$
$$x_3 \approx 11.6123$$

Stage 4: $\sin 16° = \frac{x_4}{48}$
$$x_4 = 48 \sin 16°$$
$$x_4 \approx 13.2306$$

Stage 5: $\sin 18° = \frac{x_5}{48}$
$$x_5 = 48 \sin 18°$$
$$x_5 = 14.8328$$

$$x_2 - x_1 \approx 9.9798 - 8.3351$$
$$\approx 1.6447$$
$$x_3 - x_2 \approx 11.6123 - 9.9798$$
$$\approx 1.6325$$
$$x_4 - x_3 \approx 13.2306 - 11.6123$$
$$\approx 1.6183$$
$$x_5 - x_4 \approx 14.8328 - 13.2306$$
$$\approx 1.6026$$

No, the end of the treadmill does not rise the same distance each time. The changes in the rise of the treadmill between stages are only approximately the same, about 1.6 inches.

24.

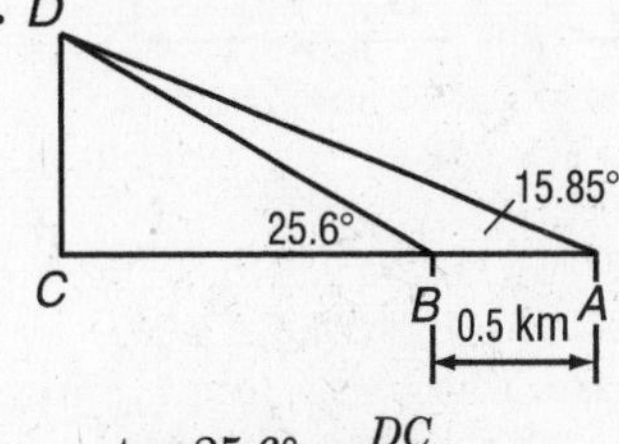

$$\tan 25.6° = \frac{DC}{BC}$$
$$BC \tan 25.6° = DC$$
$$\tan 15.85° = \frac{DC}{AC}$$

$AC \tan 15.85° = DC$
$AC \tan 15.85° = BC \tan 25.6°$
$AC = BC\left(\frac{\tan 25.6°}{\tan 15.85°}\right)$

Change AC to meters.
0.5 km = 500 m

$500 + BC = BC\left(\frac{\tan 25.6°}{\tan 15.85°}\right)$
$500 = BC\left(\frac{\tan 25.6°}{\tan 15.85°}\right) - BC$
$500 = BC\left(\frac{\tan 25.6°}{\tan 15.85°} - 1\right)$
$BC = \frac{500}{\left(\frac{\tan 25.6°}{\tan 15.85°} - 1\right)}$
$BC \approx 727$
$DC = BC \tan 25.6°$
≈ 348

Ayers Rock is about 348 meters high.

25.

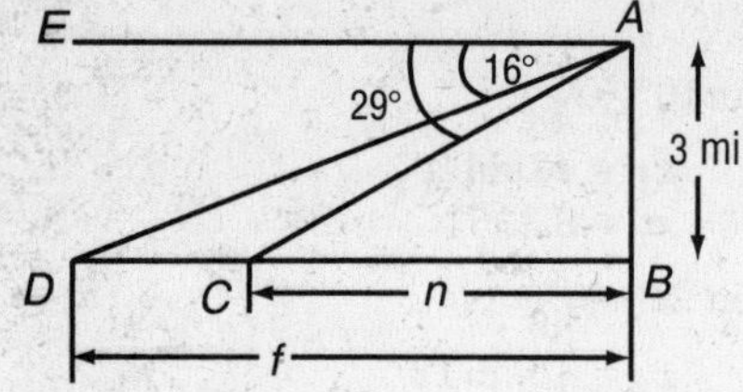

Find $f - n$.
$\overline{EA}$ is parallel to $\overline{DB}$, so $\angle EAD \cong \angle ADB$ and $\angle EAC \cong \angle ACB$ because they are alternate interior angles. Then $m\angle ADB = 16$ and $m\angle ACB = 29$.

$\tan 16° = \frac{3}{f}$
$f \tan 16° = 3$
$f = \frac{3}{\tan 16°}$
$\tan 29° = \frac{3}{n}$
$n \tan 29° = 3$
$n = \frac{3}{\tan 29°}$
$f - n = \frac{3}{\tan 16°} - \frac{3}{\tan 29°}$
$f - n \approx 5.1$

The crater is about 5.1 miles across.

26.

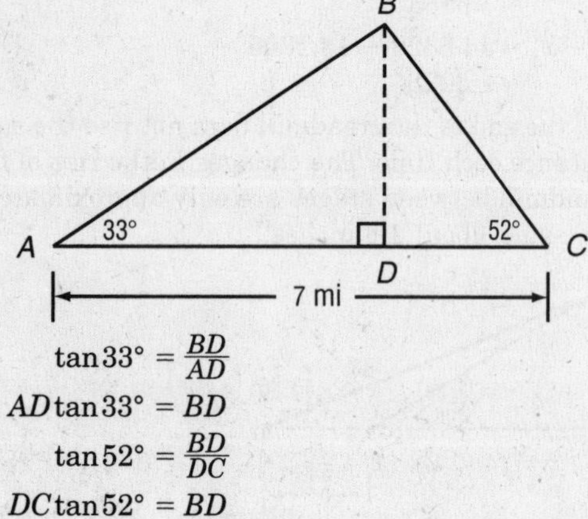

$\tan 33° = \frac{BD}{AD}$
$AD \tan 33° = BD$
$\tan 52° = \frac{BD}{DC}$
$DC \tan 52° = BD$
$AD \tan 33° = DC \tan 52°$
$AD = DC\left(\frac{\tan 52°}{\tan 33°}\right)$
$7 - DC = DC\left(\frac{\tan 52°}{\tan 33°}\right)$
$7 = DC\left(\frac{\tan 52°}{\tan 33°}\right) + DC$
$7 = DC\left(\frac{\tan 52°}{\tan 33°} + 1\right)$
$DC = \frac{7}{\frac{\tan 52°}{\tan 33°} + 1}$
$BD = \left(\frac{7}{\frac{\tan 52°}{\tan 33°} + 1}\right)\tan 52°$
$BD \approx 3.0$

The balloon is about 3.0 miles above the ground.

27. Answers should include the following.

- Pilots use angles of elevation when they are ascending and angles of depression when descending.
- Angles of elevation are formed when a person looks upward and angles of depression are formed when a person looks downward.

28. B; let x represent the distance from the ship to the foot of the tower.

$\tan 25° = \frac{120}{x}$
$x \tan 25° = 120$
$x = \frac{120}{\tan 25°}$
$x \approx 257.3$

The ship is about 257.3 meters from the tower.

29. A; $\frac{y}{28} = \frac{x}{16}$
$16y = 28x$
$16\left(\frac{1}{2}\right) = 28x$
$8 = 28x$
$\frac{8}{28} = x$
$\frac{2}{7} = x$

Page 376 Maintain Your Skills

30. $\cos A = 0.6717$
$A = \cos^{-1}(0.6717)$
$A \approx 47.8$

31. $\sin B = 0.5127$
$B = \sin^{-1}(0.5127)$
$B \approx 30.8$

32. $\tan C = 2.1758$
$C = \tan^{-1}(2.1758)$
$C \approx 65.3$

33. $\cos D = 0.3421$
$D = \cos^{-1}(0.3421)$
$D \approx 70.0$

34. $\sin E = 0.1455$
$E = \sin^{-1}(0.1455)$
$E \approx 8.4$

35. $\tan F = 0.3541$
$F = \tan^{-1}(0.3541)$
$F \approx 19.5$

36. $12 = x\sqrt{2}$

$\frac{12}{\sqrt{2}} = x$

$\frac{12}{\sqrt{2}} \cdot \frac{\sqrt{2}}{\sqrt{2}} = x$

$\frac{12\sqrt{2}}{2} = x$

$6\sqrt{2} = x$

$y = x$

$y = 6\sqrt{2}$

37. $x = 14\sqrt{3}$

$y = 2(14)$ or 28

38. $20 = 2y$

$10 = y$

$x = y\sqrt{3}$

$x = 10\sqrt{3}$

39. Let ℓ be the length of the model.

$\frac{\text{length of model}}{\text{length of plane}} = \frac{\text{wingspan of model}}{\text{wingspan of plane}}$

$\frac{\ell}{78} = \frac{36}{90}$

$90\ell = 78(36)$

$90\ell = 2808$

$\ell = 31.2$

The length of the model is 31.2 cm.

40a. $\angle 1 \cong \angle 2$

40b. AAS

40c. $\overline{FX} \cong \overline{GX}$

40d. CPCTC

40e. $\angle 4 \cong \angle 3$

40f. Isosceles Triangle Theorem

41. $\frac{x}{6} = \frac{35}{42}$

$42x = 6(35)$

$42x = 210$

$x = 5$

42. $\frac{3}{x} = \frac{5}{45}$

$3(45) = 5x$

$135 = 5x$

$27 = x$

43. $\frac{12}{17} = \frac{24}{x}$

$12x = 17(24)$

$12x = 408$

$x = 34$

44. $\frac{24}{36} = \frac{x}{15}$

$24(15) = 36x$

$360 = 36x$

$10 = x$

45. $\frac{12}{13} = \frac{48}{x}$

$12x = 13(48)$

$12x = 624$

$x = 52$

46. $\frac{x}{18} = \frac{5}{8}$

$8x = 18(5)$

$8x = 90$

$x = 11.25$

47. $\frac{28}{15} = \frac{7}{x}$

$28x = 15(7)$

$28x = 105$

$x = 3.75$

48. $\frac{x}{40} = \frac{3}{26}$

$26x = 40(3)$

$26x = 120$

$x = \frac{60}{13}$

7-6 The Law of Sines

Pages 380–381 Check for Understanding

1. Felipe; Makayla is using the definition of the sine ratio for a right triangle, but this is not a right triangle.

2.

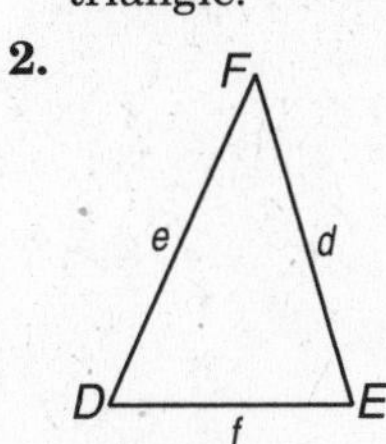

Sample answer: Let $m\angle D = 65$, $m\angle E = 73$, and $d = 15$. Then $\frac{\sin 65°}{15}$ is the fixed ratio or scale factor for the Law of Sines extended proportion. The length of e is found by using $\frac{\sin 65°}{15} = \frac{\sin 73°}{e}$. The $m\angle F$ is found by evaluating $180 - (m\angle D + m\angle E)$. In this problem $m\angle F = 42$. The length of f is found by using $\frac{\sin 65°}{15} = \frac{\sin 42°}{f}$.

3. In one case you need the measures of two sides and the measure of an angle opposite one of the sides. In the other case you need the measures of two angles and the measure of a side.

4. $\frac{\sin X}{x} = \frac{\sin Y}{y}$

$\frac{\sin 37°}{3} = \frac{\sin 68°}{y}$

$y \sin 37° = 3 \sin 68°$

$y = \frac{3 \sin 68°}{\sin 37°}$

$y \approx 4.6$

5. $m\angle X + m\angle Y + m\angle Z = 180$

$57 + m\angle Y + 72 = 180$

$m\angle Y = 51$

$\frac{\sin X}{x} = \frac{\sin Y}{y}$

$\frac{\sin 57°}{x} = \frac{\sin 51°}{12.1}$

$12.1 \sin 57° = x \sin 51°$

$\frac{12.1 \sin 57°}{\sin 51°} = x$

$13.1 \approx x$

6. $\frac{\sin Y}{y} = \frac{\sin Z}{z}$

$\frac{\sin Y}{7} = \frac{\sin 37°}{11}$

$11 \sin Y = 7 \sin 37°$

$\sin Y = \frac{7 \sin 37°}{11}$

$Y = \sin^{-1}\left(\frac{7 \sin 37°}{11}\right)$

$Y \approx 23°$

7. $\frac{\sin Y}{y} = \frac{\sin Z}{z}$

$\frac{\sin 92°}{17} = \frac{\sin Z}{14}$

$14 \sin 92° = 17 \sin Z$

$\frac{14 \sin 92°}{17} = \sin Z$

$\sin^{-1}\left(\frac{14 \sin 92°}{17}\right) = Z$

$55° \approx Z$

8. $m\angle P + m\angle Q + m\angle R = 180$

$m\angle P + 59 + 66 = 180$

$m\angle P + 125 = 180$

$m\angle P = 55$

$\frac{\sin P}{p} = \frac{\sin Q}{q}$

$\frac{\sin 55°}{72} = \frac{\sin 59°}{q}$

$q \sin 55° = 72 \sin 59°$

$q = \frac{72 \sin 59°}{\sin 55°}$

$q \approx 75.3$

$\frac{\sin P}{p} = \frac{\sin R}{r}$

$\frac{\sin 55°}{72} = \frac{\sin 66°}{r}$

$r \sin 55° = 72 \sin 66°$

$r = \frac{72 \sin 66°}{\sin 55°}$

$r \approx 80.3$

9. $\frac{\sin P}{p} = \frac{\sin R}{r}$

$\frac{\sin 105°}{32} = \frac{\sin R}{11}$

$11 \sin 105° = 32 \sin R$

$\frac{11 \sin 105°}{32} = \sin R$

$\sin^{-1}\left(\frac{11 \sin 105°}{32}\right) = R$

$19° \approx R$

$m\angle P + m\angle Q + m\angle R = 180$

$105 + m\angle Q + 19 \approx 180$

$m\angle Q + 124 \approx 180$

$m\angle Q \approx 56$

$\frac{\sin P}{p} = \frac{\sin Q}{q}$

$\frac{\sin 105°}{32} = \frac{\sin 56°}{q}$

$q \sin 105° = 32 \sin 56°$

$q = \frac{32 \sin 56°}{\sin 105°}$

$q \approx 27.5$

10. $m\angle P + m\angle Q + m\angle R = 180$

$33 + m\angle Q + 58 = 180$

$m\angle Q + 91 = 180$

$m\angle Q = 89$

$\frac{\sin P}{p} = \frac{\sin Q}{q}$

$\frac{\sin 33°}{p} = \frac{\sin 89°}{22}$

$22 \sin 33° = p \sin 89°$

$\frac{22 \sin 33°}{\sin 89°} = p$

$12.0 \approx p$

$\frac{\sin R}{r} = \frac{\sin Q}{q}$

$\frac{\sin 58°}{r} = \frac{\sin 89°}{22}$

$22 \sin 58° = r \sin 89°$

$\frac{22 \sin 58°}{\sin 89°} = r$

$18.7 \approx r$

11. $\frac{\sin P}{p} = \frac{\sin Q}{q}$

$\frac{\sin 120°}{28} = \frac{\sin Q}{22}$

$22 \sin 120° = 28 \sin Q$

$\frac{22 \sin 120°}{28} = \sin Q$

$\sin^{-1}\left(\frac{22 \sin 120°}{28}\right) = Q$

$43° \approx Q$

$m\angle P + m\angle Q + m\angle R = 180$

$120 + 43 + m\angle R \approx 180$

$163 + m\angle R \approx 180$

$m\angle R \approx 17$

$\frac{\sin P}{p} = \frac{\sin R}{r}$

$\frac{\sin 120°}{28} = \frac{\sin 17°}{r}$

$r \sin 120° = 28 \sin 17°$

$r = \frac{28 \sin 17°}{\sin 120°}$

$r \approx 9.5$

12. $m\angle P + m\angle Q + m\angle R = 180$

$50 + 65 + m\angle R = 180$

$115 + m\angle R = 180$

$m\angle R = 65$

$\frac{\sin P}{p} = \frac{\sin Q}{q}$

$\frac{\sin 50°}{12} = \frac{\sin 65°}{q}$

$q \sin 50° = 12 \sin 65°$

$q = \frac{12 \sin 65°}{\sin 50°}$

$q \approx 14.2$

$\frac{\sin P}{p} = \frac{\sin R}{r}$

$\frac{\sin 50°}{12} = \frac{\sin 65°}{r}$

$r \sin 50° = 12 \sin 65°$

$r = \frac{12 \sin 65°}{\sin 50°}$

$r \approx 14.2$

13.
$$\frac{\sin Q}{q} = \frac{\sin R}{r}$$
$$\frac{\sin 110.7°}{17.2} = \frac{\sin R}{9.8}$$
$$9.8\sin 110.7° = 17.2\sin R$$
$$\frac{9.8\sin 110.7°}{17.2} = \sin R$$
$$\sin^{-1}\left(\frac{9.8\sin 110.7°}{17.2}\right) = R$$
$$32° \approx R$$
$$m\angle P + m\angle Q + m\angle R = 180$$
$$m\angle P + 110.7 + 32 \approx 180$$
$$m\angle P + 142.7 \approx 180$$
$$m\angle P \approx 37$$
$$\frac{\sin P}{p} = \frac{\sin Q}{q}$$
$$\frac{\sin 37°}{p} = \frac{\sin 110.7°}{17.2}$$
$$17.2\sin 37° = p\sin 110.7°$$
$$\frac{17.2\sin 37°}{\sin 110.7°} = p$$
$$11.1 \approx p$$

14. $\overline{AD} \parallel \overline{BC}$, so $\angle DAC \cong \angle BCA$ because they are alternate interior angles. Thus, $m\angle DAC = 88$.
$$\frac{\sin 32°}{6} = \frac{\sin 88°}{DC}$$
$$DC\sin 32° = 6\sin 88°$$
$$DC = \frac{6\sin 88°}{\sin 32°}$$
$$DC \approx 11.3$$
$ABCD$ is a parallelogram, so $\overline{AD} \cong \overline{BC}$ and $\overline{DC} \cong \overline{AB}$. The perimeter of $ABCD$ is $2(6) + 2(11.3)$, or 34.6 units.

15. $m\angle A + m\angle B + m\angle C = 180$
$$55 + m\angle B + 62 = 180$$
$$m\angle B + 117 = 180$$
$$m\angle B = 63$$
$$\frac{\sin 63°}{240} = \frac{\sin 62°}{AB}$$
$$AB\sin 63° = 240\sin 62°$$
$$AB = \frac{240\sin 62°}{\sin 63°}$$
$$AB \approx 237.8 \text{ feet}$$

Pages 381–382 Practice and Apply

16.
$$\frac{\sin K}{k} = \frac{\sin L}{\ell}$$
$$\frac{\sin 63°}{k} = \frac{\sin 45°}{22}$$
$$22\sin 63° = k\sin 45°$$
$$\frac{22\sin 63°}{\sin 45°} = k$$
$$27.7 \approx k$$

17.
$$\frac{\sin K}{k} = \frac{\sin L}{\ell}$$
$$\frac{\sin 70°}{3.2} = \frac{\sin 52°}{\ell}$$
$$\ell\sin 70° = 3.2\sin 52°$$
$$\ell = \frac{3.2\sin 52°}{\sin 70°}$$
$$\ell \approx 2.7$$

18.
$$\frac{\sin K}{k} = \frac{\sin M}{m}$$
$$\frac{\sin 73°}{18.2} = \frac{\sin M}{10.5}$$
$$10.5\sin 73° = 18.2\sin M$$
$$\frac{10.5\sin 73°}{18.2} = \sin M$$
$$\sin^{-1}\left(\frac{10.5\sin 73°}{18.2}\right) = M$$
$$33° \approx M$$
$$m\angle M \approx 33$$

19.
$$\frac{\sin K}{k} = \frac{\sin M}{m}$$
$$\frac{\sin 96°}{10} = \frac{\sin M}{4.8}$$
$$4.8\sin 96° = 10\sin M$$
$$\frac{4.8\sin 96°}{10} = \sin M$$
$$\sin^{-1}\left(\frac{4.8\sin 96°}{10}\right) = M$$
$$29° \approx M$$
$$m\angle M \approx 29$$

20. $m\angle K + m\angle L + m\angle M = 180$
$$31 + 88 + m\angle M = 180$$
$$119 + m\angle M = 180$$
$$m\angle M = 61$$
$$\frac{\sin L}{\ell} = \frac{\sin M}{m}$$
$$\frac{\sin 88°}{\ell} = \frac{\sin 61°}{5.4}$$
$$5.4\sin 88° = \ell\sin 61°$$
$$\frac{5.4\sin 88°}{\sin 61°} = \ell$$
$$6.2 \approx \ell$$

21.
$$\frac{\sin M}{m} = \frac{\sin L}{\ell}$$
$$\frac{\sin 59°}{14.8} = \frac{\sin L}{8.3}$$
$$8.3\sin 59° = 14.8\sin L$$
$$\frac{8.3\sin 59°}{14.8} = \sin L$$
$$\sin^{-1}\left(\frac{8.3\sin 59°}{14.8}\right) = L$$
$$29° \approx L$$
$$m\angle L \approx 29$$

22.
$$\frac{\sin X}{x} = \frac{\sin Y}{y}$$
$$\frac{\sin 41°}{x} = \frac{\sin 71°}{7.4}$$
$$7.4\sin 41° = x\sin 71°$$
$$\frac{7.4\sin 41°}{\sin 71°} = x$$
$$5.1 \approx x$$
$$m\angle W + m\angle X + m\angle Y = 180$$
$$m\angle W + 41 + 71 = 180$$
$$m\angle W + 112 = 180$$
$$m\angle W = 68$$
$$\frac{\sin W}{w} = \frac{\sin Y}{y}$$
$$\frac{\sin 68°}{w} = \frac{\sin 71°}{7.4}$$
$$7.4\sin 68° = w\sin 71°$$
$$\frac{7.4\sin 68°}{\sin 71°} = w$$
$$7.3 \approx w$$

23.
$$\frac{\sin Y}{y} = \frac{\sin X}{x}$$
$$\frac{\sin 96°}{23.7} = \frac{\sin X}{10.3}$$
$$10.3 \sin 96° = 23.7 \sin X$$
$$\frac{10.3 \sin 96°}{23.7} = \sin X$$
$$\sin^{-1}\left(\frac{10.3 \sin 96°}{23.7}\right) = X$$
$$25.6° \approx X$$
$$m\angle W + m\angle X + m\angle Y = 180$$
$$m\angle W + 25.6 + 96 \approx 180$$
$$m\angle W + 121.6 \approx 180$$
$$m\angle W \approx 58.4$$
$$\frac{\sin W}{w} = \frac{\sin Y}{y}$$
$$\frac{\sin 58.4°}{w} = \frac{\sin 96°}{23.7}$$
$$23.7 \sin 58.4° = w \sin 96°$$
$$\frac{23.7 \sin 58.4°}{\sin 96°} = w$$
$$20.3 \approx w$$

24. $m\angle W + m\angle X + m\angle Y = 180$
$$52 + 25 + m\angle Y = 180$$
$$77 + m\angle Y = 180$$
$$m\angle Y = 103$$
$$\frac{\sin W}{w} = \frac{\sin Y}{y}$$
$$\frac{\sin 52°}{w} = \frac{\sin 103°}{15.6}$$
$$15.6 \sin 52° = w \sin 103°$$
$$\frac{15.6 \sin 52°}{\sin 103°} = w$$
$$12.6 \approx w$$
$$\frac{\sin X}{x} = \frac{\sin Y}{y}$$
$$\frac{\sin 25°}{x} = \frac{\sin 103°}{15.6}$$
$$15.6 \sin 25° = x \sin 103°$$
$$\frac{15.6 \sin 25°}{\sin 103°} = x$$
$$6.8 \approx x$$

25.
$$\frac{\sin X}{x} = \frac{\sin Y}{y}$$
$$\frac{\sin X}{20} = \frac{\sin 112°}{56}$$
$$56 \sin X = 20 \sin 112°$$
$$\sin X = \frac{20 \sin 112°}{56}$$
$$X = \sin^{-1}\left(\frac{20 \sin 112°}{56}\right)$$
$$X \approx 19.3°$$
$$m\angle W + m\angle X + m\angle Y = 180$$
$$m\angle W + 19.3 + 112 \approx 180$$
$$m\angle W + 131.3 \approx 180$$
$$m\angle W \approx 48.7$$
$$\frac{\sin W}{w} = \frac{\sin Y}{y}$$
$$\frac{\sin 48.7°}{w} = \frac{\sin 112°}{56}$$
$$56 \sin 48.7° = w \sin 112°$$
$$\frac{56 \sin 48.7°}{\sin 112°} = w$$
$$45.4 \approx w$$

26. $m\angle W + m\angle X + m\angle Y = 180$
$$38 + m\angle X + 115 = 180$$
$$m\angle X + 153 = 180$$
$$m\angle X = 27$$
$$\frac{\sin W}{w} = \frac{\sin X}{x}$$
$$\frac{\sin 38°}{8.5} = \frac{\sin 27°}{x}$$
$$x \sin 38° = 8.5 \sin 27°$$
$$x = \frac{8.5 \sin 27°}{\sin 38°}$$
$$x \approx 6.3$$
$$\frac{\sin W}{w} = \frac{\sin Y}{y}$$
$$\frac{\sin 38°}{8.5} = \frac{\sin 115°}{y}$$
$$y \sin 38° = 8.5 \sin 115°$$
$$y = \frac{8.5 \sin 115°}{\sin 38°}$$
$$y \approx 12.5$$

27. $m\angle W + m\angle X + m\angle Y = 180$
$$36 + m\angle X + 62 = 180$$
$$m\angle X + 98 = 180$$
$$m\angle X = 82$$
$$\frac{\sin W}{w} = \frac{\sin X}{x}$$
$$\frac{\sin 36°}{3.1} = \frac{\sin 82°}{x}$$
$$x \sin 36° = 3.1 \sin 82°$$
$$x = \frac{3.1 \sin 82°}{\sin 36°}$$
$$x \approx 5.2$$
$$\frac{\sin W}{w} = \frac{\sin Y}{y}$$
$$\frac{\sin 36°}{3.1} = \frac{\sin 62°}{y}$$
$$y \sin 36° = 3.1 \sin 62°$$
$$y = \frac{3.1 \sin 62°}{\sin 36°}$$
$$y \approx 4.7$$

28.
$$\frac{\sin W}{w} = \frac{\sin Y}{y}$$
$$\frac{\sin 107°}{30} = \frac{\sin Y}{9.5}$$
$$9.5 \sin 107° = 30 \sin Y$$
$$\frac{9.5 \sin 107°}{30} = \sin Y$$
$$\sin^{-1}\left(\frac{9.5 \sin 107°}{30}\right) = Y$$
$$17.6° \approx Y$$
$$m\angle W + m\angle X + m\angle Y = 180$$
$$107 + m\angle X + 17.6 \approx 180$$
$$m\angle X + 124.6 \approx 180$$
$$m\angle X \approx 55.4$$
$$\frac{\sin W}{w} = \frac{\sin X}{x}$$
$$\frac{\sin 107°}{30} = \frac{\sin 55.4°}{x}$$
$$x \sin 107° = 30 \sin 55.4°$$
$$x = \frac{30 \sin 55.4°}{\sin 107°}$$
$$x \approx 25.8$$

29.
$$\frac{\sin W}{w} = \frac{\sin X}{x}$$
$$\frac{\sin 88°}{21} = \frac{\sin X}{16}$$
$$16 \sin 88° = 21 \sin X$$
$$\frac{16 \sin 88°}{21} = \sin X$$
$$\sin^{-1}\left(\frac{16 \sin 88°}{21}\right) = X$$
$$49.6° \approx X$$

$m\angle W + m\angle X + m\angle Y = 180$
$88 + 49.6 + m\angle Y \approx 180$
$137.6 + m\angle Y \approx 180$
$m\angle Y \approx 42.4$

$\frac{\sin W}{w} = \frac{\sin Y}{y}$
$\frac{\sin 88°}{21} = \frac{\sin 42.4°}{y}$
$y \sin 88° = 21 \sin 42.4°$
$y = \frac{21 \sin 42.4°}{\sin 88°}$
$y \approx 14.2$

30. Let x be the measure of each base angle.
$x + x + 44 = 180$
$2x + 44 = 180$
$2x = 136$
$x = 68$
Let y be the measure of each of the congruent sides.
$\frac{\sin 44°}{46} = \frac{\sin 68°}{y}$
$y \sin 44° = 46 \sin 68°$
$y = \frac{46 \sin 68°}{\sin 44°}$
$y \approx 61.4$
The perimeter of the triangle is about $46 + 61.4 + 61.4$ or 168.8 cm.

31. $\frac{\sin 28°}{12} = \frac{\sin 40°}{AB}$
$AB \sin 28° = 12 \sin 40°$
$AB = \frac{12 \sin 40°}{\sin 28°}$
$AB \approx 16.43$
$\overline{AB} \cong \overline{DC}$ and $\overline{AD} \cong \overline{BC}$, so the perimeter of $ABCD$ is $2(16.43) + 2(12)$ or 56.9 units.

32.

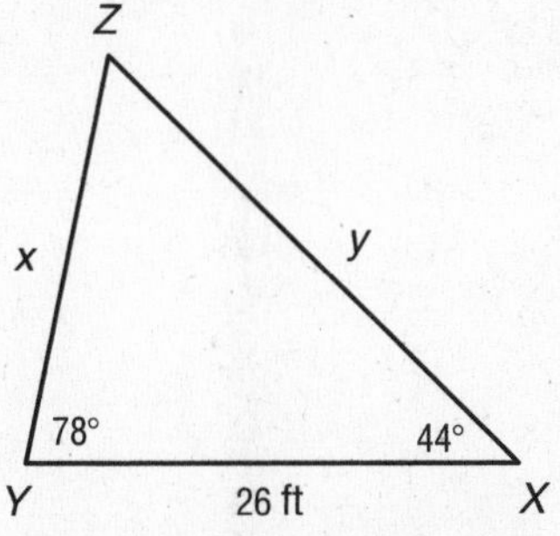

$m\angle X + m\angle Y + m\angle Z = 180$
$44 + 78 + m\angle Z = 180$
$122 + m\angle Z = 180$
$m\angle Z = 58$
$\frac{\sin 44°}{x} = \frac{\sin 58°}{26}$
$26 \sin 44° = x \sin 58°$
$\frac{26 \sin 44°}{\sin 58°} = x$
$21.3 \approx x$
$\frac{\sin 78°}{y} = \frac{\sin 58°}{26}$
$26 \sin 78° = y \sin 58°$
$\frac{26 \sin 78°}{\sin 58°} = y$
$30 \approx y$
The length of fence needed is about $30 + 21.3 + 26$ or 77.3 feet.

33.

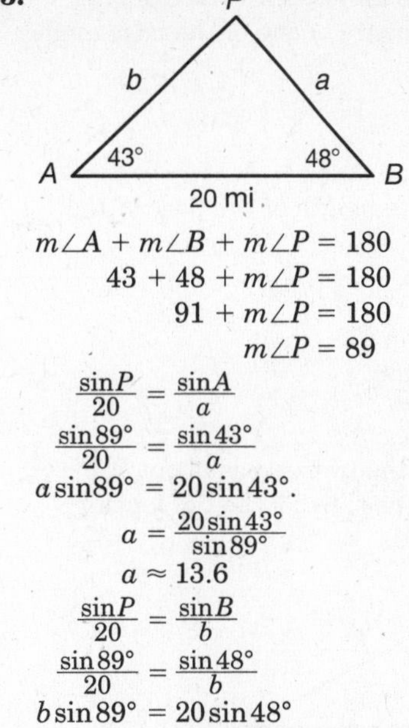

$m\angle A + m\angle B + m\angle P = 180$
$43 + 48 + m\angle P = 180$
$91 + m\angle P = 180$
$m\angle P = 89$
$\frac{\sin P}{20} = \frac{\sin A}{a}$
$\frac{\sin 89°}{20} = \frac{\sin 43°}{a}$
$a \sin 89° = 20 \sin 43°.$
$a = \frac{20 \sin 43°}{\sin 89°}$
$a \approx 13.6$
$\frac{\sin P}{20} = \frac{\sin B}{b}$
$\frac{\sin 89°}{20} = \frac{\sin 48°}{b}$
$b \sin 89° = 20 \sin 48°$
$b = \frac{20 \sin 48°}{\sin 89°}$
$b \approx 14.9$
The first station is about 14.9 miles from the plane, and the second station is about 13.6 miles from the plane.

34.

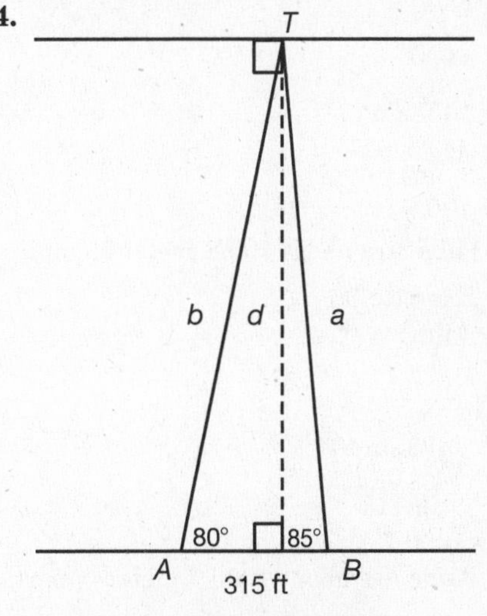

$m\angle A + m\angle B + m\angle T = 180$
$80 + 85 + m\angle T = 180$
$165 + m\angle T = 180$
$m\angle T = 15$
$\frac{\sin T}{315} = \frac{\sin A}{a}$
$\frac{\sin 15°}{315} = \frac{\sin 80°}{a}$
$a \sin 15° = 315 \sin 80°$
$a = \frac{315 \sin 80°}{\sin 15°}$
$a \approx 1198.6$
$\frac{\sin B}{d} = \frac{\sin 90°}{a}$
$\frac{\sin 85°}{d} = \frac{\sin 90°}{1198.6}$
$1198.6 \sin 85° = d \sin 90°$
$\frac{1198.6 \sin 85°}{\sin 90°} = d$
$1194.0 \approx d$
The distance across the gorge is about 1194 feet.

35. The plot of land is an isosceles triangle. Let x represent the measure of one of the base angles.

$x + x + 85 = 180$
$2x + 85 = 180$
$2x = 95$
$x = 47.5$

Let y represent the length of the base of the triangle.

$\frac{\sin 47.5°}{160} = \frac{\sin 85°}{y}$
$y \sin 47.5° = 160 \sin 85°$
$y = \frac{160 \sin 85°}{\sin 47.5°}$
$y \approx 216$

The perimeter of the property is about 160 + 160 + 216 or 536 feet, so 536 feet of fencing material is needed.

36.

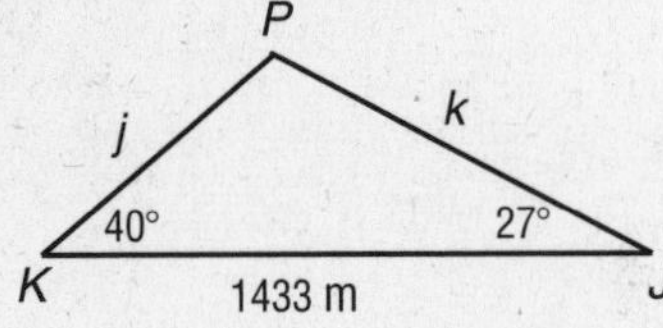

$m\angle K + m\angle J + m\angle P = 180$
$40 + 27 + m\angle P = 180$
$67 + m\angle P = 180$
$m\angle P = 113$

$\frac{\sin P}{1433} = \frac{\sin J}{j}$
$\frac{\sin 113°}{1433} = \frac{\sin 27°}{j}$
$j \sin 113° = 1433 \sin 27°$
$j = \frac{1433 \sin 27°}{\sin 113°}$
$j \approx 706.8$

Kayla and Paige are about 706.8 meters apart.

37. See art for Exercise 36.

$\frac{\sin P}{1433} = \frac{\sin K}{k}$
$\frac{\sin 113°}{1433} = \frac{\sin 40°}{k}$
$k \sin 113° = 1433 \sin 40°$
$k = \frac{1433 \sin 40°}{\sin 113°}$
$k \approx 1000.7$

Jenna and Paige are about 1000.7 meters apart.

38.

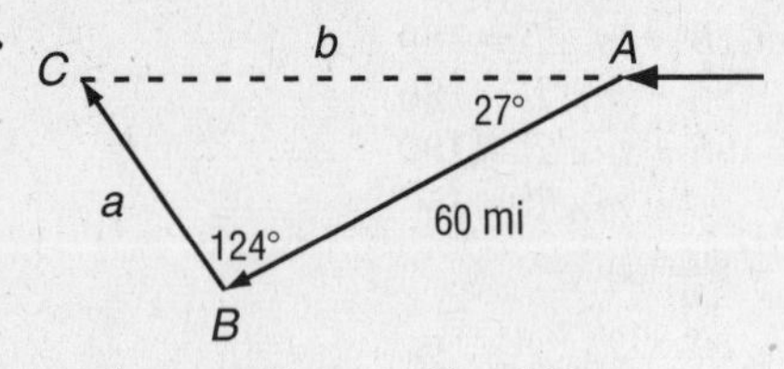

$m\angle A + m\angle B + m\angle C = 180$
$27 + 124 + m\angle C = 180$
$151 + m\angle C = 180$
$m\angle C = 29$

$\frac{\sin A}{a} = \frac{\sin C}{60}$
$\frac{\sin 27°}{a} = \frac{\sin 29°}{60}$
$60 \sin 27° = a \sin 29°$
$\frac{60 \sin 27°}{\sin 29°} = a$
$56.2 \approx a$

Keisha must fly about 56.2 miles.

39. See art for Exercise 38.

$\frac{\sin B}{b} = \frac{\sin C}{60}$
$\frac{\sin 124°}{b} = \frac{\sin 29°}{60}$
$60 \sin 124° = b \sin 29°$
$\frac{60 \sin 124°}{\sin 29°} = b$
$102.6 \approx b$
$60 + a = 60 + 56.2$ or 116.2
$116.2 - 102.6 = 13.6$

Keisha added about 13.6 miles to the flight.

40. Yes; in right $\triangle ABC$, $\frac{\sin A}{a} = \frac{\sin C}{c}$ where C is the right angle. Then $\sin A = \frac{a \sin C}{c}$. Since $m\angle C = 90$, then $\sin A = \frac{a \sin 90°}{c}$. Since $\sin 90° = 1$, then $\sin A = \frac{a}{c}$, which is the definition of the sine ratio.

41. Sample answer: Triangles are used to determine distances in space. Answers should include the following.

- The VLA is one of the world's premier astronomical radio observatories. It is used to make pictures from the radio waves emitted by astronomical objects.
- Triangles are used in the construction of the antennas.

42. $m\angle X + m\angle Y + m\angle Z = 180$
$48 + 112 + m\angle Z = 180$
$160 + m\angle Z = 180$
$m\angle Z = 20$

$\frac{\sin X}{x} = \frac{\sin Y}{y}$
$\frac{\sin 48°}{12} = \frac{\sin 112°}{y}$
$y \sin 48° = 12 \sin 112°$
$y = \frac{12 \sin 112°}{\sin 48°}$
$y \approx 15.0$

$\frac{\sin X}{x} = \frac{\sin Z}{z}$
$\frac{\sin 48°}{12} = \frac{\sin 20°}{z}$
$z \sin 48° = 12 \sin 20°$
$z = \frac{12 \sin 20°}{\sin 48°}$
$z \approx 5.5$

43. A; Metropolis Grill: $\frac{9 + 8 + 7 + 7}{4} = \frac{31}{4} = 7.75$

Le Circus: $\frac{10 + 8 + 3 + 5}{4} = \frac{26}{4} = 6.5$

Aquavent: $\frac{8 + 9 + 4 + 6}{4} = \frac{27}{4} = 6.75$

Del Blanco's: $\frac{7 + 9 + 4 + 7}{4} = \frac{27}{4} = 6.75$

Metropolis Grill has the best average rating of the four restaurant choices.

Page 383 Maintain Your Skills

44. Find x so that the angle of elevation is 73.5°.

$\tan 73.5° = \frac{6 + 1}{x}$
$x \tan 73.5° = 7$
$x = \frac{7}{\tan 73.5°}$
$x \approx 2.07$

The overhang should be about 2.07 feet long.

45. Let y represent the amount of the window that will get direct sunlight.

$$\tan 26.5° = \frac{7-y}{2.07}$$
$$2.07\tan 26.5° = 7 - y$$
$$y = 7 - 2.07\tan 26.5°$$
$$y \approx 5.97$$

About 5.97 feet of the window will get direct sunlight.

46.

$\sin J = \frac{j}{k}$	$\cos J = \frac{\ell}{k}$	$\tan J = \frac{j}{\ell}$
$\sin J = \frac{8}{17}$	$\cos J = \frac{15}{17}$	$\tan J = \frac{8}{15}$
$\sin J \approx 0.47$	$\cos J \approx 0.88$	$\tan J \approx 0.53$
$\sin L = \frac{\ell}{k}$	$\cos L = \frac{j}{k}$	$\tan L = \frac{\ell}{j}$
$\sin L = \frac{15}{17}$	$\cos L = \frac{8}{17}$	$\tan L = \frac{15}{8}$
$\sin L \approx 0.88$	$\cos L \approx 0.47$	$\tan L \approx 1.88$

47.

$\sin J = \frac{j}{k}$	$\cos J = \frac{\ell}{k}$	$\tan J = \frac{j}{\ell}$
$\sin J = \frac{20}{29}$	$\cos J = \frac{21}{29}$	$\tan J = \frac{20}{21}$
$\sin J \approx 0.69$	$\cos J \approx 0.72$	$\tan J = 0.95$
$\sin L = \frac{\ell}{k}$	$\cos L = \frac{j}{k}$	$\tan L = \frac{\ell}{j}$
$\sin L = \frac{21}{29}$	$\cos L = \frac{20}{29}$	$\tan L = \frac{21}{20}$
$\sin L \approx 0.72$	$\cos L \approx 0.69$	$\tan L = 1.05$

48.

$\sin J = \frac{j}{k}$	$\cos J = \frac{\ell}{k}$	$\tan J = \frac{j}{\ell}$
$\sin J = \frac{12}{24}$	$\cos J = \frac{12\sqrt{3}}{24}$	$\tan J = \frac{12}{12\sqrt{3}}$
$\sin J = \frac{1}{2}$	$\cos J = \frac{\sqrt{3}}{2}$	$\tan J = \frac{\sqrt{3}}{3}$
$\sin J = 0.50$	$\cos J \approx 0.87$	$\tan J \approx 0.58$
$\sin L = \frac{\ell}{k}$	$\cos L = \frac{j}{k}$	$\tan L = \frac{\ell}{j}$
$\sin L = \frac{12\sqrt{3}}{24}$	$\cos L = \frac{12}{24}$	$\tan L = \frac{12\sqrt{3}}{12}$
$\sin L = \frac{\sqrt{3}}{2}$	$\cos L = \frac{1}{2}$	$\tan L = \sqrt{3}$
$\sin L \approx 0.87$	$\cos L = 0.50$	$\tan L \approx 1.73$

49.

$\sin J = \frac{j}{k}$	$\cos J = \frac{\ell}{k}$	$\tan J = \frac{j}{\ell}$
$\sin J = \frac{7\sqrt{2}}{14}$	$\cos J = \frac{7\sqrt{2}}{14}$	$\tan J = \frac{7\sqrt{2}}{7\sqrt{2}}$
$\sin J = \frac{\sqrt{2}}{2}$	$\cos J = \frac{\sqrt{2}}{2}$	$\tan J = 1.00$
$\sin J \approx 0.71$	$\cos J \approx 0.71$	
$\sin L = \frac{\ell}{k}$	$\cos L = \frac{j}{k}$	$\tan L = \frac{\ell}{j}$
$\sin L = \frac{7\sqrt{2}}{14}$	$\cos L = \frac{7\sqrt{2}}{14}$	$\tan L = \frac{7\sqrt{2}}{7\sqrt{2}}$
$\sin L = \frac{\sqrt{2}}{2}$	$\cos L = \frac{\sqrt{2}}{2}$	$\tan L = 1.00$
$\sin L \approx 0.71$	$\cos L \approx 0.71$	

50. $m\angle 1 + 54 = 120$ Exterior Angle Theorem
$m\angle 1 = 66$

51. $m\angle 2 = 54$ alternate interior ∠s

52. $m\angle 2 + m\angle 3 + 36 = 180$ Angle Sum Theorem
$54 + m\angle 3 + 36 = 180$
$m\angle 3 + 90 = 180$
$m\angle 3 = 90$

53. $\frac{c^2 - a^2 - b^2}{-2ab} = \frac{10^2 - 7^2 - 8^2}{-2(7)(8)} = \frac{100 - 49 - 64}{-112} = \frac{-13}{-112} = \frac{13}{112}$

54. $\frac{c^2 - a^2 - b^2}{-2ab} = \frac{6^2 - 4^2 - 9^2}{-2(4)(9)} = \frac{36 - 16 - 81}{-72} = \frac{-61}{-72} = \frac{61}{72}$

55. $\frac{c^2 - a^2 - b^2}{-2ab} = \frac{10^2 - 5^2 - 8^2}{-2(5)(8)} = \frac{100 - 25 - 64}{-80} = -\frac{11}{80}$

56. $\frac{c^2 - a^2 - b^2}{-2ab} = \frac{13^2 - 16^2 - 4^2}{-2(16)(4)} = \frac{169 - 256 - 16}{-128} = \frac{-103}{-128} = \frac{103}{128}$

57. $\frac{c^2 - a^2 - b^2}{-2ab} = \frac{9^2 - 3^2 - 10^2}{-2(3)(10)} = \frac{81 - 9 - 100}{-60} = \frac{-28}{-60} = \frac{7}{15}$

58. $\frac{c^2 - a^2 - b^2}{-2ab} = \frac{11^2 - 5^2 - 7^2}{-2(5)(7)} = \frac{121 - 25 - 49}{-70} = -\frac{47}{70}$

Page 383 Practice Quiz 2

1. $\tan x° = \frac{16}{10}$
$x = \tan^{-1}\left(\frac{16}{10}\right)$
$x \approx 58.0$

2. $\cos 17° = \frac{x}{9.7}$
$9.7\cos 17° = x$
$9.3 \approx x$

3. $\cos 53° = \frac{32}{x}$
$x\cos 53° = 32$
$x = \frac{32}{\cos 53°}$
$x \approx 53.2$

4.

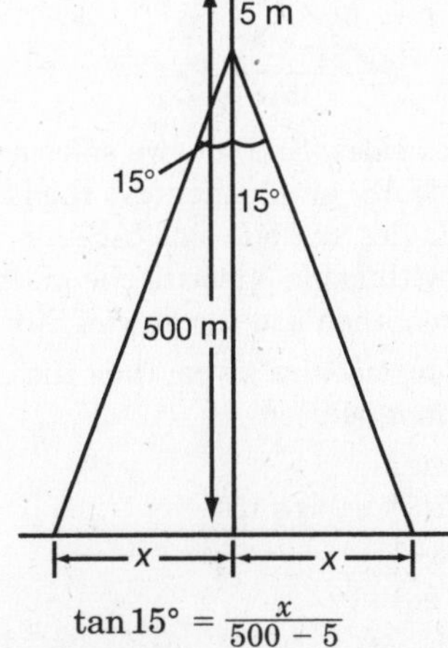

$\tan 15° = \frac{x}{500 - 5}$
$495\tan 15° = x$
$132.6 \approx x$

The distance is about 132.6 meters.

5. $\frac{\sin D}{EF} = \frac{\sin F}{DE}$

$\frac{\sin D}{8} = \frac{\sin 82°}{12}$

$12 \sin D = 8 \sin 82°$

$\sin D = \frac{8 \sin 82°}{12}$

$D = \sin^{-1}\left(\frac{8 \sin 82°}{12}\right)$

$D \approx 41°$

$m\angle D + m\angle E + m\angle F = 180$

$41 + m\angle E + 82 \approx 180$

$m\angle E + 123 \approx 180$

$m\angle E \approx 57$

$\frac{\sin E}{DF} = \frac{\sin D}{EF}$

$\frac{\sin 57°}{DF} = \frac{\sin 41°}{8}$

$8 \sin 57° = DF \sin 41°$

$\frac{8 \sin 57°}{\sin 41°} = DF$

$10.2 \approx DF$

Page 384 Geometry Software Investigation: The Ambiguous Case of the Law of Sines

1. BD, AB, and $m\angle A$
2. Sample answer: There are two different triangles.
3. Sample answer: The results are the same. In each case, two triangles are possible.
4. Sample answer: Circle B intersects $\overrightarrow{AC}$ at only one point. See students' work.
5. Yes; sample answer: There is no solution if circle B does not intersect $\overrightarrow{AC}$.

7-7 The Law of Cosines

Pages 387–388 Check for Understanding

1. Sample answer: Use the Law of Cosines when you have all three sides given (SSS) or two sides and the included angle (SAS).

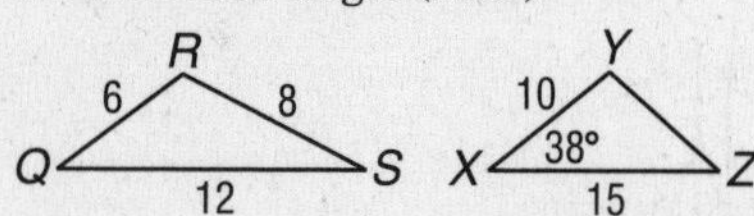

2. If you have all three sides (SSS) or two sides and the included angle (SAS) given, then use the Law of Cosines. If two angles and one side (ASA or AAS) or two sides with angle opposite one of the sides (SSA) are given, then use the Law of Sines.
3. If two angles and one side are given, then the Law of Cosines cannot be used.
4. $b^2 = a^2 + c^2 - 2ac \cos B$

$b^2 = 5^2 + (\sqrt{2})^2 - 2(5)(\sqrt{2})\cos 45°$

$b^2 = 27 - 10\sqrt{2} \cos 45°$

$b = \sqrt{27 - 10\sqrt{2} \cos 45°}$

$b \approx 4.1$

5. $a^2 = b^2 + c^2 - 2bc \cos A$

$a^2 = 107^2 + 94^2 - 2(107)(94)\cos 105°$

$a^2 = 20{,}285 - 20{,}116 \cos 105°$

$a = \sqrt{20{,}285 - 20{,}116 \cos 105°}$

$a \approx 159.7$

6. $s^2 = r^2 + t^2 - 2rt \cos S$

$65^2 = 33^2 + 56^2 - 2(33)(56)\cos S$

$4225 = 4225 - 3696 \cos S$

$0 = -3696 \cos S$

$0 = \cos S$

$S = \cos^{-1}(0)$

$S = 90°$

7. $r^2 = s^2 + t^2 - 2st \cos R$

$2.2^2 = 1.3^2 + 1.6^2 - 2(1.3)(1.6)\cos R$

$4.84 = 4.25 - 4.16 \cos R$

$0.59 = -4.16 \cos R$

$\frac{0.59}{-4.16} = \cos R$

$R = \cos^{-1}\left(\frac{0.59}{-4.16}\right)$

$R \approx 98°$

8. We know the measures of three sides (SSS), so use the Law of Cosines.

$x^2 = y^2 + z^2 - 2yz \cos X$

$5^2 = 10^2 + 13^2 - 2(10)(13)\cos X$

$25 = 269 - 260 \cos X$

$-244 = -260 \cos X$

$\frac{-244}{-260} = \cos X$

$X = \cos^{-1}\left(\frac{-244}{-260}\right)$

$X \approx 20°$

$\frac{\sin X}{x} = \frac{\sin Y}{y}$

$\frac{\sin 20°}{5} = \frac{\sin Y}{10}$

$10 \sin 20° = 5 \sin Y$

$\frac{10 \sin 20°}{5} = \sin Y$

$\sin^{-1}\left(\frac{10 \sin 20°}{5}\right) = Y$

$43° \approx Y$

$m\angle X + m\angle Y + m\angle Z = 180$

$20 + 43 + m\angle Z \approx 180$

$m\angle Z \approx 117$

9. We know the measures of two sides and the included angle (SAS), so use the Law of Cosines.

$\ell^2 = k^2 + m^2 - 2km \cos L$

$\ell^2 = 20^2 + 24^2 - 2(20)(24)\cos 47°$

$\ell^2 = 976 - 960 \cos 47°$

$\ell = \sqrt{976 - 960 \cos 47°}$

$\ell \approx 17.9$

$\frac{\sin L}{\ell} = \frac{\sin K}{k}$

$\frac{\sin 47°}{17.9} = \frac{\sin K}{20}$

$20 \sin 47° = 17.9 \sin K$

$\frac{20 \sin 47°}{17.9} = \sin K$

$\sin^{-1}\left(\frac{20 \sin 47°}{17.9}\right) \approx K$

$55° \approx K$

$m\angle K + m\angle L + m\angle M = 180$

$55 + 47 + m\angle M \approx 180$

$m\angle M \approx 78$

10. Let n, d, and q be the measures of the sides opposite $\angle N$, $\angle D$, and $\angle Q$, respectively.

$n = \frac{1}{2}(10) + \frac{1}{2}(24)$ or 17 mm

$d = \frac{1}{2}(24) + \frac{1}{2}(22)$ or 23 mm

$q = \frac{1}{2}(10) + \frac{1}{2}(22)$ or 16 mm

$$n^2 = d^2 + q^2 - 2dq\cos N$$
$$17^2 = 23^2 + 16^2 - 2(23)(16)\cos N$$
$$289 = 785 - 736\cos N$$
$$-496 = -736\cos N$$
$$\frac{-496}{-736} = \cos N$$
$$N = \cos^{-1}\left(\frac{-496}{-736}\right)$$
$$N \approx 47.6°$$
$$q^2 = n^2 + d^2 - 2nd\cos Q$$
$$16^2 = 17^2 + 23^2 - 2(17)(23)\cos Q$$
$$256 = 818 - 782\cos Q$$
$$-562 = -782\cos Q$$
$$\frac{-562}{-782} = \cos Q$$
$$Q = \cos^{-1}\left(\frac{-562}{-782}\right)$$
$$Q \approx 44.1°$$
$$m\angle Q + m\angle D + m\angle N = 180$$
$$44.1 + m\angle D + 47.6 \approx 180$$
$$m\angle D \approx 88.3$$

Pages 388–390 Practice and Apply

11. $u^2 = t^2 + v^2 - 2tv\cos U$
$$u^2 = 9.1^2 + 8.3^2 - 2(9.1)(8.3)\cos 32°$$
$$u^2 = 151.7 - 151.06\cos 32°$$
$$u = \sqrt{151.7 - 151.06\cos 32°}$$
$$u \approx 4.9$$

12. $v^2 = t^2 + u^2 - 2tu\cos V$
$$v^2 = 11^2 + 17^2 - 2(11)(17)\cos 78°$$
$$v^2 = 410 - 374\cos 78°$$
$$v = \sqrt{410 - 374\cos 78°}$$
$$v \approx 18.2$$

13. $t^2 = u^2 + v^2 - 2uv\cos T$
$$t^2 = 11^2 + 17^2 - 2(11)(17)\cos 105°$$
$$t^2 = 410 - 374\cos 105°$$
$$t = \sqrt{410 - 374\cos 105°}$$
$$t \approx 22.5$$

14. $t^2 = u^2 + v^2 - 2uv\cos T$
$$t^2 = 17^2 + 11^2 - 2(17)(11)\cos 59°$$
$$t^2 = 410 - 374\cos 59°$$
$$t = \sqrt{410 - 374\cos 59°}$$
$$t \approx 14.7$$

15. $f^2 = e^2 + g^2 - 2eg\cos F$
$$8.3^2 = 9.1^2 + 16.7^2 - 2(9.1)(16.7)\cos F$$
$$68.89 = 361.7 - 303.94\cos F$$
$$-292.81 = -303.94\cos F$$
$$\frac{-292.81}{-303.94} = \cos F$$
$$F = \cos^{-1}\left(\frac{292.81}{303.94}\right)$$
$$F \approx 16°$$

16. $e^2 = f^2 + g^2 - 2fg\cos E$
$$14^2 = 19^2 + 32^2 - 2(19)(32)\cos E$$
$$196 = 1385 - 1216\cos E$$
$$-1189 = -1216\cos E$$
$$\frac{-1189}{-1216} = \cos E$$
$$E = \cos^{-1}\left(\frac{1189}{1216}\right)$$
$$E \approx 12°$$

17. $f^2 = e^2 + g^2 - 2eg\cos F$
$$198^2 = 325^2 + 208^2 - 2(325)(208)\cos F$$
$$39{,}204 = 148{,}889 - 135{,}200\cos F$$
$$-109{,}685 = -135{,}200\cos F$$
$$\frac{-109{,}685}{-135{,}200} = \cos F$$
$$F = \cos^{-1}\left(\frac{109{,}685}{135{,}200}\right)$$
$$F \approx 36°$$

18. $g^2 = e^2 + f^2 - 2ef\cos G$
$$10^2 = 21.9^2 + 18.9^2 - 2(21.9)(18.9)\cos G$$
$$100 = 836.82 - 827.82\cos G$$
$$-736.82 = -827.82\cos G$$
$$\frac{-736.82}{-827.82} = \cos G$$
$$G = \cos^{-1}\left(\frac{736.82}{827.82}\right)$$
$$G \approx 27°$$

19. $\frac{\sin H}{h} = \frac{\sin F}{f}$
$$\frac{\sin H}{8} = \frac{\sin 40°}{10}$$
$$10\sin H = 8\sin 40°$$
$$\sin H = \frac{8\sin 40°}{10}$$
$$H = \sin^{-1}\left(\frac{8\sin 40°}{10}\right)$$
$$H \approx 31°$$
$$m\angle F + m\angle G + m\angle H = 180$$
$$40 + m\angle G + 31 \approx 180$$
$$m\angle G \approx 109$$
$$\frac{\sin G}{g} = \frac{\sin F}{f}$$
$$\frac{\sin 109°}{g} = \frac{\sin 40°}{10}$$
$$10\sin 109° = g\sin 40°$$
$$\frac{10\sin 109°}{\sin 40°} = g$$
$$14.7 \approx g$$

20. $p^2 = m^2 + q^2 - 2mq\cos P$
$$p^2 = 11^2 + 10^2 - 2(11)(10)\cos 38°$$
$$p^2 = 221 - 220\cos 38°$$
$$p = \sqrt{221 - 220\cos 38°}$$
$$p \approx 6.9$$
$$\frac{\sin M}{m} = \frac{\sin P}{p}$$
$$\frac{\sin M}{11} = \frac{\sin 38°}{6.9}$$
$$6.9\sin M = 11\sin 38°$$
$$\sin M = \frac{11\sin 38°}{6.9}$$
$$M = \sin^{-1}\left(\frac{11\sin 38°}{6.9}\right)$$
$$M \approx 79°$$
$$m\angle M + m\angle P + m\angle Q = 180$$
$$79 + 38 + m\angle Q \approx 180$$
$$m\angle Q \approx 63$$

21. $b^2 = c^2 + d^2 - 2cd\cos B$

$18^2 = 15^2 + 11^2 - 2(15)(11)\cos B$

$324 = 346 - 330\cos B$

$-22 = -330\cos B$

$\frac{-22}{-330} = \cos B$

$B = \cos^{-1}\left(\frac{22}{330}\right)$

$B \approx 86°$

$\frac{\sin B}{b} = \frac{\sin C}{c}$

$\frac{\sin 86°}{18} = \frac{\sin C}{15}$

$15\sin 86° = 18\sin C$

$\frac{15\sin 86°}{18} = \sin C$

$\sin^{-1}\left(\frac{15\sin 86°}{18}\right) = C$

$56° \approx C$

$m\angle B + m\angle C + m\angle D = 180$

$86 + 56 + m\angle D \approx 180$

$m\angle D \approx 38$

22. $\frac{\sin A}{a} = \frac{\sin C}{c}$

$\frac{\sin 42°}{a} = \frac{\sin 77°}{6}$

$6\sin 42° = a\sin 77°$

$\frac{6\sin 42°}{\sin 77°} = a$

$4.1 \approx a$

$m\angle A + m\angle B + m\angle C = 180$

$42 + m\angle B + 77 = 180$

$m\angle B = 61$

$\frac{\sin B}{b} = \frac{\sin C}{c}$

$\frac{\sin 61°}{b} = \frac{\sin 77°}{6}$

$6\sin 61° = b\sin 77°$

$\frac{6\sin 61°}{\sin 77°} = b$

$5.4 \approx b$

23. $c^2 = a^2 + b^2 - 2ab\cos C$

$c^2 = 10.3^2 + 9.5^2 - 2(10.3)(9.5)\cos 37°$

$c^2 = 196.34 - 195.7\cos 37°$

$c = \sqrt{196.34 - 195.7\cos 37°}$

$c \approx 6.3$

$\frac{\sin A}{a} = \frac{\sin C}{c}$

$\frac{\sin A}{10.3} = \frac{\sin 37°}{6.3}$

$6.3\sin A = 10.3\sin 37°$

$\sin A = \frac{10.3\sin 37°}{6.3}$

$A = \sin^{-1}\left(\frac{10.3\sin 37°}{6.3}\right)$

$A \approx 80°$

$m\angle A + m\angle B + m\angle C = 180$

$80 + m\angle B + 37 \approx 180$

$m\angle B \approx 63$

24. $a^2 = b^2 + c^2 - 2bc\cos A$

$15^2 = 19^2 + 28^2 - 2(19)(28)\cos A$

$225 = 1145 - 1064\cos A$

$-920 = -1064\cos A$

$\frac{-920}{-1064} = \cos A$

$A = \cos^{-1}\left(\frac{920}{1064}\right)$

$A \approx 30°$

$\frac{\sin A}{a} = \frac{\sin B}{b}$

$\frac{\sin 30°}{15} = \frac{\sin B}{19}$

$19\sin 30° = 15\sin B$

$\frac{19\sin 30°}{15} = \sin B$

$\sin^{-1}\left(\frac{19\sin 30°}{15}\right) = B$

$39° \approx B$

$m\angle A + m\angle B + m\angle C = 180$

$30 + 39 + m\angle C \approx 180$

$m\angle C \approx 111$

25. $\frac{\sin A}{a} = \frac{\sin C}{c}$

$\frac{\sin 53°}{a} = \frac{\sin 28°}{14.9}$

$14.9\sin 53° = a\sin 28°$

$\frac{14.9\sin 53°}{\sin 28°} = a$

$25.3 \approx a$

$m\angle A + m\angle B + m\angle C = 180$

$53 + m\angle B + 28 = 180$

$m\angle B = 99$

$\frac{\sin B}{b} = \frac{\sin C}{c}$

$\frac{\sin 99°}{b} = \frac{\sin 28°}{14.9}$

$14.9 \sin 99° = b \sin 28°$

$\frac{14.9\sin 99°}{\sin 28°} = b$

$31.3 \approx b$

26. Find $m\angle DAB$ and $m\angle BCD$.

$DB^2 = AB^2 + AD^2 - 2(AB)(AD)\cos(\angle DAB)$

$\left(7\frac{2}{3}\right)^2 = 5^2 + 5^2 - 2(5)(5)\cos(\angle DAB)$

$\frac{529}{9} = 50 - 50\cos(\angle DAB)$

$\frac{79}{9} = -50\cos(\angle DAB)$

$-\frac{1}{50}\left(\frac{79}{9}\right) = \cos(\angle DAB)$

$m\angle DAB = \cos^{-1}\left[-\frac{1}{50}\left(\frac{79}{9}\right)\right]$

$m\angle DAB \approx 100$

$BD^2 = BC^2 + DC^2 -$
$2(BC)(DC)\cos(\angle BCD)$

$\left(7\frac{2}{3}\right)^2 = 8^2 + 8^2 - 2(8)(8)\cos(\angle BCD)$

$\frac{529}{9} = 128 - 128\cos(\angle BCD)$

$\frac{-623}{9} = -128\cos(\angle BCD)$

$-\frac{1}{128}\left(-\frac{623}{9}\right) = \cos(\angle BCD)$

$m\angle BCD = \cos^{-1}\left[\frac{1}{128}\left(\frac{623}{9}\right)\right]$

$m\angle BCD \approx 57$

27. $\frac{\sin L}{\ell} = \frac{\sin M}{m}$

$\frac{\sin 23°}{54} = \frac{\sin M}{44}$

$44\sin 23° = 54\sin M$

$\frac{44\sin 23°}{54} = \sin M$

$\sin^{-1}\left(\frac{44\sin 23°}{54}\right) = M$

$18.6° \approx M$

$$m\angle L + m\angle M + m\angle N = 180$$
$$23 + 18.6 + m\angle N \approx 180$$
$$m\angle N \approx 138.4$$
$$\frac{\sin N}{n} = \frac{\sin L}{\ell}$$
$$\frac{\sin 138.4^\circ}{n} = \frac{\sin 23^\circ}{54}$$
$$54 \sin 138.4^\circ = n \sin 23^\circ$$
$$\frac{54 \sin 138.4^\circ}{\sin 23^\circ} = n$$
$$91.8 \approx n$$

28. $m^2 = \ell^2 + n^2 - 2\ell n \cos M$
$$18^2 = 24^2 + 30^2 - 2(24)(30)\cos M$$
$$324 = 1476 - 1440 \cos M$$
$$-1152 = -1440 \cos M$$
$$\frac{-1152}{-1440} = \cos M$$
$$M = \cos^{-1}\left(\frac{1152}{1440}\right)$$
$$M \approx 36.9^\circ$$
$$\frac{\sin M}{m} = \frac{\sin L}{\ell}$$
$$\frac{\sin 36.9^\circ}{18} = \frac{\sin L}{24}$$
$$24 \sin 36.9^\circ = 18 \sin L$$
$$\frac{24 \sin 36.9^\circ}{18} = \sin L$$
$$\sin^{-1}\left(\frac{24 \sin 36.9^\circ}{18}\right) = L$$
$$53.2^\circ \approx L$$
$$m\angle L + m\angle M + m\angle N = 180$$
$$53.2 + 36.9 + m\angle N \approx 180$$
$$m\angle N \approx 89.9$$

29. $\ell^2 = m^2 + n^2 - 2mn \cos L$
$$\ell^2 = 19^2 + 28^2 - 2(19)(28)\cos 49^\circ$$
$$\ell^2 = 1145 - 1064 \cos 49^\circ$$
$$\ell = \sqrt{1145 - 1064 \cos 49^\circ}$$
$$\ell \approx 21.1$$
$$\frac{\sin L}{\ell} = \frac{\sin M}{m}$$
$$\frac{\sin 49^\circ}{21.1} = \frac{\sin M}{19}$$
$$19 \sin 49^\circ = 21.1 \sin M$$
$$\frac{19 \sin 49^\circ}{21.1} = \sin M$$
$$\sin^{-1}\left(\frac{19 \sin 49^\circ}{21.1}\right) = M$$
$$42.8^\circ \approx M$$
$$m\angle L + m\angle M + m\angle N = 180$$
$$49 + 42.8 + m\angle N \approx 180$$
$$m\angle N \approx 88.2$$

30. $m\angle L + m\angle M + m\angle N = 180$
$$55 + 46 + m\angle N = 180$$
$$m\angle N = 79$$
$$\frac{\sin L}{\ell} = \frac{\sin N}{n}$$
$$\frac{\sin 55^\circ}{\ell} = \frac{\sin 79^\circ}{16}$$
$$16 \sin 55^\circ = \ell \sin 79^\circ$$
$$\frac{16 \sin 55^\circ}{\sin 79^\circ} = \ell$$
$$13.4 \approx \ell$$
$$\frac{\sin M}{m} = \frac{\sin N}{n}$$
$$\frac{\sin 46^\circ}{m} = \frac{\sin 79^\circ}{16}$$
$$16 \sin 46^\circ = m \sin 79^\circ$$
$$\frac{16 \sin 46^\circ}{\sin 79^\circ} = m$$
$$11.7 \approx m$$

31. $\ell^2 = m^2 + n^2 - 2mn \cos L$
$$423^2 = 256^2 + 288^2 - 2(256)(288)\cos L$$
$$178{,}929 = 148{,}480 - 147{,}456 \cos L$$
$$30{,}449 = -147{,}456 \cos L$$
$$\frac{30{,}449}{-147{,}456} = \cos L$$
$$L = \cos^{-1}\left(\frac{30{,}449}{-147{,}456}\right)$$
$$L \approx 101.9^\circ$$
$$\frac{\sin L}{\ell} = \frac{\sin M}{m}$$
$$\frac{\sin 101.9^\circ}{423} = \frac{\sin M}{256}$$
$$256 \sin 101.9^\circ = 423 \sin M$$
$$\frac{256 \sin 101.9^\circ}{423} = \sin M$$
$$\sin^{-1}\left(\frac{256 \sin 101.9^\circ}{423}\right) = M$$
$$36.3^\circ \approx M$$
$$m\angle L + m\angle M + m\angle N = 180$$
$$101.9 + 36.3 + m\angle N \approx 180$$
$$m\angle N \approx 41.8$$

32. $m^2 = \ell^2 + n^2 - 2\ell n \cos M$
$$m^2 = 6.3^2 + 6.7^2 - 2(6.3)(6.7)\cos 55^\circ$$
$$m^2 = 84.58 - 84.42 \cos 55^\circ$$
$$m = \sqrt{84.58 - 84.42 \cos 55^\circ}$$
$$m \approx 6.0$$
$$\frac{\sin M}{m} = \frac{\sin L}{\ell}$$
$$\frac{\sin 55^\circ}{6} = \frac{\sin L}{6.3}$$
$$6.3 \sin 55^\circ = 6 \sin L$$
$$\frac{6.3 \sin 55^\circ}{6} = \sin L$$
$$\sin^{-1}\left(\frac{6.3 \sin 55^\circ}{6}\right) = L$$
$$59.3^\circ \approx L$$
$$m\angle L + m\angle M + m\angle N = 180$$
$$59.3 + 55 + m\angle N \approx 180$$
$$m\angle N \approx 65.7$$

33. $m^2 = \ell^2 + n^2 - 2\ell n \cos M$
$$m^2 = 5^2 + 10^2 - 2(5)(10)\cos 27^\circ$$
$$m^2 = 125 - 100 \cos 27^\circ$$
$$m = \sqrt{125 - 100 \cos 27^\circ}$$
$$m \approx 6.0$$
$$\frac{\sin L}{\ell} = \frac{\sin M}{m}$$
$$\frac{\sin L}{5} = \frac{\sin 27^\circ}{6}$$
$$6 \sin L = 5 \sin 27^\circ$$
$$\sin L = \frac{5 \sin 27^\circ}{6}$$
$$L = \sin^{-1}\left(\frac{5 \sin 27^\circ}{6}\right)$$
$$L \approx 22.2^\circ$$
$$m\angle L + m\angle M + m\angle N = 180$$
$$22.2 + 27 + m\angle N \approx 180$$
$$m\angle N \approx 130.8$$

34.
$$\ell^2 = m^2 + n^2 - 2mn\cos L$$
$$14^2 = 20^2 + 17^2 - 2(20)(17)\cos L$$
$$196 = 689 - 680\cos L$$
$$-493 = -680\cos L$$
$$\frac{-493}{-680} = \cos L$$
$$L = \cos^{-1}\left(\frac{493}{680}\right)$$
$$L \approx 43.5°$$
$$\frac{\sin L}{\ell} = \frac{\sin M}{m}$$
$$\frac{\sin 43.5°}{14} = \frac{\sin M}{20}$$
$$20\sin 43.5° = 14\sin M$$
$$\frac{20\sin 43.5°}{14} = \sin M$$
$$\sin^{-1}\left(\frac{20\sin 43.5°}{14}\right) = M$$
$$79.5° \approx M$$
$$m\angle L + m\angle M + m\angle N = 180$$
$$43.5 + 79.5 + m\angle N \approx 180$$
$$m\angle N \approx 57.0$$

35.
$$m^2 = \ell^2 + n^2 - 2\ell n\cos M$$
$$m^2 = 14^2 + 21^2 - 2(14)(21)\cos 60°$$
$$m^2 = 637 - 588\cos 60°$$
$$m = \sqrt{637 - 588\cos 60°}$$
$$m \approx 18.5$$
$$\frac{\sin L}{\ell} = \frac{\sin M}{m}$$
$$\frac{\sin L}{14} = \frac{\sin 60°}{18.5}$$
$$18.5\sin L = 14\sin 60°$$
$$\sin L = \frac{14\sin 60°}{18.5}$$
$$L = \sin^{-1}\left(\frac{14\sin 60°}{18.5}\right)$$
$$L \approx 40.9°$$
$$m\angle L + m\angle M + m\angle N = 180$$
$$40.9 + 60 + m\angle N \approx 180$$
$$m\angle N \approx 79.1$$

36.
$$\ell^2 = m^2 + n^2 - 2mn\cos L$$
$$14^2 = 15^2 + 16^2 - 2(15)(16)\cos L$$
$$196 = 481 - 480\cos L$$
$$-285 = -480\cos L$$
$$\frac{-285}{-480} = \cos L$$
$$L = \cos^{-1}\left(\frac{285}{480}\right)$$
$$L \approx 53.6°$$
$$\frac{\sin L}{\ell} = \frac{\sin M}{m}$$
$$\frac{\sin 53.6°}{14} = \frac{\sin M}{15}$$
$$15\sin 53.6° = 14\sin M$$
$$\frac{15\sin 53.6°}{14} = \sin M$$
$$\sin^{-1}\left(\frac{15\sin 53.6°}{14}\right) = M$$
$$59.6° \approx M$$
$$m\angle L + m\angle M + m\angle N = 180$$
$$53.6 + 59.6 + m\angle N \approx 180$$
$$m\angle N \approx 66.8$$

37.
$$\frac{\sin L}{\ell} = \frac{\sin N}{n}$$
$$\frac{\sin 51°}{40} = \frac{\sin N}{35}$$
$$35\sin 51° = 40\sin N$$
$$\frac{35\sin 51°}{40} = \sin N$$
$$\sin^{-1}\left(\frac{35\sin 51°}{40}\right) = N$$
$$42.8° \approx N$$
$$m\angle L + m\angle M + m\angle N = 180$$
$$51 + m\angle M + 42.8 \approx 180$$
$$m\angle M \approx 86.2$$
$$\frac{\sin L}{\ell} = \frac{\sin M}{m}$$
$$\frac{\sin 51°}{40} = \frac{\sin 86.2°}{m}$$
$$m\sin 51° = 40\sin 86.2°$$
$$m = \frac{40\sin 86.2°}{\sin 51°}$$
$$m \approx 51.4$$

38.
$$\ell^2 = m^2 + n^2 - 2mn\cos L$$
$$10^2 = 11^2 + 12^2 - 2(11)(12)\cos L$$
$$100 = 265 - 264\cos L$$
$$-165 = -264\cos L$$
$$\frac{-165}{-264} = \cos L$$
$$L = \cos^{-1}\left(\frac{165}{264}\right)$$
$$L \approx 51.3°$$
$$\frac{\sin L}{\ell} = \frac{\sin M}{m}$$
$$\frac{\sin 51.3°}{10} = \frac{\sin M}{11}$$
$$11\sin 51.3° = 10\sin M$$
$$\frac{11\sin 51.3°}{10} = \sin M$$
$$\sin^{-1}\left(\frac{11\sin 51.3°}{10}\right) = M$$
$$59.1° \approx M$$
$$m\angle L + m\angle M + m\angle N = 180$$
$$51.3 + 59.1 + m\angle N \approx 180$$
$$m\angle N \approx 69.6$$

39.
$$BC^2 = BP^2 + PC^2 - 2(BP)(PC)\cos(\angle BPC)$$
$$BC^2 = \left(\frac{1}{2} \cdot 214\right)^2 + \left(\frac{1}{2} \cdot 188\right)^2 - 2\left(\frac{1}{2} \cdot 214\right)\left(\frac{1}{2} \cdot 188\right)\cos 70°$$
$$BC^2 = 20{,}285 - 20{,}116\cos 70°$$
$$BC = \sqrt{20{,}285 - 20{,}116\cos 70°}$$
$$BC \approx 115.8$$
$$AB^2 = AP^2 + BP^2 - 2(AP)(BP)\cos(\angle APB)$$
$$AB^2 = \left(\frac{1}{2} \cdot 188\right)^2 + \left(\frac{1}{2} \cdot 214\right)^2 - 2\left(\frac{1}{2} \cdot 188\right)\left(\frac{1}{2} \cdot 214\right)\cos(180 - 70)°$$
$$AB^2 = 20{,}285 - 20{,}116\cos 110°$$
$$AB = \sqrt{20{,}285 - 20{,}116\cos 110°}$$
$$AB \approx 164.8$$

$AB = DC$ and $AD = BC$, so the perimeter of $ABCD$ is $2(115.8) + 2(164.8)$ or 561.2 units.

40.
$$QS^2 = PQ^2 + PS^2 - 2(PQ)(PS)\cos P$$
$$QS^2 = 721^2 + 756^2 - 2(721)(756)\cos 58°$$
$$QS^2 = 1{,}091{,}377 - 1{,}090{,}152\cos 58°$$
$$QS = \sqrt{1{,}091{,}377 - 1{,}090{,}152\cos 58°}$$
$$QS \approx 716.7$$

$$\frac{\sin(\angle PQS)}{PS} = \frac{\sin P}{QS}$$
$$\frac{\sin(\angle PQS)}{756} = \frac{\sin 58°}{716.7}$$
$$716.7 \sin(\angle PQS) = 756 \sin 58°$$
$$\sin(\angle PQS) = \frac{756 \sin 58°}{716.7}$$
$$m\angle PQS = \sin^{-1}\left(\frac{756 \sin 58°}{716.7}\right)$$
$$m\angle PQS \approx 63.5$$
$$QS^2 = QR^2 + RS^2 - 2(QR)(RS)\cos R$$
$$716.7^2 = 547^2 + 593^2 - 2(547)(593)\cos R$$
$$513{,}658.89 = 650{,}858 - 648{,}742 \cos R$$
$$-137{,}199.11 = -648{,}742 \cos R$$
$$\frac{-137{,}199.11}{-648{,}742} = \cos R$$
$$R = \cos^{-1}\left(\frac{137{,}199.11}{648{,}742}\right)$$
$$R \approx 77.8°$$

41.

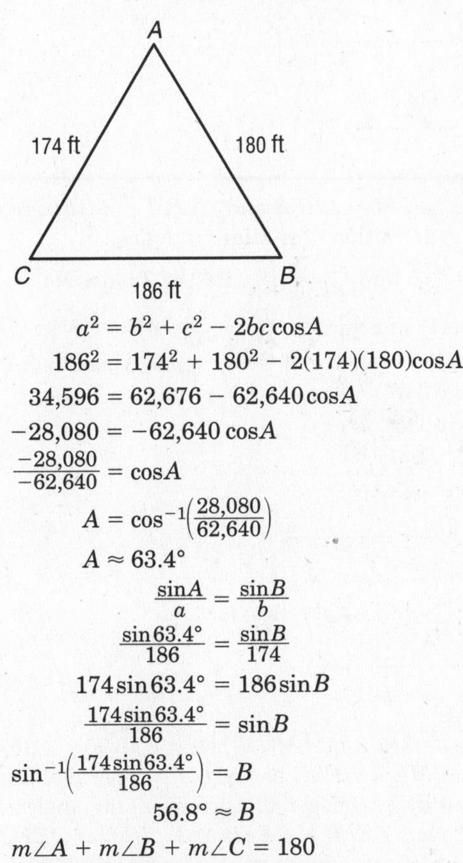

$$a^2 = b^2 + c^2 - 2bc\cos A$$
$$186^2 = 174^2 + 180^2 - 2(174)(180)\cos A$$
$$34{,}596 = 62{,}676 - 62{,}640 \cos A$$
$$-28{,}080 = -62{,}640 \cos A$$
$$\frac{-28{,}080}{-62{,}640} = \cos A$$
$$A = \cos^{-1}\left(\frac{28{,}080}{62{,}640}\right)$$
$$A \approx 63.4°$$
$$\frac{\sin A}{a} = \frac{\sin B}{b}$$
$$\frac{\sin 63.4°}{186} = \frac{\sin B}{174}$$
$$174 \sin 63.4° = 186 \sin B$$
$$\frac{174 \sin 63.4°}{186} = \sin B$$
$$\sin^{-1}\left(\frac{174 \sin 63.4°}{186}\right) = B$$
$$56.8° \approx B$$
$$m\angle A + m\angle B + m\angle C = 180$$
$$63.4 + 56.8 + m\angle C \approx 180$$
$$m\angle C \approx 59.8$$

42. Let C represent Carlos's angle and A represent Adam's angle.

$$24^2 = 40^2 + 50^2 - 2(40)(50)\cos C$$
$$576 = 4100 - 4000 \cos C$$
$$-3524 = -4000 \cos C$$
$$\frac{-3524}{-4000} = \cos C$$
$$C = \cos^{-1}\left(\frac{3524}{4000}\right)$$
$$C \approx 28.2°$$
$$24^2 = 30^2 + 22^2 - 2(30)(22)\cos A$$
$$576 = 1384 - 1320 \cos A$$
$$-808 = -1320 \cos A$$
$$\frac{-808}{-1320} = \cos A$$
$$A = \cos^{-1}\left(\frac{808}{1320}\right)$$
$$A \approx 52.3°$$

Adam has a greater angle, which is 52.3°.

43a. Pythagorean Theorem
43b. Substitution
43c. Pythagorean Theorem
43d. Substitution
43e. Def. of cosine
43f. Cross products
43g. Substitution
43h. Commutative Property

44.
$$AB = \sqrt{[10 - (-6)]^2 + [-4 - (-8)]^2}$$
$$= \sqrt{16^2 + 4^2}$$
$$= \sqrt{272}$$
$$BC = \sqrt{(6 - 10)^2 + [8 - (-4)]^2}$$
$$= \sqrt{(-4)^2 + 12^2}$$
$$= \sqrt{160}$$
$$AC = \sqrt{[6 - (-6)]^2 + [8 - (-8)]^2}$$
$$= \sqrt{12^2 + 16^2}$$
$$= \sqrt{400} \text{ or } 20$$
$$AC^2 = BC^2 + AB^2 - 2(BC)(AB)\cos B$$
$$20^2 = \left(\sqrt{160}\right)^2 + \left(\sqrt{272}\right)^2 - 2\left(\sqrt{160}\right)\left(\sqrt{272}\right)\cos B$$
$$400 = 432 - 2\sqrt{43{,}520}\cos B$$
$$-32 = -2\sqrt{43{,}520}\cos B$$
$$\frac{-32}{-2\sqrt{43{,}520}} = \cos B$$
$$B = \cos^{-1}\left(\frac{32}{2\sqrt{43{,}520}}\right)$$
$$B \approx 85.6°$$

So, $m\angle ABC \approx 85.6°$.

$$CB^2 = AB^2 + AC^2 - 2(AB)(AC)\cos A$$
$$\left(\sqrt{160}\right)^2 = \left(\sqrt{272}\right)^2 + 20^2 - 2\left(\sqrt{272}\right)(20)\cos A$$
$$160 = 672 - 40\sqrt{272}\cos A$$
$$-512 = -40\sqrt{272}\cos A$$
$$\frac{-512}{-40\sqrt{272}} = \cos A$$
$$A = \cos^{-1}\left(\frac{512}{40\sqrt{272}}\right)$$
$$A \approx 39.1°$$
$$m\angle DCA = m\angle B + m\angle A$$
$$\approx 85.6 + 39.1 \text{ or } 124.7$$

45. Sample answer: Triangles are used to build supports, walls, and foundations. Answers should include the following.

- The triangular building was more efficient with the cells around the edge.
- The Law of Sines requires two angles and a side or two sides and an angle opposite one of those sides.

46. B; $d^2 = e^2 + f^2 - 2ef\cos D$

$$d^2 = 12^2 + 15^2 - 2(12)(15)\cos 75°$$
$$d^2 = 369 - 360\cos 75°$$
$$d = \sqrt{369 - 360\cos 75°}$$
$$d \approx 16.6$$

47. C; earnings = base salary + commission

Let s represent her sales for the month.

$$4455 = 1280 + 0.125s$$
$$3175 = 0.125s$$
$$25{,}400 = s$$

Her sales that month were \$25,400.

Page 390 Maintain Your Skills

48.
$$\frac{\sin X}{x} = \frac{\sin Y}{y}$$
$$\frac{\sin 22°}{x} = \frac{\sin 49°}{4.7}$$
$$4.7\sin 22° = x\sin 49°$$
$$\frac{4.7\sin 22°}{\sin 49°} = x$$
$$2.3 \approx x$$

49.
$$\frac{\sin X}{x} = \frac{\sin Y}{y}$$
$$\frac{\sin 50°}{14} = \frac{\sin Y}{10}$$
$$10\sin 50° = 14\sin Y$$
$$\frac{10\sin 50°}{14} = \sin Y$$
$$\sin^{-1}\left(\frac{10\sin 50°}{14}\right) = Y$$
$$33° \approx Y$$

50.

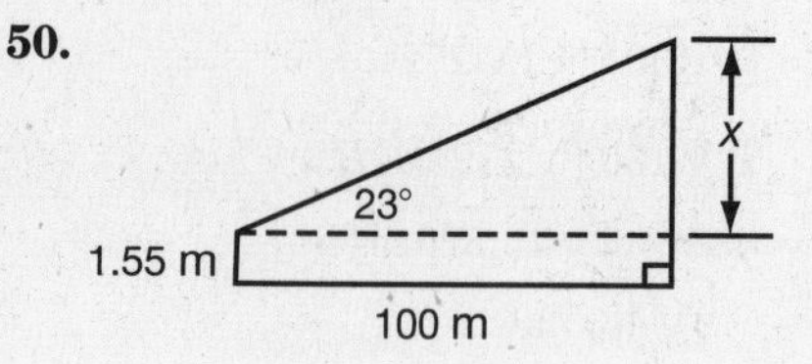

$$\tan 23° = \frac{x}{100}$$
$$100\tan 23° = x$$
$$42.45 \approx x$$
$$x + 1.55 \approx 42.45 + 1.55 \text{ or } 44.0$$

The height of the building is about 44.0 meters.

51. To show that $\overline{AB} \parallel \overline{CD}$, we must show that $\frac{AC}{CE} = \frac{BD}{DE}$.

$\frac{AC}{CE} = \frac{8.4}{6}$ or $\frac{7}{5}$, and $\frac{BD}{DE} = \frac{6.3}{4.5}$ or $\frac{7}{5}$. Thus, $\frac{AC}{CE} = \frac{BD}{DE}$. Since the sides have proportional lengths, $\overline{AB} \parallel \overline{CD}$.

52. To show that $\overline{AB} \parallel \overline{CD}$, we must show that $\frac{AC}{CE} = \frac{BD}{DE}$.

$CE = AE - AC = 15 - 7$ or 8.

So, $\frac{AC}{CE} = \frac{7}{8}$.

$DE = BE - BD = 22.5 - 10.5$ or 12.

So, $\frac{BD}{DE} = \frac{10.5}{12}$ or $\frac{7}{8}$. Thus, $\frac{AC}{CE} = \frac{BD}{DE}$. Since the sides have proportional lengths, $\overline{AB} \parallel \overline{CD}$.

53. To show that $\overline{AB} \parallel \overline{CD}$, we must show that $\frac{AB}{CD} = \frac{AE}{CE}$.

$\frac{AB}{CD} = \frac{8}{4}$ or 2, and $\frac{AE}{CE} = \frac{9}{4}$. Since the side lengths are not proportional, $\overline{AB}$ is not parallel to $\overline{CD}$.

54. To show that $\overline{AB} \parallel \overline{CD}$, we must show that $\frac{AB}{CD} = \frac{BE}{DE}$.

$\frac{AB}{CD} = \frac{5.4}{3}$ or $\frac{9}{5}$, and $\frac{BE}{DE} = \frac{18}{10}$ or $\frac{9}{5}$.

Thus, $\frac{AB}{CD} = \frac{BE}{DE}$. Since the sides have proportional lengths, $\overline{AB} \parallel \overline{CD}$.

55. Given: $\triangle JFM \sim \triangle EFB$
$\triangle LFM \sim \triangle GFB$

Prove: $\triangle JFL \sim \triangle EFG$

Proof:
Since $\triangle JFM \sim \triangle EFB$ and $\triangle LFM \sim \triangle GFB$, then by the definition of similar triangles, $\frac{JF}{EF} = \frac{MF}{BF}$ and $\frac{MF}{BF} = \frac{LF}{GF}$. By the Transitive Property of Equality, $\frac{JF}{EF} = \frac{LF}{GF}$. $\angle F \cong \angle F$ by the Reflexive Property of Congruence. Then, by SAS Similarity, $\triangle JFL \sim \triangle EFG$.

56. Given: $\overline{JM} \parallel \overline{EB}$
$\overline{LM} \parallel \overline{GB}$

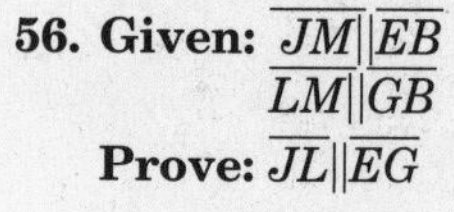

Prove: $\overline{JL} \parallel \overline{EG}$

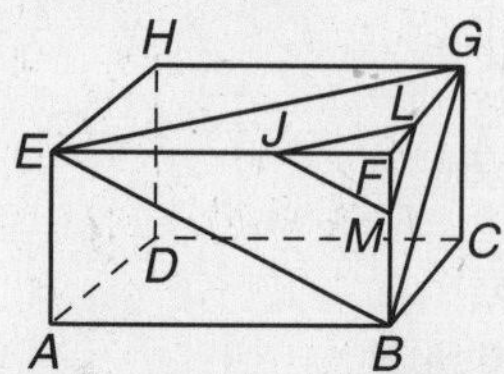

Proof:
Since $\overline{JM} \parallel \overline{EB}$ and $\overline{LM} \parallel \overline{GB}$, then $\angle MJF \cong \angle BEF$ and $\angle FML \cong \angle FBG$ because if two parallel lines are cut by a transversal, corresponding angles are congruent. $\angle EFB \cong \angle EFB$ and $\angle BFG \cong \angle BFG$ by the Reflexive Property of Congruence. Then $\triangle EFB \sim \triangle JFM$ and $\triangle FBG \sim \triangle FML$ by AA Similarity. Then $\frac{JF}{EF} = \frac{MF}{BF}$, $\frac{MF}{BF} = \frac{LF}{GF}$ by the definition of similar triangles. $\frac{JF}{EF} = \frac{LF}{GF}$ by the Transitive Porperty of Equality and $\angle EFG \cong \angle EFG$ by the Reflexive Property of Congruence. Thus, $\triangle JFL \sim \triangle EFG$ by SAS Similarity and $\angle FJL \cong \angle FEG$ by the definition of similar triangles. $\overline{JL} \parallel \overline{EG}$ because if two lines are cut by a transversal so that the corresponding angles are congruent, then the lines are parallel.

57.

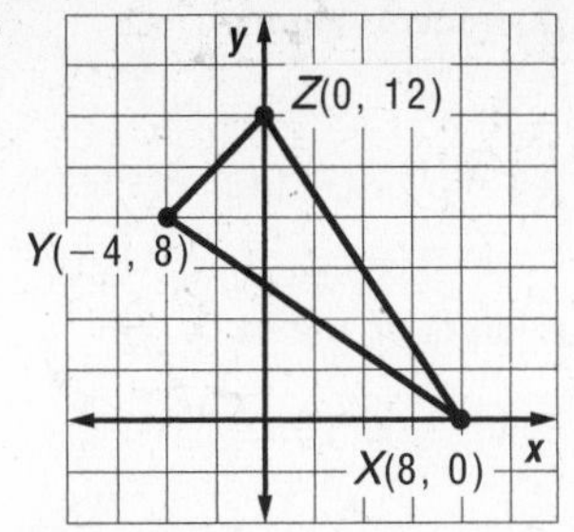

Find an equation of the altitude from X to $\overline{YZ}$.
The slope of $\overline{YZ}$ is $\frac{12-8}{0-(-4)}$ or 1, so the slope of the altitude is -1.

$y - y_1 = m(x - x_1)$
$y - 0 = -1(x - 8)$
$y = -x + 8$

Find an equation of the altitude from Z to $\overline{XY}$.
The slope of $\overline{XY}$ is $\frac{8-0}{-4-8}$ or $-\frac{2}{3}$, so the slope of the altitude is $\frac{3}{2}$.

$y - y_1 = m(x - x_1)$
$y - 12 = \frac{3}{2}(x - 0)$
$y = \frac{3}{2}x + 12$

Solve a system of equations to find the point of intersection of the altitudes.

$-x + 8 = \frac{3}{2}x + 12$
$8 = \frac{5}{2}x + 12$
$-4 = \frac{5}{2}x$
$-\frac{8}{5} = x$

Replace x with $-\frac{8}{5}$ in one of the equations to find the y-coordinate.

$y = -\left(-\frac{8}{5}\right) + 8$
$y = \frac{48}{5}$

The coordinates of the orthocenter are $\left(-\frac{8}{5}, \frac{48}{5}\right)$ or $(-1.6, 9.6)$.

58.

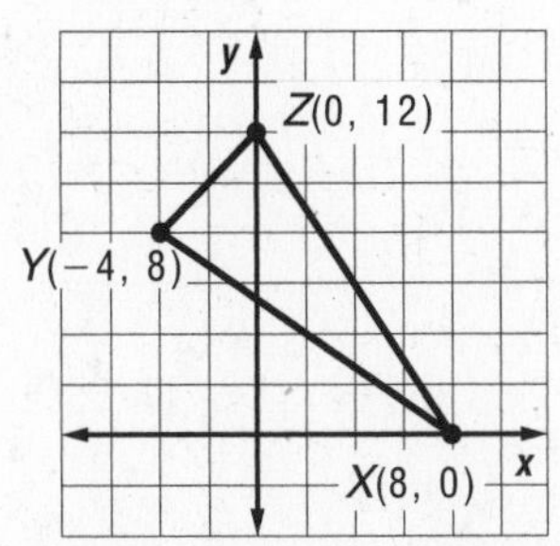

Find an equation of the median from X to $\overline{YZ}$.
The midpoint of $\overline{YZ}$ is $\left(\frac{-4+0}{2}, \frac{8+12}{2}\right)$ or $(-2, 10)$.
Then the slope of the median is $\frac{10-0}{-2-8}$ or -1.

$y - y_1 = m(x - x_1)$
$y - 0 = -1(x - 8)$
$y = -x + 8$

Find an equation of the median from Z to $\overline{XY}$.
The midpoint of $\overline{XY}$ is $\left(\frac{8+(-4)}{2}, \frac{0+8}{2}\right)$ or $(2, 4)$.
Then the slope of the median is $\frac{4-12}{2-0}$ or -4.

$y - y_1 = m(x - x_1)$
$y - 4 = -4(x - 2)$
$y - 4 = -4x + 8$
$y = -4x + 12$

Solve a system of equations to find the point of intersection of the medians.

$-x + 8 = -4x + 12$
$3x + 8 = 12$
$3x = 4$
$x = \frac{4}{3}$

Replace x with $\frac{4}{3}$ in one of the equations to find the y-coordinate.

$y = -\frac{4}{3} + 8$
$y = \frac{20}{3}$

The coordinates of the centroid are $\left(\frac{4}{3}, \frac{20}{3}\right)$ or about $(1.3, 6.7)$.

59.

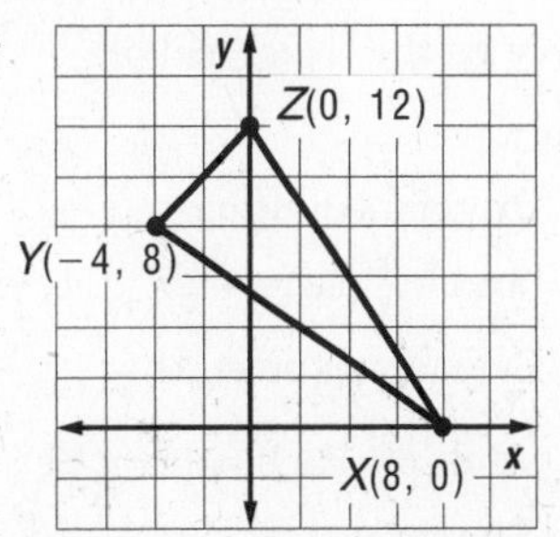

Find an equation of the perpendicular bisector of $\overline{YZ}$.
The midpoint of $\overline{YZ}$ is $\left(\frac{-4+0}{2}, \frac{8+12}{2}\right)$ or $(-2, 10)$.
The slope of $\overline{YZ}$ is $\frac{12-8}{0-(-4)}$ or 1, so the slope of the perpendicular bisector is -1.

$y - y_1 = m(x - x_1)$
$y - 10 = -1[x - (-2)]$
$y - 10 = -x - 2$
$y = -x + 8$

Find an equation of the perpendicular bisector of $\overline{XY}$.
The midpoint of $\overline{XY}$ is $\left(\frac{8+(-4)}{2}, \frac{0+8}{2}\right)$ or $(2, 4)$.
The slope of $\overline{XY}$ is $\frac{8-0}{-4-8}$ or $-\frac{2}{3}$, so the slope of the perpendicular bisector is $\frac{3}{2}$.

$y - y_1 = m(x - x_1)$
$y - 4 = \frac{3}{2}(x - 2)$
$y = \frac{3}{2}x - 3 + 4$
$y = \frac{3}{2}x + 1$

Solve a system of equations to find the point of intersection of the perpendicular bisectors.

$-x + 8 = \frac{3}{2}x + 1$
$8 = \frac{5}{2}x + 1$
$7 = \frac{5}{2}x$
$\frac{14}{5} = x$

Replace x with $\frac{14}{5}$ in one of the equations to find the y-coordinate.

$y = -\frac{14}{5} + 8$
$y = \frac{26}{5}$

The coordinates of the circumcenter are $\left(\frac{14}{5}, \frac{26}{5}\right)$ or $(2.8, 5.2)$.

Page 391 Geometry Activity: Trigonometric Identities

1. Sample answer: It is of the form $a^2 + b^2 = c^2$, where $c = 1$.

2. $\frac{1}{\cos\theta} = \sec\theta$; $\frac{1}{\tan\theta} = \cot\theta$

3. $\frac{\sin\theta}{\cos\theta} \stackrel{?}{=} \tan\theta$ Original equation

$\frac{\frac{y}{r}}{\frac{x}{r}} \stackrel{?}{=} \frac{y}{x}$ $\sin\theta = \frac{y}{r}$, $\cos\theta = \frac{x}{r}$, $\tan\theta = \frac{y}{x}$

$\left(\frac{y}{r}\right)\frac{r}{x} \stackrel{?}{=} \frac{y}{x}$ Multiply by the reciprocal of $\frac{x}{r}$.

$\frac{y}{x} = \frac{y}{x}$ ✓ Multiply.

4. $\frac{\cos\theta}{\sin\theta} \stackrel{?}{=} \cot\theta$ Original equation

$\frac{\frac{x}{r}}{\frac{y}{r}} \stackrel{?}{=} \frac{x}{y}$ $\sin\theta = \frac{y}{r}$, $\cos\theta = \frac{x}{r}$, $\cot\theta = \frac{x}{y}$

$\left(\frac{x}{r}\right)\frac{r}{y} \stackrel{?}{=} \frac{x}{y}$ Multiply by the reciprocal of $\frac{y}{r}$.

$\frac{x}{y} = \frac{x}{y}$ ✓ Multiply.

5. $\tan^2\theta + 1 \stackrel{?}{=} \sec^2\theta$ Original equation

$\left(\frac{y}{x}\right)^2 + 1 \stackrel{?}{=} \left(\frac{r}{x}\right)^2$ $\tan\theta = \frac{y}{x}$, $\sec\theta = \frac{r}{x}$

$\frac{y^2}{x^2} + 1 \stackrel{?}{=} \frac{r^2}{x^2}$ Evaluate exponents.

$x^2\left(\frac{y^2}{x^2} + 1\right) \stackrel{?}{=} (x^2)\frac{r^2}{x^2}$ Multiply each side by x^2.

$y^2 + x^2 \stackrel{?}{=} r^2$ Simplify.

$r^2 = r^2$ ✓ Substitution; $y^2 + x^2 = r^2$

6. $\cot^2\theta + 1 \stackrel{?}{=} \csc^2\theta$ Original equation

$\left(\frac{x}{y}\right)^2 + 1 \stackrel{?}{=} \left(\frac{r}{y}\right)^2$ $\cot\theta = \frac{x}{y}$, $\sec\theta = \frac{r}{y}$

$\left(\frac{x^2}{y^2} + 1\right) \stackrel{?}{=} \frac{r^2}{y^2}$ Evaluate exponents.

$y^2\left(\frac{x^2}{y^2} + 1\right) \stackrel{?}{=} (y^2)\frac{r^2}{y^2}$ Multiply each side by y^2.

$x^2 + y^2 \stackrel{?}{=} r^2$ Simplify.

$r^2 = r^2$ ✓ Substitution; $x^2 + y^2 = r^2$

Chapter 7 Study Guide and Review

Page 392 Vocabulary and Concept Check

1. true
2. false; opposite; adjacent
3. false; a right
4. true
5. true
6. false; 45°-45°-90°
7. false; depression

Pages 392–396 Lesson-by-Lesson Review

8. Let x represent the geometric mean.

$\frac{4}{x} = \frac{x}{16}$
$x^2 = 64$
$x = \sqrt{64}$
$x = 8$

9. Let x represent the geometric mean.

$\frac{4}{x} = \frac{x}{81}$
$x^2 = 324$
$x = \sqrt{324}$
$x = 18$

10. Let x represent the geometric mean.

$\frac{20}{x} = \frac{x}{35}$
$x^2 = 700$
$x = \sqrt{700}$
$x = 10\sqrt{7}$
$x \approx 26.5$

11. Let x represent the geometric mean.

$\frac{18}{x} = \frac{x}{44}$
$x^2 = 792$
$x = \sqrt{792}$
$x = 6\sqrt{22}$
$x \approx 28.1$

12. Let $x = RS$.

$\frac{PS}{RS} = \frac{RS}{QS}$
$\frac{8}{x} = \frac{x}{14}$
$x^2 = 112$
$x = \sqrt{112}$ or $4\sqrt{7}$
$x \approx 10.6$
So $RS \approx 10.6$.

13. $15^2 + 20^2 = x^2$
$225 + 400 = x^2$
$625 = x^2$
$\sqrt{625} = x$
$25 = x$

14. $x^2 + \left(\frac{5}{17}\right)^2 = \left(\frac{13}{17}\right)^2$
$x^2 + \frac{25}{289} = \frac{169}{289}$
$x^2 = \frac{144}{289}$
$x = \sqrt{\frac{144}{289}}$
$x = \frac{12}{17}$

15. $x^2 + 13^2 = 21^2$
$x^2 + 169 = 441$
$x^2 = 272$
$x = \sqrt{272}$ or $4\sqrt{17}$
$x \approx 16.5$

16. $x = 9$
$y = 9\sqrt{2}$

17. $13 = x\sqrt{2}$
$\frac{13}{\sqrt{2}} = x$
$\frac{13}{\sqrt{2}} \cdot \frac{\sqrt{2}}{\sqrt{2}} = x$
$\frac{13\sqrt{2}}{2} = x$
$y = x$
$y = \frac{13\sqrt{2}}{2}$

18. $x = 2(6)$ or 12
$y = 6\sqrt{3}$

19. $z = 18\sqrt{3}$
$a = 2z$
$a = 2(18\sqrt{3})$ or $36\sqrt{3}$

20. $14 = z\sqrt{3}$

$\frac{14}{\sqrt{3}} = z$

$\frac{14}{\sqrt{3}} \cdot \frac{\sqrt{3}}{\sqrt{3}} = z$

$\frac{14\sqrt{3}}{3} = z$

$a = 2z$

$a = 2\left(\frac{14\sqrt{3}}{3}\right)$ or $\frac{28\sqrt{3}}{3}$

$z = y\sqrt{3}$

$\frac{14\sqrt{3}}{3} = y\sqrt{3}$

$\frac{14\sqrt{3}}{3} \cdot \frac{1}{\sqrt{3}} = y$

$\frac{14}{3} = y$

$b = 2y$

$b = 2\left(\frac{14}{3}\right)$ or $\frac{28}{3}$

21. $\sin F = \frac{a}{c}$
$= \frac{9}{15}$ or $\frac{3}{5}$
$= 0.60$

$\cos F = \frac{b}{c}$
$= \frac{12}{15}$ or $\frac{4}{5}$
$= 0.80$

$\tan F = \frac{a}{b}$
$= \frac{9}{12}$ or $\frac{3}{4}$
$= 0.75$

$\sin G = \frac{b}{c}$
$= \frac{12}{15}$ or $\frac{4}{5}$
$= 0.80$

$\cos G = \frac{a}{c}$
$= \frac{9}{15}$ or $\frac{3}{5}$
$= 0.60$

$\tan G = \frac{b}{a}$
$= \frac{12}{9}$ or $\frac{4}{3}$
≈ 1.33

22. $\sin F = \frac{a}{c} = \frac{7}{25} = 0.28$ $\quad \cos F = \frac{b}{c} = \frac{24}{25} = 0.96$ $\quad \tan F = \frac{a}{b} = \frac{7}{24} \approx 0.29$

$\sin G = \frac{b}{c} = \frac{24}{25} = 0.96$ $\quad \cos G = \frac{a}{c} = \frac{7}{25} = 0.28$ $\quad \tan G = \frac{b}{a} = \frac{24}{7} \approx 3.43$

23. $\sin P = 0.4522$
$P = \sin^{-1}(0.4522)$
KEYSTROKES: [2nd] [SIN^{-1}] 0.4522 [ENTER]
$m\angle P \approx 26.9$

24. $\cos Q = 0.1673$
$Q = \cos^{-1}(0.1673)$
KEYSTROKES: [2nd] [COS^{-1}] 0.1673 [ENTER]
$m\angle Q \approx 80.4$

25. $\tan R = 0.9324$
$R = \tan^{-1}(0.9324)$
KEYSTROKES: [2nd] [TAN^{-1}] 0.9324 [ENTER]
$m\angle R \approx 43.0$

26.

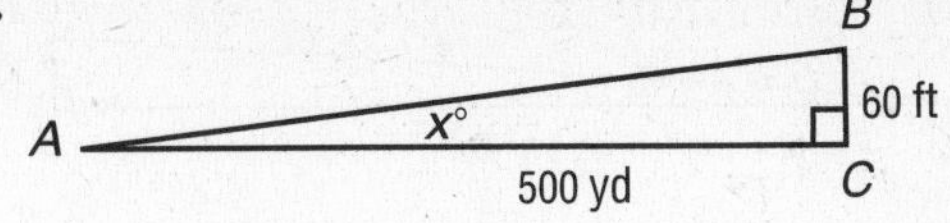

Let x represent $m\angle BAC$.

$\tan x° = \frac{60}{500 \cdot 3}$

$\tan x° = \frac{60}{1500}$

$x = \tan^{-1}\left(\frac{60}{1500}\right)$

$x \approx 2.3$

The angle of elevation must be greater than 2.3°.

27.

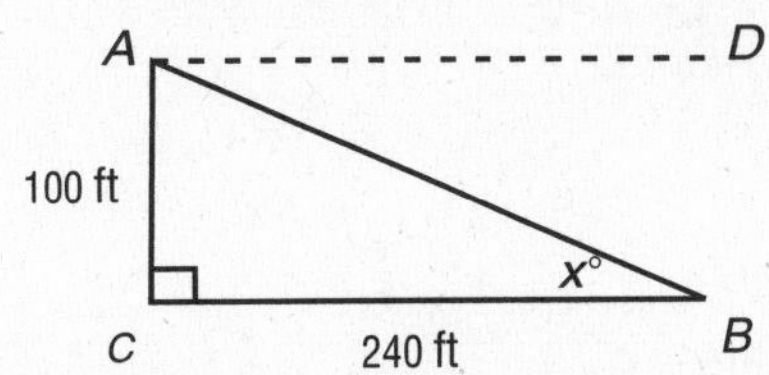

Let x represent $m\angle ABC$.
The ground and the horizontal level with the top of the escalator are parallel. Therefore, $m\angle DAB = m\angle ABC$ since they are alternate interior angles.

$\tan x° = \frac{100}{240}$

$x = \tan^{-1}\left(\frac{100}{240}\right)$

$x \approx 22.6$

The angle of depression is about 22.6°.

28.

Let x represent $m\angle ABC$.
The ground and the horizontal level with the initial point of the balloon are parallel. Therefore, $m\angle DAB = m\angle ABC$ since they are alternate interior angles.

$\tan x° = \frac{50}{1000}$

$x = \tan^{-1}\left(\frac{50}{1000}\right)$

$x \approx 2.9$

The angle of depression is about 2.9°.

29.

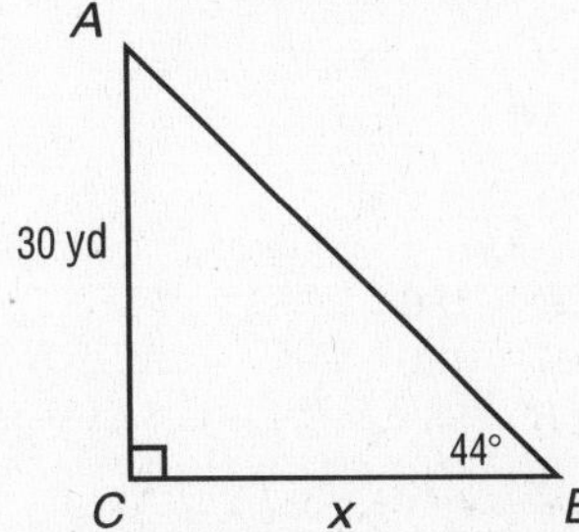

Let x represent the length of the shadow of the building, BC.

$\tan 44° = \frac{30}{x}$

$x \tan 44° = 30$

$x = \frac{30}{\tan 44°}$

$x \approx 31.1$

The shadow is about 31.1 yards long.

30.

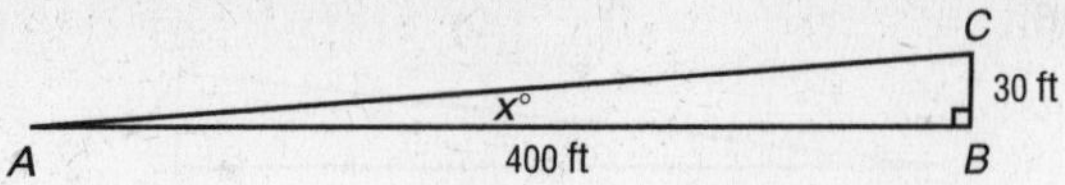

Let x represent $m\angle CAB$.

$\tan x° = \frac{30}{400}$

$x = \tan^{-1}\left(\frac{30}{400}\right)$

$x \approx 4.3$

The angle of elevation of the track is about 4.3°.

31.
$$\frac{\sin F}{f} = \frac{\sin G}{g}$$
$$\frac{\sin 82°}{f} = \frac{\sin 48°}{16}$$
$$16\sin 82° = f\sin 48°$$
$$\frac{16\sin 82°}{\sin 48°} = f$$
$$21.3 \approx f$$

32.
$$\frac{\sin H}{h} = \frac{\sin G}{g}$$
$$\frac{\sin H}{10.5} = \frac{\sin 65°}{13}$$
$$13\sin H = 10.5\sin 65°$$
$$\sin H = \frac{10.5\sin 65°}{13}$$
$$H = \sin^{-1}\left(\frac{10.5\sin 65°}{13}\right)$$
$$H \approx 47°$$

33.
$$\frac{\sin A}{a} = \frac{\sin B}{b}$$
$$\frac{\sin 64°}{15} = \frac{\sin B}{11}$$
$$11\sin 64° = 15\sin B$$
$$\frac{11\sin 64°}{15} = \sin B$$
$$\sin^{-1}\left(\frac{11\sin 64°}{15}\right) = B$$
$$41° \approx B$$
$$m\angle A + m\angle B + m\angle C = 180$$
$$64 + 41 + m\angle C \approx 180$$
$$m\angle C \approx 75$$
$$\frac{\sin C}{c} = \frac{\sin A}{a}$$
$$\frac{\sin 75°}{c} = \frac{\sin 64°}{15}$$
$$15\sin 75° = c\sin 64°$$
$$\frac{15\sin 75°}{\sin 64°} = c$$
$$16.1 \approx c$$

34.
$$\frac{\sin C}{c} = \frac{\sin A}{a}$$
$$\frac{\sin 67°}{12} = \frac{\sin 55°}{a}$$
$$a\sin 67° = 12\sin 55°$$
$$a = \frac{12\sin 55°}{\sin 67°}$$
$$a \approx 10.7$$
$$m\angle A + m\angle B + m\angle C = 180$$
$$55 + m\angle B + 67 = 180$$
$$m\angle B = 58$$
$$\frac{\sin B}{b} = \frac{\sin C}{c}$$
$$\frac{\sin 58°}{b} = \frac{\sin 67°}{12}$$
$$12\sin 58° = b\sin 67°$$
$$\frac{12\sin 58°}{\sin 67°} = b$$
$$11.1 \approx b$$

35.
$$\frac{\sin A}{a} = \frac{\sin B}{b}$$
$$\frac{\sin 29°}{4.8} = \frac{\sin B}{8.7}$$
$$8.7\sin 29° = 4.8\sin B$$
$$\frac{8.7\sin 29°}{4.8} = \sin B$$
$$\sin^{-1}\left(\frac{8.7\sin 29°}{4.8}\right) = B$$
$$61° \approx B$$
$$m\angle A + m\angle B + m\angle C = 180$$
$$29 + 61 + m\angle C \approx 180$$
$$m\angle C \approx 90$$
$$\frac{\sin C}{c} = \frac{\sin A}{a}$$
$$\frac{\sin 90°}{c} = \frac{\sin 29°}{4.8}$$
$$4.8\sin 90° = c\sin 29°$$
$$\frac{4.8\sin 90°}{\sin 29°} = c$$
$$9.9 \approx c$$

36.
$$m\angle A + m\angle B + m\angle C = 180$$
$$29 + 64 + m\angle C = 180$$
$$m\angle C = 87$$
$$\frac{\sin A}{a} = \frac{\sin B}{b}$$
$$\frac{\sin 29°}{a} = \frac{\sin 64°}{18.5}$$
$$18.5\sin 29° = a\sin 64°$$
$$\frac{18.5\sin 29°}{\sin 64°} = a$$
$$10.0 \approx a$$
$$\frac{\sin B}{b} = \frac{\sin C}{c}$$
$$\frac{\sin 64°}{18.5} = \frac{\sin 87°}{c}$$
$$c\sin 64° = 18.5\sin 87°$$
$$c = \frac{18.5\sin 87°}{\sin 64°}$$
$$c \approx 20.6$$

37.
$$z^2 = x^2 + y^2 - 2xy\cos Z$$
$$z^2 = 7.6^2 + 5.4^2 - 2(7.6)(5.4)\cos 51°$$
$$z^2 = 86.92 - 82.08\cos 51°$$
$$z = \sqrt{86.92 - 82.08\cos 51°}$$
$$z \approx 5.9$$

38.
$$y^2 = x^2 + z^2 - 2xz\cos Y$$
$$y^2 = 21^2 + 16^2 - 2(21)(16)\cos 73°$$
$$y^2 = 697 - 672\cos 73°$$
$$y = \sqrt{697 - 672\cos 73°}$$
$$y \approx 22.4$$

39.
$$a^2 = b^2 + c^2 - 2bc\cos A$$
$$a^2 = 13^2 + 18^2 - 2(13)(18)\cos 64°$$
$$a^2 = 493 - 468\cos 64°$$
$$a = \sqrt{493 - 468\cos 64°}$$
$$a \approx 17.0$$
$$\frac{\sin A}{a} = \frac{\sin B}{b}$$
$$\frac{\sin 64°}{17} = \frac{\sin B}{13}$$
$$13\sin 64° = 17\sin B$$
$$\frac{13\sin 64°}{17} = \sin B$$
$$\sin^{-1}\left(\frac{13\sin 64°}{17}\right) = B$$
$$43° \approx B$$
$$m\angle A + m\angle B + m\angle C = 180$$
$$64 + 43 + m\angle C \approx 180$$
$$m\angle C \approx 73$$

40. $\frac{\sin B}{b} = \frac{\sin C}{c}$
$\frac{\sin B}{5.2} = \frac{\sin 53°}{6.7}$
$6.7 \sin B = 5.2 \sin 53°$
$\sin B = \frac{5.2 \sin 53°}{6.7}$
$B = \sin^{-1}\left(\frac{5.2 \sin 53°}{6.7}\right)$
$B \approx 38°$
$m\angle A + m\angle B + m\angle C = 180$
$m\angle A + 38 + 53 \approx 180$
$m\angle A \approx 89$
$\frac{\sin A}{a} = \frac{\sin B}{b}$
$\frac{\sin 89°}{a} = \frac{\sin 38°}{5.2}$
$5.2 \sin 89° = a \sin 38°$
$\frac{5.2 \sin 89°}{\sin 38°} = a$
$8.4 \approx a$

Chapter 7 Practice Test

Page 397

1. $c^2 = a^2 + b^2 - 2ab \cos C$

2. Yes; two perfect squares can be written as $a \cdot a$ and $b \cdot b$. Multiplied together, we have $a \cdot a \cdot b \cdot b$. Taking the square root, we have ab, which is rational.

3. Sample answer: $2, 2\sqrt{3}, 4$

4. Let x represent the geometric mean.
$\frac{7}{x} = \frac{x}{63}$
$x^2 = 441$
$x = \sqrt{441}$ or 21

5. Let x represent the geometric mean.
$\frac{6}{x} = \frac{x}{24}$
$x^2 = 144$
$x = \sqrt{144}$ or 12

6. Let x represent the geometric mean.
$\frac{10}{x} = \frac{x}{50}$
$x^2 = 500$
$x = \sqrt{500}$ or $10\sqrt{5}$

7. $x^2 + 5^2 = 6^2$
$x^2 + 25 = 36$
$x^2 = 11$
$x = \sqrt{11}$
$x \approx 3.32$

8. $7^2 + 13^2 = x^2$
$49 + 169 = x^2$
$218 = x^2$
$\sqrt{218} = x$
$14.8 \approx x$

9. $x^2 + \left(\frac{12}{2}\right)^2 = 9^2$
$x^2 + 36 = 81$
$x^2 = 45$
$x = \sqrt{45}$
$x = 3\sqrt{5}$
$x \approx 6.7$

10. $19 = x\sqrt{2}$
$\frac{19}{\sqrt{2}} = x$
$\frac{19}{\sqrt{2}} \cdot \frac{\sqrt{2}}{\sqrt{2}} = x$
$\frac{19\sqrt{2}}{2} = x$
$y = x$
$y = \frac{19\sqrt{2}}{2}$

11. This is a 30°-60°-90° triangle with hypotenuse of length 12, shorter leg with length of y and longer leg with length of x.
$12 = 2y$
$6 = y$
$x = y\sqrt{3}$
$x = 6\sqrt{3}$

12. $x^2 + 8^2 = 16^2$
$x^2 + 64 = 256$
$x^2 = 192$
$x = \sqrt{192}$
$x = 8\sqrt{3}$
$\sin y° = \frac{8}{16}$
$y = \sin^{-1}\left(\frac{8}{16}\right)$
$y = 30$

13. $\cos B = \frac{BC}{AB}$
$= \frac{15}{21} = \frac{5}{7}$

14. $\tan A = \frac{BC}{AC}$
$= \frac{15}{16}$

15. $\sin A = \frac{BC}{AB}$
$= \frac{15}{21} = \frac{5}{7}$

16. $\frac{\sin F}{f} = \frac{\sin G}{g}$
$\frac{\sin 59°}{13} = \frac{\sin 71°}{g}$
$g \sin 59° = 13 \sin 71°$
$g = \frac{13 \sin 71°}{\sin 59°}$
$g \approx 14.3$

17. $\frac{\sin F}{f} = \frac{\sin H}{h}$
$\frac{\sin 52°}{10} = \frac{\sin H}{12.5}$
$12.5 \sin 52° = 10 \sin H$
$\frac{12.5 \sin 52°}{10} = \sin H$
$\sin^{-1}\left(\frac{12.5 \sin 52°}{10}\right) = H$
$80.1° \approx H$

18. $f^2 = g^2 + h^2 - 2gh \cos F$
$f^2 = 15^2 + 13^2 - 2(15)(13)\cos 48°$
$f^2 = 394 - 390 \cos 48°$
$f = \sqrt{394 - 390 \cos 48°}$
$f \approx 11.5$

19. $h^2 = f^2 + g^2 - 2fg \cos H$
$h^2 = 13.7^2 + 16.8^2 - 2(13.7)(16.8)\cos 71°$
$h^2 = 469.93 - 460.32 \cos 71°$
$h = \sqrt{469.93 - 460.32 \cos 71°}$
$h \approx 17.9$

20. $c^2 = a^2 + b^2 - 2ab\cos C$

$c^2 = 15^2 + 17^2 - 2(15)(17)\cos 45°$

$c^2 = 514 - 510\cos 45°$

$c = \sqrt{514 - 510\cos 45°}$

$c \approx 12.4$

$\frac{\sin C}{c} = \frac{\sin A}{a}$

$\frac{\sin 45°}{12.4} = \frac{\sin A}{15}$

$15\sin 45° = 12.4\sin A$

$\frac{15\sin 45°}{12.4} = \sin A$

$\sin^{-1}\left(\frac{15\sin 45°}{12.4}\right) = A$

$59° \approx A$

$m\angle A + m\angle B + m\angle C = 180$

$59 + m\angle B + 45 \approx 180$

$m\angle B \approx 76$

21. $\frac{\sin A}{a} = \frac{\sin B}{b}$

$\frac{\sin A}{12.2} = \frac{\sin 48°}{10.9}$

$10.9 \sin A = 12.2\sin 48°$

$\sin A = \frac{12.2\sin 48°}{10.9}$

$A = \sin^{-1}\left(\frac{12.2\sin 48°}{10.9}\right)$

$A \approx 56°$

$m\angle A + m\angle B + m\angle C = 180$

$56 + 48 + m\angle C \approx 180$

$m\angle C \approx 76$

$\frac{\sin C}{c} = \frac{\sin B}{b}$

$\frac{\sin 76°}{c} = \frac{\sin 48°}{10.9}$

$10.9\sin 76° = c\sin 48°$

$\frac{10.9\sin 76°}{\sin 48°} = c$

$14.2 \approx c$

22. $a^2 = b^2 + c^2 - 2bc\cos A$

$19^2 = 23.2^2 + 21^2 - 2(23.2)(21)\cos A$

$361 = 979.24 - 974.4\cos A$

$-618.24 = -974.4\cos A$

$\frac{-618.24}{-974.4} = \cos A$

$A = \cos^{-1}\left(\frac{618.24}{974.4}\right)$

$A \approx 51°$

$\frac{\sin A}{a} = \frac{\sin B}{b}$

$\frac{\sin 51°}{19} = \frac{\sin B}{23.2}$

$23.2\sin 51° = 19\sin B$

$\frac{23.2\sin 51°}{19} = \sin B$

$\sin^{-1}\left(\frac{23.2\sin 51°}{19}\right) = B$

$72° \approx B$

$m\angle A + m\angle B + m\angle C = 180$

$51 + 72 + m\angle C \approx 180$

$m\angle C \approx 57$

23.

A B 9° 0.5 mi C x D

The ground and the horizontal level with the plane are parallel. Therefore, $m\angle BAD = m\angle ADC$ since they are alternate interior angles.

Let x represent CD, the horizontal distance to the city.

$\tan 9° = \frac{0.5}{x}$

$x\tan 9° = 0.5$

$x = \frac{0.5}{\tan 9°}$

$x \approx 3.2$

The horizontal distance to the city is about 3.2 miles.

24.

C x A 10° 5 mi B

Let x represent CB, the height of the incline.

$\tan 10° = \frac{x}{5}$

$5\tan 10° = x$

$0.9 \approx x$

The height of the incline is about 0.9 mile.

25. D; Let y represent the unknown side length in the triangle.

$5^2 + y^2 = 13^2$

$25 + y^2 = 169$

$y^2 = 144$

$y = \sqrt{144}$

$y = 12$

$\tan X = \frac{12}{5}$

Chapter 7 Standardized Test Practice

Pages 398–399

1. C; there is no information to support choices A, B, or D. $\angle 1$ and $\angle 4$ are vertical angles, and $\angle 2$ and $\angle 3$ are vertical angles.

2. D; $AD = CD$

$3x + 5 = 5x - 1$

$5 = 2x - 1$

$6 = 2x$

$3 = x$

$AC = AD + CD$

$AC = 3x + 5 + 5x - 1$

$AC = 3(3) + 5 + 5(3) - 1$

$AC = 28$

3. B; $\frac{SR}{DC} = \frac{PT}{AE}$

$\frac{SR}{8} = \frac{6}{11}$

$11(SR) = 48$

$SR = \frac{48}{11}$ or $4\frac{4}{11}$

4. C; $\frac{AB}{AC} = \frac{AC}{AD}$

$\frac{12 + 3}{AC} = \frac{AC}{12}$

$(AC)^2 = 180$

$AC = \sqrt{180}$

$AC \approx 13.4$

5. B; $m\angle RTS = 180 - 135$ or 45.

$RT = (ST)\sqrt{2}$

$= 5\sqrt{2}$

6. D; the height of the original tower is $AB + BC$.

$$\sin 36° = \frac{60}{BC}$$
$$BC \sin 36° = 60$$
$$BC = \frac{60}{\sin 36°}$$
$$BC \approx 102$$
$$AB + BC = 60 + 102 \text{ or } 162 \text{ feet.}$$

7. C;
$$\frac{\sin R}{r} = \frac{\sin S}{s}$$
$$\frac{\sin 34°}{14} = \frac{\sin S}{21}$$
$$21 \sin 34° = 14 \sin S$$
$$\frac{21 \sin 34°}{14} = \sin S$$
$$\sin^{-1}\left(\frac{21 \sin 34°}{14}\right) = S$$
$$57° \approx S$$

8.
$$m\angle C + m\angle BDC + m\angle DBC = 180$$
$$90 + m\angle BDC + 55 = 180$$
$$m\angle BDC = 35$$
$$m\angle ADB + m\angle BDC = m\angle ADC$$
$$m\angle ADB + 35 = 61$$
$$m\angle ADB = 26$$
$$m\angle A + m\angle ABD + m\angle ADB = 180$$
$$69 + m\angle ABD + 26 = 180$$
$$m\angle ABD = 85$$
$$m\angle ABC = m\angle ABD + m\angle DBC$$
$$= 85 + 55$$
$$= 140$$

9.
$$m = \frac{y_2 - y_1}{x_2 - x_1}$$
$$= \frac{50 - 32}{10 - 0}$$
$$= \frac{18}{10} \text{ or } \frac{9}{5}$$

10.
$$y - y_1 = m(x - x_1)$$
$$y - 32 = \frac{9}{5}(x - 0)$$
$$y = \frac{9}{5}x + 32$$

11.
$$\frac{YZ}{UV} = \frac{XZ}{TV}$$
$$= \frac{6}{10} \text{ or } \frac{3}{5}$$

12. Let x represent Dee's height above the water.

$$\sin 41° = \frac{x}{500}$$
$$500 \sin 41° = x$$
$$328 \approx x$$

Dee is about 328 feet above the water.

13. Since Sasha is equidistant from Toby and Rani, $\overline{ST}$ and $\overline{SR}$ are congruent and $\triangle STR$ is an isosceles triangle. According to the Isosceles Triangle Theorem, $\angle T$ and $\angle R$ are also congruent. $\overline{SX}$ is perpendicular to $\overline{TR}$, so $\angle SXT$ and $\angle SXR$ are both right angles and congruent. Two corresponding angles and the corresponding nonincluded sides are congruent (AAS Theorem), so $\triangle STX$ and $\triangle SRX$ are congruent triangles. Since these triangles are congruent, the corresponding sides $\overline{TX}$ and $\overline{RX}$ are congruent and have equal length; therefore when Sasha is jumping at Point X she will be at the midpoint between Toby and Rani.

Chapter 8 Quadrilaterals

Page 403 Getting Started

1. The angles measuring $x°$ and $50°$ are supplementary. Find x.
$m\angle x + 50 = 180$
$m\angle x = 130$
So, x is 130.
2. $x°$ is the measure of the exterior angle of the triangle so its measure is the sum of the two remote interior angles or $25 + 20$. So, $x = 45$.
3. The measure of an internal angle of an equilateral triangle is 60. The angle measuring $x°$ is supplementary to one of the angles. Find x.
$m\angle x + 60 = 180$
$m\angle x = 120$
So, x is 120.
4. The slope is given by $m = \frac{y_2 - y_1}{x_2 - x_1}$.
$\overline{RS}$: $\frac{10-3}{-1-4} = -\frac{7}{5}$
$\overline{TS}$: $\frac{10-20}{-1-13} = \frac{-10}{-14}$
$= \frac{5}{7}$
$\overline{RS}$ and $\overline{TS}$ are perpendicular since their slopes are opposite inverses.
5. The slope is given by $m = \frac{y_2 - y_1}{x_2 - x_1}$.
$\overline{RS}$: $\frac{8-6}{3-(-9)} = \frac{2}{12}$
$= \frac{1}{6}$
$\overline{TS}$: $\frac{8-20}{3-1} = \frac{-12}{2}$
$= -6$
$\overline{RS}$ and $\overline{TS}$ are perpendicular since their slopes are opposite inverses.
6. The slope is given by $m = \frac{y_2 - y_1}{x_2 - x_1}$.
$\overline{RS}$: $\frac{3-(-1)}{5-(-6)} = \frac{4}{11}$
$\overline{TS}$: $\frac{5-3}{2-5} = -\frac{2}{3}$
$\overline{RS}$ and $\overline{TS}$ are not perpendicular since their slopes are not opposite inverses.
7. The slope is given by $m = \frac{y_2 - y_1}{x_2 - x_1}$.
$\overline{RS}$: $\frac{8-4}{-3-(-6)} = \frac{4}{3}$
$\overline{TS}$: $\frac{8-2}{-3-5} = \frac{6}{-8}$
$= -\frac{3}{4}$
$\overline{RS}$ and $\overline{TS}$ are perpendicular since their slopes are opposite inverses.
8. The slope is given by $m = \frac{y_2 - y_1}{x_2 - x_1}$.
$m = \frac{d - \frac{d}{2}}{-c - \frac{c}{2}}$
$= \frac{\frac{d}{2}}{-\frac{3}{2}c}$
$= -\frac{d}{3c}$
The slope is $-\frac{d}{3c}$.
9. The slope is given by $m = \frac{y_2 - y_1}{x_2 - x_1}$.
$m = \frac{0-a}{b-0}$
$= -\frac{a}{b}$
The slope is $-\frac{a}{b}$.
10. The slope is given by $m = \frac{y_2 - y_1}{x_2 - x_1}$.
$m = \frac{a-c}{-c-(-a)}$
$= \frac{a-c}{-c+a}$
$= \frac{a-c}{a-c}$ or 1
The slope is 1.

8-1 Angles of Polygons

Page 406 Geometry Activity: Sum of the Exterior Angles of a Polygon

1.

Polygon	number of exterior angles	sum of measure of exterior angles
triangle	3	360
quadrilateral	4	360
pentagon	5	360
hexagon	6	360
heptagon	7	360

2. The sum of the measures of the exterior angles is 360.

Page 407 Check for Understanding

1. A concave polygon has at least one obtuse angle, which means the sum will be different from the formula.
2. Yes; an irregular polygon can be separated by the diagonals into triangles so the theorems apply.
3. Sample answer:
regular quadrilateral:

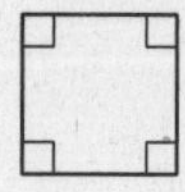

The sum of the interior angles is 360°.
quadrilateral that is not regular:

The sum of the interior angles is 360°.

4. Use the Interior Angle Sum Theorem.
$S = 180(n-2)$
$= 180(5 - 2)$
$= 540$
The sum of the measures of the interior angles of a pentagon is 540.
5. Use the Interior Angle Sum Theorem.
$S = 180(n - 2)$
$= 180(12 - 2)$
$= 1800$
The sum of the measures of the interior angles of a dodecagon is 1800.
6. Use the Interior Angle Sum Theorem to write an equation to solve for n, the number of sides.
$S = 180(n - 2)$
$(60)n = 180(n - 2)$
$60n = 180n - 360$
$0 = 120n - 360$
$360 = 120n$
$3 = n$
The polygon has 3 sides.
7. Use the Interior Angle Sum Theorem to write an equation to solve for n, the number of sides.
$S = 180(n - 2)$
$(90)n = 180(n - 2)$
$90n = 180n - 360$
$0 = 90n - 360$
$360 = 90n$
$4 = n$
The polygon has 4 sides.
8. Since $n = 4$, the sum of the measures of the interior angles is $180(4 - 2)$ or 360. Write an equation to express the sum of the measures of the interior angles of the polygon.
$360 = m\angle T + m\angle U + m\angle V + m\angle W$
$360 = x + (3x - 4) + x + (3x - 4)$
$360 = 8x - 8$
$368 = 8x$
$46 = x$
Use the value of x to find the measure of each angle.
$m\angle T = 46$, $m\angle U = 3 \cdot 46 - 8$ or 134, $m\angle V = 46$, and $m\angle W = 3 \cdot 46 - 8$ or 134.
9. Since $n = 6$, the sum of the measures of the interior angles is $180(6 - 2)$ or 720. Write an equation to express the sum of the measures of the interior angles of the polygon.
$720 = m\angle J + m\angle K + m\angle L + m\angle M + m\angle N + m\angle P$
$720 = 2x + (9x + 30) + (9x + 30) + 2x + (9x + 30) + (9x + 30)$
$720 = 40x + 120$
$600 = 40x$
$15 = x$
Use the value of x to find the measure of each angle.
$m\angle J = 30$, $m\angle K = 9 \cdot 15 + 30$ or 165, $m\angle L = 9 \cdot 15 + 30$ or 165, $m\angle M = 30$, $m\angle N = 9 \cdot 15 + 30$ or 165, and $m\angle P = 9 \cdot 15 + 30$ or 165.
10. The sum of the measures of the exterior angles is 360. There are 6 congruent exterior angles.
$6n = 360$
$n = 60$
The measure of each exterior angle is 60. Since each exterior angle and its corresponding interior angle form a linear pair, the measure of the interior angle is $180 - 60$ or 120.
11. The sum of the measures of the exterior angles is 360. There are 18 congruent exterior angles.
$18n = 360$
$n = 20$
The measure of each exterior angle is 20. Since each exterior angle and its corresponding interior angle form a linear pair, the measure of the interior angle is $180 - 20$ or 160.
12. Use the Interior Angle Sum Theorem.
$S = 180(n - 2)$
$= 180(5 - 2)$
$= 540$
The sum of the measures of the interior angles of the base of the fish tank is 540.

Pages 407–409 Practice and Apply

13. Use the Interior Angle Sum Theorem.
$S = 180(n - 2)$
$= 180(32 - 2)$
$= 5400$
The sum of the measures of the interior angles of a 32-gon is 5400.
14. Use the Interior Angle Sum Theorem.
$S = 180(n - 2)$
$= 180(18 - 2)$
$= 2880$
The sum of the measures of the interior angles of an 18-gon is 2880.
15. Use the Interior Angle Sum Theorem.
$S = 180(n - 2)$
$= 180(19 - 2)$
$= 3060$
The sum of the measures of the interior angles of a 19-gon is 3060.
16. Use the Interior Angle Sum Theorem.
$S = 180(n - 2)$
$= 180(27 - 2)$
$= 4500$
The sum of the measures of the interior angles of a 27-gon is 4500.
17. Use the Interior Angle Sum Theorem.
$S = 180(n - 2)$
$= 180(4y - 2)$
$= 720y - 360$
$= 360(2y - 1)$
The sum of the measures of the interior angles of a $4y$-gon is $360(2y - 1)$.
18. Use the Interior Angle Sum Theorem.
$S = 180(n - 2)$
$= 180(2x - 2)$
$= 360(x - 1)$
The sum of the measures of the interior angles of a $2x$-gon is $360(x - 1)$.

19. Use the Interior Angle Sum Theorem.
$S = 180(n - 2)$
$= 180(8 - 2)$
$= 1080$
The sum of the measures of the interior angles of the octagonal garden is 1080.

20. Use the Interior Angle Sum Theorem.
$S = 180(n - 2)$
$= 180(6 - 2)$
$= 720$
The sum of the measures of the interior angles of the hexagonal gazebos is 720.

21. Use the Interior Angle Sum Theorem to write an equation to solve for n, the number of sides.
$S = 180(n - 2)$
$(140)n = 180(n - 2)$
$140n = 180n - 360$
$0 = 40n - 360$
$360 = 40n$
$9 = n$
The polygon has 9 sides.

22. Use the Interior Angle Sum Theorem to write an equation to solve for n, the number of sides.
$S = 180(n - 2)$
$(170)n = 180(n - 2)$
$170n = 180n - 360$
$0 = 10n - 360$
$360 = 10n$
$36 = n$
The polygon has 36 sides.

23. Use the Interior Angle Sum Theorem to write an equation to solve for n, the number of sides.
$S = 180(n - 2)$
$(160)n = 180(n - 2)$
$160n = 180n - 360$
$0 = 20n - 360$
$360 = 20n$
$18 = n$
The polygon has 18 sides.

24. Use the Interior Angle Sum Theorem to write an equation to solve for n, the number of sides.
$S = 180(n - 2)$
$(165)n = 180(n - 2)$
$165n = 180n - 360$
$0 = 15n - 360$
$360 = 15n$
$24 = n$
The polygon has 24 sides.

25. Use the Interior Angle Sum Theorem to write an equation to solve for n, the number of sides.
$S = 180(n - 2)$
$(157.5)n = 180(n - 2)$
$157.5n = 180n - 360$
$0 = 22.5n - 360$
$360 = 22.5n$
$16 = n$
The polygon has 16 sides.

26. Use the Interior Angle Sum Theorem to write an equation to solve for n, the number of sides.
$S = 180(n - 2)$
$(176.4)n = 180(n - 2)$
$176.4n = 180n - 360$
$0 = 3.6n - 360$
$360 = 3.6n$
$100 = n$
The polygon has 100 sides.

27. Since $n = 4$, the sum of the measures of the interior angles is $180(4 - 2)$ or 360. Write an equation to express the sum of the measures of the interior angles of the polygon.
$360 = m\angle M + m\angle P + m\angle Q + m\angle R$
$360 = x + 4x + 2x + 5x$
$360 = 12x$
$30 = x$
Use the value of x to find the measure of each angle.
$m\angle M = 30$, $m\angle P = 4 \cdot 30$ or 120, $m\angle Q = 2 \cdot 30$ or 60, and $m\angle R = 5 \cdot 30$ or 150.

28. Since $n = 5$, the sum of the measures of the interior angles is $180(5 - 2)$ or 540. Write an equation to express the sum of the measures of the interior angles of the polygon.
$540 = m\angle E + m\angle F + m\angle G + m\angle H + m\angle J$
$540 = x + (x + 20) + (x + 5) + (x - 5) + (x + 10)$
$540 = 5x + 30$
$510 = 5x$
$102 = x$
Use the value of x to find the measure of each angle.
$m\angle E = 102$, $m\angle F = 102 + 20$ or 122, $m\angle G = 102 + 5$ or 107, $m\angle H = 102 - 5$ or 97, and $m\angle J = 102 + 10$ or 112.

29. Since $n = 4$, the sum of the measures of the interior angles is $180(4 - 2)$ or 360. Since a parallelogram has congruent opposite angles, the measures of angles M and P are equal, and the measures of angles N and Q are equal. Write an equation to express the sum of the measures of the interior angles of the parallelogram.
$360 = m\angle M + m\angle N + m\angle P + m\angle Q$
$360 = 10x + 20x + 10x + 20x$
$360 = 60x$
$6 = x$
Use the value of x to find the measure of each angle.
$m\angle M = 10 \cdot 6$ or 60, $m\angle N = 20 \cdot 6$ or 120, $m\angle P = 10 \cdot 6$ or 60, and $m\angle Q = 20 \cdot 6$ or 120.

30. Since $n = 4$, the sum of the measures of the interior angles is $180(4 - 2)$ or 360. Write an equation to express the sum of the measures of the interior angles of the isosceles trapezoid.
$360 = m\angle T + m\angle W + m\angle Y + m\angle Z$
$360 = 20x + 20x + 30x + 30x$
$360 = 100x$
$3.6 = x$
Use the value of x to find the measure of each angle.
$m\angle T = 20 \cdot 3.6$ or 72, $m\angle W = 20 \cdot 3.6$ or 72, $m\angle Y = 30 \cdot 3.6$ or 108, and $m\angle Z = 30 \cdot 3.6$ or 108.

31. Since $n = 10$, the sum of the measures of the interior angles is $180(10 - 2)$ or 1440. The sum of the given measures is $10x + 440$. Find x.
$1440 = 10x + 440$
$1000 = 10x$
$100 = x$
The measures of the interior angles of the decagon are 105, 110, 120, 130, 135, 140, 160, 170, 180, and 190.

32. Since $n = 5$, the sum of the measures of the interior angles is $180(5 - 2)$ or 540. Write an equation to express the sum of the measures of the interior angles of the polygon.
$540 = m\angle A + m\angle B + m\angle C + m\angle D + m\angle E$
$540 = 6x + (4x + 13) + (x + 9) + (2x - 8) + (4x - 1)$
$540 = 17x + 13$
$527 = 17x$
$31 = x$
Use the value of x to find the measure of each angle.
$m\angle A = 6 \cdot 31$ or 186, $m\angle B = 4 \cdot 31 + 13$ or 137, $m\angle C = 31 + 9$ or 40, $m\angle D = 2 \cdot 31 - 8$ or 54, and $m\angle E = 4 \cdot 31 - 1$ or 123.

33. Sample answer: Since $n = 4$, the sum of the measures of the interior angles is $180(4 - 2)$ or 360. Write an equation to express the sum of the measures of the interior angles of the quadrilateral.
$360 = x + 2x + 3x + 4x$
$360 = 10x$
$36 = x$
The measures of the interior angles of the quadrilateral are 36, $2 \cdot 36$ or 72, $3 \cdot 36$ or 108, and $4 \cdot 36$ or 144.

34. Since $n = 4$, the sum of the measures of the interior angles is $180(4 - 2)$ or 360. Write an equation to express the sum of the measures of the interior angles of the quadrilateral.
$360 = x + (x + 10) + (x + 20) + (x + 30)$
$360 = 4x + 60$
$300 = 4x$
$75 = x$
The measures of the interior angles of the quadrilateral are 75, $75 + 10$ or 85, $75 + 20$ or 95, and $75 + 30$ or 105.

35. The sum of the measures of the exterior angles is 360. A regular decagon has 10 congruent exterior angles.
$10n = 360$
$n = 36$
The measure of each exterior angle is 36. Since each exterior angle and its corresponding interior angle form a linear pair, the measure of the interior angle is $180 - 36$ or 144.

36. The sum of the measures of the exterior angles is 360. A regular hexagon has 6 congruent exterior angles.
$6n = 360$
$n = 60$
The measure of each exterior angle is 60. Since each exterior angle and its corresponding interior angle form a linear pair, the measure of the interior angle is $180 - 60$ or 120.

37. The sum of the measures of the exterior angles is 360. A regular nonagon has 9 congruent exterior angles.
$9n = 360$
$n = 40$
The measure of each exterior angle is 40. Since each exterior angle and its corresponding interior angle form a linear pair, the measure of the interior angle is $180 - 40$ or 140.

38. The sum of the measures of the exterior angles is 360. A regular octagon has 8 congruent exterior angles.
$8n = 360$
$n = 45$
The measure of each exterior angle is 45. Since each exterior angle and its corresponding interior angle form a linear pair, the measure of the interior angle is $180 - 45$ or 135.

39. Since $n = 11$, the sum of the measures of the interior angles is $180(11 - 2)$ or 1620. A regular 11-gon has 11 congruent interior angles. Let the measure of one of these angles be x.
$1620 = 11x$
$147.3 \approx x$
To the nearest tenth, the measure of each interior angle of the 11-gon is 147.3. Since each interior angle and its corresponding exterior angle form a linear pair, the measure of the exterior angle is about $180 - 147.3$ or 32.7.

40. Since $n = 7$, the sum of the measures of the interior angles is $180(7 - 2)$ or 900. A regular 7-gon has 7 congruent interior angles. Let the measure of one of these angles be x.
$900 = 7x$
$128.6 \approx x$
To the nearest tenth, the measure of each interior angle of the 7-gon is 128.6. Since each interior angle and its corresponding exterior angle form a linear pair, the measure of the exterior angle is about $180 - 128.6$ or 51.4.

41. Since $n = 12$, the sum of the measures of the interior angles is $180(12 - 2)$ or 1800. A regular 12-gon has 12 congruent interior angles. Let the measure of one of these angles be x.
$1800 = 12x$
$150 = x$
The measure of each interior angle of the 12-gon is 150. Since each interior angle and its corresponding exterior angle form a linear pair, the measure of the exterior angle is about $180 - 150$ or 30.

42. Consider the sum of the measures of the exterior angles, N, for an n-gon.
N = sum of measures of linear pairs − sum of measures of interior angles
$= 180n - 180(n - 2)$
$= 180n - 180n + 360$
$= 360$
So, the sum of the exterior angle measures is 360 for any convex polygon.

43. Since $n = 5$, the sum of the measures of the interior angles is $180(5 - 2)$ or 540. A regular pentagon has 5 congruent interior angles. Let the measure of one of these angles be x.
$540 = 5x$
$108 = x$
The measure of each interior angle of the Pentagon is 108. Since each interior angle and its corresponding exterior angle form a linear pair, the measure of the exterior angle is $180 - 108$ or 72.

44. Yes; both the dome and the architectural elements are based upon a regular octagon.
Since $n = 8$, the sum of the measures of the interior angles is $180(8 - 2)$ or 1080. A regular octagon has 8 congruent interior angles. Let the measure of one of these angles be x.
$1080 = 8x$
$135 = x$
The measure of each interior angle is 135. Since each interior angle and its corresponding exterior angle form a linear pair, the measure of the exterior angle is about $180 - 135$ or 45.

45. $\frac{180(n-2)}{n} = \frac{180n - 360}{n}$
$= \frac{180n}{n} - \frac{360}{n}$
$= 180 - \frac{360}{n}$
The two formulas are equivalent.

46. Sample answer: The outline of a scallop shell is a convex polygon that is not regular. The lines in the shell resemble diagonals drawn from one vertex of a polygon. These diagonals separate the polygon into triangles. Answers should include the following.
- The Interior Angle Sum Theorem is derived from the pattern between the number of sides in a polygon and the number of triangles. The formula is the product of the sum of the measures of the angles in a triangle, 180, and the number of triangles the polygon contains.
- The exterior angle and the interior angle of a polygon are a linear pair. So, the measure of an exterior angle is the difference between 180 and the measure of the interior angle.

47. B; since the unknown polygon is regular, its interior angles are congruent. The sum of the measures of the interior angles of the square, pentagon, and unknown regular polygon is 360. Let the measure of each interior angle of the unknown polygon be x. Find x using the fact that the measures of the interior angles of squares and regular pentagons are 90 and 108, respectively.
$360 = x + 90 + 108$
$162 = x$
The sum of the measures of the interior angles is given by $S = 180(n - 2)$, the Interior Angle Sum Theorem. This is equal to $162n$, where, in both cases, n is the number of sides. Solve for n.
$162n = 180(n - 2)$
$162n = 180n - 360$
$0 = 18n - 360$
$360 = 18n$
$20 = n$
The polygon has 20 sides.

48. Since $\frac{9y}{2x} = 9, y = 2x$. Substitute this result into the other equation relating x and y. Then solve for x.
$6x + 3y = 48$
$6x + 3(2x) = 48$
$6x + 6x = 48$
$12x = 48$
$x = 4$

Page 409 Maintain Your Skills

49. Use the Law of Cosines to find $m\angle C$ since the measures of all three sides are known.
$c^2 = a^2 + b^2 - 2ab\cos C$
$11^2 = 6^2 + 9^2 - 2(6)(9)\cos C$
$121 = 36 + 81 - 108\cos C$
$4 = -108\cos C$
$-\frac{4}{108} = \cos C$
$C = \cos^{-1}\left(-\frac{1}{27}\right)$
$C \approx 92.1$
To the nearest tenth, the measure of angle C is 92.1.

50. Use the Law of Cosines to find $m\angle B$ since the measures of all three sides are known.
$b^2 = a^2 + c^2 - 2ac\cos B$
$23.6^2 = 15.5^2 + 25.1^2 - 2(15.5)(25.1)\cos B$
$556.96 = 240.25 + 630.01 - 778.1\cos B$
$-313.3 = -778.1\cos B$
$\frac{-313.3}{-778.1} = \cos B$
$B = \cos^{-1}\left(\frac{313.3}{778.1}\right)$
$B \approx 66.3$
To the nearest tenth, the measure of angle B is 66.3.

51. Use the Law of Cosines to find $m\angle A$ since the measures of all three sides are known.
$a^2 = b^2 + c^2 - 2ab\cos A$
$47^2 = 53^2 + 56^2 - 2(53)(56)\cos A$
$2209 = 2809 + 3136 - 5936\cos A$
$-3736 = -5936\cos A$
$\frac{-3736}{-5936} = \cos A$
$A = \cos^{-1}\left(\frac{467}{742}\right)$
$A \approx 51.0$
To the nearest tenth, the measure of angle A is 51.0.

52. Use the Law of Cosines to find $m\angle C$ since the measures of all three sides are known.
$c^2 = a^2 + b^2 - 2ab\cos C$
$16^2 = 12^2 + 14^2 - 2(12)(14)\cos C$
$256 = 144 + 196 - 336\cos C$
$-84 = -336\cos C$
$\frac{-84}{-336} = \cos C$
$C = \cos^{-1}\left(\frac{1}{4}\right)$
$C \approx 75.5$
To the nearest tenth, the measure of angle C is 75.5.

53. Use the Law of Sines since we know the measures of two sides and an angle opposite one of the sides.

$\frac{\sin G}{g} = \frac{\sin F}{f}$

$\frac{\sin G}{17} = \frac{\sin 54°}{15}$

$\sin G = \frac{17 \sin 54°}{15}$

$G = \sin^{-1}\left(\frac{17 \sin 54°}{15}\right)$

$G \approx 66°$

Use the Angle Sum Theorem to find $m\angle H$.

$m\angle F + m\angle G + m\angle H = 180$

$54 + 66 + m\angle H \approx 180$

$m\angle H \approx 60$

Use the Law of Sines to find h.

$\frac{\sin F}{f} = \frac{\sin H}{h}$

$\frac{\sin 54°}{15} = \frac{\sin 60°}{h}$

$h = \frac{15 \sin 60°}{\sin 54°}$

$h \approx 16.1$

Therefore, $m\angle G \approx 66$, $m\angle H \approx 60$, and $h \approx 16.1$.

54. Use the Angle Sum Theorem to find $m\angle G$.

$m\angle F + m\angle G + m\angle H = 180$

$47 + m\angle G + 78 = 180$

$m\angle G = 55$

Use the Law of Sines to find f and h.

$\frac{\sin G}{g} = \frac{\sin F}{f}$

$\frac{\sin 55°}{31} = \frac{\sin 47°}{f}$

$f = \frac{31 \sin 47°}{\sin 55°}$

$f \approx 27.7$

$\frac{\sin G}{g} = \frac{\sin H}{h}$

$\frac{\sin 55°}{31} = \frac{\sin 78°}{h}$

$h = \frac{31 \sin 78°}{\sin 55°}$

$h \approx 37.0$

Therefore, $m\angle G = 55$, $f \approx 27.7$, and $h \approx 37.0$.

55. Use the Angle Sum Theorem to find $m\angle F$.

$m\angle F + m\angle G + m\angle H = 180$

$m\angle F + 56 + 67 = 180$

$m\angle F = 57$

Use the Law of Sines to find f and h.

$\frac{\sin G}{g} = \frac{\sin F}{f}$

$\frac{\sin 56°}{63} = \frac{\sin 57°}{f}$

$f = \frac{63 \sin 57°}{\sin 56°}$

$f \approx 63.7$

$\frac{\sin G}{g} = \frac{\sin H}{h}$

$\frac{\sin 56°}{63} = \frac{\sin 67°}{h}$

$h = \frac{63 \sin 67°}{\sin 56°}$

$h \approx 70.0$

Therefore, $m\angle F = 57$, $f \approx 63.7$, and $h \approx 70.0$.

56. Use the Law of Sines since we know the measures of two sides and an angle opposite one of the sides.

$\frac{\sin H}{h} = \frac{\sin G}{g}$

$\frac{\sin H}{32.4} = \frac{\sin 65°}{30.7}$

$\sin H = \frac{32.4 \sin 65°}{30.7}$

$H = \sin^{-1}\left(\frac{32.4 \sin 65°}{30.7}\right)$

$H \approx 73°$

Use the Angle Sum Theorem to find $m\angle F$.

$m\angle F + m\angle G + m\angle H = 180$

$m\angle F + 65 + 73 \approx 180$

$m\angle F \approx 42$

Use the Law of Sines to find f.

$\frac{\sin G}{g} = \frac{\sin F}{f}$

$\frac{\sin 65°}{30.7} = \frac{\sin 42°}{f}$

$f = \frac{30.7 \sin 42°}{\sin 65°}$

$f \approx 22.7$

Therefore, $m\angle H \approx 73$, $m\angle F \approx 42$, and $f \approx 22.7$.

57. Given: $\overline{JL} \parallel \overline{KM}$, $\overline{JK} \parallel \overline{LM}$

Prove: $\triangle JKL \cong \triangle MLK$

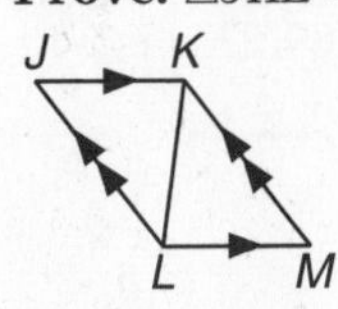

Proof:

Statements	Reasons
1. $\overline{JL} \parallel \overline{KM}$, $\overline{JK} \parallel \overline{LM}$	1. Given
2. $\angle MKL \cong \angle JLK$, $\angle JKL \cong \angle MLK$	2. Alt. int. ∠s are ≅.
3. $\overline{KL} \cong \overline{KL}$	3. Reflexive Property
4. $\triangle JKL \cong \triangle MLK$	4. ASA

58. Line b is the transversal that forms $\angle 3$ and $\angle 11$ where it intersects lines m and n. $\angle 3$ and $\angle 11$ are corresponding angles.

59. Line m is the transversal that forms $\angle 6$ and $\angle 7$ where it intersects lines b and c. $\angle 6$ and $\angle 7$ are consecutive interior angles.

60. Line c is the transversal that forms $\angle 8$ and $\angle 10$ where it intersects lines m and n. $\angle 8$ and $\angle 10$ are alternate interior angles.

61. Line n is the transversal that forms $\angle 12$ and $\angle 16$ where it intersects lines b and c. $\angle 12$ and $\angle 16$ are alternate exterior angles.

62. $\angle 1$ and $\angle 4$, $\angle 1$ and $\angle 2$, $\angle 2$ and $\angle 3$, and $\angle 3$ and $\angle 4$ are consecutive interior angles.

63. $\angle 3$ and $\angle 5$, and $\angle 2$ and $\angle 6$ are alternate interior angles.

64. $\angle 1$ and $\angle 5$, and $\angle 4$ and $\angle 6$ are corresponding angles.

65. None; there are no pairs of alternate exterior angles.

Page 410 Spreadsheet Investigation: Angles of Polygons

1. For a regular polygon, the measure of each interior angle in the polygon can be found by dividing the sum of the measures of the interior angles by the number of sides of the polygon. So, the formula to find the measure of each interior angle in the polygon is "=C2/A2".
2. For a regular polygon, the sum of the measures of the exterior angles of the polygon can be found by multiplying the number of sides by the measure of the exterior angles. So, the formula to find the sum of the measures of the exterior angles of the polygon is "=A2*E2".
3. The formula for the sum of the measures of the interior angles is "=(A2-2)*180", which gives -180 for 1 side and 0 for 2 sides.
4. No, a polygon is a closed figure formed by coplanar segments.
5. A 15-sided polygon has 13 triangles.
6. The measure of the exterior angle of a 15-sided polygon is 24.
7. The measure of the interior angle of a 110-sided polygon is about 176.7.
8. Each interior angle measures 180. This is not possible for a polygon.

8-2 Parallelograms

Pages 411–412 Geometry Activity: Properties of Parallelograms

1. $\overline{FG} \cong \overline{HJ} \cong \overline{PQ} \cong \overline{RS}$ and $\overline{FJ} \cong \overline{GH} \cong \overline{PS} \cong \overline{QR}$.
2. $\angle F \cong \angle P \cong \angle H \cong \angle R$ and $\angle J \cong \angle G \cong \angle Q \cong \angle S$.
3. Opposite angles are congruent; consecutive angles are supplementary.

Page 414 Check for Understanding

1. Opposite sides are congruent; opposite angles are congruent; consecutive angles are supplementary; and if there is one right angle, there are four right angles.
2. Diagonals bisect each other; each diagonal forms two congruent triangles in a parallelogram.
3. Sample answer:

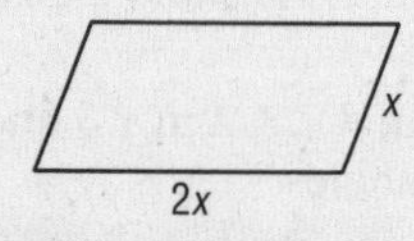

4. $\overline{SV} \cong \overline{VQ}$ because diagonals of parallelograms bisect each other.
5. Since diagonals bisect each other and opposite sides of parallelograms are congruent, $\triangle VRS \cong \triangle VTQ$ by SSS.
6. Since consecutive angles in parallelograms are supplementary, $\angle TSR$ is supplementary to $\angle STQ$ and $\angle SRQ$.
7. $\angle MJK \cong \angle KLM$ because opposite angles in a parallelogram are congruent. Find $m\angle KLM$.
   $m\angle KLM = m\angle KLR + m\angle MLR$
   $= 70 + 30 = 100$
   So, $m\angle MJK = 100$.
8. Consecutive angles in a parallelogram are supplementary. So, $m\angle JML = 180 - m\angle KLM$.
   $m\angle JML = 180 - 100$ or 80.
9. Consecutive angles in a parallelogram are supplementary. So, $m\angle JKL = 180 - m\angle KLM$.
   $m\angle JKL = 180 - 100$ or 80.
10. $\angle KJL \cong \angle JLM$ because they are alternate interior angles. The measure of $\angle JLM$ is 30. So, $m\angle KJL = 30$.
11. Opposite sides of a parallelogram are congruent, so their measures are equal.
    Find a.
    $JM = KL$
    $3a = 21$
    $a = 7$
12. Opposite sides of a parallelogram are congruent, so their measures are equal. Find b.
    $JK = ML$
    $2b + 3 = 45$
    $2b = 42$
    $b = 21$
13. **Given:** ▱$VZRQ$ and ▱$WQST$
    Prove: $\angle Z \cong \angle T$

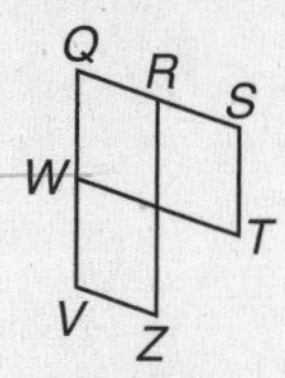

Proof:

Statements	Reasons
1. ▱$VZRQ$ and ▱$WQST$	1. Given
2. $\angle Z \cong \angle Q$, $\angle Q \cong \angle T$	2. Opp. ∠s of a ▱ are ≅.
3. $\angle Z \cong \angle T$	3. Transitive Prop.

14. **Given:** ▱$XYRZ$, $\overline{WZ} \cong \overline{WS}$
    Prove: $\angle XYR \cong \angle S$

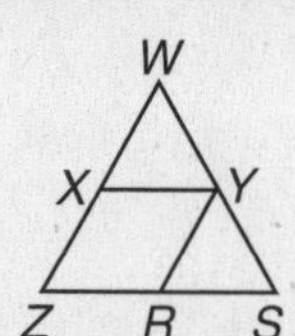

Proof: Opposite angles of a parallelogram are congruent, so $\angle Z \cong \angle XYR$. By the Isosceles Triangle Theorem, since $\overline{WZ} \cong \overline{WS}$, $\angle Z \cong \angle S$. By the Transitive Property, $\angle XYR \cong \angle S$.

15. C; $\overline{GJ}$ and $\overline{HK}$ are the diagonals. Since the diagonals of a parallelogram bisect each other, the intersection point is the midpoint of $\overline{GJ}$ and $\overline{HK}$. Find the intersection of the diagonals by finding the midpoint of $\overline{GJ}$.

$$\left(\frac{x_1+x_2}{2}, \frac{y_1+y_2}{2}\right) = \left(\frac{-3+3}{2}, \frac{4+(-5)}{2}\right)$$
$$= (0, -0.5)$$

The diagonals intersect at $(0, -0.5)$.

Pages 415–416 Practice and Apply

16. $\angle DAB \cong \angle BCD$ because opposite angles of a parallelogram are congruent.

17. $\angle ABD \cong \angle CDB$ because alternate interior angles are congruent.

18. $\overline{AB} \parallel \overline{DC}$ because opposite sides of a parallelogram are parallel.

19. $\overline{BG} \cong \overline{GD}$ because the diagonals of a parallelogram bisect each other.

20. $\triangle ABD \cong \triangle CDB$ because the diagonal $\overline{BD}$ separates the parallelogram into two congruent triangles.

21. $\angle ACD \cong \angle BAC$ because alternate interior angles are congruent.

22. $\angle RNP \cong \angle NRM$ because they are alternate interior angles. So, $m\angle RNP = 38$. Find $m\angle MNP$.

$$\begin{aligned} m\angle MNP &= m\angle MNR + m\angle RNP \\ &= 33 + 38 \\ &= 71 \end{aligned}$$

23. $\angle NRP \cong \angle MNR$ because they are alternate interior angles. So, $m\angle NRP = 33$.

24. $\angle RNP \cong \angle MRN$ because they are alternate interior angles. So, $m\angle RNP = 38$.

25. $\angle RMN$ is supplementary to $\angle MNP$. Find $m\angle RMN$.

$$\begin{aligned} m\angle RMN + m\angle MNP &= 180 \\ m\angle RMN + 71 &= 180 \\ m\angle RMN &= 109 \end{aligned}$$

26. $\angle MQN$ is supplementary to $\angle PQN$. Find $m\angle MQN$.

$$\begin{aligned} 180 &= m\angle MQN + m\angle PQN \\ 180 &= m\angle MQN + 83 \\ 97 &= m\angle MQN \end{aligned}$$

27. $\angle MQR \cong \angle PQN$ because they are vertical angles. So, $m\angle MQR = 83$.

28. Opposite sides of a parallelogram are congruent, so their measures are equal. Find x.

$$\begin{aligned} MN &= RP \\ 3x - 4 &= 20 \\ 3x &= 24 \\ x &= 8 \end{aligned}$$

29. Opposite sides of a parallelogram are congruent, so their measures are equal. Find y.

$$\begin{aligned} NP &= MR \\ 2y + 5 &= 17.9 \\ 2y &= 12.9 \\ y &= 6.45 \end{aligned}$$

30. The diagonals of a parallelogram bisect each other, so $MQ = QP$. Find w.

$$\begin{aligned} MQ &= QP \\ 4w - 3 &= 11.1 \\ 4w &= 14.1 \\ w &\approx 3.5 \end{aligned}$$

31. The diagonals of a parallelogram bisect each other, so $RQ = QN$. Find z.

$$\begin{aligned} RQ &= QN \\ 3z - 3 &= 15.4 \\ 3z &= 18.4 \\ z &\approx 6.1 \end{aligned}$$

32. The diagonals of a parallelogram bisect each other, so $EJ = JG$. Find x.

$$\begin{aligned} EJ &= JG \\ 2x + 1 &= 3x \\ 1 &= x \\ EG &= EJ + JG \\ &= [2(1) + 1] + 3(1) \\ &= 6 \end{aligned}$$

33. The diagonals of a parallelogram bisect each other, so $HJ = JF$. Find y.

$$\begin{aligned} HJ &= JF \\ \tfrac{1}{2}y + 2 &= y - \tfrac{1}{2} \\ 2 + \tfrac{1}{2} &= y - \tfrac{1}{2}y \\ \tfrac{5}{2} &= \tfrac{1}{2}y \\ 5 &= y \\ FH &= HJ + JF \\ &= \left[\tfrac{1}{2}(5) + 2\right] + \left(5 - \tfrac{1}{2}\right) \\ &= 9 \end{aligned}$$

34. Since the diagonals of a parallelogram bisect each other, the drawer pulls are at the intersection point of the diagonals.

35.

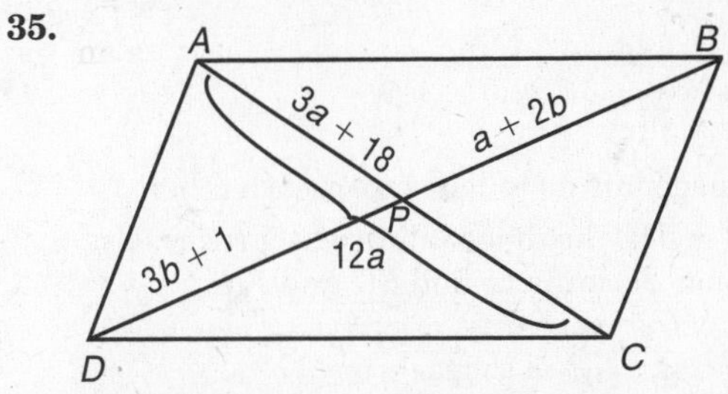

Since $ABCD$ is a parallelogram, the diagonals bisect each other. So $AC = 2(AP)$

$AC = 12a$ and $AP = 3a + 18$

$$\begin{aligned} 12a &= 2(3a + 18) \\ 12a &= 6a + 36 \\ 6a &= 36 \\ a &= 6 \end{aligned}$$

$$\begin{aligned} DP &= PB \\ 3b + 1 &= a + 2b \\ 3b + 1 &= 6 + 2b \\ b &= 5 \\ DB &= 2(3b + 1) \\ &= 2(3 \cdot 5 + 1) \\ &= 2(16) = 32 \end{aligned}$$

So, $a = 6$, $b = 5$, and $DB = 32$.

36. A $2x + 5$ B, $2y$, $120°$, $21°$, D 21 C

Opposite sides of a parallelogram are congruent, so $\overline{AB} \cong \overline{CD}$ and $AB = CD$.

Find x.

$AB = CD$
$2x + 5 = 21$
$2x = 16$
$x = 8$

Consecutive angles in a parallelogram are supplementary, so $\angle B$ is supplementary to $\angle BAD$ and $m\angle BAD = m\angle CAD + m\angle BAC$.

Find y.

$m\angle BAD + m\angle B = 180$
$m\angle CAD + m\angle BAC + m\angle B = 180$
$21 + 2y + 120 = 180$
$2y = 39$
$y = 19.5$

37. The Distance Formula is $d = \sqrt{(x_2 - x_1)^2 + (y_2 - y_1)^2}$.

The diagonals bisect each other if $EQ = QG$ and $HQ = QF$.

Find these measures.

$EQ = \sqrt{(3 - 0)^2 + (1 - 5)^2}$
$= 5$
$QG = \sqrt{(6 - 3)^2 + (-3 - 1)^2}$
$= 5$
$HQ = \sqrt{(3 - 0)^2 + [1 - (-1)]^2}$
$= \sqrt{13}$
$QF = \sqrt{(6 - 3)^2 + (3 - 1)^2}$
$= \sqrt{13}$

The diagonals do indeed bisect each other.

38. If $EG = FH$, the diagonals are congruent. Use the Distance Formula to find EG and FH.

$d = \sqrt{(x_2 - x_1)^2 + (y_2 - y_1)^2}$
$EG = \sqrt{(6 - 0)^2 + (-3 - 5)^2}$
$= 10$
$FH = \sqrt{(0 - 6)^2 + (-1 - 3)^2}$
$= \sqrt{52}$
$= 2\sqrt{13}$

No, the diagonals are not congruent, since $EG \neq FH$.

39. $\overline{EH}$ is vertical, so its slope is undefined. Find the slope of $\overline{EF}$.

$m = \frac{y_2 - y_1}{x_2 - x_1}$
$= \frac{3 - 5}{6 - 0}$
$= -\frac{1}{3}$

No, the consecutive sides are not perpendicular because the slopes of the sides are not opposite reciprocals of each other.

40. They are all congruent parallelograms. Since A, B and C are midpoints, $\overline{AC}$, $\overline{AB}$, and $\overline{BC}$ are midsegments. The midsegment is parallel to the third side and equal to half the length of the third side. So, each pair of opposite sides of $ACBX$, $ABYC$, and $ABCZ$ are parallel.

41. Given: ▱$PQRS$
Prove: $\overline{PQ} \cong \overline{RS}$
$\overline{QR} \cong \overline{SP}$

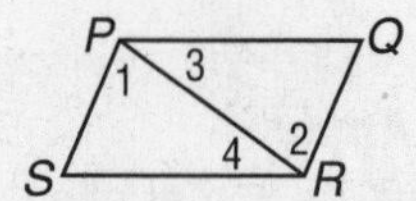

Proof:

Statements	Reasons
1. ▱$PQRS$	1. Given
2. Draw an auxiliary segment $\overline{PR}$ and label angles 1, 2, 3, and 4 as shown.	2. Diagonal of ▱$PQRS$
3. $\overline{PQ} \parallel \overline{SR}$, $\overline{PS} \parallel \overline{QR}$	3. Opp. sides of ▱ are $\parallel$.
4. $\angle 1 \cong \angle 2$, and $\angle 3 \cong \angle 4$	4. Alt. int. ∡ are $\cong$.
5. $\overline{PR} \cong \overline{PR}$	5. Reflexive Prop.
6. $\triangle QPR \cong \triangle SRP$	6. ASA
7. $\overline{PQ} \cong \overline{RS}$ and $\overline{QR} \cong \overline{SP}$	7. CPCTC

42. Given: ▱$GKLM$
Prove: $\angle G$ and $\angle K$ are supplementary.
$\angle K$ and $\angle L$ are supplementary.
$\angle L$ and $\angle M$ are supplementary.
$\angle M$ and $\angle G$ are supplementary.

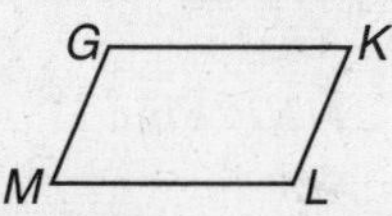

Proof:

Statements	Reasons
1. ▱$GKLM$	1. Given
2. $\overline{GK} \parallel \overline{ML}$, $\overline{GM} \parallel \overline{KL}$	2. Opp. sides of ▱ are $\parallel$.
3. $\angle G$ and $\angle K$ are supplementary. $\angle K$ and $\angle L$ are supplementary. $\angle L$ and $\angle M$ are supplementary. $\angle M$ and $\angle G$ are supplementary.	3. Cons. int. ∡ are suppl.

43. Given: $\square MNPQ$

$\angle M$ is a right angle.

Prove: $\angle N$, $\angle P$ and $\angle Q$ are right angles

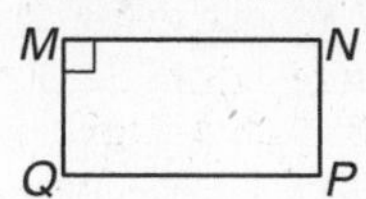

Proof: By definition of a parallelogram, $\overline{MN} \parallel \overline{QP}$. Since $\angle M$ is a right angle, $\overline{MQ} \perp \overline{MN}$. By the Perpendicular Transversal Theorem, $\overline{MQ} \perp \overline{QP}$. $\angle Q$ is a right angle, because perpendicular lines form a right angle. $\angle N \cong \angle Q$ and $\angle M \cong \angle P$ because opposite angles in a parallelogram are congruent. $\angle P$ and $\angle N$ are right angles, since all right angles are congruent.

44. Given: $ACDE$ is a parallelogram.

Prove: $\overline{EC}$ bisects $\overline{AD}$. $\overline{AD}$ bisects $\overline{EC}$.

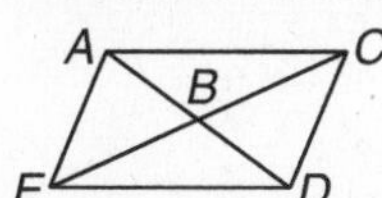

Proof: It is given that $ACDE$ is a parallelogram. Since opposite sides of a parallelogram are congruent, $\overline{EA} \cong \overline{DC}$. By definition of a parallelogram, $\overline{EA} \parallel \overline{DC}$. $\angle AEB \cong \angle DCB$ and $\angle EAB \cong \angle CDB$ because alternate interior angles are congruent. $\triangle EBA \cong \triangle CBD$ by ASA. $\overline{EB} \cong \overline{BC}$ and $\overline{AB} \cong \overline{BD}$ by CPCTC. By the definition of segment bisector, $\overline{EC}$ bisects $\overline{AD}$ and $\overline{AD}$ bisects $\overline{EC}$.

45. Given: $\square WXYZ$

Prove: $\triangle WXZ \cong \triangle YZX$

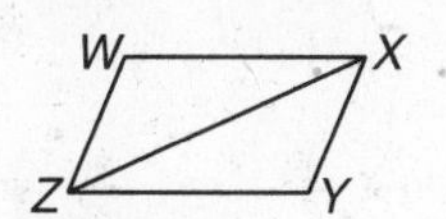

Proof:

Statements	Reasons
1. $\square WXYZ$	1. Given
2. $\overline{WX} \cong \overline{ZY}$, $\overline{WZ} \cong \overline{XY}$	2. Opp. sides of $\square$ are $\cong$.
3. $\angle ZWX \cong \angle XYZ$	3. Opp. $\angle$s of $\square$ are $\cong$.
4. $\triangle WXZ \cong \triangle YZX$	4. SAS

46. Given: $DGHK$ is a parallelogram.

$\overline{FH} \perp \overline{GD}$

$\overline{DJ} \perp \overline{HK}$

Prove: $\triangle DJK \cong \triangle HFG$

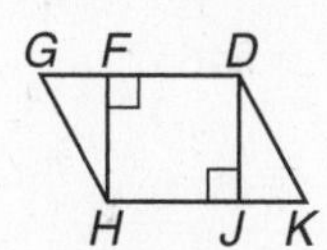

Proof:

Statements	Reasons
1. $DGHK$ is a parallelogram. $\overline{FH} \perp \overline{GD}$, $\overline{DJ} \perp \overline{HK}$	1. Given
2. $\angle G \cong \angle K$	2. Opp. $\angle$s of $\square$ are $\cong$.
3. $\overline{GH} \cong \overline{DK}$	3. Opp. sides of $\square$ are $\cong$.
4. $\angle HFG$ and $\angle DJK$ are rt. $\angle$s.	4. $\perp$ lines form four rt. $\angle$s.
5. $\triangle HFG$ and $\triangle DJK$ are rt. $\triangle$s.	5. Def. of rt. $\triangle$
6. $\triangle HFG \cong \triangle DJK$	6. HA

47. Given: $\square BCGH$, $\overline{HD} \cong \overline{FD}$

Prove: $\angle F \cong \angle GCB$

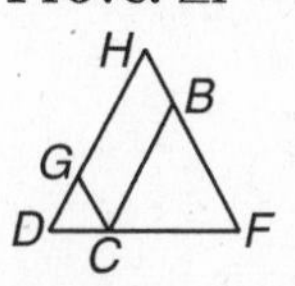

Proof:

Statements	Reasons
1. $\square BCGH$ $\overline{HD} \cong \overline{FD}$	1. Given
2. $\angle F \cong \angle H$	2. Isosceles Triangle Theorem
3. $\angle H \cong \angle GCB$	3. Opp. $\angle$s of $\square$ are $\cong$.
4. $\angle F \cong \angle GCB$	4. Congruence of angles is transitive.

48. $\angle MSR \cong \angle PST$ because they are vertical angles. $\angle NMP \cong \angle MPQ$ because they are alternate interior angles. So, $\triangle MSR$ is similar to $\triangle PST$. $\overline{MN} \cong \overline{QP}$ because opposite sides of a parallelogram are congruent. $TP = \frac{1}{2}QP$ is given, so $TP = \frac{1}{2}MN$.

Find $\frac{MS}{SP}$.

$$\frac{MS}{SP} = \frac{MR}{TP}$$
$$= \frac{\frac{1}{4}MN}{\frac{1}{2}MN}$$
$$= \frac{1}{2}$$

The ratio of MS to SP is $\frac{1}{2}$.

49. Sample answer: The graphic uses the illustration of wedges shaped like parallelograms to display the data. Answers should include the following.

- The opposite sides are parallel and congruent, the opposite angles are congruent, and the consecutive angles are supplementary.
- Sample answer:

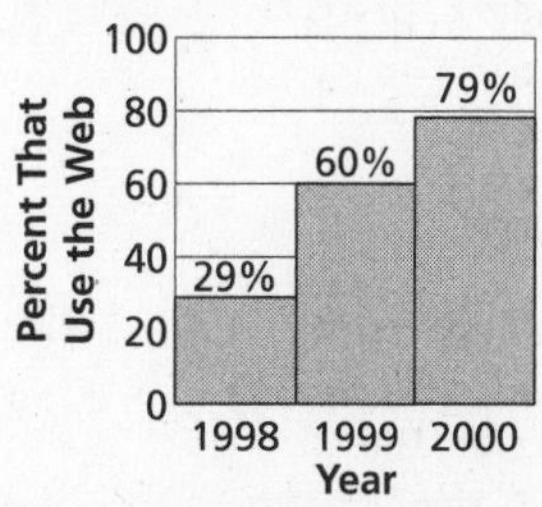

50. Consecutive angles of a parallelogram are supplementary. Find x

$$(3x + 42) + (9x - 18) = 180$$
$$12x + 24 = 180$$
$$12x = 156$$
$$x = 13$$

$3(13) + 42 = 81$

$9(13) - 18 = 99$

The measures of the angles are 81 and 99.

51. B; the perimeter p is equal to the sum of the measures of the sides. Find y.

$$2x + 2y = p$$
$$2\left(\frac{y}{5}\right) + 2y = p$$
$$\frac{2}{5}y + 2y = p$$
$$\frac{12y}{5} = p$$
$$y = \frac{5p}{12}$$

Page 416 Maintain Your Skills

52. Use the Interior Angle Sum Theorem.

$$S = 180(n - 2)$$
$$= 180(14 - 2)$$
$$= 2160$$

The sum of the measures of the interior angles of a 14-gon is 2160.

53. Use the Interior Angle Sum Theorem.

$$S = 180(n - 2)$$
$$= 180(22 - 2)$$
$$= 3600$$

The sum of the measures of the interior angles of a 22-gon is 3600.

54. Use the Interior Angle Sum Theorem.

$$S = 180(n - 2)$$
$$= 180(17 - 2)$$
$$= 2700$$

The sum of the measures of the interior angles of a 17-gon is 2700.

55. Use the Interior Angle Sum Theorem.

$$S = 180(n - 2)$$
$$= 180(36 - 2)$$
$$= 6120$$

The sum of the measures of the interior angles of a 36-gon is 6120.

56. Since the measures of two sides and the included angle are known, use the Law of Cosines.

$$a^2 = b^2 + c^2 - 2bc \cos A$$
$$a^2 = 11^2 + 13^2 - 2(11)(13) \cos 42°$$
$$a = \sqrt{11^2 + 13^2 - 2(11)(13) \cos 42°}$$
$$a \approx 8.8$$

$$b^2 = a^2 + c^2 - 2ac \cos B$$
$$11^2 \approx 8.8^2 + 13^2 - 2(8.8)(13) \cos B$$
$$-125.44 \approx -2(8.8)(13) \cos B$$
$$\frac{-125.44}{-228.8} \approx \cos B$$
$$\cos^{-1}\left(\frac{125.44}{228.8}\right) \approx B$$
$$56.8 \approx m\angle B$$

Use the Angle Sum Theorem.

$$m\angle A + m\angle B + m\angle C = 180$$
$$42 + 56.8 + m\angle C \approx 180$$
$$m\angle C \approx 81.2$$

57. Since the measures of two sides and an angle opposite one of the sides are known, use the Law of Sines.

$$\frac{\sin C}{c} = \frac{\sin B}{b}$$
$$\frac{\sin C}{14} = \frac{\sin 57°}{12.5}$$
$$\sin C = \frac{14 \sin 57°}{12.5}$$
$$C = \sin^{-1}\left(\frac{14 \sin 57°}{12.5}\right)$$
$$m\angle C \approx 69.9$$

Use the Angle Sum Theorem.

$$m\angle A + m\angle B + m\angle C = 180$$
$$m\angle A + 57 + 69.9 \approx 180$$
$$m\angle A \approx 53.1$$

$$\frac{\sin B}{b} = \frac{\sin A}{a}$$
$$\frac{\sin 57°}{12.5} = \frac{\sin 53.1°}{a}$$
$$a = \frac{12.5 \sin 53.1°}{\sin 57°}$$
$$a \approx 11.9$$

58. Since the measures of two sides and the included angle are known, use the Law of Cosines.

$$c^2 = a^2 + b^2 - 2ab \cos C$$
$$c^2 = 21^2 + 24^2 - 2(21)(24) \cos 78°$$
$$c = \sqrt{21^2 + 24^2 - 2(21)(24) \cos 78°}$$
$$c \approx 28.4$$

$$a^2 = b^2 + c^2 - 2bc \cos A$$
$$21^2 \approx 24^2 + 28.4^2 - 2(24)(28.4) \cos A$$
$$-941.56 \approx -1363.2 \cos A$$
$$\frac{-941.56}{-1363.2} \approx \cos A$$
$$\cos^{-1}\left(\frac{941.56}{1363.2}\right) \approx A$$
$$46.3 \approx m\angle A$$

Use the Angle Sum Theorem.

$$m\angle A + m\angle B + m\angle C = 180$$
$$46.3 + m\angle B + 78 \approx 180$$
$$m\angle B \approx 55.7$$

59. The numbers of the outside diagonals are all ones, so the first thirty numbers sum to 30.

60. The second diagonal consists of the natural numbers, 1, 2, 3, 4, 5, . . . , 70.

The sum is $1 + 2 + 3 + 4 + \ldots + 67 + 68 + 69 + 70$
$= (1 + 70) + (2 + 69) + (3 + 68) + \ldots + (35 + 36)$
$= 35(71)$
$= 2485$

The sum of the first 70 numbers is 2485.

61.

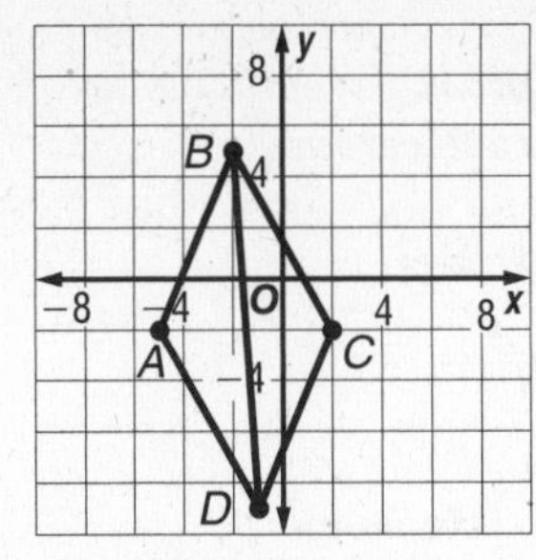

$\overline{AB}$ is a side. Find the slope of $\overline{AB}$.

$$m = \frac{y_2 - y_1}{x_2 - x_1}$$
$$= \frac{5 - (-2)}{-2 - (-5)}$$
$$= \frac{7}{3}$$

62.

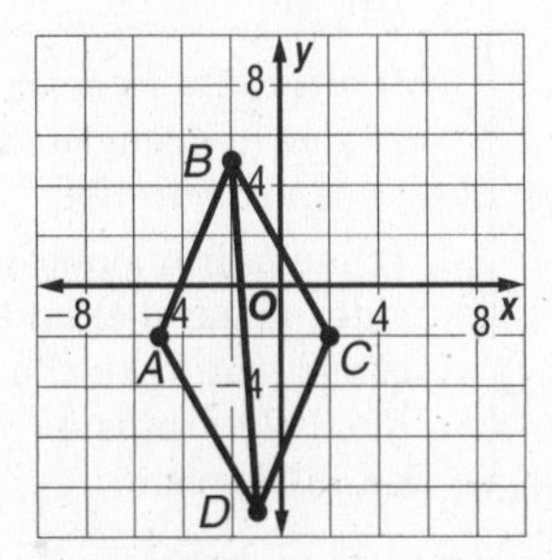

$\overline{BD}$ is a diagonal. Find the slope of $\overline{BD}$.

$$m = \frac{y_2 - y_1}{x_2 - x_1}$$
$$= \frac{-9 - 5}{-1 - (-2)}$$
$$= -14$$

63.

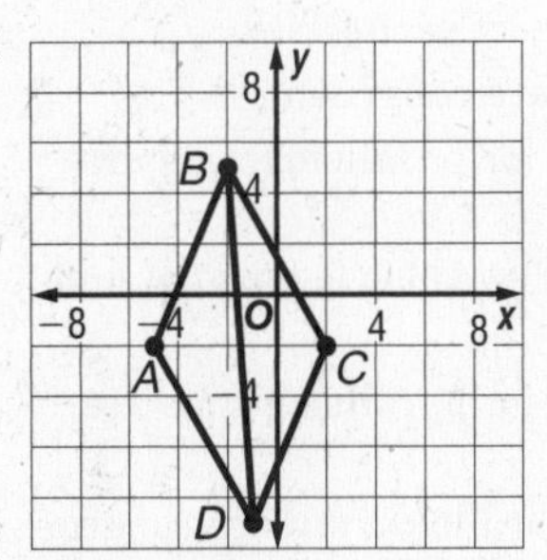

$\overline{CD}$ is a side. Find the slope of $\overline{CD}$.

$$m = \frac{y_2 - y_1}{x_2 - x_1}$$
$$= \frac{-9 - (-2)}{-1 - 2}$$
$$= \frac{7}{3}$$

8-3 Tests for Parallelograms

Page 417 Geometry Activity: Testing for a Parallelogram

1. They appear to be parallel.
2. The quadrilaterals formed are parallelograms.
3. The measures of pairs of opposite sides are equal.
4. Opposite angles are congruent, and consecutive angles are supplementary.
5. Opposite sides are parallel and congruent, opposite angles are congruent, or consecutive angles are supplementary.

Pages 420–421 Check for Understanding

1. Both pairs of opposite sides are congruent; both pairs of opposite angles are congruent; diagonals bisect each other; one pair of opposite sides is parallel and congruent.
2. Sample answer:

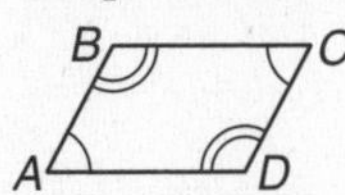

3. Shaniqua; Carter's description could result in a shape that is not a parallelogram.
4. No; one pair of opposite sides is not parallel and congruent.
5. Yes; the missing angle measure of the parallelogram is 180 − 102 or 78, so each pair of opposite angles is congruent.
6. Opposite sides of a parallelogram are congruent. Find x and y.
   $2x - 5 = 3x - 18$
   $13 = x$
   $2y + 12 = 5y$
   $12 = 3y$
   $4 = y$
7. Opposite angles of a parallelogram are congruent. Find x and y.
   $3x - 17 = 2x + 24$
   $x = 41$
   $5y - 6 = y + 58$
   $4y = 64$
   $y = 16$
8. Yes;

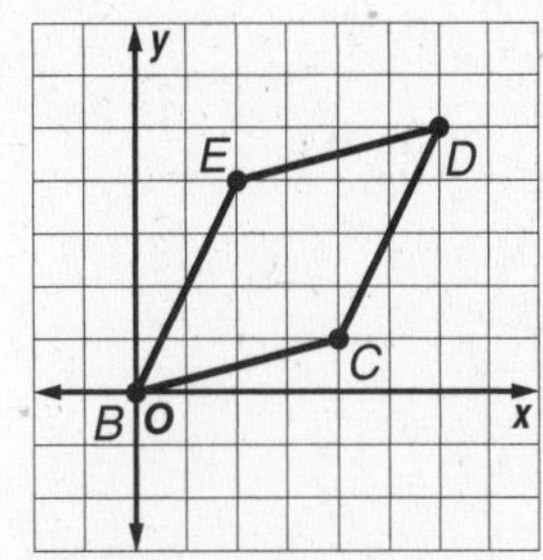

If the opposite sides of a quadrilateral are parallel, then it is a parallelogram.

$$\text{slope of } \overline{BC} = \frac{y_2 - y_1}{x_2 - x_1} = \frac{1 - 0}{4 - 0} = \frac{1}{4}$$

$$\text{slope of } \overline{DE} = \frac{4 - 5}{2 - 6} = \frac{1}{4}$$

$$\text{slope of } \overline{BE} = \frac{4 - 0}{2 - 0} = 2$$

$$\text{slope of } \overline{CD} = \frac{5 - 1}{6 - 4} = 2$$

Since opposite sides have the same slope, $\overline{BC} \parallel \overline{DE}$ and $\overline{BE} \parallel \overline{CD}$. Therefore, $BCDE$ is a parallelogram by definition.

9. Yes;

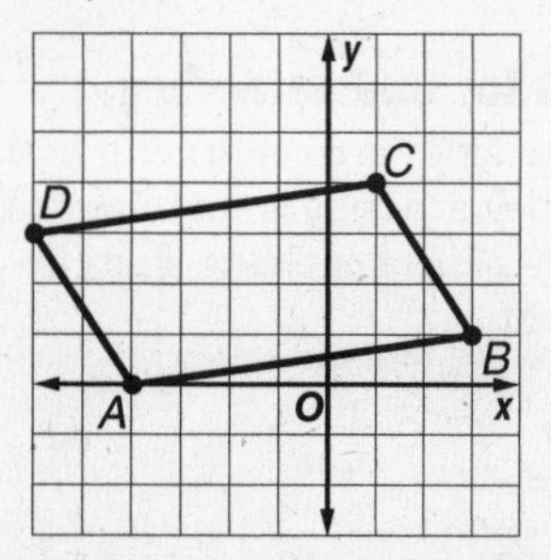

First use the Distance Formula to determine whether the opposite sides are congruent.

$$AD = \sqrt{(x_2 - x_1)^2 + (y_2 - y_1)^2}$$
$$= \sqrt{[-6 - (-4)]^2 + (3 - 0)^2}$$
$$= \sqrt{13}$$
$$BC = \sqrt{(1 - 3)^2 + (4 - 1)^2}$$
$$= \sqrt{13}$$

Since $AD = BC$, $\overline{AD} \cong \overline{BC}$.

Next, use the Slope Formula to determine whether $\overline{AD} \parallel \overline{BC}$.

$$\text{slope of } \overline{AD} = \frac{y_2 - y_1}{x_2 - x_1}$$
$$= \frac{3 - 0}{-6 - (-4)}$$
$$= -\frac{3}{2}$$
$$\text{slope of } \overline{BC} = \frac{4 - 1}{1 - 3}$$
$$= -\frac{3}{2}$$

$\overline{AD}$ and $\overline{BC}$ have the same slope, so they are parallel. Since one pair of opposite sides is congruent and parallel, $ABCD$ is a parallelogram.

10. No;

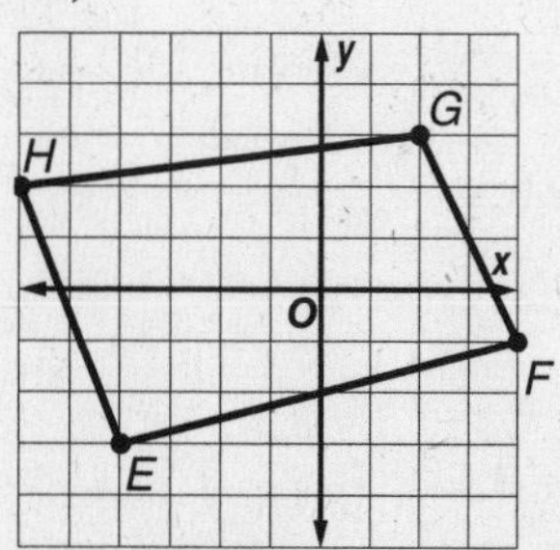

If the midpoints of the diagonals are the same, the diagonals bisect each other. If the diagonals of a quadrilateral bisect each other, then the quadrilateral is a parallelogram.

Find the midpoints of $\overline{EG}$ and $\overline{FH}$.

$$\overline{EG}: \left(\frac{x_1 + x_2}{2}, \frac{y_1 + y_2}{2}\right) = \left(\frac{-4 + 2}{2}, \frac{3 + (-3)}{2}\right)$$
$$= (-1, 0)$$
$$\overline{FH}: \left(\frac{x_1 + x_2}{2}, \frac{y_1 + y_2}{2}\right) = \left(\frac{4 + (-6)}{2}, \frac{-1 + 2}{2}\right)$$
$$= \left(-1, \frac{1}{2}\right)$$

The midpoints of $\overline{EG}$ and $\overline{FH}$ differ, so $EFGH$ is not a parallelogram.

11. Given: $\overline{PT} \cong \overline{TR}$
$\angle TSP \cong \angle TQR$
Prove: $PQRS$ is a parallelogram.

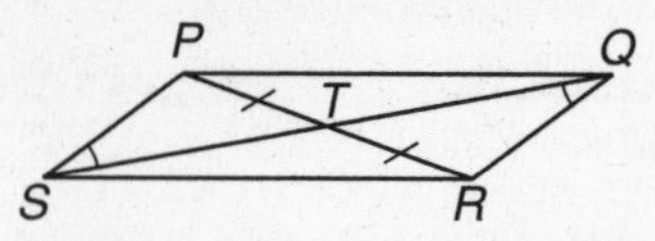

Proof:

Statements	Reasons
1. $\overline{PT} \cong \overline{TR}$, $\angle TSP \cong \angle TQR$	1. Given
2. $\angle PTS \cong \angle RTQ$	2. Vertical angles are congruent.
3. $\triangle PTS \cong \triangle RTQ$	3. AAS
4. $\overline{PS} \cong \overline{QR}$	4. CPCTC
5. $\overline{PS} \parallel \overline{QR}$	5. If alternate interior angles are congruent, lines are parallel.
6. $PQRS$ is a parallelogram.	6. If one pair of opposite sides is parallel and congruent, then the quadrilateral is a parallelogram.

12. If one pair of opposite sides is congruent and parallel, the quadrilateral is a parallelogram.

Pages 421–423 Practice and Apply

13. Yes; each pair of opposite angles is congruent.

14. Yes; the diagonals bisect each other.

15. Yes; opposite angles are congruent.

16. No; none of the tests for parallelograms are fulfilled.

17. Yes; one pair of opposite sides is parallel and congruent.

18. No; none of the tests for parallelograms are fulfilled.

19. Opposite sides of a parallelogram are congruent. Find x and y.

$$2x = 5x - 18$$
$$18 = 3x$$
$$6 = x$$
$$3y = 96 - y$$
$$4y = 96$$
$$y = 24$$

20. Diagonals of a parallelogram bisect each other. Find x and y.

$$2x + 3 = 5x$$
$$3 = 3x$$
$$1 = x$$
$$4y = 8y - 36$$
$$36 = 4y$$
$$9 = y$$

21. Opposite sides of a parallelogram are congruent. Find x and y.

$y + 2x = 4$
$y = -2x + 4$
$5y - 2x = 3y + 2x$
$2y = 4x$
$y = 2x$

Find x.

$2x = -2x + 4$
$4x = 4$
$x = 1$

So, $y = 2(1)$ or 2.

22. Since the opposite sides of a parallelogram are parallel, there are two pairs of alternate interior angles formed by the diagonal of the parallelogram. Find x and y.

$25x = 100$
$x = 4$
$10y = 40$
$y = 4$

23. Since the opposite sides of a parallelogram are parallel, there are two pairs of alternate interior angles formed by the diagonal of the parallelogram. Find y in terms of x.

$\frac{1}{2}y = x - 12$
$y = 2x - 24$

Opposite angles of a parallelogram are congruent. Find another equation for y in terms of x.

$3y - 4 = 4x - 8$
$3y = 4x - 4$
$y = \frac{4}{3}x - \frac{4}{3}$

Find x by setting the two expressions for y equal to each other

$\frac{4}{3}x - \frac{4}{3} = 2x - 24$
$\frac{68}{3} = \frac{2}{3}x$
$34 = x$

So, $y = 2(34) - 24$ or 44.

24. Diagonals of a parallelogram bisect each other. Find y in terms of x.

$3y + 4 = x$
$3y = x - 4$
$y = \frac{1}{3}x - \frac{4}{3}$
$4y = \frac{2}{3}x$
$y = \frac{1}{6}x$

Find x by setting the two expressions for y equal to each other.

$\frac{1}{6}x = \frac{1}{3}x - \frac{4}{3}$
$\frac{4}{3} = \frac{1}{6}x$
$8 = x$

So, $y = \frac{1}{6}(8)$ or $1\frac{1}{3}$.

25. Yes;

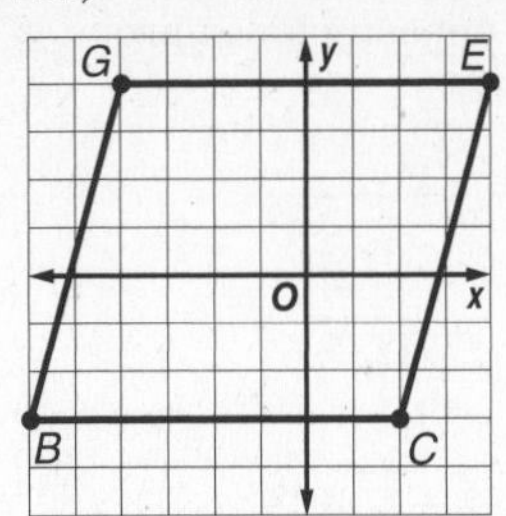

If the opposite sides of a quadrilateral are parallel, then it is a parallelogram.

slope of $\overline{BC} = \frac{y_2 - y_1}{x_2 - x_1}$
$= \frac{-3 - (-3)}{2 - (-6)}$
$= 0$

slope of $\overline{EG} = \frac{4 - 4}{-4 - 4}$
$= 0$

slope of $\overline{BG} = \frac{4 - (-3)}{-4 - (-6)}$
$= \frac{7}{2}$

slope of $\overline{CE} = \frac{4 - (-3)}{4 - 2}$
$= \frac{7}{2}$

Since opposite sides have the same slope, $\overline{BC} \parallel \overline{EG}$ and $\overline{BG} \parallel \overline{CE}$. Therefore, $BCEG$ is a parallelogram by definition.

26. No;

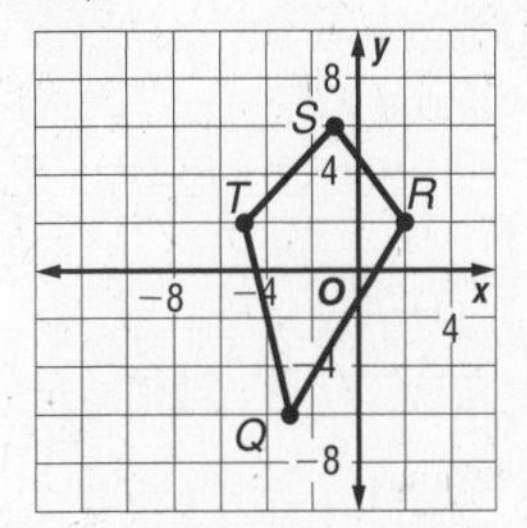

If the opposite sides of a quadrilateral are parallel, then it is a parallelogram.

slope of $\overline{QR} = \frac{y_2 - y_1}{x_2 - x_1}$
$= \frac{2 - (-6)}{2 - (-3)}$
$= \frac{8}{5}$

slope of $\overline{ST} = \frac{2 - 6}{-5 - (-1)}$
$= 1$

The opposite sides, $\overline{QR}$ and $\overline{ST}$, do not have the same slope, so they are not parallel. Therefore, $QRST$ is not a parallelogram.

27. Yes;

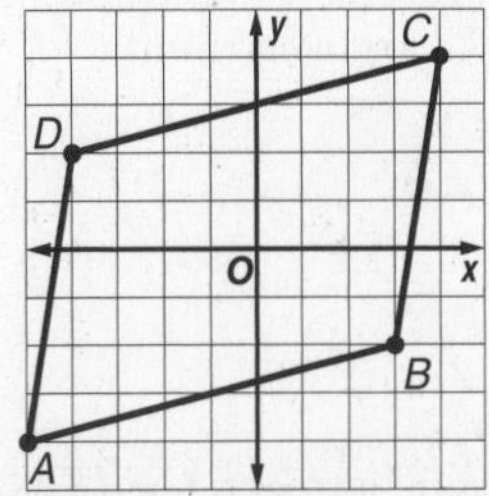

If both pairs of opposite sides of a quadrilateral are congruent, then the quadrilateral is a parallelogram. Use the Distance Formula to determine whether the opposite sides are congruent.

$AD = \sqrt{(x_2 - x_1)^2 + (y_2 - y_1)^2}$
$= \sqrt{[-4 - (-5)]^2 + [2 - (-4)]^2}$
$= \sqrt{37}$

$BC = \sqrt{(4 - 3)^2 + [4 - (-2)]^2}$
$= \sqrt{37}$

$AB = \sqrt{[3 - (-5)]^2 + [-2 - (-4)]^2}$
$= \sqrt{68}$

$CD = \sqrt{(-4 - 4)^2 + (2 - 4)^2}$
$= \sqrt{68}$

Since the measures of both pairs of opposite sides are equal, *ABCD* is a parallelogram.

28. Yes;

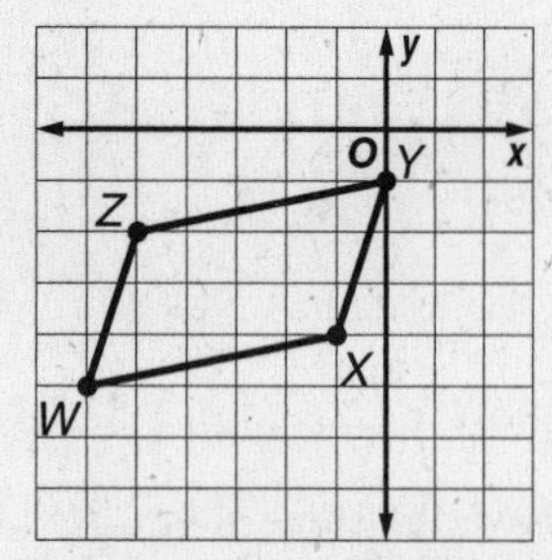

If the midpoints of the diagonals are the same, the diagonals bisect each other. If the diagonals of a quadrilateral bisect each other, then the quadrilateral is a parallelogram.

Find the midpoints of $\overline{WY}$ and $\overline{XZ}$.

$\overline{WY}$: $\left(\frac{x_1 + x_2}{2}, \frac{y_1 + y_2}{2}\right) = \left(\frac{-6 + 0}{2}, \frac{-5 + (-1)}{2}\right)$
$= (-3, -3)$

$\overline{XZ}$: $\left(\frac{x_1 + x_2}{2}, \frac{y_1 + y_2}{2}\right) = \left(\frac{-1 + (-5)}{2}, \frac{-4 + (-2)}{2}\right)$
$= (-3, -3)$

The midpoints of $\overline{WY}$ and $\overline{XZ}$ are the same, so *WXYZ* is a parallelogram.

29. No;

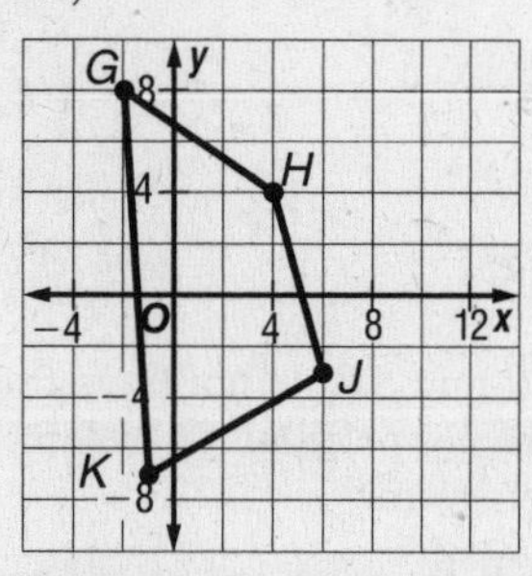

First use the Distance Formula to determine whether the opposite sides are congruent.

$GH = \sqrt{(x_2 - x_1)^2 + (y_2 - y_1)^2}$
$= \sqrt{[4 - (-2)]^2 + (4 - 8)^2}$
$= \sqrt{52}$

$JK = \sqrt{(-1 - 6)^2 + [-7 - (-3)]^2}$
$= \sqrt{65}$

Since $GH \neq JK$, $\overline{GH} \not\cong \overline{JK}$. Therefore, *GHJK* is not a parallelogram.

30. No;

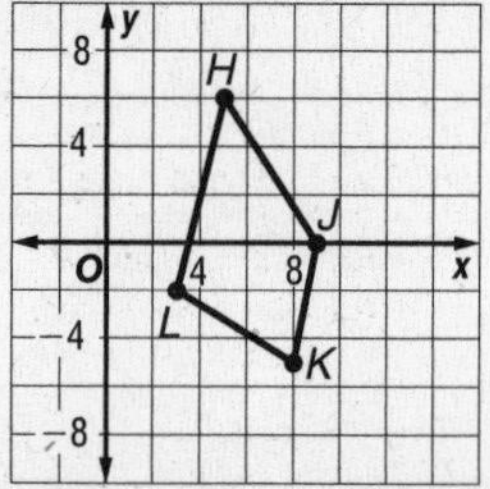

If both pairs of opposite sides of a quadrilateral are congruent, then the quadrilateral is a parallelogram. Use the Distance Formula to determine whether the opposite sides are congruent.

$HJ = \sqrt{(x_2 - x_1)^2 + (y_2 - y_1)^2}$
$= \sqrt{(9 - 5)^2 + (0 - 6)^2}$
$= \sqrt{52}$

$KL = \sqrt{(3 - 8)^2 + [-2 - (-5)]^2}$
$= \sqrt{34}$

Since $HJ \neq KL$, $\overline{HJ} \not\cong \overline{KL}$. Therefore, *HJKL* is not a parallelogram.

31. Yes;

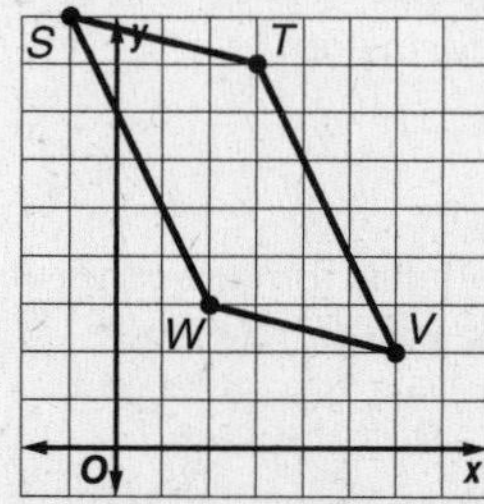

If the midpoints of the diagonals are the same, the diagonals bisect each other. If the diagonals of a quadrilateral bisect each other, then the quadrilateral is a parallelogram.

Find the midpoints of $\overline{SV}$ and $\overline{WT}$.

$\overline{SV}$: $\left(\frac{x_1 + x_2}{2}, \frac{y_1 + y_2}{2}\right) = \left(\frac{-1 + 6}{2}, \frac{9 + 2}{2}\right)$
$= \left(\frac{5}{2}, \frac{11}{2}\right)$

$\overline{WT}$: $\left(\frac{x_1 + x_2}{2}, \frac{y_1 + y_2}{2}\right) = \left(\frac{2 + 3}{2}, \frac{3 + 8}{2}\right)$
$= \left(\frac{5}{2}, \frac{11}{2}\right)$

The midpoints of $\overline{SV}$ and $\overline{WT}$ are the same, so *STVW* is a parallelogram.

32. Yes;

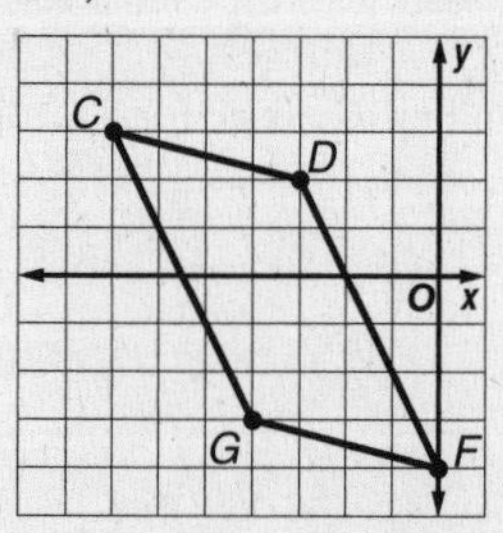

First use the Distance Formula to determine whether the opposite sides are congruent.

$CD = \sqrt{(x_2 - x_1)^2 + (y_2 - y_1)^2}$
$= \sqrt{[-3 - (-7)]^2 + (2 - 3)^2}$
$= \sqrt{17}$

$FG = \sqrt{(-4-0)^2 + [-3-(-4)]^2}$
$= \sqrt{17}$

Since $CD = FG$, $\overline{CD} \cong \overline{FG}$.

Next, use the Slope Formula to determine whether $\overline{CD} \parallel \overline{FG}$.

slope of $\overline{CD} = \frac{y_2 - y_1}{x_2 - x_1}$
$= \frac{2-3}{-3-(-7)}$
$= -\frac{1}{4}$

slope of $\overline{FG} = \frac{-3-(-4)}{-4-0}$
$= -\frac{1}{4}$

$\overline{CD}$ and $\overline{FG}$ have the same slope, so they are parallel. Since one pair of opposite sides is congruent and parallel, $CDFG$ is a parallelogram.

33.

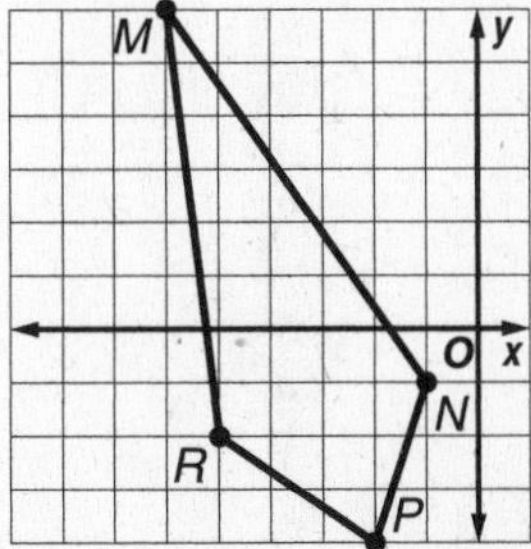

Sample answer: Hold N, P, and R fixed. Let M be $M(x, y)$. Find the slopes of $\overline{MN}$ and $\overline{MR}$ so that they equal those of $\overline{PR}$ and $\overline{NP}$, respectively.

slope of $\overline{MN}$ = slope of $\overline{PR}$

$$\frac{-1-y}{-1-x} = \frac{-2-(-4)}{-5-(-2)}$$
$$\frac{1+y}{1+x} = \frac{2}{-3}$$

If $x = -4$ and $y = 1$, then $\overline{MN} \parallel \overline{PR}$.

slope of $\overline{MR}$ = slope of $\overline{NP}$

$$\frac{-2-y}{-5-x} = \frac{-4-(-1)}{-2-(-1)}$$
$$\frac{2+y}{5+x} = \frac{3}{1}$$

If $x = -4$ and $y = 1$, then $\overline{MR} \parallel \overline{NP}$.

Move M to $(-4, 1)$ to make both pairs of opposite sides parallel. Then $MNPR$ is a parallelogram.

Perform the same process for the other vertices.

Move N: slope of $\overline{MN}$ = slope of $\overline{PR}$

$$\frac{y-6}{x-(-6)} = \frac{2}{-3}$$
$$\frac{y-6}{x+6} = \frac{-2}{3}$$

$x = -3$ and $y = 4$.

slope of $\overline{MR}$ = slope of $\overline{NP}$

$$\frac{-2-6}{-5-(-6)} = \frac{-4-y}{-2-x}$$
$$\frac{8}{-1} = \frac{4+y}{2+x}$$

Again, $x = -3$ and $y = 4$.
So, move N to $(-3, 4)$.

Move P: slope of $\overline{MN}$ = slope of $\overline{PR}$

$$\frac{-1-6}{-1-(-6)} = \frac{-2-y}{-5-x}$$
$$\frac{-7}{5} = \frac{2+y}{5+x}$$

$x = 0$ and $y = -9$.

slope of $\overline{MR}$ = slope of $\overline{NP}$

$$\frac{8}{-1} = \frac{y-(-1)}{x-(-1)}$$
$$\frac{-8}{1} = \frac{y+1}{x+1}$$

Again, $x = 0$ and $y = -9$.
So, move P to $(0, -9)$.

Move R: slope of $\overline{MN}$ = slope of $\overline{PR}$

$$\frac{-7}{5} = \frac{y-(-4)}{x-(-2)}$$
$$\frac{7}{-5} = \frac{y+4}{x+2}$$

$x = -7$ and $y = 3$.

slope of $\overline{MR}$ = slope of $\overline{NP}$

$$\frac{y-6}{x-(-6)} = \frac{3}{1}$$
$$\frac{y-6}{x+6} = \frac{-3}{-1}$$

Again, $x = -7$ and $y = 3$.
So, move R to $(-7, 3)$.

34.

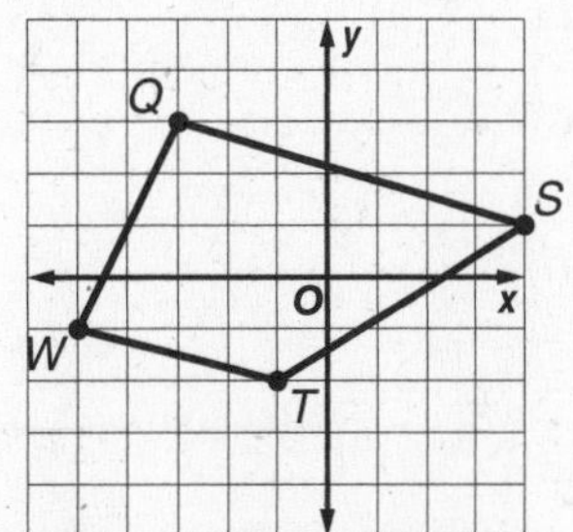

Hold S, T, and W fixed. Let Q be $Q(x, y)$. Find the slopes of $\overline{QS}$ and $\overline{QW}$ so that they equal those of $\overline{TW}$ and $\overline{ST}$, respectively.

slope of $\overline{QS}$ = slope of $\overline{TW}$

$$\frac{1-y}{4-x} = \frac{-1-(-2)}{-5-(-1)}$$
$$\frac{1-y}{4-x} = \frac{1}{-4}$$
$$\frac{1-y}{4-x} = \frac{-1}{4}$$

If $x = 0$ and $y = 2$, then $\overline{QS} \parallel \overline{TW}$.

slope of $\overline{QW}$ = slope of $\overline{ST}$

$$\frac{-1-y}{-5-x} = \frac{-2-1}{-1-4}$$
$$\frac{1+y}{5+x} = \frac{-3}{-5}$$
$$\frac{1+y}{5+x} = \frac{3}{5}$$

If $x = 0$ and $y = 2$, then $\overline{QW} \parallel \overline{ST}$. Move Q to $(0, 2)$ to make both pairs of opposite sides parallel. Then $QSTW$ is a parallelogram.

Perform the same process for the other vertices.

Move S: slope of $\overline{QS}$ = slope of $\overline{TW}$

$$\frac{y-3}{x-(-3)} = \frac{-1}{4}$$
$$\frac{y-3}{x+3} = \frac{-1}{4}$$

$x = 1$ and $y = 2$.

slope of $\overline{QW}$ = slope of $\overline{ST}$

$$\frac{-1-3}{-5-(-3)} = \frac{-2-y}{-1-x}$$
$$\frac{4}{2} = \frac{2+y}{1+x}$$

Again, $x = 1$ and $y = 2$.
So, move S to $(1, 2)$.

Move T: slope of $\overline{QS}$ = slope of $\overline{TW}$

$$\frac{1-3}{4-(-3)} = \frac{-1-y}{-5-x}$$
$$\frac{-2}{7} = \frac{1+y}{5+x}$$

$x = 2$ and $y = -3$.

slope of $\overline{QW}$ = slope of $\overline{ST}$

$$\frac{4}{2} = \frac{y-1}{x-4}$$
$$\frac{-4}{-2} = \frac{y-1}{x-4}$$

Again, $x = 2$ and $y = -3$.

So, move T to $(2, -3)$

Move W: slope of $\overline{QS}$ = slope of $\overline{TW}$

$$\frac{-2}{7} = \frac{y-(-2)}{x-(-1)}$$
$$\frac{2}{-7} = \frac{y+2}{x+1}$$

Again, $x = -8$ and $y = 0$.

slope of $\overline{QW}$ = slope of $\overline{ST}$

$$\frac{y-3}{x-(-3)} = \frac{-2-1}{-1-4}$$
$$\frac{y-3}{x+3} = \frac{-3}{-5}$$

$x = -8$ and $y = 0$.

So, move W to $(-8, 0)$.

35.

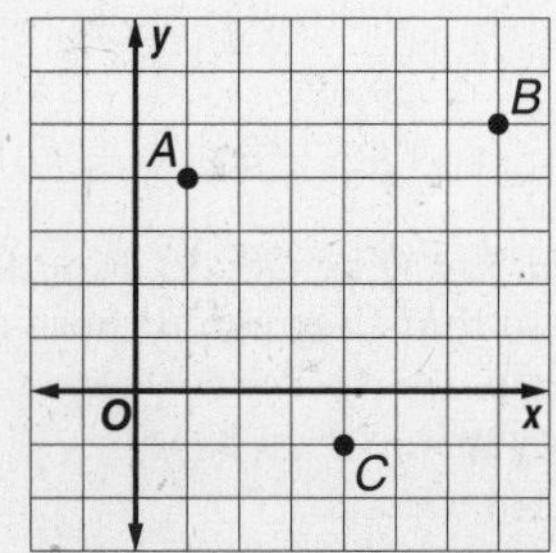

The fourth vertex can have one of three possible positions to complete the parallelogram.

Let $D(x, y)$ be the fourth vertex.

Find the slopes of $\overline{BD}$ and $\overline{CD}$ so that they equal those of $\overline{AC}$ and $\overline{AB}$, respectively.

slope of $\overline{BD}$ = slope of $\overline{AC}$

$$\frac{y-5}{x-7} = \frac{-1-4}{4-1}$$
$$\frac{y-5}{x-7} = -\frac{5}{3}$$

For $(10, 0)$ and $(4, 10)$, $\overline{BD} \parallel \overline{AC}$.

$$\text{slope of } \overline{AB} = \frac{5-4}{7-1} = \frac{1}{6}$$

Suppose D is $(10, 0)$.

$$\text{slope of } \overline{CD} = \frac{0-(-1)}{10-4} = \frac{1}{6}$$

So $\overline{AB} \parallel \overline{CD}$.

$$\text{slope of } \overline{BC} = \frac{-1-5}{4-7} = 2$$

Suppose D is $(4, 10)$.

$$\text{slope of } \overline{AD} = \frac{10-4}{4-1} = 2$$

So $\overline{AD} \parallel \overline{BC}$.

slope of $\overline{CD}$ = slope of $\overline{AB}$

$$\frac{y-(-1)}{x-4} = \frac{5-4}{7-1}$$
$$\frac{y+1}{x-4} = \frac{1}{6}$$

For $(10, 0)$ and $(-2, -2)$, $\overline{CD} \parallel \overline{AB}$.

slope of $\overline{BC} = 2$

Suppose D is $(-2, -2)$.

$$\text{slope of } \overline{DA} = \frac{4-(-2)}{1-(-2)} = 2$$

So $\overline{DA} \parallel \overline{BC}$.

So, $(-2, -2)$, $(4, 10)$, and $(10, 0)$ are the possibilities for the fourth vertex. Any of these values results in both pairs of opposite sides being parallel, and thus, the four points form a parallelogram.

36.

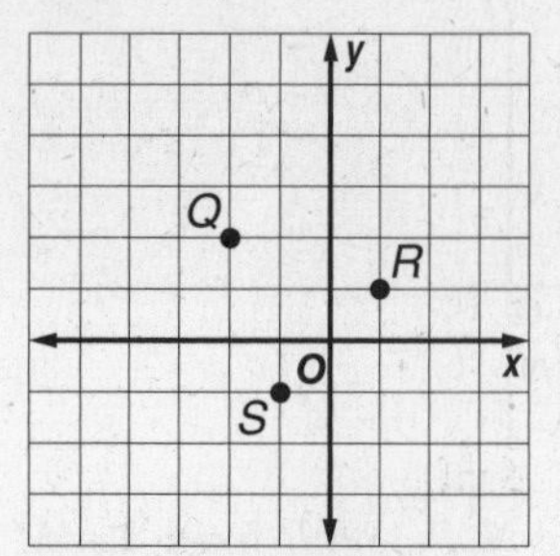

The fourth vertex can have one of three possible positions to complete the parallelogram.

Let $T(x, y)$ be the fourth vertex.

Find the slopes of $\overline{ST}$ and $\overline{RT}$ so that they equal those of $\overline{QR}$ and $\overline{QS}$, respectively.

slope of $\overline{ST}$ = slope of $\overline{QR}$

$$\frac{y-(-1)}{x-(-1)} = \frac{1-2}{1-(-2)}$$
$$\frac{y+1}{x+1} = \frac{-1}{3}$$

For $(2, -2)$ and $(-4, 0)$, $\overline{ST} \parallel \overline{QR}$.

$$\text{slope of } \overline{QS} = \frac{-1-2}{-1-(-2)} = -3$$

Suppose T is $(2, -2)$.

$$\text{slope of } \overline{RT} = \frac{-2-1}{2-1} = -3$$

So $\overline{QS} \parallel \overline{RT}$.

$$\text{slope of } \overline{RS} = \frac{-1-1}{-1-1} = 1$$

Suppose T is $(-4, 0)$.

$$\text{slope of } \overline{QT} = \frac{0-2}{-4-(-2)} = 1$$

So $\overline{QT} \parallel \overline{RS}$.

slope of $\overline{RT}$ = slope of $\overline{QS}$

$$\frac{y-1}{x-1} = \frac{-1-2}{-1-(-2)}$$
$$\frac{y-1}{x-1} = \frac{-3}{1}$$

For (2, −2) and (0, 4), $\overline{RT} \parallel \overline{QS}$.

slope of $\overline{SR} = \frac{1-(-1)}{1-(-1)}$
$= 1$

Suppose T is (0, 4).

slope of $\overline{QT} = \frac{4-2}{0-(-2)}$
$= 1$

So $\overline{SR} \parallel \overline{QT}$.

So, (2, −2), (−4, 0), and (0, 4) are the possibilities for the fourth vertex. Any of these values results in both pairs of opposite sides being parallel, and thus, the four points form a parallelogram.

37. *JKLM* is a parallelogram because $\overline{KM}$ and $\overline{JL}$ are diagonals that bisect each other.

38. If both pairs of opposite sides are parallel and congruent, then the watchbox is a parallelogram.

39. Given: $\overline{AD} \cong \overline{BC}$
$\overline{AB} \cong \overline{DC}$

Prove: *ABCD* is a parallelogram.

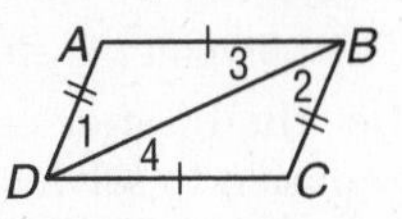

Proof:

Statements	Reasons
1. $\overline{AD} \cong \overline{BC}, \overline{AB} \cong \overline{DC}$	1. Given
2. Draw $\overline{DB}$.	2. Two points determine a line.
3. $\overline{DB} \cong \overline{DB}$	3. Reflexive Property
4. $\triangle ABD \cong \triangle CDB$	4. SSS
5. $\angle 1 \cong \angle 2, \angle 3 \cong \angle 4$	5. CPCTC
6. $\overline{AD} \parallel \overline{BC}, \overline{AB} \parallel \overline{DC}$	6. If alternate interior angles are congruent, lines are parallel.
7. *ABCD* is a parallelogram.	7. Definition of parallelogram

40. Given: $\overline{AE} \cong \overline{EC}, \overline{DE} \cong \overline{EB}$

Prove: *ABCD* is a parallelogram.

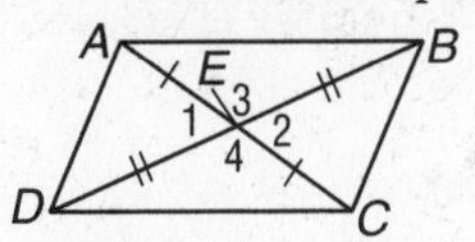

Proof:

Statements	Reasons
1. $\overline{AE} \cong \overline{EC}, \overline{DE} \cong \overline{EB}$	1. Given
2. $\angle 1 \cong \angle 2$ $\angle 3 \cong \angle 4$	2. Vertical ∠s are ≅.
3. $\triangle ABE \cong \triangle CDE$ $\triangle ADE \cong \triangle CBE$	3. SAS
4. $\overline{AB} \cong \overline{DC}$ $\overline{AD} \cong \overline{BC}$	4. CPCTC
5. *ABCD* is a parallelogram.	5. Definition of parallelogram

41. Given: $\overline{AB} \cong \overline{DC}$
$\overline{AB} \parallel \overline{DC}$

Prove: *ABCD* is a parallelogram.

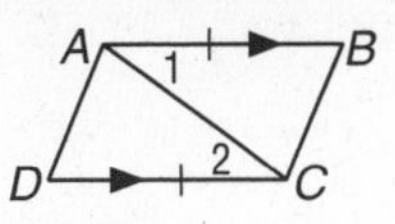

Proof:

Statements	Reasons
1. $\overline{AB} \cong \overline{DC}, \overline{AB} \parallel \overline{DC}$	1. Given
2. Draw $\overline{AC}$	2. Two points determine a line.
3. $\angle 1 \cong \angle 2$	3. If two lines are parallel, then alternate interior angles are congruent.
4. $\overline{AC} \cong \overline{AC}$	4. Reflexive Property
5. $\triangle ABC \cong \triangle CDA$	5. SAS
6. $\overline{AD} \cong \overline{BC}$	6. CPCTC
7. *ABCD* is a parallelogram.	7. If both pairs of opposite sides are congruent, then the quadrilateral is a parallelogram.

42. This theorem is not true. *ABCD* is a parallelogram with diagonal $\overline{BD}$, $\angle ABD \not\cong \angle CBD$.

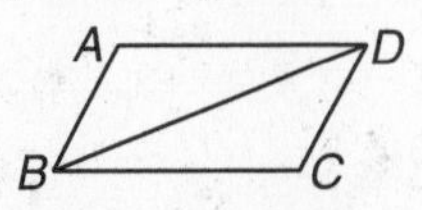

43. Given: *ABCDEF* is a regular hexagon.

Prove: *FDCA* is a parallelogram.

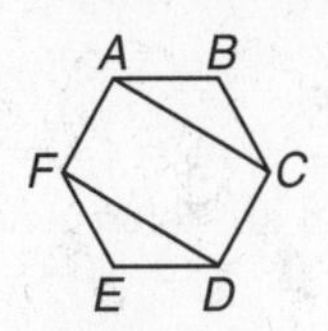

Proof:

Statements	Reasons
1. *ABCDEF* is a regular hexagon.	1. Given
2. $\overline{AB} \cong \overline{DE}, \overline{BC} \cong \overline{EF}$ $\angle E \cong \angle B, \overline{FA} \cong \overline{CD}$	2. Definition of a regular hexagon
3. $\triangle ABC \cong \triangle DEF$	3. SAS
4. $\overline{AC} \cong \overline{DF}$	4. CPCTC
5. *FDCA* is a ▱.	5. If both pairs of opposite sides are congruent, then the quadrilateral is a parallelogram.

44. Sample answer: The roofs of some covered bridges are parallelograms. The opposite sides are congruent and parallel. Answers should include the following.

- We need to know the length of the sides, or the measures of the angles formed.
- Sample answer: windows or tiles

45. B;

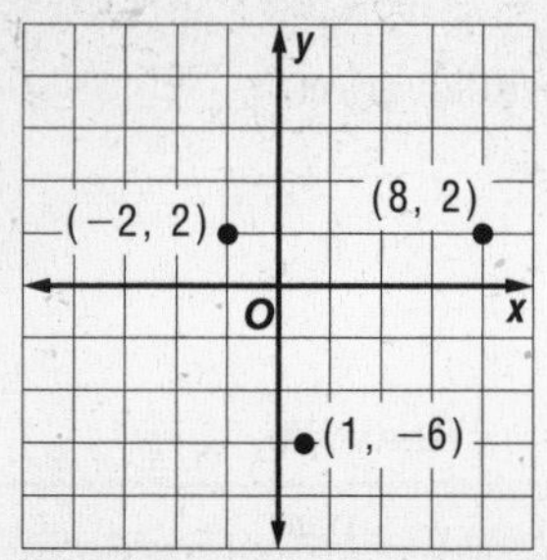

By plotting choices A, C, and D on the graph, it is obvious that they cannot be the fourth vertex. Choice B is the only possibility.
(−2, 2) and (8, 2) have the same y-component. These are the endpoints of a horizontal segment 10 units long.
(11, −6) and (1, −6) have the same y-component. These are also the endpoints of a horizontal segment 10 units long.
These two sides of the quadrilateral are congruent and parallel, so the quadrilateral is a parallelogram.

46. C; use the Distance Formula to find the distance between X and Y.

$$XY = \sqrt{(x_2 - x_1)^2 + (y_2 - y_1)^2}$$
$$= \sqrt{(-3 - 5)^2 + (-4 - 7)^2}$$
$$= \sqrt{185}$$

Page 423 Maintain Your Skills

47. The diagonals of a parallelogram bisect each other, so $ML = LQ$.
Find w.
$ML = LQ$
$w = 12$

48. Opposite sides of a parallelogram are congruent, so their measures are equal.
Find x.
$NQ = MR$
$3x + 2 = 4x - 2$
$4 = x$

49. Opposite sides of a parallelogram are congruent, so their measures are equal.
Find x.
$NQ = MR$
$3x + 2 = 4x - 2$
$4 = x$
So, NQ is $3(4) + 2$ or 14 units.

50. Opposite sides of a parallelogram are congruent, so their measures are equal.
Find y.
$QR = MN$
$3y = 2y + 5$
$y = 5$
So, QR is 3(5) or 15 units.

51. Use the Interior Angle Sum Theorem to write an equation to solve for n, the number of sides.
$S = 180(n - 2)$
$(135)n = 180(n - 2)$
$135n = 180n - 360$
$0 = 45n - 360$
$360 = 45n$
$8 = n$
The polygon has 8 sides.

52. Use the Interior Angle Sum Theorem to write an equation to solve for n, the number of sides.
$S = 180(n - 2)$
$(144)n = 180(n - 2)$
$144n = 180n - 360$
$0 = 36n - 360$
$360 = 36n$
$10 = n$
The polygon has 10 sides.

53. Use the Interior Angle Sum Theorem to write an equation to solve for n, the number of sides.
$S = 180(n - 2)$
$(168)n = 180(n - 2)$
$168n = 180n - 360$
$0 = 12n - 360$
$360 = 12n$
$30 = n$
The polygon has 30 sides.

54. Use the Interior Angle Sum Theorem to write an equation to solve for n, the number of sides.
$S = 180(n - 2)$
$(162)n = 180(n - 2)$
$162n = 180n - 360$
$0 = 18n - 360$
$360 = 18n$
$20 = n$
The polygon has 20 sides.

55. Use the Interior Angle Sum Theorem to write an equation to solve for n, the number of sides.
$S = 180(n - 2)$
$(175)n = 180(n - 2)$
$175n = 180n - 360$
$0 = 5n - 360$
$360 = 5n$
$72 = n$
The polygon has 72 sides.

56. Use the Interior Angle Sum Theorem to write an equation to solve for n, the number of sides.
$S = 180(n - 2)$
$(175.5)n = 180(n - 2)$
$175.5n = 180n - 360$
$0 = 4.5n - 360$
$360 = 4.5n$
$80 = n$
The polygon has 80 sides.

57. The legs of this right triangle have equal lengths, so it is a 45°-45°-90° triangle. The hypotenuse of a 45°-45°-90° triangle is $\sqrt{2}$ times the length of the legs. Therefore, $x = 45$ and $y = 12\sqrt{2}$.

58. The hypotenuse of this right triangle is twice the length of one of its legs, so it is a 30°-60°-90° triangle. The longer leg is $\sqrt{3}$ times the length of the shorter leg. Therefore, $x = 10\sqrt{3}$ and $y = 30$.

59. Two angles are given, 60° and 90°, so this is a 30°-60°-90° triangle. The shorter leg is half the length of the hypotenuse, and the longer leg is $\sqrt{3}$ times the length of the shorter leg. Therefore, $x = 16\sqrt{3}$ and $y = 16$.

60. Determine the slopes of $\overline{AB}$ and $\overline{BC}$.

$$\text{slope of } \overline{AB} = \frac{y_2 - y_1}{x_2 - x_1} = \frac{3 - 5}{6 - 2} = -\frac{1}{2}$$

$$\text{slope of } \overline{BC} = \frac{7 - 3}{8 - 6} = 2$$

The product of the slopes of $\overline{AB}$ and $\overline{BC}$ is -1, so $\overline{AB} \perp \overline{BC}$.

61. Determine the slopes of $\overline{AB}$ and $\overline{BC}$.

$$\text{slope of } \overline{AB} = \frac{y_2 - y_1}{x_2 - x_1} = \frac{7 - 2}{0 - (-1)} = 5$$

$$\text{slope of } \overline{BC} = \frac{1 - 7}{4 - 0} = -\frac{3}{2}$$

The product of the slopes of $\overline{AB}$ and $\overline{BC}$ is not -1, so $\overline{AB}$ is not perpendicular to $\overline{BC}$.

62. Determine the slopes of $\overline{AB}$ and $\overline{BC}$.

$$\text{slope of } \overline{AB} = \frac{y_2 - y_1}{x_2 - x_1} = \frac{7 - 4}{5 - 0} = \frac{3}{5}$$

$$\text{slope of } \overline{BC} = \frac{3 - 7}{8 - 5} = -\frac{4}{3}$$

The product of the slopes of $\overline{AB}$ and $\overline{BC}$ is not -1, so $\overline{AB}$ is not perpendicular to $\overline{BC}$.

63. Determine the slopes of $\overline{AB}$ and $\overline{BC}$.

$$\text{slope of } \overline{AB} = \frac{y_2 - y_1}{x_2 - x_1} = \frac{-3 - (-5)}{1 - (-2)} = \frac{2}{3}$$

$$\text{slope of } \overline{BC} = \frac{0 - (-3)}{-1 - 1} = -\frac{3}{2}$$

The product of the slopes of $\overline{AB}$ and $\overline{BC}$ is -1, so $\overline{AB} \perp \overline{BC}$.

Page 423 Practice Quiz 1

1. Use the Interior Angle Sum Theorem to write an equation to solve for n, the number of sides.

$$S = 180(n - 2)$$
$$\left(147\tfrac{3}{11}\right)n = 180(n - 2)$$
$$\frac{1620}{11}n = 180n - 360$$
$$0 = \frac{360}{11}n - 360$$
$$360 = \frac{360}{11}n$$
$$11 = n$$

The polygon has 11 sides.

2. Opposite sides of a parallelogram are congruent, so their measures are equal. Find x.

$$WZ = XY$$
$$x^2 = 42 + x$$
$$0 = x^2 - x - 42$$
$$0 = (x + 6)(x - 7)$$
$$0 = x + 6 \text{ or } 0 = x - 7$$
$$-6 = x \qquad 7 = x$$

$x = -6$ or 7, so $x^2 = 36$ or 49.
So, WZ is 36 or 49.

3. $\angle YXZ \cong \angle XZW$ because they are alternate interior angles. So, $m\angle YXZ$ is 54 and $m\angle WXY$ is $54 + 60$ or 114. $\angle WXY$ and $\angle XYZ$ are supplementary, so $m\angle XYZ$ is $180 - 114$ or 66.

4. Opposite angles of a parallelogram are congruent. Find x and y.

$$5x - 19 = 3x + 9$$
$$2x = 28$$
$$x = 14$$
$$6y - 57 = 3y + 36$$
$$3y = 93$$
$$y = 31$$

5. Opposite sides of a parallelogram are congruent. Find x and y.

$$2x - 4 = x + 4$$
$$x = 8$$
$$4y - 8 = 3y - 2$$
$$y = 6$$

8-4 Rectangles

Pages 427–428 Check for Understanding

1. If consecutive sides are perpendicular or diagonals are congruent, then the parallelogram is a rectangle.

2. Sample answer: sometimes

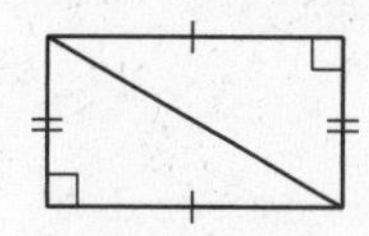

Rectangle

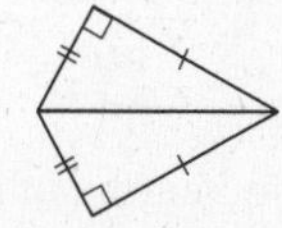

Not Rectangle

3. McKenna is correct. Consuelo's definition is correct if one pair of opposite sides is parallel and congruent.

4. The diagonals of a rectangle are congruent, so $\overline{AC} \cong \overline{BD}$.

$$\overline{AC} \cong \overline{BD}$$
$$AC = BD$$
$$30 - x = 4x - 60$$
$$90 = 5x$$
$$18 = x$$

5. The diagonals of a rectangle bisect each other and are congruent, so $NP = \frac{1}{2}NR$ and $\overline{MP} \cong \overline{NP}$.

$$NP = \tfrac{1}{2}NR$$
$$2x - 30 = \tfrac{1}{2}(2x + 10)$$
$$x = 35$$

So, $MP = 2(35) - 30$ or 40.

6. $\angle QRT \cong \angle RTS$ because they are alternate interior angles.

$$\angle QRT \cong \angle RTS$$
$$m\angle QRT = m\angle RTS$$
$$x^2 + 1 = 3x + 11$$
$$x^2 - 3x - 10 = 0$$
$$(x - 5)(x + 2) = 0$$
$$x - 5 = 0 \text{ or } x + 2 = 0$$
$$x = 5 \qquad x = -2$$

So, $x = 5$ or -2.

7. $\angle QRT$ and $\angle SRT$ are complementary.

$$m\angle QRT + m\angle SRT = 90$$

Use the values of x found in Exercise 6.

$$[(-2)^2 + 1] + m\angle SRT = 90$$
$$5 + m\angle SRT = 90$$
$$m\angle SRT = 85$$

or

$$5^2 + 1 + m\angle SRT = 90$$
$$26 + m\angle SRT = 90$$
$$m\angle SRT = 64$$

$\angle SRT \cong \angle RSQ$ and the sum of the measures of the interior angles of a triangle is 180. Find $m\angle RPS$.

$$m\angle RPS + m\angle SRT + m\angle RSQ = 180$$
$$m\angle RPS + 2m\angle SRT = 180$$
$$m\angle RPS + 2(64) = 180$$
$$m\angle RPS = 52$$

or

$$m\angle RPS + 2(85) = 180$$
$$m\angle RPS = 10$$

8. The diagonals of a rectangle are congruent. Use the Distance Formula to determine whether the diagonals of quadrilateral $EFGH$ are congruent.

$$EG = \sqrt{(x_2 - x_1)^2 + (y_2 - y_1)^2}$$
$$= \sqrt{[2 - (-4)]^2 + [3 - (-3)]^2}$$
$$= \sqrt{72}$$
$$HF = \sqrt{(x_2 - x_1)^2 + (y_2 - y_1)^2}$$
$$= \sqrt{[3 - (-5)]^2 + (-1 - 1)^2}$$
$$= \sqrt{68}$$

The lengths of the diagonals are not equal, so $EFGH$ is not a rectangle.

9. The framer can make sure that the angles measure 90 or that the diagonals are congruent.

Pages 428–430 Practice and Apply

10. The diagonals of a rectangle bisect each other and are congruent, so $\overline{NQ} \cong \overline{QM}$.

$$\overline{NQ} \cong \overline{QM}$$
$$NQ = QM$$
$$5x - 3 = 4x + 6$$
$$x = 9$$

So, $NQ = 5(9) - 3$ or 42. NK is twice NQ or 84.

11. The diagonals of a rectangle bisect each other, so $\overline{NQ} \cong \overline{QK}$.

$$\overline{NQ} \cong \overline{QK}$$
$$NQ = QK$$
$$2x + 3 = 5x - 9$$
$$12 = 3x$$
$$4 = x$$

So, $NQ = 2(4) + 3$ or 11. $\overline{JQ} \cong \overline{NQ}$, so $JQ = 11$.

12. Opposite sides of a rectangle are congruent, so $\overline{JK} \cong \overline{NM}$.

$$\overline{JK} \cong \overline{NM}$$
$$JK = NM$$
$$x^2 + 1 = 8x - 14$$
$$x^2 - 8x + 15 = 0$$
$$(x - 3)(x - 5) = 0$$
$$x - 3 = 0 \text{ or } x - 5 = 0$$
$$x = 3 \qquad x = 5$$

So, $x = 3$ or $x = 5$. $JK = 3^2 + 1$ or 10 or $JK = 5^2 + 1$ or 26.

13. The sum of the measures of $\angle NJM$ and $\angle KJM$ is 90. Find x.

$$m\angle NJM + m\angle KJM = 90$$
$$2x - 3 + x + 5 = 90$$
$$3x = 88$$
$$x = 29\tfrac{1}{3}$$

14. $\triangle KMN$ is a right triangle, so the sum of the measures of $\angle NKM$ and $\angle KNM$ is 90. Find x.

$$m\angle NKM + m\angle KNM = 90$$
$$x^2 + 4 + x + 30 = 90$$
$$x^2 + x - 56 = 0$$
$$(x + 8)(x - 7) = 0$$
$$x + 8 = 0 \text{ or } x - 7 = 0$$
$$x = -8 \qquad x = 7$$

So, $x = -8$ or $x = 7$. $m\angle KNM = -8 + 30$ or 22 or $m\angle KNM = 7 + 30$ or 37. $\angle KNM \cong \angle JKN$ because they are alternate interior angles, so $m\angle JKN =$ 22 or 37.

15. The sum of the measures of $\angle JKN$ and $\angle NKM$ is 90. Find x.

$$m\angle JKN + m\angle NKM = 90$$
$$2x^2 + 2 + 14x = 90$$
$$2x^2 + 14x - 88 = 0$$
$$x^2 + 7x - 44 = 0$$
$$(x - 4)(x + 11) = 0$$
$$x - 4 = 0 \text{ or } x + 11 = 0$$
$$x = 4 \qquad x = -11$$

Since $m\angle NKM = 14x$, and the measure of an angle must be positive, discard $x = -11$. So, $x = 4$.

16. $m\angle 1 = 30$.

17. $\angle 2$ is complementary to $\angle 1$ because consecutive sides of a rectangle are perpendicular. So, $m\angle 2 = 90 - 30$ or 60.

18. The diagonals of a rectangle are congruent and bisect each other. So, the triangles formed by the diagonals of a rectangle are isosceles. Therefore, $\angle 2 \cong \angle 3$. From Exercise 17, $m\angle 3 = 60$.

19. $\angle 4$ is complementary to $\angle 3$ because consecutive sides of a rectangle are perpendicular.

$$m\angle 4 + m\angle 3 = 90$$
$$m\angle 4 + 60 = 90$$
$$m\angle 4 = 30$$

20. $\angle 5 \cong \angle 1$ because they are alternate interior angles. So, $m\angle 5 = 30$.

21. $\triangle WXZ$ is a right triangle, so the sum of the measures of $\angle 1$ and $\angle 6$ is 90. Therefore, $m\angle 6 = 90 - 30$ or 60.

22. $\angle 7 \cong \angle 3$ because they are alternate interior angles. From Exercise 18, $m\angle 7 = 60$.

23. $\angle 8$ is complementary to $\angle 7$.

$$m\angle 8 + m\angle 7 = 90$$
$$m\angle 8 + 60 = 90$$
$$m\angle 8 = 30$$

24. The sum of the interior angles of a triangle is 180.

$$m\angle 9 + m\angle 7 + m\angle 6 = 180$$
$$m\angle 9 + 60 + 60 = 180$$
$$m\angle 9 = 60$$

25. The contractor can measure the opposite sides and the diagonals to make sure they are congruent.

26. Use the Pythagorean Theorem to find the measure of the diagonal of the television screen.

$$c = \sqrt{a^2 + b^2}$$
$$= \sqrt{21^2 + 36^2} = \sqrt{441 + 1296}$$
$$= \sqrt{1737}$$
$$\approx 42$$

The measure of the diagonal is about 42 in.

27. Opposite sides of a rectangle are parallel. Find the slopes of $\overline{DH}$ and $\overline{FG}$.

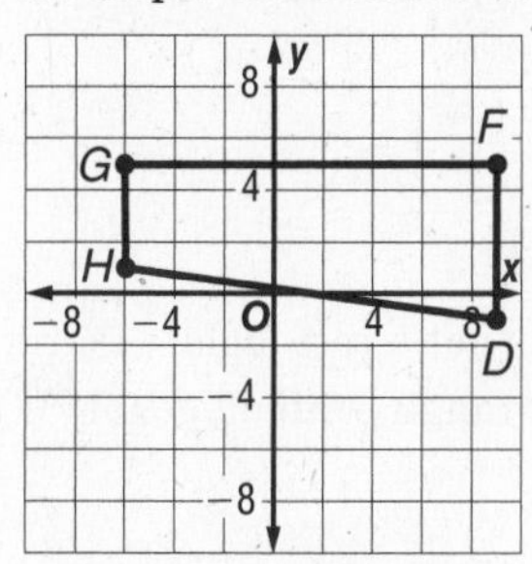

$$\text{slope of } \overline{DH} = \frac{y_2 - y_1}{x_2 - x_1}$$
$$= \frac{1 - (-1)}{-6 - 9}$$
$$= -\frac{2}{15}$$

$$\text{slope of } \overline{FG} = \frac{y_2 - y_1}{x_2 - x_1}$$
$$= \frac{5 - 5}{-6 - 9}$$
$$= 0$$

The slopes are not equal. Therefore, $\overline{DH}$ and $\overline{FG}$ are not parallel. So, $DFGH$ is not a rectangle.

28. The diagonals of a rectangle are congruent. Use the Distance Formula to determine whether the diagonals of quadrilateral $DFGH$ are congruent.

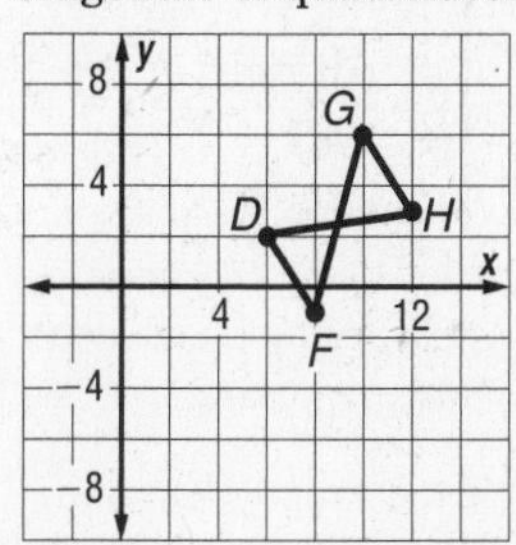

To determine if $DFGH$ is a rectangle, the points must be connected in the order given. When the points are plotted and connected, it is clear that $DFGH$ is not a quadrilateral. So $DFGH$ is not a rectangle.

29. If the opposite sides of a quadrilateral are parallel and the diagonals of the quadrilateral are congruent, then the quadrilateral is a rectangle. Find the slopes of $\overline{DH}$, $\overline{FG}$, $\overline{GH}$, and $\overline{DF}$.

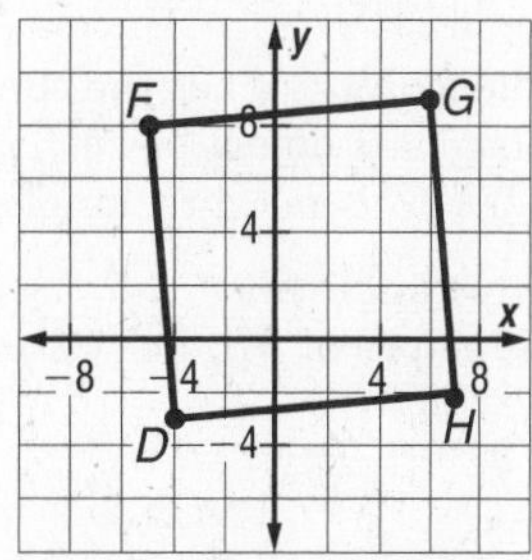

$$\text{slope of } \overline{DH} = \frac{y_2 - y_1}{x_2 - x_1}$$
$$= \frac{-2 - (-3)}{7 - (-4)}$$
$$= \frac{1}{11}$$

$$\text{slope of } \overline{FG} = \frac{9 - 8}{6 - (-5)}$$
$$= \frac{1}{11}$$

$$\text{slope of } \overline{GH} = \frac{-2 - 9}{7 - 6}$$
$$= -11$$

$$\text{slope of } \overline{DF} = \frac{8 - (-3)}{-5 - (-4)}$$
$$= -11$$

So, $\overline{DH} \parallel \overline{FG}$ and $\overline{GH} \parallel \overline{DF}$. Use the Distance Formula to determine whether the diagonals of quadrilateral $DFGH$ are congruent.

$$DG = \sqrt{(x_2 - x_1)^2 + (y_2 - y_1)^2}$$
$$= \sqrt{[6 - (-4)]^2 + [9 - (-3)]^2}$$
$$= \sqrt{244}$$

$$FH = \sqrt{(x_2 - x_1)^2 + (y_2 - y_1)^2}$$
$$= \sqrt{[7 - (-5)]^2 + (-2 - 8)^2}$$
$$= \sqrt{244}$$

So, $\overline{DG} \cong \overline{FH}$, $\overline{DH} \parallel \overline{FG}$, and $\overline{GH} \parallel \overline{DF}$. Therefore, $DFGH$ is a rectangle.

30. Use the Distance Formula to find WY and XZ.

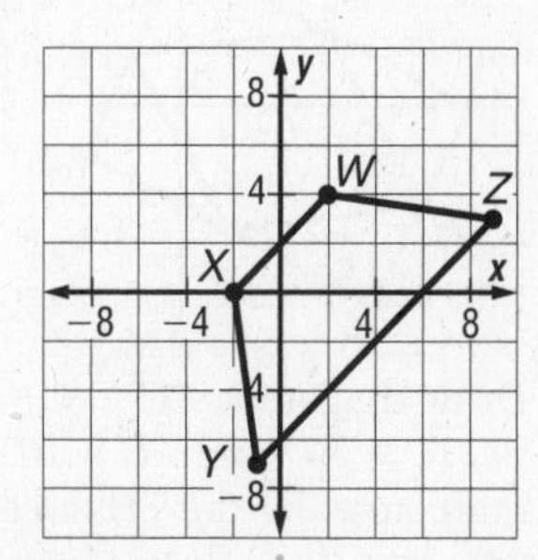

$$WY = \sqrt{(x_2 - x_1)^2 + (y_2 - y_1)^2}$$
$$= \sqrt{(-1 - 2)^2 + (-7 - 4)^2}$$
$$= \sqrt{130}$$

$$XZ = \sqrt{(x_2 - x_1)^2 + (y_2 - y_1)^2}$$
$$= \sqrt{(3 - 0)^2 + [9 - (-2)]^2}$$
$$= \sqrt{130}$$

31. See the figure in Exercise 30. Find the coordinates of the midpoints of $\overline{WY}$ and $\overline{XZ}$.

$\overline{WY}$: $\left(\frac{x_1+x_2}{2}, \frac{y_1+y_2}{2}\right) = \left(\frac{2+(-1)}{2}, \frac{4+(-7)}{2}\right)$

$= \left(\frac{1}{2}, -\frac{3}{2}\right)$

$\overline{XZ}$: $\left(\frac{x_1+x_2}{2}, \frac{y_1+y_2}{2}\right) = \left(\frac{-2+9}{2}, \frac{0+3}{2}\right)$

$= \left(\frac{7}{2}, \frac{3}{2}\right)$

32. The midpoints of the diagonals are not the same (Exercise 31), so the diagonals do not bisect each other. Therefore, $WXYZ$ is not a rectangle.

33. Consecutive sides of a rectangle are perpendicular. Find the slopes of $\overline{AB}$, $\overline{BC}$, $\overline{CD}$, and $\overline{DA}$.

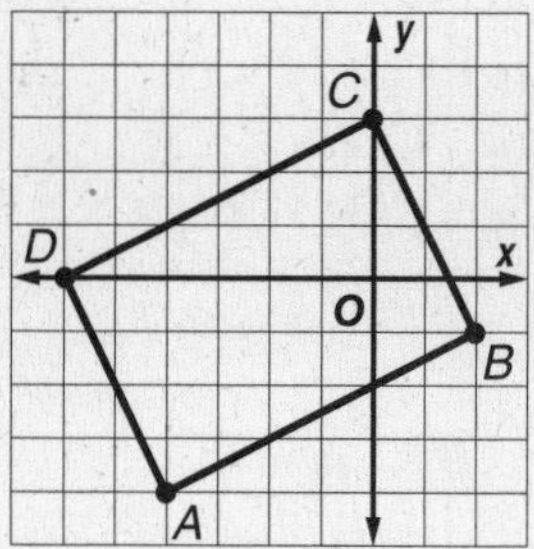

slope of $\overline{AB} = \frac{y_2 - y_1}{x_2 - x_1}$

$= \frac{-1-(-4)}{2-(-4)}$

$= \frac{1}{2}$

slope of $\overline{BC} = \frac{3-(-1)}{0-2}$

$= -2$

slope of $\overline{CD} = \frac{0-3}{-6-0}$

$= \frac{1}{2}$

slope of $\overline{DA} = \frac{-4-0}{-4-(-6)}$

$= -2$

The consecutive sides are perpendicular. Therefore, $ABCD$ is a rectangle.

34.

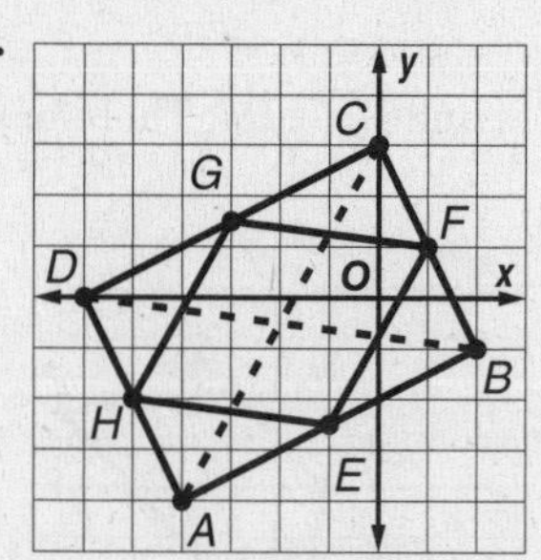

Draw diagonals $\overline{AC}$ and $\overline{BD}$. $ABCD$ is a rectangle. So $\overline{AC} \cong \overline{BD}$ and $AC = BD$. Since G and H are midpoints, by the Triangle Midsegment Theorem $\overline{HG} \parallel \overline{AC}$ and $HG = \frac{1}{2}(AC)$

Since E and F are midpoints, $\overline{EF} \parallel \overline{AC}$ and $EF = \frac{1}{2}(AC)$

So by transitivity of parallel lines and substitution, $\overline{HG} \parallel \overline{EF}$ and $HG = EF$. So $GHEF$ is a parallelogram. Since H and E are midpoints, by the Triangle Midsegment Theorem, $HE = \frac{1}{2}(BD)$ and $GF = \frac{1}{2}(BD)$. Since $AC = BD$, $HE = \frac{1}{2}(AC)$ and $GF = \frac{1}{2}(AC)$. Therefore, $HE = GH = GF = EF$. $EFGH$ is a parallelogram with all sides congruent.

35. To make the diagonals the same, either $\overline{AC}$ must be shortened or $\overline{BD}$ must be lengthened. This can be accomplished by moving L and K until the length of the diagonals is the same.

36. Find the ratio of the length to the width of the rectangle.

$\frac{19.42}{12.01} \approx 1.617$

Since 1.617 is close to 1.618, the rectangle is a golden rectangle. Use the Pythagorean Theorem to find the length of the diagonal.

$c = \sqrt{a^2 + b^2}$

$= \sqrt{19.42^2 + 12.01^2}$

≈ 22.83

The length of the diagonal is about 22.83 ft.

37. See students' work.

38. Parallelograms have opposite sides congruent and bisecting diagonals, so the minimal requirements to justify that a parallelogram is a rectangle are that diagonals are congruent or the parallelogram has one right angle.

39. Sample answer:

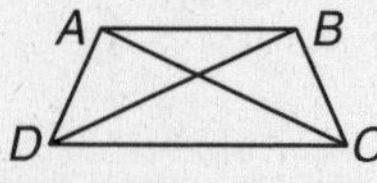

$\overline{AC} \cong \overline{BD}$ but $ABCD$ is not a rectangle.

40. Given: $WXYZ$ is a rectangle with diagonals $\overline{WY}$ and $\overline{XZ}$.

Prove: $\overline{WY} \cong \overline{XZ}$

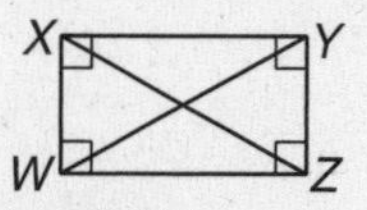

Proof:

Statements	Reasons
1. $WXYZ$ is a rectangle with diagonals $\overline{WY}$ and $\overline{XZ}$.	1. Given
2. $\overline{WX} \cong \overline{ZY}$	2. Opp. sides of ▱ are ≅.
3. $\overline{WZ} \cong \overline{WZ}$	3. Reflexive Property
4. ∠XWZ and ∠YZW are right angles	4. Def. of rectangle
5. ∠XWZ ≅ ∠YZW	5. All right ∠s are ≅.
6. △XWZ ≅ △YZW	6. SAS
7. $\overline{WY} \cong \overline{XZ}$	7. CPCTC

41. Given: ▱$WXYZ$, $\overline{WY} \cong \overline{XZ}$
Prove: $WXYZ$ is a rectangle.

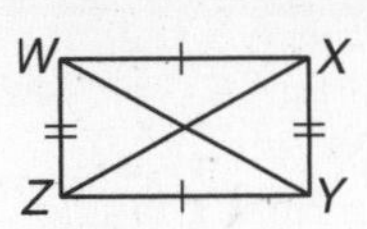

Proof:

Statements	Reasons
1. ▱$WXYZ$, $\overline{WY} \cong \overline{XZ}$	1. Given
2. $\overline{WX} \cong \overline{YZ}$, $\overline{XY} \cong \overline{WZ}$, and $\overline{WY} \cong \overline{XZ}$	2. Opposite sides of ▱ are $\cong$.
3. $\overline{WX} \cong \overline{WX}$	3. Reflexive Property
4. $\triangle WZX \cong \triangle XYW$	4. SSS
5. $\angle ZWX \cong \angle YXW$	5. CPCTC
6. $\angle ZWX$ and $\angle YXW$ are supplementary	6. Consec. ∠s of ▱ are suppl.
7. $\angle ZWX$ and $\angle YXW$ are right angles	7. If 2 ∠s are $\cong$ and suppl., each $\angle$ is a rt. $\angle$.
8. $\angle WZY$ and $\angle XYZ$ are right ∠s	8. If ▱ has 1 rt. $\angle$, it has 4 rt. ∠s.
9. $WXYZ$ is a rectangle	9. Def. of rectangle

42. Given: $PQST$ is a rectangle.
$\overline{QR} \cong \overline{VT}$
Prove: $\overline{PR} \cong \overline{VS}$

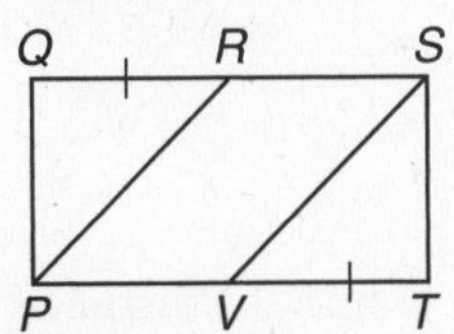

Proof:

Statements	Reasons
1. $PQST$ is a rectangle. $\overline{QR} \cong \overline{VT}$	1. Given
2. $PQST$ is a parallelogram.	2. Def. of rectangle
3. $\overline{TS} \cong \overline{PQ}$	3. Opp. sides of ▱ are $\cong$.
4. $\angle T$ and $\angle Q$ are rt. $\angle$s.	4. Definition of rectangle
5. $\angle T \cong \angle Q$	5. All rt. ∠s are $\cong$.
6. $\triangle RPQ \cong \triangle VST$	6. SAS
7. $\overline{PR} \cong \overline{VS}$	7. CPCTC

43. Given: $DEAC$ and $FEAB$ are rectangles.
$\angle GKH \cong \angle JHK$
$\overline{GJ}$ and $\overline{HK}$ intersect at L.
Prove: $GHJK$ is a parallelogram.

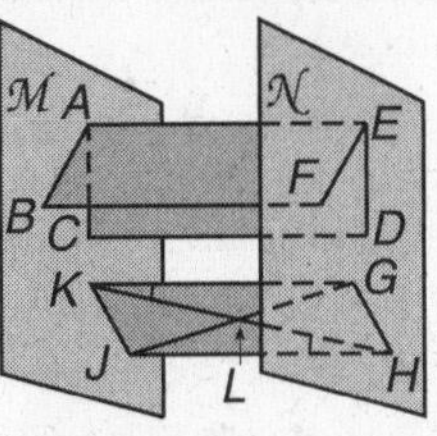

Proof:

Statements	Reasons
1. $DEAC$ and $FEAB$ are rectangles. $\angle GKH \cong \angle JHK$ $\overline{GJ}$ and $\overline{HK}$ intersect at L.	1. Given
2. $\overline{DE} \parallel \overline{AC}$ and $\overline{FE} \parallel \overline{AB}$	2. Def. of parallelogram
3. plane $\mathcal{N} \parallel$ plane $\mathcal{M}$	3. Def. of parallel plane
4. G, J, H, K, L are in the same plane.	4. Def. of intersecting lines
5. $\overline{GH} \parallel \overline{KJ}$	5. Def. of parallel lines.
6. $\overline{KG} \parallel \overline{HJ}$	6. Alt. int. ∠s are $\cong$.
7. $GHJK$ is a parallelogram.	7. Def. of parallelogram

44. Explore: We need to find the number of rectangles that can be formed using four of the twelve points as corners.
Plan: Count the number of rectangles with each possible set of dimensions. Arrange the data in a table
Solve:

Dimensions length $\times$ width	Number of rectangles
3×2	1
3×1	2
2×2	2
2×1	4
1×2	3
1×1	6

There are 18 rectangles formed by using the rows and columns as sides.
There are 2 additional rectangles formed as shown.

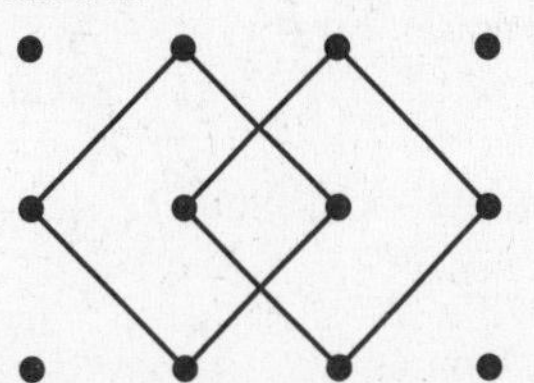

Examine: The table covers all possible dimensions. The figure shows the only rectangles that can be formed using rectangles at an angle to the rows and columns. So the answer is reasonable.

45. No; there are no parallel lines in spherical geometry.

46. AC appears to be shorter than TR, so $AC < TR$.

47. Since the sides would not be parallel in spherical geometry, a rectangle cannot exist.

48. Sample answer: The tennis court is divided into rectangular sections. The players use the rectangles to establish the playing area. Answers should include the following.
   - Not counting overlap, there are 5 rectangles on each side of a tennis court.
   - Measure each diagonal to make sure they are the same length and measure each angle to make sure they measure 90.

49. A; since $\overline{AB} \parallel \overline{CE}$, $\angle ADE \cong \angle BAD$ and $\angle BDC \cong \angle ABD$ because they are alternate interior angles. It is given that $\angle ADE \cong \angle BAD$, so $\angle BAD \cong \angle ABD$. Therefore, $\triangle ABD$ is isosceles and $DB = DA$, so DB is 6.

50. D; since s is the shorter side of the playground, $s + 10$ is the longer side. The perimeter of the fence is 80 feet, and the perimeter of a rectangle is equal to twice the width plus twice the length. Therefore, the equation to find s is $2(s + 10) + 2s = 80$.

Page 430 Maintain Your Skills

51. There are 31 parallelograms: 11 individual parallelograms, 12 using two others, 6 using three others, and 2 using four others.

52. The sum of the measures of the internal angles of a triangle is 180. Find $m\angle AFD$.

$$m\angle AFD + m\angle FDA + m\angle DAF = 180$$
$$m\angle AFD + 34 + 49 = 180$$
$$m\angle AFD = 97$$

53. $\angle ACD \cong \angle BAC$ because they are alternate interior angles. $m\angle ACD = 54$. So, $m\angle BAC = 54$. $\angle BAD$ is supplementary to $\angle ADC$. Find $m\angle CDF$.

$$m\angle CDF + m\angle FDA + m\angle DAF + m\angle BAC = 180$$
$$m\angle CDF + 34 + 49 + 54 = 180$$
$$m\angle CDF = 43$$

54. $\angle FBC \cong \angle ADF$ because they are alternate interior angles. $m\angle ADF = 34$. So, $m\angle FBC = 34$.

55. $\angle BCF \cong \angle DAF$ because they are alternate interior angles. $m\angle DAF = 49$. So, $m\angle BCF = 49$.

56. Opposite sides of a parallelogram are congruent, so their measures are equal. Find y.

$$BC = AD$$
$$3y - 4 = 29$$
$$3y = 33$$
$$y = 11$$

57. Opposite sides of a parallelogram are congruent, so their measures are equal. Find x.

$$AB = CD$$
$$5x = 25$$
$$x = 5$$

58. $\overline{ST}$ is the altitude to side $\overline{QR}$ in right triangle QSR so by Theorem 7.2, its measure is the geometric mean between the two segments of the hypotenuse. Let $x = ST$.

$$\frac{18}{x} = \frac{x}{34}$$
$$x^2 = 612$$
$$x = \sqrt{612}$$
$$x \approx 24.7$$

59. $\overline{NP}$ is the altitude to side $\overline{MO}$ in right triangle MNO so by Theorem 7.2, its measure is the geometric mean between the two segments of the hypotenuse.
Let $x = NP$.

$$\frac{11}{x} = \frac{x}{27}$$
$$x^2 = 297$$
$$x = \sqrt{297}$$
$$x \approx 17.2$$

60. The measure of the altitude to $\overline{AB}$ is the geometric mean between the two segments of the hypotenuse of $\triangle ABC$.
Let x = the measure of the altitude.

$$\frac{24}{x} = \frac{x}{14}$$
$$x^2 = 336$$
$$x = \sqrt{336}$$
$$x \approx 18.3$$

61. Use the Distance Formula to find the distance between the given points.

$$d = \sqrt{(x_2 - x_1)^2 + (y_2 - y_1)^2}$$
$$= \sqrt{(-3 - 1)^2 + [1 - (-2)]^2}$$
$$= 5$$

The distance between the points is 5 units.

62. Use the Distance Formula to find the distance between the given points.

$$d = \sqrt{(x_2 - x_1)^2 + (y_2 - y_1)^2}$$
$$= \sqrt{[5 - (-5)]^2 + (12 - 9)^2}$$
$$= \sqrt{109}$$
$$\approx 10.4$$

The distance between the points is $\sqrt{109}$ or about 10.4 units.

63. Use the Distance Formula to find the distance between the given points.

$$d = \sqrt{(x_2 - x_1)^2 + (y_2 - y_1)^2}$$
$$= \sqrt{(22 - 1)^2 + (24 - 4)^2}$$
$$= 29$$

The distance between the points is 29 units.

8-5 Rhombi and Squares

Page 434 Check for Understanding

1. Sample answer:

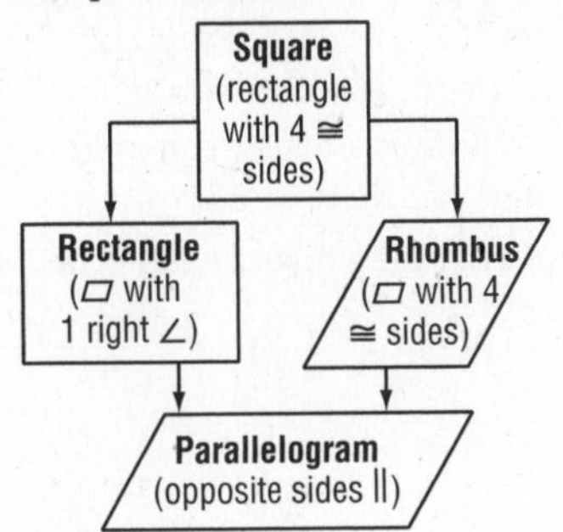

2. Sample answer:

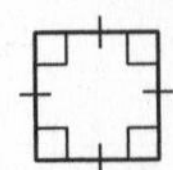

3. A square is a rectangle with all sides congruent.

4. The sides of a rhombus are congruent. Find x.
$$AB = BC$$
$$2x + 3 = 5x$$
$$3 = 3x$$
$$1 = x$$
The value of x is 1.

5. From Exercise 1, $x = 1$.
So, $BC = 5(1)$ or 5. AD is congruent to BC, so $AD = 5$.

6. The diagonals of a rhombus are perpendicular, so $m\angle AEB = 90$.

7. Consecutive angles of a rhombus are supplementary. Find $m\angle BCD$.
$$m\angle BCD + m\angle ABC = 180$$
$$m\angle BCD + 83.2 = 180$$
$$m\angle BCD = 96.8$$

8.

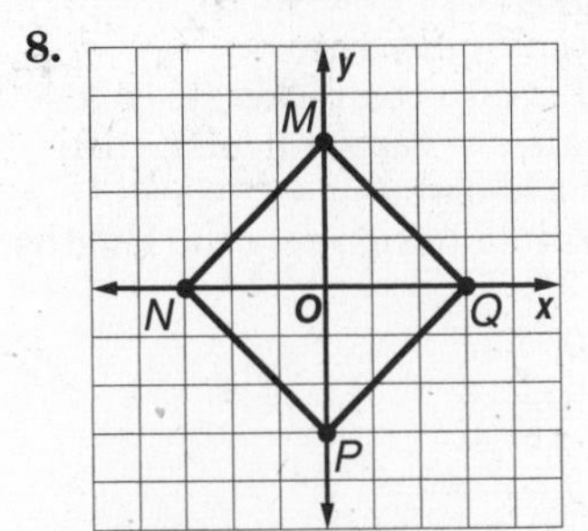

If the four sides are congruent, then parallelogram $MNPQ$ is either a rhombus or a square. If consecutive sides are perpendicular, then $MNPQ$ is a rectangle or a square.
Use the distance formula to compare the lengths of the sides.
$$MN = \sqrt{(-3-0)^2 + (0-3)^2} = \sqrt{9+9} = 3\sqrt{2}$$
$$NP = \sqrt{[0-(-3)]^2 + (-3-0)^2} = \sqrt{9+9} = 3\sqrt{2}$$
$$PQ = \sqrt{[3-0]^2 + [0-(-3)]^2} = \sqrt{9+9} = 3\sqrt{2}$$
$$QM = \sqrt{(3-0)^2 + (0-3)^2} = \sqrt{9+9} = 3\sqrt{2}$$
Use slope to determine whether the consecutive sides are perpendicular.
$$\text{slope of } \overline{MN} = \frac{0-3}{-3-0} = 1$$
$$\text{slope of } \overline{NP} = \frac{-3-0}{0-(-3)} = -1$$
$$\text{slope of } \overline{PQ} = \frac{-3-0}{0-3} = 1$$
$$\text{slope of } \overline{QM} = \frac{0-3}{3-0} = -1$$
Since the slopes of $\overline{MN}$ and $\overline{PQ}$ are opposite reciprocals of the slopes of $\overline{NP}$ and $\overline{QM}$, consecutive sides are perpendicular. The lengths of the four sides are the same, so the sides are congruent. $MNPQ$ is a rectangle, a rhombus, and a square.

9.

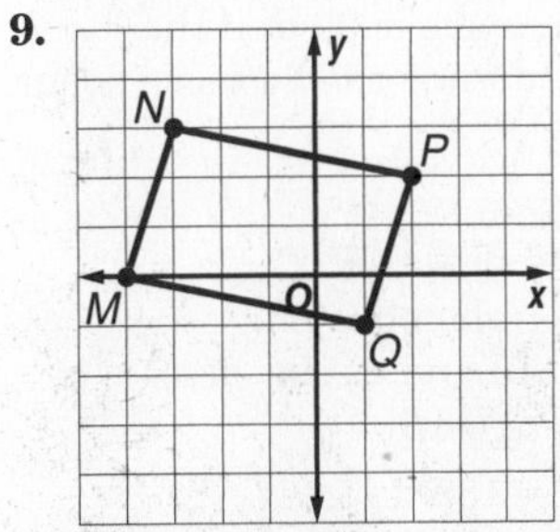

If the diagonals are congruent, then parallelogram $MNPQ$ is either a rectangle or a square. If the diagonals are perpendicular, then $MNPQ$ is a square or a rhombus.
Use the distance formula to compare the lengths of the diagonals.
$$MP = \sqrt{(-4-2)^2 + (0-2)^2} = \sqrt{36+4} = \sqrt{40}$$
$$NQ = \sqrt{(-3-1)^2 + [3-(-1)]^2} = \sqrt{16+16} = \sqrt{32}$$
Use slope to determine whether the diagonals are perpendicular.
$$\text{slope of } \overline{MP} = \frac{0-2}{-4-2} = \frac{1}{3}$$
$$\text{slope of } \overline{NQ} = \frac{-1-3}{1-(-3)} = -1$$
The diagonals are not congruent or perpendicular. $MNPQ$ is not a rhombus, a rectangle, or a square.

10. Given: $\triangle KGH$, $\triangle HJK$, $\triangle GHJ$, and $\triangle JKG$ are isosceles.

Prove: $GHJK$ is a rhombus.

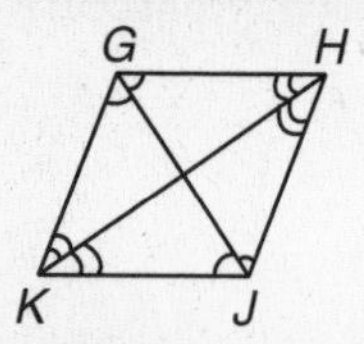

Proof:

Statements	Reasons
1. $\triangle KGH$, $\triangle HJK$, $\triangle GHJ$, and $\triangle JKG$ are isosceles.	1. Given
2. $\overline{KG} \cong \overline{GH}$, $\overline{HJ} \cong \overline{KJ}$, $\overline{GH} \cong \overline{HJ}$, $\overline{KG} \cong \overline{KJ}$	2. Def. of isosceles $\triangle$
3. $\overline{KG} \cong \overline{HJ}$, $\overline{GH} \cong \overline{KJ}$	3. Transitive Property
4. $\overline{KG} \cong \overline{GH}$, $\overline{HJ} \cong \overline{KJ}$	4. Substitution
5. $GHJK$ is a rhombus.	5. Def. of rhombus

11. If the measure of each angle is 90 or if the diagonals are congruent, then the floor is a square.

Pages 434–437 Practice and Apply

12. Consecutive angles in a rhombus are supplementary, so $\angle DAB$ and $\angle ADC$ are supplementary. Find $m\angle DAB$.

$$m\angle DAB + m\angle ADC = 180$$
$$m\angle DAB + \tfrac{1}{2}m\angle DAB = 180$$
$$\tfrac{3}{2}m\angle DAB = 180$$
$$m\angle DAB = 120$$

Opposite angles of a rhombus are congruent, so $\angle BCD \cong \angle DAB$. The diagonals of a rhombus bisect the angles, so $m\angle ACD = \frac{1}{2}m\angle BCD$ or 60.

13. Consecutive angles in a rhombus are supplementary, so $\angle DAB$ and $\angle ADC$ are supplementary. Find $m\angle DAB$.

$$m\angle DAB + m\angle ADC = 180$$
$$m\angle DAB + \tfrac{1}{2}m\angle DAB = 180$$
$$\tfrac{3}{2}m\angle DAB = 180$$
$$m\angle DAB = 120$$

14. By definition, a rhombus has four congruent sides, so $\overline{DA} \cong \overline{CB}$ and $DA = 6$.

15. Consecutive angles in a rhombus are supplementary, so $\angle DAB$ and $\angle ADC$ are supplementary. Find $m\angle ADC$.

$$m\angle DAB + m\angle ADC = 180$$
$$2(m\angle ADC) + m\angle ADC = 180$$
$$3(m\angle ADC) = 180$$
$$m\angle ADC = 60$$

The diagonals of a rhombus bisect the angles, so $m\angle ADB = \frac{1}{2}(m\angle ADC)$ or 30.

16. The diagonals of a rhombus are perpendicular, so $m\angle YVZ = 90$. The measure of the interior angles of a triangle is 180. Find $m\angle YZV$.

$$m\angle YVZ + m\angle YZV + m\angle WYZ = 180$$
$$90 + m\angle YZV + 53 = 180$$
$$m\angle YZV + 143 = 180$$
$$m\angle YZV = 37$$

17. The diagonals of a rhombus bisect the angles, so $\angle XYW \cong \angle WYZ$ and $m\angle XYW = 53$.

18. The diagonals of a rhombus bisect each other, so $\overline{XV} \cong \overline{ZV}$. Find a.

$$\overline{XV} \cong \overline{ZV}$$
$$XV = ZV$$
$$2a - 2 = \frac{5a + 1}{4}$$
$$8a - 8 = 5a + 1$$
$$3a = 9$$
$$a = 3$$

So, $XV = 2(3) - 2$ or 4 and XZ is twice XV or 8.

19. From Exercise 18, $XV = 4$. $VW = 3$.
The diagonals of a rhombus are perpendicular, so $m\angle WVX = 90$ and $\triangle WVX$ is a right triangle. $\overline{XW}$ is the hypotenuse of $\triangle WVX$. Use the Pythagorean Theorem.

$$(VW)^2 + (XV)^2 = (XW)^2$$
$$3^2 + 4^2 = (XW)^2$$
$$25 = (XW)^2$$
$$5 = XW$$

20.

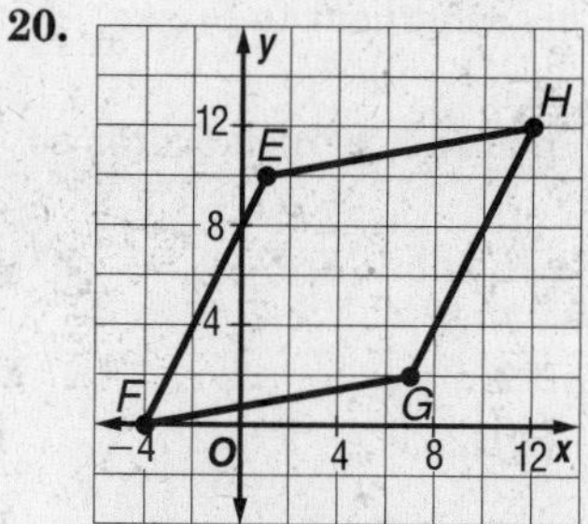

If the diagonals are congruent, then parallelogram $EFGH$ is either a rectangle or a square. If the diagonals are perpendicular, then $EFGH$ is a square or a rhombus.
Use the Distance Formula to compare the lengths of the diagonals.

$$EG = \sqrt{(1 - 7)^2 + (10 - 2)^2}$$
$$= \sqrt{36 + 64} = \sqrt{100}$$
$$= 10$$
$$FH = \sqrt{(-4 - 12)^2 + (0 - 12)^2}$$
$$= \sqrt{256 + 144} = \sqrt{400}$$
$$= 20$$

Use slope to determine whether the diagonals are perpendicular.

$$\text{slope of } \overline{EG} = \frac{10 - 2}{1 - 7}$$
$$= -\frac{4}{3}$$
$$\text{slope of } \overline{FH} = \frac{0 - 12}{-4 - 12}$$
$$= \frac{3}{4}$$

The diagonals are not congruent. Since the slopes of $\overline{EG}$ and $\overline{FH}$ are opposite reciprocals of each other, the diagonals are perpendicular. *EFGH* is a rhombus.

21. 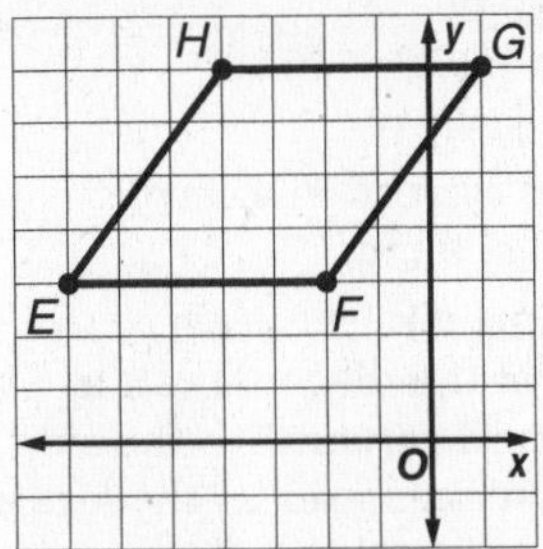

If the diagonals are congruent, then parallelogram *EFGH* is either a rectangle or a square. If the diagonals are perpendicular, then *EFGH* is a square or a rhombus.

Use the Distance Formula to compare the lengths of the diagonals.

$EG = \sqrt{(-7 - 1)^2 + (3 - 7)^2}$
$= \sqrt{64 + 16} = \sqrt{80}$
$FH = \sqrt{[-2 - (-4)]^2 + (3 - 7)^2}$
$= \sqrt{4 + 16} = \sqrt{20}$

Use slope to determine whether the diagonals are perpendicular.

slope of $\overline{EG} = \frac{3 - 7}{-7 - 1}$
$= \frac{1}{2}$
slope of $\overline{FH} = \frac{3 - 7}{-2 - (-4)}$
$= -2$

The diagonals are not congruent. Since the slopes of $\overline{EG}$ and $\overline{FH}$ are opposite reciprocals of each other, the diagonals are perpendicular. *EFGH* is a rhombus.

22. 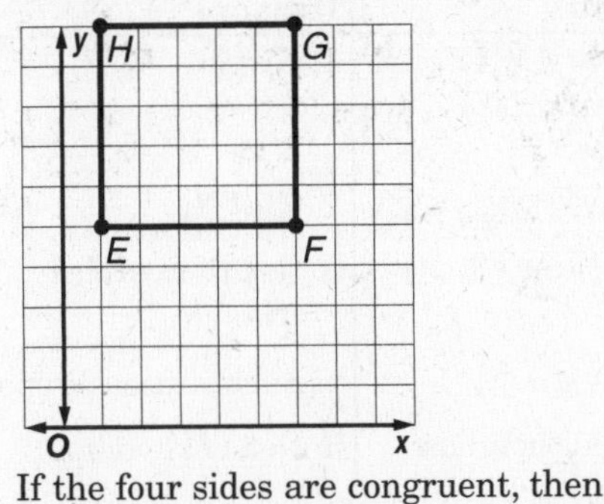

If the four sides are congruent, then parallelogram *EFGH* is either a rhombus or a square. If consecutive sides are perpendicular, then *EFGH* is a rectangle or a square.

It is obvious from the figure that each side has measure 5, so the sides are congruent. It is also obvious that $\overline{HE}$ and $\overline{GF}$ are vertical segments and $\overline{HG}$ and $\overline{EF}$ are horizontal, so consecutive sides are perpendicular. Thus, *EFGH* is a square, a rectangle, and a rhombus.

23. 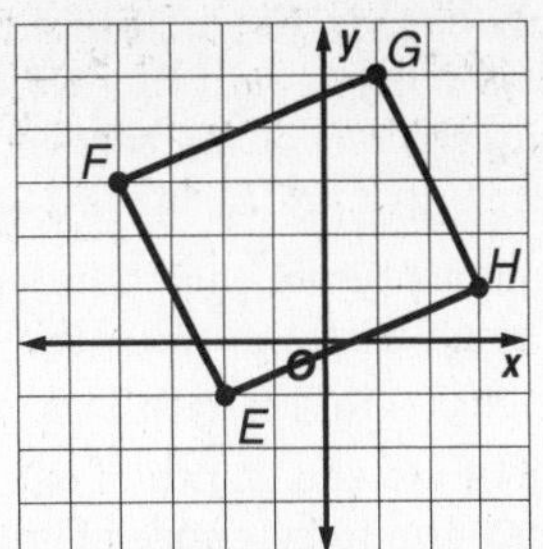

If the diagonals are congruent, then parallelogram *EFGH* is either a rectangle or a square. If the diagonals are perpendicular, then *EFGH* is a square or a rhombus.

Use the Distance Formula to compare the lengths of the diagonals.

$EG = \sqrt{(-2 - 1)^2 + (-1 - 5)^2}$
$= \sqrt{9 + 36} = \sqrt{45}$
$FH = \sqrt{(-4 - 3)^2 + (3 - 1)^2}$
$= \sqrt{49 + 4} = \sqrt{53}$

Use slope to determine whether the diagonals are perpendicular.

slope of $\overline{EG} = \frac{-1 - 5}{-2 - 1}$
$= 2$
slope of $\overline{FH} = \frac{3 - 1}{-4 - 3}$
$= -\frac{2}{7}$

The diagonals are not congruent or perpendicular. *EFGH* is not a rhombus, a rectangle, or a square.

24. Sample answer:

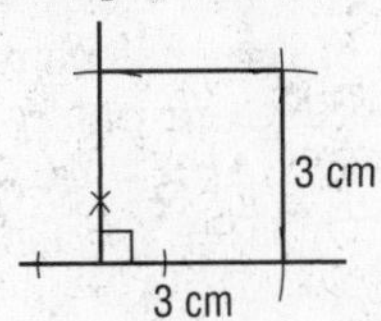

25. Sample answer:

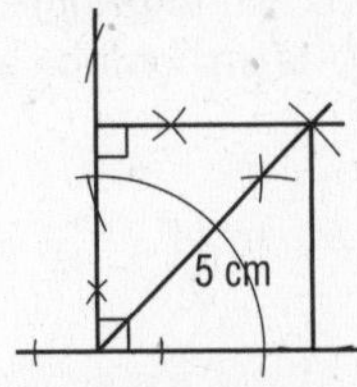

26. Every square is a parallelogram, but not every parallelogram is a square. The statement is sometimes true.

27. Every square is a rhombus. The statement is always true.

28. Every rectangle is a parallelogram. The statement is always true.

29. If a rhombus is a square, then the rhombus is also a rectangle. Otherwise the rhombus is not a rectangle. The statement is sometimes true.

30. Every square is a rhombus, but not every rhombus is a square. The statement is sometimes true.

31. Every square is a rectangle. The statement is always true.

32. The width of the square base is $15\frac{3}{4}$ in. Since the width of the smaller boxes is one half the width of the square base, the dimensions of the base of one of the smaller boxes are $7\frac{7}{8}$ in. by $7\frac{7}{8}$ in.

33. The diagonals of a rhombus bisect each other and are perpendicular, so along with the sides, they form four congruent right triangles with legs $\frac{12}{2}$ or 6 cm and $\frac{16}{2}$ or 8 cm. Find the length of one of the four congruent sides–the hypotenuse of one of the right triangles–by using the Pythagorean Theorem.

side length $= \sqrt{6^2 + 8^2}$
$= \sqrt{36 + 64} = 10$

The side length is 10 cm, so the perimeter of the rhombus is 4(10) or 40 cm.

34. $ABCD$ is a rhombus; $EFGH$ and $JKLM$ are congruent squares.

35. Given: $ABCD$ is a parallelogram.
$\overline{AC} \perp \overline{BD}$

Prove: $ABCD$ is a rhombus.

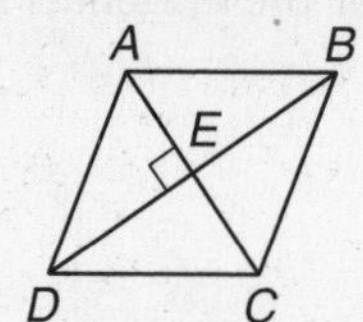

Proof: We are given that $ABCD$ is a parallelogram. The diagonals of a parallelogram bisect each other, so $\overline{AE} \cong \overline{EC}$. $\overline{BE} \cong \overline{BE}$ because congruence of segments is reflexive. We are also given that $\overline{AC} \perp \overline{BD}$. Thus, $\angle AEB$ and $\angle BEC$ are right angles by the definition of perpendicular lines. Then $\angle AEB \cong \angle BEC$ because all right angles are congruent. Therefore, $\triangle AEB \cong \triangle CEB$ by SAS. $\overline{AB} \cong \overline{CB}$ by CPCTC. Opposite sides of parallelograms are congruent, so $\overline{AB} \cong \overline{CD}$ and $\overline{BC} \cong \overline{AD}$. Then since congruence of segments is transitive, $\overline{AB} \cong \overline{CD} \cong \overline{BC} \cong \overline{AD}$. All four sides of $ABCD$ are congruent, so $ABCD$ is a rhombus by definition.

36. Given: $ABCD$ is a rhombus.

Prove: Each diagonal bisects a pair of opposite angles.

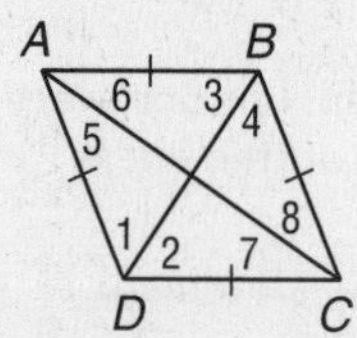

Proof: We are given that $ABCD$ is a rhombus. By definition of rhombus, $ABCD$ is a parallelogram. Opposite angles of a parallelogram are congruent, so $\angle ABC \cong \angle ADC$ and $\angle BAD \cong \angle BCD$. $\overline{AB} \cong \overline{BC} \cong \overline{CD} \cong \overline{DA}$ because all sides of a rhombus are congruent. $\triangle ABC \cong \triangle ADC$ by SAS. $\angle 5 \cong \angle 6$ and $\angle 7 \cong \angle 8$ by CPCTC. $\triangle BAD \cong \triangle BCD$ by SAS. $\angle 1 \cong \angle 2$ and $\angle 3 \cong \angle 4$ by CPCTC. By definition of angle bisector, each diagonal bisects a pair of opposite angles.

37. The width and length of the rectangular court are 6400 mm and 9750 mm, respectively. The measure of the diagonal can be found using the Pythagorean Theorem with width and length as the legs of a right triangle.

diagonal measure $= \sqrt{6400^2 + 9750^2}$
$\approx 11{,}662.9$

No; the diagram is not correct. The correct measure is about 11,662.9 mm.

38. The side length of the square service boxes is 1600 mm. The length of the diagonal can be found using the Pythagorean Theorem with the side length of the square service boxes as the legs of a right triangle.

diagonal length $= \sqrt{1600^2 + 1600^2}$
≈ 2263

The length of the diagonal of the square service boxes is about 2263 mm or 2.263 m.

39. The flag of Denmark contains four red rectangles. The flag of St. Vincent and the Grenadines contains a blue rectangle, a green rectangle, a yellow rectangle, a blue and yellow rectangle, a yellow and green rectangle, and three green rhombi. The flag of Trinidad and Tobago contains two white parallelograms and one black parallelogram.

40. Given: $\triangle WZY \cong \triangle WXY$
$\triangle WZY$ and $\triangle XYZ$ are isosceles.

Prove: $WXYZ$ is a rhombus.

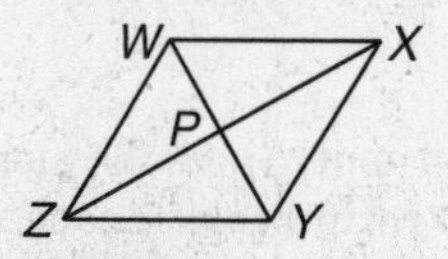

Proof:

Statements	Reasons
1. $\triangle WZY \cong \triangle WXY$ $\triangle WZY$ and $\triangle XYZ$ are isosceles.	1. Given
2. $\overline{WZ} \cong \overline{WX}$, $\overline{ZY} \cong \overline{XY}$	2. CPCTC
3. $\overline{WZ} \cong \overline{ZY}$, $\overline{WX} \cong \overline{XY}$	3. Def. of isosceles triangle
4. $\overline{WZ} \cong \overline{WX} \cong \overline{ZY} \cong \overline{XY}$	4. Substitution Property
5. $WXYZ$ is rhombus.	5. Def. of rhombus

41. Given: $\triangle TPX \cong \triangle QPX \cong \triangle QRX \cong \triangle TRX$

Prove: $TPQR$ is a rhombus.

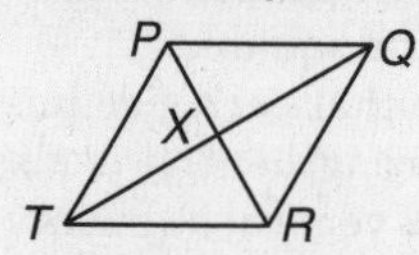

Proof:

Statements	Reasons
1. $\triangle TPX \cong \triangle QPX \cong \triangle QRX \cong \triangle TRX$	1. Given
2. $\overline{TP} \cong \overline{PQ} \cong \overline{QR} \cong \overline{TR}$	2. CPCTC
3. $TPQR$ is a rhombus.	3. Def. of rhombus

42. Given: $\triangle LGK \cong \triangle MJK$
GHJK is a parallelogram.
Prove: *GHJK* is a rhombus.

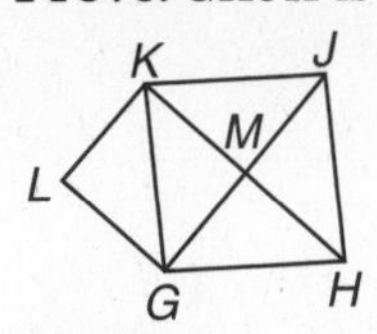

Proof:

Statements	Reasons
1. $\triangle LGK \cong \triangle MJK$ *GHJK* is a parallelogram.	1. Given
2. $\overline{KG} \cong \overline{KJ}$	2. CPCTC
3. $\overline{KJ} \cong \overline{GH}$, $\overline{KG} \cong \overline{JH}$	3. Opp. sides of $\square$ are $\cong$.
4. $\overline{KG} \cong \overline{JH} \cong \overline{GH} \cong \overline{JK}$	4. Substitution Property
5. *GHJK* is a rhombus.	5. Def. of rhombus

43. Given: *QRST* and *QRTV* are rhombi.
Prove: $\triangle QRT$ is equilateral.

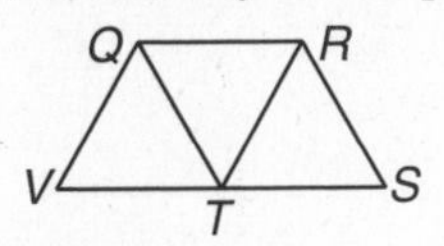

Proof:

Statements	Reasons
1. *QRST* and *QRTV* are rhombi.	1. Given
2. $\overline{QV} \cong \overline{VT} \cong \overline{TR} \cong \overline{QR}$, $\overline{QR} \cong \overline{TS} \cong \overline{RS} \cong \overline{QT}$	2. Def. of rhombus
3. $\overline{TR} \cong \overline{QR} \cong \overline{QT}$	3. Transitive Property
4. $\triangle QRT$ is equilateral.	4. Def. of equilateral triangle

44. Hexagons 1, 2, 3, and 4 have 3, 12, 27, and 48 rhombi, respectively. Note that the numbers of rhombi are $3 = 3(1) = 3(1^2)$, $12 = 3(4) = 3(2^2)$, $27 = 3(9) = 3(3^2)$, and $48 = 3(16) = 3(4^2)$. So, the number of rhombi are given by $3x^2$, where x is the hexagon number.

Hexagon	Number of rhombi
1	3
2	12
3	27
4	48
5	75
6	108
x	$3x^2$

45. Sample answer: You can ride a bicycle with square wheels over a curved road. Answers should include the following.

- Rhombi and squares both have all four sides congruent, but the diagonals of a square are congruent. A square has four right angles and rhombi have each pair of opposite angles congruent, but not all angles are necessarily congruent.
- Sample answer: Since the angles of a rhombus are not all congruent, riding over the same road would not be smooth.

46. B; the side length of the square is $\sqrt{36} = 6$ units. So, the perimeter of the square is 4(6) or 24 units. Since rectangle *ABCD* is contained within the square, which itself is a rectangle, *ABCD* cannot have a greater perimeter than the square. Therefore, the perimeter of rectangle *ABCD* is less than 24 units.

47. C; test all four values.

<x>	$\frac{1+x}{x-2}$
<0>	$-\frac{1}{2}$
<1>	-2
<3>	4
<4>	$\frac{5}{2} = 2\frac{1}{2}$

<3> has the greatest value.

Pae 437 Maintain Your Skills

48. The diagonals of a rectangle bisect each other and are congruent, so $\overline{PJ} \cong \overline{LJ}$.

$$\overline{PJ} \cong \overline{LJ}$$
$$PJ = LJ$$
$$3x - 1 = 2x + 1$$
$$x = 2$$

So, $x = 2$.

49. The interior angles of a rectangle are 90°. Find $m\angle MLK$.

$$m\angle MLK + m\angle PLM = m\angle PLK$$
$$m\angle MLK + 90 = 110$$
$$m\angle MLK = 20$$

Diagonals of a rectangle are congruent and bisect each other. So $\overline{JL} \cong \overline{JM}$. Then *LKMJ* is a rhombus because opposite sides of a parallelogram are congruent. Since each diagonal of a rhombus bisects a pair of congruent opposite angles, $\angle KML \cong \angle MLK$. The sum of the measures of the interior angles of a triangle is 180. Find $m\angle LKM$.

$$m\angle MLK + m\angle LKM + m\angle KML = 180$$
$$20 + m\angle LKM + 20 = 180$$
$$m\angle LKM = 140$$

So, $m\angle LKM = 140$.

50. $\angle MJN$ is supplementary to $\angle PJN$. Find $m\angle PJN$.

$$m\angle MJN + m\angle PJN = 180$$
$$35 + m\angle PJN = 180$$
$$m\angle PJN = 145$$

Since the diagonals of a rectangle are congruent and bisect each other, $\triangle PJN$ is isosceles with sides $\overline{PJ}$ and $\overline{JN}$ congruent. Since $\triangle PJN$ is isosceles, $\angle MPN \cong \angle PNL$. The sum of the measures of the interior angles of a triangle is 180. Find $m\angle MPN$.

$$m\angle MPN + m\angle PNL + m\angle PJN = 180$$
$$m\angle MPN + m\angle MPN + 145 = 180$$
$$2(m\angle MPN) = 35$$
$$m\angle MPN = 17.5$$

So, $m\angle MPN = 17.5$.

51. Since the diagonals of a rectangle are congruent and bisect each other, $\overline{LJ} \cong \overline{JN}$. Since the sides of a rhombus are congruent, $\overline{LJ} \cong \overline{MK}$. Therefore, $\overline{MK} \cong \overline{JN}$.

Find x.

$\overline{MK} \cong \overline{JN}$

$MK = JN$

$6x = 14 - x$

$7x = 14$

$x = 2$

Once again, the sides of a rhombus are congruent, so $\overline{MK} \cong \overline{KL}$. Find y.

$\overline{MK} \cong \overline{KL}$

$MK = KL$

$6x = 3x + 2y$

$\frac{3}{2}x = y$

Substituting the value of x from above,

$\frac{3}{2}(2) = y$

$3 = y$

So, $x = 2$ and $y = 3$.

52. Since $m\angle LMP = m\angle PMN$, the diagonal $\overline{PM}$ bisects $\angle LMN$. If the diagonals of a rectangle bisect its interior angles, the rectangle must be a square. The diagonals of a square are perpendicular, so $m\angle PJL = 90$.

53. Yes;

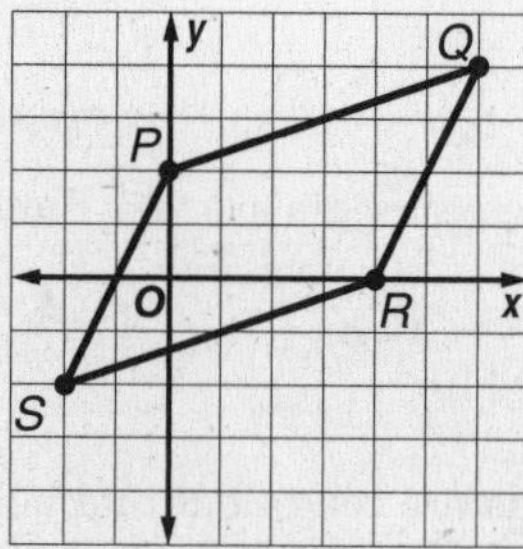

If both pairs of opposite sides of a quadrilateral are congruent, then the quadrilateral is a parallelogram. Use the Distance Formula to determine whether the opposite sides are congruent.

$PQ = \sqrt{(0 - 6)^2 + (2 - 4)^2}$

$= \sqrt{40}$

$RS = \sqrt{[4 - (-2)]^2 + [0 - (-2)]^2}$

$= \sqrt{40}$

$QR = \sqrt{(6 - 4)^2 + (4 - 0)^2}$

$= \sqrt{20}$

$PS = \sqrt{[0 - (-2)]^2 + [2 - (-2)]^2}$

$= \sqrt{20}$

Since the measures of both pairs of opposite sides are equal, $PQRS$ is a parallelogram.

54. No;

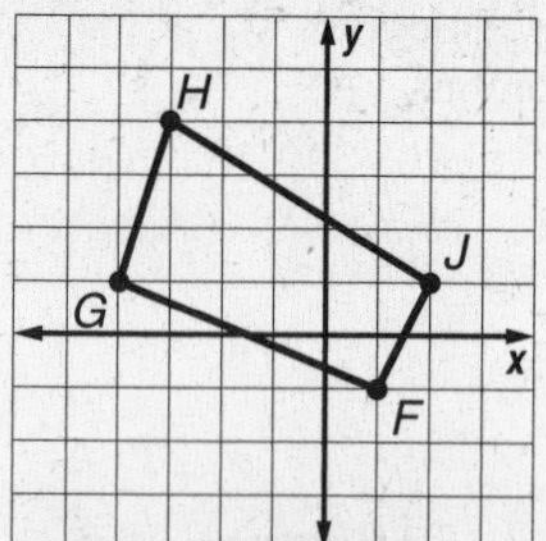

If both pairs of opposite sides of a quadrilateral are congruent, then the quadrilateral is a parallelogram. Use the Distance Formula to determine whether the opposite sides are congruent.

$HJ = \sqrt{(-3 - 2)^2 + (4 - 1)^2}$

$= \sqrt{34}$

$FG = \sqrt{[1 - (-4)]^2 + (-1 - 1)^2}$

$= \sqrt{29}$

Since $HJ \neq FG$, $\overline{HJ} \ncong \overline{FG}$. Therefore, $FGHJ$ is not a parallelogram.

55. No;

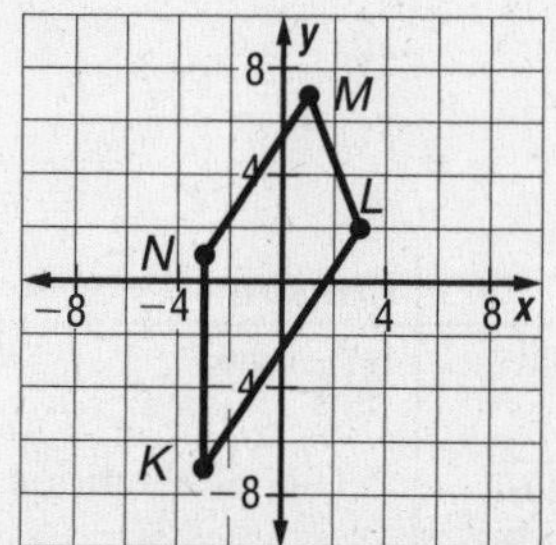

If the opposite sides of a quadrilateral are parallel, then it is a parallelogram. The slope of $\overline{KN}$ is undefined, since it is vertical. $\overline{LM}$ is clearly not vertical. The opposite sides, $\overline{KN}$ and $\overline{LM}$, do not have the same slope, so they are not parallel. Therefore, $KLMN$ is not a parallelogram.

56. Yes;

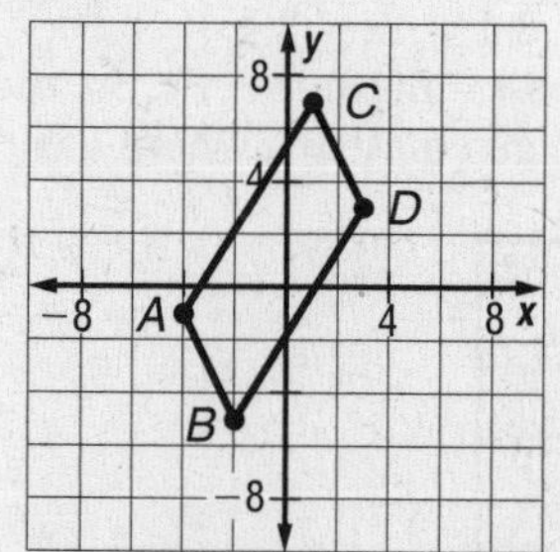

If the opposite sides of a quadrilateral are parallel, then it is a parallelogram.

slope of $\overline{AB} = \frac{-1 - (-5)}{-4 - (-2)}$

$= -2$

slope of $\overline{CD} = \frac{7 - 3}{1 - 3}$

$= -2$

slope of $\overline{AC} = \frac{-1 - 7}{-4 - 1}$

$= \frac{8}{5}$

slope of $\overline{BD} = \frac{-5 - 3}{-2 - 3}$

$= \frac{8}{5}$

Since opposite sides have the same slope, $\overline{AB} \parallel \overline{CD}$ and $\overline{AC} \parallel \overline{BD}$. Therefore, $ABDC$ is a parallelogram by definition.

57. From the Triangle Proportionality Theorem, $\frac{PS}{ST} = \frac{QP}{RT}$.
Substitute the known measures.
$$\frac{PS}{9} = \frac{24}{16}$$
$$PS(16) = 24(9)$$
$$16(PS) = 216$$
$$PS = 13.5$$

58. From the Triangle Proportionality Theorem, $\frac{PS}{PT} = \frac{QS}{QR}$.
Substitute the known measures.
$$\frac{y+2}{y-3} = \frac{16}{16-12}$$
$$4(y+2) = 16(y-3)$$
$$4y + 8 = 16y - 48$$
$$56 = 12y$$
$$4\tfrac{2}{3} = y$$

59. From the Triangle Proportionality Theorem, $\frac{TS}{PS} = \frac{RT}{QP}$.
Substitute the known measures.
$$\frac{TS}{TS+8} = \frac{15}{21}$$
$$21(TS) = 15(TS + 8)$$
$$6(TS) = 120$$
$$TS = 20$$

60. If $\overline{AG} \cong \overline{AC}$, $\triangle ACG$ is isosceles. Then $\angle AGC \cong \angle ACG$.

61. If $\overline{AJ} \cong \overline{AH}$, $\triangle AJH$ is isosceles. Then $\angle AJH \cong \angle AHJ$.

62. If $\angle AFD \cong \angle ADF$, $\triangle ADF$ is isosceles. Then $\overline{AF} \cong \overline{AD}$.

63. If $\angle AKB \cong \angle ABK$, $\triangle ABK$ is isosceles. Then $\overline{AK} \cong \overline{AB}$.

64. Solve for x.
$$\tfrac{1}{2}(8x - 6x - 7) = 5$$
$$8x - 6x - 7 = 10$$
$$2x = 17$$
$$x = 8.5$$

65. Solve for x.
$$\tfrac{1}{2}(7x + 3x + 1) = 12.5$$
$$7x + 3x + 1 = 25$$
$$10x = 24$$
$$x = 2.4$$

66. Solve for x.
$$\tfrac{1}{2}(4x + 6 + 2x + 13) = 15.5$$
$$4x + 6 + 2x + 13 = 31$$
$$6x = 12$$
$$x = 2$$

67. Solve for x.
$$\tfrac{1}{2}(7x - 2 + 3x + 3) = 25.5$$
$$7x - 2 + 3x + 3 = 51$$
$$10x = 50$$
$$x = 5$$

Page 438 Geometry Activity: Kites

1. The diagonals intersect at a right angle.
2. $\angle QRS \cong \angle QTS$
3. See students' work; $\overline{NR} \cong \overline{TN}$, but $\overline{QN} \not\cong \overline{NS}$.
4. 3 pairs: $\triangle QRN \cong \triangle QTN$, $\triangle RNS \cong \triangle TNS$, $\triangle QRS \cong \triangle QTS$
5. 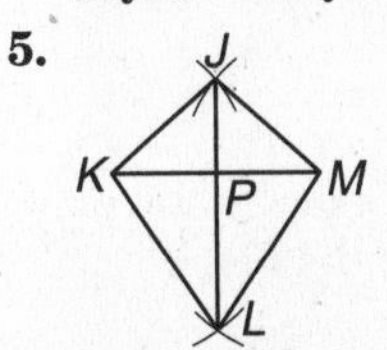

   The diagonals intersect in a right angle; $\angle JKL \cong \angle JML$; $\overline{KP} \cong \overline{PM}$, $\overline{JP} \not\cong \overline{PL}$; 3 pairs: $\triangle JPK \cong \triangle JPM$, $\triangle KPL \cong \triangle MPL$, $\triangle JKL \cong \triangle JML$.
6. One pair of opposite angles is congruent. The diagonals are perpendicular. The longer diagonal bisects the shorter diagonal. The short sides are congruent and the long sides are congruent.

8-6 Trapezoids

Page 441 Geometry Activity: Median of a Trapezoid

1. See students' work.
2. The median is the average of the lengths of the bases.
   So, $MN = \frac{1}{2}(WX + ZY)$.

Page 442 Check for Understanding

1. Exactly one pair of opposite sides is parallel.
2.

Properties	Trapezoid	Rectangle	Square	Rhombus
diagonals are congruent	only isosceles	yes	yes	no
diagonals are perpendicular	no	no	yes	yes
diagonals bisect each other	no	yes	yes	yes
diagonals bisect angles	no	no	yes	yes

3. Sample answer: The median of a trapezoid is parallel to both bases.

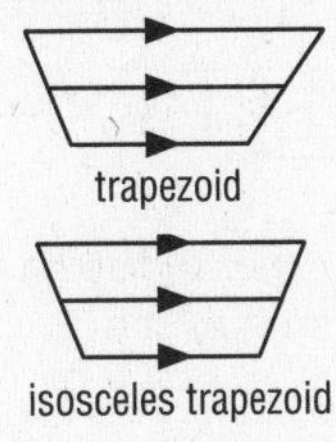

4.

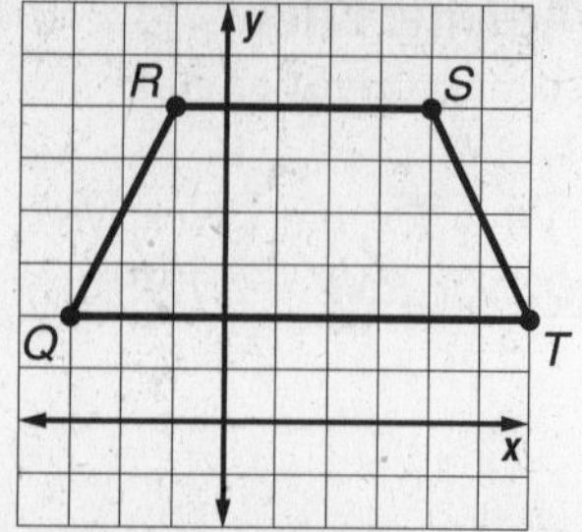

A quadrilateral is a trapezoid if exactly one pair of opposite sides is parallel. Use the Slope Formula.

slope of $\overline{QT} = \frac{2-2}{-3-6}$
$= 0$

slope of $\overline{RS} = \frac{6-6}{-1-4}$
$= 0$

slope of $\overline{QR} = \frac{2-6}{-3-(-1)}$
$= 2$

slope of $\overline{ST} = \frac{6-2}{4-6}$
$= -2$

Exactly one pair of opposite sides is parallel, $\overline{QT}$ and $\overline{RS}$. So, $QRST$ is a trapezoid.

5. See graph in Exercise 4.
Use the Distance Formula to show that the legs are congruent.

$QR = \sqrt{[-3-(-1)]^2 + (2-6)^2}$
$= \sqrt{4+16} = \sqrt{20}$

$ST = \sqrt{(4-6)^2 + (6-2)^2}$
$= \sqrt{4+16} = \sqrt{20}$

Since the legs are congruent, $QRST$ is an isosceles trapezoid.

6. Given: $CDFG$ is an isosceles trapezoid with bases $\overline{CD}$ and $\overline{FG}$.
Prove: $\angle DGF \cong \angle CFG$
Proof:

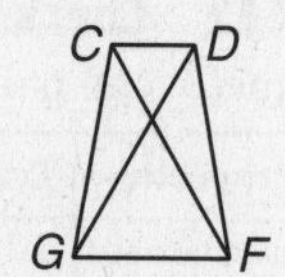

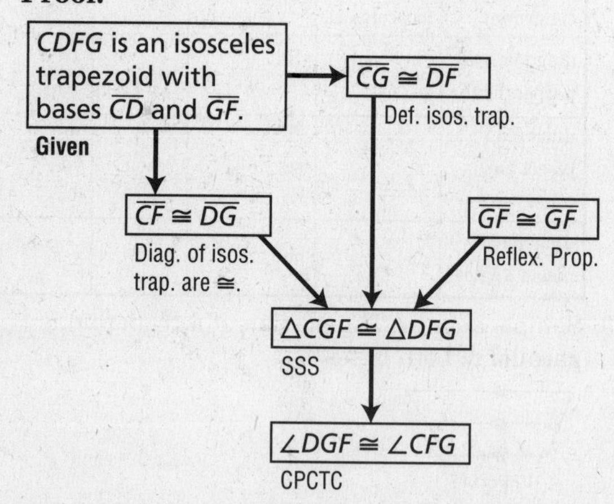

7. The median of a trapezoid is parallel to the bases, and its measure is one-half the sum of the measures of the bases. Find x.

$YZ = \frac{1}{2}(EF + HG)$

$13 = \frac{1}{2}[(3x + 8) + (4x - 10)]$

$26 = 7x - 2$

$28 = 7x$

$4 = x$

8. The perspective makes it appear that the buildings are formed by trapezoids and parallelograms.

Pages 442–445 Practice and Apply

9a.

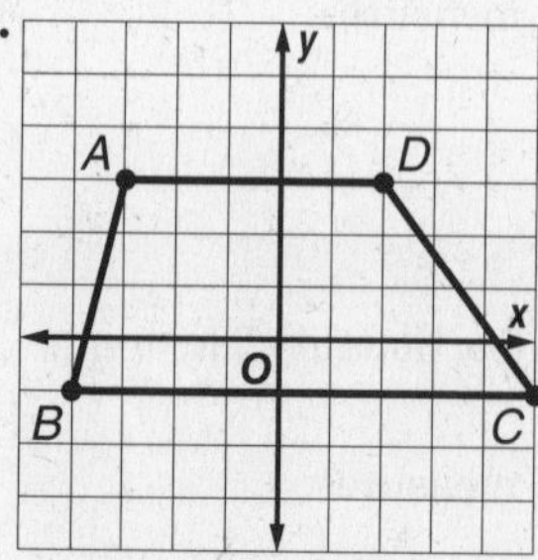

A quadrilateral is a trapezoid if exactly one pair of opposite sides is parallel. Use the Slope Formula.

slope of $\overline{AD} = \frac{3-3}{-3-2}$
$= 0$

slope of $\overline{BC} = \frac{-1-(-1)}{-4-5}$
$= 0$

slope of $\overline{AB} = \frac{3-(-1)}{-3-(-4)}$
$= 4$

slope of $\overline{CD} = \frac{-1-3}{5-2}$
$= -\frac{4}{3}$

Exactly one pair of opposite sides is parallel, $\overline{AD}$ and $\overline{BC}$. So, $ABCD$ is a trapezoid.

9b.

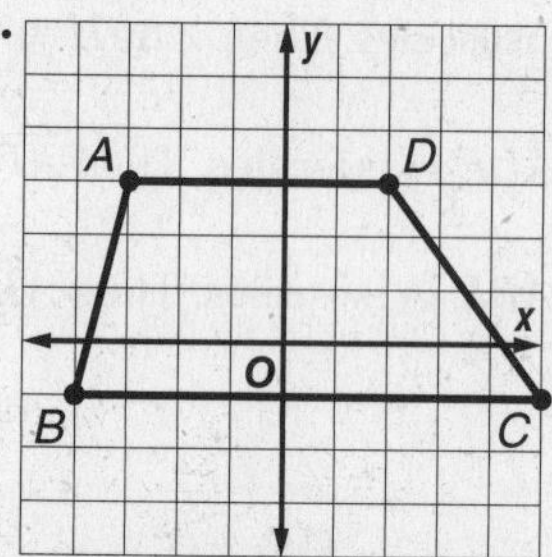

Use the Distance Formula to determine whether the legs are congruent.

$AB = \sqrt{[-3-(-4)]^2 + [3-(-1)]^2}$
$= \sqrt{1+16}$
$= \sqrt{17}$

$CD = \sqrt{(5-2)^2 + (-1-3)^2}$
$= \sqrt{9+16}$
$= 5$

Since the legs are not congruent, $ABCD$ is not an isosceles trapezoid.

10a.

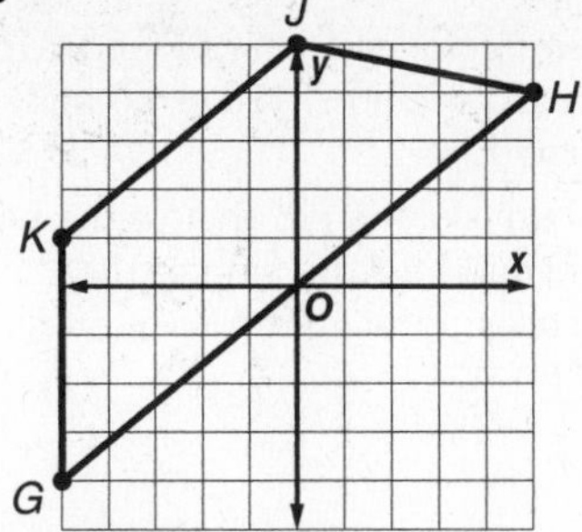

A quadrilateral is a trapezoid if exactly one pair of opposite sides is parallel. Use the Slope Formula.

slope of $\overline{GH} = \frac{-4 - 4}{-5 - 5}$

$= \frac{4}{5}$

slope of $\overline{KJ} = \frac{5 - 1}{0 - (-5)}$

$= \frac{4}{5}$

slope of $\overline{GK} = \frac{1 - (-4)}{-5 - (-5)}$

$= \frac{5}{0}$ or undefined

slope of $\overline{HJ} = \frac{4 - 5}{5 - 0}$

$= -\frac{1}{5}$

Exactly one pair of opposite sides is parallel, $\overline{KJ}$ and $\overline{GH}$. So, *GHJK* is a trapezoid.

10b.

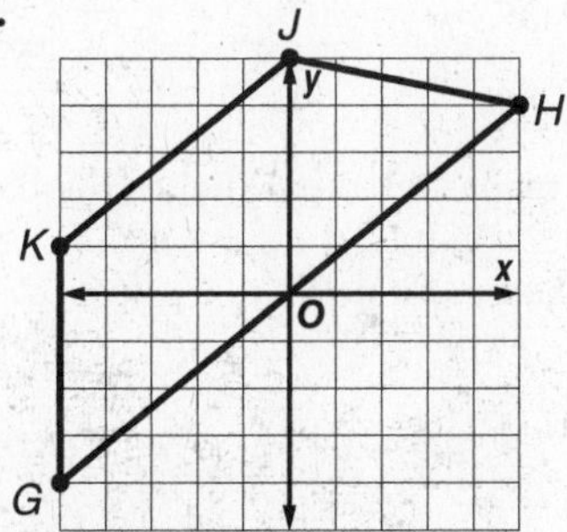

Use the Distance Formula to determine whether the legs are congruent.

$GK = \sqrt{[-5 - (-5)]^2 + (-4 - 1)^2}$

$= \sqrt{0 + 25}$

$= 5$

$JH = \sqrt{(0 - 5)^2 + (5 - 4)^2}$

$= \sqrt{25 + 1}$

$= \sqrt{26}$

Since the legs are not congruent, *GHJK* is not an isosceles trapezoid.

11a.

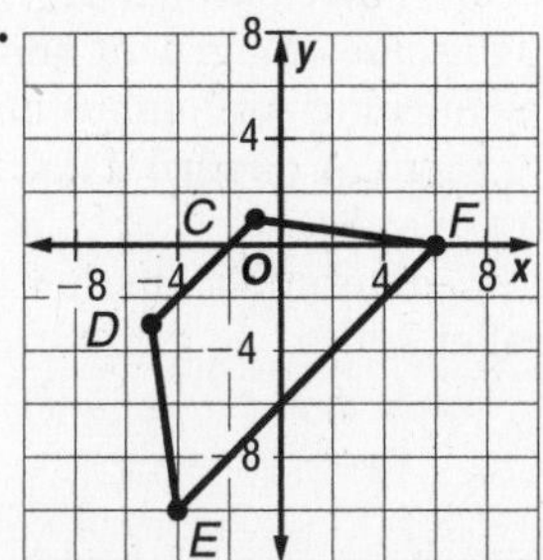

A quadrilateral is a trapezoid if exactly one pair of opposite sides is parallel. Use the Slope Formula.

slope of $\overline{DC} = \frac{-3 - 1}{-5 - (-1)}$

$= 1$

slope of $\overline{FE} = \frac{0 - (-10)}{6 - (-4)}$

$= 1$

slope of $\overline{DE} = \frac{-3 - (-10)}{-5 - (-4)}$

$= -7$

slope of $\overline{CF} = \frac{1 - 0}{-1 - 6}$

$= -\frac{1}{7}$

Exactly one pair of opposite sides is parallel, $\overline{DC}$ and $\overline{FE}$. So, *CDEF* is a trapezoid.

11b.

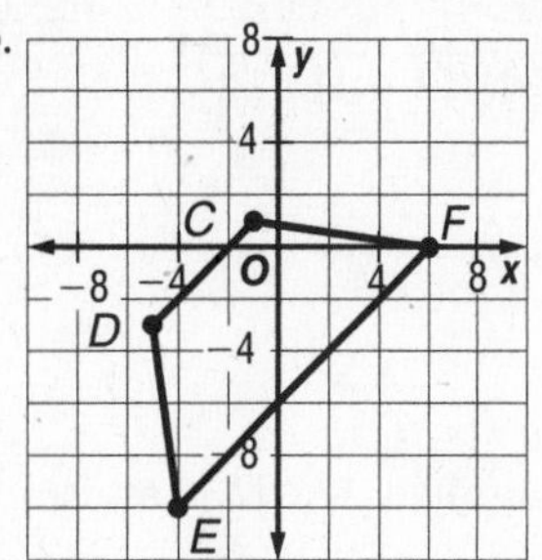

Use the Distance Formula to show that the legs are congruent.

$DE = \sqrt{[-5 - (-4)]^2 + [-3 - (-10)]^2}$

$= \sqrt{1 + 49}$

$= \sqrt{50}$

$CF = \sqrt{(-1 - 6)^2 + (1 - 0)^2}$

$= \sqrt{49 + 1}$

$= \sqrt{50}$

Since the legs are congruent, *CDEF* is an isosceles trapezoid.

12a.

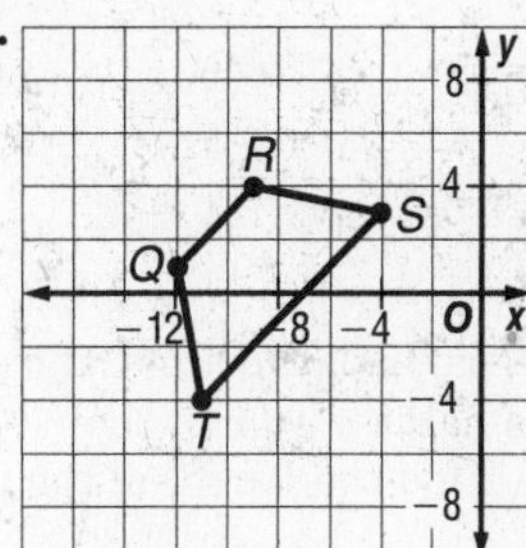

A quadrilateral is a trapezoid if exactly one pair of opposite sides is parallel. Use the Slope Formula.

slope of $\overline{QR} = \frac{1 - 4}{-12 - (-9)}$

$= 1$

slope of $\overline{TS} = \frac{-4 - 3}{-11 - (-4)}$

$= 1$

slope of $\overline{QT} = \frac{1 - (-4)}{-12 - (-11)}$

$= -5$

slope of $\overline{RS} = \frac{4 - 3}{-9 - (-4)}$

$= -\frac{1}{5}$

Exactly one pair of opposite sides is parallel, $\overline{QR}$ and $\overline{TS}$. So, *QRST* is a trapezoid.

12b. 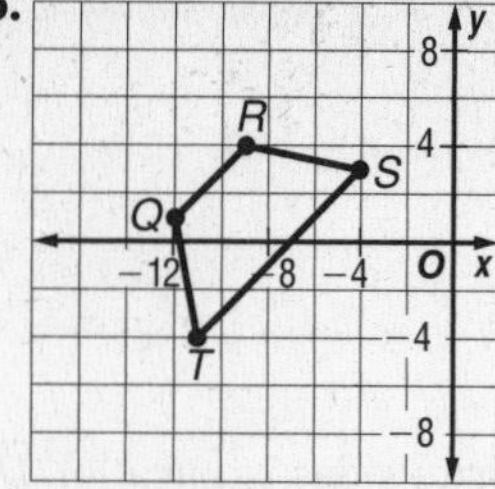

Use the Distance Formula to show that the legs are congruent.

$QT = \sqrt{[1 - (-4)]^2 + [-12 - (-11)]^2}$
$= \sqrt{25 + 1}$
$= \sqrt{26}$

$RS = \sqrt{(4 - 3)^2 + [-9 - (-4)]^2}$
$= \sqrt{1 + 25}$
$= \sqrt{26}$

Since the legs are congruent, $QRST$ is an isosceles trapezoid.

13. The median of a trapezoid is parallel to the bases, and its measure is one-half the sum of the measures of the bases. Find DE.

$XY = \frac{1}{2}(DE + HG)$
$20 = \frac{1}{2}(DE + 32)$
$40 = DE + 32$
$8 = DE$

14. The median of a trapezoid is parallel to the bases, and its measure is one-half the sum of the measures of the bases. Find VT.

$AB = \frac{1}{2}(VT + RS)$
$15 = \frac{1}{2}(VT + 26)$
$30 = VT + 26$
$4 = VT$

15. The median of a trapezoid is parallel to the bases, and its measure is one-half the sum of the measures of the bases. Find the length of the median and let it be x.

$x = \frac{1}{2}(WZ + XY)$
$= \frac{1}{2}(8 + 20)$
$= 14$

The length of the median is 14.
Both base pairs of an isosceles trapezoid are congruent, so $m\angle X = 70$ and $m\angle W = m\angle Z$. The sum of the measures of the interior angles of a quadrilateral is 360.
Find $m\angle W$ and $m\angle Z$.

$m\angle W + m\angle Z + m\angle X + m\angle Y = 360$
$m\angle W + m\angle W + 70 + 70 = 360$
$2(m\angle W) = 220$
$m\angle W = 110$
$m\angle Z = 110$

16. The median of a trapezoid is parallel to the bases, and its measure is one-half the sum of the measures of the bases. Find AB.

$AB = \frac{1}{2}(TS + QR)$
$= \frac{1}{2}(12 + 20)$
$= 16$

So, $AB = 16$.
$\angle Q$ and $\angle S$ are supplementary to $\angle T$ and $\angle R$, respectively. So, $m\angle Q = 180 - 120$ or 60 and $m\angle S = 180 - 45$ or 135.

17. The median of a trapezoid is parallel to the bases, and its measure is one-half the sum of the measures of the bases. First find AB.

$AB = \frac{1}{2}(RS + QT)$
$= \frac{1}{2}(54 + 86)$
$= 70$

So, $AB = 70$.
Find GH.

$GH = \frac{1}{2}(RS + AB)$
$= \frac{1}{2}(54 + 70)$
$= 62$

So, $GH = 62$.

18. The median of a trapezoid is parallel to the bases, and its measure is one-half the sum of the measures of the bases. Find AB.
From Exercise 17, $AB = 70$.
Find JK.

$JK = \frac{1}{2}(QT + AB)$
$= \frac{1}{2}(86 + 70)$
$= 78$

So, $JK = 78$.

19. The median of a trapezoid is parallel to the bases, and its measure is one-half the sum of the measures of the bases. Find RP.

$RP = \frac{1}{2}(JK + LM)$
$5 + x = \frac{1}{2}\left[2(x + 3) + \frac{1}{2}x - 1\right]$
$10 + 2x = 2x + 6 + \frac{1}{2}x - 1$
$5 = \frac{1}{2}x$
$10 = x$

$x = 10$, so $RP = 5 + 10$ or 15.

20. Since the two octagons are regular polygons with the same center, the quadrilaterals are trapezoids with one pair of opposite sides parallel.

21. Sample answer: triangles, quadrilaterals, trapezoids, hexagons

22. A trapezoid must have exactly one pair of opposite sides parallel. A parallelogram must have both pairs of opposite sides parallel. A square must have all four sides congruent and consecutive sides perpendicular. A rhombus must have all four sides congruent. A quadrilateral has four sides.
Use the Slope Formula to determine whether the opposite sides are parallel.

slope of $\overline{BC} = \frac{2 - 4}{1 - 4}$
$= \frac{2}{3}$

slope of $\overline{DE} = \frac{1 - (-1)}{5 - 2}$
$= \frac{2}{3}$

slope of $\overline{CD} = \frac{4 - 1}{4 - 5}$
$= -3$

slope of $\overline{BE} = \frac{2-(-1)}{1-2}$
$= -3$

Use the distance formula to compare the lengths of the sides.

$BC = \sqrt{(1-4)^2 + (2-4)^2}$
$= \sqrt{9+4}$
$= \sqrt{13}$

$DE = \sqrt{(5-2)^2 + [1-(-1)]^2}$
$= \sqrt{9+4}$
$= \sqrt{13}$

$CD = \sqrt{(4-5)^2 + (4-1)^2}$
$= \sqrt{1+9}$
$= \sqrt{10}$

$BE = \sqrt{(1-2)^2 + [2-(-1)]^2}$
$= \sqrt{1+9}$
$= \sqrt{10}$

Opposite sides are parallel. Since the slopes of consecutive sides are not negative reciprocals, consecutive sides are not perpendicular; there are no right angles. Opposite sides are congruent, but consecutive sides are not congruent. *BCDE* is a parallelogram.

23. A trapezoid must have exactly one pair of opposite sides parallel. A parallelogram must have both pairs of opposite sides parallel. A square must have all four sides congruent and consecutive sides perpendicular. A rhombus must have all four sides congruent.
A quadrilateral has four sides.
Use the Slope Formula to determine whether the opposite sides are parallel.

slope of $\overline{GH} = \frac{2-2}{-2-4}$
$= 0$

slope of $\overline{JK} = \frac{-1-(-1)}{6-(-4)}$
$= 0$

slope of $\overline{GK} = \frac{2-(-1)}{-2-(-4)}$
$= \frac{3}{2}$

slope of $\overline{HJ} = \frac{2-(-1)}{4-6}$
$= -\frac{3}{2}$

Exactly one pair of opposite sides is parallel, so *GHJK* is a trapezoid.

24. A trapezoid must have exactly one pair of opposite sides parallel. A parallelogram must have both pairs of opposite sides parallel. A square must have all four sides congruent and consecutive sides perpendicular. A rhombus must have all four sides congruent. A quadrilateral has four sides.
Use the Slope Formula to determine whether the opposite sides are parallel.

slope of $\overline{MN} = \frac{1-3}{-3-1}$
$= \frac{1}{2}$

slope of $\overline{OP} = \frac{-1-(-2)}{3-(-2)}$
$= \frac{1}{5}$

slope of $\overline{MP} = \frac{1-(-2)}{-3-(-2)}$
$= -3$

slope of $\overline{NO} = \frac{3-(-1)}{1-3}$
$= -2$

All four sides have different slopes. Therefore, opposite sides are not parallel and the figure is a quadrilateral.

25. A trapezoid must have exactly one pair of opposite sides parallel. A parallelogram must have both pairs of opposite sides parallel. A square must have all four sides congruent and consecutive sides perpendicular. A rhombus must have all four sides congruent. A quadrilateral has four sides.
Use the Slope Formula to determine whether the opposite sides are parallel.

slope of $\overline{QR} = \frac{0-3}{-3-0}$
$= 1$

slope of $\overline{ST} = \frac{0-(-3)}{3-0}$
$= 1$

slope of $\overline{QT} = \frac{0-(-3)}{-3-0}$
$= -1$

slope of $\overline{RS} = \frac{3-0}{0-3}$
$= -1$

Use the distance formula to compare the lengths of the sides.

$QR = \sqrt{(-3-0)^2 + (0-3)^2}$
$= \sqrt{9+9} = \sqrt{18}$

$ST = \sqrt{(3-0)^2 + [0-(-3)]^2}$
$= \sqrt{9+9} = \sqrt{18}$

$QT = \sqrt{(-3-0)^2 + [0-(-3)]^2}$
$= \sqrt{9+9} = \sqrt{18}$

$RS = \sqrt{(0-3)^2 + (3-0)^2}$
$= \sqrt{9+9} = \sqrt{18}$

All four sides have the same length and are congruent. Opposite sides are parallel. Because the slopes of consecutive sides are opposite reciprocals, consecutive sides are perpendicular. *QRST* is a square.

26. A quadrilateral is a trapezoid if exactly one pair of opposite sides is parallel. Use the Slope Formula.

slope of $\overline{QR} = \frac{4-1}{0-4}$
$= -\frac{3}{4}$

slope of $\overline{PS} = \frac{3-(-3)}{-4-4}$
$= -\frac{3}{4}$

slope of $\overline{PQ} = \frac{3-4}{-4-0}$
$= \frac{1}{4}$

slope of $\overline{RS} = \frac{1-(-3)}{4-4} = \frac{4}{0}$ or undefined

Exactly one pair of opposite sides is parallel, $\overline{QR}$ and $\overline{PS}$. So, *PQRS* is a trapezoid.

Use the Distance Formula to determine whether the legs are congruent.

$PQ = \sqrt{(-4-0)^2 + (3-4)^2}$
$= \sqrt{16+1}$
$= \sqrt{17}$
$RS = \sqrt{(4-4)^2 + [1-(-3)]^2}$
$= \sqrt{0+16}$
$= 4$

Since $PQ \neq RS$, $PQRS$ is not an isosceles trapezoid.

27. Use the Midpoint Formula to find the coordinates of the midpoints of $\overline{PQ}$ and $\overline{RS}$.

$\overline{PQ}$: $\left(\frac{-4+0}{2}, \frac{3+4}{2}\right) = (-2, 3.5)$

$\overline{RS}$: $\left(\frac{4+4}{2}, \frac{1+(-3)}{2}\right) = (4, -1)$

The coordinates of the midpoints of $\overline{PQ}$ and $\overline{RS}$ are $A(-2, 3.5)$ and $B(4, -1)$, respectively.

28. The median of a trapezoid is parallel to the bases, and its measure is one-half the sum of the measures of the bases. So, the length of $\overline{AB}$ is given by $AB = \frac{1}{2}(QR + PS)$. To find the lengths of the bases, recognize that the bases are the hypotenuses of two special right triangles, 3-4-5 and 6-8-10. Find AB.

$AB = \frac{1}{2}(QR + PS)$
$= \frac{1}{2}(5 + 10)$
$= 7.5$

So, $AB = 7.5$.

29. A quadrilateral is a trapezoid if exactly one pair of opposite sides is parallel. Use the Slope Formula.

slope of $\overline{DE} = \frac{2-5}{-2-5}$
$= \frac{3}{7}$

slope of $\overline{FG} = \frac{-3-(-2)}{5-(-2)}$
$= -\frac{1}{7}$

The slopes of $\overline{DG}$ and $\overline{EF}$ are undefined. They are both vertical; therefore, they are parallel.
Exactly one pair of opposite sides is parallel, $\overline{DG}$ and $\overline{EF}$. So, $DEFG$ is a trapezoid.
Use the Distance Formula to determine whether the legs are congruent.

$DE = \sqrt{(-2-5)^2 + (2-5)^2}$
$= \sqrt{49+9}$
$= \sqrt{58}$
$GF = \sqrt{[5-(-2)]^2 + [-3-(-2)]^2}$
$= \sqrt{49+1}$
$= \sqrt{50}$

Since $DE \neq GF$, $DEFG$ is not an isosceles trapezoid.

30. Use the Midpoint Formula to find the coordinates of the midpoints of $\overline{DE}$ and $\overline{GF}$.

$\overline{DE}$: $\left(\frac{-2+5}{2}, \frac{2+5}{2}\right) = (1.5, 3.5)$

$\overline{GF}$: $\left(\frac{-2+5}{2}, \frac{-2+(-3)}{2}\right) = (1.5, -2.5)$

The coordinates of the midpoints of $\overline{DE}$ and $\overline{GF}$ are $W(1.5, 3.5)$ and $V(1.5, -2.5)$, respectively.

31. The median of a trapezoid is parallel to the bases, and its measure is one-half the sum of the measures of the bases. So, the length of $\overline{WV}$ is given by $WV = \frac{1}{2}(DG + EF)$. Since $\overline{DG}$ and $\overline{EF}$ are vertical, their lengths are given by $|y_1 - y_2|$. Find WV.

$WV = \frac{1}{2}(DG + EF)$
$= \frac{1}{2}[2 - (-2) + 5 - (-3)]$
$= 6$

So, $WV = 6$.

32. Given: $\overline{HJ} \parallel \overline{GK}$, $\overline{HG} \nparallel \overline{JK}$, $\triangle HGK \cong \triangle JKG$
Prove: $GHJK$ is an isosceles trapezoid.

Proof:

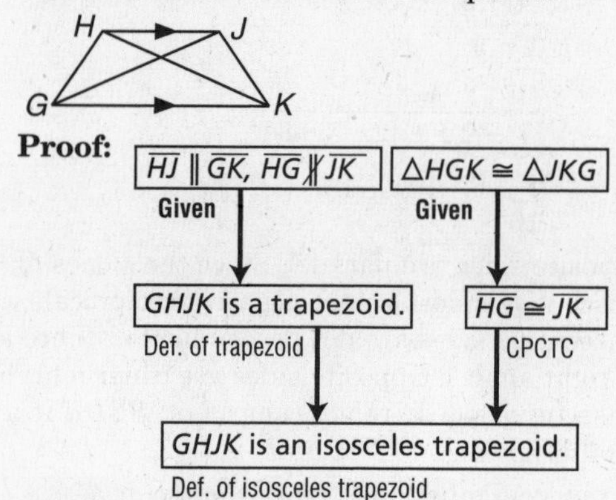

33. Given: $\triangle TZX \cong \triangle YXZ$, $\overline{WX} \nparallel \overline{ZY}$
Prove: $XYZW$ is a trapezoid.

Proof:

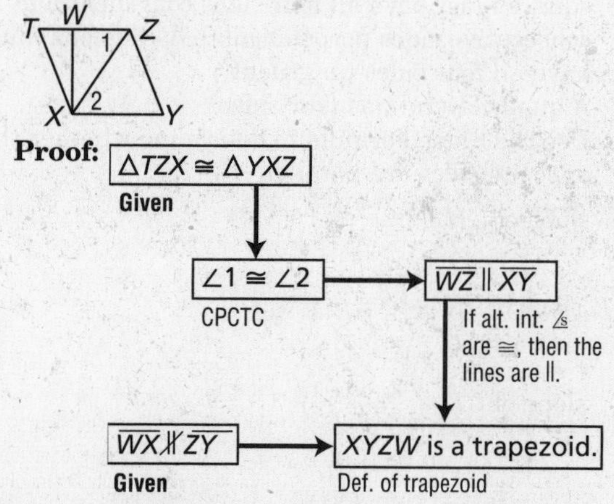

34. Given: $ZYXP$ is an isosceles trapezoid.
Prove: $\triangle PWX$ is isosceles.

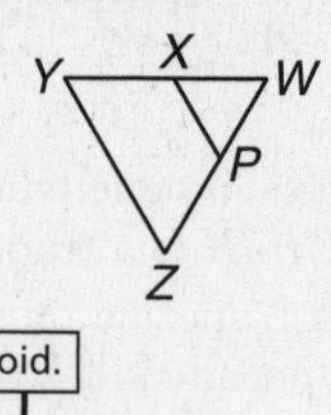

Proof:

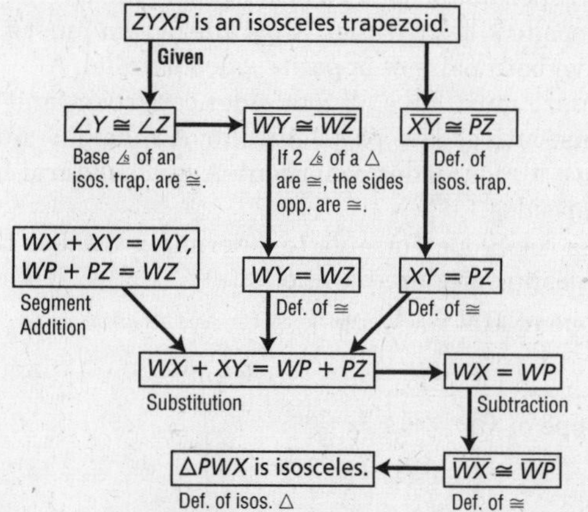

35. Given: E and C are midpoints of $\overline{AD}$ and $\overline{DB}$.
$\overline{AD} \cong \overline{DB}$
Prove: $ABCE$ is an isosceles trapezoid.

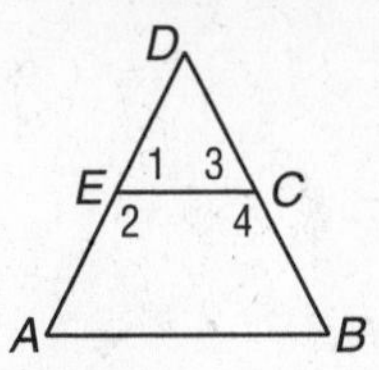

Proof:

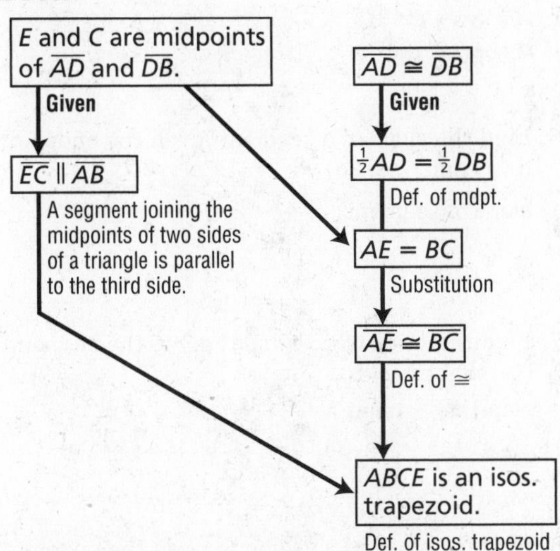

36. Given: $ABCD$ is an isosceles trapezoid.
$\overline{BC} \parallel \overline{AD}$
$\overline{AB} \cong \overline{CD}$
Prove: $\angle A \cong \angle D$
$\angle ABC \cong \angle DCB$

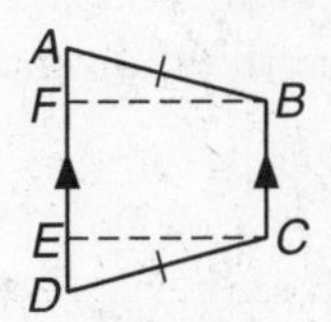

Proof: Draw auxiliary segments so that $\overline{BF} \perp \overline{AD}$ and $\overline{CE} \perp \overline{AD}$. Since $\overline{BC} \parallel \overline{AD}$ and parallel lines are everywhere equidistant, $\overline{BF} \cong \overline{CE}$. Perpendicular lines form right angles, so $\angle BFA$ and $\angle CED$ are right angles. $\triangle BFA$ and $\triangle CED$ are right triangles by definition. Therefore, $\triangle BFA \cong \triangle CED$ by HL. $\angle A \cong \angle D$ by CPCTC. Since $\angle CBF$ and $\angle BCE$ are right angles and all right angles are congruent, $\angle CBF \cong \angle BCE$. $\angle ABF \cong \angle DCE$ by CPCTC. So, $\angle ABC \cong \angle DCB$ by angle addition.

37. Sample answer:

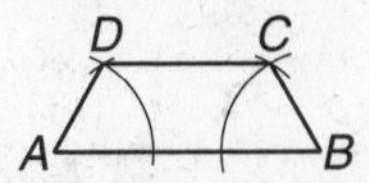

38. Sample answer:

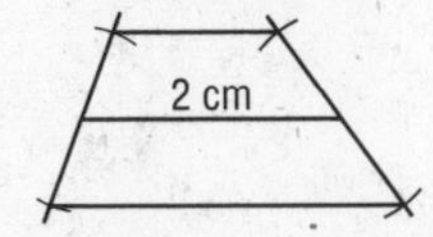

39.

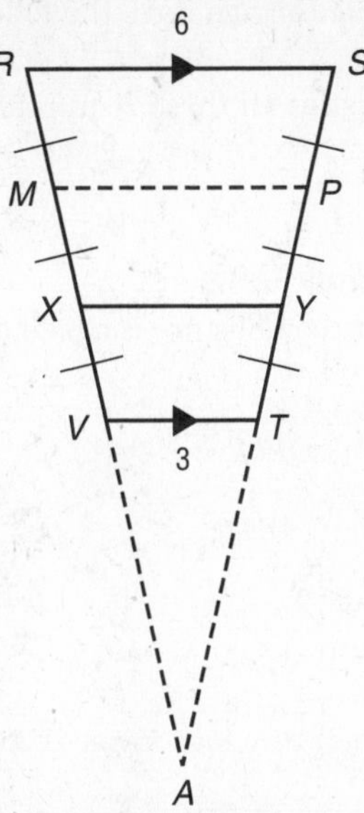

Extend $\overline{RV}$ and $\overline{ST}$ to intersect at A. Since $RSTV$ is an isoceles trapezoid, $\angle R \cong \angle S$. So $\triangle RSA$ is isosceles. Then $\overline{RA} \cong \overline{SA}$. $RA = RX + XA$ and $SA = SY + YA$. Since $RA = SA$ and $RX = SY$, $XA = YA$. By the Converse of the Triangle Proportionality Theorem, $\overline{XY} \parallel \overline{RS}$. Since $\overline{RS} \parallel \overline{VT}$, $\overline{XY} \parallel \overline{VT}$. Let M and P be the midpoints of $\overline{RX}$ and $\overline{SY}$, respectively. Draw $\overline{MP}$. $RX = 2(XV)$, so $\frac{1}{2}RX = XV$. $RM = MX = \frac{1}{2}RX$. So $RM = MX = XV$. Similarly, $SP = PY = YT$.
By an argument similar to the one above, $\overline{MP} \parallel \overline{RS}$, $\overline{MP} \parallel \overline{XY}$, $\overline{MP} \parallel \overline{VT}$. So $\overline{XY}$ is the median of isosceles trapezoid $MPTV$. And $\overline{MP}$ is the median of isosceles trapezoid $RSYX$.

So, $MP = \frac{1}{2}(RS + XY)$ and $XY = \frac{1}{2}(MP + VT)$

$MP = \frac{1}{2}(6 + XY)$ $\qquad$ $XY = \frac{1}{2}(MP + 3)$

$MP = 3 + \frac{1}{2}XY$

Substitute this expression for MP into the second equation.

$$XY = \frac{1}{2}[(3 + \frac{1}{2}XY) + 3]$$
$$XY = \frac{1}{2}(6 + \frac{1}{2}XY)$$
$$XY = 3 + \frac{1}{4}XY$$
$$\frac{3}{4}XY = 3$$
$$XY = 4$$

40. It is not possible. Since pairs of base angles of an isosceles trapezoid are congruent, if two angles are right, all four angles will be right. Then the quadrilateral would be a rectangle, not a trapezoid.

41. Sample answer: Trapezoids are used in monuments as well as other buildings. Answers should include the following.
- Trapezoids have exactly one pair of opposite sides parallel.
- Trapezoids can be used as window panes.

42. Quadrilateral $WXYZ$ is a trapezoid because exactly one pair of opposite sides is parallel, $\overline{WX}$ and $\overline{YZ}$.

43. B; points in the shaded region have the following characteristics: $x < 0, y > 0$, and $y < \frac{5}{3}x + 5$. The only choice that satisfies all three requirements is B, $(-1, 3)$.

Page 445 Maintain Your Skills

44. Opposite angles of a rhombus are congruent, so $\angle QLM \cong \angle QPM$. Find x.

$$\angle QLM \cong \angle QPM$$
$$m\angle QLM = m\angle QPM$$
$$2x^2 - 10 = 8x$$
$$2x^2 - 8x - 10 = 0$$
$$x^2 - 4x - 5 = 0$$
$$(x - 5)(x + 1) = 0$$
$$x - 5 = 0 \text{ or } x + 1 = 0$$
$$x = 5 \qquad x = -1$$

x must be greater than zero (otherwise $m\angle QPM$ is negative), so $x = 5$.

The diagonals of a rhombus bisect the angles, so $m\angle LPQ = \frac{1}{2}m\angle QPM$, and $m\angle QPM = 8(5)$ or 40. Therefore, $m\angle LPQ = 20$.

45. By definition, a rhombus has four congruent sides, so $\overline{QL} \cong \overline{MP}$ and $QL = 10$.

46. Consecutive angles in a rhombus are supplementary, so $\angle QLM$ and $\angle LQP$ are supplementary. From Exercise 44, $x = 5$. So $m\angle QLM = 2(5)^2 - 10$ or 40. Find $m\angle LQP$.

$$m\angle QLM + m\angle LQP = 180$$
$$40 + m\angle LQP = 180$$
$$m\angle LQP = 140$$

47. The diagonals of a rhombus bisect the angles, so $m\angle LQM = \frac{1}{2}m\angle LQP$. $m\angle LQM = \frac{1}{2}(140)$ or 70.

48. By definition, a rhombus has four congruent sides, so the perimeter is 4(10) or 40.

49. Use the Distance Formula to find RS and TV.

$$RS = \sqrt{(-7 - 0)^2 + (-3 - 4)^2}$$
$$= \sqrt{49 + 49} = 7\sqrt{2}$$
$$TV = \sqrt{[3 - (-4)]^2 + [1 - (-7)]^2}$$
$$= \sqrt{49 + 64} = \sqrt{113}$$

50. Find the coordinates of the midpoints of $\overline{RT}$ and $\overline{SV}$ using the Midpoint Formula.

$$\overline{RT}: \left(\frac{-7 + 3}{2}, \frac{-3 + 1}{2}\right) = (-2, -1)$$

$$\overline{SV}: \left(\frac{0 + (-4)}{2}, \frac{4 + (-7)}{2}\right) = \left(-2, -\frac{3}{2}\right)$$

51. No; $RSTV$ is not a rectangle because opposite sides are not congruent (Exercise 49) and the diagonals do not bisect each other (Exercise 50).

52. Solve the proportion for y.

$$\frac{16}{38} = \frac{24}{y}$$
$$16y = 24(38)$$
$$16y = 912$$
$$y = 57$$

53. Solve the proportion for y.

$$\frac{y}{6} = \frac{17}{30}$$
$$30y = 17(6)$$
$$30y = 102$$
$$y = \frac{17}{5}$$

54. Solve the proportion for y.

$$\frac{5}{y + 4} = \frac{20}{28}$$
$$5(28) = 20(y + 4)$$
$$140 = 20y + 80$$
$$60 = 20y$$
$$3 = y$$

55. Solve the proportion for y.

$$\frac{2y}{9} = \frac{52}{36}$$
$$36(2y) = 52(9)$$
$$72y = 468$$
$$y = \frac{13}{2}$$

56. Find the slope of a segment given the endpoints $(0, a)$ and $(-a, 2a)$.

$$\text{slope of segment} = \frac{y_2 - y_1}{x_2 - x_1}$$
$$= \frac{2a - a}{-a - 0}$$
$$= -1$$

57. Find the slope of a segment given the endpoints $(-a, b)$ and (a, b).

$$\text{slope of segment} = \frac{y_2 - y_1}{x_2 - x_1}$$
$$= \frac{b - b}{a - (-a)}$$
$$= 0$$

58. Find the slope of a segment given the endpoints (c, c) and (c, d).

$$\text{slope of segment} = \frac{y_2 - y_1}{x_2 - x_1}$$
$$= \frac{d - c}{c - c}$$
$$= \frac{d - c}{0}, \text{ which is undefined}$$

59. Find the slope of a segment given the endpoints $(a, -b)$ and $(2a, b)$.

$$\text{slope of segment} = \frac{y_2 - y_1}{x_2 - x_1}$$
$$= \frac{b - (-b)}{2a - a}$$
$$= \frac{2b}{a}$$

60. Find the slope of a segment given the endpoints $(3a, 2b)$ and $(b, -a)$.

$$\text{slope of segment} = \frac{y_2 - y_1}{x_2 - x_1}$$
$$= \frac{-a - 2b}{b - 3a}$$
$$= \frac{a + 2b}{3a - b}$$

61. Find the slope of a segment given the endpoints (b, c) and $(-b, -c)$.

$$\text{slope of segment} = \frac{y_2 - y_1}{x_2 - x_1}$$
$$= \frac{-c - c}{-b - b}$$
$$= \frac{c}{b}$$

Page 445 Practice Quiz 2

1. $\angle BAD$ is a right angle, so $\angle BAC$ and $\angle CAD$ are complementary. Find x.

$$m\angle BAC + m\angle CAD = 90$$
$$(2x + 1) + (5x + 5) = 90$$
$$7x = 84$$
$$x = 12$$

2. $\angle ABD \cong \angle BDC$ because they are alternate interior angles. Find y.

$$\angle ABD \cong \angle BDC$$
$$m\angle ABD = m\angle BDC$$
$$y^2 = 3y + 10$$
$$y^2 - 3y - 10 = 0$$
$$(y - 5)(y + 2) = 0$$
$$y - 5 = 0 \text{ or } y + 2 = 0$$
$$y = 5 \qquad y = -2$$

If $y = 5$, $m\angle ABD = 5^2$ or 25. If $y = -2$, $m\angle ABD = (-2)^2$ or 4. $\angle ABD \cong \angle BAC$ because they are base angles of an isosceles triangle, so their measures are equal.
From Question 1, $x = 12$.
So, $m\angle BAC = 2(12) + 1$ or 25, therefore, y is 5.
Reject $y = -2$ because it leads to a contradiction.

3.

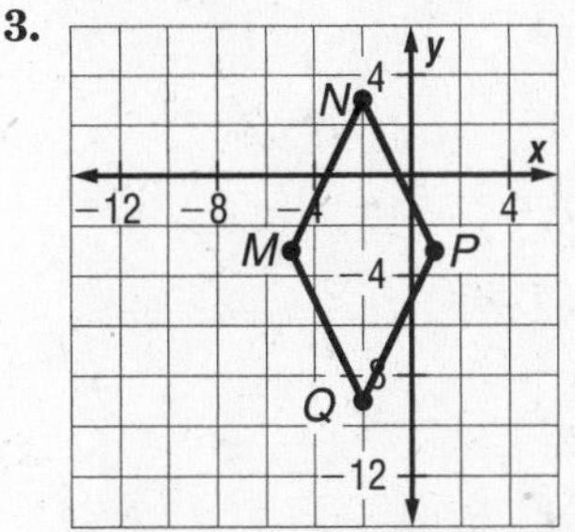

If opposite sides are parallel, then $MNPQ$ can be a rhombus, a rectangle, or a square. If diagonals are perpendicular, then $MNPQ$ can be a rhombus or a square. If consecutive sides are perpendicular, then $MNPQ$ is a rectangle or a square.
Use the Slope Formula to determine whether opposite sides are parallel and consecutive sides are perpendicular.

$$\text{slope of } \overline{MN} = \frac{-3 - 3}{-5 - (-2)} = 2$$

$$\text{slope of } \overline{QP} = \frac{-9 - (-3)}{-2 - 1} = 2$$

$$\text{slope of } \overline{NP} = \frac{3 - (-3)}{-2 - 1} = -2$$

$$\text{slope of } \overline{MQ} = \frac{-3 - (-9)}{-5 - (-2)} = -2$$

Use the Slope Formula to determine whether the diagonals are perpendicular.

$$\text{slope of } \overline{NQ} = \frac{3 - (-9)}{-2 - (-2)} = \frac{12}{0}, \text{ which is undefined}$$

$$\text{slope of } \overline{MP} = \frac{-3 - (-3)}{-5 - 1} = 0$$

Opposite sides are parallel. Since one diagonal is vertical and the other is horizontal, they are perpendicular. Since the slopes of consecutive sides are not opposite reciprocals, they are not perpendicular. So, $MNPQ$ is a rhombus.

4. The median of a trapezoid is parallel to the bases, and its measure is one-half the sum of the measures of the bases. Find MN.

$$MN = \tfrac{1}{2}(TR + VS)$$
$$= \tfrac{1}{2}(44 + 21)$$
$$= 32.5$$

5. The median of a trapezoid is parallel to the bases, and its measure is one-half the sum of the measures of the bases. Find VS.

$$MN = \tfrac{1}{2}(TR + VS)$$
$$25 = \tfrac{1}{2}(32 + VS)$$
$$50 = 32 + VS$$
$$18 = VS$$

Page 446 Reading Mathematics: Hierarchy of Polygons

1. False; all jums are mogs and some mogs are jums. Mogs is of a higher class than jums in the hierarchy.
2. False; both jebs and jums are mogs, but no jebs are jums. Jebs and jums are distinct members of the same class.
3. True; all lems are jums because every element of a class is contained within any class linked above it in a hierarchy diagram.
4. True; some wibs are jums and some wibs are jebs. Jums and jebs are members of the same class in the hierarchy. Wibs is a member of the class below and is directly linked to both jums and jebs.
5. True; all mogs are bips because every element of a class is contained within any class linked above it in a hierarchy diagram.
6. All triangles are polygons. Both isosceles and scalene triangles are triangles. All equilateral triangles are isosceles triangles.

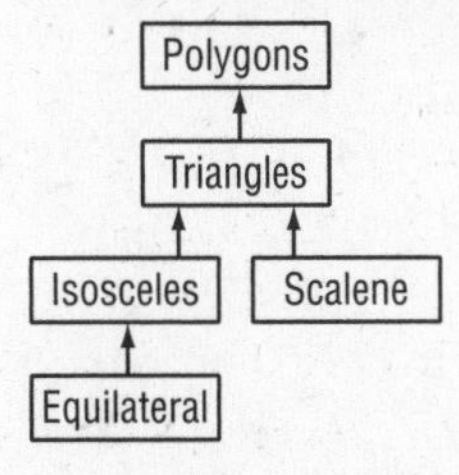

8-7 Coordinate Proof With Quadrilaterals

Page 448 Geometry Software Investigation: Quadrilaterals

1. See students' work.
2. A parallelogram is formed by the midpoints since the opposite sides are congruent.

Pages 449–450 Check for Understanding

1. Place one vertex at the origin and position the figure so another vertex lies on the positive x-axis.
2. Sample answer:

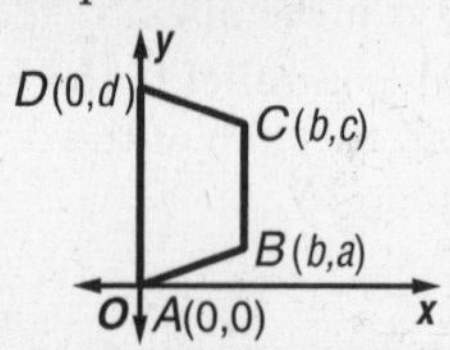

3.

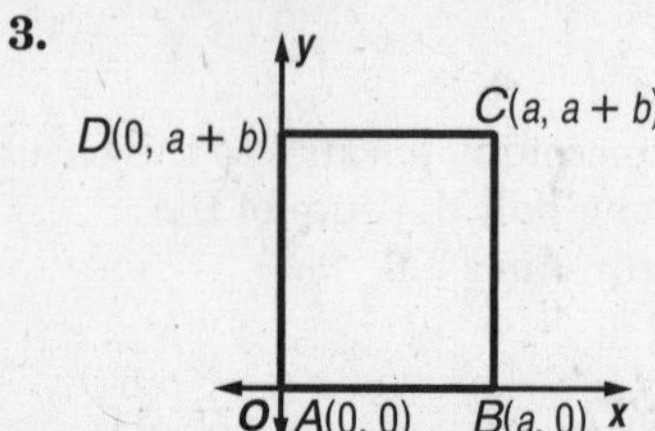

4. The quadrilateral is a square. Opposite sides of a square are congruent and parallel, and its interior angles are all 90°. So, the x-coordinate of C is a and the y-coordinate is a.
The coordinates of C are (a, a).
5. The quadrilateral is a parallelogram. Opposite sides of a parallelogram are congruent and parallel. So, the y-coordinate of D is b.
The length of $\overline{AB}$ is a, and the length of $\overline{DC}$ is a.
So, the x-coordinate of D is $(a + c) - a$ or c.
The coordinates of D are (c, b).
6. **Given:** $ABCD$ is a parallelogram.
Prove: $\overline{AC}$ and $\overline{DB}$ bisect each other.

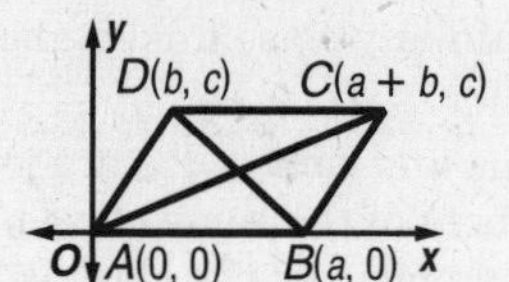

Proof:
The midpoint of $\overline{AC} = \left(\frac{0 + (a + b)}{2}, \frac{0 + c}{2}\right)$
$= \left(\frac{a + b}{2}, \frac{c}{2}\right)$
The midpoint of $\overline{DB} = \left(\frac{a + b}{2}, \frac{0 + c}{2}\right)$
$= \left(\frac{a + b}{2}, \frac{c}{2}\right)$
$\overline{AC}$ and $\overline{DB}$ bisect each other.
7. **Given:** $ABCD$ is a square.
Prove: $\overline{AC} \perp \overline{DB}$

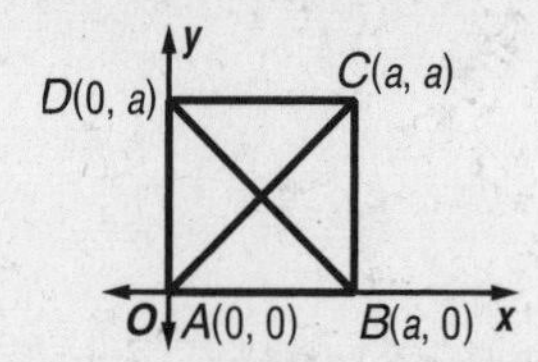

Proof:
Slope of $\overline{DB} = \frac{0 - a}{a - 0}$ or -1
Slope of $\overline{AC} = \frac{0 - a}{0 - a}$ or 1
The slope of $\overline{AC}$ is the negative reciprocal of the slope of $\overline{DB}$, so they are perpendicular.
8. **Given:** $D(195, 180)$, $E(765, 180)$, $F(533, 0)$, $G(195, 0)$
Prove: $DEFG$ is a trapezoid.

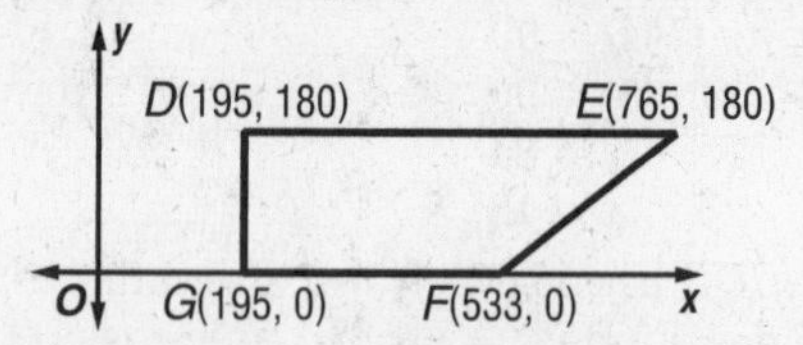

Proof:
Slope of $\overline{DE} = \frac{180 - 180}{765 - 195}$ or 0
Slope of $\overline{GF} = \frac{0 - 0}{533 - 195}$ or 0
Slope of $\overline{EF} = \frac{180 - 0}{765 - 533}$ or $\frac{45}{58}$
Slope of $\overline{DG} = \frac{180 - 0}{195 - 195}$ or undefined
$\overline{DE}$ and $\overline{GF}$ have the same slope, so exactly one pair of opposite sides are parallel. Therefore, $DEFG$ is a trapezoid.

Pages 450–451 Practice and Apply

9.

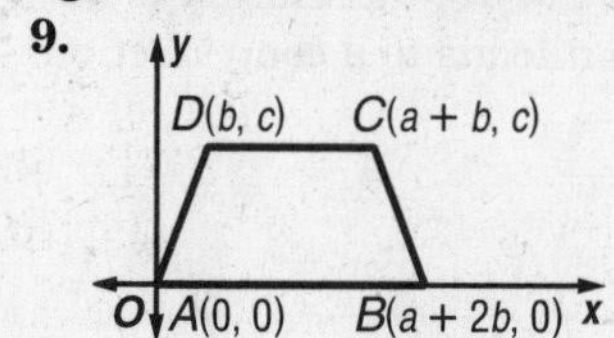

10.

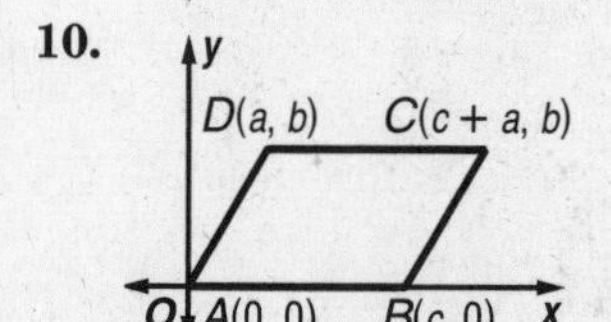

11. The quadrilateral is a parallelogram. Opposite sides of a parallelogram are congruent and parallel. So, the y-coordinate of B is c.
The length of $\overline{HG}$ is $a + b$, and the length of $\overline{BC}$ is $a + b$. So, the x-coordinate of B is $a - (a + b)$ or $-b$.
The coordinates of B are $(-b, c)$.
12. The quadrilateral is a square. Opposite sides of a square are congruent and parallel, and its interior angles are all 90°. So, the x-coordinate of A is $-b$ and the y-coordinate is b, and the x-coordinate of E is b and the y-coordinate is $-b$.
The coordinates of A and E are $(-b, b)$ and $(b, -b)$, respectively.
13. The quadrilateral is a parallelogram. Opposite sides of a parallelogram are congruent and parallel. So, the y-coordinates of E and G are c and 0, respectively.
The length of $\overline{HG}$ is a, and the length of $\overline{EF}$ is a.
So, the x-coordinate of G is a, and the x-coordinate of E is $(a - b) - a$ or $-b$.
The coordinates of G and E are $(a, 0)$ and $(-b, c)$, respectively.
14. The quadrilateral is an isosceles trapezoid. The top and bottom sides of the trapezoid are parallel, so the y-coordinate of M is c.
The length of $\overline{LM}$ is $a + 2b$, so the x-coordinate of M is $(a + b) - (a + 2b)$ or $-b$.
The coordinates of M are $(-b, c)$.

15. The quadrilateral is a rectangle. Opposite sides of a rectangle are congruent and parallel, and its interior angles are all 90°. So, the y-coordinates of T and W are c and $-c$, respectively.
The origin is at the center of the rectangle, so the x-coordinates of T and W are both $-2a$ (the opposites of the x-coordinates of U and V).
The coordinates of T and W are $(-2a, c)$ and $(-2a, -c)$, respectively.

16. The quadrilateral is an isosceles trapezoid. The right and left sides of the trapezoid are parallel, so the x-coordinates of T and S are 0 and a, respectively.
The length of $\overline{QT}$ is a, so the y-coordinate of T is $-\frac{1}{2}a$. The length of $\overline{RS}$ is $2a - 2c$, so the y-coordinate of S is $(a - c) - (2a - 2c)$ or $-a + c$.
The coordinates of T and S are $\left(0, -\frac{1}{2}a\right)$ and $(a, -a + c)$, respectively.

17. Given: $ABCD$ is a rectangle.
Prove: $\overline{AC} \cong \overline{DB}$

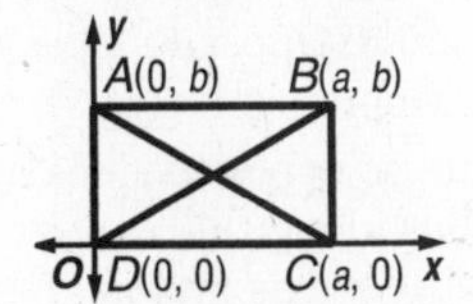

Proof: Use the Distance Formula to find $AC = \sqrt{a^2 + b^2}$ and $BD = \sqrt{a^2 + b^2}$. $\overline{AC}$ and $\overline{DB}$ have the same length, so they are congruent.

18. Given: $\square ABCD$ and $\overline{AC} \cong \overline{BD}$
Prove: $\square ABCD$ is a rectangle.

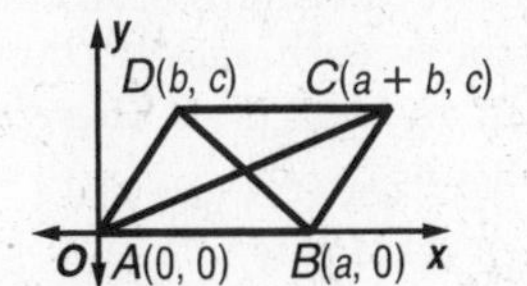

Proof:
$AC = \sqrt{(a + b - 0)^2 + (c - 0)^2}$
$BD = \sqrt{(b - a)^2 + (c - 0)^2}$
But $AC = BD$ and
$\sqrt{(a + b - 0)^2 + (c - 0)^2} = \sqrt{(b - a)^2 + (c - 0)^2}$.

$$(a + b - 0)^2 + (c - 0)^2 = (b - a)^2 + (c - 0)^2$$
$$(a + b)^2 + c^2 = (b - a)^2 + c^2$$
$$a^2 + 2ab + b^2 + c^2 = b^2 - 2ab + a^2 + c^2$$
$$2ab = -2ab$$
$$4ab = 0$$
$$a = 0 \text{ or } b = 0$$

Because A and B are different points, $a \neq 0$. Then $b = 0$. The slope of $\overline{AD}$ is undefined and the slope of $\overline{AB} = 0$. Thus $\overline{AD} \perp \overline{AB}$. $\angle DAB$ is a right angle and $ABCD$ is a rectangle.

19. Given: isosceles trapezoid $ABCD$ with $\overline{AD} \cong \overline{BC}$
Prove: $\overline{BD} \cong \overline{AC}$

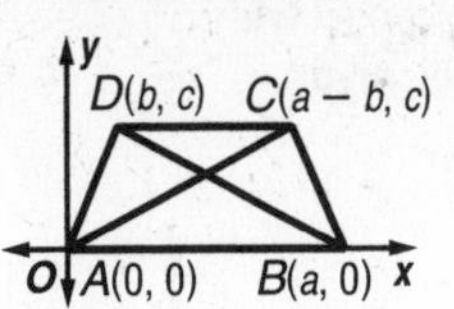

Proof:
$BD = \sqrt{(a - b)^2 + (0 - c)^2} = \sqrt{(a - b)^2 + c^2}$
$AC = \sqrt{((a - b) - 0)^2 + (c - 0)^2} = \sqrt{(a - b)^2 + c^2}$
$BD = AC$ and $\overline{BD} \cong \overline{AC}$

20. Given: $ABCD$ is an isosceles trapezoid with median $\overline{XY}$.
Prove: $\overline{XY} \parallel \overline{AB}$ and $\overline{XY} \parallel \overline{DC}$

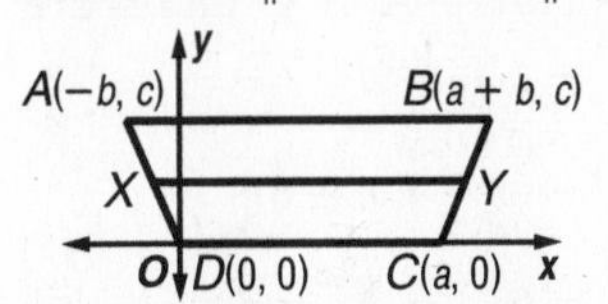

Proof: The midpoint of $\overline{AD}$ is X. The coordinates are $\left(\frac{-b}{2}, \frac{c}{2}\right)$. The midpoint of $\overline{BC}$ is $Y\left(\frac{2a + b}{2}, \frac{c}{2}\right)$. The slope of $\overline{AB} = 0$, the slope of $\overline{XY} = 0$ and the slope of $\overline{DC} = 0$. Thus, $\overline{XY} \parallel \overline{AB}$ and $\overline{XY} \parallel \overline{DC}$.

21. Given: $ABCD$ is a rectangle.
Q, R, S, and T are midpoints of their respective sides.
Prove: $QRST$ is a rhombus.

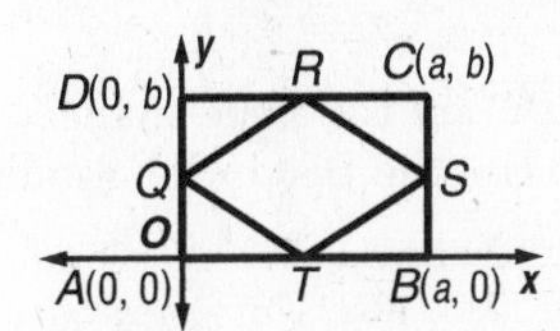

Proof:
Midpoint Q is $\left(\frac{0 + 0}{2}, \frac{b + 0}{2}\right)$ or $\left(0, \frac{b}{2}\right)$.
Midpoint R is $\left(\frac{a + 0}{2}, \frac{b + b}{2}\right)$ or $\left(\frac{a}{2}, \frac{2b}{2}\right)$ or $\left(\frac{a}{2}, b\right)$
Midpoint S is $\left(\frac{a + a}{2}, \frac{b + 0}{2}\right)$ or $\left(\frac{2a}{2}, \frac{b}{2}\right)$ or $\left(a, \frac{b}{2}\right)$.
Midpoint T is $\left(\frac{a + 0}{2}, \frac{0 + 0}{2}\right)$ or $\left(\frac{a}{2}, 0\right)$.

$$QR = \sqrt{\left(\frac{a}{2} - 0\right)^2 + \left(b - \frac{b}{2}\right)^2} = \sqrt{\left(\frac{a}{2}\right)^2 + \left(\frac{b}{2}\right)^2}$$
$$RS = \sqrt{\left(a - \frac{a}{2}\right)^2 + \left(\frac{b}{2} - b\right)^2}$$
$$= \sqrt{\left(\frac{a}{2}\right)^2 + \left(-\frac{b}{2}\right)^2} \text{ or } \sqrt{\left(\frac{a}{2}\right)^2 + \left(\frac{b}{2}\right)^2}$$
$$ST = \sqrt{\left(a - \frac{a}{2}\right)^2 + \left(\frac{b}{2} - 0\right)^2} = \sqrt{\left(\frac{a}{2}\right)^2 + \left(\frac{b}{2}\right)^2}$$
$$QT = \sqrt{\left(\frac{a}{2} - 0\right)^2 + \left(0 - \frac{b}{2}\right)^2}$$
$$= \sqrt{\left(\frac{a}{2}\right)^2 + \left(-\frac{b}{2}\right)^2} \text{ or } \sqrt{\left(\frac{a}{2}\right)^2 + \left(\frac{b}{2}\right)^2}$$

$QR = RS = ST = QT$
$\overline{QR} \cong \overline{RS} \cong \overline{ST} \cong \overline{QT}$
$QRST$ is a rhombus.

22. Given: $RSTV$ is a quadrilateral.
A, B, C, and D are midpoints of sides $\overline{RS}$, $\overline{ST}$, $\overline{TV}$, and $\overline{VR}$, respectively.
Prove: $ABCD$ is a parallelogram.

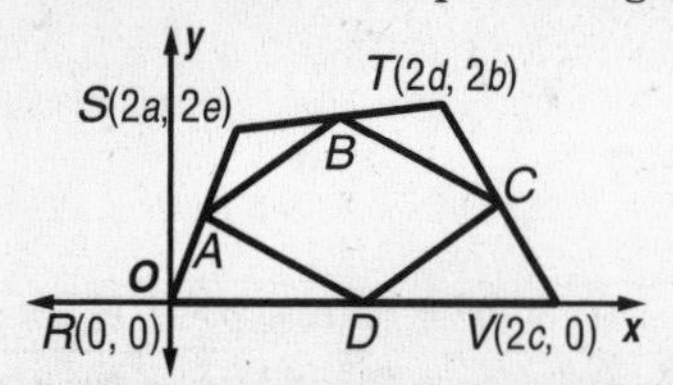

Proof: Place quadrilateral $RSTV$ on the coordinate plane and label coordinates as shown. (Using coordinates that are multiples of 2 will make the computation easier.) By the Midpoint Formula, the coordinates of A, B, C, and D are

$$A\left(\frac{2a}{2}, \frac{2e}{2}\right) = (a, e);$$
$$B\left(\frac{2d+2a}{2}, \frac{2e+2b}{2}\right) = (d+a, e+b);$$
$$C\left(\frac{2d+2c}{2}, \frac{2b}{2}\right) = (d+c, b); \text{ and}$$
$$D\left(\frac{2c}{2}, \frac{0}{2}\right) = (c, 0).$$

Find the slopes of $\overline{AB}$ and $\overline{DC}$.

Slope of $\overline{AB}$	Slope of $\overline{DC}$
$m = \frac{y_2 - y_1}{x_2 - x_1}$	$m = \frac{y_2 - y_1}{x_2 - x_1}$
$= \frac{(e+b) - e}{(d+a) - a}$	$= \frac{0 - b}{c - (d + c)}$
$= \frac{b}{d}$	$= \frac{-b}{-d}$ or $\frac{b}{d}$

The slopes of $\overline{AB}$ and $\overline{DC}$ are the same so the segments are parallel. Use the Distance Formula to find AB and DC.

$$AB = \sqrt{((d+a) - a)^2 + ((e+b) - e)^2}$$
$$= \sqrt{d^2 + b^2}$$
$$DC = \sqrt{((d+c) - c)^2 + (b - 0)^2}$$
$$= \sqrt{d^2 + b^2}$$

Thus, $\overline{AB} \cong \overline{DC}$. Therefore, $ABCD$ is a parallelogram because if one pair of opposite sides of a quadrilateral are both parallel and congruent, then the quadrilateral is a parallelogram.

23. Sample answer:

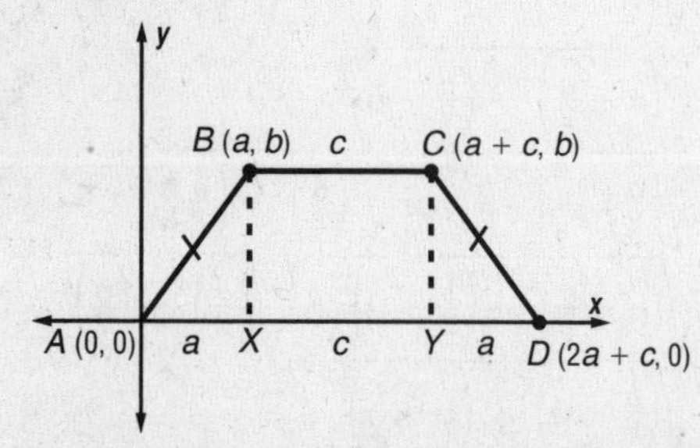

Graph points $A(0, 0)$ and $B(a, b)$. If $ABCD$ is an isosceles trapezoid, $\overline{BC} \parallel \overline{AD}$. So $\overline{BC}$ is horizontal. Let b be the y-coordinate of C. Since $ABCD$ is isosceles $AB = CD$. $AX = a$. Let $XY = c$. Then $YD = a$ because $\triangle ABX \cong \triangle DCY$ by HL.
The coordinates of C and D are $C(a + c, b)$ and $D(2a + c, 0)$.

24.

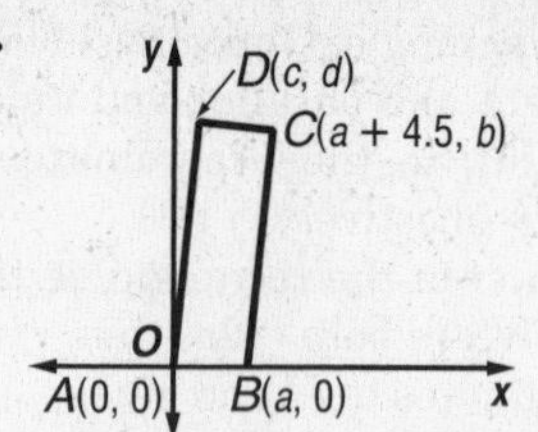

25. No, there is not enough information given to prove that the sides of the tower are parallel.

26. From the information given, we can approximate the height from the ground to the top level of the tower.

27. Sample answer: The coordinate plane is used in coordinate proofs. The Distance Formula, Midpoint Formula and Slope Formula are used to prove theorems. Answers should include the following.
- Place the figure so one of the vertices is at the origin. Place at least one side of the figure on the positive x-axis. Keep the figure in the first quadrant if possible and use coordinates that will simplify calculations.
- Sample answer: Theorem 8.3: Opposite sides of a parallelogram are congruent.

28. D; $ABCD$ is a parallelogram, so opposite sides are parallel. $\overline{BC} \parallel \overline{AD}$ and $\overline{BC}$ lies along the x-axis, so $\overline{AD}$ is parallel to the x-axis and is horizontal. Therefore, D must have the same y-coordinate as A.
Opposite sides of a parallelogram are congruent, so $\overline{AD} \cong \overline{BC}$ and $AD = BC$. According to the Distance Formula, $BC = \sqrt{(x_2 - x_1)^2 + (y_2 - y_1)^2}$ $= \sqrt{(c - b)^2 + 0} = c - b$. So, the length of AD is $c - b$. $\overline{AD}$ is horizontal, the x-coordinate of D must be greater than that of A, and $c - b > 0$, therefore, the x-coordinate of D must be $c - b$.
The coordinates of point D are $(c - b, a)$.

29. A; if $p = -5$, then $5 - p^2 - p = 5 - (-5)^2 - (-5)$ or -15.

Page 451 Maintain Your Skills

30. Given: $MNOP$ is a trapezoid with bases $\overline{MN}$ and $\overline{OP}$, $\overline{MN} \cong \overline{QO}$
Prove: $MNOQ$ is a parallelogram.

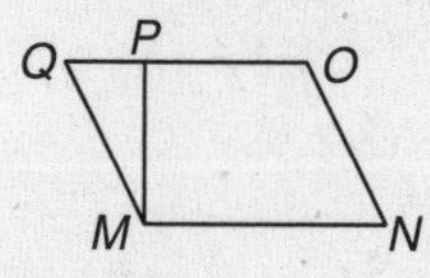

Proof:

Statements	Reasons
1. $MNOP$ is a trapezoid with bases $\overline{MN}$ and $\overline{OP}$. $\overline{MN} \cong \overline{QO}$	1. Given
2. $\overline{OP} \parallel \overline{MN}$	2. Def. of trapezoid
3. $MNOQ$ is a parallelogram.	3. If one pair of opp. sides are $\parallel$ and $\cong$, the quad. is $\square$.

31. Opposite angles of a rhombus are congruent and the diagonals of a rhombus bisect opposite angles, so $\angle RMP \cong \angle MPR$ and their measures are equal. Since $\angle RMP \cong \angle JMK$ and $m\angle JMK = 55$, $m\angle MPR = 55$.

32. $\angle KJM$ and $\angle MLK$ are right angles and $\triangle JKM$ and $\triangle KLM$ are right triangles because the interior angles of a rectangle are right angles. Opposite sides of a rectangle are parallel, so $\angle LKM \cong \angle JMK$ and their measures are equal because they are alternate interior angles. Therefore, $m\angle LKM = 55$. The sum of the measures of the interior angles of a triangle is 180. Find $m\angle KML$.

$m\angle KLM + m\angle LKM + m\angle KML = 180$
$90 + 55 + m\angle KML = 180$
$m\angle KML = 35$

33. $\angle KLM$ is a right angle because the interior angles of a rectangle are right angles. Opposite angles of a rhombus are congruent, so $\angle MLP \cong \angle MRP$ and their measures are equal. Therefore, $m\angle MLP = 70$. The measure of $\angle KLP$ is the sum of the measures of $\angle KLM$ and $\angle MLP$. Find $m\angle KLP$.

$m\angle KLP = m\angle KLM + m\angle MLP$
$m\angle KLP = 90 + 70$
$m\angle KLP = 160$

34. Let x represent the geometric mean.

$\frac{7}{x} = \frac{x}{14}$
$x^2 = 98$
$x = \sqrt{98}$
$x \approx 9.9$

The geometric mean of 7 and 14 is $\sqrt{98}$ or about 9.9.

35. Let x represent the geometric mean.

$\frac{2\sqrt{5}}{x} = \frac{x}{6\sqrt{5}}$
$x^2 = 60$
$x = \sqrt{60}$
$x \approx 7.7$

The geometric mean of $2\sqrt{5}$ and $6\sqrt{5}$ is $\sqrt{60}$ or about 7.7.

36. $\angle VXY$ is an exterior angle of $\triangle WVX$. So $m\angle WVX < m\angle VXY$.

37. VZ and XZ are equal, so $\overline{VZ} \cong \overline{XZ}$ and $\triangle VXZ$ is an isosceles triangle. The base angles of an isosceles triangle are congruent, so $\angle XVZ \cong \angle VXZ$ and $m\angle XVZ = m\angle VXZ$.

38. $\angle VXY$ and $\angle XYV$ are two interior angles of $\triangle VXY$. The side of $\triangle VXY$ opposite $\angle VXY$ measures 6 + 6.1 or 12.1. The side of $\triangle VXY$ opposite $\angle XYV$ measures 8. Therefore, $m\angle XYV < m\angle VXY$.

39. $\angle XZY$ and $\angle ZXY$ are two interior angles of $\triangle ZXY$. The side of $\triangle ZXY$ opposite $\angle XZY$ measures 9. The side of $\triangle ZXY$ opposite $\angle ZXY$ measures 6.1. According to Theorem 5.9, if one side of a triangle is longer than another side, then the angle opposite the longer side has a greater measure than the angle opposite the shorter side. Therefore, $m\angle XZY < m\angle ZXY$.

Chapter 8 Study Guide and Review

Page 452 Vocabulary and Concept Check

1. true
2. true
3. false; rectangle
4. true
5. false; trapezoid
6. false; rhombus
7. true
8. true

Pages 452–456 Lesson-by-Lesson Review

9. Use the Interior Angle Sum Theorem.

$S = 180(n - 2)$
$= 180(6 - 2) = 720$

The measure of an interior angle of a hexagon is $\frac{720}{6}$ or 120.

10. Use the Interior Angle Sum Theorem.

$S = 180(n - 2)$
$= 180(15 - 2) = 2340$

The measure of an interior angle of a regular 15-gon is $\frac{2340}{15}$ or 156.

11. Use the Interior Angle Sum Theorem.

$S = 180(n - 2)$
$= 180(4 - 2) = 360$

The measure of an interior angle of a square is $\frac{360}{4}$ or 90.

12. Use the Interior Angle Sum Theorem.

$S = 180(n - 2)$
$= 180(20 - 2) = 3240$

The measure of an interior angle of a regular 20-gon is $\frac{3240}{20}$ or 162.

13. Since $n = 4$, the sum of the measures of the interior angles is $180(4 - 2)$ or 360. Write an equation to express the sum of the measures of the interior angles of the polygon.

$360 = m\angle W + m\angle X + m\angle Y + m\angle Z$
$360 = \left(\frac{1}{2}a + 8\right) + a + (a - 28) + (a + 2)$
$360 = \frac{7}{2}a - 18$
$378 = \frac{7}{2}a$
$108 = a$

Use the value of a to find the measure of each angle.

$m\angle W = \frac{1}{2}(108) + 8$ or 62, $m\angle X = 108$, $m\angle Y = 108 - 28$ or 80, and $m\angle Z = 108 + 2$ or 110.

14. Since $n = 5$, the sum of the measures of the interior angles is $180(5 - 2)$ or 540. Write an equation to express the sum of the measures of the interior angles of the polygon.

$540 = m\angle A + m\angle B + m\angle C + m\angle D + m\angle E$
$540 = (x + 27) + (1.5x + 3) + (x + 25) + (2x - 22) + x$
$540 = 6.5x + 33$
$507 = 6.5x$
$78 = x$

Use the value of x to find the measure of each angle.
$m\angle A = 78 + 27$ or 105, $m\angle B = 1.5 \cdot 78 + 3$ or 120, $m\angle C = 78 + 25$ or 103, $m\angle D = 2 \cdot 78 - 22$ or 134, and $m\angle E = 78$.

15. $\angle BCD \cong \angle BAD$ because opposite angles in a parallelogram are congruent. Find $m\angle BAD$.
$m\angle BAD = m\angle CAD + m\angle BAC$
$= 20 + 32$
$= 52$
So, $m\angle BCD = 52$.

16. The diagonals of a parallelogram bisect each other, so AF and CF are equal. Therefore, $AF = 6.86$.

17. Consecutive angles in a parallelogram are supplementary. So, $m\angle ADC = 180 - m\angle BCD$. From Exercise 15, $m\angle BCD = 52$.
So, $m\angle ADC = 180 - 52$ or 128.
$\angle ADB \cong \angle CBD$ because they are alternate interior angles. So $m\angle ADB = 40.1$.
$m\angle BDC = m\angle ADC - m\angle ADB$
$= 128 - 40.1$
$= 87.9$
So, $m\angle BDC = 87.9$.

18. Opposite sides of a parallelogram are congruent, so their measures are equal and $BC = AD$. Therefore, $BC = 9$.

19. Opposite sides of a parallelogram are congruent, so their measures are equal and $CD = AB$. Therefore, $CD = 6$.

20. Consecutive angles in a parallelogram are supplementary. So, $m\angle ADC = 180 - m\angle BAD$. Find $m\angle BAD$.
$m\angle BAD = m\angle CAD + m\angle BAC$
$= 20 + 32$
$= 52$
So, $m\angle ADC = 180 - 52$ or 128.

21. No;

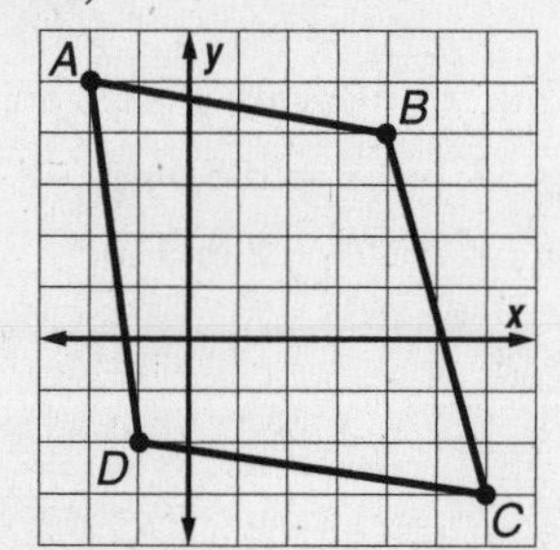

If both pairs of opposite sides of a quadrilateral are congruent, then the quadrilateral is a parallelogram. Use the Distance Formula to determine whether the opposite sides are congruent.

$AD = \sqrt{[-2 - (-1)]^2 + [5 - (-2)]^2}$
$= \sqrt{50}$
$BC = \sqrt{(4 - 6)^2 + [4 - (-3)]^2}$
$= \sqrt{53}$

$AD \neq BC$, so $ABCD$ is not a parallelogram.

22. Yes;

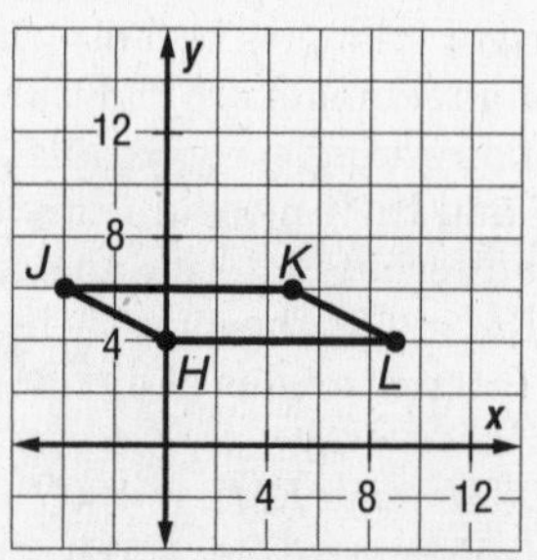

If the midpoints of the diagonals are the same, the diagonals bisect each other. If the diagonals of a quadrilateral bisect each other, then the quadrilateral is a parallelogram.
Find the midpoints of $\overline{HK}$ and $\overline{JL}$.

$\overline{HK}$: $\left(\frac{0+5}{2}, \frac{4+6}{2}\right) = \left(\frac{5}{2}, 5\right)$

$\overline{JL}$: $\left(\frac{-4+9}{2}, \frac{6+4}{2}\right) = \left(\frac{5}{2}, 5\right)$

The midpoints of $\overline{HK}$ and $\overline{JL}$ are the same, so $HKJL$ is a parallelogram.

23. Yes;

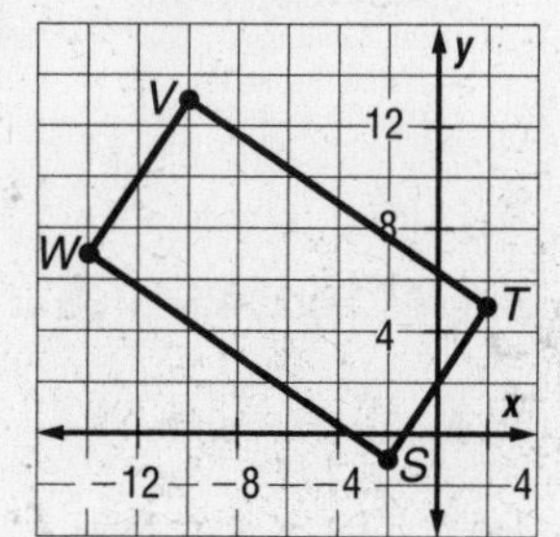

If the opposite sides of a quadrilateral are parallel, then it is a parallelogram.

slope of $\overline{ST} = \frac{-1-5}{-2-2}$
$= \frac{3}{2}$

slope of $\overline{VW} = \frac{13-7}{-10-(-14)}$
$= \frac{3}{2}$

slope of $\overline{TV} = \frac{5-13}{2-(-10)}$
$= -\frac{2}{3}$

slope of $\overline{SW} = \frac{-1-7}{-2-(-14)}$
$= -\frac{2}{3}$

Since opposite sides have the same slope, $\overline{ST} \parallel \overline{VW}$ and $\overline{TV} \parallel \overline{SW}$. Therefore, $STVW$ is a parallelogram by definition.

24. The diagonals of a rectangle bisect each other and are congruent, so $AF = \frac{1}{2}AC$.

$AF = \frac{1}{2}AC$
$2x + 7 = \frac{1}{2}(9x - 1)$
$4x + 14 = 9x - 1$
$-5x = -15$
$x = 3$

So, $AF = 2(3) + 7$ or 13.

25. The diagonals of a rectangle are congruent and bisect each other. So, the triangles formed by the diagonals of a rectangle are isosceles. Therefore, $\angle 2 \cong \angle 1$ and $m\angle 2 = m\angle 1$.

$$16x - 12 = 12x + 4$$
$$4x = 16$$
$$x = 4$$

So, $m\angle 2 = 16(4) - 12$ or 52.

26. The diagonals of a rectangle bisect each other and are congruent, so $\overline{CF} \cong \overline{DF}$. Find x.

$$\overline{CF} \cong \overline{DF}$$
$$CF = DF$$
$$4x + 1 = x + 13$$
$$3x = 12$$
$$x = 4$$

27. The interior angles of rectangles are right angles, and the sum of the measures of the interior angles of a triangle is 180. So, the sum of the measures of $\angle 2$ and $\angle 5$ is 90. Find $m\angle 5$.

$$m\angle 2 + m\angle 5 = 90$$
$$(70 - 4x) + (18x - 8) = 90$$
$$14x = 28$$
$$x = 2$$

So, $m\angle 5 = 18(2) - 8$ or 28.

28. Find the slopes of $\overline{RS}$, $\overline{ST}$, $\overline{TV}$, and $\overline{VR}$.

$$\text{slope of } \overline{RS} = \frac{-5 - (-5)}{-3 - 0} = 0$$

$$\text{slope of } \overline{ST} = \frac{-5 - 4}{0 - (-3)} = -3$$

$$\text{slope of } \overline{TV} = \frac{4 - 4}{3 - 0} = 0$$

$$\text{slope of } \overline{VR} = \frac{4 - (-5)}{0 - (-3)} = 3$$

The slopes of consecutive sides are not negative reciprocals, so consecutive sides are not perpendicular. Therefore, $RSTV$ is not a rectangle. (Note: $\overline{ST} \nparallel \overline{VR}$ so $RSTV$ is not even a parallelogram. So it is not a rectangle.)

29. If the opposite sides of a quadrilateral are parallel and the diagonals of the quadrilateral are congruent, then the quadrilateral is a rectangle. Find the slopes of $\overline{RS}$, $\overline{ST}$, $\overline{TV}$, and $\overline{VR}$.

$$\text{slope of } \overline{RS} = \frac{0 - 3}{0 - 6} = \frac{1}{2}$$

$$\text{slope of } \overline{TV} = \frac{7 - 4}{4 - (-2)} = \frac{1}{2}$$

$$\text{slope of } \overline{ST} = \frac{3 - 7}{6 - 4} = -2$$

$$\text{slope of } \overline{VR} = \frac{4 - 0}{-2 - 0} = -2$$

So, $\overline{RS} \parallel \overline{TV}$ and $\overline{ST} \parallel \overline{VR}$. Use the Distance Formula to determine whether the diagonals of quadrilateral $RSTV$ are congruent.

$$RT = \sqrt{(0 - 4)^2 + (0 - 7)^2} = \sqrt{16 + 49} = \sqrt{65}$$

$$SV = \sqrt{[6 - (-2)]^2 + (3 - 4)^2} = \sqrt{64 + 1} = \sqrt{65}$$

Since the opposite sides are parallel and the diagonals are congruent, $RSTV$ is a rectangle.

30. The diagonals of a rhombus bisect the angles, so $\angle 1 \cong \angle 2$. Find x.

$$\angle 1 \cong \angle 2$$
$$m\angle 1 = m\angle 2$$
$$2x + 20 = 5x - 4$$
$$24 = 3x$$
$$8 = x$$

31. The diagonals of a rhombus bisect each other, so $AF = \frac{1}{2}AC = \frac{1}{2}(15)$ or 7.5.

32. The diagonals of a rhombus are perpendicular, so $m\angle 3$ is 90. Find y.

$$m\angle 3 = 90$$
$$y^2 + 26 = 90$$
$$y^2 = 64$$
$$y = 8 \text{ or } -8$$

33. $\overline{BC} \parallel \overline{AD}$, so $\angle ADY$ and $\angle BCY$ are supplementary. Find $m\angle BCY$.

$$m\angle BCY + m\angle ADY = 180$$
$$m\angle BCY + 78 = 180$$
$$m\angle BCY = 102$$

Both pairs of base angles of an isosceles trapezoid are congruent, so $\angle XBC \cong \angle BCY$ and $m\angle XBC = 102$.

34. The median of a trapezoid is parallel to the bases, and its measure is one-half the sum of the measures of the bases. Find JM.

$$AB = \frac{1}{2}(KL + JM)$$
$$57 = \frac{1}{2}(21 + JM)$$
$$114 = 21 + JM$$
$$93 = JM$$

35. Given: $ABCD$ is a square.
Prove: $\overline{AC} \perp \overline{BD}$

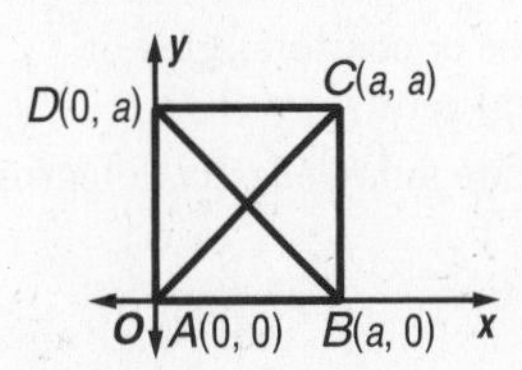

Proof:

Slope of $\overline{AC} = \frac{a - 0}{a - 0}$ or 1

Slope of $\overline{BD} = \frac{a - 0}{0 - a}$ or -1

The slope of $\overline{AC}$ is the negative reciprocal of the slope of $\overline{BD}$. Therefore, $\overline{AC} \perp \overline{BD}$.

36. Given: $ABCD$ is a parallelogram.
Prove: $\triangle ABC \cong \triangle CDA$

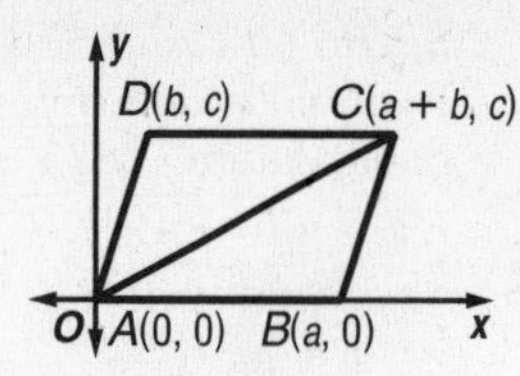

Proof:

$AB = \sqrt{(a-0)^2 + (0-0)^2} = \sqrt{a^2 + 0^2}$ or a

$DC = \sqrt{[(a+b) - b]^2 + (c-c)^2} = \sqrt{a^2 + 0^2}$ or a

$AD = \sqrt{(b-0)^2 + (c-0)^2} = \sqrt{b^2 + c^2}$

$BC = \sqrt{[(a+b) - a]^2 + (c-0)^2} = \sqrt{b^2 + c^2}$

AB and DC have the same measure, so $\overline{AB} \cong \overline{DC}$. AD and BC have the same measure, so $\overline{AD} \cong \overline{BC}$. $\overline{AC} \cong \overline{AC}$ by the Reflexive Property. Therefore, $\triangle ABC \cong \triangle CDA$ by SSS.

37. The quadrilateral is an isosceles trapezoid. The top and bottom sides of the trapezoid are parallel, so the y-coordinate of P is c.
The length of $\overline{MN}$ is $4a$, so the x-coordinate of P is $4a - a$ or $3a$.
The coordinates of P are $(3a, c)$.

38. The quadrilateral is a parallelogram. Opposite sides of a parallelogram are congruent and parallel. So, the y-coordinate of U is c.
The length of $\overline{VW}$ is $b - (-a)$ or $a + b$. So the length of $\overline{TU}$ is also $a + b$. So, the x-coordinate of U is $(a + b) - 0$ or $a + b$.
The coordinates of U are $(a + b, c)$.

Chapter 8 Practice Test

Page 457

1. true

2. false;

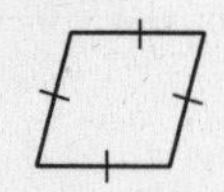

3. false;

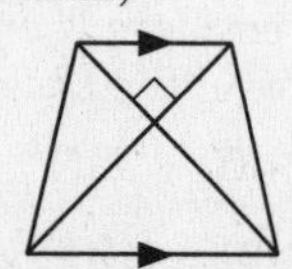

4. $\overline{HK} \cong \overline{FG}$ because opposite sides of parallelograms are congruent.

5. $\angle FKJ \cong \angle HGJ$ because alternate interior angles are congruent.

6. $\angle FKH \cong \angle FGH$ because opposite angles of parallelograms are congruent.

7. $\overline{GH} \parallel \overline{FK}$ because opposite sides of parallelograms are parallel.

8. Yes;

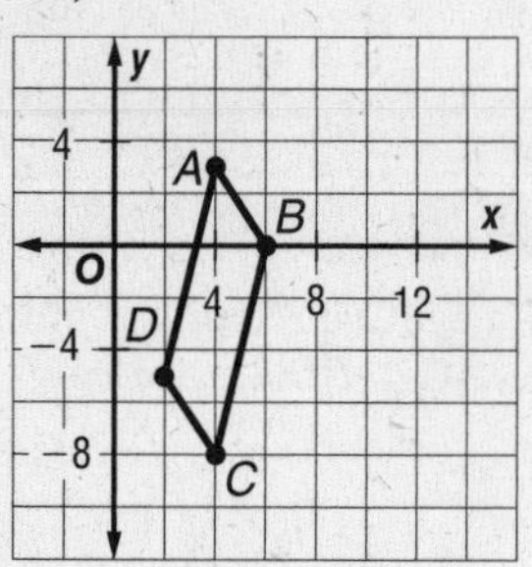

If the midpoints of the diagonals are the same, the diagonals bisect each other. If the diagonals of a quadrilateral bisect each other, then the quadrilateral is a parallelogram.
Find the midpoints of $\overline{AC}$ and $\overline{BD}$.

$\overline{AC}$: $\left(\frac{4+4}{2}, \frac{3+(-8)}{2}\right) = \left(4, -\frac{5}{2}\right)$

$\overline{BD}$: $\left(\frac{6+2}{2}, \frac{0+(-5)}{2}\right) = \left(4, -\frac{5}{2}\right)$

The midpoints of $\overline{AC}$ and $\overline{BD}$ are the same, so the diagonals bisect each other and $ABCD$ is a parallelogram.

9. Yes;

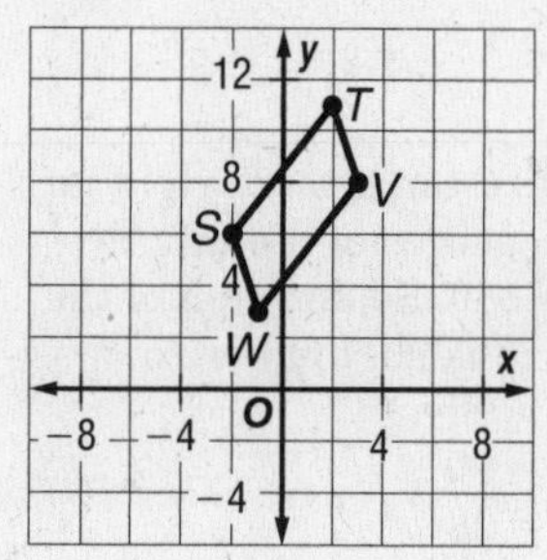

First use the Distance Formula to determine whether the opposite sides are congruent.

$ST = \sqrt{(-2-2)^2 + (6-11)^2}$
$= \sqrt{41}$

$VW = \sqrt{[3-(-1)]^2 + (8-3)^2}$
$= \sqrt{41}$

Since $ST = VW$, $\overline{ST} \cong \overline{VW}$.
Next, use the Slope Formula to determine whether $\overline{ST} \parallel \overline{VW}$.

slope of $\overline{ST} = \frac{6-11}{-2-2}$
$= \frac{5}{4}$

slope of $\overline{VW} = \frac{8-3}{3-(-1)}$
$= \frac{5}{4}$

$\overline{ST}$ and $\overline{VW}$ have the same slope, so they are parallel. Since one pair of opposite sides is congruent and parallel $STVW$ is a parallelogram.

10. No;

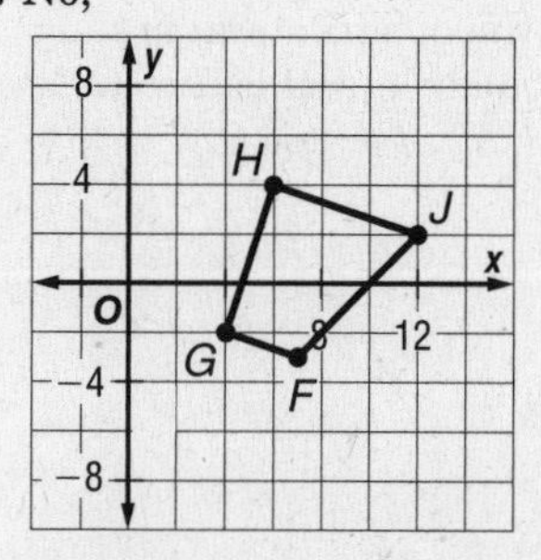

If both pairs of opposite sides of a quadrilateral are congruent, then the quadrilateral is a parallelogram. Use the Distance Formula to determine whether the opposite sides are congruent.

$FG = \sqrt{(7-4)^2 + [-3-(-2)]^2}$
$= \sqrt{10}$
$HJ = \sqrt{(6-12)^2 + (4-2)^2}$
$= \sqrt{40}$
$GH = \sqrt{(4-6)^2 + (-2-4)^2}$
$= \sqrt{40}$
$FJ = \sqrt{(7-12)^2 + (-3-2)^2}$
$= \sqrt{50}$

Since the measures of both pairs of opposite sides are not equal, they are not congruent. Therefore, *FGHJ* is not a parallelogram.

11. Yes;

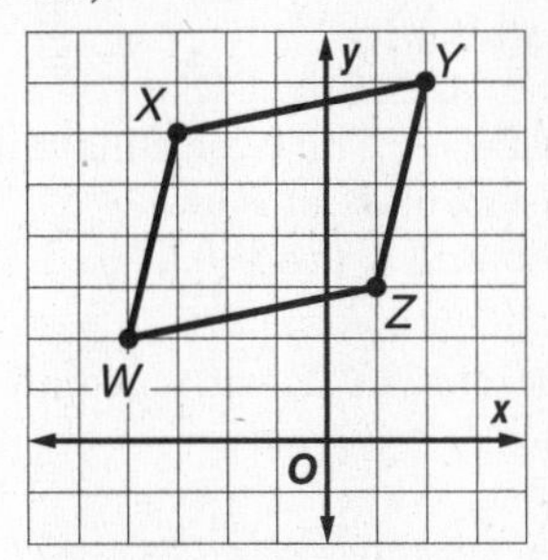

If both pairs of opposite sides of a quadrilateral are congruent, then the quadrilateral is a parallelogram. Use the Distance Formula to determine whether the opposite sides are congruent.

$WX = \sqrt{[-4-(-3)]^2 + (2-6)^2}$
$= \sqrt{17}$
$YZ = \sqrt{(2-1)^2 + (7-3)^2}$
$= \sqrt{17}$
$XY = \sqrt{(-3-2)^2 + (6-7)^2}$
$= \sqrt{26}$
$WZ = \sqrt{(-4-1)^2 + (2-3)^2}$
$= \sqrt{26}$

Since the measures of both pairs of opposite sides are equal, they are congruent. Therefore, *WXYZ* is a parallelogram.

12. The diagonals of a rectangle bisect each other, so $\overline{QP} \cong \overline{PS}$. Find x.

$\overline{QP} \cong \overline{PS}$
$QP = PS$
$3x + 11 = 4x + 8$
$3 = x$

Find QS.

$QS = QP + PS$
$= (3x + 11) + (4x + 8)$
$= 7x + 19$
$= 7(3) + 19$
$= 40$

So, $QS = 40$.

13. Opposite sides of a rectangle are parallel, so $\angle QTR \cong \angle SRT$ because they are alternate interior angles. Find x^2.

$\angle QTR \cong \angle SRT$
$m\angle QTR = m\angle SRT$
$2x^2 - 7 = x^2 + 18$
$x^2 = 25$

So, $m\angle QTR = 2(25) - 7$ or 43.

14.

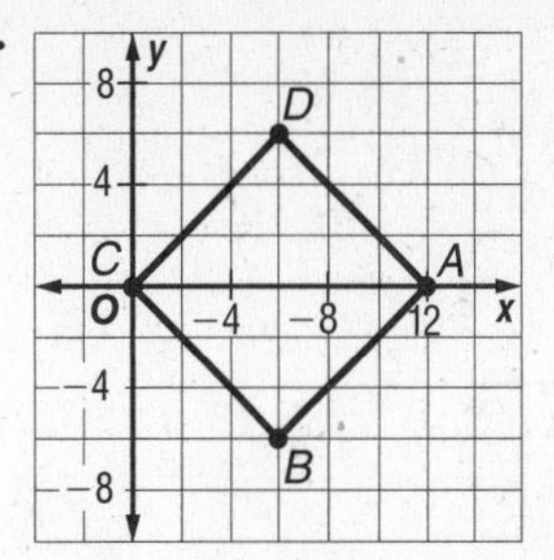

If the four sides are congruent, then parallelogram *ABCD* is either a rhombus or a square. If consecutive sides are perpendicular, then *ABCD* is a rectangle or a square.

Use the Distance Formula to compare the lengths of the sides.

$AB = \sqrt{(12-6)^2 + [0-(-6)]^2}$
$= \sqrt{36 + 36}$
$= 6\sqrt{2}$
$BC = \sqrt{(6-0)^2 + (-6-0)^2}$
$= \sqrt{36 + 36}$
$= 6\sqrt{2}$
$CD = \sqrt{(0-6)^2 + (0-6)^2}$
$= \sqrt{36 + 36}$
$= 6\sqrt{2}$
$AD = \sqrt{(12-6)^2 + (0-6)^2}$
$= \sqrt{36 + 36}$
$= 6\sqrt{2}$

Use the Slope Formula to determine whether the consecutive sides are perpendicular.

slope of $\overline{AB} = \frac{0-(-6)}{12-6}$
$= 1$

slope of $\overline{BC} = \frac{-6-0}{6-0}$
$= -1$

slope of $\overline{CD} = \frac{0-6}{0-6}$
$= 1$

slope of $\overline{AD} = \frac{0-6}{12-6}$
$= -1$

Since the slopes of $\overline{AB}$ and $\overline{CD}$ are negative reciprocals of the slopes of $\overline{BC}$ and $\overline{AD}$, consecutive sides are perpendicular. The lengths of the four sides are the same, so the sides are congruent. Therefore, *ABCD* is a rectangle, a rhombus, and a square.

15.

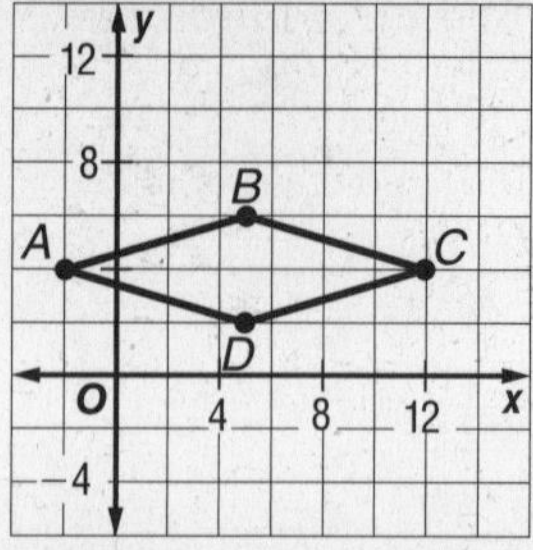

If the four sides are congruent, then parallelogram $ABCD$ is a square or a rhombus. If the diagonals are congruent, then $ABCD$ is a square or a rectangle. If the diagonals are perpendicular, then $ABCD$ is a square or a rhombus.

Use the Distance Formula to compare the lengths of the sides.

$AB = \sqrt{(-2 - 5)^2 + (4 - 6)^2}$
$= \sqrt{49 + 4} = \sqrt{53}$

$BC = \sqrt{(5 - 12)^2 + (6 - 4)^2}$
$= \sqrt{49 + 4} = \sqrt{53}$

$CD = \sqrt{(12 - 5)^2 + (4 - 2)^2}$
$= \sqrt{49 + 4} = \sqrt{53}$

$AD = \sqrt{(-2 - 5)^2 + (4 - 2)^2}$
$= \sqrt{49 + 4} = \sqrt{53}$

Use the Distance Formula to compare the lengths of the diagonals.

$AC = \sqrt{(-2 - 12)^2 + (4 - 4)^2}$
$= \sqrt{196 + 0} = 14$

$BD = \sqrt{(5 - 5)^2 + (6 - 2)^2}$
$= \sqrt{0 + 16} = 4$

Use the Slope Formula to determine whether the diagonals are perpendicular.

slope of $\overline{AC} = \frac{4 - 4}{-2 - 12}$
$= 0$

slope of $\overline{BD} = \frac{6 - 2}{5 - 5}$
$= \frac{4}{0}$, which is undefined

$\overline{AC}$ is horizontal and $\overline{BD}$ is vertical, so the diagonals are perpendicular, but not congruent since $AC \neq BD$. The lengths of the four sides are the same, so the sides are congruent. Therefore, $ABCD$ is a rhombus.

16. The quadrilateral is a parallelogram. Opposite sides of a parallelogram are congruent and parallel. So, the y-coordinate of P is c.
The length of $\overline{MQ}$ is $b - (-a)$ or $a + b$ and the length of $\overline{NP}$ is $a + b$. So, the x-coordinate of P is $(a + b) - 0$ or $a + b$.
The coordinates of P are $(a + b, c)$.

17.

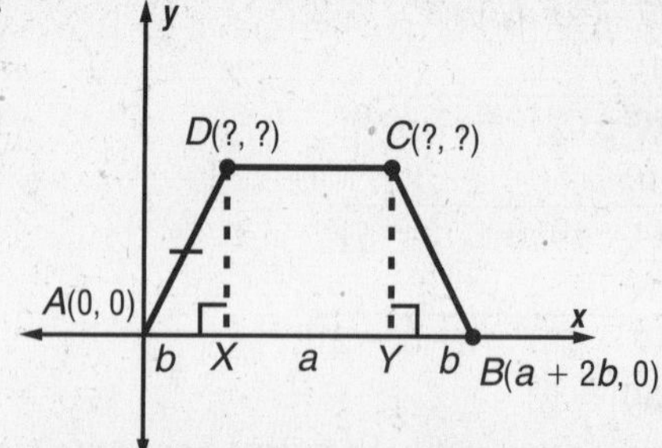

Sample answer: To find the y-coordinates of C and D notice that $\overline{CD}$ is parallel to the x-axis. So $\overline{CD}$ is a horizontal segment, and C and D both have the same y-coordinate. Call it c.
To find the x-coordinate of D, notice that $AX = b$ so the x-coordinate of D is b. The x-coordinate of c is the same as the x-coordinate of Y or $a + b$.
So the coordinates of D and C are $D(b, c)$ and $C(a + b, c)$

18. Given: isosceles trapezoid $WXYZ$ with median $\overline{ST}$
Prove: $\overline{WX} \parallel \overline{ST} \parallel \overline{YZ}$

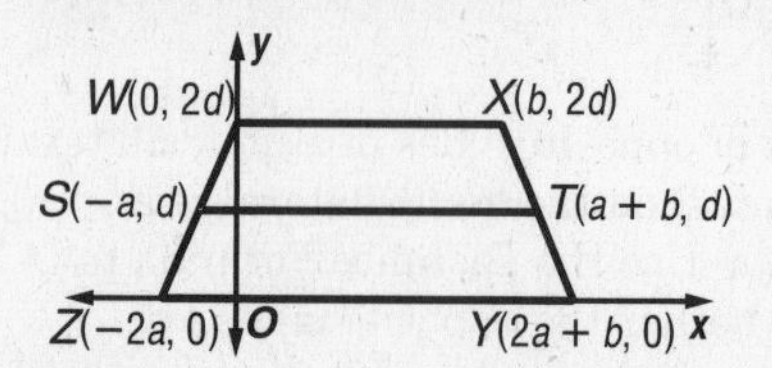

Proof:
To prove lines parallel, show their slopes equal.
The slope of WX is $\frac{2d - 2d}{b - 0}$ or 0.
The slope of ST is $\frac{d - d}{(a + b) - (-a)}$ or 0.
The slope of YZ is $\frac{0 - 0}{(2a + b) - (-2a)}$ or 0.
Since WX, ST, and YZ all have zero slope, they are parallel.

19. The median of a trapezoid is parallel to the bases, and its measure is one-half the sum of the measures of the bases. Find the length of the mid-chord of the keel.

length of mid-chord $= \frac{1}{2}$(length of root chord + length of tip chord)
$= \frac{1}{2}(9.8 + 7.4)$
$= \frac{1}{2}(17.2)$
$= 8.6$

The length of the mid-chord is 8.6 ft.

20. C; use the Interior Angle Sum Theorem to write an equation to solve for n, the number of sides.

$S = 180(n - 2)$
$(108)n = 180(n - 2)$
$108n = 180n - 360$
$0 = 72n - 360$
$360 = 72n$
$5 = n$

The polygon has 5 sides.

Chapter 8 Standardized Test Practice

Pages 458–459

1. C; the length of the ramp is the hypotenuse of a right triangle with legs measuring 3 meters and 5 meters. Use the Pythagorean Theorem to find the length of the hypotenuse.

$$c = \sqrt{a^2 + b^2}$$
$$= \sqrt{3^2 + 5^2}$$
$$= \sqrt{34}$$
$$\approx 6$$

To the nearest meter, the length of the ramp should be 6 m.

2. D; the contrapositive of the statement "If an astronaut is in orbit, then he or she is weightless" is "If an astronaut is not weightless, then he or she is not in orbit."

3. B; for the two rectangles to be similar, the measures of their corresponding sides must proportional. The ratio of the length to the width of $QRST$ is $\frac{7}{4}$ or 7 : 4. The ratios of the choices, from A to D, are: $\frac{28}{14} = 2$, $\frac{21}{12} = \frac{7}{4}$, $\frac{14}{4} = \frac{7}{2}$, and $\frac{7}{8}$. The dimensions, 21 cm by 12 cm, could be the dimensions of a rectangle similar to $QRST$.

4. C; the ladder, wall, and ground form a 30°-60°-90° right triangle. The ladder is the hypotenuse, the wall is the longer leg, and the ground is the shorter leg. In a 30°-60°-90° triangle, the length of the hypotenuse is twice the length of the shorter leg, and the length of the longer leg is $\sqrt{3}$ times the length of the shorter leg. So, the shorter leg is $\frac{24}{2}$ or 12 ft, and the longer leg is $12\sqrt{3}$ ft. The ladder reaches $12\sqrt{3}$ ft up the side of the house.

5. B; the diagonals of a rectangle are congruent, so $\overline{JL} \cong \overline{KM}$. Find x.

$$\overline{JL} \cong \overline{KM}$$
$$JL = KM$$
$$2x + 5 = 4x - 11$$
$$16 = 2x$$
$$8 = x$$

6. C; the diagonals of a rhombus bisect each other but are not necessarily congruent.

7. A; since $m\angle DAE = 74$ and $m\angle BCE = 20$, $\overline{DA}$ is not parallel to $\overline{BC}$. Since $ABCD$ is a trapezoid, and every trapezoid has exactly one pair of opposite sides that are parallel, $\overline{AB}$ is parallel to $\overline{CD}$.

8. Set y equal to zero to find the x-coordinate at which the graph crosses the x-axis.

$$y = -4x + 5$$
$$0 = -4x + 5$$
$$4x = 5$$
$$x = \frac{5}{4}$$

The point at which the graph crosses the x-axis is $\left(\frac{5}{4}, 0\right)$.

9. If one side of a triangle is longer than another side, then the angle opposite the longer side has a greater measure than the angle opposite the shorter side. The side representing the path from Candace's house to the theater is opposite a 55° angle, and the side representing the path from Julio's house to the theater is opposite a 40° angle. Since $55 > 40$, Julio's house is closer to the theater.

10. $\overline{CD}$ is the altitude of right triangle ABC so, by Theorem 7.2, its measure is the geometric mean between the segments of the hypotenuse.

Let $x = CD$

$$\frac{4}{x} = \frac{x}{25}$$

11. The sides of a rhombus are congruent, so $\overline{AC}$ separates rhombus $ABCD$ into two isosceles triangles. The base angles of an isosceles triangle are congruent, so $\angle ACD \cong \angle CAD$.

$\angle CDE$ is an exterior angle of $\triangle ACD$ so $m\angle CDE = m\angle CAD + m\angle ACD$.

Substituting $m\angle ACD$ for $m\angle CAD$,

$$116 = m\angle ACD + m\angle ACD$$
$$116 = 2(m\angle ACD)$$
$$58 = m\angle ACD$$

12a. $\angle MNR$ and $\angle PQR$ are both right angles and all right angles are congruent, so $\angle MNR \cong \angle PQR$. Since congruence of angles is reflexive, $\angle R \cong \angle R$. $\triangle MNR$ is similar to $\triangle PQR$ because two angles are congruent (AA Similarity).

12b. The ratios of corresponding sides of similar polygons are the same, so $\frac{MR}{MN} = \frac{PR}{QP}$.

The proportion is $\frac{400 + a}{126} = \frac{400}{120}$. Solve for a.

$$\frac{400 + a}{126} = \frac{400}{120}$$
$$400 + a = 126\left(\frac{400}{120}\right)$$
$$a = 420 - 400$$
$$a = 20$$

The distance across the sand trap, a, is 20 yards.

13a. Given: quadrilateral *ABCD*
Prove: *ABCD* is a parallelogram

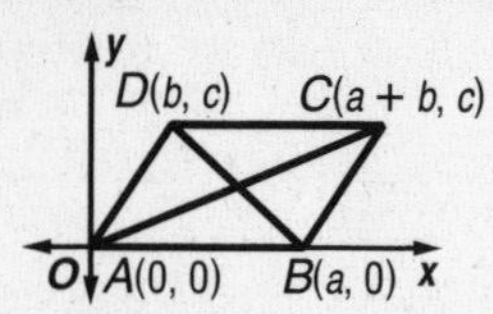

Proof:
The slope of $\overline{AD}$ is $\frac{c-0}{b-0}$ or $\frac{c}{b}$. The slope of $\overline{BC}$ is $\frac{c-0}{a+b-a}$ or $\frac{c}{b}$. $\overline{AD}$ and $\overline{BC}$ have the same slope so they are parallel.
$AD = \sqrt{(b-0)^2 + (c-0)^2} = \sqrt{b^2 + c^2}$.
$BC = \sqrt{(a+b-a)^2 + (c-0)^2} = \sqrt{b^2 + c^2}$.
Since one pair of opposite sides are parallel and congruent, *ABCD* is a parallelogram.

13b. The slope of $\overline{AC}$ is $\frac{c-0}{a+b-0}$ or $\frac{c}{a+b}$.
The slope of $\overline{BD}$ is $\frac{c-0}{b-a}$ or $\frac{c}{b-a}$.
The product of the slopes is $\frac{c}{a+b} \times \frac{c}{b-a}$ $= \frac{c^2}{b^2-a^2}$. Since $c^2 = a^2 - b^2$, the product of the slopes is $\frac{a^2-b^2}{b^2-a^2}$ or -1, so the diagonals of *ABCD* are perpendicular.

13c. Since the diagonals are perpendicular, *ABCD* is a rhombus.

Chapter 9 Transformations

Page 461 Getting Started

1.

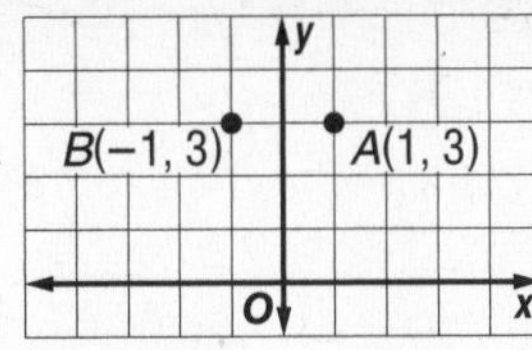

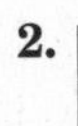
2.

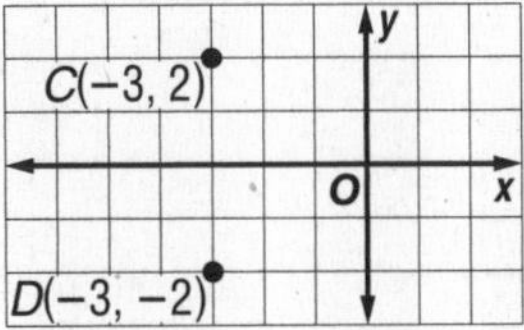

3.

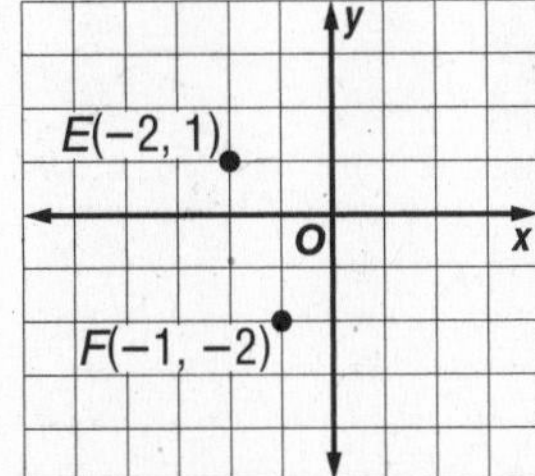

4.

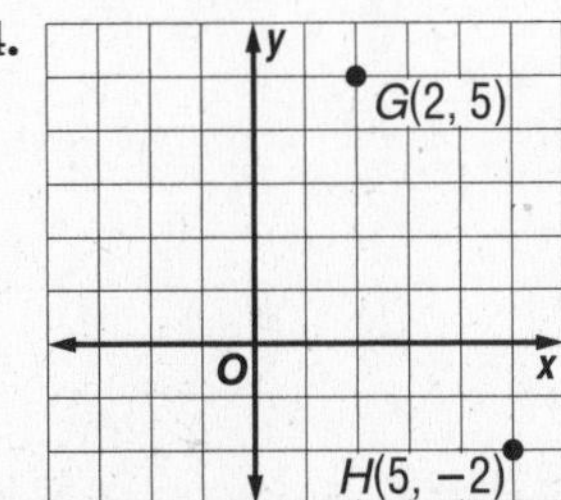

5.

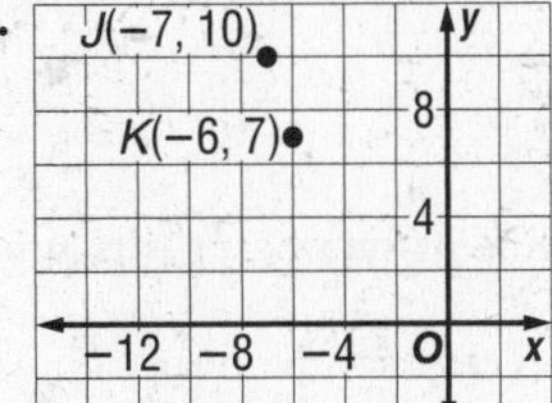

6.

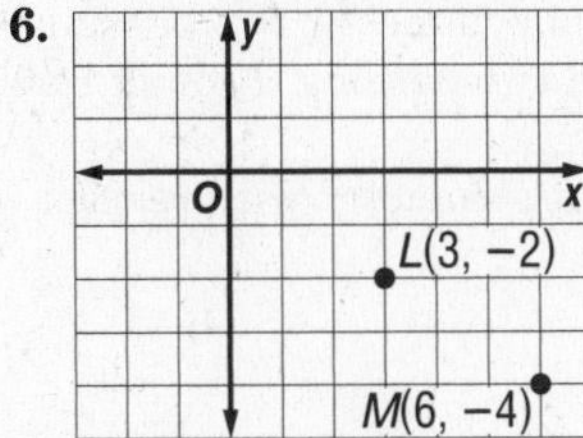

7. $\tan A = \frac{3}{4}$

$A = \tan^{-1}\left(\frac{3}{4}\right)$

$A \approx 36.9°$

The measure of angle A is approximately 36.9.

8. $\tan A = \frac{5}{8}$

$A = \tan^{-1}\left(\frac{5}{8}\right)$

$A \approx 32.0°$

The measure of angle A is approximately 32.0.

9. $\sin A = \frac{2}{3}$

$A = \sin^{-1}\left(\frac{2}{3}\right)$

$A \approx 41.8°$

The measure of angle A is approximately 41.8.

10. $\sin A = \frac{4}{5}$

$A = \sin^{-1}\left(\frac{4}{5}\right)$

$A \approx 53.1°$

The measure of angle A is approximately 53.1.

11. $\cos A = \frac{9}{12}$

$A = \cos^{-1}\left(\frac{9}{12}\right)$

$A \approx 41.4°$

The measure of angle A is approximately 41.4.

12. $\cos A = \frac{15}{17}$

$A = \cos^{-1}\left(\frac{15}{17}\right)$

$A \approx 28.1°$

The measure of angle A is approximately 28.1.

13. $\begin{bmatrix} 0 & 1 \\ 1 & -1 \end{bmatrix} \cdot \begin{bmatrix} 5 & 4 \\ -5 & -1 \end{bmatrix}$

$= \begin{bmatrix} 0(5) + 1(-5) & 0(4) + 1(-1) \\ & 1(4) + (-1)(-1) \end{bmatrix}$

$= \begin{bmatrix} -5 & -1 \\ 10 & \end{bmatrix}$

14. $\begin{bmatrix} -1 & 0 \\ 1 & 1 \end{bmatrix} \cdot \begin{bmatrix} 0 & -2 \\ -2 & 3 \end{bmatrix}$

$= \begin{bmatrix} -1(0) + 0(-2) & -1(-2) + 0(3) \\ & 1(-2) + 1(3) \end{bmatrix}$

$= \begin{bmatrix} 0 & 2 \\ -2 & 1 \end{bmatrix}$

15. $\begin{bmatrix} 0 & 1 \\ -1 & 0 \end{bmatrix} \cdot \begin{bmatrix} -3 & 4 & 5 \\ -2 & -5 & 1 \end{bmatrix}$

$= \begin{bmatrix} 0(-3)+1(-2) & 0(4)+1(-5) & 0(5)+1(1) \\ -1(-3)+0(-2) & & -1(5)+0(1) \end{bmatrix}$

$= \begin{bmatrix} -2 & -5 & 1 \\ 3 & -4 & -5 \end{bmatrix}$

16. $\begin{bmatrix} -1 & 0 \\ 0 & -1 \end{bmatrix} \begin{bmatrix} -1 & -3 & -3 & 2 \\ 3 & -1 & -2 & 1 \end{bmatrix}$

$= \begin{bmatrix} -1(-1)+0(3) & -1(-3)+0(-1) & -1(-3)+0(-2) & -1(2)+0(1) \\ 0(-1)+(-1)(3) & 0(-3)+(-1)(-1) & 0(-3)+(-1)(-2) & 0(2)+(-1)(1) \end{bmatrix}$

$\begin{bmatrix} 1 & 3 & 3 & -2 \\ -3 & 1 & 2 & -1 \end{bmatrix}$

Page 462 Geometry Activity: Transformations

1. Rotation; the figure has been turned around a point.
2. Dilation; the figure has been enlarged.
3. Reflection or rotation; the figure has either been flipped over a line or turned around a point.
4. Translation; the figure has been slid down and to the left.
5. Dilation; the figure has been reduced.
6. Reflection; the figure has been flipped over a line.
7. Translation; the figure has been slid down and to the left.

8. Reflection or rotation; the figure has either been flipped over a line or turned around a point.
9. Reflection; the figure has been flipped over a line.
10. Reflection; the figure has been flipped over a line.
11. Rotation, reflection, and translation result in an image that is congruent to its preimage. They are isometries.

9-1 Reflections

Page 467 Check for Understanding

1. Sample Answer: The centroid of an equilateral triangle is not a point of symmetry.
2. Sample answer: $W(-3, 1)$, $X(-2, 3)$, $Y(3, 3)$ and $Z(3, 1)$ with reflected image $W'(1, -3)$, $X'(3, -2)$, $Y'(3, 3)$, $Z'(1, 3)$

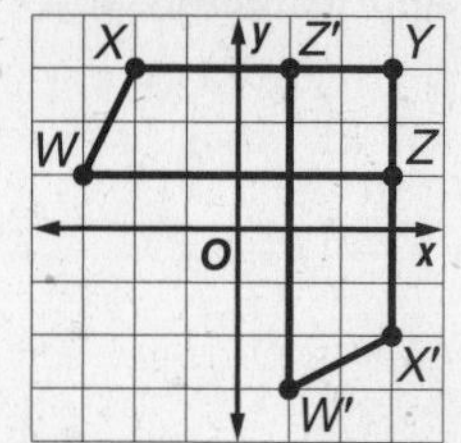

3. Angle measure, betweenness of points, collinearity, and distance are four properties that are preserved in reflections.
4.

5. 2; The figure has 2 lines of symmetry, each passing through opposite vertices (at the tips). Yes; the figure has point symmetry with respect to its center.
6. 3; The figure has 3 lines of symmetry, each passing through the center, a vertex, and the midpoint of the side opposite the vertex. No; the figure has no common point of reflection and, thus, no point symmetry.
7. 6; The figure has 3 lines of symmetry at the tips of the 5-sided figures and 3 lines of symmetry between the 5-sided figures. The figure has point symmetry with respect to its center.

8.

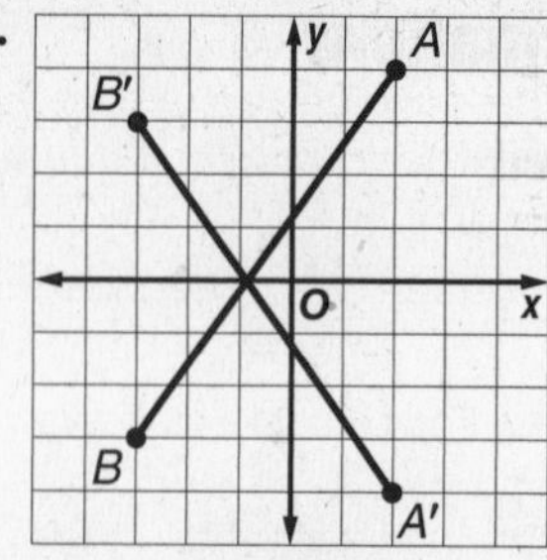

9.

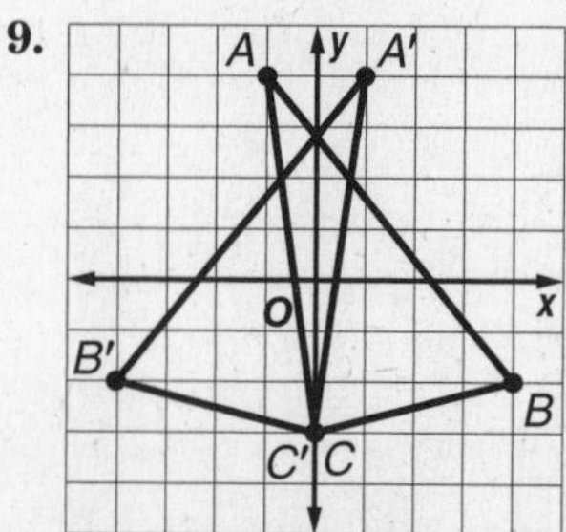

10.

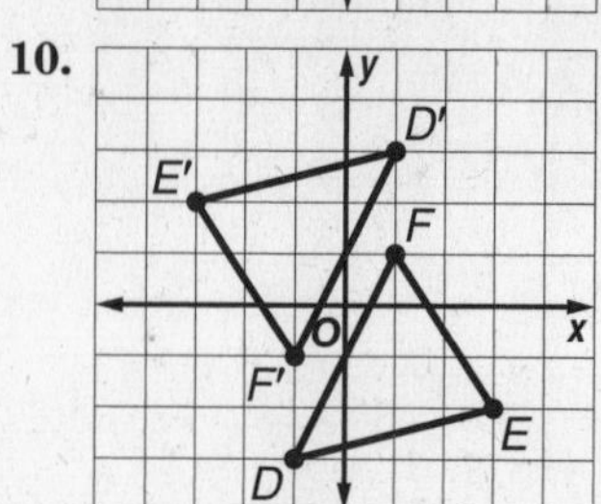

11.

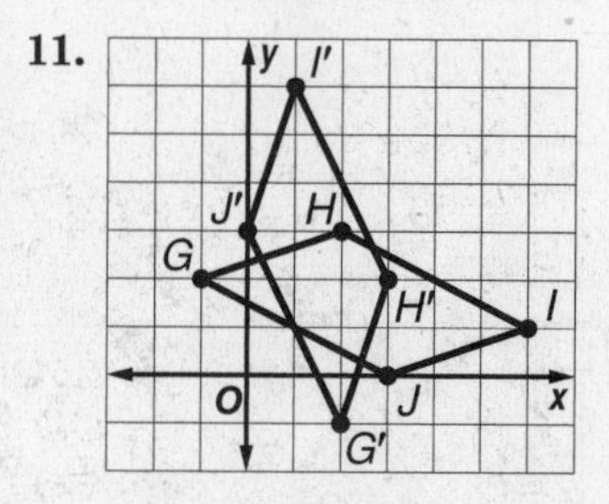

12. 1; The butterfly has 1 line of symmetry through its head and tail. No; the butterfly has no common point of reflection and, thus, no point symmetry.
13. 4; The leaf has 4 lines of symmetry: two passing between the leaves and two passing through their centers. Yes; the figure has point symmetry with respect to its center.

14. Looking at the tiger directly from the front, its face has one line of symmetry that goes down the center of the face vertically. No, the tiger face has no common point of reflection and, thus, no point symmetry.

Pages 467–469 Practice and Apply

15. X is on line ℓ, so it is its own reflection. Y is the image of W under a reflection in line ℓ. So, $\overline{YX}$ is the image of $\overline{WX}$ under a reflection in line ℓ.

16. Z is on line ℓ, so it is its own reflection. Y is the image of W under a reflection in line ℓ. So, $\overline{YZ}$ is the image of $\overline{WZ}$ under a reflection in line ℓ.

17. X and Z are on line ℓ, so they are their own reflections. W is the image of Y under a reflection in line ℓ. So, $\angle XZW$ is the image of $\angle XZY$ under a reflection in line ℓ.

18. T is on line m, so it is its own reflection.

19. U is on line m, so it is its own reflection. V is the image of Y under a reflection in line m. So, $\overline{UV}$ is the image of $\overline{UY}$ under a reflection in line m.

20. V is the image of Y, Y is the image of V, and X is the image of W under a reflection in line m. So, $\triangle VYX$ is the image of $\triangle YVW$ under a reflection in line m.

21. T is the image of U under a reflection in point Z.

22. U is the image of T, V is the image of X, and Z is its own image under a reflection in point Z. So, $\angle UVZ$ is the image of $\angle TXZ$ under a reflection in point Z.

23. W is the image of Y, T is the image of U, and Z is its own image under a reflection in point Z. So, $\triangle WTZ$ is the image of $\triangle YUZ$ under a reflection in point Z.

24. Draw perpendiculars from A, B, C, D, E, and F to line ℓ. Locate A', B', C', D', E', and F' so that line ℓ is the perpendicular bisector of $\overline{AA'}, \overline{BB'}, \overline{CC'}, \overline{DD'}, \overline{EE'}$, and $\overline{FF'}$. Points A', B', C', D', E', and F' are the respective images of A, B, C, D, E, and F. Connect vertices A', B', C', D', E', and F'.

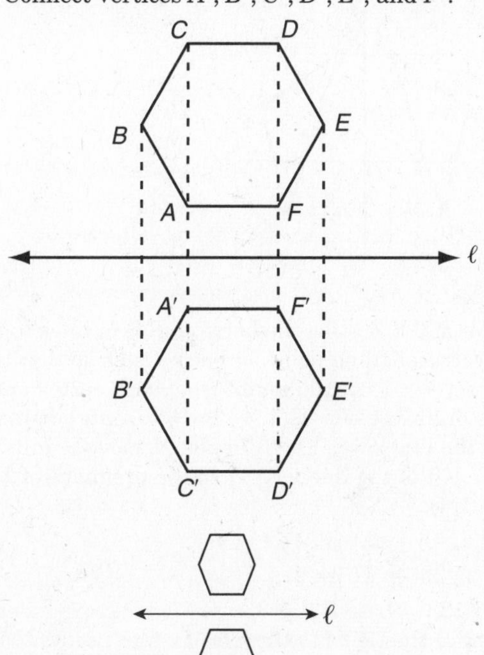

25. Draw perpendiculars from P, R, S, T, and U to line ℓ. Locate P', R', S', T', and U' so that line ℓ is the perpendicular bisector of $\overline{PP'}, \overline{RR'}, \overline{SS'}, \overline{TT'}$, and $\overline{UU'}$. P', R', S', T', and U' are the respective images of P, R, S, T, and U. Connect vertices P', R', S', T', and U'.

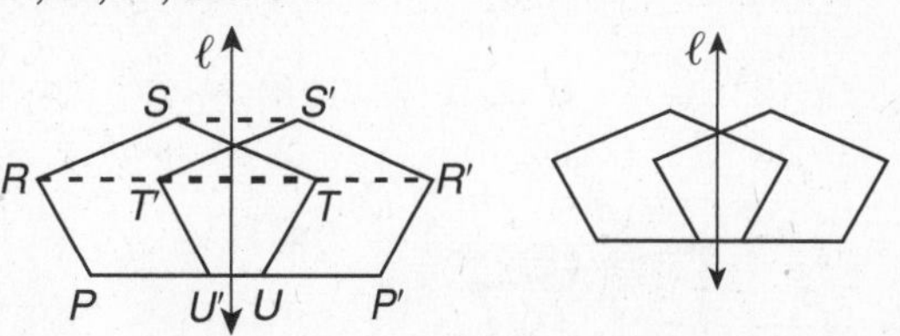

26. Since W is on line ℓ, W is its own reflection. Draw segments perpendicular to line ℓ from X, Y, and Z. Locate X', Y', and Z' so that ℓ is the perpendicular bisector of $\overline{XX'}$, $\overline{YY'}$, and $\overline{ZZ'}$. X', Y', and Z' are the respective images of X, Y, and Z. Connect vertices W, X', Y', and Z'.

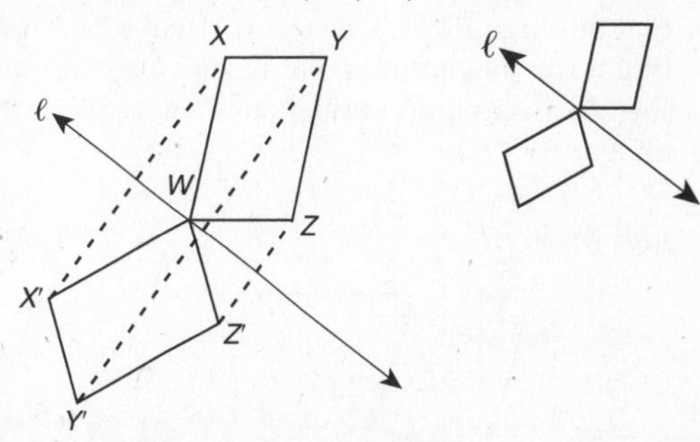

27. Plot rectangle $MNPQ$. Since $\overline{QQ'}$ passes through the origin, use the horizontal and vertical distances from Q to the origin to find Q'. From Q to the origin is 2 units to the right and 3 units down. Q' is located by repeating that pattern from the origin. Two units to the right and 3 units down yields $Q'(2, -3)$.

$Q(-2, 3) \rightarrow Q'(2, -3)$ $\quad N(2, -3) \rightarrow N'(-2, 3)$

$P(-2, -3) \rightarrow P'(2, 3)$ $\quad M(2, 3) \rightarrow M'(-2, -3)$

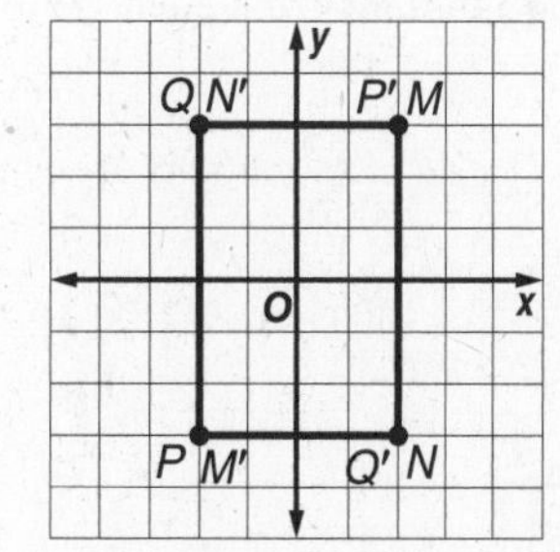

28. Plot rectangle $GHIJ$. Since $\overline{GG'}$ passes through the origin, use the horizontal and vertical distances from G to the origin to find G'. From G to the origin is 2 units up and 2 units to the right. G' is located by repeating that pattern from the origin. Two units up and 2 units to the right yields $G'(2, 2)$.

$G(-2, -2) \rightarrow G'(2, 2)$ $I(3, 3) \rightarrow I'(-3, -3)$

$H(2, 0) \rightarrow H'(-2, 0)$ $J(-2, 4) \rightarrow J'(2, -4)$

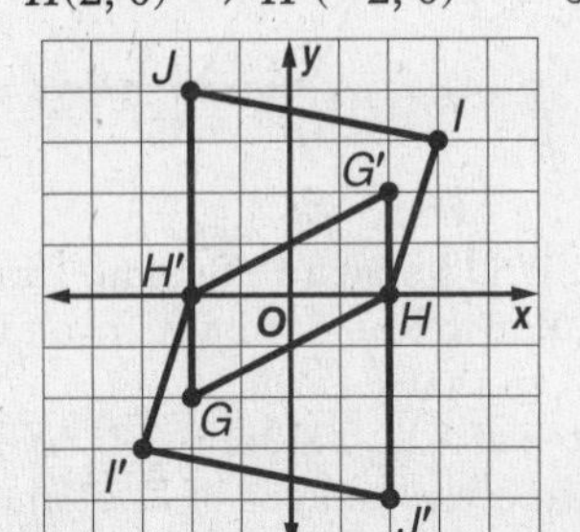

29. Plot square $QRST$. Use the vertical grid lines to find a corresponding point for each vertex so that the x-axis is equidistant from each vertex and its image.

$Q(-1, 4) \rightarrow Q'(-1, -4)$ $S(3, 2) \rightarrow S'(3, -2)$

$R(2, 5) \rightarrow R'(2, -5)$ $T(0, 1) \rightarrow T'(0, -1)$

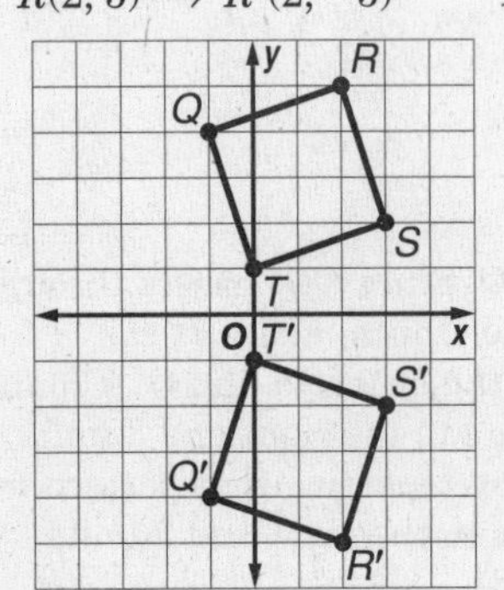

30. Plot trapezoid $DEFG$. Use the horizontal grid lines to find a corresponding point for each vertex so that the y-axis is equidistant from each vertex and its image.

$E(-2, 4) \rightarrow E'(2, 4)$ $F(-2, -1) \rightarrow F'(2, -1)$

$D(4, 0) \rightarrow D'(-4, 0)$ $G(4, -3) \rightarrow G'(-4, -3)$

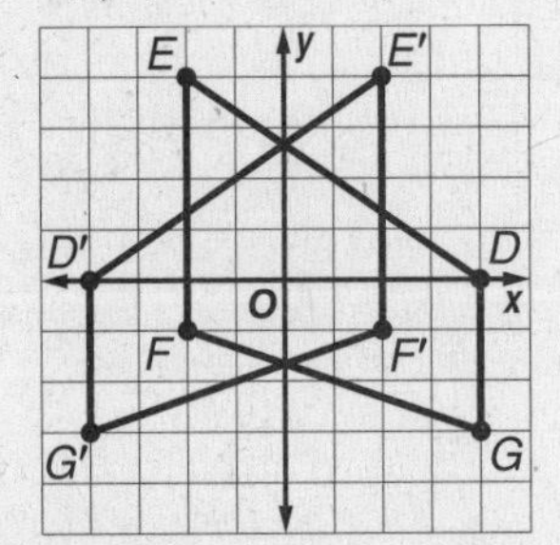

31. Plot $\triangle BCD$ and the line $y = x$. The slope of $y = x$ is 1. $\overline{BB'}$ is perpendicular to $y = x$, so its slope is -1. From B to the line $y = x$, move up 2 units and to the left 3 units. From the line $y = x$, move up 2 units and to the left 3 units to $B'(0, 5)$

$B(5, 0) \rightarrow B'(0, 5)$ $D(-2, -1) \rightarrow D'(-1, -2)$

$C(-2, 4) \rightarrow C'(4, -2)$

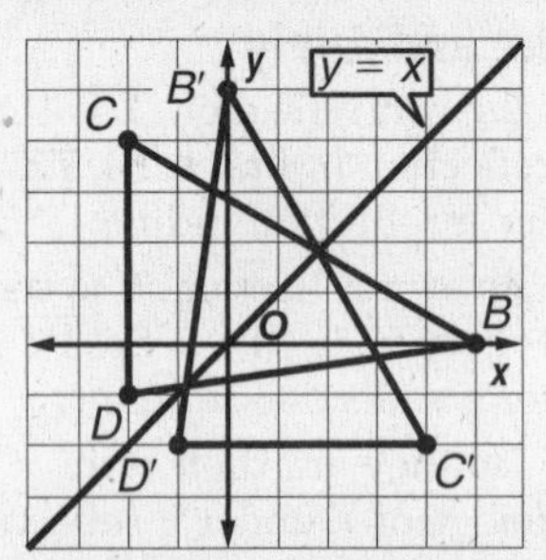

32. Plot $\triangle KLM$. Use the vertical grid lines to find a corresponding point for each vertex so that the line $y = 2$ is equidistant from each vertex and its image. For $K(4, 0)$, the vertical distance to the line $y = 2$ is 2. To plot K', move up 2 units on the vertical gridline so the image of K is $K'(4, 4)$.

$K(4, 0) \rightarrow K'(4, 4)$

$L(-2, 4) \rightarrow L'(-2, 0)$

$M(-2, 1) \rightarrow M'(-2, 3)$

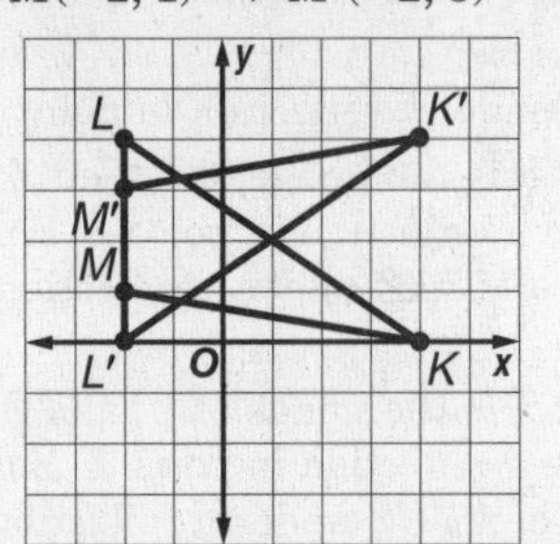

33. Plot $\triangle F'G'H'$. To find $\triangle FGH$, use the horizontal grid lines to find a corresponding point for each vertex so that the y-axis is equidistant from each vertex and its image.

$F'(1, 4) \rightarrow F(-1, 4)$ $H'(3, -2) \rightarrow H(-3, -2)$

$G'(4, 2) \rightarrow G(-4, 2)$

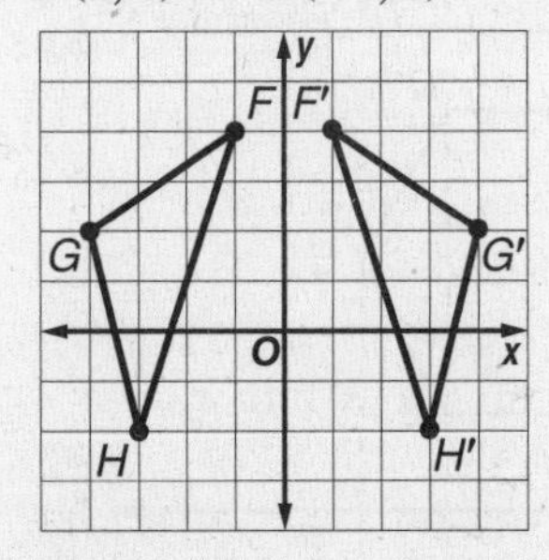

$(x, y) \rightarrow (-x, y)$

34. Plot $\triangle X'Y'Z'$. Use the horizontal grid lines to find a corresponding point for each vertex so that the line $x = -1$ is equidistant from each vertex and its preimage. For $X'(1, 4)$, the horizontal distance to the line $x = -1$ is 2. To plot X, move 2 units to the left of the line $x = -1$ so the preimage of X' is $(-3, 4)$.

$X'(1, 4) \rightarrow X(-3, 4)$

$Y'(2, 2) \rightarrow Y(-4, 2)$

$Z'(-2, -3) \rightarrow Z(0, -3)$

Notice that Z' is to the left of $x = -1$ so its preimage is to the right of the line $x = -1$.

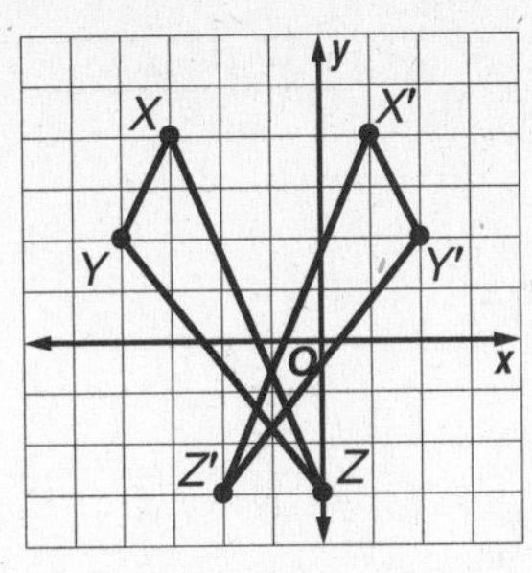

$(x, y) \rightarrow (-(x + 2), y)$

35. 2; The figure has two lines of symmetry, each passing through opposite vertices.
Yes; the figure has point symmetry with respect to its center.

36. 8; The figure has eight lines of symmetry, each passing through opposite vertices.
Yes; the figure has point symmetry with respect to its center.

37. 1; The figure has one line of symmetry, which passes horizontally through the center.
No; the figure has no common point of reflection and, thus, no point symmetry.

38. The preimage and final image have the same shape and the same orientation.

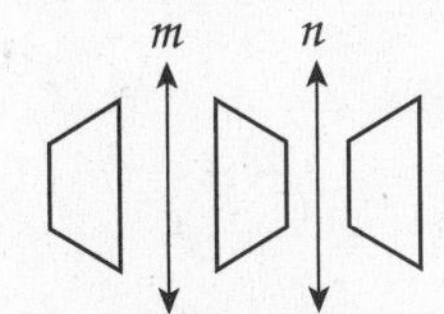

39. The preimage and final image have the same shape, but the final image is turned or rotated with respect to the preimage.

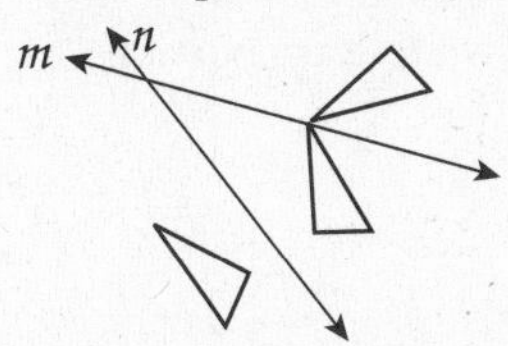

40. Apply the reflections in turn. $D(-1, 4)$, $E(2, 8)$, $F(6, 5)$, and $G(3, 1)$. Reflection in the x-axis: Multiply the y-coordinates by -1: $(a, b) \rightarrow (a, -b)$: $D'(-1, -4)$, $E'(2, -8)$, $F'(6, -5)$, and $G'(3, -1)$. Reflection in the line $y = x$: Interchange the x- and y- coordinates $(a, b) \rightarrow (b, a)$: $D''(-4, -1)$, $E''(-8, 2)$, $F''(-5, 6)$, and $G''(-1, 3)$.

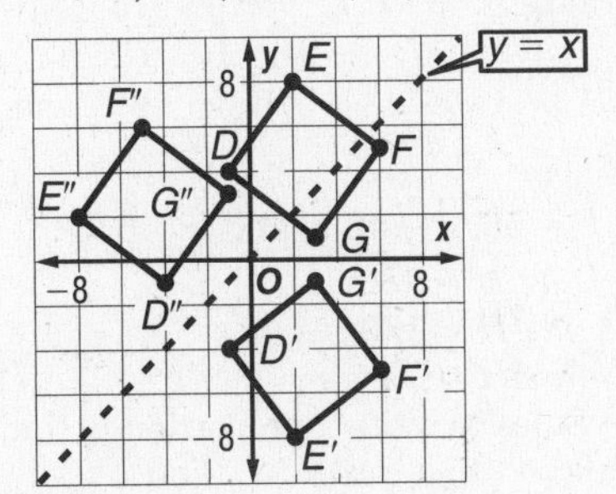

41. Undo the reflections in turn. $A'''(4, 7)$, $B'''(10, -3)$, and $C'''(-6, -8)$. Reflection in the origin: multiply both coordinates by -1: $(a, b) \rightarrow (-a, -b)$: $A''(-4, -7)$, $B''(-10, 3)$, and $C''(6, 8)$. Reflection in the y-axis: multiply the x-coordinate by -1: $(a, b) \rightarrow (-a, b)$: $A'(4, -7)$, $B'(10, 3)$, and $C'(-6, 8)$.
Reflection in the x-axis: multiply the y-coordinate by -1 $(a, b) \rightarrow (a, -b)$: $A(4, 7)$, $B(10, -3)$, and $C(-6, -8)$.
Undoing the transformations results in the triangle ABC.

42.

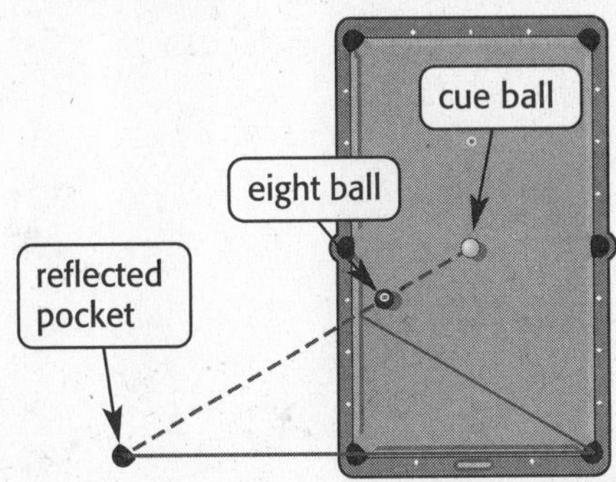

43. Consider point (a, b). Upon reflection in the origin, its image is $(-a, -b)$. Upon reflection in the x-axis and then the y-axis, its image is $(a, -b)$ and then $(-a, -b)$. The images are the same.

44. The diamond has numerous lines of symmetry, including vertical and horizontal lines of symmetry. It has a point of symmetry at the center.

45. The diamond has a vertical line of symmetry, passing through its center.

46. The diamond has a vertical line of symmetry, passing through its center.

47. The diamond has a vertical line of symmetry and a horizontal line of symmetry. It has a point of symmetry at the center.

48. Sample answer: Reflections of the surrounding vistas can be seen in bodies of water. Answers should include the following.

- Three examples of line symmetry in nature are the water's edge in a lake, the line through the middle of a pin oak leaf, and the lines of a four leaf clover.
- Each point above the water has a corresponding point in the image in the lake. The distance of a point above the water appears the same as the distance of the image below the water.

49. D; x-axis reflection: $(-2, 5) \rightarrow (-2, -5)$ and y-axis reflection: $(-2, -5) \rightarrow (2, -5)$ or reflection in the origin: $(-2, 5) \rightarrow (2, -5)$

50. B; $a \star c = 2a + b + 2c$
$25 \star 45 = 2(25) + 18 + 2(45)$
$25 \star 45 = 158$

Page 469 Maintain Your Skills

51. Given: Quadrilateral $LMNP$
X, Y, Z, and W are midpoints of their respective sides.

Prove: $\overline{YW}$ and $\overline{XZ}$ bisect each other.

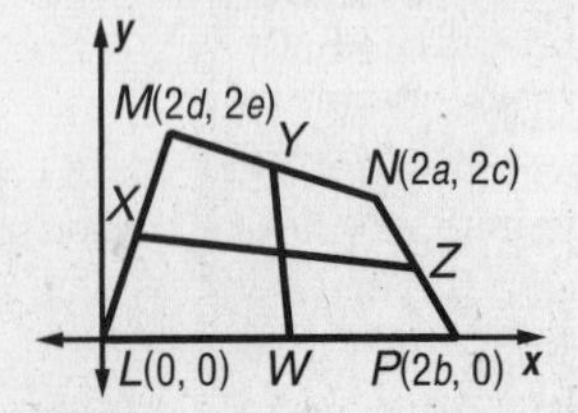

Proof:
Midpoint Y of $\overline{MN}$ is $\left(\frac{2d + 2a}{2}, \frac{2e + 2c}{2}\right)$ or $(d + a, e + c)$.
Midpoint Z of $\overline{NP}$ is $\left(\frac{2a + 2b}{2}, \frac{2c + 0}{2}\right)$ or $(a + b, c)$.
Midpoint W of $\overline{PL}$ is $\left(\frac{0 + 2b}{2}, \frac{0 + 0}{2}\right)$ or $(b, 0)$.
Midpoint X of $\overline{LM}$ is $\left(\frac{0 + 2d}{2}, \frac{0 + 2e}{2}\right)$ or (d, e).
Midpoint of $\overline{WY}$ is $\left(\frac{d + a + b}{2}, \frac{e + c + 0}{2}\right)$ or $\left(\frac{a + b + d}{2}, \frac{c + e}{2}\right)$.
Midpoint of $\overline{XZ}$ is $\left(\frac{d + a + b}{2}, \frac{e + c}{2}\right)$ or $\left(\frac{a + b + d}{2}, \frac{c + e}{2}\right)$. The midpoints of $\overline{XZ}$ and $\overline{YW}$ are the same, so $\overline{XZ}$ and $\overline{WY}$ bisect each other.

52. Given: Isosceles trapezoid
$\overline{AD} \cong \overline{BC}$
H, J, K, and G are midpoints of their respective sides.

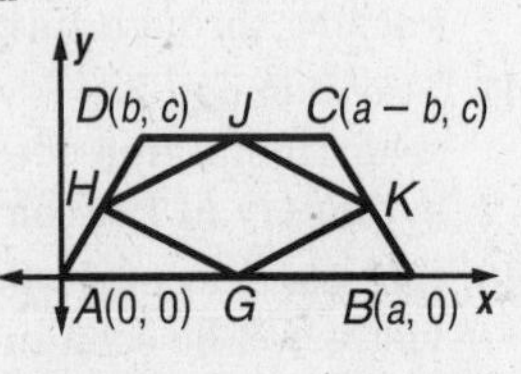

Prove: $GHJK$ is a rhombus.

Proof:
Midpoint H of $\overline{AD}$ is $\left(\frac{b + 0}{2}, \frac{c + 0}{2}\right) = \left(\frac{b}{2}, \frac{c}{2}\right)$.

Midpoint J of $\overline{DC}$ is $\left(\frac{a - b + b}{2}, \frac{c + c}{2}\right) = \left(\frac{a}{2}, c\right)$.

Midpoint K of $\overline{CB}$ is $\left(\frac{a - b + a}{2}, \frac{c + 0}{2}\right) = \left(\frac{2a - b}{2}, \frac{c}{2}\right)$.

Midpoint G of $\overline{DC}$ is $\left(\frac{a + 0}{2}, \frac{0 + 0}{2}\right) = \left(\frac{a}{2}, 0\right)$.

$$HJ = \sqrt{\left(\frac{b}{2} - \frac{a}{2}\right)^2 + \left(\frac{c}{2} - c\right)^2} = \frac{\sqrt{b^2 - 2ab + a^2 + c^2}}{2};$$

$$GK = \sqrt{\left(\frac{2a - b}{2} - \frac{a}{2}\right)^2 + \left(\frac{c}{2} - 0\right)^2} = \frac{\sqrt{b^2 - 2ab + a^2 + c^2}}{2};$$

$$HG = \sqrt{\left(\frac{b}{2} - \frac{a}{2}\right)^2 + \left(\frac{c}{2} - 0\right)^2} = \frac{\sqrt{b^2 - 2ab + a^2 + c^2}}{2};$$

$$KJ = \sqrt{\left(\frac{2a - b}{2} - \frac{a}{2}\right)^2 + \left(\frac{c}{2} - c\right)^2} = \frac{\sqrt{b^2 - 2ab + a^2 + c^2}}{2};$$

$HJ = GK = HG = KJ$, so $\overline{HJ} \cong \overline{GK} \cong \overline{HG} \cong \overline{KJ}$ and $GHJK$ is a rhombus.

53. BE is the measure of the median of the trapezoid.
$BE = \frac{AF + CD}{2}$
$= \frac{32 + 48}{2}$
$= 40$

54. $BE = \frac{AF + CD}{2}$
$= \frac{32 + 48}{2}$
$= 40$
$XY = \frac{BE + CD}{2}$
$= \frac{40 + 48}{2}$
$= 44$

55. $BE = \frac{AF + CD}{2}$
$= \frac{32 + 48}{2}$
$= 40$
$WZ = \frac{AF + BE}{2}$
$= \frac{32 + 40}{2}$
$= 36$

56. $m\angle F + m\angle G + m\angle H = 180$
$m\angle F + 53 + 71 = 180$
$m\angle F = 56$

Use the Law of Sines to write a proportion to find g.
$\frac{\sin F}{f} = \frac{\sin G}{g}$
$g = \frac{f \sin G}{\sin F}$
$g = \frac{48 \sin 53°}{\sin 56°}$
$g \approx 46.2$

Use the Law of Sines again to find the measure of the third side.
$\frac{\sin F}{f} = \frac{\sin H}{h}$
$h = \frac{f \sin H}{\sin F}$
$h = \frac{48 \sin 71°}{\sin 56°}$
$h \approx 54.7$

Therefore, $m\angle F = 56$, $g \approx 46.2$, and $h \approx 54.7$.

57. $m\angle F + m\angle G + m\angle H = 180$
$59 + 45 + m\angle H = 180$
$m\angle H = 76$

Use the Law of Sines to write a proportion to find f.
$\frac{\sin G}{g} = \frac{\sin F}{f}$
$f = \frac{g \sin F}{\sin G}$
$f = \frac{21 \sin 59°}{\sin 45°}$
$f \approx 25.5$

Use the Law of Sines again to find the measure of the third side.

$\frac{\sin G}{g} = \frac{\sin H}{h}$

$h = \frac{g \sin H}{\sin G}$

$h = \frac{21 \sin 76°}{\sin 45°}$

$h \approx 28.8$

Therefore, $m\angle H = 76$, $f \approx 25.5$, and $h \approx 28.8$.

58. We know two sides and the measure of the angle opposite one of the sides. Use the Law of Sines to find the measure of the second angle.

$\frac{\sin F}{f} = \frac{\sin H}{h}$

$\sin H = \frac{h \sin F}{f}$

$H = \sin^{-1}\left(\frac{h \sin F}{f}\right)$

$H = \sin^{-1}\left(\frac{13.2 \sin 106°}{14.5}\right)$

$H \approx 61°$

So, $m\angle H \approx 61$.

Use the Angle Sum Theorem to find the measure of angle G.

$m\angle F + m\angle G + m\angle H = 180$

$106 + m\angle G + 61 \approx 180$

$m\angle G \approx 13$

Use the Law of Sines and a proportion to find g.

$\frac{\sin H}{h} = \frac{\sin G}{g}$

$g = \frac{h \sin G}{\sin H}$

$g \approx \frac{13.2 \sin 13°}{\sin 61°}$

$g \approx 3.4$

Therefore, $m\angle H \approx 61$, $m\angle G \approx 13$, and $g \approx 3.4$.

59. $d = \sqrt{(x_2 - x_1)^2 + (y_2 - y_1)^2}$

$EF = \sqrt{(2 - 3)^2 + [0 - (-1)]^2}$

$EF = \sqrt{(-1)^2 + 1^2}$

$EF = \sqrt{2}$

60. $d = \sqrt{(x_2 - x_1)^2 + (y_2 - y_1)^2}$

$FG = \sqrt{(3 - 2)^2 + (3 - 0)^2}$

$FG = \sqrt{1^2 + 3^2}$

$FG = \sqrt{10}$

61. $d = \sqrt{(x_2 - x_1)^2 + (y_2 - y_1)^2}$

$GH = \sqrt{(5 - 3)^2 + (4 - 3)^2}$

$GH = \sqrt{2^2 + 1^2}$

$GH = \sqrt{5}$

62. $d = \sqrt{(x_2 - x_1)^2 + (y_2 - y_1)^2}$

$HE = \sqrt{(3 - 5)^2 + (-1 - 4)^2}$

$HE = \sqrt{(-2)^2 + (-5)^2}$

$HE = \sqrt{29}$

9-2 Translations

Page 472 Check for Understanding

1. Sample answer: $A(3, 5)$ and $B(-4, 7)$; start at 3, count to the left to -4, which is 7 units to the left or -7. Then count up 2 units from 5 to 7 or $+2$. The translation from A to B is $(x, y) \rightarrow (x - 7, y + 2)$.
2. The properties that are preserved include betweenness of points, collinearity, and angle and distance measure. Since translations are composites of two reflections, all translations are isometries. Thus, all properties preserved by reflections are preserved by translations.
3. Allie; counting from the point $(-2, 1)$ to $(1, -1)$ is right 3 and down 2 to the image. The reflections would be too far to the right. The image would be reversed as well.
4. Yes; $\triangle GHI$ is a translation of $\triangle ABC$. $\triangle DEF$ is the image of $\triangle ABC$ when $\triangle ABC$ is reflected in line m, and $\triangle GHI$ is the image of $\triangle DEF$ when $\triangle DEF$ is reflected in line n.
5. No; quadrilateral $WXYZ$ is oriented differently than quadrilateral $NPQR$.
6. This translation moved every point of the preimage 1 unit right and 3 units up.

   $D(-3, -4) \rightarrow D'(-3 + 1, -4 + 3)$ or $D'(-2, -1)$

   $E(4, 2) \rightarrow E'(4 + 1, 2 + 3)$ or $E'(5, 5)$

   Graph D and E and connect. Graph D' and E' and connect.

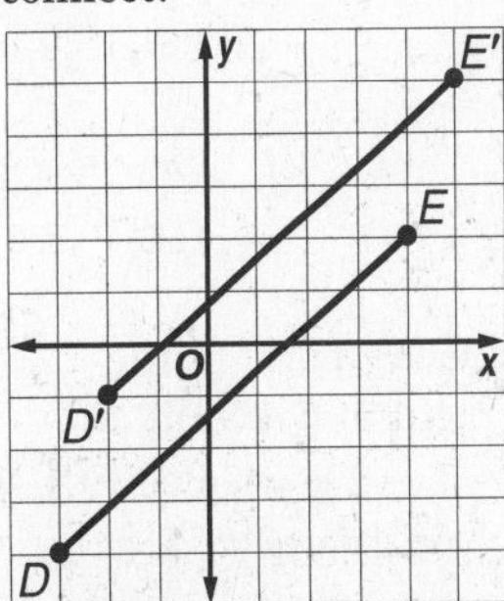

7. This translation moved every point of the preimage 3 units to the left and 4 units down.

   $K(5, -2) \rightarrow K'(5 - 3, -2 - 4)$ or $K'(2, -6)$

   $L(-3, -1) \rightarrow L'(-3 - 3, -1 - 4)$ or $L'(-6, -5)$

   $M(0, 5) \rightarrow M'(0 - 3, 5 - 4)$ or $M'(-3, 1)$

   Graph K, L, and M and connect to form $\triangle KLM$. Graph K', L', and M' to form $\triangle K'L'M'$.

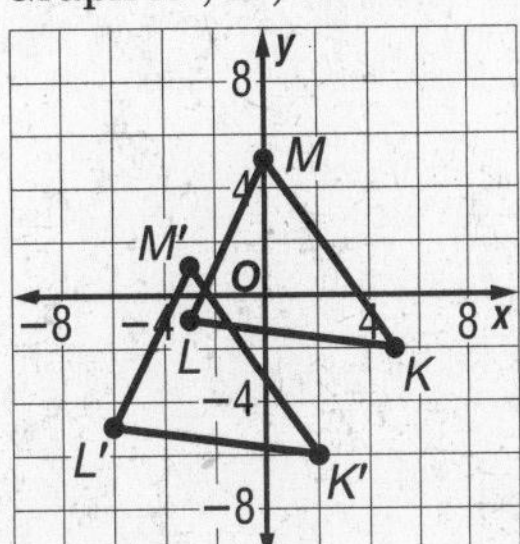

8. $1 \rightarrow 2 = (x, y) \rightarrow (x, y + 3)$
$2 \rightarrow 3 = (x, y) \rightarrow (x + 4, y)$
$3 \rightarrow 4 = (x, y) \rightarrow (x + 4, y)$

Pages 472–475 Practice and Apply

9. Yes; it is one reflection after another with respect to the two parallel lines.

10. No; it is a reflection followed by a translation.

11. No; it is a reflection followed by a rotation.

12. No; it is a reflection followed by a translation.

13. Yes; it is one reflection after another with respect to the two parallel lines.

14. No; it is a reflection followed by a translation.

15. For each endpoint of $\overline{PQ}$ move left 3 units and up 4 units to find the image. Connect P' and Q'.

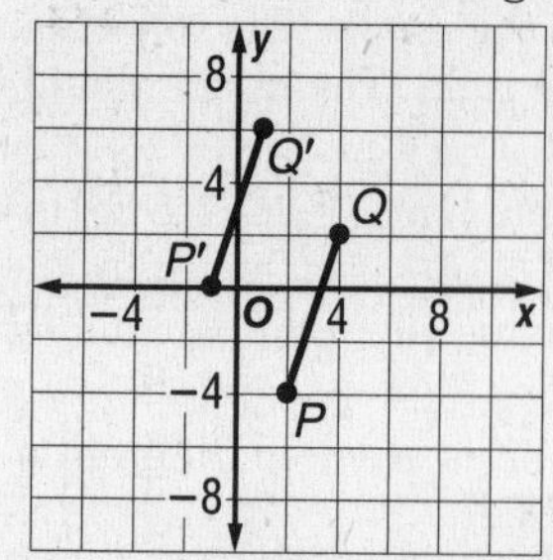

16. For each endpoint of $\overline{AB}$ move right 4 units and down 2 units. Connect A' and B'.

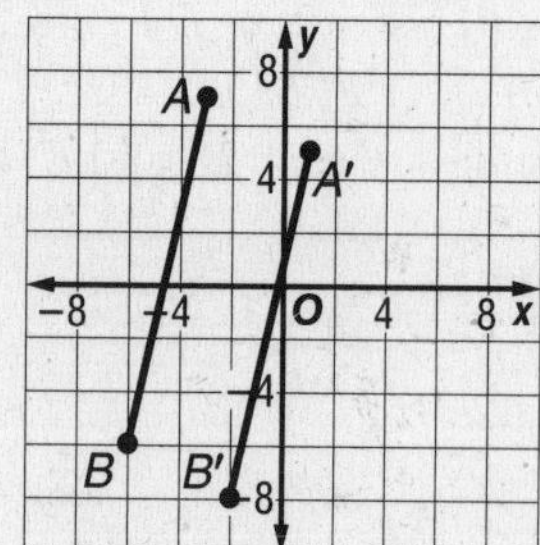

17. This translation moved every point of the preimage 1 unit to the right and 4 units up.
$M(-2, -2) \rightarrow M'(-2 + 1, -2 + 4)$ or $M'(-1, 2)$
$J(-5, 2) \rightarrow J'(-5 + 1, 2 + 4)$ or $J'(-4, 6)$
$P(0, 4) \rightarrow P'(0 + 1, 4 + 4)$ or $P'(1, 8)$
Plot the vertices of the preimage and the image and connect the respective vertices to form the preimage and the image.

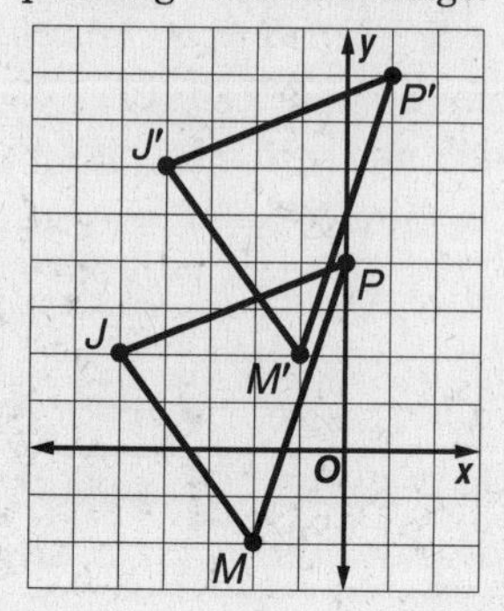

18. This translation moved every point of the preimage 2 units to the right and 1 unit down.
$E(0, -4) \rightarrow E'(0 + 2, -4 - 1)$ or $E'(2, -5)$
$F(-4, -4) \rightarrow F'(-4 + 2, -4 - 1)$ or $F'(-2, -5)$
$G(0, 2) \rightarrow G'(0 + 2, 2 - 1)$ or $G'(2, 1)$
Plot the vertices of the preimage and the image and connect the respective vertices to form the preimage and the image.

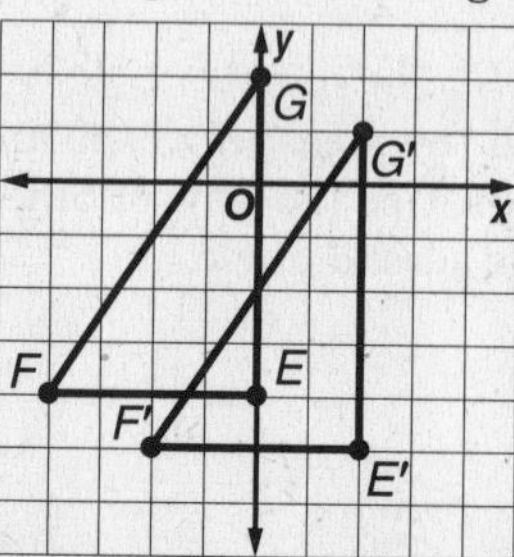

19. This translation moved every point of the preimage 5 units to the left and 3 units up.
$P(1, 4) \rightarrow P'(1 - 5, 4 + 3)$ or $P'(-4, 7)$
$Q(-1, 4) \rightarrow Q'(-1 - 5, 4 + 3)$ or $Q'(-6, 7)$
$R(-2, -4) \rightarrow R'(-2 - 5, -4 + 3)$ or $R'(-7, -1)$
$S(2, -4) \rightarrow S'(2 - 5, -4 + 3)$ or $S'(-3, -1)$
Plot the vertices of the preimage and the image and connect the respective vertices to form the preimage and the image.

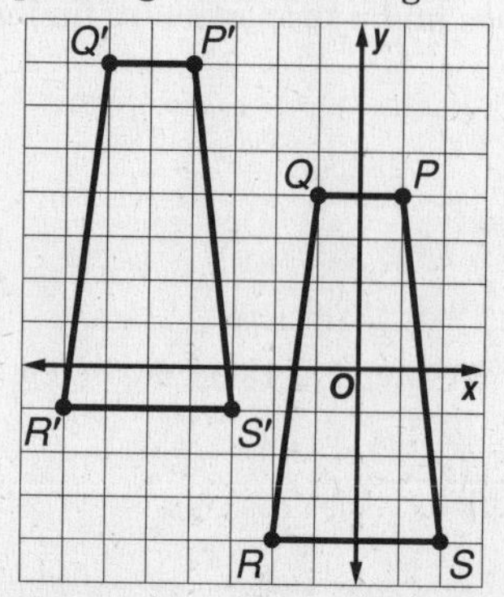

20. This translation moved every point of the preimage 4 units to the right and 3 units down.
$V(-3, 0) \rightarrow V'(-3 + 4, 0 - 3)$ or $V'(1, -3)$
$W(-3, 2) \rightarrow W'(-3 + 4, 2 - 3)$ or $W'(1, -1)$
$X(-2, 3) \rightarrow X'(-2 + 4, 3 - 3)$ or $X'(2, 0)$
$Y(0, 2) \rightarrow Y'(0 + 4, 2 - 3)$ or $Y'(4, -1)$
$Z(-1, 0) \rightarrow Z'(-1 + 4, 0 - 3)$ or $Z'(3, -3)$
Plot the vertices of the preimage and the image and connect the respective vertices to form the preimage and the image.

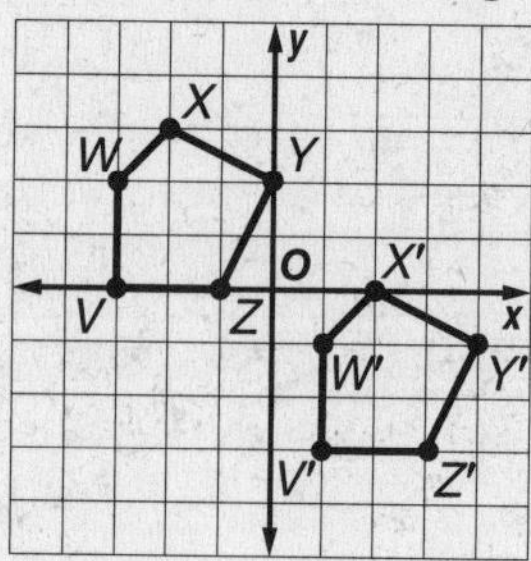

21. As a translation, the bishop moves left 3 squares and down 7 squares.

22. Sample answers: pawn: up two squares; rook: left four squares; knight: down two squares, right 1 square; bishop: up three squares, right three squares; queen: up five squares; king: right 1 square

23. Four triangle lengths is equivalent to a translation of 48 in. right.

24. Two triangle lengths left and four triangle lengths up and left (60° angle above horizontal) form one leg and the hypotenuse of a right triangle. Use the Pythagorean Theorem to find the other leg (direction up).

$$c^2 = a^2 + b^2$$
$$(4 \cdot 12)^2 = a^2 + (2 \cdot 12)^2$$
$$2304 = a^2 + 576$$
$$1728 = a^2$$
$$24\sqrt{3} = a$$
$$41.6 \approx a$$

The translation is $24\sqrt{3} \approx 41.6$ in. up and 24 in. left.

25. The red line represents a translation of six triangle lengths right and four triangle heights down. Use the Pythagorean Theorem to find the triangle height.

$$c^2 = a^2 + b^2$$
$$12^2 = a^2 + \left(\frac{12}{2}\right)^2$$
$$144 = a^2 + 36$$
$$108 = a^2$$
$$6\sqrt{3} = a$$

Six lengths: $6(12) = 72$
Four heights: $4(6\sqrt{3}) = 24\sqrt{3} \approx 41.6$
The translation is 72 in. right and $24\sqrt{3} \approx 41.6$ in. down.

26. Sample answer:
Explore: We are looking for two parallel lines such that $\triangle TWY$ is reflected over each line to result in the image $\triangle BDG$.
Plan: Once we choose any line to be the first parallel line, there is only one possible line for the second parallel line.
Solve: Choose $y = -4$ for the first parallel line. Use the vertical grid lines to determine the vertices of the image that are the same distance from $y = -4$ as the preimage. Since the y-coordinate of W is -4, W is its own image.
$T(3, -7) \rightarrow T'(3, -1)$
$Y(9, -8) \rightarrow Y'(9, 0)$
Now we need to find a line such that a reflection over that line has an image of $\triangle BDG$. We are looking for a line that is equidistant between $\triangle T'WY'$ and $\triangle BDG$. We are only looking at the y-coordinates.
$T'(3, -1) \rightarrow B(3, 3)$. The distance between the y-coordinates is $|-1 - 3| = 4$ so the parallel line that lies halfway between is 2 units above the line $y = -1$ or $y = 1$.

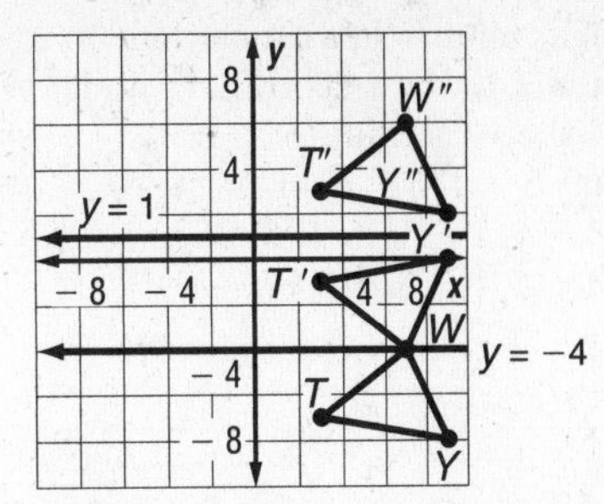

Examine: Notice that the image of $W(7, -4)$ is $W''(7, 6)$ or point D. The line $y = 1$ is equidistant from these points. The image of $Y'(9, 0)$ is $Y''(9, 2)$ or G. Again, the line $y = 1$ is equidistant from these points. So, two possible parallel lines are $y = -4$ and $y = 1$.

27. This translation moved every point of the preimage 2 units to the right and 4 units down.

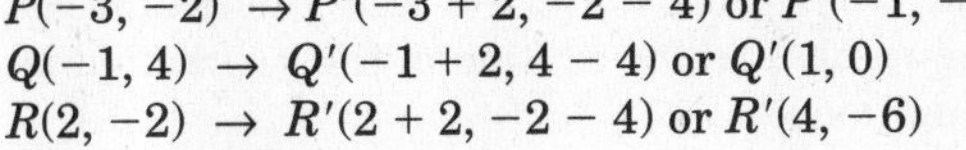

$P(-3, -2) \rightarrow P'(-3 + 2, -2 - 4)$ or $P'(-1, -6)$
$Q(-1, 4) \rightarrow Q'(-1 + 2, 4 - 4)$ or $Q'(1, 0)$
$R(2, -2) \rightarrow R'(2 + 2, -2 - 4)$ or $R'(4, -6)$

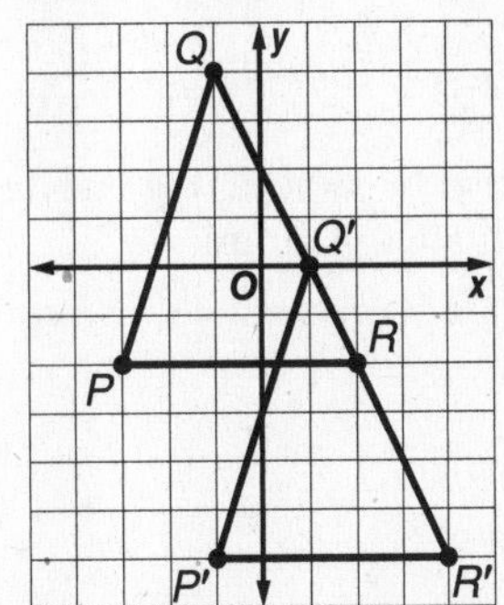

28. First reflect $\triangle RST$ in the line $y = 2$. Use the vertical grid lines to find images of the vertices that are the same distance from $y = 2$ as the preimage.
$R(-4, -1) \rightarrow R'(-4, 5)$
$S(-1, 3) \rightarrow S'(-1, 1)$
$T(-1, 1) \rightarrow T'(-1, 3)$

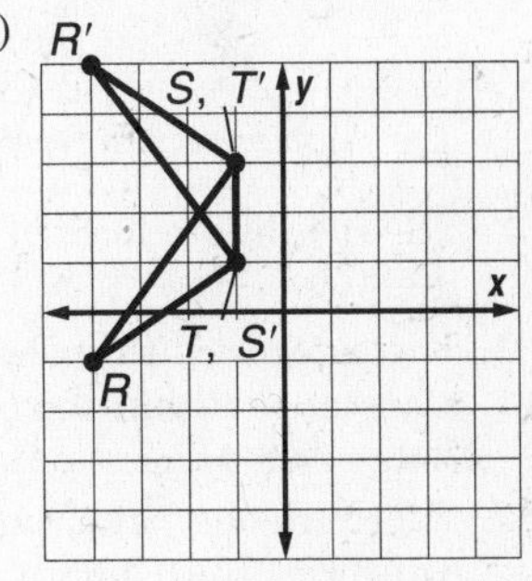

Next reflect $\triangle R'S'T'$ over the line $y = -2$ to get $\triangle R''S''T''$.
$R'(-4, 5) \rightarrow R''(-4, -9)$
$S'(-1, 1) \rightarrow S''(-1, -5)$
$T'(-1, 3) \rightarrow T''(-1, -7)$

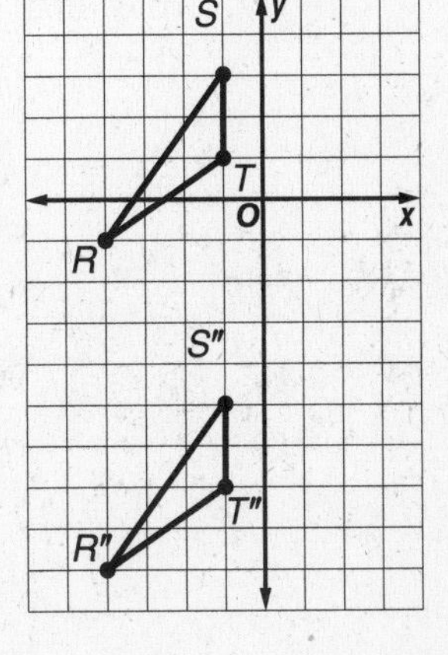

29. To find the image, "undo" the translation. To undo $(x - 4, y + 5)$, add 4 to the x-coordinate and subtract 5 from the y-coordinate.
$A'(-8, 5) \rightarrow A(-8 + 4, 5 - 5)$ or $A(-4, 0)$
$B'(2, 7) \rightarrow B(2 + 4, 7 - 5)$ or $B(6, 2)$
$C'(3, 1) \rightarrow C(3 + 4, 1 - 5)$ or $C(7, -4)$

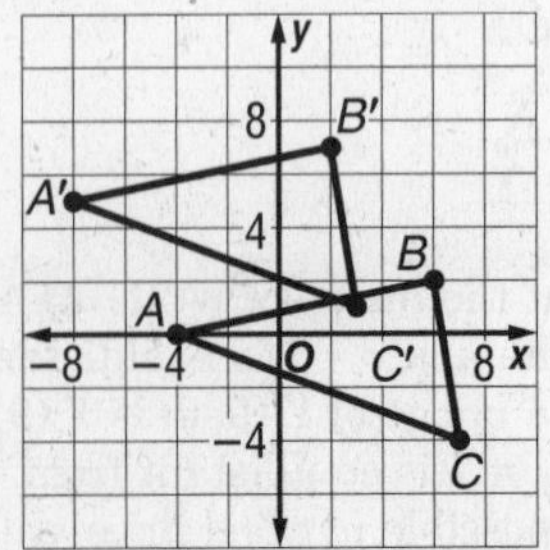

30. In order to find the coordinates of H and N we find the transformation from vertex F to vertex M.
$F(3, 9) \rightarrow M(4, 2)$
The translation in the x-direction is 1 unit to the right. The translation in the y-direction is 7 units down.
$G(-1, 4) \rightarrow N(x_1, y_1)$
$\rightarrow N(-1 + 1, 4 - 7)$ or $N(0, -3)$
$H(x_2, y_2) \rightarrow P(6, -3)$
Undo the transformation by subtracting one and adding 7. So the coordinates of H are $H(6 - 1, -3 + 7)$ or $H(5, 4)$.

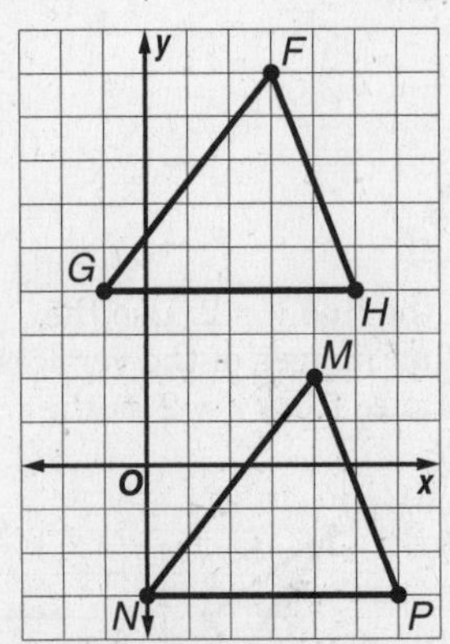

The coordinate form of the translation is $(x, y) \rightarrow (x + 1, y - 7)$.

31. The categories that show a boy-girl-boy unit translated within the bar are "more brains" and "more free time".

32. "More friends" and "more athletic ability" are the categories that show a boy-girl-boy unit reflected (about a vertical line) within the bar.

33. $\frac{80\%}{9} \approx 8.89\%$ per person
$\frac{78\%}{8} = 9.75\%$ per person
$\frac{70\%}{7} = 10\%$ per person
$\frac{62\%}{7} \approx 8.86\%$ per person
No; the percent per figure is different in each category.

34. Sample answer: Every time a band member takes a step, he or she moves a fixed amount in a certain direction. This is a translation. Answers should include the following.
- When a band member takes a step forward, backward, left, right, or on a diagonal without turning, this is a translation.
- To move in a rectangular pattern, the band member starting at (0, 0) could move to (0, 5). Moving from (0, 5) to (4, 5), from (4, 5) to (4, 0) and from (4, 0) back to the origin, the band member would have completed the rectangle.

35. Translations and reflections preserve the congruences of lengths and angles. The composition of the two transformations will preserve both congruences. Therefore, a glide reflection is an isometry.

36. First, find the vertices of the image after the translation $(x, y) \rightarrow (x, y - 2)$.
$D(4, 3) \rightarrow (4, 3 - 2)$ or $(4, 1)$
$E(2, -2) \rightarrow (2, -2 - 2)$ or $(2, -4)$
$F(0, 1) \rightarrow (0, 1 - 2)$ or $(0, -1)$
Now reflect the image in the y-axis. Use the formula $(a, b) \rightarrow (-a, b)$.
$(4, 1) \rightarrow D'(-4, 1)$
$(2, -4) \rightarrow E'(-2, -4)$
$(0, -1) \rightarrow F'(0, -1)$

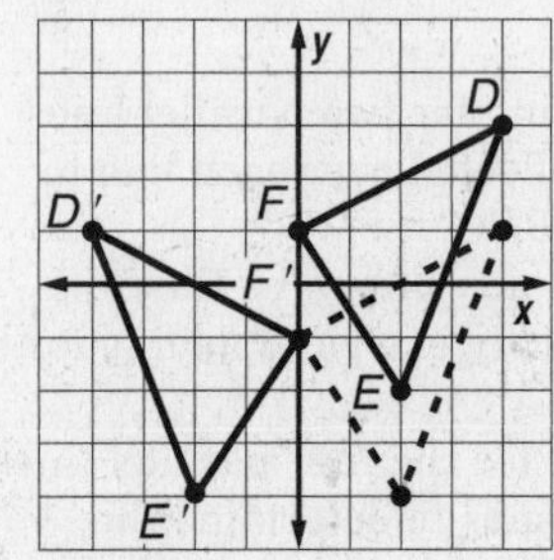

37. First, find the vertices of the image after the translation $(x, y) \rightarrow (x + 3, y)$.
$A(-3, -2) \rightarrow (-3 + 3, -2)$ or $(0, -2)$
$B(-1, -3) \rightarrow (-1 + 3, -3)$ or $(2, -3)$
$C(2, -1) \rightarrow (2 + 3, -1)$ or $(5, -1)$
Now reflect the image over the line $y = 1$.
Use the horizontal grid lines to find the vertices of $\triangle A'B'C'$ such that each vertex of the image is the same distance from the line $y = 1$ as its preimage.
$(0, -2) \rightarrow A'(0, 4)$
$(2, -3) \rightarrow B'(2, 5)$
$(5, -1) \rightarrow C'(5, 3)$

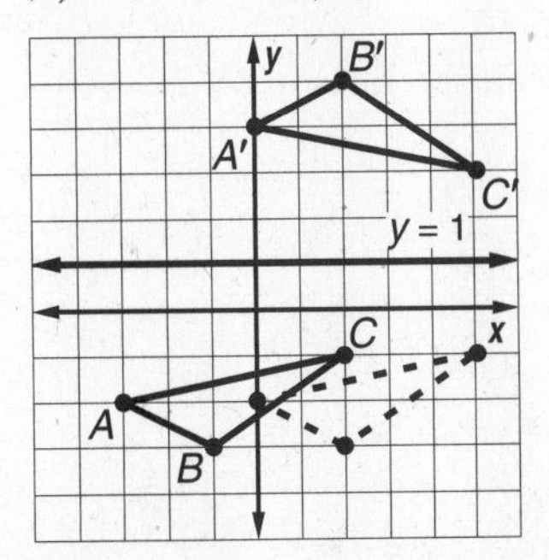

38. C; $X(5, 4) \rightarrow X'(3, 1) = X'(5 - 2, 4 - 3)$
So, $(x, y) \rightarrow (x - 2, y - 3)$.
$Y'(3 - 2, -1 - 3) = Y'(1, -4)$
$Z'(0 - 2, 2 - 3) = Z'(-2, -1)$

39. A; $m = \frac{y_2 - y_1}{x_2 - x_1}$
$= \frac{-1 - 5}{2 - (-2)}$
$= \frac{-6}{4}$
$= -\frac{3}{2}$

Page 475 Maintain Your Skills

40. Draw perpendiculars from A, B, C, D, and E to line m. Locate A', B', C', D', and E' so that line m is the perpendicular bisector of $\overline{AA'}, \overline{BB'}, \overline{CC'}, \overline{DD'}$, and $\overline{EE'}$. Points A', B', C', D', and E' are the respective images of A, B, C, D, and E. Connect vertices A', B', C', D', and E'.

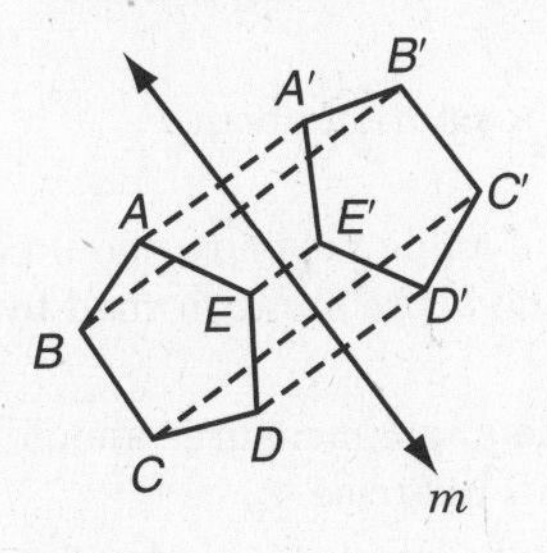

41. Draw perpendiculars from P, R, S, T, and U to line m. Locate P', R', S', T', and U' so that line m is the perpendicular bisector of $\overline{PP'}, \overline{RR'}, \overline{SS'}, \overline{TT'}$, and $\overline{UU'}$. Points P', R', S', T', and U' are the respective images of P, R, S, T, and U. Connect vertices P', R', S', T', and U'.

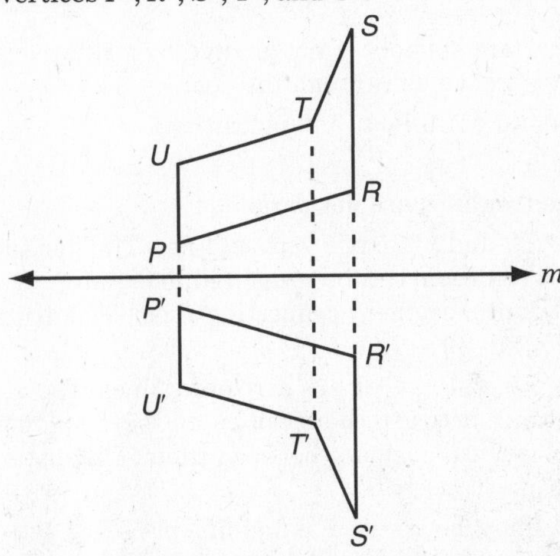

42. Since D is on line m, D is its own reflection. Draw perpendiculars from E, F, G, H, and I to line m. Locate E', F', G', H', and I' so that line m is the perpendicular bisector of $\overline{EE'}, \overline{FF'}, \overline{GG'}, \overline{HH'}$, and $\overline{II'}$. Points E', F', G', H', and I' are the respective images of E, F, G, H, and I. Connect vertices D, E', F', G', H', and I'.

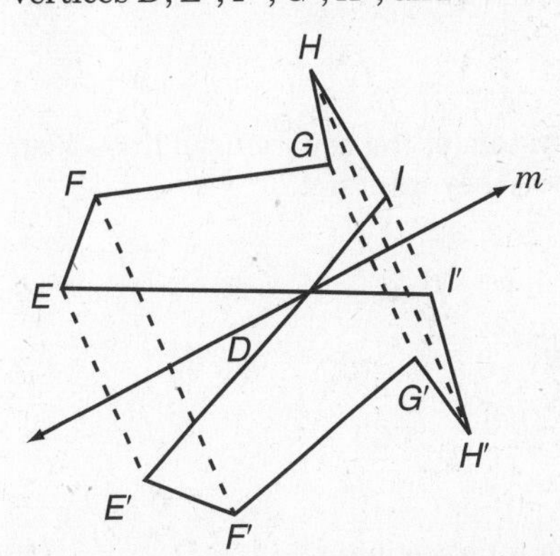

43. The top and bottom segments of the trapezoid are parallel, so the y-coordinate of Q is the same as that of R, c.
The x-coordinate of Q plus the x-coordinate of R equals the x-coordinate of S.
$x + b = a$
$x = a - b$
So, $Q(?, ?) = Q(a - b, c)$.
T is at the origin, so $T(?, ?) = T(0, 0)$.

44. Opposite sides of a parallelogram are congruent and parallel. So, the y-coordinate of B is b and that of D is 0.
The x-coordinate of B plus the x-coordinate of D equals the x-coordinate of C.
$x + d = a + d$
$x = a$
So, $B(?, ?) = B(a, b)$ and $D(d, ?) = D(d, 0)$.

45. Find the opposite side of the triangle, y.

$\tan 45° = \frac{y}{6}$
$6\tan 45° = y$
$6 = y$

Six yards is 18 feet. So, the height of the tree is $5 + 18$ or 23 ft.

46. A certain shopper is not greeted by a salesperson when he walks through the door.

47. You did not fill out an application.

48. $y > 6$

49. The two lines are not parallel.

50. $x = -2$ and $x = 5$ are vertical lines. The distance between them can be measured along any horizontal segment connecting them. The distance is $|5 - (-2)| = 7$.

51. $y = -6$ and $y = -1$ are horizontal lines. The distance between them can be measured along any vertical segment between them. The distance is $|-6 - (-1)| = 5$.

52. Let line ℓ be $y = 2x + 3$ and line m be $y = 2x - 7$. The slope of the parallel lines is 2. The slope of a line perpendicular to the parallel lines is the opposite reciprocal of 2, or $-\frac{1}{2}$. Use the y-intercept of line m, $(0, -7)$, as one of the endpoints of the perpendicular segment, P. Find P.

P: $y - y_1 = m(x - x_1)$
$y - (-7) = -\frac{1}{2}(x - 0)$
$y + 7 = -\frac{1}{2}x$
$y = -\frac{1}{2}x - 7$

Solve the system of the equations of lines P and ℓ to find where they intersect.

$-\frac{1}{2}x - 7 = 2x + 3$
$x + 14 = -4x - 6$
$5x = -20$
$x = -4$
$y = -\frac{1}{2}(-4) - 7$
$= 2 - 7$
$= -5$

The point of intersection is $(-4, -5)$.

Find the distance between the two points of intersection, $(0, -7)$ and $(-4, -5)$.

$d = \sqrt{(x_2 - x_1)^2 + (y_2 - y_1)^2}$
$= \sqrt{(-4 - 0)^2 + [-5 - (-7)]^2}$
$= \sqrt{16 + 4}$
$= 2\sqrt{5}$

The distance between the lines is $2\sqrt{5}$.

53. Let line ℓ be $y = x + 2$ and line m be $y = x - 4$. The slope of the parallel lines is 1. The slope of a line perpendicular to the parallel lines is the opposite reciprocal of 1, or -1. Use the y-intercept of line m, $(0, -4)$, as one of the endpoints of the perpendicular segment, P. Find P.

P: $y - y_1 = m(x - x_1)$
$y - (-4) = -1(x - 0)$
$y + 4 = -x$
$y = -x - 4$

Solve the system of the equations of lines P and ℓ to find where they intersect.

$-x - 4 = x + 2$
$-6 = 2x$
$-3 = x$
$y = -(-3) - 4$
$= 3 - 4$
$= -1$

The point of intersection is $(-3, -1)$.

Find the distance between the two points of intersection, $(0, -4)$ and $(-3, -1)$.

$d = \sqrt{(x_2 - x_1)^2 + (y_2 - y_1)^2}$
$= \sqrt{(-3 - 0)^2 + [-1 - (-4)]^2}$
$= \sqrt{9 + 9}$
$= 3\sqrt{2}$

The distance between the lines is $3\sqrt{2}$.

54.

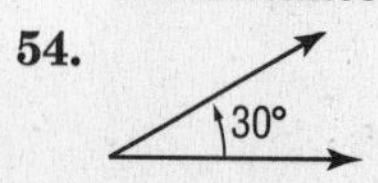

55.

56.

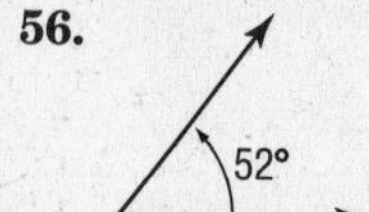

57.

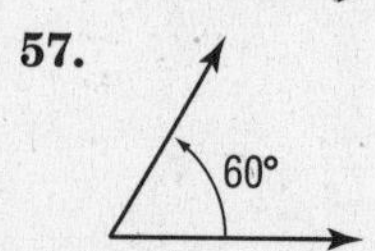

58.

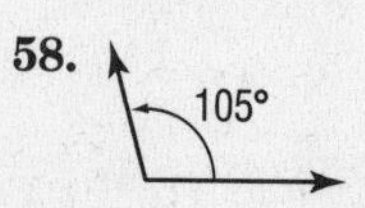

59.

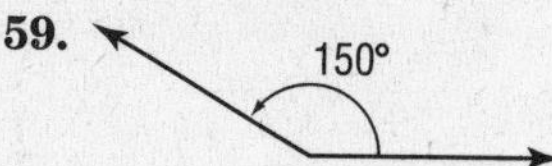

9-3 Rotations

Page 477 Geometry Software Investigation: Reflections in Intersecting Lines

1. See students' figures.
2. The transformation is a rotation about P.
3. See students' work.
4. See students' work. The angle measure should be twice the measure of the acute angle formed by the intersecting lines.
5. See students' work. The angle measures should be the same as $m\angle APA''$ in Exercise 4.
6. Sample answer: The measure of the angle of rotation is twice the measure of the acute angle formed by the intersecting lines.

Pages 478–479 Check for Understanding

1. Sample answer:

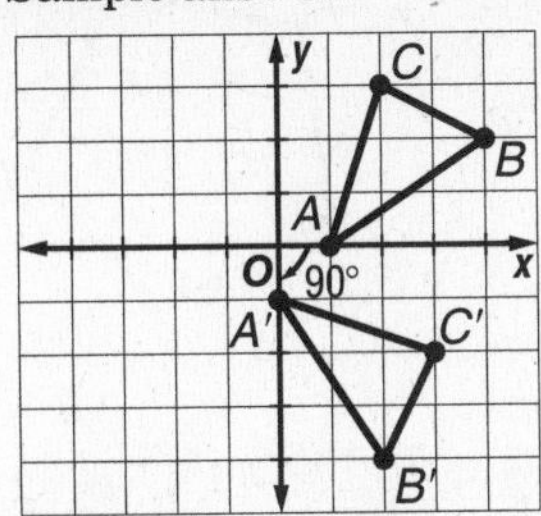

$\triangle ABC$ has vertices $A(1, 0)$, $B(4, 2)$, and $C(2, 3)$. $\triangle A'B'C'$ has vertices $A'(0, -1)$, $B'(2, -4)$, and $C'(3, -2)$.
For a clockwise rotation of 90 degrees about the origin, $(x, y) \rightarrow (y, -x)$.

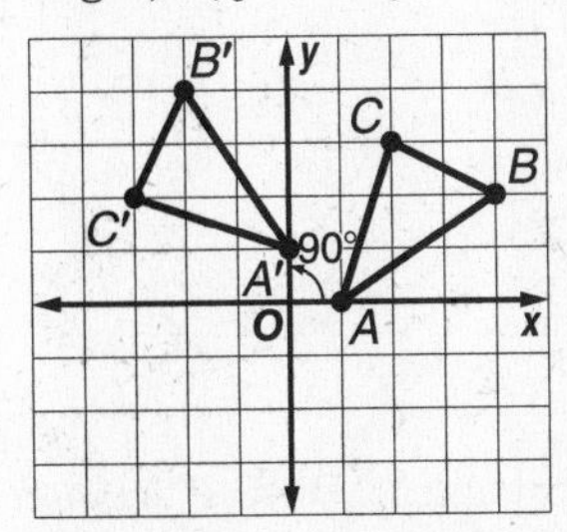

$\triangle ABC$ has vertices $A(1, 0)$, $B(4, 2)$, and $C(2, 3)$. $\triangle A'B'C'$ has vertices $A'(0, 1)$, $B'(-2, 4)$, and $C'(-3, 2)$. For a counterclockwise rotation of 90 degrees about the origin, $(x, y) \rightarrow (-y, x)$.

2. A rotation image can be found by reflecting the image in a line, then reflecting that image in a second of two intersecting lines. The second method is to rotate each point of the given figure using the angle of rotation twice the angle formed between the intersecting lines. Use the intersection point of the two lines as the point of rotation.
3. Both translations and rotations are made up of two reflections. The difference is that translations reflect across parallel lines and rotations reflect across intersecting lines.
4. Draw a segment from G to B. Use a protractor to measure a 60° angle counterclockwise with $\overline{GB}$ as one side. Draw $\overrightarrow{GX}$. Use a compass to copy $\overline{GB}$ onto $\overrightarrow{GX}$. Name the segment $\overline{GB}'$.
Repeat with points C and D. $\triangle B'C'D'$ is the image of $\triangle BCD$ under a 60° counterclockwise rotation about point G.

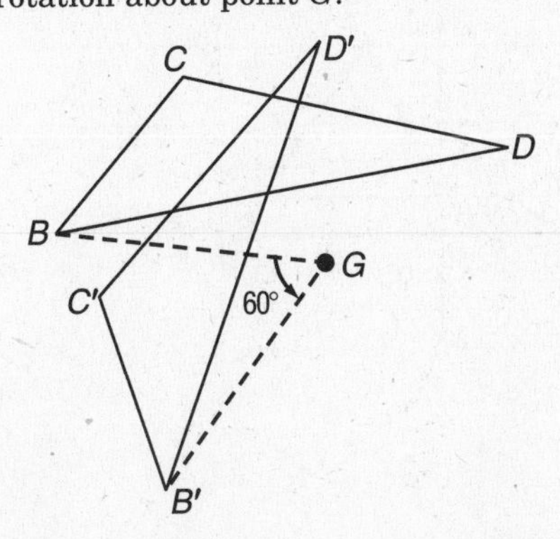

5. First reflect parallelogram $ABCD$ in line ℓ. Next, reflect the image in line m. Parallelogram $A''B''C''D''$ is the image of parallelogram $ABCD$ under reflections in lines ℓ and m.

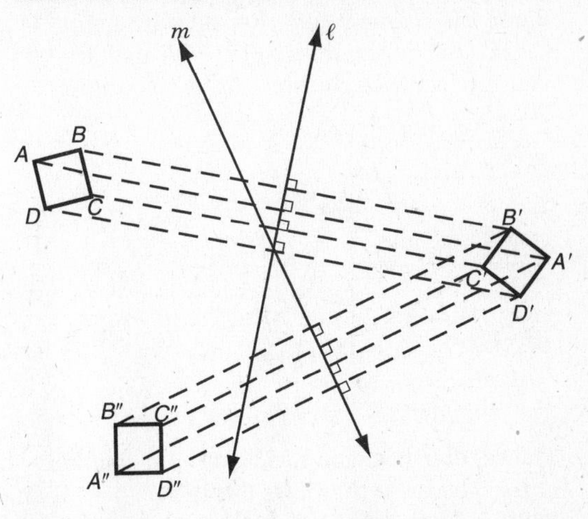

6. First reflect quadrilateral $DEFG$ in line ℓ. Since E and F are on ℓ, each point is its own image. Next, reflect the image in line m. Quadrilateral $D''E''F''G''$ is the image of quadrilateral $DEFG$ under reflections in lines ℓ and m.

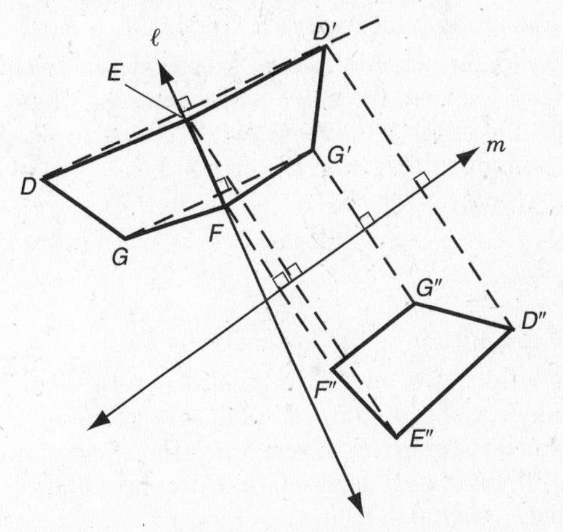

7. First graph $\overline{XY}$. Draw a segment from the origin O to point X. Use a protractor to measure a 45° angle clockwise with $\overline{OX}$ as one side. Draw $\overrightarrow{OR}$. Use a compass to copy $\overline{OX}$ onto $\overrightarrow{OR}$. Name the segment $\overline{OX'}$. Repeat with point Y. $\overline{X'Y'}$ is the image of $\overline{XY}$ under a 45° clockwise rotation about the origin.

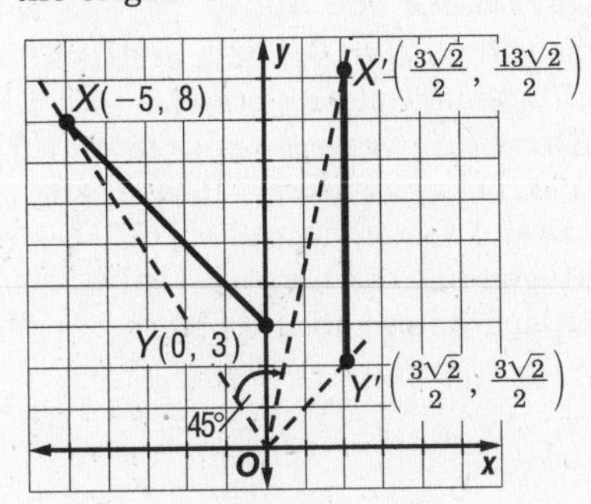

8. First graph $\triangle PQR$. Draw a segment from the origin O to point P. Use a protractor to measure a 90° angle counterclockwise with $\overline{OP}$ as one side. Draw $\overrightarrow{OX}$. Use a compass to copy $\overline{OP}$ onto $\overrightarrow{OX}$. Name the segment $\overline{OP'}$. Repeat with points Q and R. $\triangle P'Q'R'$ is the image of $\triangle PQR$ under a 90° counterclockwise rotation about the origin.

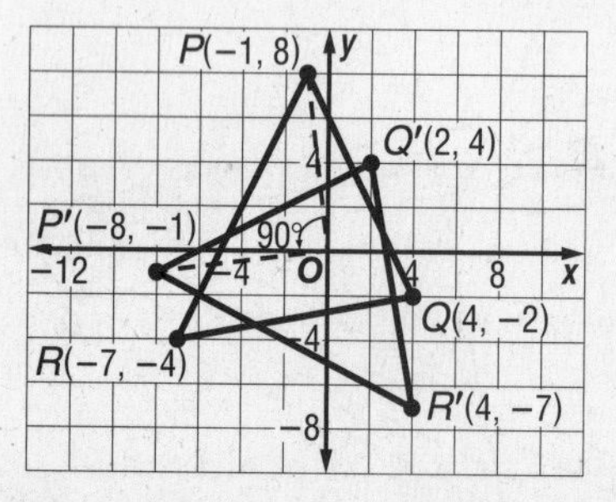

9. The regular hexagon has rotational symmetry of order 6 because there are 6 rotations less than 360° (including 0 degrees) that produce an image indistinguishable from the original.

$$\text{magnitude} = \frac{360°}{\text{order}} = \frac{360°}{6} = 60°$$

The magnitude of the symmetry is 60°.

10. The regular octagon has rotational symmetry of order 8 because there are 8 rotations less than 360° (including 0 degrees) that produce an image indistinguishable from the original.

$$\text{magnitude} = \frac{360°}{\text{order}} = \frac{360°}{8} = 45°$$

The magnitude of the symmetry is 45°.

11. The left, center, and right fans have rotational symmetry of orders 5, 4, and 3, respectively, because there are 5, 4, and 3 rotations less than 360° (including 0 degrees) that produce images indistinguishable from the originals. The magnitude of the symmetry is given by $\frac{360°}{\text{order}}$. So, the magnitudes of the symmetries for the left, center, and right fans are 72°, 90°, and 120°, respectively.

Pages 479–481 Practice and Apply

12. Draw a segment from R to B. Use a protractor to measure a 110° angle counterclockwise with $\overline{RB}$ as one side. Draw $\overrightarrow{RX}$. Use a compass to copy $\overline{RB}$ onto $\overrightarrow{RX}$. Name the segment $\overline{RB'}$. Repeat with points C, D, E, and F. Pentagon $B'C'D'E'F'$ is the image of pentagon $BCDEF$ under a 110° counterclockwise rotation about point R.

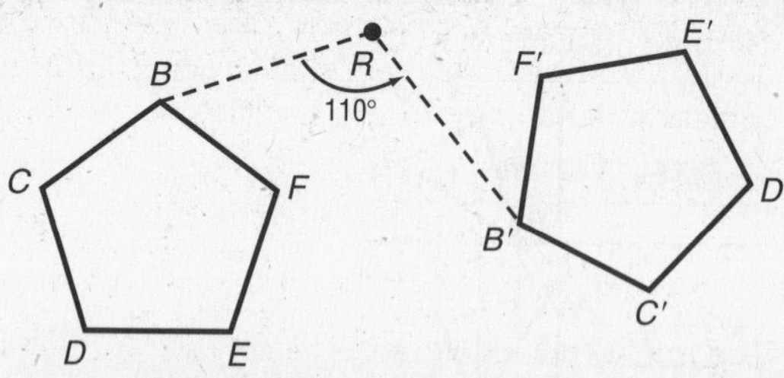

13. Draw a segment from Q to M. Use a protractor to measure a 180° angle counterclockwise with $\overline{QM}$ as one side. Draw $\overrightarrow{QX}$. Use a compass to copy $\overline{QM}$ onto $\overrightarrow{QX}$. Name the segment $\overline{QM'}$. Repeat with points N and P. $\triangle M'N'P'$ is the image of $\triangle MNP$ under a 180° counterclockwise rotation about point Q.

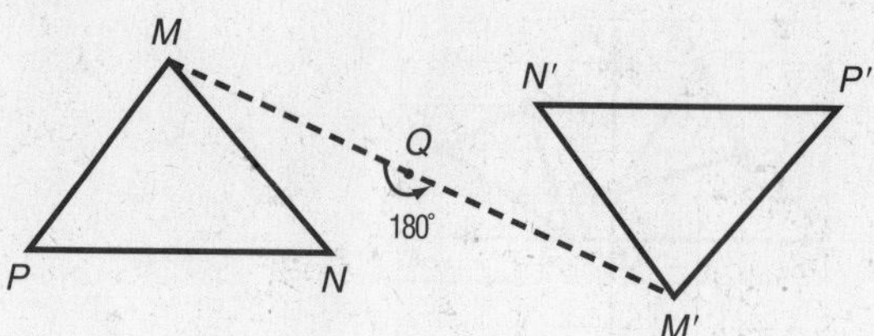

14. First graph $\triangle XYZ$ and point P. Draw a segment from point P to point X. Use a protractor to measure a 90° angle counterclockwise with $\overline{PX}$ as one side. Draw $\overrightarrow{PR}$. Use a compass to copy $\overline{PX}$ onto $\overrightarrow{PR}$. Name the segment $\overline{PX'}$. Repeat with points Y and Z. $\triangle X'Y'Z'$ is the image of $\triangle XYZ$ under a 90° counterclockwise rotation about point P.

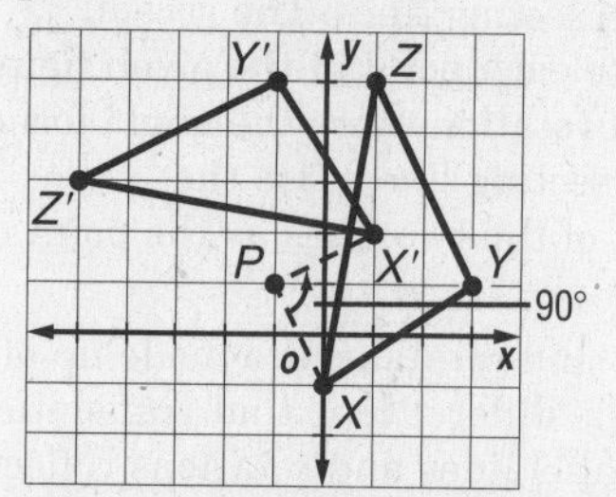

15. First graph $\triangle RST$ and point P. Draw a segment from point P to point R. Use a protractor to measure a 90° angle clockwise with $\overline{PR}$ as one side. Draw $\overrightarrow{PX}$. Use a compass to copy $\overline{PR}$ onto $\overrightarrow{PX}$. Name the segment $\overline{PR'}$. Repeat with points S and T. $\triangle R'S'T'$ is the image of $\triangle RST$ under a 90° clockwise rotation about point P.

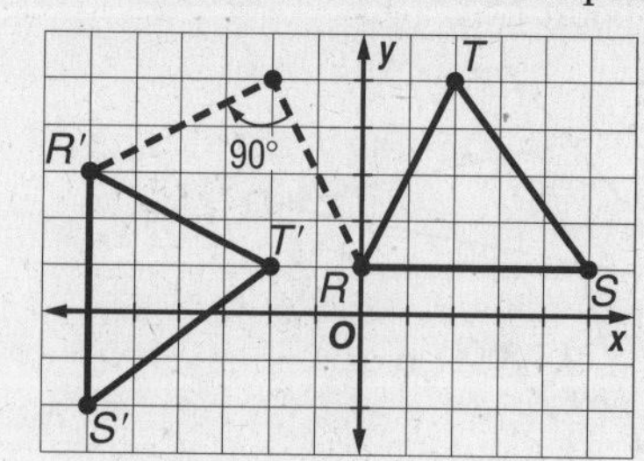

16. The Ferris wheel has rotational symmetry of order 20 because there are 20 rotations less than 360° (including 0 degrees) that produce an image indistinguishable from the original.

$$\text{magnitude} = \frac{360°}{\text{order}}$$
$$= \frac{360°}{20}$$
$$= 18°$$

The magnitude of the symmetry is 18°.

17. From Exercise 16, the magnitude of the symmetry is 18°. Seat 1 is moved 4 positions, or $4(18°) = 72°$.

18. From Exercise 16, the magnitude of the symmetry is 18°. Divide 144° by the magnitude of the rotational symmetry.

$\frac{144°}{18°} = 8$ positions

Seat 1 is moved 8 positions, or to the original position of seat 9.

19.

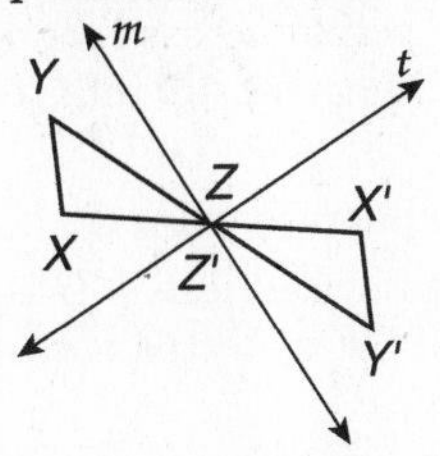

20.

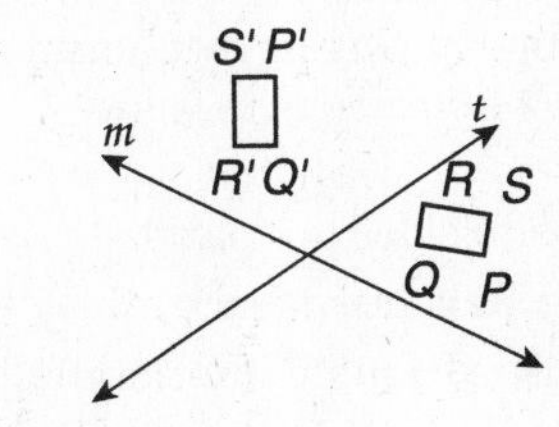

21.

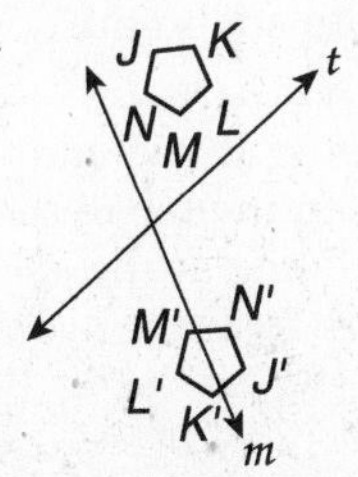

22.

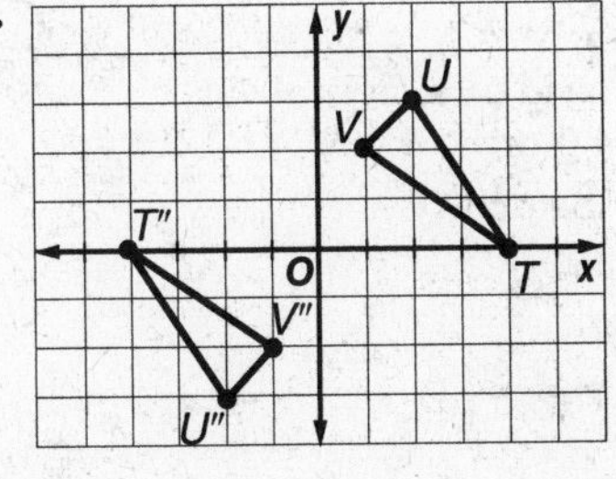

Reflection in y-axis: $T(4, 0) \rightarrow T'(-4, 0)$
$U(2, 3) \rightarrow U'(-2, 3)$
$V(1, 2) \rightarrow V'(-1, 2)$

Reflection in x-axis: $T'(-4, 0) \rightarrow T''(-4, 0)$
$U'(-2, 3) \rightarrow U''(-2, -3)$
$V'(-1, 2) \rightarrow V''(-1, -2)$

The angle of rotation is 180°.

23.

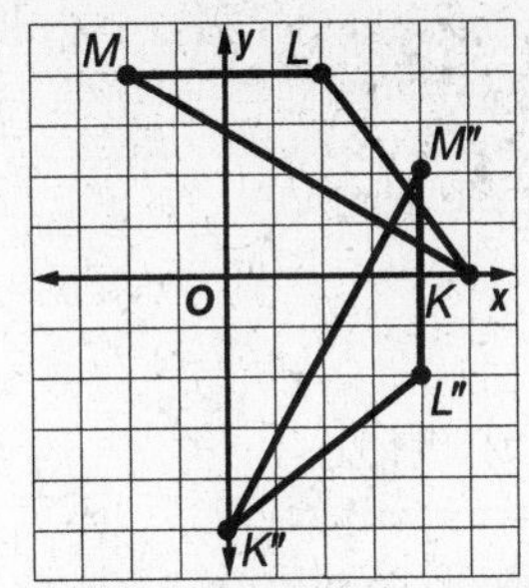

Reflection in line $y = x$: $K(5, 0) \rightarrow K'(0, 5)$
$L(2, 4) \rightarrow L'(4, 2)$
$M(-2, 4) \rightarrow M'(4, -2)$

Reflection in x-axis: $K'(0, 5) \rightarrow K''(0, -5)$
$L'(4, 2) \rightarrow L''(4, -2)$
$M'(4, -2) \rightarrow M''(4, 2)$

The angle of rotation is 90° clockwise.

24.

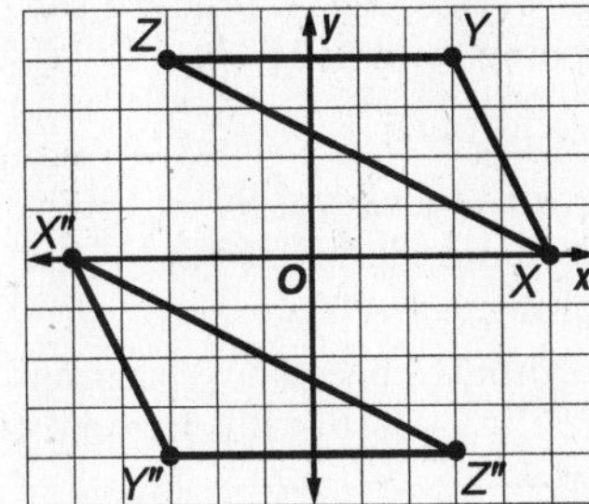

Reflection in line $y = -x$: $X(5, 0) \rightarrow X'(0, -5)$
$Y(3, 4) \rightarrow Y'(-4, -3)$
$Z(-3, 4) \rightarrow Z'(-4, 3)$

Reflection in line $y = x$: $X'(0, -5) \rightarrow X''(-5, 0)$
$Y'(-4, -3) \rightarrow Y''(-3, -4)$
$Z'(-4, 3) \rightarrow Z''(3, -4)$

The angle of rotation is 180°.

25.

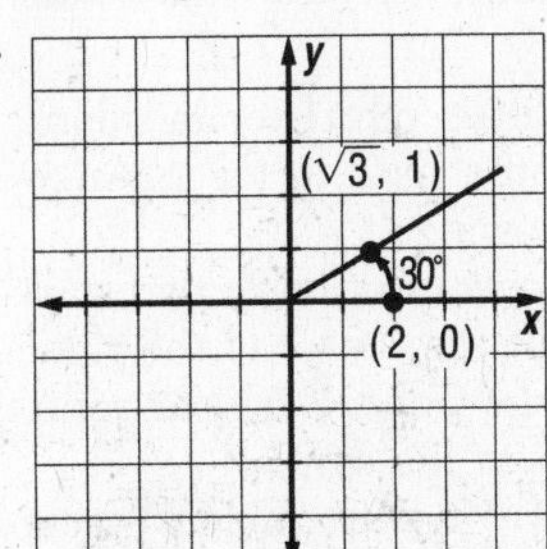

$$x = 2\cos 30°$$
$$= \sqrt{3}$$
$$y = 2\sin 30°$$
$$= 1$$

The coordinates of the image are $(\sqrt{3}, 1)$.

26. The CD changer has rotational symmetry of order 5 because there are 5 rotations less than 360° (including 0 degrees) that produce an image indistinguishable from the original.

$$\text{magnitude} = \frac{360°}{\text{order}} = \frac{360°}{5} = 72°$$

The magnitude of the symmetry is 72°.

27. Yes; it is a proper successive reflection with respect to the two intersecting lines.
28. Yes; it is a proper successive reflection with respect to the two intersecting lines.
29. Yes; the teacups are rotating.
30. Yes; the scrambler is rotating.
31. No; the roller coaster is not rotating.
32. The letters H, I, N, O, S, X, and Z produce the same letter after being rotated 180°.
33. $\frac{360°}{40°/\text{reflection}} = 9$ reflections
34. Angles of rotation with measures 90 or 180 would be easier on a coordinate plane because of the grids used in graphing.
35. In each case, y-coordinates become x-coordinates and the opposite of the x-coordinates become y-coordinates, or $(x, y) \rightarrow (y, -x)$.
36. The 80° clockwise rotation and then 150° counterclockwise rotation about the origin is equivalent to a 70° counterclockwise rotation.
37. Any point on the line of reflection is invariant.
38. The center of rotation is the only invariant point.
39. There are no invariant points. Every point is translated a units in the x-direction and b units in the y-direction.
40. Sample answer: The Tilt-A-Whirl sends riders tipping and spinning on a circular track. Answers should include the following.
    - The Tilt-A-Whirl shows rotation in two ways. The cars rotate about the center of the ride as the cars go around the track. Each car rotates around a pivot point in the car.
    - Answers will vary but the Scrambler, Teacups, and many kiddie rides use rotation.
41. B; the central angle of the octagon is $\frac{360°}{8} = 45°$. The triangle is moved three positions clockwise, or $3(45)° = 135°$.
42. D; $x = \frac{2}{5}y$ and $y = \frac{1}{3}z$, so $x = \frac{2}{5}\left(\frac{1}{3}z\right) = \frac{2}{15}z$, or $z = \frac{15}{2}x = \frac{15}{2}(6) = 45$.
43.

Transformation	reflection	translation	rotation
angle measure	yes	yes	yes
betweenness of points	yes	yes	yes
orientation	no	yes	no
collinearity	yes	yes	yes
distance measure	yes	yes	yes

44. Reflection is an indirect isometry because the image of the transformed figure cannot be found by moving it intact within the plane.
45. Translation is a direct isometry because the image of the transformed figure is found by moving it intact within the plane.
46. Rotation is a direct isometry because the image of the transformed figure is found by moving it intact within the plane.

Page 482 Maintain Your Skills

47. Yes; it is one reflection after another with respect to the two parallel lines.
48. No; it is a rotation followed by a reflection with respect to line a.
49. Yes; it is one reflection after another with respect to the two parallel lines.
50. C is the image of A in a reflection across line p. G is its own image. So, the image of $\overline{AG}$ reflected across line p is $\overline{CG}$.
51. C is the image of F reflected across point G.
52. H is the image of E reflected across line q. G is its own image. So, the image of $\overline{GE}$ reflected across line q is $\overline{GH}$.
53. A and F are the images of C and D, respectively, in a reflection across line p. G is its own image. So, the image of $\angle CGD$ reflected across line p is $\angle AGF$.
54. $\overline{QR} \parallel \overline{PS}$ because opposite sides are parallel.
55. $\overline{PT} \cong \overline{TR}$ because diagonals bisect each other.
56. $\angle SQR \cong \angle QSP$ because alternate interior angles are congruent.
57. $\angle QPS \cong \angle QRS$ because opposite angles are congruent.
58. Let y be the opposite side of the right triangle formed by the eye of the surveyor, the top of the building, and the side of the building at the eye level of the surveyor (100 meters from the eye).

$$\tan 23° = \frac{y}{100}$$
$$100\tan 23° = y$$
$$42.45 \approx y$$

So, the height of the building is about $42.45 + 1.55 = 44.0$ m.

59. Use the triangle inequality.
$$6 + 8 \stackrel{?}{>} 16$$
$$14 \not> 16 \text{ no}$$
60. Use the triangle inequality.
$$12 + 17 \stackrel{?}{>} 20$$
$$29 > 20 \text{ yes}$$
61. Use the triangle inequality.
$$22 + 23 \stackrel{?}{>} 37$$
$$45 > 37 \text{ yes}$$
62. $180a = 360$
$$a = \frac{360}{180}$$
$$a = 2$$

63. $180a + 90b = 360$
$2a + b = 4$
$b = -2a + 4$

Use a table.

a	b
0	4
1	2
2	0

Three values are (0, 4), (1, 2), and (2, 0).

64. $135a + 45b = 360$
$3a + b = 8$
$b = -3a + 8$

Use a table.

a	b
0	8
1	5
2	2

Three values are (0, 8), (1, 5), and (2, 2).

65. $120a + 30b = 360$
$4a + b = 12$
$b = -4a + 12$

Use a table.

a	b
0	12
1	8
2	4
3	0

Four values are (0, 12), (1, 8), (2, 4), and (3, 0).

66. $180a + 60b = 360$
$3a + b = 6$
$b = -3a + 6$

Use a table.

a	b
0	6
1	3
2	0

Three values are (0, 6), (1, 3), and (2, 0).

67. $180a + 30b = 360$
$6a + b = 12$
$b = -6a + 12$

Use a table.

a	b
0	12
1	6
2	0

Three values are (0, 12), (1, 6), and (2, 0).

Page 482 Practice Quiz 1

1. For a reflection in the origin, $(a, b) \rightarrow (-a, -b)$
$D(-1, 1) \rightarrow D'(1, -1)$
$E(1, 4) \rightarrow E'(-1, -4)$
$F(3, 2) \rightarrow F'(-3, -2)$

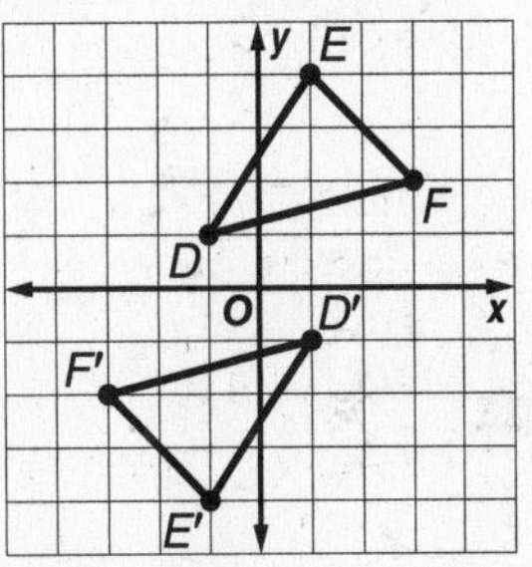

2. For a reflection in the line $y = x$, $(a, b) \rightarrow (b, a)$
$A(0, 2) \rightarrow A'(2, 0)$
$B(2, 2) \rightarrow B'(2, 2)$
$C(3, 0) \rightarrow C'(0, 3)$
$D(-1, 1) \rightarrow D'(1, -1)$

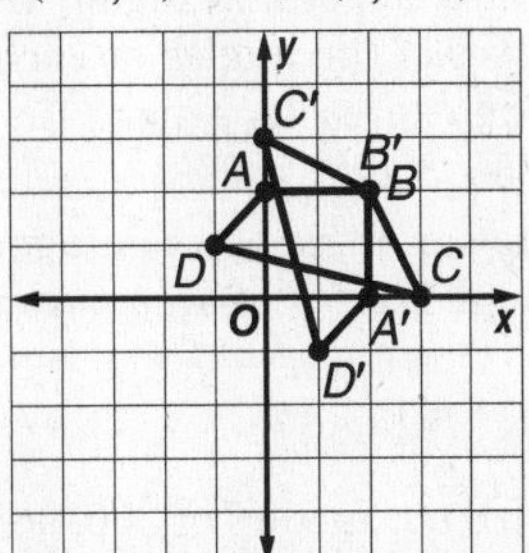

3. The translation moved each endpoint 3 units to the left and 4 units up.
$P(1, -4) \rightarrow P'(1 - 3, -4 + 4)$ or $P'(-2, 0)$
$Q(4, -1) \rightarrow Q'(4 - 3, -1 + 4)$ or $Q'(1, 3)$

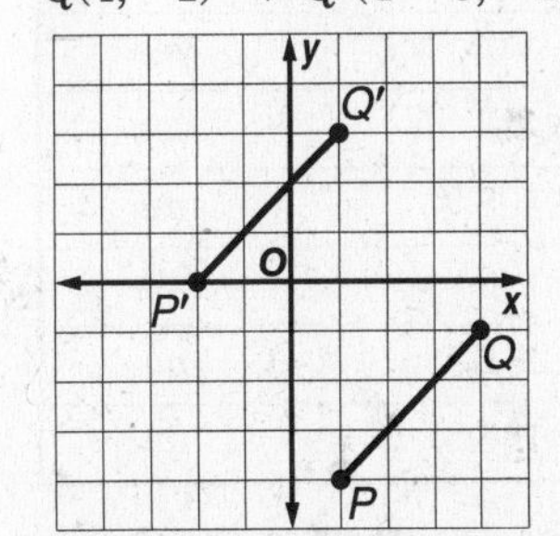

4. The translation moved each vertex 1 unit to the right and 4 units down.
$K(-2, 0) \rightarrow K'(-2 + 1, 0 - 4)$ or $K'(-1, -4)$
$L(-4, 2) \rightarrow L'(-4 + 1, 2 - 4)$ or $L'(-3, -2)$
$M(0, 4) \rightarrow M'(0 + 1, 4 - 4)$ or $M'(1, 0)$

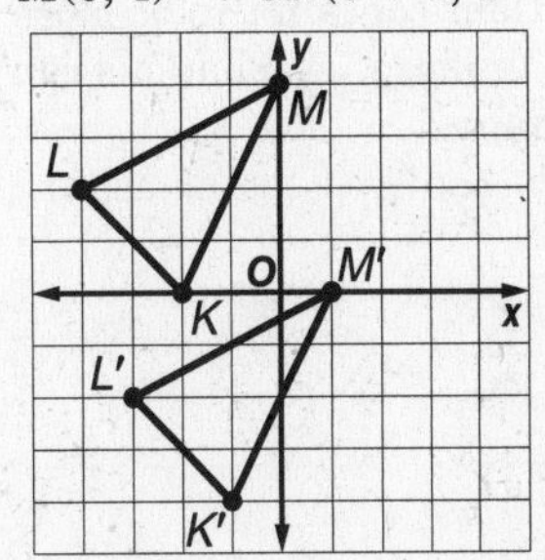

5. The 36-horse carousel has rotational symmetry of order 36 because there are 36 rotations less than 360° (including 0 degrees) that produce an image indistinguishable from the original.

$$\text{magnitude} = \frac{360°}{\text{order}} = \frac{360°}{36} = 10°$$

The magnitude of the symmetry is 10°.

9-4 Tessellations

Page 483 Geometry Activity: Tessellations of Regular Polygons

1. equilateral triangle, square, and hexagon
2. The measure of an interior angle of an equilateral triangle is 60; of a square, 90; of a hexagon, 120. The sum of the measures of the angles at each vertex must be 360. The expressions are: $6(60) = 360$; $4(90) = 360$; $3(120) = 360$.
3. The measure of an interior angle of a pentagon is 108; of a heptagon, about 128.57; of an octagon, 135.
   $\frac{360}{108} = 3\frac{1}{3}$; $\frac{360}{128.57} \approx 2.8$; $\frac{360}{135} = 2\frac{2}{3}$
   108, 128.57, and 135 are not factors of 360, so the pentagon, the heptagon, and the octagon do not tessellate the plane.

Regular Polygon	Measure of One Interior Angle	Does It Tessellate?
triangle	60	yes
square	90	yes
pentagon	108	no
hexagon	120	yes
heptagon	128.57	no
octagon	135	no

4. If a regular polygon has an interior angle with a measure that is a factor of 360, then the polygon will tessellate the plane.

Pages 485–486 Check for Understanding

1. Semi-regular tessellations contain two or more regular polygons, but uniform tessellations can be any combination of shapes.
2. Sample answer:

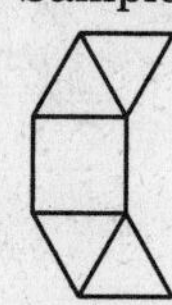

3. The figure used in the tessellation appears to be a trapezoid, which is not a regular polygon. Thus, the tessellation cannot be regular.
4. Let $m\angle 1$ represent one interior angle of the regular decagon. Use the Interior Angle Formula.

$$m\angle 1 = \frac{180(n-2)}{n} = \frac{180(10-2)}{10} = 144$$

Since 144 is not a factor of 360, a decagon will not tessellate the plane.

5. Let $m\angle 1$ represent one interior angle of the regular decagon. Use the Interior Angle Formula.

$$m\angle 1 = \frac{180(n-2)}{n} = \frac{180(30-2)}{30} = 168$$

Since 168 is not a factor of 360, a decagon will not tessellate the plane.

6. Yes; Use the algebraic method to determine whether a semi-regular tessellation can be created using squares and triangles of side length 1 unit.
   Each interior angle of a square measures 90°, and each interior angle of a triangle measures 60°.
   Find whole number values for h and t so that $90h + 60t = 360$.
   Let $h = 2$.
   $90(2) + 60t = 360$
   $180 + 60t = 360$
   $60t = 180$
   $t = 3$
   When $h = 2$ and $t = 3$, there are two squares with three triangles at each vertex.

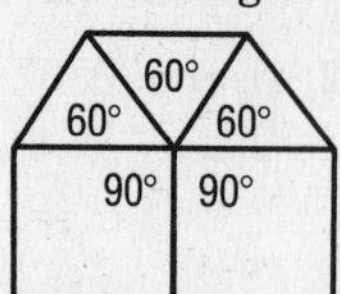

7. Yes; Use the algebraic method to determine whether a semi-regular tessellation can be created using squares and octagons of side length 1 unit.
   Each interior angle of a square measures 90°, and each interior angle of an octagon measures $\frac{180(8-2)}{8}$ or 135°.
   Find whole number values for h and t so that $90h + 135t = 360$.
   Let $h = 1$.
   $90(1) + 135t = 360$
   $90 + 135t = 360$
   $135t = 270$
   $t = 2$
   When $h = 1$ and $t = 2$, there is one square with two octagons at each vertex.

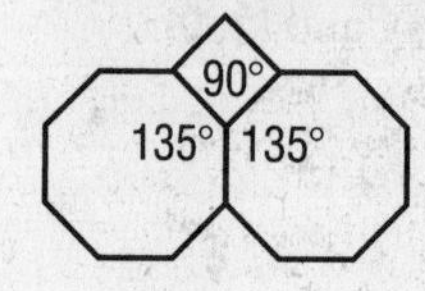

8. Yes; the pattern is a tessellation because at the different vertices the sum of the angles is 360°. The tessellation is uniform because at every vertex there is the same combination of shapes and angles.
9. Yes; the pattern is a tessellation because at the different vertices the sum of the angles is 360°. The tessellation is not uniform because the number of angles at the vertices varies.
10. Each "postage stamp" is a square that has been tessellated and 90 is a factor of 360. It is a regular tessellation since only one polygon is used.

Pages 486–487 Practice and Apply

11. No; let $m\angle 1$ represent one interior angle of the regular nonagon. Use the Interior Angle Formula.
$$m\angle 1 = \frac{180(n-2)}{n} = \frac{180(9-2)}{9} = 140$$
Since 140 is not a factor of 360, a nonagon will not tessellate the plane.
12. Yes; let $m\angle 1$ represent one interior angle of the regular nonagon. Use the Interior Angle Formula.
$$m\angle 1 = \frac{180(n-2)}{n} = \frac{180(6-2)}{6} = 120$$
Since 120 is a factor of 360, a hexagon will tessellate the plane.
13. Yes; let $m\angle 1$ represent one interior angle of the equilateral triangle. Use the Interior Angle Formula.
$$m\angle 1 = \frac{180(n-2)}{n} = \frac{180(3-2)}{3} = 60$$
Since 60 is a factor of 360, an equilateral triangle will tessellate the plane.
14. No; let $m\angle 1$ represent one interior angle of the regular dodecagon. Use the Interior Angle Formula.
$$m\angle 1 = \frac{180(n-2)}{n} = \frac{180(12-2)}{12} = 150$$
Since 150 is not a factor of 360, a dodecagon will not tessellate the plane.
15. No; let $m\angle 1$ represent one interior angle of the regular 23-gon. Use the Interior Angle Formula.
$$m\angle 1 = \frac{180(n-2)}{n} = \frac{180(23-2)}{23} \approx 164.3$$
Since 164.3 is not a factor of 360, a 23-gon will not tessellate the plane.
16. No; let $m\angle 1$ represent one interior angle of the regular 36-gon. Use the Interior Angle Formula.
$$m\angle 1 = \frac{180(n-2)}{n} = \frac{180(36-2)}{36} = 170$$
Since 170 is not a factor of 360, a 36-gon will not tessellate the plane.
17. No; A non-square rhombus is not a regular polygon, so a semi-regular tessellation cannot be created from regular octagons and non-square rhombi.
18. Yes; Use the algebraic method to determine whether a semi-regular tessellation can be created using regular dodecagons and equilateral triangles of side length 1 unit.
Each interior angle of a dodecagon measures $\frac{180(12-2)}{12}$ or 150°, and each interior angle of an equilateral triangle measures 60°.
Find whole number values for h and t so that $150h + 60t = 360$.
Let $h = 2$.
$$150(2) + 60t = 360$$
$$300 + 60t = 360$$
$$60t = 60$$
$$t = 1$$
When $h = 2$ and $t = 1$, there are two dodecagons with one triangle at each vertex.

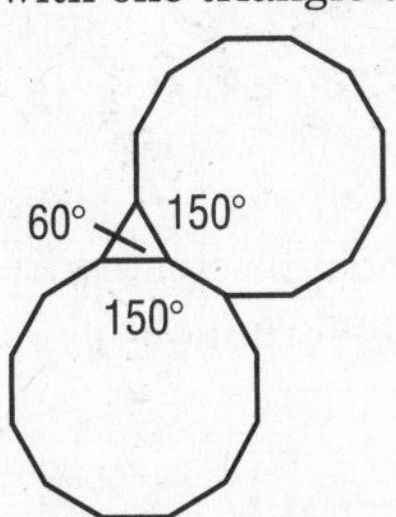

19. Yes; Use the algebraic method to determine whether a semi-regular tessellation can be created using regular dodecagons, squares, and equilateral triangles of side length 1 unit.
Each interior angle of a dodecagon measures $\frac{180(12-2)}{12}$ or 150°, and each interior angle of squares and equilateral triangles measures 90° and 60°, respectively.
Find whole number values for h, s, and t so that $150h + 90s + 60t = 360$.
Let $h = 1$ and $s = 1$.
$$150(1) + 90(1) + 60t = 360$$
$$150 + 90 + 60t = 360$$
$$60t = 120$$
$$t = 2$$
When $h = 1$, $s = 1$, and $t = 2$, there are one dodecagon, one square, and two triangles at each vertex.

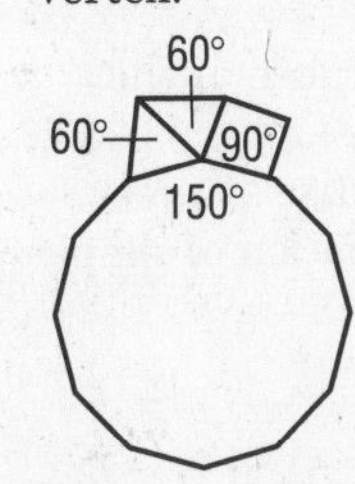

20. No; Use the algebraic method to determine whether a semi-regular tessellation can be created using regular heptagons, squares, and equilateral triangles of side length 1 unit.
Each interior angle of a heptagon measures $\frac{180(7-2)}{7} = \frac{900}{7}$ or approximately 128.6°, and each interior angle of squares and equilateral triangles measures 90° and 60°, respectively.
Find whole number values for h, s, and t so that $\frac{900}{7}h + 90s + 60t = 360$.
Let $s = 1$ and $t = 1$.

$$\frac{900}{7}h + 90(1) + 60(1) = 360$$
$$\frac{900}{7}h + 90 + 60 = 360$$
$$\frac{900}{7}h = 210$$
$$h \approx 1.63$$

Let $s = 1$ and $t = 2$.

$$\frac{900}{7}h + 90(1) + 60(2) = 360$$
$$\frac{900}{7}h + 90 + 120 = 360$$
$$\frac{900}{7}h = 150$$
$$h \approx 1.17$$

Let $s = 2$ and $t = 1$.

$$\frac{900}{7}h + 90(2) + 60(1) = 360$$
$$\frac{900}{7}h + 180 + 60 = 360$$
$$\frac{900}{7}h = 120$$
$$h \approx 0.93$$

There are no more reasonable possibilities. So, a semi-regular tessellation cannot be created from regular heptagons, squares, and equilateral triangles.

21. yes; tessellation:

The pattern is a tessellation because at the different vertices the sum of the angles is 360°. The tessellation is uniform because at every vertex there is the same combination of shapes and angles.

22. yes; tessellation:

The pattern is a tessellation because at the different vertices the sum of the angles is 360°. The tessellation is uniform because at every vertex there is the same combination of shapes and angles.

23. yes; tessellation:

The pattern is a tessellation because at the different vertices the sum of the angles is 360°. The tessellation is not uniform because the number of angles at the vertices varies.

24. No; use the algebraic method to determine whether the combination of a regular pentagon and a square with equal side length tessellates the plane.
Each interior angle of a square measures 90°, and each interior angle of a pentagon measures $\frac{180(5-2)}{5}$ or 108°.
Find whole number values for h and t so that $90h + 108t = 360$.
Let $h = 1$.

$$90(1) + 108t = 360$$
$$90 + 108t = 360$$
$$108t = 270$$
$$t = 2.5$$

Let $h = 2$.

$$90(2) + 108t = 360$$
$$180 + 108t = 360$$
$$108t = 180$$
$$t \approx 1.67$$

Let $h = 3$.

$$90(3) + 108t = 360$$
$$270 + 108t = 360$$
$$108t = 90$$
$$t \approx 0.83$$

No combination of a regular pentagon and a square with equal side length can be formed such that the total of the measures of the angles at a vertex is 360°. So, the combination does not tessellate the plane.

25. Yes; the pattern is a tessellation because at the different vertices the sum of the angles is 360°. The tessellation is not uniform because the number of angles at the vertices varies.

26. Yes; the pattern is a tessellation because at the different vertices the sum of the angles is 360°. The tessellation is not uniform because the number of angles at the vertices varies.

27. Yes; the pattern is a tessellation because at the different vertices the sum of the angles is 360°. The tessellation is uniform because at every vertex there is the same combination of shapes and angles. The tessellation is also regular since it is formed by only one type of regular polygon.

28. Yes; the pattern is a tessellation because at the different vertices the sum of the angles is 360°. The tessellation is uniform because at every vertex there is the same combination of shapes and angles. The tessellation is also semi-regular since more than one regular polygon is used.

29. The tessellation is semi-regular since more than one regular polygon is used. The tessellation is also uniform because at every vertex there is the same combination of shapes and angles.

30. Always; the sum of the measures of the interior angles of a triangle is 180°. If each angle is used twice at each vertex, the sum of the angles is 360°.

31. Never; semi-regular tessellations have the same combination of shapes and angles at each vertex like uniform tessellations. The shapes for semi-regular tessellations are just regular.

32. Sometimes; when the combination of shapes are regular polygons, then the uniform tessellation becomes semi-regular.

33. Always; the sum of the measures of the angles of a quadrilateral is 360°. So if each angle of the quadrilateral is rotated at the vertex, then that equals 360 and the tessellation is possible.

34. Never; the measure of an interior angle is $\frac{180(16-2)}{16}$ or 157.5, which is not a factor of 360.

35. yes

36. None of these; the tessellation is not uniform because the number of angles at the vertices varies. It is not regular since more than one polygon is used. It is not semi-regular since not all of the polygons are regular.

37. The tessellation is uniform because at every vertex there is the same combination of shapes and angles. The tessellation is also regular since only one regular polygon is used.

38. Sample answers:

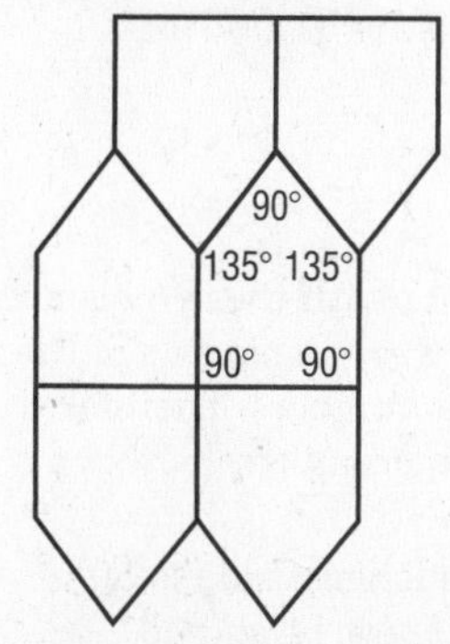

The measures are 90°, 90°, 90°, 135°, and 135°. The tessellation is not regular since the pentagons are not regular and it is not uniform since the number of angles at the vertices varies.

39. Sample answer: Tessellations can be used in art to create abstract art. Answers should include the following.
- The equilateral triangles are arranged to form hexagons, which are arranged adjacent to one another.
- Sample answers: kites, trapezoids, isosceles triangles

40. C; interior angle $= \frac{180(n-2)}{n}$

$= \frac{180(9-2)}{9}$

$= 140$

41. A; $\frac{360(12-2)}{2(12)} - \frac{180}{12} = \frac{180(10)}{12} - \frac{180}{12}$

$= \frac{1800-180}{12}$

$= \frac{1620}{12}$

$= 135$

Page 488 Maintain Your Skills

42. First graph $\triangle ABC$ and point P. Draw a segment from point P to point A. Use a protractor to measure a 90° angle counterclockwise with $\overline{PA}$ as one side. Draw $\overrightarrow{PR}$. Use a compass to copy $\overline{PA}$ onto $\overrightarrow{PR}$. Name the segment $\overline{PA'}$. Repeat with points B and C. $\triangle A'B'C'$ is the image of $\triangle ABC$ under a 90° counterclockwise rotation about point P.

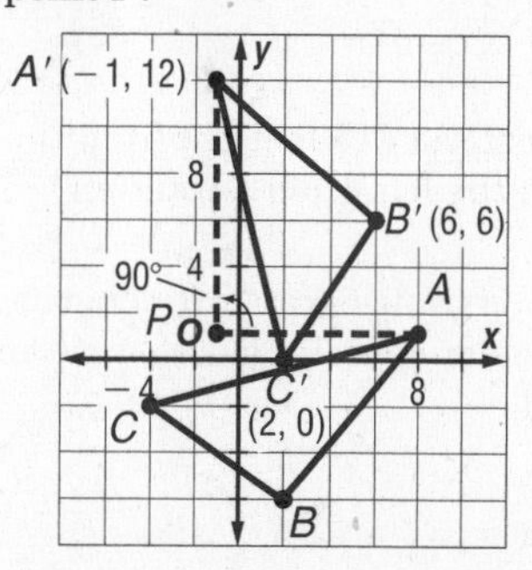

43. First graph $\triangle DEF$ and point P. Draw a segment from point P to point D. Use a protractor to measure a 90° angle clockwise with $\overline{PD}$ as one side. Draw $\overrightarrow{PR}$. Use a compass to copy $\overline{PD}$ onto $\overrightarrow{PR}$. Name the segment $\overline{PD'}$. Repeat with points E and F. $\triangle D'E'F'$ is the image of $\triangle DEF$ under a 90° clockwise rotation about point P.

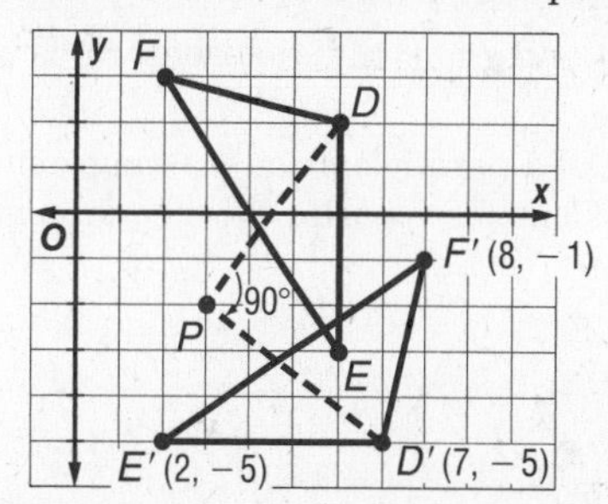

44. First graph parallelogram $GHIJ$ and point P. Draw a segment from point P to point G. Use a protractor to measure a 90° angle counterclockwise with $\overline{PG}$ as one side. Draw $\overrightarrow{PR}$. Use a compass to copy $\overline{PG}$ onto $\overrightarrow{PR}$. Name the segment $\overline{PG'}$. Repeat with points H, I, and J. Parallelogram $G'H'I'J'$ is the image of $GHIJ$ under a 90° counterclockwise rotation about point P.

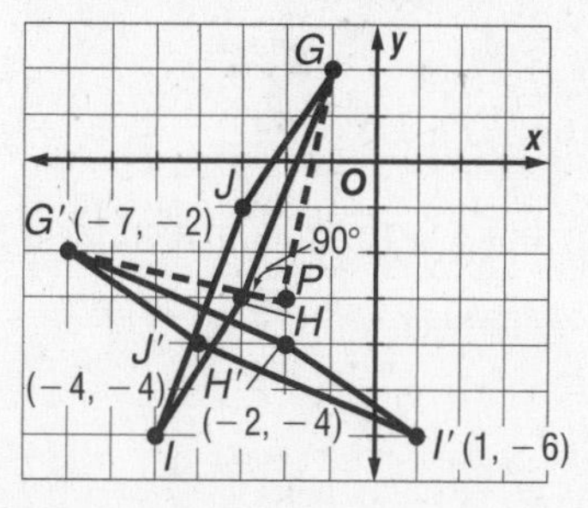

45. First graph rectangle $KLMN$ and point P. Draw a segment from point P to point K. Use a protractor to measure a 90° angle counterclockwise with $\overline{PK}$ as one side. Draw $\overrightarrow{PR}$. Use a compass to copy $\overline{PK}$ onto $\overrightarrow{PR}$. Name the segment $\overline{PK'}$. Repeat with points L, M, and N. Rectangle $K'L'M'N'$ is the image of $KLMN$ under a 90° counterclockwise rotation about point P.

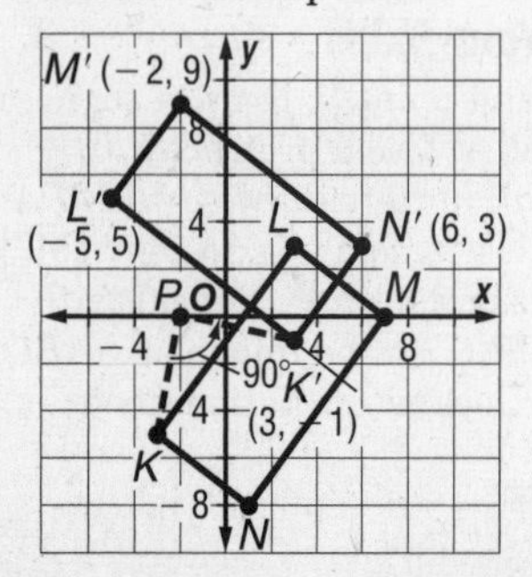

46. The move is a translation 15 feet out from the wall and 21 feet to the left, then a rotation of 90° clockwise.

47. Opposite sides of a parallelogram are congruent. Set the expressions for opposite sides equal and solve.

$y = y^2$
$0 = y^2 - y$
$0 = y(y - 1)$
$0 = y$ or $0 = y - 1$
$y = 0 \quad y = 1$

Since y represents a side length, which must be positive, $y = 1$.

$6x = 4x + 8$
$2x = 8$
$x = 4$

So, when x is 4 and y is 1, the quadrilateral is a parallelogram.

48. Opposite sides of a parallelogram are congruent. Set the expressions for opposite sides equal and solve.

$5y = 2y + 36$
$3y = 36$
$y = 12$
$6x - 2 = 64$
$6x = 66$
$x = 11$

So, when x is 11 and y is 12, the quadrilateral is a parallelogram.

49. Opposite angles of a parallelogram are congruent and the sum of its interior angles is 360°.

$2x + 8 = 120$
$2x = 112$
$x = 56$

$120 + 120 + 2(5y) = 360$
$10y = 120$
$y = 12$

So, when x is 56 and y is 12, the quadrilateral is a parallelogram.

50. 12, 16, 20

Since the measure of the longest side is 20, 20 must be c, and a and b are 12 and 16.

$a^2 + b^2 = c^2$
$12^2 + 16^2 \stackrel{?}{=} 20^2$
$144 + 256 \stackrel{?}{=} 400$
$400 = 400$

These segments form the sides of a right triangle since they satisfy the Pythagorean Theorem. The measures are whole numbers and form a Pythagorean triple.

51. 9, 10, 15

Since the measure of the longest side is 15, 15 must be c, and a and b are 9 and 10.

$a^2 + b^2 = c^2$
$9^2 + 10^2 \stackrel{?}{=} 15^2$
$81 + 100 \stackrel{?}{=} 225$
$181 \neq 225$

Since $181 \neq 225$, segments with these measures cannot form a right triangle. Therefore, they do not form a Pythagorean triple.

52. 2.5, 6, 6.5

Since the measure of the longest side is 6.5, 6.5 must be c, and a and b are 2.5 and 6.

$a^2 + b^2 = c^2$
$2.5^2 + 6^2 \stackrel{?}{=} 6.5^2$
$6.25 + 36 \stackrel{?}{=} 42.25$
$42.25 = 42.25$

Since $42.25 = 42.25$, segments with these measures form a right triangle. However, only one of the three numbers is a whole number. Therefore, they do not form a Pythagorean triple.

53. 14, $14\sqrt{3}$, 28

Since the measure of the longest side is 28, 28 must be c, and a and b are 14 and $14\sqrt{3}$.

$a^2 + b^2 = c^2$
$14^2 + (14\sqrt{3})^2 \stackrel{?}{=} 28^2$
$196 + 588 \stackrel{?}{=} 784$
$784 = 784$

Since $784 = 784$, segments with these measures form a right triangle. However, only two of the three numbers are whole numbers. Therefore, they do not form a Pythagorean triple.

54. 14, 48, 50

Since the measure of the longest side is 50, 50 must be c, and a and b are 14 and 48.

$a^2 + b^2 = c^2$
$14^2 + 48^2 \stackrel{?}{=} 50^2$
$196 + 2304 \stackrel{?}{=} 2500$
$2500 = 2500$

These segments form the sides of a right triangle since they satisfy the Pythagorean Theorem. The measures are whole numbers and form a Pythagorean triple.

55. $\frac{1}{2}, \frac{1}{3}, \frac{1}{4}$

Since the measure of the longest side is $\frac{1}{2}$, $\frac{1}{2}$ must be c, and a and b are $\frac{1}{3}$ and $\frac{1}{4}$.

$$a^2 + b^2 = c^2$$
$$\left(\frac{1}{3}\right)^2 + \left(\frac{1}{4}\right)^2 \stackrel{?}{=} \left(\frac{1}{2}\right)^2$$
$$\frac{1}{9} + \frac{1}{16} \stackrel{?}{=} \frac{1}{4}$$
$$\frac{25}{144} \neq \frac{1}{4}$$

Since $\frac{25}{144} \neq \frac{1}{4}$, segments with these measures cannot form a right triangle. Therefore, they do not form a Pythagorean triple.

56. By the Triangle Midsegment Theorem, $\overline{AB} \parallel \overline{EF}$ and $AB = \frac{1}{2}EF$, $\overline{BC} \parallel \overline{DF}$ and $BC = \frac{1}{2}DF$, and $\overline{AC} \parallel \overline{DE}$ and $AC = \frac{1}{2}DE$. So, $EF = 30$, $DF = 22$, $DE = 26$, and the perimeter is $30 + 22 + 26 = 78$.

57. By the Triangle Midsegment Theorem, $AB = FC$, $AC = \frac{1}{2}DE$, and $BC = DA$. So, $AB = 7$, $BC = 10$, and $AC = 9$.

58.

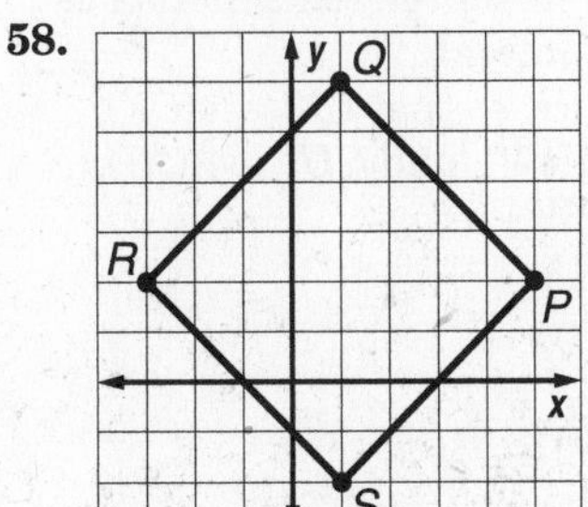

$\overline{PQ}$ and $\overline{RS}$ are opposite sides.

$$\text{slope of } \overline{PQ} = \frac{y_2 - y_1}{x_2 - x_1} = \frac{6 - 2}{1 - 5} = \frac{4}{-4} = -1$$

$$\text{slope of } \overline{RS} = \frac{-2 - 2}{1 - (-3)} = \frac{-4}{4} = -1$$

$\overline{QR}$ and $\overline{PS}$ are opposite sides.

$$\text{slope of } \overline{QR} = \frac{2 - 6}{-3 - 1} = \frac{-4}{-4} = 1$$

$$\text{slope of } \overline{PS} = \frac{-2 - 2}{1 - 5} = \frac{-4}{-4} = 1$$

The opposite sides of quadrilateral $PQRS$ have the same slopes. Therefore, the sides are parallel.

59.

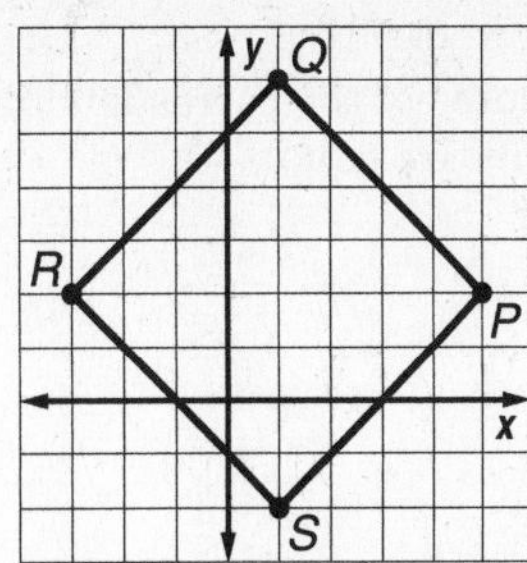

From Exercise 58, slope of $\overline{PQ} = -1$ slope of $\overline{RS} = -1$, slope of $\overline{QR} = 1$, and slope of $\overline{PS} = 1$. Since the product of the slopes of adjacent sides of quadrilateral $PQRS$ is -1, the adjacent sides are perpendicular.

60.

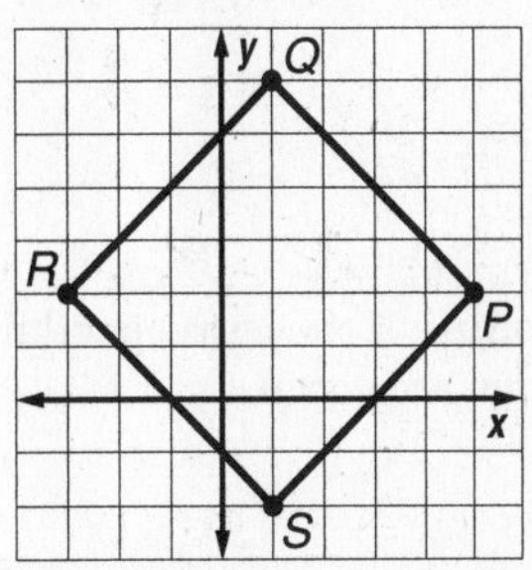

Use the distance formula to find the length of each side.

$$d = \sqrt{(x_2 - x_1)^2 + (y_2 - y_1)^2}$$
$$PQ = \sqrt{(1 - 5)^2 + (6 - 2)^2} = \sqrt{(-4)^2 + 4^2} = \sqrt{32}$$
$$QR = \sqrt{(-3 - 1)^2 + (2 - 6)^2} = \sqrt{(-4)^2 + (-4)^2} = \sqrt{32}$$
$$RS = \sqrt{[1 - (-3)]^2 + (-2 - 2)^2} = \sqrt{4^2 + (-4)^2} = \sqrt{32}$$
$$PS = \sqrt{(1 - 5)^2 + (-2 - 2)^2} = \sqrt{(-4)^2 + (-4)^2} = \sqrt{32}$$

61. The measures of the sides of quadrilateral $PQRS$ are equal. Opposite sides are parallel and adjacent sides are perpendicular. $PQRS$ is a square.

62. Corresponding sides of similar polygons are proportional.

$$\text{scale factor} = \frac{AB}{WX} = \frac{8}{12} = \frac{2}{3}$$

The scale factor is $\frac{2}{3}$.

63. The scale factor is $\frac{AB}{WX} = \frac{8}{12} = \frac{2}{3}$.
The ratios of the measures of the corresponding sides of similar polygons are equal.

$\frac{AB}{WX} = \frac{BC}{XY}$

$\frac{2}{3} = \frac{10}{XY}$

$XY = \frac{3}{2}(10)$

$XY = 15$

64. The scale factor is $\frac{AB}{WX} = \frac{8}{12} = \frac{2}{3}$.
The ratios of the measures of the corresponding sides of similar polygons are equal.

$\frac{AB}{WX} = \frac{CD}{YZ}$

$\frac{2}{3} = \frac{10}{YZ}$

$YZ = \frac{3}{2}(10)$

$YZ = 15$

65. The scale factor is $\frac{AB}{WX} = \frac{8}{12} = \frac{2}{3}$.
The ratios of the measures of the corresponding sides of similar polygons are equal.

$\frac{AB}{WX} = \frac{AD}{WZ}$

$\frac{2}{3} = \frac{15}{WZ}$

$WZ = \frac{3}{2}(15)$

$WZ = 22.5$

Page 489 Geometry Activity: Tessellations and Transformations

1. Yes; whatever space is taken out of the square is then added onto the outside of the square. The area does not change; only the shape changes.

2. Modify the bottom of the unit to be like the right side of the triangle. Erase the bottom and right original sides of the triangle.

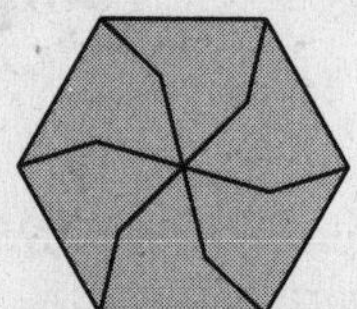

3.

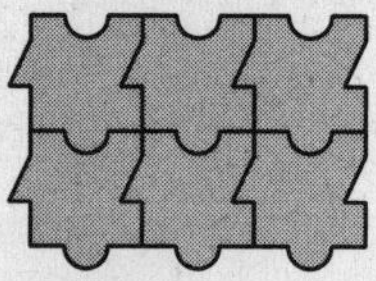

4.

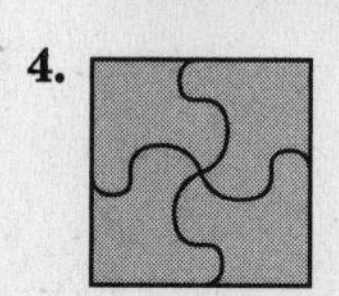

5.

9-5 Dilations

Pages 493–494 Check for Understanding

1. Dilations only preserve length if the scale factor is 1 or −1. So for any other scale factor, length is not preserved and the dilation is not an isometry.

2. Sample answer:

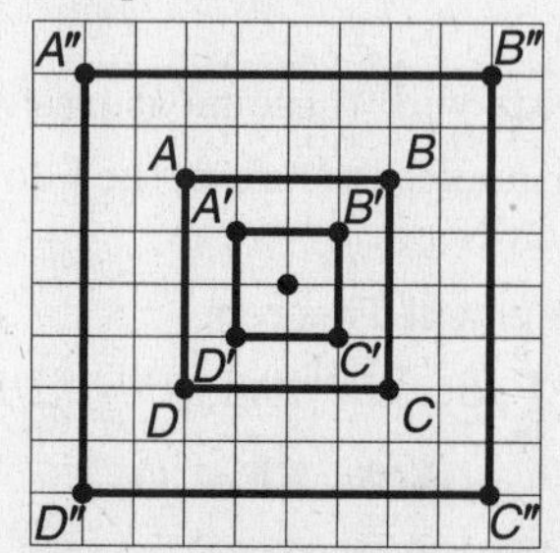

3. Trey; Desiree found the image using a positive scale factor.

4. Since $|4| > 1$, the dilation is an enlargement. Draw $\overrightarrow{CX}$, $\overrightarrow{CW}$, $\overrightarrow{CV}$, and $\overrightarrow{CU}$. Since r is positive, X', W', V', and U' will lie on the continuation of the sides of the quadrilateral.
Locate X', W', V', and U' so that $CX' = 4(CX)$, $CW' = 4(CW)$, $CV' = 4(CV)$, and $CU' = 4(CU)$.
Draw quadrilateral $X'W'V'U'$.

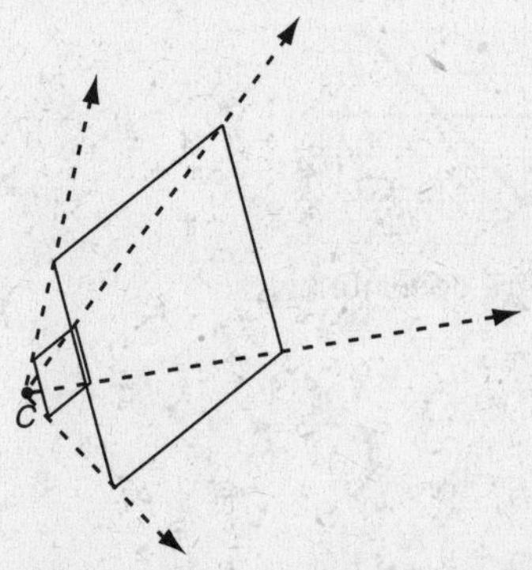

5. Since $\left|\frac{1}{5}\right| < 1$, the dilation is a reduction. Draw $\overrightarrow{CP}$, $\overrightarrow{CR}$, $\overrightarrow{CS}$, $\overrightarrow{CT}$, $\overrightarrow{CU}$, $\overrightarrow{CV}$. Locate P', R', S', T', U', and V' so that $CP' = \frac{1}{5}(CP)$, $CR' = \frac{1}{5}(CR)$, $CS' = \frac{1}{5}(CS)$, $CT' = \frac{1}{5}(CT)$, $CU' = \frac{1}{5}(CU)$, and $CV' = \frac{1}{5}(CV)$. Draw hexagon $P'R'S'T'U'V'$.

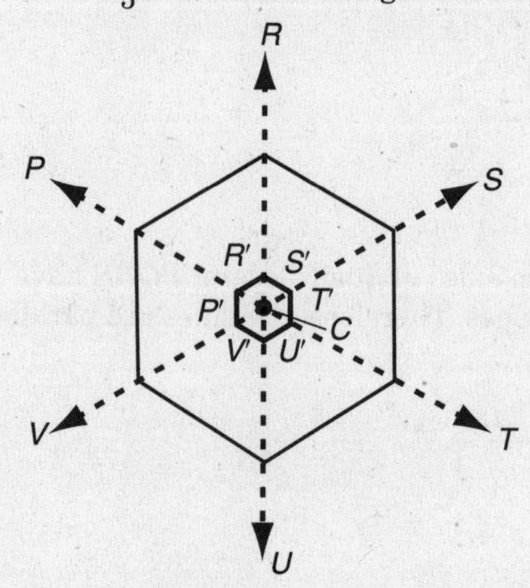

6. Since $|-2| > 1$, the dilation is an enlargement. Draw $\overline{CE}$, $\overline{CD}$, $\overline{CG}$, and $\overline{CF}$. Since r is negative, E', D', G', and F' will lie on rays that are opposite to $\overrightarrow{CE}$, $\overrightarrow{CD}$, $\overrightarrow{CG}$, and $\overrightarrow{CF}$, respectively. Locate E', D', G', and F' so that $CE' = 2(CE)$, $CD' = 2(CD)$, $CG' = 2(CG)$, and $CF' = 2(CF)$. Draw quadrilateral $E'D'G'F'$.

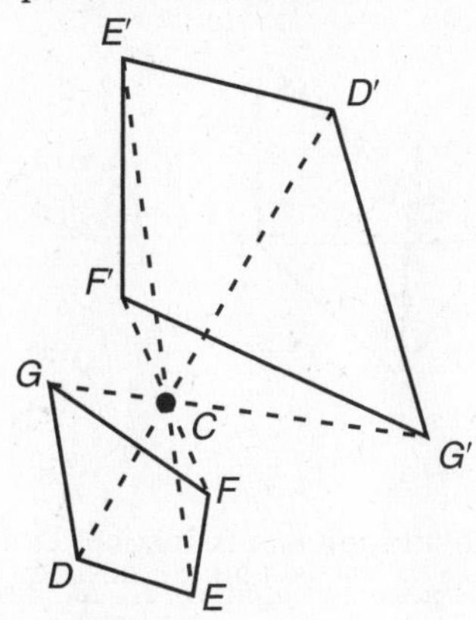

7. $AB = 3$, $r = 4$
Use the Dilation Theorem.
$A'B' = |r|(AB)$
$A'B' = (4)(3)$
$A'B' = 12$

8. $A'B' = 8$, $r = -\frac{2}{5}$
Use the Dilation Theorem.
$A'B' = |r|(AB)$
$8 = \frac{2}{5}(AB)$
$20 = AB$

9. Find P' and Q' using the scale factor, $r = \frac{1}{3}$.

Preimage (x, y)	Image $\left(\frac{x}{3}, \frac{y}{3}\right)$
$P(9, 0)$	$P'(3, 0)$
$Q(0, 6)$	$Q'(0, 2)$

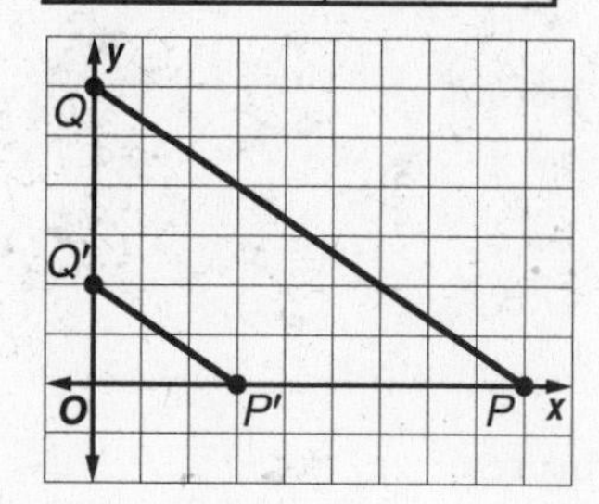

10. Find K', L', and M' using the scale factor, $r = 3$.

Preimage (x, y)	Image $(3x, 3y)$
$K(5, 8)$	$K'(15, 24)$
$L(-3, 4)$	$L'(-9, 12)$
$M(-1, -6)$	$M'(-3, -18)$

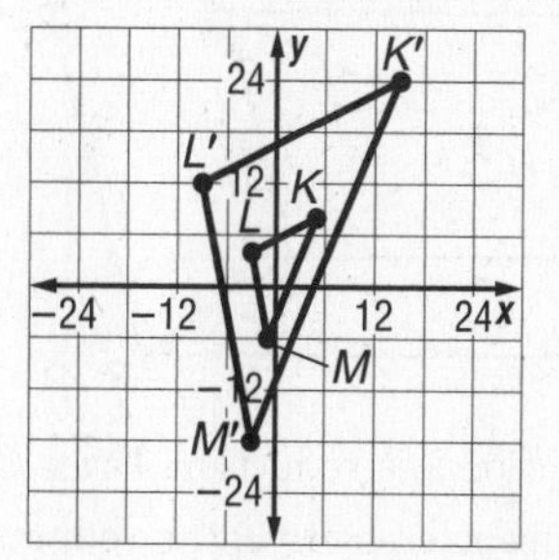

11. For ease, compare diagonals of the figures.
$$\text{scale factor} = \frac{\text{image length}}{\text{preimage length}}$$
$$r = \frac{8 \text{ units}}{4 \text{ units}}$$
$$r = 2$$
Since the scale factor is greater than 1, the dilation is an enlargement.

12. Compare the vertical sides of the triangles.
$$\text{scale factor} = \frac{\text{image length}}{\text{preimage length}}$$
$$r = \frac{4 \text{ units}}{6 \text{ units}}$$
$$r = \frac{2}{3}$$
Since $0 < |r| < 1$, the dilation is a reduction.

13. C; The drawing and the garden are similar.
$$\frac{12 \text{ ft}}{18 \text{ ft}} = \frac{x}{8 \text{ in.}}$$
$$\frac{2}{3} = \frac{x}{8 \text{ in.}}$$
$$5\tfrac{1}{3} \text{ in.} = x$$

Pages 494–496 Practice and Apply

14. Since $|3| > 1$, the dilation is an enlargement. Draw $\overline{CX}$, $\overline{CY}$, and $\overline{CZ}$. Since r is positive, X', Y', and Z' will lie on the continuation of the sides of the triangle.
Locate X', Y', and Z' so that $CX' = 3(CX)$, $CY' = 3(CY)$, and $CZ' = 3(CZ)$. Draw $\triangle X'Y'Z'$.

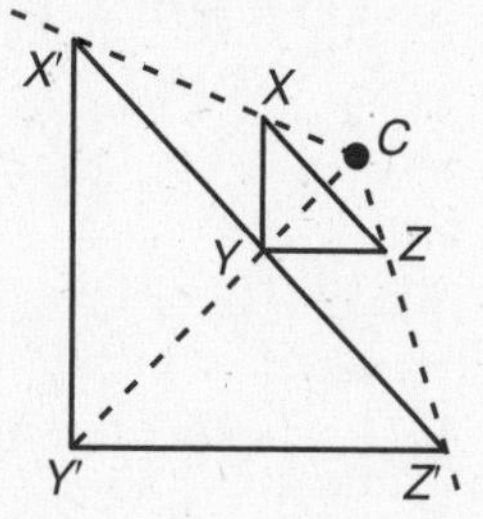

15. Since $|2| > 1$, the dilation is an enlargement. Draw $\overrightarrow{CT}$, $\overrightarrow{CS}$, $\overrightarrow{CR}$, and $\overrightarrow{CP}$. Since r is positive, T', S', R', and P' will lie on the continuations of the sides of the quadrilateral. Locate T', S', R', and P' so that $CT' = 2(CT)$, $CS' = 2(CS)$, $CR' = 2(CR)$, and $CP' = 2(CP)$. Draw quadrilateral $T'S'R'P'$.

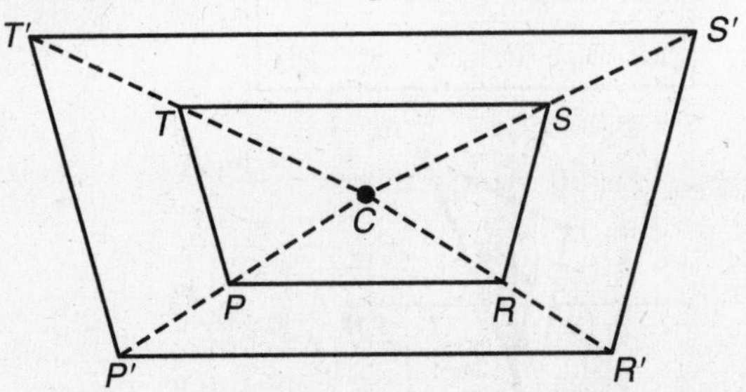

16. Since $\left|\frac{1}{2}\right| < 1$, the dilation is a reduction. Draw $\overline{CK}$, $\overline{CL}$, $\overline{CM}$, and $\overline{CN}$. Since r is positive, the reduction will have the same orientation. Locate K', L', M', and N' so that $CK' = \frac{1}{2}(CK)$, $CL' = \frac{1}{2}(CL)$, $CM' = \frac{1}{2}(CM)$, and $CN' = \frac{1}{2}(CN)$. Draw quadrilateral $K'L'M'N'$.

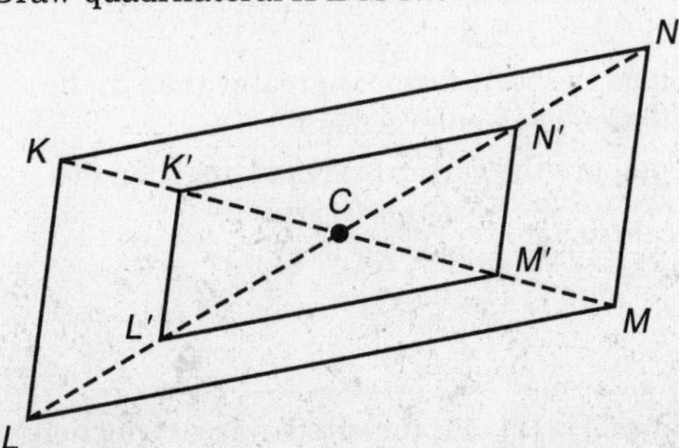

17. Since $\left|\frac{2}{5}\right| < 1$, the dilation is a reduction. Draw $\overline{CR}$, $\overline{CS}$, and $\overline{CT}$. Since r is positive, the reduction will have the same orientation. Locate R', S', and T' so that $CR' = \frac{2}{5}(CR)$, $CS' = \frac{2}{5}(CS)$, and $CT' = \frac{2}{5}(CT)$. Draw $\triangle R'S'T'$.

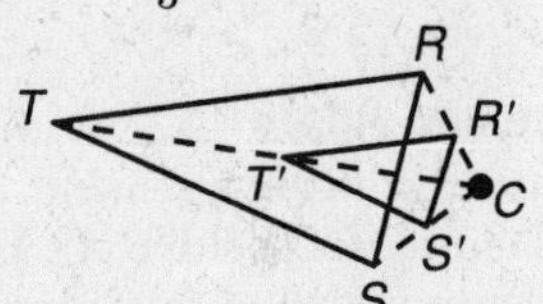

18. Since $|-1| = 1$, the dilation is a congruence transformation. Draw $\overrightarrow{CE}$, $\overrightarrow{CD}$, $\overrightarrow{CA}$, and $\overrightarrow{CB}$. Since r is negative, E', D', A', and B' will lie on rays that are opposite to $\overrightarrow{CE}$, $\overrightarrow{CD}$, $\overrightarrow{CA}$, and $\overrightarrow{CB}$. Locate E', D', A', and B' so that $CE' = CE$, $CD' = CD$, $CA' = CA$, and $CB' = CB$. Draw quadrilateral $E'D'A'B'$.

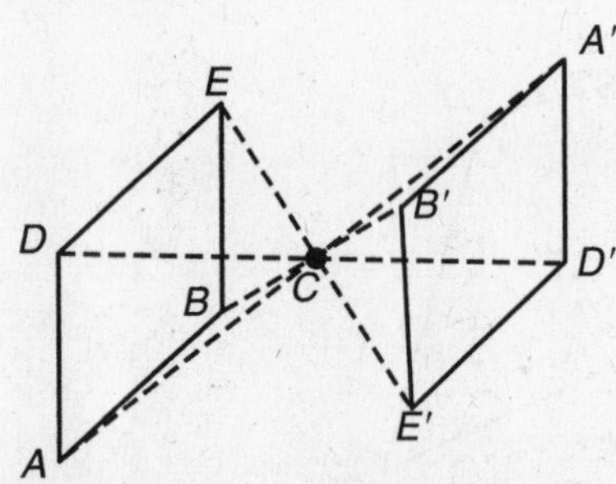

19. Since $\left|-\frac{1}{4}\right| < 1$, the dilation is a reduction. Draw $\overline{CL}$, $\overline{CM}$, and $\overline{CN}$. Since r is negative, L', M', and N' lie on rays that are opposite to $\overrightarrow{CL}$, $\overrightarrow{CM}$, and $\overrightarrow{CN}$, respectively. Locate L', M', and N' so that $CL' = \frac{1}{4}(CL)$, $CM' = \frac{1}{4}(CM)$, and $CN' = \frac{1}{4}(CN)$. Draw $\triangle L'M'N'$.

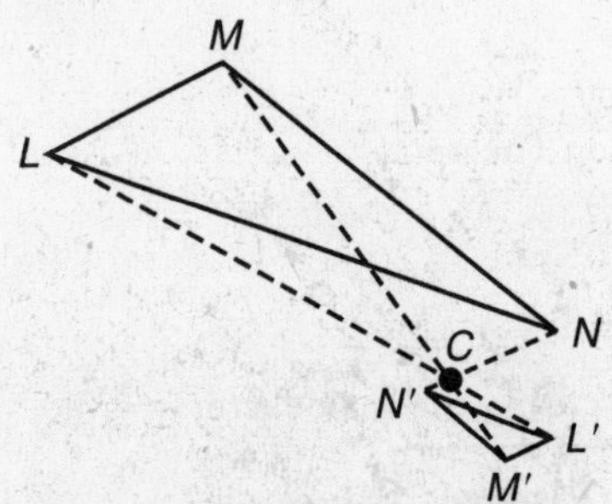

20. $ST = 6$, $r = -1$
Use the Dilation Theorem.
$S'T' = |r|(ST)$
$S'T' = (1)(6)$
$S'T' = 6$

21. $ST = \frac{4}{5}$, $r = \frac{3}{4}$
Use the Dilation Theorem.
$S'T' = |r|(ST)$
$S'T' = \left(\frac{3}{4}\right)\left(\frac{4}{5}\right)$
$S'T' = \frac{3}{5}$

22. $S'T' = 12$, $r = \frac{2}{3}$
Use the Dilation Theorem.
$S'T' = |r|(ST)$
$12 = \left(\frac{2}{3}\right)(ST)$
$18 = ST$

23. $S'T' = \frac{12}{5}$, $r = -\frac{3}{5}$
Use the Dilation Theorem.
$S'T' = |r|(ST)$
$\frac{12}{5} = \left(\frac{3}{5}\right)(ST)$
$4 = ST$

24. $ST = 32, r = -\frac{5}{4}$

Use the Dilation Theorem.

$S'T' = |r|(ST)$

$S'T' = \left(\frac{5}{4}\right)(32)$

$S'T' = 40$

25. $ST = 2.25, r = 0.4$

Use the Dilation Theorem.

$S'T' = |r|(ST)$

$S'T' = (0.4)(2.25)$

$S'T' = 0.9$

26. Find F', G', and H' using the scale factor, $r = 2$.

Preimage (x, y)	**Image** $(2x, 2y)$
$F(3, 4)$	$F'(6, 8)$
$G(6, 10)$	$G'(12, 20)$
$H(-3, 5)$	$H'(-6, 10)$

Find F'', G'', and H'' using the scale factor, $r = \frac{1}{2}$.

Preimage (x, y)	**Image** $\left(\frac{1}{2}x, \frac{1}{2}y\right)$
$F(3, 4)$	$F''\left(\frac{3}{2}, 2\right)$
$G(6, 10)$	$G''(3, 5)$
$H(-3, 5)$	$H''\left(-\frac{3}{2}, \frac{5}{2}\right)$

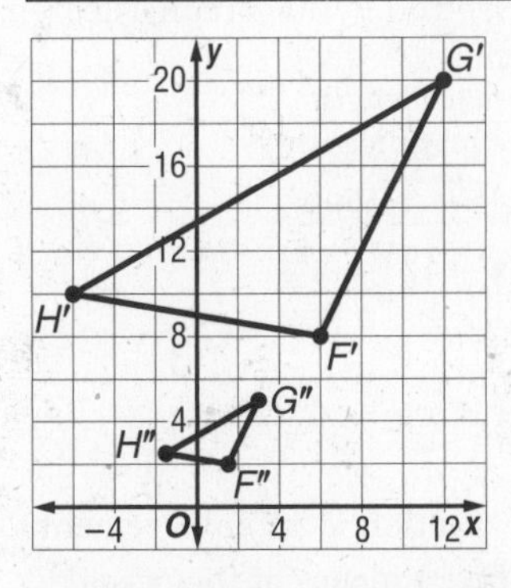

27. Find X', Y', and Z' using the scale factor, $r = 2$.

Preimage (x, y)	**Image** $(2x, 2y)$
$X(1, -2)$	$K'(2, -4)$
$Y(4, -3)$	$Y'(8, -6)$
$Z(6, -1)$	$Z'(12, -2)$

Find X'', Y'', and Z'' using the scale factor, $r = \frac{1}{2}$.

Preimage (x, y)	**Image** $\left(\frac{1}{2}x, \frac{1}{2}y\right)$
$X(1, -2)$	$X''\left(\frac{1}{2}, -1\right)$
$Y(4, -3)$	$Y''\left(2, -\frac{3}{2}\right)$
$Z(6, -1)$	$Z''\left(3, -\frac{1}{2}\right)$

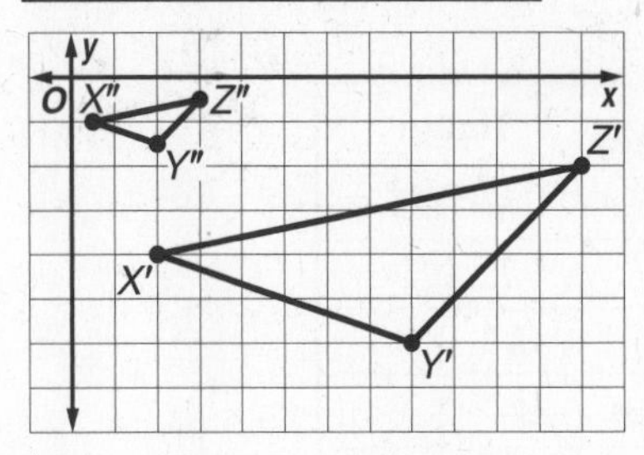

28. Find P', Q', R', and S' using the scale factor, $r = 2$.

Preimage (x, y)	**Image** $(2x, 2y)$
$P(1, 2)$	$P'(2, 4)$
$Q(3, 3)$	$Q'(6, 6)$
$R(3, 5)$	$R'(6, 10)$
$S(1, 4)$	$S'(2, 8)$

Find P'', Q'', R'', and S'' using the scale factor, $r = \frac{1}{2}$.

Preimage (x, y)	**Image** $\left(\frac{1}{2}x, \frac{1}{2}y\right)$
$P(1, 2)$	$P''\left(\frac{1}{2}, 1\right)$
$Q(3, 3)$	$Q''\left(\frac{3}{2}, \frac{3}{2}\right)$
$R(3, 5)$	$R''\left(\frac{3}{2}, \frac{5}{2}\right)$
$S(1, 4)$	$S''\left(\frac{1}{2}, 2\right)$

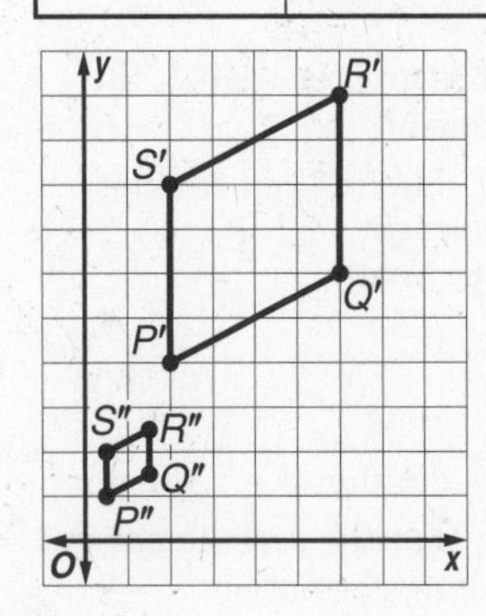

29. Find K', L', M', and N' using the scale factor, $r = 2$.

Preimage (x, y)	Image $(2x, 2y)$
$K(4, 2)$	$K'(8, 4)$
$L(-4, 6)$	$L'(-8, 12)$
$M(-6, -8)$	$M'(-12, -16)$
$N(6, -10)$	$N'(12, -20)$

Find K'', L'', M'', and N'' using the scale factor, $r = \frac{1}{2}$.

Preimage (x, y)	Image $\left(\frac{1}{2}x, y\right)$
$K(4, 2)$	$K''(2, 1)$
$L(-4, 6)$	$L''(-2, 3)$
$M(-6, -8)$	$M''(-3, -4)$
$N(6, -10)$	$N''(3, -5)$

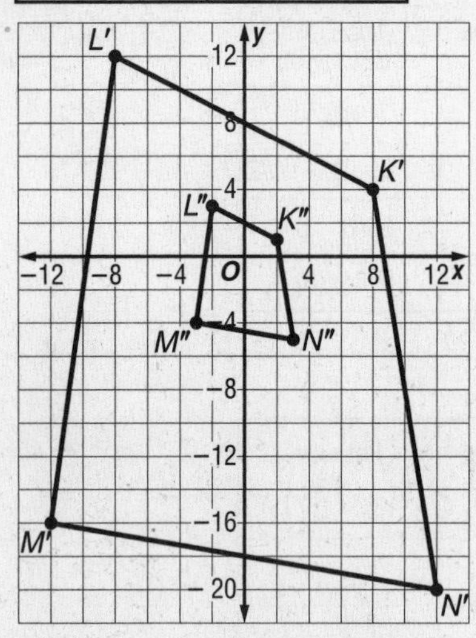

30. Compare sides of the squares.

$$\text{scale factor} = \frac{\text{image length}}{\text{preimage length}}$$

$$r = \frac{6 \text{ units}}{2 \text{ units}}$$

$$r = 3$$

Since the scale factor is greater than 1, the dilation is an enlargement.

31. Compare the vertical sides of the triangles.

$$\text{scale factor} = \frac{\text{image length}}{\text{preimage length}}$$

$$r = \frac{2 \text{ units}}{4 \text{ units}}$$

$$r = \frac{1}{2}$$

Since $0 < |r| < 1$, the dilation is a reduction.

32. By inspection, we see that the scale factor is 1. This is a congruence transformation.

33. Compare QT and $Q'T'$. Use the Pythagorean Theorem to find the side lengths.

$$\text{scale factor} = \frac{\text{image length}}{\text{preimage length}}$$

$$r = \frac{\sqrt{\left(\frac{4}{3}\right)^2 + \left(\frac{1}{3}\right)^2} \text{ units}}{\sqrt{4^2 + 1^2} \text{ units}}$$

$$r = \frac{\frac{\sqrt{17}}{3}}{\sqrt{17}}$$

$$r = \frac{1}{3}$$

Since $0 < |r| < 1$, the dilation is a reduction.

34. Compare YZ and $Y'Z'$. Note that the scale factor is negative since the image appears on the opposite side of the center with respect to the preimage. Use the Pythagorean Theorem to find the side lengths.

$$\text{scale factor} = -\frac{\text{image length}}{\text{preimage length}}$$

$$r = -\frac{\sqrt{1^2 + \left(\frac{3}{4}\right)^2} \text{ units}}{\sqrt{4^2 + 3^2} \text{ units}}$$

$$r = -\frac{\frac{\sqrt{25}}{4}}{\sqrt{25}}$$

$$r = -\frac{1}{4}$$

Since $0 < |r| < 1$, the dilation is a reduction.

35. Compare BD and $B'D'$. Note that the scale factor is negative since the image points appear on the opposite side of the center with respect to the preimage points. Use the Pythagorean Theorem to find the side lengths.

$$\text{scale factor} = -\frac{\text{image length}}{\text{preimage length}}$$

$$r = -\frac{\sqrt{4^2 + 1^2} \text{ units}}{\sqrt{2^2 + \left(\frac{1}{2}\right)^2} \text{ units}}$$

$$r = -\frac{\sqrt{17}}{\frac{\sqrt{17}}{2}}$$

$$r = -2$$

Since $|r| > 1$, the dilation is an enlargement.

36. Determine the width in inches of the actual wingspan of the SR-71.

$$(55 \text{ ft})\frac{12 \text{ in.}}{1 \text{ ft}} + 7 \text{ in.} = 660 \text{ in.} + 7 \text{ in.}$$
$$= 667 \text{ in.}$$

Find the scale factor by dividing the wingspan of the model by the wingspan of the plane.

$$\frac{14}{667} = \frac{1}{\frac{667}{14}} \approx \frac{1}{48}$$

The scale factor is about $\frac{1}{48}$.

37. Each dimension is reduced by a factor of 0.75.

$0.75(10) = 7.5$

$0.75(14) = 10.5$

The new dimensions are 7.5 in. by 10.5 in.

38. Find the ratio of the area of the image to that of the preimage.

$$\frac{\text{area of image}}{\text{area of preimage}} = \frac{0.75(10) \cdot 0.75(14)}{10 \cdot 14}$$
$$= 0.75^2$$
$$= 0.5625$$
$$= \frac{9}{16}$$

The area of the image is $\frac{9}{16}$ that of the preimage.

39. Each side of the rectangle is lengthened by a factor of 4, so the perimeter is four times the original perimeter.

40. Find the ratio of the areas.

$$\frac{\text{area of image}}{\text{area of preimage}} = \frac{(4b)(4h)}{bh}$$
$$= 16$$

The area is 16 times the original area.

41. Given: dilation with center C and scale factor r
Prove: $ED = r(AB)$

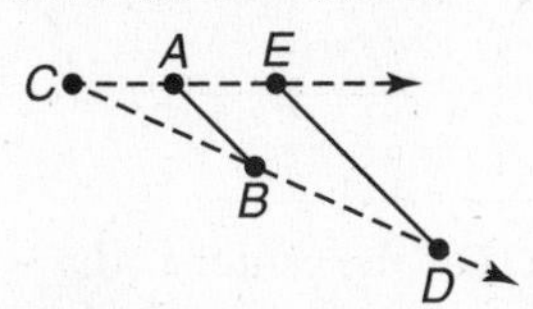

Proof:
$CE = r(CA)$ and $CD = r(CB)$ by the definition of a dilation. $\frac{CE}{CA} = r$ and $\frac{CD}{CB} = r$. So, $\frac{CE}{CA} = \frac{CD}{CB}$ by substitution. $\angle ACB \cong \angle ECD$, since congruence of angles is reflexive. Therefore, by SAS Similarity, $\triangle ACB$ is similar to $\triangle ECD$. The corresponding sides of similar triangles are proportional, so $\frac{ED}{AB} = \frac{CE}{CA}$. We know that $\frac{CE}{CA} = r$, so $\frac{ED}{AB} = r$ by substitution. Therefore, $ED = r(AB)$ by the Multiplication Property of Equality.

42. First dilation: $(x, y) \rightarrow (rx, ry)$
Second dilation: $(rx, ry) \rightarrow (r^2x, r^2y)$
Find r^2 using the x-values of A and A''.

$$3 = r^2(12)$$
$$\frac{1}{4} = r^2$$
$$\frac{1}{2} = r$$

So, the scale factor is $\frac{1}{2}$.

43. Use the distance formula to find XY and $X'Y'$.

$$d = \sqrt{(x_2 - x_1)^2 + (y_2 - y_1)^2}$$
$$XY = \sqrt{(0 - 4)^2 + (5 - 2)^2}$$
$$= \sqrt{16 + 9}$$
$$= 5$$
$$X'Y' = \sqrt{(15 - 7)^2 + (11 - 17)^2}$$
$$= \sqrt{64 + 36}$$
$$= 10$$

Find the absolute value of the scale factor.

$$|r| = \frac{\text{image length}}{\text{preimage length}}$$
$$= \frac{10}{5}$$
$$= 2$$

The absolute value of the scale factor is 2.

44. The width and height of the photograph are increased by 150%, or a factor of 1.5.
$1.5(480) = 720$
$1.5(640) = 960$
The dimensions of the image are 960 pixels by 720 pixels.

45. The original width is 640 pixels. To reduce it to 32 pixels, Dinah must use a scale factor of $\frac{32}{640} = \frac{1}{20}$.

46. The original height is 480 pixels. To enlarge it to 600 pixels, Dinah used a scale factor of $\frac{600}{480} = \frac{5}{4}$.

47. The dimensions of the photograph are 10 cm by 12 cm. The space available is 6 cm by 8 cm. $\frac{6}{10} = 0.6$ and $\frac{8}{12} \approx 0.67$, so a scale factor of 0.6 is required for the photograph to be as large as possible on the page. So, she should save the image file at 60%.

48. Use the distance formula to find the length of each side.

$$d = \sqrt{(x_2 - x_1)^2 + (y_2 - y_1)^2}$$
$$CD = \sqrt{(3 - 7)^2 + (8 - 7)^2}$$
$$= \sqrt{16 + 1}$$
$$= \sqrt{17}$$
$$BC = \sqrt{(7 - 5)^2 + [7 - (-1)]^2}$$
$$= \sqrt{4 + 64}$$
$$= 2\sqrt{17}$$
$$AB = \sqrt{[5 - (-1)]^2 + (-1 - 1)^2}$$
$$= \sqrt{36 + 4}$$
$$= 2\sqrt{10}$$
$$AD = \sqrt{[3 - (-1)]^2 + (8 - 1)^2}$$
$$= \sqrt{16 + 49}$$
$$= \sqrt{65}$$

Find the perimeter.

$$P = \sqrt{17} + 2\sqrt{17} + 2\sqrt{10} + \sqrt{65}$$
$$\approx 26.8$$

The perimeter is about 26.8 units.

49. Find A', B', C', and D' using the scale factor, $r = -2$.

Preimage (x, y)	Image $(-2x, -2y)$
$A(-1, 1)$	$A'(2, -2)$
$B(5, -1)$	$B'(-10, 2)$
$C(7, 7)$	$C'(-14, -14)$
$D(3, 8)$	$D'(-6, -16)$

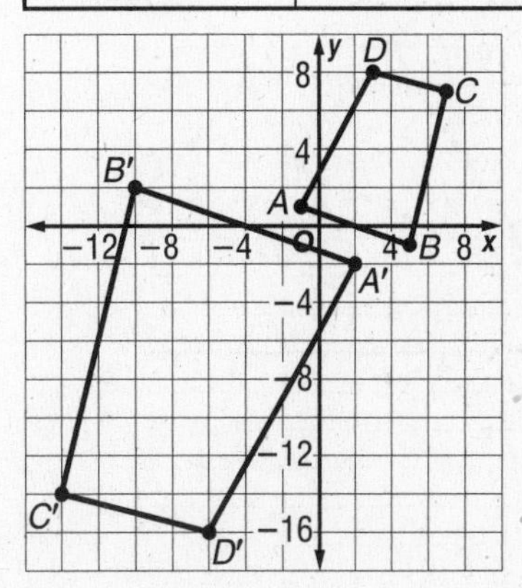

50. Use the distance formula to find the length of each side.

$d = \sqrt{(x_2 - x_1)^2 + (y_2 - y_1)^2}$

$A'B' = \sqrt{(-10 - 2)^2 + [2 - (-2)]^2}$
$= \sqrt{144 + 16}$
$= 4\sqrt{10}$

$B'C' = \sqrt{[-14 - (-10)]^2 + (-14 - 2)^2}$
$= \sqrt{16 + 256}$
$= 4\sqrt{17}$

$C'D' = \sqrt{[-6 - (-14)]^2 + [-16 - (-14)]^2}$
$= \sqrt{64 + 4}$
$= 2\sqrt{17}$

$A'D' = \sqrt{(-6 - 2)^2 + [-16 - (-2)]^2}$
$= \sqrt{64 + 196}$
$= 2\sqrt{65}$

Find the perimeter.

$P = 4\sqrt{10} + 4\sqrt{17} + 2\sqrt{17} + 2\sqrt{65}$
≈ 53.5

The perimeter of quadrilateral $A'B'C'D'$ is about 53.5 units, so it is twice the perimeter of quadrilateral $ABCD$.

51. $T(6, -5)$, $U(3, -8)$, $V(-1, -2)$

Reflection in the x-axis:

$(x, y) \rightarrow (x, -y)$
$T(6, -5) \rightarrow (6, 5)$
$U(3, -8) \rightarrow (3, 8)$
$V(-1, -2) \rightarrow (-1, 2)$

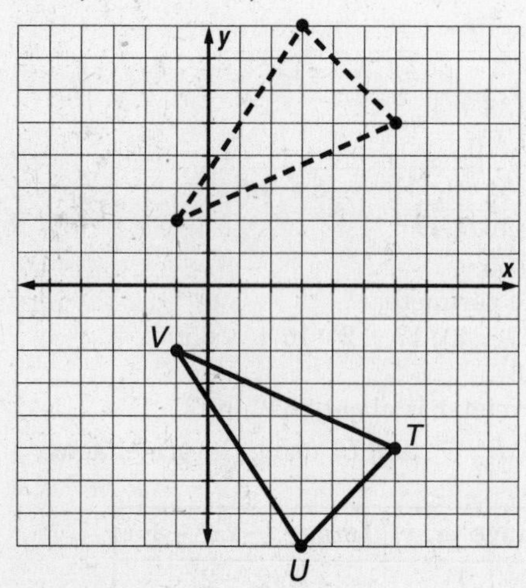

Translation with $(x, y) \rightarrow (x + 4, y - 1)$:

$(6, 5) \rightarrow (10, 4)$
$(3, 8) \rightarrow (7, 7)$
$(-1, 2) \rightarrow (3, 1)$

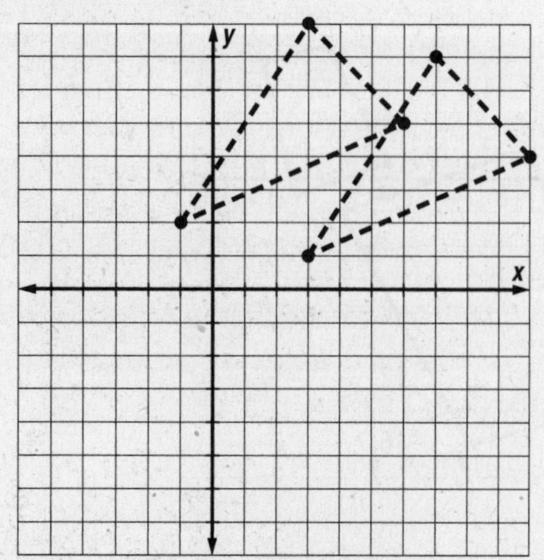

Dilation with scale factor of $\frac{1}{3}$:

$(x, y) \rightarrow \left(\frac{x}{3}, \frac{y}{3}\right)$
$(10, 4) \rightarrow T'\left(\frac{10}{3}, \frac{4}{3}\right)$
$(7, 7) \rightarrow U'\left(\frac{7}{3}, \frac{7}{3}\right)$
$(3, 1) \rightarrow V'\left(1, \frac{1}{3}\right)$

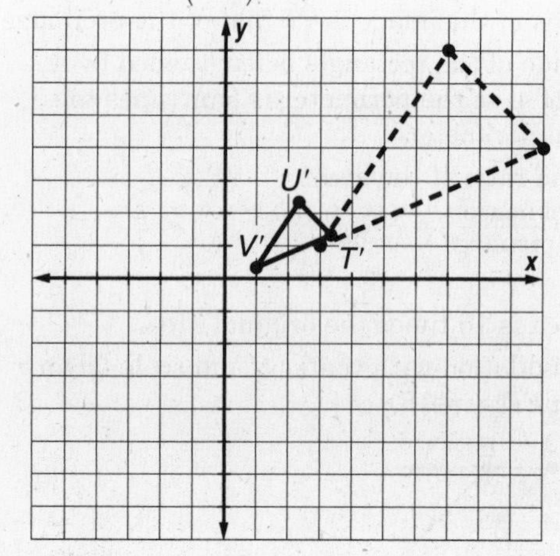

52. Translate the points so that the center is the origin.

$(x, y) \rightarrow (x - 3, y - 5)$
$G(3, 5) \rightarrow (0, 0)$
$H(7, -4) \rightarrow (4, -9)$
$I(-1, 0) \rightarrow (-4, -5)$

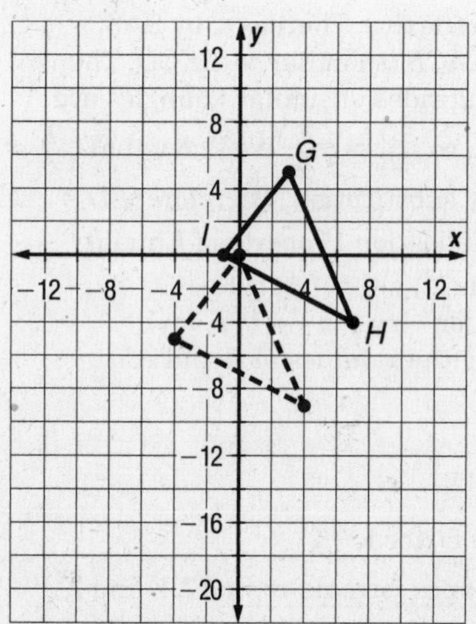

Dilate the figure using the scale factor 2.

$(x, y) \rightarrow (2x, 2y)$
$(0, 0) \rightarrow (0, 0)$
$(4, -9) \rightarrow (8, -18)$
$(-4, -5) \rightarrow (-8, -10)$

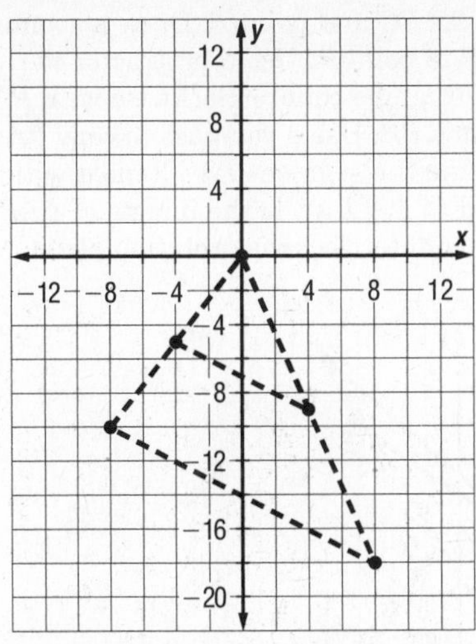

Translate the points back so that the center is at (3, 5).

$(x, y) \rightarrow (x + 3, x + 5)$

$(0, 0) \rightarrow G'(3, 5)$

$(8, -18) \rightarrow H'(11, -13)$

$(-8, -10) \rightarrow I'(-5, -5)$

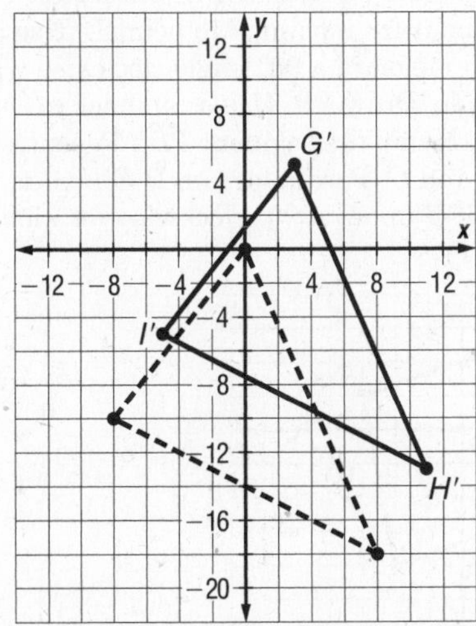

The coordinates of the vertices of the image are $G'(3, 5)$, $H'(11, -13)$ and $I'(-5, -5)$.

53. Sample answer: Yes; a cut and paste produces an image congruent to the original. Answers should include the following.

- Congruent figures are similar, so cutting and pasting is a similarity transformation.
- If you scale both horizontally and vertically by the same factor, you are creating a dilation.

54. B; the pentagons are similar. Find the scale factor by dividing the length of the radius of the larger pentagon by that of the smaller pentagon.

$$\text{Scale factor} = \frac{6+6}{6} = \frac{12}{6} = 2$$

So, the perimeter of the larger pentagon is twice that of the smaller pentagon ($5n$). The perimeter of the larger pentagon is $10n$.

55. A; find the slope of $3x + 5y = 12$.

$3x + 5y = 12$

$5y = -3x + 12$

$y = -\frac{3}{5}x + \frac{12}{5}$

The slope is $-\frac{3}{5}$.

The slopes of perpendicular lines are opposite reciprocals of each other.

$-\frac{1}{-\frac{3}{5}} = \frac{5}{3}$

The slope is $\frac{5}{3}$.

Page 497 Maintain Your Skills

56. No; use the algebraic method to determine whether a semi-regular tessellation can be created using equilateral triangles and regular pentagons of side length 1 unit.

Each interior angle of an equilateral triangle measures 60°, and each interior angle of a regular pentagon measures $\frac{180(5-2)}{5}$ or 108°.

Find whole number values for h and t so that $60h + 108t = 360$.

Let $h = 1$.

$60(1) + 108t = 360$

$60 + 108t = 360$

$108t = 300$

$t \approx 2.8$

Let $h = 2$.

$60(2) + 108t = 360$

$120 + 108t = 360$

$108t = 240$

$t \approx 2.2$

Let $h = 3$.

$60(3) + 108t = 360$

$180 + 108t = 360$

$108t = 180$

$t \approx 1.7$

Let $h = 4$.

$60(4) + 108t = 360$

$240 + 108t = 120$

$108t = 120$

$t \approx 1.1$

Let $h = 5$.

$60(5) + 108t = 360$

$300 + 108t = 360$

$108t = 60$

$t \approx 0.6$

There are no more reasonable possibilities. So, a semi-regular tessellation cannot be created from equilateral triangles and regular pentagons.

57. No; use the algebraic method to determine whether a semi-regular tessellation can be created using regular octagons and hexagons of side length 1 unit.

Each interior angle of a regular octagon measures $\frac{180(8-2)}{8}$ or 135°, and each interior angle of a regular hexagon measures $\frac{180(6-2)}{6}$ or 120°.

Find whole number values for h and t so that $135h + 120t = 360$.
Let $t = 1$.
$135h + 120(1) = 360$
$135h + 120 = 360$
$135h = 240$
$t \approx 1.8$
Let $t = 2$.
$135h + 120(2) = 360$
$135h + 240 = 360$
$135h = 120$
$t \approx 0.9$
There are no more reasonable possibilities. So, a semi-regular tessellation cannot be created from regular octagons and hexagons.

58. Yes; use the algebraic method to determine whether a semi-regular tessellation can be created using squares and equilateral triangles of side length 1 unit.
Each interior angle of a square measures 90°, and each interior angle of a triangle measures 60°.
Find whole number values for h and t so that $90h + 60t = 360$.
Let $h = 2$.
$90(2) + 60t = 360$
$180 + 60t = 360$
$60t = 180$
$t = 3$
When $h = 2$ and $t = 3$, there are two squares with three triangles at each vertex.

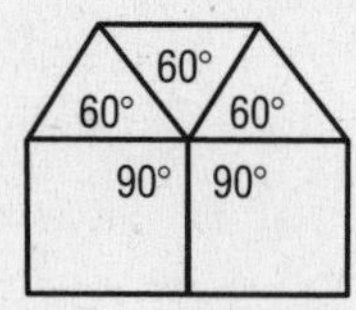

59. No; use the algebraic method to determine whether a semi-regular tessellation can be created using regular hexagons and dodecagons of side length 1 unit.
Each interior angle of a regular hexagon measures $\frac{180(6-2)}{6}$ or 120°, and each interior angle of a regular dodecagon measures $\frac{180(12-2)}{12}$ or 150°.
Find whole number values for h and t so that $120h + 150t = 360$.
Let $h = 1$.
$120(1) + 150t = 360$
$120 + 150t = 360$
$150t = 240$
$t = 1.6$
Let $h = 2$.
$120(2) + 150t = 360$
$240 + 150t = 360$
$150t = 120$
$t = 0.8$
There are no more reasonable possibilities. So, a semi-regular tessellation cannot be created from regular hexagons and dodecagons.

60. First graph $\triangle ABC$ and point P. Draw a segment from point P to point A. Use a protractor to measure a 90° angle counterclockwise with $\overline{PA}$ as one side. Draw $\overrightarrow{PR}$. Use a compass to copy $\overline{PA}$ onto $\overrightarrow{PR}$. Name the segment $\overline{PA'}$. Repeat with points B and C. $\triangle A'B'C'$ is the image of $\triangle ABC$ under a 90° counterclockwise rotation about point P.

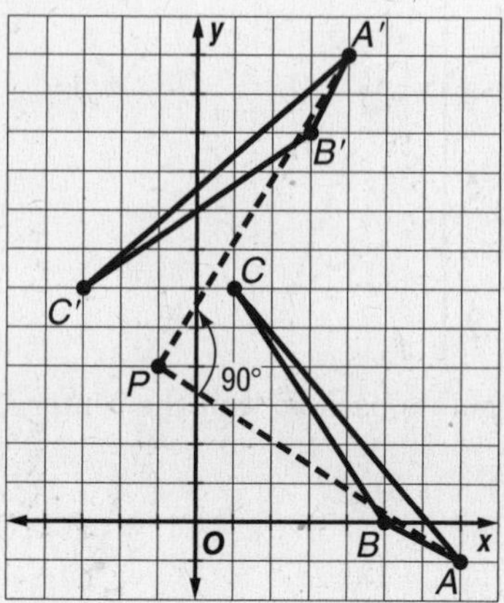

61. First graph parallelogram $DEFG$ and point P. Draw a segment from point P to point D. Use a protractor to measure a 90° angle clockwise with $\overline{PD}$ as one side. Draw $\overrightarrow{PR}$. Use a compass to copy $\overline{PD}$ onto $\overrightarrow{PR}$. Name the segment $\overline{PD'}$. Repeat with points E, F, and G. Parallelogram $D'E'F'G'$ is the image of $DEFG$ under a 90° clockwise rotation about point P.

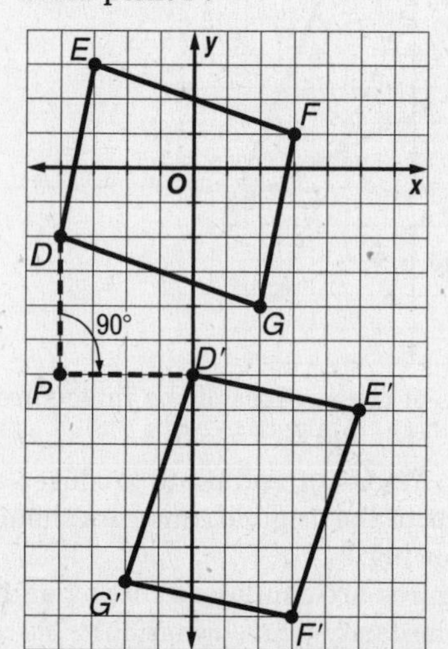

62. Yes; the opposite sides of a rectangle are congruent, and the diagonals are congruent.

63. Given: $\angle J \cong \angle L$
B is the midpoint of $\overline{JL}$.
Prove: $\triangle JHB \cong \triangle LCB$

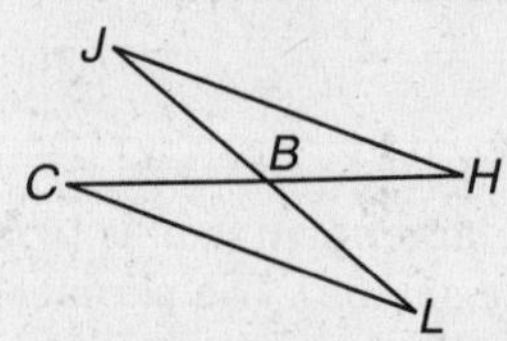

Proof: It is known that $\angle J \cong \angle L$. Since B is the midpoint of $\overline{JL}$, $\overline{JB} \cong \overline{LB}$ by the Midpoint Theorem. $\angle JBH \cong \angle LBC$ because vertical angles are congruent. Thus, $\triangle JHB \cong \triangle LCB$ by ASA.

64. Use the tangent ratio.

$\tan A = \frac{BC}{AC}$

$\tan A = \frac{2}{3}$

$A = \tan^{-1}\left(\frac{2}{3}\right)$

Use a calculator to find $m\angle A$.

$m\angle A \approx 33.7$

65. Use the tangent ratio.

$\tan A = \frac{BC}{AB}$

$\tan A = \frac{28}{7}$

$A = \tan^{-1}4$

Use a calculator to find $m\angle A$.

$m\angle A \approx 76.0$

66. Use the cosine ratio.

$\cos A = \frac{AC}{AB}$

$\cos A = \frac{20}{32}$

$A = \cos^{-1}\left(\frac{5}{8}\right)$

Use a calculator to find $m\angle A$.

$m\angle A \approx 51.3$

Page 497 Practice Quiz 2

1. Yes; the pattern is a tessellation because at the different vertices the sum of the angles is 360°. The tessellation is uniform because at every vertex there is the same combination of shapes and angles.
The tessellation is semi-regular because it is composed of more than one type of regular polygon.

2. Yes; the pattern is a tessellation because at the different vertices the sum of the angles is 360°. The tessellation is uniform because at every vertex there is the same combination of shapes and angles.

3. Since $\left|\frac{3}{4}\right| < 1$, the dilation is a reduction.
Draw $\overline{CD}$, $\overline{CF}$, and $\overline{CE}$. Locate D', F', and E' on $\overline{CD}$, $\overline{CF}$, and $\overline{CE}$ so that $CD' = \frac{3}{4}(CD)$, $CF' = \frac{3}{4}(CF)$, and $CE' = \frac{3}{4}(CE)$. Draw $\triangle D'F'E'$.

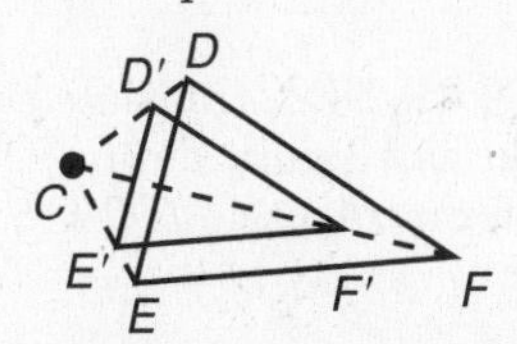

4. Since $|-2| > 1$, the dilation is an enlargement. Since r is negative, P', R', S', T', and U' will lie on rays that are opposite to $\overrightarrow{CP}$, $\overrightarrow{CR}$, $\overrightarrow{CS}$, $\overrightarrow{CT}$, and $\overrightarrow{CU}$ respectively. Locate P', R', S', T', and U' so that $CP' = 2(CP)$, $CR' = 2(CR)$, $CS', = 2(CS)$, $CT' = 2(CT)$, and $CU' = 2(CU)$. Draw $P'R'S'T'U'$.

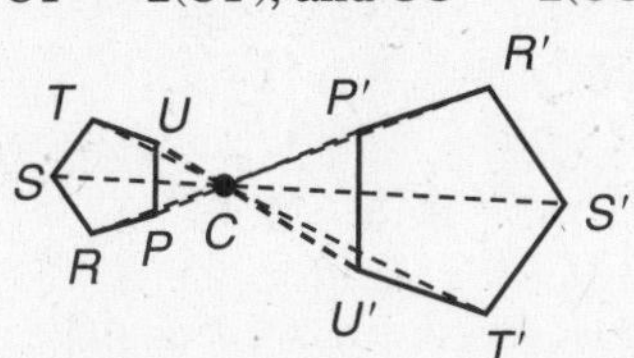

5. Find A', B', and C' using the scale factor, $r = -\frac{1}{2}$.

Preimage (x, y)	Image $\left(-\frac{x}{2}, -\frac{y}{2}\right)$
$A(10, 2)$	$A'(-5, -1)$
$B(1, 6)$	$B'(-\frac{1}{2}, -3)$
$C(-4, 4)$	$C'(2, -2)$

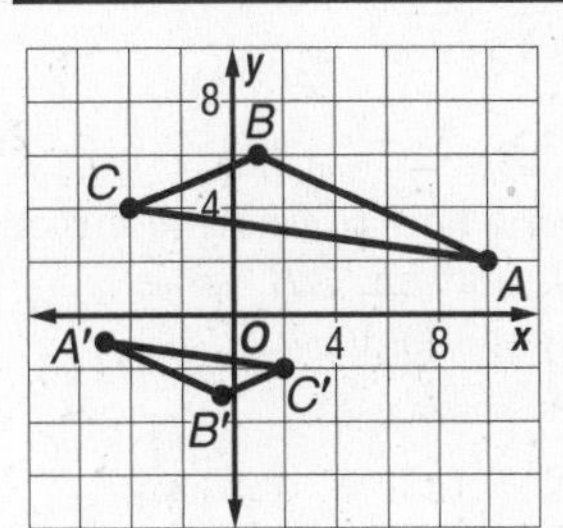

9-6 Vectors

Page 501 Geometry Activity: Comparing Magnitude and Components of Vectors

1. See students' work.

2. The components of $\vec{b}$ are twice the components of $\vec{a}$.

3. The components of $\vec{b}$ are three times the components of $\vec{a}$.

4. Sample answer: The magnitude is n times greater than the magnitude of $\langle x, y\rangle$, and the direction is the same.

Pages 502–503 Check for Understanding

1. Sample answer: $\langle 7, 7\rangle$

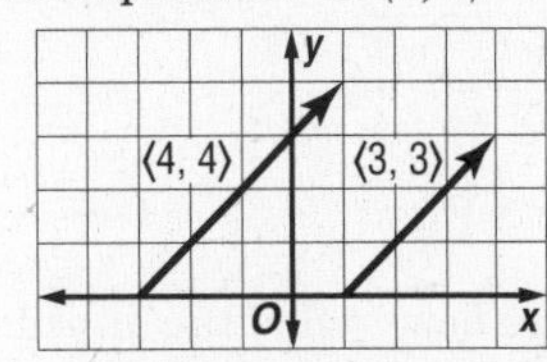

2. Two equal vectors must have the same magnitude and direction. Parallel vectors have the same direction or opposite directions. The magnitude of parallel vectors can be different.

3. Sample answer: Using a vector to translate a figure is the same as using an ordered pair because a vector has horizontal and vertical components, which can be represented by one coordinate of an ordered pair.

4. Find the change in x-values and the corresponding change in y-values.

$\overrightarrow{AB} = \langle x_2 - x_1, y_2 - y_1\rangle$

$= \langle 1 - (-4), 3 - (-3)\rangle$

$= \langle 5, 6\rangle$

5. Find the change in x-values and the corresponding change in y-values.

$\overrightarrow{CD} = \langle x_2 - x_1, y_2 - y_1 \rangle$
$= \langle 0 - (-4), 1 - 4 \rangle$
$= \langle 4, -3 \rangle$

6. Find the magnitude using the distance formula.

$|\overrightarrow{AB}| = \sqrt{(x_2 - x_1)^2 - (y_2 - y_1)^2}$
$= \sqrt{(-3 - 2)^2 + (3 - 7)^2}$
$= \sqrt{41}$
≈ 6.4

Graph $\overrightarrow{AB}$ to determine how to find the direction. Draw a right triangle that has $\overrightarrow{AB}$ as its hypotenuse and an acute angle at A.

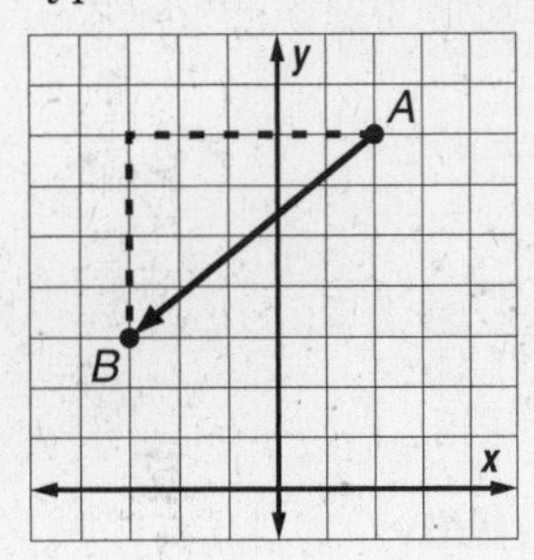

$\tan A = \frac{y_2 - y_1}{x_2 - x_1}$
$= \frac{3 - 7}{-3 - 2}$
$= \frac{4}{5}$
$m\angle A = \tan^{-1}\frac{4}{5}$
≈ 38.7

A vector in standard position that is equal to $\overrightarrow{AB}$ forms a 38.7° angle with the negative x-axis in the third quadrant. So it forms a 180 + 38.7 or 218.7° angle with the positive x-axis.
Thus, $\overrightarrow{AB}$ has a magnitude of $\sqrt{41}$ or about 6.4 units and a direction of about 218.7°.

7. Find the magnitude using the distance formula.

$|\overrightarrow{AB}| = \sqrt{(x_2 - x_1)^2 + (y_2 - y_1)^2}$
$= \sqrt{[-12 - (-6)]^2 + (-4 - 0)^2}$
$= 2\sqrt{13}$
≈ 7.2

Graph $\overrightarrow{AB}$ to determine how to find the direction. Draw a right triangle that has $\overrightarrow{AB}$ as its hypotenuse and an acute angle at A.

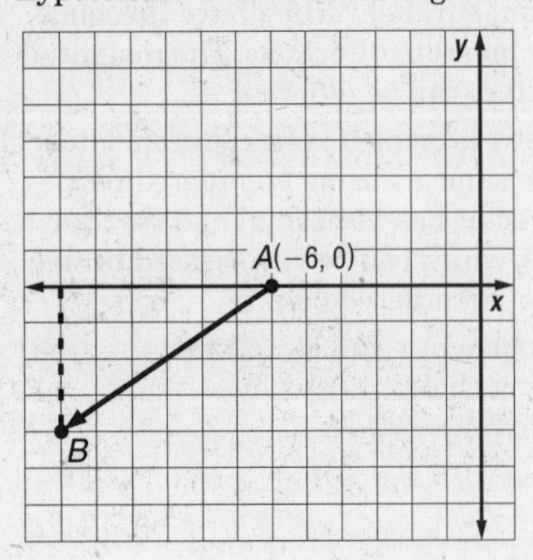

$\tan A = \frac{y_2 - y_1}{x_2 - x_1}$
$= \frac{-4 - 0}{-12 - (-6)}$
$= \frac{2}{3}$
$m\angle A = \tan^{-1}\frac{2}{3}$
≈ 33.7

A vector in standard position that is equal to $\overrightarrow{AB}$ forms a 33.7° angle with the negative x-axis in the third quadrant. So it forms a 180 + 33.7 or 213.7° angle with the positive x-axis.
Thus, $\overrightarrow{AB}$ has a magnitude of $2\sqrt{13}$ or about 7.2 units and a direction of about 213.7°.

8. Find the magnitude using the Pythagorean Theorem.

$|\vec{\mathbf{v}}| = \sqrt{x^2 + y^2}$
$= \sqrt{8^2 + (-15)^2}$
$= 17$

Vector $\vec{\mathbf{v}}$ lies in the fourth quadrant. Find the direction.

$\tan \theta = \frac{y}{x}$
$= \frac{-15}{8}$
$m\angle\theta = \tan^{-1}\left(-\frac{15}{8}\right)$
≈ -61.9

$\vec{\mathbf{v}}$ forms a 61.9° angle with the positive x-axis in the fourth quadrant. So, it forms a 360 − 61.9 or 298.1° angle with the positive x-axis.
Thus, $\vec{\mathbf{v}}$ has a magnitude of 17 units and a direction of about 298.1°.

9. First, graph $\triangle JKL$. Next, translate each vertex by $\vec{\mathbf{t}}$, 1 unit left and 9 units up. Connect the vertices to form $\triangle J'K'L'$.

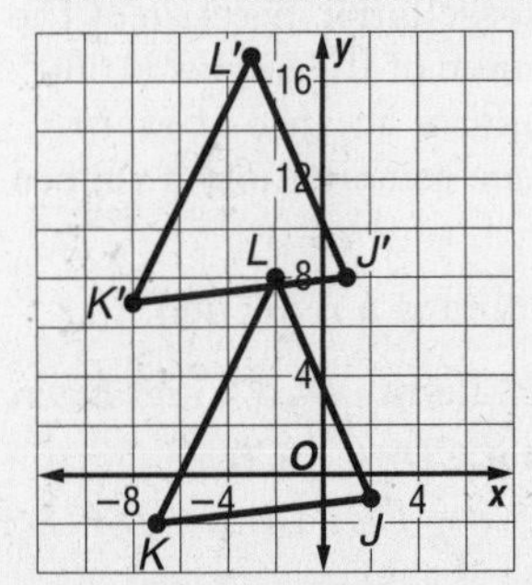

10. First, graph trapezoid $PQRS$. Next, translate each vertex by $\vec{\mathbf{u}}$, 3 units right and 3 units down. Connect the vertices to form trapezoid $P'Q'R'S'$.

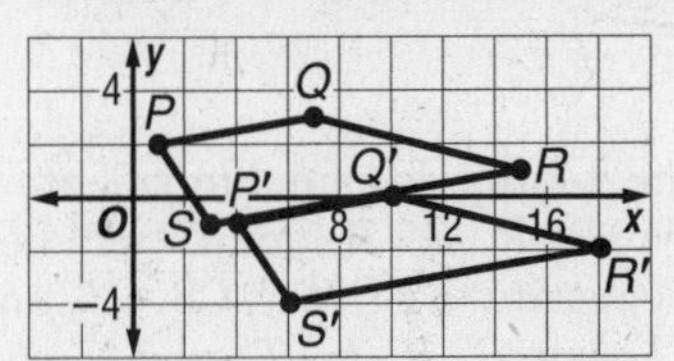

11. First, graph ▱$WXYZ$. Next, translate each vertex by $\vec{e}$, 1 unit left and 6 units up. Finally, translate each vertex by $\vec{f}$, 8 units right and 5 units down. Connect the vertices to form ▱$W'X'Y'Z'$.

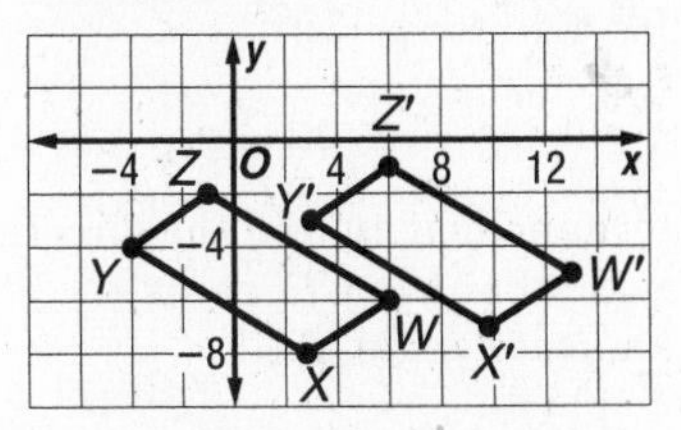

12. Find the resultant vector.

$\vec{g} + \vec{h} = \langle 4 + 0, 0 + 6\rangle$
$= \langle 4, 6\rangle$

Find the magnitude to the resultant vector.

$|\vec{g} + \vec{h}| = \sqrt{x^2 + y^2}$
$= \sqrt{4^2 + 6^2}$
$= 2\sqrt{13}$
≈ 7.2

The resultant vector lies in the first quadrant. Find the direction.

$\tan\theta = \frac{y}{x}$
$= \frac{6}{4}$
$= \frac{3}{2}$
$m\angle\theta = \tan^{-1}\left(\frac{3}{2}\right)$
≈ 56.3

The resultant vector forms a 56.3° angle with the positive x-axis in the first quadrant. Thus, $\vec{g} + \vec{h}$ has a magnitude of $2\sqrt{13}$ or about 7.2 units and a direction of about 56.3°.

13. Find the resultant vector.

$\vec{t} + \vec{u} = \langle 0 + 12, -9 - 9\rangle$
$= \langle 12, -18\rangle$

Find the magnitude of the resultant vector.

$|\vec{t} + \vec{u}| = \sqrt{x^2 + y^2}$
$= \sqrt{12^2 + (-18)^2}$
$= 6\sqrt{13}$
≈ 21.6

The resultant vector lies in the fourth quadrant. Find the direction.

$\tan\theta = \frac{y}{x}$
$= \frac{-18}{12}$
$= -\frac{3}{2}$
$m\angle\theta = \tan^{-1}\left(-\frac{3}{2}\right)$
≈ -56.3

The resultant vector forms a 56.3° angle with the positive x-axis in the fourth quadrant. So, it forms a 360 − 56.3 or 303.7° angle with the positive x-axis.
Thus, $\vec{t} + \vec{u}$ has a magnitude of $6\sqrt{13}$ or about 21.6 units and a direction of about 303.7°.

14. The initial path of the boat is due east, so a vector representing the speed of the boat lies on the positive x-axis and is 10 units long. The current is flowing 30° south of east, so a vector representing the speed of the current will be 30° below the positive x-axis 3 units long. The resultant speed can be found by adding the two vectors. Find the components of the current vector.

$\langle x, y\rangle = \langle 3\cos(-30°), 3\sin(-30°)\rangle$
$= \left\langle \frac{3\sqrt{3}}{2}, -\frac{3}{2}\right\rangle$

Find the resultant vector.

$\left\langle 10 + \frac{3\sqrt{3}}{2}, 0 + \left(-\frac{3}{2}\right)\right\rangle = \langle 10 + 1.5\sqrt{3}, -1.5\rangle$

Use the Pythagorean Theorem to find the magnitude.

$|\langle 10 + 1.5\sqrt{3}, -1.5\rangle|$
$= \sqrt{(10 + 1.5\sqrt{3})^2 + (-1.5)^2}$
≈ 12.7

The resultant vector lies in the fourth quadrant. Find the direction.

$\tan\theta = \frac{y}{x}$
$= \frac{-1.5}{10 + 1.5\sqrt{3}}$
$m\angle\theta = \tan^{-1}\left(-\frac{1.5}{10 + 1.5\sqrt{3}}\right)$
≈ -6.8

The speed of the boat is about 12.7 knots, at a direction of about 6.8° south of due east.

Pages 503–505 Practice and Apply

15. Find the change in x-values and the corresponding change in y-values.

$\overrightarrow{AB} = \langle x_2 - x_1, y_2 - y_1\rangle$
$= \langle 3 - 1, 3 - (-3)\rangle$
$= \langle 2, 6\rangle$

16. Find the change in x-values and the corresponding change in y-values.

$\overrightarrow{CD} = \langle x_2 - x_1, y_2 - y_1\rangle$
$= \langle -3 - (-2), 4 - 0\rangle$
$= \langle -1, 4\rangle$

17. Find the change in x-values and the corresponding change in y-values.

$\overrightarrow{EF} = \langle x_2 - x_1, y_2 - y_1\rangle$
$= \langle -3 - 4, -1 - 3\rangle$
$= \langle -7, -4\rangle$

18. Find the change in x-values and the corresponding change in y-values.

$\overrightarrow{GH} = \langle x_2 - x_1, y_2 - y_1\rangle$
$= \langle 2 - (-3), 4 - 4\rangle$
$= \langle 5, 0\rangle$

19. Find the change in x-values and the corresponding change in y-values.

$\overrightarrow{LM} = \langle x_2 - x_1, y_2 - y_1\rangle$
$= \langle 1 - 4, 3 - (-2)\rangle$
$= \langle -3, 5\rangle$

20. Find the change in x-values and the corresponding change in y-values.

$\overrightarrow{NP} = \langle x_2 - x_1, y_2 - y_1\rangle$
$= \langle -1 - (-4), -1 - (-3)\rangle$
$= \langle 3, 2\rangle$

21. Find the magnitude using the distance formula.

$|\overrightarrow{CD}| = \sqrt{(x_2 - x_1)^2 + (y_2 - y_1)^2}$
$= \sqrt{(9 - 4)^2 + (2 - 2)^2}$
$= 5$

Since $\overrightarrow{CD} = \langle 5, 0\rangle$, it is along the positive x-axis. Thus, $\overrightarrow{CD}$ has a magnitude of 5 units and a direction of 0°.

22. Find the magnitude using the distance formula.

$$|\overrightarrow{CD}| = \sqrt{(x_2 - x_1)^2 + (y_2 - y_1)^2}$$
$$= \sqrt{[2 - (-2)]^2 + (5 - 1)^2}$$
$$= 4\sqrt{2}$$
$$\approx 5.7$$

Graph $\overrightarrow{CD}$ to determine how to find the direction. Draw a right triangle that has $\overrightarrow{CD}$ as its hypotenuse and an acute angle at C.

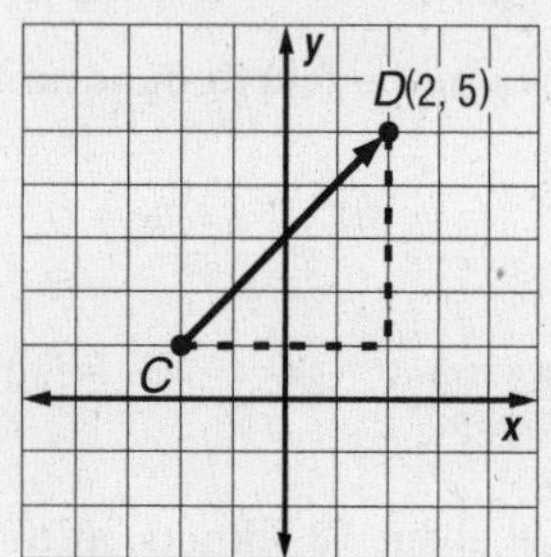

$$\tan C = \frac{y_2 - y_1}{x_2 - x_1}$$
$$= \frac{5 - 1}{2 - (-2)}$$
$$= 1$$
$$m\angle C = \tan^{-1} 1$$
$$= 45$$

A vector in standard position that is equal to $\overrightarrow{CD}$ forms a 45° angle with the positive x-axis in the first quadrant.

Thus, $\overrightarrow{CD}$ has a magnitude of $4\sqrt{2}$ or about 5.7 units and a direction of 45°.

23. Find the magnitude using the distance formula.

$$|\overrightarrow{CD}| = \sqrt{(x_2 - x_1)^2 + (y_2 - y_1)^2}$$
$$= \sqrt{[-3 - (-5)]^2 + (6 - 10)^2}$$
$$= 2\sqrt{5}$$
$$\approx 4.5$$

Graph $\overrightarrow{CD}$ to determine how to find the direction. Draw a right triangle that has $\overrightarrow{CD}$ as its hypotenuse and an acute angle at C.

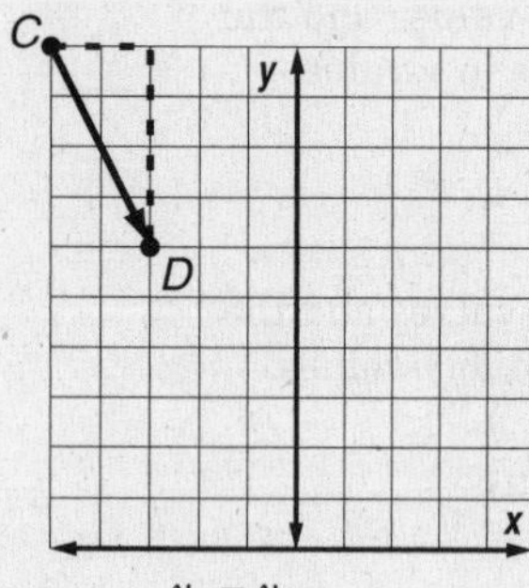

$$\tan C = \frac{y_2 - y_1}{x_2 - x_1}$$
$$= \frac{6 - 10}{-3 - (-5)}$$
$$= -2$$
$$m\angle C = \tan^{-1}(-2)$$
$$= -63.4$$

A vector in standard position that is equal to $\overrightarrow{CD}$ forms a 63.4° angle with the positive x-axis in the fourth quadrant. So it forms a 360 − 63.4 or 296.6° angle with the positive x-axis.

Thus, $\overrightarrow{CD}$ has a magnitude of $2\sqrt{5}$ or about 4.5 units and a direction of 296.6°.

24. Find the magnitude using the distance formula.

$$|\overrightarrow{CD}| = \sqrt{(x_2 - x_1)^2 + (y_2 - y_1)^2}$$
$$= \sqrt{(-2 - 0)^2 + [-4 - (-7)]^2}$$
$$= \sqrt{13}$$
$$\approx 3.6$$

Graph $\overrightarrow{CD}$ to determine how to find the direction. Draw a right triangle that has $\overrightarrow{CD}$ as its hypotenuse and an acute angle at C.

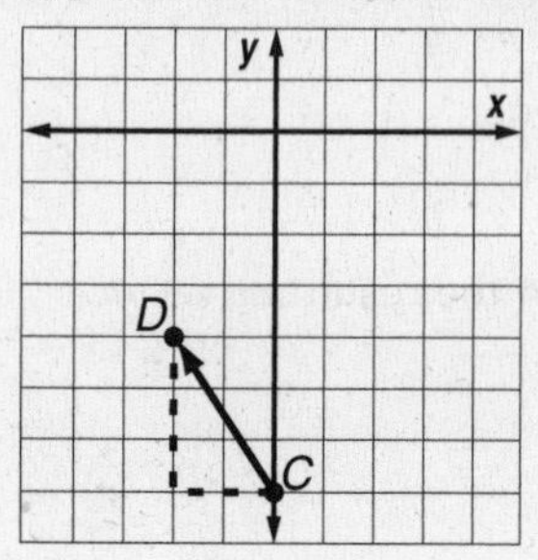

$$\tan C = \frac{y_2 - y_1}{x_2 - x_1}$$
$$= \frac{-4 - (-7)}{-2 - 0}$$
$$= -1.5$$
$$m\angle C = \tan^{-1}(-1.5)$$
$$= -56.3$$

A vector in standard position that is equal to $\overrightarrow{CD}$ forms a 56.3° angle with the negative x-axis in the second quadrant. So it forms a 180 − 56.3 or 123.7° angle with the positive x-axis.

Thus, $\overrightarrow{CD}$ has a magnitude of $\sqrt{13}$ or about 3.6 units and a direction of 123.7°.

25. Find the magnitude using the distance formula.

$$|\overrightarrow{CD}| = \sqrt{(x_2 - x_1)^2 + (y_2 - y_1)^2}$$
$$= \sqrt{[6 - (-8)]^2 + [0 - (-7)]^2}$$
$$= 7\sqrt{5}$$
$$\approx 15.7$$

Graph $\overrightarrow{CD}$ to determine how to find the direction. Draw a right triangle that has $\overrightarrow{CD}$ as its hypotenuse and an acute angle at C.

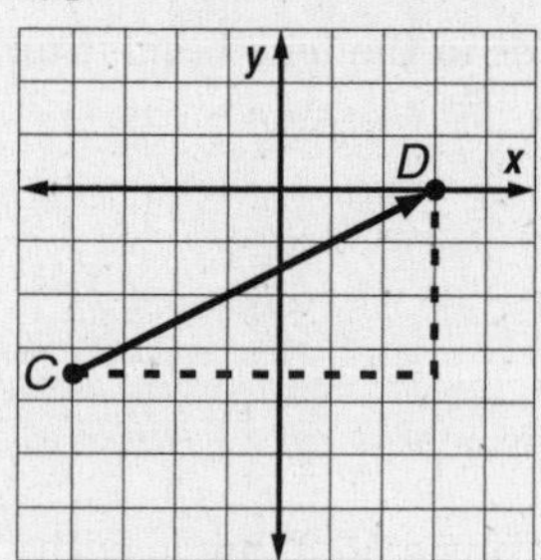

$$\tan C = \frac{y_2 - y_1}{x_2 - x_1}$$
$$= \frac{0 - (-7)}{6 - (-8)}$$
$$= 0.5$$
$$m\angle C = \tan^{-1}(0.5)$$
$$= 26.6$$

A vector in standard position that is equal to $\overrightarrow{CD}$ forms a 26.6° angle with the positive x-axis in the first quadrant.

Thus, $\overrightarrow{CD}$ has a magnitude of $7\sqrt{5}$ or about 15.7 units and a direction of 26.6°.

26. Find the magnitude using the distance formula.

$|\overrightarrow{CD}| = \sqrt{(x_2 - x_1)^2 + (y_2 - y_1)^2}$
$= \sqrt{(-2 - 10)^2 + [-2 - (-3)]^2}$
$= \sqrt{145}$
≈ 12.0

Graph $\overrightarrow{CD}$ to determine how to find the direction. Draw a right triangle that has $\overrightarrow{CD}$ as its hypotenuse and an acute angle at C.

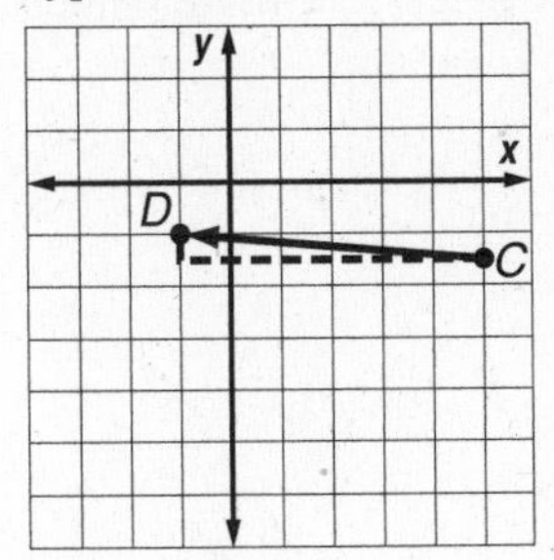

$\tan C = \frac{y_2 - y_1}{x_2 - x_1}$
$= \frac{-2 - (-3)}{-2 - 10}$
$= -\frac{1}{12}$
$m\angle C = \tan^{-1}\left(-\frac{1}{12}\right)$
$= -4.8$

A vector in standard position that is equal to $\overrightarrow{CD}$ forms a 4.8° angle with the negative x-axis in the second quadrant. So it forms a 180 − 4.8 or 175.2° angle with the positive x-axis.
Thus, $\overrightarrow{CD}$ has a magnitude of $\sqrt{145}$ or about 12.0 units and a direction of 175.2°.

27. Find the magnitude using the Pythagorean Theorem.

$|\vec{t}| = \sqrt{x^2 + y^2}$
$= \sqrt{7^2 + 24^2}$
$= 25$

Vector $\vec{t}$ lies in the first quadrant. Find the direction.

$\tan\theta = \frac{y}{x}$
$= \frac{24}{7}$
$m\angle\theta = \tan^{-1}\left(\frac{24}{7}\right)$
≈ 73.7

$\vec{v}$ forms a 73.7° angle with the positive x-axis in the first quadrant.
Thus, $\vec{v}$ has a magnitude of 25 units and a direction of about 73.7°.

28. Find the magnitude using the Pythagorean Theorem.

$|\vec{u}| = \sqrt{x^2 + y^2}$
$= \sqrt{(-12)^2 + 15^2}$
$= 3\sqrt{41}$
≈ 19.2

Vector $\vec{u}$ lies in the second quadrant. Find the direction.

$\tan\theta = \frac{y}{x}$
$= \frac{15}{-12}$
$= -\frac{5}{4}$
$m\angle\theta = \tan^{-1}\left(-\frac{5}{4}\right)$
≈ -51.3

$\vec{u}$ forms a 51.3° angle with the negative x-axis in the second quadrant. So, it forms a 180 − 51.3 or 128.7° angle with the positive x-axis.
Thus, $\vec{u}$ has a magnitude of $3\sqrt{41}$ or about 19.2 units and a direction of about 128.7°.

29. Find the magnitude using the Pythagorean Theorem.

$|\vec{v}| = \sqrt{x^2 + y^2}$
$= \sqrt{(-25)^2 + (-20)^2}$
$= 5\sqrt{41}$
≈ 32.0

Vector $\vec{v}$ lies in the third quadrant. Find the direction.

$\tan\theta = \frac{y}{x}$
$= \frac{-20}{-25}$
$= \frac{4}{5}$
$m\angle\theta = \tan^{-1}\left(\frac{4}{5}\right)$
≈ 38.7

$\vec{v}$ forms a 38.7° angle with the negative x-axis in the third quadrant. So, it forms a 180 + 38.7 or 218.7° angle with the positive x-axis.
Thus, $\vec{v}$ has a magnitude of $5\sqrt{41}$ or about 32.0 units and a direction of about 218.7°.

30. Find the magnitude using the Pythagorean Theorem.

$|\vec{w}| = \sqrt{x^2 + y^2}$
$= \sqrt{36^2 + (-15)^2}$
$= 39$

Vector $\vec{w}$ lies in the fourth quadrant. Find the direction.

$\tan\theta = \frac{y}{x}$
$= \frac{-15}{36}$
$= -\frac{5}{12}$
$m\angle\theta = \tan^{-1}\left(-\frac{5}{12}\right)$
≈ -22.6

$\vec{w}$ forms a 22.6° angle with the positive x-axis in the fourth quadrant. So, it forms a 360 − 22.6 or 337.4° angle with the positive x-axis.
Thus, $\vec{w}$ has a magnitude of 39 units and a direction of about 337.4°.

31. Find the magnitude using the Distance Formula.

$|\overrightarrow{MN}| = \sqrt{(x_2 - x_1)^2 + (y_2 - y_1)^2}$
$= \sqrt{[-9 - (-3)]^2 + (9 - 3)^2}$
$= 6\sqrt{2}$
≈ 8.5

Graph $\overrightarrow{MN}$ to determine how to find the direction. Draw a right triangle that has $\overrightarrow{MN}$ as its hypotenuse and an acute angle at M.

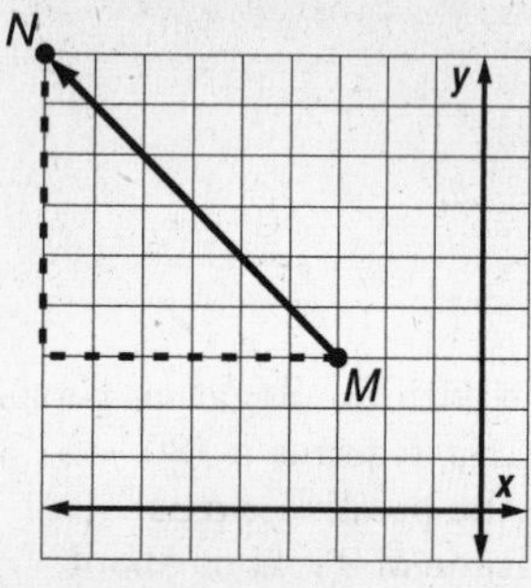

$$\tan M = \frac{y_2 - y_1}{x_2 - x_1}$$
$$= \frac{9 - 3}{-9 - (-3)}$$
$$= -1$$
$$m\angle M = \tan^{-1}(-1)$$
$$= -45$$

A vector in standard position that is equal to $\overrightarrow{MN}$ forms a 45° angle with the negative x-axis in the second quadrant. So it forms a 180 − 45 or 135° angle with the positive x-axis.
Thus, $\overrightarrow{MN}$ has a magnitude of $6\sqrt{2}$ or about 8.5 units and a direction of 135.0°.

32. Find the magnitude using the Distance Formula.

$$|\overrightarrow{MN}| = \sqrt{(x_2 - x_1)^2 + (y_2 - y_1)^2}$$
$$= \sqrt{(2 - 8)^2 + (5 - 1)^2}$$
$$= 2\sqrt{13}$$
$$\approx 7.2$$

Graph $\overrightarrow{MN}$ to determine how to find the direction. Draw a right triangle that has $\overrightarrow{MN}$ as its hypotenuse and an acute angle at M.

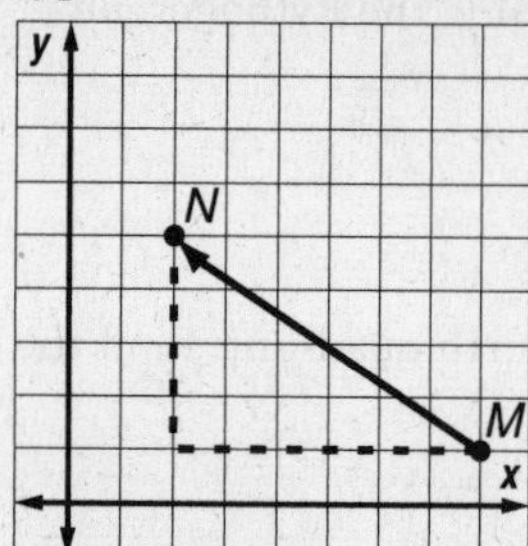

$$\tan M = \frac{y_2 - y_1}{x_2 - x_1}$$
$$= \frac{5 - 1}{2 - 8}$$
$$= -\frac{2}{3}$$
$$m\angle M = \tan^{-1}\left(-\frac{2}{3}\right)$$
$$= -33.7$$

A vector in standard position that is equal to $\overrightarrow{MN}$ forms a 33.7° angle with the negative x-axis in the second quadrant. So it forms a 180 − 33.7 or 146.3° angle with the positive x-axis.
Thus, $\overrightarrow{MN}$ has a magnitude of $2\sqrt{13}$ or about 7.2 units and a direction of 146.3°.

33. Find the magnitude using the Distance Formula.

$$|\overrightarrow{MN}| = \sqrt{(x_2 - x_1)^2 + (y_2 - y_1)^2}$$
$$= \sqrt{(-12 - 0)^2 + (-2 - 2)^2}$$
$$= 4\sqrt{10}$$
$$\approx 12.6$$

Graph $\overrightarrow{MN}$ to determine how to find the direction. Draw a right triangle that has $\overrightarrow{MN}$ as its hypotenuse and an acute angle at M.

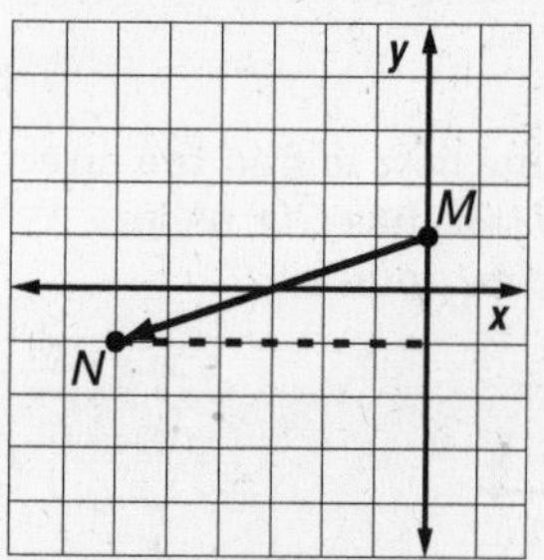

$$\tan M = \frac{y_2 - y_1}{x_2 - x_1}$$
$$= \frac{-2 - 2}{-12 - 0}$$
$$= \frac{1}{3}$$
$$m\angle M = \tan^{-1}\left(\frac{1}{3}\right)$$
$$= 18.4$$

A vector in standard position that is equal to $\overrightarrow{MN}$ forms an 18.4° angle with the negative x-axis in the third quadrant. So it forms a 180 + 18.4 or 198.4° angle with the positive x-axis.
Thus, $\overrightarrow{MN}$ has a magnitude of $4\sqrt{10}$ or about 12.6 units and a direction of 198.4°.

34. Find the magnitude using the Distance Formula.

$$|\overrightarrow{MN}| = \sqrt{(x_2 - x_1)^2 + (y_2 - y_1)^2}$$
$$= \sqrt{[6 - (-1)]^2 + (-8 - 7)^2}$$
$$= \sqrt{274}$$
$$\approx 16.6$$

Graph $\overrightarrow{MN}$ to determine how to find the direction. Draw a right triangle that has $\overrightarrow{MN}$ as its hypotenuse and an acute angle at M.

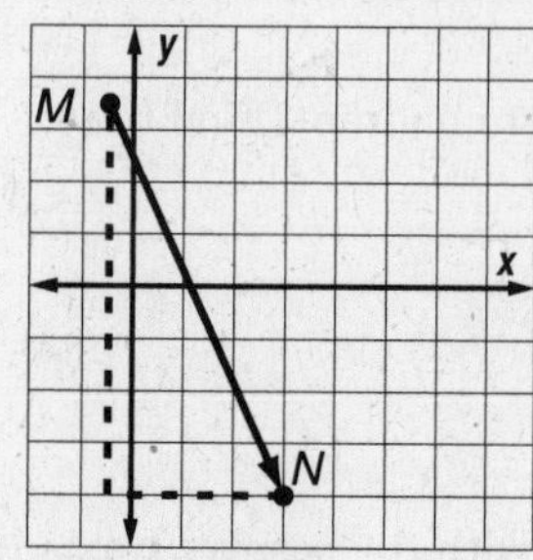

$$\tan M = \frac{y_2 - y_1}{x_2 - x_1}$$
$$= \frac{-8 - 7}{6 - (-1)}$$
$$= -\frac{15}{7}$$
$$m\angle M = \tan^{-1}\left(-\frac{15}{7}\right)$$
$$= -65.0$$

A vector in standard position that is equal to $\overrightarrow{MN}$ forms an 65.0° angle with the positive x-axis in the fourth quadrant. So it forms a 360 − 65.0 or 295.0° angle with the positive x-axis.
Thus, $\overrightarrow{MN}$ has a magnitude of $\sqrt{274}$ or about 16.6 units and a direction of 295.0°.

35. Find the magnitude using the Distance Formula.

$|\overrightarrow{MN}| = \sqrt{(x_2 - x_1)^2 + (y_2 - y_1)^2}$
$= \sqrt{[1 - (-1)]^2 + (-12 - 10)^2}$
$= 2\sqrt{122}$
≈ 22.1

Graph $\overrightarrow{MN}$ to determine how to find the direction. Draw a right triangle that has $\overrightarrow{MN}$ as its hypotenuse and an acute angle at M.

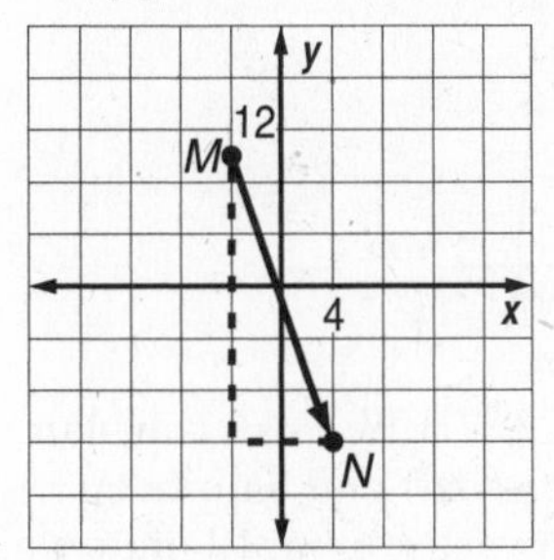

$\tan M = \frac{y_2 - y_1}{x_2 - x_1}$
$= \frac{-12 - 10}{1 - (-1)}$
$= -11$

$m\angle M = \tan^{-1}(-11)$
$= -84.8$

A vector in standard position that is equal to $\overrightarrow{MN}$ forms an 84.8° angle with the positive x-axis in the fourth quadrant. So it forms a 360 − 84.8 or 275.2° angle with the positive x-axis.3

Thus, $\overrightarrow{MN}$ has a magnitude of $2\sqrt{122}$ or about 22.1 units and a direction of 275.2°.

36. Find the magnitude using the Distance Formula.

$|\overrightarrow{MN}| = \sqrt{(x_2 - x_1)^2 + (y_2 - y_1)^2}$
$= \sqrt{[-6 - (-4)]^2 + (-4 - 0)^2}$
$= 2\sqrt{5}$
≈ 4.5

Graph $\overrightarrow{MN}$ to determine how to find the direction. Draw a right triangle that has $\overrightarrow{MN}$ as its hypotenuse and an acute angle at M.

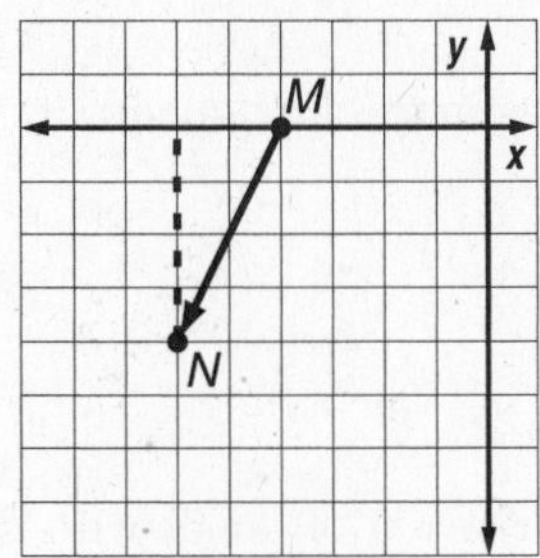

$\tan M = \frac{y_2 - y_1}{x_2 - x_1}$
$= \frac{-4 - 0}{-6 - (-4)}$
$= 2$

$m\angle M = \tan^{-1}(2)$
$= 63.4$

A vector in standard position that is equal to $\overrightarrow{MN}$ forms a 63.4° angle with the negative x-axis in the third quadrant. So it forms a 180 + 63.4 or 243.4° angle with the positive x-axis.

Thus, $\overrightarrow{MN}$ has a magnitude of $2\sqrt{5}$ or about 4.5 units and a direction of 243.4°.

37. First, graph $\triangle ABC$. Next, translate each vertex by $\vec{a}$, 6 units down. Connect the vertices to form $\triangle A'B'C'$.

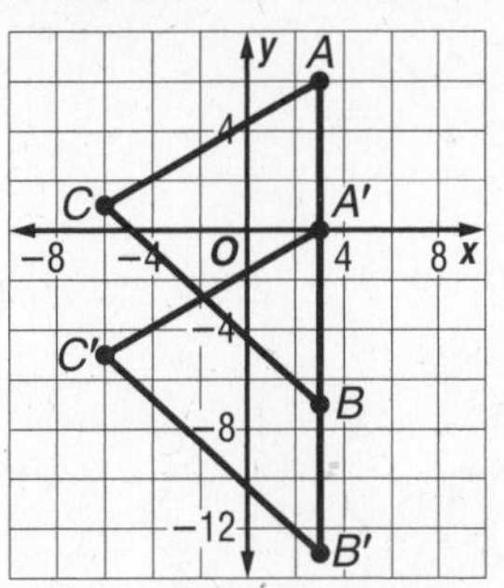

38. First, graph $\triangle DEF$. Next, translate each vertex by $\vec{b}$, 3 units left and 9 units down. Connect the vertices to form $\triangle D'E'F'$.

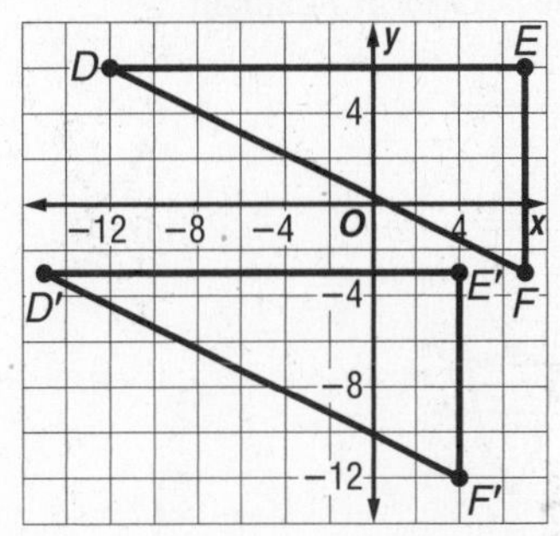

39. First, graph square $GHIJ$. Next, translate each vertex by $\vec{c}$, 3 units right and 8 units down. Connect the vertices to form square $G'H'I'J'$.

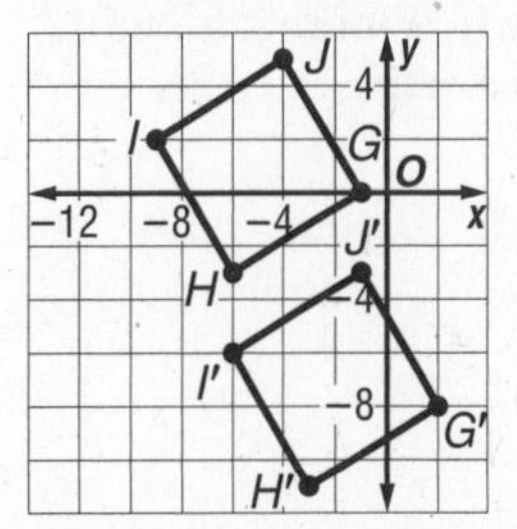

40. First, graph quadrilateral $KLMN$. Next, translate each vertex by $\vec{x}$, 10 units left and 2 units up. Connect the vertices to form quadrilateral $K'L'M'N'$.

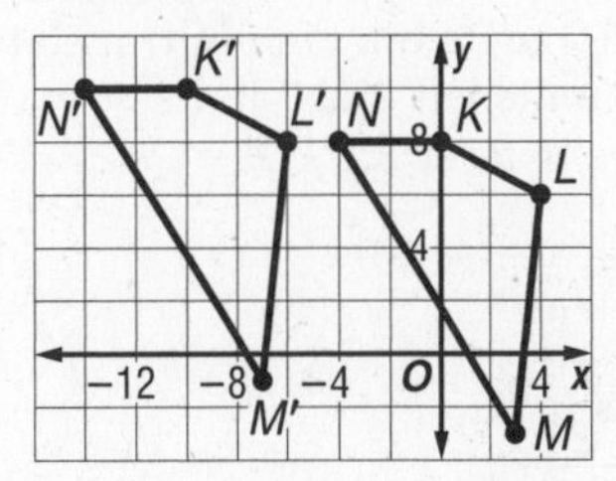

41. First, graph pentagon *OPQRS*. Next, translate each vertex by $\vec{\mathbf{y}}$, 5 units left and 11 units up. Connect the vertices to form pentagon $O'P'Q'R'S'$.

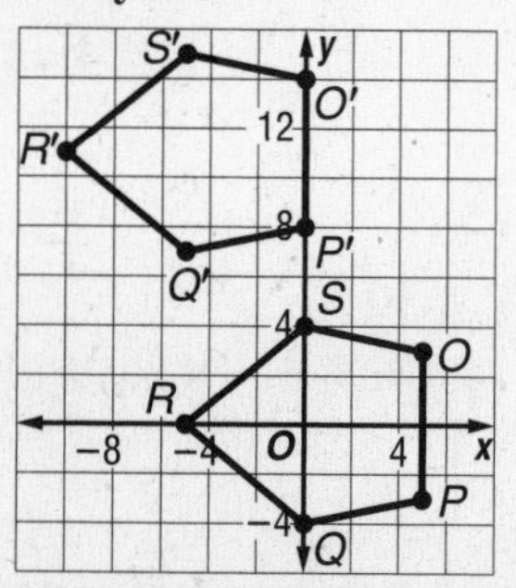

42. First, graph hexagon *TUVWXY*. Next, translate each vertex by $\vec{\mathbf{z}}$, 18 units left and 12 units up. Connect the vertices to form hexagon $T'U'V'W'X'Y'$.

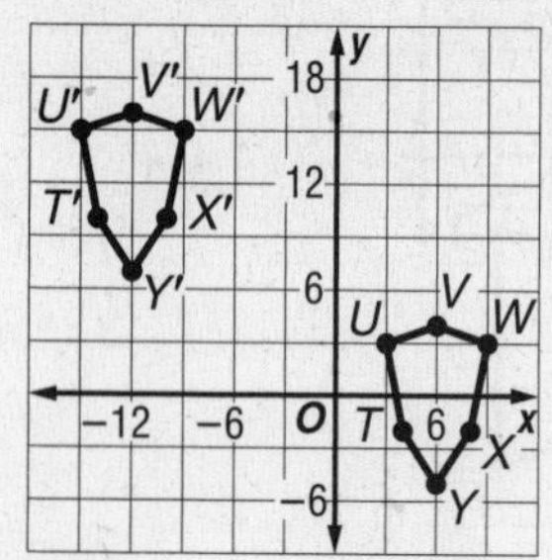

43. First, graph ▱*ABCD*. Next, translate each vertex by $\vec{\mathbf{p}}$, 11 units right and 6 units up. Finally, translate each vertex by $\vec{\mathbf{q}}$, 9 units left and 3 units down. Connect the vertices to form ▱$A'B'C'D'$.

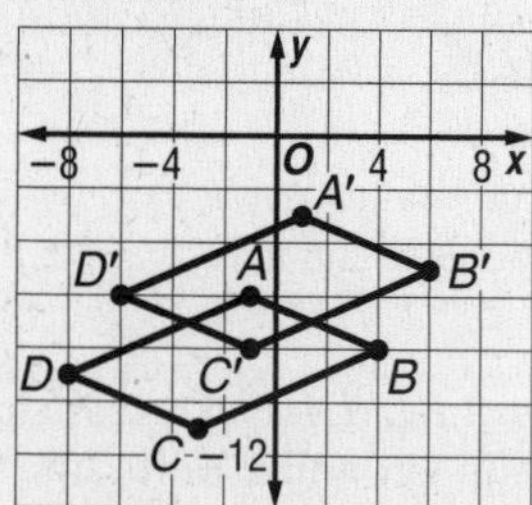

44. First, graph $\triangle XYZ$. Next, translate each vertex by $\vec{\mathbf{p}}$, 2 units right and 2 units up. Finally, translate each vertex by $\vec{\mathbf{q}}$, 4 units left and 7 units down. Connect the vertices to form $\triangle X'Y'Z'$.

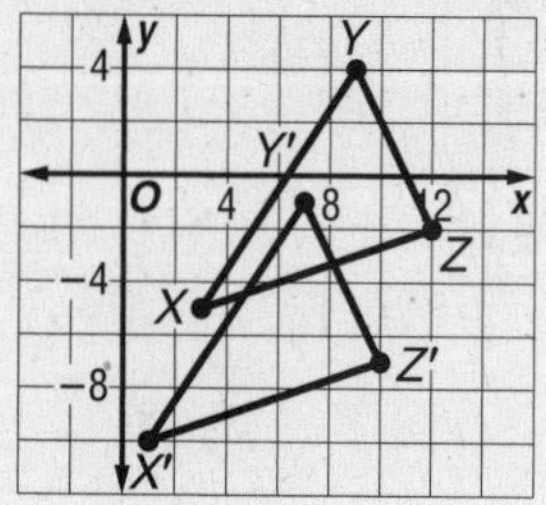

45. First, graph quadrilateral *EFGH*. Next, translate each vertex by $\vec{\mathbf{p}}$, 6 units left and 10 units up. Finally, translate each vertex by $\vec{\mathbf{q}}$, 1 unit right and 8 units down. Connect the vertices to form quadrilateral $E'F'G'H'$.

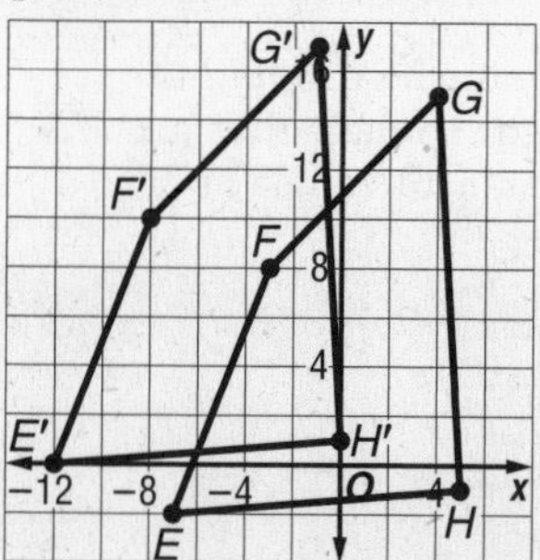

46. First, graph pentagon *STUVW*. Next, translate each vertex by $\vec{\mathbf{p}}$, 4 units left and 5 units up. Finally, translate each vertex by $\vec{\mathbf{q}}$, 12 units right and 11 units up. Connect the vertices to form pentagon $S'T'U'V'W'$.

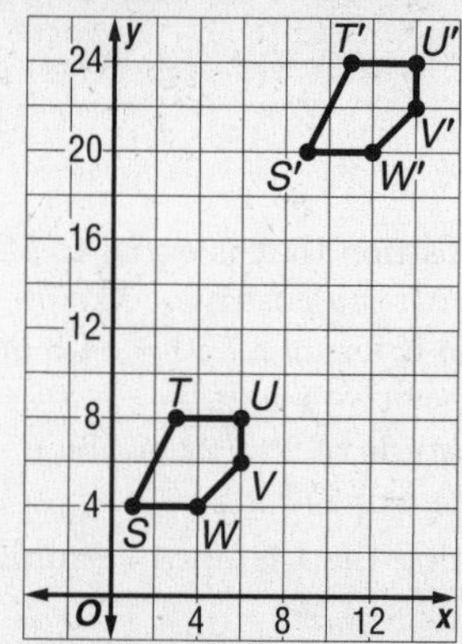

47. Find the resultant vector.

$\vec{\mathbf{a}} + \vec{\mathbf{b}} = \langle 5 + 0, 0 + 12 \rangle$

$= \langle 5, 12 \rangle$

Find the magnitude of the resultant vector.

$|\vec{\mathbf{a}} + \vec{\mathbf{b}}| = \sqrt{x^2 + y^2}$

$= \sqrt{5^2 + 12^2}$

$= 13$

The resultant vector lies in the first quadrant. Find the direction.

$\tan\theta = \frac{y}{x}$

$= \frac{12}{5}$

$m\angle\theta = \tan^{-1}\left(\frac{12}{5}\right)$

≈ 67.4

The resultant vector forms a 67.4° angle with the positive x-axis in the first quadrant.

Thus, $\vec{\mathbf{a}} + \vec{\mathbf{b}}$ has a magnitude of 13 units and a direction of about 67.4°.

48. Find the resultant vector.

$\vec{\mathbf{c}} + \vec{\mathbf{d}} = \langle 0 + (-8), -8 + 0 \rangle$

$= \langle -8, -8 \rangle$

Find the magnitude of the resultant vector.

$|\vec{\mathbf{c}} + \vec{\mathbf{d}}| = \sqrt{x^2 + y^2}$

$= \sqrt{(-8)^2 + (-8)^2}$

$= 8\sqrt{2}$

≈ 11.3

The resultant vector lies in the third quadrant.
Find the direction.

$\tan\theta = \frac{y}{x}$
$= \frac{-8}{-8}$
$= 1$
$m\angle\theta = \tan^{-1}(1)$
$= 45$

The resultant vector forms a 45° angle with the negative x-axis in the third quadrant. So it forms a 180 + 45 or 225° angle with the positive x-axis.
Thus, $\vec{\mathbf{c}} + \vec{\mathbf{d}}$ has a magnitude of $8\sqrt{2}$ or about 11.3 units and a direction of 225°.

49. Find the resultant vector.

$\vec{\mathbf{e}} + \vec{\mathbf{f}} = \langle -4 + 7, 0 + (-4)\rangle$
$= \langle 3, -4\rangle$

Find the magnitude of the resultant vector.

$|\vec{\mathbf{e}} + \vec{\mathbf{f}}| = \sqrt{x^2 + y^2}$
$= \sqrt{3^2 + (-4)^2}$
$= 5$

The resultant vector lies in the fourth quadrant.
Find the direction.

$\tan\theta = \frac{y}{x}$
$= \frac{-4}{3}$
$m\angle\theta = \tan^{-1}\left(-\frac{4}{3}\right)$
$= -53.1$

The resultant vector forms a 53.1° angle with the positive x-axis in the fourth quadrant. So it forms a 360 − 53.1 or 306.9° angle with the positive x-axis.
Thus, $\vec{\mathbf{e}} + \vec{\mathbf{f}}$ has a magnitude of 5 units and a direction of about 306.9°.

50. Find the resultant vector.

$\vec{\mathbf{u}} + \vec{\mathbf{v}} = \langle 12 + 0, 6 + 6\rangle$
$= \langle 12, 12\rangle$

Find the magnitude of the resultant vector.

$|\vec{\mathbf{u}} + \vec{\mathbf{v}}| = \sqrt{x^2 + y^2}$
$= \sqrt{12^2 + 12^2}$
$= 12\sqrt{2}$
≈ 17.0

The resultant vector lies in the first quadrant.
Find the direction.

$\tan\theta = \frac{y}{x}$
$= \frac{12}{12}$
$= 1$
$m\angle\theta = \tan^{-1}(1)$
$= 45$

The resultant vector forms a 45° angle with the positive x-axis in the first quadrant.
Thus, $\vec{\mathbf{u}} + \vec{\mathbf{v}}$ has a magnitude of $12\sqrt{2}$ or about 17.0 units and a direction of 45°.

51. Find the resultant vector.

$\vec{\mathbf{w}} + \vec{\mathbf{x}} = \langle 5 + (-1), 6 + (-4)\rangle$
$= \langle 4, 2\rangle$

Find the magnitude of the resultant vector.

$|\vec{\mathbf{w}} + \vec{\mathbf{x}}| = \sqrt{x^2 + y^2}$
$= \sqrt{4^2 + 2^2}$
$= 2\sqrt{5}$
≈ 4.5

The resultant vector lies in the first quadrant.
Find the direction.

$\tan\theta = \frac{y}{x}$
$= \frac{2}{4}$
$= \frac{1}{2}$
$m\angle\theta = \tan^{-1}\left(\frac{1}{2}\right)$
$= 26.6$

The resultant vector forms a 26.6° angle with the positive x-axis in the first quadrant.
Thus, $\vec{\mathbf{w}} + \vec{\mathbf{x}}$ has a magnitude of $2\sqrt{5}$ or about 4.5 units and a direction of about 26.6°.

52. Find the resultant vector.

$\vec{\mathbf{y}} + \vec{\mathbf{z}} = \langle 9 + (-10), -10 + (-2)\rangle$
$= \langle -1, -12\rangle$

Find the magnitude of the resultant vector.

$|\vec{\mathbf{y}} + \vec{\mathbf{z}}| = \sqrt{x^2 + y^2}$
$= \sqrt{(-1)^2 + (-12)^2}$
$= \sqrt{145}$
≈ 12.0

The resultant vector lies in the third quadrant.
Find the direction.

$\tan\theta = \frac{y}{x}$
$= \frac{-12}{-1}$
$= 12$
$m\angle\theta = \tan^{-1}(12)$
$= 85.2$

The resultant vector forms an 85.2° angle with the negative x-axis in the third quadrant. So it forms a 180 + 85.2 or 265.2° angle with the positive x-axis.
Thus, $\vec{\mathbf{y}} + \vec{\mathbf{z}}$ has a magnitude of $\sqrt{145}$ or about 12.0 units and a direction of about 265.2°.

53. The first path of the freighter is due east, so a vector representing the path lies on the positive x-axis and is 35 units long. The second path of the freighter is due south, so a vector representing this path begins at the tip of the first vector and stretches 28 units in the negative y-direction. Add the two vectors, $\langle 35, 0\rangle$ and $\langle 0, -28\rangle$.
$\langle 35 + 0, 0 + (-28)\rangle = \langle 35, -28\rangle$
Use the Pythagorean Theorem to find the magnitude.

$|\langle 35, -28\rangle| = \sqrt{35^2 + (-28)^2}$
≈ 44.8

The resultant vector lies in the fourth quadrant.
Find the direction.

$\tan\theta = \frac{y}{x}$
$= \frac{-28}{35}$
$= -\frac{4}{5}$
$m\angle\theta = \tan^{-1}\left(-\frac{4}{5}\right)$
≈ -38.7

The distance the freighter traveled is about 44.8 mi, at a direction of about 38.7° south of due east.

54. Let the initial direction of the swimmer be the positive x-direction, and the direction that the current flows by the positive y-direction. The speed and direction of the swimmer can be represented by the vector $\langle 4.5, 0\rangle$, and those of the current can be represented by the vector $\langle 0, 2\rangle$.
Add the vectors.
$\langle 4.5 + 0, 0 + 2\rangle = \langle 4.5, 2\rangle$
Use the Pythagorean Theorem to find the magnitude.
$$|\langle 4.5, 2\rangle| = \sqrt{4.5^2 + 2^2} \approx 4.9$$
The resultant vector lies in the first quadrant. Find the direction.
$$\tan\theta = \frac{y}{x} = \frac{2}{4.5} = \frac{4}{9}$$
$$m\angle\theta = \tan^{-1}\left(\frac{4}{9}\right) \approx 24$$
The swimmer is traveling about 4.9 mph at an angle of 24°.

55. Add the vectors representing the velocities of the wind and jet.
$$\langle 100, 0\rangle + \langle -450, 450\rangle = \langle 100 - 450, 0 + 450\rangle = \langle -350, 450\rangle$$
The resultant vector for the jet is $\langle -350, 450\rangle$ mph.

56. From Exercise 55, the resultant vector is $\langle -350, 450\rangle$.
Use the Pythagorean Theorem to find the magnitude.
$$|\langle -350, 450\rangle| = \sqrt{(-350)^2 + 450^2} \approx 570.1$$
The magnitude of the resultant is about 570.1 mph.

57. From Exercise 55, the resultant vector is $\langle -350, 450\rangle$.
Find the direction.
$$\tan\theta = \frac{y}{x} = \frac{450}{-350} = -\frac{9}{7}$$
$$m\angle\theta = \tan^{-1}\left(-\frac{9}{7}\right) \approx -52.1$$
The direction of the resultant is about 52.1° north of due west.

58. Sample answer: Let one vector be $\langle 1, 0\rangle$. Then the x-components of the other two vectors must sum to -1, the y-components must cancel, and the magnitudes of the other two vectors must be 1. Try $-\frac{1}{2}$ for the x-components. Find the y-components.
$$1 = \sqrt{\left(-\frac{1}{2}\right)^2 + y^2}$$
$$1 = \frac{1}{4} + y^2$$
$$\frac{3}{4} = y^2$$
$$\pm\frac{\sqrt{3}}{2} = y$$
Three vectors with equal magnitude, the sum of which is $\langle 0, 0\rangle$, are $\langle 1, 0\rangle$, $\left\langle -\frac{1}{2}, \frac{\sqrt{3}}{2}\right\rangle$, and $\left\langle -\frac{1}{2}, -\frac{\sqrt{3}}{2}\right\rangle$.

59. Sample answer: Quantities such as velocity are vectors. The velocity of the wind and the velocity of the plane together factor into the overall flight plan. Answers should include the following.
- A wind from the west would add to the velocity contributed by the plane resulting in an overall velocity with a larger magnitude.
- When traveling east, the prevailing winds add to the velocity of the plane. When traveling west, they detract from it.

60. B; find the sum of the vectors.
$$\vec{q} + \vec{r} = \langle 5, 10\rangle + \langle 3, 5\rangle = \langle 8, 15\rangle$$
Use the Pythagorean Theorem to find the magnitude of the vector sum.
$$|\vec{q} + \vec{r}| = \sqrt{8^2 + 15^2} = 17$$
The magnitude is 17.

61. D; $5^b = 125$
$5^b = 5^3$
So, $b = 3$.
$$4^b \times 3 = 4^3 \times 3 = 64 \times 3 = 192$$

Page 505 Maintain Your Skills

62. $AB = 8, r = 2$
Use the Dilation Theorem.
$A'B' = |r|(AB)$
$A'B' = (2)(8)$
$A'B' = 16$

63. $AB = 12, r = \frac{1}{2}$
Use the Dilation Theorem.
$A'B' = |r|(AB)$
$A'B' = \left(\frac{1}{2}\right)(12)$
$A'B' = 6$

64. $A'B' = 15, r = 3$
Use the Dilation Theorem.
$A'B' = |r|(AB)$
$15 = (3)(AB)$
$5 = AB$

65. $A'B' = 12, r = \frac{1}{4}$
Use the Dilation Theorem.
$A'B' = |r|(AB)$
$12 = \left(\frac{1}{4}\right)(AB)$
$48 = AB$

66. Yes; the pattern is a tessellation because at the different vertices the sum of the angles is 360°. The tessellation is uniform because at every vertex there is the same combination of shapes and angles. The tessellation is also semi-regular since more than one regular polygon is used.

67. Yes; the pattern is a tessellation because at the different vertices the sum of the angles is 360°. The tessellation is not uniform because the number of angles at the vertices varies.

68. The opposite angles of the rhombus are congruent.

$\angle WZY \cong \angle WXY$ and $\angle XYZ \cong \angle XWZ$.

$$m\angle XYZ + m\angle XWZ + m\angle WZY + m\angle WXY = 360$$
$$2m\angle XYZ + 2m\angle WZY = 360$$
$$2m\angle XYZ + 2\left(\tfrac{1}{5}m\angle XYZ\right) = 360$$
$$\tfrac{12}{5}m\angle XYZ = 360$$
$$m\angle XYZ = 150$$

69. The measures of all sides of the rhombus are equal. So, $WX = YZ = 12$.

70. $m\angle XYZ = 5m\angle WZY$

$$150 = 5m\angle WZY$$
$$30 = m\angle WZY$$
$$m\angle XZY = \tfrac{1}{2}m\angle WZY$$
$$= \tfrac{1}{2}(30)$$
$$= 15$$

71. The opposite angles of the rhombus are congruent.

$\angle WXY \cong \angle WZY$

$m\angle WXY = m\angle WZY$

$m\angle WXY = 30$ (From Exercise 70, $m\angle WZY = 30$.)

72.

30, 30, 30, 30, x, y, 25°

The diagonals bisect each other perpendicularly. They also bisect the interior angles. The lengths of the diagonals are $2x$ and $2y$.

$2x = 2(30 \cos 25°)$

≈ 54.4

$2y = 2(30 \sin 25°)$

≈ 25.4

To the nearest tenth, the lengths of the diagonals are 25.4 cm and 54.4 cm.

73. $\begin{bmatrix} -5 & 5 \\ -3 & -2 \end{bmatrix} + \begin{bmatrix} 1 & -8 \\ -7 & 6 \end{bmatrix} = \begin{bmatrix} -5+1 & 5-8 \\ -3-7 & -2+6 \end{bmatrix}$

$= \begin{bmatrix} -4 & -3 \\ -10 & 4 \end{bmatrix}$

74. $\begin{bmatrix} -2 & 2 & -2 \\ -7 & -2 & -5 \end{bmatrix} + \begin{bmatrix} -8 & -8 & -8 \\ 1 & 1 & 1 \end{bmatrix}$

$= \begin{bmatrix} -2-8 & 2-8 & -2-8 \\ -7+1 & -2+1 & -5+1 \end{bmatrix}$

$= \begin{bmatrix} -10 & -6 & -10 \\ -6 & -1 & -4 \end{bmatrix}$

75. $3\begin{bmatrix} -9 & -5 & -1 \\ 9 & 1 & 5 \end{bmatrix} = \begin{bmatrix} 3(-9) & 3(-5) & 3(-1) \\ 3(9) & 3(1) & 3(5) \end{bmatrix}$

$= \begin{bmatrix} -27 & -15 & -3 \\ 27 & 3 & 15 \end{bmatrix}$

76. $\frac{1}{2}\begin{bmatrix} -4 & -5 & 0 & 2 \\ 4 & 4 & 6 & 0 \end{bmatrix} = \begin{bmatrix} \frac{1}{2}(-4) & \frac{1}{2}(-5) & \frac{1}{2}(0) & \frac{1}{2}(2) \\ \frac{1}{2}(4) & \frac{1}{2}(4) & \frac{1}{2}(6) & \frac{1}{2}(0) \end{bmatrix}$

$= \begin{bmatrix} -2 & -2.5 & 0 & 1 \\ 2 & 2 & 3 & 0 \end{bmatrix}$

77. $\begin{bmatrix} -4 & -4 \\ 2 & 2 \end{bmatrix} + 2\begin{bmatrix} 8 & 4 \\ -3 & -7 \end{bmatrix}$

$= \begin{bmatrix} -4 & -4 \\ 2 & 2 \end{bmatrix} + \begin{bmatrix} 2(8) & 2(4) \\ 2(-3) & 2(-7) \end{bmatrix}$

$= \begin{bmatrix} -4 & -4 \\ 2 & 2 \end{bmatrix} + \begin{bmatrix} 16 & 8 \\ -6 & -14 \end{bmatrix}$

$= \begin{bmatrix} -4+16 & -4+8 \\ 2-6 & 2-14 \end{bmatrix}$

$= \begin{bmatrix} 12 & 4 \\ -4 & -12 \end{bmatrix}$

78. $\begin{bmatrix} 1 & -1 \\ -1 & 1 \end{bmatrix} \cdot \begin{bmatrix} 2 & -3 \\ -2 & -4 \end{bmatrix}$

$= \begin{bmatrix} 1(2)+(-1)(-2) & 1(-3)+(-1)(-4) \\ -1(2)+1(-2) & -1(-3)+1(-4) \end{bmatrix}$

$= \begin{bmatrix} 2+2 & -3+4 \\ -2-2 & 3-4 \end{bmatrix}$

$= \begin{bmatrix} 4 & 1 \\ -4 & -1 \end{bmatrix}$

9-7 Transformations with Matrices

Pages 508–509 Check for Understanding

1. $A(3, 3) = A'(3, 3)$

$B(4, 1)$ and $B'(1, 4)$ imply $(x, y) \rightarrow (y, x)$.

$C(-1, 1)$ and $C'(1, -1)$ imply $(x, y) \rightarrow (y, x)$.

$(x, y) \rightarrow (y, x)$ occurs with reflection in the line $y = x$.

The reflection matrix is $\begin{bmatrix} 0 & 1 \\ 1 & 0 \end{bmatrix}$.

2. Sample answer: The format used to represent the transformation is different in each method, but the result is the same. Positive values move a figure up or right, and negative values move a figure down or left.

3. Sample answer: $\begin{bmatrix} -2 & -2 & -2 & -2 \\ -1 & -1 & -1 & -1 \end{bmatrix}$

4. The vertex matrix for $\triangle ABC$ is $\begin{bmatrix} 5 & 3 & 0 \\ 4 & -1 & 2 \end{bmatrix}$.

The translation matrix is $\begin{bmatrix} -2 & -2 & -2 \\ -1 & -1 & -1 \end{bmatrix}$.

Find the vertex matrix for the image.

$$\begin{bmatrix} 5 & 3 & 0 \\ 4 & -1 & 2 \end{bmatrix} + \begin{bmatrix} -2 & -2 & -2 \\ -1 & -1 & -1 \end{bmatrix} = \begin{bmatrix} 3 & 1 & -2 \\ 3 & -2 & 1 \end{bmatrix}.$$

The coordinates of the vertices of $\triangle A'B'C'$ are $A'(3, 3)$, $B'(1, -2)$, and $C'(-2, 1)$.

5. The vertex matrix for ▱$DEFG$ is $\begin{bmatrix} -1 & 5 & 3 & -3 \\ 3 & 3 & 0 & 0 \end{bmatrix}$.

The translation matrix is $\begin{bmatrix} 0 & 0 & 0 & 0 \\ 6 & 6 & 6 & 6 \end{bmatrix}$.

Find the vertex matrix for the image.

$$\begin{bmatrix} -1 & 5 & 3 & -3 \\ 3 & 3 & 0 & 0 \end{bmatrix} + \begin{bmatrix} 0 & 0 & 0 & 0 \\ 6 & 6 & 6 & 6 \end{bmatrix} = \begin{bmatrix} -1 & 5 & 13 & -3 \\ 9 & 9 & 6 & 6 \end{bmatrix}$$

The coordinates of the vertices of the image are $D'(-1, 9)$, $E'(5, 9)$, $F'(3, 6)$, and $G'(-3, 6)$.

6. The vertex matrix for $\triangle XYZ$ is $\begin{bmatrix} 3 & 6 & -3 \\ 4 & 10 & 5 \end{bmatrix}$.

Multiply the vertex matrix by the scale factor to find the vertex matrix of the image.

$$2\begin{bmatrix} 3 & 6 & -3 \\ 4 & 10 & 5 \end{bmatrix} = \begin{bmatrix} 6 & 12 & -6 \\ 8 & 20 & 10 \end{bmatrix}$$

The coordinates of the vertices of the image are $X'(6, 8)$, $Y'(12, 20)$, and $Z'(-6, 10)$.

7. The vertex matrix for ▱$ABCD$ is $\begin{bmatrix} 1 & 3 & 3 & 1 \\ 2 & 3 & 5 & 4 \end{bmatrix}$.

Multiply the vertex matrix by the scale factor to find the vertex matrix of the image.

$$-\frac{1}{4}\begin{bmatrix} 1 & 3 & 3 & 1 \\ 2 & 3 & 5 & 4 \end{bmatrix} = \begin{bmatrix} -\frac{1}{4} & -\frac{3}{4} & -\frac{3}{4} & -\frac{1}{4} \\ -\frac{1}{2} & -\frac{3}{4} & -\frac{5}{4} & -1 \end{bmatrix}$$

The coordinates of the vertices of the image are $A'\left(-\frac{1}{4}, -\frac{1}{2}\right)$, $B'\left(-\frac{3}{4}, -\frac{3}{4}\right)$, $C'\left(-\frac{3}{4}, -\frac{5}{4}\right)$, and $D'\left(-\frac{1}{4}, -1\right)$.

8. The reflection matrix for a reflection in the x-axis is $\begin{bmatrix} 1 & 0 \\ 0 & -1 \end{bmatrix}$.

The vertex matrix for $\overline{EF}$ is $\begin{bmatrix} -2 & 5 \\ 4 & 1 \end{bmatrix}$.

Multiply the vertex matrix for $\overline{EF}$ by the reflection matrix to find the vertex matrix of the image.

$$\begin{bmatrix} 1 & 0 \\ 0 & -1 \end{bmatrix} \cdot \begin{bmatrix} -2 & 5 \\ 4 & 1 \end{bmatrix} = \begin{bmatrix} -2 & 5 \\ -4 & -1 \end{bmatrix}$$

The coordinates of the vertices of the image are $E'(-2, -4)$ and $F'(5, -1)$.

9. The reflection matrix for a reflection in the y-axis is $\begin{bmatrix} -1 & 0 \\ 0 & 1 \end{bmatrix}$.

The vertex matrix for quadrilateral $HIJK$ is $\begin{bmatrix} -5 & -1 & -3 & -7 \\ 4 & -1 & -6 & -3 \end{bmatrix}$.

Multiply the vertex matrix for quadrilateral $HIJK$ by the reflection matrix to find the vertex matrix of the image.

$$\begin{bmatrix} -1 & 0 \\ 0 & 1 \end{bmatrix} \cdot \begin{bmatrix} -5 & -1 & -3 & -7 \\ 4 & -1 & -6 & -3 \end{bmatrix} = \begin{bmatrix} 5 & 1 & 3 & 7 \\ 4 & -1 & -6 & -3 \end{bmatrix}$$

The coordinates of the vertices of the image are $H'(5, 4)$, $I'(1, -1)$, $J'(3, -6)$, and $K'(7, -3)$.

10. The rotation matrix for a counterclockwise rotation of 90° is $\begin{bmatrix} 0 & -1 \\ 1 & 0 \end{bmatrix}$.

The vertex matrix for $\overline{LM}$ is $\begin{bmatrix} -2 & 3 \\ 1 & 5 \end{bmatrix}$.

Multiply the vertex matrix for $\overline{LM}$ by the rotation matrix to find the vertex matrix of the image.

$$\begin{bmatrix} 0 & -1 \\ 1 & 0 \end{bmatrix} \cdot \begin{bmatrix} -2 & 3 \\ 1 & 5 \end{bmatrix} = \begin{bmatrix} -1 & -5 \\ -2 & 3 \end{bmatrix}$$

The coordinates of the vertices of the image are $L'(-1, -2)$ and $M'(-5, 3)$.

11. The rotation matrix for a counterclockwise rotation of 270° is $\begin{bmatrix} 0 & 1 \\ -1 & 0 \end{bmatrix}$.

The vertex matrix for $\triangle PQR$ is $\begin{bmatrix} 6 & 6 & 2 \\ 3 & 7 & 7 \end{bmatrix}$.

Multiply the vertex matrix for $\triangle PQR$ by the rotation matrix to find the vertex matrix of the image.

$$\begin{bmatrix} 0 & 1 \\ -1 & 0 \end{bmatrix} \cdot \begin{bmatrix} 6 & 6 & 2 \\ 3 & 7 & 7 \end{bmatrix} = \begin{bmatrix} 3 & 7 & 7 \\ -6 & -6 & -2 \end{bmatrix}$$

The coordinates of the vertices of the image are $P'(3, -6)$, $Q'(7, -6)$, and $R'(7, -2)$.

12. The vertex matrix for quadrilateral $STUV$ is $\begin{bmatrix} -4 & -2 & 0 & -2 \\ 1 & 2 & 1 & -2 \end{bmatrix}$.

Multiply the vertex matrix by the scale factor to find the vertex matrix of the image.

$$2\begin{bmatrix} -4 & -2 & 0 & -2 \\ 1 & 2 & 1 & -2 \end{bmatrix} = \begin{bmatrix} -8 & -4 & 0 & -4 \\ 2 & 4 & 2 & -4 \end{bmatrix}$$

The rotation matrix for a counterclockwise rotation of 90° is $\begin{bmatrix} 0 & -1 \\ 1 & 0 \end{bmatrix}$.

Multiply the vertex matrix of the image due to dilation to find the vertex matrix of the final image.

$$\begin{bmatrix} 0 & -1 \\ 1 & 0 \end{bmatrix} \cdot \begin{bmatrix} -8 & -4 & 0 & -4 \\ 2 & 4 & 2 & -4 \end{bmatrix} = \begin{bmatrix} -2 & -4 & -2 & 4 \\ -8 & -4 & 0 & -4 \end{bmatrix}$$

The coordinates of the vertices of the image are $S'(-2, -8)$, $T'(-4, -4)$, $U'(-2, 0)$ and $V'(4, -4)$.

13. The rose bed must be dilated by a scale factor of $\frac{1}{2}$.

The vertex matrix of the rose bed is $\begin{bmatrix} 3 & 7 & 5 & 1 \\ -1 & -3 & -7 & -5 \end{bmatrix}$.

Multiply the vertex matrix by the scale factor to find the vertex matrix of the new rose bed plan.

$$\frac{1}{2}\begin{bmatrix} 3 & 7 & 5 & 1 \\ -1 & -3 & -7 & -5 \end{bmatrix} = \begin{bmatrix} \frac{3}{2} & \frac{7}{2} & \frac{5}{2} & \frac{1}{2} \\ -\frac{1}{2} & -\frac{3}{2} & -\frac{7}{2} & -\frac{5}{2} \end{bmatrix}$$

The new coordinates are (1.5, −0.5), (3.5, −1.5), (2.5, −3.5), and (0.5, −2.5).

14. All dimensions have been reduced by $\frac{1}{2}$, so the coordinates of the center after the changes have been made will be $\frac{1}{2}(4, -4) = (2, -2)$.

Pages 509–511 Practice and Apply

15. The vertex matrix for $\overline{EF}$ is $\begin{bmatrix} -4 & -1 \\ 1 & 3 \end{bmatrix}$.

The translation matrix is $\begin{bmatrix} -2 & -2 \\ 5 & 5 \end{bmatrix}$.

Find the vertex matrix for the image.

$$\begin{bmatrix} -4 & -1 \\ 1 & 3 \end{bmatrix} + \begin{bmatrix} -2 & -2 \\ 5 & 5 \end{bmatrix} = \begin{bmatrix} -6 & -3 \\ 6 & 8 \end{bmatrix}$$

The coordinates of the vertices of the image are $E'(-6, 6)$ and $F'(-3, 8)$.

16. The vertex matrix for $\triangle JKL$ is $\begin{bmatrix} -3 & 4 & 7 \\ 5 & 8 & 5 \end{bmatrix}$.

The translation matrix is $\begin{bmatrix} -3 & -3 & -3 \\ -4 & -4 & -4 \end{bmatrix}$.

Find the vertex matrix for the image.

$$\begin{bmatrix} -3 & 4 & 7 \\ 5 & 8 & 5 \end{bmatrix} + \begin{bmatrix} -3 & -3 & -3 \\ -4 & -4 & -4 \end{bmatrix} = \begin{bmatrix} -6 & 1 & 4 \\ 1 & 4 & 1 \end{bmatrix}$$

The coordinates of the vertices of the image are $J'(-6, 1)$, $K'(1, 4)$, and $L'(4, 1)$.

17. The vertex matrix for $\square MNOP$ is $\begin{bmatrix} -2 & 2 & 2 & -2 \\ 7 & 9 & 7 & 5 \end{bmatrix}$.

The translation matrix is $\begin{bmatrix} 3 & 3 & 3 & 3 \\ -6 & -6 & -6 & -6 \end{bmatrix}$.

Find the vertex matrix for the image.

$$\begin{bmatrix} -2 & 2 & 2 & -2 \\ 7 & 9 & 7 & 5 \end{bmatrix} + \begin{bmatrix} 3 & 3 & 3 & 3 \\ -6 & -6 & -6 & -6 \end{bmatrix} = \begin{bmatrix} 1 & 5 & 5 & 1 \\ 1 & 3 & 1 & -1 \end{bmatrix}$$

The coordinates of the vertices of the image are $M'(1, 1)$, $N'(5, 3)$, $O'(5, 1)$, and $P'(1, -1)$.

18. The vertex matrix for trapezoid $RSTU$ is $\begin{bmatrix} 2 & 6 & 6 & -2 \\ 3 & 2 & -1 & 1 \end{bmatrix}$.

The translation matrix is $\begin{bmatrix} -6 & -6 & -6 & -6 \\ -2 & -2 & -2 & -2 \end{bmatrix}$.

Find the vertex matrix for the image.

$$\begin{bmatrix} 2 & 6 & 6 & -2 \\ 3 & 2 & -1 & 1 \end{bmatrix} + \begin{bmatrix} -6 & -6 & -6 & -6 \\ -2 & -2 & -2 & -2 \end{bmatrix} = \begin{bmatrix} -4 & 0 & 0 & -8 \\ 1 & 0 & -3 & -1 \end{bmatrix}$$

The coordinates of the vertices of the image are $R'(-4, 1)$, $S'(0, 0)$, $T'(0, -3)$, and $U'(-8, -1)$.

19. The vertex matrix for $\triangle ABC$ is $\begin{bmatrix} 6 & 4 & 3 \\ 5 & 5 & 7 \end{bmatrix}$.

Multiply the vertex matrix by the scale factor to find the vertex matrix of the image.

$$2\begin{bmatrix} 6 & 4 & 3 \\ 5 & 5 & 7 \end{bmatrix} = \begin{bmatrix} 12 & 8 & 6 \\ 10 & 10 & 14 \end{bmatrix}$$

The coordinates of the vertices of the image are $A'(12, 10)$, $B'(8, 10)$, and $C'(6, 14)$.

20. The vertex matrix for $\triangle DEF$ is $\begin{bmatrix} -1 & 0 & 2 \\ 4 & 1 & 3 \end{bmatrix}$.

Multiply the vertex matrix by the scale factor to find the vertex matrix of the image.

$$-\frac{1}{3}\begin{bmatrix} -1 & 0 & 2 \\ 4 & 1 & 3 \end{bmatrix} = \begin{bmatrix} \frac{1}{3} & 0 & -\frac{2}{3} \\ -\frac{4}{3} & -\frac{1}{3} & -1 \end{bmatrix}$$

The coordinates of the vertices of the image are $D'\left(\frac{1}{3}, -\frac{4}{3}\right)$, $E'\left(0, -\frac{1}{3}\right)$, and $F'\left(-\frac{2}{3}, -1\right)$.

21. The vertex matrix for quadrilateral $GHIJ$ is $\begin{bmatrix} 4 & -4 & -6 & 6 \\ 2 & 6 & -8 & -10 \end{bmatrix}$.

Multiply the vertex matrix by the scale factor to find the vertex matrix of the image.

$$-\frac{1}{2}\begin{bmatrix} 4 & -4 & -6 & 6 \\ 2 & 6 & -8 & -10 \end{bmatrix} = \begin{bmatrix} -2 & 2 & 3 & -3 \\ -1 & -3 & 4 & 5 \end{bmatrix}$$

The coordinates of the vertices of the image are $G'(-2, -1)$, $H'(2, -3)$, $I'(3, 4)$, and $J'(-3, 5)$.

22. The vertex matrix for pentagon $KLMNO$ is $\begin{bmatrix} 1 & 3 & 6 & 4 & 3 \\ -2 & -1 & -1 & -3 & -3 \end{bmatrix}$.

Multiply the vertex matrix by the scale factor to find the vertex matrix of the image.

$$4\begin{bmatrix} 1 & 3 & 6 & 4 & 3 \\ -2 & -1 & -1 & -3 & -3 \end{bmatrix}$$

$$= \begin{bmatrix} 4 & 12 & 24 & 16 & 12 \\ -8 & -4 & -4 & -12 & -12 \end{bmatrix}$$

The coordinates of the vertices of the image are $K'(4, -8)$, $L'(12, -4)$, $M'(24, -4)$, $N'(16, -12)$, and $O'(12, -12)$.

23. The reflection matrix for a reflection in the y-axis is $\begin{bmatrix} -1 & 0 \\ 0 & 1 \end{bmatrix}$.

The vertex matrix for $\overline{XY}$ is $\begin{bmatrix} 2 & 4 \\ 2 & -1 \end{bmatrix}$.

Multiply the vertex matrix for $\overline{XY}$ by the reflection matrix to find the vertex matrix of the image.

$$\begin{bmatrix}-1 & 0\\ 0 & 1\end{bmatrix}\cdot\begin{bmatrix}2 & 4\\ 2 & -1\end{bmatrix}=\begin{bmatrix}-2 & -4\\ 2 & -1\end{bmatrix}$$

The coordinates of the vertices of the image are $X'(-2, 2)$, and $Y'(-4, -1)$.

24. The reflection matrix for a reflection in the line $y = x$ is $\begin{bmatrix}0 & 1\\ 1 & 0\end{bmatrix}$.

The vertex matrix for $\triangle ABC$ is $\begin{bmatrix}5 & 0 & -1\\ -3 & -5 & -3\end{bmatrix}$.

Multiply the vertex matrix for $\triangle ABC$ by the reflection matrix to find the vertex matrix of the image.

$$\begin{bmatrix}0 & 1\\ 1 & 0\end{bmatrix}\cdot\begin{bmatrix}5 & 0 & -1\\ -3 & -5 & -3\end{bmatrix}=\begin{bmatrix}-3 & -5 & -3\\ 5 & 0 & -1\end{bmatrix}$$

The coordinates of the vertices of the image are $A'(-3, 5)$, $B'(-5, 0)$, and $C'(-3, -1)$.

25. The reflection matrix for a reflection in the x-axis is $\begin{bmatrix}1 & 0\\ 0 & -1\end{bmatrix}$.

The vertex matrix for quadrilateral $DEFG$ is $\begin{bmatrix}-4 & 2 & 3 & -3\\ 5 & 6 & 1 & -4\end{bmatrix}$.

Multiply the vertex matrix for quadrilateral $DEFG$ by the reflection matrix to find the vertex matrix of the image.

$$\begin{bmatrix}1 & 0\\ 0 & -1\end{bmatrix}\cdot\begin{bmatrix}-4 & 2 & 3 & -3\\ 5 & 6 & 1 & -4\end{bmatrix}=\begin{bmatrix}-4 & 2 & 3 & -3\\ -5 & -6 & -1 & 4\end{bmatrix}$$

The coordinates of the vertices of the image are $D'(-4, -5)$, $E'(2, -6)$, $F'(3, -1)$, and $G'(-3, 4)$.

26. The reflection matrix for a reflection in the y-axis is $\begin{bmatrix}-1 & 0\\ 0 & 1\end{bmatrix}$.

The vertex matrix for quadrilateral $HIJK$ is $\begin{bmatrix}9 & 2 & -4 & -2\\ -1 & -6 & -3 & 4\end{bmatrix}$.

Multiply the vertex matrix for quadrilateral $HIJK$ by the reflection matrix to find the vertex matrix of the image.

$$\begin{bmatrix}-1 & 0\\ 0 & 1\end{bmatrix}\cdot\begin{bmatrix}9 & 2 & -4 & -2\\ -1 & -6 & -3 & 4\end{bmatrix}=\begin{bmatrix}-9 & -2 & 4 & 2\\ -1 & -6 & -3 & 4\end{bmatrix}$$

The coordinates of the vertices of the image are $H'(-9, -1)$, $I'(-2, -6)$, $J'(4, -3)$, and $K'(2, 4)$.

27. The vertex matrix for $\triangle VWX$ is $\begin{bmatrix}-3 & 1 & 3\\ 3 & 3 & -2\end{bmatrix}$.

Multiply the vertex matrix by the scale factor to find the vertex matrix of the image.

$$\frac{2}{3}\begin{bmatrix}-3 & 1 & 3\\ 3 & 3 & -2\end{bmatrix}=\begin{bmatrix}-2 & \frac{2}{3} & 2\\ 2 & 2 & -\frac{4}{3}\end{bmatrix}$$

The coordinates of the vertices of the image are $V'(-2, 2)$, $W'\left(\frac{2}{3}, 2\right)$, and $X'\left(2, -\frac{4}{3}\right)$.

28. The vertex matrix for $\triangle VWX$ is $\begin{bmatrix}-3 & 1 & 3\\ 3 & 3 & -2\end{bmatrix}$.

The translation matrix is $\begin{bmatrix}-4 & -4 & -4\\ -1 & -1 & -1\end{bmatrix}$.

Find the vertex matrix for the image.

$$\begin{bmatrix}-3 & 1 & 3\\ 3 & 3 & -2\end{bmatrix}+\begin{bmatrix}-4 & -4 & -4\\ -1 & -1 & -1\end{bmatrix}=\begin{bmatrix}-7 & -3 & -1\\ 2 & 2 & -3\end{bmatrix}$$

The coordinates of the vertices of the image are $V'(-7, 2)$, $W'(-3, 2)$, and $X'(-1, -3)$.

29. The rotation matrix for a counterclockwise rotation of 90° is $\begin{bmatrix}0 & -1\\ 1 & 0\end{bmatrix}$.

The vertex matrix for $\triangle VWX$ is $\begin{bmatrix}-3 & 1 & 3\\ 3 & 3 & -2\end{bmatrix}$.

Multiply the vertex matrix for $\triangle VWX$ by the rotation matrix to find the vertex matrix of the image.

$$\begin{bmatrix}0 & -1\\ 1 & 0\end{bmatrix}\cdot\begin{bmatrix}-3 & 1 & 3\\ 3 & 3 & -2\end{bmatrix}=\begin{bmatrix}-3 & -3 & 2\\ -3 & 1 & 3\end{bmatrix}$$

The coordinates of the vertices of the image are $V'(-3, -3)$, $W'(-3, 1)$, and $X'(2, 3)$.

30. The reflection matrix for a reflection in the line $y = x$ is $\begin{bmatrix}0 & 1\\ 1 & 0\end{bmatrix}$.

The vertex matrix for $\triangle VWX$ is $\begin{bmatrix}-3 & 1 & 3\\ 3 & 3 & -2\end{bmatrix}$.

Mutiply the vertex matrix for $\triangle VWX$ by the reflection matrix to find the vertex matrix of the image.

$$\begin{bmatrix}0 & 1\\ 1 & 0\end{bmatrix}\cdot\begin{bmatrix}-3 & 1 & 3\\ 3 & 3 & -2\end{bmatrix}=\begin{bmatrix}3 & 3 & -2\\ -3 & 1 & 3\end{bmatrix}$$

The coordinates of the vertices of the image are $V'(3, -3)$, $W'(3, 1)$, and $X'(-2, 3)$.

31. The vertex matrix for polygon $PQRST$ is $\begin{bmatrix}-1 & -4 & -2 & 0 & 2\\ -1 & 1 & 4 & 4 & 1\end{bmatrix}$.

The translation matrix is $\begin{bmatrix}3 & 3 & 3 & 3 & 3\\ -2 & -2 & -2 & -2 & -2\end{bmatrix}$.

Find the vertex matrix for the image.

$$\begin{bmatrix}-1 & -4 & -2 & 0 & 2\\ -1 & 1 & 4 & 4 & 1\end{bmatrix}+\begin{bmatrix}3 & 3 & 3 & 3 & 3\\ -2 & -2 & -2 & -2 & -2\end{bmatrix}$$

$$=\begin{bmatrix}2 & -1 & 1 & 3 & 5\\ -3 & -1 & 2 & 2 & -1\end{bmatrix}$$

The coordinates of the vertices of the image are $P'(2, -3)$, $Q'(-1, -1)$, $R'(1, 2)$, $S'(3, 2)$, and $T'(5, -1)$.

32. The vertex matrix for polygon $PQRST$ is $\begin{bmatrix}-1 & -4 & -2 & 0 & 2\\ -1 & 1 & 4 & 4 & 1\end{bmatrix}$.

Multiply the vertex matrix by the scale factor to find the vertex matrix of the image.

$$-3\begin{bmatrix}-1 & -4 & -2 & 0 & 2\\ -1 & 1 & 4 & 4 & 1\end{bmatrix}=\begin{bmatrix}3 & 12 & 6 & 0 & -6\\ 3 & -3 & -12 & -12 & -3\end{bmatrix}$$

The coordinates of the vertices of the image are $P'(3, 3)$, $Q'(12, -3)$, $R'(6, -12)$, $S'(0, -12)$, and $T'(-6, -3)$.

33. The reflection matrix for a reflection in the y-axis is $\begin{bmatrix} -1 & 0 \\ 0 & 1 \end{bmatrix}$.

The vertex matrix for polygon $PQRST$ is $\begin{bmatrix} -1 & -4 & -2 & 0 & 2 \\ -1 & 1 & 4 & 4 & 1 \end{bmatrix}$.

Multiply the vertex matrix for $PQRST$ by the reflection matrix to find the vertex matrix of the image.

$$\begin{bmatrix} -1 & 0 \\ 0 & 1 \end{bmatrix} \cdot \begin{bmatrix} -1 & -4 & -2 & 0 & 2 \\ -1 & 1 & 4 & 4 & 1 \end{bmatrix}$$

$$= \begin{bmatrix} 1 & 4 & 2 & 0 & -2 \\ -1 & 1 & 4 & 4 & 1 \end{bmatrix}$$

The coordinates of the vertices of the image are $P'(1, -1)$, $Q'(4, 1)$, $R'(2, 4)$, $S'(0, 4)$, and $T'(-2, 1)$.

34. The rotation matrix for a counterclockwise rotation of 180° is $\begin{bmatrix} -1 & 0 \\ 0 & -1 \end{bmatrix}$.

The vertex matrix for polygon $PQRST$ is $\begin{bmatrix} -1 & -4 & -2 & 0 & 2 \\ -1 & 1 & 4 & 4 & 1 \end{bmatrix}$.

Multiply the vertex matrix for $PQRST$ by the rotation matrix to find the vertex matrix of the image.

$$\begin{bmatrix} -1 & 0 \\ 0 & -1 \end{bmatrix} \cdot \begin{bmatrix} -1 & -4 & -2 & 0 & 2 \\ -1 & 1 & 4 & 4 & 1 \end{bmatrix}$$

$$= \begin{bmatrix} 1 & 4 & 2 & 0 & -2 \\ 1 & -1 & -4 & -4 & -1 \end{bmatrix}$$

The coordinates of the vertices of the image are $P'(1, 1)$, $Q'(4, -1)$, $R'(2, -4)$, $S'(0, -4)$, and $T'(-2, -1)$.

35. The rotation matrix for a counterclockwise rotation of 90° is $\begin{bmatrix} 0 & -1 \\ 1 & 0 \end{bmatrix}$.

The vertex matrix for $\overline{MN}$ is $\begin{bmatrix} 12 & -3 \\ 1 & 10 \end{bmatrix}$.

Multiply the vertex matrix for $\overline{MN}$ by the rotation matrix to find the vertex matrix of the image.

$$\begin{bmatrix} 0 & -1 \\ 1 & 0 \end{bmatrix} \cdot \begin{bmatrix} 12 & -3 \\ 1 & 10 \end{bmatrix} = \begin{bmatrix} -1 & -10 \\ 12 & -3 \end{bmatrix}$$

The coordinates of the vertices of the image are $M'(-1, 12)$, and $N'(-10, -3)$.

36. The rotation matrix for a counterclockwise rotation of 180° is $\begin{bmatrix} -1 & 0 \\ 0 & -1 \end{bmatrix}$.

The vertex matrix for $\triangle PQR$ is $\begin{bmatrix} 5 & 1 & 1 \\ 1 & 2 & -4 \end{bmatrix}$.

Multiply the vertex matrix for $\triangle PQR$ by the rotation matrix to find the vertex matrix of the image.

$$\begin{bmatrix} -1 & 0 \\ 0 & -1 \end{bmatrix} \cdot \begin{bmatrix} 5 & 1 & 1 \\ 1 & 2 & -4 \end{bmatrix} = \begin{bmatrix} -5 & -1 & -1 \\ -1 & -2 & 4 \end{bmatrix}$$

The coordinates of the vertices of the image are $P'(-5, -1)$, $Q'(-1, -2)$, and $R'(-1, 4)$.

37. The rotation matrix for a counterclockwise rotation of 90° is $\begin{bmatrix} 0 & -1 \\ 1 & 0 \end{bmatrix}$.

The vertex matrix for ▱$STUV$ is $\begin{bmatrix} 2 & 6 & 5 & 1 \\ 1 & 1 & -3 & -3 \end{bmatrix}$.

Multiply the vertex matrix for ▱$STUV$ by the rotation matrix to find the vertex matrix of the iamge.

$$\begin{bmatrix} 0 & -1 \\ 1 & 0 \end{bmatrix} \cdot \begin{bmatrix} 2 & 6 & 5 & 1 \\ 1 & 1 & -3 & -3 \end{bmatrix} = \begin{bmatrix} -1 & -1 & 3 & 3 \\ 2 & 6 & 5 & 1 \end{bmatrix}$$

The coordinates of the vertices of the image are $S'(-1, 2)$, $T'(-1, 6)$, $U'(3, 5)$, and $V'(3, 1)$.

38. The rotation matrix for a counterclockwise rotation of 270° is $\begin{bmatrix} 0 & 1 \\ -1 & 0 \end{bmatrix}$.

The vertex matrix for pentagon $ABCDE$ is $\begin{bmatrix} -1 & 6 & 4 & -4 & -5 \\ 1 & 0 & -8 & -10 & -3 \end{bmatrix}$.

Multiply the vertex matrix for pentagon $ABCDE$ by the rotation matrix to find the vertex matrix of the image.

$$\begin{bmatrix} 0 & 1 \\ -1 & 0 \end{bmatrix} \cdot \begin{bmatrix} -1 & 6 & 4 & -4 & -5 \\ 1 & 0 & -8 & -10 & -3 \end{bmatrix}$$

$$= \begin{bmatrix} 1 & 0 & -8 & -10 & -3 \\ 1 & -6 & -4 & 4 & 5 \end{bmatrix}$$

The coordinates of the vertices of the image are $A'(1, 1)$, $B'(0, -6)$, $C'(-8, -4)$, $D'(-10, 4)$, and $E'(-3, 5)$.

39. The vertex matrix for polygon $ABCDEF$ is $\begin{bmatrix} -3 & -2 & 2 & 3 & 2 & -2 \\ 1 & 4 & 4 & 1 & -2 & -2 \end{bmatrix}$.

Multiply the vertex matrix by the scale factor to find the vertex matrix of the image.

$$\frac{1}{3}\begin{bmatrix} -3 & -2 & 2 & 3 & 2 & -2 \\ 1 & 4 & 4 & 1 & -2 & -2 \end{bmatrix}$$

$$= \begin{bmatrix} -1 & -\frac{2}{3} & \frac{2}{3} & 1 & \frac{2}{3} & -\frac{2}{3} \\ \frac{1}{3} & \frac{4}{3} & \frac{4}{3} & \frac{1}{3} & -\frac{2}{3} & -\frac{2}{3} \end{bmatrix}$$

The reflection matrix for a reflection in the x-axis is $\begin{bmatrix} 1 & 0 \\ 0 & -1 \end{bmatrix}$.

Multiply the vertex matrix of the image by the reflection matrix to find the vertex matrix of the final image.

$$\begin{bmatrix} 1 & 0 \\ 0 & -1 \end{bmatrix} \cdot \begin{bmatrix} -1 & -\frac{2}{3} & \frac{2}{3} & 1 & \frac{2}{3} & -\frac{2}{3} \\ \frac{1}{3} & \frac{4}{3} & \frac{4}{3} & \frac{1}{3} & -\frac{2}{3} & -\frac{2}{3} \end{bmatrix}$$

$$= \begin{bmatrix} -1 & -\frac{2}{3} & \frac{2}{3} & 1 & \frac{2}{3} & -\frac{2}{3} \\ -\frac{1}{3} & -\frac{4}{3} & -\frac{4}{3} & -\frac{1}{3} & \frac{2}{3} & \frac{2}{3} \end{bmatrix}$$

The coordinates of the vertices of the final image are $A'\left(-1, -\frac{1}{3}\right)$, $B'\left(-\frac{2}{3}, -\frac{4}{3}\right)$, $C'\left(\frac{2}{3}, -\frac{4}{3}\right)$, $D'\left(1, -\frac{1}{3}\right)$, $E'\left(\frac{2}{3}, \frac{2}{3}\right)$, and $F'\left(-\frac{2}{3}, \frac{2}{3}\right)$.

40. The vertex matrix for polygon $ABCDEF$ is

$$\begin{bmatrix} -3 & -2 & 2 & 3 & 2 & -2 \\ 1 & 4 & 4 & 1 & -2 & -2 \end{bmatrix}.$$

The translation matrix is

$$\begin{bmatrix} -5 & -5 & -5 & -5 & -5 & -5 \\ 2 & 2 & 2 & 2 & 2 & 2 \end{bmatrix}.$$

Find the vertex matrix for the image.

$$\begin{bmatrix} -3 & -2 & 2 & 3 & 2 & -2 \\ 1 & 4 & 4 & 1 & -2 & -2 \end{bmatrix} + \begin{bmatrix} -5 & -5 & -5 & -5 & -5 & -5 \\ 2 & 2 & 2 & 2 & 2 & 2 \end{bmatrix}$$

$$= \begin{bmatrix} -8 & -7 & -3 & -2 & -3 & -7 \\ 3 & 6 & 6 & 3 & 0 & 0 \end{bmatrix}$$

The rotation matrix for a counterclockwise rotation of 90° is $\begin{bmatrix} 0 & -1 \\ 1 & 0 \end{bmatrix}$.

Multiply the vertex matrix of the image by the rotation matrix to find the vertex matrix of the final image.

$$\begin{bmatrix} 0 & -1 \\ 1 & 0 \end{bmatrix} \cdot \begin{bmatrix} -8 & -7 & -3 & -2 & -3 & -7 \\ 3 & 6 & 6 & 3 & 0 & 0 \end{bmatrix}$$

$$= \begin{bmatrix} -3 & -6 & -6 & -3 & 0 & 0 \\ -8 & -7 & -3 & -2 & -3 & -7 \end{bmatrix}$$

The coordinates of the vertices of the final image are $A'(-3, -8)$, $B'(-6, -7)$, $C'(-6, -3)$, $D'(-3, -2)$, $E'(0, -3)$, and $F'(0, -7)$.

41. The reflection matrix for a reflection in the line $y = x$ is $\begin{bmatrix} 0 & 1 \\ 1 & 0 \end{bmatrix}$.

The vertex matrix for polygon $ABCDEF$ is

$$\begin{bmatrix} -3 & -2 & 2 & 3 & 2 & -2 \\ 1 & 4 & 4 & 1 & -2 & -2 \end{bmatrix}.$$

Multiply the vertex matrix for $ABCDEF$ by the reflection matrix to find the vertex matrix of the image.

$$\begin{bmatrix} 0 & 1 \\ 1 & 0 \end{bmatrix} \cdot \begin{bmatrix} -3 & -2 & 2 & 3 & 2 & -2 \\ 1 & 4 & 4 & 1 & -2 & -2 \end{bmatrix}$$

$$= \begin{bmatrix} 1 & 4 & 4 & 1 & -2 & -2 \\ -3 & -2 & 2 & 3 & 2 & -2 \end{bmatrix}$$

The translation matrix is $\begin{bmatrix} 1 & 1 & 1 & 1 & 1 & 1 \\ 4 & 4 & 4 & 4 & 4 & 4 \end{bmatrix}$.

Find the vertex matrix for the final image.

$$\begin{bmatrix} 1 & 4 & 4 & 1 & -2 & -2 \\ -3 & -2 & 2 & 3 & 2 & -2 \end{bmatrix} + \begin{bmatrix} 1 & 1 & 1 & 1 & 1 & 1 \\ 4 & 4 & 4 & 4 & 4 & 4 \end{bmatrix}$$

$$= \begin{bmatrix} 2 & 5 & 5 & 2 & -1 & -1 \\ 1 & 2 & 6 & 7 & 6 & 2 \end{bmatrix}$$

The coordinates of the vertices of the final image are $A'(2, 1)$, $B'(5, 2)$, $C'(5, 6)$, $D'(2, 7)$, $E'(-1, 6)$, and $F'(-1, 2)$.

42. The rotation matrix for a counterclockwise rotation of 180° is $\begin{bmatrix} -1 & 0 \\ 0 & -1 \end{bmatrix}$.

The vertex matrix for polygon $ABCDEF$ is

$$\begin{bmatrix} -3 & -2 & 2 & 3 & 2 & -2 \\ 1 & 4 & 4 & 1 & -2 & -2 \end{bmatrix}.$$

Multiply the vertex matrix for $ABCDEF$ by the rotation matrix to find the vertex matrix of the image.

$$\begin{bmatrix} -1 & 0 \\ 0 & -1 \end{bmatrix} \cdot \begin{bmatrix} -3 & -2 & 2 & 3 & 2 & -2 \\ 1 & 4 & 4 & 1 & -2 & -2 \end{bmatrix}$$

$$= \begin{bmatrix} 3 & 2 & -2 & -3 & -2 & 2 \\ -1 & -4 & -4 & -1 & 2 & -2 \end{bmatrix}$$

Multiply the vertex matrix of the image by the scale factor to find the vertex matrix of the final image.

$$-2\begin{bmatrix} 3 & 2 & -2 & -3 & -2 & 2 \\ -1 & -4 & -4 & -1 & 2 & 2 \end{bmatrix}$$

$$= \begin{bmatrix} -6 & -4 & 4 & 6 & 4 & -4 \\ 2 & 8 & 8 & 2 & -4 & -4 \end{bmatrix}$$

The coordinates of the vertices of the final image are A′(−6, 2), B′(−4, 8), C′(4, 8), D′(6, 2), E′(4, −4), and F′(−4, −4).

43. Each footprint is reflected in the y-axis, then translated up two units.

44. Reflect in the y-axis using the matrix $\begin{bmatrix} -1 & 0 \\ 0 & 1 \end{bmatrix}$.

Translate up 2 units using $\begin{bmatrix} 0 \\ 2 \end{bmatrix}$.

Combine the two operations into

$$\begin{bmatrix} -1 & 0 \\ 0 & 1 \end{bmatrix} \cdot \begin{bmatrix} x \\ y \end{bmatrix} + \begin{bmatrix} 0 \\ 2 \end{bmatrix}.$$

45. Imagine the y-axis in the middle of the plan. Then a reflection in the y-axis could be used to create a floor plan with the garage on the left. The reflection matrix is $\begin{bmatrix} -1 & 0 \\ 0 & 1 \end{bmatrix}$.

46. A counterclockwise rotation of 90° could be used to create a floor plan with the house facing east. The rotation matrix is $\begin{bmatrix} 0 & -1 \\ 1 & 0 \end{bmatrix}$.

47. A reflection in the line $y = -x$ transforms (x, y) into $(-y, -x)$. The matrix that performs this operation is $\begin{bmatrix} 0 & -1 \\ -1 & 0 \end{bmatrix}$.

48. Matrices make it simpler for movie makers to move figures. Answers should include the following.

- By using a succession of matrix transformations, an object will move about in a scene.
- Sample answer: programming the animation in a screen saver

49. The rotation matrix for a 90° clockwise rotation is equivalent to a 270° counterclockwise rotation, or $\begin{bmatrix} 0 & 1 \\ -1 & 0 \end{bmatrix}$.

50. B; since 26% are action movies and 14% are comedies, 60% of the movie titles are neither action movies nor comedies.
$0.60(2500) = 1500$
So, 1500 movie titles are neither action movies nor comedies.

Page 511 Maintain Your Skills

51. First, graph $\triangle ABC$. Next, translate each vertex by $\vec{v}$, 1 unit left and 5 units down. Connect the vertices to form $\triangle A'B'C'$.

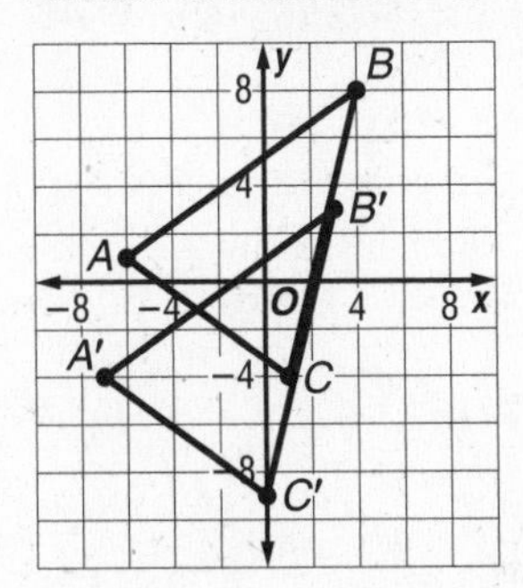

52. First, graph quadrilateral $DEFG$. Next, translate each vertex by $\overline{w}$, 7 units left and 8 units up. Connect the vertices to form quadrilateral $D'E'F'G'$.

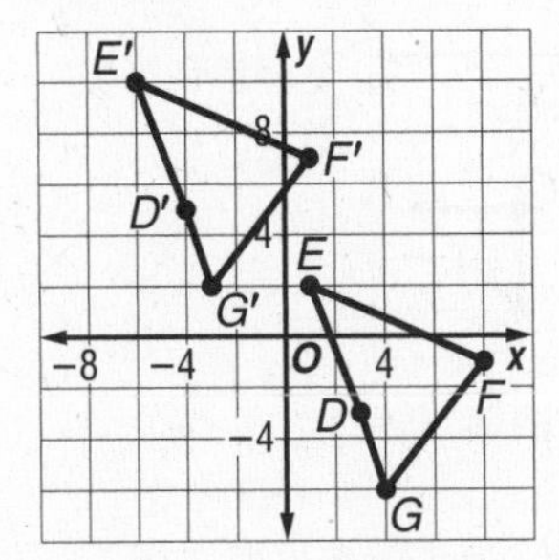

53. Compare WX and $W'X'$. Note that the scale factor is negative since the image appears on the opposite side of the center with respect to the preimage.

$$\text{scale factor} = -\frac{\text{image length}}{\text{preimage length}}$$

$$r = -\frac{2\text{units}}{4\text{units}}$$

$$r = -\frac{1}{2}$$

Since $0 < |r| < 1$, the dilation is a reduction.

54. The measure of an exterior angle of a regular polygon is given by $\frac{360}{n}$. The measure of an exterior angle of a 5-sided polygon is $\frac{360}{5}$ or 72. Use the Interior Angle Formula to find the interior angle of a regular 5-sided polygon.

$$\frac{180(n-2)}{n} = \frac{180(5-2)}{5}$$
$$= 108$$

55. The measure of an exterior angle of a regular polygon is given by $\frac{360}{n}$. The measure of an exterior angle of a 6-sided polygon is $\frac{360}{6}$ or 60. Use the Interior Angle Formula to find the interior angle of a regular 6-sided polygon.

$$\frac{180(n-2)}{n} = \frac{180(6-2)}{6}$$
$$= 120$$

56. The measure of an exterior angle of a regular polygon is given by $\frac{360}{n}$. The measure of an exterior angle of a 8-sided polygon is $\frac{360}{8}$ or 45. Use the Interior Angle Formula to find the interior angle of a regular 8-sided polygon.

$$\frac{180(n-2)}{n} = \frac{180(8-2)}{8}$$
$$= 135$$

57. The measure of an exterior angle of a regular polygon is given by $\frac{360}{n}$. The measure of an exterior angle of a 10-sided polygon is $\frac{360}{10}$ or 36. Use the Interior Angle Formula to find the interior angle of a regular 10-sided polygon.

$$\frac{180(n-2)}{n} = \frac{180(10-2)}{10}$$
$$= 144$$

58. The two right triangles are similar. So, using the definition of similar polygons, $\frac{DE}{AB} = \frac{CD}{BC}$. Solve for DE.

$$\frac{DE}{AB} = \frac{CD}{BC}$$
$$DE = (AB)\frac{CD}{BC}$$
$$DE = (1.75)\frac{34.5}{0.75}$$
$$DE = 80.5$$

The tree is 80.5 m tall.

Chapter 9 Study Guide and Review

Vocabulary and Concept Check

1. false; center
2. true
3. false; component form
4. false; magnitude
5. false; center of rotation
6. true
7. false; scale factor
8. false; resultant vector

Lesson-by-Lesson Review

9. Use the vertical grid lines to find images of each vertex of $\triangle ABC$ so that each vertex of the image is the same distance from the x-axis as the vertex of the preimage or use $(a, b) \rightarrow (a, -b)$

$A(2, 1) \rightarrow A'(2, -1)$
$B(5, 1) \rightarrow B'(5, -1)$
$C(2, 3) \rightarrow C'(2, -3)$

Draw triangles ABC and $A'B'C'$.

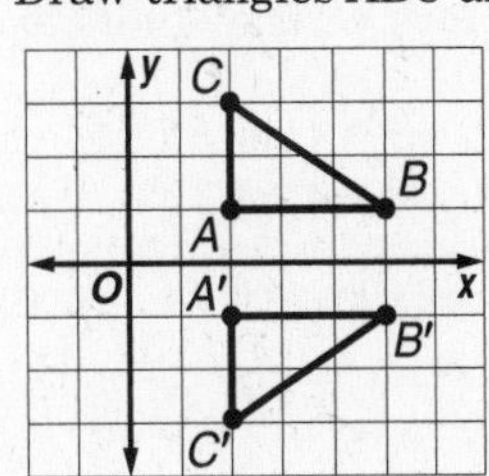

10. Use the transformation $(a, b) \rightarrow (b, a)$.
$W(-4, 5) \rightarrow W'(5, -4)$ $Y(-3, 3) \rightarrow Y'(3, -3)$
$X(-1, 5) \rightarrow X'(5, -1)$ $Z(-6, 3) \rightarrow Z'(3, -6)$
Draw parallelograms $WXYZ$ and $W'X'Y'Z'$.

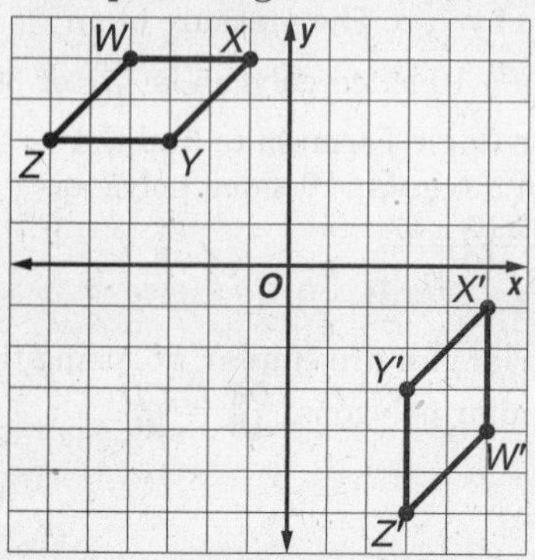

11. Use the vertical grid lines to find images of each vertex of rectangle $EFGH$ such that each vertex of the image is the same distance from the line $x = 1$ as the vertex of the preimage.
$E(-4, -2) \rightarrow E'(6, -2)$ $G(0, -4) \rightarrow G'(2, -4)$
$F(0, -2) \rightarrow F'(2, -2)$ $H(-4, -4) \rightarrow H'(6, -4)$
Draw rectangles $EFGH$ and $E'F'G'H'$.

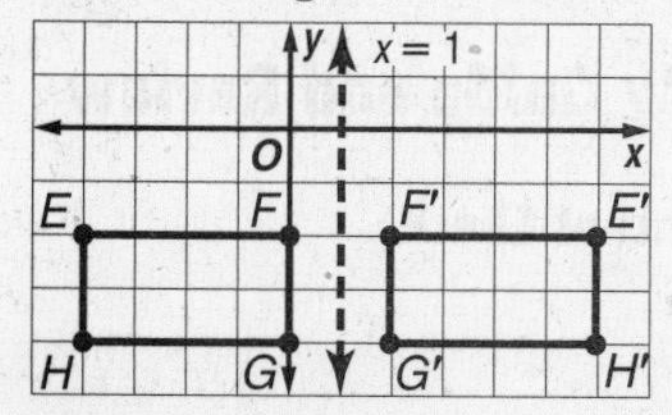

12. This translation moved each vertex 4 units to the left and 4 units down.
$E(2, 2) \rightarrow E'(2 - 4, 2 - 4)$ or $E'(-2, -2)$
$F(6, 2) \rightarrow F'(6 - 4, 2 - 4)$ or $F'(2, -2)$
$G(4, -2) \rightarrow G'(4 - 4, -2 - 4)$ or $G'(0, -6)$
$H(1, -1) \rightarrow H'(1 - 4, -1 - 4)$ or $H'(-3, -5)$
Draw quadrilaterals $EFGH$ and $E'F'G'H'$.

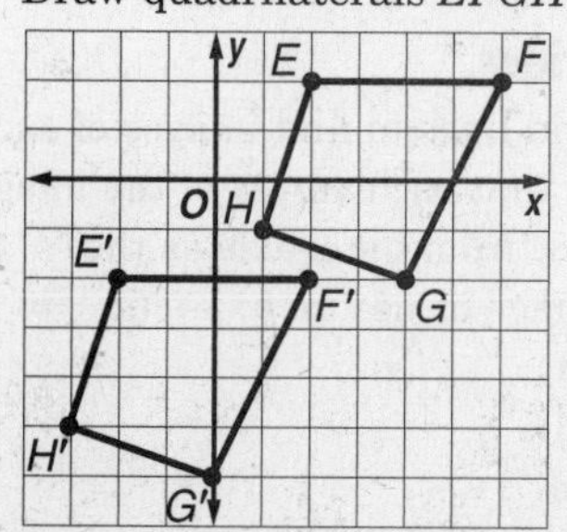

13. This translation moved each vertex 2 units to the right and 4 units up.
$S(-3, -5) \rightarrow S'(-3 + 2, -5 + 4)$ or $S'(-1, -1)$.
$T(-1, -1) \rightarrow T'(-1 + 2, -1 + 4)$ or $T'(1, 3)$
Draw $\overline{ST}$ and $\overline{S'T'}$.

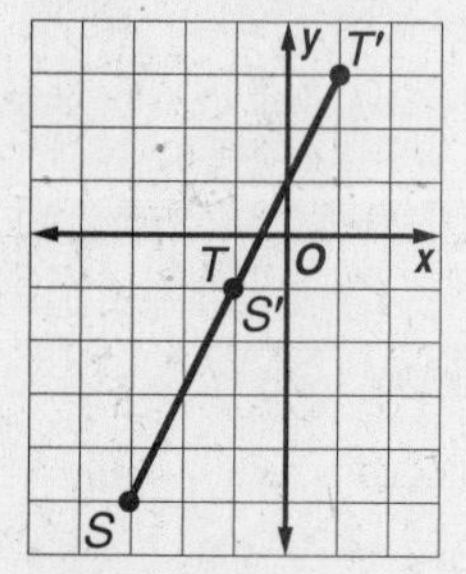

14. This translation moved each vertex 1 unit to the right and 3 units down.
$X(2, 5) \rightarrow X'(2 + 1, 5 - 3)$ or $X'(3, 2)$
$Y(1, 1) \rightarrow Y'(1 + 1, 1 - 3)$ or $Y'(2, -2)$
$Z(5, 1) \rightarrow Z'(5 + 1, 1 - 3)$ or $Z'(6, -2)$
Draw $\triangle XYZ$ and $\triangle X'Y'Z'$.

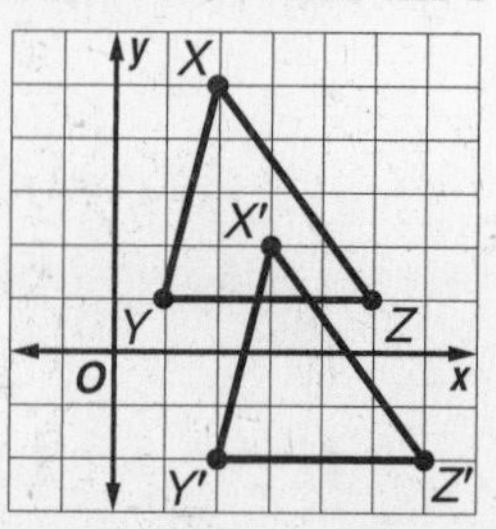

15.

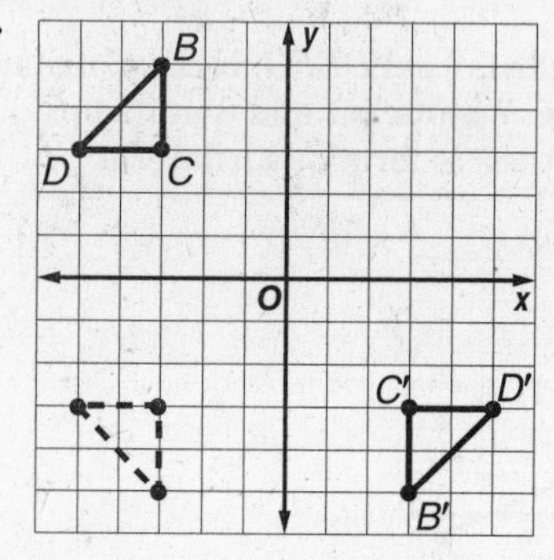

Reflection in x-axis: $B(-3, 5) \rightarrow (-3, -5)$
$C(-3, 3) \rightarrow (-3, -3)$
$D(-5, 3) \rightarrow (-5, -3)$
Reflection in the y-axis: $(-3, -5) \rightarrow B'(3, -5)$
$(-3, -3) \rightarrow C'(3, -3)$
$(-5, -3) \rightarrow D'(5, -3)$
The angle of rotation is 180°.

16.

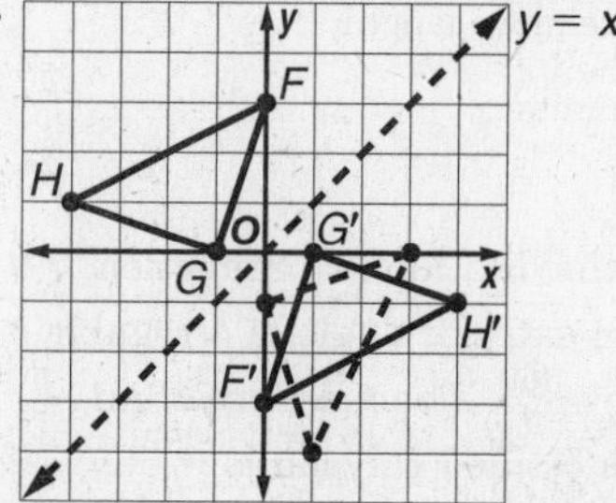

Reflection in line $y = x$: $F(0, 3) \rightarrow (3, 0)$
$G(-1, 0) \rightarrow (0, -1)$
$H(-4, 1) \rightarrow (1, -4)$
Reflection in line $y = -x$: $(3, 0) \rightarrow F'(0, -3)$
$(0, -1) \rightarrow G'(1, 0)$
$(1, -4) \rightarrow H'(4, -1)$
The angle of rotation is 180°.

17. 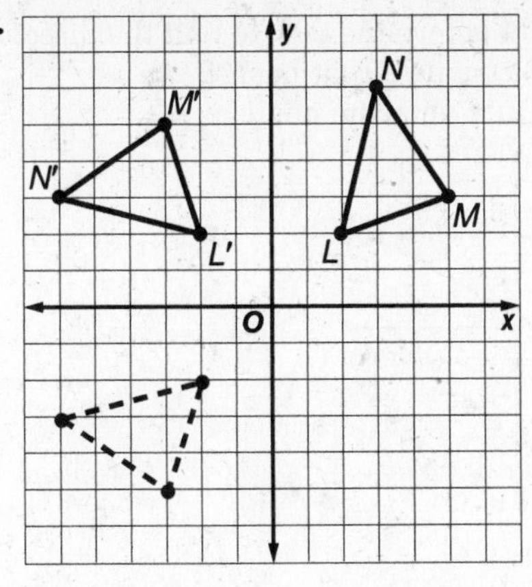

Reflection in line $y = -x$: $L(2, 2) \rightarrow (-2, -2)$
$M(5, 3) \rightarrow (-3, -5)$
$N(3, 6) \rightarrow (-6, -3)$
Reflection in the x-axis: $(-2, -2) \rightarrow L'(-2, 2)$
$(-3, -5) \rightarrow M'(-3, 5)$
$(-6, -3) \rightarrow N'(-6, 3)$
The angle of rotation is 90° counterclockwise.

18. The figure has rotational symmetry of order 9 because there are 9 rotations less than 360° (including 0 degrees) that produce an image indistinguishable from the original.
$$\text{magnitude} = \frac{360°}{\text{order}} = \frac{360°}{9} = 40°$$
The magnitude of the symmetry is 40°.

19. $$\text{magnitude} = \frac{360°}{\text{order}} = \frac{360°}{9} = 40°$$
The magnitude of the symmetry is 40°.
Vertex 2 is moved 5 positions, or 5(40°) = 200°.

20. $$\text{magnitude} = \frac{360°}{\text{order}} = \frac{360°}{9} = 40°$$
The magnitude of the symmetry is 40°.
Divide 280° by the magnitude of the rotational symmetry.
$\frac{280°}{40°} = 7$ positions
Vertex 5 is moved 7 positions, or to the original position of vertex 7.

21. Yes; the pattern is a tessellation because at the different vertices the sum of the angles is 360°. The tessellation is not uniform because the number of angles at the vertices varies.

22. Yes; the pattern is a tessellation because at the different vertices the sum of the angles is 360°. The tessellation is uniform because at every vertex there is the same combination of shapes and angles. The tessellation is also regular since it is formed by only one type of regular polygon.

23. Yes; the pattern is a tessellation because at the different vertices the sum of the angles is 360°. The tessellation is uniform because at every vertex there is the same combination of shapes and angles.

24. No; let $m\angle 1$ represent one interior angle of the regular pentagon. Use the Interior Angle Formula.
$$m\angle 1 = \frac{180(n-2)}{n} = \frac{180(5-2)}{5} = 108$$
Since 108 is not a factor of 360, a pentagon will not tessellate the plane.

25. Yes; the measure of an interior angle of an equilateral triangle is 60, which is a factor of 360, so an equilateral triangle will tessellate the plan.

26. No; let $m\angle 1$ represent one interior angle of the regular decagon. Use the Interior Angle Formula.
$$m\angle 1 = \frac{180(n-2)}{n} = \frac{180(10-2)}{10} = 144$$
Since 144 is not a factor of 360, a decagon will not tessellate the plane.

27. $CD = 8, r = 3$
Use the Dilation Theorem.
$C'D' = |r|(CD)$
$C'D' = (3)(8)$
$C'D' = 24$

28. $CD = \frac{2}{3}, r = -6$
Use the Dilation Theorem.
$C'D' = |r|(CD)$
$C'D' = (6)\left(\frac{2}{3}\right)$
$C'D' = 4$

29. $C'D' = 24, r = 6$
Use the Dilation Theorem.
$C'D' = |r|(CD)$
$24 = (6)(CD)$
$4 = CD$

30. $C'D' = 60, r = \frac{10}{3}$
Use the Dilation Theorem.
$C'D' = |r|(CD)$
$60 = \left(\frac{10}{3}\right)(CD)$
$18 = CD$

31. $CD = 12, r = -\frac{5}{6}$
Use the Dilation Theorem.
$C'D' = |r|(CD)$
$C'D' = \left(\frac{5}{6}\right)(12)$
$C'D' = 10$

32. $C'D' = \frac{55}{2}, r = \frac{5}{4}$
Use the Dilation Theorem.
$C'D' = |r|(CD)$
$\frac{55}{2} = \left(\frac{5}{4}\right)(CD)$
$22 = CD$

33. Find P', Q', and R' using the scale factor, $r = -2$.

Preimage (x, y)	Image $(-2x, -2y)$
$P(-1, 3)$	$P'(2, -6)$
$Q(2, 2)$	$Q'(-4, -4)$
$R(1, -1)$	$R'(-2, 2)$

34. Find E', F', G', and H' using the scale factor, $r = -2$.

Preimage (x, y)	Image $(-2x, -2y)$
$E(-3, 2)$	$E'(6, -4)$
$F(1, 2)$	$F'(-2, -4)$
$G(1, -2)$	$G'(-2, 4)$
$H(-3, -2)$	$H'(6, 4)$

35. Find the change in x-values and the corresponding change in y-values.

$\overrightarrow{AB} = \langle x_2 - x_1, y_2 - y_1\rangle$
$= \langle 0 - (-3), 2 - (-2)\rangle$
$= \langle 3, 4\rangle$

36. Find the change in x-values and the corresponding change in y-values.

$\overrightarrow{CD} = \langle x_2 - x_1, y_2 - y_1\rangle$
$= \langle -4 - 4, 2 - (-2)\rangle$
$= \langle -8, 4\rangle$

37. Find the change in x-values and the corresponding change in y-values.

$\overrightarrow{EF} = \langle x_2 - x_1, y_2 - y_1\rangle$
$= \langle 1 - 1, 4 - (-4)\rangle$
$= \langle 0, 8\rangle$

38. Find the magnitude using the Distance Formula.

$|\overrightarrow{AB}| = \sqrt{(x_2 - x_1)^2 + (y_2 - y_1)^2}$
$= \sqrt{[-9 - (-6)]^2 + (-3 - 4)^2}$
$= \sqrt{58}$
≈ 7.6

Graph $\overrightarrow{AB}$ to determine how to find the direction. Draw a right triangle that has $\overrightarrow{AB}$ as its hypotenuse and an acute angle at A.

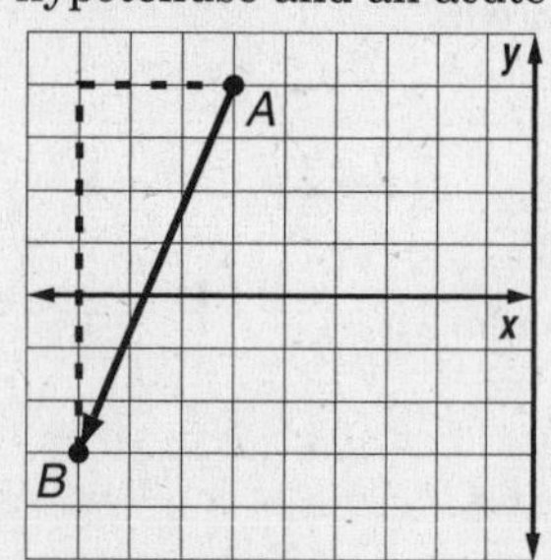

$\tan A = \frac{y_2 - y_1}{x_2 - x_1}$
$= \frac{-3 - 4}{-9 - (-6)}$
$= \frac{7}{3}$
$m\angle A = \tan^{-1}\frac{7}{3}$
≈ 66.8

A vector in standard position that is equal to $\overrightarrow{AB}$ forms a 66.8° angle with the negative x-axis in the third quadrant. So it forms a 180 + 66.8 or 246.8° angle with the positive x-axis.
Thus, $\overrightarrow{AB}$ has a magnitude of $\sqrt{58}$ or about 7.6 units and a direction of about 246.8°.

39. Find the magnitude using the Distance Formula.

$|\overrightarrow{AB}| = \sqrt{(x_2 - x_1)^2 + (y_2 - y_1)^2}$
$= \sqrt{(-5 - 8)^2 + (-2 - 5)^2}$
$= \sqrt{218}$
≈ 14.8

Graph $\overrightarrow{AB}$ to determine how to find the direction. Draw a right triangle that has $\overrightarrow{AB}$ as its hypotenuse and an acute angle at A.

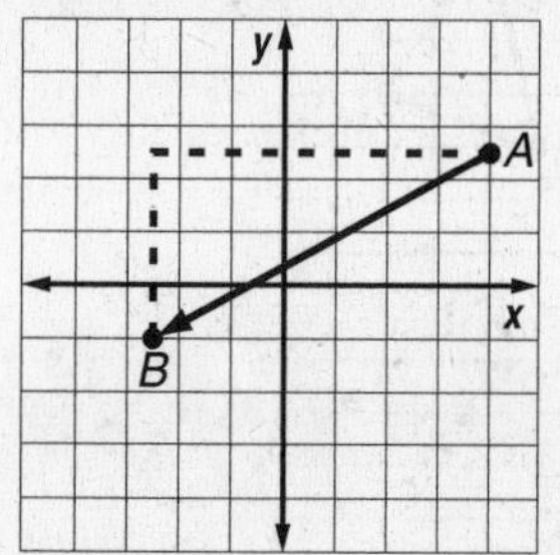

$\tan A = \frac{y_2 - y_1}{x_2 - x_1}$
$= \frac{-2 - 5}{-5 - 8}$
$= \frac{7}{13}$
$m\angle A = \tan^{-1}\frac{7}{13}$
≈ 28.3

A vector in standard position that is equal to $\overrightarrow{AB}$ forms a 28.3° angle with the negative x-axis in the third quadrant. So it forms a 180 + 28.3 or 208.3° angle with the positive x-axis.
Thus, $\overrightarrow{AB}$ has a magnitude of $\sqrt{218}$ or about 14.8 units and a direction of about 208.3°.

40. Find the magnitude using the Distance Formula.

$|\overrightarrow{AB}| = \sqrt{(x_2 - x_1)^2 + (y_2 - y_1)^2}$
$= \sqrt{[15 - (-14)]^2 + (-5 - 2)^2}$
$= \sqrt{890}$
≈ 29.8

Graph $\overrightarrow{AB}$ to determine how to find the direction. Draw a right triangle that has $\overrightarrow{AB}$ as its hypotenuse and an acute angle at A.

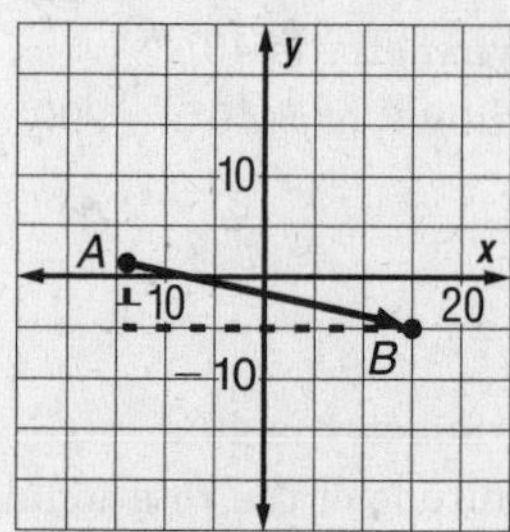

$\tan A = \frac{y_2 - y_1}{x_2 - x_1}$
$= \frac{-5 - 2}{15 - (-14)}$
$= -\frac{7}{29}$
$m\angle A = \tan^{-1}\left(-\frac{7}{29}\right)$
≈ -13.6

A vector in standard position that is equal to $\overrightarrow{AB}$ forms a −13.6° angle with the positive x-axis in the fourth quadrant. So it forms a 360 − 13.6 or 346.4° angle with the positive x-axis.
Thus, $\overrightarrow{AB}$ has a magnitude of $\sqrt{890}$ or about 29.8 units and a direction of about 346.4°.

41. Find the magnitude using the Distance Formula.

$|\overrightarrow{AB}| = \sqrt{(x_2 - x_1)^2 + (y_2 - y_1)^2}$
$= \sqrt{(-45 - 16)^2 + (0 - 40)^2}$
$= \sqrt{5321}$
≈ 72.9

Graph $\overrightarrow{AB}$ to determine how to find the direction. Draw a right triangle that has $\overrightarrow{AB}$ as its hypotenuse and an acute angle at A.

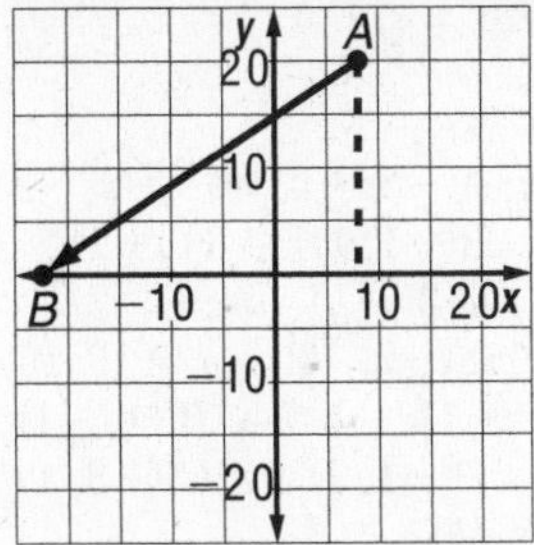

$$\tan A = \frac{y_2 - y_1}{x_2 - x_1} = \frac{0 - 40}{-45 - 16} = \frac{40}{61}$$

$$m\angle A = \tan^{-1}\frac{40}{61} \approx 33.3$$

A vector in standard position that is equal to $\overrightarrow{AB}$ forms a 33.3° angle with the negative x-axis in the third quadrant. So it forms a 180 + 33.3 or 213.3° angle with the positive x-axis.

Thus, $\overrightarrow{AB}$ has a magnitude of $\sqrt{5321}$ or about 72.9 units and a direction of about 213.3°.

42. The vertex matrix for $\triangle DEF$ is $\begin{bmatrix} -3 & 0 & 2 \\ -2 & 5 & -4 \end{bmatrix}$.

The translation matrix is $\begin{bmatrix} -3 & -3 & -3 \\ -6 & -6 & -6 \end{bmatrix}$.

Find the vertex matrix for the image.

$$\begin{bmatrix} -3 & 0 & 2 \\ -2 & 5 & -4 \end{bmatrix} + \begin{bmatrix} -3 & -3 & -3 \\ -6 & -6 & -6 \end{bmatrix} = \begin{bmatrix} -6 & -3 & -1 \\ -8 & -1 & -10 \end{bmatrix}$$

The coordinates of the vertices of the image are $D'(-6, -8)$, $E'(-3, -1)$, and $F'(-1, -10)$.

43. The vertex matrix for $\triangle DEF$ is $\begin{bmatrix} -3 & 0 & 2 \\ -2 & 5 & -4 \end{bmatrix}$.

Multiply the vertex matrix by the scale factor to find the vertex matrix of the image.

$$\frac{4}{5}\begin{bmatrix} -3 & 0 & 2 \\ -2 & 5 & -4 \end{bmatrix} = \begin{bmatrix} -\frac{12}{5} & 0 & \frac{8}{5} \\ -\frac{8}{5} & 4 & -\frac{16}{5} \end{bmatrix}$$

The coordinates of the vertices of the image are $D'\left(-\frac{12}{5}, -\frac{8}{5}\right)$, $E'(0, 4)$, and $F'\left(\frac{8}{5}, -\frac{16}{5}\right)$.

44. The reflection matrix for a reflection in the line $y = x$ is $\begin{bmatrix} 0 & 1 \\ 1 & 0 \end{bmatrix}$.

The vertex matrix for $\triangle DEF$ is $\begin{bmatrix} -3 & 0 & 2 \\ -2 & 5 & -4 \end{bmatrix}$.

Multiply the vertex matrix for $\triangle DEF$ by the reflection matrix to find the vertex matrix of the image.

$$\begin{bmatrix} 0 & 1 \\ 1 & 0 \end{bmatrix} \cdot \begin{bmatrix} -3 & 0 & 2 \\ -2 & 5 & -4 \end{bmatrix} = \begin{bmatrix} -2 & 5 & -4 \\ -3 & 0 & 2 \end{bmatrix}$$

The coordinates of the vertices of the image are $D'(-2, -3)$, $E'(5, 0)$, and $F'(-4, 2)$.

45. The rotation matrix for a counterclockwise rotation of 270° is $\begin{bmatrix} 0 & 1 \\ -1 & 0 \end{bmatrix}$.

The vertex matrix for $\triangle DEF$ is $\begin{bmatrix} -3 & 0 & 2 \\ -2 & 5 & -4 \end{bmatrix}$.

Multiply the vertex matrix for $\triangle DEF$ by the rotation matrix to find the vertex matrix of the image.

$$\begin{bmatrix} 0 & 1 \\ -1 & 0 \end{bmatrix} \cdot \begin{bmatrix} -3 & 0 & 2 \\ -2 & 5 & -4 \end{bmatrix} = \begin{bmatrix} -2 & 5 & -4 \\ 3 & 0 & -2 \end{bmatrix}$$

The coordinates of the vertices of the image are $D'(-2, 3)$, $E'(5, 0)$, and $F'(-4, -2)$.

46. The vertex matrix for $\triangle PQR$ is $\begin{bmatrix} 9 & 1 & 4 \\ 2 & -1 & 5 \end{bmatrix}$.

The translation matrix is $\begin{bmatrix} 2 & 2 & 2 \\ -5 & -5 & -5 \end{bmatrix}$.

Find the vertex matrix of the image.

$$\begin{bmatrix} 9 & 1 & 4 \\ 2 & -1 & 5 \end{bmatrix} + \begin{bmatrix} 2 & 2 & 2 \\ -5 & -5 & -5 \end{bmatrix} = \begin{bmatrix} 11 & 3 & 6 \\ -3 & -6 & 0 \end{bmatrix}$$

The reflection matrix for a reflection in the x-axis is $\begin{bmatrix} 1 & 0 \\ 0 & -1 \end{bmatrix}$.

Multiply the vertex matrix of the image by the reflection matrix to find the vertex matrix of the final image.

$$\begin{bmatrix} 1 & 0 \\ 0 & -1 \end{bmatrix} \cdot \begin{bmatrix} 11 & 3 & 6 \\ -3 & -6 & 0 \end{bmatrix} = \begin{bmatrix} 11 & 3 & 6 \\ 3 & 6 & 0 \end{bmatrix}$$

The coordinates of the vertices of the final image are $P'(11, 3)$, $Q'(3, 6)$, and $R'(6, 0)$.

47. The rotation matrix for a counterclockwise rotation of 180° is $\begin{bmatrix} -1 & 0 \\ 0 & -1 \end{bmatrix}$.

The vertex matrix for quadrilateral $WXYZ$ is $\begin{bmatrix} -8 & -2 & -1 & -6 \\ 1 & 3 & 0 & -3 \end{bmatrix}$.

Multiply the vertex matrix for quadrilateral $WXYZ$ by the rotation matrix to find the vertex matrix of the image.

$$\begin{bmatrix} -1 & 0 \\ 0 & -1 \end{bmatrix} \cdot \begin{bmatrix} -8 & -2 & -1 & -6 \\ 1 & 3 & 0 & -3 \end{bmatrix} = \begin{bmatrix} 8 & 2 & 1 & 6 \\ -1 & -3 & 0 & 3 \end{bmatrix}$$

Multiply the vertex matrix of the image by the scale factor to find the vertex matrix of the final image.

$$-2\begin{bmatrix} 8 & 2 & 1 & 6 \\ -1 & -3 & 0 & 3 \end{bmatrix} = \begin{bmatrix} -16 & -4 & -2 & -12 \\ 2 & 6 & 0 & -6 \end{bmatrix}$$

The coordinates of the vertices of the final image are $W'(-16, 2)$, $X'(-4, 6)$, $Y'(-2, 0)$, and $Z'(-12, -6)$.

Chapter 9 Practice Test

Page 517

1. isometry

2. uniform

3. scalar

4. E

5. $\overline{DC}$

6. $\triangle BCA$

7. First, graph $\triangle PQR$. Next, translate each vertex right 3 units and up 1 unit. Connect the vertices to form $\triangle P'Q'R$.
$P(-3, 5) \rightarrow P'(-3 + 3, 5 + 1)$ or $P'(0, 6)$
$Q(-2, 1) \rightarrow Q'(-2 + 3, 1 + 1)$ or $Q'(1, 2)$
$R(-4, 2) \rightarrow R'(-4 + 3, 2 + 1)$ or $R'(-1, 3)$

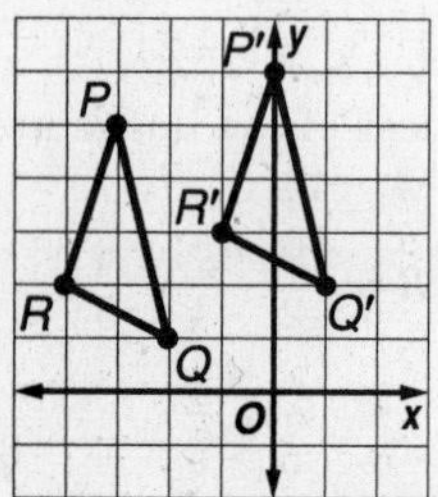

8. First, graph parallelogram $WXYZ$. Next, translate each vertex up 5 units and left 3 units. Connect the vertices to form $W'X'Y'Z$.
$W(-2, -5) \rightarrow W'(-2 - 3, -5 + 5)$ or $W'(-5, 0)$
$X(1, -5) \rightarrow X'(1 - 3, -5 + 5)$ or $X'(-2, 0)$
$Y(2, -2) \rightarrow Y'(2 - 3, -2 + 5)$ or $Y'(-1, 3)$
$Z(-1, -2) \rightarrow Z'(-1 - 3, -2 + 5)$ or $Z'(-4, 3)$

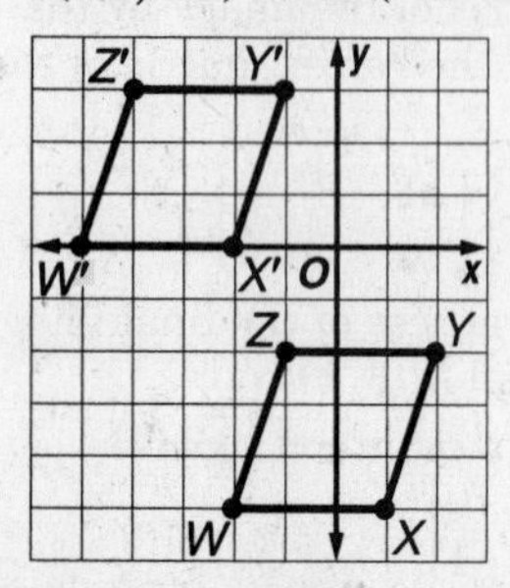

9. First, graph $\overline{FG}$. Next, translate each vertex left 4 units and down 1 unit. Connect the vertices to form $\overline{F'G'}$.
$F(3, 5) \rightarrow F'(3 - 4, 5 - 1)$ or $F'(-1, 4)$
$G(6, -1) \rightarrow G'(6 - 4, -1 - 1)$ or $G'(2, -2)$

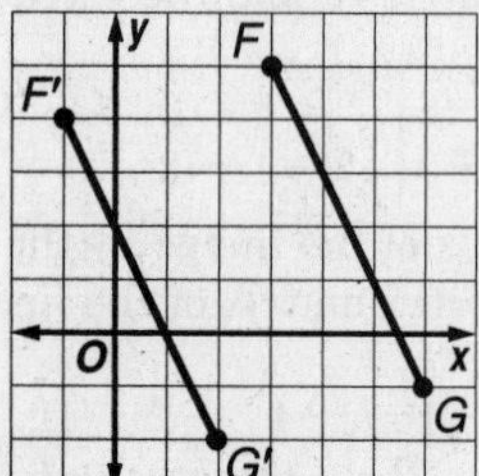

10.

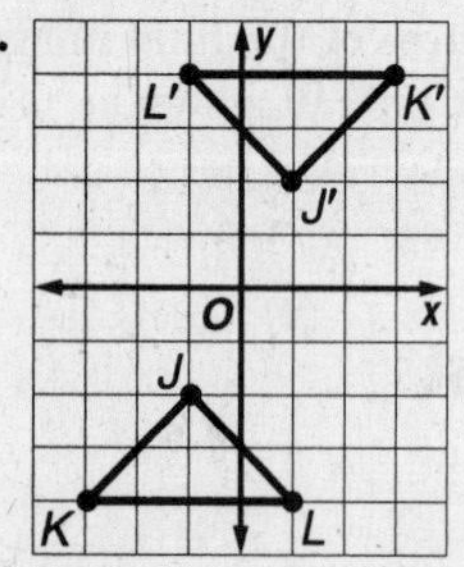

Reflection in y-axis: $J(-1, -2) \rightarrow (1, -2)$
$K(-3, -4) \rightarrow (3, -4)$
$L(1, -4) \rightarrow (-1, -4)$
Reflection in x-axis: $(1, -2) \rightarrow J'(1, 2)$
$(3, -4) \rightarrow K'(3, 4)$
$(-1, -4) \rightarrow L'(-1, 4)$
The angle of rotation is 180°.

11.

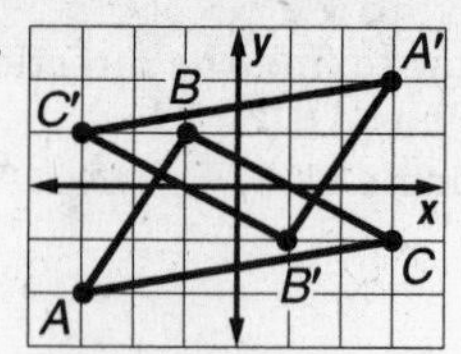

Reflection in line $y = x$: $A(-3, -2) \rightarrow (-2, -3)$
$B(-1, 1) \rightarrow (1, -1)$
$C(3, -1) \rightarrow (-1, 3)$
Reflection in line $y = -x$: $(-2, -3) \rightarrow A'(3, 2)$
$(1, -1) \rightarrow B'(1, -1)$
$(-1, 3) \rightarrow C'(-3, 1)$
The angle of rotation is 180°.

12.

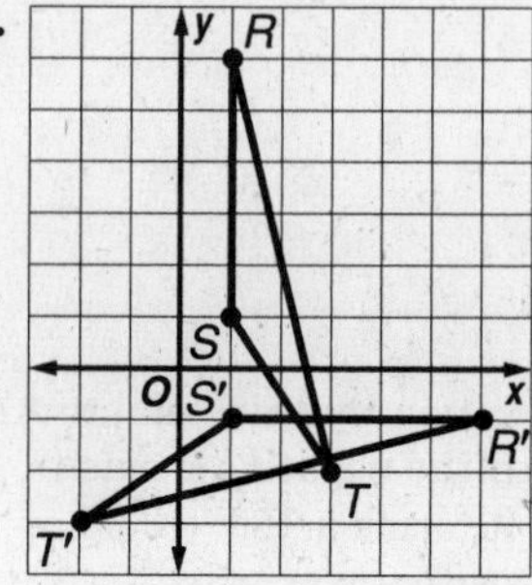

Reflection in y-axis: $R(1, 6) \rightarrow (-1, 6)$
$S(1, 1) \rightarrow (-1, 1)$
$T(3, -2) \rightarrow (-3, -2)$
Reflection in line $y = x$: $(-1, 6) \rightarrow R'(6, -1)$
$(-1, 1) \rightarrow S'(1, -1)$
$(-3, -2) \rightarrow T'(-2, -3)$
The angle of rotation is 90° clockwise or 270° counterclockwise.

13. Yes; the pattern is a tessellation because at the different vertices the sum of the angles is 360°. The tessellation is uniform because at every vertex there is the same combination of shapes and angles.

14. Yes; the pattern is a tessellation because at the different vertices the sum of the angles is 360°. The tessellation is uniform because at every vertex there is the same combination of shapes and angles. The tessellation is also semi-regular since more than one regular polygon is used.

15. Yes; the pattern is a tessellation because at the different vertices the sum of the angles is 360°. The tessellation is not uniform because the number of angles at the vertices varies.

16. $MN = 5, r = 4$
Use the Dilation Theorem.
$M'N' = |r|(MN)$
$M'N' = (4)(5)$
$M'N' = 20$

17. $MN = 8, r = \frac{1}{4}$
Use the Dilation Theorem.
$M'N' = |r|(MN)$
$M'N' = \left(\frac{1}{4}\right)(8)$
$M'N' = 2$

18. $M'N' = 36, r = 3$

Use the Dilation Theorem.

$M'N' = |r|(MN)$

$36 = (3)(MN)$

$12 = MN$

19. $MN = 9, r = -\frac{1}{5}$

Use the Dilation Theorem.

$M'N' = |r|(MN)$

$M'N' = \left(\frac{1}{5}\right)(9)$

$M'N' = \frac{9}{5}$

20. $M'N' = 20, r = \frac{2}{3}$

Use the Dilation Theorem.

$M'N' = |r|(MN)$

$20 = \left(\frac{2}{3}\right)(MN)$

$30 = MN$

21. $M'N' = \frac{29}{5}, r = -\frac{3}{5}$

Use the Dilation Theorem.

$M'N' = |r|(MN)$

$\frac{29}{5} = \left(\frac{3}{5}\right)(MN)$

$\frac{29}{3} = MN$

22. Find the magnitude of the resultant vector.

$|\vec{v}| = \sqrt{x^2 + y^2}$

$= \sqrt{(-3)^2 + 2^2}$

$= \sqrt{13}$

≈ 3.6

The resultant vector lies in the second quadrant. Find the direction.

$\tan\theta = \frac{y}{x}$

$= -\frac{2}{3}$

$m\angle\theta = \tan^{-1}\left(-\frac{2}{3}\right)$

≈ -33.7

The resultant vector forms a 33.7° angle with the negative x-axis in the second quadrant. So it forms a 180 − 33.7 or 146.3° angle with the positive x-axis.

Thus, $\vec{v}$ has a magnitude of $\sqrt{13}$ or about 3.6 units and a direction of about 146.3°.

23. Find the magnitude of the resultant vector.

$|\vec{w}| = \sqrt{x^2 + y^2}$

$= \sqrt{(-6)^2 + (-8)^2}$

$= 10$

The resultant vector lies in the third quadrant. Find the direction.

$\tan\theta = \frac{y}{x}$

$= \frac{-8}{-6}$

$= \frac{4}{3}$

$m\angle\theta = \tan^{-1}\left(\frac{4}{3}\right)$

≈ 53.1

The resultant vector forms a 53.1° angle with the negative x-axis in the third quadrant. So it forms a 180 + 53.1 or 233.1° angle with the positive x-axis.

Thus, $\vec{w}$ has a magnitude of 10 units and a direction of about 233.1°.

24. The distance Gunja must travel can be found by multiplying the distance measured on the map by the scale factor.

$2.25 \text{ in.} \times \frac{150 \text{ mi}}{1 \text{ in.}} = 337.5 \text{ mi}$

Gunja must travel 337.5 mi.

25. A; A reflection of (3, 4) in the x-axis gives (3, −4). A reflection of (3, −4) in the origin gives (−3, 4). A reflection of (3, −4) in the y-axis gives (−3, −4). A reflection of (3, 4) in the origin gives (−3, −4). Only choice A, a reflection in the x-axis, gives (3, −4).

Chapter 9 Standardized Test Practice

Pages 518–519

1. D

2. A; both $\overline{CE}$ and $\overline{DF}$ are at an angle of 70° from the line $\overleftrightarrow{AB}$, so they are parallel.

3. D; the congruence of a single pair of opposite sides is not sufficient proof that $QRST$ is a parallelogram.

4. B; in a reflection in the y-axis, x-coordinates become their opposite, so (4, 2) → (−4, 2).

5. D; parallelogram $JKLM$ can be thought of as parallelogram $ABCD$ transformed by having all of its points moved the same distance in the same direction, or translated.

6. B; congruence transformations preserve angle and distance measure, collinearity, and betweenness of points. Orientation is not necessarily preserved.

7. A; a 180° rotation of $\triangle ABC$ about point C on line m maps $\triangle ABC$ to $\triangle A'B'C'$.

8. The top and bottom sides of the hexagon must be parallel and have the same length. So, the bottom side must have length 2 and slope 0. The coordinates of the missing vertex are (1, −2).

9. The triangle is isosceles, so the missing angle is $x°$. The sum of the interior angles of a triangle is 180°. Find x.

$2x° + x° + x° = 180°$

$4x° = 180°$

$x° = 45°$

So, $x = 45$.

10. The two right triangles are similar. So, using the definition of similar polygons, $\frac{AC}{CE} = \frac{BC}{CD}$.

Use the Pythagorean Theorem to find CE.

$CE = \sqrt{(CD)^2 + (DE)^2}$

$= \sqrt{100^2 + 45^2}$

$= 5\sqrt{481}$

Find x.

$\frac{AC}{CE} = \frac{BC}{CD}$

$\frac{x}{5\sqrt{481}} = \frac{300}{100}$

$x \approx 329$

To the nearest meter, the length of the cable is 329 m.

11. Given: $\overline{AB} \cong \overline{AC}$, $\overline{AD} \cong \overline{AE}$

Prove: $\triangle ABD \cong \triangle ACE$

Proof:

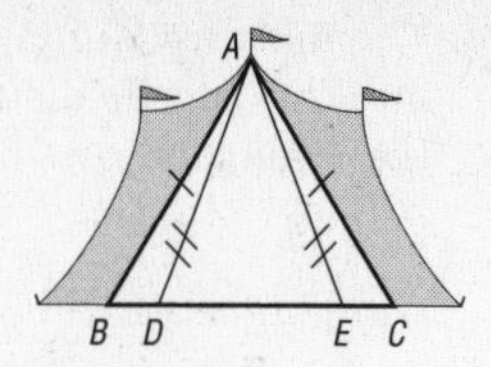

Statements	Reasons
1. $\overline{AB} \cong \overline{AC}$, $\overline{AD} \cong \overline{AE}$	1. Given
2. $\angle ABD \cong \angle ACE$ $\angle ADE \cong \angle AED$	2. Isos. $\triangle$ Thm.
3. $\angle ADB$ and $\angle ADE$ are supplementary. $\angle AEC$ and $\angle AED$ are supplementary.	3. If 2 ∠s form a linear pair, then they are suppl.
4. $\angle ADB \cong \angle AEC$	4. ∠s suppl. to ≅ ∠s are ≅.
5. $\triangle ABD \cong \triangle ACE$	5. AAS

12a.

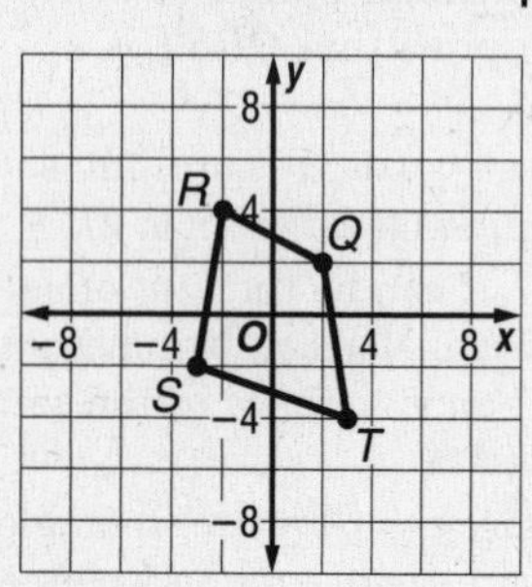

12b. Find Q', R', S', and T' using the scale factor, $r = 2$.

Preimage (x, y)	Image $(2x, 2y)$
$Q(2, 2)$	$Q'(4, 4)$
$R(-2, 4)$	$R'(-4, 8)$
$S(-3, -2)$	$S'(-6, -4)$
$T(3, -4)$	$T'(6, -8)$

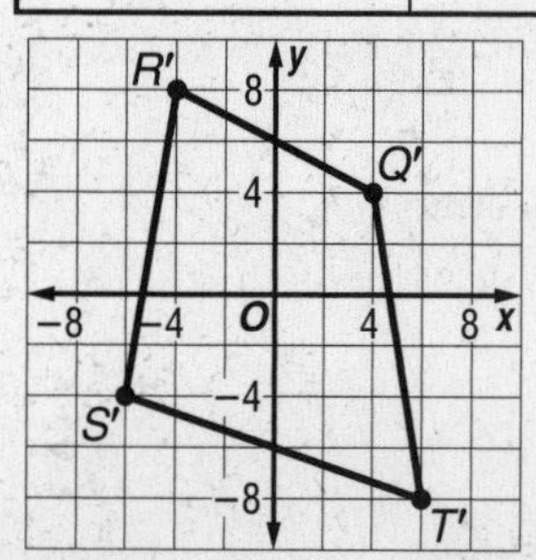

12c. Multiply the x- and y-coordinates of each vertex by the scale factor; $Q(2, 2)$ becomes $Q'(2 \times 2, 2 \times 2)$ or $Q'(4, 4)$.

12d. Sample answer: Enlargements and reductions preserve the shape of the figure. Congruence transformations preserve collinearity, betweenness of points, and angle and distance measures.

Chapter 10 Circles

Page 521 Getting Started

1. $\frac{4}{9}p = 72$
   $\frac{9}{4}\left(\frac{4}{9}p\right) = \frac{9}{4}(72)$
   $p = 162$

2. $6.3p = 15.75$
   $\frac{6.3p}{6.3} = \frac{15.75}{6.3}$
   $p = 2.5$

3. $3x + 12 = 8x$
   $3x + 12 - 3x = 8x - 3x$
   $12 = 5x$
   $\frac{12}{5} = \frac{5x}{5}$
   $\frac{12}{5} = x$ or $x = 2.4$

4. $7(x + 2) = 3(x - 6)$
   $7x + 14 = 3x - 18$
   $7x - 3x = -18 - 14$
   $4x = -32$
   $\frac{4x}{4} = \frac{-32}{4}$
   $x = -8$

5. $C = 2pr$
   $\frac{C}{2p} = \frac{2pr}{2p}$
   $\frac{C}{2p} = r$

6. $r = \frac{C}{6.28}$
   $6.28r = 6.28\left(\frac{C}{6.28}\right)$
   $6.28r = C$

7. $c^2 = a^2 + b^2$
   $17^2 = 8^2 + x^2$
   $289 = 64 + x^2$
   $225 = x^2$
   $\sqrt{225} = \sqrt{x^2}$
   $15 = x$

8. $c^2 = a^2 + b^2$
   $10^2 = 6^2 + x^2$
   $100 = 36 + x^2$
   $64 = x^2$
   $\sqrt{64} = \sqrt{x^2}$
   $8 = x$

9. $c^2 = a^2 + b^2$
   $(6x)^2 = 72^2 + 72^2$
   $36x^2 = 5184 + 5184$
   $36x^2 = 10{,}368$
   $\frac{36x^2}{36} = \frac{10{,}368}{36}$
   $x^2 = 288$
   $x = \sqrt{288}$
   $x \approx 17.0$

10. $x^2 - 4x = 10$
    $x^2 - 4x - 10 = 0$
    Use the Quadratic Formula.
    $a = 1, b = -4, c = -10$
    $x = \frac{-b \pm \sqrt{b^2 - 4ac}}{2a}$
    $x = \frac{-(-4) \pm \sqrt{(-4)^2 - 4(1)(-10)}}{2(1)}$
    $x = \frac{4 \pm \sqrt{56}}{2}$
    $x = \frac{4 \pm 2\sqrt{14}}{2}$
    $x = 2 \pm \sqrt{14}$
    $x \approx 5.7, -1.7$

11. $3x^2 - 2x - 4 = 0$
    Use the Quadratic Formula.
    $a = 3, b = -2, c = -4$
    $x = \frac{-b \pm \sqrt{b^2 - 4ac}}{2a}$
    $x = \frac{-(-2) \pm \sqrt{(-2)^2 - 4(3)(-4)}}{2(3)}$
    $x = \frac{2 \pm \sqrt{52}}{6}$
    $x = \frac{2 \pm 2\sqrt{13}}{6}$
    $x = \frac{1 \pm \sqrt{13}}{3}$
    $x \approx 1.5, -0.9$

12. $x^2 = x + 15$
    $x^2 - x - 15 = 0$
    Use the Quadratic Formula.
    $a = 1, b = -1, c = -15$
    $x = \frac{-b \pm \sqrt{b^2 - 4ac}}{2a}$
    $x = \frac{-(-1) \pm \sqrt{(-1)^2 - 4(1)(-15)}}{2(1)}$
    $x = \frac{1 \pm \sqrt{61}}{2}$
    $x \approx 4.4, -3.4$

13. $2x^2 + x = 15$
    $2x^2 + x - 15 = 0$
    Use the Quadratic Formula.
    $a = 2, b = 1, c = -15$
    $x = \frac{-b \pm \sqrt{b^2 - 4ac}}{2a}$
    $x = \frac{-1 \pm \sqrt{1^2 - 4(2)(-15)}}{2(2)}$
    $x = \frac{-1 \pm \sqrt{121}}{4}$
    $x = \frac{-1 \pm 11}{4}$
    $x = 2.5, -3$

10-1 Circles and Circumference

Page 524 Geometry Activity: Circumference Ratio

1. See students' work.
2. Each ratio should be near 3.1.
3. $C \approx 3.14d$

Pages 525–526 Check for Understanding

1. Sample answer: The value of π is calculated by dividing the circumference of a circle by the diameter.
2. $d = 2r$, $r = \frac{1}{2}d$
3. Except for a diameter, two radii and a chord of a circle can form a triangle. The Triangle Inequality Theorem states that the sum of two sides has to be greater than the third. So, $2r$ has to be greater than the measure of any chord that is not a diameter, but $2r$ is the measure of the diameter. So the diameter has to be longer than any other chord of the circle.
4. The circle has its center at E, so it is named circle E, or $\odot E$.
5. Four radii are shown: $\overline{EA}$, $\overline{EB}$, $\overline{EC}$, or $\overline{ED}$.
6. Three chords are shown: $\overline{AB}$, $\overline{AC}$, or $\overline{BD}$.
7. $\overline{AC}$ and $\overline{BD}$ are chords that go through the center, so $\overline{AC}$ and $\overline{BD}$ are diameters.
8. $r = \frac{1}{2}d$
   $r = \frac{1}{2}(12)$ or 6
   The radius is 6 mm.
9. $d = 2r$
   $d = 2(5.2)$ or 10.4
   The diameter is 10.4 in.
10. Since the radius of $\odot Z$ is 7, $XZ = 7$.
    $\overline{YZ}$ is part of radius $\overline{XZ}$.
    $XY + YZ = XZ$
    $2 + YZ = 7$
    $YZ = 5$
11. Since the radius of $\odot W$ is 4, $IW = 4$ and $WY = 4$.
    $\overline{IX}$ is part of diameter $\overline{IY}$.
    $IX + XY = IW + WY$
    $IX + 2 = 4 + 4$
    $IX = 6$
12. $IC = IW + WY + XZ + ZC - XY$
    $IC = 4 + 4 + 7 + 7 - 2$
    $IC = 20$
13. $d = 2r$
    $d = 2(5)$ or 10 m
    $C = \pi d$
    $C = \pi(10)$ or about 31.42 m
14. $C = \pi d$
    $2368 = \pi d$
    $\frac{2368}{\pi} = d$
    $753.76 \approx d$
    $d \approx 753.76$ ft
    $r = \frac{1}{2}d$
    $r \approx \frac{1}{2}(753.76)$ or 376.88 ft
15. B; $C = \pi d$
    $C = \pi(9)$ or 9π mm

Pages 526–527 Practice and Apply

16. The circle has its center at F, so it is named circle F, or $\odot F$.
17. Three radii are shown: $\overline{FA}$, $\overline{FB}$, or $\overline{FE}$.
18. Two chords are shown: $\overline{BE}$ or $\overline{CD}$.
19. $\overline{BE}$ is the only chord that goes through the center, so $\overline{BE}$ is a diameter.
20. $\overline{FA}$ is a radius not contained in a diameter.
21. The circle has its center at R, so it is named circle R, or $\odot R$.
22. Six radii are shown: $\overline{RT}$, $\overline{RU}$, $\overline{RV}$, $\overline{RW}$, $\overline{RX}$, or $\overline{RZ}$.
23. Three chords are shown: $\overline{ZV}$, $\overline{TX}$, or $\overline{WZ}$.
24. $\overline{TX}$ or $\overline{WZ}$ are the chords that go through the center, so $\overline{TX}$ and $\overline{WZ}$ are diameters.
25. $\overline{RU}$ and $\overline{RV}$ are radii not contained in a diameter.
26. $d = 2r$
    $d = 2(2)$ or 4 ft
27. $r = \frac{1}{2}d$
    $r = \frac{1}{2}(5)$ or 2.5 ft
28. $TR = \frac{1}{2}(TX)$
    $TR = \frac{1}{2}(120)$ or 60 cm
29. $ZW = 2RZ$
    $ZW = 2(32)$ or 64 in. or 5 ft 4 in.
30. $\overline{UR}$ and $\overline{RV}$ are both radii.
    $RV = UR$
    $RV = 18$ in.
31. $\overline{XT}$ is a diameter and $\overline{UR}$ is a radius.
    $UR = \frac{1}{2}XT$
    $UR = \frac{1}{2}(1.2)$ or 0.6 m
32. $AZ = CW$
    $AZ = 2$
33. $\overline{AX}$ is a radius of $\odot A$, and $\overline{ZX}$ is part of $\overline{AX}$.
    $AZ + ZX = AX$
    $2 + ZX = \frac{1}{2}(10)$
    $ZX = 3$
34. $\overline{BZ}$ is a radius of $\odot B$, and $\overline{BX}$ is part of $\overline{BZ}$.
    $ZX + BX = BZ$
    $3 + BX = \frac{1}{2}(30)$
    $3 + BX = 15$
    $BX = 12$
35. $BY = BX$
    $BY = 12$
36. $YW = ZX$
    $YW = 3$
37. $AC = AZ + ZW + WC$
    $AC = 2 + 30 + 2$ or 34
38. $FG = GH$
    $FG = 10$
39. $FH = FG + GH$
    $FH = 10 + 10$ or 20
40. $GL = GH$
    $GL = 10$
41. $\overline{GL}$ is a diameter of $\odot J$, so $GL = 10$.
    $\overline{GJ}$ is a radius of $\odot J$.
    $GJ = \frac{1}{2}(GL)$
    $GJ = \frac{1}{2}(10)$ or 5

42. $\overline{JL}$ is a radius of $\odot J$.

$JL = \frac{1}{2}(10)$ or 5

43. $\overline{JL}$ is a diameter of $\odot K$, so $JL = 5$.
$\overline{JK}$ is a radius of $\odot K$.

$JK = \frac{1}{2}JL$

$JK = \frac{1}{2}(5)$ or 2.5

44. $d = 2r$
$d = 2(7)$ or 14 mm
$C = \pi d$
$C = \pi(14)$ or about 43.98 mm

45. $r = \frac{1}{2}d$

$r = \frac{1}{2}(26.8)$ or 13.4 cm
$C = \pi d$
$C = \pi(26.8)$ or about 84.19 cm

46. $C = \pi d$
$26\pi = \pi d$
$26 = d$ or $d = 26$ mi

$r = \frac{1}{2}d$

$r = \frac{1}{2}(26)$ or 13 mi

47. $C = \pi d$
$76.4 = \pi d$
$\frac{76.4}{\pi} = d$
$24.32 \approx d$ or $d \approx 24.32$ m

$r = \frac{1}{2}d$

$r \approx \frac{1}{2}(24.32)$ or 12.16 m

48. $r = \frac{1}{2}d$

$r = \frac{1}{2}\left(12\frac{1}{2}\right)$ or $6\frac{1}{4}$ yd
$C = \pi d$
$C = \pi\left(12\frac{1}{2}\right)$ or about 39.27 yd

49. $d = 2r$

$d = 2\left(6\frac{3}{4}\right)$ or $13\frac{1}{2}$ in.
$C = \pi d$
$C = \pi\left(13\frac{1}{2}\right)$ or about 42.41 in.

50. $r = \frac{1}{2}d$

$r = \frac{1}{2}(2a)$ or a
$C = \pi d$
$C = \pi(2a)$ or about $6.28a$

51. $d = 2r$

$d = 2\left(\frac{a}{6}\right)$ or about $0.33a$
$C = 2\pi r$
$C = 2\pi\left(\frac{a}{6}\right)$ or about $1.05a$

52. The diameter of the circle is the same as the hypotenuse of the right triangle.
$d^2 = 16^2 + 30^2$
$d^2 = 1156$
$d = \sqrt{1156}$ or 34 m
$C = \pi d$
$C = \pi(34)$ or 34π m

53. The diameter of the circle is the same as the hypotenuse of the right triangle.
$d^2 = 3^2 + 4^2$
$d^2 = 25$
$d = \sqrt{25}$ or 5 ft
$C = \pi d$
$C = \pi(5)$ or 5π ft

54. The diameter of the circle is the same as the hypotenuse of the right triangle.
$d^2 = 10^2 + 10^2$
$d^2 = 200$
$d = \sqrt{200}$ or $10\sqrt{2}$ in.
$C = \pi d$
$C = \pi(10\sqrt{2})$ or $10\pi\sqrt{2}$ in.

55. The diameter of the circle is the same as the hypotenuse of the right triangle.
$d^2 = (4\sqrt{2})^2 + (4\sqrt{2})^2$
$d^2 = 64$
$d = \sqrt{64}$ or 8 cm
$C = \pi d$
$C = \pi(8)$ or 8π cm

56. 1; This description is the definition of a radius.

57. 0; The longest chord of a circle is the diameter, which contains the center.

58. $C = 2\pi r$
$C = 2\pi(800)$ or about 5026.5 ft

59. $800 - 200 = 600$
$800 - 300 = 500$
The range of values for the radius of the explosion circle is 500 to 600 ft.

60. $C = 2\pi r$
$C = 2\pi(500)$ or about 3142
$C = 2\pi(600)$ or about 3770
The least and maximum circumferences are 3142 ft and 3770 ft, respectively.

61. Let r = the radius of $\odot O$.
$x^2 + y^2 = r^2$
$p^2 + t^2 = r^2$
$x^2 + y^2 + p^2 + t^2 = 288$
Substitute r^2 for $x^2 + y^2$ and r^2 for $p^2 + t^2$.
$r^2 + r^2 = 288$
$2r^2 = 288$
$r^2 = 144$
$r = 12$
$C = 2\pi r$
$C = 2\pi(12)$ or 24π units

62. Sample answer: about 251.3 feet. Answers should include the following.

- The distance the animal travels is approximated by the circumference of the circle.
- The diameter for the circle on which the animal is located becomes $80 - 2$ or 78. The circumference of this circle is 78π. Multiply by 22 to get a total distance of $22(78\pi)$ or about 5391 feet. This is a little over a mile.

63. Let r = the radius of $\odot C$. Then $2r$ = the radius of $\odot B$, and $4r$ = the radius of $\odot A$.

sum of circumferences $= 2\pi r + 2\pi(2r) + 2\pi(4r)$

$42\pi = 2\pi r + 4\pi r + 8\pi r$

$42\pi = 14\pi r$

$3 = r$

$AC = r + 2(2r) + 4r$

$AC = 3 + 2(2 \cdot 3) + 4(3)$

$AC = 3 + 12 + 12$ or 27

64. A; $\frac{100d}{k}$% of gasoline has been pumped.

65. Small circle: $C = 2\pi r$

$C = 2\pi(5)$ or 10π

Medium circle: $r = 5 + 5$ or 10

$C = 2\pi r$

$C = 2\pi(10)$ or 20π

Large circle: $r = 5 + 5 + 5$ or 15

$C = 2\pi r$

$C = 2\pi(15)$ or 30π

The circumferences from least to greatest are 10π, 20π, and 30π.

Page 528 Maintain Your Skills

66. $|\overline{AB}| = \sqrt{1^2 + 4^2}$

$= \sqrt{17}$

≈ 4.1

$m\angle A = \tan^{-1}\frac{4}{1}$

≈ 76

The magnitude is about 4.1 and the direction is about 76°.

67. $|\vec{V}| = \sqrt{4^2 + 9^2}$

$= \sqrt{97}$

≈ 9.8

$mV = \tan^{-1}\frac{9}{4}$

≈ 66

The magnitude is about 9.8 and the direction is about 66°.

68. $|\overline{AB}| = \sqrt{(7-4)^2 + (22-2)^2}$

$= \sqrt{3^2 + 20^2}$

$= \sqrt{409}$

≈ 20.2

$\tan A = \frac{22-2}{7-4}$

$= \frac{20}{3}$

$m\angle A = \tan^{-1}\frac{20}{3}$

≈ 81

The magnitude is about 20.2 and the direction is about 81°.

69. $|\overrightarrow{CD}| = \sqrt{(40-0)^2 + (0-(-20))^2}$

$= \sqrt{40^2 + 20^2}$

$= \sqrt{2000}$

≈ 44.7

$\tan C = \frac{0-(-20)}{40-0}$

$= \frac{20}{40}$ or $\frac{1}{2}$

$m\angle C = \tan^{-1}\frac{1}{2}$

≈ 27

The magnitude is about 44.7 and the direction is about 27°.

70. $A'B' = |k|(AB)$

$A'B' = 6(5)$

$A'B' = 30$

71. $A'B' = |k|(AB)$

$A'B' = 1.5(16)$

$A'B' = 24$

72. $A'B' = |k|(AB)$

$= \frac{1}{2}\left(\frac{2}{3}\right)$

$= \frac{1}{3}$

73. Given: $\overline{RQ}$ bisects $\angle SRT$

Prove: $m\angle SQR > m\angle SRQ$

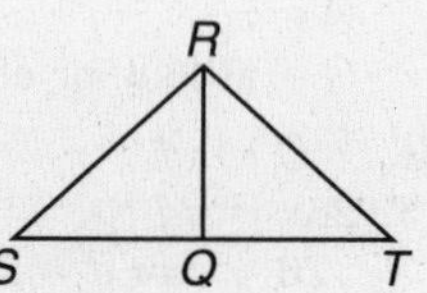

Statements	Reasons
1. $\overline{RQ}$ bisects $\angle SRT$.	1. Given
2. $\angle SRQ \cong \angle QRT$	2. Def. of $\angle$ bisector
3. $m\angle SRQ = m\angle QRT$	3. Def. of $\cong$ $\angle$s
4. $m\angle SQR = m\angle T + m\angle QRT$	4. Exterior Angle Theorem
5. $m\angle SQR > m\angle QRT$	5. Def. of Inequality
6. $m\angle SQR > m\angle SRQ$	6. Substitution

74. a is the midpoint of $\overline{OF}$. Therefore, the missing coordinates are $(2a, 0)$.

75. $x + 2x = 180$ — Linear pair

$3x = 180$

$x = 60$

76. $2x + 3x = 90$ — Lines are perpendicular.

$5x = 90$

$x = 18$

77. $(3x + x) + 2x = 180$ — Linear pair

$6x = 180$

$x = 30$

78. $3x + 5x = 180$ — Linear pair

$8x = 180$

$x = 22.5$

79. $3x = 90$ — $\perp$ lines form 2 rt $\angle$s.

$x = 30$

80. $x + x + x = 360$

$3x = 360$

$x = 120$

10-2 Angles and Arcs

Pages 532–533 Check for Understanding

1. Sample answer:

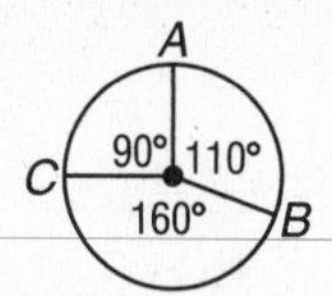

$\widehat{AB}, \widehat{BC}, \widehat{AC}, \widehat{ABC}, \widehat{BCA}, \widehat{CAB}$; $m\widehat{AB} = 110$,
$m\widehat{BC} = 160$, $m\widehat{AC} = 90$, $m\widehat{ABC} = 270$,
$m\widehat{BCA} = 250$, $m\widehat{CAB} = 200$

2. A diameter divides the circle into two congruent arcs. Without the third letter, it is impossible to know which semicircle is being referenced.
3. Sample answer: Concentric circles have the same center, but different radius measures; congruent circles usually have different centers but the same radius measure.
4. $\angle MCN$ and $\angle NCL$ are a linear pair.
$m\angle MCN + m\angle NCL = 180$
$60 + m\angle NCL = 180$
$m\angle NCL = 120$
5. $\angle MCR$ and $\angle RCL$ are a linear pair.
$m\angle MCR + m\angle RCL = 180$
$(x - 1) + (3x + 5) = 180$
$4x + 4 = 180$
$4x = 176$
$x = 44$
Use the value of x to find $m\angle RCL$.
$m\angle RCL = 3x + 5$
$= 3(44) + 5$
$= 132 + 5$ or 137
6. Use the value of x to find $m\angle RCM$.
From Exercise 5, $x = 44$.
$m\angle RCM = x - 1$
$= 44 - 1$ or 43
7. $\angle RCN$ is composed of adjacent angles, $\angle RCM$ and $\angle MCN$.
$m\angle RCN = m\angle RCM + m\angle MCN$
$= 43 + 60$ or 103
8. $\widehat{BC}$ is a minor arc, so $m\widehat{BC} = m\angle BAC$.
$\angle BAC \cong \angle EAD$
$m\angle BAC = m\angle EAD$
$m\widehat{BC} = m\angle EAD$
$m\widehat{BC} = 42$
9. $\widehat{CBE}$ is a semicircle.
$m\widehat{CBE} = 180$
10. One way to find $m\widehat{EDB}$ is by using $\widehat{EDC}$ and $\widehat{CB}$.
$\widehat{EDC}$ is a semicircle.
$m\widehat{EDB} = m\widehat{EDC} + m\widehat{CB}$
$m\widehat{EDB} = 180 + 42$ or 222
11. One way to find $m\widehat{CD}$ is by using $\widehat{CDE}$ and $\widehat{DE}$.
$\widehat{CDE}$ is a semicircle.
$m\widehat{CD} + m\widehat{DE} = m\widehat{CDE}$
$m\widehat{CD} + 42 = 180$
$m\widehat{CD} = 138$
12. $C = 2\pi r$
$C = 2\pi(12)$ or 24π
Let ℓ = arc length.
$\frac{60}{360} = \frac{\ell}{24\pi}$
$\frac{60}{360}(24\pi) = \ell$
$4\pi = \ell$
The length of $\widehat{TR}$ is 4π units or about 12.57 units.
13. Sample answer:
$25\%(360°) = 90°$, $23\%(360°) = 83°$,
$28\%(360°) = 101°$, $22\%(360°) = 79°$,
$2\%(360°) = 7°$

Pages 533–535 Practice and Apply

14. $\angle AGC$ and $\angle CGB$ are a linear pair.
$m\angle AGC + m\angle CGB = 180$
$60 + m\angle CGB = 180$
$m\angle CGB = 120$
15. $\angle AGC$ and $\angle BGE$ are vertical angles.
$m\angle BGE = m\angle AGC$
$m\angle BGE = 60$
16. $\angle AGD$ is a right angle.
$m\angle AGD = 90$
17. One way to find $m\angle DGE$ is by using $\angle AGD$ and $\angle BGE$. $\angle AGB$ is a straight angle.
$m\angle AGD + m\angle DGE + m\angle BGE = m\angle AGB$
$90 + m\angle DGE + 60 = 180$
$m\angle DGE = 30$
18. $\angle CGD$ is composed of adjacent angles, $\angle CGA$ and $\angle AGD$.
$m\angle CGD = m\angle CGA + m\angle AGD$
$m\angle CGD = 60 + 90$
$m\angle CGD = 150$
19. $\angle AGE$ is composed of adjacent angles, $\angle AGD$ and $\angle DGE$.
$m\angle AGE = m\angle AGD + m\angle DGE$
$m\angle AGE = 90 + 30$
$m\angle AGE = 120$
20. $\angle ZXV$ and $\angle YXW$ are vertical angles.
$m\angle ZXV = m\angle YXW$
$2x + 65 = 4x + 15$
$50 = 2x$
$25 = x$
Use the value of x to find $m\angle ZXV$.
$m\angle ZXV = 2x + 65$
$= 2(25) + 65$
$= 50 + 65$ or 115
21. Use the value of x to find $m\angle YXW$.
From Exercise 20, $x = 25$.
$m\angle YXW = 4x + 15$
$= 4(25) + 15$
$= 100 + 15$ or 115
22. $\angle ZXY$ and $\angle ZXV$ are a linear pair.
$m\angle ZXY + m\angle ZXV = 180$
$m\angle ZXY + 115 = 180$
$m\angle ZXY = 65$
23. $\angle ZXY$ and $\angle VXW$ are vertical angles.
$\angle VXW \cong \angle ZXY$
$m\angle VXW = m\angle ZXY$
$m\angle VXW = 65$

24. $\overarc{BC}$ is a minor arc, so $m\overarc{BC} = m\angle BOC$.
Since $\overline{AB}$ is a diameter and $\angle AOC$ is a right angle, $m\angle AOB = 180$ and $m\angle AOC = 90$.
$m\angle BOC + m\angle AOC = m\angle AOB$
$m\angle BOC + 90 = 180$
$m\angle BOC = 90$
$m\overarc{BC} = 90$

25. $\overarc{AC}$ is a minor arc, so $m\overarc{AC} = m\angle AOC$.
$\angle AOC$ is a right angle.
$m\overarc{AC} = m\angle AOC$
$m\overarc{AC} = 90$

26. $\overarc{AE}$ is a minor arc, so $m\overarc{AE} = m\angle AOE$.
$\angle AOE$ and $\angle BOC$ are vertical angles.
$\angle AOE \cong \angle BOC$
$m\angle AOE = m\angle BOC$
$m\overarc{AE} = m\angle BOC$
$m\overarc{AE} = 90$

27. $\overarc{EB}$ is a minor arc, so $m\overarc{EB} = m\angle EOB$.
$\angle EOB$ and $\angle AOC$ are vertical angles.
$\angle EOB \cong \angle AOC$
$m\angle EOB = m\angle AOC$
$m\overarc{EB} = m\angle AOC$
$m\overarc{EB} = 90$

28. Since $\overline{AB}$ is a diameter, $m\angle AOB = 180$.
$m\overarc{ACB} = m\angle AOB$
$m\overarc{ACB} = 180$

29. Since $\angle BOD \cong \angle DOE \cong \angle EOF \cong \angle FOA$,
$m\angle BOD = m\angle DOE = m\angle EOF = m\angle FOA$.
Since $m\angle AOB = 180$, each of the four angles measures $\frac{180}{4}$ or 45.
$\overarc{AD}$ is composed of adjacent arcs, $\overarc{DE}$, $\overarc{EF}$, and $\overarc{FA}$.
$m\overarc{AD} = m\overarc{DE} + m\overarc{EF} + m\overarc{FA}$
$m\overarc{AD} = m\angle DOE + m\angle EOF + m\angle FOA$
$m\overarc{AD} = 45 + 45 + 45$ or 135

30. $\overarc{CBF}$ is composed of adjacent arcs, $\overarc{CB}$, $\overarc{BD}$, $\overarc{DE}$, and $\overarc{EF}$.
$m\overarc{CBF} = m\overarc{CB} + m\overarc{BD} + m\overarc{DE} + m\overarc{EF}$
$m\overarc{CBF} = m\angle COB + m\angle BOD + m\angle DOE + m\angle EOF$
$m\overarc{CBF} = 90 + 45 + 45 + 45$ or 225

31. $m\overarc{ADC} = 360 - m\overarc{AC}$
$m\overarc{ADC} = 360 - m\angle AOC$
$m\overarc{ADC} = 360 - 90$ or 270

32. Find the value of x.
Since $\overline{VY}$ is a diameter, $\overarc{VUY}$ is a semicircle. $\overarc{VUY}$ is composed of adjacent arcs, $\overarc{VU}$ and $\overarc{UY}$. $\overarc{VU}$ is a minor arc, so $m\overarc{VU} = m\angle VZU$. $\overarc{UY}$ is a minor arc, so $m\overarc{UY} = m\angle UZY$.
$m\overarc{VU} + m\overarc{UY} = m\overarc{VUY}$
$m\angle VZU + m\angle UZY = 180$
$4x + (2x + 24) = 180$
$6x + 24 = 180$
$6x = 156$
$x = 26$
Use the value of x to find $m\overarc{UY}$.
$m\overarc{UY} = 2x + 24$
$m\overarc{UY} = 2(26) + 24$
$m\overarc{UY} = 52 + 24$ or 76

33. $\angle WZV$ and $\angle UZY$ are vertical angles.
$\angle WZV \cong \angle UZY$
$m\angle WZV = m\angle UZY$
$m\overarc{WV} = m\overarc{UY}$
$m\overarc{WV} = 76$

34. $m\overarc{WX} = m\angle WZX$
$m\overarc{WX} = 2x$
From Exercise 32, $x = 26$.
$m\overarc{WX} = 2(26)$ or 52

35. $\angle XZY \cong \angle WZX$
$m\angle XZY = m\angle WZX$
$m\overarc{XY} = m\overarc{WX}$
$m\overarc{XY} = 52$

36. $m\overarc{WUY} = 360 - m\overarc{WX} - m\overarc{XY}$
$m\overarc{WUY} = 360 - 52 - 52$ or 256

37. $m\overarc{YVW} = 360 - m\overarc{WX} - m\overarc{XY}$
$m\overarc{YVW} = 360 - 52 - 52$ or 256

38. $m\overarc{XVY} = 360 - m\overarc{XY}$
$m\overarc{XVY} = 360 - 52$ or 308

39. $m\overarc{WUX} = 360 - m\overarc{WX}$
$m\overarc{WUX} = 360 - 52$ or 308

40. $C = \pi d$
$C = \pi(32)$ or 32π
Let ℓ = arc length.
$\frac{100}{360} = \frac{\ell}{32\pi}$
$\frac{100}{360}(32\pi) = \ell$
$\frac{80\pi}{9} = \ell$
The length of $\overarc{DE}$ is $\frac{80\pi}{9}$ units or about 27.93 units.

41. $C = d\pi$
$C = 32\pi$
$m\overarc{DHE} = 360 - m\angle DCE$
$m\overarc{DHE} = 360 - 90$ or 270
Let ℓ = arc length.
$\frac{270}{360} = \frac{\ell}{32\pi}$
$\frac{270}{360}(32\pi) = \ell$
$24\pi = \ell$
The length of $\overarc{DHE}$ is 24π units or about 75.40 units.

42. $m\overarc{HDF} = 360 - m\angle HCF$
$m\overarc{HDF} = 360 - 125$ or 235
Let ℓ = arc length.
$\frac{235}{360} = \frac{\ell}{32\pi}$
$\frac{235}{360}(32\pi) = \ell$
$\frac{188\pi}{9} = \ell$
The length of $\overarc{HDF}$ is $\frac{188\pi}{9}$ units or about 65.62 units.

43. Let ℓ = arc length.
$\frac{45}{360} = \frac{\ell}{32\pi}$
$\frac{45}{360}(32\pi) = \ell$
$4\pi = \ell$
The length of $\overarc{HD}$ is 4π units or about 12.57 units.

44. Sample answer: $76\%(360°) = 273°$, $16\%(360°) = 58°$, $5\%(360°) = 18°$, $3\%(360°) = 11°$

45. The first category is a major arc, and the other three categories are minor arcs.

46.

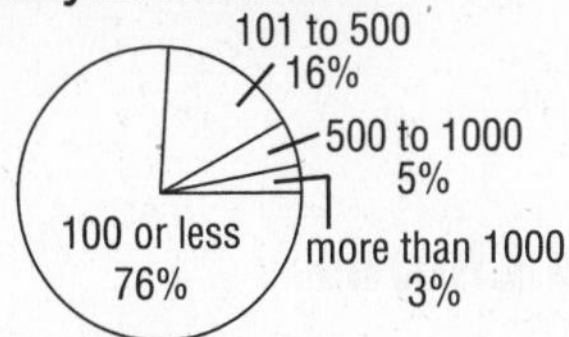

47. always

48. Sometimes; the central angle of a minor arc can be greater than 90°.

49. Never; the sum of the measures of the central angles of a circle is always 360.

50. always

51. Let $m\angle 1 = 2x$, then $m\angle 2 = 3x$ and $m\angle 3 = 4x$.

$m\angle 1 + m\angle 2 + m\angle 3 = 360$

$2x + 3x + 4x = 360$

$9x = 360$

$x = 40$

Therefore, $m\angle 1 = 2(40)$ or 80, $m\angle 2 = 3(40)$ or 120, and $m\angle 3 = 4(40)$ or 160.

52. $C = \pi d$

$C = \pi(12)$ or 12π in.

The measure of the angle from the minute hand to the hour hand at 2:00 is 60.

Let ℓ = arc length.

$\frac{60}{360} = \frac{\ell}{12\pi}$

$\frac{60}{360}(12\pi) = \ell$

$2\pi = \ell$

The arc length is 2π in. or about 6.3 in.

53. $C = 2\pi r$

$C = 2\pi(12)$ or 24π

Let ℓ = arc length.

$\frac{270}{360} = \frac{\ell}{24\pi}$

$\frac{270}{360}(24\pi) = \ell$

$18\pi = \ell$

The length of the arc is 18π ft or about 56.5 ft.

54. Given: $\angle BAC \cong \angle DAE$

Prove: $\widehat{BC} \cong \widehat{DE}$

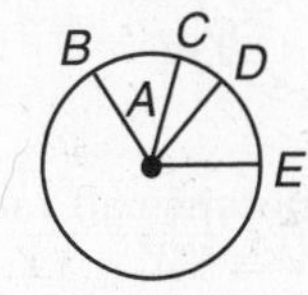

Proof:

Statements	**Reasons**
1. $\angle BAC \cong \angle DAE$	1. Given
2. $m\angle BAC = m\angle DAE$	2. Def. of ≅ ∠s
3. $m\widehat{BC} = m\widehat{DE}$	3. Def. of arc measure
4. $\widehat{BC} \cong \widehat{DE}$	4. Def. of ≅ arcs

55. No; the radii are not equal, so the proportional part of the circumferences would not be the same. Thus, the arcs would not be congruent.

56. Sample answer: The hands of the clock form central angles.

Answers should include the following.

- The hands form acute, right, and obtuse angles.
- Some times when the angles formed by the minute and hour hand are congruent are 1:00 and 11:00, 2:00 and 10:00, 3:00 and 9:00, 4:00 and 8:00, and 5:00 and 7:00. They also form congruent angles many other times of the day, such as 3:05 and 8:55.

57. B; $C = 2\pi r$ or about $6.3r$

$P = 2\ell + 2w$

$P = 2(2r) + 2r$ or $6r$

Since $6.3r > 6r$, the circumference of the circle is greater than the perimeter of the rectangle.

58. Rewrite 3:5:10 as $3x$:$5x$:$10x$ and use these measures for the measures of the central angles of the circle.

$3x + 5x + 10x = 360$

$18x = 360$

$x = 20$

The measures of the angles are 3(20) or 60, 5(20) or 100, and 10(20) or 200.

Page 535 Maintain Your Skills

59. $d = 2r$

$d = 2(10)$ or 20

$C = \pi d$

$C = \pi(20)$ or about 62.83

60. $r = \frac{1}{2}d$

$r = \frac{1}{2}(13)$ or 6.5

$C = \pi d$

$C = \pi(13)$ or about 40.84

61. $C = \pi d$

$28\pi = \pi d$

$28 = d$ or $d = 28$

$r = \frac{1}{2}d$

$r = \frac{1}{2}(28)$ or 14

62. $C = \pi d$

$75.4 = \pi d$

$\frac{75.4}{\pi} = d$

$d \approx 24.00$

$r = \frac{1}{2}d$

$r \approx \frac{1}{2}(24.00)$ or 12.00

63. magnitude = $\sqrt{(72)^2 + (45)^2}$ or about 84.9

direction = $\tan^{-1}\frac{45}{72}$ or about 32

The magnitude is about 84.9 newtons, and the direction is about 32° north of due east.

64. $\frac{12}{18 - x} = \frac{10}{x}$

$12x = 10(18 - x)$

$12x = 180 - 10x$

$22x = 180$

$x = 8\frac{2}{11}$

65. $\frac{26.2}{x} = \frac{17.3}{24.22}$
$26.2(24.22) = 17.3x$
$634.564 = 17.3x$
$36.68 = x$

66. Construct a line perpendicular to the line with the equation $y - 7 = 0$ through point $Q(6, -2)$.

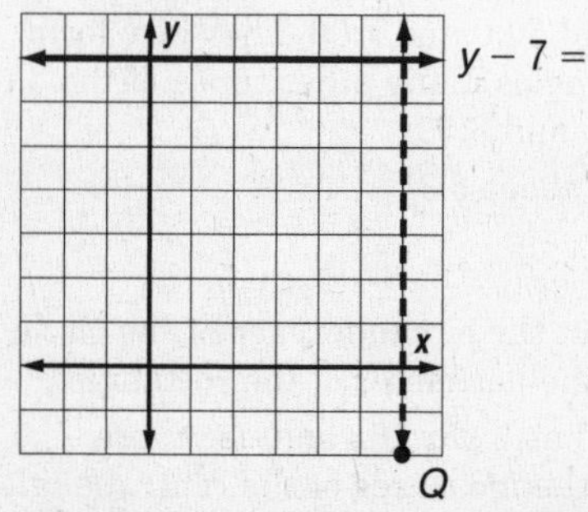

The line $y - 7 = 0$ is a horizontal line, so the line perpendicular to the given line is vertical. The desired distance is $|-2 - 7|$ or 9.
The distance is 9 units.

67. First, write the equation of a line p perpendicular to both lines. The slope of each of the given lines is 1. So the slope of p is -1. Use the y-intercept of the first line, (0, 3).
$y - y_1 = m(x - x_1)$
$y - 3 = -1(x - 0)$
$y - 3 = -x$
$y = -x + 3$
Next, find the point of intersection using the equations $y = -x + 3$ and $y = x - 4$.
$-x + 3 = x - 4$
$-2x = -7$
$x = \frac{7}{2}$
$y = \frac{7}{2} - 4$ or $y = -\frac{1}{2}$
The point of intersection is $\left(\frac{7}{2}, -\frac{1}{2}\right)$.
Finally, find the distance between (0, 3) and $\left(\frac{7}{2}, -\frac{1}{2}\right)$.
$d = \sqrt{(x_2 - x_1)^2 + (y_2 - y_1)^2}$
$= \sqrt{\left(\frac{7}{2} - 0\right)^2 + \left(-\frac{1}{2} - 3\right)^2}$
$= \sqrt{24.5}$
The distance between the lines is $\sqrt{24.5}$ units.

68. $90 - 57.5 = 32.5$
$180 - 57.5 = 122.5$
The measures of the complement and supplement are 32.5 and 122.5, respectively.

69. If ABC has three sides, then ABC is a triangle.

70. Both are true.

71. $x = 42$ since the triangle is isosceles.

72. $x + x + 30 = 180$ by the Angle Sum Theorem
$2x = 150$
$x = 75$

73. $40 + 40 + x = 180$
$x = 180 - 80$ or 100

74. $x + x + 90 = 180$
$2x = 90$
$x = 45$

75. $2x + 2x + x = 180$
$5x = 180$
$x = 36$

76. $3x = 180$
$x = 60$

10-3 Arcs and Chords

Page 538 Geometry Activity: Congruent Chords and Distance

1. $\overline{SU}$ and $\overline{SX}$ are perpendicular bisectors of $\overline{VT}$ and $\overline{WY}$, respectively.
2. $VT = WY$, $SU = SX$
3. Sample answer: When the chords are congruent, they are equidistant from the center of the circle.

Pages 539–540 Check for Understanding

1. Sample answer: An inscribed polygon has all vertices on the circle. A circumscribed circle means the circle is drawn around so that the polygon lies in its interior and all vertices lie on the circle.
2. Sample answer:

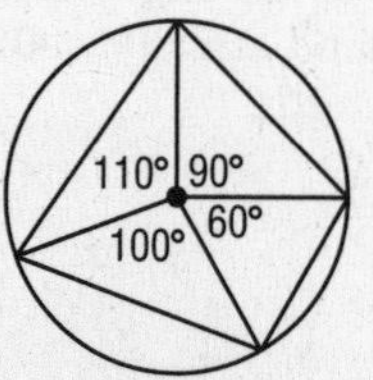

None of the sides are congruent.

3. Tokei; to bisect the chord, it must be a diameter and be perpendicular.
4. **Given:** $\odot X$, $\overline{UV} \cong \overline{WY}$
Prove: $\widehat{UV} \cong \widehat{WY}$

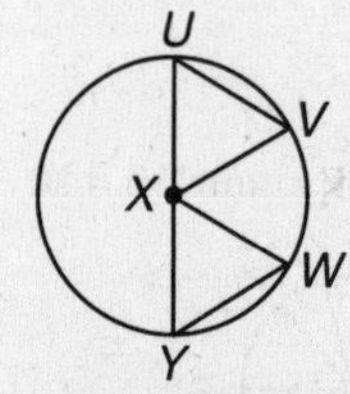

Proof: Because all radii are congruent, $\overline{XU} \cong \overline{XV} \cong \overline{XW} \cong \overline{XY}$. You are given that $\overline{UV} \cong \overline{WY}$, so $\triangle UVX \cong \triangle WYX$, by SSS. Thus, $\angle UXV \cong \angle WXY$ by CPCTC. Since the central angles have the same measure, their intercepted arcs have the same measure and are therefore, congruent. Thus, $\widehat{UV} \cong \widehat{WY}$.

5. $\overline{OY}$ bisects $\widehat{AB}$, so $m\widehat{AY} = \frac{1}{2}m\widehat{AB}$.
$m\widehat{AY} = \frac{1}{2}m\widehat{AB}$
$m\widehat{AY} = \frac{1}{2}(60)$ or 30

6. $\overline{OY}$ bisects $\overline{AB}$, so $AX = \frac{1}{2}(AB)$.
$AX = \frac{1}{2}(AB)$
$AX = \frac{1}{2}(10)$ or 5

7. Draw radius $\overline{OA}$. Radius $\overline{OA}$ is the hypotenuse of $\triangle OXA$.
$(OX)^2 + (AX)^2 = (OA)^2$
$(OX)^2 + 5^2 = 10^2$
$(OX)^2 + 25 = 100$
$(OX)^2 = 75$
$OX = \sqrt{75}$ or $5\sqrt{3}$

8. $\overline{AB}$ and $\overline{CE}$ are equidistant from P, so $\overline{AB} \cong \overline{CE}$.
$QE = \frac{1}{2}(CE)$, so $CE = 2(20)$ or 40.
$\overline{AB} \cong \overline{CE}$, so $AB = 40$.

9. $(PE)^2 = (PQ)^2 + (QE)^2$
$(PE)^2 = 10^2 + 20^2$
$(PE)^2 = 100 + 400$
$(PE)^2 = 500$
$PE = 10\sqrt{5}$ or about 22.36

10. $\widehat{AB} \cong \widehat{BC} \cong \widehat{CA}$
Each arc measures $\frac{360°}{3}$ or 120°.

Pages 540–543 Practice and Apply

11. $\overline{XY}$ bisects $\overline{AB}$, so $AM = \frac{1}{2}(AB)$.
$AM = \frac{1}{2}(AB)$
$AM = \frac{1}{2}(30)$ or 15

12. $MB = AM$
$MB = 15$

13. $\overline{XZ}$ bisects $\overline{CD}$, so $CN = \frac{1}{2}(CD)$.
$CN = \frac{1}{2}(CD)$
$CN = \frac{1}{2}(30)$ or 15

14. $ND = CN$
$ND = 15$

15. $\overline{XZ}$ bisects $\widehat{CD}$, so $m\widehat{DZ} = m\widehat{CZ}$.
$m\widehat{DZ} = m\widehat{CZ}$
$m\widehat{DZ} = 40$

16. $\overline{XZ}$ bisects $\widehat{CD}$, so $m\widehat{CZ} = \frac{1}{2}m\widehat{CD}$.
$m\widehat{CD} = 2m\widehat{CZ}$
$m\widehat{CD} = 2(40)$ or 80

17. $\overline{AB} \cong \overline{CD}$, so $m\widehat{AB} = m\widehat{CD}$.
$m\widehat{AB} = m\widehat{CD}$
$m\widehat{AB} = 80$

18. $\overline{XY}$ bisects $\widehat{AB}$, so $m\widehat{YB} = \frac{1}{2}m\widehat{AB}$.
$\widehat{YB} = \frac{1}{2}m\widehat{AB}$
$\widehat{YB} = \frac{1}{2}(80)$ or 40

19. $(QR)^2+(PR)^2 = (PQ)^2$
$(QR)^2 + 3^2 = 5^2$
$(QR)^2 + 9 = 25$
$(QR)^2 = 16$
$QR = 4$

20. $\overline{PR}$ bisects $\overline{QS}$, so $QS = 2(QR)$.
$QS = 2(QR)$
$QS = 2(4)$ or 8

21. $TV = 13 - 1$ or 12
$TX = TW$ since both are radii.
$(XV)^2 + (TV)^2 = (TX)^2$
$(XV)^2 + (12)^2 = (13)^2$
$(XV)^2 + 144 = 169$
$(XV)^2 = 25$
$XV = 5$

22. $\overline{TZ}$ bisects $\overline{XY}$, so $XV = VY$.
$XY = XV + VY$
$XY = 5 + 5$ or 10

23. $m\widehat{AB} = m\widehat{BC} = m\widehat{CD} = m\widehat{DE} = m\widehat{EF}$
$= m\widehat{FG} = m\widehat{GH} = m\widehat{HA} = \frac{360}{8}$ or 45

24. $m\widehat{LM} = m\widehat{MJ} = m\widehat{JK} = m\widehat{KL} = \frac{360}{4}$ or 90

25. $2x + x + 2x + x = 360$
$6x = 360$
$x = 60$
$2x = 120$
$m\widehat{NP} = m\widehat{RQ} = 120$;
$m\widehat{NR} = m\widehat{PQ} = 60$

26. $(LK)^2 + (FL)^2 = (FK)^2$
$(LK)^2 + 8^2 = 17^2$
$(LK)^2 + 64 = 289$
$(LK)^2 = 225$
$LK = 15$

27. $\overline{FL}$ bisects $\overline{KM}$, so $LK = LM$.
$KM = KL + LM$
$KM = 15 + 15$ or 30

28. $\overline{JG}$ and $\overline{KM}$ are equidistant from the center, so $\overline{JG} \cong \overline{KM}$.
$JG = KM$
$JG = 30$

29. $\overline{FH}$ bisects $\overline{JG}$, so $JH = \frac{1}{2}(JG)$.
$JH = \frac{1}{2}(JG)$
$JH = \frac{1}{2}(30)$ or 15

30. $\overrightarrow{DF}$ bisects $\overline{BC}$, so $FB = CF$.
$FB = CF$
$FB = 8$

31. $\overline{DF}$ bisects $\overline{BC}$, so $BC = FB + CF$.
$BC = FB + CF$
$BC = 8 + 8$ or 16

32. $\overline{AB}$ and $\overline{BC}$ are equidistant from the center, so $\overline{AB} \cong \overline{BC}$.
$AB = BC$
$AB = 16$

33. $(FD)^2 + (CF)^2 = (DC)^2$
$(FD)^2 + 8^2 = 10^2$
$(FD)^2 + 64 = 100$
$(FD)^2 = 36$
$FD = 6$
Since $ED = FD$, $ED = 6$.

34. Since $\overline{XY}$ and $\overline{ST}$ are equidistant from the center, $\overline{XY} \cong \overline{ST}$.
$XY = ST$
$4a - 5 = -5a + 13$
$9a = 18$
$a = 2$
$ST = -5a + 13$
$ST = -5(2) + 13$
$ST = -10 + 13$ or 3
$SQ = \frac{1}{2}(ST) = \frac{1}{2}(3)$ or 1.5

35. Since $AC = 20$, then $BC = \frac{1}{2}(20)$ or 10.
Since $m\angle ACE = 45$ and $m\angle BDC = 90$, then $m\angle CBD = 180 - (90 + 45)$ or 45.

Therefore $CD = 5x$.

$(CD)^2 + (BD)^2 = (BC)^2$

$(5x)^2 + (5x)^2 = 10^2$

$25x^2 + 25x^2 = 100$

$50x^2 = 100$

$x^2 = 2$

$x = \sqrt{2}$ or about 1.41

36a. Given

36b. All radii are congruent.

36c. Reflexive Property

36d. Definition of perpendicular lines

36e. $\triangle ARP \cong \triangle BRP$

36f. CPCTC

36g. If central angles are congruent, intercepted arcs are congruent.

37. Given: $\odot O$, $\overline{OS} \perp \overline{RT}$, $\overline{OV} \perp \overline{UW}$, $\overline{OS} \cong \overline{OV}$

Prove: $\overline{RT} \cong \overline{UW}$

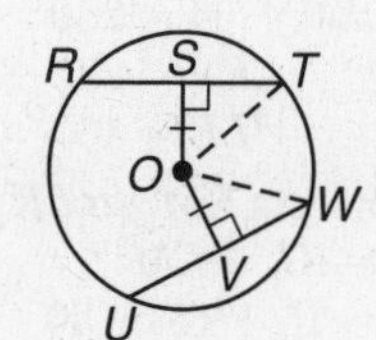

Proof:

Statements	Reasons
1. $\overline{OT} \cong \overline{OW}$	1. All radii of a $\odot$ are $\cong$.
2. $\overline{OS} \perp \overline{RT}$, $\overline{OV} \perp \overline{UW}$, $\overline{OS} \cong \overline{OV}$	2. Given
3. $\angle OST$, $\angle OVW$ are right angles.	3. Def. of $\perp$ lines
4. $\triangle STO \cong \triangle VWO$	4. HL
5. $\overline{ST} \cong \overline{VW}$	5. CPCTC
6. $ST = VW$	6. Definition of $\cong$ segments
7. $2(ST) = 2(VW)$	7. Multiplication Property
8. $\overline{OS}$ bisects $\overline{RT}$; $\overline{OV}$ bisects $\overline{UW}$.	8. Radius $\perp$ to a chord bisects the chord.
9. $RT = 2(ST)$, $UW = 2(VW)$	9. Def. of seg. bisector
10. $RT = UW$	10. Substitution
11. $\overline{RT} \cong \overline{UW}$	11. Definition of $\cong$ segments

38. Given: $\odot O$, $\overline{MN} \cong \overline{PQ}$, $\overline{ON}$ and $\overline{OQ}$ are radii. $\overline{OA} \perp \overline{MN}$; $\overline{OB} \perp \overline{PQ}$

Prove: $\overline{OA} \cong \overline{OB}$

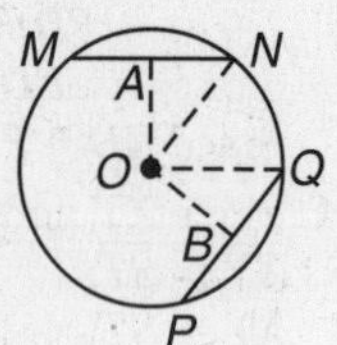

Proof:

Statements	Reasons
1. $\odot O$, $\overline{MN} \cong \overline{PQ}$, $\overline{ON}$ and $\overline{OQ}$ are radii, $\overline{OA} \perp \overline{MN}$, $\overline{OB} \perp \overline{PQ}$	1. Given
2. $\overline{OA}$ bisects $\overline{MN}$; $\overline{OB}$ bisects $\overline{PQ}$.	2. $\overline{OA}$ and $\overline{OB}$ are contained in radii. A radius $\perp$ to a chord bisects the chord.
3. $AN = \frac{1}{2}MN$; $BQ = \frac{1}{2}PQ$	3. Def. of bisector
4. $MN = PQ$	4. Def. of $\cong$ segments
5. $\frac{1}{2}MN = \frac{1}{2}PQ$	5. Mult. Prop.
6. $AN = BQ$	6. Substitution
7. $\overline{AN} \cong \overline{BQ}$	7. Def. of $\cong$ segments
8. $\overline{ON} \cong \overline{OQ}$	8. All radii of a circle are $\cong$.
9. $\triangle AON \cong \triangle BOQ$	9. HL
10. $\overline{OA} \cong \overline{OB}$	10. CPCTC

39. Let x = width of largest square. Use the Pythagorean Theorem.

$x^2 + x^2 = 4^2$

$2x^2 = 16$

$x^2 = 8$

$x = \sqrt{8}$ or about 2.82

The width is about 2.82 in.

40.

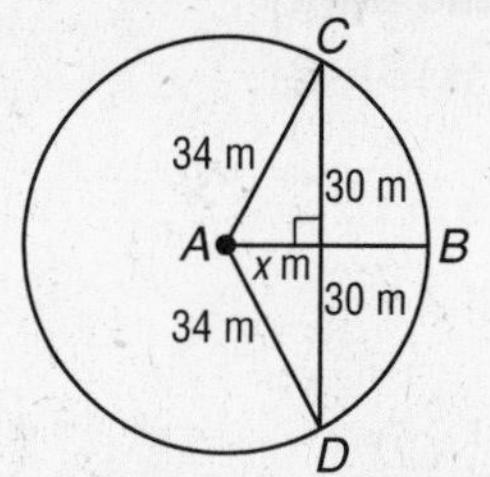

Let x = distance from the center to the chord. Use the Pythagorean Theorem.

$x^2 + 30^2 = 34^2$

$x^2 + 900 = 1156$

$x^2 = 256$

$x = 16$

The chord is 16 m from the center of the circle.

41.

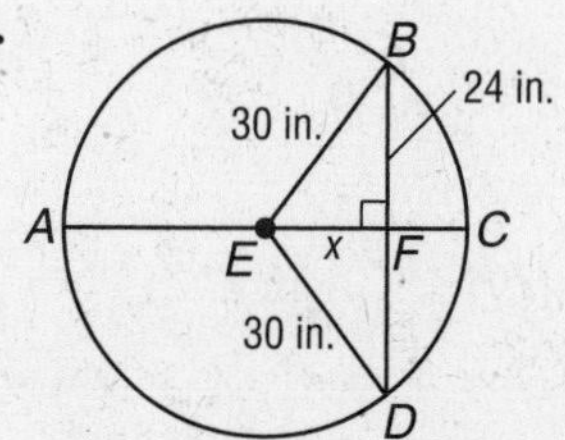

Since the diameter is 60 inches, the radius is 30 inches. Let x = distance from the center to the chord. Use the Pythagorean Theorem.

$x^2 + 24^2 = 30^2$

$x^2 + 576 = 900$

$x^2 = 324$

$x = 18$

The chord is 18 in. from the center of the circle.

42.

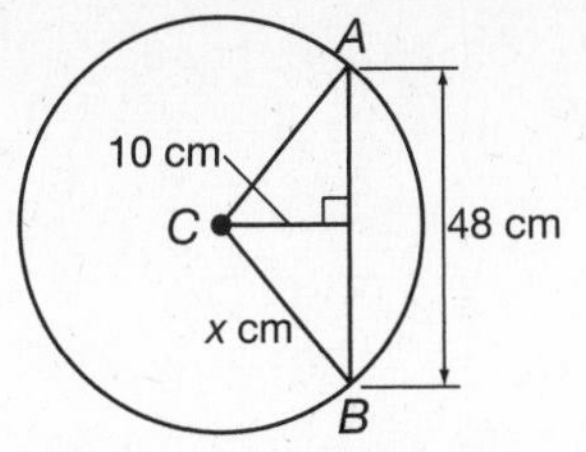

Let r = the radius of the circle. Use the Pythagorean Theorem.

$10^2 + 24^2 = r^2$
$100 + 576 = r^2$
$676 = r^2$
$26 = r$

The radius is 26 cm.

43.

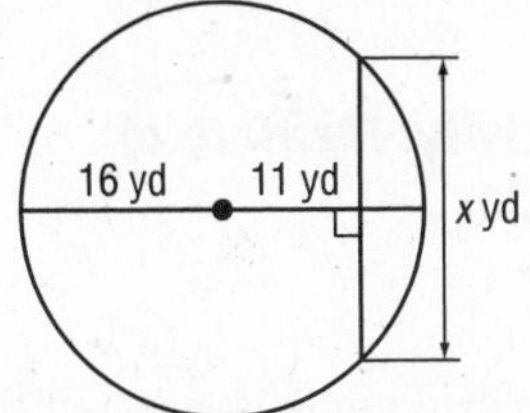

Since the diameter is 32 yards, the radius is 16 yards. Let x = half of the length of the chord. Use the Pythagorean Theorem.

$x^2 + 11^2 = 16^2$
$x^2 + 121 = 256$
$x^2 = 135$
$x = \sqrt{135}$

The length of the chord is $2\sqrt{135}$ or about 23.24 yd.

44. The line through the midpoint bisects the chord and is perpendicular to the chord, so the line is a diameter of the circle. Where two diameters meet would locate the center of the circle.

45. Let r be the radius of $\odot P$. Draw radii to points D and E to create triangles. The length DE is $r\sqrt{3}$ and $AB = 2r$; $r\sqrt{3} \neq \frac{1}{2}(2r)$.

46. Inscribed regular hexagon; the chords and the radii of the circle are congruent by construction. Thus, all triangles formed by these segments are equilateral triangles. That means each angle of the hexagon measures 120°, making all angles of the hexagon congruent and all sides congruent.

47. Inscribed equilateral triangle; the six arcs making up the circle are congruent because the chords intercepting them were congruent by construction. Each of the three chords drawn intercept two of the congruent chords. Thus, the three larger arcs are congruent. So, the three chords are congruent, making this an equilateral triangle.

48. $m\widehat{AB} = m\widehat{CD}$

49. No; congruent arcs must be in the same circle or congruent circles, but these are in concentric circles.

50. $\overline{AB} \cong \overline{CD}$; in the smaller circle, $\overline{OX} \cong \overline{OY}$ because they are radii. This means that in the larger circle, $\overline{AB}$ and $\overline{CD}$ are equidistant from the center, making them congruent chords.

51. Sample answer: The grooves of a waffle iron are chords of the circle. The ones that pass horizontally and vertically through the center are diameters. Answers should include the following.

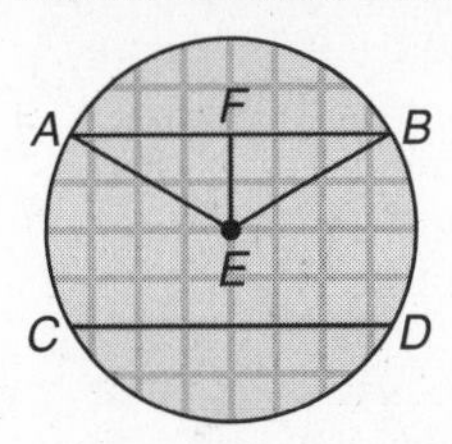

- If you know the measure of the radius and the distance the chord is from the center, you can use the Pythagorean Theorem to find the length of half of the chord and then multiply by 2.
- There are four grooves on either side of the diameter, so each groove is about 1 in. from the center. In the figure, $EF = 2$ and $EB = 4$ because the radius is half the diameter. Using the Pythagorean Theorem, you find that $FB \approx 3.464$ in. so $AB \approx 6.93$ in. Approximate lengths for other chords are 5.29 in. and 7.75 in., but exactly 8 in. for the diameter.

52. C; $\overline{DB}$ bisects $\overline{AC}$ and $OA = OC$

53. Bridgeworth population in 2010 = 1.2(204,000) or 244,800
Sutterly population in 2010 = 1.2(216,000) or 259,200
In 2010, 259,200 − 244,800 or 14,400 more people will live in Sutterly than in Bridgeworth.

Page 543 Maintain Your Skills

54. $\widehat{KTR}$ is a semicircle.
$m\widehat{KT} = m\widehat{KTR} - m\widehat{TR}$
$m\widehat{KT} = 180 - m\angle TSR$
$m\widehat{KT} = 180 - 42$ or 138

55. $\widehat{ERT}$ is a semicircle.
$m\widehat{ERT} = 180$

56. One way to find $m\widehat{KRT}$ is by using $m\widehat{KT}$.
$m\widehat{KRT} = 360 - m\widehat{KT}$
$m\widehat{KRT} = 360 - 138$ or 222

57. $\overline{SU}$ is a chord that is not a diameter.

58. MD is a radius and RI is a diameter.
$RI = 2(MD)$
$RI = 2(7)$ or 14

59. All radii are congruent: $\overline{RM}, \overline{AM}, \overline{DM}, \overline{IM}$

60. $\frac{1}{2}x = 120$
$x = 240$

61. $\frac{1}{2}x = 25$
$x = 50$

62. $2x = \frac{1}{2}(45 + 35)$
$2x = \frac{1}{2}(80)$
$2x = 40$
$x = 20$

63. $3x = \frac{1}{2}(120 - 60)$
$3x = \frac{1}{2}(60)$
$3x = 30$
$x = 10$

64. $45 = \frac{1}{2}(4x + 30)$
$45 = \frac{1}{2}(4x) + \frac{1}{2}(30)$
$45 = 2x + 15$
$30 = 2x$
$15 = x$

65. $90 = \frac{1}{2}(6x + 3x)$
$90 = \frac{1}{2}(9x)$
$90 = 4.5x$
$20 = x$

Page 543 Practice Quiz 1

1. $\overline{BC}$, $\overline{BD}$, and $\overline{BA}$ are radii.
2. $\overline{BD}$ and $\overline{CB}$ are radii, so $BD = CB$.
$BD = CB$
$3x = 7x - 3$
$-4x = -3$
$x = \frac{3}{4}$
$\overline{AC}$ is a diameter, so $AC = 2BD$
$AC = 2BD$
$AC = 2(3x)$
$AC = 6x$
$AC = 6\left(\frac{3}{4}\right)$ or 4.5
3. $\widehat{ADC}$ is a semicircle, so $m\widehat{ADC} = 180$.
$m\widehat{AD} = m\widehat{ADC} - m\widehat{CD}$
$m\widehat{AD} = 180 - m\angle CBD$
$m\widehat{AD} = 180 - 85$ or 95
4. $C = 2\pi r$
$C = 2\pi(3)$
$C = 6\pi$ or about 18.8 in.
5. The degree measure of an arc connecting two consecutive rungs is $\frac{360}{40}$ or 9.
6. $C = 2\pi r$
$C = 2\pi(3)$ or 6π
$m\widehat{CAD} = m\widehat{CA} + m\widehat{AD}$
$m\widehat{CAD} = 180 + m\angle ABD$
$m\widehat{CAD} = 180 + 150$ or 330
Let ℓ = arc length
$\frac{330}{360} = \frac{\ell}{6\pi}$
$\frac{330}{360}(6\pi) = \ell$
$5.5\pi = \ell$
The length of $\widehat{CAD}$ is 5.5π or about 17.3 units.
7. $m\angle CAM = m\angle NTM$
$m\angle CAM = 28$
8. $m\angle HMN = 180 - 2(40)$ or 100
$m\widehat{ES} = m\widehat{HN}$
$m\widehat{ES} = m\angle HMN$
$m\widehat{ES} = 100$
9. $CT = AC$
$SC = \frac{1}{2}CT$
$SC = \frac{1}{2}AC$
$SC = \frac{1}{2}(42)$ or 21
10. $\left(\frac{1}{2}x\right)^2 + 5^2 = 13^2$
$\frac{x^2}{4} + 25 = 169$
$\frac{x^2}{4} = 144$
$x^2 = 576$
$x = 24$

10-4 Inscribed Angles

Page 544 Geometry Activity: Measure of Inscribed Angles

1. See students' work.
2. $m\widehat{XZ} = 2(m\angle XYZ)$
3. The measure of an inscribed angle is one-half the measure of its intercepted arc.

Pages 548–549 Check for Understanding

1. Sample answer:

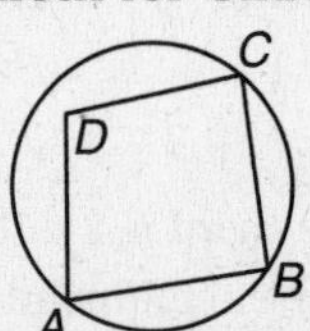

2. The measures of an inscribed angle and a central angle for the same intercepted arc can be calculated using the measure of the arc. However, the measure of the central angle equals the measure of the arc, while the measure of the inscribed angle is half the measure of the arc.
3. First find $m\widehat{QP}$ and $m\widehat{NP}$. $m\widehat{QP} = m\widehat{MN} = 120$ and $m\widehat{NP} = m\widehat{MQ} = 60$ because the central angles are congruent vertical angles.
$m\angle 1 = \frac{1}{2}m\widehat{NP} = \frac{1}{2} \cdot 60 = 30$
$m\angle 2 = \frac{1}{2}m\widehat{QP} = \frac{1}{2} \cdot 120 = 60$
$m\angle 3 = \frac{1}{2}m\widehat{MN} = \frac{1}{2} \cdot 120 = 60$
$m\angle 4 = \frac{1}{2}m\widehat{NP} = \frac{1}{2} \cdot 60 = 30$
$m\angle 5 = \frac{1}{2}\widehat{MQ} = \frac{1}{2} \cdot 60 = 30$
$m\angle 6 = \frac{1}{2}\widehat{MN} = \frac{1}{2} \cdot 120 = 60$
$m\angle 7 = \frac{1}{2}\widehat{QP} = \frac{1}{2} \cdot 120 = 60$
$m\angle 8 = \frac{1}{2}\widehat{MQ} = \frac{1}{2} \cdot 60 = 30$
4. **Given:** Quadrilateral $ABCD$ is inscribed in $\odot P$.
$m\angle C = \frac{1}{2}m\angle B$
Prove: $m\widehat{CDA} = 2(m\widehat{DAB})$

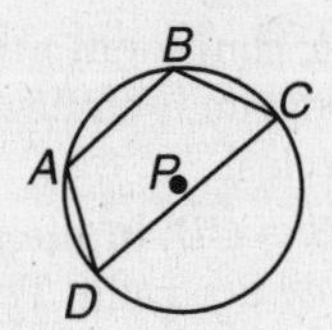

Proof: Given $m\angle C = \frac{1}{2}(m\angle B)$ means that $m\angle B = 2(m\angle C)$. Since $m\angle B = \frac{1}{2}(m\widehat{CDA})$

and $m\angle C = \frac{1}{2}(m\widehat{DAB})$, the equation becomes $\frac{1}{2}(m\widehat{CDA}) = 2\left[\frac{1}{2}(m\widehat{DAB})\right]$. Multiplying each side by 2 results in $m\widehat{CDA} = 2(m\widehat{DAB})$.

5. Angle PTS is a right angle because it intercepts a semicircle.

$$m\angle 1 + m\angle 2 + m\angle PTS = 180$$
$$(6x + 11) + (9x + 19) + 90 = 180$$
$$15x + 120 = 180$$
$$15x = 60$$
$$x = 4$$

Use the value of x to find the measures of $\angle 1$ and $\angle 2$.

$m\angle 1 = 6x + 11 = 6(4) + 11 = 35$ $\quad$ $m\angle 2 = 9x + 19 = 9(4) + 19 = 55$

Because $\widehat{PQ} \cong \widehat{RS}$, $m\widehat{PQ} = m\widehat{RS}$.

$m\angle 3 = m\angle 4$ because they are inscribed angles intercepting congruent arcs, $\widehat{PQ}$ and $\widehat{RS}$.

$$4y - 25 = 3y - 9$$
$$y = 16$$

Use the value of y to find the measures of $\angle 3$ and $\angle 4$.

$m\angle 3 = 4y - 25 = 4(16) - 25 = 39$ $\quad$ $m\angle 4 = 3y - 9 = 3(16) - 9 = 39$

6. 

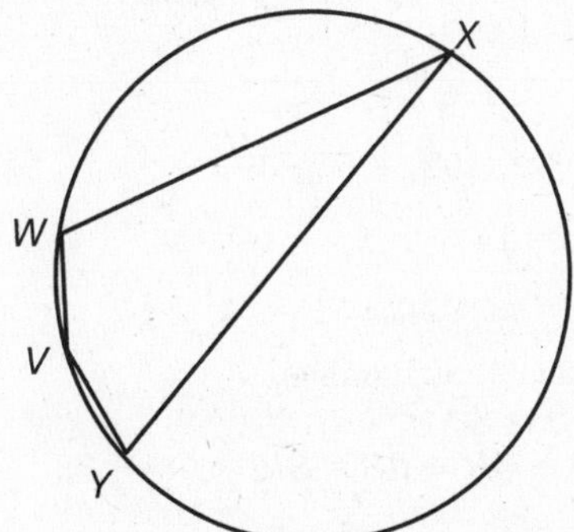

If a quadrilateral is inscribed in a circle, then its opposite angles are supplementary. $\angle V$ and $\angle X$ are opposite angles, so

$$m\angle V + m\angle X = 180$$
$$m\angle V + 28 = 180$$
$$m\angle V = 152$$

$\angle W$ and $\angle Y$ are opposite angles, so

$$m\angle W + m\angle Y = 180$$
$$110 + m\angle Y = 180$$
$$m\angle Y = 70$$

7. Since $\widehat{XZY}$ is a semicircle, $\angle XZY$ is a right angle. So, the probability is 1.

Pages 549–551 Practice and Apply

8. First find $m\widehat{PQ}$ and $m\widehat{QR}$. $m\widehat{PQ} = m\widehat{QR}$.

$$m\widehat{PS} + m\widehat{SR} + m\widehat{QR} + m\widehat{PQ} = 360$$
$$45 + 75 + m\widehat{PQ} + m\widehat{PQ} = 360$$
$$120 + 2m\widehat{PQ} = 360$$
$$2m\widehat{PQ} = 240$$
$$m\widehat{PQ} = 120$$
$$m\widehat{QR} = 120$$

$$m\angle 1 = \frac{1}{2}m\widehat{QR} = \frac{1}{2} \cdot 120 = 60$$
$$m\angle 2 = \frac{1}{2}m\widehat{PS} = \frac{1}{2} \cdot 45 = 22.5$$
$$m\angle 3 = \frac{1}{2}m\widehat{SR} = \frac{1}{2} \cdot 75 = 37.5$$
$$m\angle 4 = \frac{1}{2}m\widehat{PQ} = \frac{1}{2} \cdot 120 = 60$$
$$m\angle 5 = \frac{1}{2}m\widehat{PS} = \frac{1}{2} \cdot 45 = 22.5$$
$$m\angle 6 = \frac{1}{2}m\widehat{QR} = \frac{1}{2} \cdot 120 = 60$$
$$m\angle 7 = \frac{1}{2}m\widehat{SR} = \frac{1}{2} \cdot 75 = 37.5$$
$$m\angle 8 = \frac{1}{2}m\widehat{PQ} = \frac{1}{2} \cdot 120 = 60$$

9. $m\widehat{BC} = 2(m\angle BDC)$

$$m\widehat{BC} = 2(25) \text{ or } 50$$
$$m\widehat{AB} + m\widehat{BC} + m\widehat{CD} + m\widehat{AD} = 360$$
$$120 + 50 + 130 + m\widehat{AD} = 360$$
$$m\widehat{AD} = 60$$
$$m\angle 1 = \frac{1}{2}m\widehat{AD} = \frac{1}{2}(60) \text{ or } 30$$
$$m\angle 2 = m\angle 1 \text{ or } 30$$
$$m\angle 3 = \frac{1}{2}m\widehat{BC} = \frac{1}{2}(50) \text{ or } 25$$

10. $m\angle 1 = m\angle 6 = \frac{1}{2}m\widehat{XZ} = \frac{1}{2}(100) \text{ or } 50$

Since $\overline{XY} \perp \overline{ST}$, $\widehat{XS} \cong \widehat{SY}$, and $m\angle 8 = m\angle 11$.

$$m\angle 1 + m\angle 11 = 90$$
$$50 + m\angle 11 = 90$$
$$m\angle 11 = 40$$
$$m\angle 8 = 40$$
$$m\angle 2 + m\angle 8 = 90$$
$$m\angle 2 + 40 = 90$$
$$m\angle 2 = 50$$

Since $\overline{ZW} \perp \overline{ST}$, $\widehat{ZT} \cong \widehat{TW}$, and $m\angle 9 = m\angle 10$.

$$m\angle 10 + m\angle 6 = 90$$
$$m\angle 10 + 50 = 90$$
$$m\angle 10 = 40$$
$$m\angle 9 = 40$$
$$m\angle 9 + m\angle 4 = 90$$
$$40 + m\angle 4 = 90$$
$$m\angle 4 = 50$$

$\angle XZW$ is a right angle because it intercepts a semicircle.

$$m\angle 3 + m\angle 4 = 90$$
$$m\angle 3 + 50 = 90$$
$$m\angle 3 = 40$$

$\angle YXZ$ is a right angle because it intercepts a semicircle.

$$m\angle 2 + m\angle 5 = 90$$
$$50 + m\angle 5 = 90$$
$$m\angle 5 = 40$$
$$m\angle 5 + m\angle 3 + m\angle 7 = 180$$
$$40 + 40 + m\angle 7 = 180$$
$$m\angle 7 = 100$$

11. Given: $\widehat{AB} \cong \widehat{DE}, \widehat{AC} \cong \widehat{CE}$
Prove: $\triangle ABC \cong \triangle EDC$

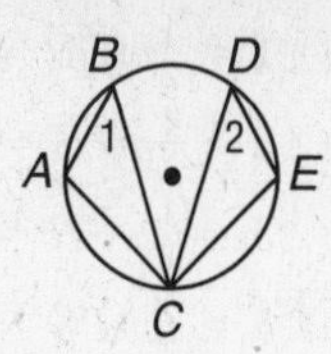

Proof:

Statements	Reasons
1. $\widehat{AB} \cong \widehat{DE}, \widehat{AC} \cong \widehat{CE}$	1. Given
2. $m\widehat{AB} = m\widehat{DE}, m\widehat{AC} = m\widehat{CE}$	2. Def. of $\cong$ arcs
3. $\frac{1}{2}m\widehat{AB} = \frac{1}{2}m\widehat{DE}$, $\frac{1}{2}m\widehat{AC} = \frac{1}{2}m\widehat{CE}$	3. Mult. Prop.
4. $m\angle ACB = \frac{1}{2}m\widehat{AB}$, $m\angle ECD = \frac{1}{2}m\widehat{DE}$, $m\angle 1 = \frac{1}{2}m\widehat{AC}, m\angle 2 = \frac{1}{2}m\widehat{CE}$	4. Inscribed $\angle$ Theorem
5. $m\angle ACB = m\angle ECD$, $m\angle 1 = m\angle 2$	5. Substitution
6. $\angle ACB \cong \angle ECD, \angle 1 \cong \angle 2$	6. Def. of $\cong$ $\angle$s
7. $\overline{AB} \cong \overline{DE}$	7. $\cong$ arcs have $\cong$ chords.
8. $\triangle ABC \cong \triangle EDC$	8. AAS

12. Given: $\odot P$
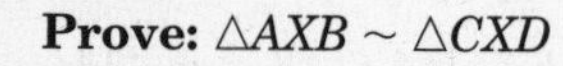
Prove: $\triangle AXB \sim \triangle CXD$

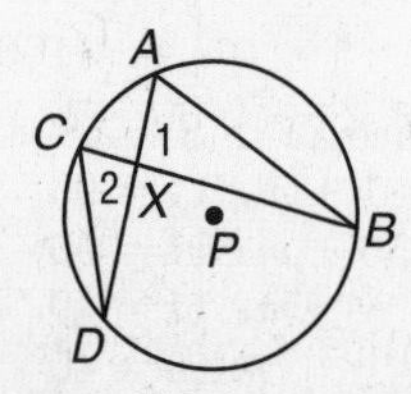

Proof:

Statements	Reasons
1. $\odot P$	1. Given
2. $\angle A \cong \angle C$	2. Inscribed $\angle$s intercepting same arc are $\cong$.
3. $\angle 1 \cong \angle 2$	3. Vertical $\angle$s are $\cong$.
4. $\triangle AXB \sim \triangle CXD$	4. AA Similarity

13. Inscribed angles that intercept the same arc are congruent.
$m\angle 1 = m\angle 2$
$x = 2x - 13$
$13 = x$
$m\angle 1 = x$ or 13
$m\angle 2 = 13$

14. $m\angle 8 = \frac{1}{2}m\widehat{AB}$
$= \frac{1}{2}(120)$ or 60
$m\angle 2 + m\angle 8 = 90$
$m\angle 2 + 60 = 90$
$m\angle 2 = 30$
$\widehat{BC} \cong \widehat{CD}$ because $\overline{BD} \perp \overline{AC}$ so
$m\angle 1 = m\angle 2$
$m\angle 1 = 30$
$m\angle 3 = m\angle 8$
$m\angle 3 = 60$
$m\widehat{AD} = m\widehat{AB} = 120$, and $m\widehat{BC} = m\widehat{CD}$.
$m\widehat{AB} + m\widehat{AD} + m\widehat{BC} + m\widehat{CD} = 360$
$120 + 120 + m\widehat{BC} + m\widehat{BC} = 360$
$2m\widehat{BC} = 120$
$m\widehat{BC} = 60$
$m\widehat{CD} = 60$
$m\angle 4 = \frac{1}{2}m\widehat{CD}$
$= \frac{1}{2}(60)$ or 30
$m\angle 7 = \frac{1}{2}m\widehat{BC}$
$= \frac{1}{2}(60)$ or 30
$m\angle 4 = 30$
$m\angle 4 + m\angle 5 = 90$
$30 + m\angle 5 = 90$
$m\angle 5 = 60$
$m\angle 6 = 60$

15. $\angle KPR$ is a right angle because $\widehat{KPR}$ is a semicircle.
$m\angle 2 = 90$
$m\angle R + m\angle K + m\angle P = 180$
$m\angle R + m\angle K + 90 = 180$
$m\angle R + m\angle K = 90$
$\frac{1}{3}x + 5 + \frac{1}{2}x = 90$
$\frac{5}{6}x = 85$
$x = 102$
$m\angle 3 = m\angle R = \frac{1}{3}x + 5$
$= \frac{1}{3}(102) + 5$ or 39
$m\angle 1 = m\angle K = \frac{1}{2}x$
$= \frac{1}{2}(102)$ or 51

16. By the definition of a rhombus, $\overline{PQ} \cong \overline{QR} \cong \overline{RS} \cong \overline{SP}$.
Therefore, $\widehat{PQ} \cong \widehat{QR} \cong \widehat{RS} \cong \widehat{SP}$.
$m\widehat{SP} = \frac{360}{4}$ or 90
$m\angle QRP = \frac{1}{2}(m\widehat{PQ})$
$= \frac{1}{2}(90)$ or 45

17. $m\angle 1 = m\angle 2$ since $\overline{DE} \cong \overline{EC}$.
$m\angle 1 + m\angle 2 = 90$
$m\angle 1 = m\angle 2 = 45$
$m\angle 3$ is a right angle because $\widehat{ABC}$ is a semicircle.
$m\angle 1 + m\angle 3 + m\angle 4 = 180$
$45 + 90 + m\angle 4 = 180$
$m\angle 4 = 45$
$m\angle 5 = \frac{1}{2}m\widehat{CF}$
$= \frac{1}{2}(60)$ or 30
$m\angle 7$ is a right angle because $\widehat{AFC}$ is a semicircle.
$m\angle 5 + m\angle 6 + m\angle 7 = 180$
$30 + m\angle 6 + 90 = 180$
$m\angle 6 = 60$
$m\widehat{AF} = 2(m\angle 6)$
$= 2(60)$ or 120

18.

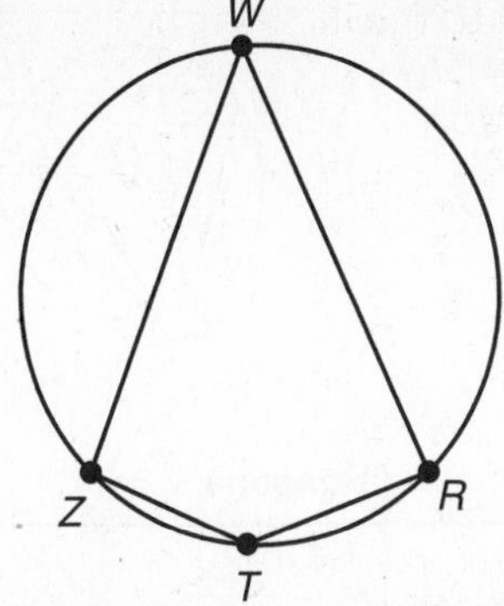

If a quadrilateral is inscribed in a circle, then its opposite angles are supplementary. $\angle W$ and $\angle T$ are opposite angles, so

$m\angle W + m\angle T = 180$
$45 + m\angle T = 180$
$m\angle T = 135$

$\angle R$ and $\angle Z$ are opposite angles, so

$m\angle R + m\angle Z = 180$
$100 + m\angle Z = 180$
$m\angle Z = 80$

19.

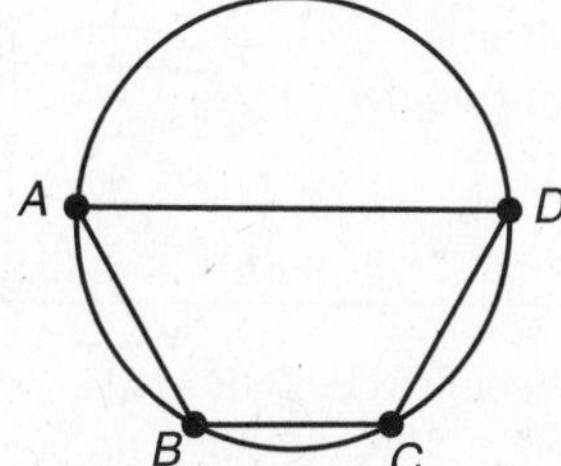

Since $ABCD$ is a trapezoid $\overline{AD} \parallel \overline{BC}$. $\angle A$ and $\angle B$ are consecutive interior angles so they are supplementary.

$m\angle A + m\angle B = 180$
$60 + m\angle B = 180$
$m\angle B = 120$

If a quadrilateral is inscribed in a circle, then its opposite angles are supplementary. $\angle A$ and $\angle C$ are opposite angles, so

$m\angle A + m\angle C = 180$
$60 + m\angle C = 180$
$m\angle C = 120$

$\angle B$ and $\angle D$ are opposite angles, so

$m\angle B + m\angle D = 180$
$120 + m\angle D = 180$
$m\angle D = 60$

20. Sample answer: $\overline{PQ}$ is a diagonal of $PDQT$ and a diameter of the circle.

21. Sample answer: $\overline{EF}$ is a diameter of the circle and a diagonal and angle bisector of $EDFG$.

22. Since pentagon $PQRST$ is equilateral, $\overline{PQ} \cong \overline{QR} \cong \overline{RS} \cong \overline{ST} \cong \overline{TP}$ and $\widehat{PQ} \cong \widehat{QR} \cong \widehat{RS} \cong \widehat{ST} \cong \widehat{TP}$.

$m\widehat{QR} = \frac{360}{5}$ or 72

23. $m\angle PSR = \frac{1}{2}(m\widehat{PQ} + m\widehat{QR})$
$= \frac{1}{2}(72 + 72)$ or 72

24. $m\angle PQR = \frac{1}{2}(m\widehat{RS} + m\widehat{ST} + m\widehat{TP})$
$= \frac{1}{2}(72 + 72 + 72)$ or 108

25. $m\widehat{PTS} = m\widehat{PT} + m\widehat{TS}$
$= 72 + 72$ or 144

26. $m\widehat{BA} = m\angle BZA$ or 104

27. $m\widehat{ADC} = 360 - (m\widehat{BA} + m\widehat{CB})$
$= 360 - (104 + 94)$ or 162

28. $m\angle BDA = \frac{1}{2}m\widehat{BA}$
$= \frac{1}{2}(104)$ or 52

29. $2(m\angle ZAC) + 180 = m\widehat{BA} + m\widehat{CB}$
$2(m\angle ZAC) + 180 = 104 + 94$
$2(m\angle ZAC) + 180 = 198$
$2(m\angle ZAC) = 18$
$m\angle ZAC = 9$

30. $m\widehat{AC} = 2m\angle ABC$
$= 2(50)$ or 100

$m\angle DEF = \frac{1}{2}m\widehat{DBF}$
$= \frac{1}{2}(128)$ or 64

31. If $m\widehat{PS} = 40$, then $m\widehat{PQS} = 360 - 40$ or 320. If T is located in $\widehat{PQS}$, then

$m\angle PTS = \frac{1}{2}m\widehat{PS}$
$= \frac{1}{2}(40)$ or 20.

The probability that $m\angle PTS = 20$ is the same as the probability that T is contained in $\widehat{PQS}$, $\frac{320}{360}$ or $\frac{8}{9}$.

32. If $m\widehat{PSR} = 110$, then $m\widehat{PQR} = 360 - 110$ or 250. If T is located in $\widehat{PQR}$, then

$m\angle PTR = \frac{1}{2}m\widehat{PSR}$
$= \frac{1}{2}(110)$ or 55.

The probability that $m\angle PTR = 55$ is the same as the probability that T is contained in $\widehat{PQR}$, $\frac{250}{360}$ or $\frac{25}{36}$.

33. No matter where T is selected, $m\angle STQ = \frac{1}{2}(180)$ or 90 because $\widehat{SPQ}$ is a semicircle. Therefore, the probability that $m\angle STQ = 90$ is 1.

34. $m\angle PTQ$ can never equal 180 since $m\widehat{PQ} \neq 360$. Therefore, the propability that $m\angle PTQ = 180$ is 0.

35. Given: T lies inside $\angle PRQ$.
$\overline{RK}$ is a diameter of $\odot T$.
Prove: $m\angle PRQ = \frac{1}{2}m\widehat{PKQ}$

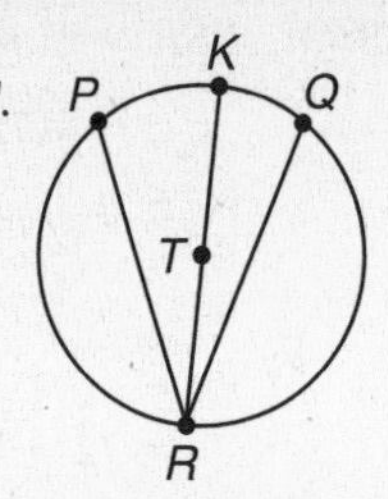

Proof:

Statements	Reasons
1. $m\angle PRQ = m\angle PRK + m\angle KRQ$	1. $\angle$ Addition Th.
2. $m\widehat{PKQ} = m\widehat{PK} + m\widehat{KQ}$	2. Arc Addition Theorem
3. $\frac{1}{2}m\widehat{PKQ} = \frac{1}{2}m\widehat{PK} + \frac{1}{2}m\widehat{KQ}$	3. Multiplication Prop.
4. $m\angle PRK = \frac{1}{2}m\widehat{PK}$, $m\angle KRQ = \frac{1}{2}m\widehat{KQ}$	4. The measure of an inscribed $\angle$ whose side is a diameter is half the measure of the intercepted arc (Case 1).
5. $\frac{1}{2}m\widehat{PKQ} = m\angle PRK + m\angle KRQ$	5. Subst. (Steps 3, 4)
6. $\frac{1}{2}m\widehat{PKQ} = m\angle PRQ$	6. Substitution (Steps 5, 1)

36. Given: T lies outside $\angle PRQ$.
$\overline{RK}$ is a diameter of $\odot T$.
Prove: $m\angle PRQ = \frac{1}{2}m\widehat{PQ}$

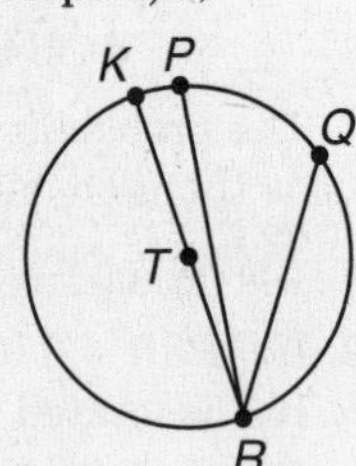

Proof:

Statements	Reasons
1. $m\angle PRQ = m\angle KRQ - m\angle PRK$	1. Angle Addition Theorem, Subtraction Property
2. $m\widehat{PQ} = m\widehat{KQ} - m\widehat{KP}$	2. Arc Addition Theorem, Subtraction Property
3. $\frac{1}{2}m\widehat{PQ} = \frac{1}{2}(m\widehat{KQ} - m\widehat{KP})$	3. Division Property
4. $m\angle PRK = \frac{1}{2}m\widehat{KP}$, $m\angle KRQ = \frac{1}{2}m\widehat{KQ}$	4. The measure of an inscribed $\angle$ whose side is a diameter is half the measure of the intercepted arc (Case 1).
5. $m\angle PRQ = \frac{1}{2}m\widehat{KQ} - \frac{1}{2}m\widehat{KP}$	5. Subst. (Steps 1, 4)
6. $m\angle PRQ = \frac{1}{2}(m\widehat{KQ} - m\widehat{KP})$	6. Distributive Property
7. $m\angle PRQ = \frac{1}{2}m\widehat{PQ}$	7. Substitution (Steps 6, 3)

37. Given: inscribed $\angle MLN$ and $\angle CED$, $\widehat{CD} \cong \widehat{MN}$
Prove: $\angle CED \cong \angle MLN$

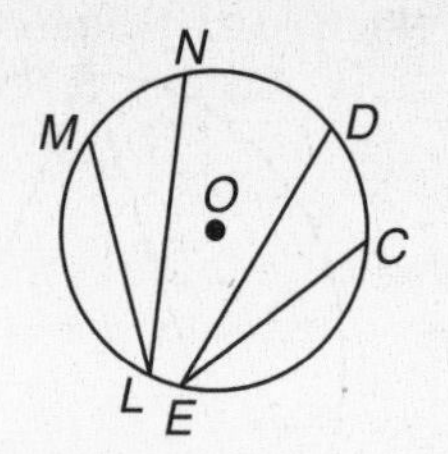

Proof:

Statements	Reasons
1. $\angle MLN$ and $\angle CED$ are inscribed; $\widehat{CD} \cong \widehat{MN}$	1. Given
2. $m\angle MLN = \frac{1}{2}m\widehat{MN}$; $m\angle CED = \frac{1}{2}m\widehat{CD}$	2. Measure of an inscribed $\angle$ = half measure of intercepted arc.
3. $m\widehat{CD} = m\widehat{MN}$	3. Def. of $\cong$ arcs
4. $\frac{1}{2}m\widehat{CD} = \frac{1}{2}m\widehat{MN}$	4. Mult. Prop.
5. $m\angle CED = m\angle MLN$	5. Substitution
6. $\angle CED \cong \angle MLN$	6. Def. of $\cong$ $\angle$s

38. Given: $\widehat{PQR}$ is a semicircle.
Prove: $\angle PQR$ is a right angle.

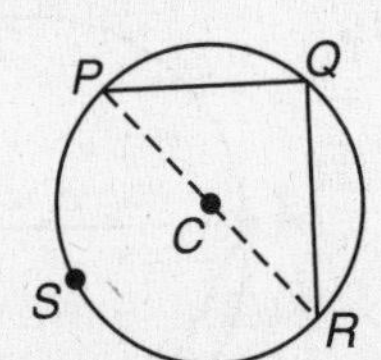

Proof: Since $\widehat{PSR}$ is a semicircle, $\widehat{PSR}$ is also a semicircle and $m\widehat{PSR} = 180$. $\angle PQR$ is an inscribed angle, and $m\angle PQR = \frac{1}{2}(m\widehat{PSR})$ or 90, making $\angle PQR$ a right angle.

39. Given: quadrilateral $ABCD$ inscribed in $\odot O$
Prove: $\angle A$ and $\angle C$ are supplementary. $\angle B$ and $\angle D$ are supplementary.

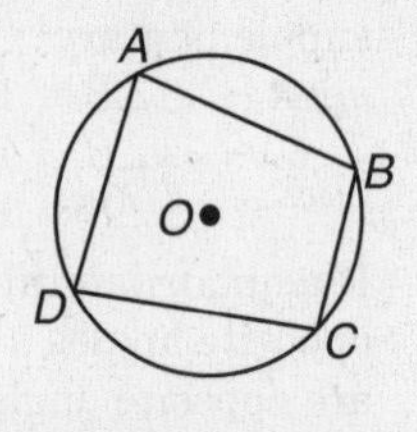

Proof: By arc addition and the definitions of arc measure and the sum of central angles, $m\widehat{DCB} + m\widehat{DAB} = 360$. Since $m\angle C = \frac{1}{2}m\widehat{DAB}$ and $m\angle A = \frac{1}{2}m\widehat{DCB}$, $m\angle C + m\angle A = \frac{1}{2}(m\widehat{DCB} + m\widehat{DAB})$, but $m\widehat{DCB} + m\widehat{DAB} = 360$, so $m\angle C + m\angle A = \frac{1}{2}(360)$ or 180. This makes $\angle C$ and $\angle A$ supplementary. Because the sum of the measures of the interior angles of a quadrilateral is 360, $m\angle A + m\angle C + m\angle B + m\angle D = 360$. But $m\angle A + m\angle C = 180$, so $m\angle B + m\angle D = 180$, making them supplementary also.

40. There are 8 congruent arcs, so each measures $\frac{360}{8}$ or 45.

41. Isosceles right triangle because sides are congruent radii making it isosceles and $\angle AOC$ is a central angle for an arc of 90°, making it a right angle.

42. Square because each angle intercepts a semicircle, making them 90° angles. Each side is a chord of congruent arcs, so the chords are congruent.

43. Square because each angle intercepts a semicircle, making them 90° angles. Each side is a chord of congruent arcs, so the chords are congruent.

44. Use the properties of trapezoids and inscribed quadrilaterals to verify that $ABCD$ is isosceles.

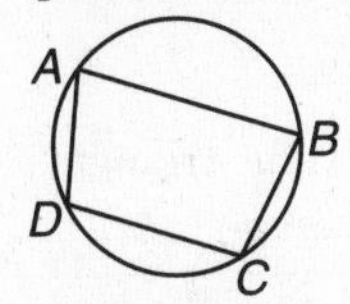

$m\angle A + m\angle D = 180$ (same side interior angles $= 180$)
$m\angle A + m\angle C = 180$ (opposite angles of inscribed quadrilaterals $= 180$)
$m\angle A + m\angle D = m\angle A + m\angle C$ (Substitution)
$m\angle D = m\angle C$ (Subtraction Property)
$\angle D \cong \angle C$ (Def. of $\cong \angle$s)
Trapezoid $ABCD$ is isosceles because the base angles are congruent.

45. Sample answer: The socket is similar to an inscribed polygon because the vertices of the hexagon can be placed on a circle that is concentric with the outer circle of the socket. Answers should include the following.

- An inscribed polygon is one in which all of its vertices are points on a circle.
- The side of the regular hexagon inscribed in a circle $\frac{3}{4}$ inch wide is $\frac{3}{8}$ inch.

46. C; $m\angle AOB = 2m\angle ACB$, so the ratio is 1 : 2.

47. There are 18 even-numbered pages and 18 odd-numbered pages, so there are $18 \cdot 6 + 18 \cdot 7$ or 234 articles.

Page 551 Maintain Your Skills

48. Since $AB = 60$, $CD = 30$. Since $DE = 48$, $FD = 24$.
$(FD)^2 + (CF)^2 = (CD)^2$
$24^2 + (CF)^2 = 30^2$
$(CF)^2 = 324$
$CF = 18$

49. Draw a line from C to E. Since $AB = 32$, $CE = 16$. Draw $\overline{CE}$. Use the Pythagorean Theorem.
$(FE)^2 + (FC)^2 = (CE)^2$
$(FE)^2 + 11^2 = (16)^2$
$FE^2 = 135$
$FE = \sqrt{135}$
$FE \approx 11.62$

50. Since $DE = 60$, $FD = 30$. Use the Pythagorean Theorem.
$(FD)^2 + (FC)^2 = (CD)^2$
$30^2 + 16^2 = (CD)^2$
$1156 = (CD)^2$
$34 = CD$
Since $AB = 2(CD)$, $AB = 2(34)$ or 68.

51. $C = 2\pi r$
$C = 2\pi(12)$ or 24π
Let $\ell =$ length of $\widehat{QR}$.
$\frac{60}{360} = \frac{\ell}{24\pi}$
$\frac{60}{360}(24\pi) = \ell$
$4\pi = \ell$
The length of $\widehat{QR}$ is 4π units.

52. $C = 2\pi r$
$C = 2\pi(16)$ or 32π
Let $\ell =$ length of $\widehat{QR}$.
$\frac{90}{360} = \frac{\ell}{32\pi}$
$\frac{90}{360}(32\pi) = \ell$
$8\pi = \ell$
The length of $\widehat{QR}$ is 8π units.

53. always

54. sometimes

55. sometimes

56. Use the converse of the Pythagorean Theorem.
$a^2 + b^2 \stackrel{?}{=} c^2$
$4^2 + 5^2 \stackrel{?}{=} 6^2$
$16 + 25 \stackrel{?}{=} 36$
$41 \neq 36$
It is not a right triangle.

57. Use the converse of the Pythagorean Theorem.
$a^2 + b^2 \stackrel{?}{=} c^2$
$3^2 + 8^2 \stackrel{?}{=} 10^2$
$9 + 64 \stackrel{?}{=} 100$
$73 \neq 100$
It is not a right triangle.

58. Use the converse of the Pythagorean Theorem.
$a^2 + b^2 \stackrel{?}{=} c^2$
$28^2 + 45^2 \stackrel{?}{=} 53^2$
$784 + 2025 \stackrel{?}{=} 2809$
$2809 = 2809$
It is a right triangle.

10-5 Tangents

Page 552 Geometry Software Investigation: Tangents and Radii

1. $\overline{WX}$ is a radius.
2. $WX < WY$
3. It doesn't, unless Y and X coincide.
4. $\overline{WX} \perp \overline{XY}$
5. Sample answer: The shortest distance from the center of a circle to the tangent is the radius of the circle, which is perpendicular to the tangent.

Page 555 Check for Understanding

1a. Two; from any point outside the circle, you can draw only two tangents.

1b. None; a line containing a point inside the circle would intersect the circle in two points. A tangent can only intersect a circle in one point.

1c. One; since a tangent intersects a circle in exactly one point, there is one tangent containing a point on the circle.

2. If the lines are tangent at the endpoints of a diameter, they are parallel and thus, not intersecting.

3. Sample answer:
polygon circumscribed about a circle

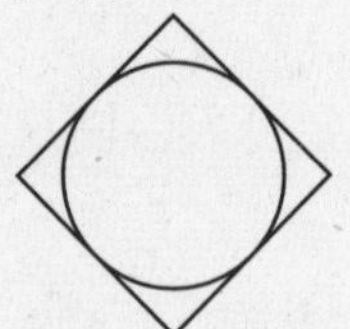

polygon inscribed in a circle

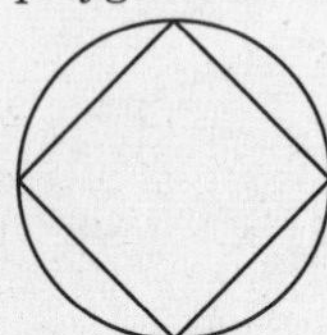

4. Triangle MPO is a right triangle with hypotenuse $\overline{MO}$. Use the Pythagorean Theorem.
$(MP)^2 + (PO)^2 = (MO)^2$
$16^2 + x^2 = 20^2$
$256 + x^2 = 400$
$x^2 = 144$
$x = 12$

5. If $\triangle PRO$ is a right triangle, then $\overline{PR}$ is tangent to $\odot O$. Use the converse of the Pythagorean Theorem.
$(PR)^2 + (PO)^2 \stackrel{?}{=} (RO)^2$
$5^2 + 12^2 \stackrel{?}{=} 13^2$
$25 + 144 \stackrel{?}{=} 169$
$169 = 169$
Yes, $\overline{PR}$ is tangent to $\odot O$.

6. $4(3) + 4x = 32$
$4x = 20$
$x = 5$

7. Each side of the square is 2(72) or 144 feet. The total length of fence is 4(144) or 576 feet.

Pages 556–558 Practice and Apply

8. Determine whether $\triangle ABC$ is a right triangle.
$(AB)^2 + (BC)^2 \stackrel{?}{=} (AC)^2$
$16^2 + 30^2 \stackrel{?}{=} 34^2$
$1156 = 1156$
Because the converse of the Pythagorean Theorem is true, $\triangle ABC$ is a right triangle with right angle ABC and $\overline{AB} \perp \overline{BC}$.
Yes; $\overline{BC}$ is tangent to $\odot A$.

9. Determine whether $\triangle DEF$ is a right triangle.
$(EF)^2 + (DF)^2 \stackrel{?}{=} (DE)^2$
$3^2 + 4^2 \stackrel{?}{=} 5^2$
$25 = 25$
Because the converse of the Pythagorean Theorem is true, $\triangle DEF$ is a right triangle. Since $\overline{DE}$ is the longest side, $\angle F$ is the right angle. $\angle E$ is not a right angle, so $\overline{DE}$ is not perpendicular to $\overline{EF}$.
No; $\overline{DE}$ is not tangent to $\odot F$.

10. Determine whether $\triangle JGH$ is a right triangle.
$(JG)^2 + (GH)^2 \stackrel{?}{=} (JH)^2$
$5^2 + 12^2 \stackrel{?}{=} 14^2$
$169 \neq 196$
Because the converse of the Pythagorean Theorem did not prove true in this case, $\triangle JGH$ is not a right triangle.
No; $\overline{GH}$ is not tangent to $\odot J$.

11. Determine whether $\triangle KLM$ is a right triangle.
$(KL)^2 + (LM)^2 \stackrel{?}{=} (KM)^2$
$10^2 + 6^2 \stackrel{?}{=} \left(\sqrt{136}\right)^2$
$136 = 136$
Because the converse of the Pythagorean Theorem is true, $\triangle KLM$ is a right triangle with right angle KLM and $\overline{LM} \perp \overline{KL}$.
Yes; $\overline{KL}$ is tangent to $\odot M$.

For Exercises 12–15, use the Pythagorean Theorem.

12. $(NO)^2 + (NP)^2 = (OP)^2$
$6^2 + x^2 = 10^2$
$36 + x^2 = 100$
$x^2 = 64$
$x = 8$

13. $(QR)^2 + (RS)^2 = (QS)^2$
$12^2 + x^2 = (12 + 8)^2$
$144 + x^2 = 400$
$x^2 = 256$
$x = 16$

14. $(WU)^2 + (UV)^2 = (WV)^2$
$12^2 + 7^2 = x^2$
$144 + 49 = x^2$
$193 = x^2$
$\sqrt{193} = x$

15. $(AC)^2 + (AB)^2 = (BC)^2$
$8^2 + x^2 = 17^2$
$64 + x^2 = 289$
$x^2 = 225$
$x = 15$

16. $DE = DF$
$x - 2 = 14$
$x = 16$

17. $HJ = HN$
$HJ = 2$
$HK = HJ + JK$
$5 = 2 + JK$
$3 = JK$
$KL = JK$
$KL = 3$

18. $RS = RQ$
$= 6$
$ST = TU$
$= 4$
$RT = RS + ST$
$x = 6 + 4$ or 10

19. $\sin B = \frac{AC}{BC}$
$\sin 30° = \frac{15}{x}$
$x = \frac{15}{\sin 30°}$ or 30

20. $(DG)^2 + (DE)^2 = (EG)^2$

$x^2 + 16^2 = (12 + x)^2$

$x^2 + 256 = 144 + 24x + x^2$

$112 = 24x$

$4\frac{2}{3} = x$

21. See students' work.

22. Given: $\ell \perp \overline{AB}$

$\overline{AB}$ is a radius of $\odot A$.

Prove: ℓ is tangent to $\odot A$.

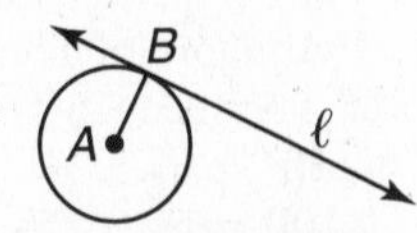

Proof: Assume ℓ is not tangent to $\odot A$. Since ℓ intersects $\odot A$ at B, it must intersect the circle in another place. Call this point C. Then $AB = AC$. But if $\overline{AB} \perp \ell$, then $\overline{AB}$ must be the shortest segment from A to ℓ. If $\overline{AB} = \overline{AC}$, then $\overline{AC}$ is the shortest segment from A to ℓ. Since B and C are two different points on ℓ, this is a contradiction. Therefore, ℓ is tangent to $\odot A$.

23. Let r = the radius of $\odot M$.

$(PL)^2 + (ML)^2 = (PM)^2$

$10^2 + r^2 = (r + 2)^2$

$100 + r^2 = r^2 + 4r + 4$

$96 = 4r$

$24 = r$

$PL + ML + MN + NP = 10 + 24 + 24 + 2$ or 60

The perimeter is 60 units.

24.

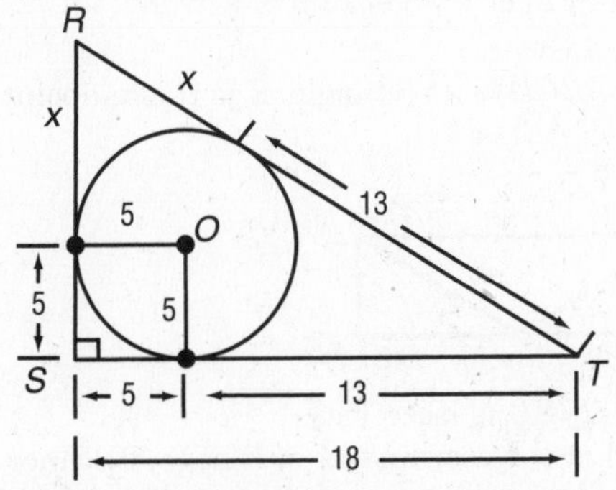

Use the Pythagorean Theorem to write an equation for $\triangle RST$.

$(x + 5)^2 + 18^2 = (x + 13)^2$

$x^2 + 10x + 25 + 324 = x^2 + 26x + 169$

$180 = 16x$

$11.25 = x$

$11.25 + 5 + 18 + 13 + 11.25 = 58.5$

The perimeter is 58.5 units.

25. $d = 5, r = 2.5, GY = EG = 2.5$. Since $\overline{CB}$ is a tangent, $\angle AEB$ is a right angle and $\triangle GEB$ is a right triangle. Use the Pythagorean Theorem.

$(EG)^2 + (EB)^2 = (GB)^2$

$(2.5)^2 + (EB)^2 = (2.5 + 2.5)^2$

$6.25 + (EB)^2 = 25$

$(EB)^2 = 18.75$

$EB = \sqrt{18.75}$

$EB = EC = DC = DA = FA = FB$

$6EB = 6\sqrt{18.75}$ or $15\sqrt{3}$

The perimeter is $15\sqrt{3}$ units.

26. $CF = CE$

$6(3 - x) = 3x$

$18 - 6x = 3x$

$18 = 9x$

$2 = x$

$CE = 3x$
$= 3(2)$ or 6

$CF = 6(3 - x)$
$= 6(3 - 2)$ or 6

$BE = BD$

$4y = 12y - 4$

$4 = 8y$

$\frac{1}{2} = y$

$BE = 4y$
$= 4\left(\frac{1}{2}\right)$ or 2

$BD = 12y - 4$
$= 12\left(\frac{1}{2}\right) - 4$ or 2

$AF = AD$

$10(z - 4) = 2z$

$10z - 40 = 2z$

$-40 = -8z$

$5 = z$

$AF = 10(z - 4)$
$= 10(5 - 4)$ or 10

$AD = 2z$
$= 2(5)$ or 10

$CE + CF + BE + BD + AF + AD = 6 + 6 + 2 + 2 + 10 + 10$ or 36

The perimeter is 36 units.

27. Given: $\overline{AB}$ is tangent to $\odot X$ at B.

$\overline{AC}$ is tangent to $\odot X$ at C.

Prove: $\overline{AB} \cong \overline{AC}$

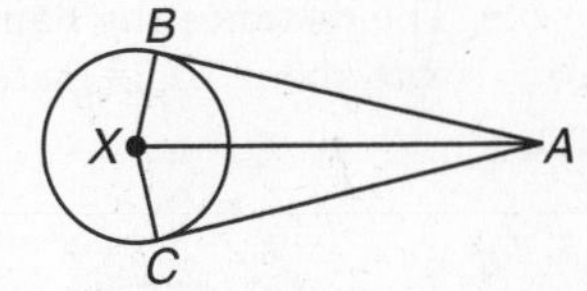

Proof:

Statements	Reasons
1. $\overline{AB}$ is tangent to $\odot X$ at B, $\overline{AC}$ is tangent to $\odot X$ at C.	1. Given
2. Draw $\overline{BX}$, $\overline{CX}$, and $\overline{AX}$.	2. Through any two points, there is one line.
3. $\overline{AB} \perp \overline{BX}$, $\overline{AC} \perp \overline{CX}$	3. Line tangent to a circle is $\perp$ to the radius at the pt. of tangency.
4. $\angle ABX$ and $\angle ACX$ are right angles.	4. Def. of $\perp$ lines
5. $\overline{BX} \cong \overline{CX}$	5. All radii of a circle are $\cong$.
6. $\overline{AX} \cong \overline{AX}$	6. Reflexive Prop.
7. $\triangle ABX \cong \triangle ACX$	7. HL
8. $\overline{AB} \cong \overline{AC}$	8. CPCTC

28. Let a = the radius of the roll of film, b = the amount of film exposed, and c = the distance from the center of the roll to the intake of the holding chamber. Since the diameter of the roll of film is 25, $a = 12.5$. Use the Pythagorean Theorem.

$a^2 + b^2 = c^2$

$12.5^2 + b^2 = 100^2$

$156.25 + b^2 = 10{,}000$

$b^2 = 9843.75$

$b \approx 99$

About 99 millimeters of film would be exposed.

29. $\overline{AE}$ and $\overline{BF}$

30. $\overline{AD}$ and $\overline{BC}$

31. 12;

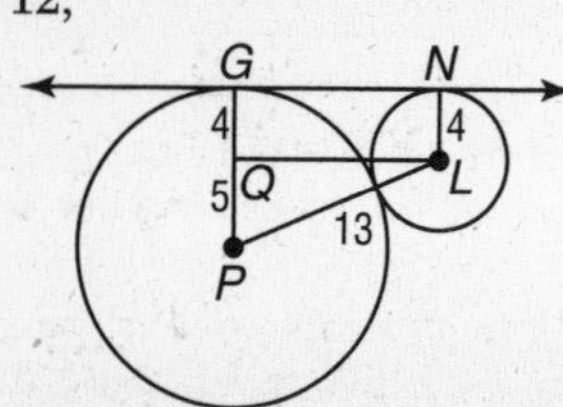

Draw $\overline{PG}$, $\overline{NL}$, and $\overline{PL}$. Construct $\overline{LQ} \perp \overline{GP}$, thus $LQGN$ is a rectangle. $GQ = NL = 4$, so $QP = 5$. Using the Pythagorean Theorem, $(QP)^2 + (QL)^2 = (PL)^2$. So, $QL = 12$. Since $GN = QL$, $GN = 12$.

32. Sample answer: Many of the field events have the athlete moving in a circular motion and releasing an object (discus, hammer, shot). The movement of the athlete models a circle and the path of the released object models a tangent.
Answers should include the following.
- The arm of the thrower, the handle, the wire, and hammer form the radius defining the circle when the hammer is spun around. The tangent is the path of the hammer when it is released.
- The distance the hammer was from the athlete was about 70.68 meters.

33.

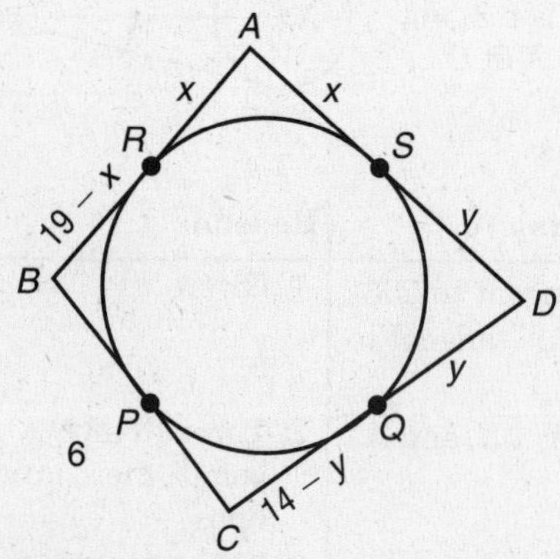

$PB = RB = 19 - x$
$CP = CQ = 14 - y$
$PB + CP = 6$
$(19 - x) + (14 - y) = 6$
$33 - x - y = 6$
$27 = x + y$
$AD = 27$

34. B; $2 + 12 + 22 + \cdots + 92 = 470$
$102 + 112 + 122 + \cdots + 192 = 1470$
$\vdots \qquad \vdots$
$902 + 912 + 922 + \cdots + 992 = 9470$
$\frac{470 + 1470 + 2470 + \cdots + 9470}{100} = \frac{49{,}700}{100}$ or 497

35. $\overleftrightarrow{AD}$ and $\overleftrightarrow{BC}$

36. $\overleftrightarrow{AE}$ and $\overleftrightarrow{BF}$

Page 558 Maintain Your Skills

37. $m\widehat{FAC} = m\widehat{FA} + m\widehat{AC}$
$180 = 90 + m\widehat{AC}$
$90 = m\widehat{AC}$
$m\angle AFC = \frac{1}{2}m\widehat{AC}$
$= \frac{1}{2}(90)$ or 45
$\widehat{BD} = \widehat{AC}$
$m\angle BED = m\angle AFC$
$m\angle BED = 45$

38. Connect L to J to form a right triangle. Use the Pythagorean Theorem.
$(LJ)^2 = x^2 + 5^2$
$10^2 = x^2 + 5^2$
$100 = x^2 + 25$
$75 = x^2$
$5\sqrt{3} = x$ or $x \approx 8.7$

39. $KB = 5 - 2$ or 3
Use the Pythagorean Theorem.
$3^2 + x^2 = 5^2$
$9 + x^2 = 25$
$x^2 = 16$
$x = 4$

40. Connect O to A to form a right triangle; $OA = 8$.
Let $x = \frac{1}{2}AP$.
Use the Pythagorean Theorem.
$4^2 + x^2 = 8^2$
$16 + x^2 = 64$
$x^2 = 48$
$x = 4\sqrt{3}$
$AP = 2(4\sqrt{3})$ or $8\sqrt{3} \approx 13.9$

41. Sample answer:
Given: $ABCD$ is a rectangle. E is the midpoint of $\overline{AB}$.

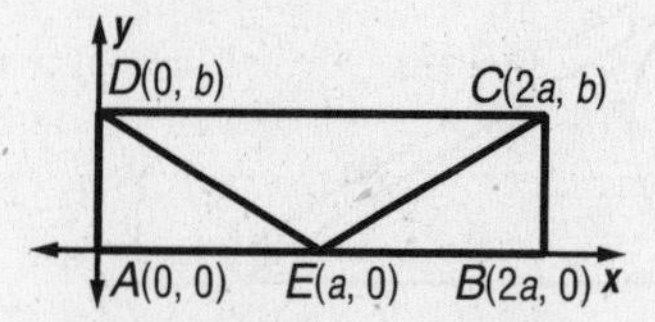

Prove: $\triangle CED$ is isosceles.
Proof: Let the coordinates of E be $(a, 0)$. Since E is the midpoint and is halfway between A and B, the coordinates of B will be $(2a, 0)$. Let the coordinates of D be $(0, b)$. The coordinates of C will be $(2a, b)$ because it is on the same horizontal as D and the same vertical as B.
$ED = \sqrt{(a - 0)^2 + (0 - b)^2}$
$= \sqrt{a^2 + b^2}$
$EC = \sqrt{(a - 2a)^2 + (0 - b)^2}$
$= \sqrt{a^2 + b^2}$
Since $ED = EC$, $\overline{ED} \cong \overline{EC}$. $\triangle DEC$ has two congruent sides, so it is isosceles.

42. $x + 3 = \frac{1}{2}[(4x + 6) - 10]$
$x + 3 = \frac{1}{2}(4x - 4)$
$x + 3 = 2x - 2$
$5 = x$

43. $2x - 5 = \frac{1}{2}[(3x + 16) - 20]$
$2x - 5 = \frac{1}{2}(3x - 4)$
$2x - 5 = \frac{3}{2}x - 2$
$\frac{1}{2}x = 3$
$x = 6$

44. $2x + 4 = \frac{1}{2}[(x + 20) - 10]$

$2x + 4 = \frac{1}{2}(x + 10)$

$2x + 4 = \frac{1}{2}x + 5$

$\frac{3}{2}x = 1$

$x = \frac{2}{3}$

45. $x + 3 = \frac{1}{2}[(4x + 10) - 45]$

$x + 3 = \frac{1}{2}(4x - 35)$

$x + 3 = 2x - \frac{35}{2}$

$\frac{41}{2} = x$ or $x = 20.5$

Page 560 Geometry Activity: Inscribed and Circumscribed Triangles

1. See students' work.
2. See students' work.
3. See students' work.
4. The incenter is equidistant from each side. The perpendicular to one side should be the same length as it is to the other two sides.
5. The incenter is equidistant from all the sides. The radius of the circle is perpendicular to the tangent sides and all radii are congruent, matching the distance from the incenter to the sides.
6. The circumcenter is equidistant from all three vertices, so the distance from the circumcenter to one vertex is the same as the distance to each of the others.
7. The circumcenter is equidistant from the vertices and all of the vertices must lie on the circle. So, this distance is the radius of the circle containing the vertices.
8. $\frac{360}{6} = 60$
9. 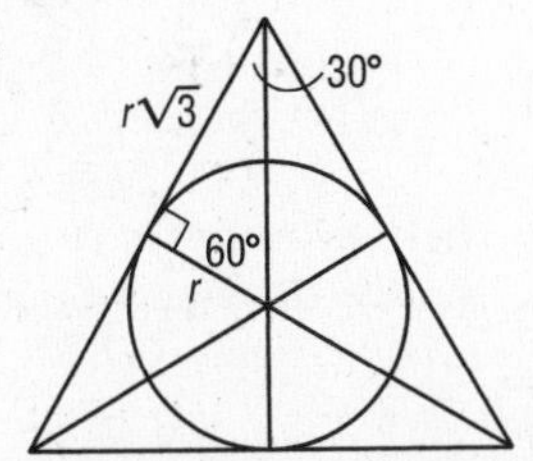

   Suppose all six radii are drawn. Each central angle measures 60°. Thus, six 30°-60°-90° triangles are formed. Each triangle has a side which is a radius r units long. Using 30°-60°-90° side ratios, the segment tangent to the circle has length $r\sqrt{3}$, making each side of the circumscribed triangle $2r\sqrt{3}$. If all three sides have the same measure, then the triangle is equilateral.
10. The incenter is the point from which you can construct a circle "in" the triangle. Circum means *around*. So the circumcenter is the point from which you can construct a circle "around" the triangle.

10-6 Secants, Tangents, and Angle Measures

Page 564 Check for Understanding

1. Sample answer: A tangent intersects the circle in only one point and no part of the tangent is in the interior of the circle. A secant intersects the circle in two points and some of its points do lie in the interior of the circle.
2. Sample answer:

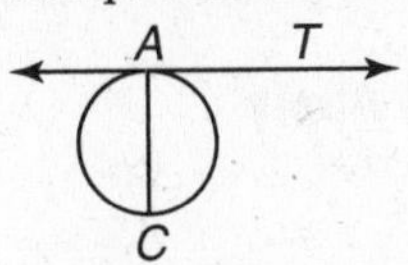

   Angle *TAC* is a right angle; There are two reasons: (1) If the point of tangency is the endpoint of a diameter, then the tangent is perpendicular to the diameter at that point. (2) The arc intercepted by the secant (diameter) and the tangent is a semicircle. Thus the measure of the angle is half of 180 or 90.

3. $m\angle 1 = 180 - \frac{1}{2}(38 + 46)$

$= 180 - \frac{1}{2}(84)$

$= 180 - 42$ or 138

4. $m\angle 2 = \frac{1}{2}(360 - 100)$

$= \frac{1}{2}(260)$ or 130

5. $x = \frac{1}{2}[84 - (180 - 84 - 52)]$

$= \frac{1}{2}(84 - 44)$

$= \frac{1}{2}(40)$ or 20

6. $x = \frac{1}{2}[128 - (360 - 148 - 128)]$

$= \frac{1}{2}(128 - 84)$

$= \frac{1}{2}(44)$ or 22

7. $55 = \frac{1}{2}[x - (360 - x)]$

$55 = \frac{1}{2}(2x - 360)$

$55 = x - 180$

$235 = x$

8. $m\angle CAS = \frac{1}{2}m\widehat{SA}$

$= \frac{1}{2}(46)$ or 23

9. $m\widehat{SAK} = 2m\angle SLK$

$= 2(78)$ or 156

$m\widehat{AK} = m\widehat{SAK} - m\widehat{SA}$

$= 156 - 46$ or 110

$m\angle QAK = \frac{1}{2}m\widehat{AK}$

$= \frac{1}{2}(110)$ or 55

10. Since $\overline{SA} \parallel \overline{LK}$, $\angle ASL$ and $\angle SLK$ are supplementary consecutive interior angles. So $m\angle ASL = 180 - 78$ or 102 and $m\widehat{AKL} = 2(m\angle ASL) = 2(102)$ or 204.

$m\widehat{AKL} = m\widehat{AK} + m\widehat{KL}$

$204 = 110 + m\widehat{KL}$ (See Exercise 9.)

$94 = m\widehat{KL}$

11. $m\widehat{SA} + m\widehat{AK} + m\widehat{KL} + m\widehat{SL} = 360$

$46 + 110 + 94 + m\widehat{SL} = 360$

$250 + m\widehat{SL} = 360$

$m\widehat{SL} = 110$

Pages 564–567 Practice and Apply

12. $m\angle 3 = \frac{1}{2}(100 + 120)$

$= \frac{1}{2}(220)$ or 110

13. $m\angle 4 = \frac{1}{2}(45 + 75)$

$= \frac{1}{2}(120)$ or 60

14. $m\angle 5 = \frac{1}{2}[360 - (110 + 150)]$

$= \frac{1}{2}(360 - 260)$

$= \frac{1}{2}(100)$ or 50

15. $5a + 3a + 6a + 4a = 360$

$18a = 360$

$a = 20$

$m\angle 6 = \frac{1}{2}(5a + 6a)$

$= \frac{1}{2}(11a)$

$= \frac{1}{2}(11 \cdot 20)$

$= \frac{1}{2}(220)$ or 110

16. $m\angle 7 = \frac{1}{2}(196)$ or 98

17. $m\angle 8 = \frac{1}{2}(180)$ or 90

18. $m\angle 9 = \frac{1}{2}(360 - 120)$

$= \frac{1}{2}(240)$ or 120

19. $m\angle 10 = \frac{1}{2}[360 - (100 + 160)]$

$= \frac{1}{2}(360 - 260)$

$= \frac{1}{2}(100)$ or 50

20. $65 = \frac{1}{2}(m\widehat{AC} + 72)$

$130 = m\widehat{AC} + 72$

$58 = m\widehat{AC}$

21. $x = \frac{1}{2}(90 - 30)$

$= \frac{1}{2}(60)$ or 30

22. $x = \frac{1}{2}[20 - (180 - 20 - 150)]$

$= \frac{1}{2}(20 - 10)$

$= \frac{1}{2}(10)$ or 5

23. $25 = \frac{1}{2}(90 - 5x)$

$50 = 90 - 5x$

$-40 = -5x$

$8 = x$

24. $360 - (160 + 34 + 106) = 60$

$x = \frac{1}{2}(60 - 34)$

$x = \frac{1}{2}(26)$ or 13

25. $x = \frac{1}{2}(7x - 20)$

$2x = 7x - 20$

$20 = 5x$

$4 = x$

26. $x = \frac{1}{2}(10x - 40)$

$x = 5x - 20$

$20 = 4x$

$5 = x$

27. $x + 2.5 = \frac{1}{2}[(4x + 5) - 50]$

$2x + 5 = 4x - 45$

$50 = 2x$

$25 = x$

28. $90 - 60 = 30$

$30 = \frac{1}{2}(105 - 5x)$

$60 = 105 - 5x$

$5x = 45$

$x = 9$

29. $50 = \frac{1}{2}[(360 - x) - x]$

$100 = 360 - 2x$

$2x = 260$

$x = 130$

30. $30 = \frac{1}{2}[x - (360 - x)]$

$60 = 2x - 360$

$420 = 2x$

$210 = x$

31. $3x = \frac{1}{2}[(4x + 50) - 30]$

$3x = \frac{1}{2}(4x + 20)$

$3x = 2x + 10$

$x = 10$

32. $40 = \frac{1}{2}[(x^2 + 12x) - (x^2 + 2x)]$

$40 = \frac{1}{2}(10x)$

$40 = 5x$

$8 = x$

33. $m\angle C = \frac{1}{2}(116 - 38)$

$= \frac{1}{2}(78)$ or 39

$m\angle C = \frac{1}{2}(m\widehat{BAH} - m\widehat{BH})$

$39 = \frac{1}{2}[(360 - m\widehat{BH}) - m\widehat{BH}]$

$39 = \frac{1}{2}(360 - 2m\widehat{BH})$

$39 = 180 - m\widehat{BH}$

$m\widehat{BH} = 141$

34. $m\overset{\frown}{BE} = 2m\angle EFB$
$= 2(30)$ or 60
$m\angle FGE = 180 - 52$ or 128
$m\angle FGE + m\angle EFB + m\angle GEF = 180$
$128 + 30 + m\angle GEF = 180$
$m\angle GEF = 22$
$m\overset{\frown}{CF} = 2m\angle GEF$
$= 2(22)$ or 44
$m\overset{\frown}{AC} = 360 - (m\overset{\frown}{AB} + m\overset{\frown}{BE} + m\overset{\frown}{FE} + m\overset{\frown}{CF})$
$= 360 - (108 + 60 + 118 + 44)$
$= 360 - 330$ or 30

35. $m\overset{\frown}{CF} = 44$ (See Exercise 34.)

36. $m\angle EDB = \frac{1}{2}(m\overset{\frown}{BE} - m\overset{\frown}{AC})$
$= \frac{1}{2}(60 - 30)$
$= \frac{1}{2}(30)$ or 15

37. $m\angle AMB = 23.5$
$m\angle AMB = \frac{1}{2}(m\overset{\frown}{BC} - m\overset{\frown}{AB})$
$23.5 = \frac{1}{2}(m\overset{\frown}{BC} - 71)$
$47 = m\overset{\frown}{BC} - 71$
$118 = m\overset{\frown}{BC}$

38. $m\overset{\frown}{ABC} = m\overset{\frown}{AB} + m\overset{\frown}{BC}$
$m\overset{\frown}{ABC} = 71 + 118$ or 189
No, its measure is 189, not 180.

39. $C = \pi d$
$C = \pi(100) \approx 314.2$ feet
$\frac{118}{360} \cdot 314.2 \approx 103$
You would walk about 103 ft.

40a. Given: $\overleftrightarrow{AC}$ and $\overleftrightarrow{AT}$ are secants to the circle.
Prove: $m\angle CAT = \frac{1}{2}(m\overset{\frown}{CT} - m\overset{\frown}{BR})$

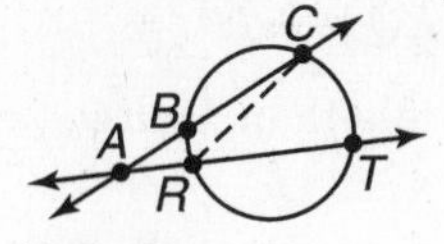

Proof:

Statements	Reasons
1. $\overleftrightarrow{AC}$ and $\overleftrightarrow{AT}$ are secants to the circle.	1. Given
2. $m\angle CRT = \frac{1}{2}m\overset{\frown}{CT}$, $m\angle ACR = \frac{1}{2}m\overset{\frown}{BR}$	2. The meas. of an inscribed $\angle = \frac{1}{2}$ the meas. of its intercepted arc.
3. $m\angle CRT = m\angle ACR + m\angle CAT$	3. Exterior ⦣ Theorem
4. $\frac{1}{2}m\overset{\frown}{CT} = \frac{1}{2}m\overset{\frown}{BR} + m\angle CAT$	4. Substitution
5. $\frac{1}{2}m\overset{\frown}{CT} - \frac{1}{2}m\overset{\frown}{BR} = m\angle CAT$	5. Subtraction Prop.
6. $\frac{1}{2}(m\overset{\frown}{CT} - m\overset{\frown}{BR}) = m\angle CAT$	6. Distributive Prop.

40b. Given: $\overleftrightarrow{DG}$ is a tangent to the circle. $\overrightarrow{DF}$ is a secant to the circle.
Prove: $m\angle FDG = \frac{1}{2}(m\overset{\frown}{FG} - m\overset{\frown}{GE})$

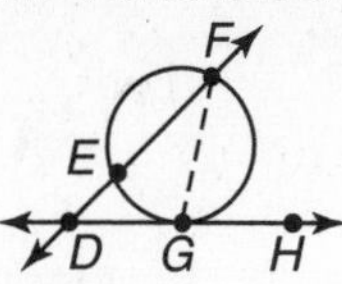

Proof:

Statements	Reasons
1. $\overleftrightarrow{DG}$ is a tangent to the circle. $\overrightarrow{DF}$ is a secant to the circle.	1. Given
2. $m\angle DFG = \frac{1}{2}m\overset{\frown}{GE}$, $m\angle FGH = \frac{1}{2}m\overset{\frown}{FG}$	2. The meas. of an inscribed $\angle = \frac{1}{2}$ the meas. of its intercepted arc.
3. $m\angle FGH = m\angle DFG + m\angle FDG$	3. Exterior ⦣ Theorem
4. $\frac{1}{2}m\overset{\frown}{FG} = \frac{1}{2}m\overset{\frown}{GE} + m\angle FDG$	4. Substitution
5. $\frac{1}{2}m\overset{\frown}{FG} - \frac{1}{2}m\overset{\frown}{GE} = m\angle FDG$	5. Subtraction Prop.
6. $\frac{1}{2}(m\overset{\frown}{FG} - m\overset{\frown}{GE}) = m\angle FDG$	6. Distributive Prop.

40c. Given: $\overrightarrow{HI}$ and $\overleftrightarrow{HJ}$ are tangents to the circle.
Prove: $m\angle IHJ = \frac{1}{2}(m\overset{\frown}{IXJ} - m\overset{\frown}{IJ})$

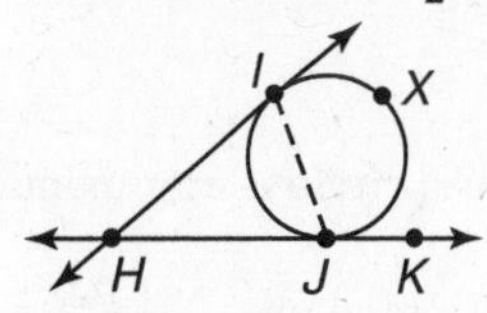

Proof:

Statements	Reasons
1. $\overrightarrow{HI}$ and $\overleftrightarrow{HJ}$ are tangents to the circle.	1. Given
2. $m\angle IJK = \frac{1}{2}m\overset{\frown}{IXJ}$, $m\angle HIJ = \frac{1}{2}m\overset{\frown}{IJ}$	2. The measure of an inscribed $\angle = \frac{1}{2}$ the measure of its intercepted arc.
3. $m\angle IJK = m\angle HIJ + m\angle IHJ$	3. Ext. ⦣ Th.
4. $\frac{1}{2}m\overset{\frown}{IXJ} = \frac{1}{2}m\overset{\frown}{IJ} + m\angle IHJ$	4. Substitution
5. $\frac{1}{2}m\overset{\frown}{IXJ} - \frac{1}{2}m\overset{\frown}{IJ} = m\angle IHJ$	5. Subtr. Prop.
6. $\frac{1}{2}(m\overset{\frown}{IXJ} - m\overset{\frown}{IJ}) = m\angle IHJ$	6. Distrib. Prop.

41. The diagonals of a rhombus are perpendicular.
Let x = side length of rhombus.

$5^2 + 12^2 = x^2$
$25 + 144 = x^2$
$169 = x^2$
$13 = x$

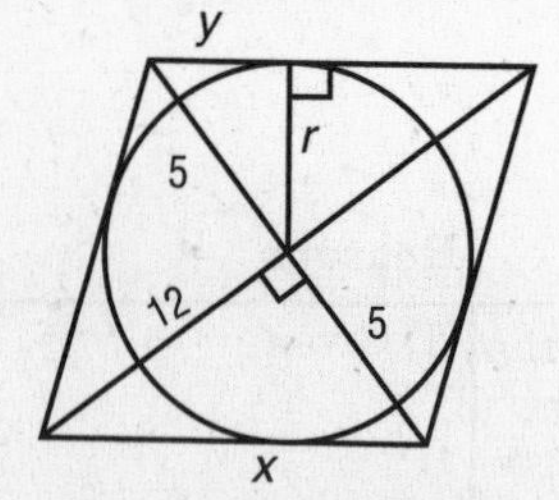

Let r = radius of inscribed circle.
The radius forms two right triangles. One triangle has a hypotenuse of 5. Let y = measure of the portion of the side of the rhombus that is a leg of this triangle. The second right triangle has a hypotenuse of length 12. Its second leg has measure $13 - y$.
Use the Pythagorean Theorem to write an equation for each triangle.

$r^2 + (13 - y)^2 = 12^2$
$r^2 + y^2 = 5^2$
$r^2 = 25 - y^2$

Substituting for r^2,

$25 - y^2 + (13 - y)^2 = 144$
$25 - y^2 + 169 - 26y + y^2 = 144$
$194 - 26y = 144$
$-26y = -50$
$y \approx 1.9$

$r^2 \approx 25 - 1.9^2$
$r^2 \approx 21.39$
$r \approx 4.6$

The radius of the inscribed circle is approximately 4.6 cm.

42. Let x = measure of intercepted arc.

$86.5 = \frac{1}{2}[(360 - x) - x]$
$173 = 360 - 2x$
$2x = 187$
$x = 93.5$

43a. Given: $\overrightarrow{AB}$ is a tangent to $\odot O$.
$\overrightarrow{AC}$ is a secant to $\odot O$. $\angle CAB$ is acute.

Prove: $m\angle CAB = \frac{1}{2}m\widehat{CA}$

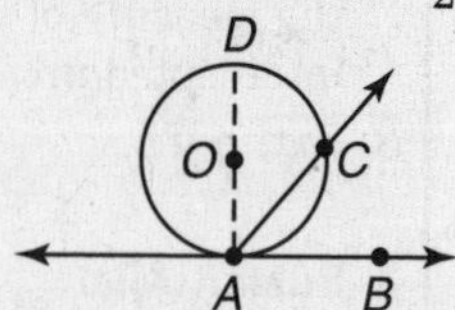

Proof: $\angle DAB$ is a right $\angle$ with measure 90, and $\widehat{DCA}$ is a semicircle with measure 180, since if a line is tangent to a $\odot$, it is $\perp$ to the radius at the point of tangency. Since $\angle CAB$ is acute, C is in the interior of $\angle DAB$, so by the Angle and Arc Addition Postulates, $m\angle DAB = m\angle DAC + m\angle CAB$ and $m\widehat{DCA} = m\widehat{DC} + m\widehat{CA}$. By substitution, $90 = m\angle DAC + m\angle CAB$ and $180 = m\widehat{DC} + m\widehat{CA}$. So, $90 = \frac{1}{2}m\widehat{DC} + \frac{1}{2}m\widehat{CA}$ by Division Prop., and $m\angle DAC + m\angle CAB = \frac{1}{2}m\widehat{DC} + \frac{1}{2}m\widehat{CA}$ by substitution. $m\angle DAC = \frac{1}{2}m\widehat{DC}$ since $\angle DAC$ is inscribed, so substitution yields $\frac{1}{2}m\widehat{DC} + m\angle CAB = \frac{1}{2}m\widehat{DC} + \frac{1}{2}m\widehat{CA}$. By Subtraction Prop., $m\angle CAB = \frac{1}{2}m\widehat{CA}$.

43b. Given: $\overrightarrow{AB}$ is a tangent to $\odot O$.
$\overrightarrow{AC}$ is a secant to $\odot O$. $\angle CAB$ is obtuse.

Prove: $m\angle CAB = \frac{1}{2}m\widehat{CDA}$

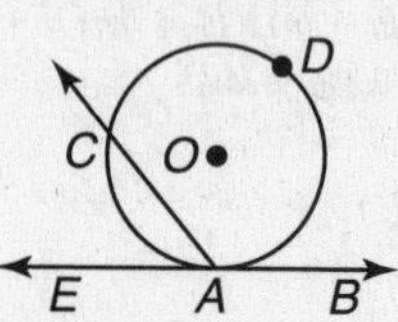

Proof: $\angle CAB$ and $\angle CAE$ form a linear pair, so $m\angle CAB + m\angle CAE = 180$. Since $\angle CAB$ is obtuse, $\angle CAE$ is acute and Case 1 applies, so $m\angle CAE = \frac{1}{2}m\widehat{CA}$. $m\widehat{CA} + m\widehat{CDA} = 360$, so $\frac{1}{2}m\widehat{CA} + \frac{1}{2}m\widehat{CDA} = 180$ by Division Prop., and $\frac{1}{2}m\angle CAE + \frac{1}{2}m\widehat{CDA} = 180$ by substitution. By the Transitive Prop., $m\angle CAB + m\angle CAE = m\angle CAE + \frac{1}{2}m\widehat{CDA}$, so by Subtraction Prop., $m\angle CAB = \frac{1}{2}m\widehat{CDA}$.

44. Let x = the measure of the angle of view.

$360 - 54 = 306$
$x = \frac{1}{2}(306 - 54)$
$x = \frac{1}{2}(252)$ or 126

45. $\angle 3, \angle 1, \angle 2$; $m\angle 3 = m\widehat{RQ}$, $m\angle 1 = \frac{1}{2}m\widehat{RQ}$ so $m\angle 3 > m\angle 1$, $m\angle 2 = \frac{1}{2}(m\widehat{RQ} - m\widehat{TP}) = \frac{1}{2}m\widehat{RQ} - \frac{1}{2}m\widehat{TP}$, which is less than $\frac{1}{2}m\widehat{RQ}$, so $m\angle 2 < m\angle 1$.

46. Sample answer: Each raindrop refracts light from the sun and sends the beam to Earth. The raindrop is actually spherical, but the angle of the light is an inscribed angle from the bent rays. Answers should include the following.

- $\angle C$ is an inscribed angle and $\angle F$ is a secant-secant angle.
- The measure of $\angle F$ can be calculated by finding the positive difference between $m\widehat{BD}$ and the measure of the small intercepted arc containing point C.

47. A; let x = the measure of the common intercepted arc of $\angle A$ and $\angle B$.

$m\angle A = \frac{1}{2}(x - 15)$
$10 = \frac{1}{2}(x - 15)$
$20 = x - 15$
$35 = x$
$m\angle B = \frac{1}{2}(95 - 35)$
$= \frac{1}{2}(60)$ or 30

48. C; the set of data can be represented by the linear equation $y = \frac{1}{2}x + 1$.

Page 568 Maintain Your Skills

49. The tangent forms a right angle with the radius. Use the Pythagorean Theorem.

$(2x)^2 + 24^2 = (16 + 24)^2$
$4x^2 + 576 = 1600$
$4x^2 = 1024$
$x^2 = 256$
$x = 16$

50. Both tangents to the circle on the left have the same measure. Both tangents to the circle on the right have the same measure. Both tangents to the circle in the middle have the same measure. Because the left tangent to the middle circle is also tangent to the left circle and the right tangent to the middle circle is also tangent to the right circle, the tangents to the left and right circles have the same measure.

$12x + 10 = 74 - 4x$
$16x = 64$
$x = 4$

51. $\angle EGN$ is an inscribed angle.

$m\angle EGN = \frac{1}{2}m\widehat{EN}$
$= \frac{1}{2}(66)$ or 33

52. $m\widehat{GEN} = m\widehat{GE} + m\widehat{EN}$
$180 = m\widehat{GE} + 66$
$114 = m\widehat{GE}$
$m\angle GME = \frac{1}{2}m\widehat{GE}$
$= \frac{1}{2}(114)$ or 57

53. $m\widehat{GM} = 89$
$\angle GNM$ is an inscribed angle that intercepts $\widehat{GM}$.
$m\angle GNM = \frac{1}{2}(m\widehat{GM})$
$= \frac{1}{2}(89)$ or 44.5

54. slope $= \frac{\text{vertical rise}}{\text{horizontal run}}$
slope $= \frac{1}{12}$
The slope is $\frac{1}{12}$.

55. 30 feet = $30 \cdot 12$ or 360 in.
Let x = the height of the ramp.
$(12x)^2 + x^2 = 360^2$
$145x^2 = 129{,}600$
$x^2 \approx \sqrt{893.8}$
$x \approx 30$
The ramp is about 30 in. high.

56. Given: $\overline{AC} \cong \overline{BF}$
Prove: $AB = CF$

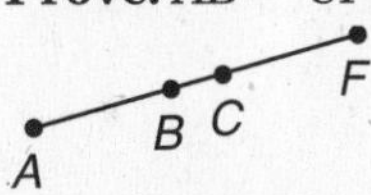

Proof: By definition of congruent segments, $AC = BF$. Using the Segment Addition Postulate, we know that $AC = AB + BC$ and $BF = BC + CF$. Since $AC = BF$, this means that $AB + BC = BC + CF$. If BC is subtracted from each side of this equation, the result is $AB = CF$.

57. $x^2 + 6x - 40 = 0$
$(x - 4)(x + 10) = 0$
$x - 4 = 0$ or $x + 10 = 0$
$x = 4$ $\quad x = -10$

58. $2x^2 + 7x - 30 = 0$
$(2x - 5)(x + 6) = 0$
$2x - 5 = 0$ or $x + 6 = 0$
$x = \frac{5}{2}$ $\quad x = -6$
$x = 2\frac{1}{2}$

59. $3x^2 - 24x + 45 = 0$
$3(x^2 - 8x + 15) = 0$
$3(x - 3)(x - 5) = 0$
$x - 3 = 0$ or $x - 5 = 0$
$x = 3$ $\quad x = 5$

Page 568 Practice Quiz 2

1. Each central angle has a measure of $\frac{360}{8}$ or 45. Therefore, the remaining 2 angles in each triangle each measure $\frac{180 - 45}{2}$ or 67.5.
2. Inscribed angles of the same arc are congruent.
   $m\angle 1 = m\angle 2 = \frac{1}{2}(68)$ or 34
3. $x = 2(6)$ or 12
4. $x = \frac{1}{2}(60 - 34)$
   $= \frac{1}{2}(26)$ or 13
5. $360 - 129 = 231$
   $x = \frac{1}{2}(231)$ or 115.5

10-7 Special Segments in a Circle

Page 569 Geometry Activity: Intersecting Chords

1. $\angle PTS \cong \angle RTQ$ (vertical $\angle$s are $\cong$);
   $\angle P \cong \angle R$ ($\angle$s intercepting same arc are $\cong$);
   $\angle S \cong \angle Q$ ($\angle$s intercepting same arc are $\cong$)
2. They are similar by AA Similarity.
3. $\frac{PT}{RT} = \frac{ST}{TQ}$ or $PT \cdot TQ = RT \cdot ST$

Pages 571–572 Check for Understanding

1. Sample answer: The product equation for secant segments equates the product of exterior segment measure and the whole segment measure for each secant. In the case of secant-tangent, the product involving the tangent segment becomes (measure of tangent segment)^2 because the exterior segment and the whole segment are the same segment.
2. Latisha; the length of the tangent segment squared equals the product of the exterior secant segment and the entire secant, not the interior secant segment.

3. Sample answer:

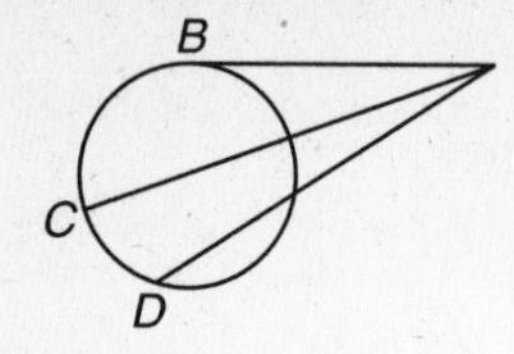

4. $9x = 3(6)$
$9x = 18$
$x = 2$

5. $31^2 = 20(20 + x)$
$961 = 400 + 20x$
$561 = 20x$
$28.1 \approx x$

6. $x(10 + x) = 3(3 + 5)$
$10x + x^2 = 24$
$x^2 + 10x - 24 = 0$
$(x - 2)(x + 12) = 0$
$x - 2 = 0$ or $x + 12 = 0$
$x = 2$ $\quad x = -12$
Since x represents a length, it must be positive. Reject the negative value. So $x = 2$.

7. Draw a model using a circle. Let x represent the unknown measure of the segment of diameter $\overline{AB}$. Use the products of the lengths of the intersecting chords to find the length of the diameter.

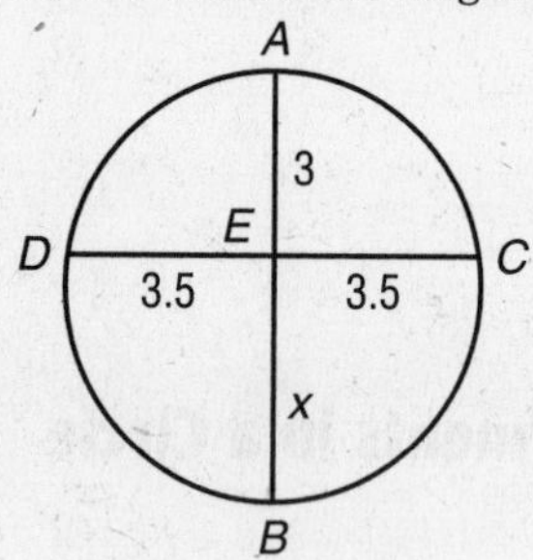

$AE \cdot EB = DE \cdot EC$
$3x = 3.5 \cdot 3.5$
$x \approx 4.08$
$AB = AE + EB$
$AB \approx 3 + 4.08$ or 7.08
The radius of the circle is about $\frac{7.08}{2}$ or 3.54.
The ratio of the arch width to the radius of the circle is about 7:3.54.

Pages 572–574 Practice and Apply

8. $2x = 4(5)$
$2x = 20$
$x = 10$

9. $6 \cdot 6 = x \cdot 9$
$36 = 9x$
$4 = x$

10. $7 \cdot 2 = 3 \cdot x$
$14 = 3x$
$\frac{14}{3} = x$ or $x \approx 4.7$

11. $x(x + 8) = 5 \cdot 4$
$x^2 + 8x = 20$
$x^2 + 8x - 20 = 0$
$(x - 2)(x + 10) = 0$
$x - 2 = 0$ or $x + 10 = 0$
$x = 2$ $\quad x = -10$
Disregard the negative value.
$x = 2$

12. $x^2 = 3(9 + 3)$
$x^2 = 36$
$x = 6$

13. $4^2 = 2(x + 2)$
$16 = 2x + 4$
$12 = 2x$
$6 = x$

14. $16^2 = x(x + x + 16)$
$256 = x(2x + 16)$
$256 = 2x^2 + 16x$
$0 = 2x^2 + 16x - 256$
$0 = x^2 + 8x - 128$
$0 = (x + 16)(x - 8)$
$0 = x + 16$ or $0 = x - 8$
$-16 = x$ $\quad 8 = x$
Since the length of a segment cannot be negative, reject $x = -16$. So $x = 8$.

15. $(9.8)^2 = 7.1(2x + 7.1)$
$96.04 = 14.2x + 50.41$
$45.63 = 14.2x$
$3.2 \approx x$

16. $4(4 + 2) = 3(3 + x)$
$24 = 3(3 + x)$
$8 = 3 + x$
$5 = x$

17. $x(5 + x) = 3(3 + 9)$
$5x + x^2 = 36$
$x^2 + 5x - 36 = 0$
$(x - 4)(x + 9) = 0$
$x - 4 = 0$ or $x + 9 = 0$
$x = 4$ $\quad x = -9$
Disregard the negative value.
$x = 4$

18. $x(x + 5 + x) = 5(5 + 5 + x)$
$x(5 + 2x) = 5(10 + x)$
$5x + 2x^2 = 50 + 5x$
$2x^2 = 50$
$x^2 = 25$
$x = 5$

19. $x(x + 3x) = 8(8 + x + 2)$
$x(4x) = 8(10 + x)$
$4x^2 = 80 + 8x$
$4x^2 - 8x - 80 = 0$
$4(x - 2x - 20) = 0$
$$x = \frac{-(-2) \pm \sqrt{(-2)^2 - 4(1)(-20)}}{2(1)}$$
$$x = \frac{2 \pm \sqrt{84}}{2}$$
$x \approx 5.6$ or $x \approx -3.6$
Disregard the negative value. So $x \approx 5.6$.

20. Let r = the radius of the circle.

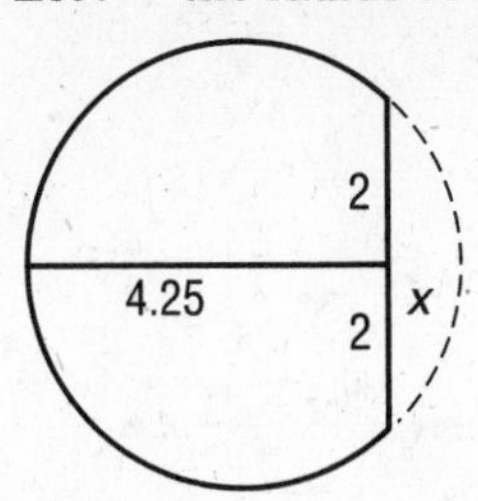

Let x represent the unknown measure of the segment of the diameter. Use the products of the lengths of the intersecting chords to find the length of the diameter.

$4.25x = 2 \cdot 2$
$x \approx 0.94$
$d \approx 4.25 + 0.94$ or 5.19 mm
$r = \frac{1}{2}d \approx \frac{1}{2}(5.19)$ or 2.6 mm

21. Given: $\overline{WY}$ and $\overline{ZX}$ intersect at T.
Prove: $WT \cdot TY = ZT \cdot TX$

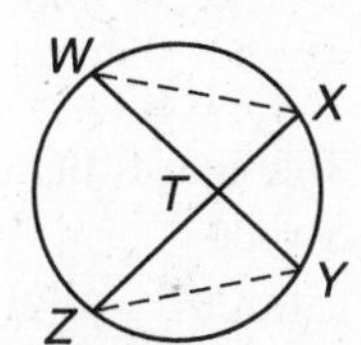

Proof:

Statements	Reasons
a. $\angle W \cong \angle Z$, $\angle X \cong \angle Y$	a. Inscribed angles that intercept the same arc are congruent.
b. $\triangle WXT \sim \triangle ZYT$	b. AA Similarity
c. $\frac{WT}{ZT} = \frac{TX}{TY}$	c. Definition of similar triangles
d. $WT \cdot TY = ZT \cdot TX$	d. Cross products

22. $4(9) = (8 + x)(8 - x)$
$36 = 64 - x^2$
$x^2 = 28$
$x \approx 5.3$

23. $2(5 + 3) = y^2$
$16 = y^2$
$4 = y$

24. $12^2 = 6(6 + y + 3)$
$144 = 54 + 6y$
$90 = 6y$
$15 = y$

$2x = 3y$
$2x = 3(15)$
$2x = 45$
$x = 22.5$

25. $10(10 + 8) = 9(9 + x)$
$180 = 81 + 9x$
$99 = 9x$
$11 = x$

26. $9^2 = x(x + 24)$
$81 = x^2 + 24x$
$0 = x^2 + 24x - 81$
$0 = (x - 3)(x + 27)$
$0 = x - 3$ or $0 = x + 27$
$3 = x \qquad -27 = x$
Disregard the negative value. So $x = 3$.

$2x(2x + 24) = y(y + 12.25)$
$2(3)(2 \cdot 3 + 24) = y^2 + 12.25y$
$180 = y^2 + 12.25y$
$0 = y^2 + 12.25y - 180$
$y = \frac{-12.25 \pm \sqrt{(12.25)^2 - 4(1)(-180)}}{2(1)}$
$y \approx 8.6$ or $y \approx -20.9$
Disregard the negative value. So $y \approx 8.6$.

27. $3(3 + x) = 4(4 + 9)$
$9 + 3x = 52$
$3x = 43$
$x \approx 14.3$

28. $10^2 = y(6 + y)$
$100 = 6y + y^2$
$0 = y^2 + 6y - 100$
$y = \frac{-6 \pm \sqrt{6^2 - 4(1)(-100)}}{2(1)}$
$y \approx 7.4$ or $y \approx -13.4$
Disregard the negative value. So $y \approx 7.4$.

29. Let r = the radius of the circle. Draw a model using a circle. Let x represent the unknown measure of the segment of the diameter. Use the products of the lengths of the intersecting chords to find the diameter.
$60x = 100 \cdot 100$
$x = 166.\overline{6}$

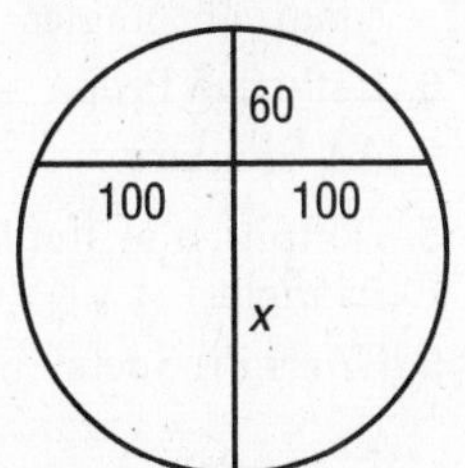

$d = 166.\overline{6} + 60$ or $226.\overline{6}$ cm
$r = \frac{1}{2}d = \frac{1}{2}(226.\overline{6})$ or $113.\overline{3}$ cm
The radius is $113.\overline{3}$ cm.

30. Given: $\overline{EC}$ and $\overline{EB}$ are secant segments.
Prove: $EA \cdot EC = ED \cdot EB$

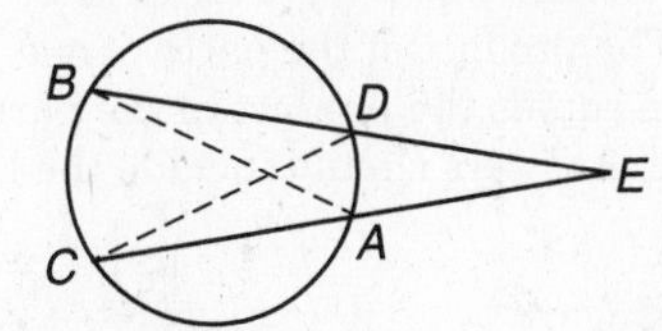

Proof:

Statements	Reasons
1. $\overline{EC}$ and $\overline{EB}$ are secant segments.	1. Given
2. $\angle DEC \cong \angle AEB$	2. They name the same angle. (Reflexive Prop.)
3. $\angle ECD \cong \angle EBA$	3. Inscribed $\measuredangle$s that intercept the same arc are $\cong$.
4. $\triangle ABE \sim \triangle DCE$	4. AA Similarity
5. $\frac{EA}{ED} = \frac{EB}{EC}$	5. Definition of similar triangles
6. $EA \cdot EC = ED \cdot EB$	6. Cross Products

31. Given: tangent $\overline{RS}$ and secant $\overline{US}$
Prove: $(RS)^2 = US \cdot TS$

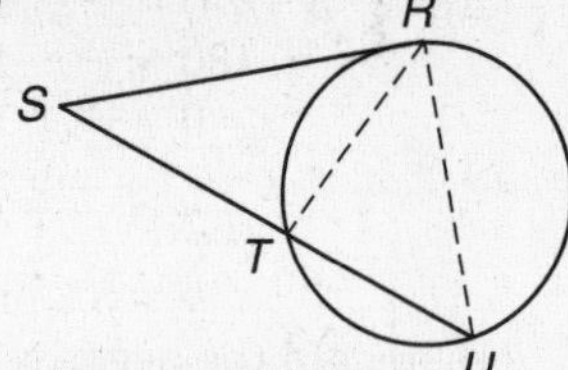

Proof:

Statements	Reasons
1. tangent $\overline{RS}$ and secant $\overline{US}$	1. Given
2. $m\angle RUT = \frac{1}{2}m\widehat{RT}$	2. The measure of an inscribed angle equals half the measure of its intercepted arc.
3. $m\angle SRT = \frac{1}{2}m\widehat{RT}$	3. The measure of an angle formed by a secant and a tangent equals half the measure of its intercepted arc.
4. $m\angle RUT = m\angle SRT$	4. Substitution
5. $\angle SUR \cong \angle SRT$	5. Definition of congruent angles
6. $\angle S \cong \angle S$	6. Reflexive Prop.
7. $\triangle SUR \sim \triangle SRT$	7. AA Similarity
8. $\frac{RS}{US} = \frac{TS}{RS}$	8. Definition of similar triangles
9. $(RS)^2 = US \cdot TS$	9. Cross Products

32. $ZY = XY$
$(WX)^2 = XY \cdot XZ$
$(WX)^2 = XY(XY + ZY)$
$(WX)^2 = XY(2XY)$
$(WX)^2 = 2(XY)^2$
$WX = \sqrt{2(XY)^2}$
$WX = \sqrt{2} \cdot XY$

33. Sample answer: The product of the parts of one intersecting chord equals the product of the parts of the other chord. Answers should include the following.
- $\overline{AF}, \overline{FD}, \overline{EF}, \overline{FB}$
- $AF \cdot FD = EF \cdot FB$

34. D; $x^2 = 20 - x$
$x^2 + x - 20 = 0$
$(x - 4)(x + 5) = 0$
$x - 4 = 0$ or $x + 5 = 0$
$x = 4$ $\quad x = -5$

35. C; let x = time working together.
$\frac{x}{15} + \frac{x}{30} = 1$
$30\left(\frac{x}{15} + \frac{x}{30}\right) = 30(1)$
$2x + x = 30$
$3x = 30$
$x = 10$
It will take them 10 minutes working together.

Page 574 Maintain Your Skills

36. $360 - 102 = 258$
$m\angle 1 = \frac{1}{2}(258)$ or 129

37. $m\angle 2 = \frac{1}{2}(85 + 230)$
$= \frac{1}{2}(315)$ or 157.5

38. $m\angle 3 = \frac{1}{2}(28 + 2 \cdot 12)$
$= \frac{1}{2}(52)$ or 26

39. $x = 7$

40. Connect the center with the point of tangency, forming a right triangle. Use the Pythagorean Theorem.
$(12 + x)^2 = 12^2 + 16^2$
$144 + 24x + x^2 = 144 + 256$
$x^2 + 24x - 256 = 0$
$(x - 8)(x + 32) = 0$
$x - 8 = 0$ or $x + 32 = 0$
$x = 8$ $\quad x = -32$
Disregard the negative value. So $x = 8$.

41. $x = 36$

42. $\tan 67° = \frac{x}{5}$
$5 \tan 67° = x$
$12 \approx x$
The distance across the stream is about 12 feet.

43. No two sides are congruent, and one angle is greater than 90. The triangle is scalene and obtuse.

44. Two sides are congruent and one angle is a right angle. The triangle is isosceles and right.

45. All of the sides are congruent and all three angles congruent. The triangle is equilateral, acute or equiangular.

46. $d = \sqrt{(x_2 - x_1)^2 + (y_2 - y_1)^2}$
$d = \sqrt{(10 - (-2))^2 + (12 - 7)^2}$
$d = \sqrt{144 + 25}$
$d = \sqrt{169}$ or 13

47. $d = \sqrt{(x_2 - x_1)^2 + (y_2 - y_1)^2}$
$d = \sqrt{(3 - 1)^2 + (4 - 7)^2}$
$d = \sqrt{4 + 9}$ or $\sqrt{13}$

48. $d = \sqrt{(x_2 - x_1)^2 + (y_2 - y_1)^2}$
$d = \sqrt{(15 - 9)^2 + (-2 - (-4))^2}$
$d = \sqrt{36 + 4}$ or $\sqrt{40}$

10-8 Equations of Circles

Pages 577–578 Check for Understanding

1. Sample answer:

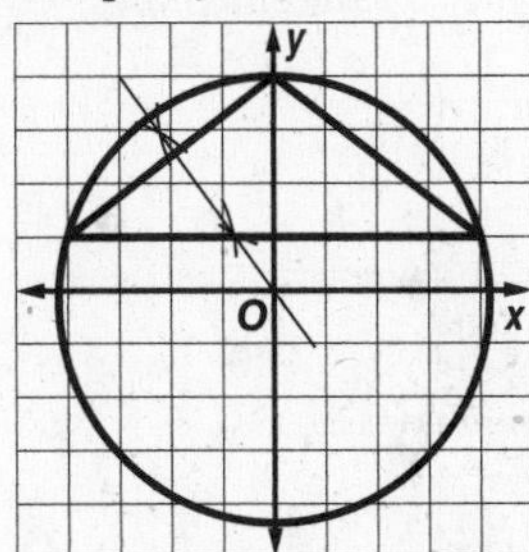

2. A circle is the locus of all points in a plane (coordinate plane) a given distance (the radius) from a given point (the center). The equation of a circle is written from knowing the location of the given point and the radius.

3. $(x - h)^2 + (y - k)^2 = r^2$
$[x - (-3)]^2 + (y - 5)^2 = 10^2$
$(x + 3)^2 + (y - 5)^2 = 100$

4. $(x - h)^2 + (y - k)^2 = r^2$
$(x - 0)^2 + (y - 0)^2 = (\sqrt{7})^2$
$x^2 + y^2 = 7$

5. First find the length of the diameter and radius.
$d = \sqrt{(x_2 - x_1)^2 + (y_2 - y_1)^2}$
$d = \sqrt{(-6 - 2)^2 + (15 - 7)^2}$
$d = \sqrt{64 + 64}$ or $8\sqrt{2}$
$r = \frac{8\sqrt{2}}{2}$ or $4\sqrt{2}$
The center is the midpoint of the diameter:
$C\left(\frac{2 + (-6)}{2}, \frac{7 + 15}{2}\right)$ or $C(-2, 11)$
$(x - h)^2 + (y - k)^2 = r^2$
$[x - (-2)]^2 + (y - 11)^2 = (4\sqrt{2})^2$
$(x + 2)^2 + (y - 11)^2 = 32$

6. $(x + 5)^2 + (y - 2)^2 = 9$
$(x - h)^2 = (x + 5)^2$ $\quad$ $(y - k)^2 = (y - 2)^2$
$x - h = x + 5$ $\quad$ $y - k = y - 2$
$-h = 5$ $\quad$ $-k = -2$
$h = -5$ $\quad$ $k = 2$
$r^2 = 9$, so $r = 3$
The center is at $(-5, 2)$, and the radius is 3.

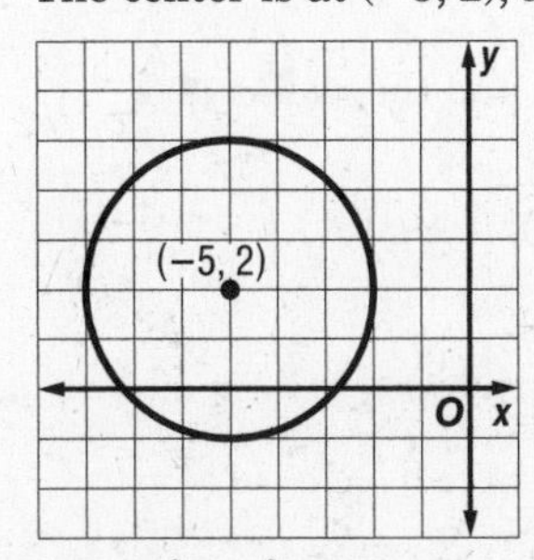

7. $(x - 3)^2 + y^2 = 16$
Write the equation in standard form.
$(x - 3)^2 + (y - 0)^2 = 4^2$
The center is at (3, 0), and the radius is 4.

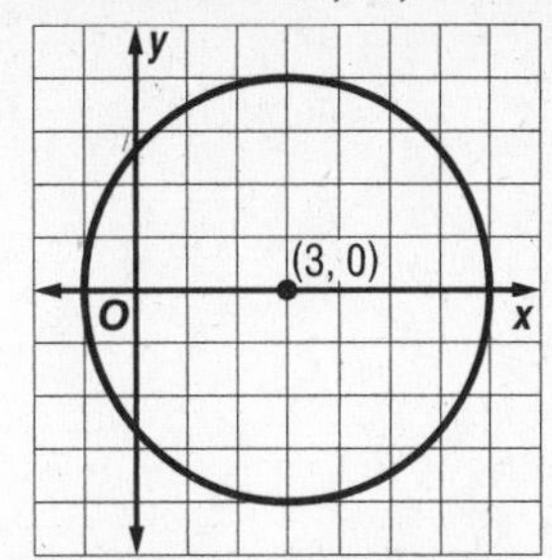

8. Explore: You are given three points that lie on a circle.
Plan: Graph $\triangle NMQ$. Construct perpendicular bisectors of two sides to locate the center. Find the length of the radius. Use the center and radius to write an equation.

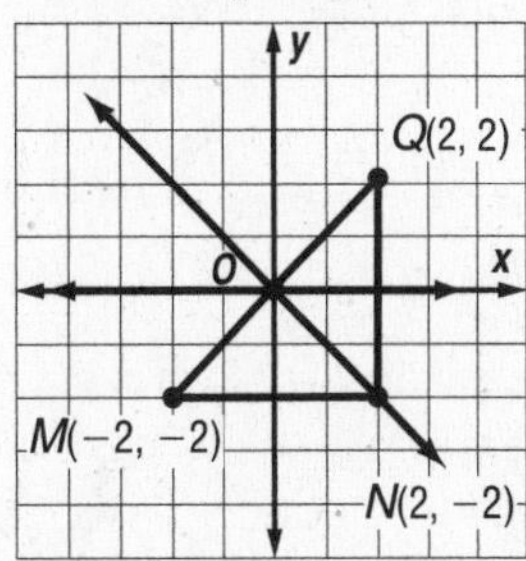

Solve: The center is at (0, 0).
$r = \sqrt{(0 - 2)^2 + [0 - (-2)]^2}$ or $\sqrt{8}$
Write an equation.
$(x - 0)^2 + (y - 0)^2 = (\sqrt{8})^2$
$x^2 + y^2 = 8$

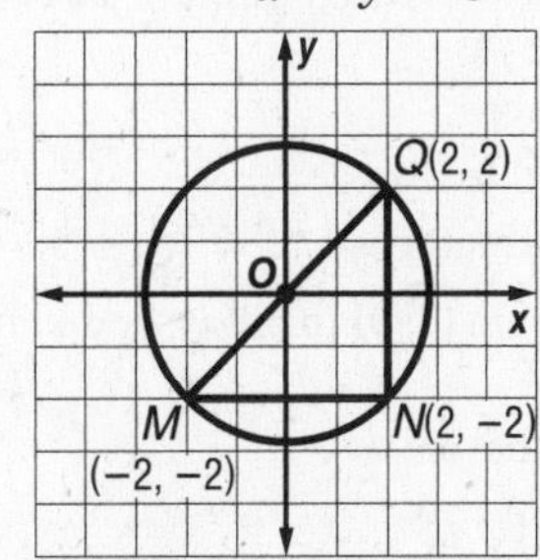

Examine: Verify the location of the center by finding the equations of the two bisectors and solving a system of equations.
⊥ bisector of $\overline{MN}$: $\overline{MN}$ is horizontal so its ⊥ bisector is vertical and goes through the midpoint of $\overline{MN}$: (0, −2).
Its equation is $x = 0$.
⊥ bisector of $\overline{QN}$: $\overline{QN}$ is vertical so its ⊥ bisector is horizontal and goes through the midpoint of $\overline{QN}$: (2, 0).
Its equation is $y = 0$.
The intersection of $x = 0$ and $y = 0$ is the point (0, 0).
So the center is correct.

9. The center is at (0, 0) and the radius is 4(10) or 40.
$(x - h)^2 + (y - k)^2 = r^2$
$(x - 0)^2 + (y - 0)^2 = 40^2$
$x^2 + y^2 = 1600$

Pages 578–580 Practice and Apply

10. $(x - h)^2 + (y - k)^2 = r^2$
$(x - 0)^2 + (y - 0)^2 = 3^2$
$x^2 + y^2 = 9$

11. $(x - h)^2 + (y - k)^2 = r^2$
$[x - (-2)]^2 + [y - (-8)]^2 = 5^2$
$(x + 2)^2 + (y + 8)^2 = 25$

12. $(x - h)^2 + (y - k)^2 = r^2$
$(x - 1)^2 + [y - (-4)]^2 = (\sqrt{17})^2$
$(x - 1)^2 + (y + 4)^2 = 17$

13. If $d = 12$, $r = 6$.
$(x - h)^2 + (y - k)^2 = r^2$
$(x - 0)^2 + (y - 0)^2 = 6^2$
$x^2 + y^2 = 36$

14. $(x - h)^2 + (y - k)^2 = r^2$
$(x - 5)^2 + (y - 10)^2 = 7^2$
$(x - 5)^2 + (y - 10)^2 = 49$

15. If $d = 20$, $r = 10$.
$(x - h)^2 + (y - k)^2 = r^2$
$(x - 0)^2 + (y - 5)^2 = 10^2$
$x^2 + (y - 5)^2 = 100$

16. If $d = 16$, $r = 8$.
$(x - h)^2 + (y - k)^2 = r^2$
$[x - (-8)]^2 + (y - 8)^2 = 8^2$
$(x + 8)^2 + (y - 8)^2 = 64$

17. If $d = 24$, $r = 12$.
$(x - h)^2 + (y - k)^2 = r^2$
$[x - (-3)]^2 + [y - (-10)]^2 = 12^2$
$(x + 3)^2 + (y + 10)^2 = 144$

18. The distance between the center and the endpoint of the radius is $r = \sqrt{[0 - (-3)]^2 + (6 - 6)^2}$ or 3.
$(x - h)^2 + (y - k)^2 = r^2$
$[x - (-3)]^2 + (y - 6)^2 = 3^2$
$(x + 3)^2 + (y - 6)^2 = 9$

19. The midpoint of the diameter is $\left(\frac{2 + (-2)}{2}, \frac{-2 + 2}{2}\right)$ or (0, 0). The distance from (0, 0) to either endpoint, say (2, −2) is
$r = \sqrt{(2 - 0)^2 + (-2 - 0)^2}$ or $\sqrt{8}$.
$(x - h)^2 + (y - k)^2 = r^2$
$(x - 0)^2 + (y - 0)^2 = (\sqrt{8})^2$
$x^2 + y^2 = 8$

20. Find the center, which is the midpoint of the diameter: $\left(\frac{-7 - 15}{2}, \frac{-2 + 6}{2}\right)$ or (−11, 2). Then find r, the distance from (−11, 2) to (−7, −2).
$r = \sqrt{[-7 - (-11)]^2 + (-2 - 2)^2}$ or $\sqrt{32}$.
$(x - h)^2 + (y - k)^2 = r^2$
$[x - (-11)]^2 + (y - 2)^2 = (\sqrt{32})^2$
$(x + 11)^2 + (y - 2)^2 = 32$

21. The distance between the center and the endpoint of the radius is $r = \sqrt{[1 - (-2)]^2 + (0 - 1)^2}$ or $\sqrt{10}$
$(x - h)^2 + (y - k)^2 = r^2$
$[x - (-2)]^2 + (y - 1)^2 = (\sqrt{10})^2$
$(x + 2)^2 + (y - 1)^2 = 10$

22. If $d = 12$, $r = 6$.
The center is at (0 − 18, 0 − 7) or (−18, −7).
$(x - h)^2 + (y - k)^2 = r^2$
$[x - (-18)]^2 + [y - (-7)]^2 = 6^2$
$(x + 18)^2 + (y + 7)^2 = 36$

23. Sketch a drawing of the two tangent lines.

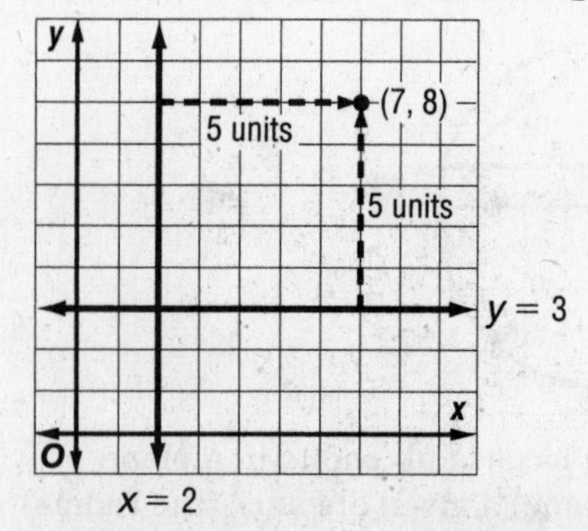

The line $x = 2$ is perpendicular to a radius. Since $x = 2$ is a vertical line, the radius lies on a horizontal line. Count 5 units to the right from $x = 2$. Find the value of h.
$h = 2 + 5$ or 7
Likewise, the radius perpendicular to the line $y = 3$ lies on a vertical line. The value of k is 5 units up from 3.
$k = 3 + 5$ or 8
The center is at (7, 8) and the radius is 5.
$(x - h)^2 + (y - k)^2 = r^2$
$(x - 7)^2 + (y - 8)^2 = 5^2$
$(x - 7)^2 + (y - 8)^2 = 25$

24. $x^2 + y^2 = 25$
Write the equation in standard form.
$(x - 0)^2 + (y - 0)^2 = 5^2$
The center is at (0, 0), and the radius is 5.

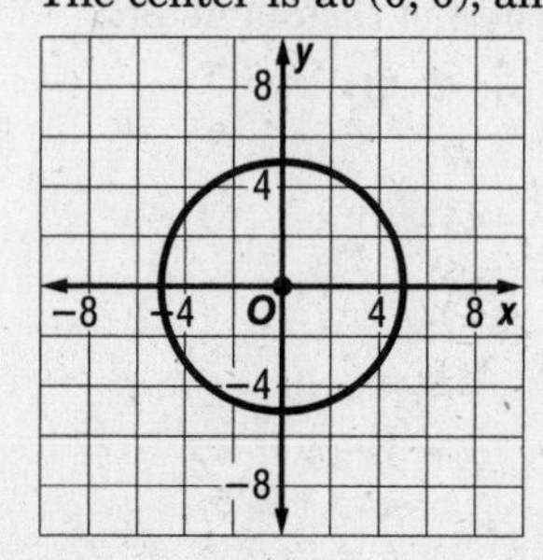

25. $x^2 + y^2 = 36$
Write the equation in standard form.
$(x - 0)^2 + (y - 0)^2 = 6^2$
The center is at (0, 0), and the radius is 6.

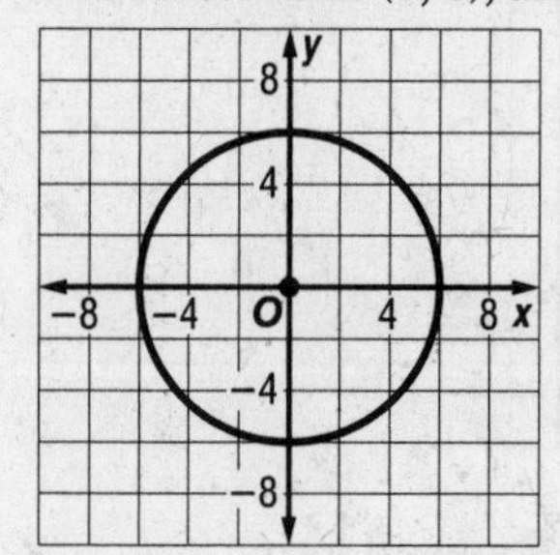

26. $x^2 + y^2 - 1 = 0$

Write the equation in standard form.

$x^2 + y^2 = 1$

$(x - 0)^2 + (y - 0)^2 = 1^2$

The center is at (0, 0), and the radius is 1.

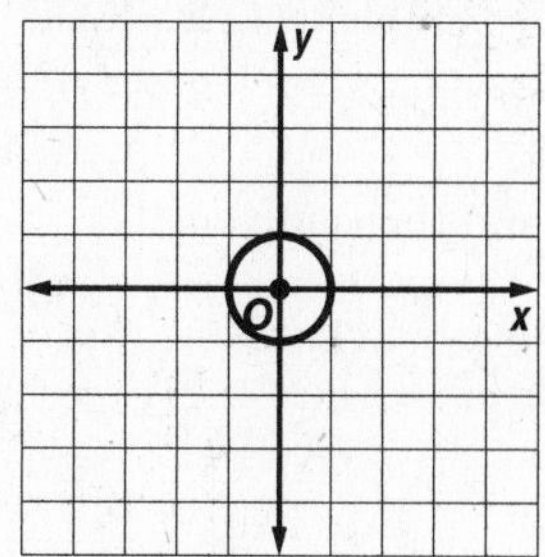

27. $x^2 + y^2 - 49 = 0$

Write the equation in standard form.

$x^2 + y^2 = 49$

$(x - 0)^2 + (y - 0)^2 = 7^2$

The center is at (0, 0), and the radius is 7.

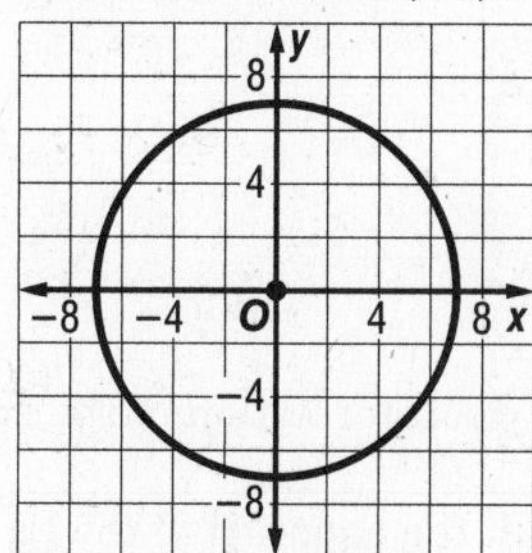

28. $(x - 2)^2 + (y - 1)^2 = 4$

Compare each expression in the equation to the standard form.

$(x - h)^2 = (x - 2)^2$ $\quad$ $(y - k)^2 = (y - 1)^2$

$x - h = x - 2$ $\quad$ $y - k = y - 1$

$h = 2$ $\quad$ $k = 1$

$r^2 = 4$, so $r = 2$.

The center is at (2, 1), and the radius is 2.

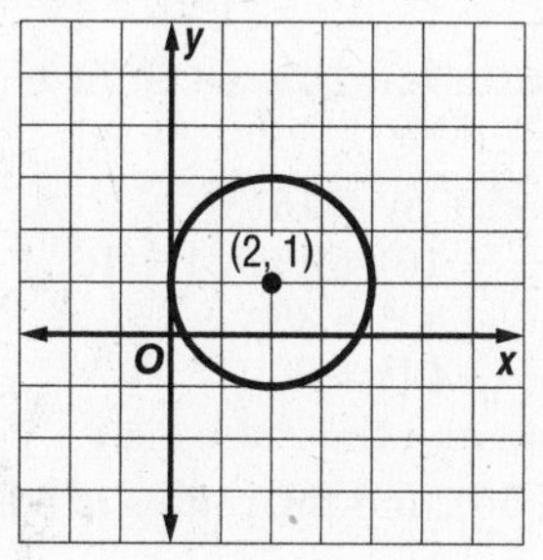

29. $(x + 1)^2 + (y + 2)^2 = 9$

Compare each expression in the equation to the standard form.

$(x - h)^2 = (x + 1)^2$ $\quad$ $(y - k)^2 = (y + 2)^2$

$x - h = x + 1$ $\quad$ $y - k = y + 2$

$-h = 1$ $\quad$ $-k = 2$

$h = -1$ $\quad$ $k = -2$

$r^2 = 9$, so $r = 3$.

The center is at (−1, −2), and the radius is 3.

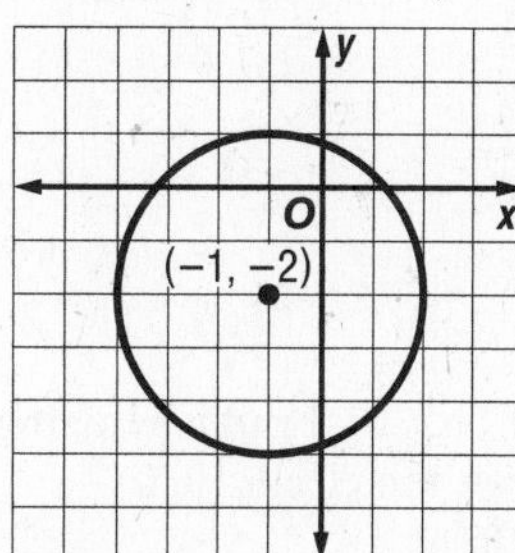

30. Explore: You are given three points that lie on a circle.

Plan: Graph $\triangle ABC$. Construct the perpendicular bisectors of two sides to locate the center. Find the length of the radius. Use the center and radius to write the equation.

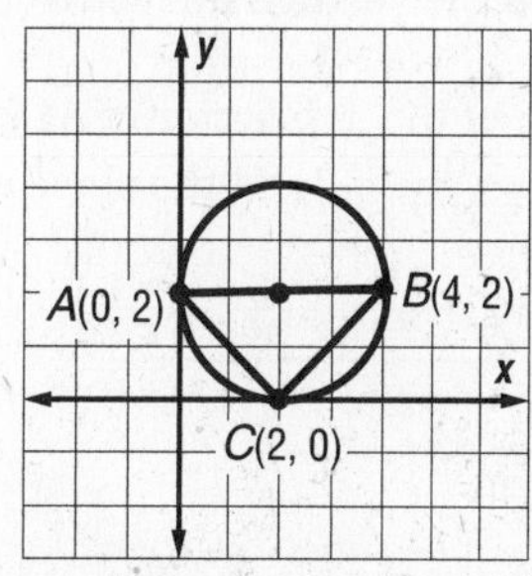

Solve: The center is at (2, 2). Find r by using the center and a point on the circle, (2, 0).

$r = \sqrt{(2 - 2)^2 + (0 - 2)^2}$ or 2

Write an equation.

$(x - 2)^2 + (y - 2)^2 = 2^2$

$(x - 2)^2 + (y - 2)^2 = 4$

Examine: Verify the location of the center by finding the equations of the two bisectors and solving a system of equations.

The perpendicular bisector of $\overline{AB}$ is the vertical line with equation $x = 2$. Next find the equation of the 1 bisector of $\overline{AC}$. The slope of $\overline{AC}$ is $\frac{0 - 2}{2 - 0} = -1$. The slope of the $\perp$ bisector is 1. The midpoint of $\overline{AC} = (1, 1)$. Use the point-slope form.

$y - 1 = 1(x - 1)$

$y = x$

Substitute $x = 2$ to find the intersection.

$y = 2$

The intersection is (2, 2) so the center is correct.

31. Explore: You are given three points that lie on a circle.

Plan: Graph $\triangle ABC$. Construct the perpendicular bisectors of two sides to locate the center. Find the length of the radius. Use the center and radius to write the equation.

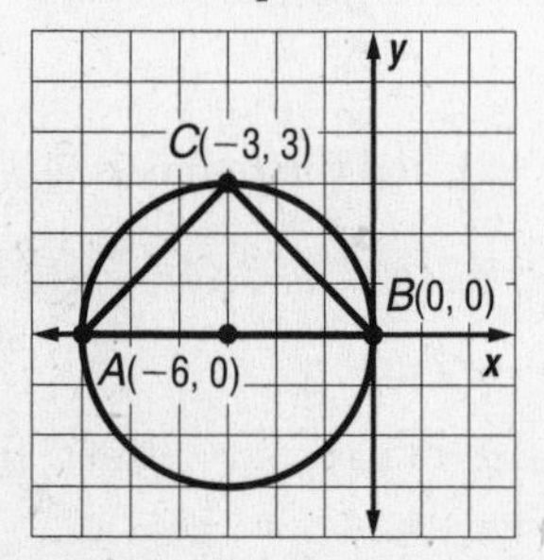

Solve: The center is at $(-3, 0)$. Find r by using the center and a point on the circle, $(0, 0)$.

$r = (0 - 0)^2 + [0 - (-3)]^2$ or 3

write an equation.

$[x - (-3)]^2 + (y - 0)^2 = 3^2$

$(x + 3)^2 + y^2 = 9$

Examine: Verify the location of the center by finding the equations of the two bisectors and solving a system of equations.

The $\perp$ bisector of $\overline{AB}$ is the vertical line with equation $x = -3$.

Next find the equation of the $\perp$ bisector of $\overline{AC}$.

The slope of $\overline{AC} = \frac{3 - 0}{-3 - (-6)} = 1$. So the slope of the $\perp$ bisector is -1.

Find the midpoint of $\overline{AC}$: $\left(\frac{-3 + (-6)}{2}, \frac{3 + 0}{2}\right)$

$= \left(-\frac{9}{2}, \frac{3}{2}\right)$

Use the point-slope form.

$y - \frac{3}{2} = -1\left(x + \frac{9}{2}\right)$

$y - \frac{3}{2} = -x - \frac{9}{2}$

$2y - 3 = -2x - 9$

$2x + 2y = -6$

$x + y = -3$

Substitute $x = -3$ to find the intersection.

$-3 + y = -3$

$y = 0$

The intersection is $(-3, 0)$ so the center is correct.

32. $(x - 2)^2 + (y - 2)^2 = r^2$

The center is at $(2, 2)$.

Find r by using the Distance Formula with the center and the point $(2, 5)$.

$r = \sqrt{(2 - 2)^2 + (5 - 2)^2}$

$= \sqrt{9}$ or 3

33. $(x - 5)^2 + (y - 3)^2 = r^2$

The center is at $(5, 3)$.

Find r by using the Distance Formula with the center and the point $(5, 1)$.

$r = \sqrt{(5 - 5)^2 + (1 - 3)^2}$

$= \sqrt{4}$ or 2

34. The slope of $\overline{AC}$ is $-\frac{1}{4}$, so the slope of its bisector is 4. The midpoint of $\overline{AC}$ is $(0, 5)$. Use the slope and the midpoint to write an equation for the bisector of $\overline{AC}$: $y = 4x + 5$. The slope of $\overline{BC}$ is $-\frac{9}{2}$, so the slope of its bisector is $\frac{2}{9}$. The midpoint of $\overline{BC}$ is $(-2, -3)$. Use the slope and the midpoint to write an equation for the bisector of $\overline{BC}$:

$y = \frac{2}{9}x - \frac{23}{9}$. Solving the system of equations,

$y = 4x + 5$ and $y = \frac{2}{9}x - \frac{23}{9}$, yields $(-2, -3)$,

which is the circumcenter. Let $(-2, -3)$ be D, then $DA = DB = DC = \sqrt{85}$.

35. The center is at $(0, 0)$ and the radius is 7.

$(x - h)^2 + (y - k)^2 = r^2$

$(x - 0)^2 + (y - 0)^2 = 7^2$

$x^2 + y^2 = 49$

36. concentric circles

37. The radius is $\frac{26}{2}$ or 13.

38. $(x - 6)^2 + (y + 2)^2 = 36$

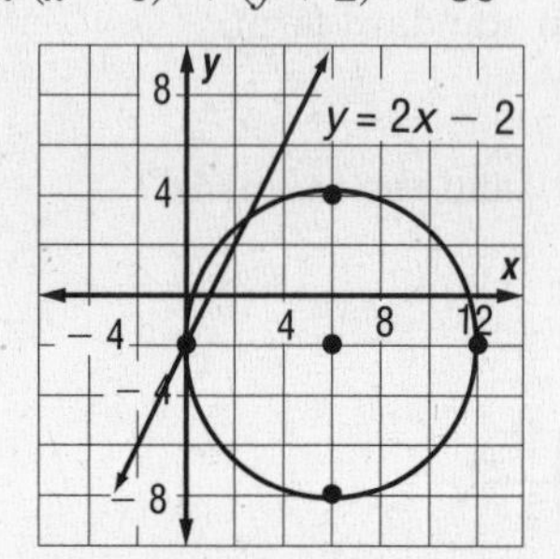

Solve algebraically for the intersection of the two graphs.

Substitute $2x - 2$ for y in the equation of the circle.

$(x - 6)^2 + (2x - 2 + 2)^2 = 36$

$x^2 - 12x + 36 + (2x)^2 = 36$

$5x^2 - 12x = 0$

$x(5x - 12) = 0$

$x = 0$ or $5x - 12 = 0$

$x = 2.4$

$y = 2x - 2$

$y = 2(0) - 2$ or $y = 2(2.4) - 2$

$= -2 \qquad = 2.8$

The line is a secant because it intersects the circle at $(0, -2)$ and $(2.4, 2.8)$.

39. $x^2 - 4x + y^2 + 8y = 16$

$(x^2 - 4x + 4) + (y^2 + 8y + 16) = 16 + 4 + 16$

$(x - 2)^2 + (y + 4)^2 = 36$

$(x - 2)^2 + [y - (-4)]^2 = 6^2$

The center is at $(2, -4)$, and the radius is 6.

40. The center is at $(-58, 55)$, and the radius is 80.

$(x - h)^2 + (y - k)^2 = r^2$

$[x - (-58)]^2 + (y - 55)^2 = 80^2$

$(x + 58)^2 + (y - 55)^2 = 6400$

41. See students' work.

42. The center is at $(0, 0)$, and the radius is $185 + 1740$ or 1925.

$(x - h)^2 + (y - k)^2 = r^2$

$(x - 0)^2 + (y - 0)^2 = (1925)^2$

$x^2 + y^2 = 3{,}705{,}625$

43a.

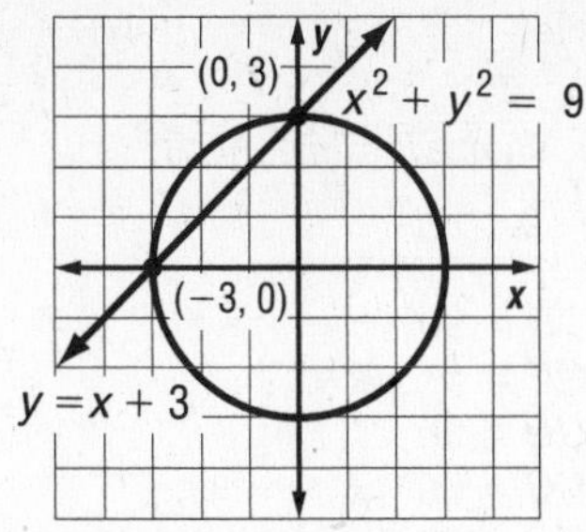

Substitute $x + 3$ for y in the equation of the circle.

$$x^2 + (x + 3)^2 = 9$$
$$x^2 + x^2 + 6x + 9 = 9$$
$$2x^2 + 6x = 0$$
$$2x(x + 3) = 0$$

$2x = 0$ or $x + 3 = 0$

$x = 0$ $\quad x = -3$

$y = x + 3$

$y = 0 + 3$ or $y = -3 + 3$

$= 3$ $\quad = 0$

The intersection points are $(0, 3)$ and $(-3, 0)$.

43b.

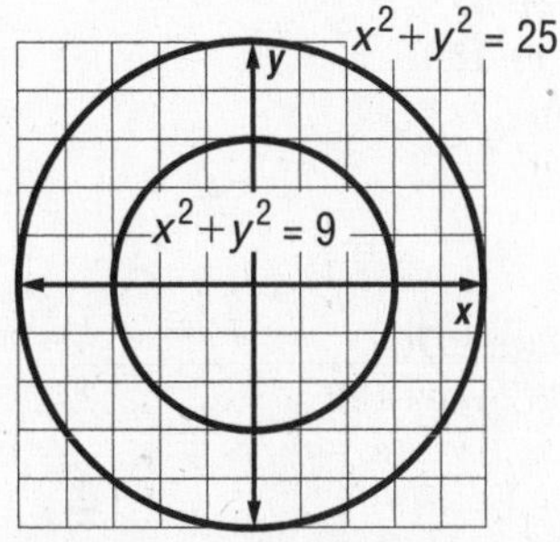

Since $x^2 + y^2 = 25$ and $x^2 + y^2 = 9$, $25 = 9$. This is never true, so there are no intersection points.

43c.

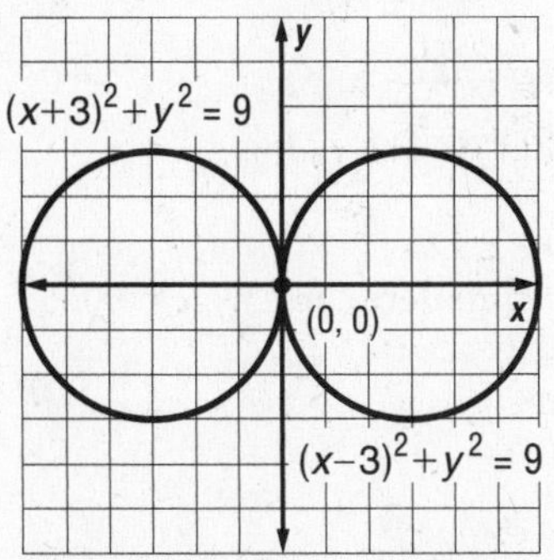

Since $(x + 3)^2 + y^2 = 9$ and $(x - 3)^2 + y^2 = 9$,

$$(x + 3)^2 + y^2 = (x - 3)^2 + y^2$$
$$(x + 3)^2 = (x - 3)^2$$
$$x^2 + 6x + 9 = x^2 - 6x + 9$$
$$6x = -6x$$
$$12x = 0$$
$$x = 0$$

Substitute $x = 0$ into the equation of either circle.

$$(0 + 3)^2 + y^2 = 9$$
$$3^2 + y^2 = 9$$
$$9 + y^2 = 9$$
$$y^2 = 0$$
$$y = 0$$

The intersection point is $(0, 0)$.

44. Sample answer: Equations of concentric circles; answers should include the following.
- $(x - h)^2 + (y - k)^2 = r^2$
- $x^2 + y^2 = 9$, $x^2 + y^2 = 36$, $x^2 + y^2 = 81$, $x^2 + y^2 = 144$, $x^2 + y^2 = 225$

45. B;
$$x^2 + y^2 + 4x - 14y + 53 = 81$$
$$(x^2 + 4x) + (y^2 - 14y) = 28$$
$$(x^2 + 4x + 4) + (y^2 - 14y + 49) = 28 + 4 + 49$$
$$(x + 2)^2 + (y - 7)^2 = 81$$
$$[x - (-2)]^2 + (y - 7)^2 = 9^2$$

The center is at $(-2, 7)$ and the diameter is $2(9)$ or 18.

46. D; there are 2 pints in one quart and 4 quarts in one gallon, so there are 8 pints in one gallon. One of the 8 pints is gone, so 7 of the 8 pints are left, or $\frac{7}{8}$.

Page 580 Maintain Your Skills

47. $AX = EX$ or 24

48. $(DX)^2 = (DE)^2 + (EX)^2$
$(DX)^2 = 7^2 + 24^2$
$(DX)^2 = 625$
$DX = 25$

49. $DX = QX + QD$
$25 = QX + 7$
$18 = QX$

50. $TX = DX + DT$
$TX = 25 + 7$ or 32

51. $360 - (117 + 125) = 118$
$x = \frac{1}{2}(118)$ or 59

52. $360 - (100 + 120 + 90) = 50$
$x = \frac{1}{2}(120 - 50)$
$= \frac{1}{2}(70)$ or 35

53. $180 - (45 + 130) = 5$
$x = \frac{1}{2}(45 - 5)$
$= \frac{1}{2}(40)$ or 20

54. $A(-3, 2) \rightarrow A'(-3 - 3, 2 + 4)$ or $A'(-6, 6)$
$B(4, -1) \rightarrow B'(4 - 3, -1 + 4)$ or $B'(1, 3)$
$C(0, -4) \rightarrow C'(0 - 3, -4 + 4)$ or $C'(-3, 0)$

55. $A(-3, 2) \rightarrow A'(3, 2)$
$B(4, -1) \rightarrow B'(-4, -1)$
$C(0, -4) \rightarrow C'(0, -4)$

56. Each child needs $2(12) + 2(10) = 44$ inches plus 1 inch overlap, or 45 inches total. The teacher needs $25 \times 45 = 1125$ inches, or $\frac{1125}{36} = 31.25$ yards.

Chapter 10 Study Guide and Review

Page 581 Vocabulary and Concept Check

1. a **2.** j
3. h **4.** i
5. b **6.** f
7. d **8.** g
9. c **10.** e

Pages 581–586 Lesson-by-Lesson Review

11. $r = \frac{1}{2}d$

$r = \frac{1}{2}(15)$ or 7.5 in.

$C = \pi d$

$C = \pi(15)$

$C \approx 47.12$ in.

12. $d = 2r$

$d = 2(6.4)$ or 12.8 m

$C = \pi d$

$C = \pi(12.8)$

$C \approx 40.21$ m

13. $C = 2\pi r$

$68 = 2\pi r$

$\frac{68}{2\pi} = r$

10.82 yd $\approx r$

$d = 2r$

$d = 2\left(\frac{68}{2\pi}\right)$

$d \approx 21.65$ yd

14. $r = \frac{1}{2}d$

$r = \frac{1}{2}(52)$ or 26 cm

$C = 2\pi r$

$C = 2\pi(26)$

$C \approx 163.36$ cm

15. $C = 2\pi r$

$138 = 2\pi r$

$\frac{138}{2\pi} = r$

21.96 ft $\approx r$

$d = 2r$

$d = 2\left(\frac{138}{2\pi}\right)$

$d \approx 43.93$ ft

16. $d = 2r$

$d = 2(11)$ or 22 mm

$C = \pi d$

$C = \pi(22)$

$C \approx 69.12$ mm

17. $m\angle BPY + m\angle YPC + m\angle CPA = 180$

$3x + (3x - 3) + (2x + 15) = 180$

$8x + 12 = 180$

$8x = 168$

$x = 21$

$\widehat{YC}$ is a minor arc, so $m\widehat{YC} = m\angle YPC$.

$m\widehat{YC} = m\angle YPC$

$m\widehat{YC} = 3x - 3$

$m\widehat{YC} = 3(21) - 3$ or 60

18. $\widehat{BC}$ is a minor arc, so $m\widehat{BC} = m\angle BPC$.

$\widehat{BC}$ is composed of adjacent arcs, $\widehat{BY}$ and $\widehat{YC}$.

$m\widehat{BC} = m\widehat{BY} + m\widehat{YC}$

$m\widehat{BC} = m\angle BPY + m\angle YPC$

$m\widehat{BC} = 3x + (3x - 3)$

$m\widehat{BC} = 6x - 3$

$m\widehat{BC} = 6(21) - 3$ or 123

19. $m\widehat{BX} = 360 - m\widehat{BCX}$

$m\widehat{BX} = 360 - (3x + 3x - 3 + 2x + 15 + 3x)$

$m\widehat{BX} = 360 - (11x + 12)$

$m\widehat{BX} = 360 - [11(21) + 12]$

$m\widehat{BX} = 360 - 243$ or 117

20. $\widehat{BCA}$ is a semicircle.

$m\widehat{BCA} = 180$

21. $\widehat{AB}$ is a minor arc, so $m\widehat{AB} = m\angle AGB$.

$m\widehat{AB} = m\angle AGB$

$m\widehat{AB} = 30$

22. $m\angle AGC = m\angle CGD = 90$

$m\angle AGB + m\angle BGC = m\angle AGC$

$30 + m\angle BGC = 90$

$m\angle BGC = 60$

$m\widehat{BC} = 60$

23. $\widehat{FD}$ is a minor arc, so $m\widehat{FD} = m\angle FGD$.

Vertical angles are congruent.

$\angle FGD \cong \angle AGB$

$m\angle FGD = m\angle AGB$

$m\widehat{FD} = m\angle AGB$

$m\widehat{FD} = 30$

24. $\widehat{CDF}$ is composed of adjacent arcs, $\widehat{CD}$ and $\widehat{DF}$.

$m\widehat{CDF} = m\widehat{CD} + m\widehat{DF}$

$m\widehat{CDF} = 90 + 30$ or 120

25. $\widehat{BCD}$ is composed of adjacent arcs, $\widehat{BC}$ and $\widehat{CD}$.

$m\widehat{BCD} = m\widehat{BC} + m\widehat{CD}$

$m\widehat{BCD} = 60 + 90$ or 150

26. $\widehat{FAB}$ is a semicircle.

$m\widehat{FAB} = 180$

27. $C = 2\pi r$

$C = 2\pi(6)$ or 12π

$m\angle DIG = 180 - 2(m\angle DGI)$

$m\angle DIG = 180 - 2(24)$ or 132

Let ℓ = arc length.

$\frac{132}{360} = \frac{\ell}{12\pi}$

$\frac{132}{360}(12\pi) = \ell$

$\frac{22}{5}\pi = \ell$

The length of $\widehat{DG}$ is $\frac{22}{5}\pi$ units.

28. $WN = IW = 5$ and IW is a radius.

$C = 2\pi r$

$C = 2\pi(5)$ or 10π

Since $\triangle IWN$ is equilateral, $m\angle WIN = 60$.

Let ℓ = arc length.

$\frac{60}{360} = \frac{\ell}{10\pi}$

$\frac{60}{360}(10\pi) = \ell$

$\frac{5}{3}\pi = \ell$

The length of $\widehat{WN}$ is $\frac{5}{3}\pi$ units.

29. $SV = \frac{1}{2}SU$

$SV = \frac{1}{2}(20)$ or 10

30. $WZ = \frac{1}{2}YW$

$WZ = \frac{1}{2}(20)$ or 10

31. $\overline{RT}$ bisects $\overline{SU}$, so $UV = SV$.

$UV = SV$

$UV = 10$

32. $\overline{RX}$ bisects $\widehat{YW}$, so $m\widehat{YW} = 2m\widehat{YX}$.

$m\widehat{YW} = 2m\widehat{YX}$

$m\widehat{YW} = 2(45)$ or 90

33. $SU = YW$ and $\overline{RX}$ bisects $\overline{YW}$, $\overline{RT}$ bisects $\overline{SU}$.

$m\widehat{ST} = m\widehat{YX}$

$m\widehat{ST} = 45$

34. Since $SU = YW$, $m\widehat{SU} = m\widehat{YW}$.
$m\widehat{SU} = m\widehat{YW}$
$m\widehat{SU} = 90$

35. $m\angle 1 = \frac{1}{2}(96)$ or 48

36. $\angle 2$ is a right angle because it intercepts a semicircle.
$m\angle 2 = 90$

37. Inscribed angles of the same arc are congruent.
$m\angle 3 = 32$

38. $m\angle 3 = \frac{1}{2}\, m\widehat{GH}$
$= \frac{1}{2}(78)$ or 39
$m\angle 2 + m\angle 3 = 90$
$m\angle 2 + 39 = 90$
$m\angle 2 = 51$
$m\angle 1 + m\angle 2 = 90$
$m\angle 1 + 51 = 90$
$m\angle 1 = 39$

39. $m\angle 2 + m\angle 3 = 90$
$2x + x = 90$
$3x = 90$
$x = 30$
$m\angle 2 = 2x$
$= 2(30)$ or 60
$m\angle 3 = x$ or 30
$m\angle 1 + m\angle 2 = 90$
$m\angle 1 + 60 = 90$
$m\angle 1 = 30$

40. $m\angle 2 = \frac{1}{2} m\widehat{JH}$
$= \frac{1}{2}(114)$ or 57
$m\angle 2 + m\angle 3 = 90$
$57 + m\angle 3 = 90$
$m\angle 3 = 33$
$m\angle 1 + m\angle 2 = 90$
$m\angle 1 + 57 = 90$
$m\angle 1 = 33$

41. $x^2 + 12^2 = 15^2$
$x^2 + 144 = 225$
$x^2 = 81$
$x = 9$

42. $x^2 + 9^2 = (6 + 9)^2$
$x^2 + 9^2 = 15^2$
$x^2 + 81 = 225$
$x^2 = 144$
$x = 12$

43. $7^2 + 24^2 = (x + 7)^2$
$49 + 576 = x^2 + 14x + 49$
$0 = x^2 + 14x - 576$
$(x - 18)(x + 32) = 0$
$x - 18 = 0$ or $x + 32 = 0$
$x = 18$ $\quad x = -32$
Disregard the negative value.
So $x = 18$.

44. $x = \frac{1}{2}(68 - 24)$
$x = \frac{1}{2}(44)$ or 22

45. $26 = \frac{1}{2}(89 - x)$
$52 = 89 - x$
$x = 37$

46. $33 = \frac{1}{2}(x - 51)$
$66 = x - 51$
$117 = x$

47. $7(7 + x) = 13^2$
$49 + 7x = 169$
$7x = 120$
$x \approx 17.1$

48. $8.1x = 10.3(17)$
$8.1x = 175.1$
$x \approx 21.6$

49. $x(15 + x) = 8(8 + 12)$
$15x + x^2 = 160$
$x^2 + 15x - 160 = 0$
$x = \frac{-b \pm \sqrt{b^2 - 4ac}}{2a}$
$x = \frac{-15 \pm \sqrt{(15)^2 - 4(1)(-160)}}{2(1)}$
$x \approx \frac{-15 \pm 29.4}{2}$
$x \approx 7.2$ or $x \approx -44.4$
Disregard the negative value. So $x \approx 7.2$.

50. $(x - h)^2 + (y - k)^2 = r^2$
$(x - 0)^2 + (y - 0)^2 = (\sqrt{5})^2$
$x^2 + y^2 = 5$

51. If $d = 6$, $r = 3$.
$(x - h)^2 + (y - k)^2 = r^2$
$[x - (-4)]^2 + (y - 8)^2 = 3^2$
$(x + 4)^2 + (y - 8)^2 = 9$

52. The center is at the midpoint of the diameter.
center $= \left(\frac{0 + 8}{2}, \frac{-4 - 4}{2}\right)$ or $(4, -4)$
The radius is the distance from the center to $(0, -4)$.
$r = \sqrt{(0 - 4)^2 + [-4 - (-4)]^2}$
$= \sqrt{16}$ or 4
Write the equation.
$(x - h)^2 + (y - k)^2 = r^2$
$(x - 4)^2 + [y - (-4)]^2 = 4^2$
$(x - 4)^2 + (y + 4)^2 = 16$

53. Since $x = 1$ is a vertical line, the radius lies on a horizontal line. Count horizontally from the point $(-1, 4)$ to the line $x = 1$ to find the radius. The radius is 2.

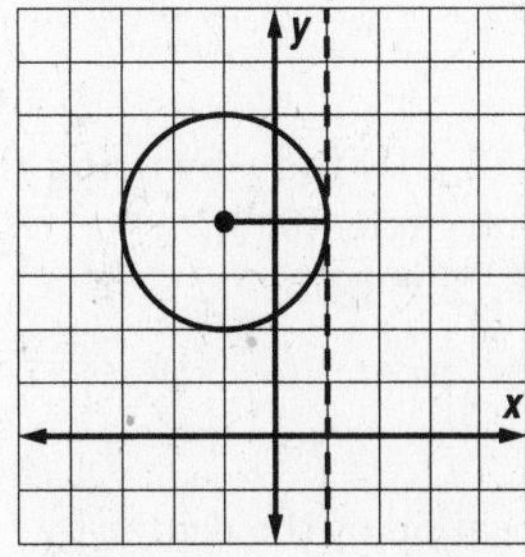

$(x - h)^2 + (y - k)^2 = r^2$
$[x - (-1)]^2 + (y - 4)^2 = 2^2$
$(x + 1)^2 + (y - 4)^2 = 4$

54. $x^2 + y^2 = 2.25$
Write the equation in standard form.
$(x - 0)^2 + (y - 0)^2 = (1.5)^2$
The center is at (0, 0), and the radius is 1.5.

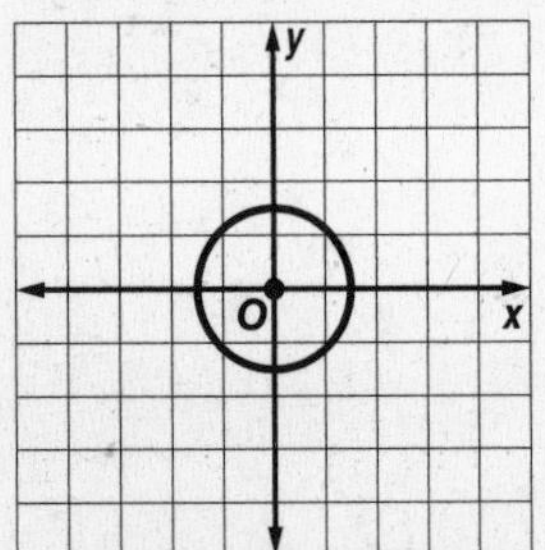

55. $(x - 4)^2 + (y + 1)^2 = 9$
Compare each expression in the equation to the standard form.

$(x - h)^2 = (x - 4)^2$ $\quad$ $(y - k)^2 = (y + 1)^2$
$x - h = x - 4$ $\quad$ $y - k = y + 1$
$-h = -4$ $\quad$ $-k = 1$
$h = 4$ $\quad$ $k = -1$

$r^2 = 9$, so $r = 3$.
The center is at (4, −1), and the radius is 3.

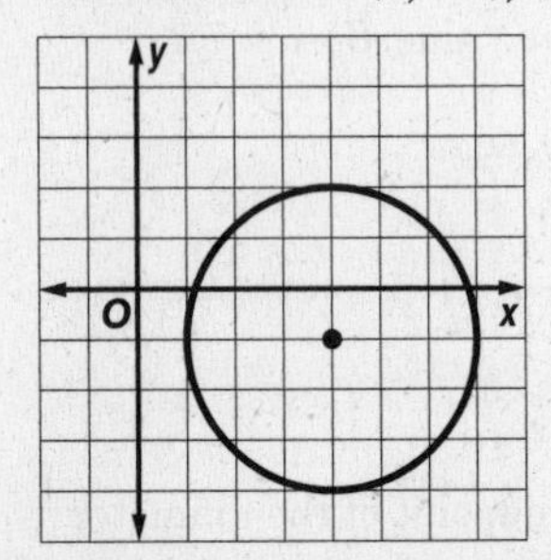

56. Explore: You are given three points that lie on a circle.
Plan: Graph $\triangle ABC$. Construct the perpendicular bisectors of two sides to locate the center. Find the length of the radius. Use the center and radius to write the equation.

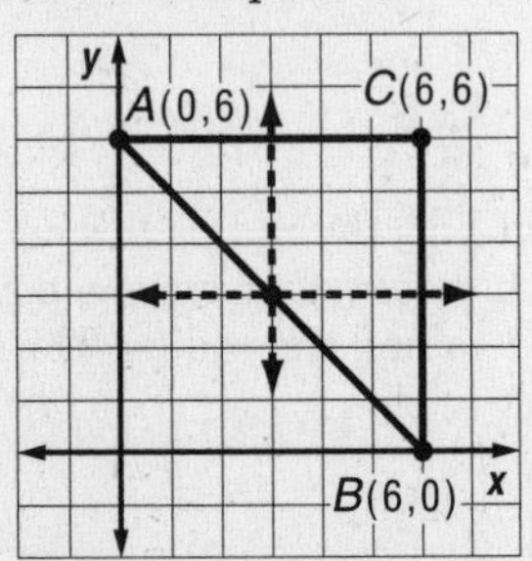

Solve: The center is at (3, 3). Find r by using the center and a point on the circle, (0, 6).
$r = \sqrt{(0 - 3)^2 + (6 - 3)^2}$ or $\sqrt{18}$
Write an equation.
$(x - 3)^2 + (y - 3)^2 = \left(\sqrt{18}\right)^2$
$(x - 3)^2 + (y - 3)^2 = 18$
Examine: Verify the location of the center by finding the equations of the two bisectors and solving a system of equations.
⊥ bisector of $\overline{AC}$: Since $\overline{AC}$ is horizontal, its ⊥ bisector is vertical. It goes through (3, 6), the midpoint of $\overline{AC}$. The equation is $x = 3$.
⊥ bisector of $\overline{BC}$: Since $\overline{BC}$ is vertical, its ⊥ bisector is horizontal. It goes through (6, 3), the midpoint of $\overline{BC}$. The equation is $y = 3$.
The intersection of $x = 3$ and $y = 3$ is the point (3, 3). This is the center we found above so the answer checks.

57.

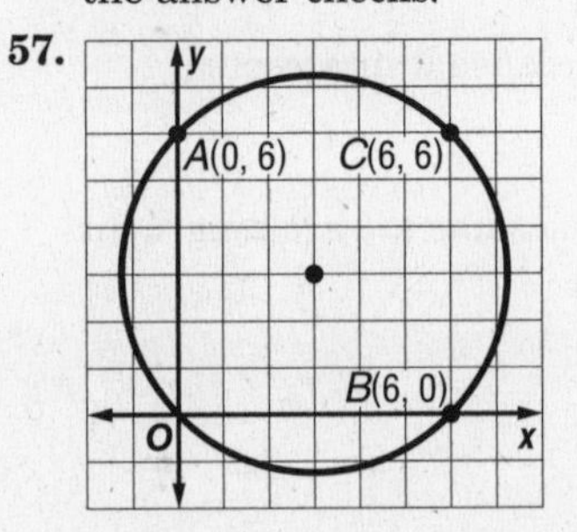

Chapter 10 Practice Test

Page 587

1. Sample answer: A chord is a segment that has its endpoints on a circle. A secant contains a chord and is a line that intersects a circle in two points. A tangent is a line that intersects a circle in exactly one point and no point of the tangent lies in the interior of the circle.
2. Find the midpoint of the diameter using the Midpoint Formula with the coordinates of the diameter's endpoints.
3. $C = 2\pi r$
   $25\pi = 2\pi r$
   $25 = 2r$
   $12.5 = r$
   The radius is 12.5 units.
4. $\overline{NA}$, $\overline{NB}$, $\overline{NC}$, and $\overline{ND}$ are radii.
5. The diameter is $\overline{AD}$, so the radius is $\frac{24}{2}$ or 12.
   $CN = 12$
6. No; diameters are the longest chords of a circle.
7. $r = AN$ or 5
   $C = 2\pi r$
   $C = 2\pi(5)$ or 10π
   The circumference is 10π meters.
8. $\widehat{BC}$ is a minor arc so $m\widehat{BC} = m\angle BNC$, $m\widehat{BC} = 20$
9. $\widehat{AD}$ is composed of adjacent arcs, $\widehat{AB}$, $\widehat{BC}$, and $\widehat{CD}$.
   $\widehat{AD}$ is a semicircle.
   $m\widehat{AD} = m\widehat{AB} + m\widehat{BC} + m\widehat{CD}$
   $180 = m\widehat{AB} + 30 + m\widehat{AB}$
   $150 = 2m\widehat{AB}$
   $75 = m\widehat{AB}$
10. $\overline{BE} \cong \overline{ED}$
    $m\widehat{BE} = m\widehat{ED}$
    $m\widehat{BE} = 120$
11. $\angle ADE$ is an inscribed angle intercepting $\widehat{AE}$.
    $m\angle ADE = \frac{1}{2} m\widehat{AE}$
    $= \frac{1}{2}(75)$ or 37.5
12. If two segments from the same exterior point are tangent to a circle, then they are congruent.
    $x = 15$

13. $6^2 = x(x + 5)$
$36 = x^2 + 5x$
$0 = x^2 + 5x - 36$
$0 = (x - 4)(x + 9)$
$x - 4 = 0$ or $x + 9 = 0$
$x = 4$ $\quad x = -9$
Discard $x = -9$. So $x = 4$.

14. $5x = 6(8)$
$5x = 48$
$x = \frac{48}{5}$ or 9.6

15. Draw the radius to the point of tangency. Use the Pythagorean Theorem.
$6^2 + 8^2 = (x + 6)^2$
$36 + 64 = x^2 + 12x + 36$
$0 = x^2 + 12x - 64$
$0 = (x - 4)(x + 16)$
$x - 4 = 0$ or $x + 16 = 0$
$x = 4$ $\quad x = -16$
Discard $x = -16$. So $x = 4$.

16. $5(5 + x) = 4(4 + 7)$
$25 + 5x = 16 + 28$
$5x = 19$
$x = \frac{19}{5}$ or 3.8

17. $35 = \frac{1}{2}[(360 - x) - x]$
$70 = 360 - 2x$
$2x = 290$
$x = 145$

18. $180 - (45 + 110) = 25$
$x = \frac{1}{2}(45 - 25)$
$= \frac{1}{2}(20)$ or 10

19. $80 = \frac{1}{2}(110 + x)$
$160 = 110 + x$
$50 = x$

20. $C = \pi d$
$C = \pi(50)$
$C \approx 157$
The circumference is about 157 ft.

21. If $d = 50$, $r = 25$.
$(x - h)^2 + (y - k)^2 = r^2$
$[x - (-2)]^2 + (y - 5)^2 = 25^2$
$(x + 2)^2 + (y - 5)^2 = 625$

22. $(x - 1)^2 + (y + 2)^2 = 4$
Compare each expression in the equation to the standard form.
$(x - h)^2 = (x - 1)^2$ $\quad (y - k)^2 = (y + 2)^2$
$x - h = x - 1$ $\quad y - k = y + 2$
$h = 1$ $\quad k = -2$
$r^2 = 4$, so $r = 2$.
The center is at $(1, -2)$, and the radius is 2.

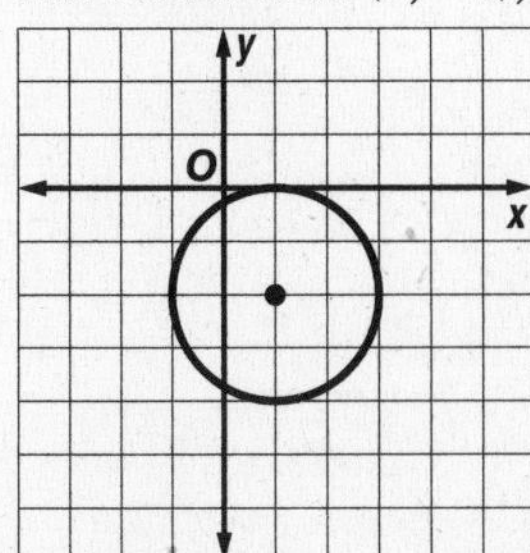

23. Sample answer:
Given: $\odot X$ with diameters $\overline{RS}$ and $\overline{TV}$
Prove: $\overset{\frown}{RT} \cong \overset{\frown}{VS}$

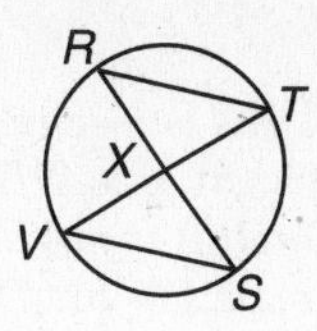

Proof:

Statements	Reasons
1. $\odot X$ with diameters $\overline{RS}$ and $\overline{TV}$	1. Given
2. $\angle RXT \cong \angle VXS$	2. Vertical ∠s are ≅.
3. $m\angle RXT = m\angle VXS$	3. Def. of ≅ ∠s
4. $m\overset{\frown}{RT} = m\angle RXT$, $m\overset{\frown}{VS} = m\angle VXS$	4. Measure of arc equals measure of its central angle.
5. $m\overset{\frown}{RT} = m\overset{\frown}{VS}$	5. Substitution
6. $\overset{\frown}{RT} \cong \overset{\frown}{VS}$	6. Def. of ≅ arcs

24. Let x = the distance from the center to the 5-inch side.
$x^2 + (2.5)^2 = 4^2$
$x^2 + 6.25 = 16$
$x^2 = 9.75$
$x \approx 3.1$
The height is about 4 + 3.1 or 7.1 in.

25. A; draw the segment from C to A.
$DB = CA$. Since diagonals of a rectangle are congruent.
$DB = r$

Chapter 10 Standardized Test Practice

Pages 588–589

1. A; $3y = 6x - 9$ is the same as $y = 2x - 3$. So the line has slope $m = 2$ and y-intercept $(0, -3)$. The graph that satisfies these conditions is A.

2. C; $m\angle 1 = m\angle 2$
$6x - 5 = 3x + 13$
$6x - 3x = 13 + 5$
$3x = 18$
$x = 6$
Use x to find the measure of $\angle 1$.
$m\angle 1 = 6x - 5$
$= 6(6) - 5$
$= 36 - 5$ or 31

3. A

4. B; $3(x + 2) = 2x + 9$
$3x + 6 = 2x + 9$
$x = 3$
Use x to find the measure of one side.
$x + 2 = 3 + 2$ or 5
Each side is 5 miles long.

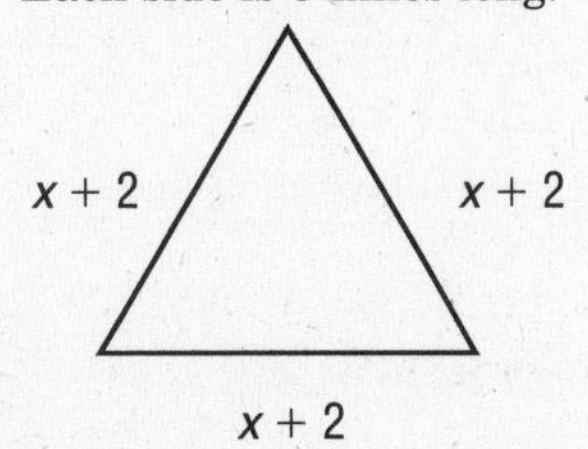

5. A

6. D

7. C; $\widehat{DEA}$ is a semicircle. $\widehat{DEA}$ is composed of adjacent arcs, $\widehat{DE}$ and $\widehat{EA}$.

$$m\widehat{DEA} = m\widehat{DE} + m\widehat{EA}$$
$$m\widehat{DEA} = m\angle DFE + m\widehat{EA}$$
$$180 = 36 + m\widehat{EA}$$
$$144 = m\widehat{EA}$$

8. *C*; the measure of a *minor* arc is the measure of its central angle.

9. D

10. $m\angle GDE + m\angle GED + m\angle DGE = 180$

$$35 + 85 + m\angle DGE = 180$$
$$120 + m\angle DGE = 180$$
$$m\angle DGE = 60$$

In $\triangle DGE$, $\overline{GE}$ is the shortest side because it is opposite the smallest angle.

$$m\angle FGE + m\angle GEF + m\angle F = 180$$
$$65 + 55 + m\angle F = 180$$
$$120 + m\angle F = 180$$
$$m\angle F = 60$$

In $\triangle GFE$, $FG < GE$ because $m\angle GEF < m\angle F$.
So FG is the shortest side of quadrilateral $DEFG$.

11. $\frac{ST}{8} = \frac{30}{12}$

$$12(ST) = 8(30)$$
$$12(ST) = 240$$
$$ST = 20$$

ST is 20 ft long.

12. Find a correspondence between $\triangle ABC$ and $\triangle DEF$ so that their sides are proportional.

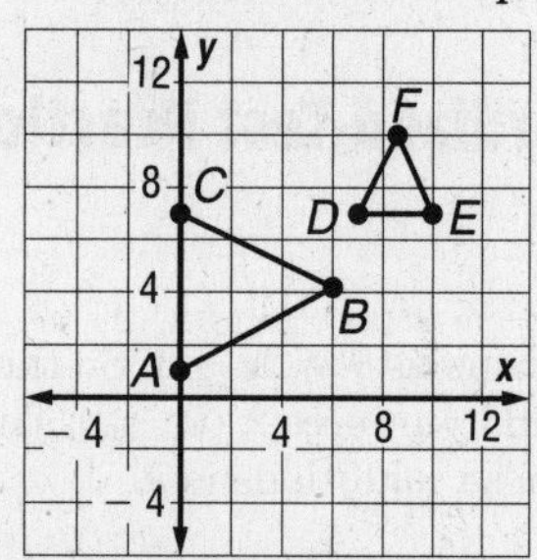

$$AC = |7 - 1| = 6$$
$$DE = |10 - 7| = 3$$
$$AB = \sqrt{(6 - 0)^2 + (4 - 1)^2}$$
$$= \sqrt{36 + 9} = \sqrt{45}$$
$$EF = \sqrt{(10 - 8.5)^2 + (7 - 10)^2}$$
$$= \sqrt{(1.5)^2 + 9} = \sqrt{11.25} = \sqrt{11\tfrac{1}{4}} = \frac{\sqrt{45}}{2}$$
$$BC = \sqrt{(6 - 0)^2 + (4 - 7)^2}$$
$$= \sqrt{45}$$
$$DF = \sqrt{(8.5 - 7)^2 + (10 - 7)^2}$$
$$= \sqrt{(1.5)^2 + 9} = \frac{\sqrt{45}}{2}$$
$$\frac{AC}{DE} = \frac{AB}{EF} = \frac{BC}{DF} = 2.$$

So the scale factor is 2.

13. $\widehat{ABC}$ is a semicircle.

$$m\angle ABC = \frac{180}{2} \text{ or } 90$$

14. $DF = AD$ or 12

$$DE = DF + FE$$
$$DE = 12 + 18 \text{ or } 30$$

Since $\overline{AE}$ is a tangent, $\angle CAE$ is a right angle and $\triangle AED$ is a right triangle. Use the Pythagorean Theorem.

$$(AE)^2 + (AD)^2 = (DE)^2$$
$$(AE)^2 + (12)^2 = 30^2$$
$$(AE)^2 + 144 = 900$$
$$(AE)^2 = 756$$
$$AE \approx 27.5$$

15. $(BK)(KC) = (KF)(JK)$

$$8(12) = 16(JK)$$
$$96 = 16(JK)$$
$$6 = JK$$

16a. $\overline{VX}$ and $\overline{WY}$

16b. Let w = width, then

$$3w + 2 = \text{length}$$
$$2(3w + 2) + 2w = 164$$
$$6w + 4 + 2w = 164$$
$$8w = 160$$
$$w = 20$$

The width is 20 in. and the length is 3(20) + 2 or 62 in.

17a.

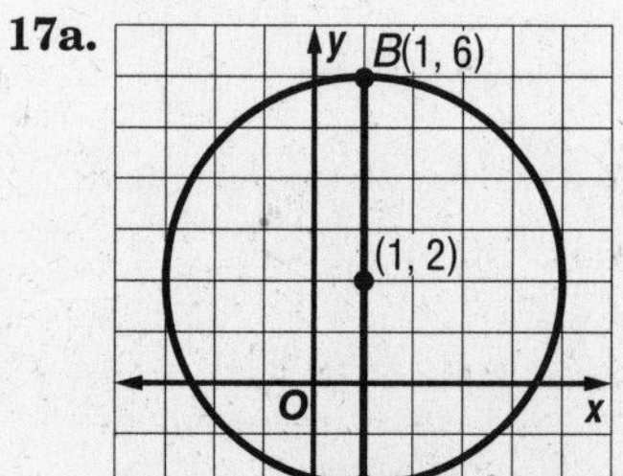

17b. The center is the midpoint of the diameter:

$$\left(\frac{1 + 1}{2}, \frac{-2 + 6}{2}\right) \text{ or } (1, 2).$$

17c. Find r by using the Distance Formula. Find the distance between the center and a point on the circle, $(1, -2)$.

$$r = \sqrt{(1 - 1)^2 + (-2 - 2)^2}$$
$$= \sqrt{16} \text{ or } 4$$

17d. $C = 2\pi r$

$$C = 2\pi(4) \text{ or } 8\pi \text{ units}$$

17e. The center is at (1, 2), and the radius is 4.

$$(x - h)^2 + (y - k)^2 = r^2$$
$$(x - 1)^2 + (y - 2)^2 = 4^2$$
$$(x - 1)^2 + (y - 2)^2 = 16$$

Chapter 11 Areas of Polygons and Circles

Page 593 Getting Started

1. $A = \ell \cdot w$
$150 = \ell \cdot 15$
$10 = \ell$

2. $A = \ell \cdot w$
$38 = \ell \cdot 19$
$2 = \ell$

3. $A = \ell \cdot w$
$21.16 = \ell \cdot 4.6$
$4.6 = \ell$

4. $A = \ell \cdot w$
$2000 = \ell \cdot 32$
$62.5 = \ell$

5. $A = \ell \cdot w$
$450 = \ell \cdot 25$
$18 = \ell$

6. $A = \ell \cdot w$
$256 = \ell \cdot 20$
$12.8 = \ell$

7. $\frac{1}{2}a(b + c) = \frac{1}{2} \cdot 6(8 + 10)$
$= \frac{1}{2} \cdot 6(18) = 54$

8. $\frac{1}{2}ab = \frac{1}{2} \cdot 6 \cdot 8 = 24$

9. $\frac{1}{2}(2b + c) = \frac{1}{2}(2 \cdot 8 + 10)$
$= \frac{1}{2}(16 + 10)$
$= \frac{1}{2} \cdot 26 = 13$

10. $\frac{1}{2}d(a + c) = \frac{1}{2} \cdot 11(6 + 10)$
$= \frac{1}{2} \cdot 11 \cdot 16$
$= 8 \cdot 11 = 88$

11. $\frac{1}{2}(b + c) = \frac{1}{2}(8 + 10)$
$= \frac{1}{2}(18) = 9$

12. $\frac{1}{2}cd = \frac{1}{2} \cdot 10 \cdot 11$
$= 55$

13. $\triangle ABD$ is a 30°-60°-90° triangle. $\overline{BD}$, or h, is the longer leg, $\overline{AD}$ is the shorter leg, and $\overline{AB}$ is the hypotenuse.
$AD = \frac{1}{2}AC$ or 6
$h = 6\sqrt{3}$

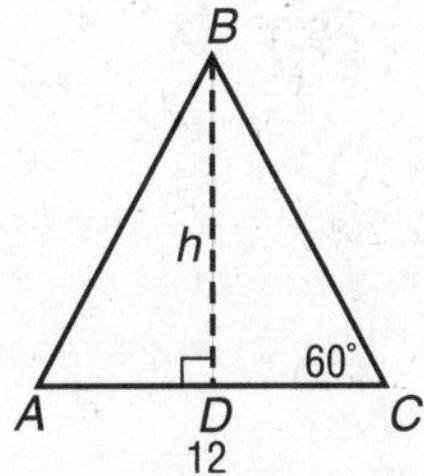

14. In the 30°-60°-90° triangle, the hypotenuse has a length of 22, so the shorter leg has a length of 11.

15. In the 45°-45°-90° triangle, the length of the hypotenuse is 15.

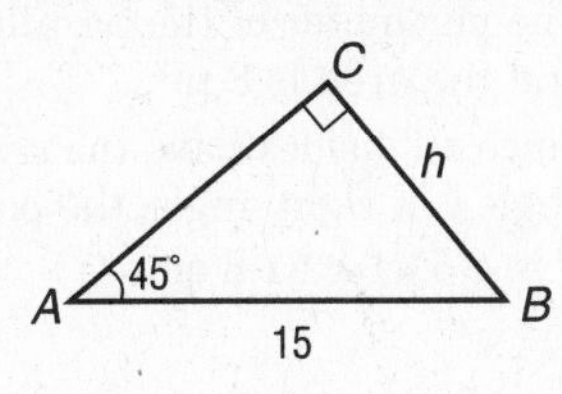

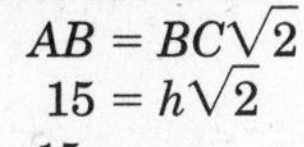

$AB = BC\sqrt{2}$
$15 = h\sqrt{2}$
$\frac{15}{\sqrt{2}} = h$
$\frac{15}{\sqrt{2}} \cdot \frac{\sqrt{2}}{\sqrt{2}} = h$
$\frac{15\sqrt{2}}{2} = h$

Page 594 Reading Mathematics

1. *bi-* 2, *sector-* a subdivision or region; divide into 2 regions
2. *poly-* many, *gon-* closed figure; closed figure with many sides
3. *equi-* equal, *lateral-* sides; having sides of equal length
4. *co-* together, *centr-* center; circles with a common center
5. *circum-* around, *scribe-* write; to write around (a geometrical figure)
6. *co-* together, *linear-* line; together on the same line
7. *circum-* around, about; *ferre-* to carry
8. Sample answers: polychromatic – multicolored, polymer – a chemical compound composed of a repeating structural unit, polysyllabic – a word with more than three syllables

11-1 Areas of Parallelograms

Page 595 Geometry Activity

1. When the parallelogram is folded, the base of the rectangle is 4 and the height is 5. So the area is 20 units^2.
2. There are 2 rectangles, one on the bottom and one formed by the folded triangles on top.
3. 40 units^2
4. The base of the parallelogram is twice of the length of the rectangle. The altitude of the parallelogram is the same width as the width of the rectangle.
5. $A = bh$

Page 598 Check for Understanding

1. The area of a rectangle is the product of the length and the width. The area of a parallelogram is the product of the base and the height. For both quadrilaterals, the measure of the length of one side is multiplied by the length of the altitude.
2. See students' work.
3. $P = 2(5) + 2(9) = 28$ ft
Use a 30°-60°-90° triangle to find the height, with x as the length of the shorter leg.
$9 = 2x$
$\frac{9}{2} = x$
height $= x\sqrt{3} = \frac{9}{2}\sqrt{3}$ ft
$A = bh$
$= 5\left(\frac{9}{2}\sqrt{3}\right)$
$= \frac{45}{2}\sqrt{3}$ or about 39.0 ft^2
The perimeter of the parallelogram is 28 ft, and the area is about 39.0 ft^2.
4. $P = 2(13) + 2(10) = 46$ yd
Use a 45°-45°-90° triangle to find the height, x.
$10 = x\sqrt{2}$
$\frac{10}{\sqrt{2}} = x$
$\frac{10}{\sqrt{2}} \cdot \frac{\sqrt{2}}{\sqrt{2}} = x$
$5\sqrt{2} = x$
height $= x = 5\sqrt{2}$ yd

$A = bh$
$= 13(5\sqrt{2})$
$= 65\sqrt{2}$ or about 91.9 yd^2

The perimeter of the parallelogram is 46 yd, and the area is about 91.9 yd^2.

5. $P = 4(3.2) = 12.8$ m
$A = s^2$
$= (3.2)^2$
$= 10.24$ or about 10.2 m^2

The perimeter of the square is 12.8 m, and the area is about 10.2 m^2.

6. 

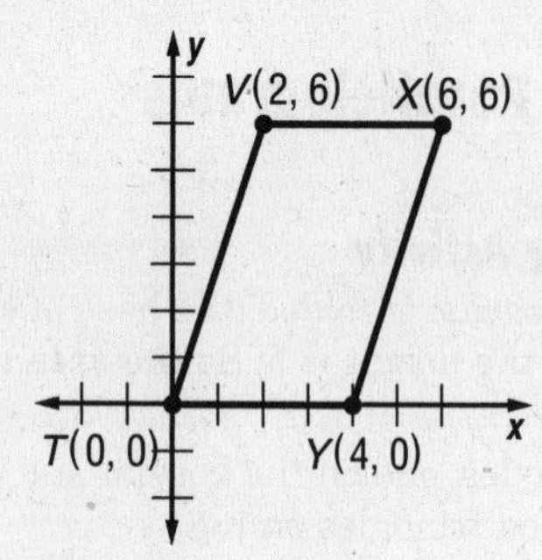

slope of $\overline{TV} = \frac{6-0}{2-0} = \frac{6}{2}$ or 3
slope of $\overline{XY} = \frac{0-6}{4-6} = \frac{-6}{-2}$ or 3
slope of $\overline{VX} = \frac{6-6}{6-2} = \frac{0}{4}$ or 0
slope of $\overline{TY} = \frac{0-0}{4-0} = \frac{0}{4}$ or 0

TVXY is a parallelogram, since opposite sides have the same slope. Slopes of consecutive sides are *not* negative reciprocals of each other, so the parallelogram is neither a square nor a rectangle.
$A = bh$
$= 4 \cdot 6$
$= 24$ units^2

7.

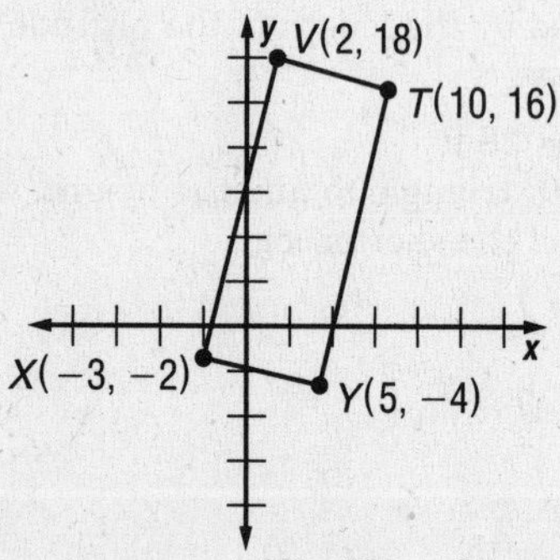

slope of $\overline{TV} = \frac{18-16}{2-10} = \frac{2}{-8}$ or $-\frac{1}{4}$
slope of $\overline{XY} = \frac{-4-(-2)}{5-(-3)} = \frac{-2}{8}$ or $-\frac{1}{4}$
slope of $\overline{VX} = \frac{-2-18}{-3-2} = \frac{-20}{-5}$ or 4
slope of $\overline{TY} = \frac{-4-16}{5-10} = \frac{-20}{-5}$ or 4

TVXY is a rectangle, since opposite sides have the same slope and the slopes of consecutive sides are negative reciprocals of each other.

Length of $\overline{XY}$ is $\sqrt{[5-(-3)]^2 + [-4-(-2)]^2}$
$= 2\sqrt{17}$.

Length of $\overline{TY}$ is $\sqrt{[10-5]^2 + [16-(-4)]^2}$
$= 5\sqrt{17}$.

$A = bh$
$= (2\sqrt{17})(5\sqrt{17}) = 10 \cdot 17$
$= 170$ units^2

8. Left Parallelogram
$A = bh$
$= 18 \cdot 15$
$= 270$ ft^2

Rectangle
$A = \ell w$
$= 35 \cdot 18$
$= 630$ ft^2

Right Parallelogram
$A = bh$
$= 18 \cdot 15$
$= 270$ ft^2

The total area is 270 + 630 + 270 or 1170 ft^2.

Pages 598–600 Practice and Apply

9. $P = 2(30) + 2(10) = 80$ in.
Use a 30°-60°-90° triangle to find the height, with x as the length of the shorter leg.
$10 = 2x$
$5 = x$
height $= x\sqrt{3} = 5\sqrt{3}$ in.
$A = bh$
$= 30(5\sqrt{3})$
$= 150\sqrt{3}$ or about 259.8 in^2.
The perimeter of the parallelogram is 80 in., and the area is about 259.8 in^2.

10. Use the 45°-45°-90° triangle with hypotenuse 4 to find the base and height of the parallelogram, both equal to x.
$4 = x\sqrt{2}$
$\frac{4}{\sqrt{2}} = x$
$\frac{4}{\sqrt{2}} \cdot \frac{\sqrt{2}}{\sqrt{2}} = x$
$2\sqrt{2} = x$
base = height $= x = 2\sqrt{2}$ m
$P = 2(2\sqrt{2}) + 2(4) = 4\sqrt{2} + 8$ or about 13.7 m
$A = bh$
$= (2\sqrt{2})(2\sqrt{2})$
$= 8$ m^2
The perimeter of the parallelgram is about 13.7 m, and the area is 8 m^2.

11. Since all 4 sides have the same measure and the angle is a right angle, the parallelogram is a square.
$P = 4(5.4) = 21.6$ cm
$A = s^2$
$= (5.4)^2$
$= 29.16$ or about 29.2 cm^2
The perimeter is 21.6 cm, and the area is about 29.2 cm^2.

12. $P = 2(15) + 2(10) = 50$ in.
Use a 45°-45°-90° triangle to find the height, x.
$10 = x\sqrt{2}$
$\frac{10}{\sqrt{2}} \cdot \frac{\sqrt{2}}{\sqrt{2}} = x$
$5\sqrt{2} = x$
height $= x = 5\sqrt{2}$ in.
$A = bh$
$= 15(5\sqrt{2})$
$= 75\sqrt{2}$ or about 106.1 in^2
The perimeter is 50 in., and the area is about 106.1 in^2.

13. $P = 2(12) + 2(10) = 44$ m

Use a 30°-60°-90° triangle to find the height, with x as the length of the shorter leg.

$10 = 2x$

$5 = x$

height $= x\sqrt{3} = 5\sqrt{3}$ m

$A = bh$

$= 12(5\sqrt{3})$

$= 60\sqrt{3}$ or about 103.9 m^2

The perimeter is 44 m, and the area is about 103.9 m^2.

14. $P = 2(5.4) + 2(4.2) = 19.2$ ft

$A = bh$

$= (5.4)(4.2)$

$= 22.68$ or about 22.7 ft^2

The perimeter is 19.2 ft, and the area is about 22.7 ft^2.

15. Rectangle

$A = \ell w$

$= 10(5\sqrt{2})$

$= 50\sqrt{2}$ mm^2

Each triangle

$A = \frac{1}{2}bh$

$= \frac{1}{2}(5)(5)$

$= \frac{25}{2}$ mm^2

The shaded area is $50\sqrt{2} - 2\left(\frac{25}{2}\right)$ or about 45.7 mm^2.

16. Rectangular outline

$A = \ell w$

$= (7 + 4 + 7)(15)$

$= 270$ cm^2

Upper cutout

$A = \ell w$

$= 12 \cdot 3$

$= 36$ cm^2

Lower cutout

$A = \ell w$

$= 4 \cdot 8$

$= 32$ cm^2

The shaded area is $270 - 36 - 32$ or 202 cm^2.

17. Big square

$A = s^2$

$= (9.2)^2$

$= 84.64$ ft^2

Small square

$A = s^2$

$= (3.1)^2$

$= 9.61$ ft^2

Rectangle

$A = \ell w$

$= (10.8)(3.1)$

$= 33.48$ ft^2

The shaded area is $84.64 + 33.48 - 9.61$ or about 108.5 ft^2.

18.

$A = bh$

$100 = (x + 15)(x)$

$100 = x^2 + 15x$

$x^2 + 15x - 100 = 0$

$(x + 20)(x - 5) = 0$

$x + 20 = 0$ or $x - 5 = 0$

$x = -20$ or $x = 5$

Reject the negative solution.

$h = 5$ units, $b = 20$ units

19.

$A = bh$

$2000 = (x + 10)(x)$

$2000 = x^2 + 10x$

$x^2 + 10x - 2000 = 0$

$(x - 40)(x + 50) = 0$

$x - 40 = 0$ or $x + 50 = 0$

$x = 40$ or $x = -50$

Reject the negative solution.

$h = 40$ units, $b = 50$ units

20.

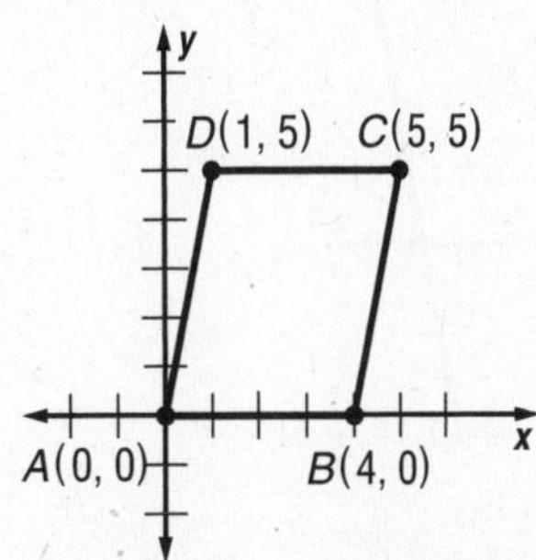

slope of $\overline{AB} = \frac{0 - 0}{4 - 0} = \frac{0}{4}$ or 0

slope of $\overline{DC} = \frac{5 - 5}{5 - 1} = \frac{0}{4}$ or 0

slope of $\overline{AD} = \frac{5 - 0}{1 - 0} = \frac{5}{1}$ or 5

slope of $\overline{BC} = \frac{5 - 0}{5 - 4} = \frac{5}{1}$ or 5

$ABCD$ is a parallelogram, since opposite sides have the same slope. Slopes of consecutive sides are *not* negative reciprocals of each other, so the parallelogram is neither a square nor a rectangle.

Base: $\overline{AB}$ lies along the x-axis and is horizontal so $AB = |4 - 0| = 4$.

Height: Since $\overline{CD}$ and $\overline{AB}$ are horizontal segments, the distance between them, or height, can be measured on any vertical segment. Reading from the graph, the height is 5.

$A = bh$

$= 4 \cdot 5$

$= 20$ units^2

21.

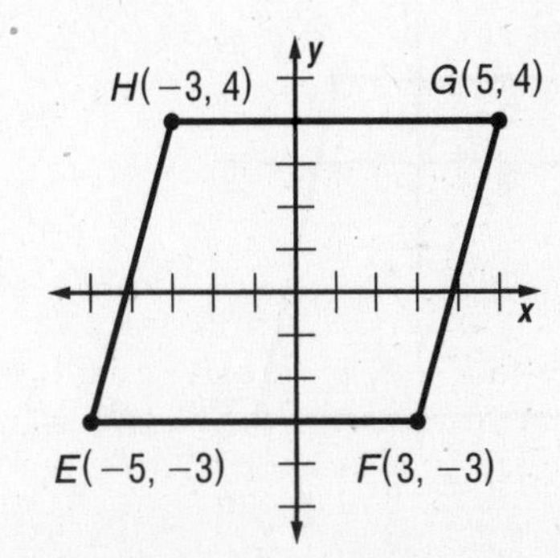

slope of $\overline{EF} = \frac{-3 - (-3)}{3 - (-5)} = \frac{0}{8}$ or 0

slope of $\overline{HG} = \frac{4 - 4}{5 - (-3)} = \frac{0}{8}$ or 0 slope of $\overline{FG} = \frac{4-(-3)}{5-3} = \frac{7}{2}$

slope of $\overline{EH} = \frac{4-(-3)}{-3-(-5)} = \frac{7}{2}$

EFGH is a parallelogram, since opposite sides have the same slope. Slopes of consecutive sides are *not* negative reciprocals of each other, so the parallelogram is neither a square nor a rectangle.

Base: $\overline{EF}$ is horizontal with length $|3-(-5)| = 8$.

Height: Since $\overline{GH}$ and $\overline{EF}$ are horizontal segments, the distance between them, or height, can be measured on any vertical segment. Reading from the graph, the height is 7.

$A = bh$
$= 8 \cdot 7$
$= 56$ units^2

22.

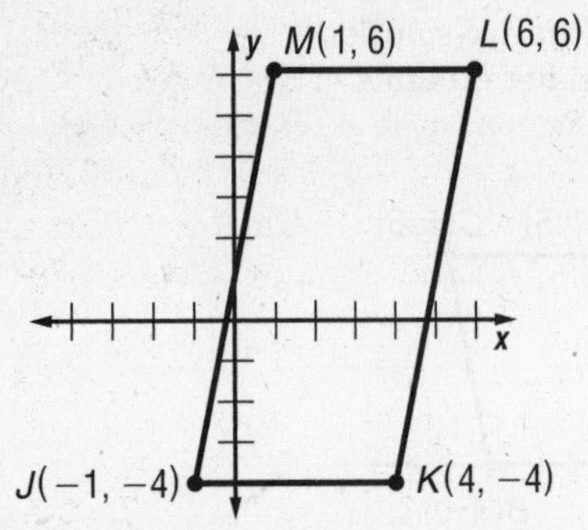

slope of $\overline{JK} = \frac{-4-(-4)}{4-(-1)} = \frac{0}{5}$ or 0

slope of $\overline{ML} = \frac{6-6}{6-1} = \frac{0}{5}$ or 0

slope of $\overline{JM} = \frac{6-(-4)}{1-(-1)} = \frac{10}{2}$ or 5

slope of $\overline{KL} = \frac{6-(-4)}{6-4} = \frac{10}{2}$ or 5

JKLM is a parallelogram, since opposite sides have the same slope. Slopes of consecutive sides are *not* negative reciprocals of each other, so the parallelogram is neither a square nor a rectangle.

Base: $\overline{JK}$ is horizontal with length $|-1-4| = 5$.

Height: Since $\overline{JK}$ and $\overline{LM}$ are horizontal segments, the distance between them, or height, can be measured on any vertical segment. Reading from the graph, the height is 10.

$A = bh$
$= 5 \cdot 10$
$= 50$ units^2

23.

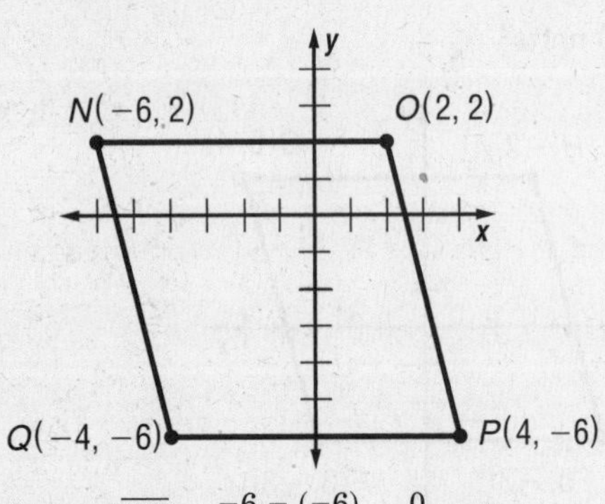

slope of $\overline{QP} = \frac{-6-(-6)}{4-(-4)} = \frac{0}{8}$ or 0

slope of $\overline{NO} = \frac{2-2}{2-(-6)} = \frac{0}{8}$ or 0

slope of $\overline{NQ} = \frac{-6-2}{-4-(-6)} = \frac{-8}{2}$ or -4

slope of $\overline{OP} = \frac{-6-2}{4-2} = \frac{-8}{2}$ or -4

NOPQ is a parallelogram, since opposite sides have the same slope. Slopes of consecutive sides are *not* negative reciprocals of each other, so the parallelogram is neither a square nor a rectangle.

Base: $\overline{PQ}$ is horizontal with length $|-4-4| = 8$.

Height: Since $\overline{PQ}$ and $\overline{ON}$ are horizontal segments, the distance between them, or height, can be measured on any vertical segment. Reading from the graph, the height is 8.

$A = bh$
$= 8 \cdot 8$
$= 64$ units^2

24.

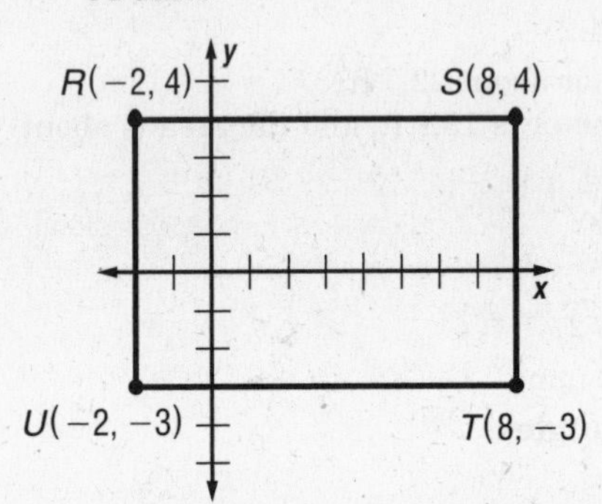

slope of $\overline{UT} = \frac{-3-(-3)}{8-(-2)} = \frac{0}{10}$ or 0

slope of $\overline{RS} = \frac{4-4}{8-(-2)} = \frac{0}{10}$ or 0

slope of $\overline{RU} = \frac{-3-4}{-2-(-2)} = \frac{-7}{0}$ is undefined

slope of $\overline{ST} = \frac{-3-4}{8-8} = \frac{-7}{0}$ is undefined

RSTU is a rectangle, since the sides are not all equal but are all horizontal or vertical.

Base: $\overline{UT}$ is horizontal with length $|-2-8| = 10$.

Height: Since $\overline{RS}$ and $\overline{UT}$ are horizontal segments, the distance between them, or height, can be measured on any vertical segment. Reading from the graph, the height is 7.

$A = bh$
$= 10 \cdot 7$
$= 70$ units^2

25.

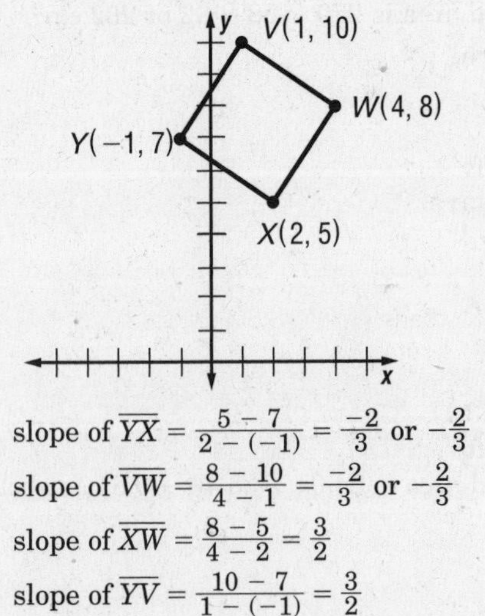

slope of $\overline{YX} = \frac{5-7}{2-(-1)} = \frac{-2}{3}$ or $-\frac{2}{3}$

slope of $\overline{VW} = \frac{8-10}{4-1} = \frac{-2}{3}$ or $-\frac{2}{3}$

slope of $\overline{XW} = \frac{8-5}{4-2} = \frac{3}{2}$

slope of $\overline{YV} = \frac{10-7}{1-(-1)} = \frac{3}{2}$

$VWXY$ is a rectangle, since opposite sides have the same slope and the slopes of consecutive sides are negative reciprocals of each other.

Length of $\overline{YX} = \sqrt{[2 - (-1)]^2 + [5 - 7]^2} = \sqrt{13}$

Length of $\overline{XW} = \sqrt{[4 - 2]^2 + [8 - 5]^2} = \sqrt{13}$

Thus $VWXY$ is in fact a square.

$A = s^2$
$= \left(\sqrt{13}\right)^2$
$= 13 \text{ units}^2$

26. Guest bedroom

$A = \ell w$
$= 20 \cdot 22$
$= 440 \text{ ft}^2$

Family room

$A = \ell w$
$= 25 \cdot 22$
$= 550 \text{ ft}^2$

Hallway

$A = \ell w$
$= 25 \cdot 3$
$= 75 \text{ ft}^2$

The Bessos need $440 + 550 + 75$ or 1065 ft^2 of carpet. Since there are 9 square feet per square yard, the family should order $1065 \div 9$ or 119 yd^2 (rounded up to the nearest yd^2).

27. The figure is composed of three 5 by 10 rectangles.

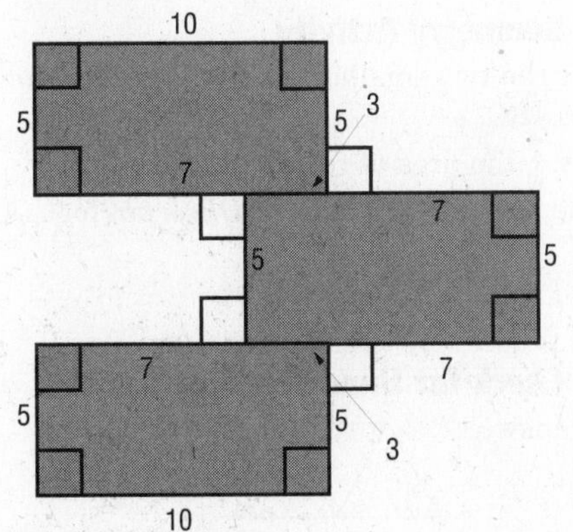

For each rectangle, $A = \ell w = 5 \cdot 10$ or 50 units^2. The total area is $3 \cdot 50$ or 150 units^2.

28. The figure can be viewed as a large 8 by 11 rectangle with two rectangular cutouts measuring 1 by 2 and 3 by 3.

Large rectangle

$A = \ell w$
$= 8 \cdot 11$
$= 88 \text{ units}^2$

Cutout 1

$A = \ell w$
$= 1 \cdot 2$
$= 2 \text{ units}^2$

Cutout 2

$A = \ell w$
$= 3 \cdot 3$
$= 9 \text{ units}^2$

The shaded area is $88 - 2 - 9$ or 77 units^2.

29. The triptych measures $3 + 5 + 2 + 12 + 2 + 5 + 3$ or 32 inches wide by $3 + 12 + 3$ or 18 inches tall. Yes, it will fit in a 45-inch by 20-inch frame.

30. By Exercise 29, the artwork measures 32 inches by 18 inches.

$A = \ell w$
$= 32 \cdot 18$
$= 576 \text{ in}^2$

31. On the crosswalk, draw a 30°-60°-90° triangle whose hypotenuse is 16 ft long and whose short side lies along the left edge of the crosswalk. That short side measures 8 ft, and the remaining side, which represents the perpendicular distance between the stripes, measures $8\sqrt{3}$ or about 13.9 ft.

32. $P = 2(8) + 2(11) = 38$ m

$A = bh$
$= 8 \cdot 10$
$= 80 \text{ m}^2$

The perimeter is 38 m, and the area is 80 m^2.

33. $P = 2(4) + 2(5.5) = 19$ m

$A = bh$
$= 4 \cdot 5$
$= 20 \text{ m}^2$

The perimeter is 19 m and the area is 80 m^2.

34. The new perimeter is half of the original perimeter. The new area is one half squared, or one fourth, the area of the original parallelogram.

35. Let the side length of one square be x. Then the side length of the other square is $\frac{48 - 4x}{4}$ or $12 - x$.

$$x^2 + (12 - x)^2 = 74$$
$$x^2 + 144 - 24x + x^2 = 74$$
$$2x^2 - 24x + 70 = 0$$
$$x^2 - 12x + 35 = 0$$
$$(x - 5)(x - 7) = 0$$

$x = 5$ or $x = 7$

The two side lengths are 5 in. and 7 in.

36. Sample answer: Area is used when designing a garden to find the total amount of materials needed. Answers should include the following.

- Find the area of one square and multiply by the number of squares in the garden.
- Knowing the area is useful when planning a stone walkway or fencing in flowers or vegetables.

37. C; the length of base $\overline{AB}$ is $\sqrt{10^2 - 6^2} = 8$, so the area is $8 \cdot 6 = 48 \text{ m}^2$.

38. D; either A, B, or C could be true, depending on whether $x > \frac{1}{2}$, $x < \frac{1}{2}$, or $x = \frac{1}{2}$.

Page 600 Maintain Your Skills

39. For the equation $(x - h)^2 + (y - k)^2 = r^2$, the center of the circle is (h, k) and the radius is r. For the equation $(x - 5)^2 + (y - 2)^2 = 49 = 7^2$, the center is $(5, 2)$ and $r = 7$.

40. For the equation $(x - h)^2 + (y - k)^2 = r^2$, the center of the circle is (h, k) and the radius is r. For the equation $(x + 3)^2 + (y + 9)^2 - 81 = 0$, which is equivalent to $[x - (-3)]^2 + [y - (-9)]^2 = 9^2$, the center is $(-3, -9)$ and $r = 9$.

41. For the equation $(x - h)^2 + (y - k)^2 = r^2$, the center of the circle is (h, k) and the radius is r. For the equation $\left(x + \frac{2}{3}\right)^2 + \left(y - \frac{1}{9}\right)^2 - \frac{4}{9} = 0$, which is equivalent to $\left[x - \left(-\frac{2}{3}\right)\right]^2 + \left[y - \frac{1}{9}\right]^2 = \left(\frac{2}{3}\right)^2$, the center is $\left(-\frac{2}{3}, \frac{1}{9}\right)$ and $r = \frac{2}{3}$.

42. For the equation $(x - h)^2 + (y - k)^2 = r^2$, the center of the circle is (h, k) and the radius is r. For the equation $(x - 2.8)^2 + (y + 7.6)^2 = 34.81$, which is equivalent to $[x - 2.8]^2 + [y - (-7.6)]^2 = 5.9^2$, the center is $(2.8, -7.6)$ and $r = 5.9$.

43. Use Theorem 10.16.

$$10(10 + 22) = 8(8 + x)$$
$$320 = 64 + 8x$$
$$256 = 8x$$
$$32 = x$$

44. Use Theorem 10.15.

$$4 \cdot 9 = x \cdot x$$
$$36 = x^2$$
$$6 = x$$

45. Use Theorem 10.17.

$$14 \cdot 14 = 7(7 + x)$$
$$196 = 49 + 7x$$
$$147 = 7x$$
$$21 = x$$

46.

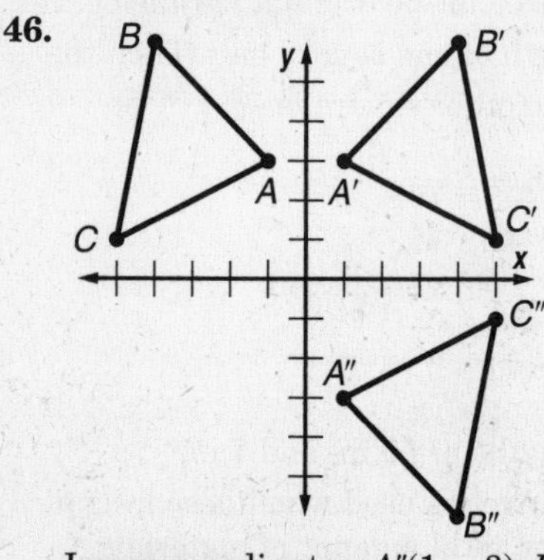

Image coordinates: $A''(1, -3)$, $B''(4, -6)$, $C''(5, -1)$
The rotation angle is 180°.

47.

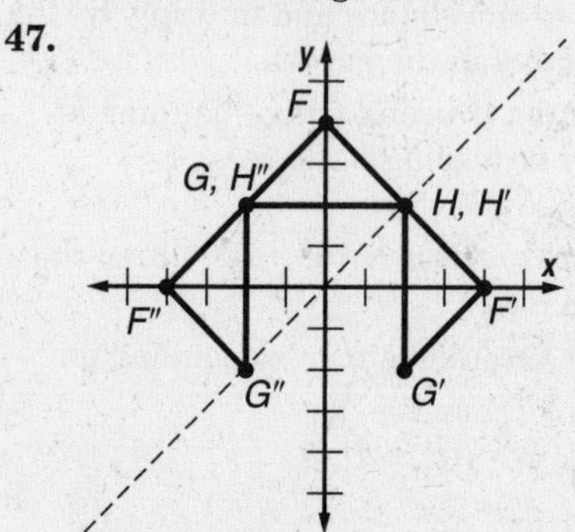

Image coordinates: $F''(-4, 0)$, $G''(-2, -2)$, $H''(-2, 2)$
The rotation angle is 90° counterclockwise.

48.

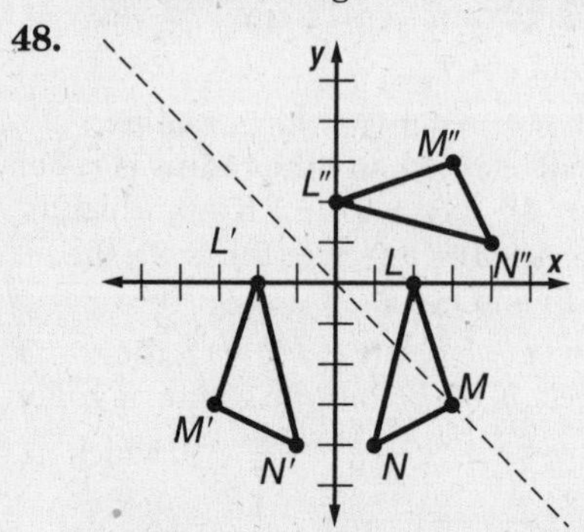

Image coordinates: $L''(0, 2)$, $M''(3, 3)$, $N''(4, 1)$
The rotation angle is 90° counterclockwise.

49. Use the Pythagorean Theorem.

$$a^2 + b^2 = c^2$$
$$5^2 + 12^2 = \ell^2$$
$$25 + 144 = \ell^2$$
$$169 = \ell^2$$
$$13 = \ell$$

The length of plywood needed is 13 ft.

50. $\frac{1}{2}(7y) = \frac{1}{2}(7 \cdot 2)$
$= 7$

51. $\frac{1}{2}wx = \frac{1}{2}(8)(4)$
$= 16$

52. $\frac{1}{2}z(x + y) = \frac{1}{2}(5)(4 + 2)$
$= 15$

53. $\frac{1}{2}x(y + w) = \frac{1}{2}(4)(2 + 8)$
$= 20$

11-2 Areas of Triangles, Trapezoids, and Rhombi

Page 601 Geometry Activity

1. Together the two smaller triangles are the same size as $\triangle ABC$.
2. $\triangle ABC$ is $\frac{1}{2}$ the area of rectangle $ACDE$.
3. Since the area of rectangle $ACDE$ is bh, for $\triangle ABC$ $A = \frac{1}{2}bh$.

Page 605 Check for Understanding

1. Sample answer:

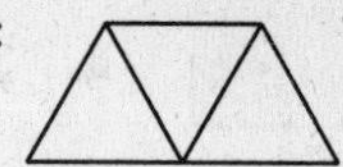

2. Kiku is correct; she simplified the formula by adding the terms in the parentheses before multiplying.
3. Sometimes; two rhombi can have different corresponding diagonal lengths and have the same area.
4. $A = \frac{1}{2}d_1d_2$
$= \frac{1}{2}(20)(24)$
$= 240 \text{ m}^2$
5. area of $FGHI$ = area of $\triangle FGH$ + area of $\triangle FHI$
$= \frac{1}{2}bh_1 + \frac{1}{2}bh_2$
$= \frac{1}{2}(37)(9) + \frac{1}{2}(37)(18)$
$= \frac{333}{2} + 333$
$= 499.5 \text{ in}^2$
6. $A = \frac{1}{2}h(b_1 + b_2)$
$= \frac{1}{2}(12)(24 + 16)$
$= 240 \text{ yd}^2$

7. $\overline{AB}$ is horizontal with length $|-5 - 2| = |-7|$ or 7. Point C lies above $\overline{AB}$ a distance of $|3 - (-3)| = |6|$ or 6 units.

$A = \frac{1}{2}bh$

$= \frac{1}{2}(7)(6)$

$= 21 \text{ units}^2$

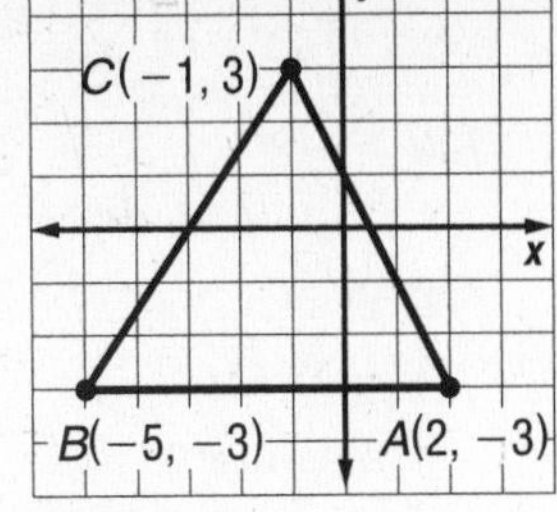

8. $\overline{FG}$ and $\overline{HJ}$ are horizontal.

$FG = |5 - (-1)|$

$= |6|$ or 6

$HJ = |1 - 3|$

$= |-2|$ or 2

$h = |8 - 4|$

$= |4|$ or 4

$A = \frac{1}{2}h(b_1 + b_2)$

$= \frac{1}{2}(4)(6 + 2)$

$= 16 \text{ units}^2$

9. $\overline{LP}$ is horizontal, $\overline{MQ}$ is vertical.

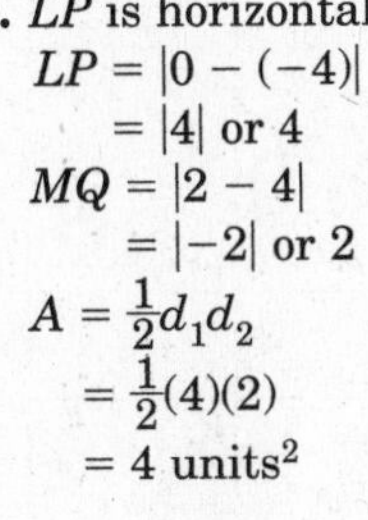

$LP = |0 - (-4)|$

$= |4|$ or 4

$MQ = |2 - 4|$

$= |-2|$ or 2

$A = \frac{1}{2}d_1d_2$

$= \frac{1}{2}(4)(2)$

$= 4 \text{ units}^2$

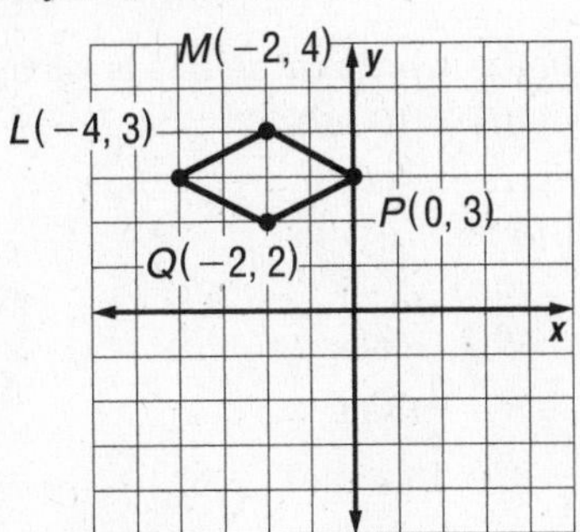

10. $A = \frac{1}{2}h(b_1 + b_2)$

Substitute the known values into the formula.

$250 = \frac{1}{2}h(20 + 30)$

$250 = \frac{1}{2}(50)h$

$250 = 25h$

$10 = h$

The height is 10 in.

11. $A = \frac{1}{2}d_1d_2$

Substitute the known values into the formula.

$675 = \frac{1}{2}(15 + 15)d_2$

$675 = 15d_2$

$45 = d_2$

$SU = 45$ m

12. From Postulate 11.1, the area of each congruent rhombus is the same, namely $82\frac{7}{8} \div 13 = 6\frac{3}{8} \text{ in}^2$. The width of one rhombus is $15 \div 5 = 3$ in. To find the other diagonal (the height), use the area formula.

$A = \frac{1}{2}d_1d_2$

$6\frac{3}{8} = \frac{1}{2}(3)d_2$

$\frac{51}{8} = \frac{3}{2}d_2$

$\frac{17}{4} = d_2$

The vertical diagonal measures $\frac{17}{4}$ or $4\frac{1}{4}$ in.

Pages 606–608 Practice and Apply

13. $A = \frac{1}{2}bh$

$= \frac{1}{2}(7.3)(3.4)$

$= 12.41$ or about 12.4 cm^2

14. $A = \frac{1}{2}bh$

$= \frac{1}{2}(10.2)(7)$

$= 35.7 \text{ ft}^2$

15. $A = \frac{1}{2}h(b_1 + b_2)$

$= \frac{1}{2}(10)(8 + 11)$

$= 95 \text{ km}^2$

16. $A = \frac{1}{2}h(b_1 + b_2)$

$= \frac{1}{2}(8.5)(8.5 + 14.2)$

$= 96.475$ or about 96.5 yd^2

17. $A = \frac{1}{2}d_1d_2$

$= \frac{1}{2}(20 + 20)(30 + 30)$

$= 1200 \text{ ft}^2$

18. $A = \frac{1}{2}d_1d_2$

$= \frac{1}{2}(17 + 17)(12 + 12)$

$= 408 \text{ cm}^2$

19. area of quadrilateral $ABCD$

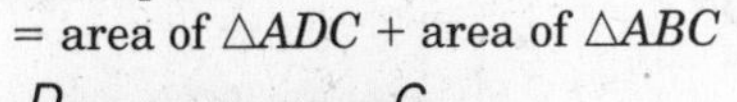

$=$ area of $\triangle ADC$ + area of $\triangle ABC$

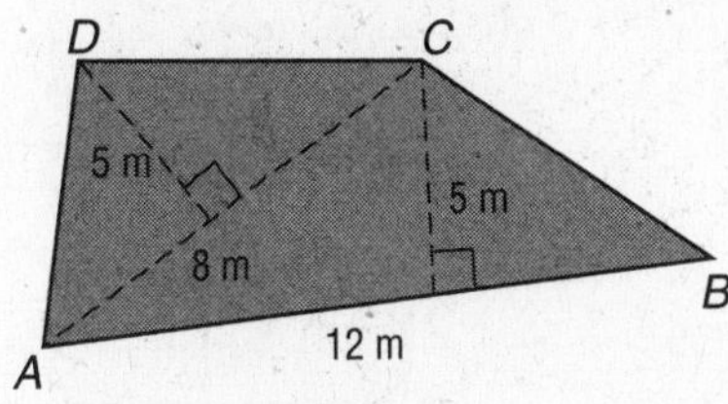

$= \frac{1}{2}b_1h_1 + \frac{1}{2}b_2h_2$

$= \frac{1}{2}(8)(5) + \frac{1}{2}(12)(5)$

$= 50 \text{ m}^2$

20. area of quadrilateral $WXYZ$

$=$ area of $\triangle WXY$ + area of $\triangle WZY$

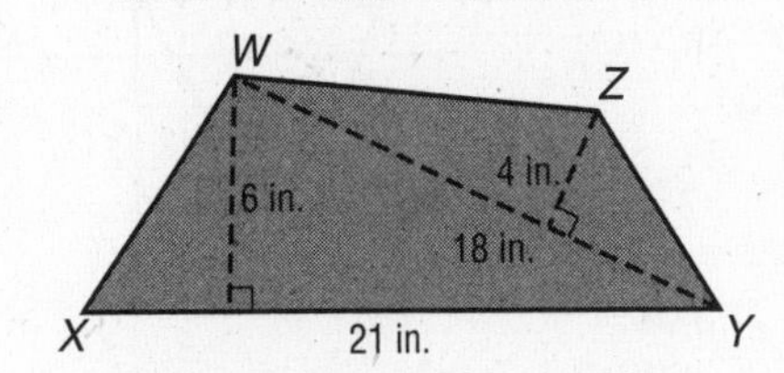

$= \frac{1}{2}b_1h_1 + \frac{1}{2}b_2h_2$

$= \frac{1}{2}(21)(6) + \frac{1}{2}(18)(4)$

$= 99 \text{ in}^2$

21. In a 30°-60°-90° right triangle, the longer leg is $\sqrt{3}$ times as long as the shorter leg. Here, then, $b = h\sqrt{3}$, so $h = \frac{b}{\sqrt{3}}$.

$A = bh$

$= 15\left(\frac{15}{\sqrt{3}}\right)$

$= 75\sqrt{3}$ or about 129.9 mm^2

22. $\overline{PT}$ and $\overline{QR}$ are horizontal.

$QR = |5 - 3|$
$= |2|$ or 2
$PT = |6 - 0|$
$= |6|$ or 6
$h = |7 - 3|$
$= |4|$ or 4
$A = \frac{1}{2}h(b_1 + b_2)$
$= \frac{1}{2}(4)(2 + 6)$
$= 16$ units^2

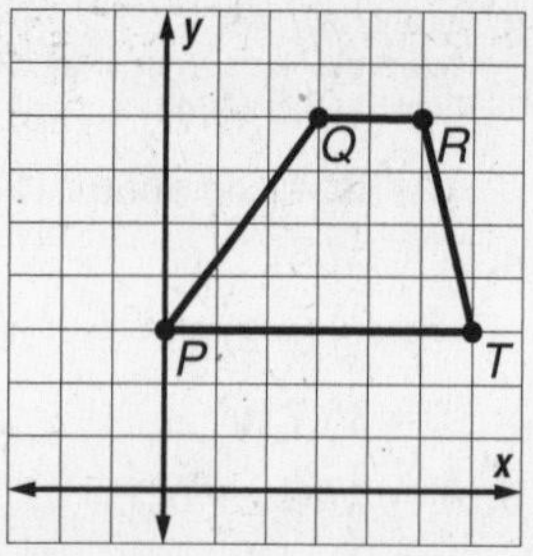

23. $\overline{RT}$ and $\overline{PQ}$ are horizontal.

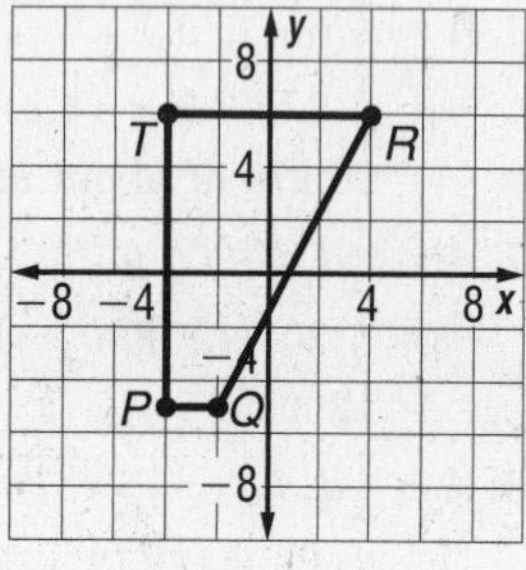

$RT = |-4 - 4|$
$= |-8|$ or 8
$PQ = |-2 - (-4)|$
$= |2|$ or 2
$h = |6 - (-5)|$
$= |11|$ or 11
$A = \frac{1}{2}h(b_1 + b_2)$
$= \frac{1}{2}(11)(8 + 2)$
$= 55$ units^2

24. $\overline{RT}$ and $\overline{PQ}$ are horizontal.

$RT = |1 - 6|$
$= |-5|$ or 5
$PQ = |6 - (-3)|$
$= |9|$ or 9
$h = |8 - 2|$
$= |6|$ or 6
$A = \frac{1}{2}h(b_1 + b_2)$
$= \frac{1}{2}(6)(5 + 9)$
$= 42$ units^2

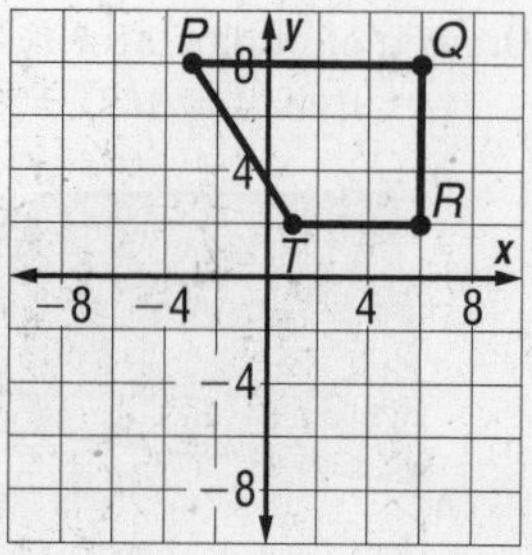

25. $\overline{RT}$ and $\overline{PQ}$ are horizontal.

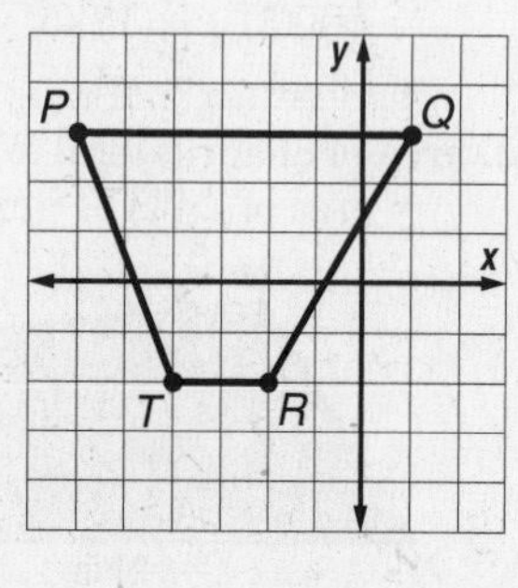

$RT = |-4 - (-2)|$
$= |-2|$ or 2
$PQ = |1 - (-6)|$
$= |7| = 7$
$h = |3 - (-2)|$
$= |5|$ or 5
$A = \frac{1}{2}h(b_1 + b_2)$
$= \frac{1}{2}(5)(2 + 7)$
$= 22.5$ units^2

26. $\overline{JL}$ is horizontal, $\overline{KM}$ is vertical.

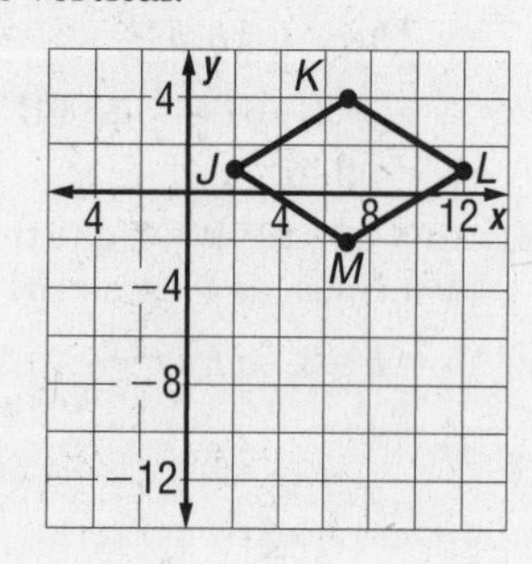

$JL = |12 - 2|$
$= |10|$ or 10
$KM = |-2 - 4|$
$= |-6|$ or 6
$A = \frac{1}{2}d_1d_2$
$= \frac{1}{2}(10)(6)$
$= 30$ units^2

27. $\overline{JL}$ is horizontal, $\overline{KM}$ is vertical.

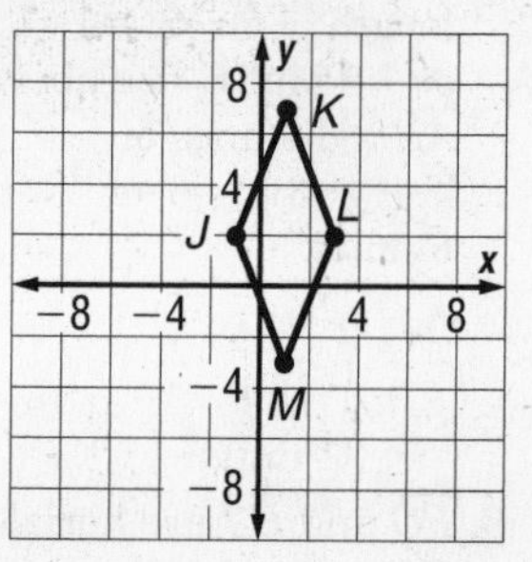

$JL = |3 - (-1)|$
$= |4|$ or 4
$KM = |-3 - 7|$
$= |-10|$ or 10
$A = \frac{1}{2}d_1d_2$
$= \frac{1}{2}(4)(10)$
$= 20$ units^2

28. $\overline{JL}$ is horizontal, $\overline{KM}$ is vertical.

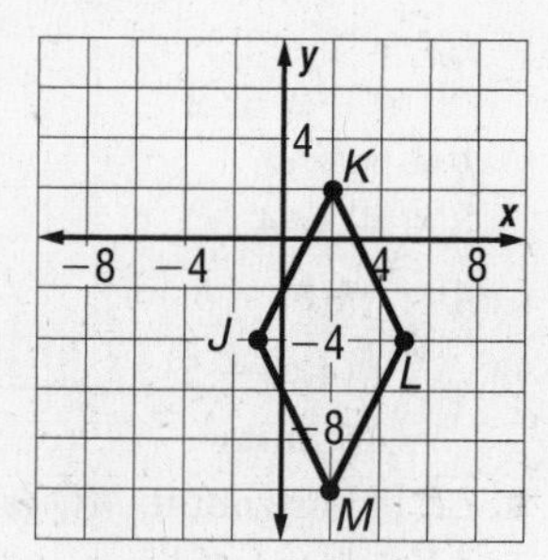

$JL = |5 - (-1)|$
$= |6|$ or 6
$KM = |-10 - 2|$
$= |-12|$ or 12
$A = \frac{1}{2}d_1d_2$
$= \frac{1}{2}(6)(12)$
$= 36$ units^2

29. $\overline{JL}$ is horizontal, $\overline{KM}$ is vertical.

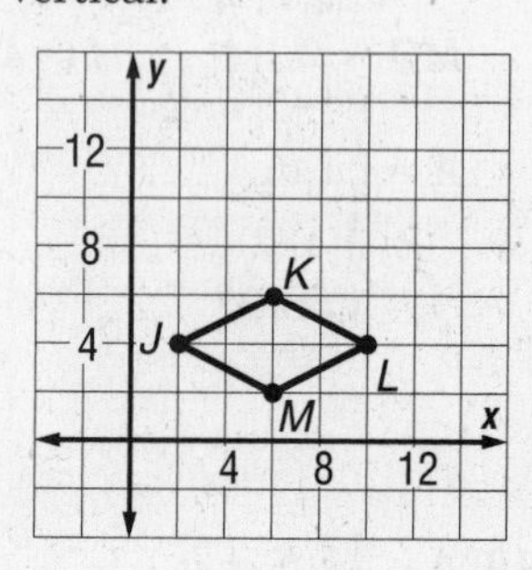

$JL = |10 - 2|$
$= |8|$ or 8
$KM = |2 - 6|$
$= |-4|$ or 4
$A = \frac{1}{2}d_1d_2$
$= \frac{1}{2}(8)(4)$
$= 16$ units^2

30. $A = \frac{1}{2}h(b_1 + b_2)$
Solve for h.
$750 = \frac{1}{2}h(35 + 25)$
$750 = \frac{1}{2}(60)h$
$750 = 30h$
$25 = h$
The height is 25 m.

31. $A = \frac{1}{2}h(b_1 + b_2)$
Solve for b_2.
$188.35 = \frac{1}{2}(8.7)(16.5 + b_2)$
$376.7 = 8.7(16.5 + b_2)$
$376.7 = 143.55 + 8.7b_2$
$233.15 = 8.7b_2$
$26.8 \approx b_2$
GK is about 26.8 ft.

32. $A = \frac{1}{2}d_1d_2$
Solve for d_2.
$375 = \frac{1}{2}(25)d_2$
$375 = 12.5d_2$
$30 = d_2$
$NQ = 30$ in.

33. $A = \frac{1}{2}d_1d_2$

Solve for d_2.

$137.9 = \frac{1}{2}(12.2)d_2$

$137.9 = 6.1d_2$

$22.6 \approx d_2$

QS is about 22.6 m.

34. $A = \frac{1}{2}bh$

Solve for b.

$248 = \frac{1}{2}b(16)$

$248 = 8b$

$31 = b$

The base measures 31 in.

35. $A = \frac{1}{2}bh$

Solve for h.

$300 = \frac{1}{2}(30)h$

$300 = 15h$

$20 = h$

The height is 20 cm.

36. Each rhombus has an area of $150 \div 2 = 75$ ft^2.

$A = \frac{1}{2}d_1d_2$

$75 = \frac{1}{2}(12)d_2$

$75 = 6d_2$

$12.5 = d_2$

Each stone walkway is 12.5 ft long.

37. From Exercise 36, $d_2 = 12.5$ ft.

$$s^2 = \left(\frac{d_1}{2}\right)^2 + \left(\frac{d_2}{2}\right)^2$$

$$s = \sqrt{\left(\frac{d_1}{2}\right)^2 + \left(\frac{d_2}{2}\right)^2}$$

$$= \sqrt{\left(\frac{12}{2}\right)^2 + \left(\frac{12.5}{2}\right)^2}$$

$$\approx 8.7$$

Each side measures about 8.7 ft.

38. $A = \frac{1}{2}h(b_1 + b_2)$

$= \frac{1}{2}(122.81)(56 + 69.7)$

≈ 7718.6 ft^2

39. $A = \frac{1}{2}h(b_1 + b_2)$

$= \frac{1}{2}(199.8)(57.8 + 75.6)$

$\approx 13{,}326.7$ ft^2

40. A side length of $20 \div 4$ or 5 m and a half-diagonal of $8 \div 2$ or 4 m implies that the other half-diagonal measures $\sqrt{5^2 - 4^2}$ or 3 m. So $d_2 = 2 \cdot 3$ or 6 m.

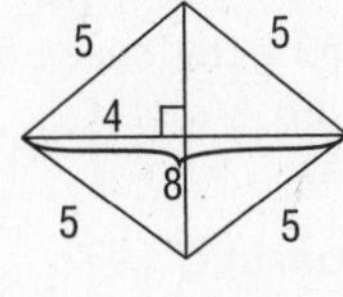

$A = \frac{1}{2}d_1d_2$

$= \frac{1}{2}(8)(6)$

$= 24$ m^2

41. A side length of $52 \div 4$ or 13 in. and a half-diagonal of $24 \div 2$ or 12 in. implies that the other half-diagonal measures $\sqrt{13^2 - 12^2}$ or 5 in. So $d_2 = 2 \cdot 5$ or 10 in.

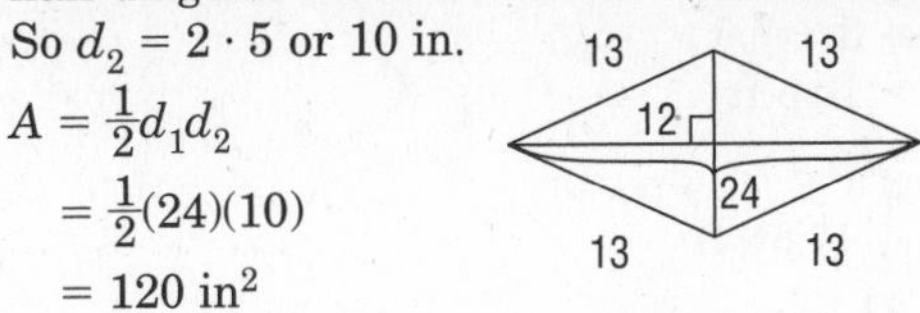

$A = \frac{1}{2}d_1d_2$

$= \frac{1}{2}(24)(10)$

$= 120$ in^2

42.

Let $b_1 = x$, $b_2 = x + 10$, $s = 2x - 3$.

$P = b_1 + b_2 + 2s$

$52 = x + (x + 10) + 2(2x - 3)$

$52 = 6x + 4$

$48 = 6x$

$8 = x$

So $b_1 = 8$, $b_2 = 18$, $s = 13$.

In order to find the area, we need to find h. Using right triangle ABC, $AB = s = 13$, and $AC = 5$.

$h^2 + 5^2 = 13^2$

$h^2 = \sqrt{144}$

$h = 12$

$A = \frac{1}{2}h(b_1 + b_2)$

$= \frac{1}{2}(12)(8 + 18)$

$= 156$ yd^2

43. $b = \frac{1}{3}P$

$= \frac{1}{3}(15)$

$= 5$ in.

Drawing the height divides the triangle into two 30°-60°-90° triangles each with base $\frac{1}{2}b$.

$h = \sqrt{3} \cdot \frac{1}{2}b$

$= \sqrt{3} \cdot \frac{1}{2}(5)$

$= \frac{5\sqrt{3}}{2}$ in.

$A = \frac{1}{2}bh$

$= \frac{1}{2}(5)\left(\frac{5\sqrt{3}}{2}\right)$

$= \frac{25\sqrt{3}}{4}$ or about 10.8 in^2

44. Because $(34.0)^2 + (81.6)^2 = (88.4)^2$, the triangle is a right triangle.

$A = \frac{1}{2}bh$

$= \frac{1}{2}(34.0)(81.6)$

$= 1387.2$ m^2

45. $LM = \sqrt{5^2 - 4^2}$ or 3 ft
$JL = \sqrt{(8.5)^2 - (4)^2}$ or 7.5 ft
$JM = JL + LM$
$= 3 + 7.5$
$= 10.5$ ft
$A = \frac{1}{2}bh$
$= \frac{1}{2}(10.5)(4)$
$= 21 \text{ ft}^2$

46. A rhombus is made up of two congruent triangles. Using d_1 and d_2 instead of b and h, its area in reference to $A = \frac{1}{2}bh$ is $2\left[\frac{1}{2}(d_1)\left(\frac{1}{2}d_2\right)\right]$ or $\frac{1}{2}d_1d_2$.

47. False; Sample answer: The area for each of these right triangles is 6 square units. The perimeter of one triangle is 12 and the perimeter of the other is $8 + \sqrt{40}$ or about 14.3.

48. Drawing the height of an equilateral triangle divides it into two 30°-60°-90° triangles each with base $\frac{1}{2}b$.

Left triangle
$h = \sqrt{3} \cdot \frac{1}{2}b$
$= \sqrt{3} \cdot \frac{1}{2}(4)$
$= 2\sqrt{3}$
$A = \frac{1}{2}bh$
$= \frac{1}{2}(4)(2\sqrt{3})$
$= 4\sqrt{3}$ or about 6.9
$P = 3(4) = 12$

Right triangle
$h = \sqrt{3} \cdot \frac{1}{2}b$
$= \sqrt{3} \cdot \frac{1}{2}(5)$
$= \frac{5}{2}\sqrt{3}$
$A = \frac{1}{2}bh$
$= \frac{1}{2}(5)\left(\frac{5}{2}\sqrt{3}\right)$
$= \frac{25}{4}\sqrt{3}$ or about 10.8
$P = 3(5) = 15$

The scale factor and ratio of perimeters is $\frac{15}{12}$ or $\frac{5}{4}$.
The ratio of areas is $\frac{25}{4}\sqrt{3} \div 4\sqrt{3}$ or $\frac{25}{16}$, which equals $\left(\frac{5}{4}\right)^2$.

49. **Left rhombus**
$A = \frac{1}{2}d_1d_2$
$= \frac{1}{2}(4)(6)$
$= 12$
$P = 4(2\sqrt{13})$
$= 8\sqrt{13}$

Right rhombus
$A = \frac{1}{2}d_1d_2$
$= \frac{1}{2}(2)(3)$
$= 3$
$P = 4\sqrt{13}$

The scale factor and ratio of perimeters is $\frac{4\sqrt{13}}{8\sqrt{13}} = \frac{1}{2}$.
The ratio of areas is $\frac{3}{12} = \frac{1}{4}$, which equals $\left(\frac{1}{2}\right)^2$.

50. The kite consists of two triangles, each with $b = 25$ in. and $h = 20 \div 2$ or 10 in.
$A = A_1 + A_2$
$= 2A_1$
$= 2(\frac{1}{2}bh)$
$= bh$
$= 25 \cdot 10$
$= 250 \text{ in}^2$

51. Comparing heights, $\frac{6}{3} = \frac{2}{1}$.

52. **Left triangle**
$P = 8 + 8.3 + 6.5$
$= 22.8$

Right triangle
$P = 4 + 4.15 + 3.25$
$= 11.4$

53. $\frac{22.8}{11.4} = \frac{2}{1}$
The ratio is the same.

54. **Left triangle**
$A = \frac{1}{2}bh$
$= \frac{1}{2}(8)(6)$
$= 24$

Right triangle
$A = \frac{1}{2}bh$
$= \frac{1}{2}(4)(3)$
$= 6$

55. $\frac{24}{6} = \frac{4}{1} = \left(\frac{2}{1}\right)^2$
The ratio of the areas is the square of the scale factor.

56. $\frac{24}{6} = \frac{4}{1} = \left(\frac{2}{1}\right)^2$
$\frac{22.8}{11.4} = \frac{2}{1}$
The ratio of the areas is the square of the ratio of the perimeters.

57. $BH = 6$ and $HA = 3$
$BA = \sqrt{6^2 + 3^2}$
$= \sqrt{45}$
area of $ABCD = \left(\sqrt{45}\right)^2$ or 45 ft^2
area of $EFGH = 9^2 = 81 \text{ ft}^2$
ratio of areas $= \frac{45}{81} = \frac{5}{9}$ or 5:9

58. Sample answer: Umbrellas have triangular panels of fabric or nylon. In order to make the panels to fit the umbrella frame, the area of the triangles is needed. Answers should include the following.
- Find the area of a triangle by multiplying the base and the height and dividing by two.
- Rhombi are composed of two congruent isosceles triangles, and trapezoids are composed of two triangles and a rectangle.

59. B; C is (0, 10), and $A = \frac{1}{2}bh = \frac{1}{2}(10)(6) = 30 \text{ units}^2$.

60. D; either $2x - 7 = 0$ or $x + 10 = 0$, so either $x = \frac{7}{2}$ or $x = -10$.

61. Let side b be the base and a be the other given side. Then $h = a\sin C$.
area $= \frac{1}{2}bh$
$= \frac{1}{2}ab\sin C$

62. $A = \frac{1}{2}ab\sin C$
$= \frac{1}{2}(4)(7)\sin 29°$
$\approx 6.79 \text{ in}^2$

63. $A = \frac{1}{2}ab\sin C$
$= \frac{1}{2}(4)(5)\sin 37°$
$\approx 6.02 \text{ cm}^2$

64. $A = \frac{1}{2}ab\sin C$
$= \frac{1}{2}(1.9)(2.3)\sin 25°$
$\approx 0.92 \text{ ft}^2$

Page 609 Maintain Your Skills

65. $A = \ell w$
$= (22)(17)$
$= 374 \text{ cm}^2$

66. Use a 30°-60°-90° triangle. The height of the parallelogram $= x\sqrt{3}$, where $x = \frac{1}{2}(10)$ or 5.
$A = bh$
$= (15)(5\sqrt{3})$
$= 75\sqrt{3}$ or about 129.9 in^2

67. area = area of large rectangle + area of "hanging" rectangle
$= b_1h_1 + b_2h_2$
$= (21)(9) + (6)(7)$
$= 231 \text{ ft}^2$

68. For the circle with center (h, k) and radius r, the equation is $(x - h)^2 + (y - k)^2 = r^2$.
Here, $(x - 1)^2 + (y - 2)^2 = 7^2$, or
$(x - 1)^2 + (y - 2)^2 = 49$.

69. For the circle with center (h, k) and radius r, the equation is $(x - h)^2 + (y - k)^2 = r^2$.
Here, $[x - (-4)]^2 + \left[y - \frac{1}{2}\right]^2 = \left(\frac{11}{2}\right)^2$,
or $(x + 4)^2 + \left(y - \frac{1}{2}\right)^2 = \frac{121}{4}$.

70. For the circle with center (h, k) and radius r, the equation is $(x - h)^2 + (y - k)^2 = r^2$.
Here, $[x - (-1.3)]^2 + [y - 5.6]^2 = 3.5^2$,
or $(x + 1.3)^2 + (y - 5.6)^2 = 12.25$.

71. Each semicircle has a radius of $\frac{1}{2}(3.5) = 1.75$ in.
total inches of trim for one flower $= 5(\pi r)$
$= 5(1.75\pi)$
≈ 27.5 in.
So she needs 10(27.5) or 275 in. to edge 10 flowers.

72. $<136 \cos 25°, 136 \sin 25°> \approx <123.3, 57.5>$

73. $<280 \cos 52°, 280 \sin 52°> \approx <172.4, 220.6>$

74. $\frac{x}{46} = \sin 73°$
$x = 46 \sin 73°$
≈ 44.0

75. $\frac{x}{30} = \sin 42°$
$x = 30 \sin 42°$
≈ 20.1

76. $\frac{1}{2}(6) = 3$
$\frac{x}{3} = \tan 58°$
$x = 3 \tan 58°$
≈ 4.8

Page 609 Practice Quiz 1

1.

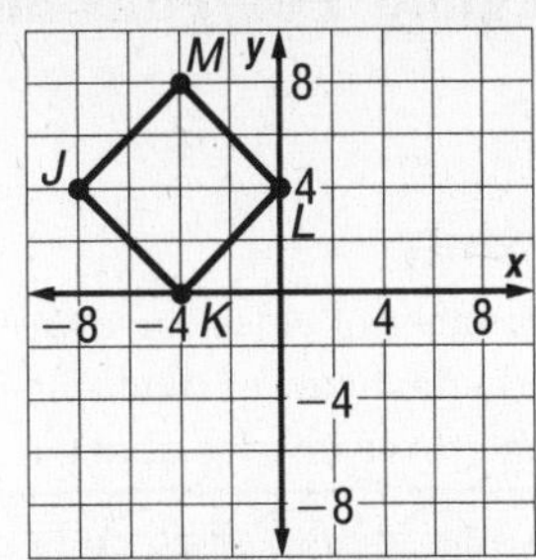

slope of $\overline{JK} = \frac{0 - 4}{-4 - (-8)} = \frac{-4}{4}$ or -1
slope of $\overline{LM} = \frac{8 - 4}{-4 - 0} = \frac{4}{-4}$ or -1
slope of $\overline{KL} = \frac{4 - 0}{0 - (-4)} = \frac{4}{4}$ or 1
slope of $\overline{JM} = \frac{8 - 4}{-4 - (-8)} = \frac{4}{4}$ or 1
Opposite sides have the same slope and slopes of consecutive sides are negative reciprocals of each other.
$JK = \sqrt{[-4 - (-8)]^2 + [0 - 4]^2} = 4\sqrt{2}$
$KL = \sqrt{[0 - (-4)]^2 + [4 - 0]^2} = 4\sqrt{2}$
Adjacent sides have the same length. $JKLM$ is a square.

2. Since $JKLM$ is a square with side length $4\sqrt{2}$,
$A = s^2$
$= (4\sqrt{2})^2$
$= 32 \text{ units}^2$

3. $\overline{NP}$ is a horizontal segment with length $|-4 - 1| = 5$ so $b_1 = 5$. $\overline{MQ}$ is a horizontal segment with length $|-6 - 7| = 13$ so $b_2 = 13$. Since $\overline{NP}$ and $\overline{MQ}$ are horizontal, the distance between them, the height, can be measured on any vertical segment. Reading from the graph, $h = 6$.
$A = \frac{1}{2}h(b_1 + b_2)$
$= \frac{1}{2}(6)(5 + 13)$
$= 54 \text{ units}^2$

4. The bases of the trapezoid are vertical segments.
$b_1 = WZ = |3 - (-1)| = 4$ and
$b_2 = XY = |7 - 1| = 6$
Since $\overline{WZ}$ and $\overline{XY}$ are vertical segments, the distance between them, or the height of the trapezoid, can be measured along any horizontal segment. Reading from the graph, $h = 5$.
$A = \frac{1}{2}h(b_1 + b_2)$
$= \frac{1}{2}(5)(4 + 6)$
$= 25 \text{ units}^2$

5. $A = \frac{1}{2}d_1d_2$
$546 = \frac{1}{2}(26)d_2$
$546 = 13d_2$
$42 = d_2$
d_2 is 42 yd long.

11-3 Areas of Regular Polygons and Circles

Page 611 Geometry Activity

1. $A = \frac{1}{2}Pa$. Since P = (number of sides)(measure of a side), each entry in the last row of the table is $\frac{1}{2}$ times the product of the three entries above it.

Number of Sides	3	5	8	10	20	50
Measure of a Side	$1.73r$	$1.18r$	$0.77r$	$0.62r$	$0.31r$	$0.126r$
Measure of Apothem	$0.5r$	$0.81r$	$0.92r$	$0.95r$	$0.99r$	$0.998r$
Area	$\mathbf{1.30r^2}$	$\mathbf{2.39r^2}$	$\mathbf{2.83r^2}$	$\mathbf{2.95r^2}$	$\mathbf{3.07r^2}$	$\mathbf{3.14r^2}$

2. The polygon appears to be a circle.
3. The areas of the polygons approach the area of the circle.
4. The formula for the area of a circle is πr^2 or about $3.14r^2$.

Page 613 Check for Understanding

1. Sample answer: Separate a hexagon inscribed in a circle into six congruent nonoverlapping isosceles triangles. The area of one triangle is one-half the product of one side of the hexagon and the apothem of the hexagon. The area of the hexagon is $6\left(\frac{1}{2}sa\right)$. The perimeter of the hexagon is $6s$, so the formula is $\frac{1}{2}Pa$.
2. Sample answer: Use the given angle measure, the given side length, and trigonometric ratios to find the missing lengths.
3. **Side length:**
Since the perimeter is 42 yards, the side length is $\frac{42}{6}$ or 7 yd.

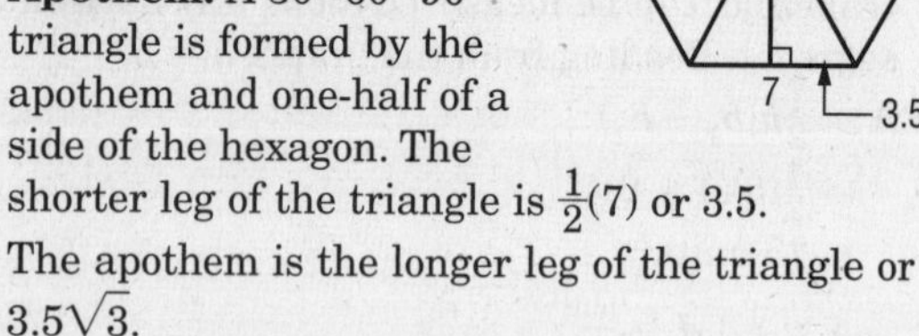

Apothem: A 30°-60°-90° triangle is formed by the apothem and one-half of a side of the hexagon. The shorter leg of the triangle is $\frac{1}{2}(7)$ or 3.5. The apothem is the longer leg of the triangle or $3.5\sqrt{3}$.

Area: $A = \frac{1}{2}Pa$
$= \frac{1}{2}(42)\ 3.5\left(\sqrt{3}\right)$
$\approx 127.3\ \text{yd}^2$

4. **Side length:**
Since the perimeter is 108 meters, the side length is $\frac{108}{9}$ or 12 m.

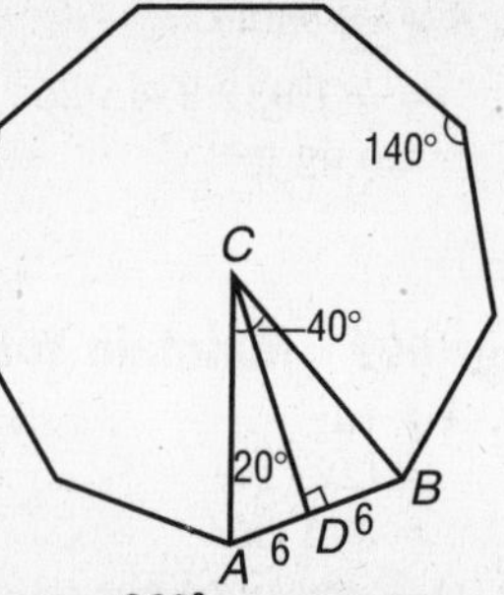

Apothem: $\overline{CD}$ is an apothem of the regular nonagon. The central angles are all congruent. The measure of each angle is $\frac{360°}{9}$ or 40°. $\overline{CD}$ bisects $\angle ACB$ so $m\angle ACD$ is 20°. $AD = 6$. Write a trigonometric ratio to find the length of $\overline{CD}$.

$\tan 20° = \frac{6}{CD}$ or $CD = \frac{6}{\tan 20°} \approx 16.485$ m

Area: $A = \frac{1}{2}Pa$
$\approx \frac{1}{2}(108)(16.485)$
$\approx 890.2\ \text{m}^2$

5. A 30°-60°-90° triangle is formed by the radius of the circle, the apothem, and half the base of the equilateral triangle. $\overline{AD}$ is the shorter leg, $\overline{BD}$ is the longer leg, and $\overline{AB}$ is the hypotenuse.

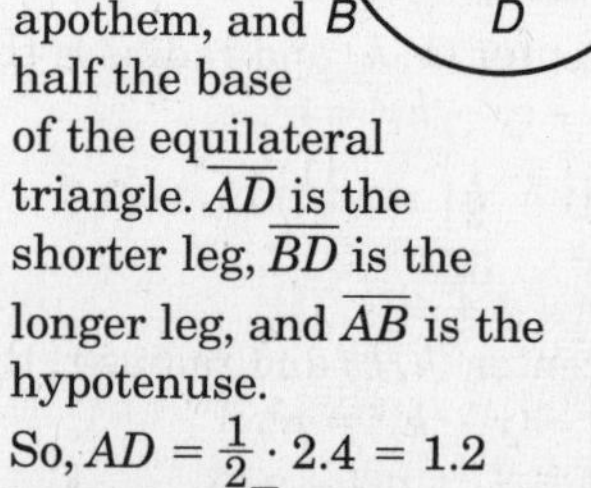

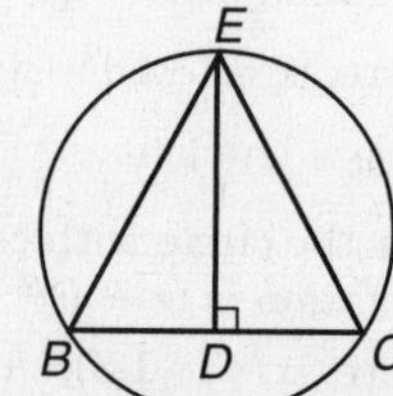

So, $AD = \frac{1}{2} \cdot 2.4 = 1.2$
$BD = 1.2\sqrt{3}$ and
$BC = 2.4\sqrt{3}$
Next, find the height of the triangle DE. Since $m\angle EBD = 60$,
$DE = \sqrt{3}\ BD = \sqrt{3}(1.2)(\sqrt{3})$
$= 3.6$
shaded area = area of circle − area of triangle
$= \pi r^2 - \frac{1}{2}bh$
$= \pi(2.4)^2 - \frac{1}{2}(2.4\sqrt{3})(3.6)$
$\approx 10.6\ \text{cm}^2$

6. $\triangle ABC$ is a 30°-60°-90° triangle. $\overline{AB}$ is the shorter leg and $\overline{AC}$ is the longer leg.

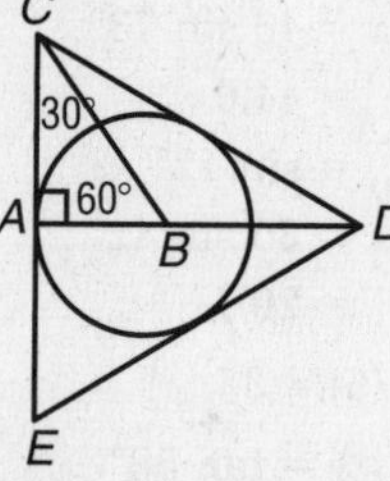

$AC = \sqrt{3}\,AB$
$= \sqrt{3} \cdot 3 = 3\sqrt{3}$
$\triangle ADC$ is also a 30°-60°-90° triangle in which $\overline{AC}$ is the shorter leg and $\overline{AD}$ is the longer leg.
So $AD = \sqrt{3}\left(3\sqrt{3}\right) = 3 \cdot 3 = 9$.
$CE = 2\,AC$ because AD is an altitude of $\triangle CDE$
$= 6\sqrt{3}$
shaded area = area of triangle − area of circle
$= \frac{1}{2}CE \cdot AD - \pi r^2$
$= \frac{1}{2}(6\sqrt{3})(9) - \pi(3)^2$
$\approx 18.5\ \text{in}^2$

7. **Small cushions**
radius = 6 in.
For cloth cover:
$r = 6 + 3 = 9$ in.
$A = \pi r^2$
$= \pi(9)^2$
$= 81\pi$

Large cushion
radius = 10 in.
For cloth cover:
$r = 10 + 3 = 13$ in.
$A = \pi r^2$
$= \pi(13)^2$
$= 169\pi$

Area of cloth to cover both sides of all cushions
$= 2(169\pi) + 14(81\pi)$
$= 1472\pi$ in^2
To convert to square yards, divide by 1296.
$\frac{1472\pi}{1296} \approx 3.6$ yd^2

Pages 613–616 Practice and Apply

8. **Side length:** $72 \div 8$ or 9 in., and $\frac{1}{2}(9) = 4.5$
Central angle: $360° \div 8$ or 45°,
and $\frac{1}{2}(45°) = 22.5°$
Apothem: $\tan 22.5° = \frac{4.5}{a}$
$a = \frac{4.5}{\tan 22.5°} \approx 10.864$ in.
$A = \frac{1}{2}Pa$
$\approx \frac{1}{2}(72)(10.864)$
≈ 391.1 in^2

9. side length $= 84\sqrt{2} \div 4$ or $21\sqrt{2}$ m
$A = s^2$
$= \left(21\sqrt{2}\right)^2$
$= 882$ m^2
(Note: This is easier than finding the central angle, the apothem, and $A = \frac{1}{2}Pa$.)

10. $\triangle ADC$ is a 45°-45°-90° triangle so $AD = 12$.
side length $= 2 \cdot 12 = 24$ cm
$A = s^2$
$= (24)^2$
$= 576$ cm^2
(Note: This is easier than finding the central angle and using $A = \frac{1}{2}Pa$.)

11. **Side length:** $m\angle ACB = 60$
so $m\angle ACD = 30$ and $\triangle ACD$ is a 30°-60°-90° triangle.
$\overline{AD}$ is the shorter leg,
so $CD = \sqrt{3}\,AD$ or
$24 = \sqrt{3}\,AD$.
$AD = \frac{24}{\sqrt{3}} \cdot \frac{\sqrt{3}}{\sqrt{3}} = \frac{24\sqrt{3}}{3} = 8\sqrt{3}$

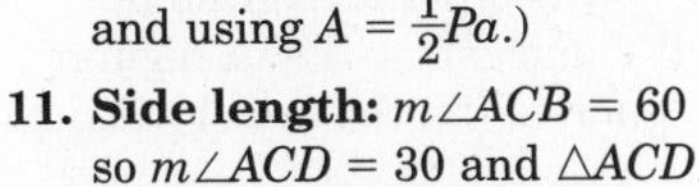

So, AB, the side length of the regular hexagon is $2(8\sqrt{3}) = 16\sqrt{3}$ in.
Perimeter: $P = 6(16\sqrt{3}) = 96\sqrt{3}$ in.
Area: $A = \frac{1}{2}Pa$
$= \frac{1}{2}(96\sqrt{3})(24)$
≈ 1995.3 in^2

12. Since we know that $\triangle ABC$ is a regular triangle, it is equilateral and $\triangle ADC$ is a 30°-60°-90° triangle. The height, $\overline{CD}$, of the triangle is the longer leg and $\overline{AD}$ is the shorter leg.
$AD = \frac{15.5}{2} = 7.75$
$CD = AD\sqrt{3} = 7.75\sqrt{3}$
The area of triangle ABC is
$\frac{1}{2}AB \cdot CD = \frac{1}{2}(15.5)(7.75\sqrt{3})$
≈ 104.0 in^2

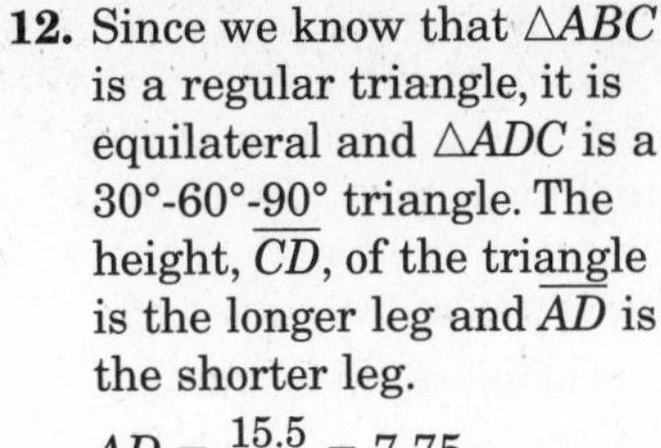

13. **Apothem:** The central angles of the octagon are all congruent so
$m\angle ACB = \frac{360}{8}$ or 45°.
$\overline{CD}$ is an apothem of the octagon. It bisects $\angle ACB$ and is a perpendicular bisector of $\overline{AB}$. So
$m\angle ACD = 22.5$. Since the side of the octagon has measure 10, $AD = 5$.
$\tan 22.5° = \frac{5}{CD}$
$CD = \frac{5}{\tan 22.5°}$
≈ 12.07
perimeter $= 10 \cdot 8 = 80$
Area: $A = \frac{1}{2}Pa$
$\approx \frac{1}{2}(80)(12.07)$
$= 482.8$ km^2

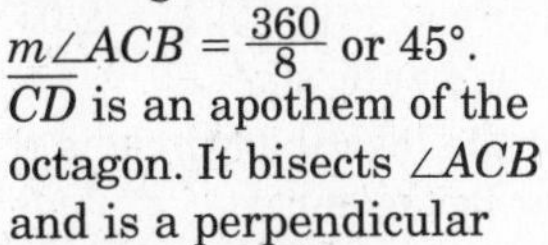

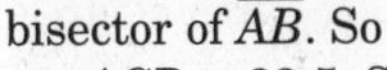

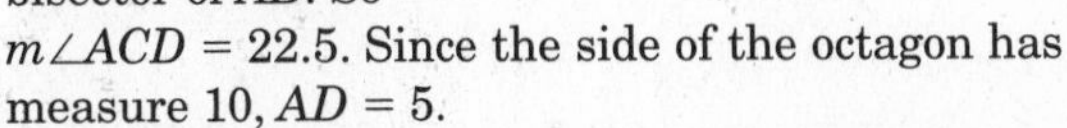

14. square side length $= 10\sqrt{2}$
shaded area = area of circle − area of square
$= \pi r^2 - s^2$
$= \pi(10)^2 - (10\sqrt{2})^2$
≈ 114.2 units^2

15. **Circle radius:** $5 \div 2 = 2.5$
shaded area = area of rectangle − area of circle
$= \ell w - \pi r^2$
$= (10)(5) - \pi(2.5)^2$
≈ 30.4 units^2

16. Use 30°-60°-90° triangles. Half the base of the equilateral triangle is $0.75\sqrt{3}$, so the base is $1.5\sqrt{3}$ and the height is $\sqrt{3}(0.75\sqrt{3})$ or 2.25.
shaded area = area of circle − area of triangle
$= \pi r^2 - \frac{1}{2}ab$
$= \pi(1.5)^2 - \frac{1}{2}(1.5\sqrt{3})(2.25)$
≈ 4.1 units^2

17. Use 30°-60°-90° triangles. Half the base of the equilateral triangle is $3.6\sqrt{3}$, so the base is $7.2\sqrt{3}$ and the height is $\sqrt{3}(3.6\sqrt{3})$ or 10.8.
shaded area = area of triangle − area of circle
$= \frac{1}{2}bh - \pi r^2$
$= \frac{1}{2}(7.2\sqrt{3})(10.8) - \pi(3.6)^2$
≈ 26.6 units^2

18. The triangle is a 30°-60°-90° right triangle whose sides measure 5, $5\sqrt{3}$, and 10. Since the hypotenuse of the triangle is a diameter of the circle, and the length of the hypotenuse is twice the length of the shorter leg, the length of the diameter is 10. So the radius of the circle is 5.

shaded area = area of circle − area of triangle

$$= \pi r^2 - \frac{1}{2}bh$$
$$= \pi(5)^2 - \frac{1}{2}(5)(5\sqrt{3})$$
$$\approx 56.9 \text{ units}^2$$

19. Use a 30°-60°-90° triangle.

Apothem: The central angle of the regular hexagon is $\frac{360°}{6} = 60°$. The apothem bisects the central angle so it forms a 30° angle with the hypotenuse of the triangle. So the apothem is the longer leg and its measure is $\sqrt{3} \cdot \frac{1}{2}(4.1) = 2.05\sqrt{3}$.

Hexagon perimeter: 6(4.1) = 24.6

Radius: The radius of the inscribed circle is the apothem of the regular hexagon.

shaded area = area of hexagon − area of circle

$$= \frac{1}{2}Pa - \pi r^2$$
$$= \frac{1}{2}(24.6)(2.05\sqrt{3}) - \pi(2.05\sqrt{3})^2$$
$$\approx 4.1 \text{ units}^2$$

20. circle radius $= \frac{1}{2}(20) = 10$ in.

Use a 30°-60° -90° triangle.

Apothem: The central angle of the regular hexagon is $\frac{360°}{6} = 60°$. The apothem bisects the central angle so it forms a 30° angle with the hypotenuse of the triangle, the radius of the circumscribed circle. So the apothem is the longer leg and has measure $\sqrt{3} \cdot \frac{1}{2}(10) = 5\sqrt{3}$ in.

Hexagon perimeter: 6(10) = 60 in.

shaded area = area of circle − area of hexagon

$$= \pi r^2 - \frac{1}{2}Pa$$
$$= \pi(10)^2 - \frac{1}{2}(60)(5\sqrt{3})$$
$$\approx 54.4 \text{ in}^2$$

21. From the solution to Exercise 14, the outer shaded area is $100\pi - 200$. The inner circle's radius is $\frac{1}{\sqrt{2}}(10) = 5\sqrt{2}$ and so its area is $\pi(5\sqrt{2})^2 = 50\pi$.

shaded area $= 100\pi - 200 + 50\pi$

$\approx 271.2 \text{ units}^2$

22. Use 30°-60°-90° triangles. Half the base of the equilateral triangle is $4\sqrt{3}$, so the base is $8\sqrt{3}$ and the height is $\sqrt{3}(4\sqrt{3})$ or 12. Also, the radius of the inner circle is 4.

shaded area = area of outer circle − area of triangle + area of inner circle

$$= \pi r_1^2 - \frac{1}{2}bh + \pi r_2^2$$
$$= \pi(8)^2 - \frac{1}{2}(8\sqrt{3})(12) + \pi(4)^2$$
$$\approx 168.2 \text{ units}^2$$

23. The radius of the larger circle is x, while the radius of the smaller circle is $\frac{1}{\sqrt{2}}x$. The ratio of areas is

$$\frac{\pi x^2}{\pi\left(\frac{1}{\sqrt{2}}x\right)^2} = \frac{\pi x^2}{\frac{1}{2}\pi x^2}$$
$$= \frac{1}{\frac{1}{2}}$$
$$= \frac{2}{1} \text{ or } 2{:}1$$

24. The scale factor is $\frac{9}{3} = \frac{3}{1}$, so the ratio of areas between a large cake and a mini-cake is $\left(\frac{3}{1}\right)^2 = \frac{9}{1}$. So when nine mini-cakes are compared to one large cake, the total area is equal. Nine mini-cakes are the same size as one 9-inch cake, but nine mini-cakes cost 9 · \$4 or \$36 while the 9-inch cake is only \$15. The 9-inch cake gives more cake for the money.

25.

16-inch pizza	**8-inch pizza**
$r = 8$	$r = 4$
$A = \pi \cdot 8^2$	$A = \pi \cdot 4^2$
$A = 64\pi$ in.^{2}	$A = 16\pi$ in.^{2}
	For 2 pizzas,
	$A = 2(16\pi)$
	$= 32\pi$ in.^{2}

One 16-inch pizza; the area of the 16-inch pizza is greater than the area of two 8-inch pizzas, so you get more pizza for the same price.

26.

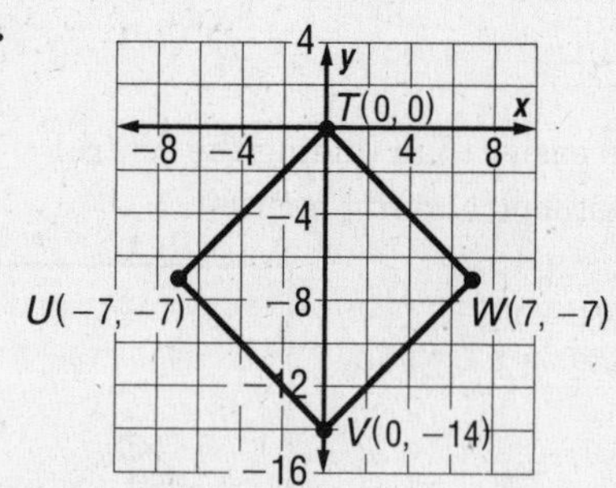

Explore: Looking at the graph, it appears that quadrilateral $TUVW$ is a square.

Plan: Show that $TUVW$ is a parallelogram by showing that $\overline{TU} \parallel \overline{WV}$ and $\overline{TU} \cong \overline{WV}$. A regular parallelogram is a square. Find the area by using the formula $A = s^2$.

Solve: slope of $\overline{TU} = \frac{-7-0}{-7-0} = \frac{-7}{-7}$ or 1

slope of $\overline{VW} = \frac{-7-(-14)}{7-0} = \frac{7}{7}$ or 1

$TU = \sqrt{[0-(-7)]^2 + [0-(-7)]^2} = \sqrt{7^2 + 7^2} = 7\sqrt{2}$

$VW = \sqrt{(0-7)^2 + [-14-(-7)]^2} = \sqrt{7^2 + 7^2} = 7\sqrt{2}$

$TU = VW$

$TUVW$ is a parallelogram.

$UV = \sqrt{(-7-0)^2 + [-7-(-14)]^2}$
$= \sqrt{7^2 + 7^2}$
$= 7\sqrt{2}$

The area of the square is $s^2 = (UV)^2 = \left(7\sqrt{2}\right)^2 = 98$ units^2.

Examine: Another way to find the area of a rhombus is using the formula $\frac{1}{2}d_1d_2$.

$A = \frac{1}{2}(14)(14) = 98$. The answer is the same.

27.

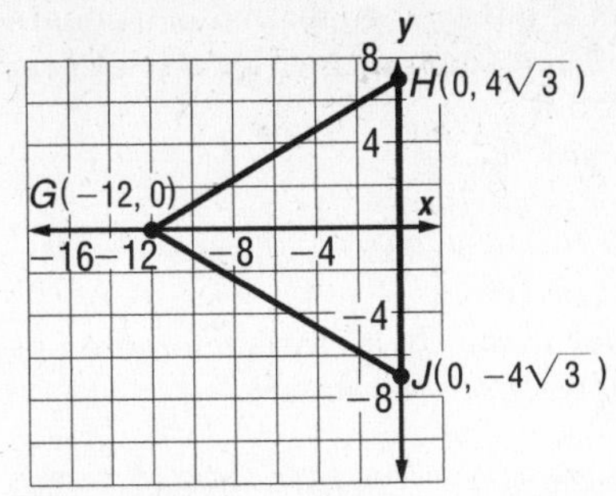

$\overline{HJ}$ is vertical.

$HJ = |-4\sqrt{3} - 4\sqrt{3}|$
$= |-8\sqrt{3}| \text{ or } 8\sqrt{3}$

$h = |0 - (-12)|$
$= |12| \text{ or } 12$

$A = \frac{1}{2}bh$
$= \frac{1}{2}(8\sqrt{3})(12)$
$= 48\sqrt{3}$ or about 83.1 units^2

28.

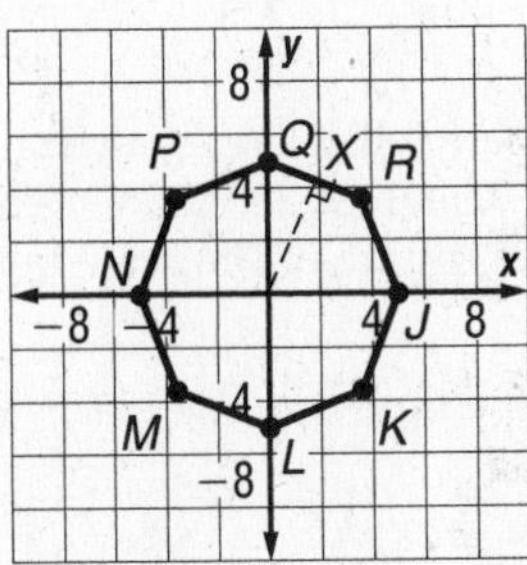

The figure is a regular octagon centered at the origin.

side length $= QR = \sqrt{(\frac{5\sqrt{2}}{2} - 0)^2 + (\frac{5\sqrt{2}}{2} - 5)^2}$
$= 5\sqrt{2 - \sqrt{2}}$, so perimeter $= 40\sqrt{2 - \sqrt{2}}$

The apothem of the regular octagon is the segment connecting the center (0, 0) with the midpoint of a side.

Use the Midpoint Formula to find the midpoint of $\overline{QR}$.

$X\left(\frac{x_1 + x_2}{2}, \frac{y_1 + y_2}{2}\right) = X\left(\frac{0 + 5\frac{\sqrt{2}}{2}}{2}, \frac{5 + 5\frac{\sqrt{2}}{2}}{2}\right)$

$= X\left(\frac{5\sqrt{2}}{4}, \frac{10 + 5\sqrt{2}}{4}\right)$

apothem length $= \left(\frac{5\sqrt{2}}{4} - 0\right)^2 + \left(\frac{10 + 5\sqrt{2}}{4}\right)$
$= \frac{5}{2}\sqrt{2+\sqrt{2}}$

$A = \frac{1}{2}Pa$
$= \frac{1}{2}\left(40\sqrt{2-\sqrt{2}}\right)\left(\frac{5}{2}\sqrt{2+\sqrt{2}}\right)$
$= 50\sqrt{2}$ or about 70.1 units^2

29.

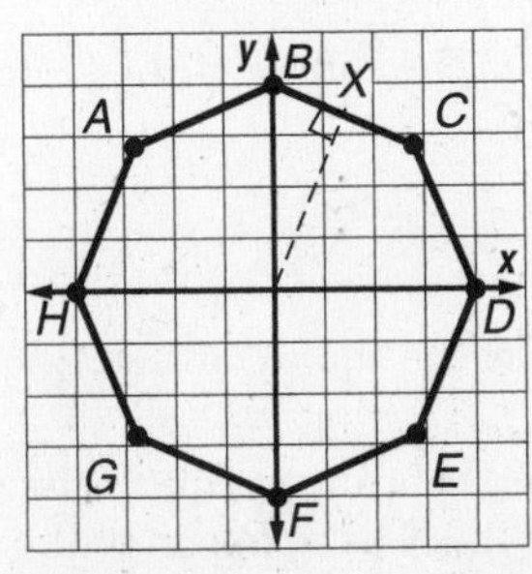

The figure is a regular octagon centered at the origin.

side length $= BC = \sqrt{(2\sqrt{2} - 0)^2 + (2\sqrt{2} - 4)^2}$
$= 4\sqrt{2 - \sqrt{2}}$

so perimeter $= 32\sqrt{2-\sqrt{2}}$

The apothem of the regular octagon is the segment connecting the center (0, 0) with the midpoint of a side.

Use the Midpoint Formula to find the midpoint of $\overline{BC}$.

$X\left(\frac{x_1 + x_2}{2}, \frac{y_1 + y_2}{2}\right) = X\left(\frac{0 + 2\sqrt{2}}{2}, \frac{4 + 2\sqrt{2}}{2}\right)$

$= X\left(\sqrt{2}, 2 + \sqrt{2}\right)$

apothem length $= \sqrt{(\sqrt{2} - 0)^2 + (2+\sqrt{2} - 0)^2}$
$= 2\sqrt{2 + \sqrt{2}}$

$A = \frac{1}{2}Pa$
$= \frac{1}{2}\left(32\sqrt{2-\sqrt{2}}\right)\left(2\sqrt{2+\sqrt{2}}\right)$
$= 32\sqrt{2}$ or about 45.3 units^2

30. $C = 2\pi r$
$r = \frac{C}{2\pi}$
$r = \frac{34\pi}{2\pi} = 17$
$A = \pi r^2$
$= \pi(17)^2$
$= 289\pi$ or about 907.9 units^2

31. $C = 2\pi r$
$r = \frac{C}{2\pi}$
$r = \frac{17\pi}{2\pi} = \frac{17}{2}$
$A = \pi r^2$
$= \pi\left(\frac{17}{2}\right)^2$
$= \frac{289}{4}\pi$ or about 227.0 units^2

32. $C = 2\pi r$
$r = \frac{C}{2\pi}$
$r = \frac{54.8}{2\pi} = \frac{27.4}{\pi}$
$A = \pi r^2$
$= \pi\left(\frac{27.4}{\pi}\right)^2$
$= \frac{750.76}{\pi}$ or about 239.0 units^2

33. $C = 2\pi r$
$r = \frac{C}{2\pi}$
$r = \frac{91.4}{2\pi} = \frac{45.7}{\pi}$
$A = \pi r^2$
$= \pi\left(\frac{45.7}{\pi}\right)^2$
$= \frac{2088.49}{\pi}$ or about 664.8 units^2

34. $A = \pi r^2$
$7850 = \pi r^2$
$r = \sqrt{\frac{7850}{\pi}} = \frac{5\sqrt{314}}{\sqrt{\pi}}$
$C = 2\pi r$
$= 2\pi \cdot \frac{5\sqrt{314}}{\sqrt{\pi}}$
$= 10\sqrt{314\pi}$ or about 314.1 ft

At 2 tiles per foot, and rounding up, that makes 629 tiles.

35. The square tiles will touch along the inner edge of the border, but there will be gaps along the outer edge. The tiles used to fill the gaps should be triangles. There will be 629 gaps between the 629 square tiles, so 629 triangular tiles will be needed.

36. $A = \pi r^2$
$= \pi(1.3)^2$
$= 1.69\pi$ or about 5.3 cm^2

37. Use radius = 11 in.
$A = \pi r^2$
$= \pi(11)^2$
$= 121\pi$ or about 380.1 in^2

38. Sample answer: Multiply the total area by 40%.

39. Use a 30°-60°-90° triangle. For the equilateral triangle, the height is opposite the 60° angle so it is the longer leg of the 30°-60°-90° triangle with hypotenuse measure of 3. So, height $= \sqrt{3} \cdot \frac{1}{2}(3) = \frac{3}{2}\sqrt{3}$.
For the circle, radius $= \frac{1}{2}(7) = 3.5$.
shaded area = area of circle − area of triangle
$= \pi r^2 - \frac{1}{2}bh$
$= \pi(3.5)^2 - \frac{1}{2}(3)\left(\frac{3}{2}\sqrt{3}\right)$
$= 12.25\pi - \frac{9}{4}\sqrt{3}$
≈ 34.6 units^2

40. circle diameter $= \sqrt{12^2 + 9^2} = 15$, and radius $= \frac{1}{2}(15) = 7.5$.
shaded area = area of circle − area of rectangle
$= \pi r^2 - \ell w$
$= \pi(7.5)^2 - (9)(12)$
$= 56.25\pi - 108$
≈ 68.7 units^2

41. r_1 = large radius = 10, r_2 = small radius = 5
shaded area = area of large circle
− area of small circles
$= \pi r_1^2 - 2\pi r_2^2$
$= \pi(10)^2 - 2\pi(5)^2$
$= 50\pi$ or about 157.1 units^2

42. circle radii = 1.5
shaded area = area of square − area of circles
$= s^2 - 4\pi r^2$
$= (6)^2 - 4\pi(1.5)^2$
$= 36 - 9\pi$
≈ 7.7 units^2

43. Flip the right half top-for-bottom, to make it a mirror image of the left half. Then there are three circles, with radii of $r_1 = 15$, $r_2 = 10$, and $r_3 = 5$.

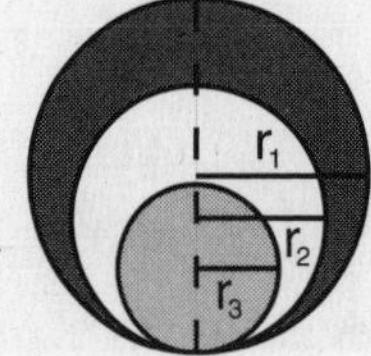

shaded area
= area of large circle
− area of medium circle + area of small circle
$= \pi r_1^2 - \pi r_2^2 + \pi r_3^2$
$= \pi(15)^2 - \pi(10)^2 + \pi(5)^2$
$= 150\pi$ or about 471.2 units^2

44. The large semicircle has radius $r_1 = 7.5$, the small semicircles have radii $r_2 = 2.5$.
shaded area = area of large semicircle
− area of small semicircles
$= \frac{1}{2}\left(\pi r_1^2\right) - 3 \cdot \frac{1}{2}\left(\pi r_2^2\right)$
$= \frac{1}{2}[\pi(7.5)^2] - 3 \cdot \frac{1}{2}[\pi(2.5)^2]$
$= 18.75\pi$ or about 58.9 units^2

45. The area of the garden equals that of a square of side length 175 ft combined with a circle of radius $\frac{1}{2}(175)$ or 87.5 ft.
area $= s^2 + \pi r^2$
$= (175)^2 + \pi(87.5)^2$
$= 7656.25\pi + 30{,}625$
$\approx 54{,}677.8$ ft^2
The perimeter is the circle's circumference plus two side lengths from the square.
perimeter $= 2\pi r + 2s$
$= 2\pi(87.5) + 2(175)$
$= 175\pi + 350$
≈ 899.8 ft

46. $\pi d_1 + \pi d_2 + \pi d_3 = 20\pi + 40\pi + 60\pi$
$= 120\pi$ or about 377.0 ft

47. The path is defined by an outer circle of radius $\frac{1}{2}(60) - 5 = 25$ and an inner circle of radius $\frac{1}{2}(40) = 20$.
path area = area of outer circle − area of inner circle
$= \pi r_1^2 - \pi r_2^2$
$= \pi(25)^2 - \pi(20)^2$
$= 225\pi$ or about 706.9 ft^2

48. No; the areas of the floors will increase by the squares of 1, 3, 5, and 7, or 1, 9, 25, and 49. The ratio of the areas is the square of the scale factor.

49. $\frac{4.2}{6.3} = \frac{2}{3}$ or 2 : 3

50. Call the perimeter of the figure on the left P_1 and the figure on the right P_2.
$P_1 = 5(4.2)$ or 21 cm
$P_2 = 5(6.3)$ or 31.5 cm

51. $\frac{21}{31.5} = \frac{2}{3}$
The ratio is the same.

52. Use the perimeter information from Exercise 50.
$\tan 36° = \frac{2.1}{a}$
$a = \frac{2.1}{\tan 36°}$
$A_1 = \frac{1}{2}Pa$
$= \frac{1}{2}(21)\left(\frac{2.1}{\tan 36°}\right)$
≈ 30.35 cm^2
$\tan 36° = \frac{3.15}{a}$
$a = \frac{3.15}{\tan 36°}$
$A_2 = \frac{1}{2}Pa$
$= \frac{1}{2}(31.5)\left(\frac{3.15}{\tan 36°}\right)$
≈ 68.29 cm^2

53. $\frac{30.24}{68.04} = \frac{4}{9} = \left(\frac{2}{3}\right)^2$
The ratio of the areas is the square of the scale factor.

54. $\frac{30.24}{68.04} = \frac{4}{9} = \left(\frac{2}{3}\right)^2$
$\frac{21}{31.5} = \frac{2}{3}$
The ratio of the areas is the square of the ratio of the perimeters.

55. The apothem of the larger hexagon equals the radius of the circle, $r = 10$. The apothem of the smaller hexagon is $\sqrt{3} \cdot \frac{1}{2}r = \frac{\sqrt{3}}{2}r$.

scale factor = $\frac{\frac{\sqrt{3}}{2}r}{r} = \frac{\sqrt{3}}{2}$

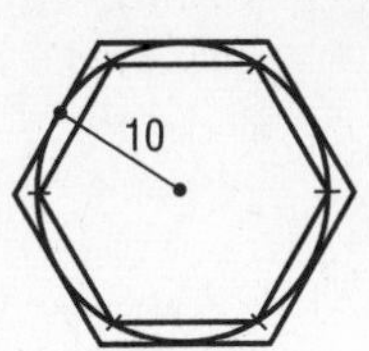

area ratio = (scale factor)^2

$= \left(\frac{\sqrt{3}}{2}\right)^2$

$= \frac{3}{4}$ or a ratio of 3 to 4

(The value of r is irrelevant.)

56. Sample answer: You can find the areas of regular polygons by finding the product of the perimeter and the apothem and then multiplying by one half. Answers should include the following.

- We need to know the length of each side and the length of the apothem.
- One method is to divide the area of the floor by the area of each tile. Since the floor is hexagonal and not rectangular, tiles of different shapes will need to be ordered to cover the floor.

57. B; for the circle, $A = \pi r^2$ so that $r = \sqrt{\frac{A}{\pi}} = \sqrt{\frac{18\pi}{\pi}}$ $= 3\sqrt{2}$. Square side length $s = \sqrt{2}r$ $= \sqrt{2}\,(3\sqrt{2}) = 6$ units.

58. B; since average = sum ÷ x, x = sum ÷ average = 90 ÷ 15 = 6.

Page 616 Maintain Your Skills

59. $A = \frac{1}{2}d_1d_2$

$= \frac{1}{2}(20)(26)$

$= 260$ cm^2

60. area = area of upper triangle + area of lower triangle

$= \frac{1}{2}bh_1 + \frac{1}{2}bh_2$

$= \frac{1}{2}(16)(7) + \frac{1}{2}(16)(6)$

$= 104$ m^2

61. Use a 30°-60°-90° triangle. Base $= \frac{1}{\sqrt{3}}(70)$ yd.

$A = bh$

$= \left(\frac{1}{\sqrt{3}}70\right)(70)$

≈ 2829.0 yd^2

62.

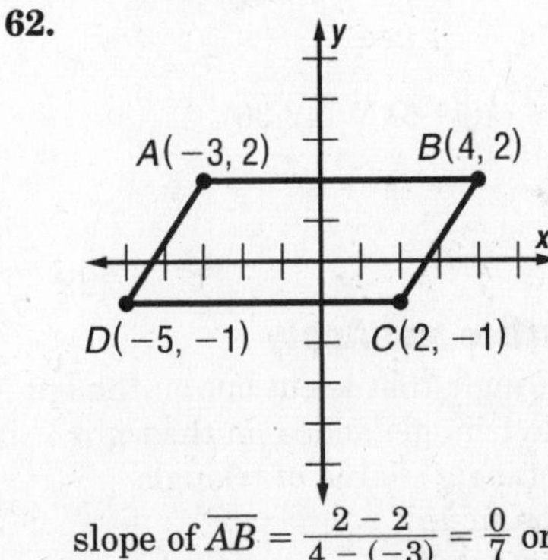

slope of $\overline{AB} = \frac{2-2}{4-(-3)} = \frac{0}{7}$ or 0

slope of $\overline{DC} = \frac{-1-(-1)}{2-(-5)} = \frac{0}{7}$ or 0

slope of $\overline{DA} = \frac{2-(-1)}{-3-(-5)} = \frac{3}{2}$

slope of $\overline{CB} = \frac{2-(-1)}{4-2} = \frac{3}{2}$

$ABCD$ is a parallelogram, since opposite sides have the same slope. Slopes of consecutive sides are *not* negative reciprocals of each other, so the parallelogram is neither a square nor a rectangle.

$A = bh$

$= (7)(3)$

$= 21$ units^2

63.

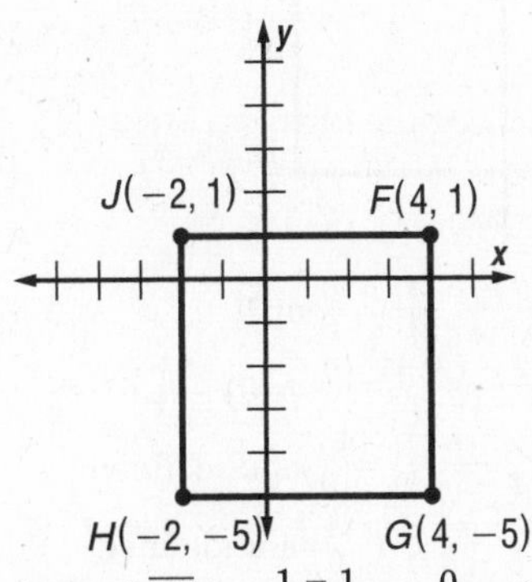

slope of $\overline{JF} = \frac{1-1}{4-(-2)} = \frac{0}{6}$ or 0

slope of $\overline{HG} = \frac{-5-(-5)}{4-(-2)} = \frac{0}{6}$ or 0

slope of $\overline{HJ} = \frac{1-(-5)}{-2-(-2)} = \frac{6}{0}$ is undefined

slope of $\overline{GF} = \frac{1-(-5)}{4-4} = \frac{6}{0}$ is undefined

$FGHJ$ is a square, since the sides are all equal and are all horizontal or vertical.

$A = s^2$

$= (6)^2$

$= 36$ units^2

64.

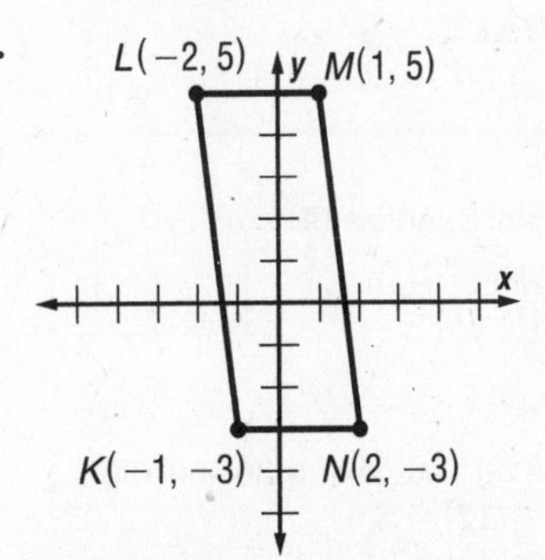

slope of $\overline{KN} = \frac{-3-(-3)}{2-(-1)} = \frac{0}{3}$ or 0

slope of $\overline{LM} = \frac{5-5}{1-(-2)} = \frac{0}{3}$ or 0

slope of $\overline{LK} = \frac{-3-5}{-1-(-2)} = \frac{-8}{1}$ or −8

slope of $\overline{MN} = \frac{-3-5}{2-1} = \frac{-8}{1}$ or −8

$KLMN$ is a parallelogram, since opposite sides have the same slope. Slopes of consecutive sides are *not* negative reciprocals of each other, so the parallelogram is neither a square nor a rectangle.

$A = bh$

$= (3)(8)$

$= 24$ units^2

65.

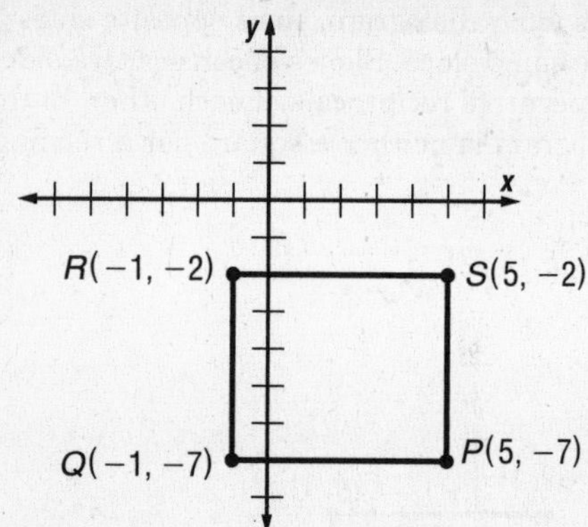

slope of $\overline{QP} = \frac{-7-(-7)}{5-(-1)} = \frac{0}{6}$ or 0

slope of $\overline{RS} = \frac{-2-(-2)}{5-(-1)} = \frac{0}{6}$ or 0

slope of $\overline{QR} = \frac{-2-(-7)}{-1-(-1)} = \frac{5}{0}$ is undefined

slope of $\overline{PS} = \frac{-2-(-7)}{5-5} = \frac{5}{0}$ is undefined

PQRS is a rectangle, since the sides are not all equal but are all horizontal or vertical.

$A = \ell w$
$= (6)(5)$
$= 30$ units^2

66. $HE = \frac{1}{2}(CD + GF)$
$38 = \frac{1}{2}(46 + GF)$
$76 = 46 + GF$
$30 = GF$

67. $WX = \frac{1}{2}(CD + HE)$
$= \frac{1}{2}(46 + 38)$
$= 42$

68. Use $GF = 30$ from Exercise 66.
$YZ = \frac{1}{2}(HE + GF)$
$= \frac{1}{2}(38 + 30)$
$= 34$

69. h is opposite the 30° angle of a 30°-60°-90° triangle, so $h = \frac{1}{2}(12) = 6$.

70. h is adjacent to the 30° angle of a 30°-60°-90° triangle, so $h = \sqrt{3} \cdot 15 = 15\sqrt{3}$.

71. h is one leg of a 45°-45°-90° triangle, so $h = \frac{1}{\sqrt{2}} \cdot 8 = 4\sqrt{2}$.

72. h is one leg of a 45°-45°-90° triangle, so $h = \frac{1}{\sqrt{2}} \cdot 21 = \frac{21\sqrt{2}}{2}$.

11-4 Areas of Irregular Figures

Page 619 Check for Understanding

1. Sample answer:
Area of irregular figure = area of triangle
+ area of rectangle
+ area of semi-circle

$= \frac{1}{2}b_1h + b_2h + \frac{1}{2}\pi r^2$
$= \frac{1}{2}(2)(4) + 2 \cdot 4 + \frac{1}{2}\pi(2)^2$
$= 4 + 8 + 2\pi$
≈ 18.3 units^2

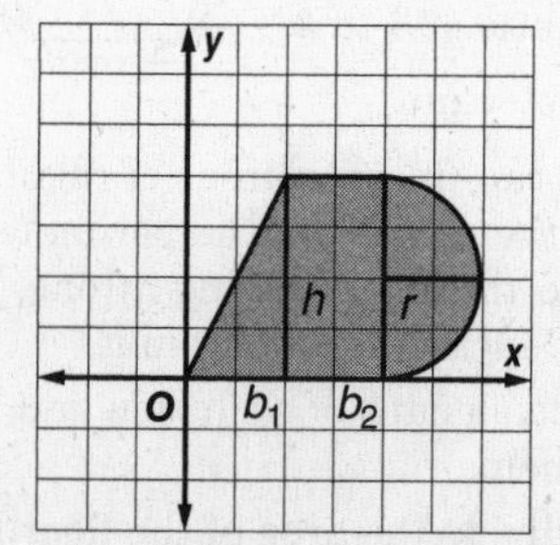

2. An irregular polygon is a polygon in which not all sides are congruent or not all angles are congruent. If a shape can be separated into semicircles or smaller circular regions, it is an irregular figure.

3. area = area of rectangle + area of triangle
$= \ell w + \frac{1}{2}bh$
$= (9.2)(3.6) + \frac{1}{2}(9.2)(8 - 3.6)$
$= 53.36$ or about 53.4 units^2

4. area = area of rectangle + area of semicircle
$= \ell w + \frac{1}{2}\pi r^2$
$= (32)(16) + \frac{1}{2}\pi(8)^2$
$= 512 + 32\pi$ or about 612.5 units^2

5. area of *MNPQ* = area of $\triangle MNQ$ + area of $\triangle QNP$
$= \frac{1}{2}bh_1 + \frac{1}{2}bh_2$
$= \frac{1}{2}(6)(5) + \frac{1}{2}(6)(3)$
$= 24$ units^2

6. area = area of rectangle + combined area of two semicircles
$= \ell w + \pi r^2$
$= (10)(4) + \pi(2)^2$
$= 40 + 4\pi$ or about 52.6 units^2

7. The height of the triangular portion is given by
$h = \sqrt{(24)^2 - \left(\frac{41.6}{2}\right)^2} = \sqrt{143.36}$.

area = area of rectangle + area of triangle
$= \ell w + \frac{1}{2}bh$
$= (41.6)(24) + \frac{1}{2}(41.6)(\sqrt{143.36})$
≈ 1247.4 in^2

Pages 619–621 Practice and Apply

8. Notice that the triangle that is cut out on the left is congruent to the triangle added on the right.
area = area of rectangle − area of triangle
+ area of triangle
= area of rectangle
$= \ell w$
$= (10)(5)$
$= 50$ units^2

9. area = area of rectangle − area of semicircle

$= \ell w - \frac{1}{2}\pi r^2$

$= (8)(12) - \frac{1}{2}\pi(4)^2$

$= 96 - 8\pi$ or about 70.9 units^2

10. area = area of rectangle + area of triangle

$= \ell w_1 + \frac{1}{2}bh$

$= (12)(25) + \frac{1}{2}(10)(8)$

$= 340$ units^2

11. area = area of rectangle + area of triangle

$= \ell w_1 + \frac{1}{2}bh$

$= (62)(54) + \frac{1}{2}(62)(27)$

$= 4185$ units^2

12. area = 2 × area of one trapezoid

$= 2 \cdot \frac{1}{2}h(b_1 + b_2)$

$= h(b_1 + b_2)$

$= 10(8 + 23)$

$= 310$ units^2

13. area = area of rectangle − combined area of two semicircles

$= \ell w - \pi r^2$

$= (22)(14) - \pi(7)^2$

$= 308 - 49\pi$ or about 154.1 units^2

14. The semicircle has radius 18 in. The perimeter is $48 + 36 + 48 + \pi \cdot 18 = 132 + 18\pi$ or about 188.5 in.

15. area = area of rectangle + area of semicircle

$= \ell w + \frac{1}{2}\pi r^2$

$= (36)(48) + \frac{1}{2}\pi(18)^2$

$= 1728 + 162\pi$ or about 2236.9 in^2

16. area = area of square + combined area of two triangles

$= s^2 + 2 \cdot \frac{1}{2}bh$

$= s^2 + bh$

$= (4)^2 + (4)(2)$

$= 24$ units^2

17. area = area of triangle + area of semicircle

$= \frac{1}{2}bh + \frac{1}{2}\pi r^2$

$= \frac{1}{2}(6)(3) + \frac{1}{2}\pi(3)^2$

$= 9 + 4.5\pi$ or about 23.1 units^2

18. area = area of rectangle + area of semicircle − area of semicircle

= area of rectangle

$= \ell w$

$= (5)(4)$

$= 20$ units^2

19.

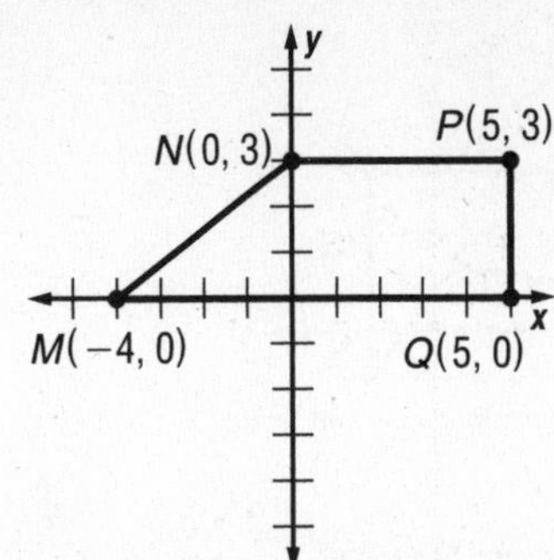

area of trapezoid $= \frac{1}{2}h(b_1 + b_2)$

$= \frac{1}{2}(3)(5 + 9)$

$= 21$ units^2

20.

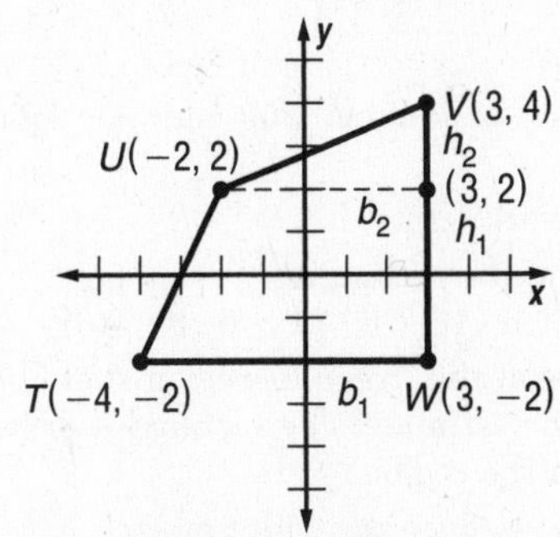

area = area of triangle + area of trapezoid

$= \frac{1}{2}b_2h_2 + \frac{1}{2}h_1(b_1 + b_2)$

$= \frac{1}{2}(5)(2) + \frac{1}{2}(4)(7 + 5)$

$= 29$ units^2

21.

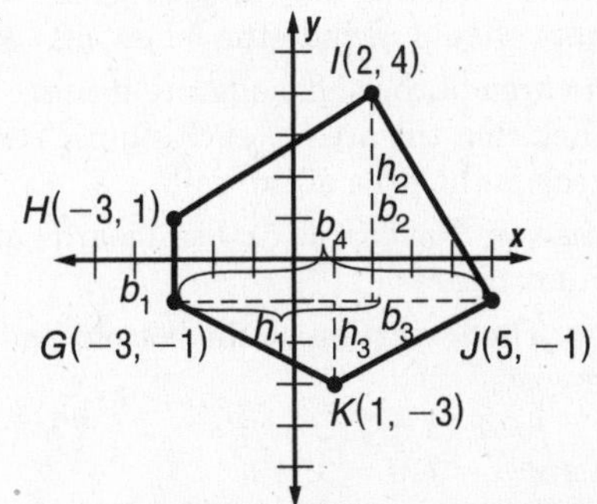

area = area of trapezoid + area of upper triangle + area of lower triangle

$= \frac{1}{2}h_1(b_1 + b_2) + \frac{1}{2}b_3h_2 + \frac{1}{2}b_4h_3$

$= \frac{1}{2}(5)(2 + 5) + \frac{1}{2}(3)(5) + \frac{1}{2}(8)(2)$

$= 33$ units^2

22.

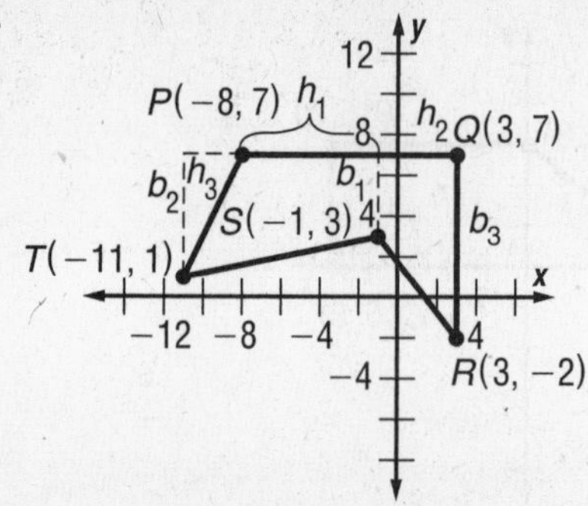

area = area of left trapezoid + area of right trapezoid − area of triangle

$= \frac{1}{2}h_1(b_1 + b_2) + \frac{1}{2}h_2(b_1 + b_3) - \frac{1}{2}b_2h_3$

$= \frac{1}{2}(10)(4 + 6) + \frac{1}{2}(4)(4 + 9) - \frac{1}{2}(6)(3)$

$= 67$ units^2

23. Sample answer: (23 squares)(2500 mi^2 per square) = 57,500 mi^2

24. See students' work.

25. Add the areas of the rectangles.
$6 \cdot 22 + 6 \cdot 20 + 6 \cdot 16 + 6 \cdot 11 + 6 \cdot 8 = 462$

26. The actual area of the irregular region should be smaller than the estimate. The rectangles drawn are larger than the region.

27. Sample answer: Reduce the width of each rectangle.

28. Let $BC = x$. Then the altitude of the triangle is $\frac{\sqrt{3}}{2}x$.

$\frac{\text{area of } \triangle ABC}{\text{area of } BCDE} = \frac{\frac{1}{2}bh}{s^2} = \frac{\frac{1}{2}(x)\left(\frac{\sqrt{3}}{2}x\right)}{x^2} = \frac{\sqrt{3}}{4}$ or $\frac{\sqrt{3}}{4} : 1$

29. Sample answer: Windsurfers use the area of the sail to catch the wind and stay afloat on the water. Answers should include the following.

- To find the area of the sail, separate it into shapes. Then find the area of each shape. The sum of areas is the area of the sail.
- Sample answer: Surfboards and sailboards are also irregular figures.

30. B; let $LM = x$. Then, starting at the bottom and moving clockwise,
$P = 7x + 4x + 4x + 2x + 2x + x + x + x = 22x$.
So $22x = 66$ and $x = LM = 3$.
area = area of A + area of B + area of C
$= s_1^2 + s_2^2 + s_3^2$
$= (12)^2 + (6)^2 + (3)^2$
$= 189$ units^2

31. C; $\sqrt{16} + 9^2 = 4 + 81 = 85$

Page 621 Maintain Your Skills

32. area = area of square − area of circle
$= s^2 - \pi r^2$
$= (14)^2 - \pi(7)^2$
$= 196 - 49\pi$ or about 42.1 units^2

33. area = area of circle − area of triangle
$= \pi r^2 - \frac{1}{2}bh$
$= \pi\left(\frac{1}{\sqrt{2}} \cdot 12\right)^2 - \frac{1}{2}(12)(12)$
$= 72\pi - 72$ or about 154.2 units^2

34. Hexagon side length = 16, and perimeter = $6 \cdot 16$ = 96. Use a 30°-60°-90° triangle to find the apothem = $x\sqrt{3}$ where $x = \frac{1}{2}(16) = 8$.
area = area of circle − area of hexagon
$= \pi r^2 - \frac{1}{2}Pa$
$= \pi(16)^2 - \frac{1}{2}(96)(8\sqrt{3})$
$= 256\pi - 384\sqrt{3}$ or about 139.1 units^2

35. Use a 30°-60°-90° triangle. Base = $\frac{1}{3}(57) = 19$ ft.
Height = $x\sqrt{3}$ where $x = \frac{1}{2}(19) = 9.5$ ft.
$A = \frac{1}{2}bh$
$= \frac{1}{2}(19)(9.5\sqrt{3})$
$= 90.25\sqrt{3}$ or about 156.3 ft^2

36. The area of the rhombus is divided into four congruent right triangles with hypotenuse

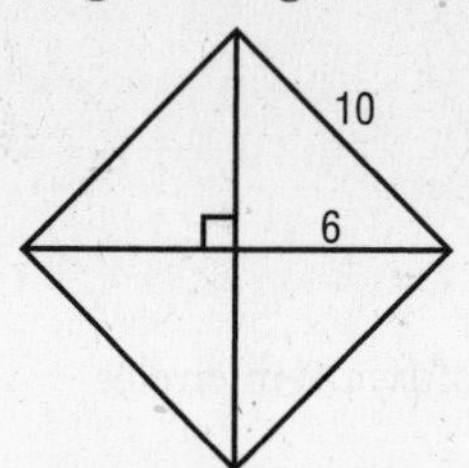

length 10 yd and one side length 6 yd. The other side length is $\sqrt{10^2 - 6^2} = 8$. So the two diagonals measure 12 yd and 2(8) = 16 yd.
$A = \frac{1}{2}d_1d_2$
$= \frac{1}{2}(12)(16)$
$= 96$ yd^2

37. Let the shorter base = $b_1 = x$. Then $b_2 = 2x - 5$ and $P = 90 = x + (2x - 5) + 2(x - 3) = 5x - 11$
$5x - 11 = 90$
$5x = 101$
$x = 20.2$
$A = \frac{1}{2}h(b_1 + b_2)$
$= \frac{1}{2} \cdot 15.43\,(x + 2x - 5)$
$= \frac{1}{2} \cdot 15.43\,(20.2 + 40.4 - 5)$
≈ 429.0 m^2

38. The image of the point lies in quadrant IV, 6 units away from the origin, at $(x, y) = \left(\frac{1}{\sqrt{2}} \cdot 6, -\frac{1}{\sqrt{2}} \cdot 6\right)$ $= (3\sqrt{2}, -3\sqrt{2})$.

39.
$$\begin{array}{r} 0.625 \\ 8\overline{)5.000} \\ \underline{4\,8} \\ 20 \\ \underline{16} \\ 40 \\ \underline{40} \\ 0 \end{array}$$

So $\frac{5}{8} \approx 0.63$.

40.
```
      0.8125
16)13.0000
    12 8
       20
       16
        40
        32
         80
         80
          0
```
So $\frac{13}{16} \approx 0.81$.

41.
```
    0.191...
47)9.000
   4 7
   4 30
   4 23
      70
      47
      23
```
So $\frac{9}{47} \approx 0.19$.

42.
```
     0.476...
21)10.000
    8 4
    1 60
    1 47
      130
      126
        4
```
So $\frac{10}{21} \approx 0.48$.

Page 621 Practice Quiz 2

1.

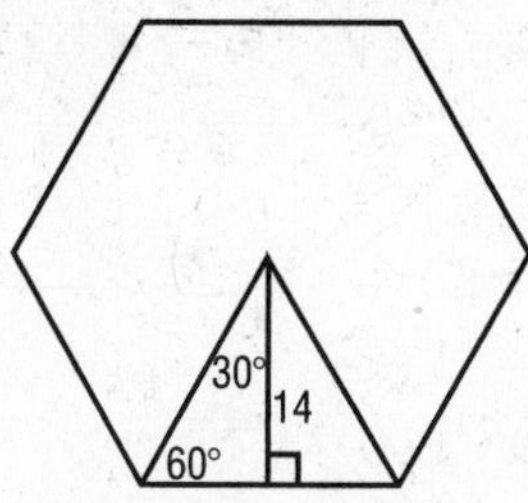

side length $= \frac{2}{\sqrt{3}} \cdot 14 = \frac{28}{\sqrt{3}}$, and

perimeter $= 6 \cdot \frac{28}{\sqrt{3}} = 56\sqrt{3}$

$A = \frac{1}{2}Pa$

$= \frac{1}{2}(56\sqrt{3})(14)$

$= 392\sqrt{3}$ or about 679.0 mm^2

2. side length = 72 ÷ 8 or 9 in., and $\frac{1}{2}(9) = 4.5$

central angle = 360° ÷ 8 or 45°, and $\frac{1}{2}(45°) = 22.5°$.

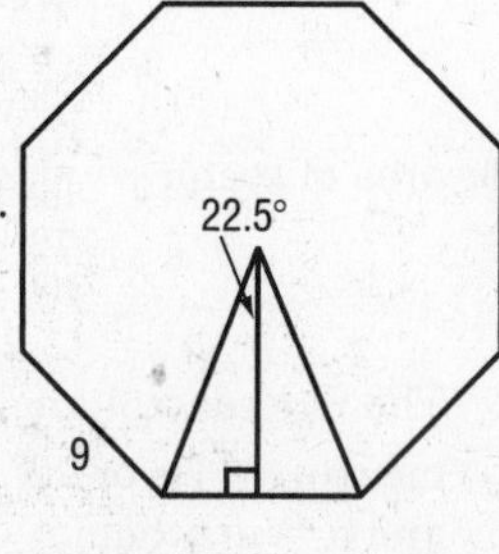

apothem $= \frac{4.5}{\tan 22.5°}$

≈ 10.864 in.

$A = \frac{1}{2}Pa$

$\approx \frac{1}{2}(72)(10.864)$

≈ 391.1 in^2

3. Use a 30°-60°-90° triangle. Half the base of the triangle is 42, so the apothem, which is also the radius of the circle is $\frac{42}{\sqrt{3}} = 14\sqrt{3}$.

Triangle perimeter = 3(84) = 252

shaded area = area of triangle − area of circle

$= \frac{1}{2}Pa - \pi r^2$

$= \frac{1}{2}(252)(14\sqrt{3}) - \pi(14\sqrt{3})^2$

$= 1764\sqrt{3} - 588\pi$

or about 1208.1 units^2

4. pentagon central angle = 360° ÷ 5 or 72°, and $\frac{1}{2}(72°) = 36°$.

half side length = 21 tan 36°, and

side length = 42 tan 36°, so

perimeter = 5(42 tan 36°) = 210 tan 36°

≈ 152.574.

shaded area = area of pentagon − area of circle

$= \frac{1}{2}Pa - \pi r^2$

$\approx \frac{1}{2}(152.574)(21) - \pi(21)^2$

≈ 216.6 units^2

5.

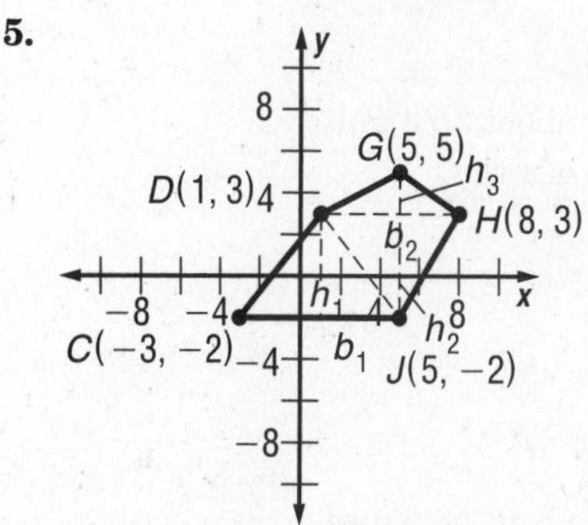

area = area of left triangle + area of middle triangle + area of upper triangle

$= \frac{1}{2}b_1h_1 + \frac{1}{2}b_2h_2 + \frac{1}{2}b_2h_3$

$= \frac{1}{2}(8)(5) + \frac{1}{2}(7)(5) + \frac{1}{2}(7)(2)$

$= 44.5$ units^2

11-5 Geometric Probability

Page 625 Check for Understanding

1. Multiply the measure of the central angle of the sector by the area of the circle and then divide the product by 360°.
2. Sample answer: darts, archery, shuffleboard
3. Rachel; Taimi did not multiply $\frac{62}{360}$ by the area of the circle.
4. $A = \frac{N}{360}\pi r^2$

$= \frac{80}{360}\pi(5^2)$

$= \frac{50}{9}\pi$ or about 17.5 units^2

$P(\text{blue}) = \frac{\text{blue area}}{\text{area of circle}}$

$= \frac{\frac{50}{9}\pi}{\pi(5^2)} = \frac{2}{9}$ or about 0.22

5. short side of one triangle = 10,
so area of square $= 4 \cdot \frac{1}{2}bh = 4 \cdot \frac{1}{2}(10)(10)$ or 200
area of circle $= \pi r^2 = \pi\left(10^2\right) = 100\pi$
blue area = area of circle − area of square
$= 100\pi - 200$ or about 114.2 units^2

$P(\text{blue}) = \frac{\text{blue area}}{\text{area of circle}}$
$= \frac{100\pi - 200}{100\pi}$
$= 1 - \frac{2}{\pi}$ or about 0.36

6. 60 out of 100 squares are shaded.
$P(\text{shaded}) = \frac{60}{100} = \frac{3}{5}$ or 0.6

Pages 625–627 Practice and Apply

7. 60 out of 100 squares are shaded.
$P(\text{shaded}) = \frac{60}{100} = 0.60$

8. 50 out of 100 squares are shaded.
$P(\text{shaded}) = \frac{50}{100} = 0.50$

9. 54 out of 100 squares are shaded.
$P(\text{shaded}) = \frac{54}{100} = 0.54$

10. $A = \frac{N}{360}\pi r^2$
$= \frac{72}{360}\pi(7.5^2)$
$= 11.25\pi$ or about 35.3 units^2

$P(\text{blue}) = \frac{\text{area of sector}}{\text{area of circle}}$
$= \frac{11.25\pi}{\pi(7.5^2)}$
$= 0.20$

11. $A = \frac{N}{360}\pi r^2$
$= \frac{60 + 60}{360}\pi\left(7.5^2\right)$
$= 18.75\pi$ or about 58.9 units^2

$P(\text{pink}) = \frac{18.75\pi}{\pi(7.5^2)}$
$= \frac{1}{3}$ or about 0.33

12. $A = \frac{N}{360}\pi r^2$
$= \frac{45 + 45 + 45}{360}\pi(7.5^2)$
≈ 66.3 units^2

$P(\text{purple}) = \frac{\text{purple area}}{\text{area of circle}}$
$= \frac{66.3}{\pi(7.5^2)}$
≈ 0.375

13. $A = \frac{N}{360}\pi r^2$
$= \frac{40}{360}\pi(7.5^2)$
$= 6.25\pi$ or about 19.6 units^2

$P(\text{red}) = \frac{\text{red area}}{\text{area of circle}}$
$= \frac{6.25\pi}{\pi(7.5^2)}$
$= \frac{1}{9}$ or about 0.11

14. $A = \frac{N}{360}\pi r^2$
$= \frac{55 + 58 + 60}{360}\pi(7.5^2)$
≈ 84.9 units^2

$P(\text{green}) = \frac{\text{green area}}{\text{area of circle}}$
$\approx \frac{84.9}{\pi(7.5^2)}$
≈ 0.48

15. $A = \frac{N}{360}\pi r^2$
$= \frac{72 + 80}{360}\pi(7.5^2)$
$= 23.75\pi$ or about 74.6 units^2

$P(\text{yellow}) = \frac{\text{yellow area}}{\text{area of circle}}$
$= \frac{23.75\pi}{\pi(7.5^2)}$
≈ 0.42

16. Using $A = \pi r^2$, we find that the smallest circle encloses π yd^2, the medium circle 4π yd^2, and the large circle 9π yd^2.
shaded area $= 9\pi - 4\pi + \pi = 6\pi$ yd^2

$P(\text{shaded}) = \frac{\text{shaded area}}{\text{overall area}}$
$= \frac{6\pi}{9\pi}$ or $\frac{2}{3}$

17. area of sector $= \frac{N}{360}\pi r^2$
$= \frac{60}{360}\pi(6^2)$
$= 6\pi$

Use a 30°-60°-90° triangle, with the length of the short leg $= \frac{1}{2}(6) = 3$ so that apothem $= 3\sqrt{3}$.
area of triangle $= \frac{1}{2}bh$
$= \frac{1}{2}(6)\left(3\sqrt{3}\right)$
$= 9\sqrt{3}$

area of segment = area of sector − area of triangle
$= 6\pi - 9\sqrt{3}$ or about 3.3 units^2

$P(\text{shaded}) = \frac{\text{area of segment}}{\text{area of circle}}$
$= \frac{6\pi - 9\sqrt{3}}{\pi(6^2)}$
$= \frac{1}{6} - \frac{\sqrt{3}}{4\pi}$ or about 0.03

18. area of sector $= \frac{N}{360}\pi r^2$
$= \frac{120}{360}\pi(8^2)$
$= \frac{64}{3}\pi$

Use a 30°-60°-90° triangle, with apothem $= \frac{1}{2}(8) = 4$.
area of $\triangle ABC = \frac{1}{2}bh$
$= \frac{1}{2}(2 \cdot 4\sqrt{3})(4)$
$= 16\sqrt{3}$

area of segment = area of sector − area of triangle
$= \frac{64}{3}\pi - 16\sqrt{3}$ or about 39.3 units^2

$P(\text{shaded}) = \frac{\text{area of segment}}{\text{area of circle}}$
$= \frac{\frac{64}{3}\pi - 16\sqrt{3}}{\pi(8^2)}$
$= \frac{1}{3} - \frac{\sqrt{3}}{4\pi}$ or about 0.20

19. area of sector $= \frac{N}{360}\pi r^2$
$= \frac{72}{360}\pi(7.5^2)$
$= 11.25\pi$

The five central angles of the pentagon each measure $\frac{360°}{5}$ or 72°. Let s = pentagon side length and a = apothem.

$7.5 \sin \frac{72°}{2} = \frac{s}{2}$
$s = 15 \sin 36°$
$7.5 \cos \frac{72°}{2} = a$
$a = 7.5 \cos 36°$

$$\begin{aligned}\text{area of triangle} &= \tfrac{1}{2}bh \\ &= \tfrac{1}{2}sa \\ &= \tfrac{1}{2}(15 \sin 36°)(7.5 \cos 36°) \\ &\approx 26.7\end{aligned}$$

$$\begin{aligned}\text{area of segment} &= \text{area of sector} \\ &\quad - \text{area of triangle} \\ &\approx 11.25\pi - 26.7 \\ &\approx 8.6\end{aligned}$$

$$\begin{aligned}\text{shaded area} &= 3(\text{area of segment}) \\ &\approx 25.8\end{aligned}$$

$$\begin{aligned}P(\text{shaded}) &= \frac{\text{shaded area}}{\text{area of circle}} \\ &\approx \frac{25.8}{\pi(7.5^2)} \\ &\approx 0.15\end{aligned}$$

In Exercises 20–23,

$$\begin{aligned}\textbf{probability} &= \frac{\textbf{area of region}}{\textbf{area of circle}} \\ &= \frac{\textbf{total angle in region}}{\textbf{360°}}.\end{aligned}$$

20. $P(\text{red}) = \frac{28.8°}{360°} = 0.08$

21. $P(\text{blue or green}) = \frac{147.6° + 97.2°}{360°} = 0.68$

22. $$\begin{aligned}P(\text{not red or blue}) &= 1 - P(\text{red or blue}) \\ &= 1 - \frac{28.8° + 147.6°}{360°} \\ &= 0.51\end{aligned}$$

23. $$\begin{aligned}P(\text{not orange or green}) &= 1 - P(\text{orange or green}) \\ &= 1 - \frac{18° + 97.2°}{360°} \\ &= 0.68\end{aligned}$$

24. $$\begin{aligned}\text{court area} &= \ell w \\ &= (39)(39) \\ &= 1521 \text{ ft}^2\end{aligned}$$

$\text{out-of-bound lane width} = \frac{39 - 27}{2} = 6 \text{ ft}$

$$\begin{aligned}\text{out-of-bounds area} &= \ell w + \ell w \\ &= 6 \cdot 39 + 6 \cdot 39 \\ &= 468 \text{ ft}^2\end{aligned}$$

$$\begin{aligned}\text{probability} &= \frac{\text{out-of-bounds area}}{\text{court area}} \\ &= \frac{468}{1521} \\ &= \frac{4}{13} \text{ or about } 0.31\end{aligned}$$

25. $\text{service box width} = \frac{27}{2} = 13.5 \text{ ft}$

$$\begin{aligned}\text{service box area} &= \ell_1 w_1 \\ &= (13.5)(21) \\ &= 283.5 \text{ ft}^2\end{aligned}$$

$$\begin{aligned}\text{court area} &= \ell_2 w_2 \\ &= (39)(39) \\ &= 1521 \text{ ft}^2\end{aligned}$$

$$\begin{aligned}\text{probability} &= \frac{\text{service box area}}{\text{court area}} \\ &= \frac{283.5}{1521} \\ &\approx 0.19\end{aligned}$$

In Exercises 26–28, the overall area is $\pi(7^2) = 49\pi$ units^2. Also,

$$\begin{aligned}\textbf{red area} &= \pi \cdot 1^2 + (\pi \cdot 4^2 - \pi \cdot 3^2) + (\pi \cdot 7^2 - \pi \cdot 6^2) \\ &= 21\pi \textbf{ units}^2\end{aligned}$$

$$\begin{aligned}\textbf{white area} &= \tfrac{1}{2}(\textbf{overall area} - \textbf{red area}) \\ &= \tfrac{1}{2}(49\pi - 21\pi) \\ &= 14\pi \textbf{ units}^2\end{aligned}$$

$$\begin{aligned}\textbf{black area} &= \textbf{white area} \\ &= 14\pi \textbf{ units}^2\end{aligned}$$

26. $$\begin{aligned}P(\text{black}) &= \frac{\text{black area}}{\text{overall area}} \\ &= \frac{14\pi}{49\pi} \\ &= \frac{2}{7} \text{ or about } 0.29\end{aligned}$$

27. $$\begin{aligned}P(\text{white}) &= \frac{\text{white area}}{\text{overall area}} \\ &= \frac{14\pi}{49\pi} \\ &= \frac{2}{7} \text{ or about } 0.29\end{aligned}$$

28. $$\begin{aligned}P(\text{red}) &= \frac{\text{red area}}{\text{overall area}} \\ &= \frac{21\pi}{49\pi} \\ &= \frac{3}{7} \text{ or about } 0.43\end{aligned}$$

29. The chances of landing on a black or white sector are the same, so they should have the same point value.

30. Of the three colors, there is the highest probability of landing on red, so red should have a lower point value than white or black.

31a. No; each colored sector has a different central angle.

31b. No; there is not an equal chance of landing on each color.

32. Sample answer: Geometric probability can help you determine the chance of a dart landing on the bullseye or high scoring sector. Answers should include the following.

- Find the area of the circles containing the red sector. Divide the difference by the area of the larger circle.
- Find the area of the center circle and divide by the area of the largest circle on the board.

33. C; $$\begin{aligned}\text{total area} &= \text{area of square} + \text{area of semicircle} \\ &= s^2 + \tfrac{1}{2}\pi r^2 \\ &= 5^2 + \tfrac{1}{2}\pi(2.5^2) \\ &\approx 34.8\end{aligned}$$

$$\begin{aligned}\text{shaded area} &= \text{area of square} - \text{area of semicircle} \\ &= 5^2 - \tfrac{1}{2}\pi(2.5^2) \\ &\approx 15.2\end{aligned}$$

$$\begin{aligned}P(\text{shaded}) &= \frac{\text{shaded area}}{\text{total area}} \\ &\approx \frac{15.2}{34.8} \\ &\approx 0.44\end{aligned}$$

34. C; $y = 16 \div 4 = 4$, so $12 \div y = 12 \div 4 = 3$.

35. area = area of rectangle + area of left triangle + area of upper triangle

$= \ell w + \frac{1}{2}b_1h_1 + \frac{1}{2}b_2h_2$

$= (28)(20) + \frac{1}{2}(16)(35) + \frac{1}{2}(28)(15)$

$= 1050$ units^2

36. area = area of rectangle − area of semicircle

$= \ell w - \frac{1}{2}\pi r^2$

$= (12)(9) - \frac{1}{2}\pi(4)^2$

$= 108 - 8\pi$ or about 82.9 ft^2

37. Side length $= \frac{1}{3}(48) = 16$ ft. Use a 30°-60°-90° triangle. Height $= x\sqrt{3}$ where $x = \frac{1}{2}(16) = 8$ ft.

$A = \frac{1}{2}bh$

$= \frac{1}{2}(16)(8\sqrt{3})$

$= 64\sqrt{3}$ or about 110.9 ft^2

38. $A = s^2$

$= (21)^2$

$= 441$ cm^2

39. Use a 30°-60°-90° triangle.

Half side length $= \frac{1}{\sqrt{3}}(8)$, side length $= \frac{1}{\sqrt{3}}(16)$,

perimeter $= 6 \cdot \frac{1}{\sqrt{3}}(16) = 32\sqrt{3}$ in.

$A = \frac{1}{2}Pa$

$= \frac{1}{2}(32\sqrt{3})(8)$

$= 128\sqrt{3}$ or about 221.7 in^2

40. $m\angle AFB + m\angle BFC = 180$

$m\angle AFB + 72 = 180$

$m\angle AFB = 108$

41. $m\angle CFD + m\angle AFD = 180$

$(4a - 1) + (2a - 5) = 180$

$6a - 6 = 180$

$6a = 186$

$a = 31$

$m\angle CFD = 4a - 1$

$= 4(31) - 1$

$= 123$

42. $m\angle CFD + m\angle AFD = 180$

$(4a - 1) + (2a - 5) = 180$

$6a - 6 = 180$

$6a = 186$

$a = 31$

$m\angle AFD = 2a - 5$

$= 2(31) - 5$

$= 57$

43. From Exercises 40 and 42, $m\angle AFB = 108$ and $m\angle AFD = 57$.

$m\angle DFB = m\angle AFB + m\angle AFD$

$= 108 + 57$

$= 165$

44. Use the Law of Cosines, with $a = p$, $b = 6.8$, $c = 11.1$, and $A = 57$.

$a^2 = b^2 + c^2 - 2bc \cos A$

$p^2 = 6.8^2 + 11.1^2 - 2(6.8)(11.1)\cos 57°$

$p^2 = 169.45 - 150.96 \cos 57°$

$p = \sqrt{169.45 - 150.96 \cos 57°}$

$p \approx 9.3$

45. Use the Law of Cosines, with $a = g$, $b = 32$, $c = 29$, and $A = 41$.

$a^2 = b^2 + c^2 - 2bc \cos A$

$g^2 = 32^2 + 29^2 - 2(32)(29) \cos 41°$

$g^2 = 1865 - 1856 \cos 41°$

$g = \sqrt{1865 - 1856 \cos 41°}$

$g \approx 21.5$

Chapter 11 Study Guide and Review

Pages 628–630

1. c

2. e

3. a

4. f

5. b

6. d

7. $P = 2(23) + 2(16) = 78$ ft

Use a 30°-60°-90° triangle to find the height, with x as the length of the shorter leg.

$16 = 2x$

$8 = x$

height $= x\sqrt{3} = 8\sqrt{3}$ ft

$A = bh$

$= (23)(8\sqrt{3})$

$= 184\sqrt{3}$ or about 318.7 ft^2

8. $P = 2(36) + 2(22) = 116$ mm

Use a 30°-60°-90° triangle to find the height, with x as the length of the shorter leg.

$22 = 2x$

$11 = x =$ height

$A = bh$

$= (36)(11)$

$= 396$ mm^2

9.

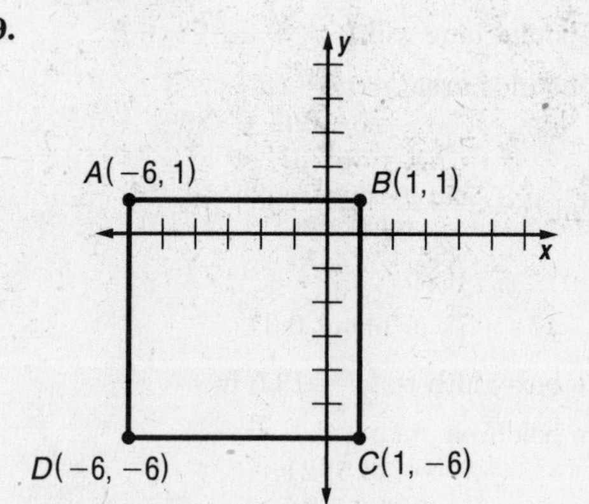

$ABCD$ is a square.

$A = s^2$

$= 7^2$

$= 49$ units^2

10.

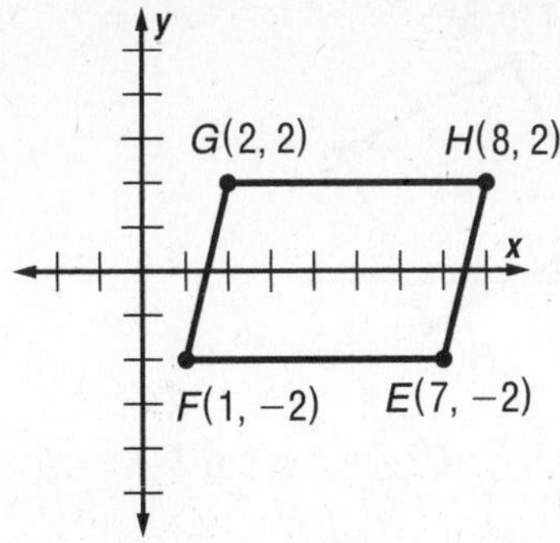

slope of $\overline{FG} = \frac{2-(-2)}{2-1} = \frac{4}{1}$ or 4

slope of $\overline{EH} = \frac{2-(-2)}{8-7} = \frac{4}{1}$ or 4

slope of $\overline{FE} = \frac{-2-(-2)}{7-1} = \frac{0}{6}$ or 0

slope of $\overline{GH} = \frac{2-2}{8-2} = \frac{0}{6}$ or 0

EFGH is a parallelogram, since opposite sides have the same slope. Slopes of consecutive sides are *not* negative reciprocals of each other, so the parallelogram is neither a square nor a rectangle.

$A = bh$
$= 6 \cdot 4$
$= 24$ units^2

11.

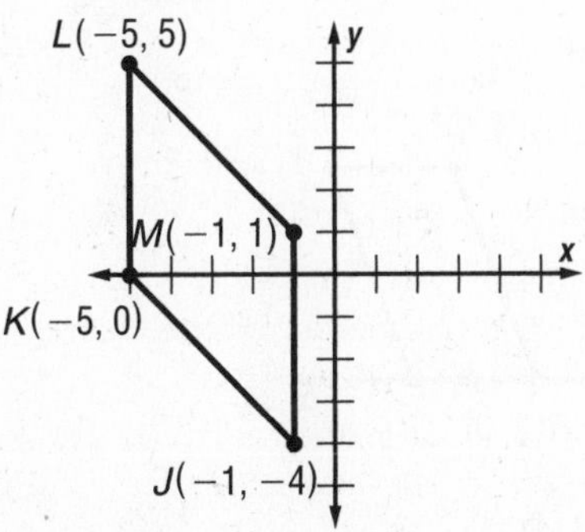

slope of $\overline{JK} = \frac{0-(-4)}{-5-(-1)} = \frac{4}{-4}$ or −1

slope of $\overline{LM} = \frac{1-5}{-1-(-5)} = \frac{-4}{4}$ or −1

slope of $\overline{JM} = \frac{1-(-4)}{-1-(-1)} = \frac{5}{0}$ undefined

slope of $\overline{KL} = \frac{5-0}{-5-(-5)} = \frac{5}{0}$ undefined

JKLM is a parallelogram, since opposite sides have the same slope. Slopes of consecutive sides are *not* negative reciprocals of each other, so the parallelogram is neither a square nor a rectangle.

$A = bh$
$= 4 \cdot 5$
$= 20$ units^2

12.

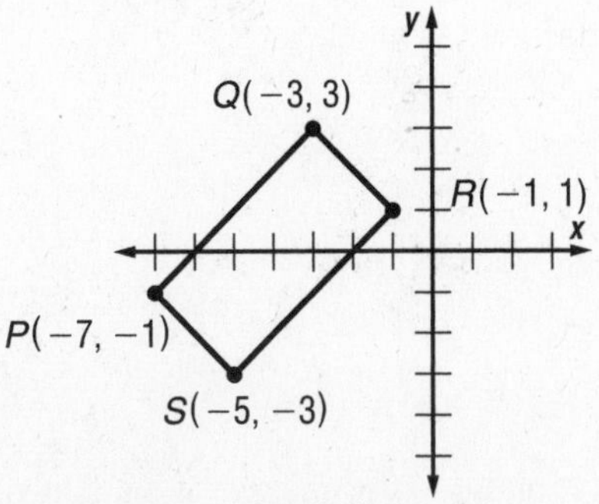

Q(−3, 3) R(−1, 1) P(−7, −1) S(−5, −3)

slope of $\overline{PQ} = \frac{3-(-1)}{-3-(-7)} = \frac{4}{4}$ or 1

slope of $\overline{RS} = \frac{-3-1}{-5-(-1)} = \frac{-4}{-4}$ or 1

slope of $\overline{PS} = \frac{-3-(-1)}{-5-(-7)} = \frac{-2}{2}$ or −1

slope of $\overline{QR} = \frac{1-3}{-1-(-3)} = \frac{-2}{2}$ or −1

PQRS is a rectangle, since opposite sides have the same slope and the slopes of consecutive sides are negative reciprocals of each other.

Length of $\overline{SR} = \sqrt{[-1-(-5)]^2 + [1-(-3)]^2}$
$= 4\sqrt{2}$

Length of $\overline{QR} = \sqrt{[-1-(-3)]^2 + [1-3]^2} = 2\sqrt{2}$

PQRS is not a square since not all sides have the same length.

$A = \ell w$
$= (4\sqrt{2})(2\sqrt{2})$
$= 16$ units^2

13. $A = \frac{1}{2}bh$
$336 = \frac{1}{2}b(24)$
$336 = 12b$
$28 = b$
base $= CE = 28$ in.

14. $A = \frac{1}{2}h(b_1 + b_2)$
$75 = \frac{1}{2}h(17 + 13)$
$75 = 15h$
$5 = h$
The height is 5 m.

15.

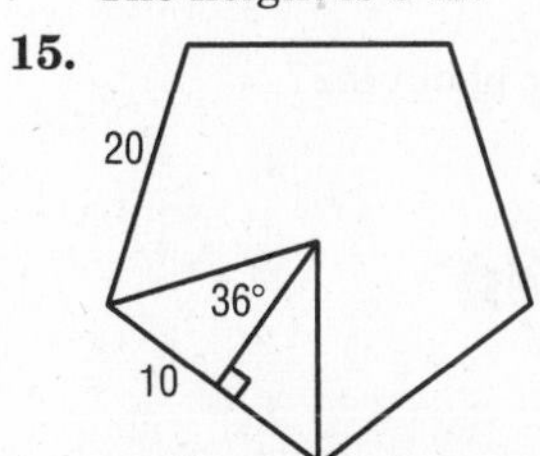

side length $= 100 \div 5 = 20$, and $\frac{1}{2}(20) = 10$

central angle $= 360 \div 5 = 72°$, and $\frac{1}{2}(72°) = 36°$

apothem $= \frac{10}{\tan 36°} \approx 13.764$

$A = \frac{1}{2}Pa$
$\approx \frac{1}{2}(100)(13.764)$
≈ 688.2 in^2

16.

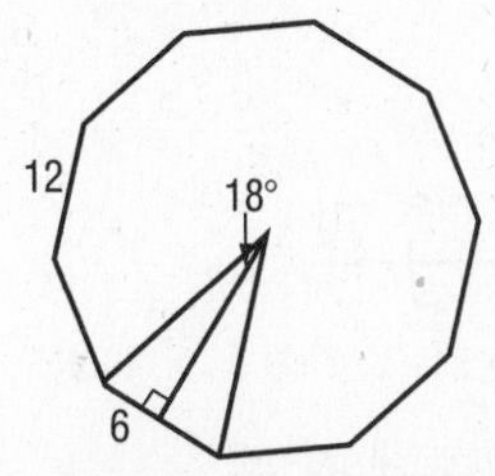

half side length $= \frac{1}{2}(12) = 6$ mm

perimeter $= 10 \cdot 12 = 120$ mm

central angle $= 360° \div 10 = 36°$, and $\frac{1}{2}(36°) = 18°$

apothem $= \frac{6}{\tan 18°} \approx 18.466$

$A = \frac{1}{2}Pa$
$\approx \frac{1}{2}(120)(18.466)$
≈ 1108.0 mm^2

17. area = area of rectangle
+ combined area of two semicircles
$= \ell w + \pi r^2$
$= (8)(3) + \pi(1.5)^2$
$= 24 + 2.25\pi$ or about 31.1 units^2

18. For the height of the trapezoid, use a 30°-60°-90° triangle: $x = 4$, so $h = 4\sqrt{3}$.
area = area of trapezoid + area of semicircle
$= \frac{1}{2}h(b_1 + b_2) + \frac{1}{2}\pi r^2$
$= \frac{1}{2}(4\sqrt{3})(8 + 10) + \frac{1}{2}\pi(4)^2$
$= 36\sqrt{3} + 8\pi$ or about 87.5 units^2

19. $A = \frac{N}{360}\pi r^2$
$= \frac{120}{360}\pi(6^2)$
$= 12\pi$

$P(\text{red}) = \frac{\text{red area}}{\text{area of circle}}$
$= \frac{12\pi}{\pi(6^2)}$
$= \frac{1}{3}$ or about 0.33

20. $A = \frac{N}{360}\pi r^2$
$= \frac{36 + 60}{360}\pi(6^2)$
$= 9.6\pi$

$P(\text{purple or green}) = \frac{\text{purple or green area}}{\text{area of circle}}$
$= \frac{9.6\pi}{\pi(6^2)}$
$= \frac{4}{15}$ or about 0.27

Chapter 11 Practice Test

Page 631

1. a

2. c

3. b

4.

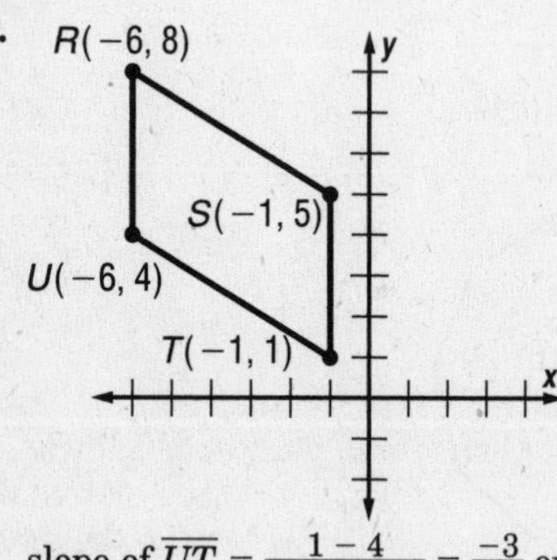

slope of $\overline{UT} = \frac{1 - 4}{-1 - (-6)} = \frac{-3}{5}$ or $-\frac{3}{5}$

slope of $\overline{RS} = \frac{5 - 8}{-1 - (-6)} = \frac{-3}{5}$ or $-\frac{3}{5}$

slope of $\overline{UR} = \frac{8 - 4}{-6 - (-6)} = \frac{4}{0}$ is undefined

slope of $\overline{TS} = \frac{5 - 1}{-1 - (-1)} = \frac{4}{0}$ is undefined

$RSTU$ is a parallelogram, since opposite sides have the same slope. Slopes of consecutive sides are *not* negative reciprocals of each other, so the parallelogram is neither a square nor a rectangle.
$A = bh$
$= (5)(4)$
$= 20$ units^2

5.

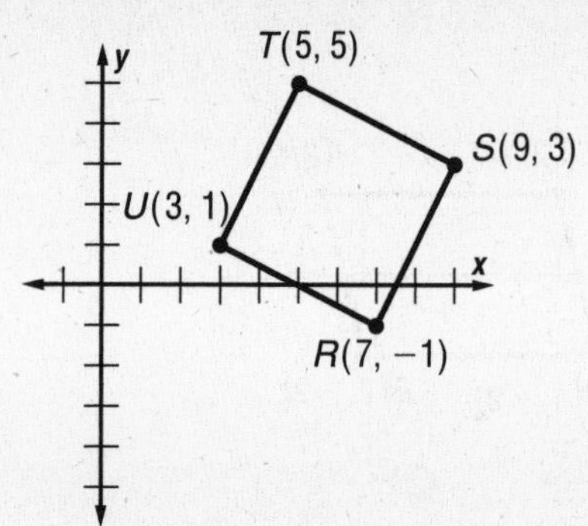

slope of $\overline{UR} = \frac{-1 - 1}{7 - 3} = \frac{-2}{4}$ or $-\frac{1}{2}$

slope of $\overline{TS} = \frac{3 - 5}{9 - 5} = \frac{-2}{4}$ or $-\frac{1}{2}$

slope of $\overline{UT} = \frac{5 - 1}{5 - 3} = \frac{4}{2}$ or 2

slope of $\overline{RS} = \frac{3 - (-1)}{9 - 7} = \frac{4}{2}$ or 2

$RSTU$ is a rectangle, since opposite sides have the same slope and the slopes of consecutive sides are negative reciprocals of each other.
Length of $\overline{UT} = \sqrt{(5 - 3)^2 + (5 - 1)^2} = 2\sqrt{5}$
Length of $\overline{UR} = \sqrt{(7 - 3)^2 + (-1 - 1)^2} = 2\sqrt{5}$
Thus $RSTU$ is in fact a square.
$A = s^2$
$= (2\sqrt{5})^2$
$= 20$ units^2

6.

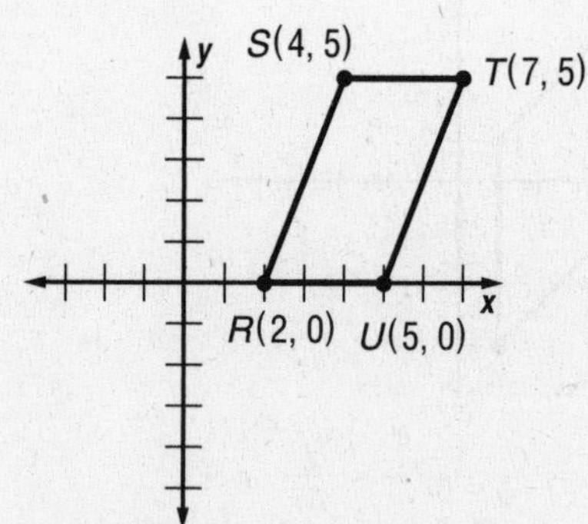

slope of $\overline{RU} = \frac{0 - 0}{5 - 2} = \frac{0}{3}$ or 0

slope of $\overline{ST} = \frac{5 - 5}{7 - 4} = \frac{0}{3}$ or 0

slope of $\overline{UT} = \frac{5 - 0}{7 - 5} = \frac{5}{2}$

slope of $\overline{RS} = \frac{5 - 0}{4 - 2} = \frac{5}{2}$

$RSTU$ is a parallelogram, since opposite sides have the same slope. Slopes of consecutive sides are *not* negative reciprocals of each other, so the parallelogram is neither a square nor a rectangle.
$A = bh$
$= (3)(5)$
$= 15$ units^2

7.

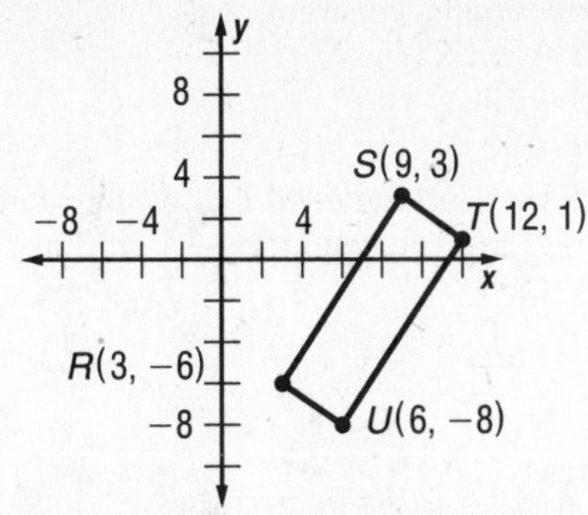

slope of $\overline{RU} = \frac{-8-(-6)}{6-3} = \frac{-2}{3}$ or $-\frac{2}{3}$

slope of $\overline{ST} = \frac{1-3}{12-9} = \frac{-2}{3}$ or $-\frac{2}{3}$

slope of $\overline{RS} = \frac{3-(-6)}{9-3} = \frac{9}{6}$ or $\frac{3}{2}$

slope of $\overline{UT} = \frac{1-(-8)}{12-6} = \frac{9}{6}$ or $\frac{3}{2}$

RSTU is a rectangle, since opposite sides have the same slope and the slopes of consecutive sides are negative reciprocals of each other.

Length of $\overline{RU} = \sqrt{[6-3]^2 + [-8-(-6)]^2} = \sqrt{13}$

Length of $\overline{RS} = \sqrt{[9-3]^2 + [3-(-6)]^2} = 3\sqrt{13}$

RSTU is not a square since not all sides have the same length.

$A = \ell w$
$= (\sqrt{13})(3\sqrt{13})$
$= 39$ units^2

8. area = area of left triangle + area of right triangle
$= \frac{1}{2}b_1h_1 + \frac{1}{2}b_2h_2$
$= \frac{1}{2}(17)(6) + \frac{1}{2}(28)(15)$
$= 261$ m^2

9. area = area of rectangle + area of triangle
$= \ell w + \frac{1}{2}bh$
$= (39)(18) + \frac{1}{2}(56 - 39)(18)$
$= 855$ yd^2

10. area $= \frac{1}{2}h(b_1 + b_2)$
$= \frac{1}{2}(22)[32 + (3 + 32 + 7)]$
$= 814$ cm^2

11. octagon central angle = 360° ÷ 8 = 45°, and $\frac{1}{2}(45°) = 22.5°$
half side length = 3 tan 22.5°, and side length = 6 tan 22.5°, so
perimeter = 8(6 tan 22.5°) = 48 tan 22.5°
≈ 19.88 ft.
$A = \frac{1}{2}Pa$
$\approx \frac{1}{2}(19.88)(3)$
≈ 29.8 ft^2

12. side length = 115 ÷ 5 = 23 cm, and $\frac{1}{2}(23) = 11.5$
central angle = 360° ÷ 5 = 72°, and $\frac{1}{2}(72°) = 36°$
apothem $= \frac{11.5}{\tan 36°} \approx 15.828$
$A = \frac{1}{2}Pa$
$\approx \frac{1}{2}(115)(15.828)$
≈ 910.1 cm^2

13. $A = \frac{N}{360}\pi r^2$
$= \frac{40 + 60}{360}\pi(6^2)$
$= 10\pi$
$P(\text{red}) = \frac{\text{red area}}{\text{area of circle}}$
$= \frac{10\pi}{\pi(6^2)}$
$= \frac{5}{18}$ or about 0.28

14. $A = \frac{N}{360}\pi r^2$
$= \frac{86}{360}\pi(6^2)$
$= 8.6\pi$
$P(\text{orange}) = \frac{\text{orange area}}{\text{area of circle}}$
$= \frac{8.6\pi}{\pi(6^2)}$
≈ 0.24

15. $A = \frac{N}{360}\pi r^2$
$= \frac{45 + 35 + 55}{360}\pi(6^2)$
$= 13.5\pi$
$P(\text{green}) = \frac{\text{green area}}{\text{area of circle}}$
$= \frac{13.5\pi}{\pi(6^2)}$
$= \frac{3}{8}$ or about 0.38

16. area = area of trapezoid + area of parallelogram
$= \frac{1}{2}h(b_1 + b_2) + b_1h_2$
$= \frac{1}{2}(8)(21 + 24) + (21)(14)$
$= 474$ units^2

17. triangle height $= x\sqrt{3}$, where $x = \frac{1}{2}(6) = 3$
area = area of rectangles + area of triangles
$= 2\ell w + 2 \cdot \frac{1}{2}bh$
$= 2(6)(5) + 2 \cdot \frac{1}{2}(6)(3\sqrt{3})$
$= 60 + 18\sqrt{3}$ or about 91.2 units^2

18. area = area of square
+ combined area of semicircles
$= s^2 + 2 \cdot \frac{1}{2}\pi r^2$
$= (7)^2 + \pi(3.5)^2$
$= 49 + 12.25\pi$ or about 87.5 units^2

19. side length = 9 ÷ 6 or 1.5 in., and $\frac{1}{2}(1.5) = 0.75$
apothem $= 0.75\sqrt{3}$
$A = \frac{1}{2}Pa$
$= \frac{1}{2}(9)(0.75\sqrt{3})$
$= 3.375\sqrt{3}$ or about 5.8 in^2

20. D; two sides are horizontal, lying on the lines $y = -1$ and $y = 4$. The lengths of the horizontal sides are $|-3 - 5| = |-1 - 7| = 8$. So the figure is a parallelogram, with height $|4 - (-1)|$ or 5.
$A = bh$
$= (8)(5)$
$= 40$ units^2

Chapter 11 Standardized Test Practice

Pages 632–633

1. B; $3\left(\frac{2x-4}{-6}\right) = 18$

$3\left(-\frac{1}{3}x + \frac{2}{3}\right) = 18$

$-x + 2 = 18$

$-x = 16$

$x = -16$

2. B; the angle of the path from the school to the baseball field is a 90° angle, and the angle of Sam's path from the library to the baseball field is greater than that. So the angle of Sam's path is obtuse.

3. A; let p, q, and r represent parts of the given statements.
p: you exercise
q: you maintain better health
r: you will live longer
Given: $p \to q$ and $q \to r$. Use the Law of Syllogism to conclude $p \to r$. That is, if you exercise, you will live longer.

4. C; $m\angle ADE = 180 - m\angle DEA - m\angle EAD$

$= 180 - 40 - 60$

$= 80$

Since $\overleftrightarrow{DE}$ is a transversal for $\overleftrightarrow{AD}$ and $\overleftrightarrow{BE}$, and corresponding angles $\angle ADE$ and $\angle BEC$ are congruent, $\overline{AD}$ and $\overline{BE}$ are parallel.

5. *D*; because the front of the tent is isosceles and the entrance is an angle bisector, the two sides of the front of the tent are congruent triangles, by SAS. Then the two bases are of equal length, namely 3 ft, and so the distance between the stakes is 3 + 3 or 6 ft.

6. D; the gazebo is a regular hexagon. The angle measuring x is an exterior angle of the hexagon. Each exterior angle of a regular hexagon measures $\frac{360}{6}$ or 60. So $x = 60$.

7. A; this is Theorem 10.5.

8. C; $A = \frac{1}{2}Pa = \frac{1}{2}(6 \cdot 9)(7.8) = 210.6 \text{ cm}^2$

9. The library coordinates are $(-4, -3)$, and the fire station coordinates are (8, 3). The post office lies at the midpoint of the segment from $(-4, -3)$ to (8, 3). So the post office coordinates are given by

$M\left(\frac{x_1 + x_2}{2}, \frac{y_1 + y_2}{2}\right) = \left(\frac{-4+8}{2}, \frac{-3+3}{2}\right)$

$= (2, 0)$

10. Put the equation in slope-intercept form.

$3x - 6y = 12$

$-6y = -3x + 12$

$y = \frac{1}{2}x - 2$

The slope of the graph of the equation is $m = \frac{1}{2}$. The slope of the perpendicular line is the negative reciprocal of $\frac{1}{2}$, or -2.

11. By the Exterior Angle Theorem,

$m\angle R + m\angle S = m\angle STP$

$m\angle R + 90 = 150$

$m\angle R = 60$

12. Since pairs of vertical angles at C are congruent, and $\angle A$ and $\angle E$ are congruent, by AA similarity $\triangle ABC \sim \triangle EDC$.

$\frac{AB}{ED} = \frac{AC}{EC}$

$\frac{AB}{250} = \frac{112}{200}$

$AB = 250\left(\frac{112}{200}\right)$

$= 140$

13. The image of $J(6, -3)$ for the translation $(x, y) \to (-x, y + 5)$ is $J'(-6, -3 + 5)$ or $J'(-6, 2)$.

14. Use trigonometry.

a. $\tan 38° = \frac{1560}{x}$

$x = \frac{1560}{\tan 38°}$

≈ 1997 ft

b. $\tan 35° = \frac{1560}{y}$

$y = \frac{1560}{\tan 35°}$

≈ 2228 ft

c. Use the results from (a) and (b).

campground width $= y - x$

$\approx 2228 - 1997$

≈ 231 ft

15.

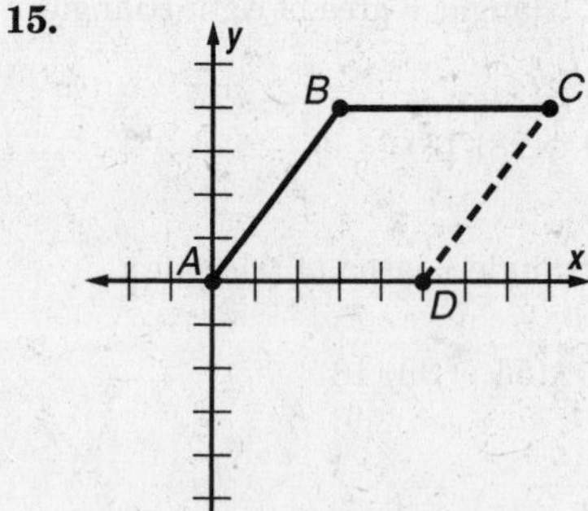

a. Since $\overline{BC}$ is horizontal, D must lie on the x-axis. And since the slope of $\overline{AB}$ is $\frac{4}{3}$, the slope of $\overline{DC}$ must be $\frac{4}{3}$. So D must be located 4 units down and 3 units left of $C(8,4)$, at $D(5,0)$.

b. $AD = 5$, and height $= 4$.

$A = bh$

$= (5)(4)$

$= 20 \text{ units}^2$

Chapter 12 Surface Area

Page 635 Getting Started

1. True; points A, C, and D lie in plane N, so $\triangle ADC$ lies in plane N.
2. False; points A, B, and C do not lie in plane K, so $\triangle ABC$ does not lie in plane K.
3. Cannot be determined; neither the given information nor the figure allow a determination of whether or not the line containing $\overline{AB}$ is parallel to plane K.
4. False; the line containing $\overline{AC}$ lies in plane N and only ℓ lies in both plane N and plane K, so the line containing $\overline{AC}$ does not lie in plane K.
5. The figure is a trapezoid with bases $b_1 = 19$ ft and $b_2 = 29$ ft and height $h = 16$ ft. The area is given by $A = \frac{1}{2}h(b_1 + b_2)$.
   $A = \frac{1}{2}(16)(19 + 29) = 384$
   The area of the figure is 384 ft^2.
6. The figure is a trapezoid with bases $b_1 = 12$ mm and $b_2 = 35$ mm and height $h = 13$ mm. The area is given by $A = \frac{1}{2}h(b_1 + b_2)$.
   $A = \frac{1}{2}(13)(12 + 35) = 305.5$
   The area of the figure is 305.5 mm^2.
7. The figure is a triangle with base $b = 1.9$ m and height $h = 1.9$ m. The area is given by $A = \frac{1}{2}bh$.
   $A = \frac{1}{2}(1.9)(1.9) = 1.805$
   Rounded to the nearest tenth, the area of the figure is 1.8 m^2.
8. $A = \pi r^2$
   $= \pi\left(\frac{d}{2}\right)^2$
   $= \frac{1}{4}\pi d^2$
   $= \frac{1}{4}\pi(19.0)^2$
   ≈ 283.5 cm^2
9. $A = \pi r^2$
   $= \pi(1.5)^2$
   ≈ 7.1 yd^2
10. $A = \pi r^2$
   $= \pi\left(\frac{d}{2}\right)^2$
   $= \frac{1}{4}\pi d^2$
   $= \frac{1}{4}\pi(10.4)^2$
   ≈ 84.9 m^2

12-1 Three-Dimensional Figures

Page 639 Check for Understanding

1. The Platonic solids are the five regular polyhedra. All of the faces are congruent, regular polygons. In other polyhedra, the bases are congruent parallel polygons, but the faces are not necessarily congruent.
2. In a square pyramid, the lateral faces are triangles. In a square prism, the faces are rectangles.
3. Sample answer:

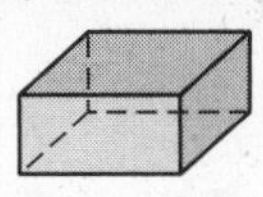

4.

back view

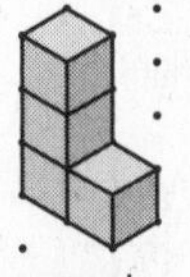

corner view

5. The base is a hexagon, and six faces meet at a point. So this solid is a hexagonal pyramid. The base is $ABCDEF$. The faces are $ABCDEF$, $\triangle AGF$, $\triangle FGE$, $\triangle EGD$, $\triangle DGC$, $\triangle CGB$, and $\triangle BGA$. The edges are $\overline{AF}$, $\overline{FE}$, $\overline{ED}$, $\overline{DC}$, $\overline{CB}$, $\overline{BA}$, $\overline{AG}$, $\overline{FG}$, $\overline{EG}$, $\overline{DG}$, $\overline{CG}$, and $\overline{BG}$. The vertices are A, B, C, D, E, F, and G.
6. The bases are squares. So this is a square prism. Sample answer: The bases are $\square KJIH$ and $\square MNOL$. The faces are $\square KJIH$, $\square MNOL$, $\square JNOI$, $\square JKMN$, $\square KHLM$, and $\square IHLO$. The edges are $\overline{KH}$, $\overline{KJ}$, $\overline{JI}$, $\overline{IH}$, $\overline{JN}$, $\overline{IO}$, $\overline{HL}$, $\overline{KM}$, $\overline{MN}$, $\overline{ML}$, $\overline{NO}$, and $\overline{LO}$. The vertices are H, K, J, I, L, M, N, and O.
7. The bases are circles. So this is a cylinder. The bases are circles P and Q.
8. To get round slices of cheese, slice the cheese parallel to the bases. To get rectangular slices, place the cheese on the slicer so the bases are perpendicular to the blade.

Pages 640–642 Practice and Apply

9.

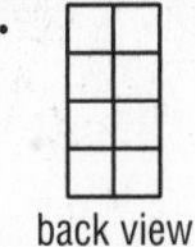

back view

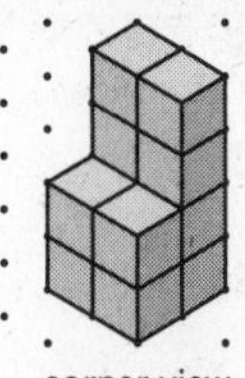

corner view

10.

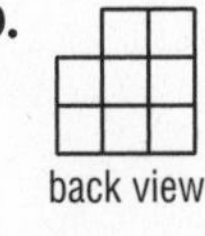

back view

corner view

11.

back view

corner view

12.

back view

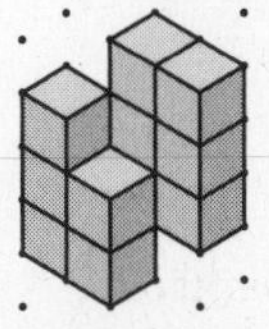

corner view

13.

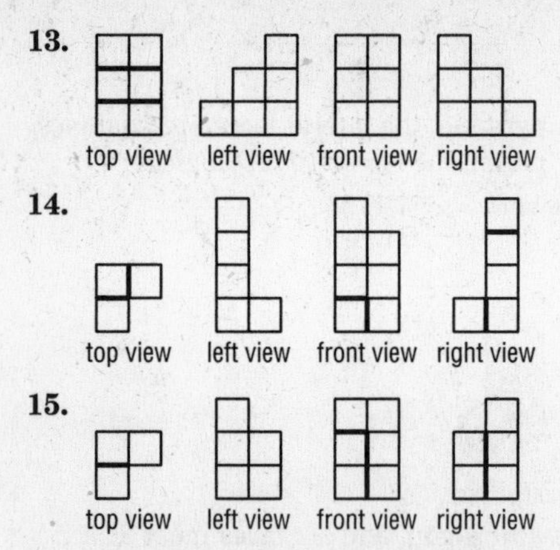

14.

15.

16. The bases are triangles. So this is a triangular prism. The bases are $\triangle MNO$, and $\triangle PQR$. The faces are $\triangle MNO$, $\triangle PQR$, $OMPR$, $ONQR$, and $PQNM$. The edges are $\overline{MN}$, $\overline{NO}$, $\overline{OM}$, $\overline{PQ}$, $\overline{QR}$, $\overline{PR}$, $\overline{NQ}$, $\overline{MP}$, and $\overline{OR}$. The vertices are M, N, O, P, Q, and R.

17. The base is a rectangle, and four faces meet in a point. So this solid is a rectangular pyramid. The base is $DEFG$. The faces are $DEFG$, $\triangle DHG$, $\triangle GHF$, $\triangle FHE$, and $\triangle DHE$. The edges are $\overline{DG}$, $\overline{GF}$, $\overline{FE}$, $\overline{ED}$, $\overline{DH}$, $\overline{EH}$, $\overline{FH}$, and $\overline{GH}$. The vertices are D, E, F, G, and H.

18. The base is a triangle, and three faces meet in a point. So this solid is a triangular pyramid. The base is $\triangle IJK$. The faces are $\triangle IJK$, $\triangle ILK$, $\triangle KLJ$, and $\triangle ILJ$. The edges are $\overline{IK}$, $\overline{KJ}$, $\overline{IJ}$, $\overline{IL}$, $\overline{KL}$, and $\overline{JL}$. The vertices are I, J, K, and L.

19. The bases are circles. The solid is a cylinder. The bases are circles S and T.

20. The solid is a sphere.

21. The solid has a circle for a base and a vertex. So it is a cone. The base is circle B. The vertex is A.

22. **16:** Yes; Euler's formula is true.

$F + V = E + 2$
$5 + 6 = 9 + 2$
$11 = 11$

17: Yes; Euler's formula is true.

$F + V = E + 2$
$5 + 5 = 8 + 2$
$10 = 10$

18: Yes; Euler's formula is true.

$F + V = E + 2$
$4 + 4 = 6 + 2$
$8 = 8$

19–21: No; these figures are not polyhedrons, so Euler's formula does not apply.

23. No, not enough information is provided by the top and front views to determine the shape.

24. Sample answer: The speaker could be shaped like a rectangular prism, or the sides could be angled.

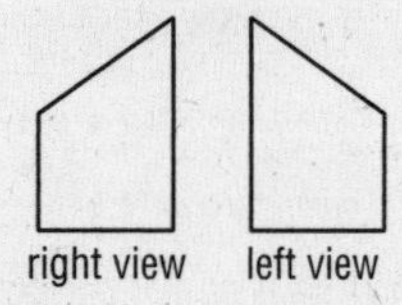

25. The resulting shape is a parabola.

26. The resulting shape is a triangle.

27. The resulting shape is a circle.

28. The resulting shape is a rectangle.

29. The resulting shape is a rectangle.

30. The resulting shape is a square.

31. intersecting three faces and parallel to base;

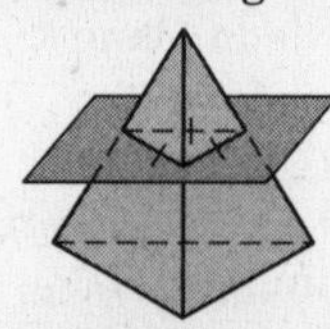

32. intersecting three faces and edges of base;

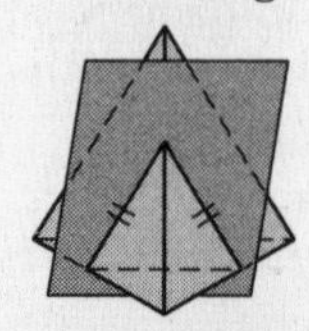

33. intersecting all four faces, not parallel to any face;

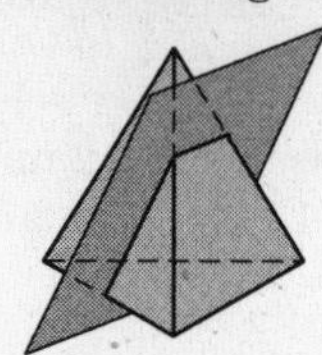

34. If the number of sides of a base of a pyramid increases infinitely, the solid that results is a cone.

35. If the number of sides of the bases of a prism increases infinitely, the solid that results is a cylinder.

36. The shapes seen in an uncut diamond are triangles and squares or rectangles.

37. The shapes seen in the emerald-cut diamond are rectangles, triangles, and quadrilaterals.

38. The shapes seen in the round-cut diamond are octagons, triangles, and quadrilaterals.

39a. 5 faces: triangular prism

39b. 6 faces: cube, rectangular prism, or hexahedron

39c. 6 faces: pentagonal pyramid

39d. 7 faces: hexagonal pyramid

39e. 8 faces: hexagonal prism

40. Yes, there is a pattern. The number of sides of the base of a prism is 2 less than the number of faces in the polyhedron. The number of sides of the base of a pyramid is 1 less than the number of faces.

41. No; the number of faces is not enough information to classify a polyhedron. A polyhedron with 6 faces could be a cube, rectangular prism, hexahedron, or a pentagonal pyramid. More information is needed to classify a polyhedron.

42.

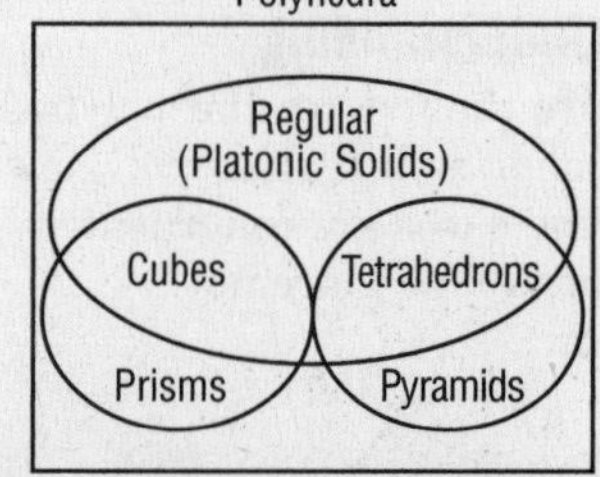

43. Sample answer: Archaeologists use two-dimensional drawings to learn more about the structure they are studying. Egyptologists can compare two-dimensional drawings of the pyramids and note similarities and any differences. Answers should include the following.
- Viewpoint drawings and corner views are types of two-dimensional drawings that show three dimensions.
- To show three dimensions in a drawing, you need to know the views from the front, top, and each side.

44. D; a circle cannot be formed by the intersection of a cube and a plane.

45. D; $\frac{x^3}{x^4} = \frac{1}{x}$

$\frac{1}{-4} > \frac{1}{-3} > \frac{1}{-2} > \frac{1}{-1}$

46. There are 6 planes of symmetry. Each plane contains the center of the tetrahedron and one edge, thus bisecting the opposite edge.

47. A cylinder has infinite planes of symmetry. Planes pass through the centers of the bases.

48. A sphere has infinite planes of symmetry. Planes pass through the center of the sphere.

Page 642 Maintain Your Skills

49. $P(\text{steak}) = \frac{87°}{360°} \approx 0.242$

50. $P(\text{not seafood}) = \frac{87° + 102° + 118°}{360°} \approx 0.853$

51. $P(\text{either pasta or chicken}) = \frac{102° + 118°}{360°} \approx 0.611$

52. $P(\text{neither pasta nor steak}) = \frac{53° + 118°}{360°} = 0.475$

53.

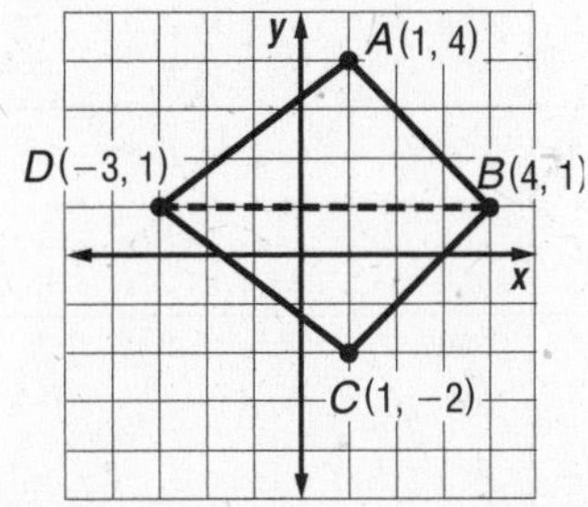

The figure is composed of two triangles, each with height 3 and base 7. Find the area.

$A = 2\left(\frac{1}{2}bh\right)$
$= bh$
$= (7)(3)$
$= 21$

The area is 21 units^2.

54.

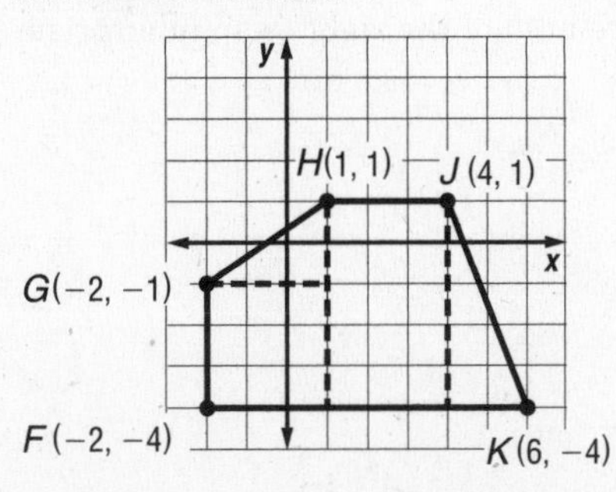

The figure is composed of two triangles, one rectangle, and one square. Find the area.

$A = \frac{1}{2}b_1h_1 + \frac{1}{2}b_2h_2 + s^2 + \ell w$
$= \frac{1}{2}(3)(2) + \frac{1}{2}(2)(5) + 3^2 + (5)(3)$
$= 32$

The area is 32 units^2.

55.

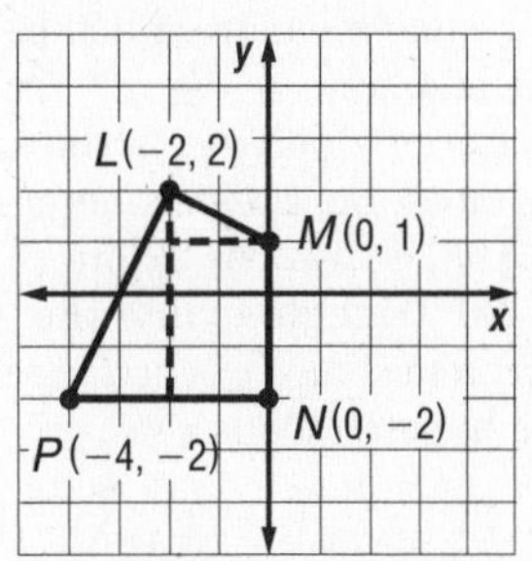

The figure is composed of two triangles and a rectangle.

$A = \frac{1}{2}b_1h_1 + \frac{1}{2}b_2h_2 + \ell w$
$= \frac{1}{2}(2)(4) + \frac{1}{2}(2)(1) + (3)(2)$
$= 11$

The area is 11 units^2.

56. Base and Sides: Each pair of opposite sides of a parallelogram has the same measure. Each base is 15 m and each side is 12 m.
Perimeter: The perimeter of a polygon is the sum of the measures of its sides. So, the perimeter of the parallelogram is 2(12) + 2(15) or 54 m.
Height: Use a 30°-60°-90° triangle to find the height. Recall that if the measure of the leg opposite the 30° angle is x, then the length of the hypotenuse is $2x$, and the length of the leg opposite the 60° angle is $x\sqrt{3}$.
$12 = 2x$
$6 = x$
So, the height of the parallelogram is $6\sqrt{3}$ meters.
Area: $A = bh$
$= (15)(6\sqrt{3})$
$= 90\sqrt{3}$
≈ 155.9
The perimeter is 54 m and the area is approximately 155.9 m^2.

57. Base and Sides: Each pair of opposite sides of a parallelogram has the same measure. Each base is 25 ft and each side is 20 ft.
Perimeter: The perimeter of a polygon is the sum of the measures of its sides. So, the perimeter of the parallelogram is 2(25) + 2(20) or 90 ft.
Height: Use a 30°-60°-90° triangle to find the height. Recall that if the measure of the leg opposite the 30° angle is x, then the length of the hypotenuse is $2x$, and the length of the leg opposite the 60° angle is $x\sqrt{3}$.
$20 = 2x$
$10 = x$
So, the height of the parallelogram is $10\sqrt{3}$ feet.

Area: $A = bh$
$= (25)(10\sqrt{3})$
$= 250\sqrt{3}$
≈ 433.0
The perimeter is 90 ft and the area is approximately 433.0 ft^2.

58. Base and Sides: Each pair of opposite sides of a parallelogram has the same measure. Each base is 68 in. and each side is 42 in.
Perimeter: The perimeter of a polygon is the sum of the measures of its sides. So, the perimeter of the parallelogram is 2(68) + 2(42) or 220 in.
Height: Use a 45°-45°-90° triangle to find the height. Recall that if the measure of each leg is x, then the length of the hypotenuse is $x\sqrt{2}$.
$42 = x\sqrt{2}$
$\frac{42}{\sqrt{2}} = x$
$21\sqrt{2} = x$
So, the height of the parallelogram is $21\sqrt{2}$ inches.
Area: $A = bh$
$= (68)(21\sqrt{2})$
$= 1428\sqrt{2}$
≈ 2019.5
The perimeter is 220 in. and the area is approximately 2019.5 in^2.

59. $A = \ell w$
$= (20)(15)$
$= 300$
The area is 300 cm^2.

60. $A = \ell w$
$= (4)(13)$
$= 52$
The area is 52 ft^2.

61. $A = \ell w$
$= (60)(72)$
$= 4320$
The area is 4320 in^2.

62. $A = s^2$
$= 1.7^2$
≈ 2.9
The area is approximately 2.9 m^2.

12-2 Nets and Surface Area

Page 645 Check for Understanding

1. Sample answer:

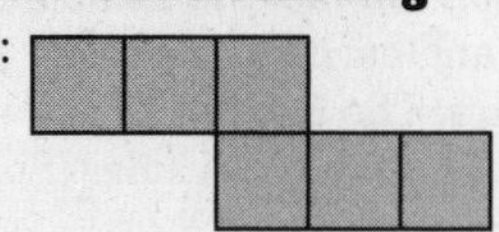

2. On isometric dot paper, the dots are arranged in triangles, which aid in drawing three-dimensional objects. On rectangular dot paper, the dots are arranged in squares, which aid in drawing the nets and orthogonal views of three-dimensional objects.

3. **4.**

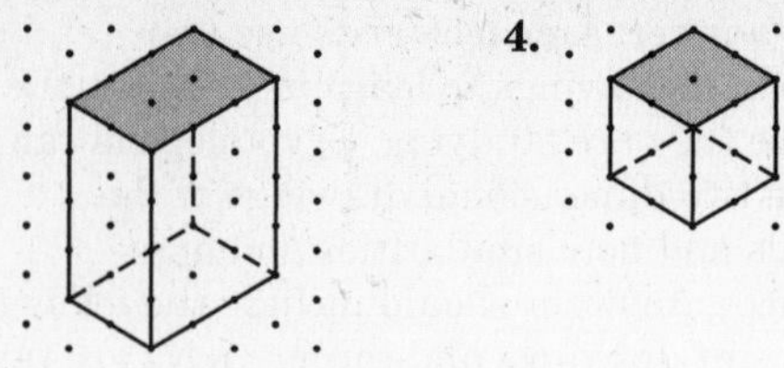

5.

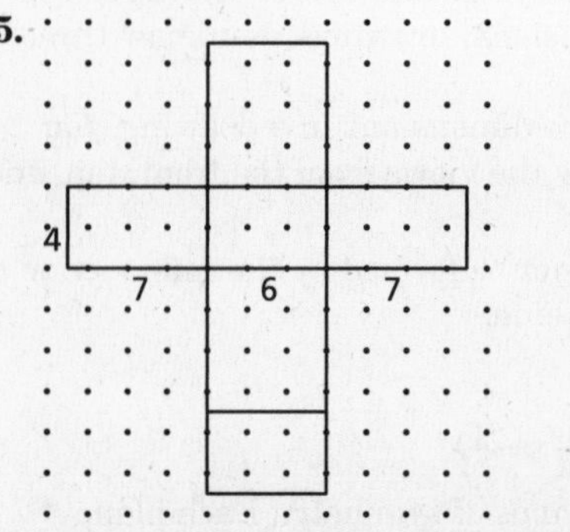

Surface area $= 2(4 \cdot 7) + 2(4 \cdot 6) + 2(6 \cdot 7)$
$= 56 + 48 + 84$
$= 188$
The surface area of the rectangular prism is 188 in^2.

6. Use the Pythagorean Theorem to find the height of the prism.
$17^2 = 8^2 + h^2$
$289 = 64 + h^2$
$225 = h^2$
$15 = h$

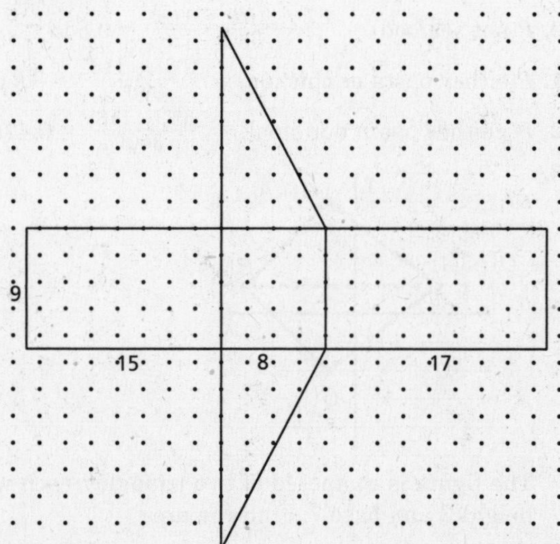

Surface area
$= 8 \cdot 9 + 9 \cdot 15 + 9 \cdot 17 + \frac{1}{2} \cdot 8 \cdot 15 + \frac{1}{2} \cdot 8 \cdot 15$
$= 72 + 135 + 153 + 60 + 60$
$= 480$
The surface area of the right triangular prism is 480 ft^2.

7.

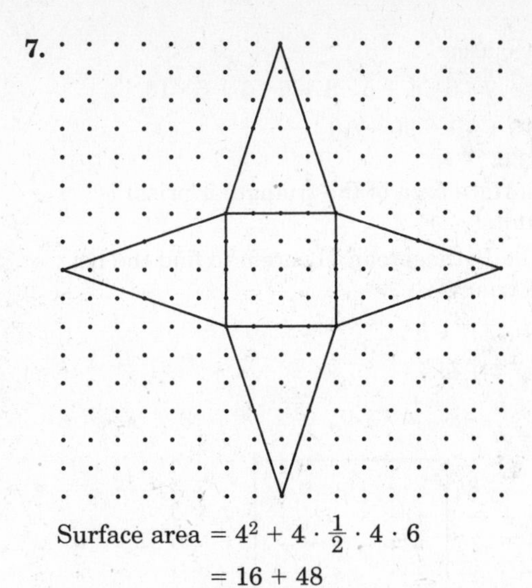

Surface area $= 4^2 + 4 \cdot \frac{1}{2} \cdot 4 \cdot 6$

$= 16 + 48$

$= 64$

The surface area of the square pyramid is 64 cm^2.

8. C; the net only forms three of the four sides. Two of the triangles overlap rather than being two different faces.

Pages 646–648 Practice and Apply

9.–12.

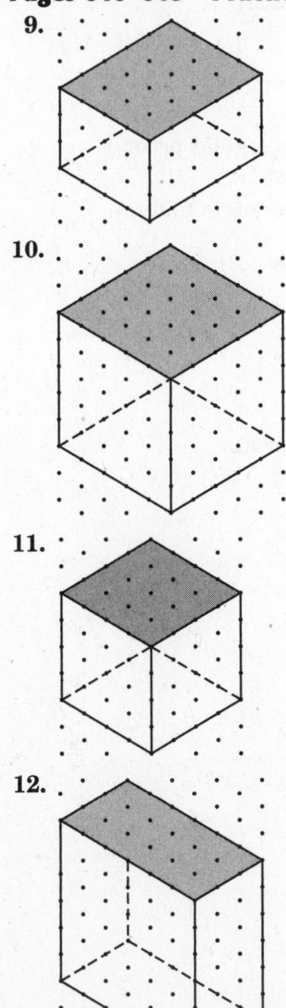

13.–15.

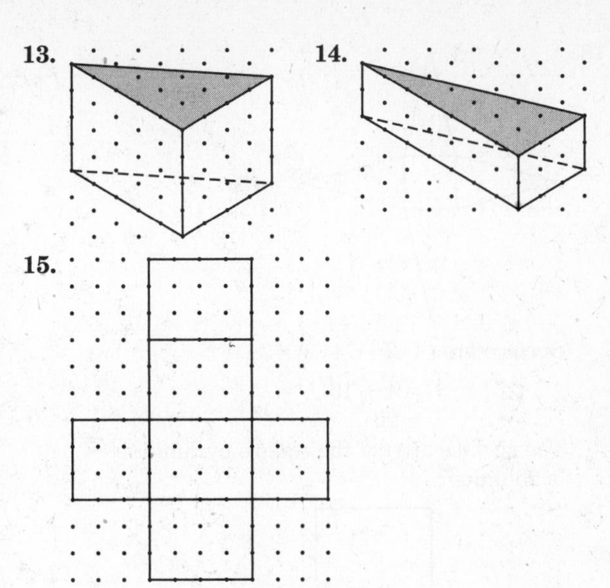

Surface area $= 2(3^2) + 4(3 \cdot 4)$

$= 18 + 48$

$= 66$

The surface area of the rectangular prism is 66 $units^2$.

16.

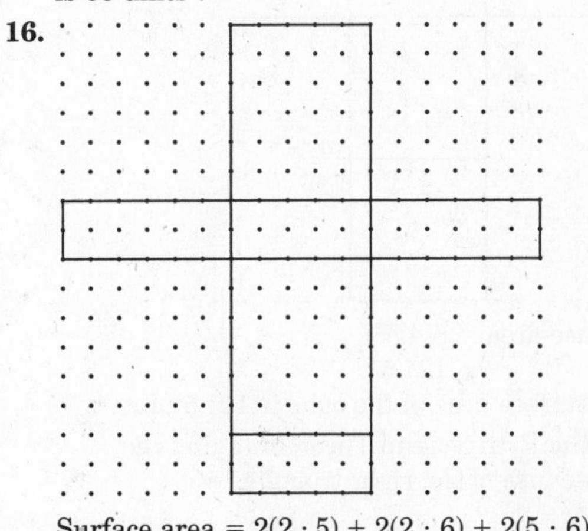

Surface area $= 2(2 \cdot 5) + 2(2 \cdot 6) + 2(5 \cdot 6)$

$= 20 + 24 + 60$

$= 104$

The surface area of the rectangular prism is 104 $units^2$.

17.

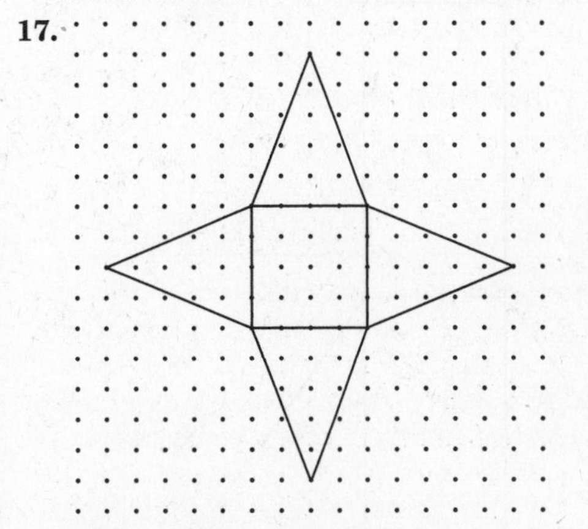

Surface area $= 4^2 + 4 \cdot \frac{1}{2} \cdot 4 \cdot 5$

$= 16 + 40$

$= 56$

The surface area of the square pyramid is 56 $units^2$.

18.

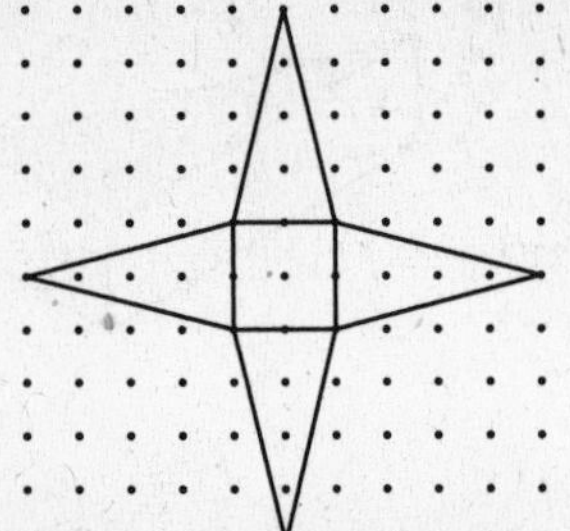

Surface area $= 2^2 + 4 \cdot \frac{1}{2} \cdot 2 \cdot 4$

$= 4 + 16$

$= 20$

The surface area of the square pyramid is 20 units^2.

19.

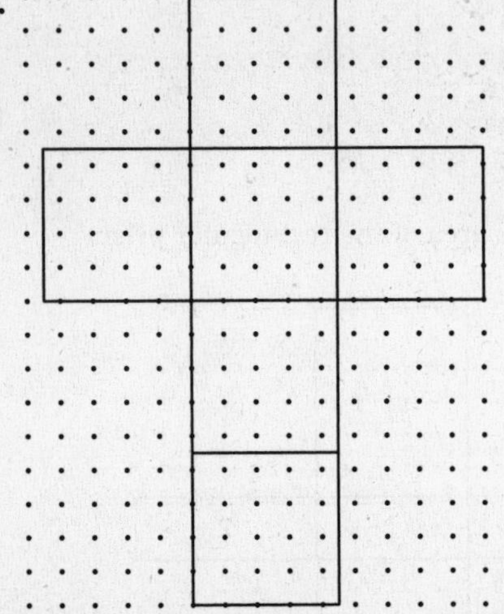

Surface area $= 6(4.5^2)$

$= 121.5$

The surface area of the cube is 121.5 units^2.

20. Use the Pythagorean Theorem to find the hypotenuse of the right triangle.

$c^2 = a^2 + b^2$

$c^2 = 6^2 + 8^2$

$c^2 = 36 + 64$

$c^2 = 100$

$c = 10$

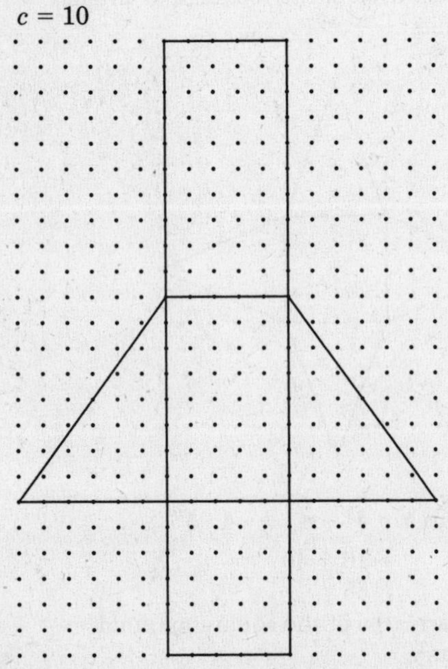

Surface area

$= 2 \cdot \frac{1}{2} \cdot 6 \cdot 8 + 5 \cdot 8 + 5 \cdot 6 + 5 \cdot 10$

$= 48 + 40 + 30 + 50$

$= 168$

The surface area of the triangular prism is 168 units^2.

21. Use the Pythagorean Theorem to find the height of the triangle.

$5^2 = \left(\frac{4}{2}\right)^2 + h^2$

$25 = 4 + h^2$

$21 = h^2$

$\sqrt{21} = h$

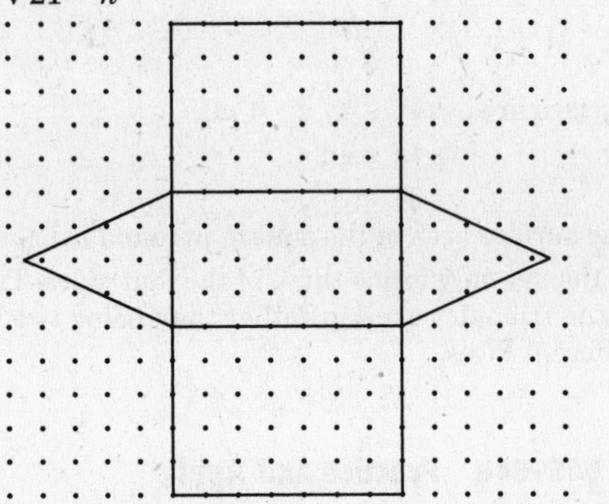

Surface area $= 2(5 \cdot 7) + 4 \cdot 7 + 2 \cdot \frac{1}{2} \cdot 4 \cdot \sqrt{21}$

$= 70 + 28 + 4\sqrt{21}$

≈ 116.3

The surface area of the triangular prism is approximately 116.3 units^2.

22. Use the Pythagorean Theorem to find the unknown edge length.

$c^2 = a^2 + b^2$

$c^2 = (8 - 6)^2 + 6^2$

$c^2 = 4 + 36$

$c^2 = 40$

$c = \sqrt{40}$

$c = 2\sqrt{10}$

Find the area of the trapezoid.

$A = \frac{1}{2}h(b_1 + b_2)$

$= \frac{1}{2}(6)(6 + 8)$

$= 42$

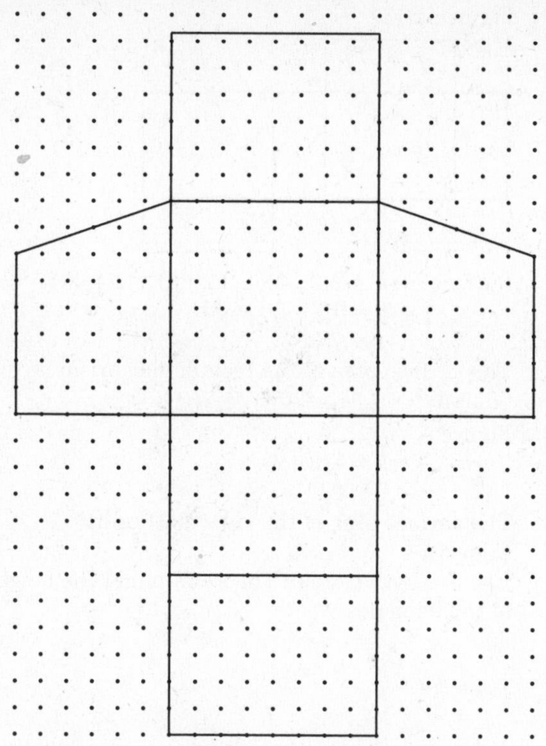

Surface area $= 2(6 \cdot 8) + 8^2 + 2\sqrt{10} \cdot 8 + 2 \cdot 42$
$= 96 + 64 + 16\sqrt{10} + 84$
≈ 294.6

The surface area of the prism is approximately 294.6 units^2.

23. Use the Pythagorean Theorem to find the unknown edge length.

$c^2 = a^2 + b^2$
$c^2 = (6 - 3)^2 + 3^2$
$c^2 = 9 + 9$
$c^2 = 18$
$c = \sqrt{18}$
$c = 3\sqrt{2}$

Find the area of the trapezoid.

$A = \frac{1}{2}h(b_1 + b_2)$
$= \frac{1}{2}(3)(3 + 6)$
$= 13.5$

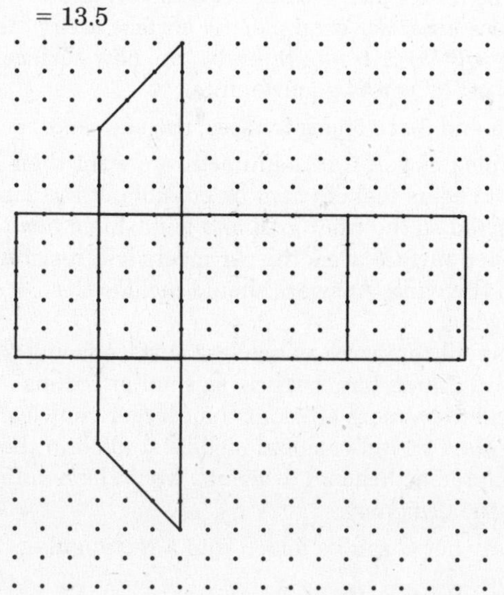

Surface area
$= 2 \cdot 13.5 + 5 \cdot 6 + 2(3 \cdot 5) + 5 \cdot 3\sqrt{2}$
$= 27 + 30 + 30 + 15\sqrt{2}$
≈ 108.2

The surface area of the prism is approximately 108.2 units^2.

24.

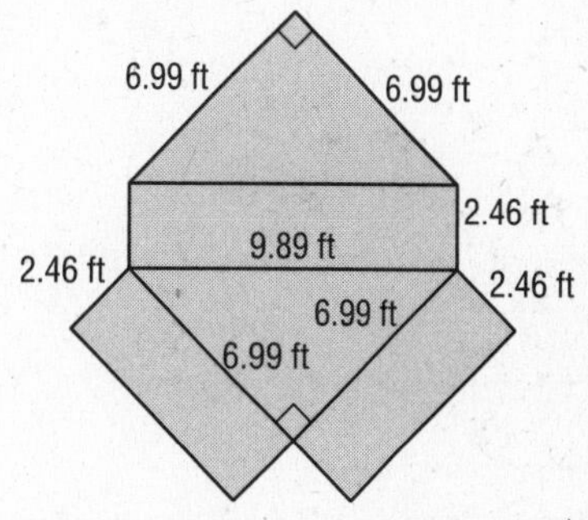

Use the Pythagorean Theorem to find the hypotenuse.

$c^2 = a^2 + b^2$
$c^2 = 6.99^2 + 6.99^2$
$c = \sqrt{97.7202}$

The thickness of the sandwich is

$8 + 13.5 + 8 = 29.5 \text{ in.} = \frac{29.5}{12} \text{ ft.}$

Surface area $= 2 \cdot \frac{1}{2} \cdot 6.99^2 + 2\left(6.99 \cdot \frac{29.5}{12}\right) + \frac{29.5}{12} \cdot \sqrt{97.7202}$
≈ 107.5

The surface area of the sandwich is approximately 107.5 ft^2.

25.

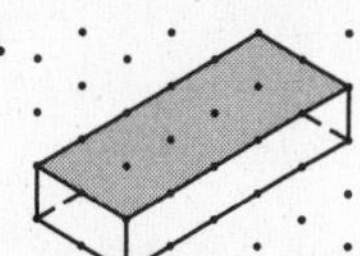

26.

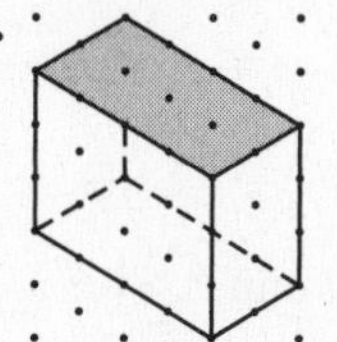

27.

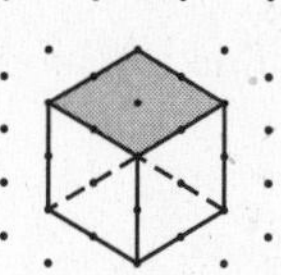

28.

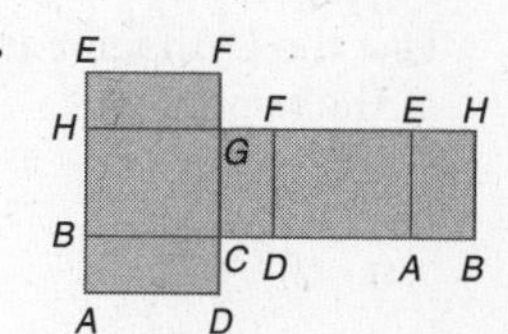

29.

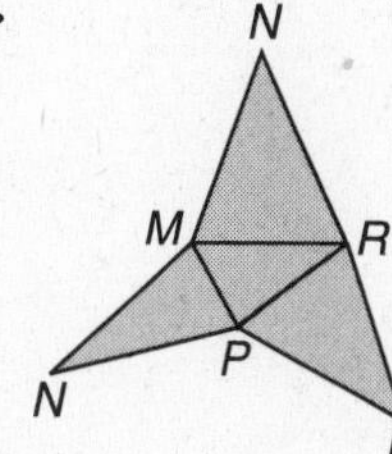

30.

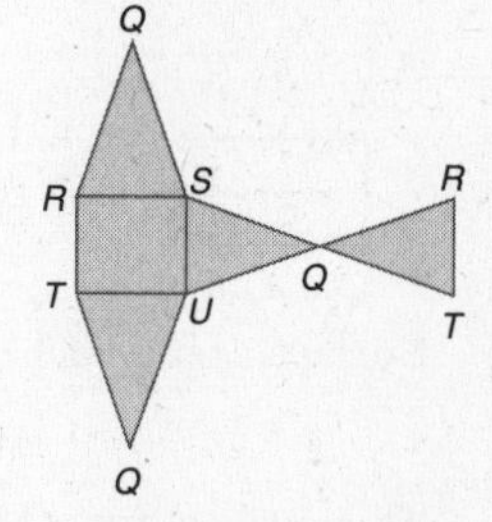

31. 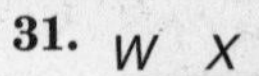

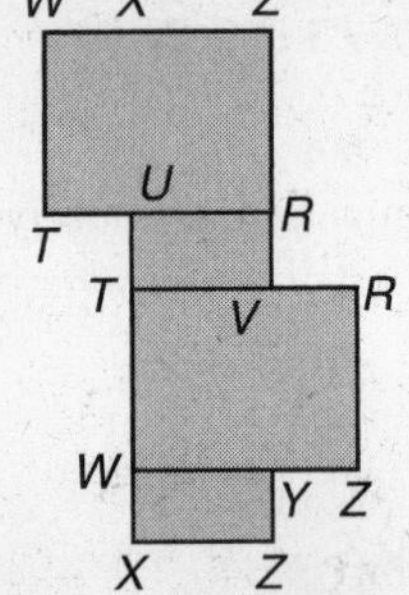

32.

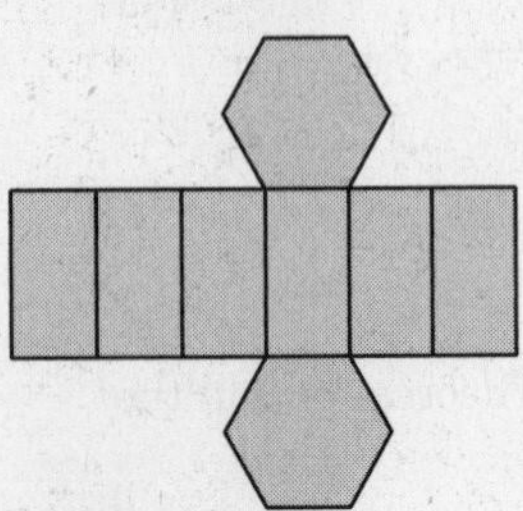

33.

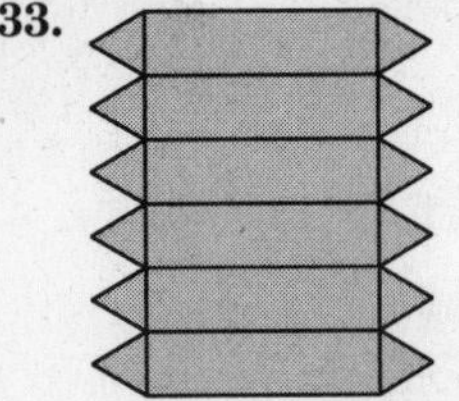

34.

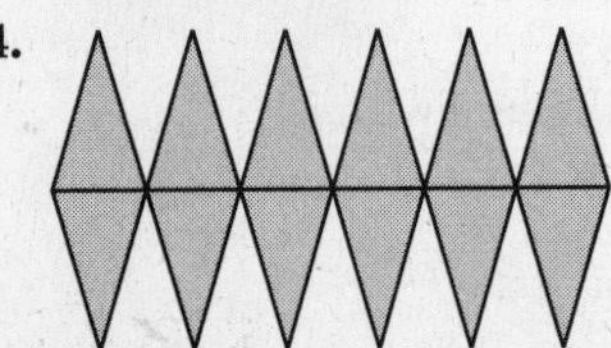

35. Figure A:

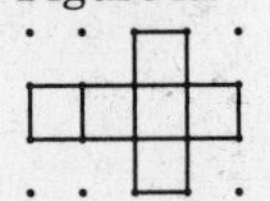

Surface area $= 6(1 \cdot 1)$
$= 6$

The surface area of the cube is 6 units^2.

Figure B:

Use the Pythagorean Theorem to find the height of the triangles.

$1^2 = h^2 + \left(\frac{1}{2}\right)^2$

$1 = h^2 + \frac{1}{4}$

$\frac{3}{4} = h^2$

$\frac{\sqrt{3}}{2} = h$

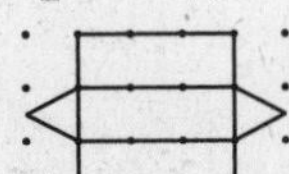

Surface area $= 3(1 \cdot 3) + 2 \cdot \frac{1}{2} \cdot 1 \cdot \frac{\sqrt{3}}{2}$
$= 9 + \frac{\sqrt{3}}{2}$

The surface area of the triangular prism is $\left(9 + \frac{\sqrt{3}}{2}\right)$ units^2.

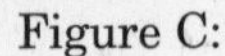

Figure C:

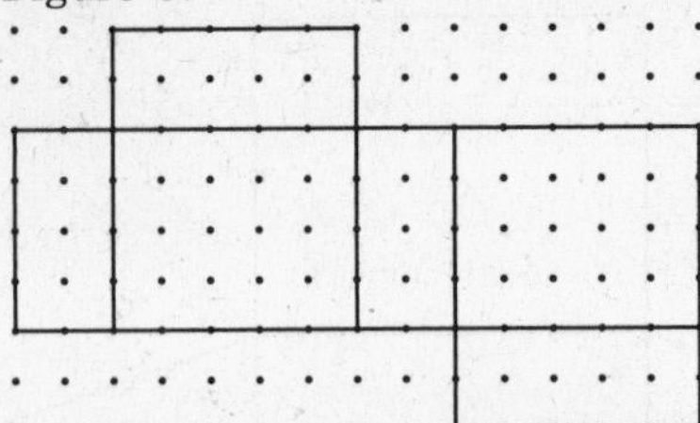

Surface area $= 2(2 \cdot 4) + 2(2 \cdot 5) + 2(4 \cdot 5)$
$= 16 + 20 + 40$
$= 76$

The surface area of the rectangular prism is 76 units^2.

36. Figure A:

Surface area $= 6(2 \cdot 2)$
$= 24$

The surface area of the cube is 24 units^2.

Figure B:

Use the Pythagorean Theorem to find the height of the triangles.

$2^2 = h^2 + \left(\frac{2}{2}\right)^2$

$4 = h^2 + 1$

$3 = h^2$

$\sqrt{3} = h$

Surface area $= 3(2 \cdot 6) + 2 \cdot \frac{1}{2} \cdot 2 \cdot \sqrt{3}$
$= 36 + 2\sqrt{3}$

The surface area of the triangular prism is $(36 + 2\sqrt{3})$ units^2.

Figure C:

Surface area $= 2(4 \cdot 8) + 2(4 \cdot 10) + 2(8 \cdot 10)$
$= 64 + 80 + 160$
$= 304$

The surface area of the rectangular prism is 304 units^2.

37. The surface area quadruples when the dimensions are doubled. For example, the surface area of the cube is $6(1^2)$ or 6 square units. When the dimensions are doubled the surface area is $6(2^2)$ or 24 square units.

38. When the dimensions are tripled, the surface area will be nine times greater than the original surface area. For example, the surface area of the cube is $6(1^2)$ or 6 square units. The new surface area is $6(3^2)$ or 54 square units.

39. No; 5 and 3 are opposite faces; the sum is 8.

40. Sample answer: Car manufacturers want their cars to be as fuel efficient as possible. If the car is designed so the front grill and windshield have a smaller surface area, the car meets less resistance from the wind. Answers should include the following.

- A small compact car has less surface facing the wind than a larger truck, so smaller sedans tend to be more efficient than larger vehicles.
- Of the two-dimensional models studied in this chapter, orthogonal drawings would be helpful to the designers.

41. C; only net *C* can be folded into a rectangular prism.

42. B; $16a^3 - 54b^3 = 2(8a^3 - 27b^3)$
$= 2[(2a)^3 - (3b)^3]$
$= 2(2a - 3b)(4a^2 + 6ab + 9b^2)$

Page 648 Maintain Your Skills

43. The resulting shape is a rectangle.
44. The resulting shape is a triangle.
45. The resulting shape is a rectangle.
46. $P(\text{butterfly in the flower bed}) = \frac{(20)(20)}{(100)(200)}$
$= 0.02$
47. $m\widehat{FLJ} = 180$ because it is a semicircle. By the Inscribed Angle Theorem,
$m\angle FHJ = \frac{1}{2}m\widehat{FLJ}$
$= \frac{1}{2}(180)$
$= 90$
48. $m\widehat{LK} = 60$ because it is $\frac{1}{6}$ the measure of a circle (360).
49. $m\widehat{GHL} = 240$ because it is $\frac{4}{6}$ the measure of a circle (360).
By the Inscribed Angle Theorem,
$m\angle LFG = \frac{1}{2}m\widehat{GHL}$
$= \frac{1}{2}(240)$
$= 120$
50. $A = bh$
$= 16 \cdot 14$
$= 224$
The area of the parallelogram is 224 ft^2.
51. $A = \frac{1}{2}h(b_1 + b_2)$
$= \frac{1}{2} \cdot 7(6 + 12)$
$= 63$
The area of the trapezoid is 63 cm^2.
52. $A = \frac{1}{2}bh$
$= \frac{1}{2} \cdot 6.5 \cdot 4$
$= 13$
The area of the triangle is 13 yd^2.
53. $A = \frac{1}{2}h(b_1 + b_2)$
$= \frac{1}{2} \cdot 10(13 + 9)$
$= 110$
The area of the trapezoid is 110 cm^2.

12-3 Surface Areas of Prisms

Page 651 Check for Understanding

1. In a right prism a lateral edge is also an altitude. In an oblique prism, the lateral edges are not perpendicular to the bases.
2. Sample answer: The bases are *ACHG* and *BDFE*. The lateral faces are *ABDC*, *GEFH*, *BEGA*, and *DFHC*. The lateral edges are $\overline{BA}$, $\overline{EG}$, $\overline{FH}$, and $\overline{DC}$.

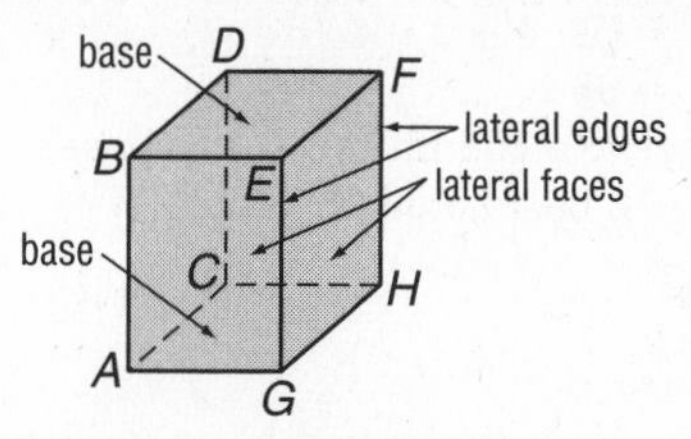

3. Use the Pythagorean Theorem to find the measure of the hypotenuse of the triangular base.
$c^2 = a^2 + b^2$
$c^2 = 8^2 + 15^2$
$c^2 = 64 + 225$
$c^2 = 289$
$c = 17$
$L = Ph$
$= (17 + 15 + 8)(21)$
$= (40)(21)$
$= 840$
$T = L + 2B$
$= 840 + 2 \cdot \frac{1}{2}bh$
$= 840 + (8)(15)$
$= 840 + 120$
$= 960$
The lateral and surface areas are 840 $units^2$ and 960 $units^2$, respectively.
4. $L(7 \times 9 \text{ base}) = Ph$
$= (2 \cdot 7 + 2 \cdot 9)(6)$
$= 192$
$L(6 \times 9 \text{ base}) = Ph$
$= (2 \cdot 6 + 2 \cdot 9)(7)$
$= 210$
$L(6 \times 7 \text{ base}) = Ph$
$= (2 \cdot 6 + 2 \cdot 7)(9)$
$= 234$
$T = L + 2B$
$= 234 + 2(6 \cdot 7)$
$= 234 + 84$
$= 318$
The lateral areas are 192 $units^2$ (7×9 base), 210 $units^2$ (6×9 base), and 234 $units^2$ (6×7 base). The surface area is 318 $units^2$.
5. The perimeter of the ceiling (base) is $2(20) + 2(15) = 40 + 30 = 70$ ft. Find the total surface area to be painted.
$T = L + B$
$= Ph + B$
$= (70)(12) + (20)(15)$
$= 840 + 300$
$= 1140$
The surface area to be painted is 1140 ft^2.

Pages 651–654 Practice and Apply

6. $L(3 \times 4 \text{ base}) = Ph$
$= (2 \cdot 3 + 2 \cdot 4)(12)$
$= 168$

$L(3 \times 12 \text{ base}) = Ph$
$= (2 \cdot 3 + 2 \cdot 12)(4)$
$= 120$
$L(4 \times 12 \text{ base}) = Ph$
$= (2 \cdot 4 + 2 \cdot 12)(3)$
$= 96$
The lateral areas are 168 units^2 (3×4 base), 120 units^2 (3×12 base), and 96 units^2 (4×12 base).

7. $L = Ph$
$= (4 + 5 + 7)(8)$
$= 128$
The lateral area is 128 units^2.

8. Use the Pythagorean Theorem to find the measure of the third side of the triangular base.
$c^2 = a^2 + b^2$
$c^2 = 7^2 + 8^2$
$c^2 = 49 + 64$
$c^2 = 113$
$c = \sqrt{113}$
$L = Ph$
$= (7 + 8 + \sqrt{113})(10)$
≈ 256.3
The lateral area is approximately 256.3 units^2.

9. $L = Ph$
$= (2 + 4 + 3 + 4 + 5)(9)$
$= 162$
The lateral area is 162 units^2.

10. $L = Ph$
$= (2 \cdot 7 + 2 \cdot 4 + 8 \cdot 2)(9)$
$= 342$
The lateral area is 342 cm^2.

11. Find the lateral area of the hollowed-out prism, including the inner faces.
$L = (\text{outer perimeter})h + (\text{inner perimeter})h$
$= (4 \cdot 4)(8) + (4 \cdot 1)(8)$
$= 160$
The lateral area is 160 units^2 (square base).
$L = 2(4^2 - 1^2) + 2(4 \cdot 8) + 4(1 \cdot 8)$
$= 2(15) + 2(32) + 4(8)$
$= 30 + 64 + 32$
$= 126$
The lateral area is 126 units^2 (rectangular base).

12. The surface area of a cube is given by $A = 6s^2$, where s is the length of a lateral edge.
$864 = 6s^2$
$144 = s^2$
$12 = s$
The length of the lateral edge is 12 in.

13. Use the Pythagorean Theorem to find the measure of the hypotenuse of the triangular base.
$c^2 = a^2 + b^2$
$c^2 = 5^2 + 12^2$
$c^2 = 169$
$c = 13$
So, $P = 5 + 12 + 13 = 30$.
$T = L + 2B$
$T = Ph + 2B$
$540 = 30(h) + 2 \cdot \frac{1}{2} \cdot 5 \cdot 12$
$540 = 30h + 60$
$480 = 30h$
$16 = h$
The height is 16 cm.

14. $L = Ph$
$156 = P(13)$
$12 = P$
The perimeter of the base must be 12 inches. There are three rectangles with integer values for the dimensions that have a perimeter of 12. The dimensions of the base could be 5×1, 4×2, or 3×3.

15. $L = Ph$
$96 = P(4)$
$24 = P$
The perimeter of the base must be 24 meters. There are six rectangles with integer values for the dimensions that have a perimeter of 24. The dimensions of the base could be 1×11, 2×10, 3×9, 4×8, 5×7, or 6×6.

16. Use the Pythagorean Theorem to find the measure of the third side of the triangular base.
$c^2 = a^2 + b^2$
$17^2 = a^2 + 8^2$
$289 = a^2 + 64$
$225 = a^2$
$15 = a$
$T = L + 2B$
$= Ph + 2B$
$= (8 + 15 + 17)(4) + 2 \cdot \frac{1}{2} \cdot 8 \cdot 15$
$= 280$
The surface area of the prism is 280 units^2.

17. $T = L + 2B$
$= Ph + 2B$
$= (3 + 3 + 3 + 3)(8) + 2(3 \cdot 3)$
$= 114$
The surface area of the prism is 114 units^2.

18. $T = L + 2B$
$= Ph + 2B$
$= (2 \cdot 4 + 2 \cdot 11)(7.5) + 2(4 \cdot 11)$
$= 313$
The surface area of the prism is 313 units^2.

19. Use the Pythagorean Theorem to find the measure of the third side of the triangular base.
$c^2 = a^2 + b^2$
$15^2 = a^2 + 9^2$
$225 = a^2 + 81$
$144 = a^2$
$12 = a$
$T = L + 2B$
$= Ph + 2B$
$= (9 + 12 + 15)(11.5) + 2 \cdot \frac{1}{2} \cdot 9 \cdot 12$
$= 522$
The surface area of the prism is 522 units^2.

20. Use trigonometry, the Pythagorean Theorem, and the figure to find the measures of the missing sides of the bases.

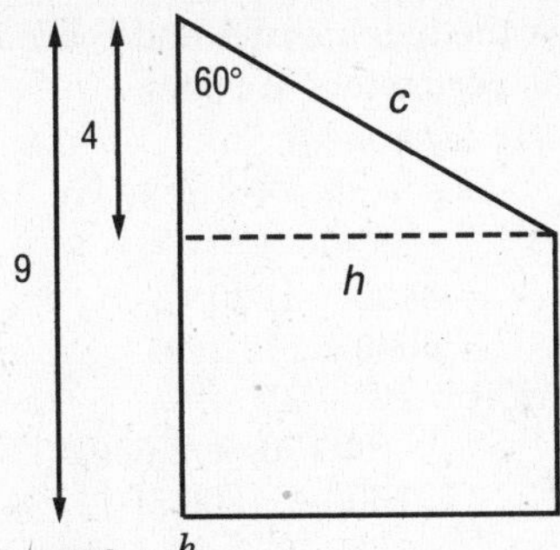

$\tan 60° = \frac{h}{4}$

$4\tan 60° = h$

$4\sqrt{3} = h$

$c^2 = a^2 + b^2$

$c^2 = 4^2 + (4\sqrt{3})^2$

$c^2 = 16 + 48$

$c^2 = 64$

$c = 8$

Find the surface area.

$T = Ph + 2B$

$= (5 + 8 + 9 + 4\sqrt{3})(11) + 2 \cdot \frac{1}{2}(4\sqrt{3})(5 + 9)$

≈ 415.2

The surface area is approximately 415.2 units^2.

21. Use trigonometry, the Pythagorean Theorem, and the figure to find the measures of the missing sides of the bases.

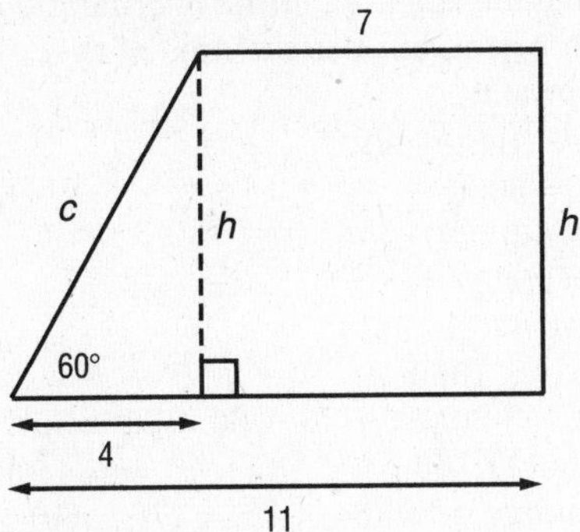

$\tan 60° = \frac{h}{4}$

$4\tan 60° = h$

$4\sqrt{3} = h$

$c^2 = a^2 + b^2$

$c^2 = 4^2 + (4\sqrt{3})^2$

$c^2 = 16 + 48$

$c^2 = 64$

$c = 8$

Find the surface area.

$T = Ph + 2B$

$= (7 + 8 + 11 + 4\sqrt{3})(10) + 2 \cdot \frac{1}{2}(4\sqrt{3})(7 + 11)$

≈ 454.0

The surface area is approximately 454.0 units^2.

22. $L = Ph$

$= (15 + 15 + 15 + 15)(10)$

$= 600$

No, the walls are 600 ft^2; 1.5 gallons will only be enough for 1 coat.

23. $L = Ph$

$= (15 + 15 + 15 + 15)(10)$

$= 600$

$\frac{600}{400} = 1.5$ gallons needed for 1 coat

So, 3 gallons are needed for 2 coats.

24. $T = L + B$

$= Ph + B$

$= (15 + 15 + 15 + 15)(10) + 15^2$

$= 825$

For two coats, the surface area is 1650 ft^2.

$\frac{1650}{400} = 4.125$

Since only whole gallons may be purchased, 5 gallons must be purchased to paint the walls and ceiling.

$5 \times \$16 = \$80$, so it will cost $80 to paint the walls and ceiling.

25. Estimate the surface area of the Corn Palace using a rectangular prism.

$L = Ph$

$= (2 \cdot 310 + 2 \cdot 185)(45)$

$= 44{,}550$

The area to be covered is estimated to be 44,550 ft^2.

26. $L = Ph$

$= (2 \cdot 310 + 2 \cdot 185)(45)$

$= 44{,}550$

$\frac{44{,}550}{15} = 2970$

It takes 2970 bushels of grain to cover the Corn Palace.

27. The actual amount needed will be higher because the area of the curved architectural elements appears to be greater than the area of the doors.

28. Use the Pythagorean Theorem to find the missing measure.

$c^2 = a^2 + b^2$

$c^2 = 2^2 + (7 - 6)^2$

$c^2 = 4 + 1$

$c = \sqrt{5}$

Find the area of the trapezoids.

$A = \frac{1}{2}h(b_1 + b_2)$

$= \frac{1}{2}(2)(6 + 7)$

$= 13$

Surface area $= 2(13) + 6^2 + 6 \cdot \sqrt{5}$

$= 26 + 36 + 6\sqrt{5}$

≈ 75.4

The surface area of the glass is approximately 75.4 ft^2.

29. Use the Pythagorean Theorem to find the measures of the third sides of the triangular bases.

Prism A:

$c^2 = a^2 + b^2$

$c^2 = 3^2 + 4^2$

$c^2 = 9 + 16$

$c^2 = 25$

$c = 5$

Prism B:
$c^2 = a^2 + b^2$
$c^2 = 6^2 + 8^2$
$c^2 = 36 + 64$
$c^2 = 100$
$c = 10$
Prism C:
$c^2 = a^2 + b^2$
$5^2 = a^2 + 3^2$
$25 = a^2 + 9$
$16 = a^2$
$4 = a$
The base of Prism $A \cong$ the base of Prism C because of the SSS Postulate.
The sides of the bases of Prism B are proportional to the sides of the bases of Prisms A and C so base of $A \sim$ base of B and base of $C \sim$ base of B.

30. Using the side lengths calculated in Exercise 29, the perimeters of the bases are as follows.
Prism A: $P = 3 + 4 + 5 = 12$
Prism B: $P = 6 + 8 + 10 = 24$
Prism C: $P = 3 + 4 + 5 = 12$
So, the ratios of the perimeters of the bases are A : B = 1 : 2, B : C = 2 : 1, and A : C = 1 : 1.

31. Prism A:
$B = \frac{1}{2}bh$
$= \frac{1}{2}(3)(4)$
$= 6$
Prism B:
$B = \frac{1}{2}bh$
$= \frac{1}{2}(6)(8)$
$= 24$
Prism C:
Use the Pythagorean Theorem to find the measure of the third side of the triangular base.
$c^2 = a^2 + b^2$
$5^2 = a^2 + 3^2$
$25 = a^2 + 9$
$16 = a^2$
$4 = a$
$B = \frac{1}{2}bh$
$= \frac{1}{2}(3)(4)$
$= 6$
So, the ratios of the areas of the bases of the prisms are A : B = 1 : 4, B : C = 4 : 1, and A : C = 1 : 1.

32. Using the values of the perimeters calculated in Exercise 30, the surface areas are as follows.
Prism A:
$T = Ph + 2B$
$= (12)(6.5) + 2 \cdot \frac{1}{2} \cdot 3 \cdot 4$
$= 90$
Prism B:
$T = Ph + 2B$
$= (24)(13) + 2 \cdot \frac{1}{2} \cdot 6 \cdot 8$
$= 360$
Prism C:
$T = Ph + 2B$
$= (12)(10) + 2 \cdot \frac{1}{2} \cdot 3 \cdot 4$
$= 132$
So, the ratios of the surface areas of the prisms are A : B = 1 : 4, B : C = 30 : 11, and A : C = 15 : 22.

33. A and B, because the heights of A and B are in the same ratio as perimeters of bases.

34. Surface area of TV $= Ph + 2B$
$= (2 \cdot 20 + 2 \cdot 30)(84) + 2(20 \cdot 30)$
$= 8400 + 1200$
$= 9600 \text{ cm}^2$
Surface area of VCR $= Ph + 2B$
$= (100)(76) + 2(600)$
$= 7600 + 1200$
$= 8800 \text{ cm}^2$
Surface area of CD $= Ph + 2B$
$= (100)(60) + 2(600)$
$= 6000 + 1200$
$= 7200 \text{ cm}^2$
Surface area of video game system
$= Ph + 2B$
$= (100)(39) + 2(600)$
$= 3900 + 1200$
$= 5100 \text{ cm}^2$
Surface area of DVD $= Ph + 2B$
$= (100)(35) + 2(600)$
$= 3500 + 1200$
$= 4700 \text{ cm}^2$

35. No, the surface area of the finished product will be the sum of the lateral areas of each prism plus the area of the bases of the TV and DVD prisms. It will also include the area of the overhang between each prism, but not the area of the overlapping prisms.

36. Area of ends $= 10(20 \cdot 30)$
$= 6000$
Area of top $= (30)(84)$
$= 2520$
Area of bottom $= (30)(35)$
$= 1050$
Area of sides $= 2(20)(84 + 76 + 60 + 39 + 35)$
$= 11{,}760$
Area of overhangs $= 30[(84 - 76) + (76 - 60) + (60 - 39) + (39 - 35)]$
$= 1470$
Total surface area
$= 6000 + 2520 + 1050 + 11{,}760 + 1470$
$= 22{,}800$
The total surface area of the finished model is $22{,}800 \text{ cm}^2$.

37. $L = Ph = 144$
$\ell = 3w$
$h = 2w$
Find the perimeter in terms of h.
$P = 2\ell + 2w$
$= 2(3w) + 2w$
$= 8w$
$= 8\left(\frac{h}{2}\right)$
$= 4h$

Find h.
$144 = Ph$
$144 = (4h)h$
$144 = 4h^2$
$36 = h^2$
$6 = h$
So, $P = (4)(6) = 24$, $w = \frac{6}{2} = 3$, and $\ell = (3)(3) = 9$.
$T = Ph + 2B$
$= (24)(6) + 2(3)(9)$
$= 144 + 54$
$= 198$
The surface area is 198 cm^2.

38. Sample answer: Brick masons use the measurements of the structure and the measurements of the bricks to find the number of bricks that will be needed. Answers should include the following.
- The lateral area is important because the sides of the brick will show. Also, depending on the project, only the lateral area of the structure may be covered with brick.
- It is important to overestimate the number of bricks ordered in case some are damaged or the calculations were inaccurate.

39. B; $121.5 = 6s^2$ where s is the length of each edge.
$s^2 = \frac{121.5}{6}$
$s = \sqrt{\frac{121.5}{6}}$
$s = 4.5$ m

40. D; $\frac{a^2 - 16}{4a - 16} = \frac{(a + 4)(a - 4)}{4(a - 4)} = \frac{a + 4}{4}, a \neq 4$

41. $L = 2 \cdot 20 \cdot 18 + 2 \cdot 16 \cdot 21$
$L = 1392\ cm^2$
$T = L + 2B$
$T = 1392 + 2(20)(16)$
$T = 2032\ cm^2$

42. $L = 4 \cdot 2.17 + 4.47 \cdot 2.17 + 1 \cdot 2.43$
$L = 20.81\ cm^2$
$T = L + 2B$
$T = 20.81 + 2 \cdot \frac{1}{2} \cdot 0.82 \cdot 4.47$
$T = 24.48\ cm^2$

43. See students' work.

Page 654 Maintain Your Skills

44. Use the Pythagorean Theorem to find the measure of the third side of the triangular base.
$c^2 = a^2 + b^2$
$c^2 = 6^2 + 8^2$
$c^2 = 36 + 64$
$c^2 = 100$
$c = 10$

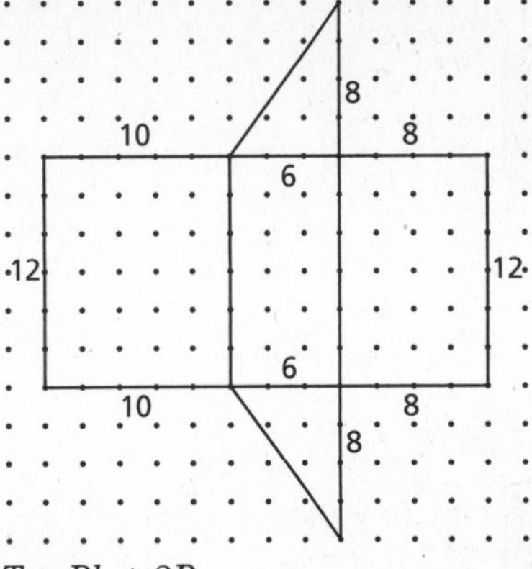

$T = Ph + 2B$
$T = (6 + 8 + 10)(12) + 2 \cdot \frac{1}{2} \cdot 6 \cdot 8$
$T = 288 + 48$
$T = 336$
The surface area of the triangular prism is 336 $units^2$.

45.

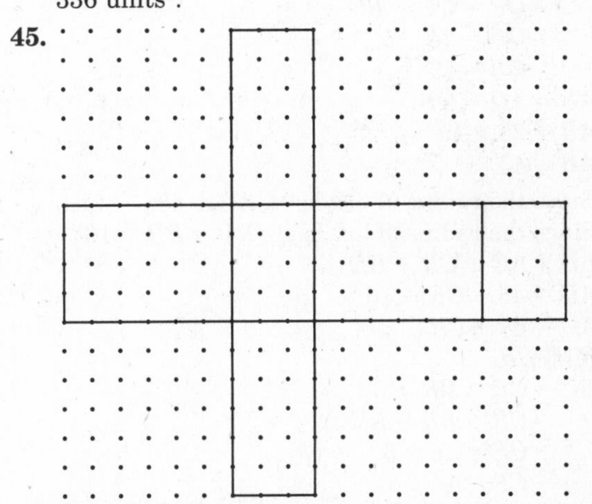

$T = Ph + 2B$
$T = (2 \cdot 3 + 2 \cdot 4)(6) + 2(3)(4)$
$T = 84 + 24$
$T = 108$
The surface area of the rectangular prism is 108 $units^2$.

46.

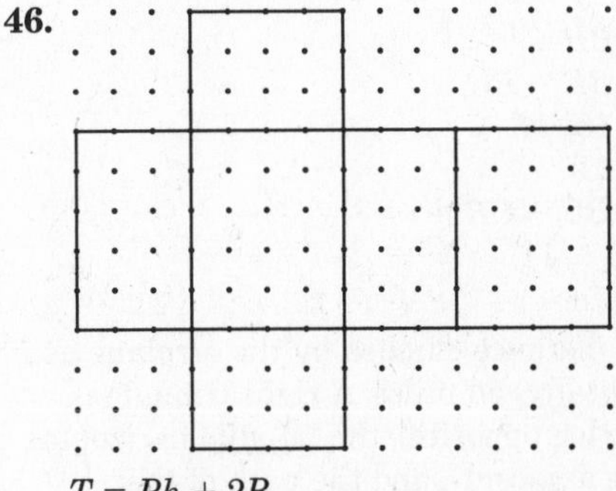

$T = Ph + 2B$
$T = (2 \cdot 4 + 2 \cdot 5)(3) + 2(4)(5)$
$T = 54 + 40$
$T = 94$
The surface area of the rectangular prism is 94 $units^2$.

47.

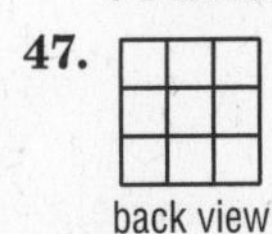

back view

48. 

49. Since the radius of $\odot Q$ is 24, $AQ = QC = 24$.
$QB + BC = QC$
$QB = QC - BC$
$QB = 24 - 5$
$QB = 19$
$AB = AQ + QB$
$AB = 24 + 19$
$AB = 43$

50. Since the radius of $\odot Q$ is 24, $AC = 48$.
Since the radius of $\odot R$ is 16, $RD = BR = 16$.
$AD = AC + CR + RD$
$AD = 48 + CR + 16$
$AD = 64 + CR$
Find CR.
$BC + CR = BR$
$CR = BR - BC$
$CR = 16 - 5$
$CR = 11$
So, $AD = 64 + 11 = 75$.

51. Since the radius of $\odot Q$ is 24, $QC = 24$.
Since the radius of $\odot R$ is 16, $BR = 16$.
$QB + BC = QC$
$QB = QC - BC$
$QB = 24 - 5$
$QB = 19$
$BC + CR = BR$
$CR = BR - BC$
$CR = 16 - 5$
$CR = 11$
$QR = QB + BC + CR$
$QR = 19 + 5 + 11$
$QR = 35$

52. Let x be the distance climbed by the airplane as it flies (horizontally) 50 miles. A right triangle is formed by x (leg opposite), the 50-mile horizontal distance (leg adjacent), and the path of the airplane (hypotenuse). Find x.
$\tan 3.5° = \frac{\text{leg opposite}}{\text{leg adjacent}}$
$\tan 3.5° = \frac{x}{50}$
$50 \tan 3.5° = x$
Use a calculator to find x.
Keystrokes: 50 [TAN] 3.5 [ENTER]
$x \approx 3.1$ mi
The total height above sea level is approximately $3 + 3.1 = 6.1$ mi.

53. To find this ratio, convert the height of the house to inches from feet, then divide the height of the drawing by the height of the house.
$\frac{\text{height of drawing in inches}}{\text{height of house in inches}} = \frac{5.5}{33 \cdot 12} = \frac{5.5}{396} = \frac{1}{72}$
The scale factor of the drawing is $\frac{1}{72}$.

54. Use a calculator.
$A = \pi r^2$
$= \pi(40)^2$
$\approx 5026.55 \text{ cm}^2$

55. Use a calculator.
$A = \pi r^2$
$= \pi\left(\frac{d}{2}\right)^2$
$= \pi\left(\frac{50}{2}\right)^2$
$= \pi(25)^2$
$\approx 1963.50 \text{ in}^2$

56. Use a calculator.
$A = \pi r^2$
$= \pi(3.5)^2$
$\approx 38.48 \text{ ft}^2$

57. Use a calculator.
$A = \pi r^2$
$= \pi(82)^2$
$\approx 21{,}124.07 \text{ mm}^2$

12-4 Surface Areas of Cylinders

Page 657 Check for Understanding

1. Multiply the circumference of the base by the height and add the area of each base.

2. Sample answer:

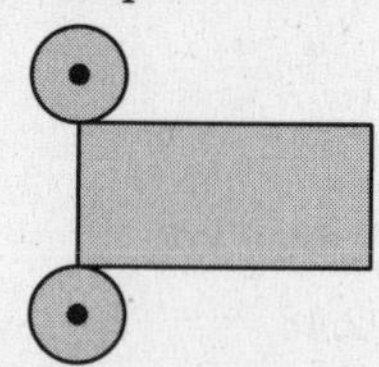

3. Jamie; since the cylinder has one base removed, the surface area will be the sum of the lateral area and one base.

4. Use a calculator.
$T = 2\pi rh + 2\pi r^2$
$= 2\pi(4)(6) + 2\pi(4)^2$
≈ 251.3
The surface area is approximately 251.3 ft^2.

5. Use a calculator.
$T = 2\pi rh + 2\pi r^2$
$= 2\pi\left(\frac{d}{2}\right)h + 2\pi\left(\frac{d}{2}\right)^2$
$= \pi dh + \frac{1}{2}\pi d^2$
$= \pi(22)(11) + \frac{1}{2}\pi(22)^2$
≈ 1520.5
The surface area is approximately 1520.5 m^2.

6. Use the formula for surface area to write and solve an equation for the radius.
$T = 2\pi rh + 2\pi r^2$
$96\pi = 2\pi r(8) + 2\pi r^2$
$96\pi = 16\pi r + 2\pi r^2$
$48 = 8r + r^2$
$0 = r^2 + 8r - 48$
$0 = (r + 12)(r - 4)$
$r = 4$ or -12
Since the radius of a circle cannot have a negative value, -12 is eliminated. So, the radius of the base is 4 cm.

7. Use the formula for surface area to write and solve an equation for the radius.
$T = 2\pi rh + 2\pi r^2$
$140\pi = 2\pi r(9) + 2\pi r^2$
$140\pi = 18\pi r + 2\pi r^2$
$70 = 9r + r^2$
$0 = r^2 + 9r - 70$
$0 = (r + 14)(r - 5)$
$r = 5$ or -14
Since the radius of a circle cannot have a negative value, -14 is eliminated. So, the radius of the base is 5 ft.

8. Find the area of one label.
$L = 2\pi rh$
$= 2\pi\left(\frac{d}{2}\right)h$
$= \pi dh$
$= \pi(2.5)(4)$
$= 10\pi$
Find the area of 3258 labels. Use a calculator.
$3258L = 3258(10\pi)$
$\approx 102{,}353.1$
The total area is approximately 102,353.1 in^2.

Pages 657–658 Practice and Apply

9. Use a calculator.
$T = 2\pi rh + 2\pi r^2$
$= 2\pi(13)(15.8) + 2\pi(13)^2$
≈ 2352.4
The surface area is approximately 2352.4 m^2.

10. Use a calculator.
$T = 2\pi rh + 2\pi r^2$
$= 2\pi\left(\frac{d}{2}\right)h + 2\pi\left(\frac{d}{2}\right)^2$
$= \pi dh + \frac{1}{2}\pi d^2$
$= \pi(13.6)(1.9) + \frac{1}{2}\pi(13.6)^2$
≈ 371.7
The surface area is approximately 371.7 ft^2.

11. Use a calculator.
$T = 2\pi rh + 2\pi r^2$
$= 2\pi\left(\frac{d}{2}\right)h + 2\pi\left(\frac{d}{2}\right)^2$
$= \pi dh + \frac{1}{2}\pi d^2$
$= \pi(14.2)(4.5) + \frac{1}{2}\pi(14.2)^2$
≈ 517.5
The surface area is approximately 517.5 in^2.

12. Use a calculator.
$T = 2\pi rh + 2\pi r^2$
$= 2\pi(14)(14) + 2\pi(14)^2$
≈ 2463.0
The surface area is approximately 2463.0 mm^2.

13. Use a calculator.
$T = 2\pi rh + 2\pi r^2$
$= 2\pi(4)(6) + 2\pi(4)^2$
≈ 251.3
The surface area is approximately 251.3 ft^2.

14. Use a calculator.
$T = 2\pi rh + 2\pi r^2$
$= 2\pi\left(\frac{d}{2}\right)h + 2\pi\left(\frac{d}{2}\right)^2$
$= \pi dh + \frac{1}{2}\pi d^2$
$= \pi(8.2)(7.2) + \frac{1}{2}\pi(8.2)^2$
≈ 291.1
The surface area is approximately 291.1 yd^2.

15. Use a calculator.
$T = 2\pi rh + 2\pi r^2$
$= 2\pi(0.9)(4.4) + 2\pi(0.9)^2$
≈ 30.0
The surface area is approximately 30.0 cm^2.

16. Use a calculator.
$T = 2\pi rh + 2\pi r^2$
$= 2\pi\left(\frac{d}{2}\right)h + 2\pi\left(\frac{d}{2}\right)^2$
$= \pi dh + \frac{1}{2}\pi d^2$
$= \pi(9.6)(3.4) + \frac{1}{2}\pi(9.6)^2$
≈ 247.3
The surface area is approximately 247.3 m^2.

17. Use the formula for surface area to write and solve an equation for the radius.
$T = 2\pi rh + 2\pi r^2$
$48\pi = 2\pi r(5) + 2\pi r^2$
$48\pi = 10\pi r + 2\pi r^2$
$24 = 5r + r^2$
$0 = r^2 + 5r - 24$
$0 = (r + 8)(r - 3)$
$r = 3$ or -8
Since the radius of a circle cannot have a negative value, -8 is eliminated. So, the radius of the base is 3 cm.

18. Use the formula for surface area to write and solve an equation for the radius.
$T = 2\pi rh + 2\pi r^2$
$340\pi = 2\pi r(7) + 2\pi r^2$
$340\pi = 14\pi r + 2\pi r^2$
$170 = 7r + r^2$
$0 = r^2 + 7r - 170$
$0 = (r + 17)(r - 10)$
$r = 10$ or -17
Since the radius of a circle cannot have a negative value, -17 is eliminated. So, the radius of the base is 10 in.

19. Use the formula for surface area to write and solve an equation for the radius.

$T = 2\pi rh + 2\pi r^2$
$320\pi = 2\pi r(12) + 2\pi r^2$
$320\pi = 24\pi r + 2\pi r^2$
$160 = 12r + r^2$
$0 = r^2 + 12r - 160$
$0 = (r + 20)(r - 8)$
$r = 8$ or -20

Since the radius of a circle cannot have a negative value, -20 is eliminated. So, the radius of the base is 8 m.

20. Use the formula for surface area to write and solve an equation for the radius.

$T = 2\pi rh + 2\pi r^2$
$425.1 = 2\pi r(6.8) + 2\pi r^2$
$425.1 = 13.6\pi r + 2\pi r^2$
$0 = 2\pi r^2 + 13.6\pi r - 425.1$

Use the quadratic formula to find r.

$a = 2\pi$
$b = 13.6\pi$
$c = -425.1$

$$r = \frac{-b \pm \sqrt{b^2 - 4ac}}{2a}$$
$$= \frac{-13.6\pi \pm \sqrt{(13.6\pi)^2 - 4(2\pi)(-425.1)}}{2(2\pi)}$$
$$\approx 5.5 \text{ or } -12.3$$

Since the radius of a circle cannot have a negative value, -12.3 is eliminated. So, the radius of the base is approximately 5.5 ft.

21. Since $L = 2\pi rh$, the lateral areas will be in the ratio 3 : 2 : 1.

$L_1 = 2\pi\left(\frac{5}{2}\right)(9)$
$= 45\pi$
$L_2 = 2\pi\left(\frac{5}{2}\right)(6)$
$= 30\pi$
$L_3 = 2\pi\left(\frac{5}{2}\right)(3)$
$= 15\pi$

The lateral areas are approximately 141.4 in^2, 94.2 in^2, and 47.1 in^2.

22. To find the amount of aluminum foil needed to cover the inside of the reflector, divide the formula for surface area of a cylinder by 2.

$T = \frac{2\pi rh + 2\pi r^2}{2}$
$= \pi rh + \pi r^2$
$= \pi\left(\frac{d}{2}\right)h + \pi\left(\frac{d}{2}\right)^2$
$= \frac{1}{2}\pi dh + \frac{1}{4}\pi d^2$
$= \frac{1}{2}\pi\left(5\frac{1}{2}\right)(18) + \frac{1}{4}\pi\left(5\frac{1}{2}\right)^2$
≈ 179.3

Approximately 179.3 in^2 of aluminum foil is needed.

23. $T = L + 2B$
$= 2\pi rh + 2\pi r^2$

Triple the height.

$T = 2\pi r(3h) + 2\pi r^2$
$= 3(2\pi rh) + 2\pi r^2$
$= 3L + 2B$

The lateral area is tripled. The surface area is increased, but not tripled.

24. $L = 2\pi rh$
$= 2\pi\left(\frac{d}{2}\right)h$
$= \pi dh$
$= \pi(5)(13)$
≈ 204.2

The lateral area of the silo is approximately 204.2 m^2.

25. From Exercise 24, $L = 65\pi$. Use the formula for lateral area to write and solve an equation for the radius.

$L = 2\pi rh$
$65\pi = 2\pi r(26)$
$1.25 = r$

The radius of the silo is 1.25 m.

26.

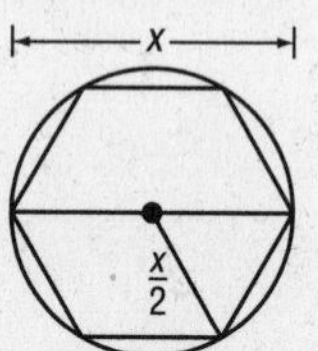

Let the diameter of the circle be x. A regular hexagon can be separated into 6 congruent nonoverlapping equilateral triangles. The sides of each triangle are $\frac{x}{2}$. The perimeter of the hexagon is $3x$. The lateral area of the hexagonal pencil is $33x$. The radius of the circle is also $\frac{x}{2}$. The circumference of the circle is $2\pi\left(\frac{x}{2}\right)$ or πx. The lateral area is approximately $34.6x$ square inches. The cylindrical pencil has the greater surface area.

27. Sample answer: Extreme sports participants use a semicylinder for a ramp. Answers should include the following.

- To find the lateral area of a semicylinder like the half-pipe, multiply the height by the circumference of the base and then divide by 2.
- A half-pipe ramp is half of a cylinder if the ramp is an equal distance from the axis of the cylinder.

28. B; $T = 2\pi rh + 2\pi r^2$
$= 2\pi\left(\frac{d}{2}\right)h + 2\pi\left(\frac{d}{2}\right)^2$
$= \pi dh + \frac{1}{2}\pi d^2$
$= \pi(8.2)(13.4) + \frac{1}{2}\pi(8.2)^2$
$\approx 450.8 \text{ cm}^2$

29. C; let x be the number of adult tickets sold. Then $200 - x$ is the number of student tickets sold.

total sales = adult sales + student sales
$500 = 5x + 2(200 - x)$
$500 = 5x + 400 - 2x$
$100 = 3x$
$33.3 \approx x$

Since the total sales were more than \$500, the minimum number of adult tickets sold was 34.

30. The locus of points 5 units from a given line is a cylinder with a radius of 5 units.

31. The locus of points equidistant from two opposite vertices of a face of a cube is a plane perpendicular to the line containing the opposite vertices of the face of the cube.

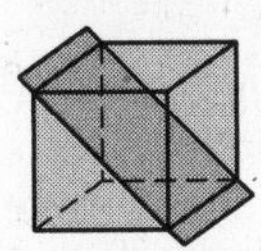

Page 659 Maintain Your Skills

32. $L(8 \times 15 \text{ base}) = Ph$
$= (2 \cdot 8 + 2 \cdot 15)(6)$
$= 276$
$L(6 \times 15 \text{ base}) = (2 \cdot 6 + 2 \cdot 15)(8)$
$= 336$
$L(8 \times 6 \text{ base}) = (2 \cdot 8 + 2 \cdot 6)(15)$
$= 420$
The lateral areas are 276 units^2 (8×15 base), 336 units^2 (6×15 base), and 420 units^2 (8×6 base).

33. Use the Pythagorean Theorem to find the measure of the third side of the triangular base.
$c^2 = a^2 + b^2$
$c^2 = 5^2 + 12^2$
$c^2 = 25 + 144$
$c^2 = 169$
$c = 13$
$L = Ph$
$= (5 + 12 + 13)(10)$
$= 300$
The lateral area is 300 units^2.

34. $L(8 \times 18 \text{ base}) = Ph$
$= (2 \cdot 8 + 2 \cdot 18)(6)$
$= 312$
$L(6 \times 18 \text{ base}) = (2 \cdot 6 + 2 \cdot 18)(8)$
$= 384$
$L(8 \times 6 \text{ base}) = (2 \cdot 8 + 2 \cdot 6)(18)$
$= 504$
The lateral areas are 312 units^2 (8×18 base), 384 units^2 (6×18 base), and 504 units^2 (8×6 base).

35.

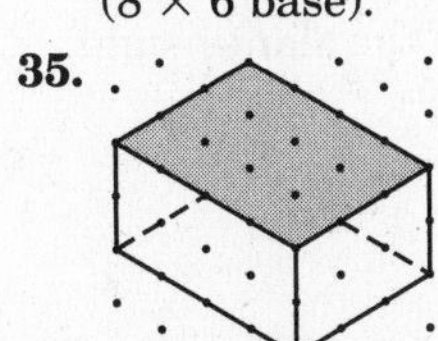

36.

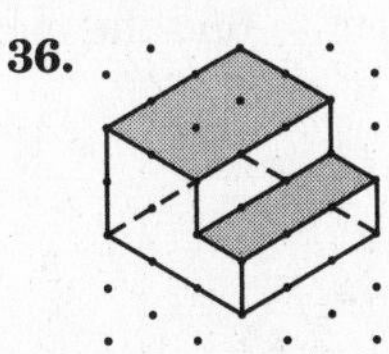

37. According to Theorem 10.11 on page 554, two segments that originate from the same exterior point and are tangent to a circle are congruent. So, $x = 27$.

38. Use the Pythagorean Theorem to find x.
$c^2 = a^2 + b^2$
$(4 + 6)^2 = x^2 + 6^2$
$100 = x^2 + 36$
$64 = x^2$
$8 = x$

39. Use the Pythagorean Theorem to find x.
$c^2 = a^2 + b^2$
$(5 + x)^2 = 5^2 + 12^2$
$25 + 10x + x^2 = 25 + 144$
$x^2 + 10x - 144 = 0$
$(x + 18)(x - 8) = 0$
$8 \text{ or } -18 = x$
Since the length of a segment cannot be negative, -18 is eliminated. So, $x = 8$.

40. Use the Law of Cosines since the measures of two sides and the included angle are known.
$a^2 = b^2 + c^2 - 2bc\cos A$
$a^2 = 6.3^2 + 7.1^2 - 2(6.3)(7.1)\cos 54°$
$a^2 = 90.1 - 89.46\cos 54°$
$a = \sqrt{90.1 - 89.46\cos 54°}$
$a \approx 6.1$
Use the Law of Sines to find $m\angle B$ and $m\angle C$.
$\frac{\sin B}{b} = \frac{\sin A}{a}$
$\sin B = \frac{b}{a}\sin A$
$B = \sin^{-1}\left(\frac{b}{a}\sin A\right)$
$B \approx \sin^{-1}\left(\frac{6.3}{6.125}\sin 54°\right)$
$B \approx 56.3°$
$\frac{\sin C}{c} = \frac{\sin A}{a}$
$C = \sin^{-1}\left(\frac{c}{a}\sin A\right)$
$C \approx \sin^{-1}\left(\frac{7.1}{6.125}\sin 54°\right)$
$C \approx 69.7°$
So, $m\angle B \approx 56.3$, $m\angle C \approx 69.7$, and $a \approx 6.1$.

41. Use the Angle Sum Theorem to find $m\angle A$.
$m\angle A + m\angle B + m\angle C = 180$
$m\angle A + 47 + 69 = 180$
$m\angle A = 64$
Use the Law of Sines to find b and c.
$\frac{\sin A}{a} = \frac{\sin B}{b}$
$b = \frac{a\sin B}{\sin A}$
$b = \frac{15\sin 47°}{\sin 64°}$
$b \approx 12.2$
$\frac{\sin A}{a} = \frac{\sin C}{c}$
$c = \frac{a\sin C}{\sin A}$
$c = \frac{15\sin 69°}{\sin 64°}$
$c \approx 15.6$
So, $m\angle A = 64$, $b \approx 12.2$, and $c \approx 15.6$.

42. $A = \frac{1}{2}bh$
$= \frac{1}{2}(20)(17)$
$= 170$
The area of the triangle is 170 in^2.

43. $A = \frac{1}{2}h(b_1 + b_2)$
$= \frac{1}{2}(6)(7 + 11)$
$= 54$
The area of the trapezoid is 54 cm^2.

44. $A = \frac{1}{2}bh$
$= \frac{1}{2}(38)(13)$
$= 247$
The area of the triangle is 247 mm^2.

Page 659 Practice Quiz 1

1. 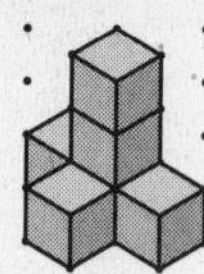

corner view

2. 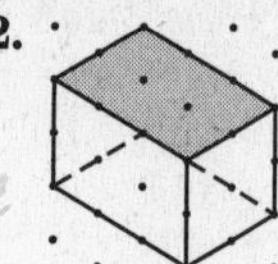

3. Use the Pythagorean Theorem to find the measure of the third side of the triangular base.

$c^2 = a^2 + b^2$
$8^2 = a^2 + 6^2$
$64 = a^2 + 36$
$28 = a^2$
$\sqrt{28} = a$
$2\sqrt{7} = a$
$L = Ph$
$= (6 + 8 + 2\sqrt{7})(12)$
≈ 231.5

The lateral area of the prism is approximately 231.5 m^2.

4. From Question 3, $L = (6 + 8 + 2\sqrt{7})(12)$

$T = L + 2B$
$= Ph + 2B$
$= (6 + 8 + 2\sqrt{7})(12) + 2 \cdot \frac{1}{2} \cdot 6 \cdot 2\sqrt{7}$
≈ 263.2

The surface area of the prism is approximately 263.2 m^2.

5. Use the formula for surface area to write and solve an equation for the radius.

$T = 2\pi rh + 2\pi r^2$
$560 = 2\pi r(11) + 2\pi r^2$
$\frac{280}{\pi} = 11r + r^2$
$0 = r^2 + 11r - \frac{280}{\pi}$

Use the quadratic formula to find r.

$a = 1$
$b = 11$
$c = -\frac{280}{\pi}$
$r = \frac{-b \pm \sqrt{b^2 - 4ac}}{2a}$
$= \frac{-11 \pm \sqrt{11^2 - 4(1)\left(-\frac{280}{\pi}\right)}}{2(1)}$
≈ 5.4 or -16.4

Since the radius of a circle cannot have a negative value, -16.4 is eliminated. So, the radius of the base is approximately 5.4 ft.

12-5 Surface Areas of Pyramids

Page 663 Check for Understanding

1. Sample answer:

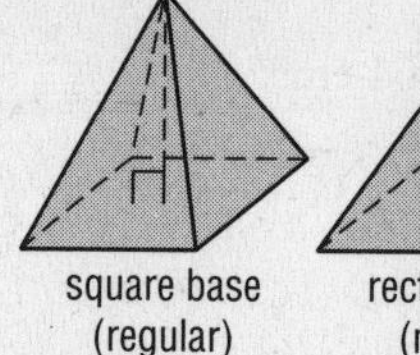

square base (regular)

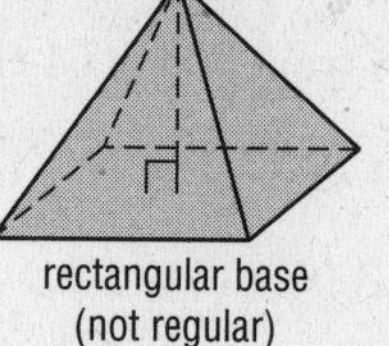

rectangular base (not regular)

2. A regular pyramid is only a regular polyhedron if all of the faces including the base are congruent regular polygons. Since the faces of a pyramid are triangles, the only regular pyramid that is also a regular polyhedron is a tetrahedron.

3. The slant height ℓ is the hypotenuse of a right triangle with legs that are the altitude and a segment with a length that is one-half the side measure of the base. Use the Pythagorean Theorem to find the slant height of the regular pyramid.

$c^2 = a^2 + b^2$
$\ell^2 = 2^2 + 7^2$
$\ell = \sqrt{53}$

Find the surface area.

$T = \frac{1}{2}P\ell + B$
$= \frac{1}{2}(4 + 4 + 4 + 4)\sqrt{53} + 4^2$
≈ 74.2

The surface area of the regular pyramid is approximately 74.2 ft^2.

4. The altitude, slant height, and apothem form a right triangle. Use the Pythagorean Theorem to find the length of the apothem. Let x represent the length of the apothem.

$c^2 = a^2 + b^2$
$(3\sqrt{2})^2 = x^2 + 3^2$
$18 = x^2 + 9$
$9 = x^2$
$3 = x$

The side measure of the base is $2x = 2(3) = 6$. Find the surface area.

$T = \frac{1}{2}P\ell + B$
$= \frac{1}{2}(6 + 6 + 6 + 6)(3\sqrt{2}) + 6^2$
≈ 86.9

The surface area of the regular pyramid is approximately 86.9 cm^2.

5. The slant height ℓ is the hypotenuse of a right triangle with legs that are a lateral edge and a segment with a length that is one-half the side measure of the base. Use the Pythagorean Theorem to find the measure of the slant height.

$c^2 = a^2 + b^2$
$13^2 = \ell^2 + \left(\frac{10}{2}\right)^2$
$169 = \ell^2 + 25$
$144 = \ell^2$
$12 = \ell$

Find the surface area.

$T = \frac{1}{2}P\ell + B$
$= \frac{1}{2}(10 + 10 + 10 + 10)(12) + 10^2$
$= 340$

The surface area of the regular pyramid is 340 cm^2.

6. To find the amount of paper used for one pyramid, find the lateral area of the pyramid.

$L = \frac{1}{2}P\ell$
$= \frac{1}{2}(2 + 2 + 2 + 2)(4)$
$= 16$

Find the total amount of paper used given that there are 6 pyramids per star.

$6L = 6(16)$
$= 96$

The amount of paper used is 16 in^2 per pyramid and 96 in^2 per star.

Pages 663–665 Practice and Apply

7. $T = \frac{1}{2}P\ell + B$
$= \frac{1}{2}(7 + 7 + 7 + 7)(5) + 7^2$
$= 119$

The surface area of the regular pyramid is 119 cm^2.

8. Find the measure of the apothem of the base, x. The central angle of the hexagon is $\frac{360°}{6} = 60°$. So, the angle formed by a radius and the apothem is $\frac{60°}{2} = 30°$.

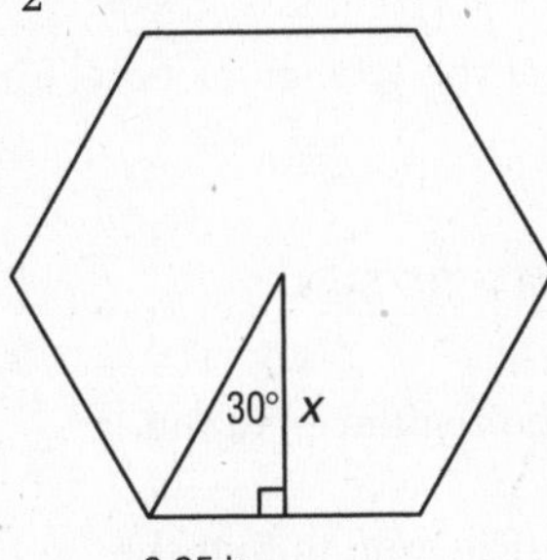

$\tan 30° = \frac{2.25}{x}$
$x = \frac{2.25}{\tan 30°}$

Find the surface area.

$T = \frac{1}{2}P\ell + \frac{1}{2}Px$
$= \frac{1}{2}P(\ell + x)$
$= \frac{1}{2}(6 \cdot 4.5)\left(6 + \frac{2.25}{\tan 30°}\right)$
≈ 133.6

The surface area of the regular pyramid is approximately 133.6 in^2.

9. Use the Pythagorean Theorem to find the height of the triangular base.

$c^2 = a^2 + b^2$
$8^2 = a^2 + \left(\frac{8}{2}\right)^2$
$64 = a^2 + 16$
$48 = a^2$
$4\sqrt{3} = a$

Find the surface area.

$T = \frac{1}{2}P\ell + B$
$= \frac{1}{2}(3 \cdot 8)(10) + \frac{1}{2}(8)(4\sqrt{3})$
≈ 147.7

The surface area of the regular pyramid is approximately 147.7 ft^2.

10. Use the Pythagorean Theorem to find the length of the apothem. Let x represent the length of the apothem.

$c^2 = a^2 + b^2$
$9^2 = x^2 + 6^2$
$81 = x^2 + 36$
$45 = x^2$
$3\sqrt{5} = x$

The side measure of the base is $2x = 2(3\sqrt{5}) = 6\sqrt{5}$. Find the surface area.

$T = \frac{1}{2}P\ell + B$
$= \frac{1}{2}(4 \cdot 6\sqrt{5})(9) + (6\sqrt{5})^2$
≈ 421.5

The surface area of the regular pyramid is approximately 421.5 cm^2.

11. Use the Pythagorean Theorem to find the slant height ℓ.

$c^2 = a^2 + b^2$
$8^2 = \left(\frac{6}{2}\right)^2 + \ell^2$
$64 = 9 + \ell^2$
$55 = \ell^2$
$\sqrt{55} = \ell$

Find the measure of the apothem of the base, x. The central angle of the pentagon is $\frac{360°}{5} = 72°$. So, the angle formed by a radius and the apothem is $\frac{72°}{2} = 36°$.

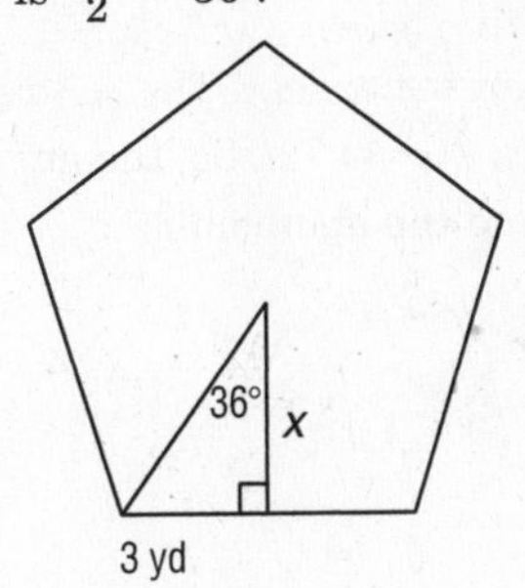

$\tan 36° = \frac{3}{x}$
$x = \frac{3}{\tan 36°}$

Find the surface area.

$T = \frac{1}{2}P\ell + \frac{1}{2}Px$
$= \frac{1}{2}P(\ell + x)$
$= \frac{1}{2}(5 \cdot 6)\left(\sqrt{55} + \frac{3}{\tan 36°}\right)$
≈ 173.2

The surface area of the regular pyramid is approximately 173.2 yd^2.

12. Use the Pythagorean Theorem to find the slant height ℓ.

$c^2 = a^2 + b^2$
$6.4^2 = \left(\frac{3.2}{2}\right)^2 + \ell^2$
$40.96 = 2.56 + \ell^2$
$38.4 = \ell^2$
$\sqrt{38.4} = \ell$

Find the measure of the apothem of the base, x. The central angle of the hexagon is $\frac{360°}{6} = 60°$. So, the angle formed by a radius and the apothem is $\frac{60°}{2} = 30°$.

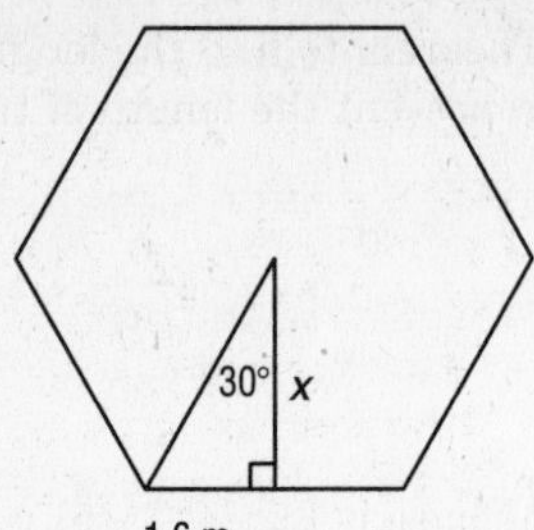

$\tan 30° = \frac{1.6}{x}$

$x = \frac{1.6}{\tan 30°}$

$x = 1.6\sqrt{3}$

Find the surface area.

$T = \frac{1}{2}P\ell + \frac{1}{2}Px$

$= \frac{1}{2}P(\ell + x)$

$= \frac{1}{2}(6 \cdot 3.2)(\sqrt{38.4} + 1.6\sqrt{3})$

≈ 86.1

The surface area of the regular pyramid is approximately 86.1 m^2.

13. Use the Pythagorean Theorem to find the apothem, x.

$c^2 = a^2 + b^2$

$13^2 = 12^2 + x^2$

$169 = 144 + x^2$

$25 = x^2$

$5 = x$

Find the side measure of the base, y. The central angle of the pentagon is $\frac{360°}{5} = 72°$. So, the angle formed by the radius and the apothem is $\frac{72°}{2} = 36°$.

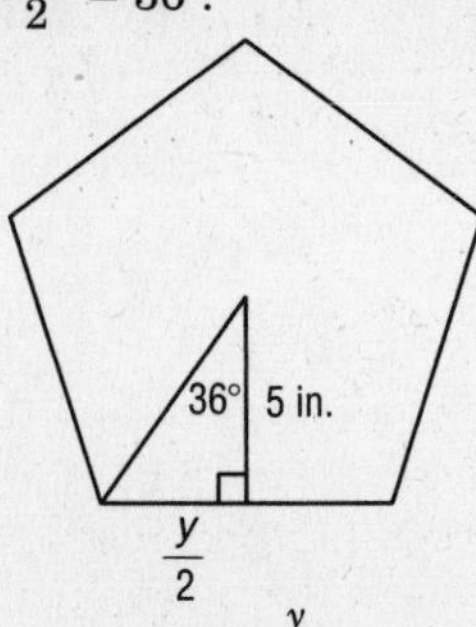

$\tan 36° = \frac{\frac{y}{2}}{5}$

$10 \tan 36° = y$

Find the surface area.

$T = \frac{1}{2}P\ell + \frac{1}{2}Px$

$= \frac{1}{2}P(\ell + x)$

$= \frac{1}{2}(5 \cdot 10 \tan 36°)(13 + 5)$

≈ 326.9

The surface area of the regular pyramid is approximately 326.9 in^2.

14. Use the Pythagorean Theorem to find the slant height ℓ.

$c^2 = a^2 + b^2$

$8^2 = \ell^2 + \left(\frac{12}{2}\right)^2$

$64 = \ell^2 + 36$

$28 = \ell^2$

$2\sqrt{7} = \ell$

Find the altitude of the triangular base.

$c^2 = a^2 + b^2$

$12^2 = a^2 + 6^2$

$144 = a^2 + 36$

$108 = a^2$

$6\sqrt{3} = a$

Find the surface area.

$T = \frac{1}{2}P\ell + B$

$= \frac{1}{2}(3 \cdot 12)(2\sqrt{7}) + \frac{1}{2}(12)(6\sqrt{3})$

≈ 157.6

The surface area of the regular pyramid is approximately 157.6 cm^2.

15. Use the Pythagorean Theorem to find the slant height ℓ.

$c^2 = a^2 + b^2$

$4^2 = \ell^2 + \left(\frac{4}{2}\right)^2$

$16 = \ell^2 + 4$

$12 = \ell^2$

$2\sqrt{3} = \ell$

ℓ is also the altitude of the triangular base. Find the surface area.

$T = \frac{1}{2}P\ell + B$

$= \frac{1}{2}(3 \cdot 4)(2\sqrt{3}) + \frac{1}{2}(4)(2\sqrt{3})$

≈ 27.7

The surface area of the regular pyramid is approximately 27.7 ft^2.

16. Use the Pythagorean Theorem to find the apothem, x.

$c^2 = a^2 + b^2$

$20^2 = 5^2 + x^2$

$400 = 25 + x^2$

$375 = x^2$

$5\sqrt{15} = x$

The side measure of the base is twice the apothem: $2x = 10\sqrt{15}$.

Find the lateral area of the roof.

$L = \frac{1}{2}P\ell$

$= \frac{1}{2}(4 \cdot 10\sqrt{15})(20)$

≈ 1549.2

The area of the roof is approximately 1549.2 ft^2.

17. Find the surface area of the first bottle.

$T = \frac{1}{2}P\ell + B$

$= \frac{1}{2}(4 \cdot 3)(4) + 3^2$

$= 33$

Find the dimensions of the base of the second bottle. Let the side measure be x.

$T = \frac{1}{2}P\ell + B$

$T = \frac{1}{2}(4x)(6) + x^2$

$33 = 12x + x^2$

$0 = x^2 + 12x - 33$

Use the quadratic formula to find x.

$a = 1$

$b = 12$

$c = -33$

$$x = \frac{-b \pm \sqrt{b^2 - 4ac}}{2a}$$
$$= \frac{-12 \pm \sqrt{12^2 - 4(1)(-33)}}{2(1)}$$
$$\approx 2.3 \text{ or } -14.3$$

The length of the side cannot be negative, so -14.3 is eliminated. The base of the second bottle is approximately 2.3 inches on each side.

18. Find the side measure of the base.

$\sqrt{360{,}000} = 600$

So, $P = 4(600) = 2400$ and the apothem is $\frac{600}{2} = 300$.

Use the Pythagorean Theorem to find the slant height ℓ.

$c^2 = a^2 + b^2$
$\ell^2 = 300^2 + 321^2$
$\ell^2 = 193{,}041$
$\ell = \sqrt{193{,}041}$

Find the lateral area.

$L = \frac{1}{2}P\ell$
$= \frac{1}{2}(2400)\sqrt{193{,}041}$
$\approx 527{,}237.2$

The lateral area of the pyramid is approximately 527,237.2 ft^2.

19. The apothem is half the length of the edge: $\frac{646}{2} = 323$. Use the Pythagorean Theorem to find the slant height ℓ.

$c^2 = a^2 + b^2$
$\ell^2 = 323^2 + 350^2$
$\ell^2 = 226{,}829$
$\ell = \sqrt{226{,}829}$

Find the lateral area to find the area of the glass.

$L = \frac{1}{2}P\ell$
$= \frac{1}{2}(4 \cdot 646)\sqrt{226{,}829}$
$\approx 615{,}335.3$

The area of the glass is approximately 615,335.3 ft^2.

20. The apothem is half the length of the side of the base: $\frac{214.5}{2} = 107.25$.

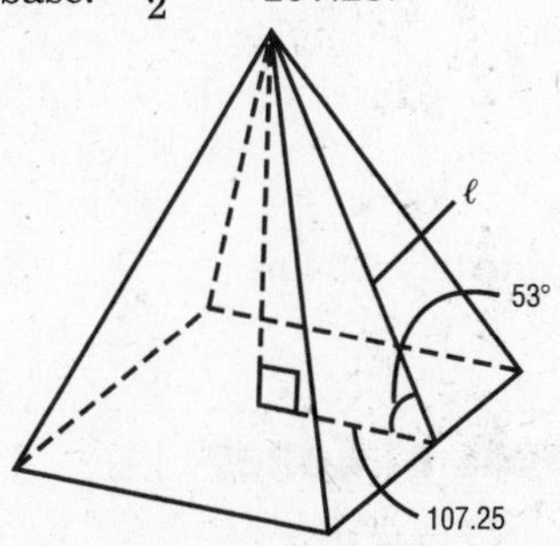

Find the slant height.

$\cos 53° = \frac{107.25}{\ell}$
$\ell = \frac{107.25}{\cos 53°}$

Find the lateral area.

$L = \frac{1}{2}P\ell$
$= \frac{1}{2}(4 \cdot 214.5)\frac{107.25}{\cos 53°}$
$\approx 76{,}452.5$

The lateral area is approximately 76,452.5 m^2.

21. Find the height of the pyramid using the Pythagorean Theorem. The apothem of the pyramid is half the length of its base, which is 12 ft: $\frac{12}{2} = 6$.

$c^2 = a^2 + b^2$
$10^2 = a^2 + 6^2$
$100 = a^2 + 36$
$64 = a^2$
$8 = a$

The height of the solid is $12 + 8 = 20$ ft.

22. Add the lateral areas of the pyramid and the cube.

$L = \frac{1}{2}P\ell + Ph$
$= \frac{1}{2}(4 \cdot 12)(10) + (4 \cdot 12)(12)$
$= 816$

The lateral area of the solid is 816 ft^2.

23. The surface area of the solid is equal to the lateral areas of the pyramid and the cube plus the area of the cube's base.

$T = \frac{1}{2}P\ell + Ph + B$
$= \frac{1}{2}(4 \cdot 12)(10) + (4 \cdot 12)(12) + 12^2$
$= 960$

The surface area of the solid is 960 ft^2.

24. Each lateral face of the frustum is a trapezoid. Find the area of one face.

$A = \frac{1}{2}h(b_1 + b_2)$
$= \frac{1}{2}\ell(b_1 + b_2)$
$= \frac{1}{2}(3)(2 + 4)$
$= 9$

The lateral area is $4A = 4(9) = 36$ yd^2.

25. Find the surface area of the truncated cube. The area of the three intact faces is 3 in^2. The truncation cuts three faces in half, leaving three triangles with a total area of 1.5 in^2. The final face is an equilateral triangle. Use the Pythagorean Theorem to find the side measure of the triangle.

$c^2 = a^2 + b^2$
$c^2 = 1^2 + 1^2$
$c^2 = 2$
$c = \sqrt{2}$

Find the altitude of the triangle.

$c^2 = a^2 + b^2$
$(\sqrt{2})^2 = a^2 + \left(\frac{\sqrt{2}}{2}\right)^2$
$2 = a^2 + \frac{1}{2}$
$1.5 = a^2$
$\sqrt{1.5} = a$

Find the area of the triangle.

$A = \frac{1}{2}bh$
$= \frac{1}{2}(\sqrt{2})\sqrt{1.5}$
$= \frac{\sqrt{3}}{2}$

The surface area is $3 + 1.5 + \frac{\sqrt{3}}{2} \approx 5.37$ in^2. The surface area of the original cube is 6 square inches. The surface area of the truncated cube is approximately 5.37 square inches. Truncating the corner of the cube reduces the surface area by about 0.63 square inch.

26. Sample answer: Pyramids are used as an alternative to rectangular prisms for the shapes of buildings. Answers should include the following.

- We need to know the dimensions of the base and slant height to find the lateral area and surface area of a pyramid.
- Sample answer: The roof of a gazebo is often a hexagonal pyramid.

27. D; $T = \frac{1}{2}P\ell + B$
$= \frac{1}{2}(20)(10) + \left(\frac{20}{4}\right)^2$
$= 125$ cm^2

28. A; $x \otimes y = \frac{1}{x - y}$
$\frac{1}{2} \otimes \frac{3}{4} = \frac{1}{\frac{1}{2} - \frac{3}{4}}$
$= \frac{1}{\frac{2}{4} - \frac{3}{4}}$
$= \frac{1}{-\frac{1}{4}}$
$= -4$

Page 665 Maintain Your Skills

29. $T = 2\pi rh + 2\pi r^2$
$= 2\pi(7)(15) + 2\pi(7)^2$
≈ 967.6
The surface area is approximately 967.6 m^2.

30. $T = 2\pi rh + 2\pi r^2$
$= 2\pi\left(\frac{d}{2}\right)h + 2\pi\left(\frac{d}{2}\right)^2$
$= \pi dh + \frac{1}{2}\pi d^2$
$= \pi(22)(14) + \frac{1}{2}\pi(22)^2$
≈ 1727.9
The surface area is approximately 1727.9 cm^2.

31. $T = 2\pi rh + 2\pi r^2$
$= 2\pi(9)(23) + 2\pi(9)^2$
≈ 1809.6
The surface area is approximately 1809.6 yd^2.

32. $T = L + 2B$
$= Ph + 2B$
$= (2 \cdot 6 + 2 \cdot 2.5)(14) + 2(6)(2.5)$
$= 268$
The surface area of the box is 268 in^2.

33. $P = 2(22) + 2(15)$
$= 44 + 30$
$= 74$
Find the height.
$\frac{h}{15} = \sin 60°$
$h = 15\sin 60°$
$h = \frac{15\sqrt{3}}{2}$
$A = bh$
$= (22)\left(\frac{15\sqrt{3}}{2}\right)$
≈ 285.8
The perimeter is 74 ft and the area is approximately 285.8 ft^2.

34. $P = 24 + 2(32) + 10 + 2(5) + 6 + (24 - 10 - 6)$
$= 122$
$A = (32)(24) + (5)(24 - 10 - 6)$
$= 808$
The perimeter is 122 m and the area is 808 m^2.

35. $P = 17 + 12 + (22 - 17) + 9 + 22 + (12 + 9 - 3 - 6) + 3(6) + 3$
$= 98$
$A = (22)(12 + 9) - (12)(22 - 17) - 6^2$
$= 366$
The perimeter is 98 m and the area is 366 m^2.

36. $\overline{FM}$ is reflected in line b, but $\overline{FM}$ lies on line b. So, the reflected image of $\overline{FM}$ is $\overline{FM}$.

37. The reflected image of $\overline{JK}$ in line a is $\overline{GF}$.

38. The reflected image of L in point M is point H.

39. The reflected image of $\overline{GM}$ in line a is $\overline{JM}$.

40. False; each pair of opposite sides must be congruent.

41. True; each pair of opposite sides is congruent.

42. $c^2 = a^2 + b^2$
$12^2 = 8^2 + b^2$
$144 = 64 + b^2$
$80 = b^2$
$4\sqrt{5} = b$
$8.9 \approx b$
The length is approximately 8.9 in.

43. $c^2 = a^2 + b^2$
$c^2 = 14^2 + 16^2$
$c^2 = 196 + 256$
$c^2 = 452$
$c = 2\sqrt{113}$
$c \approx 21.3$
The length is approximately 21.3 m.

44. $c^2 = a^2 + b^2$
$11^2 = a^2 + 6^2$
$121 = a^2 + 36$
$85 = a^2$
$\sqrt{85} = a$
$9.2 \approx a$
The length is approximately 9.2 km.

12-6 Surface Areas of Cones

Page 668 Check for Understanding

1. Sample answer:

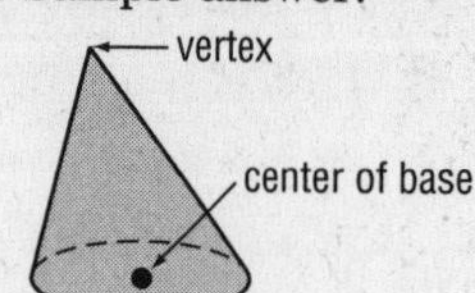

2. The formula for the lateral area is derived from the area of a sector of a circle. If the vertex of the cone is not the center of this circle, the formula is not valid.

3. $T = \pi r\ell + \pi r^2$
$= \pi(10)(17) + \pi(10)^2$
≈ 848.2
The surface area is approximately 848.2 cm^2.

4. Use the Pythagorean Theorem to find the slant height.
$c^2 = a^2 + b^2$
$\ell^2 = 12^2 + 10^2$
$\ell^2 = 144 + 100$
$\ell^2 = 244$
$\ell = 2\sqrt{61}$
$T = \pi r\ell + \pi r^2$
$= \pi(10)(2\sqrt{61}) + \pi(10)^2$
≈ 804.9
The surface area is approximately 804.9 ft^2.

5. Use the Pythagorean Theorem to find the slant height.
$c^2 = a^2 + b^2$
$\ell^2 = 8^2 + 8^2$
$\ell^2 = 128$
$\ell = 8\sqrt{2}$
$T = \pi r\ell + \pi r^2$
$= \pi(8)(8\sqrt{2}) + \pi(8)^2$
≈ 485.4
The surface area is approximately 485.4 in^2.

6. Use the Pythagorean Theorem to find the slant height.
$c^2 = a^2 + b^2$
$\ell^2 = 55^2 + \left(\frac{8.5}{2}\right)^2$
$\ell = \sqrt{3043.0625}$
$L = \pi r\ell$
$= \pi\left(\frac{8.5}{2}\right)\sqrt{3043.0625}$
≈ 736.5
The lateral area is approximately 736.5 ft^2.

Pages 668–670 Practice and Apply

7. Use the Pythagorean Theorem to find the slant height.
$c^2 = a^2 + b^2$
$\ell^2 = 5^2 + 12^2$
$\ell^2 = 169$
$\ell = 13$
$T = \pi r\ell + \pi r^2$
$= \pi(5)(13) + \pi(5)^2$
≈ 282.7
The surface area is approximately 282.7 cm^2.

8. Use the Pythagorean Theorem to find the radius.
$c^2 = a^2 + b^2$
$10^2 = r^2 + 8^2$
$100 = r^2 + 64$
$36 = r^2$
$6 = r$
$T = \pi r\ell + \pi r^2$
$= \pi(6)(10) + \pi(6)^2$
≈ 301.6
The surface area is approximately 301.6 ft^2.

9. Use the Pythagorean Theorem to find the slant height.
$c^2 = a^2 + b^2$
$\ell^2 = 9^2 + 9^2$
$\ell^2 = 162$
$\ell = 9\sqrt{2}$
$T = \pi r\ell + \pi r^2$
$= \pi(9)(9\sqrt{2}) + \pi(9)^2$
≈ 614.3
The surface area is approximately 614.3 in^2.

10. Use the Pythagorean Theorem to find the radius.
$c^2 = a^2 + b^2$
$17^2 = r^2 + 15^2$
$289 = r^2 + 225$
$64 = r^2$
$8 = r$
$T = \pi r\ell + \pi r^2$
$= \pi(8)(17) + \pi(8)^2$
≈ 628.3
The surface area is approximately 628.3 ft^2.

11. Use the Pythagorean Theorem to find the radius.
$c^2 = a^2 + b^2$
$12^2 = r^2 + 7.5^2$
$144 = r^2 + 56.25$
$87.75 = r^2$
$\sqrt{87.75} = r$
$T = \pi r\ell + \pi r^2$
$= \pi(\sqrt{87.75})(12) + \pi(\sqrt{87.75})^2$
≈ 628.8
The surface area is approximately 628.8 m^2.

12. Use the Pythagorean Theorem to find the slant height.
$c^2 = a^2 + b^2$
$\ell^2 = 2.6^2 + 6.4^2$
$\ell^2 = 47.72$
$\ell = \sqrt{47.72}$
$T = \pi r\ell + \pi r^2$
$= \pi(2.6)\sqrt{47.72} + \pi(2.6)^2$
≈ 77.7
The surface area is approximately 77.7 yd^2.

13. Use the Pythagorean Theorem to find the radius.
$c^2 = a^2 + b^2$
$18^2 = r^2 + 16^2$
$324 = r^2 + 256$
$68 = r^2$
$2\sqrt{17} = r$
$T = \pi r\ell + \pi r^2$
$= \pi(2\sqrt{17})(18) + \pi(2\sqrt{17})^2$
≈ 679.9
The surface area is approximately 679.9 in^2.

14. Use the Pythagorean Theorem to find the radius.
$c^2 = a^2 + b^2$
$19.1^2 = r^2 + 8.7^2$
$364.81 = r^2 + 75.69$
$289.12 = r^2$
$\sqrt{289.12} = r$
$T = \pi r\ell + \pi r^2$
$= \pi(\sqrt{289.12})(19.1) + \pi(\sqrt{289.12})^2$
≈ 1928.6
The surface area is approximately 1928.6 m^2.

15. Solve the surface area equation for the slant height.

$$T = \pi r\ell + \pi r^2$$
$$T - \pi r^2 = \pi r\ell$$
$$\frac{T - \pi r^2}{\pi r} = \ell$$
$$\frac{1020 - \pi(14.5)^2}{\pi(14.5)} = \ell$$
$$7.9 \approx \ell$$

The slant height is approximately 7.9 m.

16. Solve the surface area equation for the slant height.

$$T = \pi r\ell + \pi r^2$$
$$T - \pi r^2 = \pi r\ell$$
$$\frac{T - \pi r^2}{\pi r} = \ell$$
$$\frac{293.2 - \pi(6.1)^2}{\pi(6.1)} = \ell$$
$$9.2 \approx \ell$$

The slant height is approximately 9.2 ft.

17. Use the surface area equation to solve for the radius.

$$T = \pi r\ell + \pi r^2$$
$$359 = \pi r(15) + \pi r^2$$
$$0 = r^2 + 15r - \frac{359}{\pi}$$

Use the Quadratic Formula to solve for r.

$a = 1$

$b = 15$

$c = -\frac{359}{\pi}$

$$r = \frac{-b \pm \sqrt{b^2 - 4ac}}{2a}$$
$$= \frac{-15 \pm \sqrt{15^2 - 4(1)\left(-\frac{359}{\pi}\right)}}{2(1)}$$
$$\approx 5.6 \text{ or } -20.6$$

Since the radius of a circle cannot be negative, -20.6 is eliminated. So, the radius of the cone is approximately 5.6 ft.

18. Use the surface area equation to solve for the radius.

$$T = \pi r\ell + \pi r^2$$
$$523 = \pi r(12.1) + \pi r^2$$
$$0 = r^2 + 12.1r - \frac{523}{\pi}$$

Use the Quadratic Formula to solve for r.

$a = 1$

$b = 12.1$

$c = -\frac{523}{\pi}$

$$r = \frac{-b \pm \sqrt{b^2 - 4ac}}{2a}$$
$$= \frac{-12.1 \pm \sqrt{12.1^2 - 4(1)\left(-\frac{523}{\pi}\right)}}{2(1)}$$
$$\approx 8.2 \text{ or } -20.3$$

Since the radius of a circle cannot be negative, -20.3 is eliminated. So, the radius of the cone is approximately 8.2 m.

19. Use the Pythagorean Theorem to find the slant height of the cone.

$$c^2 = a^2 + b^2$$
$$\ell^2 = 4^2 + 6^2$$
$$\ell^2 = 16 + 36$$
$$\ell = 2\sqrt{13}$$
$$T = \pi r\ell + \pi r^2 + 2\pi rh$$
$$= \pi r(\ell + r + 2h)$$
$$= \pi(6)[2\sqrt{13} + 6 + 2(6)]$$
$$\approx 475.2$$

The surface area is approximately 475.2 in^2.

20.
$$T = \pi r\ell + \pi r^2 + 2\pi rh$$
$$= \pi r(\ell + r + 2h)$$
$$= \pi(3)[5 + 3 + 2(5)]$$
$$\approx 169.6$$

The surface area is approximately 169.6 ft^2.

21. Use the Pythagorean Theorem to find the slant height of the cone.

$$c^2 = a^2 + b^2$$
$$\ell^2 = 14^2 + 6.2^2$$
$$\ell = \sqrt{234.44}$$
$$T = \pi r\ell + 2\pi rh + \pi r^2$$
$$= \pi r(\ell + 2h + r)$$
$$= \pi(6.2)[\sqrt{234.44} + 2(28) + 6.2]$$
$$\approx 1509.8$$

The surface area is approximately 1509.8 m^2.

22.
$$T = \pi r\ell$$
$$= \pi\left(\frac{d}{2}\right)\ell$$
$$= \frac{1}{2}\pi(42)(47.9)$$
$$\approx 3160.1$$

The area of the canvas used is approximately 3160.1 ft^2.

23. Find the radius.

$$C = 2\pi r$$
$$\frac{C}{2\pi} = r$$
$$\frac{22}{2\pi} = r$$
$$\frac{11}{\pi} = r$$

Use the Pythagorean Theorem to find the slant height.

$$c^2 = a^2 + b^2$$
$$\ell^2 = \left(\frac{11}{\pi}\right)^2 + 18^2$$
$$\ell = \sqrt{\frac{121}{\pi^2} + 324}$$

Find the lateral area of all eight hats.

$$8L = 8\pi r\ell$$
$$= 8\pi\left(\frac{11}{\pi}\right)\sqrt{\frac{121}{\pi^2} + 324}$$
$$\approx 1613.7$$

She will use approximately 1613.7 in^2 of material.

24. Use the Pythagorean Theorem to find the slant height.

$$c^2 = a^2 + b^2$$
$$\ell^2 = 24^2 + \left(\frac{45}{2}\right)^2$$
$$\ell = \sqrt{1082.25}$$
$$L = \pi r\ell$$
$$= \pi\left(\frac{45}{2}\right)\sqrt{1082.25}$$
$$\approx 2325.4$$

The lateral area is approximately 2325.4 ft^2.

25. Use the equation for surface area to solve for the diameter.

$$T = \pi r\ell + \pi r^2$$
$$T = \pi\left(\frac{d}{2}\right)\ell + \pi\left(\frac{d}{2}\right)^2$$
$$T = \frac{1}{2}\pi d\ell + \frac{1}{4}\pi d^2$$
$$500 = \frac{1}{2}\pi d(20) + \frac{1}{4}\pi d^2$$
$$\frac{2000}{\pi} = 40d + d^2$$
$$0 = d^2 + 40d - \frac{2000}{\pi}$$

Use the Quadratic Formula to solve for d.

$a = 1$
$b = 40$
$c = -\frac{2000}{\pi}$

$$d = \frac{-b \pm \sqrt{b^2 - 4ac}}{2a}$$
$$= \frac{-40 \pm \sqrt{40^2 - 4(1)\left(-\frac{2000}{\pi}\right)}}{2(1)}$$
$$\approx 12 \text{ or } -52$$

Since the diameter of a circle cannot be negative, -52 is eliminated. So, the diameter of light on stage is approximately 12 ft.

26. Use the Pythagorean Theorem to find the slant height.

$c^2 = a^2 + b^2$
$\ell^2 = 7^2 + 4^2$
$\ell = \sqrt{65}$
$\ell \approx 8.062257748$

Use the store feature of the calculator to save ℓ.

$L = \pi r\ell$
$\approx \pi(4)(8.062257748)$
≈ 101.3133

The lateral area of the cone is approximately 101.3133 in^2.

27. Use the Pythagorean Theorem to find the slant height.

$c^2 = a^2 + b^2$
$\ell^2 = 7^2 + 4^2$
$\ell = \sqrt{65}$
$\ell \approx 8.1$
$L = \pi r\ell$
$= \pi(4)(8.1)$
≈ 101.7876

The slant height and lateral area of the cone are approximately 8.1 in. and 101.7876 in^2, respectively.

28. Use the Pythagorean Theorem to find the slant height.

$c^2 = a^2 + b^2$
$\ell^2 = 7^2 + 4^2$
$\ell = \sqrt{65}$
$\ell \approx 8.06$
$L = \pi r\ell$
$= \pi(4)(8.06)$
≈ 101.2849

The slant height and lateral area of the cone are approximately 8.06 in. and 101.2849 in^2, respectively.

29. Using the store feature on the calculator is the most accurate technique to find the lateral area. Rounding the slant height to either the tenths place or hundredths place changes the value of the slant height, which affects the final computation of the lateral area.

30. Never; the pyramid could be inscribed in the cone.

31. Sometimes; only when the heights are in the same ratio as the radii of the bases.

32. As the altitude approaches zero, the slant height of the cone approaches the radius of the base. The lateral area approaches the area of the base. The surface area approaches twice the area of the base.

33. Sample answer: Tepees are conical shaped structures. Lateral area is used because the ground may not always be covered in circular canvas. Answers should include the following.

- We need to know the circumference of the base or the radius of the base and the slant height of the cone.
- The open top reduces the lateral area of canvas needed to cover the sides. To find the actual lateral area, subtract the lateral area of the conical opening from the lateral area of the structure.

34. B; $L = \pi r\ell$
$91.5\pi = \pi r(15)$
$6.1 \text{ ft} = r$

35. D; let the odd integers be $x - 4$, $x - 2$, and x.
$3(x - 4) = 3 + 2x$
$3x - 12 = 3 + 2x$
$x = 15$

Page 670 Maintain Your Skills

36. Use the Pythagorean Theorem to find the slant height. The apothem is half the length of the base's side.

$c^2 = a^2 + b^2$
$\ell^2 = \left(\frac{149}{2}\right)^2 + 853^2$
$\ell = \sqrt{733{,}159.25}$
$L = \frac{1}{2}P\ell$
$= \frac{1}{2}(4 \cdot 149)\sqrt{733{,}159.25}$
$\approx 255{,}161.7$

The lateral area is approximately 255,161.7 ft^2.

37. $T = 2\pi rh + 2\pi r^2$
$563 = 2\pi r(9.5) + 2\pi r^2$
$0 = r^2 + 9.5r - \frac{563}{2\pi}$

Use the quadratic formula to solve for r.

$a = 1$
$b = 9.5$
$c = -\frac{563}{2\pi}$

$$r = \frac{-b \pm \sqrt{b^2 - 4ac}}{2a}$$
$$= \frac{-9.5 \pm \sqrt{9.5^2 - 4(1)\left(-\frac{563}{2\pi}\right)}}{2(1)}$$
$$\approx 5.8 \text{ or } -15.3$$

Since the radius of a circle cannot be negative, -15.3 is eliminated. So, the radius is approximately 5.8 ft.

38. $T = 2\pi rh + 2\pi r^2$
$185 = 2\pi r(11) + 2\pi r^2$
$0 = r^2 + 11r - \frac{185}{2\pi}$
Use the Quadratic Formula to solve for r.
$a = 1$
$b = 11$
$c = -\frac{185}{2\pi}$
$r = \frac{-b \pm \sqrt{b^2 - 4ac}}{2a}$
$= \frac{-11 \pm \sqrt{11^2 - 4(1)\left(-\frac{185}{2\pi}\right)}}{2(1)}$
≈ 2.2 or -13.2
Since the radius of a circle cannot be negative, -13.2 is eliminated. So, the radius is approximately 2.2 m.

39. $T = 2\pi rh + 2\pi r^2$
$470 = 2\pi r(6.5) + 2\pi r^2$
$0 = r^2 + 6.5r - \frac{470}{2\pi}$
Use the Quadratic Formula to solve for r.
$a = 1$
$b = 6.5$
$c = -\frac{470}{2\pi}$
$r = \frac{-b \pm \sqrt{b^2 - 4ac}}{2a}$
$= \frac{-6.5 \pm \sqrt{6.5^2 - 4(1)\left(-\frac{470}{2\pi}\right)}}{2(1)}$
≈ 6.0 or -12.5
Since the radius of a circle cannot be negative, -12.5 is eliminated. So, the radius is approximately 6.0 yd.

40. $T = 2\pi rh + 2\pi r^2$
$951 = 2\pi r(14) + 2\pi r^2$
$0 = r^2 + 14r - \frac{951}{2\pi}$
Use the Quadratic Formula to solve for r.
$a = 1$
$b = 14$
$c = -\frac{951}{2\pi}$
$r = \frac{-b \pm \sqrt{b^2 - 4ac}}{2a}$
$= \frac{-14 \pm \sqrt{14^2 - 4(1)\left(-\frac{951}{2\pi}\right)}}{2(1)}$
≈ 7.2 or -21.2
Since the radius of a circle cannot be negative, -21.2 is eliminated. So, the radius is approximately 7.2 cm.

41. Since the radius $\overline{MK}$ is perpendicular to the chord $\overline{FG}$, it bisects the chord. So, FG is twice FL, or 48.

42. Since the radius $\overline{MP}$ is perpendicular to the chord $\overline{HJ}$, it bisects the chord. So, NJ is half of HJ, or 24.

43. Since the radius $\overline{MP}$ is perpendicular to the chord $\overline{HJ}$, it bisects the chord. So, HN is half of HJ, or 24.

44. Since the radius $\overline{MK}$ is perpendicular to the chord $\overline{FG}$, it bisects the chord. So, LG is equal to FL, or 24.

45. Since the radius $\overline{MP}$ is perpendicular to the chord $\overline{HJ}$, it bisects the chord and its arc. So, $m\widehat{PJ}$ is equal to $m\widehat{HP}$, or 45.

46. Since the radius $\overline{MP}$ is perpendicular to the chord $\overline{HJ}$, it bisects the chord and its arc. So, $m\widehat{HJ}$ is twice $m\widehat{HP}$, or 90.

47. Let x represent the geometric mean.
$\frac{7}{x} = \frac{x}{63}$
$x^2 = 441$
$x = 21$

48. Let x represent the geometric mean.
$\frac{8}{x} = \frac{x}{18}$
$x^2 = 144$
$x = 12$

49. Let x represent the geometric mean.
$\frac{16}{x} = \frac{x}{44}$
$x^2 = 704$
$x = 8\sqrt{11} \approx 26.5$

50. $C = 2\pi r$
$= 2\pi(6)$
≈ 37.7

51. $C = \pi d$
$= \pi(8)$
≈ 25.1

52. $C = \pi d$
$= \pi(18)$
≈ 56.5

53. $C = 2\pi r$
$= 2\pi(8.2)$
≈ 51.5

54. $C = \pi d$
$= \pi(19.8)$
≈ 62.2

55. $C = 2\pi r$
$= 2\pi(4.1)$
≈ 25.8

Page 670 Practice Quiz 2

1. Use the Pythagorean Theorem to find the slant height. The apothem is equal to half the base's edge.
$c^2 = a^2 + b^2$
$\ell^2 = 6^2 + 10^2$
$\ell = 2\sqrt{34}$
$T = \frac{1}{2}P\ell + B$
$= \frac{1}{2}(4 \cdot 12)(2\sqrt{34}) + 12^2$
≈ 423.9
The surface area is approximately 423.9 cm^2.

2. Find the apothem of the base, x. The central angle of the hexagon is $\frac{360°}{6} = 60°$. So, the angle formed by the radius and the apothem is $\frac{60°}{2} = 30°$.

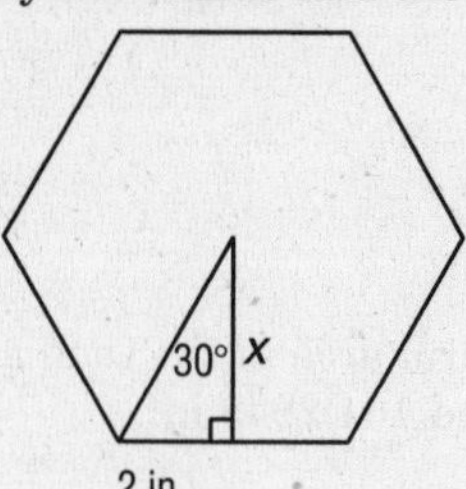

$\tan 30° = \frac{2}{x}$
$x = \frac{2}{\tan 30°}$
$x = 2\sqrt{3}$

Find the surface area.

$T = \frac{1}{2}P\ell + \frac{1}{2}Px$
$= \frac{1}{2}P(\ell + x)$
$= \frac{1}{2}(6 \cdot 4)(11 + 2\sqrt{3})$
≈ 173.6

The surface area is approximately 173.6 in^2.

3. Use the Pythagorean Theorem to find the slant height.
$c^2 = a^2 + b^2$
$\ell^2 = 3^2 + 12^2$
$\ell = 3\sqrt{17}$
$T = \pi r\ell + \pi r^2$
$= \pi(3)(3\sqrt{17}) + \pi(3)^2$
≈ 144.9
The surface area is approximately 144.9 ft^2.

4. Use the Pythagorean Theorem to find the slant height.
$c^2 = a^2 + b^2$
$\ell^2 = 6^2 + 2^2$
$\ell = 2\sqrt{10}$
$T = \pi r\ell + \pi r^2$
$= \pi(6)(2\sqrt{10}) + \pi(6)^2$
≈ 232.3
The surface area is approximately 232.3 m^2.

5. Use the equation for the lateral area to solve for the slant height.
$L = \pi r\ell$
$\ell = \frac{L}{\pi r}$
$\ell = \frac{123}{\pi(10)}$
$\ell \approx 3.9$
The slant height is approximately 3.9 in.

12-7 Surface Areas of Spheres

Page 672 Geometry Activity: Surface Area of a Sphere

1. $\frac{1}{4}$

2. πr^2

3. The surface area of a sphere is 4 times the area of the great circle.

Page 674 Check for Understanding

1. Sample answer:

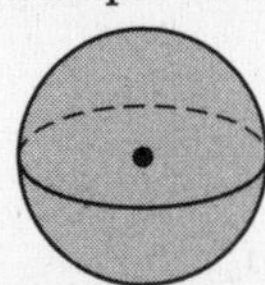

2. Tim; the surface area of a hemisphere is half of the surface area of the sphere plus the area of the great circle.

3. Use the Pythagorean Theorem to find AB.
$AB^2 = AC^2 + BC^2$
$AB^2 = 9^2 + 12^2$
$AB^2 = 81 + 144$
$AB^2 = 225$
$AB = 15$

4. Use the Pythagorean Theorem to find AC.
$AB^2 = AC^2 + BC^2$
$15^2 = AC^2 + 10^2$
$225 = AC^2 + 100$
$125 = AC^2$
$11.2 \approx AC$

5. If Q is a point on $\odot C$, then AQ is equal to the radius of the sphere, and thus, AB. So, $AQ = 18$.

6. $T = 4\pi r^2$
$= 4\pi(6.8)^2$
≈ 581.1
The surface area is approximately 581.1 in^2.

7. Find the radius.
$C = 2\pi r$
$8\pi = 2\pi r$
$4 = r$
$T = \frac{1}{2}(4\pi r^2) + \pi r^2$
$= 2\pi(4)^2 + \pi(4)^2$
≈ 150.8
The surface area is approximately 150.8 cm^2.

8. Find the radius.
$A = \pi r^2$
$18.1 = \pi r^2$
$\sqrt{\frac{18.1}{\pi}} = r$
$T = 4\pi r^2$
$= 4\pi\left(\sqrt{\frac{18.1}{\pi}}\right)^2$
$= 72.4$
The surface area is approximately 72.4 m^2.

9. $T = 4\pi r^2$
$= 4\pi(4.75)^2$
≈ 283.5
The surface area is approximately 283.5 in^2.

Pages 674–676 Practice and Apply

10. Use the Pythagorean Theorem to find PR.
$PR^2 = PT^2 + RT^2$
$PR^2 = 4^2 + 3^2$
$PR^2 = 16 + 9$
$PR^2 = 25$
$PR = 5$

11. Use the Pythagorean Theorem to find PR.
$PR^2 = PT^2 + RT^2$
$PR^2 = 3^2 + 8^2$
$PR^2 = 9 + 64$
$PR^2 = 73$
$PR \approx 8.5$

12. Use the Pythagorean Theorem to find PT.
$PR^2 = PT^2 + RT^2$
$13^2 = PT^2 + 12^2$
$169 = PT^2 + 144$
$25 = PT^2$
$5 = PT$

13. Use the Pythagorean Theorem to find PT.

$$PR^2 = PT^2 + RT^2$$
$$17^2 = PT^2 + 15^2$$
$$289 = PT^2 + 225$$
$$64 = PT^2$$
$$8 = PT$$

14. If X is a point on $\odot T$, then PX is equal to the radius of the sphere, and thus, PR. So, $PX = 9.4$.

15. If Y is a point on $\odot T$, then PY is equal to the radius of the sphere, and thus, PR. So, $PY = 12.8$.

16. Use the Pythagorean Theorem to find the radius of the charcoal rack.

$$c^2 = a^2 + b^2$$
$$11^2 = r^2 + 5^2$$
$$121 = r^2 + 25$$
$$96 = r^2$$

Find the difference in the areas.

$$\pi(11)^2 - \pi r^2 = 121\pi - 96\pi$$
$$= 25\pi$$
$$\approx 78.5$$

The difference in the areas is $25\pi \approx 78.5$ in^2.

17. $T = 4\pi r^2$

$$= 4\pi(25)^2$$
$$\approx 7854.0$$

The surface area is approximately 7854.0 in^2.

18. $T = 4\pi r^2$

$$= 4\pi(14.5)^2$$
$$\approx 2642.1$$

The surface area is approximately 2642.1 cm^2.

19. $T = 4\pi r^2$

$$= 4\pi\left(\frac{d}{2}\right)^2$$
$$= \pi d^2$$
$$= \pi(450)^2$$
$$\approx 636{,}172.5$$

The surface area is approximately 636,172.5 m^2.

20. $T = 4\pi r^2$

$$= 4\pi\left(\frac{d}{2}\right)^2$$
$$= \pi d^2$$
$$= \pi(3.4)^2$$
$$\approx 36.3$$

The surface area is approximately 36.3 ft^2.

21. Find the radius.

$$C = 2\pi r$$
$$40.8 = 2\pi r$$
$$\frac{20.4}{\pi} = r$$
$$T = \frac{1}{2}(4\pi r^2) + \pi r^2$$
$$= 3\pi r^2$$
$$= 3\pi\left(\frac{20.4}{\pi}\right)^2$$
$$\approx 397.4$$

The surface area is approximately 397.4 in^2.

22. Find the radius.

$$C = 2\pi r$$
$$30.2 = 2\pi r$$
$$\frac{15.1}{\pi} = r$$
$$T = 4\pi r^2$$
$$= 4\pi\left(\frac{15.1}{\pi}\right)^2$$
$$\approx 290.3$$

The surface area is approximately 290.3 ft^2.

23. Find the radius.

$$A = \pi r^2$$
$$814.3 = \pi r^2$$
$$\sqrt{\frac{814.3}{\pi}} = r$$
$$T = 4\pi r^2$$
$$= 4\pi\left(\sqrt{\frac{814.3}{\pi}}\right)^2$$
$$= 3257.2$$

The surface area is 3257.2 m^2.

24. Find the radius.

$$A = \pi r^2$$
$$227.0 = \pi r^2$$
$$\sqrt{\frac{227.0}{\pi}} = r$$
$$T = \frac{1}{2}(4\pi r^2) + \pi r^2$$
$$= 3\pi r^2$$
$$= 3\pi\left(\sqrt{\frac{227.0}{\pi}}\right)^2$$
$$= 681.0$$

The surface area is 681.0 km^2.

25. True; a great circle is formed by the intersection of a plane with a sphere such that the plane contains the center of the sphere, so they have the same center and radii.

26. False; two great circles will intersect at two points.

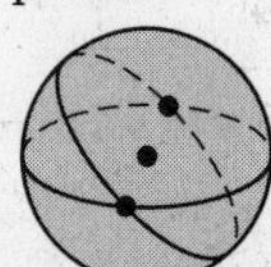

27. True; two spheres can intersect in a point or a circle, regardless of the lengths of their radii. So the statement is true.

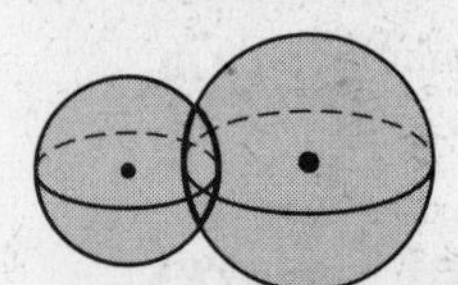

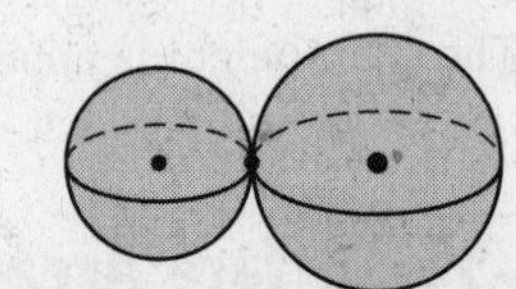

28. True; a chord that contains the center of a sphere is a diameter of a sphere. The diameter is the width of a sphere, and thus, is the longest chord.

29. True; when two spheres are tangent they intersect in one point.

30. Pole to pole: $T = 4\pi r^2$

$$= 4\pi\left(\frac{d}{2}\right)^2$$
$$= \pi d^2$$
$$= \pi(7899.83)^2$$
$$\approx 196{,}058{,}359.3 \text{ mi}^2$$

Equator: $T = \pi(7926.41)^2$

$$\approx 197{,}379{,}906.2 \text{ mi}^2$$

31. Find the mean diameter.

$\frac{7899.83 + 7926.41}{2} = 7913.12$

The diameter of the atmosphere is 200 miles longer than the mean value: 8113.12 mi.

$T = 4\pi r^2$

$= 4\pi\left(\frac{d}{2}\right)^2$

$= \pi d^2$

$= \pi(8113.12)^2$

$\approx 206{,}788{,}161.4 \text{ mi}^2$

32. Find the mean diameter.

$\frac{7899.83 + 7926.41}{2} = 7913.12$

$T = 4\pi r^2$, so $0.75T = 0.75(4\pi r^2) = 0.75\pi d^2$ is the surface area of the water.

$0.75\pi(7913.12)^2 \approx 147{,}538{,}933.4 \text{ mi}^2$

33. $T = \frac{1}{2}(4\pi r^2) + \pi r^2$

$= 2\pi\left(\frac{d}{2}\right)^2 + \pi\left(\frac{d}{2}\right)^2$

$= \frac{1}{2}\pi d^2 + \frac{1}{4}\pi d^2$

$= \frac{3}{4}\pi(13)^2$

$= 398.2$

The surface area is approximately 398.2 ft^2.

34. $T = 4\pi r^2$, so if the radius is twice as large, $4\pi(2r)^2 = 16\pi r^2$, and the ratio is

$\frac{16\pi r^2}{4\pi r^2} = 4$, or 4 : 1.

35. Let $T_2 = \frac{1}{2}T_1$. Then

$4\pi r_2^2 = \frac{1}{2}(4\pi r_1^2)$

$r_2^2 = \frac{1}{2}r_1^2$

$r_2 = \frac{r_1}{\sqrt{2}}$

$r_2 = \frac{\sqrt{2}}{2}r_1$

The ratio is $\frac{\sqrt{2}}{2}$: 1.

36. $T = 4\pi r^2$, so if the radius is three times as large, $4\pi(3r)^2 = 36\pi r^2$, and the ratio is

$\frac{36\pi r^2}{4\pi r^2} = 9$, or 9 : 1.

37. $T = 4\pi r^2$

$= 4\pi\left(\frac{d}{2}\right)^2$

$= \pi d^2$

12 mi : $\pi d^2 = \pi(12)^2$

$= 144\pi$

≈ 452.4

20 mi : $\pi d^2 = \pi(20)^2$

$= 400\pi$

≈ 1256.6

The surface area can range from about 452.4 to about 1256.6 mi^2.

38. $T = 4\pi r^2$

$= 4\pi\left(\frac{d}{2}\right)^2$

$= \pi d^2$

$= \pi(7)^2$

≈ 153.9

The surface area is approximately 153.9 mi^2.

39. The side length of the cube is equal to the diameter of the sphere, so the radius of the sphere is half the side of the cube.

40. The distance between opposite corners of the cube is equal to the diameter of the sphere. Use the Pythagorean Theorem to find the diagonal of a cube face if the side is x.

$c^2 = a^2 + b^2$

$c^2 = x^2 + x^2$

$c = x\sqrt{2}$

Now, find the opposite corner distance.

$c^2 = a^2 + b^2$

$c^2 = x^2 + (x\sqrt{2})^2$

$c^2 = 3x^2$

$c = x\sqrt{3}$

The radius is half this, so $r = \frac{x\sqrt{3}}{2}$, where x is the length of each side of the cube.

41. None; every line (great circle) that passes through X will also intersect g. All great circles intersect.

42. Sample answer: Sports equipment manufacturers use the surface area of spheres to determine the amount of material to cover the balls for different sports. Answers should include the following.

- The surface area of a sphere is four times the area of the great circle of the sphere.
- Racquetball and basketball are other sports that use balls.

43. A; the distance between opposite corners of the rectangular solid is equal to the diameter of the sphere.

Use the Pythagorean Theorem to find the diagonal of the 4 × 5 face.

$c^2 = a^2 + b^2$

$c^2 = 4^2 + 5^2$

$c = \sqrt{41}$

Now, find the opposite corner distance.

$c^2 = a^2 + b^2$

$c^2 = 7^2 + (\sqrt{41})^2$

$c^2 = 90$

$c = 3\sqrt{10}$

The radius is half this, so

$r = \frac{3\sqrt{10}}{2} \approx 4.74$ in.

44. C; $\sqrt{x^2 + 7} - 2 = x - 1$

$\sqrt{x^2 + 7} = x + 1$

$x^2 + 7 = (x + 1)^2$

$x^2 + 7 = x^2 + 2x + 1$

$6 = 2x$

$3 = x$

Page 676 Maintain Your Skills

45. Use the Pythagorean Theorem to find the radius of the base.

$c^2 = a^2 + b^2$

$19^2 = r^2 + 13^2$

$192 = r^2$

$8\sqrt{3} = r$

$T = \pi r\ell + \pi r^2$

$= \pi(8\sqrt{3})(19) + \pi(8\sqrt{3})^2$

≈ 1430.3

The surface area is approximately 1430.3 in^2.

46. Use the Pythagorean Theorem to find the slant height.
$c^2 = a^2 + b^2$
$\ell^2 = 7^2 + 10^2$
$\ell = \sqrt{149}$
$T = \pi r\ell + \pi r^2$
$= \pi(7)\sqrt{149} + \pi(7)^2$
≈ 422.4
The surface area is approximately 422.4 m^2.

47. $T = \pi r\ell + \pi r^2$
$= \pi(4.2)(15.1) + \pi(4.2)^2$
≈ 254.7
The surface area is approximately 254.7 cm^2.

48. Use the Pythagorean Theorem to find the slant height.
$c^2 = a^2 + b^2$
$\ell^2 = 7.4^2 + \left(\frac{11.2}{2}\right)^2$
$\ell = \sqrt{86.12}$
$T = \pi r\ell + \pi r^2$
$= \pi\left(\frac{d}{2}\right)\ell + \pi\left(\frac{d}{2}\right)^2$
$= \frac{1}{2}\pi d\ell + \frac{1}{4}\pi d^2$
$= \frac{1}{2}\pi(11.2)\sqrt{86.12} + \frac{1}{4}\pi(11.2)^2$
≈ 261.8
The surface area is approximately 261.8 ft^2.

49. $T = \frac{1}{2}P\ell + B$
$= \frac{1}{2}(4 \cdot 19)(16) + 19^2$
$= 969$
The surface area is 969 yd^2.

50. The apothem is half the base, or 6 feet.
Use the Pythagorean Theorem to find the slant height.
$c^2 = a^2 + b^2$
$\ell^2 = 6^2 + (13)^2$
$\ell = \sqrt{205}$
$T = \frac{1}{2}P\ell + B$
$= \frac{1}{2}(4 \cdot 12)\sqrt{205} + 12^2$
≈ 487.6
The surface area is approximately 487.6 ft^2.

51. $T = \frac{1}{2}P\ell + B$
$= \frac{1}{2}(4 \cdot 11)(24) + 11^2$
$= 649$
The surface area is 649 cm^2.

52. The diameter of the fabric required is 9 + 2(3) = 15 in.
$A = \pi r^2$
$= \pi\left(\frac{d}{2}\right)^2$
$= \frac{1}{4}\pi(15)^2$
≈ 176.7
The area of fabric needed is approximately 176.7 in^2.

53. Find the radius squared.
$(x - h)^2 + (y - k)^2 = r^2$
$[3 - (-2)]^2 + (2 - 7)^2 = r^2$
$5^2 + (-5)^2 = r^2$
$25 + 25 = r^2$
$50 = r^2$
The equation of the circle is $(x + 2)^2 + (y - 7)^2 = 50$.

54. Find the center.
$h = \frac{6 + 2}{2} = 4$
$k = \frac{-8 + 5}{2} = -\frac{3}{2}$
Find the radius squared.
$(2 - 4)^2 + \left[5 - \left(-\frac{3}{2}\right)\right]^2 = r^2$
$(-2)^2 + \left(\frac{13}{2}\right)^2 = r^2$
$\frac{185}{4} = r^2$
The equation of the circle is
$(x - 4)^2 + \left(y + \frac{3}{2}\right)^2 = \frac{185}{4}$.

Page 677 Geometry Activity: Locus and Spheres

1.

The locus of all points in space at a specific distance from a given point is a sphere. Thus, for this problem, the locus of points is two spheres each with a radius of 5 units with centers that are endpoints of the given line segment.

2. Yes, the radii are congruent.

3. Each sphere has a radius of 5 units and a diameter of 10 units.

4. The segment is 25 units long and the radii of the spheres are 5 units. So, the spheres are 15 units apart on the given segment.

5. The spheres intersect at a plane. The intersection of a plane and a sphere is a circle or a point. So, the intersection is a circle.

6. A circle is a locus of points on a plane.

7. The intersection is the set of all points equidistant from the midpoint of the given line segment in the plane containing the perpendicular bisector of the given line segment.

8. The particles from an explosion disperse in a spherical pattern. Since the explosion is at ground level, the locus of points describing the dispersion of particles is a hemisphere with a radius of 300 ft.

Chapter 12 Study Guide and Review

Page 678 Vocabulary and Concept Check

1. d	**2.** i	**3.** b	**4.** h
5. a	**6.** j	**7.** e	
8. g	**9.** c	**10.** f	

Pages 678–682 Lesson-by-Lesson Review

11. The solid is a cylinder.
Bases: $\odot F$ and $\odot G$
There are no faces, edges, or vertices.

12. Sample answer: The solid is a rectangular prism.
Bases: rectangle $WXYZ$ and rectangle $STUV$
Faces: rectangles $WXYZ$, $STUV$, $WXTS$, $XTUY$, $YUVZ$, and $WZVS$
Edges: $\overline{WX}$, $\overline{XY}$, $\overline{YZ}$, $\overline{ZW}$, $\overline{ST}$, $\overline{TU}$, $\overline{UV}$, $\overline{VS}$, $\overline{WS}$, $\overline{XT}$, $\overline{YU}$, and $\overline{ZV}$
Vertices: S, T, U, V, W, X, Y, and Z

13. Sample answer: The solid is a triangular prism.
Base: $\triangle BCD$
Faces: $\triangle ABC$, $\triangle ABD$, $\triangle ACD$, and $\triangle BCD$
Edges: $\overline{AB}$, $\overline{BC}$, $\overline{AC}$, $\overline{AD}$, $\overline{BD}$, and $\overline{CD}$
Vertices: A, B, C, and D

14. Use the Pythagorean Theorem to find the hypotenuse of the triangular base.
$c^2 = a^2 + b^2$
$c^2 = 3^2 + 4^2$
$c^2 = 25$
$c = 5$

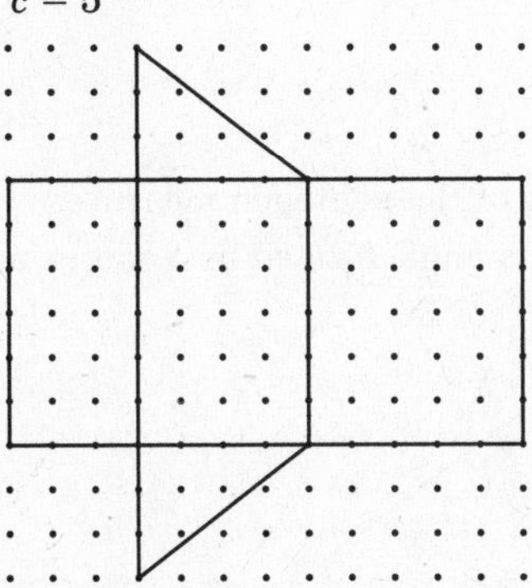

$T = Ph + 2B$
$= (3 + 4 + 5)(6) + 2 \cdot \frac{1}{2} \cdot 3 \cdot 4$
$= 84$
The surface area is 84 units^2.

15. Use the Pythagorean Theorem to find the slant height.
$c^2 = a^2 + b^2$
$13^2 = \ell^2 + \left(\frac{10}{2}\right)^2$
$169 = \ell^2 + 25$
$144 = \ell^2$
$12 = \ell$

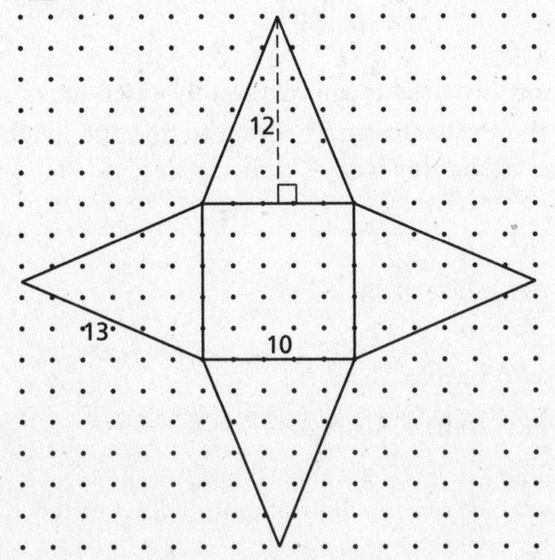

$T = \frac{1}{2}P\ell + B$
$= \frac{1}{2}(4 \cdot 10)(12) + 10^2$
$= 340$
The surface area is 340 units^2.

16.

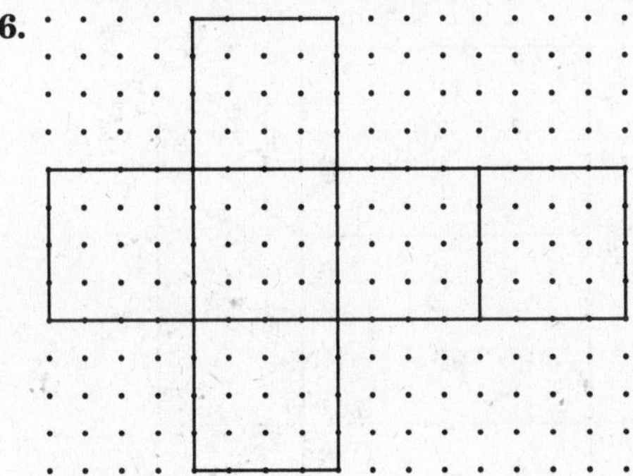

$T = 6s^2$
$= 6(4)^2$
$= 96$
The surface area is 96 units^2.

17. Use the Pythagorean Theorem to find the measure of the third side of the triangular base.
$c^2 = a^2 + b^2$
$6^2 = 4^2 + b^2$
$36 = 16 + b^2$
$20 = b^2$
$2\sqrt{5} = b$

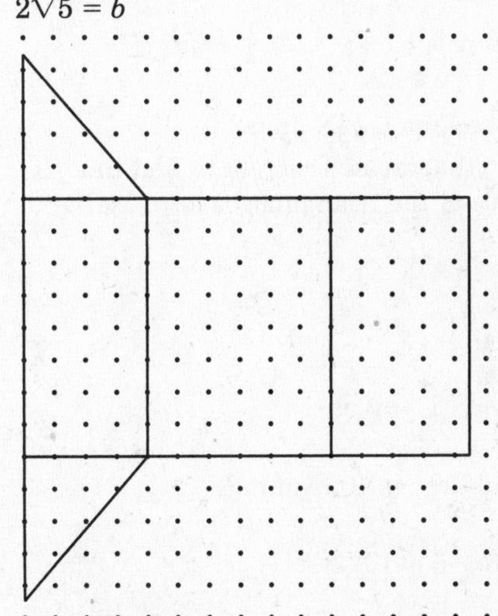

$T = Ph + 2B$
$= (4 + 6 + 2\sqrt{5})(8) + 2 \cdot \frac{1}{2} \cdot 4 \cdot 2\sqrt{5}$
≈ 133.7
The surface area is approximately 133.7 units^2.

18.

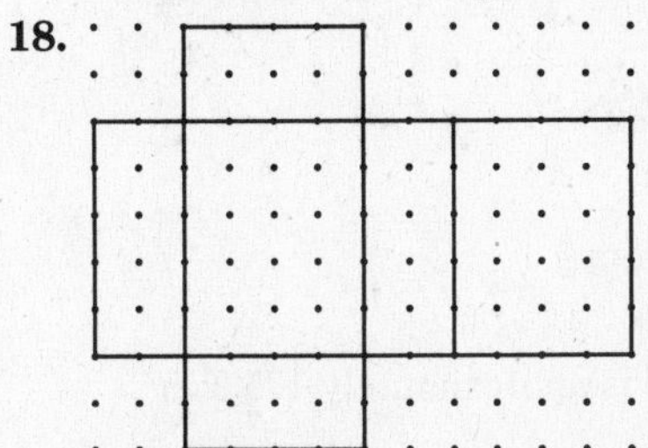

$T = Ph + 2B$
$= (2 \cdot 2 + 2 \cdot 4)(5) + 2(2)(4)$
$= 76$
The surface area is approximately 76 units^2.

19. Use the Pythagorean Theorem to find the measure of the fourth side of the trapezoidal base.

$c^2 = a^2 + b^2$
$c^2 = 4^2 + (8 - 5)^2$
$c^2 = 25$
$c = 5$

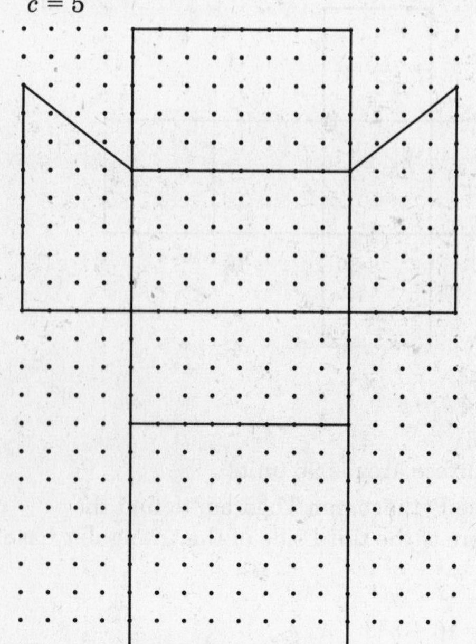

$T = Ph + 2B$
$= (4 + 5 + 5 + 8)(8) + 2 \cdot \frac{1}{2}(4)(5 + 8)$
$= 228$

The surface area is 228 units^2.

20. Use the Pythagorean Theorem to find the hypotenuse of the triangular base.

$c^2 = a^2 + b^2$
$c^2 = 15^2 + 20^2$
$c^2 = 625$
$c = 25$
$L = Ph$
$= (15 + 20 + 25)(18)$
$= 1080$

The lateral area is 1080 units^2.

21. $L = Ph$
$= (3 + 5 + 6 + 10)(3)$
$= 72$

The lateral area is 72 units^2.

22. $L = Ph$
$= (3 + 8 + 7 + 5)(4)$
$= 92$

The lateral area is 92 units^2.

23. $T = 2\pi rh + 2\pi r^2$
$= 2\pi\left(\frac{d}{2}\right)h + 2\pi\left(\frac{d}{2}\right)^2$
$= \pi dh + \frac{1}{2}\pi d^2$
$= \pi(4)(12) + \frac{1}{2}\pi(4)^2$
≈ 175.9

The surface area is approximately 175.9 in^2.

24. $T = 2\pi rh + 2\pi r^2$
$= 2\pi(6)(8) + 2\pi(6)^2$
≈ 527.8

The surface area is approximately 527.8 ft^2.

25. $T = 2\pi rh + 2\pi r^2$
$= 2\pi(4)(58) + 2\pi(4)^2$
≈ 1558.2

The surface area is approximately 1558.2 mm^2.

26. $T = 2\pi rh + 2\pi r^2$
$= 2\pi\left(\frac{d}{2}\right)h + 2\pi\left(\frac{d}{2}\right)^2$
$= \pi dh + \frac{1}{2}\pi d^2$
$= \pi(4)(8) + \frac{1}{2}\pi(4)^2$
≈ 125.7

The surface area is approximately 125.7 km^2.

27. $T = \frac{1}{2}P\ell + B$
$= \frac{1}{2}(4 \cdot 8)(15) + 8^2$
$= 304$

The surface area is 304 units^2.

28. Use the Pythagorean Theorem to find the slant height.

$c^2 = a^2 + b^2$
$13^2 = \ell^2 + \left(\frac{10}{2}\right)^2$
$169 = \ell^2 + 25$
$144 = \ell^2$
$12 = \ell$

The central angle of the pentagon measures $\frac{360°}{5}$ or 72°. So, the angle formed by a radius and the apothem is $\frac{72°}{2}$ or 36°.

Find the apothem, x.

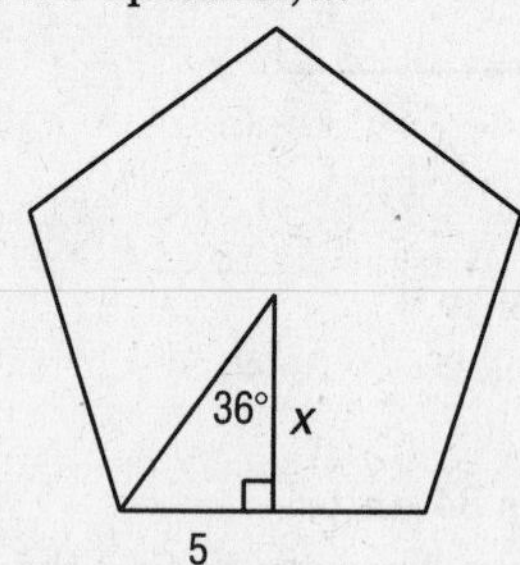

$\tan 36° = \frac{5}{x}$
$x = \frac{5}{\tan 36°}$
$T = \frac{1}{2}P\ell + \frac{1}{2}Px$
$= \frac{1}{2}P(\ell + x)$
$= \frac{1}{2}(5 \cdot 10)\left(12 + \frac{5}{\tan 36°}\right)$
≈ 472.0

The surface area is approximately 472.0 units^2.

29. Use the Pythagorean Theorem to find the height of the triangular base.

$c^2 = a^2 + b^2$
$5^2 = h^2 + \left(\frac{5}{2}\right)^2$
$25 = h^2 + 6.25$
$\sqrt{18.75} = h$
$T = \frac{1}{2}P\ell + \frac{1}{2}bh$
$= \frac{1}{2}(3 \cdot 5)(3) + \frac{1}{2}(5)\sqrt{18.75}$
≈ 33.3

The surface area is approximately 33.3 units^2.

30. $T = \pi r\ell + \pi r^2$
$= \pi(5)(18) + \pi(5)^2$
≈ 361.3
The surface area is approximately 361.3 mm^2.

31. Use the Pythagorean Theorem to find the radius of the base.
$c^2 = a^2 + b^2$
$5^2 = 4^2 + r^2$
$25 = 16 + r^2$
$9 = r^2$
$3 = r$
$T = \pi r\ell + \pi r^2$
$= \pi(3)(5) + \pi(3)^2$
≈ 75.4
The surface area is approximately 75.4 yd^2.

32. Use the Pythagorean Theorem to find the slant height.
$c^2 = a^2 + b^2$
$\ell^2 = 3^2 + 7^2$
$\ell^2 = 9 + 49$
$\ell = \sqrt{58}$
$T = \pi r\ell + \pi r^2$
$= \pi(3)\sqrt{58} + \pi(3)^2$
≈ 100.1
The surface area is approximately 100.1 in^2.

33. $T = 4\pi r^2$
$= 4\pi\left(\frac{d}{2}\right)^2$
$= \pi d^2$
$= \pi(18.2)^2$
≈ 1040.6
The surface area is approximately 1040.6 ft^2.

34. $T = \frac{1}{2}(4\pi r^2) + \pi r^2$
$= 3\pi r^2$
$= 3\pi(3.9)^2$
≈ 143.4
The surface area is approximately 143.4 cm^2.

35. Find the radius squared.
$A = \pi r^2$
$121 = \pi r^2$
$\frac{121}{\pi} = r^2$
$T = \frac{1}{2}(4\pi r^2) + \pi r^2$
$= 3\pi r^2$
$= 3\pi\left(\frac{121}{\pi}\right)$
$= 363$
The surface area is 363 mm^2.

36. Find the radius squared.
$A = \pi r^2$
$218 = \pi r^2$
$\frac{218}{\pi} = r^2$
$T = 4\pi r^2$
$= 4\pi\left(\frac{218}{\pi}\right)$
$= 872$
The surface area is 872 in^2.

37. $T = \frac{1}{2}(4\pi r^2) + \pi r^2$
$= 3\pi r^2$
$= 3\pi(16)^2$
≈ 2412.7
The surface area is approximately 2412.7 ft^2.

38. $T = 4\pi r^2$
$= 4\pi\left(\frac{d}{2}\right)^2$
$= \pi d^2$
$= \pi(5)^2$
≈ 78.5
The surface area is approximately 78.5 m^2.

39. Find the radius squared.
$A = \pi r^2$
$220 = \pi r^2$
$\frac{220}{\pi} = r^2$
$T = 4\pi r^2$
$= 4\pi\left(\frac{220}{\pi}\right)$
$= 880$
The surface area is 880 ft^2.

40. Find the radius squared.
$A = \pi r^2$
$30 = \pi r^2$
$\frac{30}{\pi} = r^2$
$T = \frac{1}{2}(4\pi r^2) + \pi r^2$
$= 3\pi r^2$
$= 3\pi\left(\frac{30}{\pi}\right)$
$= 90$
The surface area is 90 cm^2.

Chapter 12 Practice Test

Page 683

1. c **2.** a **3.** b

4. Sample answer: The solid is a rectangular prism.
Bases: rectangles *PQRS* and *TUVW*
Faces: rectangles *PQRS*, *TUVW*, *SPUT*, *QRWV*, *STWR*, and *PUVQ*
Edges: $\overline{PS}$, $\overline{QR}$, $\overline{VW}$, $\overline{UT}$, $\overline{PU}$, $\overline{ST}$, $\overline{RW}$, $\overline{QV}$, $\overline{PQ}$, $\overline{SR}$, $\overline{UV}$, and $\overline{TW}$
Vertices: *P*, *Q*, *R*, *S*, *T*, *U*, *V*, and *W*

5. The solid is a sphere.

6. The solid is a cone.
Base: $\odot F$
Vertex: *H*

7.

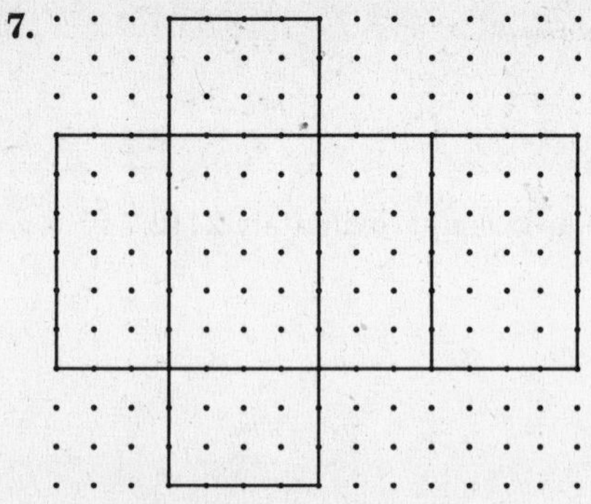

$T = Ph + 2B$
$= (2 \cdot 4 + 2 \cdot 6)(3) + 2(4)(6)$
$= 108$
The surface area is 108 units^2.

8. Use the Pythagorean Theorem to find the slant height.

$$c^2 = a^2 + b^2$$
$$(2\sqrt{26})^2 = \ell^2 + \left(\frac{4}{2}\right)^2$$
$$104 = \ell^2 + 4$$
$$100 = \ell^2$$
$$10 = \ell$$

$2\sqrt{26}$

4

$T = \frac{1}{2}P\ell + B$
$= \frac{1}{2}(4 \cdot 4)(10) + 4^2$
$= 96$
The surface area is 96 units^2.

9. $L(3 \times 5 \text{ base}) = Ph$
$= (2 \cdot 3 + 2 \cdot 5)(6)$
$= 96$
$L(3 \times 6 \text{ base}) = (2 \cdot 3 + 2 \cdot 6)(5)$
$= 90$
$L(6 \times 5 \text{ base}) = (2 \cdot 6 + 2 \cdot 5)(3)$
$= 66$
The lateral areas are 96 units^2 (3×5 base), 90 units^2 (3×6 base), and 66 units^2 (6×5 base).

10. Use the Pythagorean Theorem to find the hypotenuse.
$c^2 = a^2 + b^2$
$c^2 = 15^2 + 20^2$
$c^2 = 625$
$c = 25$
$L = Ph$
$= (15 + 20 + 25)(8)$
$= 480$
The lateral area is 480 units^2.

11. $L = Ph$
$= (5 + 3 + 4 + 7 + 4)(8)$
$= 184$
The lateral area is 184 units^2.

12. $T = 2\pi rh + 2\pi r^2$
$= 2\pi(8)(22) + 2\pi(8)^2$
≈ 1508.0
The surface area is approximately 1508.0 ft^2.

13. $T = 2\pi rh + 2\pi r^2$
$= 2\pi(3)(2) + 2\pi(3)^2$
≈ 94.2
The surface area is approximately 94.2 mm^2.

14. $T = 2\pi rh + 2\pi r^2$
$= 2\pi(78)(100) + 2\pi(78)^2$
$\approx 87{,}235.7$
The surface area is approximately 87,235.7 m^2.

15. Use the Pythagorean Theorem to find the slant height of the tetrahedron.
$c^2 = a^2 + b^2$
$6^2 = \ell^2 + \left(\frac{6}{2}\right)^2$
$36 = \ell^2 + 9$
$27 = \ell^2$
$3\sqrt{3} = \ell$
The central angle of the triangular base is $\frac{360°}{3}$ or 120°. So, the angle formed by a radius and the apothem is $\frac{120°}{2}$ or 60°. Find the apothem, x.

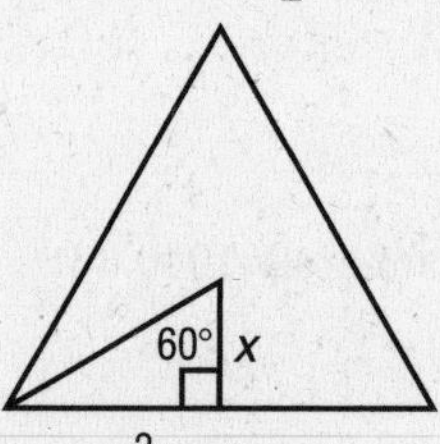

$\tan 60° = \frac{3}{x}$
$x = \frac{3}{\tan 60°}$
$x = \sqrt{3}$
Use the Pythagorean Theorem with the apothem and slant height to find the height of the tetrahedron.
$c^2 = a^2 + b^2$
$(3\sqrt{3})^2 = (\sqrt{3})^2 + b^2$
$27 = 3 + b^2$
$24 = b^2$
$4.9 \approx b$
The height is approximately $4.9 + 8 = 12.9$ units.

16. Use the Pythagorean Theorem to find the slant height of the tetrahedron.
$c^2 = a^2 + b^2$
$6^2 = \ell^2 + \left(\frac{6}{2}\right)^2$
$36 = \ell^2 + 9$
$27 = \ell^2$
$3\sqrt{3} = \ell$
$L = L_{\text{prism}} + L_{\text{tetrahedron}}$
$= Ph + \frac{1}{2}P\ell$
$= (3 \cdot 6)(8) + \frac{1}{2}(3 \cdot 6)(3\sqrt{3})$
≈ 190.8
The lateral area is approximately 190.8 units^2.

17. Use the Pythagorean Theorem to find the slant height of the tetrahedron. The slant height is the same as the altitude of the base.

$c^2 = a^2 + b^2$

$6^2 = \ell^2 + \left(\frac{6}{2}\right)^2$

$36 = \ell^2 + 9$

$27 = \ell^2$

$3\sqrt{3} = \ell$

$T = Ph + \frac{1}{2}P\ell + B$

$= (3 \cdot 6)(8) + \frac{1}{2}(3 \cdot 6)(3\sqrt{3}) + \frac{1}{2} \cdot 6 \cdot 3\sqrt{3}$

≈ 206.4

The surface area is approximately 206.4 units^2.

18. Use the Pythagorean Theorem to find the slant height.

$c^2 = a^2 + b^2$

$\ell^2 = 24^2 + 7^2$

$\ell^2 = 625$

$\ell = 25$

$T = \pi r\ell + \pi r^2$

$= \pi(7)(25) + \pi(7)^2$

≈ 703.7

The surface area is approximately 703.7 units^2.

19. Use the Pythagorean Theorem to find the radius.

$c^2 = a^2 + b^2$

$4^2 = 3^2 + r^2$

$16 = 9 + r^2$

$7 = r^2$

$\sqrt{7} = r$

$T = \pi r\ell + \pi r^2$

$= \pi(\sqrt{7})(4) + \pi(\sqrt{7})^2$

≈ 55.2

The surface area is approximately 55.2 units^2.

20. $T = \pi r\ell + \pi r^2$

$= \pi(7)(12) + \pi(7)^2$

≈ 417.8

The surface area is approximately 417.8 units^2.

21. $T = 4\pi r^2$

$= 4\pi(15)^2$

≈ 2827.4

The surface area is approximately 2827.4 in^2.

22. $T = 4\pi r^2$

$= 4\pi\left(\frac{d}{2}\right)^2$

$= \pi d^2$

$= \pi(14)^2$

≈ 615.8

The surface area is approximately 615.8 m^2.

23. Find the radius squared.

$A = \pi r^2$

$116 = \pi r^2$

$\frac{116}{\pi} = r^2$

$T = 4\pi r^2$

$= 4\pi\left(\frac{116}{\pi}\right)$

$= 464$

The surface area is 464 ft^2.

24. The area of the plastic is equal to the area of the prism minus the area of the 12 ft by 25 ft rectangular face.
Use the Pythagorean Theorem to find the height of the triangular part of the base.

$c^2 = a^2 + b^2$

$10^2 = a^2 + \left(\frac{12}{2}\right)^2$

$100 = a^2 + 36$

$64 = a^2$

$8 = a$

$T = Ph + 2B - (12)(25)$

$= (2 \cdot 10 + 2 \cdot 8 + 12)(25) + 2\left[(8)(12) + \frac{1}{2} \cdot 12 \cdot 8\right] - 300$

$= 1188$

The amount of plastic needed is 1188 ft^2.

25. D; $T = 6s^2$

$150 = 6s^2$

$25 = s^2$

$5 = s$

The length of each edge is 5 cm.

Chapter 12 Standardized Test Practice

Pages 684–685

1. D; $3x - 16 = 2x + 9$

$x = 25$

$[3(25) - 16]° = 59°$

2. D; the change in x is $6 - 0 = 6$ units.
The change in y is $9 - 1 = 8$ units.
So, the other endpoint has coordinates $(10 - 6, 6 - 8) = (4, -2)$ or $(10 + 6, 6 + 8) = (16, 14)$.

3. C

4. B; the relative length of each side corresponds to the relative measure of its opposite angle.
$m\angle C = 180 - 70 - 48 = 62$
$\overline{AC}$ is longest, $\overline{AB}$ is in the middle, and $\overline{BC}$ is shortest.

5. B; use the Pythagorean Theorem to find the length of $\overline{AB}$.

$AC^2 = BC^2 + AB^2$

$12^2 = 5^2 + AB^2$

$144 = 25 + AB^2$

$119 = AB^2$

$10.9 \approx AB$

The length of $\overline{AB}$ is approximately 10.9 in.

6. B; $A = \frac{N}{360}\pi r^2$

$= \frac{N}{360}\pi\left(\frac{d}{2}\right)^2$

$= \frac{N}{1440}\pi d^2$

$= \frac{360 - 120}{1440}\pi(18)^2$

$= 54\pi$

7. C; for example, a tetrahedron is a Platonic Solid but it is not a prism.

8. B; use the Pythagorean Theorem to find the height of the triangular bases.

$c^2 = a^2 + b^2$

$2^2 = a^2 + \left(\frac{2}{2}\right)^2$

$4 = a^2 + 1$

$3 = a^2$

$\sqrt{3} = a$

$T = Ph + 2B$

$= (3 \cdot 2)(4) + 2 \cdot \frac{1}{2} \cdot 2 \cdot \sqrt{3}$

≈ 27

The surface area is approximately 27 cm^2.

9. A; $T = 4\pi r^2$

$= 4\pi\left(\frac{d}{2}\right)^2$

$= \pi d^2$

$= \pi(4)^2$

≈ 50

The surface area is approximately 50 ft^2.

10. Let y be the height of the opposite side of the right triangle.

$\tan 58° = \frac{y}{47}$

$47\tan 58° = y$

$75 \approx y$

So, the height of the tree is approximately $75 + 5 = 80$ ft.

11. $A = \frac{1}{2}bh$

$= \frac{1}{2}(12x - 2x)(8x - 2x)$

$= \frac{1}{2}(10x)(6x)$

$= 30x^2$

The area is $30x^2$ $units^2$.

12. area of deck

= area of rectangle − area of semicircle

$= \ell w - \frac{1}{2}\pi r^2$

$= (26)(16) - \frac{1}{2}\pi\left(\frac{26-6}{2}\right)^2$

≈ 259

The area of the deck is approximately 259 ft^2.

13. Find the apothem, x.

The central angle of the pentagonal base is $\frac{360°}{5}$ or 72°. So, the angle formed by a radius and the apothem is $\frac{72°}{2}$ or 36°.

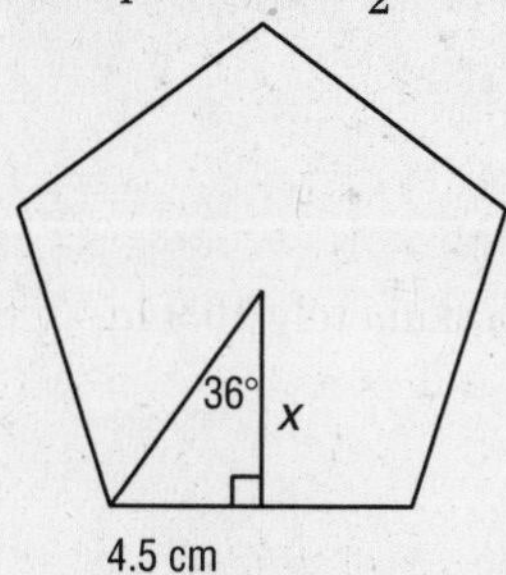

$\tan 36° = \frac{4.5}{x}$

$x = \frac{4.5}{\tan 36°}$

$T = \frac{1}{2}P\ell + \frac{1}{2}Px$

$= \frac{1}{2}(5 \cdot 9)(15) + \frac{1}{2}(5 \cdot 9)\left(\frac{4.5}{\tan 36°}\right)$

≈ 476.9

The surface area is approximately 476.9 cm^2.

14a. Sample answer:

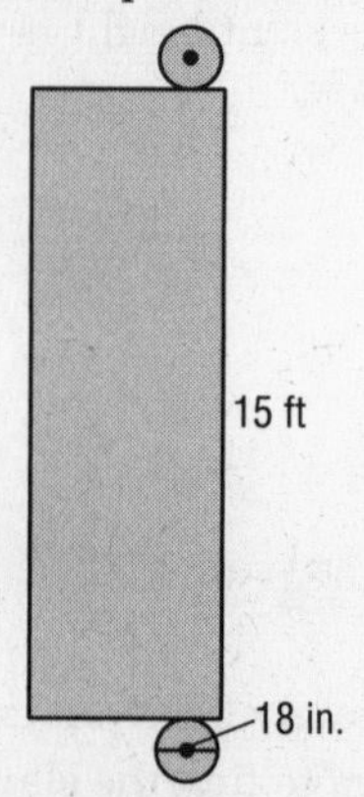

14b. Change the diameter from inches to feet. There are 12 inches in 1 foot, so 18 inches $= \frac{18}{12}$ or 1.5 feet.

$L = 2\pi rh$

$= 2\pi\left(\frac{d}{2}\right)h$

$= \pi dh$

$= \pi(1.5)(15)$

≈ 71

The lateral area is approximately 71 ft^2.

14c. $T = 2\pi rh + 2\pi r^2$

$= 2\pi\left(\frac{d}{2}\right)h + 2\pi\left(\frac{d}{2}\right)^2$

$= \pi dh + \frac{1}{2}\pi d^2$

$= \pi(1.5)(15) + \frac{1}{2}\pi(1.5)^2$

≈ 74

The surface area is approximately 74 ft^2.

15a. $T = \pi r\ell + \pi r^2$

$= \pi\left(\frac{d}{2}\right)\ell + \pi\left(\frac{d}{2}\right)^2$

$= \frac{1}{2}\pi d\ell + \frac{1}{4}\pi d^2$

$= \frac{1}{2}\pi(8)(7) + \frac{1}{4}\pi(8)^2$

≈ 138

The surface area is approximately 138 in^2.

15b. $T = 2\pi rh + 2\pi r^2$

$= 2\pi\left(\frac{d}{2}\right)h + 2\pi\left(\frac{d}{2}\right)^2$

$= \pi dh + \frac{1}{2}\pi d^2$

$= \pi(8)(22) + \frac{1}{2}\pi(8)^2$

≈ 653

The surface area is approximately 653 in^2.

15c. $T = \pi r\ell + 2\pi rh + \pi r^2$

$= \pi\left(\frac{d}{2}\right)\ell + 2\pi\left(\frac{d}{2}\right)h + \pi\left(\frac{d}{2}\right)^2$

$= \frac{1}{2}\pi d\ell + \pi dh + \frac{1}{4}\pi d^2$

$= \frac{1}{2}\pi(8)(7) + \pi(8)(22) + \frac{1}{4}\pi(8)^2$

≈ 691

The surface area is approximately 691 in^2.

Chapter 13 Volume

Page 687 Getting Started

1. $a^2 + 12^2 = 13^2$

Solve for a.

$a^2 + 144 = 169$

$a^2 = 25$

$a = \pm\sqrt{25}$

$a = \pm 5$

2. $\left(4\sqrt{3}\right)^2 + b^2 = 8^2$

Solve for b.

$48 + b^2 = 64$

$b^2 = 16$

$b = \pm\sqrt{16}$

$b = \pm 4$

3. $a^2 + a^2 = \left(3\sqrt{2}\right)^2$

Solve for a.

$2a^2 = 18$

$a^2 = 9$

$a = \pm\sqrt{9}$

$a = \pm 3$

4. $b^2 + 3b^2 = 192$

Solve for b.

$4b^2 = 192$

$b^2 = 48$

$b = \pm\sqrt{48}$

$b = \pm 4\sqrt{3}$

5. $256 + 7^2 = c^2$

Solve for c.

$256 + 49 = c^2$

$305 = c^2$

$c = \pm\sqrt{305}$

6. $144 + 12^2 = c^2$

Solve for c.

$144 + 144 = c^2$

$288 = c^2$

$\pm\sqrt{288} = c$

$c = \pm 12\sqrt{2}$

7.

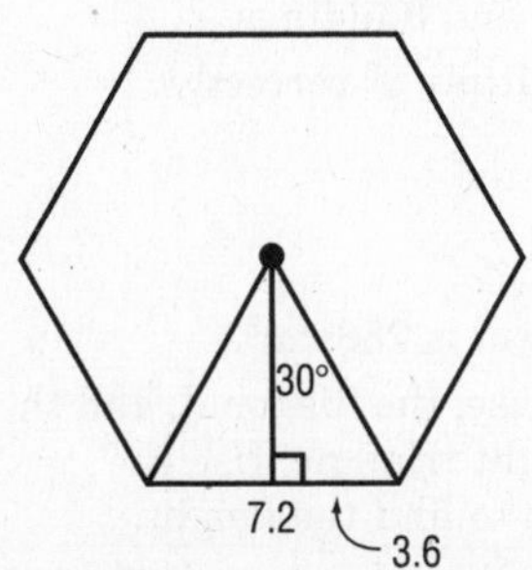

Apothem: A 30°-60°-90° triangle is formed by the apothem and one-half of a side of the hexagon. The shorter leg of the triangle is $\frac{1}{2}(7.2)$ or 3.6. The apothem is the longer leg of the triangle or $3.6\sqrt{3}$.

perimeter = (7.2)(6) = 43.2

Area: $A = \frac{1}{2}Pa$

$= \frac{1}{2}(43.2)\left(3.6\sqrt{3}\right)$

≈ 134.7

The area is approximately 134.7 cm^2.

8.

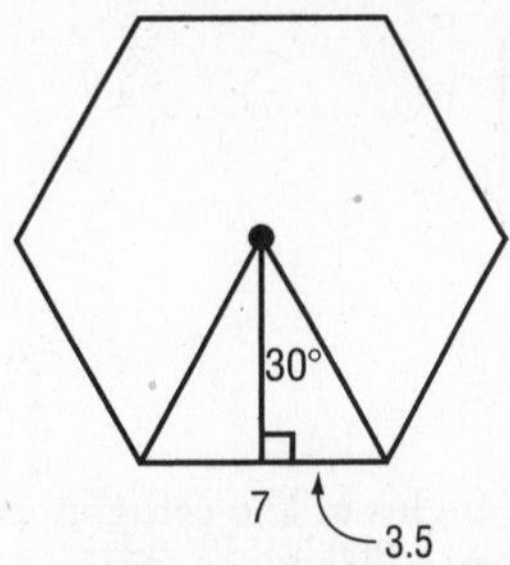

Apothem: A 30°-60°-90° triangle is formed by the apothem and one-half of a side of the hexagon. The shorter leg of the triangle is $\frac{1}{2}(7)$ or 3.5. The apothem is the longer leg of the triangle or $3.5\sqrt{3}$.

perimeter = $7 \cdot 6 = 42$

Area: $A = \frac{1}{2}Pa$

$= \frac{1}{2}(42)\left(3.5\sqrt{3}\right)$

≈ 127.3

The area is approximately 127.3 ft^2.

9.

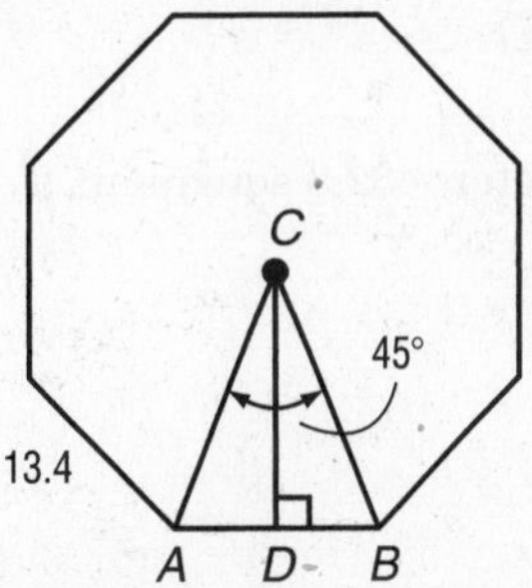

Apothem: The central angles of the octagon are all congruent, so $m\angle ACB = \frac{360}{8}$ or 45. $\overline{CD}$ is an apothem of the octagon. It bisects $\angle ACB$ and is a perpendicular bisector of $\overline{AB}$. So $m\angle ACD = 22.5$. Since the side of the octagon has measure 13.4, $AD = 6.7$.

$\tan 22.5° = \frac{6.7}{CD}$

$CD = \frac{6.7}{\tan 22.5°}$

≈ 16.175

perimeter = (13.4)(8) = 107.2

Area: $A = \frac{1}{2}Pa$

$\approx \frac{1}{2}(107.2)(16.175)$

≈ 867.0

The area is approximately 867.0 mm^2.

10.

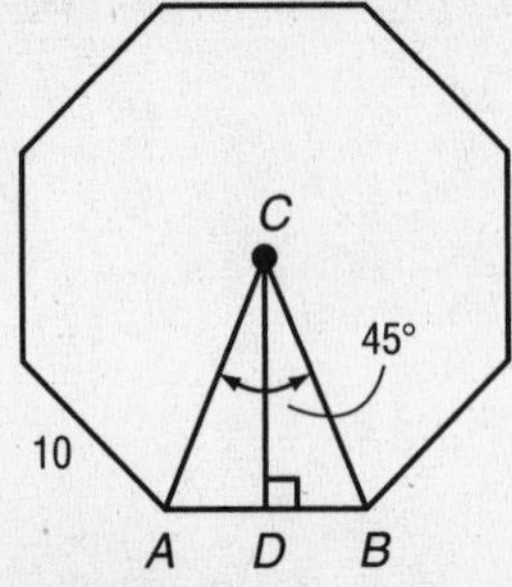

Apothem: The central angles of the octagon are all congruent, so $m\angle ACB = \frac{360}{8}$ or 45. $\overline{CD}$ is an apothem of the octagon. It bisects $\angle ACB$ and is a perpendicular bisector of $\overline{AB}$. So $m\angle ACB = 22.5$. Since the side of the octagon has measure 10, $AD = 5$.

$$\tan 22.5° = \frac{5}{CD}$$
$$CD = \frac{5}{\tan 22.5°}$$
$$\approx 12.07$$

perimeter = (10)(8) = 80

Area: $A = \frac{1}{2}Pa$
$$\approx \frac{1}{2}(80)(12.07)$$
$$\approx 482.8$$

The area is approximately 482.8 square in^2.

11. $(5b)^2 = (5b)(5b)$
$= (5)(5)(b)(b)$
$= 25b^2$

12. $\left(\frac{n}{4}\right)^2 = \left(\frac{n}{4}\right)\left(\frac{n}{4}\right)$
$= \frac{n \cdot n}{4 \cdot 4}$
$= \frac{n^2}{16}$

13. $\left(\frac{3x}{4y}\right)^2 = \left(\frac{3x}{4y}\right)\left(\frac{3x}{4y}\right)$
$= \frac{3x \cdot 3x}{4y \cdot 4y}$
$= \frac{3 \cdot 3 \cdot x \cdot x}{4 \cdot 4 \cdot y \cdot y}$
$= \frac{9x^2}{16y^2}$

14. $\left(\frac{4y}{7}\right)^2 = \left(\frac{4y}{7}\right)\left(\frac{4y}{7}\right)$
$= \frac{4y \cdot 4y}{7 \cdot 7}$
$= \frac{4 \cdot 4 \cdot y \cdot y}{7 \cdot 7}$
$= \frac{16y^2}{49}$

15. Let A be (x_1, y_1) and B be (x_2, y_2). The coordinates of the midpoint W are
$\left(\frac{x_1 + x_2}{2}, \frac{y_1 + y_2}{2}\right) = \left(\frac{0 + (-5)}{2}, \frac{-1 + 4}{2}\right)$
$= \left(-\frac{5}{2}, \frac{3}{2}\right)$ or $(-2.5, 1.5)$

16. Let A be (x_1, y_1) and B be (x_2, y_2). The coordinates of the midpoint W are
$\left(\frac{x_1 + x_2}{2}, \frac{y_1 + y_2}{2}\right) = \left(\frac{5 + (-3)}{2}, \frac{0 + 6}{2}\right)$
$= (1, 3)$

17. Let A be (x_1, y_1) in the Midpoint Formula.
$W(10, 10) = W\left(\frac{1 + x_2}{2}, \frac{-1 + y_2}{2}\right)$
Write two equations to find the coordinates of B.

$10 = \frac{1 + x_2}{2}$ $\quad$ $10 = \frac{-1 + y_2}{2}$
$20 = 1 + x_2$ $\quad$ $20 = -1 + y_2$
$19 = x_2$ $\quad$ $21 = y_2$

The coordinates of B are (19, 21).

18. Let B be (x_2, y_2) in the Midpoint Formula.
$W(0, 0) = W\left(\frac{x_1 + (-2)}{2}, \frac{y_1 + 2}{2}\right)$
Write two equations to find the coordinates of A.

$0 = \frac{x_1 + (-2)}{2}$ $\quad$ $0 = \frac{y_1 + 2}{2}$
$0 = x_1 - 2$ $\quad$ $0 = y_1 + 2$
$2 = x_1$ $\quad$ $-2 = y_1$

The coordinates of A are $(2, -2)$.

13-1 Volumes of Prisms and Cylinders

Page 688 Geometry Activity: Volume of a Rectangular Prism

1. 12 cubes in top layer +12 cubes in bottom layer = 24 cubes
2. The prism is 4 cubes long, 3 cubes wide, and 2 cubes high. $4 \times 3 \times 2 = 24$.
3. They are the same.
4. See students' work.
5. $V = \ell wh$

Page 691 Check for Understanding

1. Sample answers: cans, roll of paper towels, and chalk; boxes, crystals, and buildings
2. Julia; Che did not multiply 3^3 correctly.
3. $V = Bh$
   $= \frac{1}{2}(8)(12)(6)$
   $= 288$
   The volume of the prism is 288 cm^3.
4. The diameter of the base, the diagonal, and the lateral edge form a right triangle. Use the Pythagorean Theorem to find the height.
   $a^2 + b^2 = c^2$
   $h^2 + 8^2 = 17^2$
   $h^2 + 64 = 289$
   $h^2 = 225$
   $h = 15$
   Now find the volume.
   $V = \pi r^2 h$
   $= \pi(4^2)(15)$
   ≈ 754.0
   The volume is approximately 754.0 in^3.
5. $V = \pi r^2 h$
   $= \pi(7.5^2)(18)$
   ≈ 3180.9
   The volume is approximately 3180.9 mm^3.

6. Use the formula for the volume of a rectangular prism.
$V = Bh$
$= (12)(12)(14)$
$= 2016$
The volume of the prism is 2016 ft^3.

Pages 692–694 Practice and Apply

7. The diameter of the base, the diagonal, and the lateral edge form a right triangle. Use the Pythagorean Theorem to find the height.
$a^2 + b^2 = c^2$
$h^2 + 9^2 = 15^2$
$h^2 + 81 = 225$
$h^2 = 144$
$h = 12$
Now find the volume.
$V = \pi r^2 h$
$= \pi(4.5^2)(12)$
≈ 763.4
The volume is approximately 763.4 cm^3.

8. $V = Bh$
$= (18.7)(12.2)(3.6)$
≈ 821.3
The volume is approximately 821.3 in^3.

9. Use the Pythagorean Theorem to find the height of the prism's triangular base.
Let $b = \frac{1}{2}$(base of triangle) = 3 cm
$a^2 + b^2 = c^2$
$a^2 + 3^2 = 8^2$
$a^2 + 9 = 64$
$a^2 = 55$
$a = \sqrt{55}$
Now find the volume of the prism.
$V = Bh$
$= \frac{1}{2}(6)(\sqrt{55})(12)$
≈ 267.0
The volume is approximately 267.0 cm^3.

10. The radius of the base is $\frac{1}{2}(18)$ or 9.
$V = \pi r^2 h$
$= \pi(9^2)(12.4)$
≈ 3155.4
The volume is approximately 3155.4 m^3.

11. $V = Bh$
$= (15)(10)(5)$
$= 750$
The volume of the prism is 750 in^3.

12. Find the area, B, of the base using the formula for the area of a trapezoid.
$B = \frac{1}{2}h(b_1 + b_2)$
$= \frac{1}{2}(4)(6 + 10)$
$= 32$
Now find the volume of the prism.
$V = Bh$
$= (32)(18)$
$= 576$
The volume of the prism is 576 in^3.

13. Find the volume of the oblique prism using the formula for a rectangular prism.
$V = Bh$
$= (2.5)(3.5)(3.2)$
$= 28$
The volume of the oblique prism is 28 ft^3.

14. Find the volume of the oblique prism using the formula for a rectangular prism.
$V = Bh$
$= (55)(35)(30)$
$= 57{,}750$
The volume of the oblique prism is 57,750 m^3.

15. Find the volume of the oblique cylinder using the formula for a right cylinder.
$V = \pi r^2 h$
$= \pi(13.2^2)(27.6)$
$\approx 15{,}108.0$
The volume is approximately 15,108.0 mm^3.

16. Find the volume of the oblique cylinder using the formula for a right cylinder.
$V = \pi r^2 h$
$= \pi(2.6^2)(7.8)$
≈ 165.6
The volume is approximately 165.6 yd^3.

17. $V = \pi r^2 h$
Solve for r.
$615.8 = \pi r^2(4)$
$49 \approx r^2$
$7 \approx r$
The diameter is about 2(7) or 14 m.

18. $V = Bh$
$1152 = 64h$
$18 = h$
The lateral edge length is 18 in.

19. The solid is a rectangular prism with $\ell = 4$, $w = 3$, $h = 2$.
$V = Bh$
$= (4)(3)(2)$
$= 24$
The volume is 24 $units^3$.

20. The solid is a triangular prism. Its height is $h = 5$ and its base has area $B = \frac{1}{2}(1)(1)$.
$V = Bh$
$= \frac{1}{2}(1)(1)(5)$
$= 2.5$
The volume is 2.5 $units^3$.

21. The solid is a cylinder with $r = 2.1$, $h = 3.5$.
$V = \pi r^2 h$
$= \pi(2.1^2)(3.5)$
≈ 48.5
The volume is approximately 48.5 mm^3.

22. Treat the solid as a large rectangular prism with a smaller one attached underneath.

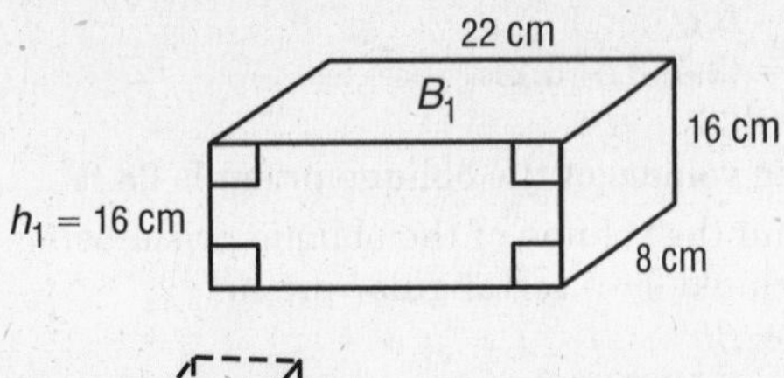

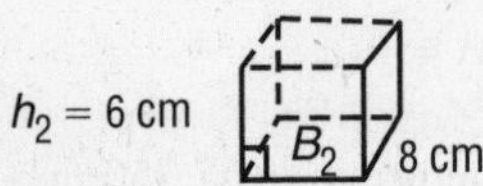

volume = volume of large rectangular prism + volume of smaller rectangular prism

$= B_1h_1 + B_2h_2$

$= (22)(8)(16) + (8)(6)(6)$

$= 3104$

The volume of the solid is 3104 cm^3.

23. volume = volume of rectangular prism − volume of cylinder

$= Bh - \pi r^2h$

$= (6)(6)(6) - \pi(1.5^2)(6)$

≈ 173.6

The volume is approximately 173.6 ft^3.

24. The solid covers 360° − 120° = 240° of the 360° in a cylinder. So the solid is $\frac{240}{360}$ of a cylinder.

volume = $\frac{240}{360}$ × volume of cylinder

$= \frac{2}{3}\pi r^2h$

$= \frac{2}{3}\pi(4^2)(10)$

≈ 335.1

The volume is approximately 335.1 ft^3.

25. The holder can be modeled as a cylindrical solid with a smaller cylindrical solid cut out of it. The smaller solid has the same radius as a can, and its height is 11.5 − 1 = 10.5. The larger solid has a radius of $\frac{1}{2}(8.5)$ or 4.25.

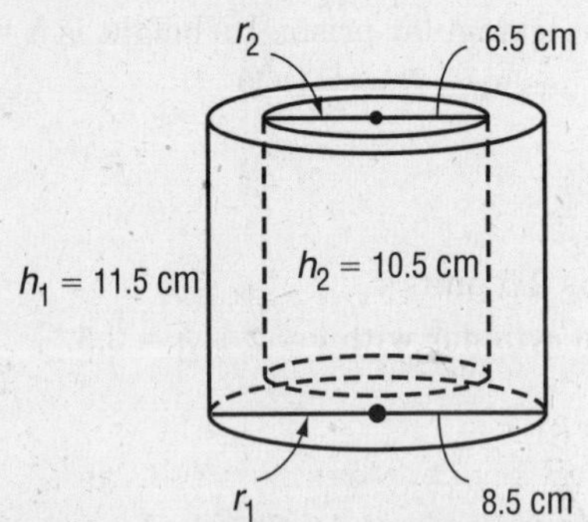

volume = volume of holder and "cut out" − volume of "cut out"

$= \pi {r_1}^2h_1 - \pi {r_2}^2h_2$

$= \pi(4.25)^2(11.5) - \pi(3.25)^2(10.5)$

≈ 304.1

The volume is approximately 304.1 cm^3.

26. The core radius is $\frac{1}{2}(35)$ or 17.5. The core height is 586 + 40 or 626.

$V = \pi r^2h$

$= \pi(17.5^2)(626)$

$\approx 602{,}282.6$

The volume is approximately 602,282.6 ft^3.

27. The volume in gallons equals the volume πr^2h in cubic feet multiplied by a conversion factor of $7\frac{1}{2}$ gal/ft^3.

$V = \pi r^2h\left(7\frac{1}{2}\right)$

$200{,}000 = \pi r^2h(7.5)$

Solve for r.

$26{,}667 \approx \pi r^2h$

$26{,}667 \approx \pi r^2(23)$

$r^2 \approx 369.1$

$r \approx 19.2$

The radius is approximately 19.2 ft.

28. Consider the water to be added as a rectangular prism, with a height of 3 − 0.3 or 2.7.

$V = Bh$

$= (50)(25)(2.7)$

$= 3375\ m^3$

Since 1 cubic meter equals 1000 liters, the volume of water to be added is 3375(1000) or 3,375,000 L.

29. Begin by finding the area of the hexagon.

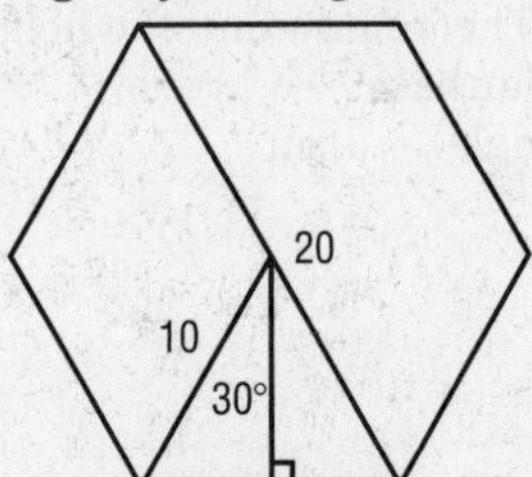

Apothem: A 30°-60°-90° triangle is formed by the apothem and one-half of a side of the hexagon. The hypotenuse is $\frac{1}{2}(20)$ or 10. So the shorter leg is $\frac{1}{2}(10) = 5$ and the longer leg is $5\sqrt{3}$. This is the length of the apothem. One side of the hexagon measures 10 mm, so the perimeter is 10 · 6 or 60.

volume of brass block

= volume of rectangular prism − volume of hexagonal prism

$= B_1h_1 - B_2h_2$

$= \ell wh_1 - \frac{1}{2}Pah_2$

$= (50)(40)(60) - \frac{1}{2}(60)(5\sqrt{3})(60)$

$\approx 104{,}411.5$

The volume is approximately 104,411.5 mm^3.

30. Set the density equal to mass ÷ volume, then use 8.0 g/cm^3 as the density. For the volume, use 104,411.5 mm^3 = 104.4115 cm^3 from Exercise 29.

$d = \frac{m}{V}$

$8.0 = \frac{m}{104.4115}$

$m = (8.0)(104.4115)$

≈ 835.3

The mass is approximately 835.3 g.

31. Start by finding the area of the base.

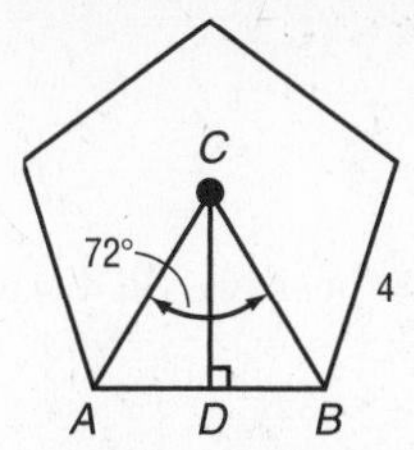

Apotem: The central angles of the pentagon are all congruent, so $m\angle ACB = \frac{360}{5}$ or 72. $\overline{CD}$ is an apothem of the pentagon. It bisects $\angle ACB$ and is a perpendicular bisector of $\overline{AB}$. So, $m\angle ACD = 36$. Since the perimeter is 20, $AB = \frac{20}{5} = 4$ and $AD = \frac{1}{2}(4) = 2$.

$\tan 36° = \frac{2}{CD}$

$CD = \frac{2}{\tan 36°}$

≈ 2.7528

Area: $A = \frac{1}{2}Pa$

$\approx \frac{1}{2}(20)(2.7528)$

Now find the volume of the prism.

$V = Bh$

$\approx \frac{1}{2}(20)(2.7528)(5)$

≈ 137.6

The volume is approximately 137.6 ft^3.

32. Sample answer: Cartoonists use mathematical concepts or terms in comics because of the difficulty that many people have had with math. Answers should include the following.

- A volume means a book.
- In mathematics, volume refers to the amount of space that a figure encloses.

33. A; use the formula for the volume of a rectangular prism.

$V = Bh$

$16{,}320 = (85)w(8)$

$16{,}320 = 680w$

$w = 24$ ft

34. B; $\pi r^2h - 2\pi rh = \pi rh(r) - \pi rh(2)$

$= \pi rh(r - 2)$

Page 694 Maintain Your Skills

35. radius $= \frac{1}{2}(12) = 6$

$T = 4\pi r^2$

$= 4\pi(6^2)$

≈ 452.4

The surface area is approximately 452.4 ft^2.

36. $T = 4\pi r^2$

$= 4\pi(41^2)$

$\approx 21{,}124.1$

The surface area is approximately 21,124.1 cm^2.

37. radius $= \frac{1}{2}(18) = 9$

$T = 4\pi r^2$

$= 4\pi(9^2)$

≈ 1017.9

The surface area is approximately 1017.9 m^2.

38. $T = 4\pi r^2$

$= 4\pi(8.5)^2$

≈ 907.9

The surface area is approximately 907.9 in^2.

39. $T = \pi r\ell + \pi r^2$

$= \pi(6)(11) + \pi(6^2)$

≈ 320.4

The surface area is approximately 320.4 m^2.

40. radius $= \frac{1}{2}(16) = 8$

$T = \pi r\ell + \pi r^2$

$= \pi(8)(13.5) + \pi(8^2)$

≈ 540.4

The surface area is approximately 540.4 cm^2.

41. Use the Pythagorean Theorem to find the slant height.

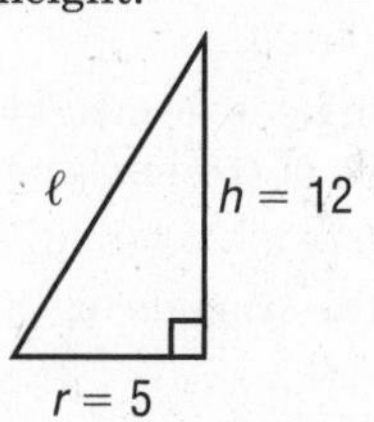

$\ell^2 = 5^2 + 12^2$

$\ell^2 = 169$

$\ell = 13$

Now find the surface area of the cone.

$T = \pi r\ell + \pi r^2$

$= \pi(5)(13) + \pi(5^2)$

≈ 282.7

The surface area is approximately 282.7 in^2.

42. radius $= \frac{1}{2}(14) = 7$

Use the Pythagorean Theorem to find the slant height.

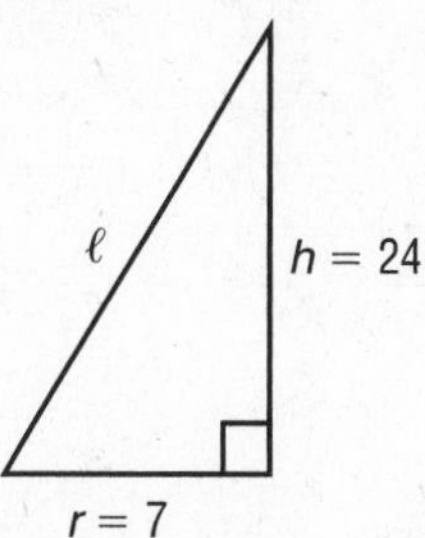

$\ell^2 = 7^2 + 24^2$

$\ell^2 = 625$

$\ell = 25$

Now find the surface area of the cone.

$T = \pi r\ell + \pi r^2$

$= \pi(7)(25) + \pi(7^2)$

≈ 703.7

The surface area is approximately 703.7 in^2.

43. Treat the floor plan as a large rectangle of width $12 + 9 = 21$ and length $13 + 8 + 7 = 28$, plus a 9×9 square.

total area = area of large rectangle + area of square

$= \ell w + s^2$

$= (28)(21) + (9)^2$

$= 669$

shaded area $= \ell_1w_1 + \ell_2w_2 + s^2$

$= (9)(13) + (12)(7) + (9)^2$

$= 282$

$P(\text{shaded}) = \frac{\text{shaded area}}{\text{total area}}$

$= \frac{282}{669}$

≈ 0.42

The probability is about 0.42, or 42%.

44.

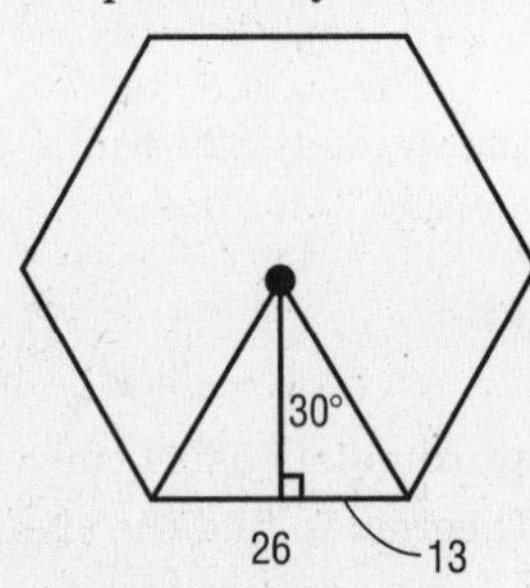

The length of one side is $\frac{156}{6}$ or 26.

Apothem: A 30°-60°-90° triangle is formed by the apothem and one-half of a side of the hexagon. The shorter leg of the triangle is $\frac{1}{2}(26)$ or 13. The apothem is the longer leg of the triangle or $13\sqrt{3}$.

Area: $A = \frac{1}{2}Pa$

$= \frac{1}{2}(156)(13\sqrt{3})$

≈ 1756.3

The area is approximately 1756.3 in^2.

45. perimeter = (6.2)(8) = 49.6

$A = \frac{1}{2}Pa$

$= \frac{1}{2}(49.6)(7.5)$

$= 186$

The area of the octagon is 186 m^2.

46. Use Theorem 10.17 about a tangent segment and a secant segment.

$13^2 = x(x + 8)$

$169 = x^2 + 8x$

$0 = x^2 + 8x - 169$

Use the Quadratic Formula.

$x = \frac{-b \pm \sqrt{b^2 - 4ac}}{2a}$

$= \frac{-8 \pm \sqrt{8^2 - 4(1)(-169)}}{2(1)}$

$= \frac{-8 + \sqrt{740}}{2}$ or $\frac{-8 - \sqrt{740}}{2}$

Since x is a length, it must be positive. So discard $\frac{-8 - \sqrt{740}}{2}$.

$x \approx 9.6$

47. Use Theorem 10.16 about 2 secant segments.

$8(8 + 6) = x(x + 4)$

$112 = x^2 + 4x$

$0 = x^2 + 4x - 112$

Use the Quadratic Formula.

$x = \frac{-b \pm \sqrt{b^2 - 4ac}}{2a}$

$= \frac{-4 \pm \sqrt{4^2 - 4(1)(-112)}}{2(1)}$

$= \frac{-4 + 4\sqrt{29}}{2}$ or $\frac{-4 - 4\sqrt{29}}{2}$

Since x is a length, it must be positive. So discard $\frac{-4 - 4\sqrt{29}}{2}$.

$x \approx 8.8$

48. Use Theorem 10.17, about a tangent segment and a secant segment.

$x^2 = 12(12 + 9.5)$

$x^2 = 258$

$x = \sqrt{258}$ or $-\sqrt{258}$

Since x is a length, it must be positive. So discard $-\sqrt{258}$.

$x \approx 16.1$

49. Drawing the height divides the triangle into two 30°-60°-90° triangles each with base $\frac{1}{2}(7)$ or 3.5. The height is then $3.5\sqrt{3}$. For the whole equilateral triangle,

$A = \frac{1}{2}bh$

$= \frac{1}{2}(7)(3.5\sqrt{3})$

≈ 21.22

The area is approximately 21.22 in^2.

50.

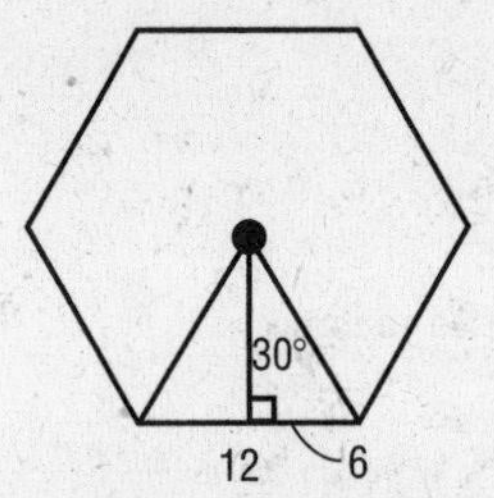

Apothem: A 30°-60°-90° triangle is formed by the apothem and one-half of a side of the hexagon. The shorter leg of the triangle is $\frac{1}{2}(12)$ or 6. The apothem is the longer leg of the triangle or $6\sqrt{3}$.

perimeter = $12 \cdot 6 = 72$

Area: $A = \frac{1}{2}Pa$

$= \frac{1}{2}(72)(6\sqrt{3})$

≈ 374.12

The area is approximately 374.12 cm^2.

51.

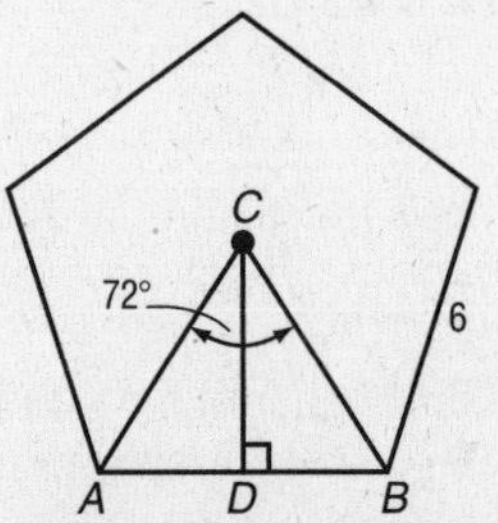

Apothem: The central angles of the pentagon are all congruent, so $m\angle ACB = \frac{360}{5}$ or 72. $\overline{CD}$ is an apothem of the pentagon. It bisects $\angle ACB$ and is a perpendicular bisector of $\overline{AB}$. So, $m\angle ACD = 36$. Since the side of the pentagon has measure 6, $AD = 3$.

$\tan 36° = \frac{3}{CD}$

$CD = \frac{3}{\tan 36°}$

≈ 4.129

perimeter = $5 \cdot 6 = 30$

Area: $A = \frac{1}{2}Pa$

$\approx \frac{1}{2}(30)(4.129)$

≈ 61.94

The area is approximately 61.94 m^2.

52.

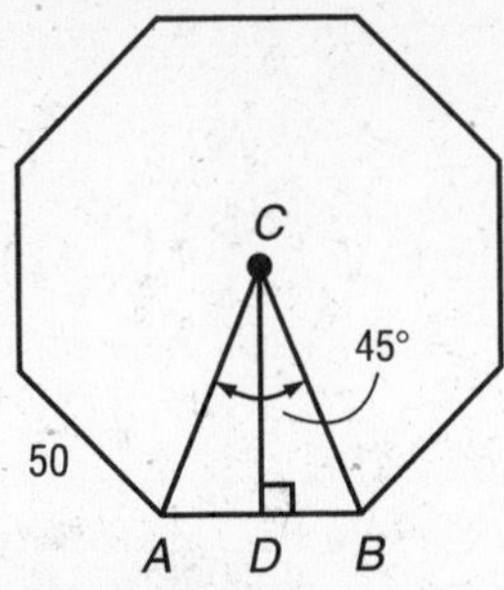

Apothem: The central angles of the octagon are all congruent, so $m\angle ACB = \frac{360}{8}$ or 45. $\overline{CD}$ is an apothem of the octagon. It bisects $\angle ACB$ and is a perpendicular bisector of $\overline{AB}$. So, $m\angle ACD = 22.5$. Since the side of the octagon has measure 50, $AD = 25$.

$\tan 22.5° = \frac{25}{CD}$

$CD = \frac{25}{\tan 22.5°}$

≈ 60.35534

perimeter $= 50 \cdot 8 = 400$

Area: $A = \frac{1}{2}Pa$

$\approx \frac{1}{2}(400)(60.35534)$

$\approx 12{,}071.07$

The area is approximately 12,071.07 ft^2.

Page 695 Spread Sheet Investigation: Prisms

Use the formula $A = 2(\ell + w)h + 2\ell w$ in column E.
Use the formula $V = \ell wh$ in column F.

Prism	Length	Width	Height	Surface Area	Volume
1	1	2	3	22	6
2	2	4	6	88	48
3	3	6	9	198	162
4	4	8	12	352	384
5	8	16	24	1408	3072

1. For each pair of prisms, the dimensions of the second prism are two times the dimensions of the first prism.
2. The surface areas of prisms 2, 4, and 5 are four times the respective surface areas of prisms 1, 2, and 4.
3. The volumes of prisms 2, 4, and 5 are eight times the respective volumes of prisms 1, 2, and 4.
4. When the dimensions are doubled, the surface area is multiplied by 4, or 2^2, and the volume is multiplied by 8, or 2^3.

13-2 Volumes of Pyramids and Cones

Page 696 Geometry Activity: Investigating the Volume of a Pyramid

1. It took 3 pyramids of rice.
2. The areas of the bases are the same.
3. The heights are the same.
4. $V = \frac{1}{3}Bh$

Pages 698–699 Check for Understanding

1. Consider first the base of the cone or pyramid. When a circle with area πr^2 is doubled in size, the new area is $\pi(2r)^2 = 4(\pi r^2)$, or 4 times the original area. Similarly, when a square with area s^2 is doubled in size, the new area is $(2s)^2 = 4s^2$, or 4 times the original area.
   When, for a cone or pyramid, the height is also doubled, the new volume is
   $V = \frac{1}{3}(4B)(2h)$
   $= 8\left(\frac{1}{3}Bh\right)$
   So in each case the volume is 8 times the original volume.
2. The volume of a pyramid is one-third the volume of a prism of the same height as the pyramid and with bases congruent to the base of the pyramid.
3. Sample answer:

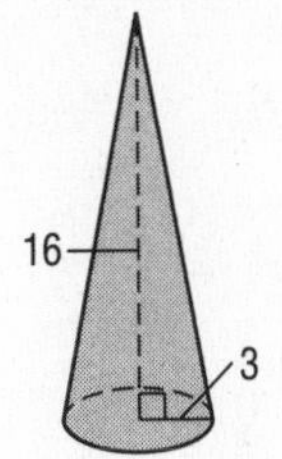

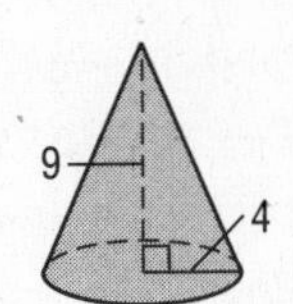

$V = \frac{1}{3}\pi(3^2)(16)$ $\quad$ $V = \frac{1}{3}\pi(4^2)(9)$
$V = 48\pi$ $\quad$ $V = 48\pi$

4. $V = \frac{1}{3}Bh$
   $= \frac{1}{3}(16)(10)(12)$
   $= 640$
   The volume of the pyramid is 640 in^3.
5. Use the 30°-60°-90° triangle to find the radius of the base. The height of the cone is the longer leg and the radius of the cone is the shorter leg.
   $r = \frac{1}{\sqrt{3}} \cdot 12 = 4\sqrt{3}$
   $V = \frac{1}{3}\pi r^2 h$
   $= \frac{1}{3}\pi(4\sqrt{3})^2(12)$
   ≈ 603.2
   The volume is approximately 603.2 mm^3.
6. $V = \frac{1}{3}Bh$
   $= \frac{1}{3}\pi r^2 h$
   $= \frac{1}{3}\pi(8^2)(20)$
   ≈ 1340.4
   The volume is approximately 1340.4 ft^3.
7. Use the formula for the volume of a cone.
   $V = \frac{1}{3}Bh$
   $= \frac{1}{3}(38{,}000)(77)$
   $\approx 975{,}333.3$
   The volume is approximately 975,333.3 ft^3.

Pages 699–701 Practice and Apply

8. Find the area of the base first.

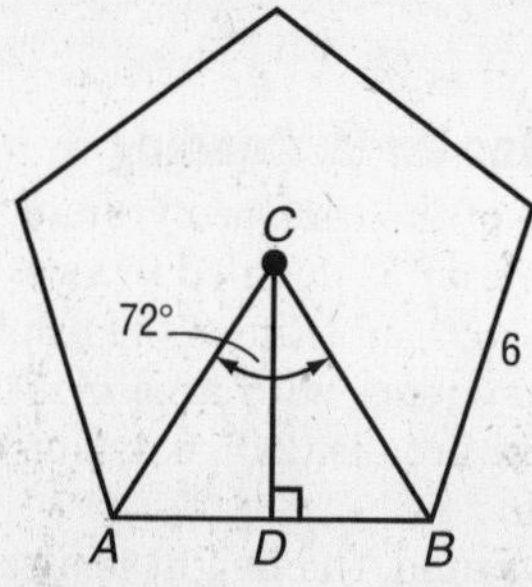

Apothem: The central angles of the pentagon are all congruent, so $m\angle ACB = \frac{360}{5}$ or 72. $\overline{CD}$ is an apothem of the pentagon. It bisects $\angle ACB$ and is a perpendicular bisector of $\overline{AB}$. So, $m\angle ACD = 36$. Since the side of the pentagon has measure 6, $AD = 3$.

$$\tan 36° = \frac{3}{CD}$$
$$CD = \frac{3}{\tan 36°}$$
$$\approx 4.13$$

perimeter $= 5(6) = 30$ cm

Base area: $A = \frac{1}{2}Pa$
$$\approx \frac{1}{2}(30)(4.13)$$
$$\approx 61.95$$

Now find the volume of the pyramid.

$$V = \frac{1}{3}Bh$$
$$\approx \frac{1}{3}(61.95)(10)$$
$$\approx 206.5$$

The volume of the pyramid is approximately 206.5 cm^3.

9. Find the height of the pyramid by first finding the height of one of the 20-20-20 triangular faces (a pyramid slant height).

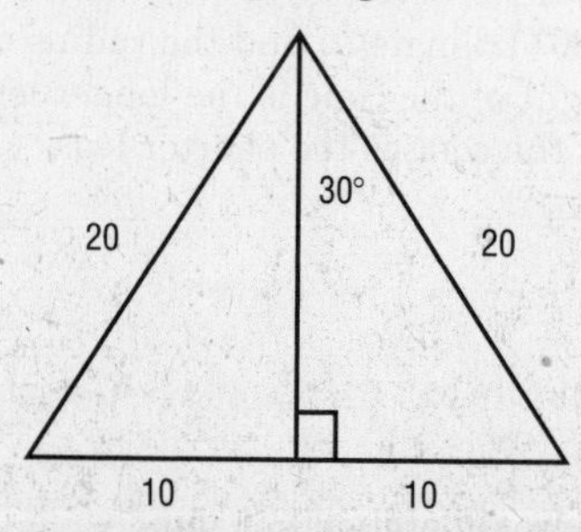

The height of this triangle is $10\sqrt{3}$.
Now use the Pythagorean Theorem with h = pyramid height.
The distance from the center of the base of the pyramid to one of the 20-in. edges is $\frac{1}{2}(15)$ or 7.5.

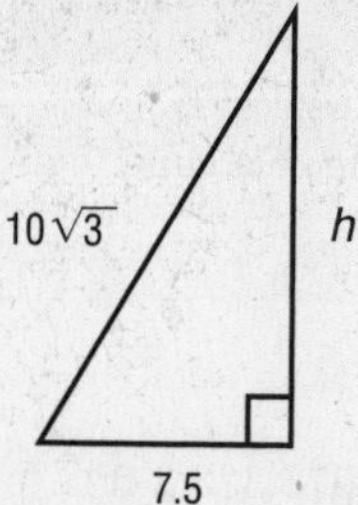

$$a^2 + b^2 = c^2$$
$$7.5^2 + h^2 = (10\sqrt{3})^2$$
$$56.25 + h^2 = 300$$
$$h^2 = 243.75$$
$$h \approx 15.612$$

For the pyramid,

$$V = \frac{1}{3}Bh$$
$$\approx \frac{1}{3}(20)(15)(15.612)$$
$$\approx 1561.2$$

The volume of the pyramid is approximately 1561.2 ft^3.

10. Use the Pythagorean Theorem with h = pyramid height.

$$a^2 + b^2 = c^2$$
$$9^2 + h^2 = 15^2$$
$$81 + h^2 = 225$$
$$h^2 = 144$$
$$h = 12$$

For the pyramid,

$$V = \frac{1}{3}Bh$$
$$= \frac{1}{3}(24)(18)(12)$$
$$= 1728$$

The volume of the pyramid is 1728 in^3.

11. Use the Pythagorean Theorem with h = cone height.

$$a^2 + b^2 = c^2$$
$$18^2 + h^2 = 30^2$$
$$324 + h^2 = 900$$
$$h^2 = 576$$
$$h = 24$$

For the cone,

$$V = \frac{1}{3}\pi r^2 h$$
$$= \frac{1}{3}\pi(18^2)(24)$$
$$\approx 8143.0$$

The volume is approximately 8143.0 mm^3.

12. Use the 45°-45°-90° triangle to find the radius of the base: $r = \frac{1}{\sqrt{2}} \cdot 10 = 5\sqrt{2}$. In the same way, find the height: $h = \frac{1}{\sqrt{2}} \cdot 10 = 5\sqrt{2}$. For the pyramid,

$$V = \frac{1}{3}\pi r^2 h$$
$$= \frac{1}{3}\pi(5\sqrt{2})^2(5\sqrt{2})$$
$$\approx 370.2$$

The volume is approximately 370.2 in^3

13.

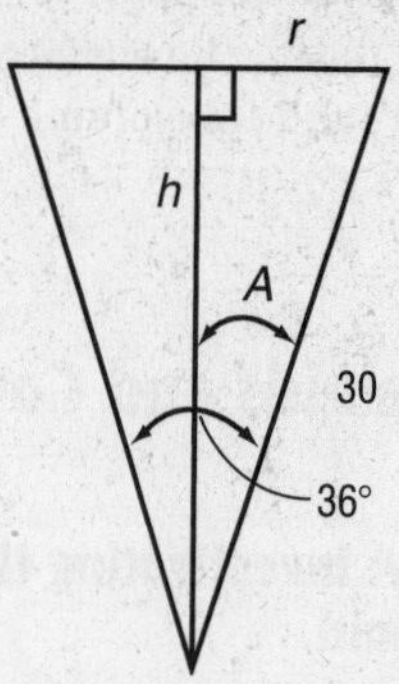

r
h
A
30
36°

Use trigonometry to find the radius of the base.

$\frac{1}{2}(36°) = 18°$

$\sin A = \frac{\text{opposite}}{\text{hypotenuse}}$

$\sin 18° = \frac{r}{30}$

$r = 30 \sin 18°$

$r \approx 9.2705$

Similarly, find the height.

$\cos A = \frac{\text{adjacent}}{\text{hypotenuse}}$

$\cos 18° = \frac{h}{30}$

$h = 30 \cos 18°$

$h \approx 28.5317$

Now find the volume.

$V = \frac{1}{3}\pi r^2 h$

$\approx \frac{1}{3}\pi(9.2705)^2(28.5317)$

≈ 2567.8

The volume of the cone is approximately 2567.8 m^3.

14. Calculate the area of the base, B, first.

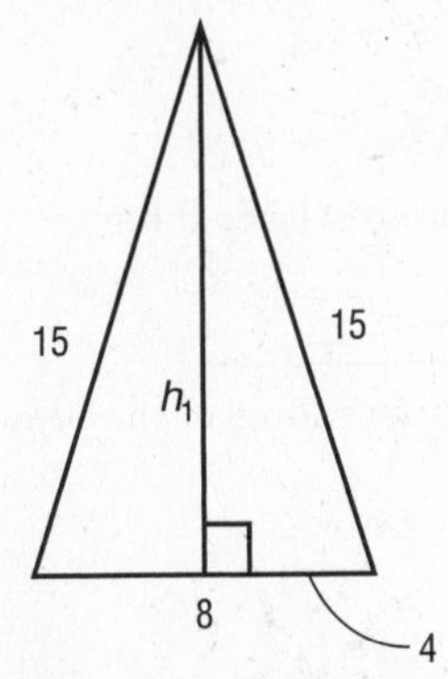

$\frac{1}{2}(8) = 4$. Now use the Pythagorean Theorem to find the height of the triangular base.

$a^2 + b^2 = c^2$

$4^2 + h_1^2 = 15^2$

$16 + h_1^2 = 225$

$h_1^2 = 209$

$h_1 = \sqrt{209}$

Now calculate the area of the base.

$B = \frac{1}{2}bh_1$

$= \frac{1}{2}(8)(\sqrt{209})$

$= 4\sqrt{209}$

Now find the height of the pyramid.

$a^2 + b^2 = c^2$

$15^2 + h_2^2 = 17^2$

$225 + h_2^2 = 289$

$h_2^2 = 64$

$h_2 = 8$

For the pyramid,

$V = \frac{1}{3}Bh_2$

$= \frac{1}{3}(4\sqrt{209})(8)$

≈ 154.2

The volume is approximately 154.2 m^3.

15. Use the Pythagorean Theorem with d = diameter of base.

$a^2 + b^2 = c^2$

$d^2 + 5^2 = 13^2$

$d^2 + 25 = 169$

$d^2 = 144$

$d = 12$

For the cone, the radius is $\frac{1}{2}(12)$ or 6. The volume is

$V = \frac{1}{3}Bh$

$= \frac{1}{3}\pi r^2 h$

$= \frac{1}{3}\pi(6^2)(5)$

≈ 188.5

The volume is approximately 188.5 cm^3.

16. The radius of the base is $r = \frac{1}{2}(24) = 12$. Use the 30°-60°-90° triangle to find the height, which is the shorter leg: $h = \frac{1}{2}(15) = 7.5$.

For the cone,

$V = \frac{1}{3}\pi r^2 h$

$= \frac{1}{3}\pi(12^2)(7.5)$

≈ 1131.0

The volume is approximately 1131.0 ft^3.

17. Find the height of the pyramid portion by first finding the height of one of the triangular sides (the pyramid slant height). Use the Pythagorean Theorem.

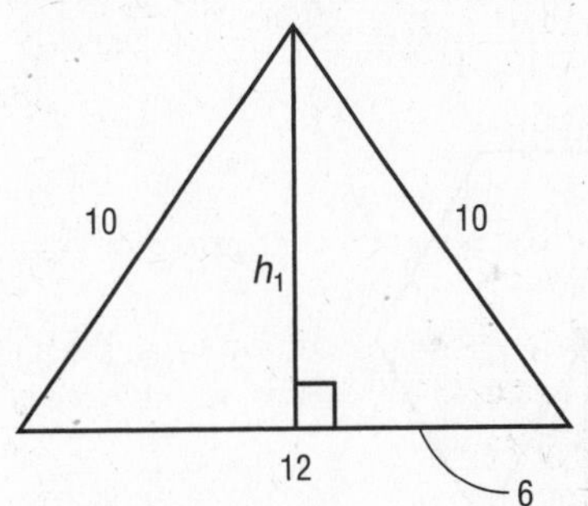

$a^2 + b^2 = c^2$

$6^2 + h_1^2 = 10^2$

$36 + h_1^2 = 100$

$h_1^2 = 64$

$h_1 = 8$

Now find the height of the pyramid, again using the Pythagorean Theorem.

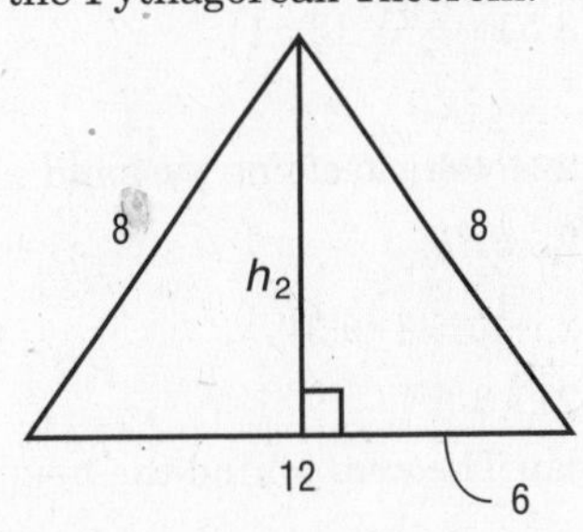

$$a^2 + b^2 = c^2$$
$$6^2 + h_2^{\,2} = 8^2$$
$$36 + h_2^{\,2} = 64$$
$$h_2^{\,2} = 28$$
$$h_2 = 2\sqrt{7}$$

volume of solid = volume of cube + volume of pyramid

$$= s^3 + \tfrac{1}{3}Bh_2$$
$$= 12^3 + \tfrac{1}{3}(12)(12)(2\sqrt{7})$$
$$\approx 1982.0$$

The volume is approximately 1982.0 mm^3.

18. Calculate the area of the hexagon first. The center-to-vertex distance, s, which in a regular hexagon equals the side length, is found using the Pythagorean Theorem.

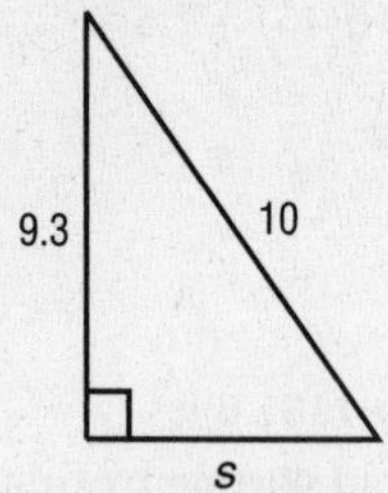

$$a^2 + b^2 = c^2$$
$$s^2 + 9.3^2 = 10^2$$
$$s^2 + 86.49 = 100$$
$$s^2 = 13.51$$
$$s = \sqrt{13.51}$$

Apothem:

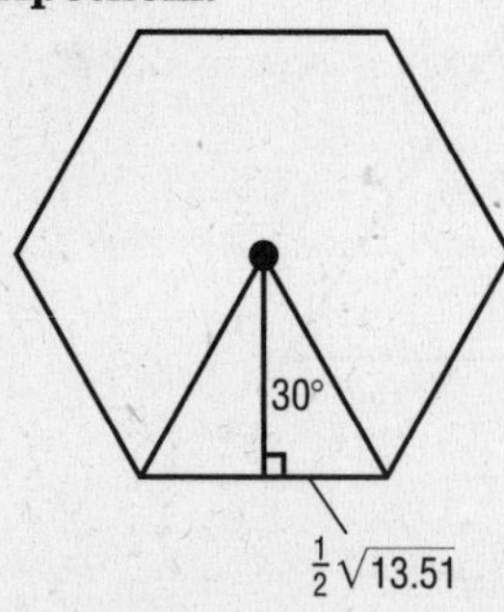

The apothem is $\sqrt{3} \cdot \frac{1}{2}\sqrt{13.51}$.

perimeter = $6\sqrt{13.51}$

Area: $A = \frac{1}{2}Pa$

$$= \tfrac{1}{2}(6\sqrt{13.51})\left(\tfrac{\sqrt{3}}{2}\sqrt{13.51}\right)$$
$$\approx 35.1$$

volume of solid = 2 × volume of one pyramid

$$= 2 \cdot \tfrac{1}{3}Bh$$
$$\approx 2 \cdot \tfrac{1}{3}(35.1)(9.3)$$
$$\approx 217.6 \text{ ft}^3$$

19. Use the Pythagorean Theorem to find the height of the frustum.

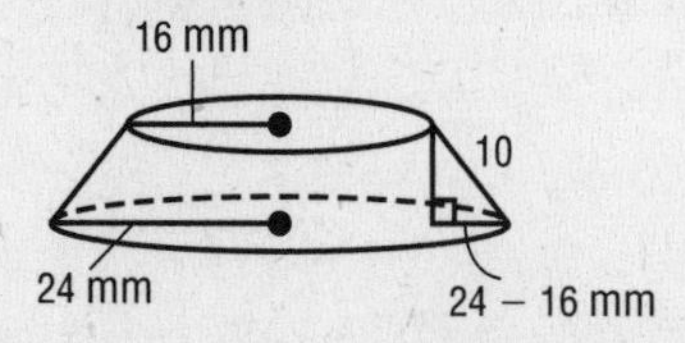

$$a^2 + b^2 = c^2$$
$$(24 - 16)^2 + h^2 = 10^2$$
$$64 + h^2 = 100$$
$$h^2 = 36$$
$$h = 6 \text{ cm}$$

Let x equal the height of the "missing" cone. By using similar triangles,

$$\frac{16}{24} = \frac{x}{x + 6}$$
$$\frac{2}{3} = \frac{x}{x + 6}$$
$$2(x + 6) = 3x$$
$$2x + 12 = 3x$$
$$x = 12 \text{ cm}$$

volume of frustum = volume of large cone − volume of "missing" cone

$$= \tfrac{1}{3}\pi r^2(x + h) - \tfrac{1}{3}\pi r^2 x$$
$$= \tfrac{1}{3}\pi(24^2)(12 + 6) - \tfrac{1}{3}\pi(16^2)(12)$$
$$\approx 7640.4 \text{ cm}^3$$

20. $r = \frac{1}{2}(103) = 51.5$ km, $h = 4.17$ km

$$V = \tfrac{1}{3}\pi r^2 h$$
$$= \tfrac{1}{3}\pi(51.5^2)(4.17)$$
$$\approx 11{,}581.9$$

The volume is approximately 11,581.9 km^3.

21.

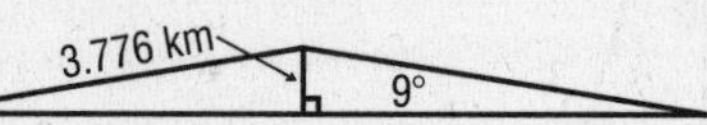

Use trigonometry to find the radius of the base.

$$\tan 9° = \frac{3.776}{r}$$
$$r = \frac{3.776}{\tan 9°}$$
$$\approx 23.840726$$
$$V = \tfrac{1}{3}\pi r^2 h$$
$$\approx \tfrac{1}{3}\pi(23.840726^2)(3.776)$$
$$\approx 2247.5$$

The volume is approximately 2247.5 km^3.

22.

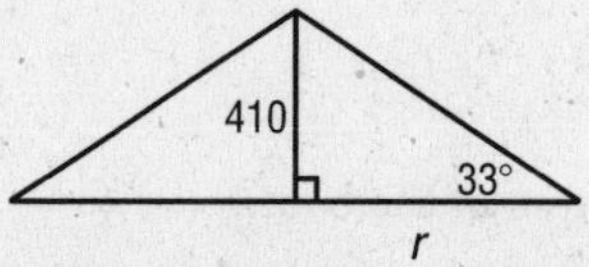

Use trigonometry to find the radius of the base.

$$\tan 33° = \frac{410}{r}$$
$$r = \frac{410}{\tan 33°}$$
$$\approx 631.344635$$
$$V = \tfrac{1}{3}\pi r^2 h$$
$$\approx \tfrac{1}{3}\pi(631.344635^2)(410)$$
$$\approx 171{,}137{,}610.4$$

The volume is approximately 171,137,610.4 m^3.

23. $r = \frac{1}{2}(22.3) = 11.15$

$V = \frac{1}{3}\pi r^2 h$

$= \frac{1}{3}\pi(11.15^2)(1.22)$

≈ 158.8

The volume is approximately 158.8 km^3.

24. $\frac{2}{3}$; The volume of each pyramid that makes up the solid on the left is $\frac{1}{3}$ of the volume of the prism, so the total volume of the solid on the left is $\frac{1}{3} + \frac{1}{3}$ or $\frac{2}{3}$ of the volume of the prism.

25. Use the formula for the volume of a pyramid.

$V = \frac{1}{3}Bh$

$= \frac{1}{3}(755^2)(481)$

$\approx 91{,}394{,}008.3$

The original volume of the pyramid is approximately 91,394,008.3 ft^3.

26. Use the formula for the volume of a pyramid.

$V = \frac{1}{3}Bh$

$= \frac{1}{3}(755^2)(449)$

$= 85{,}313{,}741.7$

The present day volume of the pyramid is approximately 85,313,741.7 ft^3.

27. $91{,}394{,}008.33 - 85{,}313{,}741.67 \approx 6{,}080{,}266.7$

Approximately 6,080,266.7 cubic feet have been lost.

28. The volume of the cone is $\frac{1}{3}$ of the volume of the cylinder, so

probability $= 1 - \frac{1}{3}$

$= \frac{2}{3}$

29. Find the height of the pyramid using the Pythagorean Theorem.

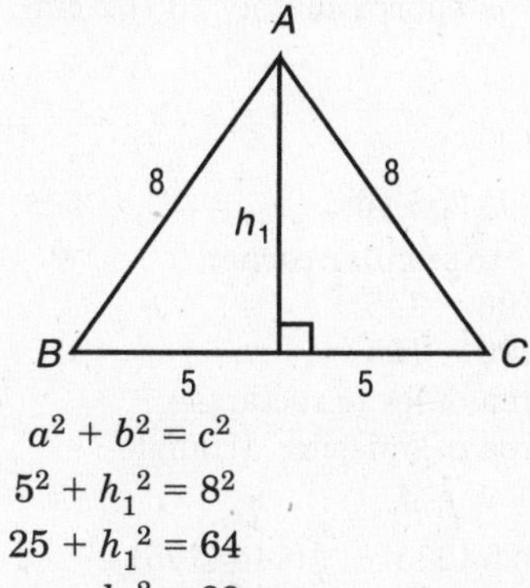

$a^2 + b^2 = c^2$

$5^2 + {h_1}^2 = 8^2$

$25 + {h_1}^2 = 64$

${h_1}^2 = 39$

$h_1 = \sqrt{39}$

For the cone, radius $= \frac{1}{2}(10)$ or 5.

volume of solid = volume of oblique pyramid + volume of cone

$= \frac{1}{3}Bh_1 + \frac{1}{3}\pi r^2 h_2$

$= \frac{1}{3}(10^2)(\sqrt{39}) + \frac{1}{3}\pi(5^2)(12)$

≈ 522.3

The volume of the solid is approximately 522.3 $units^3$.

30. Side view of tower with "missing portion":

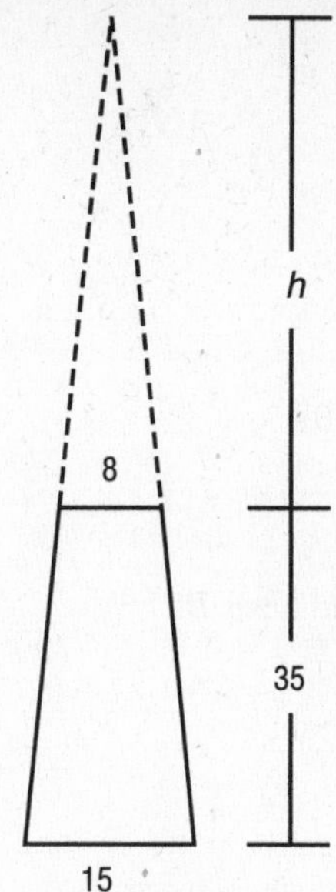

By similar triangles,

$\frac{8}{15} = \frac{h}{h + 35}$

$8h + 280 = 15h$

$280 = 7h$

$h = 40$ ft

volume of frustum = volume of large pyramid − volume of "missing" portion

$= \frac{1}{3}B_1h_1 - \frac{1}{3}B_2h_2$

$= \frac{1}{3}(15^2)(35 + 40) - \frac{1}{3}(8^2)(40)$

≈ 4771.7

The volume of the frustum is approximately 4771.7 ft^3.

31. Calculate the slant height, using the fact that each face is an equilateral triangle.

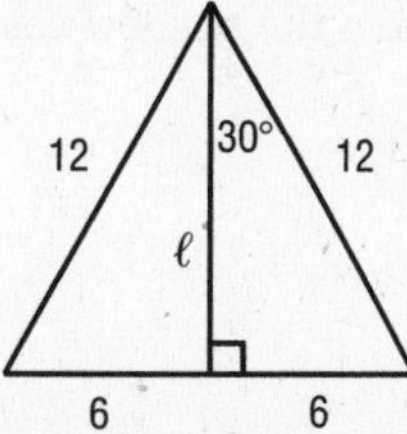

The slant height ℓ equals $6\sqrt{3}$ because it is the measure of the longer leg of the 30°-60°-90° right triangle. Now find the apothem of the base.

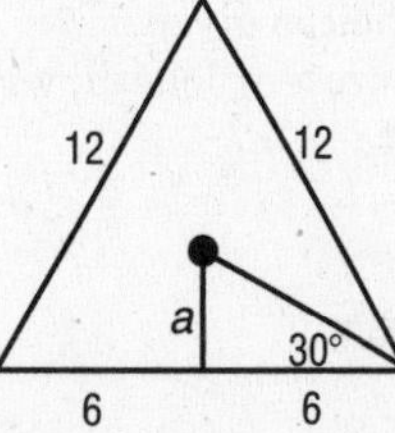

The apothem a equals $\frac{1}{\sqrt{3}} \cdot 6 = 2\sqrt{3}$. Now find the height of the solid, using the Pythagorean Theorem.

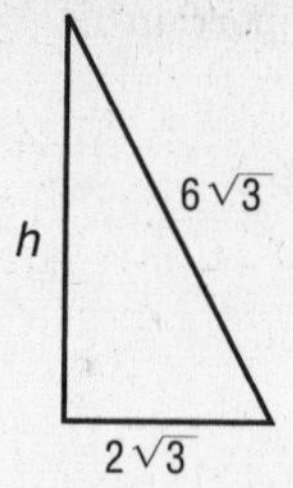

$$a^2 + b^2 = c^2$$
$$h^2 + (2\sqrt{3})^2 = (6\sqrt{3})^2$$
$$h^2 + 12 = 108$$
$$h^2 = 96$$
$$h = 4\sqrt{6}$$

Find the area of the base, B.

$$B = \frac{1}{2}Pa$$
$$= \frac{1}{2}(36)(2\sqrt{3})$$
$$= 36\sqrt{3}$$

For the pyramid,

$$V = \frac{1}{3}Bh$$
$$= \frac{1}{3}(36\sqrt{3})(4\sqrt{6})$$
$$\approx 203.6$$

The volume is approximately 203.6 in^3.

32. Sample answer: Architects use geometry to design buildings that meet the needs of their clients. Answers should include the following.

- The surface area at the top of a pyramid is much smaller than the surface area of the base. There is less office space at the top, than on the first floor.
- The silhouette of a pyramid-shaped building is smaller than the silhouette of a rectangular prism with the same height. If the light conditions are the same, the shadow cast by the pyramid is smaller than the shadow cast by the rectangular prism.

33. B; $V = \frac{1}{3}Bh$

$$= \frac{1}{3}b^2h$$
$$= \frac{1}{3}(2h)^2h$$
$$= \frac{4h^3}{3}$$

34. A; $x^3 \pm 9x = x(x^2 \pm 9)$

$$= x(x^2 - 9) \text{ or } x(x^2 + 9)$$
$$= x(x + 3)(x - 3) \text{ or } x(x^2 + 9)$$

$x(x^2 + 9)$ is not one of the choices given. So the factors that could represent length, width, and height are x, $x + 3$, and $x - 3$.

Page 701 Maintain Your Skills

35. $V = Bh$

$$= (14)(12)(6)$$
$$= 1008$$

The volume of the prism is 1008 in^3.

36. $V = \pi r^2h$

$$= \pi(8^2)(17)$$
$$\approx 3418.1$$

The volume is approximately 3418.1 m^3.

37. Use the Pythagorean Theorem to find the height of the triangle. One leg of the right triangle is $\frac{1}{2}(10)$ or 5 ft.

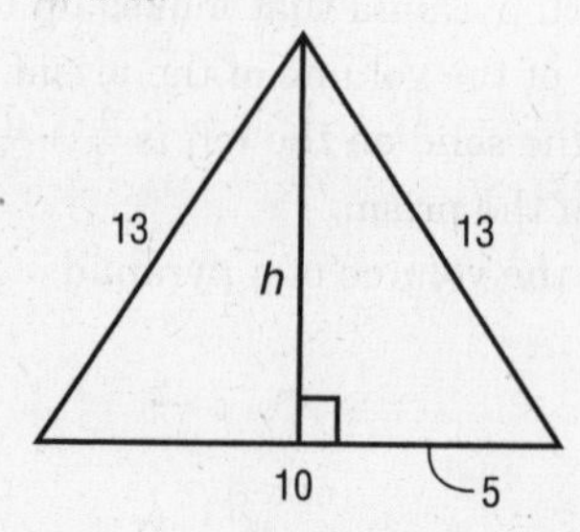

$$h^2 + 5^2 = 13^2$$
$$h^2 + 25 = 169$$
$$h^2 = 144$$
$$h = 12$$

For the prism, taking the triangles as bases,

$$V = Bh$$
$$= \frac{1}{2}(10)(12)(19)$$
$$= 1140$$

The volume of the prism is 1140 ft^3.

38. Use the circumference to find the radius.

$$C = 2\pi r$$
$$86 = 2\pi r$$
$$\frac{43}{\pi} = r$$

Now find the area.

$$T = 4\pi r^2$$
$$= 4\pi\left(\frac{43}{\pi}\right)^2$$
$$\approx 2354.2$$

The surface area is approximately 2354.2 cm^2.

39. $T = 4\pi r^2$

$$= 4(64.5)$$
$$= 258$$

The surface area is 258 yd^2.

40. For the "missing" triangular corner,

base = 335 − 190 = 145

height = 325 − 220 = 105

area of field = area of large rectangle − area of "missing" triangle

$$= \ell w - \frac{1}{2}bh$$
$$= (325)(335) - \frac{1}{2}(145)(105)$$
$$= 101{,}262.5$$

The total area is 101,262.5 ft^2.

41. $4\pi r^2 = 4\pi(3.4^2)$

$$\approx 145.27$$

42. $\frac{4}{3}\pi r^3 = \frac{4}{3}\pi(7^3)$

$$\approx 1436.76$$

43. $4\pi r^2 = 4\pi(12^2)$

$$\approx 1809.56$$

Page 701 Practice Quiz 1

1. Use the formula for the volume of a cylinder. The radius is $r = \frac{1}{2}(4)$ or 2.

$V = \pi r^2 h$
$= \pi(2^2)(10)$
≈ 125.7

The volume is approximately 125.7 in^3.

2. Use the formula for the volume of a cylinder. The radius is $r = \frac{1}{2}(12)$ or 6.

$V = \pi r^2 h$
$= \pi(6^2)(15)$
≈ 1696.5

The volume is approximately 1696.5 m^3.

3. First find the area of the hexagonal base.

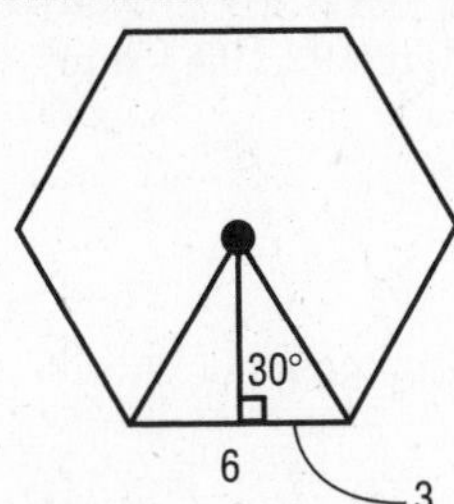

Apothem: A 30°-60°-90° triangle is formed by the apothem and one-half of a side of the hexagon. The shorter leg of the triangle is $\frac{1}{2}(6)$ or 3. The apothem is the longer leg of the triangle or $3\sqrt{3}$.

perimeter = $6 \cdot 6 = 36$

The area of the hexagonal base is

$A = \frac{1}{2}Pa$
$= \frac{1}{2}(36)(3\sqrt{3})$
$= 54\sqrt{3}$

For the prism,

$V = Bh$
$= (54\sqrt{3})(10)$
≈ 935.3

The volume is approximately 935.3 cm^3.

4.

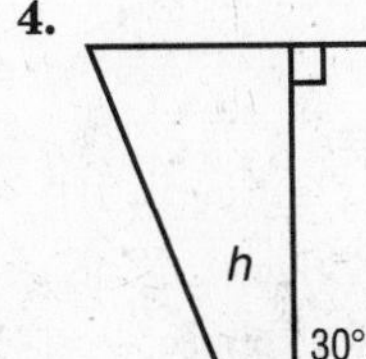

Use a 30°-60°-90° triangle to find the radius of the base and the height.

$r = \frac{1}{2}(20)$ or 10 ft, and $h = 10\sqrt{3}$ ft.

For the cone,

$V = \frac{1}{3}\pi r^2 h$
$= \frac{1}{3}\pi(10^2)(10\sqrt{3})$
≈ 1813.8

The volume is approximately 1813.8 ft^3.

5. Use the Pythagorean Theorem to find the height of the triangular base.

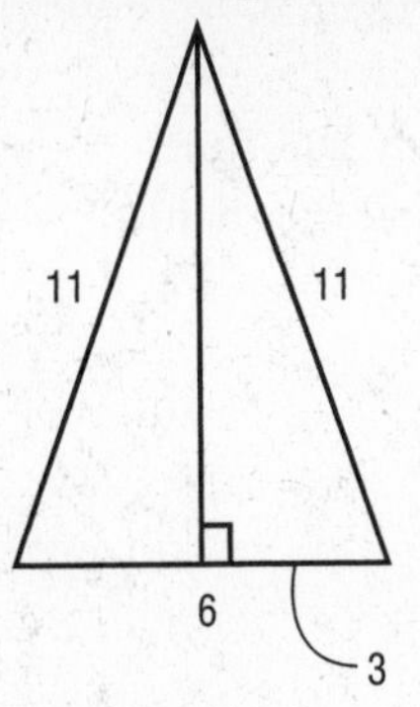

$a^2 + b^2 = c^2$
$3^2 + h^2 = 11^2$
$9 + h^2 = 121$
$h^2 = 112$
$h = 4\sqrt{7}$

Now find the area of the base, B.

$B = \frac{1}{2}bh$
$= \frac{1}{2}(6)(4\sqrt{7})$
$= 12\sqrt{7}$

For the pyramid,

$V = \frac{1}{3}Bh$
$= \frac{1}{3}(12\sqrt{7})(4)$
≈ 42.3

The volume is approximately 42.3 in^3.

13-3 Volumes of Spheres

Page 704 Check for Understanding

1. The volume of a sphere was generated by adding the volumes of an infinite number of small pyramids. Each pyramid has its base on the surface of the sphere and its height from the base to the center of the sphere.

2. Kenji; Winona divided the 12 by 3 before raising the result to the third power. Thus the order of operations was not followed correctly.

3. $V = \frac{4}{3}\pi r^3$
$= \frac{4}{3}\pi(13^3)$
≈ 9202.8

The volume is approximately 9202.8 in^3.

4. The radius is $\frac{1}{2}(12.5) = 6.25$.

$V = \frac{4}{3}\pi r^3$
$= \frac{4}{3}\pi(6.25^3)$
≈ 1022.7

The volume is approximately 1022.7 cm^3.

5. $V = \frac{4}{3}\pi r^3$
$= \frac{4}{3}\pi(4^3)$
≈ 268.1
The volume is approximately 268.1 in^3.

6. Use the circumference to find the radius.
$C = 2\pi r$
$18 = 2\pi r$
$\frac{9}{\pi} = r$
$V = \frac{4}{3}\pi r^3$
$\approx \frac{4}{3}\pi\left(\frac{9}{\pi}\right)^3$
≈ 98.5
The volume is approximately 98.5 cm^3.

7. The radius is $\frac{1}{2}(8.4) = 4.2$.
$V = \frac{1}{2}\left(\frac{4}{3}\pi r^3\right)$
$= \frac{2}{3}\pi(4.2^3)$
≈ 155.2
The volume is approximately 155.2 m^3.

8. For the sphere,
$V = \frac{4}{3}\pi r^3$
$= \frac{4}{3}\pi(5^3)$
$= \frac{500}{3}\pi$
For the cone,
$V = \frac{1}{3}\pi r^2 h$
$= \frac{1}{3}\pi(5^2)(20)$
$= \frac{500}{3}\pi$
The two volumes are equal.

Pages 704–706 Practice and Apply

9. $V = \frac{4}{3}\pi r^3$
$= \frac{4}{3}\pi(7.62^3)$
≈ 1853.3
The volume is approximately 1853.3 m^3.

10. radius $= \frac{1}{2}(33) = 16.5$
$V = \frac{4}{3}\pi r^3$
$= \frac{4}{3}\pi(16.5^3)$
$\approx 18{,}816.6$
The volume is approximately 18,816.6 in^3.

11. radius $= \frac{1}{2}(18.4) = 9.2$
$V = \frac{4}{3}\pi r^3$
$= \frac{4}{3}\pi(9.2^3)$
≈ 3261.8
The volume is approximately 3261.8 ft^3.

12. $V = \frac{4}{3}\pi r^3$
$= \frac{4}{3}\pi\left(\frac{\sqrt{3}}{2}\right)^3$
≈ 2.7
The volume is approximately 2.7 cm^3.

13. Use the circumference to find the radius.
$C = 2\pi r$
$24 = 2\pi r$
$\frac{12}{\pi} = r$
Now find the volume.
$V = \frac{4}{3}\pi r^3$
$= \frac{4}{3}\pi\left(\frac{12}{\pi}\right)^3$
≈ 233.4
The volume is approximately 233.4 in^3.

14. $V = \frac{4}{3}\pi r^3$
$= \frac{4}{3}\pi(35.8^3)$
$\approx 192{,}193.1$
The volume is approximately 192,193.1 mm^3.

15. $V = \frac{1}{2}\left(\frac{4}{3}\pi r^3\right)$
$= \frac{2}{3}\pi(3.2^3)$
≈ 68.6
The volume is approximately 68.6 m^3.

16. radius $= \frac{1}{2}(28) = 14$
$V = \frac{1}{2}\left(\frac{4}{3}\pi r^3\right)$
$= \frac{2}{3}\pi(14^3)$
≈ 5747.0
The volume is approximately 5747.0 ft^3.

17. $V = \frac{4}{3}\pi r^3$
$= \frac{4}{3}\pi(12^3)$
≈ 7238.2
The volume is approximately 7238.2 in^3.

18. Find the radius.
$C = 2\pi r$
$48 = 2\pi r$
$\frac{24}{\pi} = r$
Now find the volume.
$V = \frac{4}{3}\pi r^3$
$= \frac{4}{3}\pi\left(\frac{24}{\pi}\right)^3$
≈ 1867.6
The volume is approximately 1867.6 cm^3.

19. Use the formula for the volume of a sphere.
radius $= \frac{1}{2}(3476) = 1738$
$V = \frac{4}{3}\pi r^3$
$= \frac{4}{3}\pi(1738^3)$
$\approx 21{,}990{,}642{,}871$
The volume is approximately 21,990,642,871 km^3.

20. For the golf ball, $r = \frac{1}{2}(4.3) = 2.15$.
$V = \frac{4}{3}\pi r^3$
$= \frac{4}{3}\pi(2.15^3)$
≈ 41.63

For the tennis ball, $r = \frac{1}{2}(6.9) = 3.45$.

$V = \frac{4}{3}\pi r^3$

$= \frac{4}{3}\pi(3.45^3)$

≈ 172.01

The difference is about 172.01 − 41.63 or 130.4 cm^3.

21. For the cone, $r = \frac{1}{2}(4) = 2$.

$V = \frac{1}{3}\pi r^2 h$

$= \frac{1}{3}\pi(2^2)(10)$

$\approx 41.9 \text{ cm}^3$

For the ice cream, $r = \frac{1}{2}(4) = 2$.

$V = \frac{4}{3}\pi r^3$

$= \frac{4}{3}\pi(2^3)$

$\approx 33.5 \text{ cm}^3$

Since the volume of the ice cream is less than the volume of the cone, the cone will not overflow.

22. $\frac{\text{volume of ice cream}}{\text{volume of cone}} \approx \frac{33.5}{41.9}$

≈ 0.80

The cone will be about 80% filled.

23. $V = \frac{4}{3}\pi r^3$

$= \frac{4}{3}\pi(17^3)$

$\approx 20{,}579.5$

The volume is approximately 20,579.5 mm^3.

24. total volume × "just right" percentage

$\approx 20{,}579.5 \times 0.59$

$\approx 12{,}141.9$

The "just right" volume is approximately 12,141.9 mm^3.

25. $T = 4\pi r^2$

$= 4\pi(17^2)$

≈ 3631.68

total area × "wish for more" percentage

$\approx 3631.7 \times 0.32$

≈ 1162.1

The "wish for more" surface area is approximately 1162.1 mm^2.

26. $A = \pi r^2$

$= \pi(17^2)$

$\approx 907.9 \text{ mm}^2$

total area × "wish for less" percentage

$\approx 907.9 \times 0.09$

≈ 81.7

The area of the "wish for less" sector is approximately 81.7 mm^2.

27. For the sphere,

$V = \frac{4}{3}\pi r^3$

$= \frac{4}{3}\pi(6^3)$

$= 288\pi \text{ cm}^3$

For the cylinder,

$V = \pi r^2 h$

$= \pi(6^2)(12)$

$= 432\pi$

$\frac{\text{volume of sphere}}{\text{volume of cylinder}} = \frac{288\pi}{432\pi} = \frac{2}{3}$

28. For the cylinder, $r = \frac{1}{2}(2.5) = 1.25$.

$V = \pi r^2 h$

$= \pi(1.25^2)(7.5)$

≈ 36.82

For the three balls, $r = \frac{1}{2}(2.5) = 1.25$.

$V = 3\left(\frac{4}{3}\pi r^3\right)$

$= 4\pi(1.25^3)$

≈ 24.54

volume of empty space $\approx 36.82 - 24.54$

$\approx 12.3 \text{ in}^3$

29. For the cube,

$V = s^3$

$216 = s^3$

$6 = s$

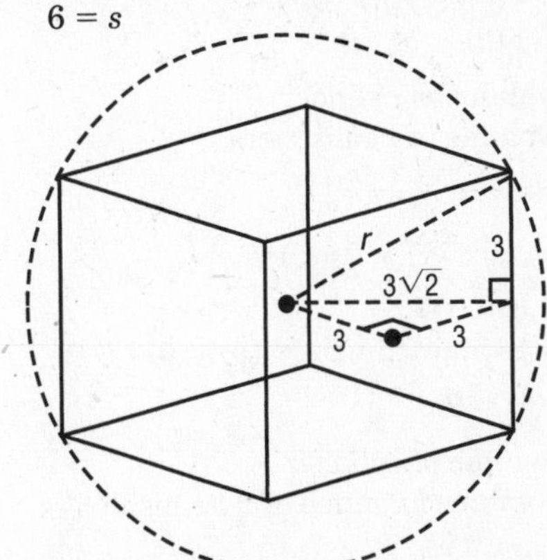

The radius of the sphere appears as r in the figure. It is found by two applications of the Pythagorean Theorem. First, for the horizontal right triangle:

$a^2 + b^2 = c^2$

$3^2 + 3^2 = c^2$

$18 = c^2$

$\sqrt{18} = c$

$3\sqrt{2} = c$

Now for the vertical right triangle:

$a^2 + b^2 = c^2$

$(3\sqrt{2})^2 + 3^2 = r^2$

$27 = r^2$

$\sqrt{27} = r$

$3\sqrt{3} = r$

For the sphere,

$V = \frac{4}{3}\pi r^3$

$= \frac{4}{3}\pi(3\sqrt{3})^3$

≈ 587.7

The volume of the sphere is approximately 587.7 in^3.

30. Find the radius.

$T = 4\pi r^2$

$784\pi = 4\pi r^2$

$196 = r^2$

$14 = r$

Now find the volume.

$V = \frac{4}{3}\pi r^3$

$= \frac{4}{3}\pi(14^3)$

$\approx 11{,}494.0$

The surface area is approximately 11,494.0 in^3.

31. total area = area of curved half-sphere + area of flat bottom

$T = \frac{1}{2}(4\pi r^2) + \pi r^2$

Find the radius.

$18.75\pi = 3\pi r^2$

$6.25 = r^2$

$2.5 = r$

Now find the volume.

$V = \frac{1}{2}\left(\frac{4}{3}\pi r^3\right)$

$= \frac{2}{3}\pi(2.5^3)$

≈ 32.7

The volume is approximately 32.7 m^3.

32. radius $= \frac{1}{2}(142) = 71$

volume = volume of cylinder + volume of hemisphere

$= \pi r^2 h + \frac{1}{2}\left(\frac{4}{3}\pi r^3\right)$

$= \pi(71^2)(71) + \frac{2}{3}\pi(71^3)$

$\approx 1{,}874{,}017.6$

The volume is approximately 1,874,017.6 ft^3.

33. radius $= \frac{1}{2}(4) = 2$

volume = volume of cylinder + combined volume of 2 hemispheres

$= \pi r^2 h + \frac{4}{3}\pi r^3$

$= \pi(2^2)(12) + \frac{4}{3}\pi(2^3)$

≈ 184

The volume is approximately 184 mm^3.

34. Sample answer: If a student knows the circumference of a sphere, then the volume can be found. Answers should include the following.

- One needs to know the radius of the Earth.
- The radius of Earth is about 6366.2 km and the volume is about 1.1×10^{12} km^3.

35. See student's work.

36. A; the ratio of the volumes is the cube of the ratio of the radii, namely $\left(\frac{3}{5}\right)^3 = 0.216$ or 21.6%.

37. A; $\frac{1}{2}(4\pi r^2) + \pi r^2 h + \frac{1}{2}(4\pi r^2) = 2\pi r^2 + \pi r^2 h + 2\pi r^2$

$= \pi r^2(2 + h + 2)$

$= \pi r^2(4 + h)$

Page 706 Maintain Your Skills

38. $V = \frac{1}{3}\pi r^2 h$

$= \frac{1}{3}\pi(6^2)(9.5)$

≈ 358.1

The volume is approximately 358.1 m^3.

39. radius $= \frac{1}{2}(15) = 7.5$

$V = \frac{1}{3}\pi r^2 h$

$= \frac{1}{3}\pi(7.5^2)(7)$

≈ 412.3

The volume is approximately 412.3 m^3.

40. Use the formula for the volume of a rectangular prism.

$V = Bh$

$25.9 = \ell(2.4)(5.0)$

$25.9 = 12\ell$

$2.2 \approx \ell$

The depth is approximately 2.2 ft.

41. $(x - h)^2 + (y - k)^2 = r^2$

$(x - 2)^2 + [y - (-1)]^2 = 8^2$

$(x - 2)^2 + (y + 1)^2 = 64$

42. $(x - h)^2 + (y - k)^2 = r^2$

$[x - (-4)]^2 + [y - (-3)]^2 = (\sqrt{19})^2$

$(x + 4)^2 + (y + 3)^2 = 19$

43. Find the center of the circle.

$M\left(\frac{x_1 + x_2}{2}, \frac{y_1 + y_2}{2}\right) = \left(\frac{5 + (-1)}{2}, \frac{-4 + 6}{2}\right)$

$= (2, 1)$

Find the radius.

$\frac{1}{2} \times \text{diameter} = \frac{1}{2}\sqrt{[-1 - 5]^2 + [6 - (-4)]^2}$

$= \sqrt{34}$

$(x - h)^2 + (y - k)^2 = r^2$

$(x - 2)^2 + (y - 1)^2 = (\sqrt{34})^2$

$(x - 2)^2 + (y - 1)^2 = 34$

44. $(2a)^2 = (2a)(2a)$

$= 2 \cdot 2 \cdot a \cdot a$

$= 4a^2$

45. $(3x)^3 = (3x)(3x)(3x)$

$= 3 \cdot 3 \cdot 3 \cdot x \cdot x \cdot x$

$= 27x^3$

46. $\left(\frac{5a}{b}\right)^2 = \left(\frac{5a}{b}\right)\left(\frac{5a}{b}\right)$

$= \frac{5a \cdot 5a}{b \cdot b}$

$= \frac{5 \cdot 5 \cdot a \cdot a}{b \cdot b}$

$= \frac{25a^2}{b^2}$

47. $\left(\frac{2k}{5}\right)^3 = \left(\frac{2k}{5}\right)\left(\frac{2k}{5}\right)\left(\frac{2k}{5}\right)$

$= \frac{2k \cdot 2k \cdot 2k}{5 \cdot 5 \cdot 5}$

$= \frac{2 \cdot 2 \cdot 2 \cdot k \cdot k \cdot k}{5 \cdot 5 \cdot 5}$

$= \frac{8k^3}{125}$

13-4 Congruent and Similar Solids

Page 709 Spreadsheet Investigation: Explore Similar Solids

1. If a number in row 6 is a, then row 7 contains a^2, and row 8 contains a^3.

2. The ratio of the surface areas is $a^2 : b^2$.

3. The ratio of the volumes is $a^3 : b^3$.

Page 710 Check for Understanding

1. Sample answer:

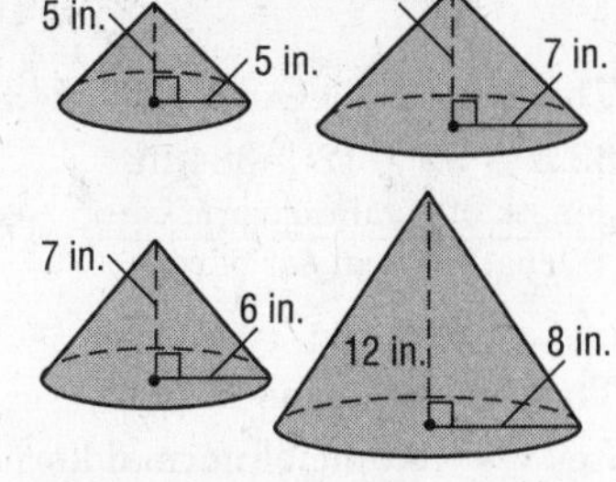

2. If two solids are similar with a scale factor of $a : b$, then the surface areas have a ratio of $a^2 : b^2$ and the volumes have a ratio of $a^3 : b^3$.

3. The two cones are of exactly the same shape and size, so they are congruent.

4. $\frac{\text{height of larger cylinder}}{\text{height of smaller cylinder}} = \frac{30}{20} = \frac{3}{2}$

$\frac{\text{diameter of larger cylinder}}{\text{diameter of smaller cylinder}} = \frac{22.5}{15} = \frac{3}{2}$

The two solids are similar. Since the scale factor is not 1, they are not congruent.

5. $\frac{\text{height of larger pyramid}}{\text{height of smaller pyramid}} = \frac{24}{18} = \frac{4}{3}$

The scale factor is 4 : 3.

6. $\frac{\text{surface area of larger pyramid}}{\text{surface area of smaller pyramid}} = \frac{a^2}{b^2} = \frac{4^2}{3^2} = \frac{16}{9}$

The ratio of the surface areas is 16 : 9.

7. $\frac{\text{volume of larger pyramid}}{\text{volume of smaller pyramid}} = \frac{a^3}{b^3} = \frac{4^3}{3^3} = \frac{64}{27}$

The ratio of the volumes is 64 : 27.

8. $\frac{\text{diameter of smaller ball}}{\text{diameter of larger ball}} = \frac{2}{16} = \frac{1}{8}$

The scale factor is 1 : 8.

9. $\frac{\text{surface area of smaller ball}}{\text{surface area of larger ball}} = \frac{a^2}{b^2} = \frac{1^2}{8^2} = \frac{1}{64}$

The ratio of the surface areas is 1 : 64.

10. $\frac{\text{volume of smaller ball}}{\text{volume of larger ball}} = \frac{a^3}{b^3} = \frac{1^3}{8^3} = \frac{1}{512}$

The ratio of the volumes is 1 : 512.

Pages 711–713 Practice and Apply

11. The bases have different shapes, so the two pyramids are neither congruent nor similar.

12. The spheres are identical in shape but not in size (unless $a = b$), so they are considered similar.

13. Use the Pythagorean Theorem to find the height of the second cylinder.

$a^2 + b^2 = c^2$
$12^2 + h^2 = 20^2$
$144 + h^2 = 400$
$h^2 = 256$
$h = 16$ in.

Since the two cylinders have the same height and diameter $(2 \cdot 6 = 12)$, they are congruent.

14. $\frac{\text{base edge of larger pyramid}}{\text{base edge of smaller pyramid}} = \frac{12\sqrt{6}}{4\sqrt{3}} = 3\sqrt{2}$

$\frac{\text{height of larger pyramid}}{\text{height of smaller pyramid}} = \frac{36\sqrt{6}}{12\sqrt{2}} = 3\sqrt{3}$

Since the ratios are not the same, the pyramids are neither congruent nor similar.

15. $\frac{\text{length of smaller prism}}{\text{length of larger prism}} = \frac{10}{15} = \frac{2}{3}$

$\frac{\text{width of smaller prism}}{\text{width of larger prism}} = \frac{1}{3}$

Since the ratios are not the same, the pyramids are neither congruent nor similar.

16. The cubes are identical in shape but not in size, so they are similar (but not congruent).

17. $26 \times \frac{5}{1} = 130$ m high

$49 \times \frac{5}{1} = 245$ m wide

$93 \times \frac{5}{1} = 465$ m long

18. Always; spheres have only one measure to compare.

19. Always; congruent solids have equal dimensions.

20. Sometimes; if the solids have a scale factor of 1, the volumes will be equal.

21. Never; different types of solids cannot be similar.

22. Never; different types of solids cannot be similar.

23. Sometimes; solids that are not similar can have the same surface area.

24. $15 \times \frac{1}{1000} = 0.015$

The Micro-Car door handle is 0.015 cm long.

25. $x \times \left(\frac{1000}{1}\right)^2 = x \times \frac{1{,}000{,}000}{1} = 1{,}000{,}000x$

The full-sized car's surface area is $1{,}000{,}000x$ cm^2.

26. $15 \times \frac{1}{18} = \frac{15}{18} = \frac{5}{6}$

The miniature door handle would be $\frac{5}{6}$ or about 0.83 cm long.

27. $\dfrac{\text{perimeter of smaller prism}}{\text{perimeter of larger prism}}$

$= \dfrac{\text{height of smaller prism's base}}{\text{height of larger prism's base}}$

$= \dfrac{4}{10}$

$= \dfrac{2}{5}$

The ratio of the perimeters of the bases is 2 : 5.

28. $\dfrac{\text{surface area of smaller prism}}{\text{surface area of larger prism}} = \dfrac{a^2}{b^2}$

$= \dfrac{2^2}{5^2}$

$= \dfrac{4}{25}$

The ratio of the surface areas is 4 : 25.

29. $\dfrac{\text{volume of smaller prism}}{\text{volume of larger prism}} = \dfrac{a^3}{b^3}$

$= \dfrac{2^3}{5^3}$

$= \dfrac{8}{125}$

The ratio of the volumes is 8 : 125.

30. Let V = volume of larger prism.

$\dfrac{48}{V} = \dfrac{8}{125}$

$6000 = 8V$

$V = 750$

The volume of the larger prism is 750 in^3.

31. scale factor $= \dfrac{5}{6}$

Let V = volume of larger cone.

$\dfrac{125\pi}{V} = \dfrac{a^3}{b^3}$

$\dfrac{125\pi}{V} = \dfrac{5^3}{6^3}$

$\dfrac{125\pi}{V} = \dfrac{125}{216}$

$27{,}000\pi = 125V$

$V = 216\pi$

Since $V = \frac{1}{3}\pi r^2 h$,

$216\pi = \frac{1}{3}\pi(6^2)h$

$18 = h$

The height of the larger cone is 18 cm.

32. 5 ft $= 5 \times 12 = 60$ in.

$\dfrac{\text{diameter of normal pie}}{\text{diameter of large pie}} = \dfrac{8}{60}$

$= \dfrac{2}{15}$

$\dfrac{\text{volume of normal pie}}{\text{volume of large pie}} = \dfrac{a^3}{b^3}$

$= \dfrac{2^3}{15^3}$

$= \dfrac{8}{3375}$

The ratio of the volumes is 8 : 3375.

33. $\dfrac{\text{diameter of smaller ball}}{\text{diameter of larger ball}} = \dfrac{29}{30}$

The scale factor is 29 : 30.

34. $\dfrac{\text{surface area of smaller ball}}{\text{surface area of larger ball}} = \dfrac{a^2}{b^2}$

$= \dfrac{29^2}{30^2}$

$= \dfrac{841}{900}$

The ratio of the surface areas is 841 : 900.

35. $\dfrac{\text{volume of smaller ball}}{\text{volume of larger ball}} = \dfrac{29^3}{30^3}$

$= \dfrac{24{,}389}{27{,}000}$

The ratio of the volumes is 24,389 : 27,000.

36. 32 ft $= 32 \times 12 = 384$ in.

$\dfrac{\text{length of gigantic ear of corn}}{\text{length of real ear of corn}} = \dfrac{384}{10}$

$= \dfrac{192}{5}$

The scale factor is 192 : 5.

37. Let V = volume of normal kernel.

$\dfrac{231}{V} = \dfrac{a^3}{b^3}$

$= \dfrac{192^3}{5^3}$

$= \dfrac{7{,}077{,}888}{125}$

$28{,}875 = 7{,}077{,}888V$

$V \approx 0.004$

The volume of a normal kernel is approximately 0.004 in^3.

38. scale factor $= \dfrac{5}{10}$

$= \dfrac{1}{2}$

$\dfrac{\text{volume of smaller cone}}{\text{volume of original cone}} = \dfrac{a^3}{b^3}$

$= \dfrac{1^3}{2^3}$

$= \dfrac{1}{8}$

Since the smaller cone has $\frac{1}{8}$ the volume of the original cone, the frustum has $1 - \frac{1}{8} = \frac{7}{8}$ the volume of the original cone.

$\dfrac{\text{volume of frustum}}{\text{volume of original cone}} = \dfrac{\frac{7}{8}}{1} = \dfrac{7}{8}$ or 7:8

$\dfrac{\text{volume of frustum}}{\text{volume of smaller cone}} = \dfrac{\frac{7}{8}}{\frac{1}{8}} = \dfrac{7}{1}$ or 7:1

39. Let r = the radius of the upper circle. Since the scale factor $= \dfrac{5}{10}$ or $\dfrac{1}{2}$,

$\dfrac{\text{radius of upper circle}}{\text{radius of lower circle}} = \dfrac{1}{2}$

$\dfrac{r}{\text{radius of lower circle}} = \dfrac{1}{2}$

radius of lower circle $= 2r$

The lateral area of the smaller cone is $L = \pi r(5) = 5\pi r$.

The lateral area of the larger cone is $L = \pi(2r)(10) = 20\pi r$.

lateral area of the frustum

= lateral area of the larger cone − lateral area of the smaller cone

$= 20\pi r - 5\pi r$

$= 15\pi r$

$\dfrac{\text{lateral area of frustum}}{\text{lateral area of original cone}} = \dfrac{15\pi r}{20\pi r} = \dfrac{3}{4}$ or 3 : 4

$\dfrac{\text{lateral area of frustum}}{\text{lateral area of smaller cone}} = \dfrac{15\pi r}{5\pi r} = \dfrac{3}{1}$ or 3 : 1

40. Yes, both cones have congruent radii. If the heights are the same measure, the cones are congruent.

41. The volume of the cone on the right is equal to the sum of the volumes of the cones inside the cylinder. Justification: Call h the height of both solids. The volume of the cone on the right is $\frac{1}{3}\pi r^2 h$. If the height of one cone inside the cylinder is c, then the height of the other one is $h - c$. Therefore, the sum of the volumes of the two cones is: $\frac{1}{3}\pi r^2 c + \frac{1}{3}\pi r^2(h - c)$ or $\frac{1}{3}\pi r^2(c + h - c)$ or $\frac{1}{3}\pi r^2 h$.

42. Sample answer: Scale factors relate the actual object to the miniatures. Answers should include the following.

- The scale factors that are commonly used are 1 : 24, 1 : 32, 1 : 43, and 1 : 64.
- The actual object is 108 in. long.

43. C; $\frac{\text{small area}}{\text{large area}} = \frac{a^2}{b^2} = \frac{4}{9} = \frac{2^2}{3^2}$

scale factor $= \frac{2}{3}$

$\frac{\text{small volume}}{\text{large volume}} = \frac{a^3}{b^3} = \frac{2^3}{3^3} = \frac{8}{27}$ or 8 : 27

44. D; $\frac{x}{y} = \frac{x}{y} \cdot \frac{yz}{yz} = \frac{xyz}{y^2z}$

$= \frac{4}{5}$ or 0.8

Page 713 Maintain Your Skills

45. radius $= \frac{1}{2}(8) = 4$

$V = \frac{4}{3}\pi r^3$

$= \frac{4}{3}\pi(4^3)$

≈ 268.1

The volume is approximately 268.1 ft^3.

46. $V = \frac{4}{3}\pi r^3$

$= \frac{4}{3}\pi(9.5^3)$

≈ 3591.4

The volume is approximately 3591.4 m^3.

47. $V = \frac{4}{3}\pi r^3$

$= \frac{4}{3}\pi(15.1^3)$

$\approx 14{,}421.8$

The volume is approximately 14,421.8 cm^3.

48. radius $= \frac{1}{2}(23) = 11.5$

$V = \frac{4}{3}\pi r^3$

$= \frac{4}{3}\pi(11.5^3)$

≈ 6370.6

The volume is approximately 6370.6 in^3.

49.

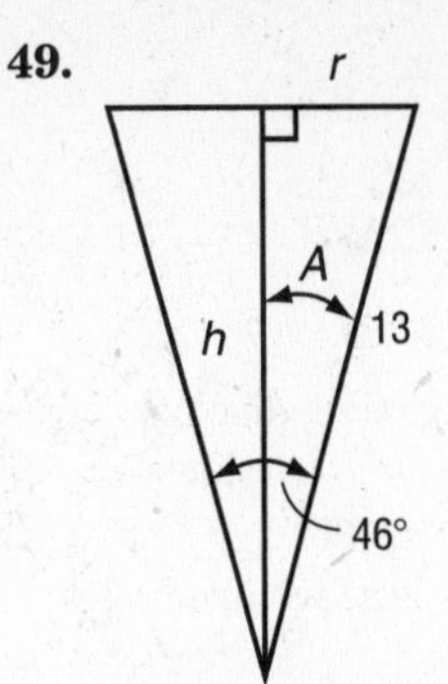

Use trigonometry to find the radius of the base.

$\frac{1}{2}(46°) = 23°$

$\sin A = \frac{\text{opposite}}{\text{hypotenuse}}$

$\sin 23° = \frac{r}{13}$

$r = 13 \sin 23°$

≈ 5.0795

Similarly, find the height.

$\cos A = \frac{\text{adjacent}}{\text{hypotenuse}}$

$\cos 23° = \frac{h}{13}$

$h = 13 \cos 23°$

≈ 11.967

Now find the volume.

$V = \frac{1}{3}\pi r^2 h$

$\approx \frac{1}{3}\pi(5.0795)^2(11.967)$

≈ 323.3

The volume is approximately 323.3 in^3.

50. $V = \frac{1}{3}Bh$

$= \frac{1}{3}(11)(7)(15)$

$= 385$

The volume of the pyramid is 385 m^3.

51. Use trigonometry to find the radius of the base.

$\tan A = \frac{\text{opposite}}{\text{adjacent}}$

$\tan 62° = \frac{21}{r}$

$r = \frac{21}{\tan 62°}$

≈ 11.166

Now find the volume.

$V = \frac{1}{3}\pi r^2 h$

$\approx \frac{1}{3}\pi(11.166)^2(21)$

≈ 2741.8

The volume is approximately 2741.8 ft^3.

52. Start with the formula for the surface area of a cylinder.

$T = 2\pi rh + 2\pi r^2$

$430 = 2\pi r(7.4) + 2\pi r^2$

Solve for r.

$2\pi r^2 + 14.8\pi r - 430 = 0$

Use the Quadratic Formula.

$r = \frac{-b \pm \sqrt{b^2 - 4ac}}{2a}$

$= \frac{-14.8\pi \pm \sqrt{(14.8\pi)^2 - 4(2\pi)(-430)}}{2(2\pi)}$

≈ 5.4 or -12.8

Since the radius cannot be negative, discard -12.8.

The radius is approximately 5.4 cm.

53. Start with the formula for the surface area of a cylinder.

$T = 2\pi rh + 2\pi r^2$

$224.7 = 2\pi r(10) + 2\pi r^2$

Solve for r.

$2\pi r^2 + 20\pi r - 224.7 = 0$

Use the Quadratic Formula.

$r = \frac{-b \pm \sqrt{b^2 - 4ac}}{2a}$

$= \frac{-20\pi \pm \sqrt{(20\pi)^2 - 4(2\pi)(-224.7)}}{2(2\pi)}$

≈ 2.8 or -12.8

Since the radius cannot be negative, discard -12.8.

The radius is approximately 2.8 yd.

54. Use the Pythagorean Theorem to find the height of the triangle.

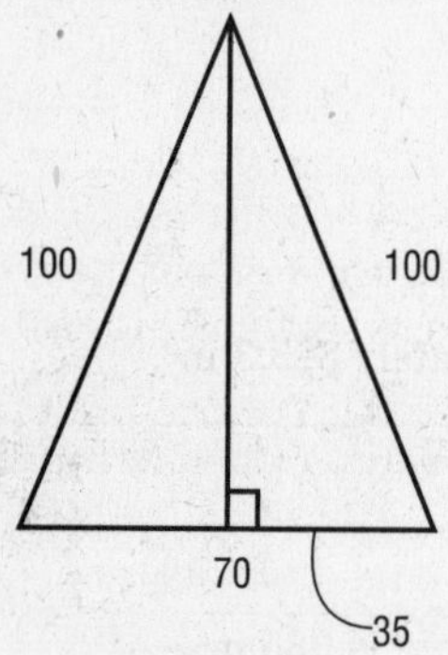

$a^2 + b^2 = c^2$

$35^2 + h^2 = 100^2$

$1225 + h^2 = 10{,}000$

$h^2 = 8775$

$h = 15\sqrt{39}$

For the triangle,

$A = \frac{1}{2}bh$

$= \frac{1}{2}(70)(15\sqrt{39})$

≈ 3279

The area is approximately 3279 yd^2.

55. Use the formula for the area of a rectangle.

$A = \ell w$

$= (12)(3)$

$= 36$

The area of the rowboat is 36 ft^2.

56. $3279 \text{ yd}^2 = 3279 \times 9 = 29{,}511 \text{ ft}^2$

$\text{probability} = \frac{\text{area of rowboat}}{\text{shaded area}}$

$\approx \frac{36}{29{,}511}$

≈ 0.0012

57. $y = 3x + 5$

Use $x = 4, y = 17$.

$17 \stackrel{?}{=} 3(4) + 5$

$17 \stackrel{?}{=} 12 + 5$

$17 = 17$

Yes, the ordered pair is on the graph.

58. $y = -4x + 1$

Use $x = -2, y = 9$.

$9 \stackrel{?}{=} -4(-2) + 1$

$9 \stackrel{?}{=} 8 + 1$

$9 = 9$

Yes, the ordered pair is on the graph.

59. $y = 7x - 4$

Use $x = -1, y = 3$.

$3 \stackrel{?}{=} 7(-1) - 4$

$3 \stackrel{?}{=} -7 - 4$

$3 \neq -11$

No, the ordered pair is not on the graph.

Page 713 Practice Quiz 2

1. $V = \frac{4}{3}\pi r^3$

$= \frac{4}{3}\pi(25.3)^3$

$\approx 67{,}834.4$

The volume is approximately 67,834.4 ft^3.

2. $\text{radius} = \frac{1}{2}(36.8) = 18.4$

$V = \frac{4}{3}\pi r^3$

$= \frac{4}{3}\pi(18.4)^3$

$\approx 26{,}094.1$

The volume is approximately 26,094.1 cm^3.

3. $\frac{\text{base side of left pyramid}}{\text{base side of right pyramid}} = \frac{7}{5}$

The scale factor is 7 : 5.

4. $\frac{\text{surface area of left pyramid}}{\text{surface area of right pyramid}} = \frac{a^2}{b^2}$

$= \frac{7^2}{5^2}$

$= \frac{49}{25}$

The ratio of the surface areas is 49 : 25.

5. $\frac{\text{volume of left pyramid}}{\text{volume of right pyramid}} = \frac{a^3}{b^3}$

$= \frac{7^3}{5^3}$

$= \frac{343}{125}$

The ratio of the volumes is 343 : 125.

13-5 Coordinates in Space

Page 717 Check for Understanding

1. The coordinate plane has 4 regions or quadrants with 4 possible combinations of signs for the ordered pairs. Three-dimensional space is the intersection of 3 planes that create 8 regions with 8 possible combinations of signs for the ordered triples.

2. Sample answer: Use the point at (2, 3, 4); $A(2, 3, 4)$, $B(2, 0, 4)$, $C(0, 0, 4)$, $D(0, 3, 4)$, $E(2, 3, 0)$, $F(2, 0, 0)$, $G(0, 0, 0)$, and $H(0, 3, 0)$.

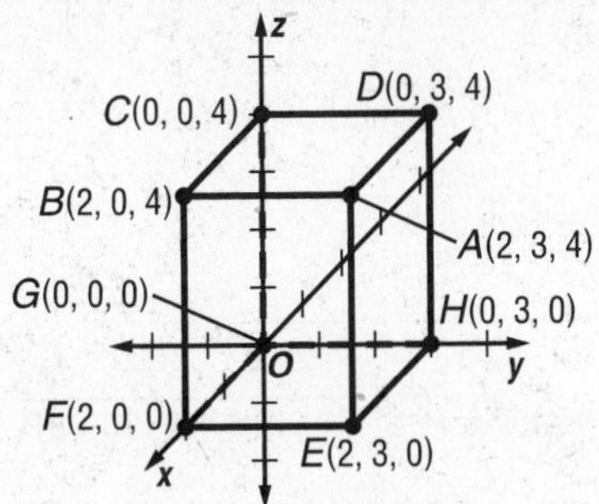

3. A dilation of a rectangular prism will provide a similar figure, but not a congruent one unless $r = 1$ or $r = -1$.

4.

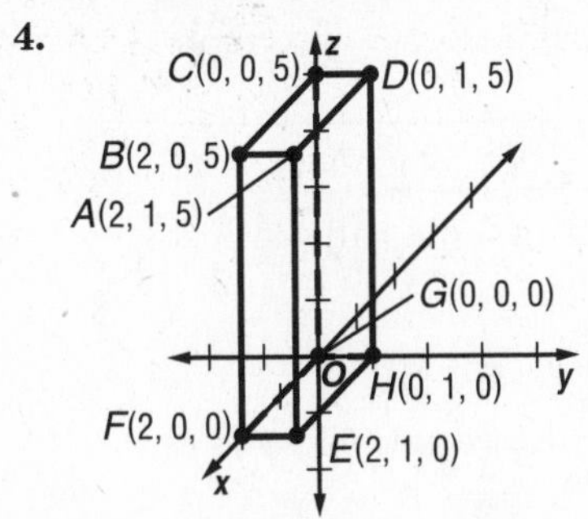

5.

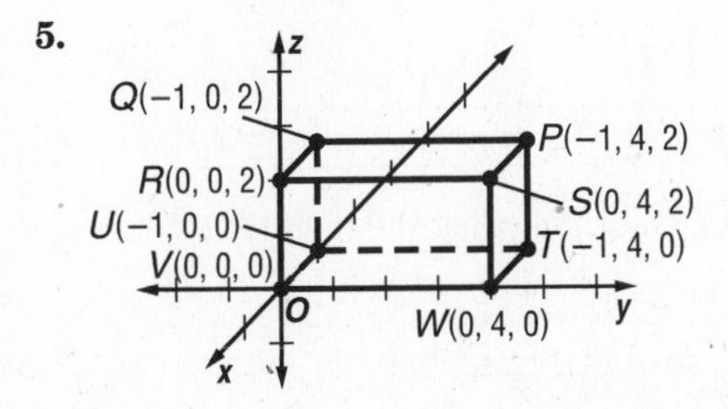

6. $DE = \sqrt{(x_2 - x_1)^2 + (y_2 - y_1)^2 + (z_2 - z_1)^2}$

$= \sqrt{(1 - 0)^2 + (5 - 0)^2 + (7 - 0)^2}$

$= \sqrt{75}$ or $5\sqrt{3}$

$M = \left(\frac{x_1 + x_2}{2}, \frac{y_1 + y_2}{2}, \frac{z_1 + z_2}{2}\right)$

$= \left(\frac{0 + 1}{2}, \frac{0 + 5}{2}, \frac{0 + 7}{2}\right)$

$= \left(\frac{1}{2}, \frac{5}{2}, \frac{7}{2}\right)$

7. $GH = \sqrt{(x_2 - x_1)^2 + (y_2 - y_1)^2 + (z_2 - z_1)^2}$

$= \sqrt{[5 - (-3)]^2 + [-3 - (-4)]^2 + (-5 - 6)^2}$

$= \sqrt{186}$

$M = \left(\frac{x_1 + x_2}{2}, \frac{y_1 + y_2}{2}, \frac{z_1 + z_2}{2}\right)$

$= \left(\frac{-3 + 5}{2}, \frac{-4 - 3}{2}, \frac{6 - 5}{2}\right)$

$= \left(1, -\frac{7}{2}, \frac{1}{2}\right)$

8. First, write a vertex matrix.

$$\begin{array}{c} \\ x \\ y \\ z \end{array}\begin{array}{c} \begin{matrix} M & N & P & Q & R & S & T & V \end{matrix} \\ \begin{bmatrix} 0 & -3 & -3 & 0 & 0 & 0 & -3 & -3 \\ 0 & 0 & 4 & 4 & 0 & 4 & 4 & 0 \\ 0 & 0 & 0 & 0 & 2 & 2 & 2 & 2 \end{bmatrix} \end{array}$$

Next, multiply each element by the scale factor, 2.

$$2\begin{bmatrix} 0 & -3 & -3 & 0 & 0 & 0 & -3 & -3 \\ 0 & 0 & 4 & 4 & 0 & 4 & 4 & 0 \\ 0 & 0 & 0 & 0 & 2 & 2 & 2 & 2 \end{bmatrix}$$

$$= \begin{array}{c} \begin{matrix} M' & N' & P' & Q' & R' & S' & T' & V' \end{matrix} \\ \begin{bmatrix} 0 & -6 & -6 & 0 & 0 & 0 & -6 & -6 \\ 0 & 0 & 8 & 8 & 0 & 8 & 8 & 0 \\ 0 & 0 & 0 & 0 & 4 & 4 & 4 & 4 \end{bmatrix} \end{array}$$

The coordinates of the vertices of the dilated image are $M'(0, 0, 0)$, $N'(-6, 0, 0)$, $P'(-6, 8, 0)$, $Q'(0, 8, 0)$, $R'(0, 0, 4)$, $S'(0, 8, 4)$, $T'(-6, 8, 4)$, and $V'(-6, 0, 4)$.

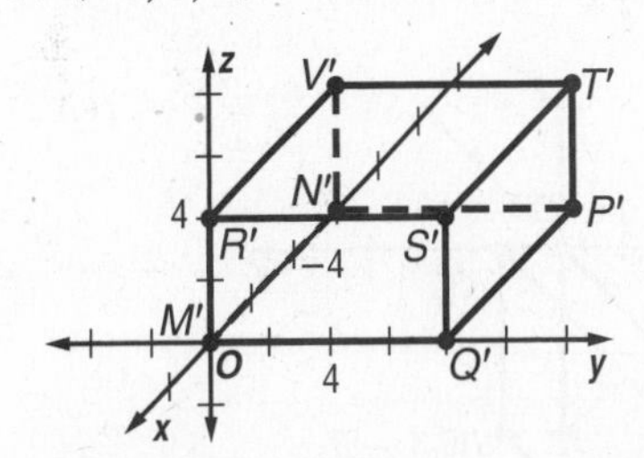

9. Write the coordinates of each corner. Then use the translation equation $(x, y, z) \rightarrow (x - 48, y, z + 16)$ to find the coordinates of each vertex of the rectangular prism that represents the storage container.

Coordinates of the vertices, (x, y, z) Preimage	Translated coordinates, $(x - 48, y, z + 16)$ Image
(12, 8, 8)	(−36, 8, 24)
(12, 0, 8)	(−36, 0, 24)
(0, 0, 8)	(−48, 0, 24)
(0, 8, 8)	(−48, 8, 24)
(12, 8, 0)	(−36, 8, 16)
(12, 0, 0)	(−36, 0, 16)
(0, 0, 0)	(−48, 0, 16)
(0, 8, 0)	(−48, 8, 16)

Pages 717–719 Practice and Apply

10.

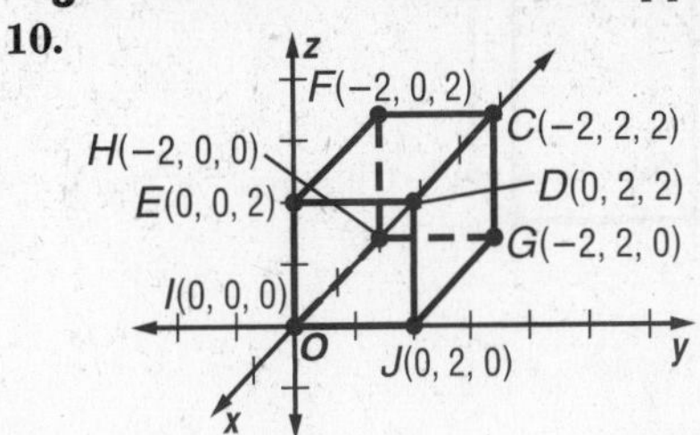

11.

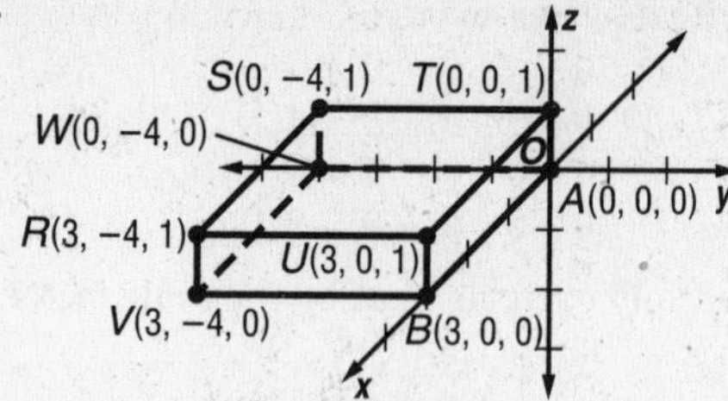

12.

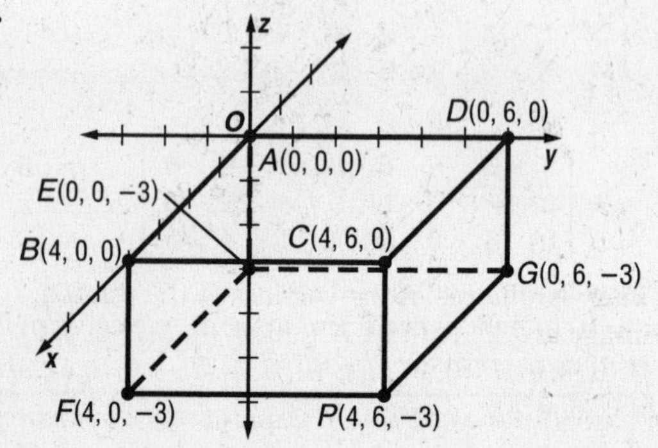

13.

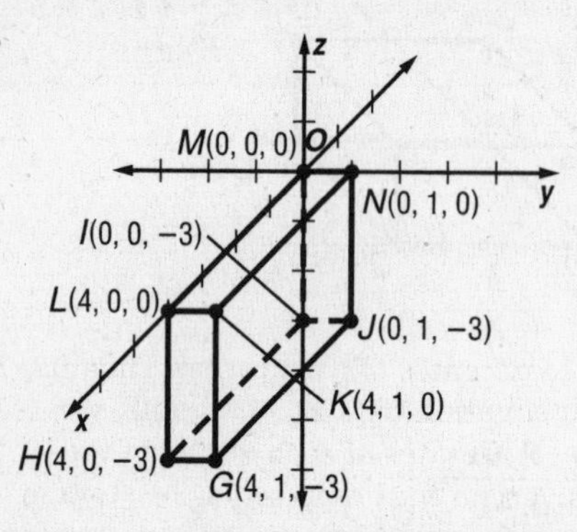

14.

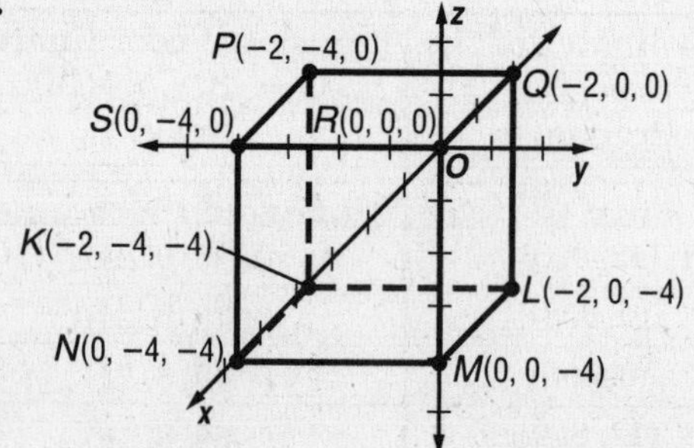

15.

z
R(−1, −3, 0)
Q(−1, 0, 0)
O
S(0, −3, 0)
y
P(0, 0, 0)
W(−1, −3, −6)
x
V(−1, 0, −6)
T(0, −3, −6)
U(0, 0, −6)

16. $KL = \sqrt{(x_2 - x_1)^2 + (y_2 - y_1)^2 + (z_2 - z_1)^2}$

$= \sqrt{(-2-2)^2 + (-2-2)^2 + (0-0)^2}$

$= \sqrt{32}$ or $4\sqrt{2}$

$M = \left(\frac{x_1 + x_2}{2}, \frac{y_1 + y_2}{2}, \frac{z_1 + z_2}{2}\right)$

$= \left(\frac{2 + (-2)}{2}, \frac{2 + (-2)}{2}, \frac{0 + 0}{2}\right)$

$= (0, 0, 0)$

17. $PQ = \sqrt{(x_2 - x_1)^2 + (y_2 - y_1)^2 + (z_2 - z_1)^2}$

$= \sqrt{[3 - (-2)]^2 + [-2 - (-5)]^2 + (-1 - 8)^2}$

$= \sqrt{115}$

$M = \left(\frac{x_1 + x_2}{2}, \frac{y_1 + y_2}{2}, \frac{z_1 + z_2}{2}\right)$

$= \left(\frac{-2 + 3}{2}, \frac{-5 + (-2)}{2}, \frac{8 + (-1)}{2}\right)$

$= \left(\frac{1}{2}, -\frac{7}{2}, \frac{7}{2}\right)$

18. $FG = \sqrt{(x_2 - x_1)^2 + (y_2 - y_1)^2 + (z_2 - z_1)^2}$

$= \sqrt{\left(0 - \frac{3}{5}\right)^2 + (3 - 0)^2 + \left(0 - \frac{4}{5}\right)^2}$

$= \sqrt{10}$

$M = \left(\frac{x_1 + x_2}{2}, \frac{y_1 + y_2}{2}, \frac{z_1 + z_2}{2}\right)$

$= \left(\frac{\frac{3}{5} + 0}{2}, \frac{0 + 3}{2}, \frac{\frac{4}{5} + 0}{2}\right)$

$= \left(\frac{3}{10}, \frac{3}{2}, \frac{2}{5}\right)$

19. $GH = \sqrt{(x_2 - x_1)^2 + (y_2 - y_1)^2 + (z_2 - z_1)^2}$

$= \sqrt{\left(\frac{1}{5} - 1\right)^2 + \left[-\frac{2}{5} - (-1)\right]^2 + (2 - 6)^2}$

$= \sqrt{17}$

$M = \left(\frac{x_1 + x_2}{2}, \frac{y_1 + y_2}{2}, \frac{z_1 + z_2}{2}\right)$

$= \left(\frac{1 + \frac{1}{5}}{2}, \frac{-1 + \left(-\frac{2}{5}\right)}{2}, \frac{6 + 2}{2}\right)$

$= \left(\frac{3}{5}, -\frac{7}{10}, 4\right)$

20. $ST = \sqrt{(x_2 - x_1)^2 + (y_2 - y_1)^2 + (z_2 - z_1)^2}$

$= \sqrt{(4\sqrt{3} - 6\sqrt{3})^2 + (5 - 4)^2 + (\sqrt{2} - 4\sqrt{2})^2}$

$= \sqrt{31}$

$M = \left(\frac{x_1 + x_2}{2}, \frac{y_1 + y_2}{2}, \frac{z_1 + z_2}{2}\right)$

$= \left(\frac{6\sqrt{3} + 4\sqrt{3}}{2}, \frac{4 + 5}{2}, \frac{4\sqrt{2} + \sqrt{2}}{2}\right)$

$= \left(5\sqrt{3}, \frac{9}{2}, \frac{5\sqrt{2}}{2}\right)$

21. $BC = \sqrt{(x_2 - x_1)^2 + (y_2 - y_1)^2 + (z_2 - z_1)^2}$

$= \sqrt{(-2\sqrt{3} - \sqrt{3})^2 + (4 - 2)^2 + (4\sqrt{2} - 2\sqrt{2})^2}$

$= \sqrt{39}$

$M = \left(\frac{x_1 + x_2}{2}, \frac{y_1 + y_2}{2}, \frac{z_1 + z_2}{2}\right)$

$= \left(\frac{\sqrt{3} - 2\sqrt{3}}{2}, \frac{2 + 4}{2}, \frac{2\sqrt{2} + 4\sqrt{2}}{2}\right)$

$= \left(-\frac{\sqrt{3}}{2}, 3, 3\sqrt{2}\right)$

22. distance

$$= \sqrt{(x_2 - x_1)^2 + (y_2 - y_1)^2 + (z_2 - z_1)^2}$$
$$= \sqrt{(-240 - 50)^2 + (140 - 100)^2 + (2.5 - 2)^2}$$
$$\approx 292.7$$

The distance is approximately 292.7 miles.

23. First, write a vertex matrix.

$$\begin{array}{c} \\ x \\ y \\ z \end{array}\begin{array}{c} \begin{array}{cccccccc} A & B & C & D & E & H & G & F \end{array} \\ \begin{bmatrix} 1 & 0 & 0 & 1 & 1 & 0 & 0 & 1 \\ 0 & 0 & 1 & 1 & 0 & 0 & 1 & 1 \\ -1 & -1 & -1 & -1 & 1 & 1 & 1 & 1 \end{bmatrix} \end{array}$$

Next, multiply each element by the scale factor, 3.

$$3\begin{bmatrix} 1 & 0 & 0 & 1 & 1 & 0 & 0 & 1 \\ 0 & 0 & 1 & 1 & 0 & 0 & 1 & 1 \\ -1 & -1 & -1 & -1 & 1 & 1 & 1 & 1 \end{bmatrix}$$

$$= \begin{array}{c} \begin{array}{cccccccc} A' & B' & C' & D' & E' & H' & G' & F' \end{array} \\ \begin{bmatrix} 3 & 0 & 0 & 3 & 3 & 0 & 0 & 3 \\ 0 & 0 & 3 & 3 & 0 & 0 & 3 & 3 \\ -3 & -3 & -3 & -3 & 3 & 3 & 3 & 3 \end{bmatrix} \end{array}$$

The coordinates of the dilated image are $A'(3, 0, -3)$, $B'(0, 0, -3)$, $C'(0, 3, -3)$, $D'(3, 3, -3)$, $E'(3, 0, 3)$, $H'(0, 0, 3)$, $G'(0, 3, 3)$, and $F'(3, 3, 3)$.

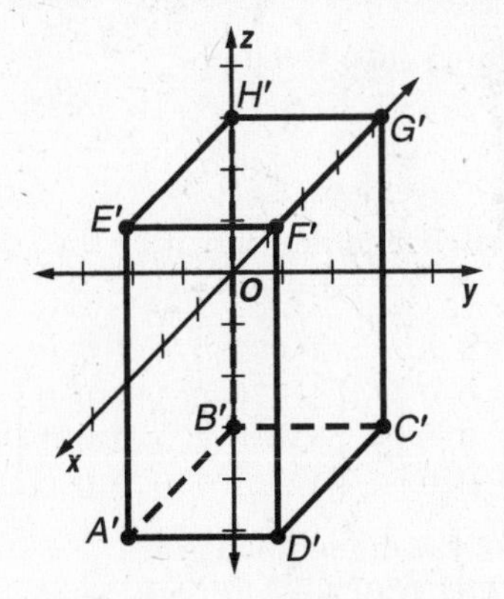

24. First, write a vertex matrix.

$$\begin{array}{c} \\ x \\ y \\ z \end{array}\begin{array}{c} \begin{array}{cccccccc} G & H & J & K & L & M & N & P \end{array} \\ \begin{bmatrix} 4 & 4 & 0 & 0 & 4 & 0 & 0 & 4 \\ -3 & 0 & 0 & -3 & -3 & -3 & 0 & 0 \\ 2 & 2 & 2 & 2 & 0 & 0 & 0 & 0 \end{bmatrix} \end{array}$$

Next, multiply each element by the scale factor, 2.

$$2\begin{bmatrix} 4 & 4 & 0 & 0 & 4 & 0 & 0 & 4 \\ -3 & 0 & 0 & -3 & -3 & -3 & 0 & 0 \\ 2 & 2 & 2 & 2 & 0 & 0 & 0 & 0 \end{bmatrix}$$

$$= \begin{array}{c} \begin{array}{cccccccc} G' & H' & J' & K' & L' & M' & N' & P' \end{array} \\ \begin{bmatrix} 8 & 8 & 0 & 0 & 8 & 0 & 0 & 8 \\ -6 & 0 & 0 & -6 & -6 & -6 & 0 & 0 \\ 4 & 4 & 4 & 4 & 0 & 0 & 0 & 0 \end{bmatrix} \end{array}.$$

The coordinates of the dilated image are $G'(8, -6, 4)$, $H'(8, 0, 4)$, $J'(0, 0, 4)$, $K'(0, -6, 4)$, $L'(8, -6, 0)$, $M'(0, -6, 0)$, $N'(0, 0, 0)$, $P'(8, 0, 0)$.

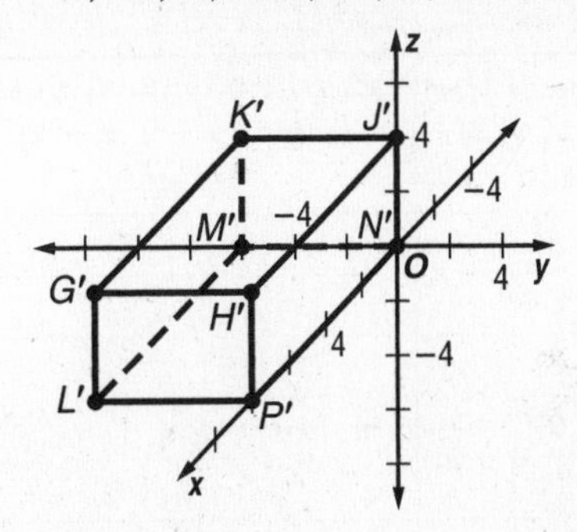

25. Write the coordinates of each corner. Then use the translation equation $(x, y, z) \rightarrow (x + 2, y + 5, z - 5)$ to find the coordinates of each vertex of the rectangular prism.

Coordinates of the vertices, (x, y, z) Preimage	**Translated coordinates, $(x + 2, y + 5, z - 5)$ Image**
$P(-2, -3, 3)$	$P'(0, 2, -2)$
$Q(-2, 0, 3)$	$Q'(0, 5, -2)$
$R(0, 0, 3)$	$R'(2, 5, -2)$
$S(0, -3, 3)$	$S'(2, 2, -2)$
$T(-2, 0, 0)$	$T'(0, 5, -5)$
$U(-2, -3, 0)$	$U'(0, 2, -5)$
$V(0, -3, 0)$	$V'(2, 2, -5)$
$W(0, 0, 0)$	$W'(2, 5, -5)$

26. Write the coordinates of each corner. Then use the translation equation $(x, y, z) \rightarrow (x - 2, y + 1, z - 1)$ to find the coordinates of each vertex of the rectangular prism.

Coordinates of the vertices, (x, y, z) Preimage	**Translated coordinates, $(x - 2, y + 1, z - 1)$ Image**
$A(2, 0, 1)$	$A'(0, 1, 0)$
$B(2, 0, 0)$	$B'(0, 1, -1)$
$C(2, 1, 0)$	$C'(0, 2, -1)$
$D(2, 1, 1)$	$D'(0, 2, 0)$
$E(0, 0, 1)$	$E'(-2, 1, 0)$
$F(0, 1, 1)$	$F'(-2, 2, 0)$
$G(0, 1, 0)$	$G'(-2, 2, -1)$
$H(0, 0, 0)$	$H'(-2, 1, -1)$

27. Write the coordinates of each corner. Then use the translation equation $(x, y, z) \rightarrow (x + 1, y + 2, z - 2)$ to find the coordinates of each vertex of the cube.

Coordinates of the vertices, (x, y, z) Preimage	Translated coordinates, $(x + 1, y + 2, z - 2)$ Image
$A(3, 3, 3)$	$A'(4, 5, 1)$
$B(3, 0, 3)$	$B'(4, 2, 1)$
$C(0, 0, 3)$	$C'(1, 2, 1)$
$D(0, 3, 3)$	$D'(1, 5, 1)$
$E(3, 3, 0)$	$E'(4, 5, -2)$
$F(3, 0, 0)$	$F'(4, 2, -2)$
$G(0, 0, 0)$	$G'(1, 2, -2)$
$H(0, 3, 0)$	$H'(1, 5, -2)$

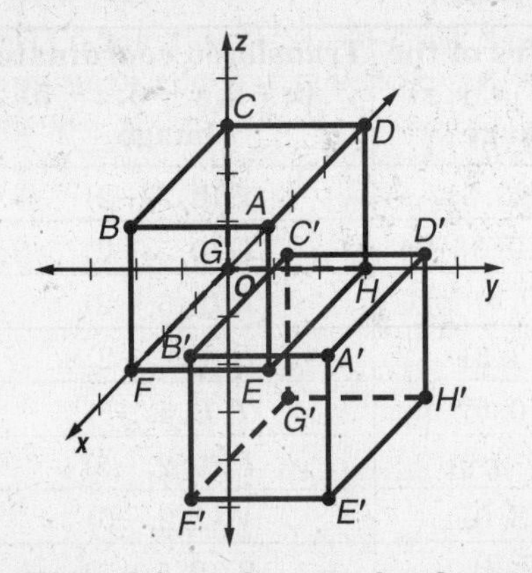

28. Write the coordinates of each corner. Then use the translation equation $(x, y, z) \rightarrow (x - 2, y - 3, z + 2)$ to find the coordinates of each vertex of the cube.

Coordinates of the vertices, (x, y, z) Preimage	Translated coordinates, $(x - 2, y - 3, z + 2)$ Image
$A(3, 3, 3)$	$A'(1, 0, 5)$
$B(3, 0, 3)$	$B'(1, -3, 5)$
$C(0, 0, 3)$	$C'(-2, -3, 5)$
$D(0, 3, 3)$	$D'(-2, 0, 5)$
$E(3, 3, 0)$	$E'(1, 0, 2)$
$F(3, 0, 0)$	$F'(1, -3, 2)$
$G(0, 0, 0)$	$G'(-2, -3, 2)$
$H(0, 3, 0)$	$H'(-2, 0, 2)$

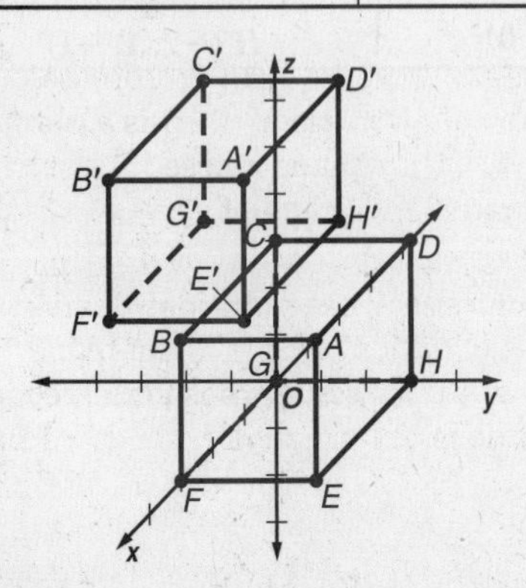

29. Write a vertex matrix and multiply it by the scale factor, 2.

$$2\begin{bmatrix} \overset{A}{3} & \overset{B}{3} & \overset{C}{0} & \overset{D}{0} & \overset{E}{3} & \overset{F}{3} & \overset{G}{0} & \overset{H}{0} \\ 3 & 0 & 0 & 3 & 3 & 0 & 0 & 3 \\ 3 & 3 & 3 & 3 & 0 & 0 & 0 & 0 \end{bmatrix} = \begin{bmatrix} \overset{A'}{6} & \overset{B'}{6} & \overset{C'}{0} & \overset{D'}{0} & \overset{E'}{6} & \overset{F'}{6} & \overset{G'}{0} & \overset{H'}{0} \\ 6 & 0 & 0 & 6 & 6 & 0 & 0 & 6 \\ 6 & 6 & 6 & 6 & 0 & 0 & 0 & 0 \end{bmatrix}$$

The coordinates of the dilated image are $A'(6, 6, 6)$, $B'(6, 0, 6)$, $C'(0, 0, 6)$, $D'(0, 6, 6)$, $E'(6, 6, 0)$, $F'(6, 0, 0)$, $G'(0, 0, 0)$, and $H'(0, 6, 0)$.

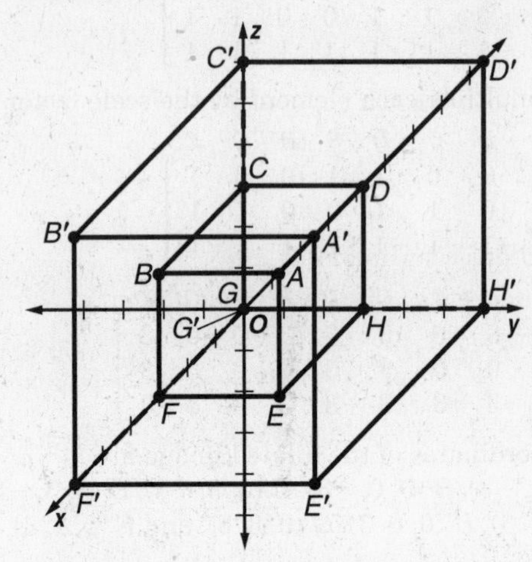

side length of dilated cube = 6 units

$V = s^3$

$= 6^3$

$= 216 \text{ units}^3$

30. Write a vertex matrix and multiply it by the scale factor $\frac{1}{3}$.

$$\frac{1}{3}\begin{bmatrix} \overset{A}{3} & \overset{B}{3} & \overset{C}{0} & \overset{D}{0} & \overset{E}{3} & \overset{F}{3} & \overset{G}{0} & \overset{H}{0} \\ 3 & 0 & 0 & 3 & 3 & 0 & 0 & 3 \\ 3 & 3 & 3 & 3 & 0 & 0 & 0 & 0 \end{bmatrix} = \begin{bmatrix} \overset{A'}{1} & \overset{B'}{1} & \overset{C'}{0} & \overset{D'}{0} & \overset{E'}{1} & \overset{F'}{1} & \overset{G'}{0} & \overset{H'}{0} \\ 1 & 0 & 0 & 1 & 1 & 0 & 0 & 1 \\ 1 & 1 & 1 & 1 & 0 & 0 & 0 & 0 \end{bmatrix}$$

The coordinates of the dilated image are $A'(1, 1, 1)$, $B'(1, 0, 1)$, $C'(0, 0, 1)$, $D'(0, 1, 1)$, $E'(1, 1, 0)$, $F'(1, 0, 0)$, $G'(0, 0, 0)$, and $H'(0, 1, 0)$.

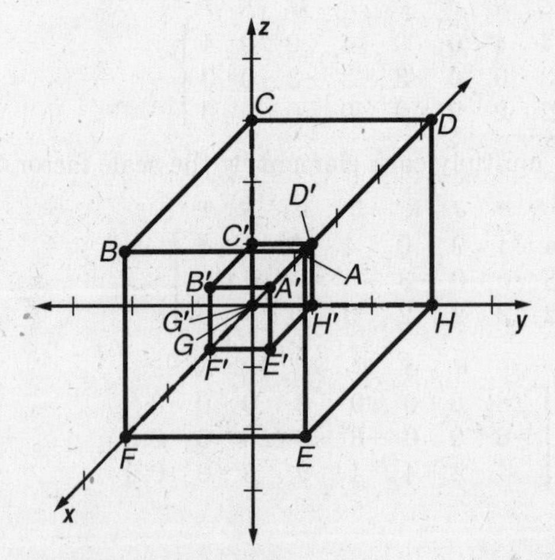

The scale factor is $\frac{1}{3}$, so the ratio of the volumes for these two cubes is

$$\frac{\text{volume of new cube}}{\text{volume of original cube}} = \frac{1^3}{3^3} = \frac{1}{27}.$$

31. first balloon location: $(-12, -12, 0.4)$
second balloon location: $(-4, -10, 0.3)$

$$\text{distance} = \sqrt{(x_2 - x_1)^2 + (y_2 - y_1)^2 + (z_2 - z_1)^2}$$
$$= \sqrt{[-4 - (-12)]^2 + [-10 - (-12)]^2 + [0.3 - 0.4]^2}$$
$$\approx 8.2$$

The distance between the balloons is approximately 8.2 miles.

32.
$$M = \left(\frac{x_1 + x_2}{2}, \frac{y_1 + y_2}{2}, \frac{z_1 + z_2}{2}\right)$$
$$(5, 1, 2) = \left(\frac{2 + x_2}{2}, \frac{4 + y_2}{2}, \frac{7 + z_2}{2}\right)$$

$5 = \frac{2 + x_2}{2}$ $\quad 1 = \frac{4 + y_2}{2}$ $\quad 2 = \frac{7 + z_2}{2}$

$10 = 2 + x_2$ $\quad 2 = 4 + y_2$ $\quad 4 = 7 + z_3$

$8 = x_2$ $\quad -2 = y_2$ $\quad -3 = z_3$

Point B has coordinates $(8, -2, -3)$.

33. The center of the sphere is the midpoint of the diameter.
$$M = \left(\frac{x_1 + x_2}{2}, \frac{y_1 + y_2}{2}, \frac{z_1 + z_2}{2}\right)$$
$$(4, -2, 6) = \left(\frac{8 + x_2}{2}, \frac{10 + y_2}{2}, \frac{-2 + z_2}{2}\right)$$

$4 = \frac{8 + x_2}{2}$ $\quad -2 = \frac{10 + y_2}{2}$ $\quad 6 = \frac{-2 + z_2}{2}$

$8 = 8 + x_2$ $\quad -4 = 10 + y_2$ $\quad 12 = -2 + z_2$

$0 = x_2$ $\quad -14 = y_2$ $\quad 14 = z_2$

The other endpoint has coordinates $(0, -14, 14)$.

34. Use the midpoint formula with the endpoints of the diameter.
$$M = \left(\frac{x_1 + x_2}{2}, \frac{y_1 + y_2}{2}, \frac{z_1 + z_2}{2}\right)$$
$$\text{center} = \left(\frac{-12 + 14}{2}, \frac{10 - 8}{2}, \frac{12 + 2}{2}\right)$$
$$= (1, 1, 7)$$

The radius is the distance from the center $(1, 1, 7)$ to $(14, -8, 2)$. Let the center be the point (x_c, y_c, z_c).
$$\text{radius} = \sqrt{(x_2 - x_c)^2 + (y_2 - y_c)^2 + (z_2 - z_c)^2}$$
$$= \sqrt{(14 - 1)^2 + (-8 - 1)^2 + (2 - 7)^2}$$
$$= \sqrt{275} \text{ or } 5\sqrt{11}$$

35. The prism has moved down 5 units, right 3 units, and forward 2 units.
$(x, y, z) \rightarrow (x + 2, y + 3, z - 5)$

36. The cube extends from $x = 2 - 4 = -2$ to $x = 2 + 4 = 6$, from $y = 4 - 4 = 0$ to $y = 4 + 4 = 8$, and from $z = 6 - 4 = 2$ to $z = 6 + 4 = 10$. So the coordinates of the vertices are $A(-2, 0, 2)$, $B(6, 0, 2)$, $C(6, 8, 2)$, $D(-2, 8, 2)$, $E(-2, 8, 10)$, $F(6, 8, 10)$, $G(6, 0, 10)$, and $H(-2, 0, 10)$.

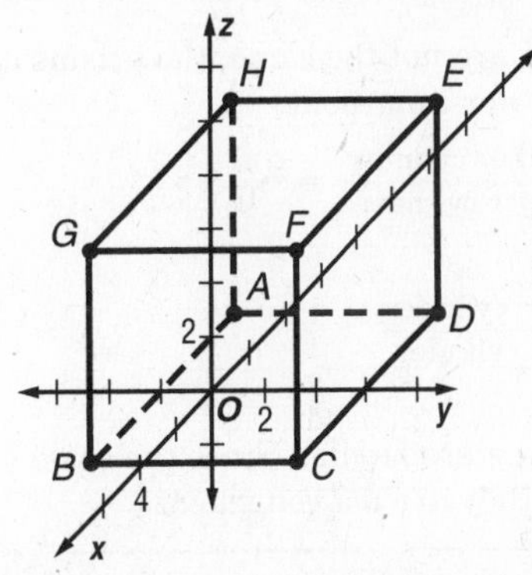

37. Sample answer: Three-dimensional graphing is used in computer animation to render images and allow them to move realistically. Answers should include the following.

- Ordered triples are a method of locating and naming points in space. An ordered triple is unique to one point.
- Applying transformations to points in space would allow an animator to create realistic movement in animation.

38. C;
$$M = \left(\frac{x_1 + x_2}{2}, \frac{y_1 + y_2}{2}, \frac{z_1 + z_2}{2}\right)$$
$$(4, -5, 3) = \left(\frac{5 + x_2}{2}, \frac{-4 + y_2}{2}, \frac{-2 + z_2}{2}\right)$$

$4 = \frac{5 + x_2}{2}$ $\quad -5 = \frac{-4 + y_2}{2}$ $\quad 3 = \frac{-2 + z_2}{2}$

$8 = 5 + x_2$ $\quad -10 = -4 + y_2$ $\quad 6 = -2 + z_2$

$3 = x_2$ $\quad -6 = y_2$ $\quad 8 = z_2$

The other endpoint has coordinates $(3, -6, 8)$.

39. B; $\sqrt{x + 1} = x - 1$
$$x + 1 = (x - 1)^2$$
$$x + 1 = x^2 - 2x + 1$$
$$0 = x^2 - 3x$$
$$0 = x(x - 3)$$
$$x = 0 \text{ or } x = 3$$

Check if each value satisfies the original equation.

$\sqrt{0 + 1} \stackrel{?}{=} 0 - 1$ $\quad \sqrt{3 + 1} \stackrel{?}{=} 3 - 1$

$\sqrt{1} \stackrel{?}{=} -1$ $\quad \sqrt{4} \stackrel{?}{=} 2$

$1 \neq -1$ $\quad 2 \stackrel{\checkmark}{=} 2$

So the solution is $x = 3$.

40. The locus of points in space with coordinates that satisfy the equation $x + y = -5$ is a plane perpendicular to the xy-plane whose intersection with the xy-plane is the graph of $y = -x - 5$ in the xy-plane.

41. The locus of points in space with coordinates that satisfy the equation $x + z = 4$ is a plane perpendicular to the xz-plane whose intersection with the xz-plane is the graph of $z = -x + 4$ in the xz-plane.

Page 719 Maintain Your Skills

42. $\frac{\text{width of smaller prism}}{\text{width of larger prism}} = \frac{9}{18}$

$= \frac{1}{2}$

$\frac{\text{height of smaller prism}}{\text{height of larger prism}} = \frac{7}{13}$

Since the ratios are not the same, the prisms are neither similar nor congruent.

43. $\frac{\text{diameter of smaller cylinder}}{\text{diameter of larger cylinder}} = \frac{2 \times 5}{15}$

$= \frac{2}{3}$

$\frac{\text{height of smaller cylinder}}{\text{height of larger cylinder}} = \frac{12}{18}$

$= \frac{2}{3}$

The two cylinders are similar. Since the scale factor is not 1, they are not congruent.

44. $V = \frac{4}{3}\pi r^3$

$= \frac{4}{3}\pi(10^3)$

≈ 4188.8

The volume is approximately 4188.8 cm^3.

45. radius $= \frac{1}{2}(13) = 6.5$ yd

$V = \frac{4}{3}\pi r^3$

$= \frac{4}{3}\pi(6.5^3)$

≈ 1150.3

The volume is approximately 1150.3 yd^3.

46. $V = \frac{4}{3}\pi r^3$

$= \frac{4}{3}\pi(17.2^3)$

$\approx 21{,}314.4$

The volume is approximately 21,314.4 m^3.

47. radius $= \frac{1}{2}(29) = 14.5$ ft

$V = \frac{4}{3}\pi r^3$

$= \frac{4}{3}\pi(14.5^3)$

$\approx 12{,}770.1$

The volume is approximately 12,770.1 ft^3.

Chapter 13 Study Guide and Review

Page 720 Vocabulary and Concept Check

1. pyramid
2. Congruent
3. an ordered triple
4. cylinder
5. similar
6. prism
7. the Distance Formula in Space
8. sphere
9. Cavalieri's Principle
10. cone

Pages 720–722 Lesson-by-Lesson Review

11. $V = Bh$

$= (18)(7)(4)$

$= 504$

The volume of the prism is 504 in^3.

12. $V = \pi r^2 h$

$= \pi(3^2)(11)$

≈ 311.0

The volume is approximately 311.0 m^3.

13. The 15-ft diagonal forms a right triangle with the height and width. Use the Pythagorean Theorem to find the width.

$a^2 + b^2 = c^2$

$w^2 + 3^2 = 15^2$

$w^2 + 9 = 225$

$w^2 = 216$

$w = 6\sqrt{6}$ ft

Now find the volume.

$V = Bh$

$= (17)(6\sqrt{6})(3)$

≈ 749.5

The volume is approximately 749.5 ft^3.

14. First find the area of the hexagonal base.

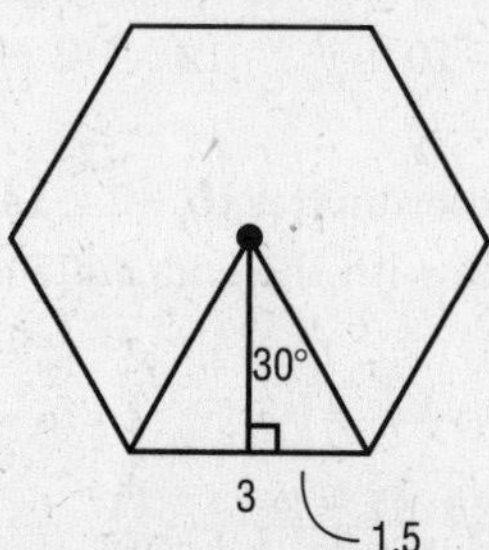

Apothem: A 30°-60°-90° triangle is formed by the apothem and one-half of a side of the hexagon. The shorter leg of the triangle is $\frac{1}{2}(3)$ or 1.5. The apothem is the longer leg of the triangle or $1.5\sqrt{3}$.

perimeter $= 3 \cdot 6 = 18$

Area: $A = \frac{1}{2}Pa$

$= \frac{1}{2}(18)(1.5\sqrt{3})$

$= 13.5\sqrt{3}$

Now find the volume of the pyramid.

$V = \frac{1}{3}Bh$

$= \frac{1}{3}(13.5\sqrt{3})(14)$

≈ 109.1

The volume is approximately 109.1 cm^3.

15. radius $= \frac{1}{2}(15) = 7.5$

Use the Pythagorean Theorem to find the height.

$a^2 + b^2 = c^2$

$7.5^2 + h^2 = 26^2$

$56.25 + h^2 = 676$

$h^2 = 619.75$

$h = \sqrt{619.75}$

For the cone,

$V = \frac{1}{3}\pi r^2 h$

$= \frac{1}{3}\pi(7.5^2)(\sqrt{619.75})$

≈ 1466.4

The volume is approximately 1466.4 ft^3.

16. $V = \frac{1}{3}Bh$

$= \frac{1}{3}(17)(5)(13)$

≈ 368.3

The volume is approximately 368.3 m^3.

17. $V = \frac{4}{3}\pi r^3$

$= \frac{4}{3}\pi(2^3)$

≈ 33.5

The volume is approximately 33.5 ft^3.

18. radius $= \frac{1}{2}(4) = 2$

$V = \frac{4}{3}\pi r^3$

$= \frac{4}{3}\pi(2^3)$

≈ 33.5

The volume is approximately 33.5 ft^3.

19. Find the radius.

$C = 2\pi r$

$65 = 2\pi r$

$\frac{65}{2\pi} = r$

Now find the volume.

$V = \frac{4}{3}\pi r^3$

$= \frac{4}{3}\pi\left(\frac{65}{2\pi}\right)^3$

≈ 4637.6

The volume is approximately 4637.6 mm^3.

20. Find the radius.

$T = 4\pi r^2$

$126 = 4\pi r^2$

$\frac{63}{2\pi} = r^2$

$\sqrt{\frac{63}{2\pi}} = r$

Now find the volume.

$V = \frac{4}{3}\pi r^3$

$= \frac{4}{3}\pi\left(\sqrt{\frac{63}{2\pi}}\right)^3$

≈ 133.0

The volume is approximately 133.0 cm^3.

21. Find the radius.

$A = \pi r^2$

$25\pi = \pi r^2$

$25 = r^2$

$5 = r$

Now find the volume.

$V = \frac{4}{3}\pi r^3$

$= \frac{4}{3}\pi(5^3)$

≈ 523.6

The volume is approximately 523.6 units^3.

22. For the left solid, the surface area is

$T = Ph + 2B$

$232 = 2(\ell + 7)(4) + 2(\ell \cdot 7)$

Now solve for ℓ.

$232 = 22\ell + 56$

$176 = 22\ell$

$8 = \ell$

For the right solid, the surface area is

$T = Ph + 2B$

$232 = 2(8 + 7)(h) + 2(8 \cdot 7)$

Now solve for h.

$232 = 30h + 112$

$120 = h$

$4 = h$

The solids have the same dimensions, so they are congruent.

23. Two spheres with different radii are similar, though not congruent.

24. $AB = \sqrt{(x_2 - x_1)^2 + (y_2 - y_1)^2 + (z_2 - z_1)^2}$

$= \sqrt{[3 - (-5)]^2 + [-8 - (-8)]^2 + [4 - (-2)]^2}$

$= \sqrt{100} = 10$

$M = \left(\frac{x_1 + x_2}{2}, \frac{y_1 + y_2}{2}, \frac{z_1 + z_2}{2}\right)$

$= \left(\frac{-5 + 3}{2}, \frac{-8 + (-8)}{2}, \frac{-2 + 4}{2}\right)$

$= (-1, -8, 1)$

25. $CD = \sqrt{(x_2 - x_1)^2 + (y_2 - y_1)^2 + (z_2 - z_1)^2}$

$= \sqrt{[-9 - (-9)]^2 + [9 - 2]^2 + [7 - 4]^2}$

$= \sqrt{58}$

$M = \left(\frac{x_1 + x_2}{2}, \frac{y_1 + y_2}{2}, \frac{z_1 + z_2}{2}\right)$

$= \left(\frac{-9 + (-9)}{2}, \frac{2 + 9}{2}, \frac{4 + 7}{2}\right)$

$= (-9, 5.5, 5.5)$

26. $EO = \sqrt{(x_2 - x_1)^2 + (y_2 - y_1)^2 + (z_2 - z_1)^2}$

$= \sqrt{(-4 - 0)^2 + (5 - 0)^2 + (5 - 0)^2}$

$= \sqrt{66}$

$M = \left(\frac{x_1 + x_2}{2}, \frac{y_1 + y_2}{2}, \frac{z_1 + z_2}{2}\right)$

$= \left(\frac{-4 + 0}{2}, \frac{5 + 0}{2}, \frac{5 + 0}{2}\right)$

$= (-2, 2.5, 2.5)$

27. $FG = \sqrt{(x_2 - x_1)^2 + (y_2 - y_1)^2 + (z_2 - z_1)^2}$

$= \sqrt{(-2\sqrt{2} - 5\sqrt{2})^2 + (3\sqrt{7} - 3\sqrt{7})^2 + (-12 - 6)^2}$

$= \sqrt{422}$

$M = \left(\frac{x_1 + x_2}{2}, \frac{y_1 + y_2}{2}, \frac{z_1 + z_2}{2}\right)$

$= \left(\frac{5\sqrt{2} - 2\sqrt{2}}{2}, \frac{3\sqrt{7} + 3\sqrt{7}}{2}, \frac{6 + (-12)}{2}\right)$

$= (1.5\sqrt{2}, 3\sqrt{7}, -3)$

Chapter 13 Practice Test

Page 723

1. b

2. c

3. a

4. The diameter of the base, the diagonal, and the lateral edge form a right triangle. Find the diameter using the Pythagorean Theorem.

$a^2 + b^2 = c^2$
$8^2 + d^2 = 10^2$
$64 + d^2 = 100$
$d^2 = 36$
$d = 6$ yd

radius $= \frac{1}{2}(6) = 3$

Now find the volume, using the formula for a cylinder.

$V = \pi r^2 h$
$= \pi(3^2)(8)$
≈ 226.2

The volume is approximately 226.2 yd^3.

5. Use the formula for a rectangular prism.

$V = Bh$
$= (6)(14)(10)$
$= 840$

The volume of the prism is 840 mm^3.

6. Find the width using the Pythagorean Theorem.

$a^2 + b^2 = c^2$
$7^2 + w^2 = (\sqrt{74})^2$
$49 + w^2 = 74$
$w^2 = 25$
$w = 5$

Now find the volume, using the formula for a rectangular prism.

$V = Bh$
$= (7)(5)(2)$
$= 70$

The volume of the prism is 70 km^3.

7. Use the formula for a pyramid.

$V = \frac{1}{3}Bh$
$= \frac{1}{3}(5)(5)(3)$
$= 25$

The volume of the pyramid is 25 ft^3.

8. First find the area of the hexagonal base.

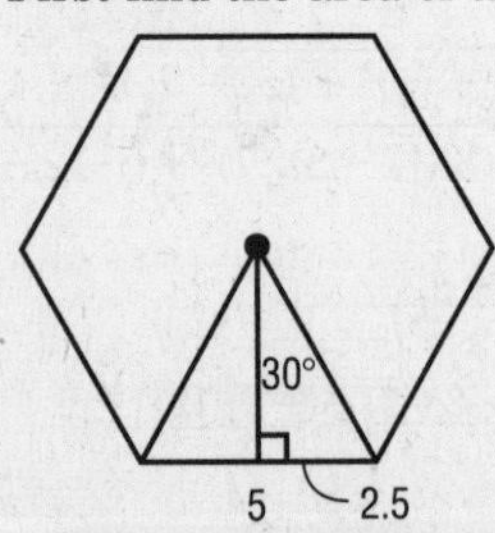

Apothem: A 30°-60°-90° triangle is formed by the apothem and one-half of a side of the hexagon. The shorter leg of the triangle is $\frac{1}{2}(5)$ or 2.5. The apothem is the longer leg of the triangle or $2.5\sqrt{3}$.

perimeter $= 5 \cdot 6 = 30$

Area: $A = \frac{1}{2}Pa$
$= \frac{1}{2}(30)(2.5\sqrt{3})$
$= 37.5\sqrt{3}$

Find the height using the Pythagorean Theorem and the fact that for a regular hexagon the distance from the center to a vertex is the same as the side length.

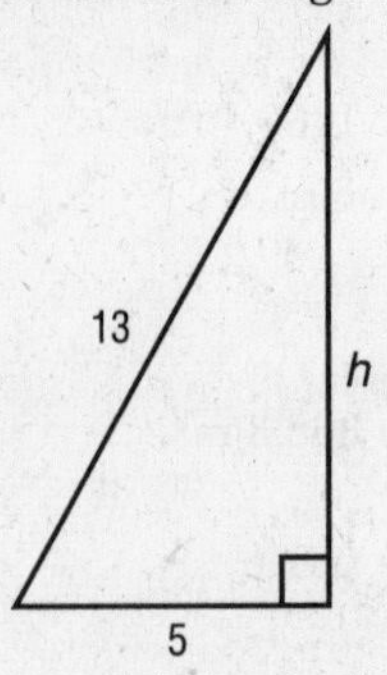

$a^2 + b^2 = c^2$
$5^2 + h^2 = 13^2$
$25 + h^2 = 169$
$h^2 = 144$
$h = 12$ m

Now find the volume.

$V = \frac{1}{3}Bh$
$= \frac{1}{3}(37.5\sqrt{3})(12)$
≈ 259.8

The volume is approximately 259.8 m^3.

9. Use the formula for a cone.

radius $= \frac{1}{2}(8.2) = 4.1$

$V = \frac{1}{3}\pi r^2 h$
$= \frac{1}{3}\pi(4.1^2)(6.8)$
≈ 119.7

The volume is approximately 119.7 cm^3.

10. Use the formula for an oblique cone.
First find the radius.

$C = 2\pi r$
$22\pi = 2\pi r$
$11 = r$

Now find the volume.

$V = \frac{1}{3}Bh$
$= \frac{1}{3}\pi r^2 h$
$= \frac{1}{3}\pi(11^2)(9)$
≈ 1140.4

The volume is approximately 1140.4 in^3.

11. The length, the width, and the diagonal form a right triangle. Use the Pythagorean Theorem to find the width w.

$a^2 + b^2 = c^2$
$78^2 + w^2 = 110.3^2$
$6084 + w^2 = 12{,}166.09$
$w^2 = 6082.09$
$w \approx 77.98776$ ft

Consider the water as a rectangular prism. Now find the volume in cubic feet, multiplied by a conversion factor of $7\frac{1}{2}$ gallons per cubic foot.

$V = Bh\left(7\frac{1}{2}\right)$

$\approx (78)(77.98776)(17)(7.5)$

$\approx 775{,}588$

The volume of water is approximately 775,588 gal.

12. $V = \frac{4}{3}\pi r^3$

$= \frac{4}{3}\pi(3^3)$

≈ 113.1

The volume is approximately 113.1 cm^3.

13. Find the radius.

$C = 2\pi r$

$34 = 2\pi r$

$\frac{17}{\pi} = r$

Now find the volume.

$V = \frac{4}{3}\pi r^3$

$= \frac{4}{3}\pi\left(\frac{17}{\pi}\right)^3$

≈ 663.7

The volume is approximately 663.7 ft^3.

14. Find the radius.

$T = 4\pi r^2$

$184 = 4\pi r^2$

$\frac{46}{\pi} = r^2$

$r \approx 3.8265$

Now find the volume.

$V = \frac{4}{3}\pi r^3$

$\approx \frac{4}{3}\pi(3.8265^3)$

≈ 234.7

The volume is approximately 234.7 in^3.

15. Find the radius.

$A = \pi r^2$

$157 = \pi r^2$

$\frac{157}{\pi} = r^2$

$r \approx 7.06928$

Now find the volume.

$V = \frac{4}{3}\pi r^3$

$\approx \frac{4}{3}\pi(7.06928^3)$

≈ 1479.8

The volume is approximately 1479.8 mm^3.

16. $\frac{\text{radius of larger cylinder}}{\text{radius of smaller cylinder}}$

$= \frac{\text{height of larger cylinder}}{\text{height of smaller cylinder}}$

$= \frac{15}{10}$

$= \frac{3}{2}$

The ratio of the radii is 3 : 2.

17. $\frac{\text{surface area of larger cylinder}}{\text{surface area of smaller cylinder}} = \frac{a^2}{b^2}$

$= \left(\frac{15}{10}\right)^2$

$= \left(\frac{3}{2}\right)^2$

$= \frac{3^2}{2^2}$

$= \frac{9}{4}$

The ratio of the surface areas is 9 : 4.

18. $\frac{\text{volume of larger cylinder}}{\text{volume of smaller cylinder}} = \frac{a^3}{b^3}$

$= \left(\frac{15}{10}\right)^3$

$= \left(\frac{3}{2}\right)^3$

$= \frac{3^3}{2^3}$

$= \frac{27}{8}$

The ratio of the volumes is 27 : 8.

19. $d = \sqrt{(x_2 - x_1)^2 + (y_2 - y_1)^2 + (z_2 - z_1)^2}$

$= \sqrt{(0 - 0)^2 + (-3 - 0)^2 + (5 - 0)^2}$

$= \sqrt{34}$

$M = \left(\frac{x_1 + x_2}{2}, \frac{y_1 + y_2}{2}, \frac{z_1 + z_2}{2}\right)$

$= \left(\frac{0 + 0}{2}, \frac{0 + (-3)}{2}, \frac{0 + 5}{2}\right)$

$= (0, -1.5, 2.5)$

20. $d = \sqrt{(x_2 - x_1)^2 + (y_2 - y_1)^2 + (z_2 - z_1)^2}$

$= \sqrt{(-1 - 0)^2 + (10 - 0)^2 + (-5 - 0)^2}$

$= \sqrt{126}$ or $3\sqrt{14}$

$M = \left(\frac{x_1 + x_2}{2}, \frac{y_1 + y_2}{2}, \frac{z_1 + z_2}{2}\right)$

$= \left(\frac{0 + (-1)}{2}, \frac{0 + 10}{2}, \frac{0 + (-5)}{2}\right)$

$= (-0.5, 5, -2.5)$

21. $d = \sqrt{(x_2 - x_1)^2 + (y_2 - y_1)^2 + (z_2 - z_1)^2}$

$= \sqrt{(9 - 0)^2 + (5 - 0)^2 + (-7 - 0)^2}$

$= \sqrt{155}$

$M = \left(\frac{x_1 + x_2}{2}, \frac{y_1 + y_2}{2}, \frac{z_1 + z_2}{2}\right)$

$= \left(\frac{0 + 9}{2}, \frac{0 + 5}{2}, \frac{0 + (-7)}{2}\right)$

$= (4.5, 2.5, -3.5)$

22. $d = \sqrt{(x_2 - x_1)^2 + (y_2 - y_1)^2 + (z_2 - z_1)^2}$

$= \sqrt{[-3 - (-2)]^2 + [-5 - 2]^2 + [-4 - 2]^2}$

$= \sqrt{86}$

$M = \left(\frac{x_1 + x_2}{2}, \frac{y_1 + y_2}{2}, \frac{z_1 + z_2}{2}\right)$

$= \left(\frac{-2 + (-3)}{2}, \frac{2 + (-5)}{2}, \frac{2 + (-4)}{2}\right)$

$= (-2.5, -1.5, -1)$

23. $d = \sqrt{(x_2 - x_1)^2 + (y_2 - y_1)^2 + (z_2 - z_1)^2}$

$= \sqrt{(-9 - 9)^2 + (-7 - 3)^2 + (6 - 4)^2}$

$= \sqrt{428}$ or $2\sqrt{107}$

$M = \left(\frac{x_1 + x_2}{2}, \frac{y_1 + y_2}{2}, \frac{z_1 + z_2}{2}\right)$

$= \left(\frac{9 - 9}{2}, \frac{3 - 7}{2}, \frac{4 + 6}{2}\right)$

$= (0, -2, 5)$

$= \sqrt{(-3 - 8)^2 + [5 - (-6)]^2 + (10 - 1)^2}$
$= \sqrt{323}$

$M = \left(\frac{x_1 + x_2}{2}, \frac{y_1 + y_2}{2}, \frac{z_1 + z_2}{2}\right)$
$= \left(\frac{8 + (-3)}{2}, \frac{-6 + 5}{2}, \frac{1 + 10}{2}\right)$
$= (2.5, -0.5, 5.5)$

25. C; $V = Bh$
$360 = (15)w(2)$
$360 = 30w$
$12 = w$

Chapter 13 Standardized Test Practice

Pages 724–725

1. A; $\angle ACD$ and $\angle ACB$ together form the right angle, $\angle BCD$.

2. B; $5x° = x° + 90°$
$4x° = 90°$
$x = 22.5$
and since $x + m\angle DEF = 90$,
$22.5 + m\angle DEF = 90$
$m\angle DEF = 67.5$

3. C; the third side length must be greater than $21 - 13 = 8$ and less than $21 + 13 = 33$.

4. C; from the statement $\triangle QRS$ is similar to $\triangle TUV$, we know that $\angle Q$ and $\angle T$ are corresponding angles.

5. B; volume of drilled block = volume of prism − volume of cylinder
$= Bh - \pi r^2h$
$= (11)(5)(8) - \pi(2^2)(8)$
$\approx 339.5 \text{ cm}^3$

6. B; $C = 2\pi r$
$25 = 2\pi r$
$\frac{25}{2\pi} = r$
$V = \frac{4}{3}\pi r^3$
$= \frac{4}{3}\pi\left(\frac{25}{2\pi}\right)^3$
$\approx 264 \text{ in}^3$

7. C; $\frac{\text{height of larger cylinder}}{\text{height of smaller cylinder}} = \frac{16}{12}$
$= \frac{4}{3}$

$\frac{r}{4.5} = \frac{4}{3}$
$3r = 18$
$r = 6$
$V = \pi r^2h$
$= \pi(6^2)(16)$
$\approx 1809.6 \text{ cm}^3$

8. D; $r = \sqrt{(x_2 - x_1)^2 + (y_2 - y_1)^2 + (z_2 - z_1)^2}$
$= \sqrt{(9 - 3)^2 + (-2 - 1)^2 + (-2 - 4)^2}$
$= 9$

9. $\frac{12z^5 + 27z^2 - 6z}{3z} = \frac{12z^5}{3z} + \frac{27z^2}{3z} - \frac{6z}{3z}$
$= 4z^4 + 9z - 2$

10. Sierra: $p \rightarrow q$
Carlos: $\sim q \rightarrow \sim p$
Carlos formed the contrapositive.

11. If the measures of the corresponding sides are the same, the triangles are congruent.

12. From Theorem 8.2, the sum of eight exterior angles is 360, so that one exterior angle measures $\frac{360}{8} = 45$.

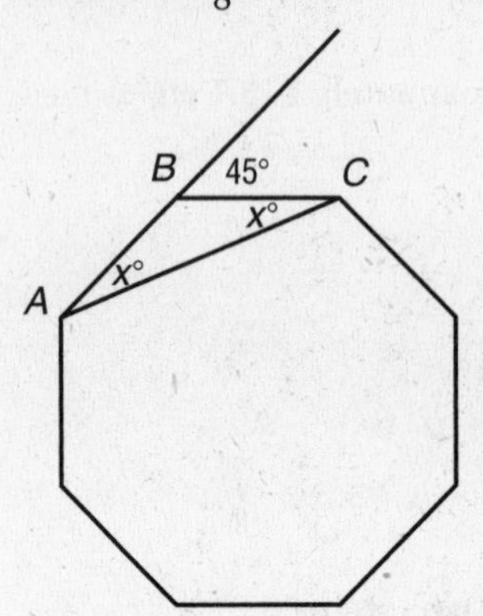

Because the octagon is regular, $AB = CB$, and $\triangle ABC$ is an isosceles triangle, so $m\angle A = m\angle C$. By the Exterior Angle Theorem,
$m\angle A + m\angle C = 45$
$x + x = 45$
$2x = 45$
$x = 22.5$

13. Point A lies on the x-axis, b units to the left of $D(0, c)$ just as B is b units to the right of C. The coordinates of A are $(-b, 0)$.

14. $V = \frac{1}{3}\pi r^2h$
$= \frac{1}{3}\pi(10^2)(18)$
$= 600\pi \text{ cm}^3$

15a. The surface area of the small can is $54\pi \text{ in}^2$ and the surface area of the large can is $90\pi \text{ in}^2$. When the height is doubled, the lateral area of the cylinder is doubled, but the area of the bases remains the same. The surface area increases by a factor of $1\frac{2}{3}$ times.

15b. The volume of the small can is $54\pi \text{ in}^3$ and the volume of the larger can is $108\pi \text{ in}^3$. The volume increases by a factor of 2.

16. volume of tank = volume of cone + volume of cylinder + volume of hemisphere
$= \frac{1}{3}\pi r^2h_1 + \pi r^2h_2 + \frac{1}{2}\left(\frac{4}{3}\pi r^3\right)$
$= \frac{1}{3}\pi(5^2)(15) + \pi(5^2)(45) + \frac{2}{3}\pi(5^3)$
≈ 4188.8

To the nearest cubic meter, the volume is 4189 m^3.